T0181807

Lecture Notes in Computer Science 12360

More information about this series at http://www.springer.com/series/7412

Andrea Vedaldi · Horst Bischof ·
Thomas Brox · Jan-Michael Frahm (Eds.)

Computer Vision – ECCV 2020

16th European Conference
Glasgow, UK, August 23–28, 2020
Proceedings, Part XV

 Springer

Editors
Andrea Vedaldi 📵
University of Oxford
Oxford, UK

Horst Bischof 📵
Graz University of Technology
Graz, Austria

Thomas Brox 📵
University of Freiburg
Freiburg im Breisgau, Germany

Jan-Michael Frahm
University of North Carolina at Chapel Hill
Chapel Hill, NC, USA

ISSN 0302-9743   ISSN 1611-3349   (electronic)
Lecture Notes in Computer Science
ISBN 978-3-030-58554-9   ISBN 978-3-030-58555-6   (eBook)
https://doi.org/10.1007/978-3-030-58555-6

LNCS Sublibrary: SL6 – Image Processing, Computer Vision, Pattern Recognition, and Graphics

This Springer imprint is published by the registered company Springer Nature Switzerland AG
The registered company address is: Gewerbestrasse 11, 6330 Cham, Switzerland

Foreword

Hosting the European Conference on Computer Vision (ECCV 2020) was certainly an exciting journey. From the 2016 plan to hold it at the Edinburgh International Conference Centre (hosting 1,800 delegates) to the 2018 plan to hold it at Glasgow's Scottish Exhibition Centre (up to 6,000 delegates), we finally ended with moving online because of the COVID-19 outbreak. While possibly having fewer delegates than expected because of the online format, ECCV 2020 still had over 3,100 registered participants.

Although online, the conference delivered most of the activities expected at a face-to-face conference: peer-reviewed papers, industrial exhibitors, demonstrations, and messaging between delegates. In addition to the main technical sessions, the conference included a strong program of satellite events with 16 tutorials and 44 workshops.

Furthermore, the online conference format enabled new conference features. Every paper had an associated teaser video and a longer full presentation video. Along with the papers and slides from the videos, all these materials were available the week before the conference. This allowed delegates to become familiar with the paper content and be ready for the live interaction with the authors during the conference week. The live event consisted of brief presentations by the oral and spotlight authors and industrial sponsors. Question and answer sessions for all papers were timed to occur twice so delegates from around the world had convenient access to the authors.

As with ECCV 2018, authors' draft versions of the papers appeared online with open access, now on both the Computer Vision Foundation (CVF) and the European Computer Vision Association (ECVA) websites. An archival publication arrangement was put in place with the cooperation of Springer. SpringerLink hosts the final version of the papers with further improvements, such as activating reference links and supplementary materials. These two approaches benefit all potential readers: a version available freely for all researchers, and an authoritative and citable version with additional benefits for SpringerLink subscribers. We thank Alfred Hofmann and Aliaksandr Birukou from Springer for helping to negotiate this agreement, which we expect will continue for future versions of ECCV.

August 2020

Vittorio Ferrari
Bob Fisher
Cordelia Schmid
Emanuele Trucco

Preface

Welcome to the proceedings of the European Conference on Computer Vision (ECCV 2020). This is a unique edition of ECCV in many ways. Due to the COVID-19 pandemic, this is the first time the conference was held online, in a virtual format. This was also the first time the conference relied exclusively on the Open Review platform to manage the review process. Despite these challenges ECCV is thriving. The conference received 5,150 valid paper submissions, of which 1,360 were accepted for publication (27%) and, of those, 160 were presented as spotlights (3%) and 104 as orals (2%). This amounts to more than twice the number of submissions to ECCV 2018 (2,439). Furthermore, CVPR, the largest conference on computer vision, received 5,850 submissions this year, meaning that ECCV is now 87% the size of CVPR in terms of submissions. By comparison, in 2018 the size of ECCV was only 73% of CVPR.

The review model was similar to previous editions of ECCV; in particular, it was double blind in the sense that the authors did not know the name of the reviewers and vice versa. Furthermore, each conference submission was held confidentially, and was only publicly revealed if and once accepted for publication. Each paper received at least three reviews, totalling more than 15,000 reviews. Handling the review process at this scale was a significant challenge. In order to ensure that each submission received as fair and high-quality reviews as possible, we recruited 2,830 reviewers (a 130% increase with reference to 2018) and 207 area chairs (a 60% increase). The area chairs were selected based on their technical expertise and reputation, largely among people that served as area chair in previous top computer vision and machine learning conferences (ECCV, ICCV, CVPR, NeurIPS, etc.). Reviewers were similarly invited from previous conferences. We also encouraged experienced area chairs to suggest additional chairs and reviewers in the initial phase of recruiting.

Despite doubling the number of submissions, the reviewer load was slightly reduced from 2018, from a maximum of 8 papers down to 7 (with some reviewers offering to handle 6 papers plus an emergency review). The area chair load increased slightly, from 18 papers on average to 22 papers on average.

Conflicts of interest between authors, area chairs, and reviewers were handled largely automatically by the Open Review platform via their curated list of user profiles. Many authors submitting to ECCV already had a profile in Open Review. We set a paper registration deadline one week before the paper submission deadline in order to encourage all missing authors to register and create their Open Review profiles well on time (in practice, we allowed authors to create/change papers arbitrarily until the submission deadline). Except for minor issues with users creating duplicate profiles, this allowed us to easily and quickly identify institutional conflicts, and avoid them, while matching papers to area chairs and reviewers.

Papers were matched to area chairs based on: an affinity score computed by the Open Review platform, which is based on paper titles and abstracts, and an affinity score computed by the Toronto Paper Matching System (TPMS), which is based on the paper's full text, the area chair bids for individual papers, load balancing, and conflict avoidance. Open Review provides the program chairs a convenient web interface to experiment with different configurations of the matching algorithm. The chosen configuration resulted in about 50% of the assigned papers to be highly ranked by the area chair bids, and 50% to be ranked in the middle, with very few low bids assigned.

Assignments to reviewers were similar, with two differences. First, there was a maximum of 7 papers assigned to each reviewer. Second, area chairs recommended up to seven reviewers per paper, providing another highly-weighed term to the affinity scores used for matching.

The assignment of papers to area chairs was smooth. However, it was more difficult to find suitable reviewers for all papers. Having a ratio of 5.6 papers per reviewer with a maximum load of 7 (due to emergency reviewer commitment), which did not allow for much wiggle room in order to also satisfy conflict and expertise constraints. We received some complaints from reviewers who did not feel qualified to review specific papers and we reassigned them wherever possible. However, the large scale of the conference, the many constraints, and the fact that a large fraction of such complaints arrived very late in the review process made this process very difficult and not all complaints could be addressed.

Reviewers had six weeks to complete their assignments. Possibly due to COVID-19 or the fact that the NeurIPS deadline was moved closer to the review deadline, a record 30% of the reviews were still missing after the deadline. By comparison, ECCV 2018 experienced only 10% missing reviews at this stage of the process. In the subsequent week, area chairs chased the missing reviews intensely, found replacement reviewers in their own team, and managed to reach 10% missing reviews. Eventually, we could provide almost all reviews (more than 99.9%) with a delay of only a couple of days on the initial schedule by a significant use of emergency reviews. If this trend is confirmed, it might be a major challenge to run a smooth review process in future editions of ECCV. The community must reconsider prioritization of the time spent on paper writing (the number of submissions increased a lot despite COVID-19) and time spent on paper reviewing (the number of reviews delivered in time decreased a lot presumably due to COVID-19 or NeurIPS deadline). With this imbalance the peer-review system that ensures the quality of our top conferences may break soon.

Reviewers submitted their reviews independently. In the reviews, they had the opportunity to ask questions to the authors to be addressed in the rebuttal. However, reviewers were told not to request any significant new experiment. Using the Open Review interface, authors could provide an answer to each individual review, but were also allowed to cross-reference reviews and responses in their answers. Rather than PDF files, we allowed the use of formatted text for the rebuttal. The rebuttal and initial reviews were then made visible to all reviewers and the primary area chair for a given paper. The area chair encouraged and moderated the reviewer discussion. During the discussions, reviewers were invited to reach a consensus and possibly adjust their ratings as a result of the discussion and of the evidence in the rebuttal.

After the discussion period ended, most reviewers entered a final rating and recommendation, although in many cases this did not differ from their initial recommendation. Based on the updated reviews and discussion, the primary area chair then made a preliminary decision to accept or reject the paper and wrote a justification for it (meta-review). Except for cases where the outcome of this process was absolutely clear (as indicated by the three reviewers and primary area chairs all recommending clear rejection), the decision was then examined and potentially challenged by a secondary area chair. This led to further discussion and overturning a small number of preliminary decisions. Needless to say, there was no in-person area chair meeting, which would have been impossible due to COVID-19.

Area chairs were invited to observe the consensus of the reviewers whenever possible and use extreme caution in overturning a clear consensus to accept or reject a paper. If an area chair still decided to do so, she/he was asked to clearly justify it in the meta-review and to explicitly obtain the agreement of the secondary area chair. In practice, very few papers were rejected after being confidently accepted by the reviewers.

This was the first time Open Review was used as the main platform to run ECCV. In 2018, the program chairs used CMT3 for the user-facing interface and Open Review internally, for matching and conflict resolution. Since it is clearly preferable to only use a single platform, this year we switched to using Open Review in full. The experience was largely positive. The platform is highly-configurable, scalable, and open source. Being written in Python, it is easy to write scripts to extract data programmatically. The paper matching and conflict resolution algorithms and interfaces are top-notch, also due to the excellent author profiles in the platform. Naturally, there were a few kinks along the way due to the fact that the ECCV Open Review configuration was created from scratch for this event and it differs in substantial ways from many other Open Review conferences. However, the Open Review development and support team did a fantastic job in helping us to get the configuration right and to address issues in a timely manner as they unavoidably occurred. We cannot thank them enough for the tremendous effort they put into this project.

Finally, we would like to thank everyone involved in making ECCV 2020 possible in these very strange and difficult times. This starts with our authors, followed by the area chairs and reviewers, who ran the review process at an unprecedented scale. The whole Open Review team (and in particular Melisa Bok, Mohit Unyal, Carlos Mondragon Chapa, and Celeste Martinez Gomez) worked incredibly hard for the entire duration of the process. We would also like to thank René Vidal for contributing to the adoption of Open Review. Our thanks also go to Laurent Charling for TPMS and to the program chairs of ICML, ICLR, and NeurIPS for cross checking double submissions. We thank the website chair, Giovanni Farinella, and the CPI team (in particular Ashley Cook, Miriam Verdon, Nicola McGrane, and Sharon Kerr) for promptly adding material to the website as needed in the various phases of the process. Finally, we thank the publication chairs, Albert Ali Salah, Hamdi Dibeklioglu, Metehan Doyran, Henry Howard-Jenkins, Victor Prisacariu, Siyu Tang, and Gul Varol, who managed to compile these substantial proceedings in an exceedingly compressed schedule. We express our thanks to the ECVA team, in particular Kristina Scherbaum for allowing open access of the proceedings. We thank Alfred Hofmann from Springer who again serve as the publisher. Finally, we thank the other chairs of ECCV 2020, including in particular the general chairs for very useful feedback with the handling of the program.

August 2020

Andrea Vedaldi
Horst Bischof
Thomas Brox
Jan-Michael Frahm

Organization

General Chairs

Vittorio Ferrari   Google Research, Switzerland
Bob Fisher   University of Edinburgh, UK
Cordelia Schmid   Google and Inria, France
Emanuele Trucco   University of Dundee, UK

Program Chairs

Andrea Vedaldi   University of Oxford, UK
Horst Bischof   Graz University of Technology, Austria
Thomas Brox   University of Freiburg, Germany
Jan-Michael Frahm   University of North Carolina, USA

Industrial Liaison Chairs

Jim Ashe   University of Edinburgh, UK
Helmut Grabner   Zurich University of Applied Sciences, Switzerland
Diane Larlus   NAVER LABS Europe, France
Cristian Novotny   University of Edinburgh, UK

Local Arrangement Chairs

Yvan Petillot   Heriot-Watt University, UK
Paul Siebert   University of Glasgow, UK

Academic Demonstration Chair

Thomas Mensink   Google Research and University of Amsterdam,
The Netherlands

Poster Chair

Stephen Mckenna   University of Dundee, UK

Technology Chair

Gerardo Aragon Camarasa   University of Glasgow, UK

Tutorial Chairs

Carlo Colombo   University of Florence, Italy
Sotirios Tsaftaris   University of Edinburgh, UK

Publication Chairs

Albert Ali Salah   Utrecht University, The Netherlands
Hamdi Dibeklioglu   Bilkent University, Turkey
Metehan Doyran   Utrecht University, The Netherlands
Henry Howard-Jenkins   University of Oxford, UK
Victor Adrian Prisacariu   University of Oxford, UK
Siyu Tang   ETH Zurich, Switzerland
Gul Varol   University of Oxford, UK

Website Chair

Giovanni Maria Farinella   University of Catania, Italy

Workshops Chairs

Adrien Bartoli   University of Clermont Auvergne, France
Andrea Fusiello   University of Udine, Italy

Area Chairs

Lourdes Agapito   University College London, UK
Zeynep Akata   University of Tübingen, Germany
Karteek Alahari   Inria, France
Antonis Argyros   University of Crete, Greece
Hossein Azizpour   KTH Royal Institute of Technology, Sweden
Joao P. Barreto   Universidade de Coimbra, Portugal
Alexander C. Berg   University of North Carolina at Chapel Hill, USA
Matthew B. Blaschko   KU Leuven, Belgium
Lubomir D. Bourdev   WaveOne, Inc., USA
Edmond Boyer   Inria, France
Yuri Boykov   University of Waterloo, Canada
Gabriel Brostow   University College London, UK
Michael S. Brown   National University of Singapore, Singapore
Jianfei Cai   Monash University, Australia
Barbara Caputo   Politecnico di Torino, Italy
Ayan Chakrabarti   Washington University, St. Louis, USA
Tat-Jen Cham   Nanyang Technological University, Singapore
Manmohan Chandraker   University of California, San Diego, USA
Rama Chellappa   Johns Hopkins University, USA
Liang-Chieh Chen   Google, USA

Timothy Hospedales	University of Edinburgh and Samsung, UK
Gang Hua	Wormpex AI Research, USA
Slobodan Ilic	Siemens AG, Germany
Hiroshi Ishikawa	Waseda University, Japan
Jiaya Jia	The Chinese University of Hong Kong, SAR China
Hailin Jin	Adobe Research, USA
Justin Johnson	University of Michigan, USA
Frederic Jurie	University of Caen Normandie, France
Fredrik Kahl	Chalmers University, Sweden
Sing Bing Kang	Zillow, USA
Gunhee Kim	Seoul National University, South Korea
Junmo Kim	Korea Advanced Institute of Science and Technology, South Korea
Tae-Kyun Kim	Imperial College London, UK
Ron Kimmel	Technion-Israel Institute of Technology, Israel
Alexander Kirillov	Facebook AI Research, USA
Kris Kitani	Carnegie Mellon University, USA
Iasonas Kokkinos	Ariel AI, UK
Vladlen Koltun	Intel Labs, USA
Nikos Komodakis	Ecole des Ponts ParisTech, France
Piotr Koniusz	Australian National University, Australia
M. Pawan Kumar	University of Oxford, UK
Kyros Kutulakos	University of Toronto, Canada
Christoph Lampert	IST Austria, Austria
Ivan Laptev	Inria, France
Diane Larlus	NAVER LABS Europe, France
Laura Leal-Taixe	Technical University Munich, Germany
Honglak Lee	Google and University of Michigan, USA
Joon-Young Lee	Adobe Research, USA
Kyoung Mu Lee	Seoul National University, South Korea
Seungyong Lee	POSTECH, South Korea
Yong Jae Lee	University of California, Davis, USA
Bastian Leibe	RWTH Aachen University, Germany
Victor Lempitsky	Samsung, Russia
Ales Leonardis	University of Birmingham, UK
Marius Leordeanu	Institute of Mathematics of the Romanian Academy, Romania
Vincent Lepetit	ENPC ParisTech, France
Hongdong Li	The Australian National University, Australia
Xi Li	Zhejiang University, China
Yin Li	University of Wisconsin-Madison, USA
Zicheng Liao	Zhejiang University, China
Jongwoo Lim	Hanyang University, South Korea
Stephen Lin	Microsoft Research Asia, China
Yen-Yu Lin	National Chiao Tung University, Taiwan, China
Zhe Lin	Adobe Research, USA

Haibin Ling	Stony Brooks, State University of New York, USA
Jiaying Liu	Peking University, China
Ming-Yu Liu	NVIDIA, USA
Si Liu	Beihang University, China
Xiaoming Liu	Michigan State University, USA
Huchuan Lu	Dalian University of Technology, China
Simon Lucey	Carnegie Mellon University, USA
Jiebo Luo	University of Rochester, USA
Julien Mairal	Inria, France
Michael Maire	University of Chicago, USA
Subhransu Maji	University of Massachusetts, Amherst, USA
Yasushi Makihara	Osaka University, Japan
Jiri Matas	Czech Technical University in Prague, Czech Republic
Yasuyuki Matsushita	Osaka University, Japan
Philippos Mordohai	Stevens Institute of Technology, USA
Vittorio Murino	University of Verona, Italy
Naila Murray	NAVER LABS Europe, France
Hajime Nagahara	Osaka University, Japan
P. J. Narayanan	International Institute of Information Technology (IIIT), Hyderabad, India
Nassir Navab	Technical University of Munich, Germany
Natalia Neverova	Facebook AI Research, France
Matthias Niessner	Technical University of Munich, Germany
Jean-Marc Odobez	Idiap Research Institute and Swiss Federal Institute of Technology Lausanne, Switzerland
Francesca Odone	Università di Genova, Italy
Takeshi Oishi	The University of Tokyo, Tokyo Institute of Technology, Japan
Vicente Ordonez	University of Virginia, USA
Manohar Paluri	Facebook AI Research, USA
Maja Pantic	Imperial College London, UK
In Kyu Park	Inha University, South Korea
Ioannis Patras	Queen Mary University of London, UK
Patrick Perez	Valeo, France
Bryan A. Plummer	Boston University, USA
Thomas Pock	Graz University of Technology, Austria
Marc Pollefeys	ETH Zurich and Microsoft MR & AI Zurich Lab, Switzerland
Jean Ponce	Inria, France
Gerard Pons-Moll	MPII, Saarland Informatics Campus, Germany
Jordi Pont-Tuset	Google, Switzerland
James Matthew Rehg	Georgia Institute of Technology, USA
Ian Reid	University of Adelaide, Australia
Olaf Ronneberger	DeepMind London, UK
Stefan Roth	TU Darmstadt, Germany
Bryan Russell	Adobe Research, USA

Mathieu Salzmann	EPFL, Switzerland
Dimitris Samaras	Stony Brook University, USA
Imari Sato	National Institute of Informatics (NII), Japan
Yoichi Sato	The University of Tokyo, Japan
Torsten Sattler	Czech Technical University in Prague, Czech Republic
Daniel Scharstein	Middlebury College, USA
Bernt Schiele	MPII, Saarland Informatics Campus, Germany
Julia A. Schnabel	King's College London, UK
Nicu Sebe	University of Trento, Italy
Greg Shakhnarovich	Toyota Technological Institute at Chicago, USA
Humphrey Shi	University of Oregon, USA
Jianbo Shi	University of Pennsylvania, USA
Jianping Shi	SenseTime, China
Leonid Sigal	University of British Columbia, Canada
Cees Snoek	University of Amsterdam, The Netherlands
Richard Souvenir	Temple University, USA
Hao Su	University of California, San Diego, USA
Akihiro Sugimoto	National Institute of Informatics (NII), Japan
Jian Sun	Megvii Technology, China
Jian Sun	Xi'an Jiaotong University, China
Chris Sweeney	Facebook Reality Labs, USA
Yu-wing Tai	Kuaishou Technology, China
Chi-Keung Tang	The Hong Kong University of Science and Technology, SAR China
Radu Timofte	ETH Zurich, Switzerland
Sinisa Todorovic	Oregon State University, USA
Giorgos Tolias	Czech Technical University in Prague, Czech Republic
Carlo Tomasi	Duke University, USA
Tatiana Tommasi	Politecnico di Torino, Italy
Lorenzo Torresani	Facebook AI Research and Dartmouth College, USA
Alexander Toshev	Google, USA
Zhuowen Tu	University of California, San Diego, USA
Tinne Tuytelaars	KU Leuven, Belgium
Jasper Uijlings	Google, Switzerland
Nuno Vasconcelos	University of California, San Diego, USA
Olga Veksler	University of Waterloo, Canada
Rene Vidal	Johns Hopkins University, USA
Gang Wang	Alibaba Group, China
Jingdong Wang	Microsoft Research Asia, China
Yizhou Wang	Peking University, China
Lior Wolf	Facebook AI Research and Tel Aviv University, Israel
Jianxin Wu	Nanjing University, China
Tao Xiang	University of Surrey, UK
Saining Xie	Facebook AI Research, USA
Ming-Hsuan Yang	University of California at Merced and Google, USA
Ruigang Yang	University of Kentucky, USA

Kwang Moo Yi   University of Victoria, Canada
Zhaozheng Yin   Stony Brook, State University of New York, USA
Chang D. Yoo   Korea Advanced Institute of Science and Technology,
                                      South Korea
Shaodi You   University of Amsterdam, The Netherlands
Jingyi Yu   ShanghaiTech University, China
Stella Yu   University of California, Berkeley, and ICSI, USA
Stefanos Zafeiriou   Imperial College London, UK
Hongbin Zha   Peking University, China
Tianzhu Zhang   University of Science and Technology of China, China
Liang Zheng   Australian National University, Australia
Todd E. Zickler   Harvard University, USA
Andrew Zisserman   University of Oxford, UK

Technical Program Committee

Sathyanarayanan	Samuel Albanie	Pablo Arbelaez
N. Aakur	Shadi Albarqouni	Shervin Ardeshir
Wael Abd Almgaeed	Cenek Albl	Sercan O. Arik
Abdelrahman	Hassan Abu Alhaija	Anil Armagan
Abdelhamed	Daniel Aliaga	Anurag Arnab
Abdullah Abuolaim	Mohammad	Chetan Arora
Supreeth Achar	S. Aliakbarian	Federica Arrigoni
Hanno Ackermann	Rahaf Aljundi	Mathieu Aubry
Ehsan Adeli	Thiemo Alldieck	Shai Avidan
Triantafyllos Afouras	Jon Almazan	Angelica I. Aviles-Rivero
Sameer Agarwal	Jose M. Alvarez	Yannis Avrithis
Aishwarya Agrawal	Senjian An	Ismail Ben Ayed
Harsh Agrawal	Saket Anand	Shekoofeh Azizi
Pulkit Agrawal	Codruta Ancuti	Ioan Andrei Bârsan
Antonio Agudo	Cosmin Ancuti	Artem Babenko
Eirikur Agustsson	Peter Anderson	Deepak Babu Sam
Karim Ahmed	Juan Andrade-Cetto	Seung-Hwan Baek
Byeongjoo Ahn	Alexander Andreopoulos	Seungryul Baek
Unaiza Ahsan	Misha Andriluka	Andrew D. Bagdanov
Thalaiyasingam Ajanthan	Dragomir Anguelov	Shai Bagon
Kenan E. Ak	Rushil Anirudh	Yuval Bahat
Emre Akbas	Michel Antunes	Junjie Bai
Naveed Akhtar	Oisin Mac Aodha	Song Bai
Derya Akkaynak	Srikar Appalaraju	Xiang Bai
Yagiz Aksoy	Relja Arandjelovic	Yalong Bai
Ziad Al-Halah	Nikita Araslanov	Yancheng Bai
Xavier Alameda-Pineda	Andre Araujo	Peter Bajcsy
Jean-Baptiste Alayrac	Helder Araujo	Slawomir Bak

Mahsa Baktashmotlagh
Kavita Bala
Yogesh Balaji
Guha Balakrishnan
V. N. Balasubramanian
Federico Baldassarre
Vassileios Balntas
Shurjo Banerjee
Aayush Bansal
Ankan Bansal
Jianmin Bao
Linchao Bao
Wenbo Bao
Yingze Bao
Akash Bapat
Md Jawadul Hasan Bappy
Fabien Baradel
Lorenzo Baraldi
Daniel Barath
Adrian Barbu
Kobus Barnard
Nick Barnes
Francisco Barranco
Jonathan T. Barron
Arslan Basharat
Chaim Baskin
Anil S. Baslamisli
Jorge Batista
Kayhan Batmanghelich
Konstantinos Batsos
David Bau
Luis Baumela
Christoph Baur
Eduardo
   Bayro-Corrochano
Paul Beardsley
Jan Bednavr'ik
Oscar Beijbom
Philippe Bekaert
Esube Bekele
Vasileios Belagiannis
Ohad Ben-Shahar
Abhijit Bendale
Róger Bermúdez-Chacón
Maxim Berman
Jesus Bermudez-cameo

Florian Bernard
Stefano Berretti
Marcelo Bertalmio
Gedas Bertasius
Cigdem Beyan
Lucas Beyer
Vijayakumar Bhagavatula
Arjun Nitin Bhagoji
Apratim Bhattacharyya
Binod Bhattarai
Sai Bi
Jia-Wang Bian
Simone Bianco
Adel Bibi
Tolga Birdal
Tom Bishop
Soma Biswas
Mårten Björkman
Volker Blanz
Vishnu Boddeti
Navaneeth Bodla
Simion-Vlad Bogolin
Xavier Boix
Piotr Bojanowski
Timo Bolkart
Guido Borghi
Larbi Boubchir
Guillaume Bourmaud
Adrien Bousseau
Thierry Bouwmans
Richard Bowden
Hakan Boyraz
Mathieu Brédif
Samarth Brahmbhatt
Steve Branson
Nikolas Brasch
Biagio Brattoli
Ernesto Brau
Toby P. Breckon
Francois Bremond
Jesus Briales
Sofia Broomé
Marcus A. Brubaker
Luc Brun
Silvia Bucci
Shyamal Buch

Pradeep Buddharaju
Uta Buechler
Mai Bui
Tu Bui
Adrian Bulat
Giedrius T. Burachas
Elena Burceanu
Xavier P. Burgos-Artizzu
Kaylee Burns
Andrei Bursuc
Benjamin Busam
Wonmin Byeon
Zoya Bylinskii
Sergi Caelles
Jianrui Cai
Minjie Cai
Yujun Cai
Zhaowei Cai
Zhipeng Cai
Juan C. Caicedo
Simone Calderara
Necati Cihan Camgoz
Dylan Campbell
Octavia Camps
Jiale Cao
Kaidi Cao
Liangliang Cao
Xiangyong Cao
Xiaochun Cao
Yang Cao
Yu Cao
Yue Cao
Zhangjie Cao
Luca Carlone
Mathilde Caron
Dan Casas
Thomas J. Cashman
Umberto Castellani
Lluis Castrejon
Jacopo Cavazza
Fabio Cermelli
Hakan Cevikalp
Menglei Chai
Ishani Chakraborty
Rudrasis Chakraborty
Antoni B. Chan

Kwok-Ping Chan
Siddhartha Chandra
Sharat Chandran
Arjun Chandrasekaran
Angel X. Chang
Che-Han Chang
Hong Chang
Hyun Sung Chang
Hyung Jin Chang
Jianlong Chang
Ju Yong Chang
Ming-Ching Chang
Simyung Chang
Xiaojun Chang
Yu-Wei Chao
Devendra S. Chaplot
Arslan Chaudhry
Rizwan A. Chaudhry
Can Chen
Chang Chen
Chao Chen
Chen Chen
Chu-Song Chen
Dapeng Chen
Dong Chen
Dongdong Chen
Guanying Chen
Hongge Chen
Hsin-yi Chen
Huaijin Chen
Hwann-Tzong Chen
Jianbo Chen
Jianhui Chen
Jiansheng Chen
Jiaxin Chen
Jie Chen
Jun-Cheng Chen
Kan Chen
Kevin Chen
Lin Chen
Long Chen
Min-Hung Chen
Qifeng Chen
Shi Chen
Shixing Chen
Tianshui Chen

Weifeng Chen
Weikai Chen
Xi Chen
Xiaohan Chen
Xiaozhi Chen
Xilin Chen
Xingyu Chen
Xinlei Chen
Xinyun Chen
Yi-Ting Chen
Yilun Chen
Ying-Cong Chen
Yinpeng Chen
Yiran Chen
Yu Chen
Yu-Sheng Chen
Yuhua Chen
Yun-Chun Chen
Yunpeng Chen
Yuntao Chen
Zhuoyuan Chen
Zitian Chen
Anchieh Cheng
Bowen Cheng
Erkang Cheng
Gong Cheng
Guangliang Cheng
Jingchun Cheng
Jun Cheng
Li Cheng
Ming-Ming Cheng
Yu Cheng
Ziang Cheng
Anoop Cherian
Dmitry Chetverikov
Ngai-man Cheung
William Cheung
Ajad Chhatkuli
Naoki Chiba
Benjamin Chidester
Han-pang Chiu
Mang Tik Chiu
Wei-Chen Chiu
Donghyeon Cho
Hojin Cho
Minsu Cho

Nam Ik Cho
Tim Cho
Tae Eun Choe
Chiho Choi
Edward Choi
Inchang Choi
Jinsoo Choi
Jonghyun Choi
Jongwon Choi
Yukyung Choi
Hisham Cholakkal
Eunji Chong
Jaegul Choo
Christopher Choy
Hang Chu
Peng Chu
Wen-Sheng Chu
Albert Chung
Joon Son Chung
Hai Ci
Safa Cicek
Ramazan G. Cinbis
Arridhana Ciptadi
Javier Civera
James J. Clark
Ronald Clark
Felipe Codevilla
Michael Cogswell
Andrea Cohen
Maxwell D. Collins
Carlo Colombo
Yang Cong
Adria R. Continente
Marcella Cornia
John Richard Corring
Darren Cosker
Dragos Costea
Garrison W. Cottrell
Florent Couzinie-Devy
Marco Cristani
Ioana Croitoru
James L. Crowley
Jiequan Cui
Zhaopeng Cui
Ross Cutler
Antonio D'Innocente

Rozenn Dahyot
Bo Dai
Dengxin Dai
Hang Dai
Longquan Dai
Shuyang Dai
Xiyang Dai
Yuchao Dai
Adrian V. Dalca
Dima Damen
Bharath B. Damodaran
Kristin Dana
Martin Danelljan
Zheng Dang
Zachary Alan Daniels
Donald G. Dansereau
Abhishek Das
Samyak Datta
Achal Dave
Titas De
Rodrigo de Bem
Teo de Campos
Raoul de Charette
Shalini De Mello
Joseph DeGol
Herve Delingette
Haowen Deng
Jiankang Deng
Weijian Deng
Zhiwei Deng
Joachim Denzler
Konstantinos G. Derpanis
Aditya Deshpande
Frederic Devernay
Somdip Dey
Arturo Deza
Abhinav Dhall
Helisa Dhamo
Vikas Dhiman
Fillipe Dias Moreira
  de Souza
Ali Diba
Ferran Diego
Guiguang Ding
Henghui Ding
Jian Ding

Mingyu Ding
Xinghao Ding
Zhengming Ding
Robert DiPietro
Cosimo Distante
Ajay Divakaran
Mandar Dixit
Abdelaziz Djelouah
Thanh-Toan Do
Jose Dolz
Bo Dong
Chao Dong
Jiangxin Dong
Weiming Dong
Weisheng Dong
Xingping Dong
Xuanyi Dong
Yinpeng Dong
Gianfranco Doretto
Hazel Doughty
Hassen Drira
Bertram Drost
Dawei Du
Ye Duan
Yueqi Duan
Abhimanyu Dubey
Anastasia Dubrovina
Stefan Duffner
Chi Nhan Duong
Thibaut Durand
Zoran Duric
Iulia Duta
Debidatta Dwibedi
Benjamin Eckart
Marc Eder
Marzieh Edraki
Alexei A. Efros
Kiana Ehsani
Hazm Kemal Ekenel
James H. Elder
Mohamed Elgharib
Shireen Elhabian
Ehsan Elhamifar
Mohamed Elhoseiny
Ian Endres
N. Benjamin Erichson

Jan Ernst
Sergio Escalera
Francisco Escolano
Victor Escorcia
Carlos Esteves
Francisco J. Estrada
Bin Fan
Chenyou Fan
Deng-Ping Fan
Haoqi Fan
Hehe Fan
Heng Fan
Kai Fan
Lijie Fan
Linxi Fan
Quanfu Fan
Shaojing Fan
Xiaochuan Fan
Xin Fan
Yuchen Fan
Sean Fanello
Hao-Shu Fang
Haoyang Fang
Kuan Fang
Yi Fang
Yuming Fang
Azade Farshad
Alireza Fathi
Raanan Fattal
Joao Fayad
Xiaohan Fei
Christoph Feichtenhofer
Michael Felsberg
Chen Feng
Jiashi Feng
Junyi Feng
Mengyang Feng
Qianli Feng
Zhenhua Feng
Michele Fenzi
Andras Ferencz
Martin Fergie
Basura Fernando
Ethan Fetaya
Michael Firman
John W. Fisher

Matthew Fisher
Boris Flach
Corneliu Florea
Wolfgang Foerstner
David Fofi
Gian Luca Foresti
Per-Erik Forssen
David Fouhey
Katerina Fragkiadaki
Victor Fragoso
Jean-Sébastien Franco
Ohad Fried
Iuri Frosio
Cheng-Yang Fu
Huazhu Fu
Jianlong Fu
Jingjing Fu
Xueyang Fu
Yanwei Fu
Ying Fu
Yun Fu
Olac Fuentes
Kent Fujiwara
Takuya Funatomi
Christopher Funk
Thomas Funkhouser
Antonino Furnari
Ryo Furukawa
Erik Gärtner
Raghudeep Gadde
Matheus Gadelha
Vandit Gajjar
Trevor Gale
Juergen Gall
Mathias Gallardo
Guillermo Gallego
Orazio Gallo
Chuang Gan
Zhe Gan
Madan Ravi Ganesh
Aditya Ganeshan
Siddha Ganju
Bin-Bin Gao
Changxin Gao
Feng Gao
Hongchang Gao

Jin Gao
Jiyang Gao
Junbin Gao
Katelyn Gao
Lin Gao
Mingfei Gao
Ruiqi Gao
Ruohan Gao
Shenghua Gao
Yuan Gao
Yue Gao
Noa Garcia
Alberto Garcia-Garcia
Guillermo
    Garcia-Hernando
Jacob R. Gardner
Animesh Garg
Kshitiz Garg
Rahul Garg
Ravi Garg
Philip N. Garner
Kirill Gavrilyuk
Paul Gay
Shiming Ge
Weifeng Ge
Baris Gecer
Xin Geng
Kyle Genova
Stamatios Georgoulis
Bernard Ghanem
Michael Gharbi
Kamran Ghasedi
Golnaz Ghiasi
Arnab Ghosh
Partha Ghosh
Silvio Giancola
Andrew Gilbert
Rohit Girdhar
Xavier Giro-i-Nieto
Thomas Gittings
Ioannis Gkioulekas
Clement Godard
Vaibhava Goel
Bastian Goldluecke
Lluis Gomez
Nuno Gonçalves

Dong Gong
Ke Gong
Mingming Gong
Abel Gonzalez-Garcia
Ariel Gordon
Daniel Gordon
Paulo Gotardo
Venu Madhav Govindu
Ankit Goyal
Priya Goyal
Raghav Goyal
Benjamin Graham
Douglas Gray
Brent A. Griffin
Etienne Grossmann
David Gu
Jiayuan Gu
Jiuxiang Gu
Lin Gu
Qiao Gu
Shuhang Gu
Jose J. Guerrero
Paul Guerrero
Jie Gui
Jean-Yves Guillemaut
Riza Alp Guler
Erhan Gundogdu
Fatma Guney
Guodong Guo
Kaiwen Guo
Qi Guo
Sheng Guo
Shi Guo
Tiantong Guo
Xiaojie Guo
Yijie Guo
Yiluan Guo
Yuanfang Guo
Yulan Guo
Agrim Gupta
Ankush Gupta
Mohit Gupta
Saurabh Gupta
Tanmay Gupta
Danna Gurari
Abner Guzman-Rivera

JunYoung Gwak
Michael Gygli
Jung-Woo Ha
Simon Hadfield
Isma Hadji
Bjoern Haefner
Taeyoung Hahn
Levente Hajder
Peter Hall
Emanuela Haller
Stefan Haller
Bumsub Ham
Abdullah Hamdi
Dongyoon Han
Hu Han
Jungong Han
Junwei Han
Kai Han
Tian Han
Xiaoguang Han
Xintong Han
Yahong Han
Ankur Handa
Zekun Hao
Albert Haque
Tatsuya Harada
Mehrtash Harandi
Adam W. Harley
Mahmudul Hasan
Atsushi Hashimoto
Ali Hatamizadeh
Munawar Hayat
Dongliang He
Jingrui He
Junfeng He
Kaiming He
Kun He
Lei He
Pan He
Ran He
Shengfeng He
Tong He
Weipeng He
Xuming He
Yang He
Yihui He

Zhihai He
Chinmay Hegde
Janne Heikkila
Mattias P. Heinrich
Stéphane Herbin
Alexander Hermans
Luis Herranz
John R. Hershey
Aaron Hertzmann
Roei Herzig
Anders Heyden
Steven Hickson
Otmar Hilliges
Tomas Hodan
Judy Hoffman
Michael Hofmann
Yannick Hold-Geoffroy
Namdar Homayounfar
Sina Honari
Richang Hong
Seunghoon Hong
Xiaopeng Hong
Yi Hong
Hidekata Hontani
Anthony Hoogs
Yedid Hoshen
Mir Rayat Imtiaz Hossain
Junhui Hou
Le Hou
Lu Hou
Tingbo Hou
Wei-Lin Hsiao
Cheng-Chun Hsu
Gee-Sern Jison Hsu
Kuang-jui Hsu
Changbo Hu
Di Hu
Guosheng Hu
Han Hu
Hao Hu
Hexiang Hu
Hou-Ning Hu
Jie Hu
Junlin Hu
Nan Hu
Ping Hu

Ronghang Hu
Xiaowei Hu
Yinlin Hu
Yuan-Ting Hu
Zhe Hu
Binh-Son Hua
Yang Hua
Bingyao Huang
Di Huang
Dong Huang
Fay Huang
Haibin Huang
Haozhi Huang
Heng Huang
Huaibo Huang
Jia-Bin Huang
Jing Huang
Jingwei Huang
Kaizhu Huang
Lei Huang
Qiangui Huang
Qiaoying Huang
Qingqiu Huang
Qixing Huang
Shaoli Huang
Sheng Huang
Siyuan Huang
Weilin Huang
Wenbing Huang
Xiangru Huang
Xun Huang
Yan Huang
Yifei Huang
Yue Huang
Zhiwu Huang
Zilong Huang
Minyoung Huh
Zhuo Hui
Matthias B. Hullin
Martin Humenberger
Wei-Chih Hung
Zhouyuan Huo
Junhwa Hur
Noureldien Hussein
Jyh-Jing Hwang
Seong Jae Hwang

Sung Ju Hwang
Ichiro Ide
Ivo Ihrke
Daiki Ikami
Satoshi Ikehata
Nazli Ikizler-Cinbis
Sunghoon Im
Yani Ioannou
Radu Tudor Ionescu
Umar Iqbal
Go Irie
Ahmet Iscen
Md Amirul Islam
Vamsi Ithapu
Nathan Jacobs
Arpit Jain
Himalaya Jain
Suyog Jain
Stuart James
Won-Dong Jang
Yunseok Jang
Ronnachai Jaroensri
Dinesh Jayaraman
Sadeep Jayasumana
Suren Jayasuriya
Herve Jegou
Simon Jenni
Hae-Gon Jeon
Yunho Jeon
Koteswar R. Jerripothula
Hueihan Jhuang
I-hong Jhuo
Dinghuang Ji
Hui Ji
Jingwei Ji
Pan Ji
Yanli Ji
Baoxiong Jia
Kui Jia
Xu Jia
Chiyu Max Jiang
Haiyong Jiang
Hao Jiang
Huaizu Jiang
Huajie Jiang
Ke Jiang

Lai Jiang
Li Jiang
Lu Jiang
Ming Jiang
Peng Jiang
Shuqiang Jiang
Wei Jiang
Xudong Jiang
Zhuolin Jiang
Jianbo Jiao
Zequn Jie
Dakai Jin
Kyong Hwan Jin
Lianwen Jin
SouYoung Jin
Xiaojie Jin
Xin Jin
Nebojsa Jojic
Alexis Joly
Michael Jeffrey Jones
Hanbyul Joo
Jungseock Joo
Kyungdon Joo
Ajjen Joshi
Shantanu H. Joshi
Da-Cheng Juan
Marco Körner
Kevin Köser
Asim Kadav
Christine Kaeser-Chen
Kushal Kafle
Dagmar Kainmueller
Ioannis A. Kakadiaris
Zdenek Kalal
Nima Kalantari
Yannis Kalantidis
Mahdi M. Kalayeh
Anmol Kalia
Sinan Kalkan
Vicky Kalogeiton
Ashwin Kalyan
Joni-kristian Kamarainen
Gerda Kamberova
Chandra Kambhamettu
Martin Kampel
Meina Kan

Christopher Kanan
Kenichi Kanatani
Angjoo Kanazawa
Atsushi Kanehira
Takuhiro Kaneko
Asako Kanezaki
Bingyi Kang
Di Kang
Sunghun Kang
Zhao Kang
Vadim Kantorov
Abhishek Kar
Amlan Kar
Theofanis Karaletsos
Leonid Karlinsky
Kevin Karsch
Angelos Katharopoulos
Isinsu Katircioglu
Hiroharu Kato
Zoltan Kato
Dotan Kaufman
Jan Kautz
Rei Kawakami
Qiuhong Ke
Wadim Kehl
Petr Kellnhofer
Aniruddha Kembhavi
Cem Keskin
Margret Keuper
Daniel Keysers
Ashkan Khakzar
Fahad Khan
Naeemullah Khan
Salman Khan
Siddhesh Khandelwal
Rawal Khirodkar
Anna Khoreva
Tejas Khot
Parmeshwar Khurd
Hadi Kiapour
Joe Kileel
Chanho Kim
Dahun Kim
Edward Kim
Eunwoo Kim
Han-ul Kim

Hansung Kim
Heewon Kim
Hyo Jin Kim
Hyunwoo J. Kim
Jinkyu Kim
Jiwon Kim
Jongmin Kim
Junsik Kim
Junyeong Kim
Min H. Kim
Namil Kim
Pyojin Kim
Seon Joo Kim
Seong Tae Kim
Seungryong Kim
Sungwoong Kim
Tae Hyun Kim
Vladimir Kim
Won Hwa Kim
Yonghyun Kim
Benjamin Kimia
Akisato Kimura
Pieter-Jan Kindermans
Zsolt Kira
Itaru Kitahara
Hedvig Kjellstrom
Jan Knopp
Takumi Kobayashi
Erich Kobler
Parker Koch
Reinhard Koch
Elyor Kodirov
Amir Kolaman
Nicholas Kolkin
Dimitrios Kollias
Stefanos Kollias
Soheil Kolouri
Adams Wai-Kin Kong
Naejin Kong
Shu Kong
Tao Kong
Yu Kong
Yoshinori Konishi
Daniil Kononenko
Theodora Kontogianni
Simon Korman

Adam Kortylewski
Jana Kosecka
Jean Kossaifi
Satwik Kottur
Rigas Kouskouridas
Adriana Kovashka
Rama Kovvuri
Adarsh Kowdle
Jedrzej Kozerawski
Mateusz Kozinski
Philipp Kraehenbuehl
Gregory Kramida
Josip Krapac
Dmitry Kravchenko
Ranjay Krishna
Pavel Krsek
Alexander Krull
Jakob Kruse
Hiroyuki Kubo
Hilde Kuehne
Jason Kuen
Andreas Kuhn
Arjan Kuijper
Zuzana Kukelova
Ajay Kumar
Amit Kumar
Avinash Kumar
Suryansh Kumar
Vijay Kumar
Kaustav Kundu
Weicheng Kuo
Nojun Kwak
Suha Kwak
Junseok Kwon
Nikolaos Kyriazis
Zorah Lähner
Ankit Laddha
Florent Lafarge
Jean Lahoud
Kevin Lai
Shang-Hong Lai
Wei-Sheng Lai
Yu-Kun Lai
Iro Laina
Antony Lam
John Wheatley Lambert

Xiangyuan lan
Xu Lan
Charis Lanaras
Georg Langs
Oswald Lanz
Dong Lao
Yizhen Lao
Agata Lapedriza
Gustav Larsson
Viktor Larsson
Katrin Lasinger
Christoph Lassner
Longin Jan Latecki
Stéphane Lathuilière
Rynson Lau
Hei Law
Justin Lazarow
Svetlana Lazebnik
Hieu Le
Huu Le
Ngan Hoang Le
Trung-Nghia Le
Vuong Le
Colin Lea
Erik Learned-Miller
Chen-Yu Lee
Gim Hee Lee
Hsin-Ying Lee
Hyungtae Lee
Jae-Han Lee
Jimmy Addison Lee
Joonseok Lee
Kibok Lee
Kuang-Huei Lee
Kwonjoon Lee
Minsik Lee
Sang-chul Lee
Seungkyu Lee
Soochan Lee
Stefan Lee
Taehee Lee
Andreas Lehrmann
Jie Lei
Peng Lei
Matthew Joseph Leotta
Wee Kheng Leow

Gil Levi
Evgeny Levinkov
Aviad Levis
Jose Lezama
Ang Li
Bin Li
Bing Li
Boyi Li
Changsheng Li
Chao Li
Chen Li
Cheng Li
Chenglong Li
Chi Li
Chun-Guang Li
Chun-Liang Li
Chunyuan Li
Dong Li
Guanbin Li
Hao Li
Haoxiang Li
Hongsheng Li
Hongyang Li
Houqiang Li
Huibin Li
Jia Li
Jianan Li
Jianguo Li
Junnan Li
Junxuan Li
Kai Li
Ke Li
Kejie Li
Kunpeng Li
Lerenhan Li
Li Erran Li
Mengtian Li
Mu Li
Peihua Li
Peiyi Li
Ping Li
Qi Li
Qing Li
Ruiyu Li
Ruoteng Li
Shaozi Li

Sheng Li
Shiwei Li
Shuang Li
Siyang Li
Stan Z. Li
Tianye Li
Wei Li
Weixin Li
Wen Li
Wenbo Li
Xiaomeng Li
Xin Li
Xiu Li
Xuelong Li
Xueting Li
Yan Li
Yandong Li
Yanghao Li
Yehao Li
Yi Li
Yijun Li
Yikang LI
Yining Li
Yongjie Li
Yu Li
Yu-Jhe Li
Yunpeng Li
Yunsheng Li
Yunzhu Li
Zhe Li
Zhen Li
Zhengqi Li
Zhenyang Li
Zhuwen Li
Dongze Lian
Xiaochen Lian
Zhouhui Lian
Chen Liang
Jie Liang
Ming Liang
Paul Pu Liang
Pengpeng Liang
Shu Liang
Wei Liang
Jing Liao
Minghui Liao

Renjie Liao
Shengcai Liao
Shuai Liao
Yiyi Liao
Ser-Nam Lim
Chen-Hsuan Lin
Chung-Ching Lin
Dahua Lin
Ji Lin
Kevin Lin
Tianwei Lin
Tsung-Yi Lin
Tsung-Yu Lin
Wei-An Lin
Weiyao Lin
Yen-Chen Lin
Yuewei Lin
David B. Lindell
Drew Linsley
Krzysztof Lis
Roee Litman
Jim Little
An-An Liu
Bo Liu
Buyu Liu
Chao Liu
Chen Liu
Cheng-lin Liu
Chenxi Liu
Dong Liu
Feng Liu
Guilin Liu
Haomiao Liu
Heshan Liu
Hong Liu
Ji Liu
Jingen Liu
Jun Liu
Lanlan Liu
Li Liu
Liu Liu
Mengyuan Liu
Miaomiao Liu
Nian Liu
Ping Liu
Risheng Liu

Sheng Liu
Shu Liu
Shuaicheng Liu
Sifei Liu
Siqi Liu
Siying Liu
Songtao Liu
Ting Liu
Tongliang Liu
Tyng-Luh Liu
Wanquan Liu
Wei Liu
Weiyang Liu
Weizhe Liu
Wenyu Liu
Wu Liu
Xialei Liu
Xianglong Liu
Xiaodong Liu
Xiaofeng Liu
Xihui Liu
Xingyu Liu
Xinwang Liu
Xuanqing Liu
Xuebo Liu
Yang Liu
Yaojie Liu
Yebin Liu
Yen-Cheng Liu
Yiming Liu
Yu Liu
Yu-Shen Liu
Yufan Liu
Yun Liu
Zheng Liu
Zhijian Liu
Zhuang Liu
Zichuan Liu
Ziwei Liu
Zongyi Liu
Stephan Liwicki
Liliana Lo Presti
Chengjiang Long
Fuchen Long
Mingsheng Long
Xiang Long

Yang Long
Charles T. Loop
Antonio Lopez
Roberto J. Lopez-Sastre
Javier Lorenzo-Navarro
Manolis Lourakis
Boyu Lu
Canyi Lu
Feng Lu
Guoyu Lu
Hongtao Lu
Jiajun Lu
Jiasen Lu
Jiwen Lu
Kaiyue Lu
Le Lu
Shao-Ping Lu
Shijian Lu
Xiankai Lu
Xin Lu
Yao Lu
Yiping Lu
Yongxi Lu
Yongyi Lu
Zhiwu Lu
Fujun Luan
Benjamin E. Lundell
Hao Luo
Jian-Hao Luo
Ruotian Luo
Weixin Luo
Wenhan Luo
Wenjie Luo
Yan Luo
Zelun Luo
Zixin Luo
Khoa Luu
Zhaoyang Lv
Pengyuan Lyu
Thomas Möllenhoff
Matthias Müller
Bingpeng Ma
Chih-Yao Ma
Chongyang Ma
Huimin Ma
Jiayi Ma

K. T. Ma
Ke Ma
Lin Ma
Liqian Ma
Shugao Ma
Wei-Chiu Ma
Xiaojian Ma
Xingjun Ma
Zhanyu Ma
Zheng Ma
Radek Jakob Mackowiak
Ludovic Magerand
Shweta Mahajan
Siddharth Mahendran
Long Mai
Ameesh Makadia
Oscar Mendez Maldonado
Mateusz Malinowski
Yury Malkov
Arun Mallya
Dipu Manandhar
Massimiliano Mancini
Fabian Manhardt
Kevis-kokitsi Maninis
Varun Manjunatha
Junhua Mao
Xudong Mao
Alina Marcu
Edgar Margffoy-Tuay
Dmitrii Marin
Manuel J. Marin-Jimenez
Kenneth Marino
Niki Martinel
Julieta Martinez
Jonathan Masci
Tomohiro Mashita
Iacopo Masi
David Masip
Daniela Massiceti
Stefan Mathe
Yusuke Matsui
Tetsu Matsukawa
Iain A. Matthews
Kevin James Matzen
Bruce Allen Maxwell
Stephen Maybank

Helmut Mayer
Amir Mazaheri
David McAllester
Steven McDonagh
Stephen J. Mckenna
Roey Mechrez
Prakhar Mehrotra
Christopher Mei
Xue Mei
Paulo R. S. Mendonca
Lili Meng
Zibo Meng
Thomas Mensink
Bjoern Menze
Michele Merler
Kourosh Meshgi
Pascal Mettes
Christopher Metzler
Liang Mi
Qiguang Miao
Xin Miao
Tomer Michaeli
Frank Michel
Antoine Miech
Krystian Mikolajczyk
Peyman Milanfar
Ben Mildenhall
Gregor Miller
Fausto Milletari
Dongbo Min
Kyle Min
Pedro Miraldo
Dmytro Mishkin
Anand Mishra
Ashish Mishra
Ishan Misra
Niluthpol C. Mithun
Kaushik Mitra
Niloy Mitra
Anton Mitrokhin
Ikuhisa Mitsugami
Anurag Mittal
Kaichun Mo
Zhipeng Mo
Davide Modolo
Michael Moeller

Pritish Mohapatra
Pavlo Molchanov
Davide Moltisanti
Pascal Monasse
Mathew Monfort
Aron Monszpart
Sean Moran
Vlad I. Morariu
Francesc Moreno-Noguer
Pietro Morerio
Stylianos Moschoglou
Yael Moses
Roozbeh Mottaghi
Pierre Moulon
Arsalan Mousavian
Yadong Mu
Yasuhiro Mukaigawa
Lopamudra Mukherjee
Yusuke Mukuta
Ravi Teja Mullapudi
Mario Enrique Munich
Zachary Murez
Ana C. Murillo
J. Krishna Murthy
Damien Muselet
Armin Mustafa
Siva Karthik Mustikovela
Carlo Dal Mutto
Moin Nabi
Varun K. Nagaraja
Tushar Nagarajan
Arsha Nagrani
Seungjun Nah
Nikhil Naik
Yoshikatsu Nakajima
Yuta Nakashima
Atsushi Nakazawa
Seonghyeon Nam
Vinay P. Namboodiri
Medhini Narasimhan
Srinivasa Narasimhan
Sanath Narayan
Erickson Rangel
    Nascimento
Jacinto Nascimento
Tayyab Naseer

Lakshmanan Nataraj
Neda Nategh
Nelson Isao Nauata
Fernando Navarro
Shah Nawaz
Lukas Neumann
Ram Nevatia
Alejandro Newell
Shawn Newsam
Joe Yue-Hei Ng
Trung Thanh Ngo
Duc Thanh Nguyen
Lam M. Nguyen
Phuc Xuan Nguyen
Thuong Nguyen Canh
Mihalis Nicolaou
Andrei Liviu Nicolicioiu
Xuecheng Nie
Michael Niemeyer
Simon Niklaus
Christophoros Nikou
David Nilsson
Jifeng Ning
Yuval Nirkin
Li Niu
Yuzhen Niu
Zhenxing Niu
Shohei Nobuhara
Nicoletta Noceti
Hyeonwoo Noh
Junhyug Noh
Mehdi Noroozi
Sotiris Nousias
Valsamis Ntouskos
Matthew O'Toole
Peter Ochs
Ferda Ofli
Seong Joon Oh
Seoung Wug Oh
Iason Oikonomidis
Utkarsh Ojha
Takahiro Okabe
Takayuki Okatani
Fumio Okura
Aude Oliva
Kyle Olszewski

Björn Ommer
Mohamed Omran
Elisabeta Oneata
Michael Opitz
Jose Oramas
Tribhuvanesh Orekondy
Shaul Oron
Sergio Orts-Escolano
Ivan Oseledets
Aljosa Osep
Magnus Oskarsson
Anton Osokin
Martin R. Oswald
Wanli Ouyang
Andrew Owens
Mete Ozay
Mustafa Ozuysal
Eduardo Pérez-Pellitero
Gautam Pai
Dipan Kumar Pal
P. H. Pamplona Savarese
Jinshan Pan
Junting Pan
Xingang Pan
Yingwei Pan
Yannis Panagakis
Rameswar Panda
Guan Pang
Jiahao Pang
Jiangmiao Pang
Tianyu Pang
Sharath Pankanti
Nicolas Papadakis
Dim Papadopoulos
George Papandreou
Toufiq Parag
Shaifali Parashar
Sarah Parisot
Eunhyeok Park
Hyun Soo Park
Jaesik Park
Min-Gyu Park
Taesung Park
Alvaro Parra
C. Alejandro Parraga
Despoina Paschalidou

Nikolaos Passalis
Vishal Patel
Viorica Patraucean
Badri Narayana Patro
Danda Pani Paudel
Sujoy Paul
Georgios Pavlakos
Ioannis Pavlidis
Vladimir Pavlovic
Nick Pears
Kim Steenstrup Pedersen
Selen Pehlivan
Shmuel Peleg
Chao Peng
Houwen Peng
Wen-Hsiao Peng
Xi Peng
Xiaojiang Peng
Xingchao Peng
Yuxin Peng
Federico Perazzi
Juan Camilo Perez
Vishwanath Peri
Federico Pernici
Luca Del Pero
Florent Perronnin
Stavros Petridis
Henning Petzka
Patrick Peursum
Michael Pfeiffer
Hanspeter Pfister
Roman Pflugfelder
Minh Tri Pham
Yongri Piao
David Picard
Tomasz Pieciak
A. J. Piergiovanni
Andrea Pilzer
Pedro O. Pinheiro
Silvia Laura Pintea
Lerrel Pinto
Axel Pinz
Robinson Piramuthu
Fiora Pirri
Leonid Pishchulin
Francesco Pittaluga

Daniel Pizarro
Tobias Plötz
Mirco Planamente
Matteo Poggi
Moacir A. Ponti
Parita Pooj
Fatih Porikli
Horst Possegger
Omid Poursaeed
Ameya Prabhu
Viraj Uday Prabhu
Dilip Prasad
Brian L. Price
True Price
Maria Priisalu
Veronique Prinet
Victor Adrian Prisacariu
Jan Prokaj
Sergey Prokudin
Nicolas Pugeault
Xavier Puig
Albert Pumarola
Pulak Purkait
Senthil Purushwalkam
Charles R. Qi
Hang Qi
Haozhi Qi
Lu Qi
Mengshi Qi
Siyuan Qi
Xiaojuan Qi
Yuankai Qi
Shengju Qian
Xuelin Qian
Siyuan Qiao
Yu Qiao
Jie Qin
Qiang Qiu
Weichao Qiu
Zhaofan Qiu
Kha Gia Quach
Yuhui Quan
Yvain Queau
Julian Quiroga
Faisal Qureshi
Mahdi Rad

Filip Radenovic
Petia Radeva
Venkatesh
  B. Radhakrishnan
Ilija Radosavovic
Noha Radwan
Rahul Raguram
Tanzila Rahman
Amit Raj
Ajit Rajwade
Kandan Ramakrishnan
Santhosh
  K. Ramakrishnan
Srikumar Ramalingam
Ravi Ramamoorthi
Vasili Ramanishka
Ramprasaath R. Selvaraju
Francois Rameau
Visvanathan Ramesh
Santu Rana
Rene Ranftl
Anand Rangarajan
Anurag Ranjan
Viresh Ranjan
Yongming Rao
Carolina Raposo
Vivek Rathod
Sathya N. Ravi
Avinash Ravichandran
Tammy Riklin Raviv
Daniel Rebain
Sylvestre-Alvise Rebuffi
N. Dinesh Reddy
Timo Rehfeld
Paolo Remagnino
Konstantinos Rematas
Edoardo Remelli
Dongwei Ren
Haibing Ren
Jian Ren
Jimmy Ren
Mengye Ren
Weihong Ren
Wenqi Ren
Zhile Ren
Zhongzheng Ren

Zhou Ren
Vijay Rengarajan
Md A. Reza
Farzaneh Rezaeianaran
Hamed R. Tavakoli
Nicholas Rhinehart
Helge Rhodin
Elisa Ricci
Alexander Richard
Eitan Richardson
Elad Richardson
Christian Richardt
Stephan Richter
Gernot Riegler
Daniel Ritchie
Tobias Ritschel
Samuel Rivera
Yong Man Ro
Richard Roberts
Joseph Robinson
Ignacio Rocco
Mrigank Rochan
Emanuele Rodolà
Mikel D. Rodriguez
Giorgio Roffo
Grégory Rogez
Gemma Roig
Javier Romero
Xuejian Rong
Yu Rong
Amir Rosenfeld
Bodo Rosenhahn
Guy Rosman
Arun Ross
Paolo Rota
Peter M. Roth
Anastasios Roussos
Anirban Roy
Sebastien Roy
Aruni RoyChowdhury
Artem Rozantsev
Ognjen Rudovic
Daniel Rueckert
Adria Ruiz
Javier Ruiz-del-solar
Christian Rupprecht

Chris Russell
Dan Ruta
Jongbin Ryu
Ömer Sümer
Alexandre Sablayrolles
Faraz Saeedan
Ryusuke Sagawa
Christos Sagonas
Tonmoy Saikia
Hideo Saito
Kuniaki Saito
Shunsuke Saito
Shunta Saito
Ken Sakurada
Joaquin Salas
Fatemeh Sadat Saleh
Mahdi Saleh
Pouya Samangouei
Leo Sampaio
  Ferraz Ribeiro
Artsiom Olegovich
  Sanakoyeu
Enrique Sanchez
Patsorn Sangkloy
Anush Sankaran
Aswin Sankaranarayanan
Swami Sankaranarayanan
Rodrigo Santa Cruz
Amartya Sanyal
Archana Sapkota
Nikolaos Sarafianos
Jun Sato
Shin'ichi Satoh
Hosnieh Sattar
Arman Savran
Manolis Savva
Alexander Sax
Hanno Scharr
Simone Schaub-Meyer
Konrad Schindler
Dmitrij Schlesinger
Uwe Schmidt
Dirk Schnieders
Björn Schuller
Samuel Schulter
Idan Schwartz

William Robson Schwartz
Alex Schwing
Sinisa Segvic
Lorenzo Seidenari
Pradeep Sen
Ozan Sener
Soumyadip Sengupta
Arda Senocak
Mojtaba Seyedhosseini
Shishir Shah
Shital Shah
Sohil Atul Shah
Tamar Rott Shaham
Huasong Shan
Qi Shan
Shiguang Shan
Jing Shao
Roman Shapovalov
Gaurav Sharma
Vivek Sharma
Viktoriia Sharmanska
Dongyu She
Sumit Shekhar
Evan Shelhamer
Chengyao Shen
Chunhua Shen
Falong Shen
Jie Shen
Li Shen
Liyue Shen
Shuhan Shen
Tianwei Shen
Wei Shen
William B. Shen
Yantao Shen
Ying Shen
Yiru Shen
Yujun Shen
Yuming Shen
Zhiqiang Shen
Ziyi Shen
Lu Sheng
Yu Sheng
Rakshith Shetty
Baoguang Shi
Guangming Shi

Hailin Shi
Miaojing Shi
Yemin Shi
Zhenmei Shi
Zhiyuan Shi
Kevin Jonathan Shih
Shiliang Shiliang
Hyunjung Shim
Atsushi Shimada
Nobutaka Shimada
Daeyun Shin
Young Min Shin
Koichi Shinoda
Konstantin Shmelkov
Michael Zheng Shou
Abhinav Shrivastava
Tianmin Shu
Zhixin Shu
Hong-Han Shuai
Pushkar Shukla
Christian Siagian
Mennatullah M. Siam
Kaleem Siddiqi
Karan Sikka
Jae-Young Sim
Christian Simon
Martin Simonovsky
Dheeraj Singaraju
Bharat Singh
Gurkirt Singh
Krishna Kumar Singh
Maneesh Kumar Singh
Richa Singh
Saurabh Singh
Suriya Singh
Vikas Singh
Sudipta N. Sinha
Vincent Sitzmann
Josef Sivic
Gregory Slabaugh
Miroslava Slavcheva
Ron Slossberg
Brandon Smith
Kevin Smith
Vladimir Smutny
Noah Snavely

Roger
    D. Soberanis-Mukul
Kihyuk Sohn
Francesco Solera
Eric Sommerlade
Sanghyun Son
Byung Cheol Song
Chunfeng Song
Dongjin Song
Jiaming Song
Jie Song
Jifei Song
Jingkuan Song
Mingli Song
Shiyu Song
Shuran Song
Xiao Song
Yafei Song
Yale Song
Yang Song
Yi-Zhe Song
Yibing Song
Humberto Sossa
Cesar de Souza
Adrian Spurr
Srinath Sridhar
Suraj Srinivas
Pratul P. Srinivasan
Anuj Srivastava
Tania Stathaki
Christopher Stauffer
Simon Stent
Rainer Stiefelhagen
Pierre Stock
Julian Straub
Jonathan C. Stroud
Joerg Stueckler
Jan Stuehmer
David Stutz
Chi Su
Hang Su
Jong-Chyi Su
Shuochen Su
Yu-Chuan Su
Ramanathan Subramanian
Yusuke Sugano

Masanori Suganuma
Yumin Suh
Mohammed Suhail
Yao Sui
Heung-Il Suk
Josephine Sullivan
Baochen Sun
Chen Sun
Chong Sun
Deqing Sun
Jin Sun
Liang Sun
Lin Sun
Qianru Sun
Shao-Hua Sun
Shuyang Sun
Weiwei Sun
Wenxiu Sun
Xiaoshuai Sun
Xiaoxiao Sun
Xingyuan Sun
Yifan Sun
Zhun Sun
Sabine Susstrunk
David Suter
Supasorn Suwajanakorn
Tomas Svoboda
Eran Swears
Paul Swoboda
Attila Szabo
Richard Szeliski
Duy-Nguyen Ta
Andrea Tagliasacchi
Yuichi Taguchi
Ying Tai
Keita Takahashi
Kouske Takahashi
Jun Takamatsu
Hugues Talbot
Toru Tamaki
Chaowei Tan
Fuwen Tan
Mingkui Tan
Mingxing Tan
Qingyang Tan
Robby T. Tan

Xiaoyang Tan
Kenichiro Tanaka
Masayuki Tanaka
Chang Tang
Chengzhou Tang
Danhang Tang
Ming Tang
Peng Tang
Qingming Tang
Wei Tang
Xu Tang
Yansong Tang
Youbao Tang
Yuxing Tang
Zhiqiang Tang
Tatsunori Taniai
Junli Tao
Xin Tao
Makarand Tapaswi
Jean-Philippe Tarel
Lyne Tchapmi
Zachary Teed
Bugra Tekin
Damien Teney
Ayush Tewari
Christian Theobalt
Christopher Thomas
Diego Thomas
Jim Thomas
Rajat Mani Thomas
Xinmei Tian
Yapeng Tian
Yingli Tian
Yonglong Tian
Zhi Tian
Zhuotao Tian
Kinh Tieu
Joseph Tighe
Massimo Tistarelli
Matthew Toews
Carl Toft
Pavel Tokmakov
Federico Tombari
Chetan Tonde
Yan Tong
Alessio Tonioni

Andrea Torsello
Fabio Tosi
Du Tran
Luan Tran
Ngoc-Trung Tran
Quan Hung Tran
Truyen Tran
Rudolph Triebel
Martin Trimmel
Shashank Tripathi
Subarna Tripathi
Leonardo Trujillo
Eduard Trulls
Tomasz Trzcinski
Sam Tsai
Yi-Hsuan Tsai
Hung-Yu Tseng
Stavros Tsogkas
Aggeliki Tsoli
Devis Tuia
Shubham Tulsiani
Sergey Tulyakov
Frederick Tung
Tony Tung
Daniyar Turmukhambetov
Ambrish Tyagi
Radim Tylecek
Christos Tzelepis
Georgios Tzimiropoulos
Dimitrios Tzionas
Seiichi Uchida
Norimichi Ukita
Dmitry Ulyanov
Martin Urschler
Yoshitaka Ushiku
Ben Usman
Alexander Vakhitov
Julien P. C. Valentin
Jack Valmadre
Ernest Valveny
Joost van de Weijer
Jan van Gemert
Koen Van Leemput
Gul Varol
Sebastiano Vascon
M. Alex O. Vasilescu

Subeesh Vasu
Mayank Vatsa
David Vazquez
Javier Vazquez-Corral
Ashok Veeraraghavan
Erik Velasco-Salido
Raviteja Vemulapalli
Jonathan Ventura
Manisha Verma
Roberto Vezzani
Ruben Villegas
Minh Vo
MinhDuc Vo
Nam Vo
Michele Volpi
Riccardo Volpi
Carl Vondrick
Konstantinos Vougioukas
Tuan-Hung Vu
Sven Wachsmuth
Neal Wadhwa
Catherine Wah
Jacob C. Walker
Thomas S. A. Wallis
Chengde Wan
Jun Wan
Liang Wan
Renjie Wan
Baoyuan Wang
Boyu Wang
Cheng Wang
Chu Wang
Chuan Wang
Chunyu Wang
Dequan Wang
Di Wang
Dilin Wang
Dong Wang
Fang Wang
Guanzhi Wang
Guoyin Wang
Hanzi Wang
Hao Wang
He Wang
Heng Wang
Hongcheng Wang

Hongxing Wang
Hua Wang
Jian Wang
Jingbo Wang
Jinglu Wang
Jingya Wang
Jinjun Wang
Jinqiao Wang
Jue Wang
Ke Wang
Keze Wang
Le Wang
Lei Wang
Lezi Wang
Li Wang
Liang Wang
Lijun Wang
Limin Wang
Linwei Wang
Lizhi Wang
Mengjiao Wang
Mingzhe Wang
Minsi Wang
Naiyan Wang
Nannan Wang
Ning Wang
Oliver Wang
Pei Wang
Peng Wang
Pichao Wang
Qi Wang
Qian Wang
Qiaosong Wang
Qifei Wang
Qilong Wang
Qing Wang
Qingzhong Wang
Quan Wang
Rui Wang
Ruiping Wang
Ruixing Wang
Shangfei Wang
Shenlong Wang
Shiyao Wang
Shuhui Wang
Song Wang

Tao Wang
Tianlu Wang
Tiantian Wang
Ting-chun Wang
Tingwu Wang
Wei Wang
Weiyue Wang
Wenguan Wang
Wenlin Wang
Wenqi Wang
Xiang Wang
Xiaobo Wang
Xiaofang Wang
Xiaoling Wang
Xiaolong Wang
Xiaosong Wang
Xiaoyu Wang
Xin Eric Wang
Xinchao Wang
Xinggang Wang
Xintao Wang
Yali Wang
Yan Wang
Yang Wang
Yangang Wang
Yaxing Wang
Yi Wang
Yida Wang
Yilin Wang
Yiming Wang
Yisen Wang
Yongtao Wang
Yu-Xiong Wang
Yue Wang
Yujiang Wang
Yunbo Wang
Yunhe Wang
Zengmao Wang
Zhangyang Wang
Zhaowen Wang
Zhe Wang
Zhecan Wang
Zheng Wang
Zhixiang Wang
Zilei Wang
Jianqiao Wangni

Anne S. Wannenwetsch
Jan Dirk Wegner
Scott Wehrwein
Donglai Wei
Kaixuan Wei
Longhui Wei
Pengxu Wei
Ping Wei
Qi Wei
Shih-En Wei
Xing Wei
Yunchao Wei
Zijun Wei
Jerod Weinman
Michael Weinmann
Philippe Weinzaepfel
Yair Weiss
Bihan Wen
Longyin Wen
Wei Wen
Junwu Weng
Tsui-Wei Weng
Xinshuo Weng
Eric Wengrowski
Tomas Werner
Gordon Wetzstein
Tobias Weyand
Patrick Wieschollek
Maggie Wigness
Erik Wijmans
Richard Wildes
Olivia Wiles
Chris Williams
Williem Williem
Kyle Wilson
Calden Wloka
Nicolai Wojke
Christian Wolf
Yongkang Wong
Sanghyun Woo
Scott Workman
Baoyuan Wu
Bichen Wu
Chao-Yuan Wu
Huikai Wu
Jiajun Wu

Jialin Wu
Jiaxiang Wu
Jiqing Wu
Jonathan Wu
Lifang Wu
Qi Wu
Qiang Wu
Ruizheng Wu
Shangzhe Wu
Shun-Cheng Wu
Tianfu Wu
Wayne Wu
Wenxuan Wu
Xiao Wu
Xiaohe Wu
Xinxiao Wu
Yang Wu
Yi Wu
Yiming Wu
Ying Nian Wu
Yue Wu
Zheng Wu
Zhenyu Wu
Zhirong Wu
Zuxuan Wu
Stefanie Wuhrer
Jonas Wulff
Changqun Xia
Fangting Xia
Fei Xia
Gui-Song Xia
Lu Xia
Xide Xia
Yin Xia
Yingce Xia
Yongqin Xian
Lei Xiang
Shiming Xiang
Bin Xiao
Fanyi Xiao
Guobao Xiao
Huaxin Xiao
Taihong Xiao
Tete Xiao
Tong Xiao
Wang Xiao

Yang Xiao
Cihang Xie
Guosen Xie
Jianwen Xie
Lingxi Xie
Sirui Xie
Weidi Xie
Wenxuan Xie
Xiaohua Xie
Fuyong Xing
Jun Xing
Junliang Xing
Bo Xiong
Peixi Xiong
Yu Xiong
Yuanjun Xiong
Zhiwei Xiong
Chang Xu
Chenliang Xu
Dan Xu
Danfei Xu
Hang Xu
Hongteng Xu
Huijuan Xu
Jingwei Xu
Jun Xu
Kai Xu
Mengmeng Xu
Mingze Xu
Qianqian Xu
Ran Xu
Weijian Xu
Xiangyu Xu
Xiaogang Xu
Xing Xu
Xun Xu
Yanyu Xu
Yichao Xu
Yong Xu
Yongchao Xu
Yuanlu Xu
Zenglin Xu
Zheng Xu
Chuhui Xue
Jia Xue
Nan Xue

Tianfan Xue
Xiangyang Xue
Abhay Yadav
Yasushi Yagi
I. Zeki Yalniz
Kota Yamaguchi
Toshihiko Yamasaki
Takayoshi Yamashita
Junchi Yan
Ke Yan
Qingan Yan
Sijie Yan
Xinchen Yan
Yan Yan
Yichao Yan
Zhicheng Yan
Keiji Yanai
Bin Yang
Ceyuan Yang
Dawei Yang
Dong Yang
Fan Yang
Guandao Yang
Guorun Yang
Haichuan Yang
Hao Yang
Jianwei Yang
Jiaolong Yang
Jie Yang
Jing Yang
Kaiyu Yang
Linjie Yang
Meng Yang
Michael Ying Yang
Nan Yang
Shuai Yang
Shuo Yang
Tianyu Yang
Tien-Ju Yang
Tsun-Yi Yang
Wei Yang
Wenhan Yang
Xiao Yang
Xiaodong Yang
Xin Yang
Yan Yang

Yanchao Yang
Yee Hong Yang
Yezhou Yang
Zhenheng Yang
Anbang Yao
Angela Yao
Cong Yao
Jian Yao
Li Yao
Ting Yao
Yao Yao
Zhewei Yao
Chengxi Ye
Jianbo Ye
Keren Ye
Linwei Ye
Mang Ye
Mao Ye
Qi Ye
Qixiang Ye
Mei-Chen Yeh
Raymond Yeh
Yu-Ying Yeh
Sai-Kit Yeung
Serena Yeung
Kwang Moo Yi
Li Yi
Renjiao Yi
Alper Yilmaz
Junho Yim
Lijun Yin
Weidong Yin
Xi Yin
Zhichao Yin
Tatsuya Yokota
Ryo Yonetani
Donggeun Yoo
Jae Shin Yoon
Ju Hong Yoon
Sung-eui Yoon
Laurent Younes
Changqian Yu
Fisher Yu
Gang Yu
Jiahui Yu
Kaicheng Yu

Ke Yu
Lequan Yu
Ning Yu
Qian Yu
Ronald Yu
Ruichi Yu
Shoou-I Yu
Tao Yu
Tianshu Yu
Xiang Yu
Xin Yu
Xiyu Yu
Youngjae Yu
Yu Yu
Zhiding Yu
Chunfeng Yuan
Ganzhao Yuan
Jinwei Yuan
Lu Yuan
Quan Yuan
Shanxin Yuan
Tongtong Yuan
Wenjia Yuan
Ye Yuan
Yuan Yuan
Yuhui Yuan
Huanjing Yue
Xiangyu Yue
Ersin Yumer
Sergey Zagoruyko
Egor Zakharov
Amir Zamir
Andrei Zanfir
Mihai Zanfir
Pablo Zegers
Bernhard Zeisl
John S. Zelek
Niclas Zeller
Huayi Zeng
Jiabei Zeng
Wenjun Zeng
Yu Zeng
Xiaohua Zhai
Fangneng Zhan
Huangying Zhan
Kun Zhan

Xiaohang Zhan
Baochang Zhang
Bowen Zhang
Cecilia Zhang
Changqing Zhang
Chao Zhang
Chengquan Zhang
Chi Zhang
Chongyang Zhang
Dingwen Zhang
Dong Zhang
Feihu Zhang
Hang Zhang
Hanwang Zhang
Hao Zhang
He Zhang
Hongguang Zhang
Hua Zhang
Ji Zhang
Jianguo Zhang
Jianming Zhang
Jiawei Zhang
Jie Zhang
Jing Zhang
Juyong Zhang
Kai Zhang
Kaipeng Zhang
Ke Zhang
Le Zhang
Lei Zhang
Li Zhang
Lihe Zhang
Linguang Zhang
Lu Zhang
Mi Zhang
Mingda Zhang
Peng Zhang
Pingping Zhang
Qian Zhang
Qilin Zhang
Quanshi Zhang
Richard Zhang
Rui Zhang
Runze Zhang
Shengping Zhang
Shifeng Zhang

Shuai Zhang
Songyang Zhang
Tao Zhang
Ting Zhang
Tong Zhang
Wayne Zhang
Wei Zhang
Weizhong Zhang
Wenwei Zhang
Xiangyu Zhang
Xiaolin Zhang
Xiaopeng Zhang
Xiaoqin Zhang
Xiuming Zhang
Ya Zhang
Yang Zhang
Yimin Zhang
Yinda Zhang
Ying Zhang
Yongfei Zhang
Yu Zhang
Yulun Zhang
Yunhua Zhang
Yuting Zhang
Zhanpeng Zhang
Zhao Zhang
Zhaoxiang Zhang
Zhen Zhang
Zheng Zhang
Zhifei Zhang
Zhijin Zhang
Zhishuai Zhang
Ziming Zhang
Bo Zhao
Chen Zhao
Fang Zhao
Haiyu Zhao
Han Zhao
Hang Zhao
Hengshuang Zhao
Jian Zhao
Kai Zhao
Liang Zhao
Long Zhao
Qian Zhao
Qibin Zhao

Qijun Zhao
Rui Zhao
Shenglin Zhao
Sicheng Zhao
Tianyi Zhao
Wenda Zhao
Xiangyun Zhao
Xin Zhao
Yang Zhao
Yue Zhao
Zhichen Zhao
Zijing Zhao
Xiantong Zhen
Chuanxia Zheng
Feng Zheng
Haiyong Zheng
Jia Zheng
Kang Zheng
Shuai Kyle Zheng
Wei-Shi Zheng
Yinqiang Zheng
Zerong Zheng
Zhedong Zheng
Zilong Zheng
Bineng Zhong
Fangwei Zhong
Guangyu Zhong
Yiran Zhong
Yujie Zhong
Zhun Zhong
Chunluan Zhou
Huiyu Zhou
Jiahuan Zhou
Jun Zhou
Lei Zhou
Luowei Zhou
Luping Zhou
Mo Zhou
Ning Zhou
Pan Zhou
Peng Zhou
Qianyi Zhou
S. Kevin Zhou
Sanping Zhou
Wengang Zhou
Xingyi Zhou

Yanzhao Zhou
Yi Zhou
Yin Zhou
Yipin Zhou
Yuyin Zhou
Zihan Zhou
Alex Zihao Zhu
Chenchen Zhu
Feng Zhu
Guangming Zhu
Ji Zhu
Jun-Yan Zhu
Lei Zhu
Linchao Zhu
Rui Zhu
Shizhan Zhu
Tyler Lixuan Zhu

Wei Zhu
Xiangyu Zhu
Xinge Zhu
Xizhou Zhu
Yanjun Zhu
Yi Zhu
Yixin Zhu
Yizhe Zhu
Yousong Zhu
Zhe Zhu
Zhen Zhu
Zheng Zhu
Zhenyao Zhu
Zhihui Zhu
Zhuotun Zhu
Bingbing Zhuang
Wei Zhuo

Christian Zimmermann
Karel Zimmermann
Larry Zitnick
Mohammadreza
    Zolfaghari
Maria Zontak
Daniel Zoran
Changqing Zou
Chuhang Zou
Danping Zou
Qi Zou
Yang Zou
Yuliang Zou
Georgios Zoumpourlis
Wangmeng Zuo
Xinxin Zuo

Additional Reviewers

Victoria Fernandez
    Abrevaya
Maya Aghaei
Allam Allam
Christine
    Allen-Blanchette
Nicolas Aziere
Assia Benbihi
Neha Bhargava
Bharat Lal Bhatnagar
Joanna Bitton
Judy Borowski
Amine Bourki
Romain Brégier
Tali Brayer
Sebastian Bujwid
Andrea Burns
Yun-Hao Cao
Yuning Chai
Xiaojun Chang
Bo Chen
Shuo Chen
Zhixiang Chen
Junsuk Choe
Hung-Kuo Chu

Jonathan P. Crall
Kenan Dai
Lucas Deecke
Karan Desai
Prithviraj Dhar
Jing Dong
Wei Dong
Turan Kaan Elgin
Francis Engelmann
Erik Englesson
Fartash Faghri
Zicong Fan
Yang Fu
Risheek Garrepalli
Yifan Ge
Marco Godi
Helmut Grabner
Shuxuan Guo
Jianfeng He
Zhezhi He
Samitha Herath
Chih-Hui Ho
Yicong Hong
Vincent Tao Hu
Julio Hurtado

Jaedong Hwang
Andrey Ignatov
Muhammad
    Abdullah Jamal
Saumya Jetley
Meiguang Jin
Jeff Johnson
Minsoo Kang
Saeed Khorram
Mohammad Rami Koujan
Nilesh Kulkarni
Sudhakar Kumawat
Abdelhak Lemkhenter
Alexander Levine
Jiachen Li
Jing Li
Jun Li
Yi Li
Liang Liao
Ruochen Liao
Tzu-Heng Lin
Phillip Lippe
Bao-di Liu
Bo Liu
Fangchen Liu

Hanxiao Liu
Hongyu Liu
Huidong Liu
Miao Liu
Xinxin Liu
Yongfei Liu
Yu-Lun Liu
Amir Livne
Tiange Luo
Wei Ma
Xiaoxuan Ma
Ioannis Marras
Georg Martius
Effrosyni Mavroudi
Tim Meinhardt
Givi Meishvili
Meng Meng
Zihang Meng
Zhongqi Miao
Gyeongsik Moon
Khoi Nguyen
Yung-Kyun Noh
Antonio Norelli
Jaeyoo Park
Alexander Pashevich
Mandela Patrick
Mary Phuong
Bingqiao Qian
Yu Qiao
Zhen Qiao
Sai Saketh Rambhatla
Aniket Roy
Amelie Royer
Parikshit Vishwas
    Sakurikar
Mark Sandler
Mert Bülent Sarıyıldız
Tanner Schmidt
Anshul B. Shah

Ketul Shah
Rajvi Shah
Hengcan Shi
Xiangxi Shi
Yujiao Shi
William A. P. Smith
Guoxian Song
Robin Strudel
Abby Stylianou
Xinwei Sun
Reuben Tan
Qingyi Tao
Kedar S. Tatwawadi
Anh Tuan Tran
Son Dinh Tran
Eleni Triantafillou
Aristeidis Tsitiridis
Md Zasim Uddin
Andrea Vedaldi
Evangelos Ververas
Vidit Vidit
Paul Voigtlaender
Bo Wan
Huanyu Wang
Huiyu Wang
Junqiu Wang
Pengxiao Wang
Tai Wang
Xinyao Wang
Tomoki Watanabe
Mark Weber
Xi Wei
Botong Wu
James Wu
Jiamin Wu
Rujie Wu
Yu Wu
Rongchang Xie
Wei Xiong

Yunyang Xiong
An Xu
Chi Xu
Yinghao Xu
Fei Xue
Tingyun Yan
Zike Yan
Chao Yang
Heran Yang
Ren Yang
Wenfei Yang
Xu Yang
Rajeev Yasarla
Shaokai Ye
Yufei Ye
Kun Yi
Haichao Yu
Hanchao Yu
Ruixuan Yu
Liangzhe Yuan
Chen-Lin Zhang
Fandong Zhang
Tianyi Zhang
Yang Zhang
Yiyi Zhang
Yongshun Zhang
Yu Zhang
Zhiwei Zhang
Jiaojiao Zhao
Yipu Zhao
Xingjian Zhen
Haizhong Zheng
Tiancheng Zhi
Chengju Zhou
Hao Zhou
Hao Zhu
Alexander Zimin

Contents – Part XV

ReDro: Efficiently Learning Large-Sized SPD Visual Representation

Saimunur Rahman[1,2]®, Lei Wang[1(✉)]®, Changming Sun[2]®,
and Luping Zhou[3]®

[1] VILA, School of Computing and Information Technology,
University of Wollongong, Wollongong, NSW 2522, Australia
sr801@uowmail.edu.au, leiw@uow.edu.au
[2] CSIRO Data61, PO Box 76, Epping, NSW 1710, Australia
changming.sun@csiro.au
[3] School of Electrical and Information Engineering, University of Sydney,
Sydney, NSW 2006, Australia
luping.zhou@sydney.edu.au

Abstract. Symmetric positive definite (SPD) matrix has recently been used as an effective visual representation. When learning this representation in deep networks, eigen-decomposition of covariance matrix is usually needed for a key step called matrix normalisation. This could result in significant computational cost, especially when facing the increasing number of channels in recent advanced deep networks.

This work proposes a novel scheme called Relation Dropout (ReDro). It is inspired by the fact that eigen-decomposition of a *block diagonal* matrix can be efficiently obtained by decomposing each of its diagonal square matrices, which are of smaller sizes. Instead of using a full covariance matrix as in the literature, we generate a block diagonal one by randomly grouping the channels and only considering the covariance within the same group. We insert ReDro as an additional layer before the step of matrix normalisation and make its random grouping transparent to all subsequent layers. Additionally, we can view the ReDro scheme as a dropout-like regularisation, which drops the channel relationship across groups. As experimentally demonstrated, for the SPD methods typically involving the matrix normalisation step, ReDro can effectively help them reduce computational cost in learning large-sized SPD visual representation and also help to improve image recognition performance.

Keywords: Block diagonal matrix · Covariance ·
Eigen-decomposition · SPD representation · Fine-grained image recognition

Electronic supplementary material The online version of this chapter (https://doi.org/10.1007/978-3-030-58555-6_1) contains supplementary material, which is available to authorized users.

1   Introduction

Learning good visual representation remains a central issue in computer vision. Representing images with local descriptors and pooling them to a global representation has been effective. Among the pooling methods, covariance matrix based pooling has gained substantial interest by exploiting the second-order information of features. Since covariance matrix is symmetric positive definite (SPD), the resulting representation is often called SPD visual representation. It has shown promising performance in various tasks, including fine-grained image classification [21], image segmentation [11], generic image classification [12,19], image set classification [33], activity and action recognition [14,39] and few-shot learning [36,37], to name a few. With the advent of deep learning, several pieces of pioneering work have integrated SPD representation into convolutional neural networks (CNNs) and investigated a range of important issues such as matrix function back-propagation [11], compact matrix estimation [7], matrix normalisation [15,16,20] and kernel-based extension [6]. These progresses bring forth effective SPD visual representations and improve image recognition performance.

Despite the successes, the end-to-end learning of SPD representation in CNNs poses a computational challenge. This is because i) the size of covariance matrix increases quadratically with the channel number in a convolutional feature map and ii) eigen-decomposition is often needed to normalise the covariance matrix in back-propagation for each training sample. This results in significant computation, especially considering that many channels are deployed in recent advanced deep networks. Although a dimension reduction layer could always be used to reduce the channel number beforehand, we are curious about if this computational challenge can be mitigated from another orthogonal perspective.

This work is inspired by the following fact: the eigen-decomposition of a *block diagonal* matrix can be obtained by simply assembling the eigenvectors and eigenvalues of its diagonal square matrices [23]. Each diagonal matrix is smaller in size and the eigen-decomposition needs less computation. Motivated by this, we propose to replace a full covariance matrix with a block diagonal one. To achieve this, all channels must be partitioned into mutually exclusive groups and the covariance of the channels in different groups shall be omitted (i.e., set as zero). A question that may arise is how to optimally partition the channels to minimise the loss of covariance information or maximise the final recognition performance. Although this optimum could be pursued by redesigning network architecture or loss function (e.g., considering the idea in [38]), it will alter the original SPD methods that use matrix normalisation and potentially increase the complexity of network training.

To realise a block diagonal covariance matrix with the minimal alteration of the original methods, negligible extra computation and no extra parameters to learn, we resort to a *random* partitioning of channels. This can be trivially implemented, with some housekeeping operation to make the randomness transparent to all the network layers after the matrix normalisation step. To carry out the end-to-end training via back-propagation, we derive the relevant matrix gradients in the presence of this randomisation. The saving on computation and the intensity of changing the random partitioning pattern are also discussed.

In addition, we conceptually link the proposed random omission of relationship (i.e., covariance) of channels to the dropout scheme commonly used in deep learning [27]. We call our scheme "Relation Dropout (ReDro)" for short. It is found that besides serving the goal of mitigating the computational challenge, the proposed scheme could bring forth an additional advantage of improving the network training and image recognition performance, which is consistent with the spirit of the extensively used dropout techniques.

The main contributions of this paper are summarised as follows.

- To mitigate the computational issue in learning large-sized SPD representation for the methods using matrix normalisation, this paper proposes a scheme called ReDro to take advantage of the eigen-decomposition efficiency of a block diagonal matrix. To the best of our survey, such a random partition based scheme is new for the deep learning of SPD visual representation.
- Via the randomisation mechanism, the ReDro scheme maintains the minimal change to the original network design and negligible computational overhead. This work derives the forward and backward propagations in the presence of the proposed scheme and discusses its properties.
- Conceptually viewing the ReDro scheme as a kind of dropout, we investigate its regularisation effect and find that it could additionally help improving network training efficiency and image recognition performance.

Extensive experiments are conducted on one scene dataset and three fine-grained image datasets to verify the effectiveness of the proposed scheme.

2   Related Work

Learning SPD Visual Representation. SPD visual representation can be traced back to covariance region descriptor in object detection, classification and tracking [25, 29, 30]. The advent of deep learning provides powerful image features and further exhibits the potential of SPD visual representation. After early attempts which compute covariance matrix on pre-extracted deep features [3], research along this line quickly enters the end-to-end learning paradigm and thrives. Covariance matrix is embedded into CNNs as a special layer and jointly learned with network weights to obtain the best possible SPD visual representation.

DeepO$_2$P [11] and Bilinear CNN (BCNN) [21] are two pieces of pioneering work that learn SPD visual representation in an end-to-end manner. The framework of DeepO$_2$P is largely followed and continuously improved by subsequent works. It generally consists of three parts. The first part feeds an image into a CNN backbone and processes it till the last layer of 3D convolutional feature maps, with width w, height h and channel number d. Viewing this map as a set of $w \times h$ local descriptors of d dimensions, the second part computes a (normalised, which will be detailed shortly) $d \times d$ covariance matrix to characterise the channel correlation. The last part is routine, usually consisting of fully connected layers and the softmax layer for prediction.

Matrix Normalisation. The step of matrix normalisation in the second part above plays a crucial role. It is widely seen in the recent work to learn SPD visual representation due to three motivations: i) Covariance matrix resides on a Riemannian manifold, whose geometric structure needs to be considered; ii) Normalisation is required to battle the "burstiness" phenomenon—a visual pattern usually occurs more times once it appears in an image; iii) Normalisation helps to achieve robust covariance estimation against small sample.

After element-wise normalisation, the recent work turns to matrix-logarithm or matrix-power normalisation because they usually produce better SPD representation.[1] Nevertheless, both of them involve the eigen-decomposition of covariance matrix, whose computational complexity can be up to $\mathcal{O}(d^3)$. This operation has to be applied for each training sample in every forward and backward propagations. The step of matrix normalisation becomes a computational bottleneck in the end-to-end learning of SPD visual representation.

The recent literature has made an effort to reduce the computation of matrix normalisation. They consider a special case of matrix power normalisation, that is, matrix square-root normalisation. In the work of [20], this is approximately calculated by applying Newton-Schulz iteration for root finding. It makes forward propagation computationally more efficient since only matrix multiplications are involved. The backward propagation still needs to solve a Lyapunov equation, which has the complexity at the same level of eigen-decomposition. After that, the work in [18] solves matrix square-rooting more efficiently. It proposes a sandwiched Newton-Schulz iteration and implements it via a set of layers with loop-embedded directed graph structure to obtain an approximate matrix square-root. It can be used for both forward and backward propagations.

Although the work in [18,20] achieves computational advantage and promising performance by using matrix square-root, their methods do not generalise to matrix normalisation with other power value p or other normalisation function f. In addition, the applicability of iterative Newton-Schulz equation to large-sized covariance matrix is unclear since only smaller-sized covariance matrices (i.e., of size 256×256) are used in that two works. These motivate us to mitigate the computational issue of matrix normalisation from other perspectives.

Finally, there are also research works to address the complexity of learning large-sized SPD representations by focusing on the parts other than matrix normalisation. Compact, low-rank and group bilinear pooling methods [7,13,38] address the high dimensionality of the feature representation after vectorising covariance matrix, and group convolution is used in [32] for the similar purpose. Linear transformation is designed in [1,10,24,32,35] to project large covariance matrices to more discriminative, smaller ones. To efficiently capture higher-order feature interactions, kernel pooling is developed [4]. Furthermore, the work in

---

[1] Let $\mathbf{C}$ be a SPD matrix and its eigen-decomposition is $\mathbf{C} = \mathbf{U}\mathbf{D}\mathbf{U}^{\top}$. The columns of $\mathbf{U}$ are eigenvectors while the diagonal of the diagonal matrix $\mathbf{D}$ consists of eigenvalues. Matrix normalisation with a function f is defined as $f(\mathbf{C}) = \mathbf{U}f(\mathbf{D})\mathbf{U}^{\top}$, where $f(\mathbf{D})$ means f is applied to each diagonal entry of $\mathbf{D}$. Matrix-logarithm and matrix-power based normalisations correspond to $f(x) = \log(x)$ and $f(x) = x^p$.

[2] develops a new learning framework to directly process manifold-valued data including covariance matrix. For our work, instead of competing with these works, it complements them and could be jointly used to mitigate the computational issue of eigen-decomposition of large SPD matrices when needed. In this work, we focus on the frameworks in [6,18–20] which typically employ matrix normalisation as an important step in learning SPD representation.

Dropout Schemes. Dropout [27] is a common regularisation technique that randomly drops neuron units from fully connected layers to improve generalisation. Several new schemes have extended this idea to convolutional layers. Spatial-Dropout [28] randomly drops feature channels. DropBlock [8] drops a block of pixels from convolutional feature maps. Weighted channel dropout [9] randomly drops feature channels with lower activations. Conceptually, the proposed "Relation Dropout (ReDro)" can be viewed as another scheme. Unlike the above ones, it randomly drops the covariance relationship of the channels across groups. It yields block-diagonal variants of a covariance matrix, and could produce dropout-like regularisation effect in training, as will be experimentally demonstrated.

3 The Proposed Relation Dropout (ReDro)

An overview of ReDro is in Fig. 1. From the left end, an image is fed into a CNN backbone, and the last convolutional feature map of d channels is routinely obtained. ReDro firstly conducts a random permutation of these channels and then evenly partitions them into k groups by following the channel number. Restricted to group i ($i = 1, 2, \cdots, k$), a smaller-sized covariance matrix $\mathbf{C}_i$ is computed and its eigen-decomposition is conducted as $\mathbf{C}_i = \mathbf{U}_i \mathbf{D}_i \mathbf{U}_i^\top$. The eigenvectors in $\mathbf{U}_1, \mathbf{U}_2, \cdots, \mathbf{U}_k$ are then arranged to form a larger, block-diagonal matrix $\mathbf{U}_b$. The similar procedure applies to the eigenvalues in $\mathbf{D}_1, \mathbf{D}_2, \cdots, \mathbf{D}_k$ to form $\mathbf{D}_b$. Note that $\mathbf{U}_b$ and $\mathbf{D}_b$ are just the eigen-decomposition of the $d \times d$ block-diagonal covariance matrix $\mathbf{C}_b = \mathrm{diag}(\mathbf{C}_1, \mathbf{C}_2, \cdots, \mathbf{C}_k)$. At the last step of ReDro, $\mathbf{U}_b$ and $\mathbf{D}_b$ is *permuted back* to the original order of the channels in the last convolutional feature map. This is important because it makes the random permutation *transparent* to subsequent network layers. This completes the proposed ReDro scheme.

In doing so, the eigen-decomposition of a covariance matrix, with part of the entries dropped, can be more efficiently obtained by taking advantage of the block-diagonal structure of $\mathbf{C}_b$. Then matrix normalisation can be readily conducted with any valid normalisation function.

3.1 Forward Propagation in the Presence of ReDro

Let $\mathbf{X}_{d \times (wh)}$ be a data matrix consisting of the d-dimensional local descriptors in the last convolutional feature map. Recall that d is the channel number while w and h are the width and height of the feature map. A random partitioning of

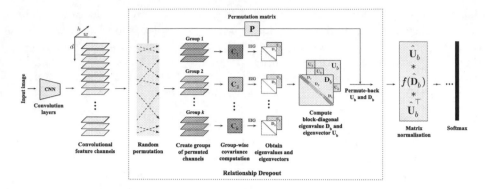

Fig. 1. Proposed relation dropout (ReDro) scheme for mitigating the computational issue of matrix normalisation in learning large-sized SPD visual representation.

the d channels can be represented by k index sets, $\mathcal{G}_1, \mathcal{G}_2, \cdots, \mathcal{G}_k$, which contain the IDs of the channels in each group, respectively.

Let a roster of all the IDs in these k index sets be $\{r_1, r_2, \cdots, r_d\}$. It is a permutation of the original channel IDs $\{1, 2, \cdots, d\}$, and therefore induces a permutation matrix $\mathbf{P}_{d \times d}$.[2] The ith row of $\mathbf{P}$ is $\mathbf{e}_{r_i}^\top = (0, \cdots, 0, 1, 0, \cdots, 0)$, which is a standard basis vector with "1" at its r_ith entry and zeros elsewhere. The effect of $\mathbf{P}$ can be intuitively interpreted. By left-multiplying $\mathbf{P}$ to $\mathbf{X}$, the rows of $\mathbf{X}$ will be permuted. A more intuitive interpretation, which will be used later, is that it permutes the d axes of an original coordinate frame $\mathcal{F}$ to form another new frame $\mathcal{F}'$. For a quantity (e.g., eigenvector) represented in $\mathcal{F}'$, we can inverse the permutation by left-multiplying $\mathbf{P}^{-1}$ to it. It is known in matrix analysis that for any permutation matrix $\mathbf{P}$, it satisfies $\mathbf{P}\mathbf{P}^\top = \mathbf{I}$. Therefore, $\mathbf{P}^{-1}$ can be trivially obtained as $\mathbf{P}^\top$. This result will be used shortly.

Now, for the channels within each group $\mathcal{G}_i$ $(i = 1, \cdots, k)$, we compute a covariance matrix $\mathbf{C}_i$. Collectively, they form a $d \times d$ block-diagonal matrix

$$\mathbf{C}_b = \mathrm{diag}(\mathbf{C}_1, \mathbf{C}_2, \cdots, \mathbf{C}_k). \tag{1}$$

Let the eigen-decomposition of $\mathbf{C}_i$ be $\mathbf{C}_i = \mathbf{U}_i \mathbf{D}_i \mathbf{U}_i^\top$. It is well-known by matrix analysis that the eigen-decomposition of $\mathbf{C}_b$ can be expressed as [23]

$$\mathbf{C}_b = \mathbf{U}_b \mathbf{D}_b \mathbf{U}_b^\top, \tag{2}$$

$$\mathbf{U}_b = \mathrm{diag}(\mathbf{U}_1, \mathbf{U}_2, \cdots, \mathbf{U}_k) \quad \text{and} \quad \mathbf{D}_b = \mathrm{diag}(\mathbf{D}_1, \mathbf{D}_2, \cdots, \mathbf{D}_k). \tag{3}$$

Note that the eigenvectors in $\mathbf{U}_b$ are obtained in the new coordinate frame $\mathcal{F}'$. To retrieve their counterparts, $\hat{\mathbf{U}}_b$, in the original frame $\mathcal{F}$ (i.e., corresponding to the original order of the d channels), we apply the inverse permutation as

$$\hat{\mathbf{U}}_b = \mathbf{P}^{-1} \mathbf{U}_b = \mathbf{P}^\top \mathbf{U}_b. \tag{4}$$

---

[2] In matrix analysis, a permutation matrix $\mathbf{P}$ is a square binary matrix. It has one and only one "1" entry in each row and each column, with all the remainder being "0". It is easy to verify that $\mathbf{P}\mathbf{P}^\top = \mathbf{P}^\top\mathbf{P} = \mathbf{I}$, where $\mathbf{I}$ is an identity matrix.

The eigenvalue matrix $\mathbf{D}_b$ does not need to be permuted back because an eigenvalue represents the data variance along the corresponding eigenvector. It is not affected by the permutation of the coordinate axes.

In this way, we obtain the eigen-decomposition of $\mathbf{C}_b$ *corresponding to the original order of the channels (i.e, $1, 2, \cdots, d$)* as

$$\hat{\mathbf{U}}_b = \mathbf{P}^\top \cdot \mathrm{diag}(\mathbf{U}_1, \mathbf{U}_2, \cdots, \mathbf{U}_k); \quad \hat{\mathbf{D}}_b = \mathrm{diag}(\mathbf{D}_1, \mathbf{D}_2, \cdots, \mathbf{D}_k). \quad (5)$$

With this result, matrix normalisation with any valid function f can now be applied. Algorithm 1 summarises the steps of the proposed ReDro scheme.

Algorithm 1. Relation Dropout (ReDro)

Input : Convolutional feature map $\mathbf{X}_{d \times (wh)}$; The number of groups k.
1. Randomly partition the d channels to groups $\mathcal{G}_1, \mathcal{G}_2, \cdots, \mathcal{G}_k$;
2. Create the permutation matrix $\mathbf{P}$ accordingly;
3. **foreach** *group $\mathcal{G}_i$* **do**
  | 1. Compute the covariance matrix $\mathbf{C}_i$;
  | 2. Calculate its eigen-decomposition $\mathbf{C}_i = \mathbf{U}_i \mathbf{D}_i \mathbf{U}_i^\top$;
  end
4. Form the eigenvectors and eigenvalues of the block-diagonal matrix $\mathbf{C}_b$:
$\mathbf{U}_b = \mathrm{diag}(\mathbf{U}_1, \mathbf{U}_2, \cdots, \mathbf{U}_k)$, $\mathbf{D}_b = \mathrm{diag}(\mathbf{D}_1, \mathbf{D}_2, \cdots, \mathbf{D}_k)$;
5. By permuting back, the eigen-decomposition of $\mathbf{C}_b$ *corresponding to the original order of the d channels* are $\hat{\mathbf{U}}_b = \mathbf{P}^\top \mathbf{U}_b$ and $\hat{\mathbf{D}}_b = \mathbf{D}_b$.

3.2 Backward Propagation in the Presence of ReDro

To derive the gradients for back-propagation, the composition of functions from the feature map $\mathbf{X}$ to the objective function $J(\mathbf{X})$ is illustrated as follows.

$$\mathbf{X} \rightarrow \underbrace{\mathbf{PX}}_{\mathbf{Y}} \rightarrow \underbrace{(\mathbf{YY}^\top) \circ \mathbf{S}}_{C_b} \rightarrow \underbrace{(\mathbf{P}^\top \mathbf{C}_b \mathbf{P})}_{\mathbf{A}(\text{Auxiliary})} \rightarrow \underbrace{f(\mathbf{A})}_{\mathbf{Z}} \rightarrow \cdots \text{layers} \cdots \rightarrow \underbrace{J(\mathbf{X})}_{\text{Objective}} \quad (6)$$

$\mathbf{P}$ is the permutation matrix. $\mathbf{Y}$ is the feature map with its channels permutated. $\mathbf{S} = \mathrm{diag}(\mathbf{1}_1, \mathbf{1}_2, \cdots, \mathbf{1}_k)$ is a block-diagonal binary matrix, where $\mathbf{1}_i$ is a square matrix of all "1"s and its size is the same as that of channel group $\mathcal{G}_i$. Noting that "$\circ$" denotes element-wise multiplication, $\mathbf{S}$ represents the selection of $\mathbf{C}_b$ out of the full covariance matrix $\mathbf{YY}^\top$. The letter under each term is used to assist derivation. $\mathbf{A}$ is an auxiliary variable and not computed in practice. $f(\mathbf{A})$ is the step of matrix normalisation. Its result $\mathbf{Z}$ is used by the subsequent fully connected and softmax layers. Our goal is to work out $\frac{\partial J}{\partial \mathbf{X}}$. Once obtained, all gradients before the convolutional feature map $\mathbf{X}$ can be routinely obtained.

In the literature, the seminal work in [11] and other work such as [6] demonstrate the rules of differentiation for matrix-valued functions. By these rules, we derive the following results (details are provided in the supplement).

$$\frac{\partial J}{\partial \mathbf{A}} = \hat{\mathbf{U}}_b \left(\mathbf{G} \circ \left(\hat{\mathbf{U}}_b^\top \frac{\partial J}{\partial \mathbf{Z}} \hat{\mathbf{U}}_b \right) \right) \hat{\mathbf{U}}_b^\top; \quad \frac{\partial J}{\partial \mathbf{C}_b} = \mathbf{P} \frac{\partial J}{\partial \mathbf{A}} \mathbf{P}^\top;$$

$$\frac{\partial J}{\partial \mathbf{Y}} = \left(\mathbf{S} \circ \left(\frac{\partial J}{\partial \mathbf{C}_b} + \left(\frac{\partial J}{\partial \mathbf{C}_b} \right)^\top \right) \right) \mathbf{Y}; \quad \frac{\partial J}{\partial \mathbf{X}} = \mathbf{P}^\top \frac{\partial J}{\partial \mathbf{Y}}, \qquad (7)$$

where $\mathbf{G}$ is a $d \times d$ matrix defined based on $\hat{\mathbf{D}}_b$.[3] As seen, given $\frac{\partial J}{\partial \mathbf{Z}}$, we can work out $\frac{\partial J}{\partial \mathbf{A}}$, $\frac{\partial J}{\partial \mathbf{C}_b}$, $\frac{\partial J}{\partial \mathbf{Y}}$ and then $\frac{\partial J}{\partial \mathbf{X}}$. In the whole course, only $\hat{\mathbf{U}}_b$, $\hat{\mathbf{D}}_b$, $\mathbf{P}$ and $\mathbf{S}$ are needed. The first two have been efficiently obtained via the proposed scheme ReDro in Sect. 3.1, while the latter two are known once the random grouping is done. Now, we have all the essential results for back-propagation. An end-to-end learning with ReDro can be readily implemented.

3.3   Discussion

Computational Savings. As aforementioned, a computational bottleneck in SPD representation learning is the eigen-decomposition of a $d \times d$ full covariance matrix. Generally, its complexity is at the order of $\mathcal{O}(d^3)$.[4] Without loss of generality, assuming that d can exactly be divided by the number of groups k, the size of each group will be d/k. The complexity incurred by ReDro, which conducts eigen-decomposition of the k smaller-sized covariance matrices, will be $\mathcal{O}(d^3/k^2)$. Therefore, in the theoretical sense, the proposed ReDro scheme can save the computation up to k^2 times.

The implementation of ReDro only needs a random permutation of the IDs of the d channels. For the gradient computation in Eq. (7), it appears that compared with the case of using a full covariance matrix, ReDro incurs extra computation involving the multiplication of $\mathbf{P}$ or $\mathbf{S}$ in $\frac{\partial J}{\partial \mathbf{C}_b}$, $\frac{\partial J}{\partial \mathbf{Y}}$ and $\frac{\partial J}{\partial \mathbf{X}}$ (Note that $\frac{\partial J}{\partial \mathbf{A}}$ needs to be computed for any eigen-decomposition based matrix normalisation, even if ReDro is not used). Nevertheless, both $\mathbf{P}$ and $\mathbf{S}$ are binary matrices simply induced by the random permutation. Their multiplication with other variables can be trivially implemented, incurring little computational overhead.

Two Key Parameters. One key parameter is the number of groups, k. Since the improvement on computational efficiency increases quadratically with k, a larger k would be preferred. Meanwhile, it is easy to see that the percentage of the entries dropped by ReDro is $(1 - \frac{1}{k}) \times 100\%$. A larger k will incur more significant loss of information. As a result, a value of k balancing these two aspects shall be used, which will be demonstrated in the experimental study.

When k is given, the other key parameter is the "intensity" of conducting the random permutation. Doing it for every training sample leads to the most intensive change of dropout pattern. As will be shown, this could make the

---

[3] For the matrix $\mathbf{G}$, its (i,j)th entry g_{ij} is defined as $\frac{f(\lambda_i) - f(\lambda_j)}{\lambda_i - \lambda_j}$ if $\lambda_i \neq \lambda_j$ and $f'(\lambda_i)$ otherwise, where λ_i is the ith diagonal element of $\hat{\mathbf{D}}_b$.

[4] For a symmetric matrix, the complexity of eigen-decomposition could be improved up to the order of $\mathcal{O}(d^{2.38})$ by more sophisticated algorithms though [5].

objective function fluctuate violently, affecting the convergence. To show the impact of this intensity, we will experiment the random permutation at three levels, namely, epoch-level (EL), batch-level (BL) and sample-level (SL) and hold it for various intervals. For example, batch-level with interval of 2 (i.e., "BL-2") uses the same random permutation for two consecutive batches before refreshed. Similarly, SL-1 conducts the random permutation for every training sample.

In addition, ReDro could bring less biased eigenvalue estimate. As known, when eigen-decomposition is applied to a large, full covariance matrix, eigenvalue estimation will be considerably biased (i.e., larger/smaller eigenvalues are estimated to be over-large/over-small) when samples are not sufficient. When ReDro is used, eigenvalues are estimated from each block sub-matrix on the diagonal. Because they are smaller in size, eigenvalue estimate could become less biased. This property can be regarded as a by-product of the ReDro scheme.

4   Experimental Result

We conduct extensive experiments on scene classification and fine-grained image classification to investigate the proposed ReDro scheme. For scene classification, *MIT Indoor* dataset [26] is used. For fine-grained image classification, the commonly used *Birds* [34], *Airplanes* [22] and *Cars* [17] datasets are tested. For all datasets, the original training and testing protocols are followed, and we do not utilise any bounding box or part annotations. Following the literature [18,20], we resize all images to 448×448 during training and testing. More details on datasets and implementation of ReDro are provided in the supplement material.

This experiment consists of three main parts, with an additional ablation study. Section 4.1 shows the computational advantage brought by ReDro. Section 4.2 investigates the efficiency of ReDro versus the intensity level at which it is applied. Section 4.3 validates the performance of ReDro via multiple typical SPD representation learning methods that explicitly use matrix normalisation.

4.1   On the Computational Advantage of ReDro

This part compares the computational cost between the case without ReDro and the case using ReDro. The former means a full covariance matrix is used.

Specifically, we compare with four typical SPD representation learning methods, namely, MPN-COV [19], DeepCOV [6], Improved BCNN (IBCNN) [20] and iSQRT-COV [18]. The first two methods conduct matrix normalisation via eigen-decomposition, while the last two realise normalisation more efficiently by matrix square-rooting. These four methods represent the case without using ReDro. To compare with them, we implement ReDro within the DeepCOV method with ResNet-50 network as backbone. When doing this, we only modify its COV layers with the proposed ReDro scheme, leaving all the other settings unchanged.

Since the layers of covariance computation and matrix normalisation in these methods are often tightly coupled, it is difficult to exclude the normalisation step to exactly compare with ReDro for computational cost. To be fair, the total

Table 1. Comparison of computational time (in second) for covariance estimation and matrix normalisation by using or not using the proposed ReDro scheme. The reported time is the sum of forward and backward propagation (individual propagation time is given in the supplement). The four methods to the left represent the case not using ReDro. The case using ReDro is implemented upon DeepCOV [6] with various k. The boldface shows that ReDro saves computational time

Matrix dimension	No ReDro used				DeepCOV [6] using ReDro			
	MPN -COV [19]	Deep -COV [6]	IBCNN [20]	iSQRT -COV [18]	with $k=2$	with $k=4$	with $k=8$	with $k=16$
128×128	0.004	0.004	0.006	0.001	0.007	0.009	0.011	0.015
256×256	0.013	0.013	0.014	0.006	0.011	0.011	0.013	0.020
512×512	0.031	0.031	0.032	0.030	0.030	**0.022**	**0.023**	0.030
1024×1024	0.097	0.097	0.121	0.097	**0.090**	**0.076**	**0.056**	**0.062**

time taken by both the steps of ReDro and matrix normalisation is used for comparison. To test the computational cost over covariance matrix of various sizes, we follow the literature to apply an additional 1×1 convolutional layer to reduce the number of channels when necessary. The comparison is conducted on a computer with a Tesla P100 GPU, 12-core CPU and 12 GB RAM. ReDro is implemented with MatConvNet library [31] on Matlab 2019a.

Table 1 shows the timing result. Firstly, along the size of covariance matrix, we can observe that i) when the size is relatively smaller (i.e., 256×256), ReDro is unnecessary. This is because the eigen-decomposition does not incur significant computation. Rigidly using ReDro in this case complicates the procedure, adding computational overhead; ii) however, when the matrix size increases to 512×512, the advantage of ReDro emerges, and becomes more pronounced for 1024×1024. In these cases, ReDro is computationally more efficient by 20%–50% than the no-ReDro counterparts (see the results in bold). In addition, in terms of the total training time, ReDro can significantly shorten the time from 52.4 h down to 24.3 h on Birds dataset, with 1.3% point improvement on classification accuracy, as will be detailed in Fig. 2.

Secondly, we test different group number k in ReDro. As seen, ReDro shows higher computational efficiency when $k = 4$ or 8. For $k = 2$, we can expect from the previous complexity analysis that its efficiency shall be lower. For $k = 16$, the overhead for processing more matrices, although in smaller sizes, becomes non-trivial. Since $k = 4$ gives rise to a high computational efficiency (and overall good classification, as will be shown later), the following two experiments will focus on this setting, with more discussion on k left to the ablation study.

4.2   On the Efficiency of ReDro Versus Its Intensity Level

Based on the findings in the last experiment, we focus on testing with 512×512-and 1024×1024-sized matrices in this experiment.

Table 2. Results using ReDro with $k = 4$ at different intensity levels on DeepCOV [6] (left) and DeepCOV-ResNet (right). The results higher than the baseline are shown in bold. The highest results for each dataset are marked by asterisks

		DeepCOV [6] (512 × 512)				DeepCOV-ResNet (1024 × 1024)			
		MIT	Airplane	Cars	Birds	MIT	Airplane	Cars	Birds
No ReDro		79.2	88.7	91.7	85.4	83.4	83.9	85.0	86.0
ReDro with $k=4$	EL-1	**80.5***	**89.1**	**92.2***	**86.5**	**84.0***	**85.4**	**88.9**	**86.2***
	EL-2	**80.5***	**88.9**	**92.1**	**86.5**	–	–	–	–
	EL-5	**80.1**	**89.0**	91.7	**86.0**	**83.6**	**85.8***	**89.8***	86.0
	EL-10	**80.3**	88.6	88.9	**85.5**	–	–	–	–
	EL-20	**79.9**	88.6	91.3	85.0	81.5	**85.3**	**87.9**	85.9
	EL-50	**79.4**	**89.0**	91.0	**85.8**	–	–	–	–
	EL-100	**80.2**	**89.2***	90.8	**86.1**	–	–	–	–
	BL-1	76.8	**88.8**	87.5	**86.6**	**84.0***	**85.2**	**87.9**	**86.2***
	BL-2	76.9	**89.1**	88.0	**86.4**	–	–	–	–
	BL-5	76.6	**89.1**	87.8	**86.6**	83.8	**85.2**	**88.0**	**86.1**
	BL-10	76.8	**89.0**	88.0	**86.5**	–	–	–	–
	BL-20	78.6	88.7	87.5	**86.5**	**84.0***	**85.5**	**87.9**	**86.1**
	SL-1	77.7	**88.9**	90.0	**86.6**	83.8	**85.1**	84.8	86.0
	SL-3	78.1	**89.0**	89.8	**86.5**	–	–	–	–
	SL-6	79.1	**89.2***	91.1	**86.7***	83.8	**85.2**	**86.2**	**86.1**

This experiment investigates the impact of intensity level, at which ReDro is applied, to classification improvement. Again, we implement ReDro with Deep-COV [6]. Specifically, we train DeepCOV by applying ReDro with 15 intensity levels, i.e., EL-$\{100, 50, 20, 10, 5, 2, 1\}$, BL-$\{20, 10, 5, 2, 1\}$ and SL-$\{6, 3, 1\}$, whose meaning is explained in Sect. 3.3. The obtained classification accuracy is compared with those of the original DeepCOV reported in [6].

The left part of Table 2 shows the results when the covariance matrix size is 512×512. Firstly, along the intensity level, we can observe that, i) the efficiency of ReDro indeed varies with the intensity level; ii) multiple intensity levels lead to improved classification over the baseline "No ReDro." In particular, EL-1 and EL-2 work overall better than others across all datasets; iii) also, EL-1 works well with MIT and Cars, and SL-6 works well with Airplane and Birds though. This suggests that for some datasets, less intensive change of random permutation, i.e., at epoch level (EL), is preferred. We observe that in this case, applying more intensive change, i.e., at sample level (SL), could cause objective function to fluctuate violently and affect convergence.

Secondly, along the datasets, we can observe that DeepCOV trained with ReDro improves classification over its baseline on all datasets, with the magnitude of 0.5–1.3%. For example, on Birds with the intensity level of SL-6, ReDro improves the accuracy from the baseline 85.4% to 86.7%.

Besides improving classification, ReDro at various intensity levels also shortens the total network training time. At each epoch of training DeepCOV, ReDro

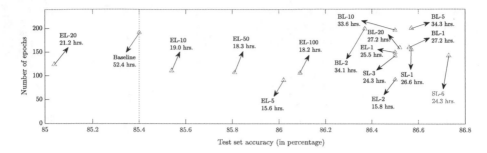

Fig. 2. Comparison of the classification error, network training time, and the total number of epochs obtained by using ReDro at various intensity levels. Each small triangle in this figure represents a ReDro case. The arrow points to its intensity level and the network training time taken to achieve its highest test accuracy, as reflected by the x-axis. The result in red indicates the best performer. Birds dataset is used. (Color figure online)

saves about 40% of the GPU time. We find that it also helps network to converge faster. Figure 2 shows that DeepCOV with ReDro achieves lower classification error with a smaller number of epochs than the baseline "No ReDro" on Birds. As seen, i) when ReDro is used, training the network, including the one achieving the highest accuracy, generally takes about half of the time of the baseline (indicated by the vertical dotted line), regardless of the intensity level; ii) with ReDro, to train a network to achieve comparable performance with the baseline, it only takes about 36% (i.e, down from 52.4 h to 19.0 h) of the baseline training time. The similar observation is seen from other datasets besides Birds.

The right part of Table 2 shows the result when the covariance matrix size is 1024×1024. Note that no existing networks have ever tried such a large matrix. We use DeepCOV network with ResNet-50 as the backbone, and apply 1×1 convolution to reduce the number of feature channels from 2048 to 1024. We call this network "DeepCOV-ResNet." Due to the longer training period caused by the larger covariance matrix, we sample eight out of the 15 intensity levels in the table and test them. From the results, we can observe that, i) except for a few intensity levels, DeepCOV-ResNet with ReDro outperforms the baseline consistently across all datasets; ii) the improvement varies between 0.7–4.8% (This largest improvement is achieved by ReDro with EL-5 on Cars). The improvement could be attributed to two factors: a) The dropout-like regularisation effect in ReDro helps the network learn better features, and b) ReDro estimates smaller-sized covariance matrices, instead of a full $d \times d$ one, from the local descriptors in the convolutional feature map. This helps to mitigate the bias issue in the estimation of the eigenvalues of covariance matrix.

4.3   On the Performance of ReDro with Typical Methods

The above experiments verify the efficiency of ReDro by integrating it into the DeepCOV method. Now, we further integrate ReDro into other typical SPD representation methods that use matrix normalisation as an important step, namely,

Improved BCNN [20], MPN-COV [19] and iSQRT-COV [18]. Since iSQRT-COV uses matrix square-root normalisation without involving eigen-decomposition, we investigate the regularisation effect of ReDro for it. This also applies to BCNN [21] which does not have the matrix normalisation step. Since most of these methods are originally proposed for fine-grained image classification, we focus on the datasets of Airplane, Cars and Birds.

Table 3 shows the results. In total, six scenarios are implemented with these methods, using differently sized covariance matrix. For each scenario, ReDro at the same intensity level is used. We compare the baseline "No ReDro" to the cases in which ReDro is integrated. As seen, i) except iSQRT-COV, networks trained with ReDro generally outperform their baseline counterparts. ReDro-based iSQRT-COV is comparable with the baseline; ii) overall, the improvement is consistent across all datasets. The improvement higher than 1% can be commonly seen, with the maximum one being 4.8%; iii) For the results from ReDro with $k = 2$ and 4, they are overall comparable or the latter is slightly better (e.g., higher in 10 out of the total 18 results). Taking the computational saving into account, $k = 4$ is a better option.

Additionally, we provide some explanation to the performance of iSQRT-COV. All the methods except iSQRT-COV use a backbone model that is pretrained *without* incorporating the SPD representation layers. These layers are only incorporated later and then fine-tuned with a fine-grained image dataset. Differently, the pretrained backbone model in iSQRT-COV has incorporated the SPD representation layers that are further fine-tuned with the fine-grained image dataset. This helps iSQRT-COV achieve more promising performance. Meanwhile, as noted previously, iSQRT-COV [18] utilises a special matrix normalisation without involving eigen-decomposition. Our ReDro could provide iSQRT-COV with extra potential in efficiently utilising the eigen-decomposition based matrix normalisation to access more normalisation functions.

Table 3. Results using ReDro in typical SPD visual representation methods. The results higher than the baseline (indicated with "No ReDro") are shown in bold. The IBCNN [20], BCNN [21] and iSQRT-COV [18] networks are trained (including the baseline) with the settings in their original papers. As for MPN-COV [19], it is trained with the same pretrained network as IBCNN, BCNN and DeepCOV for consistency. Note that to ensure a fair of comparison, we report the classification result obtained by softmax predictions and do not utilise the additional step of training a separate SVM classifier. The highest results on each method are marked by asterisks.

Dataset	Training mode	DeepCOV -ResNet (1024 × 1024)	DeepCOV (512 × 512)	IBCNN (512 × 512)	BCNN (512 × 512)	MPN -COV (256 × 256)	iSQRT -COV (256 × 256)
Airp.	No ReDro	83.9	88.7	87.0	85.3	86.1	91.1
	ReDro ($k = 2$)	**85.3**	**89.3***	**88.8***	**86.6***	**87.4**	90.6
	ReDro ($k = 4$)	**85.8***	**89.2**	**88.6**	**86.6***	**88.2***	91.1
Cars	No ReDro	85.0	91.7	90.6	89.1	89.8	92.6
	ReDro ($k = 2$)	**88.2**	90.8	**92.6***	**90.9***	**91.4**	92.3
	ReDro ($k = 4$)	**89.8***	**92.2***	**91.2**	**90.5**	**91.7***	92.6
Birds	No ReDro	86.0	85.4	85.4	84.1	82.9	88.5
	ReDro ($k = 2$)	85.9	**86.5**	**85.5***	**84.6***	**83.5***	88.0
	ReDro ($k = 4$)	**86.2***	**86.7***	84.6	83.9	83.2	**88.6***

Table 4. Impact of group number k in ReDro. The Birds dataset is used

Matrix size	No ReDro	Value of k for ReDro				
		2	3	4	8	16
1024×1024	86.0	85.9	**86.2***	**86.2***	85.3	82.9
512×512	85.6	**85.8**	**85.8**	**85.9***	85.5	83.2
256×256	85.4	**85.6**	**85.7***	**85.5**	83.9	81.5

4.4 Ablation Study on the Group Number k

To gain more understanding on the group number k in ReDro, the following experiment is conducted. With DeepCOV-ResNet, ReDro using different values of k is implemented. Table 4 shows the results. As seen, i) for all matrix sizes, $k = \{3, 4\}$ yields overall better performance than the baseline; ii) larger values of k, i.e., 8 and 16, produce inferior results because they drop a significant amount of information from the covariance matrix.

5 Conclusion

This paper proposes a novel scheme called Relation Dropout (ReDro) for reducing the computational cost in learning large-sized SPD visual representation by deep neural networks. Focusing on the step of matrix normalisation, it utilises the nice property of a block-diagonal matrix to facilitate the eigen-decomposition often required in this process. A detailed description of the proposed scheme including its forward and backward propagations is provided. Extensive experiments are conducted on multiple benchmark datasets with various settings. The result demonstrates the improvement brought by the proposed scheme on both computational efficiency and classification accuracy, when working with the SPD visual representation learning methods typically involving matrix normalisation. This ReDro scheme can be readily inserted into a network to function, without the need to alter the network architecture or the loss function.

In the future work, we will investigate the effectiveness of the ReDro scheme for more methods designed for learning SPD visual representation and gain more insight into its regularisation effect.

Acknowledgement. This work was supported by the CSIRO Data61 Scholarship; the University of Wollongong Australia IPTA scholarship; the Australian Research Council (grant number DP200101289); and the Multi-modal Australian ScienceS Imaging and Visualisation Environment (MASSIVE).

References

1. Brooks, D., Schwander, O., Barbaresco, F., Schneider, J.Y., Cord, M.: Riemannian batch normalization for SPD neural networks. In: Advances in Neural Information Processing Systems, pp. 15489–15500 (2019)

2. Chakraborty, R., Bouza, J., Manton, J., Vemuri, B.C.: A deep neural network for manifold-valued data with applications to neuroimaging. In: Chung, A.C.S., Gee, J.C., Yushkevich, P.A., Bao, S. (eds.) IPMI 2019. LNCS, vol. 11492, pp. 112–124. Springer, Cham (2019). https://doi.org/10.1007/978-3-030-20351-1_9
3. Cimpoi, M., Maji, S., Vedaldi, A.: Deep filter banks for texture recognition and segmentation. In: Proceedings of the International Conference on Computer Vision and Pattern Recognition, pp. 3828–3836. IEEE (2015)
4. Cui, Y., Zhou, F., Wang, J., Liu, X., Lin, Y., Belongie, S.: Kernel pooling for convolutional neural networks. In: Proceedings of the IEEE Conference on Computer Vision and Pattern Recognition, pp. 2921–2930 (2017)
5. Demmel, J., Dumitriu, I., Holtz, O.: Fast linear algebra is stable. Numer. Math. **108**(1), 59–91 (2007)
6. Engin, M., Wang, L., Zhou, L., Liu, X.: DeepKSPD: learning kernel-matrix-based SPD representation for fine-grained image recognition. In: Ferrari, V., Hebert, M., Sminchisescu, C., Weiss, Y. (eds.) ECCV 2018. LNCS, vol. 11206, pp. 629–645. Springer, Cham (2018). https://doi.org/10.1007/978-3-030-01216-8_38
7. Gao, Y., Beijbom, O., Zhang, N., Darrell, T.: Compact bilinear pooling. In: Proceedings of the Conference on Computer Vision and Pattern Recognition, pp. 317–326. IEEE (2016)
8. Ghiasi, G., Lin, T.Y., Le, Q.V.: Dropblock: a regularization method for convolutional networks. In: Proceedings of the Advances in Neural Information Processing Systems, pp. 10727–10737 (2018)
9. Hou, S., Wang, Z.: Weighted channel dropout for regularization of deep convolutional neural network. Proc. AAAI Conf. Artif. Intell. **33**, 8425–8432 (2019)
10. Huang, Z., Van Gool, L.: A Riemannian network for SPD matrix learning. In: Thirty-First AAAI Conference on Artificial Intelligence (2017)
11. Ionescu, C., Vantzos, O., Sminchisescu, C.: Matrix backpropagation for deep networks with structured layers. In: Proceedings of the International Conference on Computer Vision, pp. 2965–2973. IEEE (2015)
12. Jayasumana, S., Hartley, R., Salzmann, M., Li, H., Harandi, M.: Kernel methods on Riemannian manifolds with Gaussian RBF kernels. IEEE Trans. Pattern Anal. Mach. Intell. **37**(12), 2464–2477 (2015)
13. Kong, S., Fowlkes, C.: Low-rank bilinear pooling for fine-grained classification. In: Proceedings of the IEEE Conference on Computer Vision and Pattern Recognition, pp. 365–374 (2017)
14. Koniusz, P., Wang, L., Cherian, A.: Tensor representations for action recognition. IEEE Trans. Pattern Anal. Mach. Intell. (2020)
15. Koniusz, P., Zhang, H.: Power normalizations in fine-grained image, few-shot image and graph classification. IEEE Trans. Pattern Anal. Mach. Intell. (2020)
16. Koniusz, P., Zhang, H., Porikli, F.: A deeper look at power normalizations. In: 2018 IEEE Conference on Computer Vision and Pattern Recognition, CVPR 2018, Salt Lake City, UT, USA, 18–22 June 2018, pp. 5774–5783. IEEE Computer Society (2018). https://doi.org/10.1109/CVPR.2018.00605
17. Krause, J., Stark, M., Deng, J., Fei-Fei, L.: 3D object representations for fine-grained categorization. In: Proceedings of the International Conference on Computer Vision Workshops, pp. 554–561. IEEE (2013)
18. Li, P., Xie, J., Wang, Q., Gao, Z.: Towards faster training of global covariance pooling networks by iterative matrix square root normalization. In: Proceedings of the Conference on Computer Vision and Pattern Recognition, pp. 947–955. IEEE (2018)

19. Li, P., Xie, J., Wang, Q., Zuo, W.: Is second-order information helpful for large-scale visual recognition? In: Proceedings of the International Conference on Computer Vision, pp. 2070–2078. IEEE (2017)
20. Lin, T.Y., Maji, S.: Improved Bilinear Pooling with CNNs. arXiv preprint arXiv:1707.06772 (2017)
21. Lin, T.Y., RoyChowdhury, A., Maji, S.: Bilinear CNN models for fine-grained visual recognition. In: Proceedings of the International Conference on Computer Vision, pp. 1449–1457. IEEE (2015)
22. Maji, S., Rahtu, E., Kannala, J., Blaschko, M., Vedaldi, A.: Fine-grained visual classification of aircraft. arXiv preprint arXiv:1306.5151 (2013)
23. Ng, A.Y., Jordan, M.I., Weiss, Y.: On spectral clustering: analysis and an algorithm. In: Proceedings of the Advances in Neural Information Processing Systems, pp. 849–856 (2002)
24. Nguyen, X.S., Brun, L., Lézoray, O., Bougleux, S.: A neural network based on SPD manifold learning for skeleton-based hand gesture recognition. In: Proceedings of the IEEE Conference on Computer Vision and Pattern Recognition, pp. 12036–12045 (2019)
25. Porikli, F., Tuzel, O., Meer, P.: Covariance tracking using model update based on lie algebra. In: Proceedings of the Conference on Computer Vision and Pattern Recognition, vol. 1, pp. 728–735. IEEE (2006)
26. Quattoni, A., Torralba, A.: Recognizing indoor scenes. In: Proceedings of the Conference on Computer Vision and Pattern Recognition, pp. 413–420. IEEE (2009)
27. Srivastava, N., Hinton, G., Krizhevsky, A., Sutskever, I., Salakhutdinov, R.: Dropout: a simple way to prevent neural networks from overfitting. J. Mach. Learn. Res. 15(1), 1929–1958 (2014)
28. Tompson, J., Goroshin, R., Jain, A., LeCun, Y., Bregler, C.: Efficient object localization using convolutional networks. In: Proceedings of the Conference on Computer Vision and Pattern Recognition, pp. 648–656. IEEE (2015)
29. Tuzel, O., Porikli, F., Meer, P.: Region covariance: a fast descriptor for detection and classification. In: Leonardis, A., Bischof, H., Pinz, A. (eds.) ECCV 2006. LNCS, vol. 3952, pp. 589–600. Springer, Heidelberg (2006). https://doi.org/10.1007/11744047_45
30. Tuzel, O., Porikli, F., Meer, P.: Human detection via classification on riemannian manifolds. In: Proceedings of the Conference on Computer Vision and Pattern Recognition, pp. 1–8. IEEE (2007)
31. Vedaldi, A., Lenc, K.: MatConvNet: convolutional neural networks for Matlab. In: Proceedings of the 23rd ACM International Conference on Multimedia, pp. 689–692 (2015)
32. Wang, Q., Xie, J., Zuo, W., Zhang, L., Li, P.: Deep CNNs meet global covariance pooling: better representation and generalization. IEEE Trans. Pattern Anal. Mach. Intell. (2020)
33. Wang, R., Guo, H., Davis, L.S., Dai, Q.: Covariance discriminative learning: A natural and efficient approach to image set classification. In: Proceedings of the Conference on Computer Vision and Pattern Recognition, pp. 2496–2503. IEEE (2012)
34. Welinder, P., et al.: Caltech-UCSD birds 200. Technical report CNS-TR-2010-001, California Institute of Technology (2010)
35. Yu, K., Salzmann, M.: Second-order convolutional neural networks. arXiv preprint arXiv:1703.06817 (2017)

36. Zhang, H., Zhang, J., Koniusz, P.: Few-shot learning via saliency-guided hallu-
    cination of samples. In: Proceedings of the Conference on Computer Vision and
    Pattern Recognition, pp. 2770–2779. IEEE (2019)
37. Zhang, H., Zhang, L., Qui, X., Li, H., Torr, P.H.S., Koniusz, P.: Few-shot action
    recognition with permutation-invariant attention. In: Proceedings of the European
    Conference on Computer Vision (ECCV) (2020)
38. Zheng, H., Fu, J., Zha, Z.J., Luo, J.: Learning deep bilinear transformation for
    fine-grained image representation. In: Proceedings of the Advances in Neural Infor-
    mation Processing Systems, pp. 4279–4288 (2019)
39. Zhu, X., Xu, C., Hui, L., Lu, C., Tao, D.: Approximated bilinear modules for
    temporal modeling. In: Proceedings of the International Conference on Computer
    Vision, pp. 3494–3503. IEEE (2019)

Graph-Based Social Relation Reasoning

Wanhua Li[1,2,3], Yueqi Duan[1,2,3,5], Jiwen Lu[1,2,3]([✉]), Jianjiang Feng[1,2,3], and Jie Zhou[1,2,3,4]

[1] Department of Automation, Tsinghua University, Beijing, China
li-wh17@mails.tsinghua.edu.cn, {lujiwen,jfeng,jzhou}@tsinghua.edu.cn
[2] State Key Lab of Intelligent Technologies and Systems, Beijing, China
[3] Beijing National Research Center for Information Science and Technology, Beijing, China
[4] Tsinghua Shenzhen International Graduate School, Tsinghua University, Beijing, China
[5] Stanford University, Stanford, USA
duanyq19@stanford.edu

Abstract. Human beings are fundamentally sociable—that we generally organize our social lives in terms of relations with other people. Understanding social relations from an image has great potential for intelligent systems such as social chatbots and personal assistants. In this paper, we propose a simpler, faster, and more accurate method named graph relational reasoning network (GR^{2}N) for social relation recognition. Different from existing methods which process all social relations on an image independently, our method considers the paradigm of jointly inferring the relations by constructing a social relation graph. Furthermore, the proposed GR^{2}N constructs several virtual relation graphs to explicitly grasp the strong logical constraints among different types of social relations. Experimental results illustrate that our method generates a reasonable and consistent social relation graph and improves the performance in both accuracy and efficiency.

Keywords: Social relation reasoning · Paradigm shift · Graph neural networks · Social relation graph

1 Introduction

Social relations are the theme and basis of human life, where most human behaviors occur in the context of the relationships between individuals and others [32]. The social relation is derived from human social behaviors and defined as the association between individual persons. Social relation recognition from images

---

Y. Duan—Work done while at Tsinghua University.

Electronic supplementary material The online version of this chapter (https://doi.org/10.1007/978-3-030-58555-6_2) contains supplementary material, which is available to authorized users.

© Springer Nature Switzerland AG 2020
A. Vedaldi et al. (Eds.): ECCV 2020, LNCS 12360, pp. 18–34, 2020.
https://doi.org/10.1007/978-3-030-58555-6_2

Fig. 1. Examples of how the relations on the same image help each other in reasoning. We observe that social relations on an image usually follow strong *logical constraints*.

is in the ascendant in the computer vision community [20,41], while social relationships have been studied in social psychology for decades [2,6,10]. There has been a growing interest in understanding relations among persons in a given still image due to the broad applications such as potential privacy risk warning [41], intelligent and autonomous systems [52], and group activity analysis [19].

In a social scene, there are usually many people appearing at the same time, which contains various social relations. Most existing methods recognize social relations between person pairs separately [13,20,41,52], where each forward pass only processes a pair of bounding boxes on an image. However, as social relations usually form a reasonable social scene, they are not independent of each other, but highly correlated instead. Independently predicting the relations on the same image suffers from the high locality in social scenes, which may result in an unreasonable and contradictory social relation graph. To this end, we consider that jointly inferring all relations for each image helps construct a reasonable and consistent social relation graph with a thorough understanding of the social scene. Moreover, as social relations on the same image usually follow strong *logical constraints*, simultaneously considering all relations can effectively exploit the consistency of them. Two examples are shown in Fig. 1. In the first example, when we know that the relation between A and C is mother-child and that between B and C is father-child, we easily infer that A and B are lovers. In the second example, we can quickly understand the social scene through the relations among A, B, and C, and infer the relations between D and others, even if D is heavily occluded. Clearly, the relations on the same image help each other in reasoning, which is not exploited in existing methods as an important cue.

In this paper, we propose a graph relational reasoning network (GR^{2}N) to jointly infer all relations on an image with the paradigm of treating images as independent samples rather than person-pairs. In fact, the image-based paradigm is closer to the nature of human perception of social relations since human beings do not perceive the social relations of a pair of people in isolation. Many variants of graph neural networks (GNNs) can collaborate with the image-based paradigm by building a graph for each image, where the nodes represent the persons and the edges represent the relations. However, for social relation recognition, different kinds of social relations have strong logical constraints as shown in Fig. 1. Most existing GNNs' methods simply exploit contextual information via message passing, which fails to explicitly grasp the logical constraints among different types of social relations. To exploit the strong logical constraints, the proposed GR^{2}N constructs different virtual relation graphs for different relation types with shared node representations. Our method learns type-specific messages on each virtual relation graph and updates the node representations by aggregating all neighbor messages across all virtual relation graphs. In the end, the final representations of nodes are utilized to predict the relations of all pairs of nodes on the graph.

To summarize, the contributions of this paper are three-fold:

- To the best of our knowledge, our method is the first attempt to jointly reason all relations on the same image for social relation recognition. Experimental results verify the superiority of the paradigm shift from person pair-based to image-based.
- We proposed a *simper* method named GR^{2}N to collaborate with the image-based paradigm, which constructs several virtual relation graphs to explicitly model the logical constraints among different types of social relations. GR^{2}N only uses *one input image patch* and *one convolutional neural net* while the existing approaches usually use many more patches and nets.
- The proposed GR^{2}N is *faster* and *more accurate*. Unlike existing methods that only handle the relation of one person pair at a time, our method processes all relations on an image at the same time with a single forward pass, which makes our method computationally efficient. Finally, our method is 2×–7× faster than other methods due to the paradigm shift. The proposed GR^{2}N effectively exploits the information of relations on the same image to generate a reasonable and consistent social relation graph. Extensive experimental results on the PIPA and PISC datasets demonstrate that our approach outperforms the state-of-the-art methods.

2   Related Work

Social Relationship Recognition. Human face and body images contain rich information [8,22,26,27,45]. In recent years, social relationship recognition from an image has attracted increasing interest in the computer vision community [13,20,41,52]. Some early efforts to mine social information include kinship recognition [23,25,28,44], social role recognition [29,38] and occupation recognition [37,39]. Lu *et al.* [28] proposed a neighborhood repulsed metric learning (NRML) method for kinship verification. Wang *et al.* [44] explored using familial social relationships as a context for recognizing people and the social relationships between pairs of people. Ramanathan *et al.* [29] presented a conditional random field method to recognize social roles played by people in an event. Personal attributes and contextual information were usually used in occupation recognition [37]. Zhang *et al.* [51] devised an effective multitask network to learn social relation traits from face images in the wild. Sun *et al.* [41] extended the PIPA database by 26,915 social relation annotations based on the social domain theory [2]. Li *et al.* [20] collected a new social relation database and proposed a dual-glance model, which explored useful and complementary visual cues. Wang *et al.* [52] proposed a graph reasoning model to incorporate common sense knowledge of the correlation between social relationships and semantic contextual cues. Arushi *et al.* [13] used memory cells like GRUs to combine human attributes and contextual cues. However, these methods deal with all the relationships on the same image independently, which ignore the strong logical constraints among the social relationships on an image.

Graph Neural Networks. The notation of GNNs was firstly introduced by Gori *et al.* [14], and further elaborated in [35]. Scarselli *et al.* trained the GNNs via the Almeida-Pineda algorithm. Li *et al.* [24] modified the previous GNNs to use gated recurrent units and modern optimization techniques. Inspired by the huge success of convolutional networks in the computer vision domain, many efforts have been devoted to re-defining the notation of convolution for graph data [1,7,18]. One of the most famous works is GCN proposed by Kipf *et al.* [18]. Li *et al.* [21] developed deeper insights into the GCN model and argued that if a GCN was deep with many convolutional layers, the output features might be over-smoothed. Gilmer *et al.* [11] reformulated the existing GNN models into a single common framework called message passing neural networks (MPNNs). Veličković *et al.* [43] presented graph attention networks leveraging masked self-attentional layers. Xu *et al.* [48] characterized the discriminative power of popular GNN variants and developed a simple architecture that was provably the most expressive among the class of GNNs. Hamilton *et al.* [15] presented Graph-SAGE to efficiently generate node embeddings for previously unseen data. GNNs have been proven to be an effective tool for relational reasoning [3,17,36,40,42]. Schlichtkrull *et al.* [36] proposed R-GCNs and applied them to knowledge base completion tasks. Chen *et al.* [4] proposed a GloRe unit for reasoning globally.

3   Approach

In this section, we first illustrate the importance of the paradigm shift. Then we present the details of the graph building and the proposed GR^{2}N. Finally, we provide comparisons between the proposed approach and other GNNs' methods.

3.1   Revisiting the Paradigm of Social Relation Recognition

Formally, the problem of social relation recognition can be formulated as a probability function: given an input image I, bounding box values b_i and queries x:

$$x = \{x_{i,j} | i = 1, 2, ..., N, j = 1, 2, ..., N\} \tag{1}$$

where $x_{i,j}$ is the social relation between the person i and j, and N is the total number of people in the image I. The goal of social relation recognition is to find an optimal value of x:

$$x^* = \arg\max_x P(x|I, b_1, b_2, ..., b_N) \tag{2}$$

As far as we know, all existing methods of social relation recognition organize data as person pair-based, and each sample of these methods consists of an input image I and a target pair of people highlighted by bounding boxes b_i, b_j. Therefore, the actual optimization objective of these methods is as follows:

$$x_{i,j}^* = \arg\max_{x_{i,j}} P(x_{i,j}|I, b_i, b_j) \tag{3}$$

The optimization objective in (3) is consistent with that in (2) if and only if the following equation holds:

$$P(x|I, b_1, b_2, ..., b_N) = \prod_{1 \leq i,j \leq N} P(x_{i,j}|I, b_i, b_j) \tag{4}$$

It is known that (4) holds if and only if all relations on the input image I are independent of each other. Unfortunately, as we discussed earlier, this is not true and the relations on an image are usually highly related. Let y be the ground truths corresponding to queries x: $y = \{y_{i,j}|i = 1, 2, ..., N, j = 1, 2, ..., N\}$. Then the paradigm of existing methods treats quadruplets $<I, b_i, b_j, y_{i,j}>$ as independent samples to optimize the objective in (3), which can not lead to the optimal value of that in (2). On the contrary, we reorganize the data as image-based to directly optimize the objective formulated in (2) with a thorough understanding of the social scene. Our method formulates each sample as an $(N+2)$-tuple $<I, b_1, b_2, ..., b_N, y>$ and jointly infers the relations for each image.

3.2   From Image to Graph

We model each person on an image as a node in a graph, and the edge between two nodes represents the relation between the corresponding two people. Then graph operations are applied to perform joint reasoning. So the first step is to create nodes in the graph using the input image I and bounding boxes $b_1, b_2, ..., b_N$. As commonly used in detection [12,30,31,33], a convolutional neural network takes the input image I as input, and the features of object proposals are extracted directly from the last convolutional feature map H. More specifically, we first map the bounding boxes $b_1, b_2, ..., b_N$ to $b_1', b_2', ..., b_N'$ according to the transformation from I to H. Then the feature p_i of person i is obtained by employing an RoI pooling layer on the feature map H: $p_i = f_{RoIPooling}(H, b_i')$. In this way, we obtain all the features of people on the image I: $p = \{p_1, p_2, ..., p_N\}$, $p_i \in \mathbb{R}^F$, where F is the feature dimension for each person. Subsequently, these features are set as the initial feature representations of nodes in the graph: $h^0 = \{h_1^0, h_2^0, ..., h_N^0\}$, where $h_i^0 = p_i$.

3.3   Graph Relational Reasoning Network

Having created the nodes in the graph, we propose a graph relational reasoning network to perform relational reasoning on the graph. We begin with a brief description of the overall framework of our GR^{2}N. Let $\mathcal{G} = (\mathcal{V}, \mathcal{E})$ denotes a graph with node feature vectors $\boldsymbol{h}_v$ for $v \in \mathcal{V}$ and edge features $\boldsymbol{e}_{vw}$ for $(v, w) \in \mathcal{E}$. The framework of GR^{2}N can be formulated as a message passing phase and a readout phase following [11]. The message passing phase, which is defined in terms of message function M_t and vertex update function U_t, runs for T time steps. During the message passing phase, the aggregated messages $\boldsymbol{m}_v^{t+1}$ are used to update the hidden states $\boldsymbol{h}_v^t$:

$$\boldsymbol{m}_v^{t+1} = \sum_{w \in N(v)} M_t(\boldsymbol{h}_v^t, \boldsymbol{h}_w^t, \boldsymbol{e}_{vw}) \tag{5}$$

$$\boldsymbol{h}_v^{t+1} = U_t(\boldsymbol{h}_v^t, \boldsymbol{m}_v^{t+1}) \tag{6}$$

where $N(v)$ denotes the neighbors of v in graph $\mathcal{G}$. The readout phase computes feature vectors for edges using the readout function R. The message function M_t, vertex update function U_t, and readout function R are all differentiable functions.

In each time step of the message passing phase, the state of the node is updated according to the messages sent by its neighbor nodes, which requires us to know the topology of the entire graph. In social relation recognition, we use edges to represent social relations, and the edge type represents the category of the relation. However, the task of social relation recognition is to predict the existence of edges and the type of edges, which means that we do not know the topology of the graph at the beginning. To address this issue, the proposed GR^{2}N first models the edges in the graph.

The social relations on an image have strong *logical constraints* and our GR^{2}N aims to grasp these constraints to generate a reasonable and consistent social relation graph. As each kind of social relations has its specific logical constraints, we let GR^{2}N propagate different types of messages for different types of edges. Specifically, we construct a virtual relation graph for each kind of relationship, in which the edge represents the existence of this kind of relationship between two nodes. To achieve mutual reasoning and information fusion among different relationships, the feature representations of nodes are shared across all created virtual relation graphs.

Mathematically, we use K to denote the number of social relationship categories, and K virtual relation graphs $(\mathcal{V}, \mathcal{E}^1), (\mathcal{V}, \mathcal{E}^2), ..., (\mathcal{V}, \mathcal{E}^K)$ are created to model the K social relationships separately. Assuming that we have obtained the node feature representations at time step t: $\boldsymbol{h}^t = \{\boldsymbol{h}_1^t, \boldsymbol{h}_2^t, ..., \boldsymbol{h}_N^t\}$, we first model the edge embedding in the virtual relation graph:

$$\boldsymbol{e}_{i,j}^{k,t+1} = f_{r_k}(\boldsymbol{h}_i^t, \boldsymbol{h}_j^t) = [\boldsymbol{W}_{r_k}\boldsymbol{h}_i^t \| \boldsymbol{W}_{r_k}\boldsymbol{h}_j^t] \tag{7}$$

where $\boldsymbol{e}_{i,j}^{k,t+1}$ denotes the embedding of edge (i, j) in virtual relation graph $(\mathcal{V}, \mathcal{E}^k)$ at time $t + 1$ and $\|$ is the concatenation operation. $f_{r_k}(\cdot)$ is parameterized by a weight matrix $\boldsymbol{W}_{r_k} \in \mathbb{R}^{F \times F}$. In standard GNNs, the existence of edges is binary and deterministic. In this paper, we consider rather a probabilistic soft edge, *i.e.* edge (i, j) in the virtual relation graph $(\mathcal{V}, \mathcal{E}^k)$ exists according to the probability:

$$\alpha_{i,j}^{k,t+1} = \sigma(\boldsymbol{a}_{r_k}^\top \boldsymbol{e}_{i,j}^{k,t+1}) \tag{8}$$

where $\boldsymbol{a}_{r_k} \in \mathbb{R}^{2F}$ is a weight vector and $\cdot^\top$ denotes transposition. The sigmoid function $\sigma(\cdot)$ is employed as the activation function to normalize the values between 0 and 1. In the end, the $\mathcal{E}^k$ can be represented as the set $\{\alpha_{i,j}^k | 1 \leq i, j \leq N\}$ if we ignore the time step.

Once obtained, we propagate messages on each virtual relation graph according to the soft edges and aggregate messages across all virtual relation graphs:

$$\boldsymbol{m}_i^{t+1} = \sum_{j \in N(i)} \sum_{k=1}^{K} \alpha_{i,j}^{k,t+1} f_m(\boldsymbol{h}_i^t, \boldsymbol{h}_j^t, \boldsymbol{e}_{i,j}^{k,t+1}) \tag{9}$$

For simplicity, we reuse the previous weight $\boldsymbol{W}_{r_k}$ and let $f_m(\boldsymbol{h}_i^t, \boldsymbol{h}_j^t, \boldsymbol{e}_{i,j}^{k,t+1}) = \boldsymbol{W}_{r_k} \boldsymbol{h}_j^t$. So we reformulate $\boldsymbol{m}_i^{t+1}$ as:

$$\boldsymbol{m}_i^{t+1} = \sum_{j \in N(i)} M_t(\boldsymbol{h}_i^t, \boldsymbol{h}_j^t, \boldsymbol{e}_{i,j}^{k,t+1}) = \sum_{j \in N(i)} (\sum_{k=1}^{K} \alpha_{i,j}^{k,t+1} \boldsymbol{W}_{r_k} \boldsymbol{h}_j^t) \tag{10}$$

Finally, we use the aggregated messages to update the hidden state of node:

$$\boldsymbol{h}_i^{t+1} = U_t(\boldsymbol{h}_i^t, \boldsymbol{m}_i^{t+1}) = f_{GRU}(\boldsymbol{h}_i^t, \boldsymbol{m}_i^{t+1}) \tag{11}$$

where $f_{GRU}(\cdot)$ is the Gated Recurrent Unit update function introduced in [5].

Having repeated the above process for T time steps, we obtain the final node feature representations: $\boldsymbol{h}^T = \{\boldsymbol{h}_1^T, \boldsymbol{h}_2^T, ..., \boldsymbol{h}_N^T\}$. Then the readout function R is applied to the features $\boldsymbol{h}^T$ for social relation recognition. As mentioned above, the task of social relation recognition is to predict the existence and type of edges in the graph, which is exactly what virtual social graphs accomplish. So the readout function R simply reuses the functions that create the virtual social graphs:

$$\hat{x}_{i,j}^k = R(\boldsymbol{h}_i^T, \boldsymbol{h}_j^T, k) = \sigma(\boldsymbol{a}_{r_k}^\top [\boldsymbol{W}_{r_k} \boldsymbol{h}_i^T \| \boldsymbol{W}_{r_k} \boldsymbol{h}_j^T]) \tag{12}$$

where $\hat{x}_{i,j}^k$ indicates the probability of the existence of the k-th social relation between person i and j. During the test stage, the social relation between person i and j is chosen as the category with the greatest probability: $\arg\max_k \hat{x}_{i,j}^k$.

The overall framework including the backbone CNN and the proposed GR^{2}N is jointly optimized end-to-end. If we extend the ground truths $\boldsymbol{y}$ to one hot, then the loss function for the sample $<\boldsymbol{I}, \boldsymbol{b}_1, \boldsymbol{b}_2, ..., \boldsymbol{b}_N, \boldsymbol{y}>$ is:

$$\mathcal{L} = \sum_{1 \leq i,j \leq N} \sum_{k=1}^{K} -[y_{i,j}^k \log(\hat{x}_{i,j}^k) + (1 - y_{i,j}^k) \log(1 - \hat{x}_{i,j}^k)] \tag{13}$$

where $y_{i,j}^k$ is the k-th element of one hot form of $y_{i,j}$.

3.4   Discussion

Another intuitive way to jointly model all the social relationships among people on an image is to use some common variants of GNNs, such as GGNN [24] and GCN [18]. They can be utilized to perform graph operations on the graphs to replace the proposed GR^{2}N. In addition, GNNs are used in some other tasks such as group activity recognition [9], and scene graph generation [47] to model the relationships between objects. In the following, we will elaborate on the key difference between the proposed GR^{2}N and those GNNs' methods.

Different from group activity recognition, scene graph recognition, and other tasks, different kinds of social relations have *strong logical constraints* in the social relation recognition task. For example, knowing that A-C is mother-child and B-C is father-child, we can directly infer that A-B is spouses even without checking the relation A-B itself. However, most existing methods of jointly reasoning all relations such as GGNN [24], GCN [18], Structure Inference Machines [9] and Iterative Message Passing [47] simply exploit *contextual information* via message passing, which fails to *explicitly* grasp the *logical constraints* among different types of social relations. Instead, our GR^{2}N constructs a virtual social relation graph for each social relation type, where each graph learns a *type-specific* message passing matrix W_{r_k}. As each type of social relation has its *specific logical constraints*, the proposed GR^{2}N is aware of *relation types* and better exploits the *type-specific logical constraints* in social relation recognition. In this way, our method can better perform relational reasoning and generate a reasonable and consistent social relation graph.

4   Experiments

In this section, we first introduce the datasets and present some implementation details of our approach. Then we evaluate the performance of our method using quantitative and qualitative analysis.

4.1   Datasets

We evaluated the GR^{2}N and existing competing methods on the People in Photo Album (PIPA) [41] database and People in Social Context (PISC) database [20].

PIPA: The PIPA database is collected from Flickr photo albums for the task of person recognition [50]. Then the dataset is extended by Sun *et al.* with 26,915 person pair annotations based on the social domain theory [2]. The PIPA dataset involves two-level recognition tasks: 1) social domain recognition focuses on five categories of social domain, i.e., attachment domain, reciprocity domain, mating domain, hierarchical power domain, and coalitional group domain; 2) social relationship recognition focuses on 16 finer categories of relationship, such as friends, classmates, father-child, leader-subordinate, band members, and so on. For fair comparisons, we adopt the standard train/val/test split released by Sun *et al.* [41], which uses 13,672 domain/relation instances in 5,857 images for training, 709 domain/relation instances in 261 images for validation, and 5,106 domain/relation instances in 2,452 images for testing. The top-1 classification accuracy is used for evaluation.

PISC: The PISC database collects 22,670 images and is annotated following the relational model theory [10]. It has hierarchical social relationship categories: coarse-grained relationships (intimate, not-intimate and no relation) and fine-grained relationships (commercial, couple, family, friends, professional and no-relation). For coarse-grained relationship, 13,142 images with 49,017 relationship instances are used for training, 4,000 images with 14,536 instances are used for validation, 4,000 images with 15,497 instances are used for testing. For fine-grained relationship, the train/val/test set has 16,828 images and 55,400 instances, 500 images and 1,505 instances, 1250 images and 3,961 instances, respectively. We follow the commonly used metrics [13,20,52] on the PISC dataset and report the per-class recall for each relationship and the mean average precision (mAP) over all relationships.

4.2 Implementation Details

For fair comparisons, we employed ResNet-101 as the backbone CNN following [20,52]. The ResNet-101 was initialized with the pre-trained weights on ImageNet [34]. The shape of the output region of the RoI pooling layer was set to be the same as the shape of the last convolution feature map H, and then a global average pooling layer was used to obtain a 2048-D feature for each person. The time step T was set to 1, which achieved the best result. The model was trained by Adam optimization [16] with a learning rate of 10^{-5}. We trained our method for 10 epochs with a batch size of 32. During training, images were horizontally flipped with probability 0.5 and randomly cropped for data augmentation.

In a mini-batch, different images might have different numbers of people, which made the size of the graphs variable. To deal with this problem, we set the number of people N on an image to the maximum possible persons N_{max}. If the number of people on an image was less than N_{max}, then we set the missing nodes as empty ones and there would be no soft edges between the empty nodes and the real nodes. For those relationships that were not labeled, no loss would occur from them in (13) to avoid having an impact on network training. Another issue was that the social relationships on the PISC dataset were highly imbalanced, especially for the fine-level relationships. To address this, we adopted the reweighting strategy for fine-grained relationship recognition on the PISC dataset. Specifically, the samples were reweighted inversely proportionally to the class frequencies.

4.3 Results and Analysis

Comparisons with the State-of-the-Art Methods. We compare the proposed approach with several existing state-of-the-art methods. The details of these competing methods are as follows:

Table 1. Comparisons of the accuracy between our GR^{2}N and other state-of-the-art methods on the PIPA dataset.

Methods	# of input Image patches	# of CNNs	Domain	Relation
Finetuned CNN + SVM [41]	2	2	63.2%	48.6%
All attributes + SVM [41]	4	10	67.8%	57.2%
Pair CNN [20]	2	1	65.9%	58.0%
Dual-Glance [20]	4	3	–	59.6%
SRG-GN [13]	3	5	-	53.6%
GRM [52]	4	3	–	62.3%
MGR [49]	4	5	–	**64.4%**
GR^{2}N-224	1	1	69.3%	61.3%
GR^{2}N-448	1	1	**72.3%**	**64.3%**

Finetuned CNN + SVM [41]. Double-stream CaffeNet is used to extract features, then the extracted features are utilized to train a linear SVM.

All attributes + SVM [41]. Many semantic attribute categories including age, gender, location, and activity are used in this method. Then all attribute features are concatenated to train a linear SVM.

Pair CNN [20]. This model consists of two CNNs (ResNet-101) with shared weights. The input is two cropped image patches for the two individuals and the extracted features are concatenated for social relation recognition.

Dual-Glance [20]. Two cropped individual patches and the union region of them are sent to CNNs to extract features. The introduced second glance exploits surrounding proposals as contextual information to refine the predictions.

SRG-GN [13]. Five CNNs are utilized to extract scene and attribute context information (*e.g.*, age, gender, and activity). Then these features are utilized to update the state of memory cells like GRUs.

MGR [49]. It employ five CNNs to extract the global, middle, and fine granularity features to comprehensively capture the multi-granularity semantics.

GRM [52]. It replaces the second glance in Dual-Glance with a pre-trained Faster-RCNN detector [33] and a Gated Graph Neural Network [24] to model the interaction between the contextual objects and the persons of interest.

All of the above methods are person pair-based, which means that they consider the social relations on the same image separately. It is worth noting that although the SRG-GN mentions that they can generate a social relationship graph, they actually still process each relationship independently during the training and testing phases. These methods usually crop the image patches for interested individuals and resize them to 224 × 224 pixels while our approach

Table 2. We present the per-class recall for each relationship and the mAP over all relationships (in %) on the PISC dataset. The first and second best scores are highlighted in red and blue colors, respectively. (Int: Intimate, Non: Non-Intimate, NoR: No Relation, Fri: Friends, Fam: Family, Cou: Couple, Pro: Professional, Com: Commercial)

Methods	Coarse relationships				Fine relationships						
	Int	Non	NoR	mAP	Fri	Fam	Cou	Pro	Com	NoR	mAP
Pair CNN [20]	70.3	80.5	38.8	65.1	30.2	59.1	69.4	57.5	41.9	34.2	48.2
Dual-Glance [20]	73.1	84.2	59.6	79.7	35.4	68.1	76.3	70.3	57.6	60.9	63.2
GRM [52]	81.7	73.4	65.5	82.8	59.6	64.4	58.6	76.6	39.5	67.7	68.7
MGR [49]	–	–	–	–	64.6	67.8	60.5	76.8	34.7	70.4	70.0
SRG-GN [13]	–	–	–	–	25.2	80.0	100.0	78.4	83.3	62.5	71.6
GR^{2}N-224	76.3	65.3	74.0	72.2	54.7	68.5	78.1	78.0	49.7	57.5	68.6
GR^{2}N-448	81.6	74.3	70.8	**83.1**	60.8	65.9	84.8	73.0	51.7	70.4	**72.7**

takes the entire image as input. In order to make a fair comparison, we should choose the appropriate entire input image size to ensure that the personal area of interest is roughly equal to 224×224. In the PIPA and PISC dataset, the area of the bounding boxes is on average 1/4 and 1/5 of the area of the images, respectively, so we resize the original image to 448×448 pixels, which is denoted as GR^{2}N-448. Although this is still unfavorable to our approach on the PISC datasets, our method can still achieve promising performance. Besides, we also report the performance of GR^{2}N-224 to show the effectiveness of our method, which resizes the original image to 224×224 pixels.

The experimental results of social domain recognition and social relationship recognition on the PIPA database are shown in Table 1. We first compare our method with a simple baseline Pair CNN, which, like our method, does not use any scene and attribute context cues, so we can see the benefits brought by the paradigm shift. We observe that our method significantly outperforms Pair CNN. Specifically, the GR^{2}N-448 achieves an accuracy of 64.3% for social relation recognition and 72.3% for social domain recognition, improving the Pair CNN by 6.3% and 6.4% respectively. What's more, even the GR^{2}N-224 with inferior input image size outperforms the Pair CNN by 3.3% and 3.4% for social relation recognition and social domain recognition respectively, which demonstrates that the paradigm shift and GR^{2}N's superior relational reasoning ability bring significant performance gains.

Next, we compare our method with state-of-the-art methods. In terms of social relation recognition, the proposed GR^{2}N-448 improves GRM by 2.0%. What is worth mentioning is that GRM uses three individual CNNs to extract features from four image patches including one 448×448 entire image and three 224×224 person patches to exploit key contextual cues, while the GR^{2}N only uses one input image and one convolutional neural network. Our method also achieves competitive results with MGR whereas MGR uses 5 CNNs and

Table 3. Comparisons of the runtime (seconds/image) under different batch sizes for social relation recognition on the PIPA dataset.

Methods	Batch size			
	1	2	4	8
GRM	0.294	0.171	0.089	*
Pair CNN	0.077	0.045	0.039	0.037
GR^{2}N-448	**0.046**	**0.025**	**0.021**	**0.021**

Table 4. Comparisons of the runtime (seconds/image) under different batch sizes for social relation recognition on the PIPA dataset.

Methods	Accuracy
Pair CNN	58.0%
GCN	59.3%
GGNN	59.8%
GR^{2}N	**61.3%**

4 patches. For social domain recognition, our method GR^{2}N-448 achieves an accuracy of 72.3%, which improves the performance of All attributes + SVM by 4.5% without using any semantic attribute categories. Clearly, these results illustrate the superiority of our method.

Table 2 shows the experimental comparison with the recent state-of-the-art methods on the PISC database. We observe that our method achieves an mAP of 83.1% for the coarse-level recognition and 72.7% for the fine-level recognition, improving the simple baseline Pair CNN by a large margin: 18.0% for coarse relationships and 24.5% for fine relationships, which further validates the advantage of the paradigm shift. Our approach outperforms SRG-GN by 1.1% with a much simpler framework (1 neural net vs. 5 neural nets, 1 image patch vs. 3 image patches). Compared with GRM, GR^{2}N-448 achieves competitive performance for coarse relationship recognition and superior performance for fine relationship recognition with only one image patch and one neural net, while the GRM uses four image patches and three neural nets. It is worth noting that the experimental results of our approach on the PISC dataset are achieved with inferior input image size, which further illustrates the effectiveness of our GR^{2}N.

Runtime Analysis. In addition to being *simpler* and *more accurate*, another advantage of our GR^{2}N is that it is *faster*. Since our method handles all relationships on an image at the same time, the GR^{2}N is computationally efficient. We conduct experiments under different batch sizes to compare the speed of different methods using a GeForce GTX 1080Ti GPU on the PIPA dataset. To be fair, the version of GR^{2}N used here is GR^{2}N-448. The results of the forward runtime (seconds/image) are reported in Table 3. The * in the table represents memory overflow. Compared with Pair CNN, our method achieves about 2× speed-ups, which shows the benefits of the paradigm shift. The GR^{2}N-448 is 4×–7× faster than the state-of-the-art method GRM, which further demonstrates the *efficiency* of our approach. Since the average number of bounding boxes per image on the PIPA datasets is only 2.5, we expect higher speedup when there are more individuals on an image.

Ablation Study. To validate the effectiveness of the proposed GR^{2}N, we compare it with two commonly used GNN variants: GGNN [24] and GCN [18]. To apply GGNN and GCN, we connect the nodes together into a fully connected

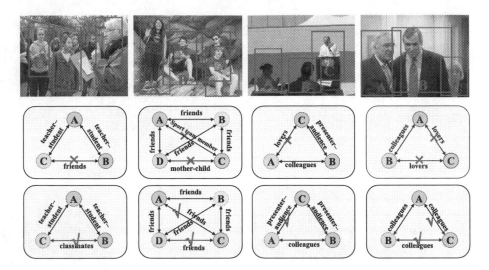

Fig. 2. Some examples of qualitative evaluation. The first row shows the input images, the second row is the corresponding predicted results of Pair CNN, and the third row is the corresponding predicted results of our method. The red cross mark indicates a prediction error, while the green check mark indicates that the corresponding error is corrected. We see that our method corrects the prediction errors by performing graph-based relational reasoning. (Color figure online)

graph. Following [46], we also add a residual connection in every layer of GCN, which observes an improvement in performance in the experiments. For a fair comparison, all three methods use an image of size 224×224 as input. The readout function of GGNN and GCN is a two-layer neural network and the edge embedding is obtained by concatenating the features of the corresponding two nodes. The cross-entropy loss function is used to train GGNN and GCN.

The comparisons on the PIPA dataset by accuracy are listed in Table 4. We observe that our proposed GR^2N outperforms GCN and GGNN, which demonstrates that the proposed GR^2N effectively performs relational reasoning and exploits the strong logical constraints of social relations on the same image. On the other hand, all three methods adopt the image-based paradigm and we see that they all outperform Pair CNN. In fact, these three methods achieve superior or competitive performance compared with some of the state-of-the-art methods which adopt the person pair-based paradigm, such as Dual-Glance and SRG-GN. Note that these GNNs' methods only use one input patch and one convolutional neural net, which illustrates the superiority of the image-based paradigm over the person pair-based paradigm.

Qualitative Evaluation. In this part, we present some examples to illustrate how our GR^2N improves performance by performing relational reasoning and exploiting the strong logical constraints of social relations on a graph in Fig. 2. For the first example, Pair CNN independently processes the relations on the

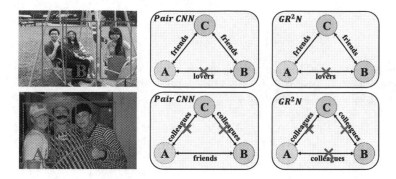

Fig. 3. Some failure cases of our method on the PIPA dataset. The first column shows the input images, the second column is the corresponding predicted results of Pair CNN, and the third column is the corresponding predicted results of our method. The red cross mark indicates a prediction error.

image and predicts that the relation between B and C is friends. On the contrary, our method jointly infers all relations for each image. We easily infer that the relation between B and C is classmates when we know that the relation between A and B and the relation between A and C both are teacher-student. This is also the case in the rest of the examples. When the relationships on an image are processed independently, some obviously unreasonable and contradictory relationships may occur simultaneously, and our method is capable of correcting them through relational reasoning with a thorough understanding of the social scene. Clearly, our method better models the interactions among people on the graph, and generates a reasonable and consistent social relation graph. We also present some failure cases in Fig. 3. We see that our method can not correct the prediction error in the first case since the error does not cause any obvious conflict. Although the ground truth of the relationship between person A and B is friends, predicting it as lovers does not lead to a contradictory social relation graph. Our method may also fail when most predictions are wrong and our model performs relational reasoning based on these false predictions, as shown in the second case.

5   Conclusion

In this paper, we have presented the GR^{2}N for social relation recognition, which is a *simpler*, *faster*, and *more accurate* method. Unlike existing methods, we simultaneously reason all relations for each image and treat images as independent samples rather than person-pairs. Our method constructs a graph for each image to model the interactions among people and perform relational reasoning on this graph with fully exploiting the strong logical constraints of relations. Experimental results demonstrate that our method generates a reasonable and consistent social relation graph. Moreover, GR^{2}N achieves better performance with less time cost compared with the state-of-the-art methods, which further illustrates the *effectiveness* and *efficiency* of our method.

Acknowledgments. This work was supported in part by the National Key Research and Development Program of China under Grant 2017YFA0700802, in part by the National Natural Science Foundation of China under Grant 61822603, Grant U1813218, Grant U1713214, and Grant 61672306, in part by Beijing Natural Science Foundation under Grant No. L172051, in part by Beijing Academy of Artificial Intelligence (BAAI), in part by a grant from the Institute for Guo Qiang, Tsinghua University, in part by the Shenzhen Fundamental Research Fund (Subject Arrangement) under Grant JCYJ20170412170602564, and in part by Tsinghua University Initiative Scientific Research Program.

References

1. Bruna, J., Zaremba, W., Szlam, A., LeCun, Y.: Spectral networks and locally connected networks on graphs. arXiv preprint arXiv:1312.6203 (2013)
2. Bugental, D.B.: Acquisition of the algorithms of social life: a domain-based approach. Psychol. Bull. **126**(2), 187 (2000)
3. Chen, S., Abhinav, S., Carl, V., Rahul, S., Kevin, M., Cordelia, S.: Relational action forecasting. In: CVPR (2019)
4. Chen, Y., Rohrbach, M., Yan, Z., Shuicheng, Y., Feng, J., Kalantidis, Y.: Graph-based global reasoning networks. In: CVPR, pp. 433–442 (2019)
5. Cho, K., et al.: Learning phrase representations using RNN encoder-decoder for statistical machine translation. In: EMNLP (2014)
6. Conte, H.R., Plutchik, R.: A circumplex model for interpersonal personality traits. J. Pers. Soc. Psychol. **40**(4), 701 (1981)
7. Defferrard, M., Bresson, X., Vandergheynst, P.: Convolutional neural networks on graphs with fast localized spectral filtering. In: NeurIPS, pp. 3844–3852 (2016)
8. Deng, J., Guo, J., Xue, N., Zafeiriou, S.: Arcface: additive angular margin loss for deep face recognition. In: CVPR, pp. 4690–4699 (2019)
9. Deng, Z., Vahdat, A., Hu, H., Mori, G.: Structure inference machines: recurrent neural networks for analyzing relations in group activity recognition. In: CVPR, pp. 4772–4781 (2016)
10. Fiske, A.P.: The four elementary forms of sociality: framework for a unified theory of social relations. Psychol. Rev. **99**(4), 689 (1992)
11. Gilmer, J., Schoenholz, S.S., Riley, P.F., Vinyals, O., Dahl, G.E.: Neural message passing for quantum chemistry. In: ICML, pp. 1263–1272 (2017)
12. Girshick, R.: Fast R-CNN. In: ICCV, pp. 1440–1448 (2015)
13. Goel, A., Ma, K.T., Tan, C.: An end-to-end network for generating social relationship graphs. In: CVPR, pp. 11186–11195 (2019)
14. Gori, M., Monfardini, G., Scarselli, F.: A new model for learning in graph domains. In: IJCNN, pp. 729–734 (2005)
15. Hamilton, W., Ying, Z., Leskovec, J.: Inductive representation learning on large graphs. In: NeurIPS, pp. 1024–1034 (2017)
16. Kingma, D.P., Ba, J.: Adam: a method for stochastic optimization. In: ICLR (2015)
17. Kipf, T., Fetaya, E., Wang, K.C., Welling, M., Zemel, R.: Neural relational inference for interacting systems. In: ICML (2018)

18. Kipf, T.N., Welling, M.: Semi-supervised classification with graph convolutional networks. In: ICLR (2017)
19. Lan, T., Sigal, L., Mori, G.: Social roles in hierarchical models for human activity recognition. In: CVPR, pp. 1354–1361 (2012)
20. Li, J., Wong, Y., Zhao, Q., Kankanhalli, M.S.: Dual-glance model for deciphering social relationships. In: ICCV, pp. 2650–2659 (2017)
21. Li, Q., Han, Z., Wu, X.M.: Deeper insights into graph convolutional networks for semi-supervised learning. In: AAAI (2018)
22. Li, W., Lu, J., Feng, J., Xu, C., Zhou, J., Tian, Q.: Bridgenet: a continuity-aware probabilistic network for age estimation. In: CVPR, pp. 1145–1154 (2019)
23. Li, W., Zhang, Y., Lv, K., Lu, J., Feng, J., Zhou, J.: Graph-based kinship reasoning network. In: ICME, pp. 1–6 (2020)
24. Li, Y., Tarlow, D., Brockschmidt, M., Zemel, R.: Gated graph sequence neural networks. In: ICLR (2016)
25. Lu, J., Hu, J., Tan, Y.P.: Discriminative deep metric learning for face and kinship verification. TIP **26**(9), 4269–4282 (2017)
26. Lu, J., Tan, Y.P.: Cost-sensitive subspace analysis and extensions for face recognition. TIFS **8**(3), 510–519 (2013)
27. Lu, J., Wang, G., Deng, W., Jia, K.: Reconstruction-based metric learning for unconstrained face verification. TIFS **10**(1), 79–89 (2014)
28. Lu, J., Zhou, X., Tan, Y.P., Shang, Y., Zhou, J.: Neighborhood repulsed metric learning for kinship verification. TPAMI **36**(2), 331–345 (2013)
29. Ramanathan, V., Yao, B., Fei-Fei, L.: Social role discovery in human events. In: CVPR, pp. 2475–2482 (2013)
30. Redmon, J., Divvala, S., Girshick, R., Farhadi, A.: You only look once: unified, real-time object detection. In: CVPR, pp. 779–788 (2016)
31. Redmon, J., Farhadi, A.: Yolo9000: better, faster, stronger. In: CVPR, pp. 7263–7271 (2017)
32. Reis, H.T., Collins, W.A., Berscheid, E.: The relationship context of human behavior and development. Psychol. Bull. **126**(6), 844 (2000)
33. Ren, S., He, K., Girshick, R., Sun, J.: Faster r-cnn: towards real-time object detection with region proposal networks. In: NeurIPS, pp. 91–99 (2015)
34. Russakovsky, O., et al.: Imagenet large scale visual recognition challenge. IJCV **115**(3), 211–252 (2015)
35. Scarselli, F., Gori, M., Tsoi, A.C., Hagenbuchner, M., Monfardini, G.: The graph neural network model. TNNLS **20**(1), 61–80 (2008)
36. Schlichtkrull, M., Kipf, T.N., Bloem, P., Van Den Berg, R., Titov, I., Welling, M.: Modeling relational data with graph convolutional networks. In: ESWC, pp. 593–607 (2018)
37. Shao, M., Li, L., Fu, Y.: What do you do? occupation recognition in a photo via social context. In: ICCV, pp. 3631–3638 (2013)
38. Shu, T., Xie, D., Rothrock, B., Todorovic, S., Chun Zhu, S.: Joint inference of groups, events and human roles in aerial videos. In: CVPR, pp. 4576–4584 (2015)
39. Song, Z., Wang, M., Hua, X.s., Yan, S.: Predicting occupation via human clothing and contexts. In: ICCV, pp. 1084–1091 (2011)
40. Sun, C., Karlsson, P., Wu, J., Tenenbaum, J.B., Murphy, K.: Stochastic prediction of multi-agent interactions from partial observations. In: ICLR (2019)
41. Sun, Q., Schiele, B., Fritz, M.: A domain based approach to social relation recognition. In: CVPR, pp. 3481–3490 (2017)
42. Tacchetti, A., et al.: Relational forward models for multi-agent learning. In: ICLR (2019)

43. Veličković, P., Cucurull, G., Casanova, A., Romero, A., Liò, P., Bengio, Y.: Graph attention networks. In: ICLR (2018)
44. Wang, G., Gallagher, A., Luo, J., Forsyth, D.: Seeing people in social context: recognizing people and social relationships. In: ECCV, pp. 169–182 (2010)
45. Wang, T., Gong, S., Zhu, X., Wang, S.: Person re-identification by video ranking. In: Fleet, D., Pajdla, T., Schiele, B., Tuytelaars, T. (eds.) ECCV 2014. LNCS, vol. 8692, pp. 688–703. Springer, Cham (2014). https://doi.org/10.1007/978-3-319-10593-2_45
46. Wang, X., Gupta, A.: Videos as space-time region graphs. In: ECCV, pp. 399–417 (2018)
47. Xu, D., Zhu, Y., Choy, C.B., Fei-Fei, L.: Scene graph generation by iterative message passing. In: CVPR, pp. 5410–5419 (2017)
48. Xu, K., Hu, W., Leskovec, J., Jegelka, S.: How powerful are graph neural networks? In: ICLR (2019)
49. Zhang, M., Liu, X., Liu, W., Zhou, A., Ma, H., Mei, T.: Multi-granularity reasoning for social relation recognition from images. In: ICME, pp. 1618–1623 (2019)
50. Zhang, N., Paluri, M., Taigman, Y., Fergus, R., Bourdev, L.: Beyond frontal faces: improving person recognition using multiple cues. In: CVPR, pp. 4804–4813 (2015)
51. Zhang, Z., Luo, P., Loy, C.C., Tang, X.: From facial expression recognition to interpersonal relation prediction. IJCV **126**(5), 550–569 (2018)
52. Zhouxia, W., Tianshui, C., Jimmy, R., Weihao, Y., Hui, C., Liang, L.: Deep reasoning with knowledge graph for social relationship understanding. In: IJCAI (2018)

EPNet: Enhancing Point Features with Image Semantics for 3D Object Detection

Tengteng Huang, Zhe Liu, Xiwu Chen, and Xiang Bai$^{(\boxtimes)}$

Huazhong University of Science and Technology, Wuhan, China
{huangtengtng,zheliu1994,xiwuchen,xbai}@hust.edu.cn

Abstract. In this paper, we aim at addressing two critical issues in the 3D detection task, including the exploitation of multiple sensors (namely LiDAR point cloud and camera image), as well as the inconsistency between the localization and classification confidence. To this end, we propose a novel fusion module to enhance the point features with semantic image features in a point-wise manner without any image annotations. Besides, a consistency enforcing loss is employed to explicitly encourage the consistency of both the localization and classification confidence. We design an end-to-end learnable framework named EPNet to integrate these two components. Extensive experiments on the KITTI and SUN-RGBD datasets demonstrate the superiority of EPNet over the state-of-the-art methods. Codes and models are available at: https://github.com/happinesslz/EPNet.

Keywords: 3D object detection · Point cloud · Multiple sensors

1 Introduction

The last decade has witnessed significant progress in the 3D object detection task via different types of sensors, such as monocular images [1,36], stereo cameras [2], and LiDAR point clouds [22,39,43]. Camera images usually contain plenty of semantic features (*e.g.*, color, texture) while suffering from the lack of depth information. LiDAR points provide depth and geometric structure information, which are quite helpful for understanding 3D scenes. However, LiDAR points are usually sparse, unordered, and unevenly distributed. Figure 1(a) illustrates a typical example of leveraging the camera image to improve the 3D detection task. It is challenging to distinguish between the closely packed white and yellow chairs by only the LiDAR point cloud due to their similar geometric

---

T. Huang and Z. Liu—Equal contribution.

Electronic supplementary material The online version of this chapter (https://doi.org/10.1007/978-3-030-58555-6_3) contains supplementary material, which is available to authorized users.

© Springer Nature Switzerland AG 2020
A. Vedaldi et al. (Eds.): ECCV 2020, LNCS 12360, pp. 35–52, 2020.
https://doi.org/10.1007/978-3-030-58555-6_3

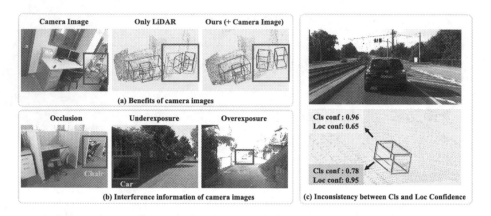

Fig. 1. Illustration of (a) the benefit and (b) potential interference information of the camera image. (c) demonstrates the inconsistency of the classification confidence and localization confidence. Green box denotes the ground truth. Blue and yellow boxes are predicted bounding boxes. (Color figure online)

structure, resulting in chaotically distributed bounding boxes. In this case, utilizing the color information is crucial to locate them precisely. This motivates us to design an effective module to fuse different sensors for a more accurate 3D object detector.

However, fusing the representations of LiDAR and camera image is a non-trivial task for two reasons. On the one hand, they possess highly different data characteristics. On the other hand, the camera image is sensitive to illumination, occlusion, *etc.* (see Fig. 1(b)), and thus may introduce interfering information that is harmful to the 3D object detection task. Previous works usually fuse these two sensors with the aid of image annotations (namely 2D bounding boxes). According to different ways of utilizing the sensors, we summarize previous works into two main categories, including 1) cascading approaches using different sensors in different stages [27,37,42], and 2) fusion methods that jointly reason over multi-sensor inputs [17,18]. Although effective, these methods have several limitations. Cascading approaches cannot leverage the complementarity among different sensors, and their performance is bounded by each stage. Fusion methods [17,18] need to generate BEV data through perspective projection and voxelization, leading to information loss inevitably. Besides, they can only approximately establish a relatively coarse correspondence between the voxel features and semantic image features. We propose a LiDAR-guided Image Fusion (LI-Fusion) module to address both the two issues mentioned above. LI-Fusion module establishes the correspondence between raw point cloud data and the camera image in a point-wise manner, and adaptively estimate the importance of the image semantic features. In this way, useful image features are utilized to enhance the point features while interfering image features are suppressed. Comparing with previous method, our solution possesses four main advantages, including 1) achieving fine-grained point-wise correspondence between LiDAR and camera image data through a simpler pipeline without complicated procedure for BEV data generation; 2) keeping the original geometric structure without information loss; 3) addressing the issue of the interference information that may be brought by the camera image; 4) free of image annotations, namely 2D bounding box annotations, as opposed to previous works [18,27].

Besides multi-sensor fusion, we observe the issue of the inconsistency between the classification confidence and localization confidence, which represent whether an object exists in a bounding box and how much overlap it shares with the ground truth. As shown in Fig. 1(c), the bounding box with higher classification confidence possesses lower localization confidence instead. This inconsistency will lead to degraded detection performance since the Non-Maximum Suppression (NMS) procedure automatically filters out boxes with large overlaps but low classification confidence. However, this problem is rarely discussed in the 3D detection task. Jiang *et al.* [9] attempt to alleviate this problem by improving the NMS procedure. They introduce a new branch to predict the localization confidence and replace the threshold for the NMS process as a multiplication of both the classification and localization confidences. Though effective to some extent, there is no explicit constraint to force the consistency of these two confidences. Different from [9], we present a consistency enforcing loss (CE loss) to guarantee the consistency of these two confidences explicitly. With its aid, boxes with high classification confidence are encouraged to possess large overlaps with the ground truth, and vice versa. This approach owns two advantages. First, our solution is easy to implement without any modifications to the architecture of the detection network. Second, our solution is entirely free of learnable parameters and extra inference time overhead.

Our key contributions are as follows:

1. Our LI-Fusion module operates on LiDAR point and camera image directly and effectively enhances the point features with corresponding semantic image features in a point-wise manner without image annotations.
2. We propose a CE loss to encourage the consistency between the classification and localization confidence, leading to more accurate detection results.
3. We integrate the LI-Fusion module and CE loss into a new framework named EPNet, which achieves state-of-the-art results on two common 3D object detection benchmark datasets, *i.e.*, the KITTI dataset [6] and SUN-RGBD dataset [33].

2   Related Work

3D Object Detection Based on Camera Images. Recent 3D object detection methods pay much attention to camera images, such as monocular [12,15,20,23,29] and stereo images [16,35]. Chen *et al.* [1] obtain 2D bounding boxes with a CNN-based object detector and infer their corresponding 3D bounding boxes with semantic, context, and shape information. Mousavian *et al.* [25] estimate localization and orientation from 2D bounding boxes of objects by exploiting the constraint of projective geometry. However, methods based on the camera image have difficulty in generating accurate 3D bounding boxes due to the lack of depth information.

3D Object Detection Based on LiDAR. Many LiDAR-based methods [24,39,40] are proposed in recent years. VoxelNet [43] divides a point cloud into voxels and employs stacked voxel feature encoding layers to extract voxel features. SECOND [38] introduces a sparse convolution operation to improve the computational efficiency of [43]. PointPillars [14] converts the point cloud to a pseudo-image and gets rid of time-consuming 3D convolution operations. PointRCNN [31] is a pioneering two-stage detector, which consists of a region proposal network (RPN) and a refinement network. The RPN network predicts the foreground points and outputs coarse bounding boxes which are then refined by the refinement network. However, LiDAR data is usually extremely sparse, posing a challenge for accurate localization.

3D Object Detection Based on Multiple Sensors. Recently, much progress has been made in exploiting multiple sensors, such as camera image and LiDAR. Qi *et al.* [27] propose a cascading approach F-PointNet, which first produces 2D proposals from camera images and then generates corresponding 3D boxes based on LiDAR point clouds. However, cascading methods need extra 2D annotations, and their performance is bounded by the 2D detector. Many methods attempt to reason over camera images and BEV jointly. MV3D [3] and AVOD [11] refine the detection box by fusing BEV and camera feature maps for each ROI region. ConFuse [18] proposes a novel continuous fusion layer that achieves the voxel-wise alignment between BEV and image feature maps. Different from previous works, our LI-Fusion module operates on LiDAR data directly and establishes a finer point-wise correspondence between the LiDAR and camera image features.

3   Method

Exploiting the complementary information of multiple sensors is important for accurate 3D object detection. Besides, it is also valuable to resolve the performance bottleneck caused by the inconsistency between the localization and classification confidence.

In this paper, we propose a new framework named EPNet to improve the 3D detection performance from these two aspects. EPNet consists of a two-stream RPN for proposal generation and a refinement network for bounding box refining, which can be trained end-to-end. The two-stream RPN effectively combines the LiDAR point feature and semantic image feature via the proposed LI-Fusion module. Besides, we provide a consistency enforcing loss (CE loss) to improve the consistency between the classification and localization confidence. In the following, we present the details of our two-steam RPN and refinement network in Subsect. 3.1 and Subsect. 3.2, respectively. Then we elaborate our CE loss and the overall loss function in Subsect. 3.4.

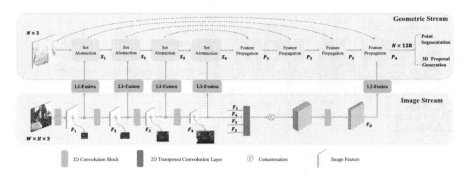

Fig. 2. Illustration of the architecture of the two-stream RPN which is composed of a geometric stream and an image stream. We employ several LI-Fusion modules to enhance the LiDAR point features with corresponding semantic image features in multiple scales. N represents the number of LiDAR points. H and W denote the height and width of the input camera image, respectively. (Color figure online)

3.1 Two-Stream RPN

Our two-stream RPN is composed of a geometric stream and an image stream. As shown in Fig. 2, the geometric stream and the image stream produce the point features and semantic image features, respectively. We employ multiple LI-Fusion modules to enhance the point features with corresponding semantic image features in different scales, leading to more discriminative feature representations.

Image Stream. The image stream takes camera images as input and extracts the semantic image information with a set of convolution operations. We adopt an especially simple architecture composed of four light-weighted convolutional blocks. Each convolutional block consists of two 3×3 convolution layers followed by a batch normalization layer [8] and a ReLU activation function. We set the second convolution layer in each block with stride 2 to enlarge the receptive field and save GPU memory. F_i ($i = 1,2,3,4$) denotes the outputs of these four convolutional blocks. As illustrated in Fig. 2, F_i provides sufficient semantic image information to enrich the LiDAR point features in different scales. We further employ four parallel transposed convolution layers with different strides to recover the image resolution, leading to feature maps with the same size as the original image. We combine them in a concatenation manner and obtain a more representative feature map F_U containing rich semantic image information with different receptive fields. As is shown later, the feature map F_U is also employed to enhance the LiDAR point features to generate more accurate proposals.

Geometric Stream. The geometric stream takes LiDAR point cloud as input and generates the 3D proposals. The geometric stream comprises four paired Set Abstraction (SA) [28] and Feature Propagation (FP) [28] layers for feature extraction. For the convenience of description, the outputs of SA and FP layers are denoted as S_i and P_i ($i = 1,2,3,4$), respectively. As shown in Fig. 2, we

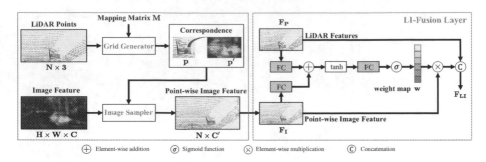

Fig. 3. Illustration of the LI-Fusion module, which consists of a grid generator, an image sampler, and a LI-Fusion layer.

combine the point features S_i with the semantic image features F_i with the aid of our LI-Fusion module. Besides, The point feature P_4 is further enriched by the multi-scale image feature F_U to obtain a compact and discriminative feature representation, which is then fed to the detection heads for foreground point segmentation and 3D proposal generation.

LI-Fusion Module. The LiDAR-guided image fusion module consists of a grid generator, an image sampler, and a LI-Fusion layer. As illustrated in Fig. 3, the LI-Fusion module involves two parts, *i.e.*, point-wise correspondence generation and LiDAR-guided fusion. Concretely, we project the LiDAR points onto the camera image and denote the mapping matrix as M. The grid generator takes a LiDAR point cloud and a mapping matrix M as inputs, and outputs the point-wise correspondence between the LiDAR points and the camera image under different resolutions. In more detail, for a particular point $p(x, y, z)$ in the point cloud, we can get its corresponding position $p'(x', y')$ in the camera image, which can be written as:

$$p' = M \times p, \tag{1}$$

where M is of size 3×4. Note that we convert p' and p into 3-dimensional and 4-dimensional vector in homogeneous coordinates in the projection process formula (1).

After establishing the correspondence, we propose to use an image sampler to get the semantic feature representation for each point. Specifically, our image sampler takes the sampling position p' and the image feature map F as inputs to produce a point-wise image feature representation V for each sampling position. Considering that the sampling position may fall between adjacent pixels, we use bilinear interpolation to get the image feature at the continuous coordinates, which can be formularized as follows:

$$V^{(p)} = \mathcal{K}(F^{(\mathcal{N}(p'))}), \tag{2}$$

where $V^{(p)}$ is the corresponding image feature for point p, $\mathcal{K}$ denotes the bilinear interpolation function, and $F^{(\mathcal{N}(p'))}$ represents the image features of the neighboring pixels for the sampling position p'.

Fusing the LiDAR feature and the point-wise image feature is non-trivial since the camera image is challenged by many factors, including illumination, occlusion, etc. In these cases, the point-wise image feature will introduce interfering information. To address this issue, we adopt a LiDAR-guided fusion layer, which utilizes the LiDAR feature to adaptively estimate the importance of the image feature in a point-wise manner. As illustrated in Fig. 3, we first feed the LiDAR feature F_P and the point-wise feature F_I into a fully connected layer and map them into the same channel. Then we add them together to form a compact feature representation, which is then compressed into a weight map $\mathbf{w}$ with a single channel through another fully connected layer. We use a sigmoid activation function to normalize the weight map $\mathbf{w}$ into the range of $[0, 1]$.

$$\mathbf{w} = \sigma(\mathcal{W}\tanh(\mathcal{U}F_P \mid \mathcal{V}F_I)) \tag{3}$$

where $\mathcal{W}, \mathcal{U}, \mathcal{V}$ denote the learnable weight matrices in our LI-Fusion layer. σ represents the sigmoid activation function.

After obtaining the weight map $\mathbf{w}$, we combine the LiDAR feature F_P and the semantic image feature F_I in a concatenation manner, which can be formularized as follows:

$$F_{LI} = F_P \parallel \mathbf{w}F_I \tag{4}$$

3.2   Refinement Network

We employ the NMS procedure to keep the high-quality proposals and feed them into the refinement network. For each input proposal, we generate its feature descriptor by randomly selecting 512 points in the corresponding bounding box on top of the last SA layer of our two-stream RPN. For those proposals with less than 512 points, we simply pad the descriptor with zeros. The refinement network consists of three SA layers to extract a compact global descriptor, and two subnetworks with two cascaded 1×1 convolution layers for the classification and regression, respectively.

3.3   Consistency Enforcing Loss

Common 3D object detectors usually generate much more bounding boxes than the number of the real objects in the scene. It poses a great challenge of how to select the high-quality bounding boxes. NMS attempts to filter unsatisfying bounding boxes according to their classification confidence. In this case, it is assumed that the classification confidence can serve as an agent for the real IoU between the bounding and the ground truth, *i.e.*, the localization confidence. However, the classification confidence and the localization confidence is often inconsistent, leading to sub-optimal performance.

This motivates us to introduce a consistency enforcing loss to ensure the consistency between the localization and classification confidence so that boxes with high localization confidence possess high classification confidence, and vice versa. The consistency enforcing loss can be written as follows:

$$L_{ce} = -log(c \times \frac{Area(D \cap G)}{Area(D \cup G)}) \qquad (5)$$

where D and G represents the predicted bounding box and the ground truth. c denotes the classification confidence for D. Towards optimizing this loss function, the classification confidence and localization confidence (*i.e.*, the IoU) are encouraged to be as high as possible jointly. Hence, boxes with large overlaps will possess high classification possibilities and be kept in the NMS procedure.

Relation to IoU Loss. Our CE loss is similar to the IoU loss [41] in the formula, but completely different in the motivation and the function. The IoU loss attempts to generate more precise regression through optimizing the IoU metric, while CE loss aims at ensuring the consistency between the localization and classification confidence to assist the NMS procedure to keep more accurate bounding boxes. Although with a simple formula, quantitative results and analyses in Sect. 4.3 demonstrates the effectiveness of our CE loss in ensuring the consistency and improving the 3D detection performance.

3.4 Overall Loss Function

We utilize a multi-task loss function for jointly optimizing the two-stream RPN and the refinement network. The total loss can be formulated as:

$$L_{total} = L_{rpn} + L_{rcnn}, \qquad (6)$$

where L_{rpn} and L_{rcnn} denote the training objective for the two-stream RPN and the refinement network, both of which adopt a similar optimizing goal, including a classification loss, a regression loss and a CE loss. We adopt the focal loss [19] as our classification loss to balance the positive and negative samples with the setting of $\alpha = 0.25$ and $\gamma = 2.0$. For a bounding box, the network needs to regress its center point (x, y, z), size (l, h, w), and orientation θ.

Since the range of the Y-axis (the vertical axis) is relatively small, we directly calculate its offset to the ground truth with a smooth L1 loss [7]. Similarly, the size of the bounding box (h, w, l) is also optimized with a smooth L1 loss. As for the X-axis, the Z-axis and the orientation θ, we adopt a bin-based regression loss [27,31]. For each foreground point, we split its neighboring area into several bins. The bin-based loss first predicts which bin b_u the center point falls in, and then regress the residual offset r_u within the bin. We formulate the loss functions as follows:

$$L_{rpn} = L_{cls} + L_{reg} + \lambda L_{cf} \qquad (7)$$

$$L_{cls} = -\alpha(1 - c_t)^\gamma \log c_t \qquad (8)$$

$$L_{reg} = \sum_{u \in x,z,\theta} E(b_u, \hat{b_u}) + \sum_{u \in x,y,z,h,w,l,\theta} S(r_u, \hat{r_u}) \qquad (9)$$

where E and S denote the cross entropy loss and the smooth L1 loss, respectively. c_t is the probability of the point in consideration belong to the ground truth category. $\hat{b}_u$ and $\hat{r}_u$ denote the ground truth of the bins and the residual offsets.

4 Experiments

We evaluate our method on two common 3D object detection datasets, including the KITTI dataset [6] and the SUN-RGBD dataset [33]. KITTI is an outdoor dataset, while SUN-RGBD focuses on the indoor scenes. In the following, we first present a brief introduction to these datasets in Subsect. 4.1. Then we provide the implementation details in Subsect. 4.2. Comprehensive analyses of the LI-Fusion module and the CE loss are elaborated in Subsect. 4.3. Finally, we exhibit the comparisons with state-of-the-art methods on the KITTI dataset and the SUN-RGBD dataset in Subsect. 4.4 and Subsect. 4.5, respectively.

4.1 Datasets and Evaluation Metric

KITTI Dataset is a standard benchmark dataset for autonomous driving, which consists of 7,481 training frames and 7,518 testing frames. Following the same dataset split protocol as [27,31], the 7,481 frames are further split into 3,712 frames for training and 3,769 frames for validation. In our experiments, we provide the results on both the validation and the testing set for all the three difficulty levels, *i.e.*, Easy, Moderate, and Hard. Objects are classified into different difficulty levels according to the size, occlusion, and truncation.

SUN-RGBD Dataset is an indoor benchmark dataset for 3D object detection. The dataset is composed of 10,335 images with 700 annotated object categories, including 5,285 images for training and 5,050 images for testing. We report results on the testing set for ten main object categories following previous works [27,37] since objects of these categories are relatively large.

Metrics. We adopt the Average Precision (AP) as the metric following the official evaluation protocol of the KITTI dataset and the SUN-RGBD dataset. Recently, the KITTI dataset applies a new evaluation protocol [32] which uses 40 recall positions instead of the 11 recall positions as before. Thus it is a fairer evaluation protocol. We compare our methods with state-of-the-art methods under this new evaluation protocol.

4.2 Implementation Details

Network Settings. The two-stream RPN takes both the LiDAR point cloud and the camera image as inputs. For each 3D scene, the range of LiDAR point cloud is $[-40, 40]$, $[-1, 3]$, $[0, 70.4]$ meters along the X (right), Y (down), Z (forward) axis in camera coordinate, respectively. And the orientation of θ is in the range of $[-\pi, \pi]$. We subsample 16,384 points from the raw LiDAR point cloud as the input for the geometric stream, which is same with PointRCNN [31]. And the image stream takes images with a resolution of 1280 × 384 as input. We employ four set abstraction layers to subsample the input LiDAR point cloud with the size of 4096, 1024, 256, and 64, respectively. Four feature propagation layers are used to recover the size of the point cloud for the foreground segmentation and 3D proposal generation. Similarly, we use four convolution block with stride 2 to downsample the input image. Besides, we employ four parallel transposed convolution with stride 2, 4, 8, 16 to recover the resolution from feature maps in different scales. In the NMS process, we select the top 8000 boxes generated by the two-stream RPN according to their classification confidence. After that, we filter redundant boxes with the NMS threshold of 0.8 and obtain 64 positive candidate boxes which will be refined by the refinement network. For both datasets, we utilize similar architecture design for the two-stream RPN as discussed above.

The Training Scheme. Our two-stream RPN and refinement network are end-to-end trainable. In the training phase, the regression loss L_{reg} and the CE loss are only applied to positive proposals, *i.e.*, proposals generated by foreground points for the RPN stage, and proposals sharing IoU larger than 0.55 with the ground truth for RCNN stage.

Parameter Optimization. The Adaptive Moment Estimation (Adam) [10] is adopted to optimize our network. The initial learning rate, weight decay, and momentum factor are set to 0.002, 0.001, and 0.9, respectively. We train the model for around 50 epochs on four Titan XP GPUs with a batch size of 12 in an end-to-end manner. The balancing weights λ in the loss function are set to 5.

Data Augmentation. Three common data augmentation strategies are adopted to prevent over-fitting, including rotation, flipping, and scale transformations. First, we randomly rotate the point cloud along the vertical axis within the range of $[-\pi/18, \pi/18]$. Then, the point cloud is randomly flipped along the forward axis. Besides, each ground truth box is randomly scaled following the uniform distribution of $[0.95, 1.05]$. Many LiDAR-based methods sample ground truth boxes from the whole dataset and place them into the raw 3D frames to simulate real scenes with crowded objects following [38,43]. Although effective, this data augmentation needs the prior information of road plane which is usually difficult to acquire for kinds of real scenes. Hence, we do not utilize this augmentation mechanism in our framework for the applicability and generality.

4.3   Ablation Study

We conduct extensive experiments on the KITTI validation dataset to evaluate the effectiveness of our LI-Fusion module and CE loss.

Analysis of the Fusion Architecture. We remove all the LI-Fusion modules to verify the effectiveness of our LI-Fusion module. As is shown in Table 1, adding LI-Fusion module yields an improvement of 1.73% in terms of 3D mAP, demonstrating its effectiveness in combining the point features and semantic image

Fig. 4. Visualization of the learned semantic image feature. The image stream mainly focuses on the foreground objects (cars). The red arrow marks the region under bad illumination, which show a distinct feature representation to its neighboring region. (Color figure online)

features. We further present comparisons with two alternative fusion solutions in Table 2. One alternative is simple concatenation (SC). We modify the input of the geometric stream as the combination of the raw camera image and LiDAR point cloud instead of their feature representations. Concretely, we append the RGB channels of camera images to the spatial coordinate channels of LiDAR point cloud in a concatenation fashion. It should be noted that no image stream is employed for SC. The other alternative is the single scale (SS) fusion, which shares a similar architecture as our two-stream RPN. The difference is that we remove all the LI-Fusion modules in the set abstraction layers and only keep the LI-Fusion module in the last feature propagation layer (see Fig. 2). As shown in Table 2, SC yields a decreasement of 3D mAP 0.28% over the baseline, indicating that simple combination in the input level cannot provide sufficient guidance information. Besides, our method outperforms SS by 3D mAP 1.31%. It suggests the effectiveness of applying the LI-Fusion modules in multiple scales.

Table 1. Ablation experiments on the KITTI val dataset.

LI-Fusion	CE	Easy	Moderate	Hard	3D mAP	Gain
×	×	86.34	77.52	75.96	79.94	–
✓	×	89.44	78.84	76.73	81.67	↑ 1.73
×	✓	90.87	81.15	79.59	83.87	↑ 3.93
✓	✓	**92.28**	**82.59**	**80.14**	**85.00**	↑ 5.06

Table 2. Analysis of different fusion mechanism on the KITTI val dataset.

SC	SS	Ours	Easy	Moderate	Hard	3D mAP	Gain
×	×	×	86.34	77.52	75.96	79.94	–
✓	×	×	85.97	77.37	75.65	79.66	↓ 0.28
×	✓	×	87.46	78.27	75.35	80.36	↑ 0.42
×	×	✓	**89.44**	**78.84**	**76.73**	**81.67**	↑ 1.73

Visualization of Learned Semantic Image Features. It should be noted that we do not add explicit supervision information (*e.g.*, annotations of 2D detection boxes) to the image stream of our two-stream RPN. The image stream is optimized together with the geometric stream with the supervision information of 3D boxes from the end of the two-stream RPN. Considering the distinct data characteristics of the camera image and LiDAR point cloud, we visualize the semantic image features to figure out what the image stream learns, as presented in Fig. 4. Although no explicit supervision is applied, surprisingly, the image stream learns well to differentiate the foreground objects from the background and extracts rich semantic features from camera images, demonstrating that the LI-Fusion module accurately establishes the correspondence between

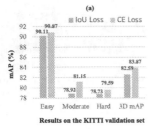 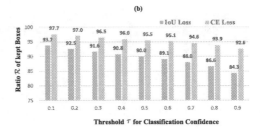

Fig. 5. Illustration of the ratio of kept positive candidate boxes varying with different classification confidence threshold. The CE loss leads to significantly larger ratios than those of the IoU loss, suggesting its effectiveness in improving the consistency of localization and classification confidence.

LiDAR point cloud and camera image, thus can provide the complementary semantic image information to the point features. It is also worth noting that the image stream mainly focuses on the representative region of the foreground objects and that the region under poor illumination demonstrates very distinct features to neighboring region, as marked by the red arrow. It indicates that it is necessary to adaptively estimate the importance of the semantic image feature since the variance of the illumination condition may introduce harmful interference information. Hence, we further provide the analysis of the weight map **w** for the semantic image feature in the following.

Analysis of the Weight Map in the LI-Fusion Layer. In a real scene, the camera image is usually disturbed by the illumination, suffering from underexposure and overexposure. To verify the effectiveness of the weight map **w** in alleviating the interference information brought by the unsatisfying camera image, we simulate the real environment by changing the illumination of the camera image. For each image in the KITTI dataset, we simulate the illumination variance through the transformation $y = a * x + b$, where x and y denote the original and transformed RGB value for a pixel. a and b represent the coefficient and the offset, respectively. We randomly lighten up (resp. darken) the camera images in the KITTI dataset by setting a to 3 (resp. 0.3) and b to 5. The quantitative results are presented in Table 3. For comparison, we remove the image stream and use our model based on only the LiDAR as the baseline, which yields a 3D mAP of 83.87%. We also provide the results of simply concatenating the RGB and LiDAR coordinates in the input level (denoted by SC), which leads to an obvious performance decreasement of 1.08% and demonstrates that images under poor quality is harmful for the 3D detection task. Besides, our method without estimating the weight map **w** also results in a decreasement of 0.69%. However, with the guidance of the weight map **w**, our method yields an improvement of 0.65% compared to the baseline. It means that introducing the weight map can adaptively select the beneficial features and ignore those harmful features.

Table 3. Comparison between the results our LI-Fusion module with and without estimating the weight map **w**.

Method	Easy	Moderate	Hard	3D mAP	Gain
Only LiDAR	90.87	81.15	79.59	83.87	–
SC	90.73	79.93	77.70	82.79	↓ 1.08
Ours (without **w**)	91.52	80.08	77.95	83.18	↓ 0.69
Ours	**91.65**	**81.77**	**80.13**	**84.52**	↑ 0.65

Table 4. The results of our approach on three benchmarks of the the KITTI validation set (*Cars*).

Benchmark	Easy	Moderate	Hard	mAP
3D Detection	92.28	82.59	80.14	85.00
Bird's Eye View	95.51	88.76	88.36	90.88
Orientation	98.48	91.74	91.16	93.79

Analysis of the CE Loss. As shown in Table 1, adding the CE loss yields a significant improvement of 3.93% over the baseline. We further present a quantitative comparison with the IoU loss to verify the superiority of our CE loss in improving the 3D detection performance. As shown in Fig. 5(a), the CE loss leads to an improvement of 3D mAP of 1.28% over the IoU loss, which indicates the benefits of ensuring the consistency of the classification and localization confidence in the 3D detection task.

To figure out how the consistency between these two confidences is improved, we give a thorough analysis of the CE loss. For the convenience of description, we denote predicted boxes possessing overlaps larger than a predefined IoU threshold τ as positive candidate boxes. Moreover, we adopt another threshold of v to filter positive candidate boxes with smaller classification confidence. Hence, the consistency can be evaluated by the ratio of $\mathcal{R}$ of how many positive candidate boxes are kept, which can be written as follows:

$$\mathcal{R} = \frac{\mathcal{N}(\mathbf{b}|\mathbf{b} \in \mathcal{B} \text{ and } \mathbf{c_b} > v)}{\mathcal{N}(\mathcal{B})}, \qquad (10)$$

where $\mathcal{B}$ represents the set of positive candidate boxes. $\mathbf{c_b}$ denotes the classification confidence of the box **b**. $\mathcal{N}(\cdot)$ calculates the number of boxes. It should be noted that all the boxes in $\mathcal{B}$ possess an overlap larger than τ with the corresponding ground truth box.

We provide evaluation results on two different settings, *i.e.*, the model trained with IoU loss and that trained with CE loss. For each frame in the KITTI validation dataset, the model generates 64 boxes without NMS procedure employed. Then we get the positive candidate boxes by calculating the overlaps with the ground truth boxes. We set τ to 0.7 following the evaluation protocol of 3D detection metric. v is varied from 0.1 to 0.9 to evaluate the consistency under different classification confidence thresholds. As is shown in Fig. 5(b), the model trained with CE loss demonstrates better consistency than that trained with IoU loss in all the different settings of classification confidence threshold v.

Table 5. Comparisons with state-of-the-art methods on the testing set of the KITTI dataset (*Cars*). L and I represent the LiDAR point cloud and the camera image.

Method	Modality	3D Detection				Bird's Eye View				Orientation			
		Easy	Moderate	Hard	3D mAP	Easy	Moderate	Hard	BEV mAP	Easy	Moderate	Hard	Ori mAP
SECOND [38]	L	83.34	72.55	65.82	73.90	89.39	83.77	78.59	83.92	90.93	82.55	73.62	82.37
PointPillars [14]	L	82.58	74.31	68.99	75.29	90.07	86.56	82.81	86.48	93.84	90.70	87.47	90.67
TANet [21]	L	84.39	75.94	68.82	76.38	91.58	86.54	81.19	86.44	93.52	90.11	84.61	89.41
PointRCNN [31]	L	86.96	75.64	70.70	77.77	92.13	87.39	82.72	87.41	95.90	91.77	86.92	91.53
Fast Point R-CNN [4]	L	85.29	77.40	70.24	77.64	90.87	87.84	80.52	86.41	-	-	-	-
F-PointNet [27]	L+I	82.19	69.79	60.59	70.86	91.17	84.67	74.77	83.54	-	-	-	-
MV3D [3]	L+I	74.97	63.63	54.00	64.20	86.62	78.93	69.80	78.45	-	-	-	-
AVOD [11]	L+I	76.39	66.47	60.23	67.70	89.75	84.95	78.32	84.34	94.98	89.22	82.14	88.78
AVOD-FPN [11]	L+I	83.07	71.76	65.73	73.52	90.99	84.82	79.62	85.14	94.65	88.61	83.71	88.99
ContFuse [18]	L+I	83.68	68.78	61.67	71.38	94.07	85.35	75.88	85.10	-	-	-	-
PC-CNN [5]	L+I	85.57	73.79	65.65	75.00	91.19	87.40	79.35	85.98	-	-	-	-
MMF [17]	L+I	88.40	77.43	70.22	78.68	93.67	88.21	81.99	87.96	-	-	-	-
Ours	L+I	**89.81**	**79.28**	**74.59**	**81.23**	**94.22**	**88.47**	**83.69**	**88.79**	**96.13**	**94.22**	**89.68**	**93.34**

Table 6. Quantitative comparisons with the state-of-the-art methods on the SUN-RGBD test set. P and I represent the point cloud and the camera image.

Method	Modality	Bathtub	Bed	Bookshelf	Chair	Desk	Dresser	Nightstand	Sofa	Table	Toilet	3D mAP
DSS [34]	P + I	44.2	78.8	11.9	61.2	20.5	6.4	15.4	53.5	50.3	78.9	42.1
2d-driven [13]	P + I	43.5	64.5	31.4	48.3	27.9	25.9	41.9	50.4	37.0	80.4	45.1
COG [30]	P + I	58.3	63.7	31.8	62.2	**45.2**	15.5	27.4	51.0	**51.3**	70.1	47.6
PointFusion [37]	P + I	37.3	68.6	**37.7**	55.1	17.2	24.0	32.3	53.8	31.0	83.8	44.1
F-PointNet [27]	P + I	43.3	81.1	33.3	64.2	24.7	32.0	58.1	61.1	51.1	**90.9**	54.0
VoteNet [26]	P	74.4	83.0	28.8	**75.3**	22.0	29.8	62.2	64.0	47.3	90.1	57.7
Ours	P + I	**75.4**	**85.2**	35.4	75.0	26.1	31.3	62.0	**67.2**	52.1	88.2	**59.8**

4.4   Experiments on KITTI Dataset

Table 5 presents quantitative results on the KITTI test set. The proposed method outperforms multi-sensor based methods F-PointNet [27], MV3D [3], AVOD-FPN [11], PC-CNN [5], ContFuse [18], and MMF [17] by 10.37%, 17.03%, 7.71%, 6.23%, 9.85% and 2.55% in terms of 3D mAP. It should be noted that MMF [17] exploits multiple auxiliary tasks (*e.g.*, 2D detection, ground estimation, and depth completion) to boost the 3D detection performance, which requires many extra annotations. These experiments consistently reveal the superiority of our method over the cascading approach [27], as well as fusion approaches based on RoIs [3,5,11] and voxels [17,18].

We also provide the quantitative results on the KITTI validation split in the Table 4 for the convenience of comparison with future work. Besides, we present the qualitative results on the KITTI validation dataset in the supplementary materials.

4.5   Experiments on SUN-RGBD Dataset

We further conduct experiments on the SUN-RGBD dataset to verify the effectiveness of our approach in the indoor scenes. Table 6 demonstrates the results compared with the state-of-the-art methods. Our EPNet achieves superior detection performance, outperforming PointFusion [37] by 15.7%, COG [30] by 12.2%,

F-PointNet [27] by 5.8% and VoteNet [26] by 2.1% in terms of 3D mAP. The comparisons with multi-sensor based methods PointFusion [37] and F-PointNet [27] are especially valuable. Both of them first generate 2D bounding boxes from camera images using 2D detectors and then outputs the 3D boxes in a cascading manner. Specifically, F-PointNet utilizes only the LiDAR data to predict the 3D boxes. PointFusion combines global image features and points features in a concatenation fashion. Different from them, our method explicitly establishes the correspondence between point features and camera image features, thus providing finer and more discriminative representations. Besides, we provide the qualitative results on the SUN-RGBD dataset in the supplementary materials.

5 Conclusion

We have presented a new 3D object detector named EPNet, which consists of a two-stream RPN and a refinement network. The two-stream RPN reasons about different sensors (*i.e.*, LiDAR point cloud and camera image) jointly and enhances the point features with semantic image features effectively by using the proposed LI-Fusion module. Besides, we address the issue of inconsistency between the classification and localization confidence by the proposed CE loss, which explicitly guarantees the consistency between the localization and classification confidence. Extensive experiments have validated the effectiveness of the LI-Fusion module and the CE loss. In the future, we are going to explore how to enhance the image feature representation with depth information of the LiDAR point cloud instead, and its application in 2D detection tasks.

Acknowledgement. This work was supported by National Key R&D Program of China (No. 2018YFB 1004600), Xiang Bai was supported by the National Program for Support of Top-notch Young Professionals and the Program for HUST Academic Frontier Youth Team 2017QYTD08.

References

1. Chen, X., Kundu, K., Zhang, Z., Ma, H., Fidler, S., Urtasun, R.: Monocular 3D object detection for autonomous driving. In: Proceedings of IEEE International Conference on Computer Vision and Pattern Recognition (2016)
2. Chen, X., Kundu, K., Zhu, Y., Ma, H., Fidler, S., Urtasun, R.: 3D object proposals using stereo imagery for accurate object class detection. IEEE Trans. Pattern Anal. Mach. Intell. **40**(5), 1259–1272 (2017)
3. Chen, X., Ma, H., Wan, J., Li, B., Xia, T.: Multi-view 3D object detection network for autonomous driving. In: Proceedings of IEEE International Conference on Computer Vision and Pattern Recognition (2017)
4. Chen, Y., Liu, S., Shen, X., Jia, J.: Fast point R-CNN. In: Proceedings of IEEE International Conference on Computer Vision (2019)
5. Du, X., Ang, M.H., Karaman, S., Rus, D.: A general pipeline for 3D detection of vehicles. In: 2018 IEEE International Conference on Robotics and Automation (ICRA), pp. 3194–3200 (2018). https://doi.org/10.1109/ICRA.2018.8461232

6. Geiger, A., Lenz, P., Urtasun, R.: Are we ready for autonomous driving. In: Proceedings of IEEE International Conference on Computer Vision and Pattern Recognition

7. Girshick, R.: Fast R-CNN. In: Proceedings of the IEEE International Conference on Computer Vision, pp. 1440–1448 (2015)

8. Ioffe, S., Szegedy, C.: Batch normalization: accelerating deep network training by reducing internal covariate shift. In: Proceedings of International Conference on Machine Learning (2015)

9. Jiang, B., Luo, R., Mao, J., Xiao, T., Jiang, Y.: Acquisition of localization confidence for accurate object detection. In: Ferrari, V., Hebert, M., Sminchisescu, C., Weiss, Y. (eds.) Computer Vision – ECCV 2018. LNCS, vol. 11218, pp. 816–832. Springer, Cham (2018). https://doi.org/10.1007/978-3-030-01264-9_48

10. Kingma, D.P., Ba, J.: Adam: a method for stochastic optimization. In: ICLR (2014)

11. Ku, J., Mozifian, M., Lee, J., Harakeh, A., Waslander, S.L.: Joint 3D proposal generation and object detection from view aggregation. In: IROS, pp. 1–8. IEEE (2018)

12. Ku, J., Pon, A.D., Waslander, S.L.: Monocular 3D object detection leveraging accurate proposals and shape reconstruction. In: CVPR (2019)

13. Lahoud, J., Ghanem, B.: 2D-driven 3D object detection in RGB-D images. In: Proceedings of IEEE International Conference on Computer Vision (2017)

14. Lang, A.H., Vora, S., Caesar, H., Zhou, L., Yang, J., Beijbom, O.: Pointpillars: fast encoders for object detection from point clouds. In: Proceedings of IEEE International Conference on Computer Vision and Pattern Recognition (2019)

15. Li, B., Ouyang, W., Sheng, L., Zeng, X., Wang, X.: GS3D: an efficient 3D object detection framework for autonomous driving. In: IEEE Conference on Computer Vision and Pattern Recognition (CVPR) (2019)

16. Li, P., Chen, X., Shen, S.: Stereo R-CNN based 3D object detection for autonomous driving. In: CVPR (2019)

17. Liang, M., Yang, B., Chen, Y., Hu, R., Urtasun, R.: Multi-task multi-sensor fusion for 3D object detection. In: Proceedings of IEEE International Conference on Computer Vision and Pattern Recognition (2019)

18. Liang, M., Yang, B., Wang, S., Urtasun, R.: Deep continuous fusion for multi-sensor 3D object detection. In: Ferrari, V., Hebert, M., Sminchisescu, C., Weiss, Y. (eds.) ECCV 2018. LNCS, vol. 11220, pp. 663–678. Springer, Cham (2018). https://doi.org/10.1007/978-3-030-01270-0_39

19. Lin, T.Y., Goyal, P., Girshick, R., He, K., Dollár, P.: Focal loss for dense object detection. In: Proceedings of IEEE International Conference on Computer Vision (2017)

20. Liu, L., Lu, J., Xu, C., Tian, Q., Zhou, J.: Deep fitting degree scoring network for monocular 3D object detection. In: Proceedings of the IEEE Conference on Computer Vision and Pattern Recognition, pp. 1057–1066 (2019)

21. Liu, Z., Zhao, X., Huang, T., Hu, R., Zhou, Y., Bai, X.: Tanet: robust 3D object detection from point clouds with triple attention. In: AAAI, pp. 11677–11684 (2020)

22. Luo, W., Yang, B., Urtasun, R.: Fast and furious: real time end-to-end 3D detection, tracking and motion forecasting with a single convolutional net. In: Proceedings of IEEE International Conference on Computer Vision and Pattern Recognition (2018)

23. Ma, X., Wang, Z., Li, H., Zhang, P., Ouyang, W., Fan, X.: Accurate monocular object detection via color-embedded 3D reconstruction for autonomous driving. In: Proceedings of the IEEE International Conference on Computer Vision (ICCV) (2019)
24. Meyer, G.P., Laddha, A., Kee, E., Vallespi-Gonzalez, C., Wellington, C.K.: Laser-Net: an efficient probabilistic 3D object detector for autonomous driving. In: Proceedings of the IEEE Conference on Computer Vision and Pattern Recognition (CVPR) (2019)
25. Mousavian, A., Anguelov, D., Flynn, J., Kosecka, J.: 3D bounding box estimation using deep learning and geometry. In: Proceedings of IEEE International Conference on Computer Vision and Pattern Recognition (2017)
26. Qi, C.R., Litany, O., He, K., Guibas, L.J.: Deep hough voting for 3D object detection in point clouds. In: Proceedings of IEEE International Conference on Computer Vision (2019)
27. Qi, C.R., Liu, W., Wu, C., Su, H., Guibas, L.J.: Frustum PointNets for 3D object detection from RGB-D data. In: Proceedings of IEEE International Conference on Computer Vision and Pattern Recognition (2018)
28. Qi, C.R., Yi, L., Su, H., Guibas, L.J.: PointNet++: deep hierarchical feature learning on point sets in a metric space. In: Advances in Neural Information Processing Systems, pp. 5099–5108 (2017)
29. Qin, Z., Wang, J., Lu, Y.: MonoGRNet: a geometric reasoning network for 3D object localization. In: The Thirty-Third AAAI Conference on Artificial Intelligence (AAAI-19) (2019)
30. Ren, Z., Sudderth, E.B.: Three-dimensional object detection and layout prediction using clouds of oriented gradients. In: Proceedings of IEEE International Conference on Computer Vision and Pattern Recognition (2016)
31. Shi, S., Wang, X., Li, H.: PointRCNN: 3D object proposal generation and detection from point cloud. In: Proceedings of IEEE International Conference on Computer Vision and Pattern Recognition (2019)
32. Simonelli, A., Bulò, S.R.R., Porzi, L., López-Antequera, M., Kontschieder, P.: Disentangling monocular 3D object detection. arXiv preprint arXiv:1905.12365 (2019)
33. Song, S., Lichtenberg, S.P., Xiao, J.: Sun RGB-D: a RGB-D scene understanding benchmark suite. In: Proceedings of IEEE International Conference on Computer Vision and Pattern Recognition (2015)
34. Song, S., Xiao, J.: Deep sliding shapes for amodal 3D object detection in RGB-D images. In: Proceedings of IEEE International Conference on Computer Vision and Pattern Recognition (2016)
35. Wang, Y., Chao, W.L., Garg, D., Hariharan, B., Campbell, M., Weinberger, K.: Pseudo-LIDAR from visual depth estimation: bridging the gap in 3D object detection for autonomous driving. In: CVPR (2019)
36. Xu, B., Chen, Z.: Multi-level fusion based 3D object detection from monocular images. In: Proceedings of IEEE International Conference on Computer Vision and Pattern Recognition (2018)
37. Xu, D., Anguelov, D., Jain, A.: PointFusion: deep sensor fusion for 3D bounding box estimation. In: Proceedings of IEEE International Conference on Computer Vision and Pattern Recognition (2018)
38. Yan, Y., Mao, Y., Li, B.: Second: sparsely embedded convolutional detection. Sensors 18(10), 3337 (2018)
39. Yang, B., Luo, W., Urtasun, R.: PIXOR: real-time 3D object detection from point clouds. In: Proceedings of IEEE International Conference on Computer Vision and Pattern Recognition (2018)

40. Yang, Z., Sun, Y., Liu, S., Shen, X., Jia, J.: STD: sparse-to-dense 3D object detector for point cloud. In: ICCV (2019). http://arxiv.org/abs/1907.10471
41. Yu, J., Jiang, Y., Wang, Z., Cao, Z., Huang, T.: UnitBox: an advanced object detection network. In: Proceedings of the 24th ACM International Conference on Multimedia (2016)
42. Zhao, X., Liu, Z., Hu, R., Huang, K.: 3D object detection using scale invariant and feature reweighting networks. In: Proceedings of the AAAI Conference on Artificial Intelligence, vol. 33, pp. 9267–9274 (2019)
43. Zhou, Y., Tuzel, O.: VoxelNet: end-to-end learning for point cloud based 3D object detection. In: Proceedings of IEEE International Conference on Computer Vision and Pattern Recognition (2018)

Self-Supervised Monocular 3D Face Reconstruction by Occlusion-Aware Multi-view Geometry Consistency

Jiaxiang Shang[1]([⊠]) [iD], Tianwei Shen[1] [iD], Shiwei Li[1] [iD], Lei Zhou[1] [iD],
Mingmin Zhen[1] [iD], Tian Fang[2] [iD], and Long Quan[1] [iD]

[1] Hong Kong University of Science and Technology, Kowloon, Hong Kong
{jshang,tshenaa,lzhouai,mzhen,quan}@cse.ust.hk
[2] Everest Innovation Technology, Kowloon, Hong Kong
{sli,fangtian}@altizure.com

Abstract. Recent learning-based approaches, in which models are trained by single-view images have shown promising results for monocular 3D face reconstruction, but they suffer from the ill-posed face pose and depth ambiguity issue. In contrast to previous works that only enforce 2D feature constraints, we propose a self-supervised training architecture by leveraging the multi-view geometry consistency, which provides reliable constraints on face pose and depth estimation. We first propose an occlusion-aware view synthesis method to apply multi-view geometry consistency to self-supervised learning. Then we design three novel loss functions for multi-view consistency, including the pixel consistency loss, the depth consistency loss, and the facial landmark-based epipolar loss. Our method is accurate and robust, especially under large variations of expressions, poses, and illumination conditions. Comprehensive experiments on the face alignment and 3D face reconstruction benchmarks have demonstrated superiority over state-of-the-art methods. Our code and model are released in https://github.com/jiaxiangshang/MGCNet.

Keywords: 3D face reconstruction · Multi-view geometry consistency

1 Introduction

3D face reconstruction is extensively studied in the computer vision community. Traditional optimization-based methods [2,4,10,27,41–43] formulate the 3D Morphable Model (3DMM) [7] parameters into a cost minimization problem, which is usually solved by expensive iterative nonlinear optimization. The supervised CNN-based methods [16–18,24,31,32,39,49,53,59] require abundant 3D face scans and corresponding RGB images, which are limited in amount

Electronic supplementary material The online version of this chapter (https://doi.org/10.1007/978-3-030-58555-6_4) contains supplementary material, which is available to authorized users.

A. Vedaldi et al. (Eds.): ECCV 2020, LNCS 12360, pp. 53–70, 2020.
https://doi.org/10.1007/978-3-030-58555-6_4

and expensive to acquire. Methods that focus on face detail reconstruction [12,21,49,53,54,62] need even high-quality 3D faces scans. To address the insufficiency of scanned 3D face datasets, some unsupervised or self-supervised methods are proposed [15,22,40,50–52,54,55,61,66], which employ the 2D facial *landmark loss* between inferred 2D landmarks projected from 3DMM and the ground truth 2D landmarks from images, as well as the *render loss* between the rendered images from 3DMM and original images. One critical drawback of existing unsupervised methods is that both *landmark loss* and *render loss* are measured in projected 2D image space and do not penalize incorrect face pose and depth value of 3DMM, resulting in the ambiguity issue of the 3DMM in the face pose and depth estimation.

To address this issue, we resort to the multi-view geometry consistency. Multi-view images not only contain 2D landmarks and pixel features but also they form the multi-view geometry constraints. Such training data is publicly available and efficient to acquire (*e.g.*,videos). Fortunately, a series of multi-view 3D reconstruction techniques named view synthesis [13,14,19] help to formulate self-supervised learning architecture based on multi-view geometry. View synthesis is a classic task that estimates proxy 3D geometry and establishes pixel correspondences among multi-view input images. Then they generate $N-1$ synthetic target view images by compositing image patches from the other $N-1$ input view images. View synthesis is commonly used in Monocular Depth Estimation (MDE) task [11,34,63,65]. However, MDE only predicts depth map and relative poses between views without inferring camera intrinsics. The geometry of MDE is incomplete as MDE loses the relationship from 3D to 2D, and they can not reconstruct a full model in scene. MDE also suffers from erroneous penalization due to self-occlusion.

Inspired by multi-view geometry consistency, we propose a self-supervised Multi-view Geometry Consistency based 3D Face Reconstruction framework (MGCNet). The workflow of MGCNet is shown in Fig. 1. I_t is always considered to be the **target** of multi-view data. To simplify the following formulation, we denote all $N-1$ views adjacent to the **target view** as the **source views**. To build up the multi-view consistency in the training process via view synthesis, we first design a covisible map that stores the mask of covisible pixels for each target-source view pair to solve self-occlusion, as the large and extreme face pose cases is common in the real world, and the self-occlusion always happens in such profile face pose cases. Secondly, we feed the 3DMM coefficients and face poses to the differentiable rendering module [22], producing the rendered image, depth map, and covisible map for each view. Thirdly, pixel consistency loss and depth consistency loss are formulated by input images and rendered depth maps in covisible regions, which ensures the consistency of 3DMM parameters in the multi-view training process. Finally, we introduce the facial epipolar loss, which formulates the epipolar error of 2D facial landmarks via the relative pose of two views, as facial landmarks is robust to illumination changes, scale ambiguity, and calibration errors. With these multi-view supervised losses, we are able to achieve accurate 3D face reconstruction and face alignment result on multiple datasets

[3, 36, 60, 64, 67]. We conduct ablation experiments to validate the effectiveness of covisible map and multi-view supervised losses.

To summarize, this paper makes the following main contributions:

- We propose an end-to-end self-supervised architecture MGCNet for face alignment and monocular 3D face reconstruction tasks. To our best knowledge, we are the first to leverage multi-view geometry consistency to mitigate the ambiguity from monocular face pose estimation and depth reconstruction in the training process.
- We build a differentiable covisible map for general view synthesis, which can mitigate the self-occlusion crux of view synthesis. Based on view synthesis, three differentiable multi-view geometry consistency loss functions are proposed as pixel consistency loss, depth consistency loss, and facial epipolar loss.
- Our MGCNet result on the face alignment benchmark [67] shows that we achieve more than a **12%** improvement over other state-of-the-art methods, especially in large and extreme face pose cases. Comparison on the challenging 3D Face Reconstruction datasets [3, 36, 60, 64] shows that MGCNet outperforms the other methods with the largest margin of **17%**.

2   Related Work

2.1   Single-View Method

Recent CNN methods [12, 17, 18, 21, 24, 28, 32, 53, 54, 56, 59, 62, 67] train the CNN network supervised by 3D face scan ground truth and achieve impressive results. [17, 24, 39] generate synthetic rendered face images with real 3D scans. [18, 28, 32, 53, 56, 59] propose their deep neural networks trained using fitted 3D shapes by traditional methods as substitute labels. Lack of realistic training data is still a great hindrance.

Recently, some self-supervised or weak-supervised methods are proposed [22, 40, 45, 51, 52, 54, 55, 61, 66] to solve the lack of high-quality 3D face scans with robust testing result. Tewari *et al.* [52] propose an differentiable rendering process to build unsupervised face autoencoder based on pixel loss. Genova *et al.* [22] train a regression network mainly focus on identity loss that compares the features of the predicted face and the input photograph. Nevertheless, face pose and depth ambiguity originated from only monocular images still a limitation.

2.2   Multi-view or Video Based Method

There are established toolchains of 3D reconstruction [20, 46, 47, 57], aimming at recovering 3D geometry from multi-view images. One related operation is view synthesis [13, 19, 48], and the goal is to synthesize the appearance of the scene from novel camera viewpoints.

Several unsupervised approaches [15, 45, 50, 58] are proposed recently to address the 3D face reconstruction from multiple images or videos.

Deng *et al.* [15] perform multi-image face reconstruction from different images by shape aggregation. Sanyal *et al.* [45] take multiple images of the same and different person, then enforce shape consistency between the same subjects and shape inconsistency between the different subjects. Wu *et al.* [58] design an impressive multi-view framework (MVFNet), which is view-consistent by design, and photometric consistency is used to generate consistent texture across views. However, MVFNet is not able to generate results via a single input since it relies on multi-view aggregation during inference. Our MGCNet explicitly exploit multi-view consistency (both geometric and photometric) to constrain the network to produce view-consistent face geometry from **a single input**, which provides better supervision than 2D information only. Therefore, MGCNet improves the performance of face alignment and 3D face reconstruction as Sect. 4.

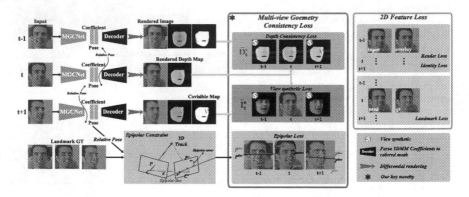

Fig. 1. The training flow of our MGCNet architecture, which is annotated in Sect. 3.1. The 2D feature loss part is our **baseline** in Sect. 3.3. Our novel multi-view geometry consistency loss functions are highlighted as ∗ in Sect. 3.

3   Method

3.1   Overview

The inference process of MGCNet takes a single image as input, while in the training process the input is N-view images (e.g. I_{t-1}, I_t, I_{t+1} for $N = 3$) of the same face and the corresponding ground-truth 2D landmarks $q_{t-1}^{gt}, q_t^{gt}, q_{t+1}^{gt}$. Then, the MGCNet estimates the 3DMM coefficients and face poses, whose notations are introduced in Sect. 3.2 in detail. $\widetilde{\mathcal{I}}_t^{(s)}, \widetilde{\mathcal{D}}_t^{(s)}$ represent the synthesized target images and the depth maps from the source views.

3.2   Model

Face Model. 3D Morphable Model (3DMM) proposed by [7] is the face prior model of the MGCNet. Specifically, the 3DMM encodes both face shape and texture as

$$S = S(\alpha, \beta) = S_{mean} + A_{id}\alpha + B_{exp}\beta$$
$$T = T(\gamma) = T_{mean} + T_{id}\gamma, \tag{1}$$

where S_{mean} and T_{mean} denote the mean shape and the mean albedo respectively. A_{id}, B_{exp} and T_{id} are the PCA bases of identity, expression and texture. $\alpha, \beta \in R^{80}$ and $\gamma \in R^{64}$ are corresponding coefficient vectors to be estimated follow [15,51,52]. We use S_{mean}, T_{mean}, A_{id} and T_{id} provided by the Basel Face Model (BFM) [7,35], and B_{exp} from FaceWarehouse [9]. We exclude the ear and neck region as [15], and our final face model contains $\sim$36K vertices.

Camera Model. The pinhole camera model is employed to define the 3D-2D projection. We assume the camera is calibrated. The face pose P is represented by an euler angle rotation $R \in SO(3)$ and translation $t \in R^3$. The relative poses $P_{t \to s}^{rel} \in SE(3)$ from the **target view** to $N-1$ **source views** are defined as $P_{t \to s}^{rel} = \begin{bmatrix} R_t^{-1}R_s & R_t^{-1}(t_s - t_t) \\ 0 & 1 \end{bmatrix}$.

Illumination Model. To acquire realistic rendered face images, we model the scene illumination by Spherical Harmonics (SH) [37,38] as $SH(N_{re}, T|\theta) = T * \sum_{b=1}^{B^2} \theta_b H_b$, where N_{re} is the normal of the face mesh, $\theta \in R^{27}$ is the coefficient. The $H_b : R^3 \to R$ are SH basis functions and the $B^2 = 9$ (B = 3 bands) parameterizes the colored illumination in red, green and blue channels.

Finally, we concatenate all the parameters together into a $(\alpha, \beta, \gamma, R, t, \theta)$ 257-dimensional vector. All 257 parameters encode the 3DMM coefficients and the face pose, which are abbreviated as *coefficient* and *pose* in Fig. 1.

3.3   2D Feature Loss

Inspired by recent related works, we leverage preliminary 2D feature loss functions in our framework.

Render Loss. The render loss aims to minimize the difference between the input face image and the rendered image as $\mathcal{L}_{render} = \frac{1}{M}\sum_{i=1}^{M} w_{skin}^i ||I^i - I_{re}^i||$, where I_{re} is the rendered image, I is the input image, and M is the number of all 2D pixels in the projected 3D face region. w_{skin}^i is the skin confidence of ith pixel as in [15]. Render loss mainly contributes to the albedo of 3DMM.

Landmark Loss. To improve the accuracy of face alignment, we employ the 2D landmark loss which defines the distance between predicted landmarks and ground truth landmarks as $\mathcal{L}_{lm} = \sum_{i=1}^{N} c_{lm}^i (q_{gt}^i - q^i)^2$, where N is the number of landmarks, and q the projection of the 3D landmarks picked from our face model. It is noted that the landmarks have different levels of importance, denoted as the confidence c_{lm} for each landmark. We set the confidence to 10 only for the nose and inner mouth landmarks, and to 1 else wise.

Identity Loss. The fidelity of the reconstructed face is an important criterion. We use the identity loss as in [22], which is the cosine distance between deep features of the input images and rendered images as $\mathcal{L}_{id} = \frac{\eta_1 \circ \eta_2}{|\eta_1||\eta_2|}$, where $\circ$ means the element-wise multiplication. η_1 and η_2 are deep features of input images and rendered images.

Regularization Loss. To prevent the face shape and texture parameters from diverging, regularization loss of 3DMM is used as $\mathcal{L}_{reg} = w_{id} \sum_{i=1}^{N_\alpha} \alpha^2 + w_{exp} \sum_{i=1}^{N_\beta} \beta^2 + w_{tex} \sum_{i=1}^{N_\gamma} \gamma^2$ where $w_{id}, w_{exp}, w_{shape}$ are trade-off parameters for 3DMM coefficients regularization ($1.0, 0.8, 3e{-}3$ by default). $N_\alpha, N_\beta, N_\gamma$ are the length of 3DMM parameters α, β, γ.

Final 2D Feature Loss. The combined 2D feature loss function $\mathcal{L}_{2D}$ is defined as $\mathcal{L}_{2D} = w_{render}\mathcal{L}_{render} + w_{lm}\mathcal{L}_{lm} + w_{id}\mathcal{L}_{id} + w_{reg}\mathcal{L}_{reg}$, where the trade-off parameters for 2D feature losses are set empirically $w_{render} = 1.9, w_{lm} = 1e{-}3, w_{id} = 0.2, w_{reg} = 1e{-}4$. We regard the **baseline** approach in the later experiement as the model trained by only 2D feature losses. In the followings, we present key ingredients of our contributions.

3.4   Occlusion-Aware View Synthesis

The key of our idea is to enforce multi-view geometry consistency, so as to achieve the self-supervised training. This could be done via view synthesis, which establishes dense pixel correspondences across multi-view input images. However, the view synthesis for face reconstruction is very easily affected by self-occlusion, as large and extreme face pose cases are common in read world applications. As shown in Fig. 2, assuming a pixel p_t is visible in the left cheek as shown in Fig. 2(a), the correspondence pixel p_s could not be found in Fig. 2(b) due to nose occlusion. Self-occlusion leads to redundant pixel consistency loss and depth consistency loss. Furthermore the related gradient of self-occlusion pixels will be highly affected by the salient redundant error as red part in Fig. 3 (Pixel Consistency Loss subfigure), which makes the training more difficult. For more practical and useful navigation in real scenarios, self-occlusion is worth to solve.

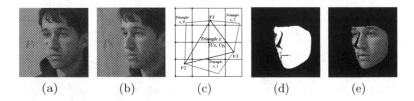

(a)   (b)   (c)   (d)   (e)

Fig. 2. (a) and (b) are the target view and source view pair; (c) is the covisible points and triangles; (d) is the covisible map; and (e) is the synthetic target view

We introduce the covisible maps C_s to account for the self-occlusion. Covisible map is a binary mask indicating the pixels which are visible in both

source and target views. During the rendering process of the MGCNet, ras-
terization builds the correspondence between vertices of a triangle and image
pixels $(V_{1,2,3} \sim U_x, U_y)$, as shown in Fig. 2(c). The common vertices visible in
two views (i.e., vertices that contribute to pixel rendering) are called covisible
points. Then we define all triangles adjacent to covisible points as covisible triangles. Finally, we project covisible triangles of the 3D face from the target view
to image space, as shown in Fig. 2(d), where the white region is covisible region.
The improvement brings from covisible maps is elaborated in Fig. 3, pixels are
not covisible in the left of the nose in target view (red in Fig. 3), which result in
redundant error. The quantitative improvements are discussed in Sect. 4.5.

To generate the synthetic target RGB images from source RGB images, we
first formulate the pixel correspondences between view pairs (I_s, I_t). Given a pair
correspondence pixel coordinate p_t, p_s in I_t, I_s, the pixel value p_s is computed
by bilinear-sampling [29], and the pixel coordinate p_s is defined as

$$p_s \sim \mathbf{K}_s [\mathbf{P}_{t \to s}^{rel}] \mathbf{D}_t(p_t) \mathbf{K}_t^{-1} p_t, \tag{2}$$

where $\sim$ represents the equality in the homogeneous coordinates, $\mathbf{K}_s$ and $\mathbf{K}_t$
are the intrinsics for the input image pairs, $\mathcal{D}_t$ is the rendered depth map of the
target view, and $\mathbf{D}_t(p_t)$ is the depth for this particular pixel p_t in $\mathcal{D}_t$.

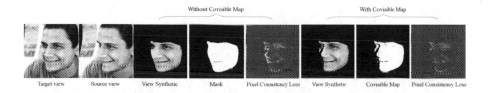

Fig. 3. The view synthesis results with and without covisible map. Without covisiable
map, the pixel consistency loss is highly affected by self-occlusion. (Color figure online)

3.5   Pixel Consistency Loss

We generate the synthesized target images by view synthesis, then we minimize
the pixel error between the target view and the synthesized target views from
the source views as

$$\mathcal{L}_{pixel} = \frac{1}{|C_s|} \sum_{i=1}^{|C_s|} C_s^i * \left| \widetilde{\mathcal{I}}_t^s(i) - \mathcal{I}_t(i) \right|, \tag{3}$$

where $\widetilde{\mathcal{I}}_t^s$ represents the synthesized target views from the source views. $\mathcal{I}_t(i)$
is the $i - th$ pixel value . Concretely, the first term $\widetilde{\mathcal{I}}_t^s$ is the bilinear-sampling
operation, which computes the corresponding pixel coordinates using the relative
pose $P_{t \to s}^{rel}$ and target depth map D_t^{re}. C_s is covisible map and $|C_s|$ denotes the
total number of covisible pixels.

3.6 Dense Depth Consistency Loss

Compared to RGB images, depth maps are less adversely affected by the gradient locality issue [5]. Thus, we propose a dense depth consistency loss function which contributes to solving depth ambiguity more explicitly, enforcing the multi-view consistency upon depth maps. Similarly, we synthesize the target depth maps $\widetilde{\mathbf{D}}_t^s$ from the source views via bilinear interpolation, and compute the consistency against the target depth map $\mathbf{D}_t$.

One critical issue is that the face region is cropped in the face detection data-preprocessing stage, making the depth value up to scale. To tackle this issue, we compute a ratio of two depth maps $\mathcal{S}_{depth}$ and rectify the scale of depth maps. Therefore, we define the dense depth consistency loss as

$$
\begin{aligned}
\mathcal{S}_{depth} &= \frac{\sum_{i=1}^{|C_s|} \mathbf{D}_t(i) C_s(i)}{\sum_{i=1}^{|C_s|} \widetilde{\mathbf{D}}_t^s(i) C_s(i)} \\
\mathcal{L}_{depth} &= \frac{1}{|C_s|} \sum_{i=1}^{|C_s|} \left| \mathcal{S}_{depth} \cdot \widetilde{\mathbf{D}}_t^s(i) - \mathbf{D}_t(i) \right|,
\end{aligned}
\tag{4}
$$

where $\mathcal{C}_s(i), \mathcal{D}_t(i)$ are the $i - th$ covisible and depth value. $\mathcal{S}_{depth}$ is the depth scale ratio. Our experiment shows that the multi-view geometry supervisory signals significantly improve the accuracy of the 3D face shape.

(a) Baseline   (b) Multi-view

Fig. 4. Eipipolar error from target to source views, (a) is the epipolar visualization from baseline network trained by 2D feature loss; and (b) is epipolar visualization of the MGCNet. The red, green and blue point means left ear bound, lower jaw and nose tip landmarks. The epipolar error of baseline is significantly larger than MGCNet. (Color figure online)

3.7 Facial Epipolar Loss

We use facial landmarks to build the epipolar consistency, as our epipolar loss is based on sparse ground-truth 2D facial landmarks, which is less likely to be affected by radiometric or illumination changes compared to pixel consistency or depth consistency losses, The epipolar loss in the *symmetric epipolar distance* [25] form between source and target 2D landmark $q_{t \leftrightarrow s} = \{\mathbf{p} \leftrightarrow \mathbf{p}'\}$ is defined as

$$
\mathcal{L}_{epi}(q|\mathbf{R}, \mathbf{t}) = \sum_{\forall (\mathbf{p}, \mathbf{p}') \in q} \frac{\mathbf{p}'^{\mathbf{T}} \mathbf{E} \mathbf{p}}{\sqrt{(\mathbf{E}\mathbf{p})_{(1)}^2 + (\mathbf{E}\mathbf{p})_{(2)}^2}}
\tag{5}
$$

where $\mathbf{E}$ being the essential matrix computed by $\mathbf{E} = [\mathbf{t}]_\times \mathbf{R}$, $[\cdot]_\times$ is the matrix representation of the cross product with $\mathbf{t}$. We simply omit the subindices for conciseness (q for $q_{t \leftrightarrow s}$, $\mathbf{R}$ for $\mathbf{R}_{t \to s}$, $\mathbf{t}$ for $\mathbf{t}_{t \to s}$).

3.8   Combined Loss

The final loss function $\mathcal{L}$ for our MGCNet is the combination of 2D feature loss and multi-view geometry loss. Training the network by only 2D feature losses leads to face pose and depth ambiguity, which is reflected in geometry inconsistency as the large epipolar error shown in Fig. 4(a). Our MGCNet trained with pixel consistency loss, dense depth consistency, and facial epipolar loss shows remarkable improvement in Fig. 4(b), which outstands our novel multi-view geometry consistency based self supervised training pipeline. Finally, the combined loss function is defined as

$$\mathcal{L} = w_{2D} * \mathcal{L}_{2D} + w_{mul} * [w_{pixel} * \mathcal{L}_{pixel} + w_{depth} * \mathcal{L}_{depth} + w_{epi} * \mathcal{L}_{epi}], \quad (6)$$

where $w_{2D} = 1.0$ and $w_{mul} = 1.0$ balance the weights between the 2D feature loss for each view and the multi-view geometry consistency loss. The trade-off parameters to take into account are $w_{pixel} = 0.15, w_{depth} = 1e{-}4, w_{epi} = 1e{-}3$.

4   Experiment

We evaluate the performance of our MGCNet on the face alignment and 3D face reconstruction tasks which compared with the most recent state-of-the-art methods [6,8,15,18,22,45,50–52,56,66,67] on diverse test datasets including AFLW20003D [67], MICC Florence [3], Binghamton University 3D Facial Expression (BU-3DFE) [60,64], and FRGC v2.0 [36].

4.1   Implementation Details

Data. 300W-LP [67] has multi-view face images with fitted 3DMM model, the model is widely used as ground truth in [18,28,56,59], such multi-view images provide better supervision than only 2D features. Multi-PIE [23] are introduced to provide multi-view face images that help solve face pose and depth ambiguity. As multi-view face datasets are always captured indoor, and thus cannot provide diversified illumination and background for training, CelebA [33] and LS3D [8] are used as part of training data, which only contribute to 2D feature losses. Detail data process can be found in the *suppl.* material.

Network. We use the ResNet50 [26] network as the backbone of our MGC-Net, we only convert the last fully-connected layer to 257 neurons to match the dimension of 3DMM coefficients. The pre-trained model from ImageNet [44] is used as an initialization. We only use $N = 3$ views in practice, as $N = 5$ views lead to a large pose gap between the first view and the last view. We implement our approach by Tensorflow [1]. The training process is based on Adam optimizer [30] with a batch size of 5. The learning rate is set to $1e{-}4$, and there are $400\,\mathrm{K}$ total iterations for the whole training process.

Fig. 5. A few results on a full range of lighting, pose, including large expressions. Each image pair is input image (left) and reconstruction result overlay (right). Further detail result (shape, albedo, and lighting) can be found in the *suppl.* material.

4.2 Qualitative Result

Result in Different Situations. Our MGCNet allows for high-quality reconstruction of facial geometry, reflectance and incident illumination as Fig. 5, under full range of lighting, pose, and expressions situations.

Geometry. We evaluate the qualitative results of our MGCNet on AFLW20003D [67]. First, we compare our MGCNet with 3DDFA [67], RingNet [45], PRN [18], and Deng *et al.* [15] on front view samples, as Row 1 and Row 2 in Fig. 6. Our predicted 3DMM coefficients produce more accurate results than the most methods, and we get comparable results with Deng *et al.* [15].

For these large and extreme pose cases as Row 3–6 in Fig. 6, our MGCNet has better face alignment and face geometry than other methods. We have more vivid emotion in Row 4 of Fig. 6, and the mouths of our result in Row 3,5 have obviously better shape than 3DDFA [67], RingNet [45], PRN [18], and Deng *et al.* [15]. Besides, the face alignment results from Row 3 to Row 6 support that we achieve better face pose estimation, especially in large and extreme pose cases.

Texture, Illumination Shadings. We also visualize our result under geometry, texture, illumination shadings, and notice that our approach performs better than Tewari18 *et al.* [51] and Tewari19 *et al.* [50], where the overlay result is very similar to the input image as Fig. 7(a). Further result and analysis about the result can be found in the *suppl.* material.

MGCNet does not focus on the appearance of 3DMM as [12,21,49–51,53,54, 62], which is only constrained by render loss. However, our multi-view geometry supervision can help render loss maximize the potential during training by accurate face alignment and depth value estimation. This makes MGCNet able to handle 3DMM texture and illumination robustly.

4.3 2D Face Alignment

The quantitative comparison of 6Dof pose is not conducted due to different camera intrinsic assumptions of different methods. Therefore, to validate that our MGCNet can mitigate the ambiguity of monocular face pose estimation,

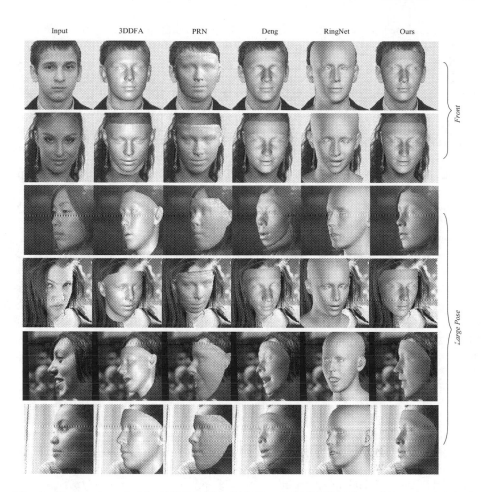

Fig. 6. Comparisons with 3DDFA [67], RingNet [45], PRN [18], and Deng *et al.* [15] on ALFW20003D.

Table 1. (a) Performance comparison on AFLW2000-3D (68 landmarks). The normalized mean error (NME) for 2D landmarks with different yaw angles is reported. The first best result is highlighted in bold. (b) Average and standard deviation root mean squared error (RMSE) with *mm* in three environments of MICC Florence.

(a)

Method	0 to 30	30 to 60	60 to 90	Mean
3DDFA [67]	3.78	4.54	7.93	5.42
3D-FAN [8]	3.61	4.34	6.87	4.94
3DSTN [6]	3.15	4.33	5.98	4.49
CMD [66]	–	–	–	3.98
PRN [18]	2.75	3.55	5.11	3.62
Ours+BL	**2.75**	**3.28**	**4.31**	**3.45**
Ours+MGCNet	**2.72**	**3.12**	**3.76**	**3.20**

(b)

Method	Cooperative	Indoor	Outdoor
Zhu *et al.* [67]	2.69 ± 0.64	2.23 ± 0.49	2.22 ± 0.56
Sanyal *et al.* [45]	2.33 ± 0.43	2.19 ± 0.43	2.07 ± 0.45
Feng *et al.* [18]	2.30 ± 0.54	2.02 ± 0.50	2.10 ± 0.60
Tran *et al.* [56]	2.00 ± 0.55	2.05 ± 0.51	1.95 ± 0.51
Genova *et al.* [22]	1.87 ± 0.61	1.86 ± 0.60	1.87 ± 0.57
Deng *et al.* [15]	1.83 ± 0.59	1.78 ± 0.53	1.78 ± 0.59
Ours	**1.73 ± 0.48**	**1.78 ± 0.47**	**1.75 ± 0.47**

we evaluate our method on AFLW2000-3D, and compare our result with Zhu
et al. [67] (3DDFA), Bulat and Tzimiropoulos [8] (3D-FAN), Bhagavatula
et al. [6] (3DSTN), Zhou *et al.* [66] (CMD), and Feng *et al.* [18] (PRN). Normalized mean error (NME) is used as the evaluation metric, and the bounding
box size of ground truth landmarks is deemed as the normalization factor. As
shown in Table 1(a) Column 5, our result outperforms the best method with a
large margin of **12%** improvement. Qualitative results can be found in the *suppl.*
material.

Learning face pose from 2D features of monocular images leads to face pose
ambiguity, the results of large and extreme face pose test samples suffer from
this heavily. As the supervision of large and extreme face pose case is even less,
which is not enough for training monocular face pose regressor. Our MGCNet
provides further robust and dense supervision by multi-view geometry for face
alignment in both frontal and profile face pose situations. The comparison in
Table 1(a) corroborates our point that the compared methods [6,8,18,66,67]
obviously degrades when the yaw angles increase from $(30, 60)$ to $(60, 90)$ in
Column 4 of Table 1(a). We also conduct an ablation study that our MGCNet
outperforms the baseline, especially on large and extreme pose case.

4.4   3D Face Reconstruction

MICC Florence with Video. MICC Florence provides videos of each subject in
cooperative, indoor and outdoor scenarios. For a fair comparison with Genova
et al. [22], Trans *et al.* [56] and Deng *et al.* [15], we calculate error with the
average shape for each video in different scenarios. Following [22], we crop the
ground truth mesh to $95mm$ around the nose tip and run iterative closest point
(ICP) algorithm for rigid alignment. The results of [56] only contain part of
the forehead region. For a fair comparison, we process the ground-truth meshes
similarly. We use the point-to-plane root mean squared error(RMSE) as the
evaluation metric. We compare with the methods of Zhu *et al.* [67] (3DDFA),
Sanyal *et al.* [45] (RingNet), Feng *et al.* [18] (PRN), Genova *et al.* [22], Trans
et al. [56] and Deng *et al.* [15]. Table 1(b) shows that our method outperforms
state-of-the-art methods [15,18,22,45,56,67] on all three scenarios.

MICC Florence with Rendered Images. Several current methods [15,18,
28,59] also generate rendered images as test input. Following [15,18,28,59], we
render face images of each subject with 20 poses: a pitch of -15, 20 and $25°$,
yaw angles of -80, -60, 0, 60 and $80°$, and 5 random poses. We use the point-
to-plane RMSE as the evaluation metric, and we process the ground truth mesh
as above. Figure 7(b) shows that our method achieves a significant improvement
of **17%** higher than the state-of-the-art methods.

The plot also shows that our MGCNet performs obvious improvement on
the extreme pose setting $x - axis[-80, 80]$ in Fig. 7(b). As we mitigate both
pose and depth ambiguity by multi-view geometry consistency in the training
process. Extreme pose sample benefits from this more significantly, since the
extreme pose input images have even less 2D features. Profile face case contains
more pronounced depth info (eg. bridge of the nose), where large error happen.

FRGC v2.0 Dataset. FRGC v2.0 is a large-scale benchmark includes 4007 scans. We random pick 1335 scans as test samples, then we crop the ground truth mesh to 95 mm around the nose tip. We first use 3D landmark as correspondence to align the predict and ground truth result, then ICP algorithm is used as fine alignment. Finally, point-to-point mean average error (MAE) is used as the evaluation metric. We compare with the methods of Galteri *et al.* [21] (D3R), 3DDFA [67], RingNet [45], PRN [18], and Deng *et al.* [15]. Table 2(a) shows that our method outperforms these state-of-the-art methods. Our MGCNet performs higher fidelity and accurate result on both frontal and profile face pose view.

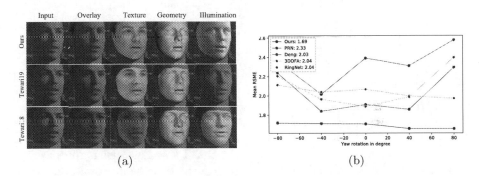

Fig. 7. (a) Comparison to Tewari18 *et al.* [51] and Tewari19 *et al.* [50]. Our MGC-Net trained by multi-view consistency loss outperforms Tewari's results in face pose, illumination and geometry. Further result can be found in the *suppl.* material. (b) Comparison with 3DDFA [67], RingNet [45], PRN [18], and Deng *et al.* [15] on MICC Florence rendered images.

Table 2. (a) Comparison with D3R [21], 3DDFA [67], RingNet [45], PRN [18], and Deng *et al.* [15] with MAE of *mm* on FRGC v2.0 dataset. (b) Mean and standard deviation point-to-point RMSE with *mm* on the BU-3DFE dataset [60,64] compared with Tewari17 *et al.* [52], Tewari18 *et al.* [51], Tewari19 *et al.* [50], Deng *et al.* [15].

<table>
<tr><td colspan="7" align="center">(a)</td><td colspan="7" align="center">(b)</td></tr>
<tr><td>Method</td><td>[21]</td><td>[18]</td><td>[45]</td><td>[67]</td><td>[15]</td><td>Ours</td><td>Method</td><td>[52]</td><td>[51] Fine</td><td>[51] Coarse</td><td>[50]</td><td>[15]</td><td>Ours</td></tr>
<tr><td>MAE</td><td>3.63</td><td>2.33</td><td>2.22</td><td>2.21</td><td>2.18</td><td>**1.93**</td><td>Mean</td><td>3.22</td><td>1.83</td><td>1.81</td><td>1.79</td><td>1.63</td><td>**1.55**</td></tr>
<tr><td>Time</td><td>–</td><td></td><td>9.8 ms</td><td>2.7 ms</td><td>75.7 ms</td><td>20 ms</td><td>20 ms</td><td>Std</td><td>0.77</td><td>0.39</td><td>0.47</td><td>0.45</td><td>0.33</td><td>**0.31**</td></tr>
</table>

BU-3DFE Dataset. We evaluate our method on the BU-3DFE dataset following [50]. Following [50], a pre-computed dense correspondence map is used to calculate a similarity transformation from predict mesh to the original ground-truth 3D mesh, and help to calculate the point-to-point RMSE. From Table 2(b), the reconstruction error of our method is lower than the current state-of-art methods [15,50–52]. Our MGCNet achieves better performance by using the multi-view geometry consistency loss functions in the training phase. Qualitative results can be found in the *suppl.* material.

Table 3. Evaluation of different training loss configurations. The ablation study performance of MGCNet is evaluated on MICC Florence 3D Face dataset [3] by RMSE.

Loss configuration					MICC florence video		
2D feature	Pixel consistency	Depth consistency	Epipolar	Covisible map	Cooperative	Indoor	Outdoor
✓	–	–	–	–	1.83	1.82	1.81
✓	✓	–	–	✓	1.80	1.80	1.80
✓	–	✓	–	✓	1.77	1.79	1.80
✓	–	–	✓	–	1.79	1.81	1.77
✓	✓	✓	–	✓	1.76	1.81	1.81
✓	✓	✓	✓	–	1.80	1.81	1.82
✓	✓	✓	✓	✓	**1.73**	**1.78**	**1.75**

4.5   Ablation Study

To validate the efficiency of our multi-view geometry consistency loss functions. We conduct ablation studies for each component on the MICC Florence dataset [3], as shown in Table 3. The ablation study mainly focuses on the proposed multi-view geometry consistency loss functions. Firstly, we deem the baseline method as the model trained with only 2D feature losses, as in Row 1. Secondly, we add our pixel consistency loss, dense depth consistency loss, and epipolar loss to the baseline in Row 2, Row 3 and Row 4. It shows that these losses help produce lower reconstruction errors than the baseline, even when they are used separately. Thirdly, comparing from Row 5 to Row 7, we combine multiple multi-view geometry loss functions and achieve state-of-the-art results, which demonstrates the effectiveness of the proposed self-supervised learning pipeline. Finally, comparing from Row 6 to Row 7, we prove that our novel covisible map to solve self-occlusion in view synthesis algorithm can help training a more accurate model. The qualitative ablation study is in the *suppl.* material.

5   Conclusion

We have presented a self-supervised pipeline MGCNet for monocular 3D Face reconstruction and demonstrated the advantages of exploiting multi-view geometry consistency to provide more reliable constraint on face pose and depth estimation. We emphasize on the occlusion-aware view synthesis and multi-view losses to make the result more robust and reliable. Our MGCNet profoundly reveals the capability of multi-view geometry consistency self-supervised learning in capturing both high-level cues and feature correspondences with geometry reasoning. The results compared to other methods indicate that our MGCNet can achieve the outstanding result without costly labeled data. Our further investigations will focus on multi-view or video-based 3D face reconstruction.

Acknowledgements. This work is supported by Hong Kong RGC GRF 16206819 & 16203518 and T22-603/15N.

References

1. Abadi, M., et al.: TensorFlow: a system for large-scale machine learning. In: 12th USENIX Symposium on Operating Systems Design and Implementation (OSDI 2016), pp. 265–283 (2016)
2. Aldrian, O., Smith, W.A.: Inverse rendering of faces with a 3D morphable model. IEEE Trans. Pattern Anal. Mach. Intell. **35**(5), 1080–1093 (2012)
3. Bagdanov, A.D., Del Bimbo, A., Masi, I.: The florence 2D/3D hybrid face dataset. In: Proceedings of the 2011 Joint ACM Workshop on Human Gesture and Behavior Understanding, J-HGBU 2011, pp. 79–80. ACM, New York (2011). https://doi.org/10.1145/2072572.2072597
4. Bas, A., Smith, W.A.P., Bolkart, T., Wuhrer, S.: Fitting a 3D morphable model to edges: a comparison between hard and soft correspondences. In: Chen, C.-S., Lu, J., Ma, K.-K. (eds.) ACCV 2016. LNCS, vol. 10117, pp. 377–391. Springer, Cham (2017). https://doi.org/10.1007/978-3-319-54427-4_28
5. Bergen, J.R., Anandan, P., Hanna, K.J., Hingorani, R.: Hierarchical model-based motion estimation. In: Sandini, G. (ed.) ECCV 1992. LNCS, vol. 588, pp. 237–252. Springer, Heidelberg (1992). https://doi.org/10.1007/3-540-55426-2_27
6. Bhagavatula, C., Zhu, C., Luu, K., Savvides, M.: Faster than real-time facial alignment: a 3D spatial transformer network approach in unconstrained poses. In: Proceedings of the IEEE International Conference on Computer Vision, pp. 3980–3989 (2017)
7. Blanz, V., Vetter, T., et al.: A morphable model for the synthesis of 3d faces. In: SIGGRAPH, vol. 99, pp. 187–194 (1999)
8. Bulat, A., Tzimiropoulos, G.: How far are we from solving the 2D & 3D face alignment problem? (And a dataset of 230,000 3D facial landmarks). In: Proceedings of the IEEE International Conference on Computer Vision, pp. 1021–1030 (2017)
9. Cao, C., Weng, Y., Zhou, S., Tong, Y., Zhou, K.: Facewarehouse: A 3D facial expression database for visual computing. IEEE Trans. Visual Comput. Graph. **20**(3), 413–425 (2013)
10. Cao, C., Wu, H., Weng, Y., Shao, T., Zhou, K.: Real-time facial animation with image-based dynamic avatars. ACM Trans. Graph. **35**(4) (2016)
11. Casser, V., Pirk, S., Mahjourian, R., Angelova, A.: Depth prediction without the sensors: leveraging structure for unsupervised learning from monocular videos. In: Proceedings of the AAAI Conference on Artificial Intelligence, vol. 33, pp. 8001–8008 (2019)
12. Chen, A., Chen, Z., Zhang, G., Mitchell, K., Yu, J.: Photo-realistic facial details synthesis from single image. In: Proceedings of the IEEE International Conference on Computer Vision, pp. 9429–9439 (2019)
13. Chen, S.E., Williams, L.: View interpolation for image synthesis. In: Proceedings of the 20th Annual Conference on Computer Graphics and Interactive Techniques, pp. 279–288. ACM (1993)
14. Debevec, P.E., Taylor, C.J., Malik, J.: Modeling and rendering architecture from photographs. University of California, Berkeley (1996)
15. Deng, Y., Yang, J., Xu, S., Chen, D., Jia, Y., Tong, X.: Accurate 3D face reconstruction with weakly-supervised learning: from single image to image set. In: Proceedings of the IEEE Conference on Computer Vision and Pattern Recognition Workshops (2019)
16. Dou, P., Kakadiaris, I.A.: Multi-view 3D face reconstruction with deep recurrent neural networks. Image Vis. Comput. **80**, 80–91 (2018)

17. Dou, P., Shah, S.K., Kakadiaris, I.A.: End-to-end 3D face reconstruction with deep neural networks. In: Proceedings of the IEEE Conference on Computer Vision and Pattern Recognition, pp. 5908–5917 (2017)
18. Feng, Y., Wu, F., Shao, X., Wang, Y., Zhou, X.: Joint 3D face reconstruction and dense alignment with position map regression network. In: Ferrari, V., Hebert, M., Sminchisescu, C., Weiss, Y. (eds.) Computer Vision – ECCV 2018. LNCS, vol. 11218, pp. 557–574. Springer, Cham (2018). https://doi.org/10.1007/978-3-030-01264-9_33
19. Fitzgibbon, A., Wexler, Y., Zisserman, A.: Image-based rendering using image-based priors. Int. J. Comput. Vision **63**(2), 141–151 (2005). https://doi.org/10.1007/s11263-005-6643-9
20. Furukawa, Y., Curless, B., Seitz, S.M., Szeliski, R.: Towards internet-scale multi-view stereo. In: 2010 IEEE Computer Society Conference on Computer Vision and Pattern Recognition, pp. 1434–1441. IEEE (2010)
21. Galteri, L., Ferrari, C., Lisanti, G., Berretti, S., Del Bimbo, A.: Deep 3D morphable model refinement via progressive growing of conditional generative adversarial networks. Comput. Vis. Image Underst. **185**, 31–42 (2019)
22. Genova, K., Cole, F., Maschinot, A., Sarna, A., Vlasic, D., Freeman, W.T.: Unsupervised training for 3d morphable model regression. In: Proceedings of the IEEE Conference on Computer Vision and Pattern Recognition, pp. 8377–8386 (2018)
23. Gross, R., Matthews, I., Cohn, J., Kanade, T., Baker, S.: Multi-pie. Image Vis. Comput. **28**(5), 807–813 (2010)
24. Guo, Y., Cai, J., Jiang, B., Zheng, J., et al.: CNN-based real-time dense face reconstruction with inverse-rendered photo-realistic face images. IEEE Trans. Pattern Anal. Mach. Intell. **41**(6), 1294–1307 (2018)
25. Hartley, R., Zisserman, A.: Multiple View Geometry in Computer Vision. Cambridge University Press, Cambridge (2003)
26. He, K., Zhang, X., Ren, S., Sun, J.: Deep residual learning for image recognition. In: Proceedings of the IEEE Conference on Computer Vision and Pattern Recognition, pp. 770–778 (2016)
27. Hu, L.: Avatar digitization from a single image for real-time rendering. ACM Trans. Graph. (TOG) **36**(6), 195 (2017)
28. Jackson, A.S., Bulat, A., Argyriou, V., Tzimiropoulos, G.: Large pose 3D face reconstruction from a single image via direct volumetric CNN regression. In: Proceedings of the IEEE International Conference on Computer Vision, pp. 1031–1039 (2017)
29. Jaderberg, M., Simonyan, K., Zisserman, A., et al.: Spatial transformer networks. In: Advances in Neural Information Processing Systems, pp. 2017–2025 (2015)
30. Kingma, D.P., Ba, J.: Adam: a method for stochastic optimization. arXiv preprint arXiv:1412.6980 (2014)
31. Liu, F., Zeng, D., Zhao, Q., Liu, X.: Joint face alignment and 3D face reconstruction. In: Leibe, B., Matas, J., Sebe, N., Welling, M. (eds.) ECCV 2016. LNCS, vol. 9909, pp. 545–560. Springer, Cham (2016). https://doi.org/10.1007/978-3-319-46454-1_33
32. Liu, F., Zhu, R., Zeng, D., Zhao, Q., Liu, X.: Disentangling features in 3D face shapes for joint face reconstruction and recognition. In: Proceedings of the IEEE Conference on Computer Vision and Pattern Recognition, pp. 5216–5225 (2018)
33. Liu, Z., Luo, P., Wang, X., Tang, X.: Deep learning face attributes in the wild. In: Proceedings of the IEEE International Conference on Computer Vision, pp. 3730–3738 (2015)

34. Mahjourian, R., Wicke, M., Angelova, A.: Unsupervised learning of depth and ego-motion from monocular video using 3D geometric constraints. In: Proceedings of the IEEE Conference on Computer Vision and Pattern Recognition, pp. 5667–5675 (2018)
35. Paysan, P., Knothe, R., Amberg, B., Romdhani, S., Vetter, T.: A 3D face model for pose and illumination invariant face recognition. In: 2009 Sixth IEEE International Conference on Advanced Video and Signal Based Surveillance, pp. 296–301. IEEE (2009)
36. Phillips, P.J., et al.: Overview of the face recognition grand challenge. In: 2005 IEEE Computer Society Conference on Computer Vision and Pattern Recognition (CVPR 2005), vol. 1, pp. 947–954. IEEE (2005)
37. Ramamoorthi, R., Hanrahan, P.: An efficient representation for irradiance environment maps. In: Proceedings of the 28th Annual Conference on Computer Graphics and Interactive Techniques, pp. 497–500. ACM (2001)
38. Ramamoorthi, R., Hanrahan, P.: A signal-processing framework for inverse rendering. In: Proceedings of the 28th Annual Conference on Computer Graphics and Interactive Techniques, pp. 117–128. ACM (2001)
39. Richardson, E., Sela, M., Kimmel, R.: 3D face reconstruction by learning from synthetic data. In: 2016 Fourth International Conference on 3D Vision (3DV), pp. 460–469. IEEE (2016)
40. Richardson, E., Sela, M., Or-El, R., Kimmel, R.: Learning detailed face reconstruction from a single image. In: Proceedings of the IEEE Conference on Computer Vision and Pattern Recognition, pp. 1259–1268 (2017)
41. Romdhani, S., Vetter, T.: Estimating 3D shape and texture using pixel intensity, edges, specular highlights, texture constraints and a prior. In: 2005 IEEE Computer Society Conference on Computer Vision and Pattern Recognition (CVPR 2005), vol. 2, pp. 986–993. IEEE (2005)
42. Roth, J., Tong, Y., Liu, X.: Unconstrained 3D face reconstruction. In: Proceedings of the IEEE Conference on Computer Vision and Pattern Recognition, pp. 2606–2615 (2015)
43. Roth, J., Tong, Y., Liu, X.: Adaptive 3D face reconstruction from unconstrained photo collections. In: Proceedings of the IEEE Conference on Computer Vision and Pattern Recognition, pp. 4197–4206 (2016)
44. Russakovsky, O., et al.: Imagenet large scale visual recognition challenge. Int. J. Comput. Vis. **115**(3), 211–252 (2015). https://doi.org/10.1007/s11263-015-0816-y
45. Sanyal, S., Bolkart, T., Feng, H., Black, M.J.: Learning to regress 3d face shape and expression from an image without 3D supervision. In: Proceedings of the IEEE Conference on Computer Vision and Pattern Recognition, pp. 7763–7772 (2019)
46. Schönberger, J.L., Frahm, J.M.: Structure-from-motion revisited. In: Conference on Computer Vision and Pattern Recognition (CVPR) (2016)
47. Schönberger, J.L., Zheng, E., Frahm, J.-M., Pollefeys, M.: Pixelwise view selection for unstructured multi-view stereo. In: Leibe, B., Matas, J., Sebe, N., Welling, M. (eds.) ECCV 2016. LNCS, vol. 9907, pp. 501–518. Springer, Cham (2016). https://doi.org/10.1007/978-3-319-46487-9_31
48. Seitz, S.M., Dyer, C.R.: View morphing. In: Proceedings of the 23rd Annual Conference on Computer Graphics and Interactive Techniques, pp. 21–30. ACM (1996)
49. Sela, M., Richardson, E., Kimmel, R.: Unrestricted facial geometry reconstruction using image-to-image translation. In: Proceedings of the IEEE International Conference on Computer Vision, pp. 1576–1585 (2017)
50. Tewari, A., et al.: FML: face model learning from videos. In: Proceedings of the IEEE Conference on Computer Vision and Pattern Recognition, pp. 10812–10822 (2019)

51. Tewari, A., et al.: Self-supervised multi-level face model learning for monocular reconstruction at over 250 HZ. In: Proceedings of the IEEE Conference on Computer Vision and Pattern Recognition, pp. 2549–2559 (2018)
52. Tewari, A., et al.: MoFA: model-based deep convolutional face autoencoder for unsupervised monocular reconstruction. In: Proceedings of the IEEE International Conference on Computer Vision, pp. 1274–1283 (2017)
53. Tran, A.T., Hassner, T., Masi, I., Paz, E., Nirkin, Y., Medioni, G.G.: Extreme 3D face reconstruction: seeing through occlusions. In: CVPR, pp. 3935–3944 (2018)
54. Tran, L., Liu, F., Liu, X.: Towards high-fidelity nonlinear 3D face morphable model. In: Proceedings of the IEEE Conference on Computer Vision and Pattern Recognition, pp. 1126–1135 (2019)
55. Tran, L., Liu, X.: Nonlinear 3D face morphable model. In: Proceedings of the IEEE Conference on Computer Vision and Pattern Recognition, pp. 7346–7355 (2018)
56. Tuan Tran, A., Hassner, T., Masi, I., Medioni, G.: Regressing robust and discriminative 3D morphable models with a very deep neural network. In: Proceedings of the IEEE Conference on Computer Vision and Pattern Recognition, pp. 5163–5172 (2017)
57. Wu, C., et al.: VisualSFM: a visual structure from motion system (2011)
58. Wu, F., et al.: MVF-Net: multi-view 3D face morphable model regression. In: Proceedings of the IEEE Conference on Computer Vision and Pattern Recognition, pp. 959–968 (2019)
59. Yi, H., et al.: MMFace: a multi-metric regression network for unconstrained face reconstruction. In: Proceedings of the IEEE Conference on Computer Vision and Pattern Recognition, pp. 7663–7672 (2019)
60. Yin, L., Wei, X., Sun, Y., Wang, J., Rosato, M.J.: A 3D facial expression database for facial behavior research. In: 7th International Conference on Automatic Face and Gesture Recognition (FGR06), pp. 211–216. IEEE (2006)
61. Yoon, J.S., Shiratori, T., Yu, S.I., Park, H.S.: Self-supervised adaptation of high-fidelity face models for monocular performance tracking. In: Proceedings of the IEEE Conference on Computer Vision and Pattern Recognition, pp. 4601–4609 (2019)
62. Zeng, X., Peng, X., Qiao, Y.: DF2Net: a dense-fine-finer network for detailed 3D face reconstruction. In: Proceedings of the IEEE International Conference on Computer Vision, pp. 2315–2324 (2019)
63. Zhan, H., Garg, R., Saroj Weerasekera, C., Li, K., Agarwal, H., Reid, I.: Unsupervised learning of monocular depth estimation and visual odometry with deep feature reconstruction. In: Proceedings of the IEEE Conference on Computer Vision and Pattern Recognition, pp. 340–349 (2018)
64. Zhang, X., et al.: A high-resolution spontaneous 3D dynamic facial expression database. In: 2013 10th IEEE International Conference and Workshops on Automatic Face and Gesture Recognition (FG), pp. 1–6. IEEE (2013)
65. Zhou, T., Brown, M., Snavely, N., Lowe, D.G.: Unsupervised learning of depth and ego-motion from video. In: Proceedings of the IEEE Conference on Computer Vision and Pattern Recognition, pp. 1851–1858 (2017)
66. Zhou, Y., Deng, J., Kotsia, I., Zafeiriou, S.: Dense 3D face decoding over 2500 FPS: joint texture & shape convolutional mesh decoders. In: Proceedings of the IEEE Conference on Computer Vision and Pattern Recognition, pp. 1097–1106 (2019)
67. Zhu, X., Lei, Z., Liu, X., Shi, H., Li, S.Z.: Face alignment across large poses: a 3D solution. In: Proceedings of the IEEE Conference on Computer Vision and Pattern Recognition, pp. 146–155 (2016)

Asynchronous Interaction Aggregation for Action Detection

Jiajun Tang, Jin Xia, Xinzhi Mu, Bo Pang, and Cewu Lu$^{(\boxtimes)}$

Shanghai Jiao Tong University, Shanghai, China
{yelantingfeng,draconids,pangbo,lucewu}@sjtu.edu.cn, ga.xiajin@gmail.com

Abstract. Understanding interaction is an essential part of video action detection. We propose the Asynchronous Interaction Aggregation network (AIA) that leverages different interactions to boost action detection. There are two key designs in it: one is the Interaction Aggregation structure (IA) adopting a uniform paradigm to model and integrate multiple types of interaction; the other is the Asynchronous Memory Update algorithm (AMU) that enables us to achieve better performance by modeling very long-term interaction dynamically without huge computation cost. We provide empirical evidence to show that our network can gain notable accuracy from the integrative interactions and is easy to train end-to-end. Our method reports the new state-of-the-art performance on AVA dataset, with *3.7 mAP* gain (12.6% relative improvement) on validation split comparing to our strong baseline. The results on datasets UCF101-24 and EPIC-Kitchens further illustrate the effectiveness of our approach. Source code will be made public at: https://github.com/MVIG-SJTU/AlphAction.

Keywords: Action detection · Video understanding · Interaction · Memory

1 Introduction

The task of action detection (spatio-temporal action localization) aims at detecting and recognizing actions in space and time. As an essential task of video understanding, it has a variety of applications such as abnormal behavior detection and autonomous driving. On top of spatial representation and temporal features [3,10,21,27], the interaction relationships [13,29,39,47] are crucial for understanding actions. Take Fig. 1 for example. The appearance of the man, the

---

J. Tang and J. Xia—Both authors contributed equally to this work.
C. Lu is a member of Qing Yuan Research Institute and MoE Key Lab of Artificial Intelligence, AI Institute, Shanghai Jiao Tong University, China.

Electronic supplementary material The online version of this chapter (https://doi.org/10.1007/978-3-030-58555-6_5) contains supplementary material, which is available to authorized users.

A. Vedaldi et al. (Eds.): ECCV 2020, LNCS 12360, pp. 71–87, 2020.
https://doi.org/10.1007/978-3-030-58555-6_5

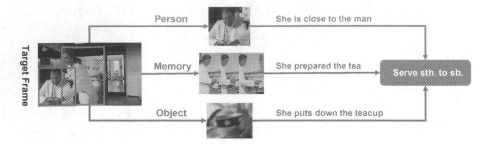

Fig. 1. Interaction Aggregation. In this target frame, we can tell that the women is serving tea to the man with following clues: (1) She is close to the man. (2) She puts down the tea cup before the man. (3) She prepared the tea a few seconds ago. These three clues correspond respectively the person-person, person-object and temporal interactions

tea cup as well as the previous movement of the woman help to predict the action of the woman. In this paper, we propose a new framework which emphasizes on the interactions for action detection.

Interactions can be briefly considered as the relationship between the target person and context. Many existing works try to explore interactions in videos, but there are two problems in the current methods: (1) Previous methods such as [13,15] focus on a single type of interaction (eg. person-object). They can only boost one specific kind of actions. Methods such as [46] intend to merge different interactions, but they model them separately. Information of one interaction can't contribute to another interaction modeling. How to find interactions correctly in video and use them for action detection remains challenging. (2) The long-term temporal interaction is important but hard to track. Methods which use temporal convolution [10,21,27] have very limited temporal reception due to the resource challenge. Methods such as [41] require a duplicated feature extracting pre-process which is not practical in reality.

In this work, we propose a new framework, the Asynchronous Interaction Aggregation network (AIA), who explores three kinds of interactions (person-person, person-object, and temporal interaction) that cover nearly all kinds of person-context interactions in the video. As a first try, AIA makes them work cooperatively in a hierarchical structure to capture higher level spatial-temporal features and more precise attentions. There are two main designs in our network: the Interaction Aggregation (IA) structure and the Asynchronous Memory Update (AMU) algorithm.

The former design, IA structure, explores and integrates all three types of interaction in a deep structure. More specifically, it consists of multiple elemental interaction blocks, of each enhances the target features with one type of interaction. These three types of interaction blocks are nested along the depth of IA structure. One block may use the result of previous interactions blocks. Thus, IA structure is able to model interactions precisely using information across different types.

Jointly training with long memory features is infeasible due to the large size of video data. The AMU algorithm is therefore proposed to estimate intractable features during training. We adopt a memory-like structure to store the spatial features and propose a series of write-read algorithm to update the content in memory: features extracted from target clips at each iteration are written to a memory pool and they can be retrieved in subsequent iterations to model temporal interaction. This effective strategy enables us to train the whole network in an end-to-end manner and the computational complexity doesn't increase linearly with the length of temporal memory features. In comparison to previous solution [41] that extracted features in advance, the AMU is much simpler and achieves better performance.

In summary, our key contributions are: (1) A deep IA structure that integrates a diversity of person-context interactions for robust action detection and (2) an AMU algorithm to estimate the memory features dynamically. We perform an extensive ablation study on the AVA [17] dataset for spatio-temporal action localization task. Our method shows a huge boost on performance, which yields the new state-of-the-art on both validation and test set. We also test our method on dataset UCF101-24 [32] and a segment level action recognition dataset EPIC-Kitchens [6]. Results further validate its generality.

2   Related Works

Video Classification. Various 3D CNN models [21,33,34,36] have been developed to handle video input. To leverage the huge image dataset, I3D [3] has been proposed to benefit from ImageNet[7] pre-training. In [4,8,27,35,44], the 3D kernels in above models are simulated by temporal filters and spatial filters which can significantly decrease the model size.

Previous two-stream methods [11,30] use optical flow to extract motion information, while recent work SlowFast [10] manages to do so using only RGB frames with different sample rates.

Spatio-temporal Action Detection. Action detection is more difficult than action classification because the model needs to not only predict the action labels but also localize the action in time and space. Most of the recent approaches [10,12,17,19,42] follow the object detection frameworks [14,28] by classifying the features generated by the detected bounding boxes. In contrast to our method, their results depend only on the cropped features. While all the other information is discarded and contributes nothing to the final prediction.

Attention Mechanism for Videos. The transformer [37] consists of several stacked self-attention layers and fully connected layers. Non-Local [38] concludes that the previous self-attention model can be viewed as a form of classical computer vision method of non-local means [2]. Hence a generic non-local block[38] is introduced. This structure enables models to compute the response by relating the features at different time or space, which makes the attention mechanism applicable for video-related tasks like action classification. The non-local block

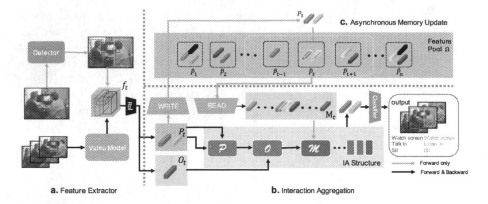

Fig. 2. Pipeline of the proposed AIA. **a.** We crop features of persons and objects from the extracted video features. **b.** Person features, object features and memory features from the feature pool Ω in **c** are fed to IA in order to integrate multiple interactions. The output of IA is passed to the final classifier for predictions. **c.** Our AMU algorithm reads memory features from feature pool and writes fresh person features to it

also plays an important role in [41] where the model references information from the long-term feature bank via a non-local feature bank operator.

3   Proposed Method

In this section, we will describe our method that localizes actions in space and time. Our approach aims at modeling and aggregating various interactions to achieve better action detection performance. In Sect. 3.1, we describe two important types of instance level features in short clips and the memory features in long videos. In Sect. 3.2, the Interaction Aggregation structure (IA) is explored to gather knowledge of interactions. In Sect. 3.3, we introduce the Asynchronous Memory Update algorithm (AMU) to alleviate the problem of heavy computation and memory consumption in temporal interaction modeling. The overall pipeline of our method is demonstrated in Fig. 2.

3.1   Instance Level and Temporal Memory Features

To model interactions in video, we need to find correctly what the queried person is interacted with. Previous works such as [38] calculate the interactions among all the pixels in feature map. Being computational expensive, these brute-force methods struggle to learn interactions among pixels due to the limited size of video dataset. Thus we go down to consider how to obtain concentrated interacted features. We observe that persons are often interacting with concrete objects and other persons. Therefore, we extract object and person embedding as the instance level features. In addition, video frames are usually highly correlated, thus we keep the long-term person features as the memory features.

Instance level features are cropped from the video features. Since computing the whole long video is impossible, we split it to consecutive short video clips $[v_1, v_2, \ldots, v_T]$. The d-dimensional features of the tth clip v_t are extracted using a video backbone model: $f_t = \mathcal{F}(v_t, \phi_\mathcal{F})$ where $\phi_\mathcal{F}$ is the parameters.

A detector is applied on the middle frame of v_t to get person boxes and object boxes. Based on the detected bounding boxes, we apply RoIAlign [18] to crop the person and object features out from extracted features f_t. The person and object features in v_t are denoted respectively as P_t and O_t.

One clip is only a short session and misses the temporal global semantics. In order to model the temporal interaction, we keep tracks of memory features. The memory features consist of person features in consecutive clips: $M_t = [P_{t-L}, \ldots, P_t, \ldots, P_{t+L}]$, where $(2L + 1)$ is the size of clip-wise reception field. In practice, a certain number of persons are sampled from each neighbor clip.

The three features above have semantic meaning and contain concentrated information to recognize actions. With these three features, we are now able to model semantic interactions explicitly.

3.2   Interaction Modeling and Aggregation

How do we leverage these extracted features? For a target person, there are multiple detected objects and persons. The main challenge is how to correctly pay more attention to the objects or the persons that the target person is interacted with. In this section, we introduce first our Interaction Block that can adaptively model each type of interactions in a uniform structure. Then we describe our Interaction Aggregation (IA) structure that aggregates multiple interactions.

Overview. Given different human P_t, object O_t and memory features M_t, the proposed IA structure outputs action features $A_t = \mathcal{E}(P_t, O_t, M_t, \phi_\mathcal{E})$, where $\phi_\mathcal{E}$ denotes the parameters in the IA structure. A_t is then passed to the final classifier for final predictions.

The hierarchical IA structure consists of multiple interaction blocks. Each of them is tailored for a single type of interactions. The interaction blocks are deep nested with other blocks to efficiently integrate different interactions for higher level features and more precise attentions.

Interaction Block. The structure of interaction block is adapted from Transformer Block originally proposed in [37] whose specific design basically follows [38,41]. Briefly speaking, one of the two inputs is used as the query and the other is mapped to key and value. Through the dot-product attention, which is the output of the softmax layer in Fig. 3 **a**, the block is able to select value features that are highly activated to the query features and merge them to enhance the query features. There are three types of interaction blocks in our design, which are P-Block, O-Block and M-Block.

–P-Block: P-Block models person-person interaction in the same clip. It is helpful for recognizing actions like listening and talking. Since the query input

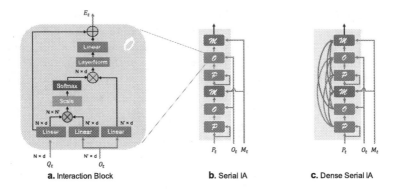

Fig. 3. Interaction Block and IA structure. **a.** The O-Block: the query input is the feature of the target person and the key/value input is the feature of objects. The P-Block and M-Block are similar. **b.** Serial IA. **c.** Dense Serial IA

is already the person features or the enhanced person features, we take the key/value input the same as the query input.

–O-Block: In O-Block, we aim to distill person-object interactions such as pushing and carrying an object. Our key/value input is the detected object features O_t. In the case where detected objects are too many, we sample based on detection scores. Figure 3a is an illustration of O-Block.

–M-Block: Some actions have strong logical connections along the temporal dimension like opening and closing. We model this type of interaction as temporal interactions. To operate this type, we take memory features M_t as key/value input of an M-Block.

Interaction Aggregation Structure. The Interaction Blocks extract three types of interaction. We now propose two IA structures to integrate these different interactions. The proposed IA structures are the naive parallel IA, the serial IA and the dense serial IA. For clarity, we use $\mathcal{P}$, $\mathcal{O}$, and $\mathcal{M}$ to represent the P-Block, O-Block, and M-Block respectively.

–Parallel IA: A naive approach is to model different interactions separately and merge them at last. As displayed in Fig. 4a, each branch follows similar structure to [13] that treats one type of interactions without the knowledge of other interactions. We argue that the parallel structure struggles to find interaction precisely. We illustrate the attention of the last P-Block in Fig. 4c by displaying the output of the softmax layer for different persons. As we can see, the target person is apparently watching and listening to the man in red. However, the P-block pays similar attention to two men.

–Serial IA: The knowledge across different interactions is helpful for recognizing interactions. We propose the serial IA to aggregate different types of interactions. As shown in Fig. 3b, different types of interaction blocks are stacked in sequence. The queried features are enhanced in one interaction block and then passed to an interaction block of a different type. Figure 4f

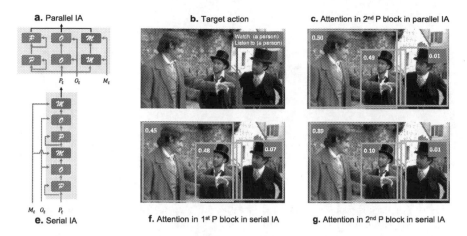

Fig. 4. We visualize attention by displaying the output of the softmax layer in P-Block. The original output contains the attention to zero padding person. We remove those meaningless attention and normalize the rest attention to 1

and **4g** demonstrate the advantage of serial IA: The first P-block can not differ the importance of the man in left and the man in middle. After gaining knowledge from O-block and M-block, the second P-block is able to pay more attention to man in left who is talking to the target person. Comparing to the attention in parallel IA (Fig. 4c), our serial IA is better in finding interactions. *–Dense Serial IA:* In above structures, the connections between interaction blocks are totally manually designed and the input of an interaction block is simply the output of another one. We expect the model to further learn which interaction features to take by itself. With this in mind, we propose the Dense Serial IA extension. In Dense Serial IA, each interaction block takes all the outputs of previous blocks and aggregates them using a learnable weight. Formally, the query of the ith block can be represent as

$$Q_{t,i} = \sum_{j \in \mathbf{C}} W_j \odot E_{t,j}, \tag{1}$$

where $\odot$ denotes the element-wise multiplication, $\mathbf{C}$ is the set of indices of previous blocks, W_j is a learnable d-dimenional vector normalized with a Softmax function among $\mathbf{C}$, $E_{t,j}$ is the enhanced output features from the jth block. Dense Serial IA is illustrated in Fig. 3c.

3.3   Asynchronous Memory Update Algorithm

Long-term memory features can provide useful temporal semantics to aid recognizing actions. Imagine a scene where a person opens the bottle cap, drinks water, and finally closes the cap, it could be hard to detect opening and closing

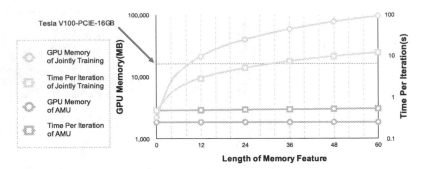

Fig. 5. Joint training with memory features is restricted by limited hardware resource. In this minor experiment, we take a 32-frame video clip with 256×340 resolution as input. The backbone is ResNet-50. During joint training (yellow line), rapidly growing GPU memory and computation time restricted the length of memory features to be very small value (8 in this experiment). With larger input or deeper backbone, this problem will be more serious. Our method (cyan line) doesn't have such problem. (Color figure online)

with subtle movements. But knowing the context of drinking water, things get much easier.

Resource Challenge. To capture more temporal information, we hope our M_t can gather features from enough number of clips, however, using more clips will increase the computation and memory consumption dramatically. Depicted with Fig. 5, when jointly training, the memory usage and computation consumption increase rapidly as the temporal length of M_t grows. To train on one target person, we must propagate forward and backward $(2L + 1)$ video clips at one time, which consumes much more time, and even worse, cannot make full use of enough long-term information due to limited GPU memory.

Insight. In the previous work [41], they pre-train another duplicated backbone to extract memory features to avoid this problem. However, this method makes use of frozen memory features, whose representation power can not be enhanced as model training goes. We expect the memory features can be updated dynamically and benefit from the improvement from parameter update in training process. Therefore, we propose the asynchronous memory update method which can generate effective dynamic long-term memory features and make the training process more lightweight. The details of training process with this algorithm are presented in Algorithm 1.

A naive design could be: pass forward all clips to get memory features and propagate current clip backward to calculate the gradients. This method alleviates the memory issue but is still slow in the training speed. We could also try to utilize the memory features like in Transformer-XL [5], but this requires training along the sequence direction and is thus unable to access future information.

Inspired by [40], our algorithm is composed of a memory component, the memory pool Ω and two basic operations, *READ* and *WRITE*. The memory

---

Algorithm 1. Training with asynchronous memory update

Input: Video dataset $\mathbf{V} = \{v^{(1)}, v^{(2)}, \ldots, v^{(|\mathbf{V}|)}\}$ with $v^{(i)} = [v_1^{(i)}, v_2^{(i)}, \ldots, v_{T_i}^{(i)}]$;
    The whole network $\mathcal{N}$, with its parameter $\phi_{\mathcal{N}}$;
Output: Optimized network $\mathcal{N}$ with $\phi_{\mathcal{N}}$ for inference.
    // Initialization :
1:   $\Omega = \{(\hat{P}_t^{(i)} \leftarrow \text{zero vectors}, \delta_t^{(i)} \leftarrow 0) \mid \forall t, i\}$.
2:   $err \leftarrow \infty$.
    // Training Process:
3: **for** $iter = 1$ to $iter_{max}$ **do**
4:     Sample a video clip $v_t^{(i)}$ from dataset $\mathbf{V}$.
5:     **for** $t' = t - L$ to $t + L$ **do**
6:       **if** $t' \neq t$ **then**
7:         **READ** $\hat{P}_{t'}^{(i)}$ and $\delta_{t'}^{(i)}$ from memory pool Ω.
8:         $w_{t'}^{(i)} = \min\{err/\delta_{t'}^{(i)}, \delta_{t'}^{(i)}/err\}$.
9:         Impose penalty: $\hat{P}_{t'}^{(i)} \leftarrow w_{t'}^{(i)} \hat{P}_{t'}^{(i)}$.
10:      **end if**
11:     **end for**
12:     Extract $P_t^{(i)}$ and $O_t^{(i)}$ with the backbone in $\mathcal{N}$.
13:     Estimated memory features: $\hat{M}_t^{(i)} \leftarrow [\hat{P}_{t-L}^{(i)}, \ldots, \hat{P}_{t-1}^{(i)}, P_t^{(i)}, \hat{P}_{t+1}^{(i)}, \ldots, \hat{P}_{t+L}^{(i)}]$.
14:     Forward $(P_t^{(i)}, O_t^{(i)}, M_t^{(i)})$ with the head in $\mathcal{N}$ and backward to optimize $\phi_{\mathcal{N}}$.
15:     Update err as the output of current loss function.
16:     **WRITE** $\hat{P}_t^{(i)} \leftarrow P_t^{(i)}, \delta_t^{(i)} \leftarrow err$ back to Ω.
17: **end for**
18: **return** $\mathcal{N}, \phi_{\mathcal{N}}$

---

pool Ω records memory features. Each feature $\hat{P}_t^{(i)}$ in this pool is an estimated value and tagged with a loss value $\delta_t^{(i)}$. This loss value $\delta_t^{(i)}$ logs the convergence state of the whole network. Two basic operations are invoked at each iteration of training:

    –*READ:* At the beginning of each iteration, given a video clip $v_t^{(i)}$ from the ith video, estimated memory features around the target clip are read from the memory pool Ω, which are $[\hat{P}_{t-L}^{(i)}, \ldots, \hat{P}_{t-1}^{(i)}]$ and $[\hat{P}_{t+1}^{(i)}, \ldots, \hat{P}_{t+L}^{(i)}]$ specifically.
    –*WRITE:* At the end of each iteration, personal features for the target clip $P_t^{(i)}$ are written back to the memory pool Ω as estimated memory features $\hat{P}_t^{(i)}$, tagged with current loss value.
    –*Reweighting:* The features we *READ* are written at different training steps. Therefore, some early written features are extracted from the model whose parameters are much different from current ones. Therefore, we impact a penalty factor $w_{t'}^{(i)}$ to discard badly estimated features. We design a simple yet effective way to compute such penalty factor by using loss tag. The difference between the loss tag $\delta_{t'}^{(i)}$ and current loss value is expressed as,

$$w_{t'}^{(i)} = \min\{err/\delta_{t'}^{(i)}, \delta_{t'}^{(i)}/err\}, \qquad (2)$$

which should be very close to 1 when the difference is small. As the network converges, the estimated features in the memory pool are expected to be closer and closer to the precise features and $w_{t'}^{(i)}$ approaches to 1.

As shown in Fig. 5, the consumption of our algorithm has no obvious increase in both GPU memory and computation as the length of memory features grows, and thus we can use long enough memory features on current common devices. With dynamic updating, the asynchronous memory features can be better exploited than frozen ones.

4   Experiments on AVA

The Atomic Visual Actions (AVA) [17] dataset is built for spatio-temporal action localization. In this dataset, each person is annotated with a bounding box and multiple action labels at 1 FPS. There are 80 atomic action classes which cover pose actions, person-person interactions and person-object interactions. This dataset contains 235 training movie videos and 64 validation movie videos.

Since our method is originally designed for spatio-temporal action detection, we use AVA dataset as the main benchmark to conduct detailed ablation experiments. The performances are evaluated with official metric frame level mean average precision(mAP) at spatial IoU ≥ 0.5 and only the top 60 most common action classes are used for evaluation, according to [17].

4.1   Implementation Details

Instance Detector. We apply Faster R-CNN [28] framework to detect persons and objects on the key frames of each clip. A model with ResNeXt-101-FPN [23,43] backbone from maskrcnn-benchmark [26] is adopted for object detection. It is firstly pre-trained on ImageNet [7] and then fine-tuned on MSCOCO [25] dataset. For human detection, we further fine-tune the model on AVA for higher detection precision.

Backbone. Our method can be easily applied to any kind of 3D CNN backbone. We select state-of-the-art backbone SlowFast [10] network with ResNet-50 structure as our baseline model. Basically following the recipe in [10], our backbone is pre-trained on Kinetics-700 [3] dataset for action classification task. This pre-trained backbone produces 66.34% top-1 and 86.66% top-5 accuracy on the Kinetics-700 validation set.

Training and Inference. Initialized from Kinetics pre-trained weights, we then fine-tune the whole model with focal loss [24] on AVA dataset. The inputs of our network are 32 RGB frames, sampled from a 64-frame raw clip with one frame interval. Clips are scaled such that the shortest side becomes 256, and then fed into the fully convolution backbone. We use only the ground-truth human boxes for training and the randomly jitter them for data augmentation. For the object boxes, we set the detection threshold to 0.5 in order to have higher recall. During inference, detected human boxes with a confidence score larger than 0.8 are used. We set $L = 30$ for memory features in our experiments. We train our network using the SGD algorithm with batch size 64 on 16 GPU (4 clips per device). BatchNorm(BN) [20] statistics are set frozen. We train for 27.5k iterations with

Table 1. Ablation Experiments. We use a ResNet-50 SlowFast backbone to perform our ablation study. Models are trained on the AVA (v2.2) training set and evaluated on the validation set. The evaluation metric mAP is shown in %

(a) **3 Interactions**

$\mathcal{P}$	$\mathcal{O}$	$\mathcal{M}$	mAP
			26.54
✓	✓		28.04
✓		✓	28.86
	✓	✓	28.92
✓	✓	✓	**29.26**

(b) **Num of I-Blocks**

blocks	mAP
$1 \times \{\mathcal{P}, \mathcal{M}, \mathcal{O}\}$	29.26
$2 \times \{\mathcal{P}, \mathcal{M}, \mathcal{O}\}$	**29.64**
$3 \times \{\mathcal{P}, \mathcal{M}, \mathcal{O}\}$	29.61

(c) **Interaction Order**

order	mAP	order	mAP
$\mathcal{M} \to \mathcal{O} \to \mathcal{P}$	29.48	$\mathcal{M} \to \mathcal{P} \to \mathcal{O}$	29.46
$\mathcal{O} \to \mathcal{P} \to \mathcal{M}$	29.51	$\mathcal{O} \to \mathcal{M} \to \mathcal{P}$	29.53
$\mathcal{P} \to \mathcal{M} \to \mathcal{O}$	29.44	$\mathcal{P} \to \mathcal{O} \to \mathcal{M}$	**29.64**

(d) **IA Structure**

structure	mAP
Parallel	28.85
Serial	29.64
Dense Serial	**29.80**

(e) **Asynchronous Memory Update**

model	params	FLOPs	mAP
Baseline	1.00×	1.00×	26.54
LFB(w/o AMU)	2.18×	2.12×	27.02
LFB(w/ AMU)	1.18×	1.12×	28.57
IA(w/o AMU)	2.35×	2.15×	28.07
IA(w/ AMU)	1.35×	1.15×	29.64

(f) **Compare to NL**

model	mAP
Baseline	26.54
+NL	26.85
+IA(w/o $\mathcal{M}$)	**28.23**

base learning rate 0.004 and the learning rate is reduced by a factor 10 at 17.5k and 22.5k iteration. A linear warm-up [16] scheduler is applied for the first 2k iterations.

4.2 Ablation Experiments

Three Interactions. We first study the importance of three kinds of interactions. For each interaction type, we use at most one block in the experiment. These blocks are then stacked in serial. To evaluate the importance of person-object interaction, we remove the O-Block in the structure. Other interactions are evaluated in the same way. Table 1a compares the model performance, where used interaction types are marked with "✓". A backbone baseline without any interaction is also listed in this table. Overall we observe that removing any of these three type interactions results in a significant performance decrease, which confirms that all these three interactions are important for action detection.

Number of Interaction Blocks. We then experiment with different settings for the number of interaction blocks in our IA structure. The interaction blocks are nested in serial structure in this experiment. In Table 1b, $N \times \{\mathcal{P}, \mathcal{M}, \mathcal{O}\}$ denotes N blocks are used for each interaction type, with the total number as $3N$. We find that with the setting $N = 2$ our method can achieve the best performance, so we use this as our default configuration.

Interaction Order. In our serial IA, different type of interactions are alternately integrated in sequential. We investigate effect of different interaction order design in Table 1c. As shown in this experiment, the performance with different

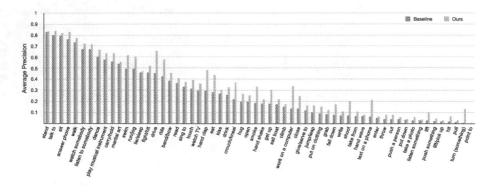

Fig. 6. Per category results comparison on the validation set of AVA v2.2

order are quite similar, we thus choose the slightly better one $\mathcal{P} \to \mathcal{O} \to \mathcal{M}$ as our default setting.

Interaction Aggregation Structure. We analyze different IA structure in this part. Parallel IA, serial IA and the dense serial IA extension are compared in Table 1d. As we expect, the parallel IA performs much worse than serial structure. With dense connections between blocks, our model is able to learn more knowledge of interactions, which further boosts the performance.

Asynchronous Memory Update. In the previous work LFB [41], the memory features are extracted with another backbone, which is frozen during training. In this experiment we compare our asynchronous memory features with the frozen ones. For fair comparison, we re-implement LFB with SlowFast backbone, and also apply our AMU algorithm to LFB. In Table 1d, we find that our asynchronous memory features can gain much better performance than the frozen method with nearly half of the parameters and computation cost. We argue that this is because our dynamic features can provide better representation.

Comparison to Non-local Attention. Finally we compare our interaction aggregation method with prior work non-local block [38] (NL). Following [10], we augment the backbone with a non-local branch, where attention is computed between the person features and global pooled features. Since there is no long-term features in this branch, we eliminate $\mathcal{M}$ in this experiment. In Table 1f, we see that our serial IA works significantly better than NL block. This confirms that our method can better learn to find potential interactions than NL block.

4.3   Main Results

Finally, we compare our results on AVA v2.1 and v2.2 with previous methods in Table 2. Our method surpasses all previous works on both versions.

The AVA v2.2 dataset, is the newer benchmark used in ActivityNet challenge 2019 [9]. On the validation set, our method reports a new state-of-the-art *33.11 mAP* with one single model, which outperforms the strong baseline SlowFast by

Table 2. Main results on AVA. Here, we display our best results with both ResNet50(R50) and ResNet101(R101). "*" indicates multi-scale testing. The input sizes are shown in frame number and sample rate. SlowFast R101 backbone models re-implemented in this work are also displayed as "ours" for comparison.

(a) **Comparison on AVA v2.1.**

model	input	pretrain	val
SlowFast [10]	32 × 2	K400	26.3
LFB [41]	32 × 2	K400	27.7
I3D [12]	64 × 1	K600	21.9
SlowFast+NL [10]	32 × 2	K600	28.2
SlowFast (ours)	32 × 2	K600	-
SlowFast (ours)	32 × 2	K700	28.1
AIA R50	32 × 2	K700	28.9
AIA R101	32 × 2	K700	**31.2**

(b) **Comparison on AVA v2.2.**

model	input	pretrain	val	test
SlowFast+NL [10]	32 × 2	K600	29.1	-
SlowFast+NL [10]	64 × 2	K600	29.4	-
SlowFast*, 7 ens. [10]	-	K600	-	34.25
SlowFast (ours)	32 × 2	K600	-	-
SlowFast (ours)	32 × 2	K700	29.3	-
AIA R50	32 × 2	K700	29.80	-
AIA R101	32 × 2	K700	32.26	-
AIA R101*	32 × 2	K700	**33.11**	**32.25**
AIA R101*, 3 ens.	-	K700	-	**34.42**

Table 3. Results on UCF101-24 Split1

Method	mAP	Method	mAP	Method	mAP	Method	mAP
ACT [22]	69.5	Gu et al. [17]	76.3	C2D (ours)	75.5	I3D (ours)	79.6
STEP [45]	75.0	Zhang et al. [46]	77.9	C2D+AIA	78.8	I3D+AIA	**81.7**

3.7 mAP. On the test split, we train our model on both training and validation splits and use a relative longer scheduler. With an ensemble of three models with different learning rates and aggregation structures, our method achieves better performance than the winning entry of AVA challenge 2019 (an ensemble with 7 SlowFast [10] networks). The per category results for our method and SlowFast baseline is illustrated in Fig. 6. We can observe the performance gain for each category, especially for those who contain interactions with video context.

As shown in Table 2, we pre-train the backbone model with a new larger Kinetics-700 for better performance. However, it is worth noting that we do not use non-local block in our backbone model and there are some other slight differences between our implementation and the official one [10]. As a result, our K700 backbone model has a similar performance to the official K600 one. That is to say, very most of the performance advantages benefit from our proposed method instead of the backbone.

5   Experiments on UCF101-24

UCF101-24 [32] is an action detection set with 24 action categories. We conduct experiments on the first split of this dataset following previous works and use the corrected annotations provided by Singh et al. [31].

We experiment two different backbone models, C2D and I3D. Both of them are pre-trained on the Kinetics-400 dataset. Other settings are basically the

Table 4. EPIC-Kitchens validation results

	Verbs		Nouns		Actions	
	top-1	top-5	top-1	top-5	top-1	top-5
Baradel [1]	40.9	–	–	–	–	–
LFB NL [41]	52.4	80.8	29.3	54.9	20.8	39.8
SlowFast (ours)	56.8	82.8	32.3	56.7	24.1	42.0
AIA-Parallel	57.6	83.9	36.3	63.0	26.4	47.4
AIA-Serial	59.2	84.2	**37.2**	**63.2**	**27.7**	**48.0**
AIA-Dense-Serial	**60.0**	**84.6**	**37.2**	62.1	27.1	47.8

same as AVA experiments. More implementation details are provided in Supplementary Material. Table 3 shows the result on UCF101-24 test split in terms of Frame-mAP with 0.5 IOU threshold. As we can see in the table, AIA achieves 3.3% and 2.1% improvement over two different backbones. Moreover, with a relative weak 2D backbone, our method still achieves very competitive results.

6   Experiments on EPIC-Kitchens

To demonstrate the generalizability of AIA, we evaluate our method on the segment level dataset EPIC-Kitchens [6]. In EPIC Kitchens, each segment is annotated with one verb and one noun. The action is defined by their combination.

For both verb model and noun model, we use the extracted segment features (global average pooling of f_t) as query input for IA structure. Hand features and object features are cropped and then fed into IA to model person-person and person-object interactions. For verb model, the memory features are the segment features. For noun model, the memory features are the object features extracted from object detector feature map, thus the AMU algorithm is only applied to the verb model. More details are available in Supplementary Material. From Table 4, we observe a significant gain for all three tasks. All the variants of AIA outperform the SlowFast baseline. Among them, the dense serial IA achieves the best performance for the verbs test, leading to 3.2% improvement on top-1 score. The serial IA results in 4.9% for the nouns test and 3.6% for the action test.

7   Conclusion

In this paper, we present the Asynchronous Interaction Aggregation network and its performance in action detection. Our method reports the new start-of-the-art on AVA dataset. Nevertheless, the performance of action detection and the interaction recognition is far from perfect. The poor performance is probably due to the limited video dataset. Transferring the knowledge of action and interaction from image could be a further improvement for AIA network.

Acknowledgements. This work is supported in part by the National Key R&D Program of China, No. 2017YFA0700800, National Natural Science Foundation of China under Grants 61772332 and Shanghai Qi Zhi Institute.

References

1. Baradel, F., Neverova, N., Wolf, C., Mille, J., Mori, G.: Object level visual reasoning in videos. In: Ferrari, V., Hebert, M., Sminchisescu, C., Weiss, Y. (eds.) ECCV 2018. LNCS, vol. 11217, pp. 106–122. Springer, Cham (2018). https://doi.org/10.1007/978-3-030-01261-8_7
2. Buades, A., Coll, B., Morel, J.M.: A non-local algorithm for image denoising. In: 2005 IEEE Computer Society Conference on Computer Vision and Pattern Recognition (CVPR 2005), vol. 2, pp. 60–65. IEEE (2005)
3. Carreira, J., Zisserman, A.: Quo Vadis, action recognition? A new model and the kinetics dataset. In: proceedings of the IEEE Conference on Computer Vision and Pattern Recognition, pp. 6299–6308 (2017)
4. Christoph, R., Pinz, F.A.: Spatiotemporal residual networks for video action recognition. In: Advances in Neural Information Processing Systems, pp. 3468–3476 (2016)
5. Dai, Z., Yang, Z., Yang, Y., Carbonell, J., Le, Q.V., Salakhutdinov, R.: Transformer-xl: Attentive language models beyond a fixed-length context. arXiv preprint arXiv:1901.02860 (2019)
6. Damen, D., et al.: Scaling egocentric vision: the epic-kitchens dataset. In: Ferrari, V., Hebert, M., Sminchisescu, C., Weiss, Y. (eds.) ECCV 2018. LNCS, vol. 11208, pp. 753–771. Springer, Cham (2018). https://doi.org/10.1007/978-3-030-01225-0_44
7. Deng, J., Dong, W., Socher, R., Li, L.J., Li, K., Fei-Fei, L.: Imagenet: a large-scale hierarchical image database. In: 2009 IEEE Conference on Computer Vision and Pattern Recognition, pp. 248–255. IEEE (2009)
8. Diba, A., et al.: Spatio-temporal channel correlation networks for action classification. In: Ferrari, V., Hebert, M., Sminchisescu, C., Weiss, Y. (eds.) ECCV 2018. LNCS, vol. 11208, pp. 299–315. Springer, Cham (2018). https://doi.org/10.1007/978-3-030-01225-0_18
9. Caba Heilbron, F., Victor Escorcia, B.G., Niebles, J.C.: Activitynet: a large-scale video benchmark for human activity understanding. In: Proceedings of the IEEE Conference on Computer Vision and Pattern Recognition, pp. 961–970 (2015)
10. Feichtenhofer, C., Fan, H., Malik, J., He, K.: Slowfast networks for video recognition. In: Proceedings of the IEEE International Conference on Computer Vision, pp. 6202–6211 (2019)
11. Feichtenhofer, C., Pinz, A., Zisserman, A.: Convolutional two-stream network fusion for video action recognition. In: Proceedings of the IEEE Conference on Computer Vision and Pattern Recognition, pp. 1933–1941 (2016)
12. Girdhar, R., Carreira, J., Doersch, C., Zisserman, A.: A better baseline for AVA. CoRR abs/1807.10066 (2018). http://arxiv.org/abs/1807.10066
13. Girdhar, R., Carreira, J., Doersch, C., Zisserman, A.: Video action transformer network. In: Proceedings of the IEEE Conference on Computer Vision and Pattern Recognition, pp. 244–253 (2019)
14. Girshick, R.: Fast R-CNN. In: Proceedings of the IEEE International Conference on Computer Vision, pp. 1440–1448 (2015)

15. Gkioxari, G., Girshick, R., Dollár, P., He, K.: Detecting and recognizing human-object interactions. In: Proceedings of the IEEE Conference on Computer Vision and Pattern Recognition, pp. 8359–8367 (2018)
16. Goyal, P., et al.: Accurate, large minibatch SGD: training imagenet in 1 hour. arXiv preprint arXiv:1706.02677 (2017)
17. Gu, C., et al.: Ava: a video dataset of spatio-temporally localized atomic visual actions. In: Proceedings of the IEEE Conference on Computer Vision and Pattern Recognition, pp. 6047–6056 (2018)
18. He, K., Gkioxari, G., Dollár, P., Girshick, R.: Mask R-CNN. In: Proceedings of the IEEE International Conference on Computer Vision, pp. 2961–2969 (2017)
19. Hou, R., Chen, C., Shah, M.: Tube convolutional neural network (T-CNN) for action detection in videos. In: Proceedings of the IEEE International Conference on Computer Vision, pp. 5822–5831 (2017)
20. Ioffe, S., Szegedy, C.: Batch normalization: accelerating deep network training by reducing internal covariate shift. arXiv preprint arXiv:1502.03167 (2015)
21. Ji, S., Xu, W., Yang, M., Yu, K.: 3D convolutional neural networks for human action recognition. IEEE Trans. Pattern Anal. Mach. Intell. **35**(1), 221–231 (2012)
22. Kalogeiton, V., Weinzaepfel, P., Ferrari, V., Schmid, C.: Action tubelet detector for spatio-temporal action localization. In: Proceedings of the IEEE International Conference on Computer Vision, pp. 4405–4413 (2017)
23. Lin, T.Y., Dollár, P., Girshick, R., He, K., Hariharan, B., Belongie, S.: Feature pyramid networks for object detection. In: Proceedings of the IEEE Conference on Computer Vision and Pattern Recognition, pp. 2117–2125 (2017)
24. Lin, T.Y., Goyal, P., Girshick, R., He, K., Dollár, P.: Focal loss for dense object detection. In: Proceedings of the IEEE International Conference on Computer Vision, pp. 2980–2988 (2017)
25. Lin, T.-Y.: Microsoft COCO: common objects in context. In: Fleet, D., Pajdla, T., Schiele, B., Tuytelaars, T. (eds.) ECCV 2014. LNCS, vol. 8693, pp. 740–755. Springer, Cham (2014). https://doi.org/10.1007/978-3-319-10602-1_48
26. Massa, F., Girshick, R.: Mask R-CNN-benchmark: fast, modular reference implementation of Instance Segmentation and Object Detection algorithms in PyTorch (2018). https://github.com/facebookresearch/maskrcnn-benchmark. Accessed 29 Feb 2020
27. Qiu, Z., Yao, T., Mei, T.: Learning spatio-temporal representation with pseudo-3D residual networks. In: Proceedings of the IEEE International Conference on Computer Vision, pp. 5533–5541 (2017)
28. Ren, S., He, K., Girshick, R., Sun, J.: Faster R-CNN: towards real-time object detection with region proposal networks. In: Advances in Neural Information Processing Systems, pp. 91–99 (2015)
29. Sigurdsson, G.A., Divvala, S., Farhadi, A., Gupta, A.: Asynchronous temporal fields for action recognition. In: The IEEE Conference on Computer Vision and Pattern Recognition (CVPR) (2017)
30. Simonyan, K., Zisserman, A.: Two-stream convolutional networks for action recognition in videos. In: Advances in Neural Information Processing Systems, pp. 568–576 (2014)
31. Singh, G., Saha, S., Sapienza, M., Torr, P.H., Cuzzolin, F.: Online real-time multiple spatiotemporal action localisation and prediction. In: Proceedings of the IEEE International Conference on Computer Vision, pp. 3637–3646 (2017)
32. Soomro, K., Zamir, A.R., Shah, M.: UCF101: a dataset of 101 human actions classes from videos in the wild. arXiv preprint arXiv:1212.0402 (2012)

33. Taylor, G.W., Fergus, R., LeCun, Y., Bregler, C.: Convolutional learning of spatio-temporal features. In: Daniilidis, K., Maragos, P., Paragios, N. (eds.) ECCV 2010. LNCS, vol. 6316, pp. 140–153. Springer, Heidelberg (2010). https://doi.org/10.1007/978-3-642-15567-3_11

34. Tran, D., Bourdev, L.D., Fergus, R., Torresani, L., Paluri, M.: C3D: generic features for video analysis. CoRR, abs/1412.0767, vol. 2, no. 7, p. 8 (2014)

35. Tran, D., Wang, H., Torresani, L., Ray, J., LeCun, Y., Paluri, M.: A closer look at spatiotemporal convolutions for action recognition. In: Proceedings of the IEEE conference on Computer Vision and Pattern Recognition, pp. 6450–6459 (2018)

36. Varol, G., Laptev, I., Schmid, C.: Long-term temporal convolutions for action recognition. IEEE Trans. Pattern Anal. Mach. Intell. **40**(6), 1510–1517 (2017)

37. Vaswani, A., et al.: Attention is all you need. In: Guyon, I., et al. (eds.) Advances in Neural Information Processing Systems, vol. 30, pp. 5998–6008. Curran Associates, Inc. (2017). http://papers.nips.cc/paper/7181-attention-is-all-you-need.pdf

38. Wang, X., Girshick, R., Gupta, A., He, K.: Non-local neural networks. In: Proceedings of the IEEE Conference on Computer Vision and Pattern Recognition, pp. 7794–7803 (2018)

39. Wang, X., Gupta, A.: Videos as space-time region graphs. In: Ferrari, V., Hebert, M., Sminchisescu, C., Weiss, Y. (eds.) ECCV 2018. LNCS, vol. 11209, pp. 413–431. Springer, Cham (2018). https://doi.org/10.1007/978-3-030-01228-1_25

40. Weston, J., Chopra, S., Bordes, A.: Memory networks. arXiv preprint arXiv:1410.3916 (2014)

41. Wu, C.Y., Feichtenhofer, C., Fan, H., He, K., Krahenbuhl, P., Girshick, R.: Long-term feature banks for detailed video understanding. In: The IEEE Conference on Computer Vision and Pattern Recognition (CVPR) (2019)

42. Xia, J., Tang, J., Lu, C.: Three branches: detecting actions with richer features. arXiv preprint arXiv:1908.04519 (2019)

43. Xie, S., Girshick, R., Dollár, P., Tu, Z., He, K.: Aggregated residual transformations for deep neural networks. In: Proceedings of the IEEE Conference on Computer Vision and Pattern Recognition, pp. 1492–1500 (2017)

44. Xie, S., Sun, C., Huang, J., Tu, Z., Murphy, K.: Rethinking spatiotemporal feature learning for video understanding. arXiv preprint arXiv:1712.04851, vol. 1, no. 2, p. 5 (2017)

45. Yang, X., Yang, X., Liu, M.Y., Xiao, F., Davis, L.S., Kautz, J.: Step: spatio-temporal progressive learning for video action detection. In: Proceedings of the IEEE Conference on Computer Vision and Pattern Recognition, pp. 264–272 (2019)

46. Zhang, Y., Tokmakov, P., Hebert, M., Schmid, C.: A structured model for action detection. In: Proceedings of the IEEE Conference on Computer Vision and Pattern Recognition, pp. 9975–9984 (2019)

47. Zhou, B., Andonian, A., Oliva, A., Torralba, A.: Temporal relational reasoning in videos. In: Ferrari, V., Hebert, M., Sminchisescu, C., Weiss, Y. (eds.) ECCV 2018. LNCS, vol. 11205, pp. 831–846. Springer, Cham (2018). https://doi.org/10.1007/978-3-030-01246-5_49

Shape and Viewpoint Without Keypoints

Shubham Goel$^{(\boxtimes)}$, Angjoo Kanazawa, and Jitendra Malik

UC Berkeley, Berkeley, USA
shubham-goel@berkeley.edu

Abstract. We present a learning framework that learns to recover the 3D shape, pose and texture from a single image, trained on an image collection without any ground truth 3D shape, multi-view, camera viewpoints or keypoint supervision. We approach this highly under-constrained problem in a "analysis by synthesis" framework where the goal is to predict the likely shape, texture and camera viewpoint that could produce the image with various learned category-specific priors. Our particular contribution in this paper is a representation of the distribution over cameras, which we call "camera-multiplex". Instead of picking a point estimate, we maintain a set of camera hypotheses that are optimized during training to best explain the image given the current shape and texture. We call our approach Unsupervised Category-Specific Mesh Reconstruction (U-CMR), and present qualitative and quantitative results on CUB, Pascal 3D and new web-scraped datasets. We obtain state-of-the-art camera prediction results and show that we can learn to predict diverse shapes and textures across objects using an image collection without any keypoint annotations or 3D ground truth. Project page: https://shubham-goel.github.io/ucmr.

1 Introduction

There has been much progress in recent years in training deep networks to infer 3D shape from 2D images. These approaches fall into two major families based on the supervisory signal used (a) 3D models, as available in collection of CAD models such as ShapeNet or (b) multiple views of the same object which permit deep learning counterparts of classical techniques such as shape carving or structure-from-motion. But do we need all this supervision?

It is easy to find on the internet large collections of images of objects belonging to particular categories, such as birds or cars or chairs. Let us focus on birds, for which datasets like CUB [35], shown in Fig. 1 (left) exist. Note that this is a "Multiple Instance Single View" setting. For each bird instance we have only a single view, and every bird is a slightly different shape, even though the multiple instances share a family resemblance. Compared to classical SFM, where we have "Single Instance Multiple Views", our goal is to "3Dfy" these birds. From a single

Electronic supplementary material The online version of this chapter (https:// doi.org/10.1007/978-3-030-58555-6_6) contains supplementary material, which is available to authorized users.

© Springer Nature Switzerland AG 2020
A. Vedaldi et al. (Eds.): ECCV 2020, LNCS 12360, pp. 88–104, 2020.
https://doi.org/10.1007/978-3-030-58555-6_6

Fig. 1. Given an image collection of an object category, like birds, we propose a computational framework that given a single image of an object, predicts its 3D shape, viewpoint and texture, without using any 3D shape, viewpoints or keypoint supervision during training. On the right we show the input image and the results obtained by our method, shown from multiple views.

image, create a 3D model and its texture map, which can then be rendered from different camera viewpoints as shown in the rows of Fig. 1 (right). This particular formulation was presented in the "Category-Specific Mesh Reconstruction" work of Kanazawa *et al.* [15], and their algorithm (CMR) is an inspiration for our work. Even earlier the work of Cashman and Fitzgibbon [3] working on analyzing images of dolphins showed how an "analysis by synthesis" paradigm with a deformable template model of a shape category could enable one to infer 3D shapes in an optimization framework.

It is under-appreciated that these approaches, while pioneering, do exploit some supervisory information. This includes (1) knowledge of a mean shape for the category (2) silhouettes for each instance (3) marked keypoints on the various instances (e.g. beak tip for each bird). Of these, the need for labeled keypoints is the most troublesome. Only one mean shape is needed for the entire category and sometimes a very generic initialization such as a sphere is good enough. Silhouettes could presumably be marked (perhaps in a category-agnostic way) by an instance segmentation network like Mask R-CNN. But keypoints are tedious to mark on every instance. This effort can be justified for a single important category like humans, but we cannot afford to do for the thousands of categories which we might want to "3Dfy". Yes, we could have the keypoints be marked by a network but to train that would require keypoint labels! There have been recent efforts in unsupervised keypoint-detection [13], however, so far, these methods learn viewpoint-dependent keypoints that often get mixed-up in the presence of 180° rotations.

In this paper we present an approach, U-CMR (for Unsupervised CMR) which enables us to train a function which can factor an image into a 3D shape, texture and camera, in a roughly similar setting to CMR [15], except that we replace the need for keypoint annotation with a single 3D template shape for the entire category. It turns out that keypoint annotations are needed for recovering cameras (using SFM like techniques) and if we don't have keypoint annotations, we have to solve for the camera simultaneously with shape and texture. Extending the "analysis by synthesis" paradigm to also recover cameras is unfortunately rather hard. Intuitively speaking, this is because of the discontinuous and multi-modal nature of the space of possible camera viewpoints. While shape and texture have a smooth, well behaved optimization surface that is amenable to gradient descent optimization, this does not hold for the space of possible cameras. The two most likely camera explanations might lie on the opposite sides of the viewing sphere, where it is not possible to approach the optimal camera in an iterative manner as done when optimizing the energy landscape of a deep network. This typically causes the camera prediction to be stuck in bad local minima. Our solution to this problem is to maintain a set of possible camera hypotheses for each training instance that we call a **camera-multiplex**. This is reminiscent of particle filtering approaches with the idea being to maintain a distribution rather than prematurely pick a point estimate. We can iteratively refine the shape, texture as well as the camera-multiplex. More details are presented in Sect. 3.

We evaluate our approach on 3D shape, pose, and texture reconstruction on 4 categories: CUB-200-2011 birds [35], PASCAL-3D [38] cars, motorcycles and a new dataset of shoes we scraped from the Internet. We show that naively predicting shape, camera, and texture directly results in a degenerate solution where the shape is flat and the camera collapses to a single mode. We quantitatively evaluate the final camera prediction where the proposed camera-multiplex approach obtains the state of the art results under this weakly-supervised setting of no keypoint annotations. We show that despite the lack of viewpoints and keypoints, we can learn a reasonable 3D shape space, approaching that of the shapes obtained by a previous method that uses keypoint supervision [15].

2   Related Work

Recent deep learning based 3D reconstruction methods can be categorized by the required supervisory signal and the output of the system. This is illustrated in Table 1. Earlier methods formulate the problem assuming full 3D shape supervision for an image [5,7,9,10,31], which is enabled by synthetic datasets such as ShapeNet [37] and SunCG [27]. Some approaches generate synthetic datasets using data gathered from the real world to train their models [4,33,41]. However, requiring 3D supervision severely restricts these approaches, since ground truth 3D shape is costly or not possible to acquire, especially at a large scale. As such, follow up methods explore more natural forms of supervision, where multiple-views of the same object are available. Some of these approaches assume known viewpoints [16,32,39] akin to the setting of traditional MVS or visual hull. Other approaches explore the problem with unknown viewpoint setting [8,12,30]. These approaches assume that multi-view silhouettes, images, and/or depth images are available, and train their models such that the predicted 3D shapes reconstruct the images after projection or differentiable rendering. A variety of differentiable rendering mechanisms have been explored [18,20,21].

While multi-view images may be obtained in the real world, the vast amount of available visual data corresponds to the setting of unconstrained collection of single-view images, where no simultaneous multiple views of the same instance are available. This is also the natural setting for non-rigid objects where the shape may change over time. The traditional non-rigid structure from motion [29] also falls under this category, where the input is a tracked set of corresponding points [6,29] or 2D keypoints [22,34]. Earlier approaches fit a deformable 3D model [1,3,14,17] to 2D keypoints and silhouettes. Kanazawa et al.[15] propose CMR, a learning based framework where 3D shape, texture, and camera pose are predicted from a single image, trained under this setting of single-view image collections with known mask and keypoint annotations. While this is a step in the right direction, the requirement of keypoint annotation is still restrictive. More recently, Kulkarni et al. [19] bypass this requirement of keypoints to learn a dense canonical surface mapping of objects from a set of image collections with mask supervision and a template 3D shape. They focus on predicting the surface correspondences on images and learn to predict the camera viewpoints during the training, but do not learn to predict the 3D shape. While we tackle a different problem, we operate under the same required supervision. As such we quantitatively compare with CSM on the quality of the camera predicted, where our approach obtains considerably better camera predictions. Note that there are several recent approaches that explore disentangling images into 2.5D surface properties, camera, and texture of the visible regions from a collection of monocular images, without any masks [26,28,36]. However these approaches are mainly demonstrated on faces. In this work we recover a full 3D representation and texture from a single image.

Table 1. A comparison of different approaches highlighting the differences between the input (during training) and the output (during inference). Our approach (U-CMR) uses only silhouette supervision but predicts full 3D shape, camera and texture. *MeshRCNN predicts shape in camera-coordinates instead of a canonical frame.

Approach	Required supervision per image					Output			
	3D Shape	Multi-view	Cam	Keypoints	Mask	3D Shape	2.5D	Cam	Texture
MeshRCNN* [10]	✗					✓			
DeepSDF [23]	✗					✓			
Smalst [41]	✗		✗	✗	✗	✓		✓	✓
PTN [39]		✗	✗				✓		
MVC [30]		✗				✓	✓		
CMR [15]			✗	✗	✗	✓		✓	✓
CSM [19]					✗			✓	
Wu et al. [36]							✓	✓	✓
U-CMR					✗	✓		✓	✓

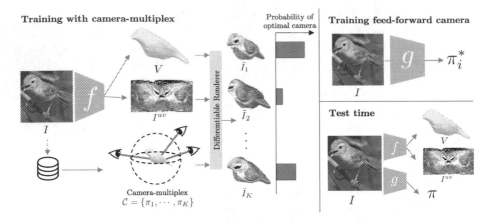

Fig. 2. Overview of the proposed framework. We first train a shape and texture predictor f, while simultaneously optimizing the camera-multiplex $\mathcal{C}$, a set of K possible camera hypotheses maintained for every image. We render the predicted shape and texture from every camera in the multiplex, compute a per-camera reconstruction loss, treat it as negative log-likelihood of the camera, and update the f against the expected loss. We also update each camera in the multiplex against the loss incurred by it. After training f, we train a feed-forward model g to predict the best camera π^* in the multiplex from an image. As such, at test time our approach is able to predict all shape, texture, and camera from a single image.

3   Approach

3.1   Preliminaries

Shape Representation. We represent 3D shape as a mesh $M \equiv (V, F)$ with vertices $V \in \mathbb{R}^{|V| \times 3}$ and faces F. The set of faces F defines the connectivity of vertices in the mesh and we assume it remains fixed. We choose a mesh topology that is homeomorphic to a sphere. We model the vertex positions of a deformable object as $V = \Delta_V + \bar{V}$, the summation of an instance-specific deformation Δ_V that is predicted from an image to a learned instance-independent mean shape $\bar{V}$ [15]. We initialize the mean shape with the template 3D mesh. This parameterization allows the model to learn the space of possible deformations for each category.

Texture Representation. As the topology of our mesh is fixed, we can use a UV image I^{uv} representation to model the texture. The values in a UV image get mapped onto the surface via a fixed UV mapping. The UV mapping is either a spherical projection akin to unrolling a globe into a flat map [11], or when that is not good enough, is a distortion-minimizing unwrap of a template mesh along manually defined seams computed using blender [2].

Camera Projection. We assume a weak-perspective camera projection, parametrized by scale $\mathbf{s} \in \mathbb{R}$, translation $\mathbf{t} \in \mathbb{R}^2$ and rotation $\mathbf{R}$ (captured as Euler angles azimuth, elevation, cyclo-rotation $[az, el, cr] \in \mathbb{R}^3$). We use $\pi(P)$ to denote the projection of a set of 3D points P onto the image coordinates via the weak-perspective projection defined by $\pi \equiv (s, \mathbf{t}, \mathbf{R})$. We denote the image rendered by composing all three factors as $\tilde{I} = \mathcal{R}(V, I^{uv}, \pi)$ and silhouette rendered just from the shape and camera as $\tilde{S} = \mathcal{R}(V, \pi)$ where $\mathcal{R}(\cdot)$ is a differentiable renderer. We denote a set of camera hypotheses kept for each image, a camera-multiplex $\mathcal{C} = \{\pi_1, \cdots, \pi_K\}$. We describe its training details below.

3.2   Our Method

Figure 2 shows an overview of our approach. During training, we learn a function $f(I)$ to predict the 3D shape and texture of the object underlying image I. We optimize over the camera-multiplex for each instance in the training dataset instead of making a deterministic prediction. For every shape and texture prediction, we compute the loss from every camera in the camera-multiplex, which induces a distribution on the camera poses inside a multiplex. We then use the expected loss over the camera-multiplex to update $f(I)$. When the training of $f(\cdot)$ converges, we identify the optimal camera for each training example in the camera-multiplex. We then train a function $g(I)$ that predicts the optimal camera from a single image, such that at test time we can infer all shape, texture, and camera from a single image. We provide the details for the training process below.

For each training instance I, let $\mathcal{C} = \{\pi_1, \cdots, \pi_K\}$ denote its camera-multiplex with K cameras and S its silhouette. Note that while we omit the subscript on training instances for brevity, *every* instance maintains its own $\mathcal{C}$ independently. For every predicted shape $V = \bar{V} + \Delta V$ and texture I^{uv}, we compute the silhouette and image reconstruction loss against each camera π_k:

$$L_{\text{mask},k} = ||S - \tilde{S}_k||_2^2 + \text{dt}(S) * \tilde{S}_k, \tag{1}$$

$$L_{\text{pixel},k} = \text{dist}(\tilde{I}_k \odot S, I \odot S,) \tag{2}$$

$L_{\text{mask},k}$ is the silhouette loss where $\tilde{S}_k = \mathcal{R}(V, \pi_k)$ is the silhouette rendered from camera π_k, and $\text{dt}(S)$ is the uni-directional distance transform of the ground truth silhouette. $L_{\text{pixel},k}$ is the image reconstruction loss computed over the foreground regions where $\mathcal{R}(V, I^{uv}, \pi_k)$ is the rendered image from camera π_k. For this, we use the perceptual distance metric of Zhang *et al.* [40]. To exploit the bilateral symmetry and to ensure symmetric texture prediction, we also render the mesh under a bilaterally symmetric second camera, and compute $L_{\text{pixel},k}$ as the average pixel loss from the two cameras.

In addition to these losses, we employ a graph-laplacian smoothness prior on our shape $L_{\text{lap}} = ||V_i - \frac{1}{|N(i)|} \sum_{j \in N(i)} V_j||^2$ that penalizes vertices i that are far away from the centroid of their adjacent vertices $N(i)$. For cars, motorcycles and shoes, we empirically observe better results using $L_{\text{lap}} = ||LV||_2$ where L

is the discrete Laplace-Beltrami operator that minimizes mean curvature [25]. For this, we construct L once using the template mesh at the start of training. Following [1,3,17], we also find it beneficial to regularize the deformations as it discourages arbitrarily large deformations and helps learn a meaningful mean shape. The corresponding energy term is expressed as $L_{\text{def}} = ||\Delta_V||_2$.

Model Update. For iteratively refining the camera-multiplex, we use the summation of the silhouette and image reconstruction loss $L_{\pi_k} = L_{\text{mask},k} + L_{\text{pixel},k}$ as the loss for each camera π_k in the camera-multiplex. We optimize each camera to minimize L_k every time the training instance is encountered during the training. For updating the shape and the texture, we use the resulting losses over the cameras as a distribution over the most likely camera pose in the camera-multiplex, and minimize the expected loss over all the cameras. Specifically, we compute the probability of π_k being the optimal camera through a softmin function $p_k = \frac{e^{-L_k/\sigma}}{\sum_j e^{-L_j/\sigma}}$ and train the shape and texture prediction modules with the final loss:

$$L_{\text{total}} = \sum_k p_k (L_{\text{mask},k} + L_{\text{pixel},k}) + L_{\text{def}} + L_{\text{lap}}. \tag{3}$$

In practice, the temperature σ changes dynamically while computing p_k by linearly normalizing L_k to have a fixed range to standardize the peakiness of the probability distribution. We do not backpropagate through p_k. In summary, we iteratively refine the cameras in the multiplex against loss L_{π_k} and update the parameters of f through L_{total} for every training sample.

We implement the camera multiplex for each image $\mathcal{C}_i$ as a variable stored in a dictionary. Every time an image is encountered during training, the corresponding camera multiplex is fetched from this dictionary of variables and used as if it were an input to the rest of the training pipeline. Most modern deep learning frameworks such as PyTorch [24] support having such a dictionary of variables.

Training a Feed-Forward Camera Predictor. When the training of f converges, for each training image we select the optimal camera to be the camera that minimizes the silhouette and image reconstruction losses. We then train a new camera prediction module $g(I)$ in a supervised manner such that at inference time our model can predict all 3D shape, texture, and camera at the same time.

Approach at Test Time. Given a novel image I at test time, we can use the learnt modules f and g to predict the 3D shape, texture and camera-viewpoint of the object underlying image I. $f(I)$ predicts shape $V = \bar{V} + \Delta_V$ and texture I^{uv} while $g(I)$ predicts the camera-viewpoint π. This is illustrated in Fig. 2.

4   Experiments

In this section we provide quantitative and qualitative evaluation of our approach that learns to predict 3D shape, texture, and camera from a single image without using any keypoint annotations during training. We explore our approach on four object categories: birds, cars, motorcycles and shoes.

4.1   Experimental Detail

Datasets. We primarily use the CUB-200-2011 dataset [35], which has 6000 training and test images of 200 species of birds. In addition to this, we train and evaluate U-CMR on multiple categories: car, motorcycles from the Pascal3D+ dataset and shoes scraped from zappos.com. For CUB and Pascal3D, we use the same train-test splits as CMR [15]. For the initial meshes for birds and cars, we use the 3D template meshes used by Kulkarni *et al.* [19]. For others, we download freely available online models, homogenize to a sphere and simplify to reduce the number of vertices. We symmetrize all meshes to exploit bilateral symmetry. We compute masks for the zappos shoes dataset, which contains white background images, via simple threshold-based background subtraction and hole-filling.

Architecture. For all but texture, we use the same architecture as that of CMR [15] and pass Resnet18 features into two modules - one each for predicting shape and texture. The shape prediction module is a set of 2 fully connected layers with $\mathbb{R}^{3|V|}$ outputs that are reshaped into Δ_V following [15]. For the texture prediction, prior work predicted flow, where the final output is an offset that indicates where to sample pixels from. In this work we directly predict the pixel values of the UV image through a decoder. The texture head is a set of upconvolutional layers that takes the output of Resnet18 preserving the spatial dimensions. We find that this results in a more stable camera, as the decoder network is able to learn a spatial prior over the UV image. We use SoftRas [20] as our renderer. Please see the supplementary material for details and ablation studies.

Camera-Multiplex Implementation. We use $K = 40$ for camera-multiplex. We initialize the camera multiplex $\mathcal{C}$ for every image in the training set, to a set of K points whose azimuth and elevation are uniformly spaced on the viewing sphere. For cars and motorcycles, we use $K = 8$ cameras - all initialized to zero elevation. We optimize each camera in the multiplex using the silhouette loss $L_{mask,k}$ before training shape and texture. To reduce compute time while training shape and texture, we reduce K from 40 to 4 after 20 epochs by pruning the camera-multiplex and keeping the top 4 cameras. Note that naive data augmentation that scales and/or translates the image without adjusting the camera-multiplex will result in the rendered shape being pixel-unaligned. We handle random crop and scale data augmentation during training by adjusting the scale and translation in the stored camera multiplex with a deterministic affine transformation before using it for rendering the shape.

Baselines. As no other approach predicts 3D shape, texture and pose without relying on keypoints or known camera or multi-view cues during training, as baseline we compare with ablations of our approach that do not use the camera-multiplex. This can be thought of CMR without keypoints, which simultaneously predicts shape, camera and texture and only supervises rendered silhouette and texture. We call this approach CMR-nokp and ensure that the experimental setup is comparable to U-CMR. Additionally, in the supplementary, we compare to two variants of CMR [15] that have more supervision than our setting. For camera prediction, we compare with CMR [15] (which uses additional keypoint supervision), CSM [19] and U-CMR without texture prediction (U-CMR-noTex).

4.2 Qualitative Evaluation

The problem when no keypoints and viewpoints are available is that there always exists a planar shape and texture that explains the image and silhouette for any arbitrary camera pose. We first demonstrate this point using CMR-nokp. We observe that, as expected, CMR-nokp results in a degenerate solution shown in Fig. 3 where the recovered shape explains the image silhouette well, but when seen from a different viewpoint the shape is planar.

In Fig. 6, we visualize U-CMR predictions on unseen images from the CUB test set. Our approach, despite not using any viewpoint or keypoint supervision is able to recover a full, plausible 3D shape of the birds and learns to predict their texture from a single image. Our approach captures various types of bird shapes, including shapes of water birds and songbirds. We are able to recover sharp long tails and some protrusion of legs and beaks. Please see supplementary for more results of random samples from test set and comparisons to CMR.

Fig. 3. CMR without keypoints. CMR-nokp, which directly predicts shape, texture, and camera without keypoint supervision or the proposed camera-multiplex, obtains degenerate solutions. The shape from predicted camera viewpoint (centre) explains the silhouette well but an alternate view (right) reveals that the model has learned to output a planar, flat bird shape.

We further analyze the shape space that we learn in Fig. 4, where we run principal component analysis on all the shapes obtained on the train set. We find directions that capture changes in the body type, the head shapes, and the tail shapes. In Fig. 4, we also show that the final mean shape deviates significantly from the template mesh it was initialized to, by becoming thinner and developing a more prominent tail. Please see the supplemental for more results.

4.3 Quantitative Evaluation

We conduct quantitative evaluation on the camera poses obtained from our approach, since there are no 3D ground truth shapes on this dataset. For camera evaluation, we compare our approach to CSM [19], which learns to out-

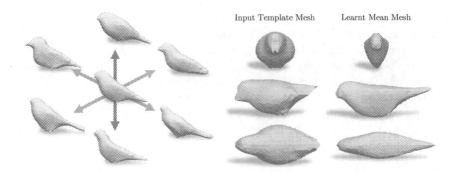

Input Template Mesh   Learnt Mean Mesh

Fig. 4. Learned Shape. On the left, we visualize the space of learned shapes by running PCA. We see that the model has learned to output changes in body type, head shape, and tail types. On the right, we compare the template shape to the final learnt mean mesh. See text for discussion.

Table 2. Quantitative evaluation of camera pose predictions on the test dataset. We plot rotation error as the geodesic distance from the ground-truth, the entropy of the azimuth-elevation distribution and the wasserstein distance of marginal Az/El w.r.t.ground-truth. U-CMR outperforms all methods in the absence of keypoints.

	Rotation error ↓	Entropy (nats) ↑	Wasserstein Dist ↓	
			Azimuth	Elevation
GT	–	7.44	–	–
CMR [15]	22.94°	7.25	6.03°	4.33°
CMR-nokp	87.52°	5.73	64.66°	12.39°
CSM [19]	61.93°	5.83	27.34°	11.28°
U-CMR (noTex)	61.82°	**7.36**	16.08°	7.90°
U-CMR	**45.52°**	7.26	**8.66°**	**6.50°**

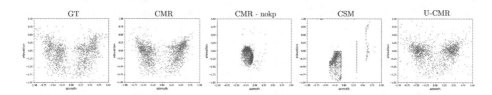

Fig. 5. Camera Pose Distributions on CUB Test. We show azimuth-elevation scatter plots over the entire CUB test set for different approaches. From left to right, we show (i) pseudo ground-truth cameras computed via running SfM on keypoints, (ii) the cameras predicted by CMR which uses the SfM cameras as supervision, (iii) CMR without viewpoint and keypoint supervision (CMR-nokp), (iv) CSM [19] and (v) our approach U-CMR. The last three approaches are weakly-supervised and do not use any keypoint annotation. Notice how the camera pose collapses in CMR-nokp and CSM, while U-CMR with camera-multiplex is able to obtain distribution similar to the ground truth cameras.

put dense correspondences of the image against a template 3D shape as well as the camera poses from image collections without keypoints. Note that they do not learn to predict 3D shapes. We used the same 3D template mesh as CSM and therefore are comparable to CSM. We evaluate cameras from different approaches on metrics measuring their accuracy and collapse. We compare our predicted cameras to the pseudo ground-truth cameras computed in CMR [15] using SFM on keypoints. For evaluating accuracy, we compute the rotation error $\mathrm{err}_R = \arccos\left(\frac{\mathrm{Tr}(\tilde{R}^T R^*)-1}{2}\right)$ as the geodesic distance between the predicted camera rotation $\tilde{R}$ and the pseudo ground-truth camera rotation R^*. We report the average rotation error (in degrees) over the entire test dataset. To measure collapse, we analyze the azimuth-rotation distribution and report (i) its entropy (in nats) and (ii) it's Wasserstein distance to the pseudo ground truth azimuth-elevation distribution. Because of computational ease, we only report the Wasserstein distance on the azimuth and elevation marginals. We primarily focus on azimuth and rotation because changes in scale, translation and cyclo-rotation of a camera only warp the image in 2D and don't constitute a "novel viewpoint".

Table 2 reports the numbers on all metrics for supervised (CMR) and weakly-supervised (CMR-nokp, CSM, U-CMR) methods. Observe that all the weakly-supervised baselines incur significant camera pose collapse - as can be seen by the entropy of their distributions. In contrast, U-CMR, despite being weakly-supervised, achieves an entropy that is slightly better than CMR - the supervised baseline. We automatically learn a camera distribution that is almost as close to the ground-truth distribution (in Wasserstein distance) as the supervised baseline (CMR). U-CMR is more accurate than CMR-nokp and CSM and achieves an average rotation error at least 15° better than them. This table also suggests that the texture loss helps with refining camera poses to make them more accurate as U-CMR (noTex) is well-distributed with a very high entropy but is not as accurate as U-CMR.

Figure 5 visualizes the azimuth-elevation distributions of different approaches. This figure illustrates that while CSM prevents an extreme mode collapse of the camera, their camera pose distribution still collapses into certain modes. For making this figure, we employ CSM's public model, trained on the same CUB dataset with a fixed template shape. This empirical evidence exemplifies the fact that U-CMR's weakly-supervised camera-multiplex optimization approach learns a camera pose distribution that's much better than other weakly-supervised baselines and almost as good as supervised methods.

4.4   Evaluations on Other Categories

While our primary evaluation is on the CUB Birds dataset, we also run our approach on cars and motorcycles from the Pascal3D+ dataset, and the shoe images we scraped from zappos.com. We show qualitative visualizations of predicted shape, texture and camera for all categories. For Pascal3D cars, we also compare IoU of predicted shapes to CMR, previous deformable model fitting-

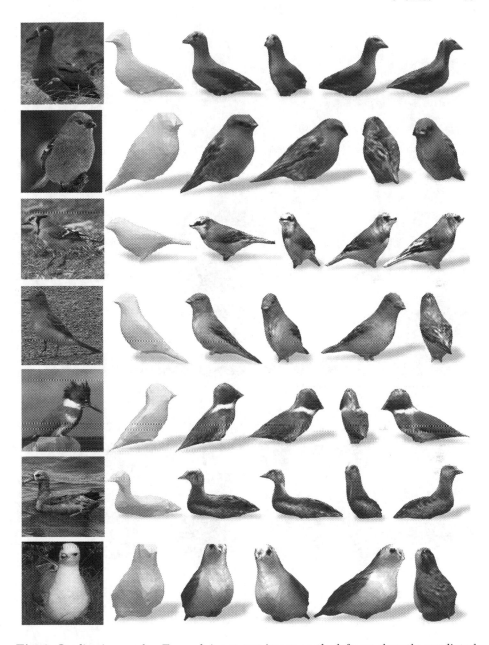

Fig. 6. Qualitative results. For each input test image on the left, we show the predicted mesh, the textured mesh, and the textured mesh from multiple views.

based [17] and volumetric prediction [32] methods. All three of these approaches leverage segmentation masks and cameras and keypoints to learn 3D shape inference.

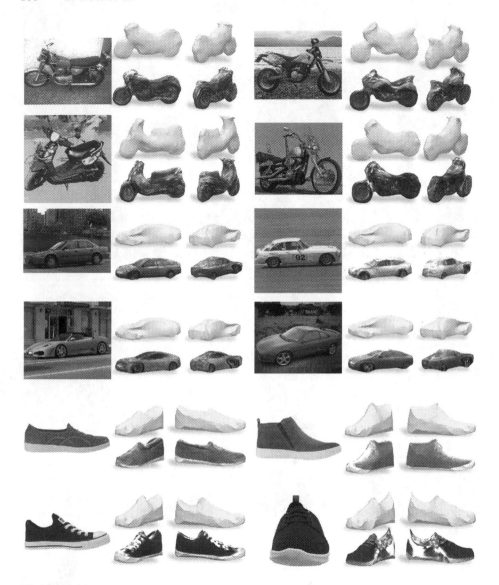

Fig. 7. Qualitative results on cars, motorcycles and shoes. For each image, we show the predicted 3D shape and texture from two viewpoints.

Figure 7 shows qualitative results on selected images from their respective test set. U-CMR learns accurate camera poses and diverse yet plausible shapes for cars and motorcycles. For shoes, U-CMR shapes are reasonable but not as diverse because of biases in the underlying dataset. We observe some artifacts where the sides of the cars have concave indentations and some parts of the shoes are tapered and pointy. These issues stem from using weak-perspective projections and limitations of the regularization, which is not spatially adaptive.

Please see the supplemental for each category's PCA visualizations and the initial template.

We report the mean IoU on the test set in Table 3 and observe that U-CMR performs comparably to alternate methods that require more supervision.

Table 3. Reconstruction evaluation using PASCAL 3D+. We report the mean intersection over union (IoU) on PASCAL 3D+ to benchmark the obtained 3D reconstructions (higher is better). We compare to CMR [15], a deformable model fitting approach (CSDM [17]) and a volumetric prediction approach (DRC [32]). CSDM and DRC use image collection supervision in addition to keypoints/cameras.

Category	CSDM [17]	DRC [32]	CMR [15]	U-CMR
Car	0.60	0.67	0.640	0.646

Fig. 8. Failure Modes. The columns, from left to right, show the input image, the predicted shape and texture from the predicted camera, and finally a different view of the textured mesh. See the text for discussion.

4.5   Limitations

While U-CMR shows promising results in the direction of weakly supervised 3D shape understanding, it has some limitations. Foremost limitation is that we do not model articulation and expect to fail in cases with significant articulation. In Fig. 8, we demonstrate various modes of failure for shape, texture and camera-viewpoint prediction. Our approach struggles when the bird shape is significantly different from the template mesh and undergoes large articulation, such as the case with flying birds. It is challenging to identify correct camera poses when there's a large deformation like this without keypoints. The data imbalance between flying and not flying birds also exacerbates this problem. The two examples in the top row of the figure show how our shape prediction fails when the bird in the image is flying, or has it's wings open. The example in the bottom right shows an articulated bird with it's head twisted back. Due to the lack of an articulation model, these failure cases are expected. We also fail at predicting good texture sometimes - especially for parts of the object that are not visible. The example on the bottom left of Fig. 8 and bottom right of Fig. 7 shows how background colours may leak into the predicted texture.

5   Conclusion

In this work, we present a learning framework that can decompose an image of a deformable object into its 3D shape, texture, and camera viewpoint. In order to solve this highly under-constrained problem, we propose a representation for maintaining a distribution over possible camera viewpoints called camera-multiplex. This allows the model to maintain a possible set of camera hypothesis, avoiding the learning process from getting stuck in a bad local minima. We show our approach on four categories, where we show that U-CMR can recover reasonable 3D shape and texture without viewpoints and keypoints.

Acknowledgements. We thank Jasmine Collins for scraping the zappos shoes dataset and members of the BAIR community for helpful discussions. This work was supported in-part by eBay, Stanford MURI and the DARPA MCS program.

References

1. Blanz, V., Vetter, T.: A morphable model for the synthesis of 3D faces. In: SIG-GRAPH (1999)
2. Blender Online Community: Blender - a 3D modelling and rendering package. Blender Institute, Amsterdam (2019). http://www.blender.org
3. Cashman, T.J., Fitzgibbon, A.W.: What shape are dolphins? Building 3D morphable models from 2D images. TPAMI **35**, 232–244 (2013)
4. Chen, W., et al.: Synthesizing training images for boosting human 3D pose estimation. In: 3DV (2016)
5. Choy, C.B., Xu, D., Gwak, J.Y., Chen, K., Savarese, S.: 3D-R2N2: a unified approach for single and multi-view 3D object reconstruction. In: Leibe, B., Matas, J., Sebe, N., Welling, M. (eds.) ECCV 2016. LNCS, vol. 9912, pp. 628–644. Springer, Cham (2016). https://doi.org/10.1007/978-3-319-46484-8_38
6. Dai, Y., Li, H., He, M.: A simple prior-free method for non-rigid structure-from-motion factorization. IJCV **107**, 101–122 (2014). https://doi.org/10.1007/s11263-013-0684-2
7. Fan, H., Su, H., Guibas, L.J.: A point set generation network for 3D object reconstruction from a single image. In: CVPR (2017)
8. Gadelha, M., Maji, S., Wang, R.: 3D shape induction from 2D views of multiple objects. In: 3DV (2017)
9. Girdhar, R., Fouhey, D.F., Rodriguez, M., Gupta, A.: Learning a predictable and generative vector representation for objects. In: Leibe, B., Matas, J., Sebe, N., Welling, M. (eds.) ECCV 2016. LNCS, vol. 9910, pp. 484–499. Springer, Cham (2016). https://doi.org/10.1007/978-3-319-46466-4_29
10. Gkioxari, G., Malik, J., Johnson, J.: Mesh R-CNN. In: ICCV (2019)
11. Hughes, J.F., Foley, J.D.: Computer Graphics: Principles and Practice. Pearson Education, London (2014)
12. Insafutdinov, E., Dosovitskiy, A.: Unsupervised learning of shape and pose with differentiable point clouds. In: NeurIPS (2018)
13. Jakab, T., Gupta, A., Bilen, H., Vedaldi, A.: Unsupervised learning of object landmarks through conditional image generation. In: NeurIPS (2018)

14. Kanazawa, A., Kovalsky, S., Basri, R., Jacobs, D.: Learning 3D deformation of animals from 2D images. In: Computer Graphics Forum. Wiley Online Library (2016)
15. Kanazawa, A., Tulsiani, S., Efros, A.A., Malik, J.: Learning category-specific mesh reconstruction from image collections. In: Ferrari, V., Hebert, M., Sminchisescu, C., Weiss, Y. (eds.) ECCV 2018. LNCS, vol. 11219, pp. 386–402. Springer, Cham (2018). https://doi.org/10.1007/978-3-030-01267-0_23
16. Kar, A., Häne, C., Malik, J.: Learning a multi-view stereo machine. In: NeurIPS (2017)
17. Kar, A., Tulsiani, S., Carreira, J., Malik, J.: Category-specific object reconstruction from a single image. In: CVPR (2015)
18. Kato, H., Ushiku, Y., Harada, T.: Neural 3D mesh renderer. In: CVPR (2018)
19. Kulkarni, N., Gupta, A., Tulsiani, S.: Canonical surface mapping via geometric cycle consistency. In: ICCV (2019)
20. Liu, S., Li, T., Chen, W., Li, H.: Soft rasterizer: a differentiable renderer for image-based 3D reasoning. In: ICCV (2019)
21. Loper, M.M., Black, M.J.: OpenDR: an approximate differentiable renderer. In: Fleet, D., Pajdla, T., Schiele, B., Tuytelaars, T. (eds.) ECCV 2014. LNCS, vol. 8695, pp. 154–169. Springer, Cham (2014). https://doi.org/10.1007/978-3-319-10584-0_11
22. Novotny, D., Ravi, N., Graham, B., Neverova, N., Vedaldi, A.: C3DPO: canonical 3D pose networks for non-rigid structure from motion. In: ICCV (2019)
23. Park, J.J., Florence, P., Straub, J., Newcombe, R., Lovegrove, S.: DeepSDF: learning continuous signed distance functions for shape representation. In: CVPR (2019)
24. Paszke, A., et al.: Pytorch: an imperative style, high-performance deep learning library. In: NeurIPS (2019)
25. Pinkall, U., Polthier, K.: Computing discrete minimal surfaces and their conjugates. Exp. Math. 2, 15–36 (1993)
26. Shu, Z., Sahasrabudhe, M., Alp Güler, R., Samaras, D., Paragios, N., Kokkinos, I.: Deforming autoencoders: unsupervised disentangling of shape and appearance. In: Ferrari, V., Hebert, M., Sminchisescu, C., Weiss, Y. (eds.) ECCV 2018. LNCS, vol. 11214, pp. 664–680. Springer, Cham (2018). https://doi.org/10.1007/978-3-030-01249-6_40
27. Song, S., Yu, F., Zeng, A., Chang, A.X., Savva, M., Funkhouser, T.: Semantic scene completion from a single depth image. In: CVPR (2017)
28. Thewlis, J., Bilen, H., Vedaldi, A.: Unsupervised learning of object frames by dense equivariant image labelling. In: NeurIPS (2017)
29. Torresani, L., Hertzmann, A., Bregler, C.: Nonrigid structure-from-motion: estimating shape and motion with hierarchical priors. TPAMI 30, 878–892 (2008)
30. Tulsiani, S., Efros, A.A., Malik, J.: Multi-view consistency as supervisory signal for learning shape and pose prediction. In: CVPR (2018)
31. Tulsiani, S., Gupta, S., Fouhey, D., Efros, A.A., Malik, J.: Factoring shape, pose, and layout from the 2D image of a 3D scene. In: CVPR (2018)
32. Tulsiani, S., Zhou, T., Efros, A.A., Malik, J.: Multi-view supervision for single-view reconstruction via differentiable ray consistency. In: CVPR (2017)
33. Varol, G., et al.: Learning from synthetic humans. In: CVPR (2017)
34. Vicente, S., Carreira, J., Agapito, L., Batista, J.: Reconstructing PASCAL VOC. In: CVPR (2014)
35. Wah, C., Branson, S., Welinder, P., Perona, P., Belongie, S.: The Caltech-UCSD Birds-200-2011 Dataset. Technical report CNS-TR-2011-001, California Institute of Technology (2011)

36. Wu, S., Rupprecht, C., Vedaldi, A.: Unsupervised learning of probably symmetric deformable 3D objects from images in the wild. arXiv preprint arXiv:1911.11130 (2019)
37. Wu, Z., et al.: 3D ShapeNets: a deep representation for volumetric shapes. In: CVPR (2015)
38. Xiang, Y., Mottaghi, R., Savarese, S.: Beyond pascal: a benchmark for 3D object detection in the wild. In: WACV (2014)
39. Yan, X., Yang, J., Yumer, E., Guo, Y., Lee, H.: Perspective transformer nets: learning single-view 3D object reconstruction without 3D supervision. In: NeurIPS (2016)
40. Zhang, R., Isola, P., Efros, A.A., Shechtman, E., Wang, O.: The unreasonable effectiveness of deep networks as a perceptual metric. In: CVPR (2018)
41. Zuffi, S., Kanazawa, A., Berger-Wolf, T., Black, M.J.: Three-D safari: learning to estimate zebra pose, shape, and texture from images "in the wild". In: ICCV (2019)

Learning Attentive and Hierarchical Representations for 3D Shape Recognition

Jiaxin Chen[1], Jie Qin[1(✉)], Yuming Shen[3], Li Liu[1], Fan Zhu[1], and Ling Shao[1,2]

[1] Inception Institute of Artificial Intelligence, Abu Dhabi, UAE
qinjiebuaa@gmail.com
[2] Mohamed bin Zayed University of Artificial Intelligence, Abu Dhabi, UAE
[3] eBay, Shanghai, China

Abstract. This paper proposes a novel method for 3D shape representation learning, namely Hyperbolic Embedded Attentive Representation (HEAR). Different from existing multi-view based methods, HEAR develops a unified framework to address both multi-view redundancy and single-view incompleteness. Specifically, HEAR firstly employs a hybrid attention (HA) module, which consists of a view-agnostic attention (VAA) block and a view-specific attention (VSA) block. These two blocks jointly explore distinct but complementary spatial saliency of local features for each single-view image. Subsequently, a multi-granular view pooling (MVP) module is introduced to aggregate the multi-view features with different granularities in a coarse-to-fine manner. The resulting feature set implicitly has hierarchical relations, which are therefore projected into a Hyperbolic space by adopting the Hyperbolic embedding. A hierarchical representation is learned by Hyperbolic multi-class logistic regression based on the Hyperbolic geometry. Experimental results clearly show that HEAR outperforms the state-of-the-art approaches on three 3D shape recognition tasks including generic 3D shape retrieval, 3D shape classification and sketch-based 3D shape retrieval.

Keywords: 3D shape recognition · View-agnostic/specific attentions · Multi-granularity view aggregation · Hyperbolic neural networks

1 Introduction

Recently, 3D shape analysis [8,18,24,33,45–47,70–72] has emerged as a hot research topic in computer vision, due to the increasing demand from real applications in virtual reality, autonomous driving, 3D printing and gaming. Learning 3D shape representations for downstream tasks, *e.g.*, 3D shape classification/retrieval, is a fundamental problem for 3D shape analysis. However, this problem is very challenging, considering the varying modalities, complicated geometries and variability of 3D shapes.

© Springer Nature Switzerland AG 2020
A. Vedaldi et al. (Eds.): ECCV 2020, LNCS 12360, pp. 105–122, 2020.
https://doi.org/10.1007/978-3-030-58555-6_7

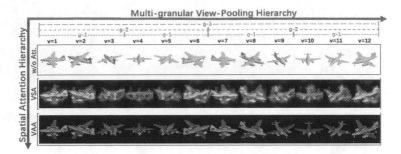

Fig. 1. Illustration of the spatial attention hierarchy as well as the multi-granular view-pooling hierarchy.

A variety of methods have been proposed to learn 3D shape representations, which can generally be divided into the following two categories: 1) 3D model-based methods, learning representations directly from the raw format of 3D shapes, such as point cloud [32,45,47], voxel [3,42,46] and mesh [14]; 2) multi-view based approaches [1,2,6,13,21,24,25,51,56,71,72], which first represent a 3D object by a set of rendered 2D images to extract individual features, and then aggregate the features to a global descriptor. Benefiting from the success of CNN in 2D image representation learning, the multi-view based approaches have surpassed their model-based counterparts in most cases. However, it remains difficult to effectively aggregate multi-view data because of their following characteristics: **(a)** *Single-view incompleteness.* As shown in Fig. 1, a 2D image rendered from a single view only captures partial appearance and geometry structures of a 3D object, due to the self-occlusion and information loss by 2D projection during the rendering procedure; **(b)** *Multi-view redundancy.* Multiple images rendered from a sequence of over-completely sampled views contain a large amount of redundant information, since images from neighboring views often capture similar geometric structures of the 3D object and many 3D shapes are geometrically symmetric. This kind of redundancy suppresses the effects of discriminative local regions, which will deteriorate the final performance.

Most of the existing works focus on addressing problem (a) by developing various view-wise pooling strategies [55,65], exploring view importance [73] or modeling multi-view data by sequence [10,22,24], while they improperly neglect problem (b). In our work, we take into account both problems (a) and (b) and propose a unified framework, namely **Hyperbolic Embedded Attentive Representation (HEAR)**, as illustrated in Fig. 2.

On the one hand, HEAR develops a hybrid attention (HA) module to extensively explore the spatial attentions of local features for each single-view image. Specifically, HA consists of two blocks, *i.e.*, the View Agnostic Attention (VAA) block and the View Specific Attention (VSA) block. Basically, VAA attempts to learn high-level spatial attentions by adopting a trainable spatial attention network shared across different views. In contrast, the parameter-free VSA aims to explore low-level view-specific spatial attentions by calculating the maximal accumulated top-M correlations with local features from other views. As shown in Fig. 1, VAA and VSA capture complementary spatial attentions that correspond to discriminative local parts of a 3D shape. Accordingly, HEAR imposes large weights on salient local features, whilst suppressing less salient ones. In this way, HEAR alleviates the negative effect caused by the multi-view redundancy.

On the other hand, HEAR employs a multi-granular view-pooling (MVP) module to aggregate multi-view features. Concretely, as shown in Fig. 1, MVP evenly partitions the 12 views into 1, 2, 4 non-overlapped segments, in each of which the views are ensembled by average/max pooling. In this manner, MVP can preserve more visual details by using this coarse-to-fine view aggregation strategy, and thus can mitigate the single-view incompleteness, as mentioned in problem (a). Based on HA and MVP, a 3D shape can be represented by a set of features, which encode distinct spatial attentions and view-pooling granularities. As observed in Fig. 1, these features implicitly have hierarchical relations w.r.t. the spatial attention and multi-granular view-pooling. We therefore employ a Hyperbolic embedding, to endow the feature space with a Hyperbolic geometry, which has recently been successfully applied to represent hierarchical structured data [4,19,20,30,49]. Accordingly, the Hyperbolic multi-class logistic regression (MLR) is applied to accomplish classification/retrieval in the Hyperbolic space.

Our main contributions are summarized as follows:

– We simultaneously address the problems of single-view incompleteness and multi-view redundancy for 3D shape representation learning by a unified framework, namely Hyperbolic Embedded Attentive Representation (HEAR).
– We propose a hybrid attention module to explore view-agnostic and view-specific attentions, which capture distinct but complementary spatial saliency.
– We present a multi-granular view-pooling mechanism to aggregate multi-view features in a hierarchical manner, which are subsequently encoded into a hierarchical representation space by employing the Hyperbolic embedding.

2   Related Work

Model-Based Methods. Several recent works learn representations from raw 3D shape data, which can be divided into the following categories. 1) Voxel-based models such as 3DShapeNet [67], VoxelNet [42], Subvolume Net [46] and VRN [3]. They directly apply the 3D convolution neural networks to learn the representation based on voxelized shapes. However, these approaches are usually computationally costly, and severely affected by the low resolution caused by the data sparsity. 2) Point cloud-based methods. Point cloud is a set of unordered 3D points, which has attracted increasing interests due to its wide applications. Qi *et al.* propose the seminal work, *i.e.*, PointNet [45] by building deep neural networks on the point sets. Afterwards, a large amount of approaches, such as PointNet++ [47], Kd-Networks [32], SO-Net [39], KPConv [62], IntepCNN [41], DPAM [40]), have been proposed to improve PointNet [45] by modeling fine-grained local patterns. 3) Mesh-based methods. A majority of CAD models are stored as meshes, which consist of 3D vertices, edges and faces. [14] presents

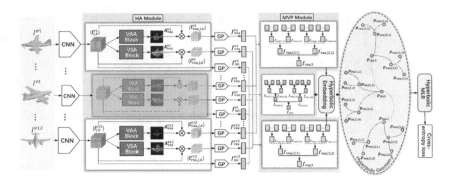

Fig. 2. Framework of Hyperbolic Embedded Attentive Representation (HEAR). HEAR firstly adopts base CNN networks to extract local feature maps, based on which a hybrid attention (HA) module is employed to explore distinct attentions, *i.e.*, the View-Agnostic Attention (VAA) and View-Specific Attention (VSA). The local features re-weighted by each attention are transformed to global features through global pooling (GP). A Multi-granular View Pooling (MVP) is subsequently adopted to aggregate the multi-view global features in a hierarchical manner. The resulting feature set is further endowed with a Hyperbolic geometry through Hyperbolic embedding, and used for classification or retrieval by the Hyperbolic multi-class logistic regression (MLR).

the MeshNet to solve the complexity and irregularity problems of meshes, and achieves comparable performance with methods using other types of data.

Multi-view Based Methods. The multi-view based method represents a 3D object by a set of 2D images, rendered from a sequence of views. This kind of approaches leverage the well-studied 2D convolutional neural networks, and thus performs better than the model-based ones. In [55], Su *et al.* develops a multi-view convolutional neural network (MVCNN) which extracts features from each single-view image followed by a view-pooling strategy to aggregate the multi-view data into the 3D descriptor. Based on MVCNN, various methods have been proposed by developing different view-wise pooling strategies [65], exploring the view importance [73] and modeling the multi-view data by sequence [10,22,24] or graphs [12]. Among these methods, VDN [35] is closely related to our work, which also explores locality attentions. However, VDN mainly focuses on single-view images, and fails to capture cross-view attention patterns. In contrast, our method explores the inter-view correlations by the VSA block. Relation-Net [73] also learns the cross-view relations of local features. Nevertheless, our work employs hybrid attentions, as well as considering the multi-granular view-pooling.

3   Proposed Method

3.1   Framework

As shown in Fig. 2, the proposed method mainly consists of three modules: the hybrid attention (HA) module, the multi-granular view pooling (MVP) module and the Hyperbolic neural networks with Hyperbolic embedding (HE).

Specifically, suppose that $\boldsymbol{\mathcal{T}} = \{\mathcal{O}^i; y^i\}_{i=1}^N$ is a training set of 3D shapes, where $\mathcal{O}^i$ is the i-th 3D shape and $y^i \in \{1, \cdots, C\}$ refers to the class label. We firstly represent the 3D shape $\mathcal{O}^i$ as a group of gray-scale images by rendering with the Phong reflection model [43] from N_v views, which are evenly placed in a plane around the 3D shape. The resulting multi-view representation is denoted by $\boldsymbol{I}^i = \{I^{v,i}\}_{v=1}^{N_v}$, where $I^{v,i}$ is a 2D rendered image of $\mathcal{O}^i$ from the v-th view. In this paper, we use $N_v = 12$ views.

Subsequently, we adopt a base convolutional network (e.g., VGG-A [53], VGG-19 [53] and ResNet-50 [23]) $\mathcal{F}_{\boldsymbol{\theta}}(\cdot)$, parameterized by $\boldsymbol{\theta}$, to extract an initial feature map $\boldsymbol{L}_{\mathrm{ini}}^{v,i} = [\boldsymbol{l}_{j,k}^{v,i}]_{1 \leq j \leq H, 1 \leq k \leq W} \in \mathbb{R}^{H \times W \times d}$ for each image $I^{v,i}$, where H, W and d denote the height, width and number of channels of the feature map, respectively. $\boldsymbol{l}_{j,k}^{v,i}$ refers to the d-dimensional local feature at the (j,k)-th location. Thereafter, a *view-agnostic attention* (VAA) block $\mathcal{VAA}_\phi(\cdot)$ together with a *view-specific attention* (VSA) block $\mathcal{VSA}(\cdot)$ are proposed to learn two different kinds of attention weights for each local feature, which we denote by $\boldsymbol{A}_{\mathrm{vaa}}^{v,i} = [\alpha_{\mathrm{vaa},j,k}^{v,i}] \in \mathbb{R}^{H \times W}$ and $\boldsymbol{A}_{\mathrm{vsa}}^{v,i} = [\alpha_{\mathrm{vsa},j,k}^{v,i}] \in \mathbb{R}^{H \times W}$, respectively. Here, ϕ refers to learnable parameters of $\mathcal{VAA}_\phi(\cdot)$. Accordingly, we can obtain three local feature maps for a single image $I^{v,i}$: the initial feature map $\boldsymbol{L}_{\mathrm{ini}}^{v,i}$ without attentions, the VAA induced feature map $\boldsymbol{L}_{\mathrm{vaa}}^{v,i} = \left[\alpha_{\mathrm{vaa},j,k}^{v,i} \cdot \boldsymbol{l}_{j,k}^{v,i}\right]_{j,k} \in \mathbb{R}^{H \times W \times d}$, as well as the VSA induced feature map $\boldsymbol{L}_{\mathrm{vsa}}^{v,i} = \left[\alpha_{\mathrm{vsa},j,k}^{v,i} \cdot \boldsymbol{l}_{j,k}^{v,i}\right]_{j,k} \in \mathbb{R}^{H \times W \times d}$.
By passing through a global pooling module $GP(\cdot)$, the local feature maps are successively aggregated into three global features $\boldsymbol{f}_{\mathrm{ini}}^{v,i} \in \mathbb{R}^D$, $\boldsymbol{f}_{\mathrm{vaa}}^{v,i} \in \mathbb{R}^D$ and $\boldsymbol{f}_{\mathrm{vsa}}^{v,i} \in \mathbb{R}^D$. For N_v rendering views, we therefore obtain three sets of global features $\boldsymbol{F}_{\mathrm{ini}}^i = [\boldsymbol{f}_{\mathrm{ini}}^{1,i}, \cdots, \boldsymbol{f}_{\mathrm{ini}}^{v,i}, \cdots, \boldsymbol{f}_{\mathrm{ini}}^{N_v,i}]$, $\boldsymbol{F}_{\mathrm{vaa}}^i = [\boldsymbol{f}_{\mathrm{vaa}}^{1,i}, \cdots \boldsymbol{f}_{\mathrm{vaa}}^{v,i}, \cdots, \boldsymbol{f}_{\mathrm{vaa}}^{N_v,i}]$, and $\boldsymbol{F}_{\mathrm{vsa}}^i = [\boldsymbol{f}_{\mathrm{vsa}}^{1,i}, \cdots, \boldsymbol{f}_{\mathrm{vsa}}^{v,i}, \cdots, \boldsymbol{f}_{\mathrm{vsa}}^{N_v,i}] \in \mathbb{R}^{D \times N_v}$. Subsequently, the *multi-granular view pooling* (MVP) $\mathcal{MVP}(\cdot)$ is proposed to aggregate multi-view features $\boldsymbol{F}_{\mathrm{ini}}^i$, $\boldsymbol{F}_{\mathrm{vaa}}^i$ and $\boldsymbol{F}_{\mathrm{vsa}}^i$ in a multi-granular manner. The aggregated features implicitly have hierarchical relations. We therefore employ a Hyperbolic embedding $\mathcal{HE}(\cdot)$ to project them into a Hyperbolic space, and learn hierarchical representations by the Hyperbolic multi-class logistic regression (MLR) with parameters $\boldsymbol{\psi}$. The cross-entropy loss is adopted to train the overall network.

3.2 Hybrid Attentions

In this section, we will elaborate the view-agnostic and view-specific attentions. Without loss of generality, we omit the index i for a more neat description.

View-Agnostic Attention. Basically, the VAA block is a variant of the Squeeze-and-Excitation network [26]. It firstly applies a 1×1 convolutional layer $conv_{1 \times 1}(\cdot)$ to squeeze the local feature map $\boldsymbol{L}_{\mathrm{ini}}^v$ to an $H \times W$ matrix $\boldsymbol{E}_{\mathrm{ini}}^v$, which is subsequently flattened into an $HW \times 1$ vector $\overrightarrow{\boldsymbol{E}_{\mathrm{ini}}^v}$. Specifically, the spatial attention map $\boldsymbol{A}_{\mathrm{vaa}}^v$ is computed by:

$$\boldsymbol{A}_{\mathrm{vaa}}^v = Reshape\left(\sigma\left(W_2 \cdot ReLU\left(W_1 \cdot \overrightarrow{\boldsymbol{E}_{\mathrm{ini}}^v}\right)\right)\right), \tag{1}$$

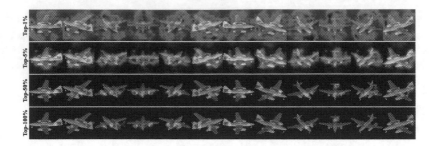

Fig. 3. Visualization of the view-specific attentions by selecting top-M cross-view responses with $M = 1\%/5\%/50\%/100\% \times H \times W$, respectively.

where $W_1 \in \mathbb{R}^{\frac{HW}{r} \times HW}$ and $W_2 \in \mathbb{R}^{HW \times \frac{HW}{r}}$ are learnable parameter matrices, $ReLU(\cdot)$ denotes the ReLU activation function, $\sigma(\cdot)$ is the Softmax activation function, $Reshape(\cdot)$ indicates the operation of reshaping an $HW \times 1$ vector to an $H \times W$ matrix, and r refers to the ratio of dimension reduction.

Note that the parameter matrices W_1 and W_2 of $\mathcal{VAA}_\psi(\cdot)$ are shared across different rendering views, *i.e.*, for all $v \in \{1, \cdots, N_v\}$. As a consequence, $\mathcal{VAA}_\psi(\cdot)$ are encouraged to pay more attention to view-independent salient local regions. In this regard, we call $\mathcal{VAA}_\psi(\cdot)$ the view-agnostic attention block.

By encoding the view-agnostic spatial attention $\boldsymbol{A}_{\text{vaa}}^v$ to the initial feature map, we can obtain the VAA induced feature map $\boldsymbol{L}_{\text{vaa}}^v = \boldsymbol{L}_{\text{ini}}^v \odot \boldsymbol{A}_{\text{vaa}}^v$, where $\odot$ indicates the element-wise production along the channel.

View-Specific Attention. Despite that $\mathcal{VAA}(\cdot)$ is view-agnostic, it tends to neglect some view-dependent local details (as shown in Fig. 1), which are discriminative for distinguishing 3D shapes. Therefore, it is reasonable to explore the view-specific attention as a complement to VAA. To this end, we propose a parameter-free view-specific attention block $\mathcal{VSA}(\cdot)$.

Given a feature map $\boldsymbol{L}_{\text{ini}}^v$ from the v-th view, $\mathcal{VSA}(\cdot)$ aims to compute the spatial attentions by exploring its saliency in feature maps from the rest $N_v - 1$ views, *i.e.*, $\{\boldsymbol{L}_{\text{ini}}^w : w \in \{1, \cdots, N_v\}; w \neq v\}$. Specifically, $\mathcal{VSA}(\cdot)$ first densely computes the response $\gamma_{j,k}^v(p, q, w)$ at location (p, q) in $\boldsymbol{L}_{\text{ini}}^w$ w.r.t. $\boldsymbol{l}_{j,k}^v$:

$$\gamma_{j,k}^v(p, q, w) = \frac{\left(\boldsymbol{l}_{j,k}^v\right)^T \cdot \boldsymbol{l}_{p,q}^w}{\|\boldsymbol{l}_{j,k}^v\|_2 \cdot \|\boldsymbol{l}_{p,q}^w\|_2}, \tag{2}$$

where $\boldsymbol{l}_{p,q}^w$ is the local feature at location (p, q) of $\boldsymbol{L}_{\text{ini}}^w$.

As shown in Eq. (2), the response $\gamma_{j,k}^v(p, q, w)$ is actually the Cosine distance between local features, implying that a large $\gamma_{j,k}^v(p, q, w)$ corresponds to a high visual similarity, and vice versa. Subsequently, we select the subset $\tilde{R}_{j,k}^v(w; M)$ with the top-M largest responses from $R_{j,k}^v(w) = \{\gamma_{j,k}^v(p, q, w)\}_{1 \leq p \leq H, 1 \leq q \leq W}$:

$$\tilde{R}_{j,k}^v(w; M) = \text{argmax}_{R' \subset R, |R'| = M} \sum_{\gamma \in R'} \gamma, \tag{3}$$

where $|R'|$ indicates the number of elements in R'.

The unnormalized view-specific spatial attention at location (j, k) for the v-th view is then formulated as:

$$\tilde{\alpha}_{\text{vsa},j,k}^{v} = \max_{w \in \{1, \cdots, N_v; w \neq v\}} \sum_{\gamma \in \tilde{R}_{j,k}^{v}(w;M)} \gamma. \tag{4}$$

The normalized view-specific attention A_{vsa}^{v} can be obtained by applying the Softmax function as follows: $A_{\text{vsa}}^{v} = \text{Softmax}\left([\tilde{\alpha}_{\text{vsa},j,k}^{v}]\right)$.

From Eqs. (3) and (4), we can observe that A_{vsa}^{v} firstly explores the non-local saliency of $l_{j,k}^{v}$ in each cross-view feature map L_{ini}^{w}. Different from the standard non-local mean operation [69] used in non-local attentions, we adopt the top-M largest responses. As illustrated in Fig. 3, for a large M, the local patch $l_{j,k}^{v}$ with high non-local co-occurrence appearance will have a large value. And the attentions are concentrated on an extremely small number of local parts, which may lose some discriminative local details. For a small M, $l_{j,k}^{v}$ with high local co-occurrence will have a large value. In this case, the attentions become more diverse, but will be more sensitive to noise (e.g., backgrounds). In order to simultaneously maintain more local details and remove outliers, we set M to a mediate value, which is fixed to $5\% \times H \times W$ in our work.

Similar to VAA, we encode the view-specific attention A_{vsa}^{v} to the initial feature map, and attain the VSA induced feature map as $L_{\text{vsa}}^{v} = L_{\text{ini}}^{v} \odot A_{\text{vsa}}^{v}$.

By using the global pooling in base networks (e.g., the global average pooling in ResNet-50), the original local feature map $L_{\text{ini}}^{v,i}$, the VAA induced local feature map $L_{\text{vaa}}^{v,i}$ and the VSA induced local feature map $L_{\text{vsa}}^{v,i}$ of $\mathcal{O}^i$ are aggregated into three sets of global features $\{f_{\text{ini}}^{v,i}\}_{v=1}^{N_v}$, $\{f_{\text{vaa}}^{v,i}\}_{v=1}^{N_v}$ and $\{f_{\text{vsa}}^{v,i}\}_{v=1}^{N_v}$, respectively.

3.3   Hierarchical Representation Learning

Multi-granular View Pooling. As shown in Fig. 1, the rendered 2D images $\{I^{v,i}\}$ of a 3D shape $\mathcal{O}^i$ from different views capture *distinct* but *incomplete* spatial and visual structures of $\mathcal{O}^i$. Conventional methods aggregate the multi-view features from N_v views by using view-level average/max pooling [7,25,55, 70,72], exploring view attentions [24], adopting the sequence modeling model such as the recurrent neural networks [22], or using 3D convolutions [34] as well as graph neural networks [12].

In our work, we develop a multi-granular view pooling (MVP) module to aggregate the multi-view features $\{f_t^{v,i}\}_{v=1}^{N_v}$ ($t \in \{\text{ini}, \text{vaa}, \text{vsa}\}$) based on the following three levels of granularity. 1) **Granularity-1 (g-1).** The N_v rendering views are sequentially divided into four groups, each of which consists of $\frac{N_v}{4}$ views. The features in each group are aggregated into one single vector by average pooling, and finally resulting in four vectors $\{f_{t,(1,g_1)}^{i}\}_{g_1=1,\cdots,4}$. 2) **Granularity-2 (g-2).** Similar to Granularity-1, the N_v rendering views are divided into two groups, each of them having $\frac{N_v}{2}$ views and therefore outputting two feature vectors $\{f_{t,(2,g_2)}^{i}\}_{g_2=1,2}$. 3) **Granularity-3 (g-3).** All N_v features are aggregated into the averaged vector $f_{t,3}^{i}$.

The above three sets of aggregated features, *i.e.*, $\boldsymbol{f}^i_{t,3}$, $\{\boldsymbol{f}^i_{t,(2,g_2)}\}_{g_2=1,2}$ and $\{\boldsymbol{f}^i_{t,(1,g_1)}\}_{g_1=1,\cdots,4}$, capture different view-dependent visual details of $\mathcal{O}^i$ in a coarse-to-fine granularity. In this way, we desire to mitigate the single-view incompleteness problem as aforementioned.

Based on the hybrid attentions and multi-granular view pooling, a 3D object $\mathcal{O}^i$ can be represented by a feature set $F^i = \{\{\boldsymbol{f}^i_{t,(1,g_1)}\}, \{\boldsymbol{f}^i_{t,(2,g_2)}\}, \boldsymbol{f}^i_{t,3} : t \in \{\text{ini, vaa, vsa}\}\}$. We note that F^i has the following kinds of hierarchical relations:

1) *Spatial Attention Hierarchy.* As shown in Fig. 1, the global feature $\boldsymbol{f}^{v,i}_{\text{ini}}$ is pooled from the original feature map $\boldsymbol{L}^{v,i}_{\text{ini}}$, which equally treats each local feature. Therefore, $\boldsymbol{f}^{v,i}_{\text{ini}}$ represents the most diversified but less salient visual information. In contrast, $\boldsymbol{f}^v_{\text{vaa}}$, which is pooled from the VAA induced feature map $\boldsymbol{L}^v_{\text{vaa}}$, encodes extremely concentrated but salient local details. $\boldsymbol{f}^v_{\text{vsa}}$, derived from $\boldsymbol{L}^v_{\text{vaa}}$, alternatively makes a trade-off, and intermediately keep the diversity and saliency controlled by K. In this manner, $\boldsymbol{f}^{v,i}_{\text{ini}}$, $\boldsymbol{f}^{v,i}_{\text{ini}}$ and $\boldsymbol{f}^{v,i}_{\text{ini}}$ have hierarchical relations in regard to the diversity and saliency of spatial attentions.

2) *Pooling-view Hierarchy.* As described above, $\boldsymbol{f}^i_{t,3}$, $\{\boldsymbol{f}^i_{t,(2,g_2)}\}_{g_2=1,2}$ and $\{\boldsymbol{f}^i_{t,(1,g_1)}\}_{g_1=1,\cdots,4}$ aggregate multi-view features using the full Nv, partially $\frac{Nv}{2}$ and $\frac{Nv}{4}$ views, respectively. As a consequence, they naturally have hierarchical relations in terms of the aggregation granularity.

Based on the above two observations, we therefore leverage the Hyperbolic geometry to learn the embedding of F^i, due to their intrinsic capability of representing hierarchies, such as the tree graphs, taxonomies and linguistic ontology in natural language processing (NLP) [49,50].

Hyperbolic Space. Formally, we denote a D-dimensional Hyperbolic space by $\mathbb{H}^D$, which is defined as a simply connected n-dimensional Riemannian manifold of constant negative sectional curvature. Basically, there exist many distinct but isomorphic models of the Hyperbolic geometry. In our work, we adopt the Poincaré ball model, considering its preveiling applications in NLP and its numerical stability as well.

Specifically, a Poincaré ball is a manifold $\mathbb{P}^D_c = \{\boldsymbol{x} \in \mathbb{R}^D : c\|\boldsymbol{x}\|^2 < 1, c \geq 0\}$ endowed with the Riemannian metric $g^{\mathbb{P}}(\boldsymbol{x}) = (\lambda^c_{\boldsymbol{x}})^2 g^{\mathbb{E}}$, where $\lambda^c_{\boldsymbol{x}} = \frac{2}{1-c\|\boldsymbol{x}\|^2}$ is the conformal factor and $g^{\mathbb{E}}$ is the Euclidean metric tensor, *i.e.*, $g^{\mathbb{E}} = \mathbf{I}_D$. Note that in a standard definition of the Poincaré ball, c equals to 1. We follow [19] and introduce the hyperparameter c to represent the radius of the Poincaré ball. Actually, c allows one to make a balance between the Hyperbolic and Euclidean geometry, considering that $\mathbb{P}^D_c$ converges to the Euclidean space $\mathbb{R}^D$ as $c \to 0$.

Basic Operations in $\mathbb{P}^D_c$. To formulate our method, we introduce some basic arithmetic operations in the Hyperbolic space.

Möbius addition. Given two points $\boldsymbol{p}, \boldsymbol{q} \in \mathbb{P}^D_c$, $\boldsymbol{p} \oplus_c \boldsymbol{q}$ refers to the Möbius addition, which is defined as the following:

$$\frac{(1 + 2c <\boldsymbol{p}, \boldsymbol{q}> +c\|\boldsymbol{q}\|^2) \cdot \boldsymbol{p} + (1 - c\|\boldsymbol{q}\|^2) \cdot \boldsymbol{q}}{1 + 2c <\boldsymbol{p}, \boldsymbol{q}> +c^2\|\boldsymbol{p}\|^2\|\boldsymbol{q}\|^2}, \tag{5}$$

where $< \cdot >$ refers the Euclidean inner product, and $\|\|$ is the l–2 vector norm.

Geodesic Distance. Based on $\oplus_c$, the geodesic distance is formuated as:

$$d_c(\boldsymbol{p}, \boldsymbol{q}) = \frac{2}{\sqrt{c}} \text{arctanh}(\sqrt{c} \cdot \| - \boldsymbol{p} \oplus_c \boldsymbol{q} \|). \tag{6}$$

Möbius Matrix-Vector Product. Suppose we have a standard Euclidean linear matrix $\boldsymbol{M} \in \mathbb{R}^{d \times D}$, the Möbius matrix-vector product $\boldsymbol{M}^{\otimes_c}(\boldsymbol{p})$ between $\boldsymbol{M}$ and $\boldsymbol{p}$ is defined as follows:

$$\boldsymbol{M}^{\otimes_c}(\boldsymbol{p}) = \frac{1}{\sqrt{c}} \tanh \left(\frac{\|\boldsymbol{M}\boldsymbol{p}\|}{\|\boldsymbol{p}\|} \text{arctanh}(\sqrt{c}\|\boldsymbol{p}\|) \right) \frac{\boldsymbol{M}\boldsymbol{p}}{\|\boldsymbol{M}\boldsymbol{p}\|}, \tag{7}$$

if $\boldsymbol{M}\boldsymbol{p} \neq \boldsymbol{0}$, and otherwise $\boldsymbol{M}^{\otimes_c}(\boldsymbol{p}) = \boldsymbol{0}$.

Hyperbolic Embedding. Usually, the feature set F^i is not located in $\mathbb{P}_c^D$. We utilize the *exponential map* $\exp_0^c$ at $\boldsymbol{0}$ as the projection from $\mathbb{R}^D$ to $\mathbb{P}_c^D$:

$$\boldsymbol{0} \oplus_c \left(\tanh \left(\sqrt{c} \cdot \frac{\lambda_0^c \cdot \|\boldsymbol{f}\|}{2} \right) \frac{\boldsymbol{f}}{\sqrt{c} \cdot \|\boldsymbol{f}\|} \right). \tag{8}$$

The inverse projection *logarithmic map* $\log_0^c(\boldsymbol{y})$ from $\mathbb{P}_c^D$ to $\mathbb{R}^D$ is defined by:

$$\frac{2}{\sqrt{c} \cdot \lambda_0^c} \text{arctanh}(\sqrt{c} \cdot \| - \boldsymbol{0} \oplus_c \boldsymbol{y} \|) \frac{-\boldsymbol{0} \oplus_c \boldsymbol{y}}{\| - \boldsymbol{0} \oplus_c \boldsymbol{y} \|}. \tag{9}$$

As a result, F^i is projected to the Hyperbolic space and turned to

$$P^i = \left\{ \{\boldsymbol{p}_{t,(1,g_1)}^i\}, \{\boldsymbol{p}_{t,(1,g_2)}^i\}, \boldsymbol{p}_{t,g_3}^i \right\},$$

where $\boldsymbol{p}_{t,(1,g_1)}^i = \exp_0^c(\boldsymbol{f}_{t,(1,g_1)}^i)$, $\boldsymbol{p}_{t,(1,g_2)}^i = \exp_0^c(\boldsymbol{f}_{t,(1,g_2)}^i)$ and $\boldsymbol{p}_{t,g_3}^i = \exp_0^c(\boldsymbol{f}_{t,g_3}^i)$. We can finally obtain the 3D representation $\boldsymbol{p}^i \in \mathbb{P}_c^{d'}$ by using the following vector concatenation in P^i:

$$\mathcal{HE}(F^i) = \boldsymbol{M}^{\oplus_c}(\boldsymbol{p}_{\text{ini},(1,g_1)}^i) \oplus_c \cdots \oplus_c \boldsymbol{M}^{\oplus_c}(\boldsymbol{p}_{\text{vsa},g_3}^i),$$

where $\boldsymbol{M} \in \mathbb{R}^{d' \times D}$ is the parameter matrix of $\mathcal{HE}(\cdot)$, and d' indicates the dimension of the concatenated feature vector.

Hyperbolic Neural Networks. In order to perform 3D shape classification and retrieval, we leverage the generalized multi-class logistic regression (MLR) to Hyperbolic spaces [19]. The basic idea lies in the following observation: the logits of MLR in the Euclidean space can be represented as the distances to certain hyperplanes, where each hyperplane can be specified with a point of origin and a normal vector. This observation can be extended to the Poincaré ball $\mathbb{P}_c^n$. Specifically, suppose C points $\{\boldsymbol{h}_k \in \mathbb{P}_c^n\}_{k=1}^C$ and normal vectors $\{\boldsymbol{a}_k \in T_{\boldsymbol{h}_k} \mathcal{P}_c^n \backslash \{\boldsymbol{0}\}\}_{k=1}^C$ are learnable parameters, where $T_{\boldsymbol{h}_k}$ stands for the tangent space at $\boldsymbol{h}_k$. Given a feature $\boldsymbol{p} \in \mathbb{P}_c^n$, the Hyperbolic MLR $\mathcal{H}_\psi(\cdot)$ for C classes is thereafter formulated as follows:

$$p_k(\boldsymbol{p}) = p(y = k|\boldsymbol{p}) \propto \exp \left(\frac{\lambda_{\boldsymbol{h}_k}^c \|\boldsymbol{a}_k\|}{\sqrt{c}} \text{arcsinh} \left(\frac{2\sqrt{c}\langle -\boldsymbol{h}_k \oplus_c \boldsymbol{p}, \boldsymbol{a}_k \rangle}{(1-c\|-\boldsymbol{h}_k \oplus_c \boldsymbol{p}\|^2)\|\cdot\boldsymbol{a}_k\|} \right) \right). \tag{10}$$

Based on Eq. (10), we then apply the cross-entropy loss for the concatenated feature $\boldsymbol{p}^i$ as well as for all the individual features $\{\boldsymbol{p}^i_{\text{ini},(1,g_1)}, \cdots, \boldsymbol{p}^i_{\text{vsa},g_3}\}$:

$$\mathcal{L}_{xent} = -\frac{1}{N} \sum_i^N \sum_{k=1}^C \sum_{\boldsymbol{p} \in \mathscr{P}^i \cup \{\boldsymbol{p}^i\}} y_k^i \cdot \log(p_k(\boldsymbol{p})). \tag{11}$$

Optimization. As shown in Eq. (11), the parameters $\{\boldsymbol{h}_k\}_{k=1}^C$ of the Hyperbolic MLR $\mathcal{H}_\psi(\cdot)$ are located inside the Poincaré ball. One way to optimize $\mathcal{H}_\psi(\cdot)$ is using the Riemannian Adam optimizer [17] with pre-conditioners [63,64]. However, as suggested in [30], we utilize a more efficient yet effective solution, *i.e.*, first optimizing $\{\boldsymbol{h}_k\}_{k=1}^C$ via the standard Adam optimizer, and then mapping them to their Hyperbolic counterparts with the exponential map $\exp_0^c(\cdot)$.

4 Experimental Results and Analysis

4.1 3D Shape Classification and Retrieval

Datasets. For 3D shape classification and retrieval, we conduct experiments on two widely used datasets: **ModelNet10** and **ModelNet40**, both of which are subsets of ModelNet [67] with 151,128 3D CAD models from 660 categories. ModelNet10 includes 4,899 3D shapes belonging to 10 classes. We follow the 3,991/908 training/test split as commonly used in literature [73]. ModelNet40 contains 12,311 3D shapes from 40 categories. For 3D shape retrieval, most existing works select 80/20 objects per class for training/testing [67] and [55]. In regard to 3D shape classification, more recent works use the full split [12,45–47,73], which has 9,843/2,468 training/test 3D models. Therefore, we adopt the 80/20 split for 3D shape retrieval, and the full split for 3D shape classification.

Evaluation Metrics. As for classification, we follow previous works and report both *per instance accuracy* and *per class accuracy*, regarding the class-imbalance problem in the ModelNet40 dataset. Concretely, the per instance accuracy is the percentage of correctly classified 3D models among all the whole test set, and the per class accuracy refers to the averaged accuracy per class. To evaluate the retrieval performance, we report the widely used *mean Average Precision (mAP)* and *Area Under Curve (AUC)* of the precision-recall curve.

Implementation Details. Following the identical rendering protocol as MVCNN [55], we render a 3D object to a set of 2D 224 × 224 greyscale images by placing virtual cameras around the 3D model every 30°. Each 3D shape is then represented by 12 view images.

As suggested in MVCNN [56], we train our model by two stages. In the first stage, we adopt the CNN backbone network pre-trained on ImageNet [48], and fine-tune it on the training set by training as a single-view image classification task. In the second stage, we initialize the convolutional layers with the model fine-tuned in stage 1, and train the full model in Fig. 2, by removing the fully-connected classifier. We adopt the Adam optimizer [31], and set the learning rate

Table 1. Comparison results on 3D shape classification. (Best results in **bold**.)

Method	Reference	Input Modality	ModelNet40		ModelNet10	
			Per instance (%)	Per class (%)	Per instance (%)	Per class (%)
SPH [29]	SPG2003	Hand-crafted	-	68.2	-	-
LFD [6]	CGF2003	Hand-crafted	-	75.5	-	40.9
Subvolume Net [46]	CVPR2016	Volume	89.2	86.0	-	-
Voxception-ResNet [3]	NIPS2016	Volume	91.3	-	93.6	-
PointNet++ [47]	NIPS2017	Points	91.9	-	-	-
SO-Net [39]	CVPR 2018	Points	93.4	90.8	95.7	95.5
DensePoint [40]	ICCV 2019	Points	93.2	-	96.6	-
MeshNet [14]	AAAI 2019	Mesh	-	-	93.1	-
MVCNN^{V-M} [55]	CVPR2015	Multi-view (#Views=12)	92.1	89.9	-	-
MVCNN-MultiRes^{V-M} [46]	CVPR2016	Multi-resolution Views	93.8	91.4	-	-
MVCN-New^{R-50} [56]	ECCVW2018	Multi View (#Views=12)	95.5	94.0	-	-
Pairwise Network^{V-M} [27]	CVPR2016	Multi-view (#Views=12)	-	91.1	-	93.2
GVCNN^{G} [13]	CVPR2018	Multi-view	93.1	-	-	-
RotationNet^{R-50} [28]	ICCV2017	Multi-view (#Views=12)	-	-	94.8	-
MHBN^{V-M} [75]	CVPR2018	Multi-view	94.1	92.2	94.9	94.9
HGNN [73]	AAAI2019	Multi-view (#Views=12)	**96.7**	-	-	-
RelationNet^{V-M} [73]	ICCV2019	Multi-view (#Views=12)	94.3	92.3	95.3	95.1
HEAR^{V-M}	Ours	Multi-view (#Views=12)	95.5	94.2	98.2	98.1
HEAR^{R-50}	Ours	Multi-view (#Views=12)	**96.7**	**95.2**	**98.6**	**98.5**

$(V - M/G/R - 50$ indicate **VGG-M [5]/GoogLeNet [58]/ResNet-50 [23]**.$)$

to 5×10^{-5} and 1×10^{-5} for the first and second stages, respectively. For both stages, the model is trained within 30 epochs with weight decay 0.001. As to the Hyperbolic embedding, we set the hyper-parameter c and the dimension d' of concatenated features as 5×10^{-5} and 1,024, respectively. All the experiments are conducted on a Telsa V100 GPU.

As summarized in Table 1, HEAR achieves the best performance on Model-Net40 and ModelNet10, when using the same base networks. For instance, the per instance/class accuracy of HEAR is 1.6%/2.6% higher than the second best one on ModelNet10, when using VGG-M. With the ResNet-50 backbone, the performance of HEAR can be further improved. Note that HGNN achieve the same per instance accuracy, *i.e.*, 96.7%, as ours on ModelNet40. However, it combines multiple types of deep features including GVCNN and MVCNN, while HEAR only requires one backbone. In [28], a higher result is reported for Rota-tionNet, by extensively exploring the rendering view coordinates. In contrast, our method uses the standard 12 fixed views. For a fair comparison, we only report the averaged accuracy of RotationNet.

Results on 3D Shape Classification. We compare our method with the hand-crafted [6,29], voxel-based [3,46], points-based [39,40,47], mesh-based [14], and multi-view based approaches [12,13,27,28,46,55,56,73,75]. As for the multi-view based methods, different base networks are utilized, such as VGG-M [5],

Table 2. Comparison results on 3D shape retrieval. (Best results in **bold**.)

Method	Reference	ModelNet40		ModelNet10	
		AUC	MAP	AUC	MAP
SPH [29]	SPG2003	34.5	33.3	46.0	44.1
3DShapeNet [68]	CVPR2015	49.9	49.2	69.3	68.3
DLAN[16]	BMVC2016	-	85.0	-	90.6
MVCNN^{V-M} [55]	CVPR2015	-	80.2	-	-
MVCNN^{V-A} [55]	CVPR2015	73.7	72.9	80.8	80.1
GIFT^{V-S} [1]	CVPR2016	83.1	81.9	92.4	91.1
RED $^{R-50}$ [2]	ICCV2017	87.0	86.3	93.2	92.2
GVCNN ^{G} [13]	CVPR2018	-	85.7	-	-
TCL^{V-A} [25]	CVPR2018	89.0	88.0	-	-
SeqViews^{V-19} [22]	TIP2018	-	89.1	-	91.4
VDN^G [35]	TVCG2018	87.6	86.6	93.6	93.2
Batch-wise [72]	CVPR2019	-	83.8	-	87.5
VNN $^{V-A}$ [24]	ICCV2019	89.6	88.9	93.5	92.8
VNN $^{V-19}$ [24]	ICCV2019	90.2	89.3	-	-
NCENet ^{G} [71]	ICCV2019	88.0	87.1	-	-
HEAR^{V-A}	Ours	91.8	91.1	95.0	94.2
HEAR^{V-19}	Ours	92.5	91.8	95.3	94.4
HEAR^{R-50}	Ours	**92.8**	**92.0**	**95.5**	**94.7**

($V - S$/$V - A$/$V - 19$ **indicate VGG-S [53]/VGG-A [53]/VGG-19 [53].**)

GoogLeNet [58] and ResNet-50 [23]. We adopt the VGG-M and ResNet-50 to make a fair comparison, which are used by most existing works.

Results on 3D Shape Retrieval. We compare HEAR with the state-of-the-art approaches including model-based [16,29,67] and multi-view based ones [1, 2,13,21,24,25,35,56,71,72]. We report the results of our method based on three commonly-used backbones (*i.e.*, VGG-A, VGG-19 and ResNet-50). In addition, we use the 1,024-dimensional concatenated vector after Hyperbolic embedding as the representation. Note that we **do not employ the triplet loss** as commonly used in 3D shape retrieval, and only use the cross-entropy loss for training.

The comparison results are summarized in Table 2. As shown, our method remarkably outperforms the state-of-the-art methods, and achieves 2.7% and 1.5% improvement w.r.t. mAP on ModelNet40 and ModelNet10, respectively. The improvement is consistent, regardless of the choice of backbone networks.

4.2 Sketch-Based 3D Shape Retrieval

Datasets. SHREC'13 [36] contains 7,200 human-drawn sketches, and 1,258 shapes from 90 classes, which are collected from the Princeton Shape Benchmark (PSB) [52]. There are a total of 80 sketches per class, 50 of which are selected

Table 3. Comparison results on sketch-based 3D shape retrieval. (Best results in **bold**.)

Method	Reference	SHREC'13						SHREC'14					
		NN	FT	ST	E	DCG	mAP	NN	FT	ST	E	DCG	mAP
CDMR [15]	ICW2013	27.9	20.3	29.6	16.6	45.8	25.0	10.9	5.7	8.9	4.1	32.8	5.4
SBR-VC [36]	SHREC13'track	16.4	9.7	14.9	8.5	34.8	11.4	9.5	5.0	8.1	3.7	31.9	5.0
SP [54]	JVLC2010	1.7	1.6	3.1	1.8	24.0	2.6	-	-	-	-	-	-
FDC [36]	SHREC13'track	11.0	6.9	10.7	6.1	30.7	8.6	-	-	-	-	-	-
DB-VLAT [61]	SIPAASC2012	-	-	-	-	-	-	16.0	11.5	17.0	7.9	37.6	13.1
CAT-DTW [74]	VC2017	23.5	13.5	19.8	10.9	39.2	14.1	13.7	6.8	10.2	5.0	33.8	6.0
Siamese [66]	CVPR2015	40.5	40.3	54.8	28.7	60.7	46.9	23.9	21.2	31.6	14.0	49.6	22.8
KECNN [59]	NC2017	32.0	31.9	39.7	23.6	48.9		-	-	-	-	-	-
DCML [11]	AAAI2017	65.0	63.4	71.9	34.8	76.6	67.4	27.2	27.5	34.5	17.1	49.8	28.6
DCHML [9]	TIP2018	73.0	71.5	77.3	36.8	81.6	74.4	40.3	32.9	39.4	20.1	54.4	33.6
LWBR [70]	CVPR2017	71.2	72.5	78.5	36.9	81.4	75.2	40.3	37.8	45.5	23.6	58.1	40.1
DCML^{R-50} [11]	TIP2017	74.0	75.2	79.7	36.5	82.9	77.4	57.8	59.1	64.7	72.3	35.1	61.5
LWBR^{R-50} [70]	CVPR2017	73.5	74.5	78.4	35.9	82.5	76.7	62.1	64.1	69.1	76.0	36.1	66.5
Shape2Vec [60]	TOG2016	-	-	-	-	-	-	71.4	69.7	74.8	36.0	81.1	72.0
DCA^{R-50} [7]	ECCV2018	78.3	79.6	82.9	37.6	85.6	81.3	77.0	78.9	82.3	39.8	85.9	80.3
Semantic^{IR} [44]	BMCV2018	82.3	82.8	86.0	40.3	88.4	84.3	80.4	74.9	81.3	39.5	87.0	78.0
DSSH^{R-50} [8]	CVPR2019	79.9	81.4	86.0	40.4	87.3	83.1	77.5	78.8	83.1	40.4	87.0	80.6
DSSH^{IR} [8]	CVPR2019	83.1	84.4	88.6	41.1	89.3	85.8	79.6	81.3	85.1	41.2	88.1	82.6
HEAR^{R-50}	Ours	82.1	83.7	87.8	40.9	88.8	85.4	79.2	80.7	84.6	40.9	87.8	82.2
HEAR^{IR}	Ours	**84.2**	**85.6**	**88.8**	**41.3**	**90.0**	**86.9**	**80.9**	**82.6**	**86.3**	**41.4**	**89.0**	**83.6**

(IR represents using Inception-ResNet-v2 [57] as the base network.)

for training and the rest for test. **SHREC'14** [38] consists of 13,680 sketches and 8,987 3D shapes belonging to 171 classes. There are 80 sketches, and around 53 3D shapes on average per class. The sketches are split into 8,550 image for training and 5,130 for testing.

Evaluation Metrics. We utilize the following widely-adopted metrics [11,37, 70] for sketch-based 3D shape retrieval: *nearest neighbor (NN)*, *first tier (FT)*, *second tier (ST)*, *E-measure (E)*, *discounted cumulated gain (DCG)* as well as the *mean average precision (mAP)*.

Implementation Details. We employ the ResNet-50 and Inception-ResNet-v2 as the base network, similar to the state-of-the-art methods [7,8,44]. We follow the same 'two branch' architecture as depicted in [44], *i.e.*, one branch for sketches and the other one for 3D shapes. The same batch-hard triplet loss and cross-entropy loss in [44] are utilized for training. The only difference between our method and [44] lies in the designed 3D shape branch. The learning rate is set to 3×10^{-5} with decay rate 0.9 for every 20,000 steps.

Experimental Results. We compare HEAR with the state-of-the-art methods for sketch-based 3D shape retrieval, including hand-crafted [15,36,36,54,61,74] and deep learning based ones [7–9,11,11,44,59,60,66,70,70].

As summarized in Table 3, our method achieves the best performance on both SHREC'13 and SHREC'14. For instance, by using the same ResNet-50 base model, HEAR improves the mAP of DCML, LWBR, DCA and DSSH by

Table 4. Ablation study of HEAR w.r.t. mAP by using the VGG-A backbone.

Method	ModelNet40	ModelNet10
Baseline_MVCNN	72.9	80.1
Baseline_MVCNN+VAA	86.3	88.7
Baseline_MVCNN+VSA	87.5	90.2
Baseline_MVCNN+HA	89.7	92.8
Baseline_MVCNN+HA+MVP	90.0	93.0
Baseline_MVCNN+HA+MVP+HNet (**HEAR**)	91.1	94.2

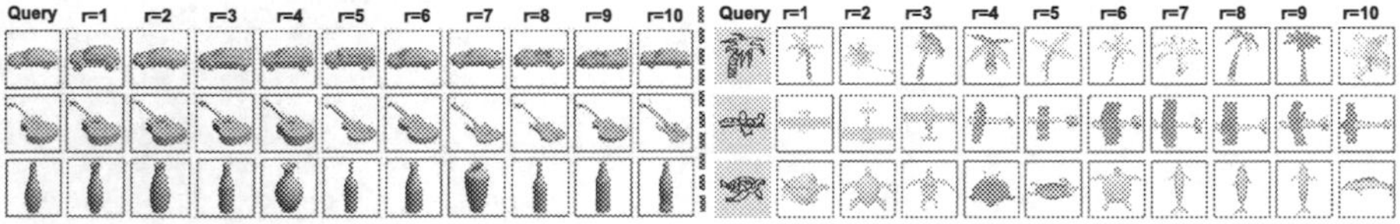

Fig. 4. Retrieval results by using HEAR on ModelNet40 (Left) and SHREC'13 (Right). Images with yellow backgrounds and blue/green/red bounding boxes indicate query 2D sketches, query 3D shapes/correct matches/false matches, respectively. (Color figure online)

8.0%, 10.2%, 4.1% and 2.3% on SHREC'13, respectively. Similar improvements can be seen on SHREC'14 and by using the Inception-ResNet-v2 backbone network. It also can be seen that the performance margin between HEAR and Semantic^{IR} [44] is significant, though the identical learning objective is applied. This suggests that the proposed network learns more descriptive 3D shape patterns for the respective task. In addition, HEAR works well with different learning objectives, further endorsing its ability to learn compact 3D representations.

4.3   Ablation Study

To evaluate each component of our method, *i.e.*, the Hybrid Attention (HA) module consisting of VAA and VSA, the Multi-granular View Pooling (MVP) module as well as the Hyperbolic Neural Networks with Hyperbolic Embedding (HNet), we conduct ablation studies on ModelNet10 and ModelNet40 for the 3D retrieval task. Specifically, we choose MVCNN with VGG-A network structure as the baseline, denoted by *Baseline_MVCNN*. We then successively add VAA/VSA, HA, MVP and HNet to validate their influences on the performance of HEAR. Note that *Baseline_MVCNN+HA+MVP* uses the concatenation and the standard linear classifier in the Euclidean space, instead of adopting the concatenation with the Hyperbolic embedding and the Hyperbolic MLR.

Table 4 summarizes the mAP of the baselines with different combinations of the components involved. We can observe that both VAA and VSA significantly improve the baseline by exploring the spatial saliency. After combining VAA and VSA, the hybrid attention (HA) can further boost the performance.

By employing MVP, HEAR can be slightly improved. The view shift of a 3D object is literally continuous throughout different view points. MVP provides a non-parametric way to perceive this via fusing multi-view data with minimal information wastage. The Hyperbolic embedding and the Hyperbolic MLR can further promote the mAP of HEAR, by endowing and modeling the hierarchical structures in the Hyperbolic space. Without the hyperbolic projection, the proposed model reduces to a conventional representation learning scheme, which is not able to fully acknowledge the structured conceptual similarities.

In addition, we qualitatively show some retrieval results by HEAR in Fig. 4.

5   Conclusion

This paper proposed a novel 3D shape representation method, namely Hyperbolic Embedded Attentive Representation (HEAR). HEAR developed a hybrid attention to explore distinct yet complementary spatial attentions. A multi-granular view-pooling module was subsequently employed to aggregate features from multi-views in a coarse-to-fine hierarchy. The resulting feature set was finally encoded into a hierarchical representation by the Hyperbolic geometry. Experiments on various tasks revealed the superiority of the proposed method.

References

1. Bai, S., Bai, X., Zhou, Z., Zhang, Z., Jan Latecki, L.: GIFT: a real-time and scalable 3D shape search engine. In: CVPR (2016)
2. Bai, S., Zhou, Z., Wang, J., Bai, X., Jan Latecki, L., Tian, Q.: Ensemble diffusion for retrieval. In: ICCV (2017)
3. Brock, A., Lim, T., Ritchie, J.M., Weston, N.: Generative and discriminative voxel modeling with convolutional neural networks. In: NeurIPS (2016)
4. Chami, I., Ying, Z., Ré, C., Leskovec, J.: Hyperbolic graph convolutional neural networks. In: NeurIPS (2019)
5. Chatfield, K., Simonyan, K., Vedaldi, A., Zisserman, A.: Return of the devil in the details: delving deep into convolutional nets. arXiv preprint arXiv:1405.3531 (2014)
6. Chen, D.Y., Tian, X.P., Shen, Y.T., Ouhyoung, M.: On visual similarity based 3D model retrieval. In: Computer Graphics Forum, vol. 22, pp. 223–232. Wiley Online Library (2003)
7. Chen, J., Fang, Y.: Deep cross-modality adaptation via semantics preserving adversarial learning for sketch-based 3D shape retrieval. In: ECCV (2018)
8. Chen, J., et al.: Deep sketch-shape hashing with segmented 3D stochastic viewing. In: CVPR (2019)
9. Dai, G., Xie, J., Fang, Y.: Deep correlated holistic metric learning for sketch-based 3D shape retrieval. IEEE Trans. Image Process. **27**, 3374–3386 (2018)
10. Dai, G., Xie, J., Fang, Y.: Siamese CNN-BiLSTM architecture for 3D shape representation learning. In: Proceedings of the 27th International Joint Conference on Artificial Intelligence, IJCAI 2018, pp. 670–676 (2018)
11. Dai, G., Xie, J., Zhu, F., Fang, Y.: Deep correlated metric learning for sketch-based 3D shape retrieval. In: AAAI (2017)

12. Feng, Y., You, H., Zhang, Z., Ji, R., Gao, Y.: Hypergraph neural networks. In: AAAI (2019)
13. Feng, Y., Zhang, Z., Zhao, X., Ji, R., Gao, Y.: GVCNN: group-view convolutional neural networks for 3D shape recognition. In: CVPR (2018)
14. Feng, Y., Feng, Y., You, H., Zhao, X., Gao, Y.: MeshNet: mesh neural network for 3D shape representation. In: AAAI 2019 (2018)
15. Furuya, T., Ohbuchi, R.: Ranking on cross-domain manifold for sketch-based 3D model retrieval. In: International Conference on Cyberworlds (2013)
16. Furuya, T., Ohbuchi, R.: Deep aggregation of local 3D geometric features for 3D model retrieval. In: BMVC (2016)
17. Bécigneul, G., Ganea, O.E.: Riemannian adaptive optimization methods (2019)
18. Gabeur, V., Franco, J.S., Martin, X., Schmid, C., Rogez, G.: Moulding humans: non-parametric 3D human shape estimation from single images. In: The IEEE International Conference on Computer Vision (ICCV) (2019)
19. Gulcehre, C., et al.: Hyperbolic neural networks. In: NeurIPS (2018)
20. Gulcehre, C., et al.: Hyperbolic attention networks. In: ICLR (2019)
21. Han, Z., et al.: 3D2SeqViews: aggregating sequential views for 3D global feature learning by CNN with hierarchical attention aggregation. IEEE Trans. Image Process. **28**(8), 3986–3999 (2019)
22. Han, Z., et al.: SeqViews2SeqLabels: learning 3D global features via aggregating sequential views by RNN with attention. IEEE Trans. Image Process. **28**(2), 658–672 (2018)
23. He, K., Zhang, X., Ren, S., Sun, J.: Deep residual learning for image recognition. In: CVPR (2016)
24. He, X., Huang, T., Bai, S., Bai, X.: View n-gram network for 3D object retrieval. In: ICCV (2019)
25. He, X., Zhou, Y., Zhou, Z., Bai, S., Bai, X.: Triplet-center loss for multi-view 3D object retrieval. In: CVPR (2018)
26. Hu, J., Shen, L., Sun, G.: Squeeze-and-excitation networks. In: CVPR (2018)
27. Johns, E., Leutenegger, S., Davision, A.J.: Pairwise decomposition of image sequences for active multiview recognition. In: CVPR (2016)
28. Kanezaki, A., Matsushita, Y., Nishida, Y.: RotationNet: joint object categorization and pose estimation using multiviews from unsupervised viewpoints. In: CVPR (2018)
29. Kazhdan, M., Funkhouser, T., Rusinkiewicz, S.: Rotation invariant spherical harmonic representation of 3D shape descriptors. In: Symposium on Geometry Processing, vol. 6, pp. 156–164 (2003)
30. Khrulkov, V., Mirvakhabova, L., Ustinova, E., Oseledets, I., Lempitsky, V.: Hyperbolic image embeddings. arXiv preprint arXiv:1904.02239 (2019)
31. Kingma, D.P., Ba, J.: Adam: a method for stochastic optimization. In: ICLR (2015)
32. Klokov, R., Lempitsky, V.: Escape from cells: deep Kd-networks for the recognition of 3D point cloud models. In: CVPR (2017)
33. Kolotouros, N., Pavlakos, G., Black, M.J., Daniilidis, K.: Learning to reconstruct 3D human pose and shape via model-fitting in the loop. In: The IEEE International Conference on Computer Vision (ICCV) (2019)
34. Kumawat, S., Raman, S.: LP-3DCNN: unveiling local phase in 3D convolutional neural networks. In: CVPR (2019)
35. Leng, B., Zhang, C., Zhou, X., Xu, C., Xu, K.: Learning discriminative 3D shape representations by view discerning networks. IEEE Trans. Visual. Comput. Graph. **25**, 2896–2909 (2018)

36. Li, B., et al.: SHREC13 track: large scale sketch-based 3D shape retrieval (2013)
37. Li, B., et al.: A comparison of methods for sketch-based 3D shape retrieval. CVIU **119**, 57–80 (2014)
38. Li, B., et al.: SHREC14 track: extended large scale sketch-based 3D shape retrieval. In: Eurographics Workshop on 3D Object Retrieval (2014)
39. Li, J., Chen, B., Hee, L.G.: SO-Net: self-organizing network for point cloud analysis. In: CVPR (2018)
40. Liu, Y., Fan, B., Meng, G., Lu, J., Xiang, S., Pan, C.: DensePoint: learning densely contextual representation for efficient point cloud processing. In: The IEEE International Conference on Computer Vision (ICCV) (2019)
41. Mao, J., Wang, X., Li, H.: Interpolated convolutional networks for 3D point cloud understanding. In: The IEEE International Conference on Computer Vision (ICCV) (2019)
42. Maturana, D., Scherer, S.: Multi-view harmonized bilinear network for 3D object recognition. In: IROS (2015)
43. Phong, B.T.: Illumination for computer generated pictures. Commun. ACM **18**(6), 311–317 (1975)
44. Qi, A., Song, Y., Xiang, T.: Semantic embedding for sketch-based 3D shape retrieval. In: BMVC (2018)
45. Qi, C.R., Su, H., Mo, K., Guibas, L.J.: PointNet: deep learning on point sets for 3D classification and segmentation. In: CVPR (2017)
46. Qi, C.R., Su, H., Niebner, M., Dai, A., Yan, M.: Volumetric and multi-view CNNs for object classification on 3D data. In: CVPR (2016)
47. Qi, C.R., Yi, L., Su, H., Guibas, L.J.: PointNet++: deep hierarchical feature learning on point sets in a metric space. In: NeurIPS (2017)
48. Russakovsky, O., et al.: ImageNet large scale visual recognition challenge. Int. J. Comput. Vis. **115**(3), 211–252 (2015)
49. Sala, F., De Sa, C., Gu, A., Ré, C.: Representation tradeoffs for hyperbolic embeddings. In: ICML (2019)
50. Sarkar, R.: Low distortion delaunay embedding of trees in hyperbolic plane. In: van Kreveld, M., Speckmann, B. (eds.) GD 2011. LNCS, vol. 7034, pp. 355–366. Springer, Heidelberg (2012). https://doi.org/10.1007/978-3-642-25878-7_34
51. Shi, B., Bai, S., Zhou, Z., Bai, X.: DeepPano: deep panoramic representation for 3D shape recognition. IEEE Signal Process. Lett. **22**(12), 2339–2343 (2015)
52. Shilane, P., Min, P., Kazhdan, M., Funkhouser, T.: The Princeton shape benchmark. In: Shape Modeling Applications (2004)
53. Simonyan, K., Zisserman, A.: Very deep convolutional networks for large-scale image recognition. arXiv preprint arXiv:1409.1556 (2014)
54. Sousa, P., Fonseca, M.J.: Sketch-based retrieval of drawings using spatial proximity. J. Vis. Lang. Comput. **21**(2), 69–80 (2010)
55. Su, H., Maji, S., Kalogerakis, E., Learned-Miller, E.: Multi-view convolutional neural networks for 3D shape recognition. In: ICCV (2015)
56. Su, J.C., Gadelha, M., Wang, R., Maji, S.: A deeper look at 3D shape classifiers. In: ECCV (2018)
57. Szegedy, C., Ioffe, S., Vanhoucke, V., Alemi, A.A.: Inception-v4, inception-ResNet and the impact of residual connections on learning. In: AAAI (2017)
58. Szegedy, C., et al.: Going deeper with convolutions. In: CVPR (2015)
59. Tabia, H., Laga, H.: Learning shape retrieval from different modalities. Neurocomputing **253**, 24–33 (2017)
60. Tasse, F.P., Dodgson, N.: Shape2Vec: semantic-based descriptors for 3D shapes, sketches and images. ACM Trans. Graph. **35**(6), 208 (2016)

61. Tatsuma, A., Koyanagi, H., Aono, M.: A large-scale shape benchmark for 3D object retrieval: Toyohashi shape benchmark. In: Asia-Pacific Signal & Information Processing Association Annual Summit and Conference (2012)
62. Thomas, H., Qi, C.R., Deschaud, J.E., Marcotegui, B., Goulette, F., Guibas, L.J.: KPConv: flexible and deformable convolution for point clouds. In: The IEEE International Conference on Computer Vision (ICCV) (2019)
63. Wang, C., Li, H., Zhao, D.: Preconditioning Toeplitz-plus-diagonal linear systems using the Sherman-Morrison-Woodbury formula. J. Comput. Appl. Math. **309**, 312–319 (2017)
64. Wang, C., Li, H., Zhao, D.: Improved block preconditioners for linear systems arising from half-quadratic image restoration. Appl. Math. Comput. **363**, 124614 (2019)
65. Wang, C., Pelillo, M., Siddiqi, K.: Dominant set clustering and pooling for multi-view 3D object recognition. In: BMVC (2017)
66. Wang, F., Kang, L., Li, Y.: Sketch-based 3D shape retrieval using convolutional neural networks. In: CVPR (2015)
67. Wu, Z., et al.: RotationNet: joint object categorization and pose estimation using multiviews from unsupervised viewpoints. In: CVPR (2015)
68. Wu, Z., et al.: 3D ShapeNets: a deep representation for volumetric shapes. In: CVPR (2015)
69. Wang, X., Girshick, R., Gupta, A., He, K.: Non-local neural networks. In: CVPR (2018)
70. Xie, J., Dai, G., Zhu, F., Fang, Y.: Learning barycentric representations of 3D shapes for sketch-based 3D shape retrieval. In: CVPR (2017)
71. Xu, C., Li, Z., Qiu, Q., Leng, B., Jiang, J.: Enhancing 2D representation via adjacent views for 3D shape retrieval. In: ICCV (2019)
72. Xu, L., Sun, H., Liu, Y.: Learning with batch-wise optimal transport loss for 3D shape recognition. In: CVPR (2019)
73. Yang, Z., Wang, L.: Learning relationships for multi-view 3D object recognition. In: ICCV (2019)
74. Yasseen, Z., Verroust-Blondet, A., Nasri, A.: View selection for sketch-based 3D model retrieval using visual part shape description. Vis. Comput. **33**(5), 565–583 (2017)
75. Yu, T., Meng, J., Yuan, J.: Multi-view harmonized bilinear network for 3D object recognition. In: CVPR (2018)

TF-NAS: Rethinking Three Search Freedoms of Latency-Constrained Differentiable Neural Architecture Search

Yibo Hu[1,2], Xiang Wu[1], and Ran He[1(✉)]

[1] CRIPAC & NLPR, CASIA, Beijing, China
huyibo871079699@gmail.com, alfredxiangwu@gmail.com, rhe@nlpr.ia.ac.cn
[2] JD AI Research, Beijing, China

Abstract. With the flourish of differentiable neural architecture search (NAS), automatically searching latency-constrained architectures gives a new perspective to reduce human labor and expertise. However, the searched architectures are usually suboptimal in accuracy and may have large jitters around the target latency. In this paper, we rethink three freedoms of differentiable NAS, i.e. operation-level, depth-level and width-level, and propose a novel method, named Three-Freedom NAS (TF-NAS), to achieve both good classification accuracy and precise latency constraint. For the operation-level, we present a **bi-sampling** search algorithm to moderate the operation collapse. For the depth-level, we introduce a **sink-connecting** search space to ensure the mutual exclusion between skip and other candidate operations, as well as eliminate the architecture redundancy. For the width-level, we propose an **elasticity-scaling** strategy that achieves precise latency constraint in a progressively fine-grained manner. Experiments on ImageNet demonstrate the effectiveness of TF-NAS. Particularly, our searched TF-NAS-A obtains 76.9% top-1 accuracy, achieving state-of-the-art results with less latency. Code is available at https://github.com/AberHu/TF-NAS.

Keywords: Differentiable NAS · Latency-constrained · Three Freedoms

1 Introduction

With the rapid developments of deep learning, ConvNets have been the *de facto* method for various computer vision tasks. It takes a long time and substantial effort to devise many useful models [14,18,19,23,28,29], boosting significant improvements in accuracy. However, instead of accuracy improvement, designing efficient ConvNets with specific resource constraints (e.g. FLOPs, latency,

Electronic supplementary material The online version of this chapter (https://doi.org/10.1007/978-3-030-58555-6_8) contains supplementary material, which is available to authorized users.

© Springer Nature Switzerland AG 2020
A. Vedaldi et al. (Eds.): ECCV 2020, LNCS 12360, pp. 123–139, 2020.
https://doi.org/10.1007/978-3-030-58555-6_8

energy) is more important in practice. Manual design requires a huge number of exploratory experiments, which is time-consuming and labor intensive. Recently, Neural Architecture Search (NAS) has attracted lots of attentions [21,22,26,33,39]. It learns to automatically discover resource-constrained architectures, which can achieve better performance than hand-craft architectures.

Most NAS methods are based on reinforcement learning (RL) [30,39,40] or evolutionary algorithms (EA) [6,8,26], leading to expensive or even unaffordable computing resources. Differentiable NAS [4,22,33] couples architecture sampling and training into a supernet to reduce huge resource overhead. This supernet supports the whole search space with three freedoms, including the operation-level, the depth-level and the width-level freedoms. However, due to the various combinations of search freedoms and the coarse-grained discreteness of search space, differentiable NAS often makes the searched architectures suboptimal with specific resource constraints. For example, setting the GPU latency constraint to 15ms and carefully tuning the trade-off parameters, we search for architectures based on the latency objective from ProxylessNAS [4]. The searched architecture has 15.76ms GPU latency, exceeding the target by a large margin. More analyses are presented in Sect. 4.5.

To address the above issue, in this paper, we first rethink the operation-level, the depth-level and the width-level search freedoms, tracing back to the source of search instability. For the operation-level, we observe operation collapse phenomenon, where the search procedure falls into some fixed operations. To alleviate such collapse, we propose a bi-sampling search algorithm. For the depth-level, we analyze the special role of skip operation and explain the mutual exclusion between skip and other operations. Furthermore, we also illustrate architecture redundancy by a simple case study in Fig. 3. To address these phenomena, we design a sink-connecting search space for NAS. For the width-level, we explore that due to the coarse-grained discreteness of search space, it is hard to search target architectures with precise resource constraints (e.g. latency). Accordingly, we present an elasticity-scaling strategy that progressively refines the coarse-grained search space by shrinking and expanding the model width, to precisely ensure the latency constraint. Combining the above components, we propose Three-Freedom Neural Architecture Search (TF-NAS) to search accurate latency-constrained architectures. To summarize, our main contributions lie in four-folds:

- Motivated by rethinking the operation-level, the depth-level and the width-level search freedoms, a novel TF-NAS is proposed to search accurate architectures with latency constraint.
- We introduce a simple bi-sampling search algorithm to moderate operation collapse phenomenon. Besides, the mutual exclusion between skip and other candidate operations, as well as the architecture redundancy, are first considered to design a new sink-connecting search space. Both of them ensure the search flexibility and stability.

- By investigating the coarse-grained discreteness of search space, we propose an elasticity-scaling strategy that progressively shrinks and expands the model width to ensure the latency constraint in a fine-grained manner.
- Our TF-NAS can search architectures with precise latency on target devices, achieving state-of-the-art performance on ImageNet classification task. Particularly, our searched TF-NAS-A achieves 76.9% top-1 accuracy with only 1.8 GPU days of search time.

2 Related Work

Micro Search focuses on finding robust cells [25–27,34,40] and stacking many copies of them to design the network architecture. AmoebaNet [26] and NAS-Net [40], which are based on Evolutionary Algorithm (EA) and Reinforcement Learning (RL) respectively, are the pioneers of micro search algorithms. However, these approaches take an expensive computational overhead, i.e. over 2,000 GPU days, for searching. DARTS [22] achieves a remarkable efficiency improvement (about 1 GPU day) by formulating the neural architecture search tasks in a differentiable manner. Following gradient based optimization in DARTS, GDAS [11] is proposed to sample one sub-graph from the whole directed acyclic graph (DAG) in one iteration, accelerating the search procedure. Xu et al. [36] randomly sample a proportion of channels for operation search in cells, leading to both faster search speed and higher training stability. P-DARTS [5] allows the depth of architecture to grow progressively in the search procedure, to alleviate memory/computational overheads and weak search instability. Comparing with accuracy, it is obvious that micro search algorithms are unfriendly to constrain the number of parameters, FLOPs and latency for neural architecture search.

Macro Search aims to search the entire neural architecture [4,6,30,31,33, 39], which is more flexible to obtain efficient networks. Baker et al. [1] introduce MetaQNN to sequentially choose CNN layers using Q-learning with an ϵ-greedy exploration strategy. MNASNet [30] and FBNet [33] are proposed to search efficient architectures with higher accuracy but lower latency. One-shot architecture search [2] designs a good search space and incorporates path drop when training the over-parameterized network. Since it suffers from the large memory usage to train an over-parameterized network, Cai et al. [4] propose ProxylessNAS to provide a new path-level pruning perspective for NAS. Different from the previous neural architecture search, EfficientNet [31] proposes three model scaling factors including width, depth and resolution for network design-ment. Benefiting from compounding scales, they achieve state-of-the-art performance on various computer vision tasks. Inspired by EfficientNet [31], in order to search for flexible architectures, we rethink three search freedoms, including **operation-level**, **depth-level** and **width-level**, for latency-constrained differentiable neural architecture search.

3   Our Method

3.1   Review of Differentiable NAS

In this paper, we focus on differentiable neural architecture search to search accurate macro architectures constrained by various inference latencies. Similar with [11,22,33], the search problem is formulated as a bi-level optimization:

$$\min_{\alpha \in A} L_{\text{val}}\left(\omega^*, \alpha\right) + \lambda C\left(LAT(\alpha)\right) \tag{1}$$

$$\text{s.t.} \quad \omega^* = \arg\min_{\omega} L_{\text{train}}\left(\omega, \alpha\right) \tag{2}$$

where ω and α are the supernet weights and the architecture distribution parameters, respectively. Given a supernet A, we aim to search a subnet $\alpha^* \in A$ that minimizes the validation loss $L_{val}\left(\omega^*, \alpha\right)$ and the latency constraint $C\left(LAT(\alpha)\right)$, where the weights ω^* of supernet are obtained by minimizing the training loss $L_{train}\left(\omega, \alpha\right)$ and λ is a trade-off hyperparameter.

Different from RL-based [30,39,40] or EA-based [6,8,26] NAS, where the outer objective Eq. (1) is treated as reward or fitness, differentiable NAS optimizes Eq. (1) by gradient descent. Sampling a subnet from supernet A is a non-differentiable process w.r.t. the architecture distribution parameters α. Therefore, a continuous relaxation is needed to allow back-propagation. Assuming there are N operations to be searched in each layer, we define op_i^l and α_i^l as the i-th operation in layer l and its architecture distribution parameter, respectively. Let x^l present the input feature map of layer l. A commonly used continuous relaxation is based on Gumbel Softmax trick [11,33]:

$$x^{l+1} = \sum_i u_i^l \cdot \text{op}_i^l\left(x^l\right), u_i^l = \frac{\exp\left((\alpha_i^l + g_i^l)/\tau\right)}{\sum_j \exp\left((\alpha_j^l + g_j^l)/\tau\right)} \tag{3}$$

$$LAT(\alpha) = \sum_l LAT\left(\alpha^l\right) = \sum_l \sum_i u_i^l \cdot LAT\left(\text{op}_i^l\right) \tag{4}$$

where τ is the temperature parameter, g_i^l is a random variable i.i.d sampled from $Gumbel(0, 1)$, $LAT\left(\alpha^l\right)$ is the latency of layer l and $LAT\left(\text{op}_i^l\right)$ is indexed from a pre-built latency lookup table. The superiority of Gumbel Softmax relaxation is to save GPU memory by approximate N times and to reduce search time. That is because only one operation with max u_i^l is chosen during forward pass. And the gradients of all the α_i^l can be back-propagated through Eq. (3).

3.2   The Search Space

In this paper, we focus on latency-constrained macro search. Inspired by EfficientNet [31], we build a layer-wise search space, which is depicted in Fig. 1 and Table 1. The input shapes and the channel numbers are the same as EfficientNet-B0 [31]. Different from EfficientNet-B0, we use ReLU in the first three stages.

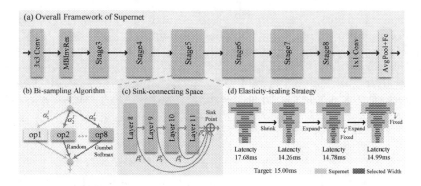

Fig. 1. The search space of TF-NAS. It contains (b) operation-level, (c) depth-level and (d) width-level search freedoms.

Table 1. Left: Macro architecture of the supernet. "OPS" denotes the operations to be searched. "MBInvRes" is the basic block in [28]. "C_{out}" means the output channels. "Act" denotes the activation function used in a stage. "L" is the number of layers in a stage, where $[a, b]$ is a discrete interval. If necessary, the down-sampling occurs at the first operation of a stage. **Right**: Candidate operations to be searched. "Expansion" defines the width of an operation and $[a, b]$ is a continuous interval. "SE Expansion" determines the width of the SE module.

Stage	Input	Operation	C_{out}	Act	L
1	$224^2 \times 3$	3×3 Conv	32	ReLU	1
2	$112^3 \times 32$	MBInvRes	16	ReLU	1
3	$112^2 \times 16$	OPS	24	ReLU	[1, 2]
4	$56^2 \times 24$	OPS	40	Swish	[1, 3]
5	$28^2 \times 40$	OPS	80	Swish	[1, 4]
6	$14^2 \times 80$	OPS	112	Swish	[1, 4]
7	$14^2 \times 112$	OPS	192	Swish	[1, 4]
8	$7^2 \times 192$	OPS	320	Swish	1
9	$7^2 \times 320$	1×1 Conv	1280	Swish	1
10	$7^2 \times 1280$	AvgPool	1280	-	1
11	1280	Fc	1000	-	1

OPS	Kernel	Expansion	SE Expansion
$k3\_e3$	3	[2, 4]	-
$k3\_e3\_e_{se}1$	3	[2, 4]	1
$k5\_e3$	5	[2, 4]	-
$k5\_e3\_e_{se}1$	5	[2, 4]	1
$k3\_e6$	3	[4, 8]	-
$k3\_e6\_e_{se}2$	3	[4, 8]	2
$k5\_e6$	5	[4, 8]	-
$k5\_e6\_e_{se}2$	5	[4, 8]	2

The reason is that the large resolutions of the early inputs mainly dominate the inference latency, leading to worse optimization during architecture searching.

Layers from stage 3 to stage 8 are searchable, and each layer can choose an operation to form the operation-level search space. The basic units of the candidate operations are MBInvRes (the basic block in MobileNetV2 [28]) with or without Squeeze-and-Excitation (SE) module, which are illustrated in Supp. 1. In our experiments, there are 8 candidate operations to be searched in each searchable layer. The detailed configurations are listed in Table 1. Each candidate operation has a kernel size $k = 3$ or $k = 5$ for the depthwise convolution, and a continuous expansion ratio $e \in [2, 4]$ or $e \in [4, 8]$, which constitutes to the width-level search space. Considering the operations with SE module, the SE expansion ratio is $e_{se} = 1$ or $e_{se} = 2$. In Table 1, the ratio of e_{se} to e for all the candidate operations lies in $[0.25, 0.5]$. $e3$ or $e6$ in the first column of Table 1 defines the expansion ratio is 3 or 6 at the beginning of searching, and e can vary in $[2, 4]$ or $[4, 8]$ during searching. Following the same naming schema, MBInvRes at stage 2 has a fixed configuration of $k3\_e1\_e_{se}0.25$. Besides, we also construct a depth-level search space based on a new sink-connecting schema. As shown in Fig. 1(c), during searching, the outputs of all the layers in a stage are connected to a sink point, which is the input to the next stage. After searching, only one connection, i.e. depth, is chosen in each stage.

3.3   Three-Freedom NAS

In this section, we investigate the operation-level, depth-level and width-level search freedoms, respectively, and accordingly make considerable improvements of the search flexibility and stability. Finally, our Three-Freedom NAS is summarized at the end of section.

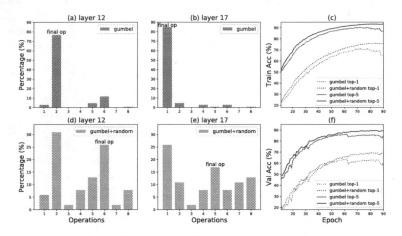

Fig. 2. (a)-(b): The counting percentages of the derived operations during searching by Gumbel Softmax relaxation. (d)-(e): The counting percentages of the derived operations during searching by bi-sampling algorithm. (c): Training accuracy of the supernet. (f): Validating accuracy of the supernet. Zoom in for better view.

Rethinking Operation-level Freedom. As demonstrated in Sect. 3.1, NAS based on Gumbel Softmax relaxation samples one operation per layer during forward pass. It means when optimizing the inner objective Eq. (2), only one path is chosen and updated by gradient descent. However, due to the alternative update between ω and α in the bi-level optimization, one path sampling strategy may focus on some specific operations and update their parameters more frequently than others. Then the architecture distribution parameters of these operations will get better when optimizing Eq. (1). Accordingly, the same operation is more likely to be selected in the next sampling. This phenomenon may cause the search procedure to fall into the specific operations at some layers, leading to suboptimal architectures. We call it operation collapse. Although there is

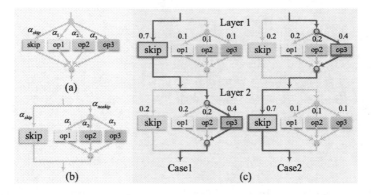

Fig. 3. (a)-(b): The mutual exclusion between skip and other operations. (c): A case study for architecture redundancy.

a temperature parameter τ to control the sampling, we find that the operation collapse still occurs in practice. We conduct an experiment based on our search space with the Gumbel Softmax relaxation, where τ linearly decreases from 5.0 to 0.2. The results are shown in Fig. 2(a)-(b), where we count the derived operations for layer 12 and 17 during searching (after each search epoch). It can be observed that almost 80% architecture derivations fall into specific operations in both layer 12 and 17, illustrating the occurrence of operation collapse.

To remedy the operation collapse, a straightforward method is early stopping [20,35]. However, it may lead to suboptimal architectures due to incomplete supernet training and operation exploration (Supp. 5). In this paper, we propose a simple bi-sampling search algorithm, where two independent paths are sampled for each time. In this way, when optimizing Eq. (2), two different paths are chosen and updated in a mini-batch. We implement it by conducting two times forward but one time backward. The second path is used to enhance the competitiveness of other operations against the one operation sampling in Gumbel Softmax. In Sect. 4.3, we conduct several experiments to explore various sampling strategies for the second path and find random sampling is the best one. Similarly, we also conduct an experiment based on our bi-sampling search algorithm and present the results in Fig. 2(d)–(e). Compared with Gumbel Softmax based sampling, our bi-sampling strategy is able to explore more operations during searching. Furthermore, as shown in Fig. 2(c) and Fig. 2(f), our bi-sampling strategy is superior to the Gumbel Softmax based sampling in both the supernet accuracy on the training and the validating set.

Rethinking Depth-level Freedom. In order to search for flexible architectures, an important component of differentiable NAS is depth-level search. Previous works [6,33] usually add a skip operation in the candidates and search them together (Fig. 3(a)). In this case, skip has equal importance to other operations and the probability of $op2$ is $P(\alpha_2)$. However, it makes the search unstable, where the derived architecture is relatively shallow and the depth has a large jitter, especially in the early search phase, as shown in orange line in

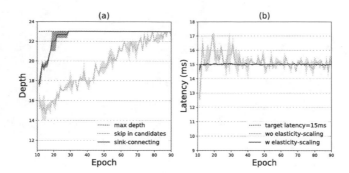

Fig. 4. (a): The searched depth for different depth-level search spaces. (b): The searched latency w/wo elasticity-scaling. All the search procedures are repeated 5 times, and we plot the mean, the maximum and the minimum. Zoom in for better view. (Color figure online)

Fig. 4(a). We argue that it is because the skip has higher priority to rule out other operations during searching, since it has no parameter. Therefore, the skip operation should be independent of other candidates, as depicted in Fig. 3(b). We call it as the mutual exclusion between skip and other candidate operations. In this case, skip competes with all the other operations and the probability of $op2$ is $P(\alpha_2, \alpha_{noskip})$. However, directly applying such a scheme will lead to architecture redundancy. Assuming there are two searchable layers in Fig. 3(c). Case 1: we choose skip in layer 1 and $op3$ in layer 2. Case 2: we choose $op3$ in layer 1 and skip in layer 2. Both cases have the same derived architectures $op3$ but quite different architecture distributions. As the number of searchable layers increases, such architecture redundancy will be more serious.

To address the above issue, we introduce a new sink-connecting search space to ensure the mutual exclusion between skip and other candidate operations, as well as eliminate the architecture redundancy. The basic framework is illustrated in Fig. 1(c), where the outputs of all the layers in a stage are connected to a sink point. During searching, the weighted sum of the output feature maps is calculated at the sink point, which is the input to the next stage. When deriving architectures, only one connection, i.e. depth, is chosen in each stage. Obviously, our sink-connecting search space makes the skip operation independent of the other candidates and has no architecture redundancy, because if a layer is skipped, then all the following layers in the same stage are also skipped. Let β_l^s be the architecture distribution parameter of l-th connection in stage s. We employ a Softmax function as the continuous relaxation:

$$x^{s+1} = \sum_{l \in s} v_l^s \cdot x^l, \ v_l^s = \frac{\exp(\beta_l^s)}{\sum_k \exp(\beta_k^s)} \tag{5}$$

$$Lat(\alpha, \beta) = \sum_s \sum_{l \in s} v_l^s \cdot Lat(\alpha^l) = \sum_s \sum_{l \in s} \sum_i v_l^s \cdot u_i^l \cdot Lat(op_i^l) \tag{6}$$

Blue line in Fig. 4(a) shows the search on sink-connecting search space. It is obvious that the search procedure is stable and the derived depth converges quickly. We do not sample for depth-level search, because if bi-sampling for β, we must independently sample 2 paths of depth and operation, respectively, leading to 4 times forward, which notably increases GPU memory and search time.

Rethinking Width-level Freedom. Due to the coarse-grained discreteness of search space, current NAS methods cannot satisfy the precise latency constraints. Each searchable layer has a fixed number of channels for the candidate operations, which means each layer has a fixed number of latency options. Furthermore, in each stage, all the layers excluding the first one have the same input and output shapes, so the latency options of these layers are all the same. Although the search space of NAS is huge (e.g. it is 10^{21} for FBNet [33]), the statuses of architectures with different latencies are finite and discrete. Due to the coarse-grained search space for latency, some target latency cannot be precisely satisfied, leading to instability during architecture searching. For example, setting the target latency to be 15 ms, we search two architectures: one is 14.32 ms and the other is 15.76 ms. Both of them have around 0.7 ms gaps for the target latency. More analyses are presented in Sect. 4.5.

In order to refine the coarse-grained search space for latency, previous works [12,33] introduce a global scaling factor or add additional candidate operations for width-level search. However, these methods are not flexible. Inspired by MorphNet [13], we propose an elasticity-scaling approach that adaptively shrinks and expands the model width to precisely satisfy the latency constraint in a progressively fine-grained manner. Our approach does not increase additional GPU memory and is insensitive to hyperparameter settings.

Given a supernet, we derive a discrete seed network (sn) based on the current architecture distribution parameters, where the strongest operation in each layer and the strongest depth in each stage are chosen. We can multiply sn by a scaling factor γ to control the width. Let $\gamma \cdot sn_{i:j}$ be a network whose layer width from stage i to stage j is multiplied by γ. Our elasticity-scaling strategy is presented in Algorithm 1, including a global scaling ($i = 3$) and a series of progressively fine-grained scaling ($i = 4 \ldots 8$). Note that the searchable stages are from stage 3 to stage 8 in our search space. More implementation details can be found in Supp. 3. In Fig. 4 (b), we observe that our elasticity-scaling strategy is effective in stabilizing the architecture search with the precise latency constraint.

Algorithm 1. Elasticity-scaling Strategy

1: Derive a seed network sn from the supernet A.
2: **for** $i = 3, \ldots, 8$ **do**
3:    Find the largest γ such that $LAT(\gamma \cdot sn_{i:8}) \leq lat_{\text{target}}$.
4:    Set $sn = \gamma \cdot sn_{i:8}$.
5: **end for**
6: Put sn back to the supernet A.
7: **return** A;

Overall Algorithm. Our Three-Freedom NAS (TF-NAS) contains all above components: the bi-sampling search algorithm, the sink-connecting search space and the elasticity-scaling strategy. It finds latency-constrained architectures from the supernet (Table 1) by solving the following bi-level problem:

$$\min_{\alpha,\beta} L_{\text{val}}\left(\omega^*, \alpha, \beta\right) + \lambda C\left(LAT(\alpha, \beta)\right) \tag{7}$$

$$\text{s.t. } \omega^* = \arg\min_{\omega} L_{\text{t\_g}}\left(\omega, \alpha, \beta\right) + L_{\text{t\_r}}\left(\omega, \alpha, \beta\right) \tag{8}$$

where $L_{\text{t\_g}}$ and $L_{\text{t\_r}}$ denote the training losses for Gumbel Softmax based sampling and random sampling, respectively. The latency-constrained objectives in [4,33] do not employ the target latency, leading to imprecise latency compared with the target one. Therefore, we introduce a new objective that explicitly contains the target latency lat_{target}:

$$C\left(LAT(\alpha, \beta)\right) = \max\left(\frac{LAT(\alpha, \beta)}{lat_{\text{target}}} - 1, 0\right) \tag{9}$$

The continuous relaxations of α and β are based on Eq. (3)–(4) and Eq. (5)–(6), respectively. We employ elasticity-scaling after each searching epoch, making it barely increase the search time. After searching, the best architecture is derived from the supernet based on α and β, where the strongest operation in each layer and the strongest depth in each stage are chosen.

4   Experiments

4.1   Dataset and Settings

All the experiments are conducted on ImageNet [9] under the mobile setting. Similar with [3], the latency is measured with a batch size of 32 on a Titan RTX GPU. We set the number of threads for OpenMP to 1 and use Pytorch1.1+cuDNN7.6.0 to measure the latency. Before searching, we pre-build a latency look up table as described in [3,33]. To reduce the search time, we choose 100 classes from the original 1000 classes to train our supernet. The supernet is trained for 90 epochs, where the first 10 epochs do not update the architecture distribution parameters. This procedure takes about 1.8 days on 1 Titan RTX GPU. After searching, the derived architecture is trained from scratch on the whole ImageNet training set. For fair comparison, we train it for 250 epochs with standard data augmentation [4], in which no auto-augmentation or mixup is used. More experimental details are provided in Supp. 2.

4.2   Comparisons with Current SOTA

We compare TF-NAS with various manually designed and automatically searched architectures. According to the latency, we divide them into four groups. For each group, we set a target latency and search an architecture. Totally, there are four latency settings, including 18ms, 15ms, 12ms and 10ms, and the final architectures are named as TF-NAS-A, TF-NAS-B, TF-NAS-C and TF-NAS-D, respectively. The comparisons are presented in Table 2. There is a slight latency error for each model. As shown in [4], the error mainly comes from the slight difference between the pre-built lookup table and the actual inference latency.

As shown in Table 2, our TF-NAS-A achieves 76.9% top-1 accuracy, which is better than NASNet-A [40] (+2.9%), PC-DARTS [36] (+1.1%), MixNet-S [32] (+1.1%) and EfficientNet-B0 [31] (+0.6%). For the GPU latency, TF-NAS-A is 6.2ms, 2.15ms, 1.83ms and 1.23ms better than NASNet-A, MdeNAS, PC-DARTS, MixNet-S and EfficientNet-B0, respectively. In the second group, our TF-NAS-B obtains 76.3% top-1 accuracy with 15.06ms. It exceeds the micro search methods (DARTS [22], DGAS [11], SETN [10], CARS-I [37]) by an average of 2.1%, and the macro search methods (SCARLET-C [6], DenseNAS-Large [12]) by an average of 0.5%. For the 12ms latency group, our TF-NAS-C is superior to ShuffleNetV1 2.0x [38], AtomNAS-A [24], FBNet-C [33] and Proxy-lessNAS (GPU) [4] both in accuracy and latency. Besides, it is comparable with MobileNetV3 [16] and MnasNet-A1 [30]. Note that MnasNet-A1 is trained for more epochs than our TF-NAS-C (350 vs 250). Obviously, training longer makes an architecture generalize better [15]. In the last group, our TF-NAS-D achieve 74.2% top-1 accuracy, outperforming MobileNetV1 [17] (+3.6%), ShuffleNetV1 1.5x [38] (+2.6%) and FPNASNet [7] (+0.9%) by large margins.

Further to investigate the impact of the SE module, we remove SE from our candidate operations and search new architectures based on the four latency settings. The result architectures are marked as TF-NAS-A-wose, TF-NAS-B-wose, TF-NAS-C-wose and TF-NAS-D-wose. As shown in Table 2, they obtain 76.5%, 76.0%, 75.0% and 74.0% top-1 accuracy, respectively, which are competitive with or even superior to the previous state-of-the-arts. Due to the page limitation, more results are presented in our supplementary materials.

4.3 Analyses of Bi-sampling Search Algorithm

As described in Sect. 3.3, our bi-sampling algorithm samples two paths in the forward pass. One path is based on Gumbel Softmax trick and the other is selected from the remaining paths. In this subsection, we set the target latency to 15ms and employ four types of sampling methods for the second path, including the Gumbel Softmax (Gumbel), the minimum architecture distribution parameter ($\min \alpha^l$), the maximum architecture distribution parameter ($\max \alpha^l$) and the random sampling (Random). As shown in Table 3, compared with other methods, random sampling achieves the best top-1 accuracy. As a consequence, we employ random sampling in our bi-sampling search algorithm. Another interesting observation is that *Gumbel+Gumbel* and *Gumbel*+$\max \alpha^l$ are inferior to one path *Gumbel* sampling strategy. This is due to the fact that both *Gumbel+Gumbel* and *Gumbel*+$\max \alpha^l$ will exacerbate the operation collapse phenomenon, leading to inferior architectures. Compared with one path *Gumbel* sampling, our bi-sampling algorithm increases the search time by 0.3 GPU day, but makes a significant improvement in top-1 accuracy (76.3% vs 75.8%).

Table 2. Comparisons with state-of-the-art architectures on the ImageNet classification task. For the competitors, we directly cite the FLOPs, the training epochs, the search time and the top-1 accuracy from their original papers or official codes. For the GPU latency, we measure it with a batch size of 32 on a Titan RTX GPU.

Architecture	Top-1 Acc(%)	GPU latency	FLOPs (M)	Training epochs	Search time (GPU days)	Venue
NASNet-A [40]	74.0	24.23ms	564	-	2,000	CVPR'18
PC-DARTS [36]	75.8	20.18ms	597	250	3.8	ICLR'20
MixNet-S [32]	75.8	19.86ms	256	-	-	BMVC'19
EfficientNet-B0 [31]	76.3	19.26ms	390	350	-	ICML'19
TF-NAS-A-wose (Ours)	76.5	18.07ms	504	250	1.8	-
TF-NAS-A (Ours)	**76.9**	**18.03ms**	457	250	1.8	-
DARTS [22]	73.3	17.53ms	574	250	4	ICLR'19
DGAS [11]	74.0	17.23ms	581	250	0.21	CVPR'19
SETN [10]	74.3	17.42ms	600	250	1.8	ICCV'19
MobileNetV2 1.4x [28]	74.7	16.18ms	585	-	-	CVPR'18
CARS-I [37]	75.2	17.80ms	591	250	0.4	CVPR'20
SCARLET-C [6]	75.6	15.09ms	280	-	12	ArXiv'19
DenseNAS-Large [12]	76.1	15.71ms	479	240	2.67	CVPR'20
TF-NAS-B-wose (Ours)	76.0	15.09ms	433	250	1.8	-
TF-NAS-B (Ours)	**76.3**	**15.06ms**	361	250	1.8	-
ShuffleNetV1 2.0x [38]	74.1	14.82ms	524	240	-	CVPR'18
AtomNAS-A [24]	74.6	12.21ms	258	350	-	ICLR'20
FBNet-C [33]	74.9	12.86ms	375	360	9	CVPR'19
ProxylessNAS (GPU) [4]	75.1	12.02ms	465	300	8.3	ICLR'18
MobileNetV3 [16]	75.2	12.36ms	219	-	-	ICCV'19
MnasNet-A1 [30]	75.2	11.98ms	312	350	288	CVPR'18
TF-NAS-C-wose (Ours)	75.0	12.06ms	315	250	1.8	-
TF-NAS-C (Ours)	**75.2**	**11.95ms**	284	250	1.8	-
MobileNetV1 [17]	70.6	**9.73ms**	569	-	-	ArXiv'17
ShuffleNetV1 1.5x [38]	71.6	10.84ms	292	240	-	CVPR'18
FPNASNet [7]	73.3	11.60ms	300	-	0.83	ICCV'19
TF-NAS-D-wose (Ours)	74.0	10.10ms	286	250	1.8	-
TF-NAS-D (Ours)	**74.2**	10.08ms	219	250	1.8	-

4.4   Analyses of Sink-connecting Search Space

As mentioned in Sect. 3.3, skip operation has a special role in depth-level search. Ensuring the mutual exclusion between skip and other candidate operations, as well as eliminating the architecture redundancy are important stability factors for the architecture search procedure. In this subsection, we set the target latency to 15ms and compare our sink-connecting search space with the other two depth-level search spaces. The results are presented in Table 4, where "skip in candidates" means adding the skip operation in the candidates (Fig. 3(a)), and "skip out candidates" denotes putting the skip operation independent of the candidates (Fig. 3(b)). Obviously, our "sink-connecting" achieves the best top-1 accuracy, demonstrating its effectiveness in finding accurate architectures

Table 3. Comparisons with different sampling methods for the second path in bi-sampling search algorithm.

Sampling	Top-1 Acc(%)	GPU latency	FLOPs(M)	Search time
Gumbel	75.8	15.05ms	374	1.5 days
Gumbel+Gumbel	75.7	15.04ms	371	1.8 days
Gumbel+$\min \alpha^l$	76.0	15.11ms	368	1.8 days
Gumbel+$\max \alpha^l$	75.5	14.92ms	354	1.8 days
Gumbel+Random	76.3	15.06ms	361	1.8 days

during searching. The "skip out candidates" beats the "skip in candidates" by about 0.5% top-1 accuracy, and the "sink-connecting" is 0.2% higher than the "skip out candidates". The former achieves more improvement than the later, indicating that the mutual exclusion between skip and other operations is more important than the architecture redundancy.

Table 4. Comparisons with different depth-level search spaces.

Method	Mutual exclusion	Architecture redundancy	Top-1 Acc(%)	GPU latency	FLOPs(M)
Skip in candidates	×	×	75.6	15.10ms	384
Skip out candidates	√	×	76.1	15.07ms	376
Sink-connecting	√	√	76.3	15.06ms	361

4.5   Analyses of Elasticity-scaling Strategy

The key to search latency-constrained architectures is the differentiable latency objective $C\left(LAT(\alpha, \beta)\right)$ in Eq. (7). Previous methods [4,33] employ diverse latency objectives with one or two hyperparameters. We list them in Table 5 and name them as C1 and C2, respectively. By tuning the hyperparameters, both C1 and C2 can be trade-off between the accuracy and the latency. We set the target latency to 15ms and directly employ C1 and C2 (without elasticity-scaling strategy) to search architectures. We try our best to fine-tune the hyperparameters in C1 and C2, so that the searched architectures conform to the latency constraint as much as possible. The search procedure is repeated 5 times for each latency objective, and we plot the average latencies of the derived architectures during searching (orange lines in Fig. 5(a)-(b)). It is obvious that both C1 and C2 cannot reach the target latency before the first 50 epochs. After that, the architecture searched by C1 fluctuates down and up around the target latency, but the architecture searched by C2 always exceeds the target latency. We also plot the results of our proposed latency objective Eq. (9) (orange line in Fig. 4(b)) and find it is more precise than C1 and C2 after the first 30 epochs. The reason is that the target latency term is explicitly employed in our latency objective.

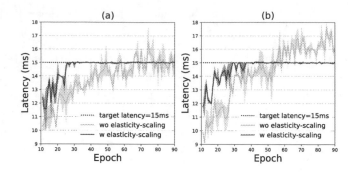

Fig. 5. (a): The searched latency by C1 w/wo elasticity-scaling. (b): The searched latency by C2 w/wo elasticity-scaling. All the search procedures are repeated 5 times, and we plot the mean, the maximum and the minimum. Zoom in for better view. (Color figure online)

The proposed elasticity-scaling strategy is the vital component in our TF-NAS to ensure the searched architectures precisely satisfy the target latency. By employing it, all the objectives are able to quickly search latency-satisfied architectures (blue lines in Fig. 4(b), Fig. 5(a) and Fig. 5(b)), demonstrating the effectiveness and the versatility of our elasticity-scaling strategy. Furthermore, we also evaluate the searched architectures based on C1, C2 and our proposed objective with and without elasticity-scaling. As shown in Table 5, our method achieves the best top-1 accuracy at 15ms latency constraint, which is slightly superior to C2 and beats C1 by a large margin no matter with or without elasticity-scaling. Therefore, explicitly introducing the target latency into the latency-constrained objective not only stabilizes large latency changes but also facilitates more accurate architecture discovery. Another observation is that under the similar backbone, the searched architectures with less/greater latencies than the target usually obtain lower/higher top-1 accuracies, especially when the latency gap is large. For example, C1 with elasticity-scaling achieves 75.9% top-1/15.05ms, which beats its counterpart without elasticity-scaling (75.6% top-1/14.32ms) by 0.3% top-1 accuracy and the latency gap is approximate 0.7ms.

Table 5. Comparisons with different latency objectives w/wo elasticity-scaling.

Name	Formulation	Elasticity-scaling	Top-1 Acc(%)	GPU Latency
C1 [33]	$\lambda_1 \log \left[(LAT(\alpha,\beta)) \right]^{\lambda_2}$	×	75.6	14.32ms
		√	75.9	15.05ms
C2 [4]	$\lambda_1 \left(LAT(\alpha,\beta) \right)$	×	76.2	15.76ms
		√	76.1	15.08ms
Ours	$\lambda_1 \max \left(\frac{LAT(\alpha,\beta)}{lat_{\text{target}}} - 1, 0 \right)$	×	76.3	15.28ms
		√	76.3	15.06ms

5   Conclusion

In this paper, we have proposed Three-Freedom NAS (TF-NAS) to seek an architecture with good accuracy as well as precise latency on the target devices. For operation-level, the proposed bi-sample search algorithm moderates the operation collapse in Gumbel Softmax relaxation. For depth-level, a novel sink-connecting search space is defined to address the mutual exclusion between skip operation and other candidate operations, as well as architecture redundancy. For width-level, an elasticity-scaling strategy progressively shrinks or expands the width of operations, contributing to precise latency constraint in a fine-grained manner. Benefiting from investigating the three freedoms of differentiable NAS, our TF-NAS achieves state-of-the-art performance on ImageNet classification task. Particularly, the searched TF-NAS-A achieves 76.9% top-1 accuracy with less latency and training epochs.

Acknowledgement. This work is partially funded by Beijing Natural Science Foundation (Grant No. JQ18017) and Youth Innovation Promotion Association CAS (Grant No. Y201929).

References

1. Baker, B., Gupta, O., Naik, N., Raskar, R.: Designing neural network architectures using reinforcement learning. arXiv preprint arXiv:1611.02167 (2017)
2. Bender, G., Kindermans, P., Zoph, B., Vasudevan, V., Le, Q.V.: Understanding and simplifying one-shot architecture search. In: International Conference on Machine Learning, pp. 550–559 (2018)
3. Cai, H., Gan, C., Han, S.: Once for all: train one network and specialize it for efficient deployment. arXiv preprint arXiv:1908.09791 (2019)
4. Cai, H., Zhu, L., Han, S.: Proxylessnas: direct neural architecture search on target task and hardware. arXiv preprint arXiv:1812.00332 (2019)
5. Chen, X., Xie, L., Wu, J., Tian, Q.: Progressive differentiable architecture search: bridging the depth gap between search and evaluation. In: Proceedings of the IEEE International Conference on Computer Vision, pp. 1294–1303 (2019)
6. Chu, X., Zhang, B., Li, J., Li, Q., Xu, R.: Scarletnas: bridging the gap between scalability and fairness in neural architecture search. arXiv preprint arXiv:1908.06022 (2019)
7. Cui, J., Chen, P., Li, R., Liu, S., Shen, X., Jia, J.: Fast and practical neural architecture search. In: Proceedings of the IEEE International Conference on Computer Vision, pp. 6509–6518 (2019)
8. Dai, X., et al.: Chamnet: towards efficient network design through platform-aware model adaptation. In: Proceedings of the IEEE Conference on Computer Vision and Pattern Recognition, pp. 11398–11407 (2019)
9. Deng, J., Dong, W., Socher, R., Li, L., Li, K., Li, F.: Imagenet: a large-scale hierarchical image database. In: 2009 IEEE Conference on Computer Vision and Pattern Recognition, pp. 248–255. IEEE (2009)
10. Dong, X., Yang, Y.: One-shot neural architecture search via self-evaluated template network. In: Proceedings of the IEEE International Conference on Computer Vision, pp. 3681–3690 (2019)

11. Dong, X., Yang, Y.: Searching for a robust neural architecture in four GPU hours. In: Proceedings of the IEEE Conference on Computer Vision and Pattern Recognition, pp. 1761–1770 (2019)
12. Fang, J., Sun, Y., Zhang, Q., Li, Y., Liu, W., Wang, X.: Densely connected search space for more flexible neural architecture search. In: Proceedings of the IEEE/CVF Conference on Computer Vision and Pattern Recognition, pp. 10628–10637 (2020)
13. Gordon, A., et al.: Morphnet: fast & simple resource-constrained structure learning of deep networks. In: Proceedings of the IEEE Conference on Computer Vision and Pattern Recognition, pp. 1586–1595 (2018)
14. He, K., Zhang, X., Ren, S., Sun, J.: Deep residual learning for image recognition. In: Proceedings of the IEEE Conference on Computer Vision and Pattern Recognition, pp. 770–778 (2016)
15. Hoffer, E., Hubara, I., Soudry, D.: Train longer, generalize better: closing the generalization gap in large batch training of neural networks. In: Advances in Neural Information Processing Systems, pp. 1731–1741(2017)
16. Howard, A., et al.: Searching for mobilenetv3. In: Proceedings of the IEEE International Conference on Computer Vision, pp. 1314–1324 (2019)
17. Howard, A.G., et al.: Mobilenets: efficient convolutional neural networks for mobile vision applications. arXiv preprint arXiv:1704.04861 (2017)
18. Hu, J., Shen, L., Sun, G.: Squeeze-and-excitation networks. In: Proceedings of the IEEE Conference on Computer Vision and Pattern Recognition, pp. 7132–7141 (2018)
19. Huang, G., Liu, Z., van der Maaten, L., Weinberger, K.Q.: Densely connected convolutional networks. In: Proceedings of the IEEE Conference on Computer Vision and Pattern Recognition, pp. 4700–4708 (2017)
20. Liang, H., et al.: DARTS+: improved differentiable architecture search with early stopping. arXiv preprint arXiv:1909.06035 (2019)
21. Liu, C., et al.: Progressive neural architecture search. In: Proceedings of the European Conference on Computer Vision (ECCV), pp. 19–34 (2018)
22. Liu, H., Simonyan, K., Yang, Y.: Darts: differentiable architecture search. arXiv preprint arXiv:1806.09055 (2019)
23. Ma, N., Zhang, X., Zheng, H., Sun, J.: Shufflenet V2: practical guidelines for efficient CNN architecture design. In: Proceedings of the European Conference on Computer Vision (ECCV), pp. 116–131 (2018)
24. Mei, J., et al.: Atomnas: fine-grained end-to-end neural architecture search. arXiv preprint arXiv:1912.09640 (2020)
25. Pham, H., Guan, M.Y., Zoph, B., Le, Q.V., Dean, J.: Efficient neural architecture search via parameter sharing. arXiv preprint arXiv:1802.03268 (2018)
26. Real, E., Aggarwal, A., Huang, Y., Le, Q.V.: Regularized evolution for image classifier architecture search. In: Proceedings of the AAAI Conference on Artificial Intelligence. **33**, 4780-4789 (2019)
27. Real, E., et al.: Large-scale evolution of image classifiers. arXiv preprint arXiv:1703.01041 (2017)
28. Sandler, M., Howard, A.G., Zhu, M., Zhmoginov, A., Chen, L.: Mobilenetv 2: inverted residuals and linear bottlenecks. In: Proceedings of the IEEE Conference on Computer Vision and Pattern Recognition, pp. 4510–4520 (2018)
29. Szegedy, C., Vanhoucke, V., Ioffe, S., Shlens, J., Wojna, Z.: Rethinking the inception architecture for computer vision. In: Proceedings of the IEEE Conference on Computer Vision and Pattern Recognition, pp. 2818–2826 (2016)

30. Tan, M., et al.: Mnasnet: platform-aware neural architecture search for mobile. In: Proceedings of the IEEE Conference on Computer Vision and Pattern Recognition, pp. 2820–2828 (2019)
31. Tan, M., Le, Q.V.: Efficientnet: rethinking model scaling for convolutional neural networks. arXiv preprint arXiv:1905.11946 (2019)
32. Tan, M., Le, Q.V.: Mixconv: mixed depthwise convolutional kernels. arXiv preprint arXiv:1907.09595 (2019)
33. Wu, B., et al.: FBNet: hardware-aware efficient convnet design via differentiable neural architecture search. In: Proceedings of the IEEE Conference on Computer Vision and Pattern Recognition, pp. 10734–10742 (2019)
34. Xie, S., Zheng, H., Liu, C., Lin, L.: SNAS: stochastic neural architecture search. arXiv preprint arXiv:1812.09926 (2019)
35. Xiong, Y., Mehta, R., Singh, V.: Resource constrained neural network architecture search: will a submodularity assumption help? In: Proceedings of the IEEE International Conference on Computer Vision, pp. 1901–1910 (2019)
36. Xu, Y., et al.: PC-DARTS: partial channel connections for memory-efficient differentiable architecture search. arXiv preprint arXiv:1907.05737 (2020)
37. Yang, Z., et al.: Cars: continuous evolution for efficient neural architecture search. In: Proceedings of the IEEE/CVF Conference on Computer Vision and Pattern Recognition, pp. 1829–1838 (2020)
38. Zhang, X., Zhou, X., Lin, M., Sun, J.: Shufflenet: an extremely efficient convolutional neural network for mobile devices. In: Proceedings of the IEEE Conference on Computer Vision and Pattern Recognition, pp. 6848–6856 (2018)
39. Zoph, B., Le, Q.V.: Neural architecture search with reinforcement learning. arXiv preprint arXiv:1611.01578 (2017)
40. Zoph, B., Vasudevan, V., Shlens, J., Le, Q.V.: Learning transferable architectures for scalable image recognition. In: Proceedings of the IEEE Conference on Computer Vision and Pattern Recognition, pp. 8697–8710 (2018)

Associative3D: Volumetric Reconstruction from Sparse Views

Shengyi Qian$^{(\boxtimes)}$ (ID), Linyi Jin(ID), and David F. Fouhey(ID)

University of Michigan, Ann Arbor, USA
{syqian,jinlinyi,fouhey}@umich.edu

Abstract. This paper studies the problem of 3D volumetric reconstruction from two views of a scene with an unknown camera. While seemingly easy for humans, this problem poses many challenges for computers since it requires simultaneously reconstructing objects in the two views while also figuring out their relationship. We propose a new approach that estimates reconstructions, distributions over the camera/object and camera/camera transformations, as well as an inter-view object affinity matrix. This information is then jointly reasoned over to produce the most likely explanation of the scene. We train and test our approach on a dataset of indoor scenes, and rigorously evaluate the merits of our joint reasoning approach. Our experiments show that it is able to recover reasonable scenes from sparse views, while the problem is still challenging. Project site: https://jasonqsy.github.io/Associative3D.

Keyword: 3D reconstruction

1 Introduction

How would you make sense of the scene in Fig. 1? After rapidly understanding the individual pictures, one can fairly quickly attempt to match the objects in each: the TV on the left in image A must go with the TV on the right in image B, and similarly with the couch. Therefore, the two chairs, while similar, are not actually the same object. Having pieced this together, we can then reason that the two images depict the same scene, but seen with a 180° change of view and infer the 3D structure of the scene. Humans have an amazing ability to reason about the 3D structure of scenes, even with as little as two sparse views with an unknown relationship. We routinely use this ability to understand images taken at an event, look for a new apartment on the Internet, or evaluate possible hotels (e.g., for ECCV). The goal of this paper is to give the same ability to computers.

Unfortunately, current techniques are not up to this challenge of volumetric reconstruction given two views from unknown cameras: this approach requires both reconstruction and pose estimation. Classic methods based on correspondence [9,20] require many more views in practice and cannot make inferences

---

S. Qian and L. Jin – Equal contribution.

© Springer Nature Switzerland AG 2020
A. Vedaldi et al. (Eds.): ECCV 2020, LNCS 12360, pp. 140–157, 2020.
https://doi.org/10.1007/978-3-030-58555-6_9

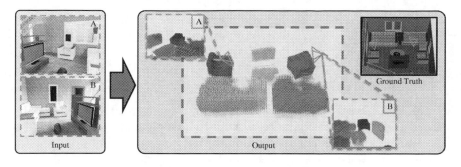

Fig. 1. Given two views from unknown cameras, we aim to extract a coherent 3D space in terms of a set of volumetric objects placed in the scene. We represent the scene with a factored representation [49] that splits the scene into per-object voxel grids with a scale and pose.

about unseen parts of the scene (i.e., what the chair looks like from behind) since this requires some form of learning. While there has been success in learning-based techniques for this sort of object reconstructions [7,17,27,49], it is unknown how to reliably stitch together the set of reconstructions into a single coherent story. Certainly there are systems that can identify pose with respect to a fixed scene [26] or a pair of views [15]; these approaches, however cannot reconstruct.

This paper presents a learning-based approach to this problem, whose results are shown in Fig. 1. The system can take two views with unknown relationship, and produce a 3D scene reconstruction for both images jointly. This 3D scene reconstruction comprises a set of per-object reconstructions rigidly placed in the scene with a pose as in [27,30,49]. Since the 3D scene reconstruction is the union of the posed objects, getting the *3D scene reconstruction* correct requires getting both the *3D object reconstruction* right as well as correctly identifying *3D object pose*. Our key insight is that jointly reasoning about objects and poses improves the results. Our method, described in Sect. 3, predicts evidence including: (a) voxel reconstructions for each object; (b) distributions over rigid body transformations between cameras and objects; and (c) an inter-object affinity for stitching. Given this evidence, our system can stitch them together to find the most likely reconstruction. As we empirically demonstrate in Sect. 4, this joint reasoning is crucial – understanding each image independently and then estimating a relative pose performs substantially worse compared to our approach. These are conducted on a challenging and large dataset of indoor scenes. We also show some common failure modes and demonstrate transfer to NYUv2 [44] dataset.

Our primary contributions are: (1) Introducing a novel problem – volumetric scene reconstruction from two unknown sparse views; (2) Learning an inter-view object affinity to find correspondence between images; (3) Our joint system, including the stitching stage, is better than adding individual components.

2   Related Work

The goal of this work is to take two views from cameras related by an unknown transformation and produce a single volumetric understanding of the scene. This touches on a number of important problems in computer vision ranging from the estimation of the pose of objects and cameras, full shape of objects, and correspondence across views. Our approach deliberately builds heavily on these works and, as we show empirically, our success depends crucially on their fusion.

This problem poses severe challenges for classic correspondence-based approaches [20]. From a purely geometric perspective, we are totally out of luck: even if we can identify the position of the camera via epipolar geometry and wide baseline stereo [36,39], we have no correspondence for most objects in Fig. 1 that would permit depth given known baseline, let alone another view that would help lead to the understanding of the full shape of the chair.

Recent work has tackled identifying this full volumetric reconstruction via learning. Learning-based 3D has made significant progress recently, including 2.5D representations [5,14,29,51], single object reconstruction [8,19,41,52,55], and scene understanding [6,12,23,32,33]. Especially, researchers have developed increasingly detailed volumetric reconstructions beginning with objects [7,17,18] and then moving to scenes [27,30,37,49] as a composition of object reconstructions that have a pose with respect to the camera. Focusing on full volumetric reconstruction, our approach builds on this progression, and creates an understanding that is built upon jointly reasoning over parses of two scenes, affinities, and relative poses; as we empirically show, this produces improvements in results. Of these works, we are most inspired by Kulkarni et al. [27] in that it also reasons over a series of relative poses; our work builds on top of this as a base inference unit and handles multiple images. We note that while we build on a particular approach to scenes [27] and objects [17], our approach is general.

While much of this reconstruction work is single-image, some is multiview, although usually in the case of an isolated object [7,24,25] or with hundreds of views [22]. Our work aims at the particular task of as little as two views, and reasons over multiple objects. While traditional local features [34] are insufficient to support reasoning over objects, semantic features are useful [2,13,50].

At the same time, there has been considerable progress in identifying the relative pose from images [1,15,26,35], RGB-D Scans [53,54] or video sequences [42,46,57]. Of these, our work is most related to learning-based approaches to identifying relative pose from RGB images, and semantic Structure-from-Motion [1] and SLAM [42], which make use of semantic elements to improve the estimation of camera pose. We build upon this work in our approach, especially work like RPNet [15] that directly predicts relative pose, although we do so with a regression-by-classification formulation that provides uncertainty. As we show

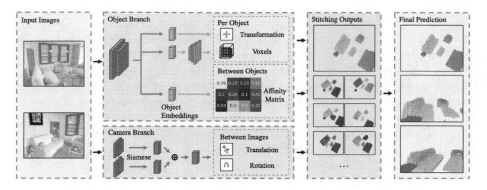

Fig. 2. Our approach. We pass the two RGB image inputs into two branches that extract evidence, which is then fused together to stitch a final result. Our first network, object branch, is a detection network following [27] that produces a set of objects in terms of voxels and a transformation into the scene. We also predict an object embedding which we can use to form an affinity matrix between objects across images. Our second network, camera branch, is a siamese network that predicts a distribution over translations and rotations between the cameras. Finally our **stitching stage** examines the evidence from the networks and produces a final prediction.

empirically, propagating this uncertainty forward lets us reason about objects and produce superior results to only focusing on pose.

3   Approach

The goal of the system is to map a pair of sparse views of a room to a full 3D reconstruction. As input, we assume a pair of images of a room. As output, we produce a set of objects represented as voxels, which are rigidly transformed and anisotropically scaled into the scene in a single coordinate frame. We achieve this with an approach, summarized in Fig. 2, that consists of three main parts: an object branch, a camera branch, and a stitching stage.

The output space is a factored representation of a 3D scene, similar to [27, 30,49]. Specifically, in contrast to using a single voxel-grid or mesh, the scene is represented as a set of per-object voxel-grids with a scale and pose that are placed in the scene. These can be converted to a single 3D reconstruction by taking their union, and so improving the 3D reconstruction can be done by either improving the per-object voxel grid or improving its placement in the scene.

The first two parts of our approach are two neural networks. An **object branch** examines each image and detects and produces single-view 3D reconstructions for objects in the camera's coordinate frame, as well as a per-object embedding that helps find the object in the other image. At the same time, an

camera branch predicts relative pose between images, represented as a *distribution* over a discrete set of rigid transformations between the cameras. These networks are trained separately to minimize complexity.

The final step, a **stitching stage**, combines these together. The output of the two networks gives: a collection of objects per image in the image's coordinate frame; a cross-image affinity which predicts object correspondence in two views; and a set of likely transformations from one camera to the other. The stitching stage aims to select a final set of predictions minimizing an objective function that aims to ensure that similar objects are in the same location, the camera pose is likely, etc. Unlike the first two stages, this is an optimization rather than a feedforward network.

3.1 Object Branch

The goal of our object branch is to take an image and produce a set of reconstructed objects in the camera's coordinate frame as well as an embedding that lets us match across views. We achieve this by extending 3D-RelNet [27] and adjust it as little as possible to ensure fair comparisons. We refer the reader for a fuller explanation in [27,49], but briefly, these networks act akin to an object detector like Faster-RCNN [40] with additional outputs. As input, 3D-RelNet takes as input an image and a set of 2D bounding box proposals, and maps the image through convolutional layers to a feature map, from which it extracts per-bounding box convolutional features. These features pass through fully connected layers to predict: a detection score (to suppress bad proposals), voxels (to represent the object), and a transformation to the world frame (represented by rotation, scale, and translation and calculated via both per-object and pairwise poses). We extend this to also produce an n-dimensional embedding $\mathbf{e} \in \mathbb{R}^n$ on the unit sphere (i.e., $||\mathbf{e}||_2^2 = 1$) that helps associate objects across images.

We use and train the embedding by creating a cross-image affinity matrix between objects. Suppose the first and second images have N and M objects each with embeddings $\mathbf{e}_i$ and $\mathbf{e}'_j$ respectively. We then define our affinity matrix $\mathbf{A} \in \mathbb{R}^{N \times M}$ as

$$\mathbf{A}_{i,j} = \sigma(k\mathbf{e}_i^T \mathbf{e}'_j) \tag{1}$$

where σ is the sigmoid/logistic function and where $k = 5$ scales the output. Ideally, $A_{i,j}$ should indicate whether objects i and j are the same object seen from a different view, where $A_{i,j}$ is high if this is true and low otherwise.

We train this embedding network using ground-truth bounding box proposals so that we can easily calculate a ground-truth affinity matrix $\hat{\mathbf{A}}$. We then minimize L_{aff}, a balanced mean-square loss between $\mathbf{A}$ and $\hat{\mathbf{A}}$: if all positive labels are $(i,j) \in \mathcal{P}$, and all negative labels are $(i,j) \in \mathcal{N}$, then the loss is

$$L_{\text{aff}} = \frac{1}{|\mathcal{P}|} \sum_{(i,j)\in\mathcal{P}} (A_{ij} - \hat{A}_{ij})^2 + \frac{1}{|\mathcal{N}|} \sum_{(i,j)\in\mathcal{N}} (A_{ij} - \hat{A}_{ij})^2. \tag{2}$$

which balances positive and negative labels (since affinity is imbalanced).

3.2   Camera Branch

Our camera branch aims to identify or narrow down the possible relationship between the two images. We approach this by building a siamese network [3] that predicts the relative camera pose T_c between the two images. We use ResNet-50 [21] to extract features from two input images. We concatenate the output features and then use two linear layers to predict the translation and rotation.

We formulate prediction of rotation and translation as a classification problem to help manage the uncertainty in the problem. We found that propagating uncertainty (via top predictions) was helpful: a single feedforward network suggests likely rotations and a subsequent stage can make a more detailed assessment in light of the object branch's predictions. Additionally, even if we care about only one output, we found regression-by-classification to be helpful since the output tended to have multiple modes (e.g., being fairly certain of the rotation modulo 90° by recognizing that both images depict a cuboidal room). Regression tends to split the difference, producing predictions which satisfy neither mode, while classification picks one, as observed in [28,49].

We cluster the rotation and translation vectors into 30 and 60 bins respectively, and predict two multinomial distributions over them. Then we minimize the cross entropy loss. At test time, we select the cartesian product of the top 3 most likely bins for rotation and top 10 most likely bins for translation as the final prediction results. The results are treated as proposals in the next section.

3.3   Stitching Object and Camera Branches

Once we have run the object and camera branches, our goal is to then produce a single stitched result. As input to this step, our object branch gives: for view 1, with N objects, the voxels $V_1, \ldots, V_N$ and transformations $T_1, \ldots, T_N$; and similarly, for M objects in view 2, the voxels $V'_1, \ldots, V'_M$ and transformations $T'_1, \ldots, T'_M$; and a cross-view affinity matrix $\mathbf{A} \in [0,1]^{N \times M}$. Additionally, we have a set of potential camera transformations $P_1, \ldots, P_F$ between two views.

The goal of this section is to integrate this evidence to find a final cross-camera pose P and correspondence $\mathbf{C} \in \{0,1\}^{M \times N}$ from view 1 to view 2. This correspondence is one-to-one and has the option to ignore an object (i.e., $\mathbf{C}_{i,j} = 1$ if and only if i and j are in correspondence and for all i, $\sum_j \mathbf{C}_{i,j} \leq 1$, and similarly for $\mathbf{C}^T$).

We cast this as a minimization problem over P and $\mathbf{C}$ including terms in the objective function that incorporate the above evidence. The cornerstone term is one that integrates all the evidence to examine the quality of the stitch, akin to trying and seeing how well things match up under a camera hypothesis. We implement this by computing the distance $\mathcal{L}_D$ between corresponding object voxels according to $\mathbf{C}$, once the transformations are applied, or:

$$\mathcal{L}_D = \frac{1}{|\mathbf{C}|_1} \sum_{(i,j) \text{ s.t. } \mathbf{C}_{i,j}=1} D(P(T_i(V_i)), T'_j(V'_j)). \tag{3}$$

Here, D is the chamfer distance between points on the edges of each shape, as defined in [38,43], or for two point clouds X and Y:

$$D(X,Y) = \frac{1}{|X|} \sum_{x \in X} \min_{y \in Y} ||x - y||_2^2 + \frac{1}{|Y|} \sum_{y \in Y} \min_{x \in X} ||x - y||_2^2. \qquad (4)$$

Additionally, we have terms that reward making $\mathbf{C}$ likely according to our object and image networks, or: the sum of similarities between corresponding objects according to the affinity matrix $\mathbf{A}$, $\mathcal{L}_S = \sum_{(i,j),\mathbf{C}_{i,j}=1}(1 - A_{i,j})$; as well as the probability of the camera pose transformation P from the image network $\mathcal{L}_P = (1 - Pr(P))$. Finally, to preclude trivial solutions, we include a term rewarding minimizing the number of un-matched objects, or $\mathcal{L}_U = \min(M, N) - |\mathbf{C}|_1$. In total, our objective function is the sum of these terms, or:

$$\min_{P,\mathbf{C}} \quad \mathcal{L}_D + \lambda_P \mathcal{L}_P + \lambda_S \mathcal{L}_S + \lambda_U \mathcal{L}_U. \qquad (5)$$

The search space is intractably large, so we optimize the objective function by RANSAC-like search over the top hypotheses for P and feasible object correspondences. For each top hypothesis of P, we randomly sample K object correspondence proposals. Here we use $K = 128$. It is generally sufficient since the correspondence between two objects is feasible only if the similarity of them is higher than a threshold according to the affinity matrix. We use random search over object correspondences because the search space increases factorially between the number of objects in correspondence. Once complete, we average the translation and scale, and randomly pick one rotation and shape from corresponding objects. Averaging performs poorly for rotation since there are typically multiple rotation modes that cannot be averaged: a symmetric table is correct at either $0°$ or $180°$ but not at $90°$. Averaging voxel grids does not make sense since there are partially observable objects. We therefore pick one mode at random for rotation and shape. Details are available in the appendix.

4   Experiments

We now describe a set of experiments that aim to address the following questions: (1) how well does the proposed method work and are there simpler approaches that would solve the problem? and (2) how does the method solve the problem? We first address question (1) by evaluating the proposed approach compared to alternate approaches both qualitatively and by evaluating the full reconstruction quantitatively. We then address question (2) by evaluating individual components of the system. We focus on what the affinity matrix learns and whether the stitching stage can jointly improve object correspondence and relative camera pose estimation. Throughout, we test our approach on the SUNCG dataset [45,56], following previous work [27,30,49,53,56]. To demonstrate transfer to other data, we also show qualitative results on NYUv2 [44].

| Input Images | | Camera 1 | | Camera 2 | | Birdview | |
| Image 1 | Image 2 | Prediction | GT | Prediction | GT | Prediction | GT |

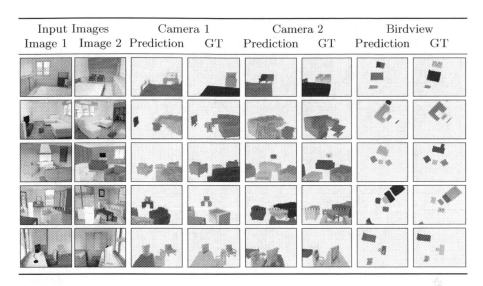

Fig. 3. Qualitative results on the SUNCG test set [45]. The final 3D predictions are shown in three different camera poses (1) the same camera as image 1; (2) the same camera as image 2; (3) a bird view to see all the objects in the whole scene. In the prediction, red/orange objects are from the left image, blue objects are from the right image, green/yellow objects are stitched. (Color figure online)

4.1 Experimental Setup

We train and do extensive evaluation on SUNCG [45] since it provides 3D scene ground truth including voxel representation of objects. There are realistic datasets such as ScanNet [10] and Matterport3D [4], but they only provide non-watertight meshes. Producing filled object voxel representation from non-watertight meshes remains an open problem. For example, Pix3D [47] aligns IKEA furniture models with images, but not all objects are labeled.

Datasets. We follow the 70%/10%/20% training, validation and test split of houses from [27]. For each house, we randomly sample up to ten rooms; for each room, we randomly sample one pair of views. Furthermore, we filter the validation and test set: we eliminate pairs where there is no overlapping object between views, and pairs in which all of one image's objects are in the other view (i.e., one is a proper subset of the other). We do not filter the training set since learning relative pose requires a large and diverse training set. Overall, we have 247532/1970/2964 image pairs for training, validation and testing, respectively. Following [49], we use six object classes - bed, chair, desk, sofa, table and tv.

Full-Scene Evaluation: Our output is a full-scene reconstruction, represented as a set of per-object voxel grids that are posed and scaled in the scene. A scene prediction can be totally wrong if one of the objects has correct shape while its translation is off by 2 m. Therefore, we quantify performance by treating the problem as a 3D detection problem in which we predict a series of 3D boxes and

Image 1	Image 2	Feedforward	NMS	Raw Affinity	**Ours**	GT

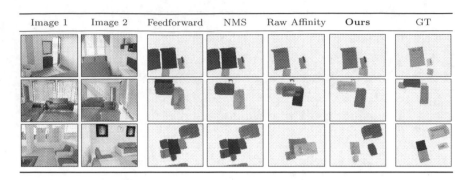

Fig. 4. Comparison between Associative3D and alternative approaches. **Row 1:** Associative3D fixes the incorrect top-1 relative camera pose in light of a single bed in the room. **Row 2**: NMS works when the relative camera pose is accurate. **Row 3:** Associative3D outperforms all alternative approaches in finding correspondence in object clutter.

voxel grids. This lets us evaluate which aspect of the problem currently hold methods back. Similar to [27], for each object, we define error metrics as follows:

- Translation ($\mathbf{t}$): Euclidean distance, or $\delta_t = ||t - \hat{t}||_2$, thresholded at $\delta_t = 1$m.
- Scale ($\mathbf{s}$): Average log difference in scaling factors, or $\delta_s = \frac{1}{3}\sum_{i=1}^{3}|\log_2(s_1^i) - \log_2(s_2^i)|$, thresholded at $\delta_s = 0.2$.
- Rotation ($\mathbf{R}$): Geodesic rotation distance, or $\delta_q = (2)^{-1/2}||\log(\mathbf{R}^T\hat{\mathbf{R}})||_F$, thresholded at $\delta_q = 30°$.
- Shape ($\mathbf{V}$): Following [48], we use F-score@0.05 to measure the difference between prediction and ground truth, thresholded at $\delta_V = 0.25$.

A prediction is a true positive only if all errors are lower than our thresholds. We calculate the precision-recall curve based on that and report average precision (AP). We also report AP for each single error metric.

Baselines. Since there is no prior work on this task, our experiments compare to ablations and alternate forms of our method. We use the following baseline methods, each of which tests a concrete hypothesis. (**Feedforward**): This method uses the object branch to recover single-view 3D scenes, and our camera branch to estimate the relative pose between different views. We ignore the affinity matrix and pick the top-1 relative pose predicted by the camera branch. There can be many duplicate objects in the output of this approach. This tests if a simple feedforward method is sufficient. (**NMS**): In addition to the feedforward approach, we perform non-maximum suppression on the final predictions. If two objects are close to each other, we merge them. This tests if a simple policy to merge objects would work. (**Raw Affinity**): Here, we use the predicted affinity matrix to merge objects based on top-1 similarity from the affinity matrix. This tests whether our stitching stage is necessary. (**Associative3D**): This is our complete method. We optimize the objective function by searching possible rotations, translations and object correspondence.

Table 1. We report the average precision (AP) in evaluation of the 3D detection setting. **All** means a prediction is a true positive only if all of translation, scale, rotation and shape are correct. **Shape**, **Translation**, **Rotation**, and **Scale** mean a prediction is a true positive when a single error is lower than thresholds. We include results on the whole test set, and top 25%, 50% and 75% examples ranked by single-view predictions.

Methods	All Examples					Top 25%	Top 50%	Top 75%
	All	Shape	Trans	Rot	Scale	All	All	All
Feedforward	21.2	22.5	31.7	28.5	26.9	41.6	34.6	28.6
NMS	21.1	23.5	31.9	29.0	27.2	42.0	34.7	28.7
Raw Affinity	15.0	24.4	26.3	28.2	25.9	28.6	23.5	18.9
Associative3D	**23.3**	**24.5**	**38.4**	**29.5**	**27.3**	**48.3**	**38.8**	**31.4**

4.2   Full Scene Evaluation

We begin by evaluating our full scene reconstruction. Our output is a set of per-object voxels that are posed and scaled in the scene. The quality of reconstruction of a single object is decided by both the voxel grids and the object pose.

First, we show qualitative examples from the proposed method in Fig. 3 as well as a comparison with alternate approaches in Fig. 4 on the SUNCG test set. The Feedforward approach tends to have duplicate objects since it does not know object correspondence. However, figuring out the camera pose and common objects is a non-trivial task. Raw Affinity does not work since it may merge objects based on their similarity, regardless of possible global conflicts. NMS works when the relative camera pose is accurate but cannot work when many objects are close to each other. Instead, Associative3D demonstrates the ability to jointly reason over reconstructions, object pose and camera pose to produce a reasonable explanation of the scene. More qualitative examples are available in the supplementary material.

We then evaluate our proposed approach quantitatively. In a factored representation [49], both *object* poses and shapes are equally important to the full *scene* reconstruction. For instance, the voxel reconstruction of a scene may have no overlap if all the shapes are right, but they are in the wrong place. Therefore, we formulate it as a 3D detection problem, as a prediction is a true positive only if all of translation, scale, rotation and shape are correct. However, 3D detection is a very strict metric. If the whole scene is slightly off in one aspect, we may have a very low AP. But the predicted scene may still be reasonable. We mainly use it quantify our performance.

Table 1 shows our performance compared with all three baseline methods. Our approach outperforms all of them, which verifies what we see in the qualitative examples. Moreover, the improvement mainly comes from that on translation. The translation-only AP is around 7 points better than Feedforward. Meanwhile, the improvement of NMS over Feedforward is limited. As we see in qualitative examples, it cannot work when many objects are close to each other. Finally, raw affinity is even worse than Feedforward, since raw affinity

Table 2. AUROC and rank correlation between the affinity matrix and *category*, *model*, *shape*, and *instance*, respectively. **Model | Category** means the ability of the affinity matrix to distinguish different models given the same category/semantic label.

	Category	Model | Category	Shape | Category	Instance | Model
AUROC	0.92	0.73	–	0.59
Correlation	0.72	0.33	0.34	0.14

may merge objects incorrectly. We will discuss why the affinity is informative, but top-1 similarity is not a good choice in Sect. 4.3.

We notice our performance gain over Feedforward and NMS is especially large when single-view predictions are reasonable. On top 25% examples which single-view prediction does a good job, Associative3D outperforms Feedforward and NMS by over 6 points. On top 50% examples, the improvement is around 4 points. It is still significant but slightly lower than that of top 25% examples. When single-view prediction is bad, our performance gain is limited since Associative3D is built upon it. We will discuss this in Sect. 4.5 as failure cases.

4.3   Inter-view Object Affinity Matrix

We then turn to evaluating how the method works by analyzing individual components. We start with the affinity matrix and study what it learns.

We have three non-mutually exclusive hypotheses: (1) **Semantic labels.** The affinity is essentially doing object recognition. After detecting the category of the object, it simply matches objects with the same category. (2) **Object shapes.** The affinity matches objects with similar shapes since it is constructed from the embedding vectors which are also used to generate shape voxels and the object pose. (3) **Correspondence.** Ideally, the affinity matrix should give us ground truth correspondence. It is challenging given duplicate objects in the scene. For example, people can have three identical chairs in their office. These hypotheses are three different levels the affinity matrix may learn, but they are not in conflict. Learning semantic labels do not mean the affinity does not learn anything about shapes.

We study this by examining a large number of pairs of objects and testing the relationship between affinity and known relationships (e.g., categories, model ids) using ground truth bounding boxes. We specifically construct three binary labels (same *category*, same *model*, same *instance*) and a continuous label shape similarity (namely F-score@0.05 [48]). When we evaluate shape similarity, we condition on the category to test if affinity distinguishes between different models of the same category, (e.g. chair). Similarly, we condition on the model when we evaluate instance similarity.

We compute two metrics: a binary classification metric that treats the affinity as a predictor of the label as well as a correlation that tests if a monotonic relationship exists between the affinity and the label. For binary classification, we

| Before | After | Before | After | Before | After |

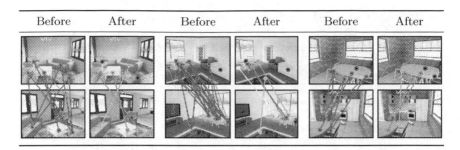

Fig. 5. Visualization of the stitching stage. The affinity matrix generates proposals of corresponding objects, and then the stitching stage removes outliers by inferring the most likely explanation of the scene.

use AUROC to evaluate the performance since it is invariant to class imbalance and has a natural interpretation. For correlation, we compute Spearman's rank correlation coefficient [58] between the affinity predictors and labels. This tests how well the relationship between affinity and each label (e.g., shape overlap) fits a monotonic function (1 is perfect agreement, 0 no agreement).

The results are shown in Table 2. Both the binary classification and the rank correlation show that the affinity matrix is able to distinguish different categories and objects of different shapes, but is sub-optimal in distinguishing the same instance. These results justify our stitching stage, which addresses the problem based on joint reasoning. It also explains why Raw Affinity underperforms all other baselines by a large margin in the full-scene evaluation. Additionally, the ability to distinguish categories and shapes provides important guidance to the stitching stage. For example, a sofa and bed are similar in 3D shapes. It is infeasible to distinguish them by simply looking at the chamfer distance, which can be distinguished by the affinity matrix.

4.4 Stitching Stage

We evaluate the stitching stage by studying two questions: (1) How well can it predict object correspondence? (2) Can it improve relative camera pose estimation? For example, if the top-1 relative pose is incorrect, could the stitching stage fix it by considering common objects in two views?

Object Correspondence. To answer the first question, we begin with qualitative examples in Fig. 5, which illustrate object correspondence before and after the stitching stage. Before our stitching stage, our affinity matrix has generated correspondence proposals based on their similarity. However, there are outliers since the affinity is sub-optimal in distinguishing the same instance. The stitching stage removes these outliers.

We evaluate object correspondence in the same setting as Sect. 4.3. Suppose the first and second images have N and M objects respectively. We then have $N \times M$ pairs. The pair is a positive example if and only if they are corresponding.

Table 3. Evaluation of object correspondence with and without the stitching stage.

	All Negative	Affinity	Affinity Top1	**Associative3D**
AP	10.1	38.8	49.4	**60.0**

Table 4. Evaluation of relative camera pose from the camera branch and picked by the stitching.

Method	Translation (meters)			Rotation (degrees)		
	Median	Mean	(Err $\leq$ 1m)%	Median	Mean	(Err $\leq 30°$)%
Top-1	1.24	1.80	41.26	**6.96**	29.90	77.56
Associative3D	**0.88**	**1.44**	**54.89**	6.97	**29.02**	**78.31**

We use average precision (AP) to measure the performance since AP pays more attention to the low recall [11,16]. For i^{th} object in view 1 and j^{th} object in view 2, we produce a confidence score by γA_{ij} where $\gamma = 1$ if the pair is predicted to be corresponding and $\gamma = 0.5$ otherwise. This γ term updates the confidence based on stitching stage to penalize pairs which have a high affinity score but are not corresponding.

We compare Associative3D with 3 baselines. (**All Negative**): The prediction is always negative (the most frequent label). This serves as a lower bound. (**Affinity**): This simply uses the affinity matrix as the confidence. (**Affinity Top1**): Rather than using the raw affinity matrix, it uses affinity top-1 similarity as the correspondence and the same strategy to decide confidence as Associative3D. Table 3 shows that our stitching stage improves AP by 10% compared to using the affinity matrix only as correspondence.

Relative Camera Pose Estimation. We next evaluate the performance of relative camera pose (i.e., camera translation and rotation) estimation and see if the stitching stage improves the relative camera pose jointly. We compare the camera pose picked by the stitching stage and top-1 camera pose predicted by the camera branch. We follow the rotation and translation metrics in our full-scene evaluation to measure the error of our predicted camera poses. We summarize results in Table 4. There is a substantial improvement in translations, with the percentage of camera poses within 1m of the ground truth being boosted from 41.3% to 54.9%. The improvement in rotation is smaller and we believe this is because the network already starts out working well and can exploit the fact that scenes tend to have three orthogonal directions. In conclusion, the stitching stage can mainly improve the prediction of camera translation.

4.5   Failure Cases

To understand the problem of reconstruction from sparse views better, we identify some representative failure cases and show them in Fig. 6. While our method

| Input Images | | Camera 1 | | Camera 2 | | Birdview | |
| Image 1 | Image 2 | Prediction | GT | Prediction | GT | Prediction | GT |

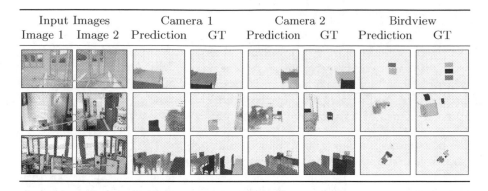

Fig. 6. Representative failure cases on the SUNCG test set [45]. **Row 1:** The input images are ambiguous. There can be two or three beds in the scene. **Row 2:** The single-view backbone does not produce a reasonable prediction. **Row 3:** This is challenging because all chairs are the same.

| Image 1 | Image 2 | Sideview | Birdview | Image 1 | Image 2 | Sideview | Birdview |

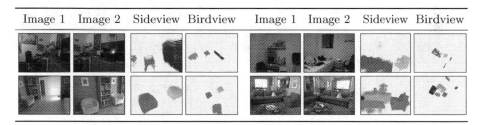

Fig. 7. Qualitative results on NYUv2 dataset [44]. Sideview corresponds to the camera view slightly transformed from the image 2 camera position.

is able to generate reasonable results on SUNCG, it cannot solve some common failure cases: (1) The image pair is ambiguous. (2) The single-view backbone does not produce reasonable predictions as we discuss in Sect. 4.2. (3) There are too many similar objects in the scene. The affinity matrix is then not able to distinguish them since it is sub-optimal in distinguishing the same instance. Our stitching stage is also limited by the random search over object correspondence. Due to factorial growth of search space, we cannot search all possible correspondences. The balancing of our sub-losses can also be sensitive.

4.6 Results on NYU Dataset

To test generalization, we also test our approach on images from NYUv2 [44]. Our only change is using proposals from Faster-RCNN [40] trained on COCO [31], since Faster-RCNN trained on SUNCG cannot generalize to NYUv2 well. We do not finetune any models and show qualitative results in Fig. 7. Despite training on synthetic data, our model can often obtain a reasonable interpretation.

5   Conclusion

We have presented Associative3D, which explores 3D volumetric reconstruction from sparse views. While the output is reasonable, failure modes indicate the problem is challenging to current techniques. Directions for future work include joint learning of object affinity and relative camera pose, and extending the approach to many views and more natural datasets other than SUNCG.

Acknowledgments. We thank Nilesh Kulkarni and Shubham Tulsiani for their help of 3D-RelNet; Zhengyuan Dong for his help of visualization; Tianning Zhu for his help of video; Richard Higgins, Dandan Shan, Chris Rockwell and Tongan Cai for their feedback on the draft. Toyota Research Institute ("TRI") provided funds to assist the authors with their research but this article solely reflects the opinions and conclusions of its authors and not TRI or any other Toyota entity.

References

1. Bao, S.Y., Bagra, M., Chao, Y.W., Savarese, S.: Semantic structure from motion with points, regions, and objects. In: 2012 IEEE Conference on Computer Vision and Pattern Recognition, pp. 2703–2710. IEEE (2012)
2. Bowman, S.L., Atanasov, N., Daniilidis, K., Pappas, G.J.: Probabilistic data association for semantic slam. In: 2017 IEEE international conference on robotics and automation (ICRA), pp. 1722–1729. IEEE (2017)
3. Bromley, J., Guyon, I., LeCun, Y., Säckinger, E., Shah, R.: Signature verification using a "siamese" time delay neural network. In: Advances in neural information processing systems, pp. 737–744 (1994)
4. Chang, A., et al.: Matterport3D: learning from RGB-D data in indoor environments. arXiv preprint arXiv:1709.06158 (2017)
5. Chen, W., Qian, S., Deng, J.: Learning single-image depth from videos using quality assessment networks. In: Proceedings of the IEEE Conference on Computer Vision and Pattern Recognition, pp. 5604–5613 (2019)
6. Chen, Y., Huang, S., Yuan, T., Qi, S., Zhu, Y., Zhu, S.C.: Holistic++ scene understanding: single-view 3D holistic scene parsing and human pose estimation with human-object interaction and physical commonsense. In: Proceedings of the IEEE International Conference on Computer Vision (ICCV), pp. 8648–8657 (2019)
7. Choy, C.B., Gwak, J., Savarese, S., Chandraker, M.: Universal correspondence network. In: Advances in Neural Information Processing Systems, pp. 2414–2422 (2016)
8. Choy, C.B., Xu, D., Gwak, J.Y., Chen, K., Savarese, S.: 3D-R2N2: a unified approach for single and multi-view 3D object reconstruction. In: Leibe, B., Matas, J., Sebe, N., Welling, M. (eds.) ECCV 2016. LNCS, vol. 9912, pp. 628–644. Springer, Cham (2016). https://doi.org/10.1007/978-3-319-46484-8_38
9. Crandall, D., Owens, A., Snavely, N., Huttenlocher, D.: SfM with MRFs: discrete-continuous optimization for large-scale structure from motion. IEEE Trans. Pattern Anal. Mach. Intell. (PAMI) 35(12), 2841–2853 (2013)
10. Dai, A., Chang, A.X., Savva, M., Halber, M., Funkhouser, T., Nießner, M.: Scannet: richly-annotated 3D reconstructions of indoor scenes. In: Proceedings of the IEEE Conference on Computer Vision and Pattern Recognition, pp. 5828–5839 (2017)

11. Davis, J., Goadrich, M.: The relationship between Precision-Recall and ROC curves. In: Proceedings of the 23rd International Conference on Machine learning, pp. 233–240 (2006)
12. Du, Y., et al.: Learning to exploit stability for 3D scene parsing. In: Advances in Neural Information Processing Systems, pp. 1726–1736 (2018)
13. Duggal, S., Wang, S., Ma, W.C., Hu, R., Urtasun, R.: DeepPruner: learning efficient stereo matching via differentiable patchmatch. In: Proceedings of the IEEE International Conference on Computer Vision, pp. 4384–4393 (2019)
14. Eigen, D., Fergus, R.: Predicting depth, surface normals and semantic labels with a common multi-scale convolutional architecture. In: Proceedings of the IEEE international conference on computer vision, pp. 2650–2658 (2015)
15. En, S., Lechervy, A., Jurie, F.: RPNet: an end-to-end network for relative camera pose estimation. In: Leal-Taixé, L., Roth, S. (eds.) ECCV 2018. LNCS, vol. 11129, pp. 738–745. Springer, Cham (2019). https://doi.org/10.1007/978-3-030-11009-3_46
16. Everingham, M., Van Gool, L., Williams, C.K., Winn, J., Zisserman, A.: The pascal visual object classes (voc) challenge. Int. J. Comput. Vision **88**(2), 303–338 (2010)
17. Girdhar, R., Fouhey, D.F., Rodriguez, M., Gupta, A.: Learning a predictable and generative vector representation for objects. In: Leibe, B., Matas, J., Sebe, N., Welling, M. (eds.) ECCV 2016. LNCS, vol. 9910, pp. 484–499. Springer, Cham (2016). https://doi.org/10.1007/978-3-319-46466-4_29
18. Gkioxari, G., Malik, J., Johnson, J.: Mesh r-cnn. In: Proceedings of the IEEE International Conference on Computer Vision, pp. 9785–9795 (2019)
19. Groueix, T., Fisher, M., Kim, V.G., Russell, B., Aubry, M.: AtlasNet: a papier-mâché approach to learning 3D surface generation. In: CVPR 2018 (2018)
20. Hartley, R.I., Zisserman, A.: Multiple View Geometry in Computer Vision. Cambridge University Press, ISBN: 0521540518, 2nd edn. (2004)
21. He, K., Zhang, X., Ren, S., Sun, J.: Deep residual learning for image recognition. In: Proceedings of the IEEE Conference on Computer Vision and Pattern Recognition, pp. 770–778 (2016)
22. Huang, P.H., Matzen, K., Kopf, J., Ahuja, N., Huang, J.B.: DeepMVS: learning multi-view stereopsis. In: Proceedings of the IEEE Conference on Computer Vision and Pattern Recognition, pp. 2821–2830 (2018)
23. Huang, S., Qi, S., Zhu, Y., Xiao, Y., Xu, Y., Zhu, S.C.: Holistic 3D scene parsing and reconstruction from a single rgb image. In: Proceedings of the European Conference on Computer Vision (ECCV), pp. 187–203 (2018)
24. Huang, Z., et al.: Deep volumetric video from very sparse multi-view performance capture. In: Proceedings of the European Conference on Computer Vision (ECCV), pp. 336–354 (2018)
25. Kar, A., Häne, C., Malik, J.: Learning a multi-view stereo machine. In: Advances in Neural Information Processing Systems, pp. 365–376 (2017)
26. Kendall, A., Grimes, M., Cipolla, R.: Posenet: a convolutional network for real-time 6-dof camera relocalization. In: Proceedings of the IEEE International Conference on Computer Vision, pp. 2938–2946 (2015)
27. Kulkarni, N., Misra, I., Tulsiani, S., Gupta, A.: 3D-RelNet: joint object and relational network for 3D prediction. In: Proceedings of the IEEE International Conference on Computer Vision, pp. 2212–2221 (2019)
28. Ladický, L., Zeisl, B., Pollefeys, M.: Discriminatively trained dense surface normal estimation. In: Fleet, D., Pajdla, T., Schiele, B., Tuytelaars, T. (eds.) ECCV 2014. LNCS, vol. 8693, pp. 468–484. Springer, Cham (2014). https://doi.org/10.1007/978-3-319-10602-1_31

29. Lasinger, K., Ranftl, R., Schindler, K., Koltun, V.: Towards robust monocular depth estimation: mixing datasets for zero-shot cross-dataset transfer. arXiv preprint arXiv:1907.01341 (2019)
30. Li, L., Khan, S., Barnes, N.: Silhouette-assisted 3D object instance reconstruction from a cluttered scene. In: Proceedings of the IEEE International Conference on Computer Vision Workshops (2019)
31. Lin, T.Y., et al.: Microsoft COCO: common objects in context. In: Fleet, D., Pajdla, T., Schiele, B., Tuytelaars, T. (eds.) ECCV 2014. LNCS, vol. 8693, pp. 740–755. Springer, Cham (2014). https://doi.org/10.1007/978-3-319-10602-1_48
32. Liu, C., Kim, K., Gu, J., Furukawa, Y., Kautz, J.: PlaneRCNN: 3D plane detection and reconstruction from a single image. In: Proceedings of the IEEE Conference on Computer Vision and Pattern Recognition, pp. 4450–4459 (2019)
33. Liu, C., Wu, J., Furukawa, Y.: Floornet: a unified framework for floorplan reconstruction from 3D scans. In: Proceedings of the European Conference on Computer Vision (ECCV), pp. 201–217 (2018)
34. Lowe, D.G.: Distinctive image features from scale-invariant keypoints. Int. J. Comput. Vision $60(2)$, 91–110 (2004)
35. Melekhov, I., Ylioinas, J., Kannala, J., Rahtu, E.: Relative camera pose estimation using convolutional neural networks. In: Blanc-Talon, J., Penne, R., Philips, W., Popescu, D., Scheunders, P. (eds.) ACIVS 2017. LNCS, vol. 10617, pp. 675–687. Springer, Cham (2017). https://doi.org/10.1007/978-3-319-70353-4_57
36. Mishkin, D., Perdoch, M., Matas, J.: Mods: fast and robust method for two-view matching. Comput. Vis. Image Underst. $1(141)$, 81–93 (2015)
37. Nie, Y., Han, X., Guo, S., Zheng, Y., Chang, J., Zhang, J.J.: Total3Dunderstanding: joint layout, object pose and mesh reconstruction for indoor scenes from a single image. In: Proceedings of the IEEE/CVF Conference on Computer Vision and Pattern Recognition, pp. 55–64 (2020)
38. Price, A., Jin, L., Berenson, D.: Inferring occluded geometry improves performance when retrieving an object from dense clutter. arXiv preprint arXiv:1907.08770 (2019)
39. Pritchett, P., Zisserman, A.: Wide baseline stereo matching. In: Sixth International Conference on Computer Vision (IEEE Cat. No. 98CH36271), pp. 754–760. IEEE (1998)
40. Ren, S., He, K., Girshick, R., Sun, J.: Faster r-cnn: towards real-time object detection with region proposal networks. In: Advances in neural information processing systems, pp. 91–99 (2015)
41. Richter, S.R., Roth, S.: Matryoshka networks: predicting 3D geometry via nested shape layers. In: Proceedings of the IEEE conference on computer vision and pattern recognition, pp. 1936–1944 (2018)
42. Salas-Moreno, R.F., Newcombe, R.A., Strasdat, H., Kelly, P.H., Davison, A.J.: Slam++: simultaneous localisation and mapping at the level of objects. In: Proceedings of the IEEE conference on computer vision and pattern recognition, pp. 1352–1359 (2013)
43. Sharma, G., Goyal, R., Liu, D., Kalogerakis, E., Maji, S.: Csgnet: neural shape parser for constructive solid geometry. In: Proceedings of the IEEE Conference on Computer Vision and Pattern Recognition, pp. 5515–5523 (2018)
44. Silberman, N., Hoiem, D., Kohli, P., Fergus, R.: Indoor segmentation and support inference from RGBD images. In: Fitzgibbon, A., Lazebnik, S., Perona, P., Sato, Y., Schmid, C. (eds.) ECCV 2012. LNCS, vol. 7576, pp. 746–760. Springer, Heidelberg (2012). https://doi.org/10.1007/978-3-642-33715-4_54

45. Song, S., Yu, F., Zeng, A., Chang, A.X., Savva, M., Funkhouser, T.: Semantic scene completion from a single depth image. In: Proceedings of the IEEE Conference on Computer Vision and Pattern Recognition, pp. 1746–1754 (2017)
46. Sui, Z., Chang, H., Xu, N., Jenkins, O.C.: Geofusion: geometric consistency informed scene estimation in dense clutter. arXiv:2003.12610 (2020)
47. Sun, X., et al.: Pix3d: dataset and methods for single-image 3D shape modeling. In: Proceedings of the IEEE Conference on Computer Vision and Pattern Recognition, pp. 2974–2983 (2018)
48. Tatarchenko, M., Richter, S.R., Ranftl, R., Li, Z., Koltun, V., Brox, T.: What do single-view 3D reconstruction networks learn? In: Proceedings of the IEEE Conference on Computer Vision and Pattern Recognition, pp. 3405–3414 (2019)
49. Tulsiani, S., Gupta, S., Fouhey, D.F., Efros, A.A., Malik, J.: Factoring shape, pose, and layout from the 2D image of a 3D scene. In: Proceedings of the IEEE Conference on Computer Vision and Pattern Recognition, pp. 302–310 (2018)
50. Wang, Q., Zhou, X., Daniilidis, K.: Multi-image semantic matching by mining consistent features. In: Proceedings of the IEEE Conference on Computer Vision and Pattern Recognition, pp. 685–694 (2018)
51. Wang, X., Fouhey, D., Gupta, A.: Designing deep networks for surface normal estimation. In: Proceedings of the IEEE Conference on Computer Vision and Pattern Recognition, pp. 539–547 (2015)
52. Wu, J., Wang, Y., Xue, T., Sun, X., Freeman, B., Tenenbaum, J.: Marrnet: 3D shape reconstruction via 2.5D sketches. In: Advances in neural information processing systems, pp. 540–550 (2017)
53. Yang, Z., Pan, J.Z., Luo, L., Zhou, X., Grauman, K., Huang, Q.: Extreme relative pose estimation for rgb-d scans via scene completion. In: Proceedings of the IEEE Conference on Computer Vision and Pattern Recognition, pp. 4531–4540 (2019)
54. Yang, Z., Yan, S., Huang, Q.: Extreme relative pose network under hybrid representations. In: Proceedings of the IEEE/CVF Conference on Computer Vision and Pattern Recognition, pp. 2455–2464 (2020)
55. Zhang, X., Zhang, Z., Zhang, C., Tenenbaum, J., Freeman, B., Wu, J.: Learning to reconstruct shapes from unseen classes. In: Advances in Neural Information Processing Systems, pp. 2257–2268 (2018)
56. Zhang, Y., et al.: Physically-based rendering for indoor scene understanding using convolutional neural networks. In: Proceedings of the IEEE Conference on Computer Vision and Pattern Recognition, pp. 5287–5295 (2017)
57. Zhou, T., Brown, M., Snavely, N., Lowe, D.G.: Unsupervised learning of depth and ego-motion from video. In: Proceedings of the IEEE Conference on Computer Vision and Pattern Recognition, pp. 1851–1858 (2017)
58. Zwillinger, D., Kokoska, S.: CRC Standard Probability and Statistics Tables and Formulae. Crc Press (1999)

PlugNet: Degradation Aware Scene Text Recognition Supervised by a Pluggable Super-Resolution Unit

Yongqiang Mou[1(✉)], Lei Tan[2], Hui Yang[1], Jingying Chen[2], Leyuan Liu[2], Rui Yan[1], and Yaohong Huang[1]

[1] AI -Labs, GuangZhou Image Data Technology Co., Ltd., Guangzhou, China
`yongqiang.mou@gmail.com`, `huiyang865@hotmail.com`, `reeyree@163.com`,
`hyh362@me.com`
[2] Nercel, Central China Normal University, Wuhan, China
`lei.tan@mails.ccnu.edu.cn`, {`chenjy,lyliu`}`@mail.ccnu.edu.cn`

Abstract. In this paper, we address the problem of recognizing degradation images that are suffering from high blur or low-resolution. We propose a novel degradation aware scene text recognizer with a pluggable super-resolution unit (PlugNet) to recognize low-quality scene text to solve this task from the feature-level. The whole networks can be trained end-to-end with a pluggable super-resolution unit (PSU) and the PSU will be removed after training so that it brings no extra computation. The PSU aims to obtain a more robust feature representation for recognizing low-quality text images. Moreover, to further improve the feature quality, we introduce two types of feature enhancement strategies: Feature Squeeze Module (FSM) which aims to reduce the loss of spatial acuity and Feature Enhance Module (FEM) which combines the feature maps from low to high to provide diversity semantics. As a consequence, the PlugNet achieves state-of-the-art performance on various widely used text recognition benchmarks like IIIT5K, SVT, SVTP, ICDAR15 and etc.

Keywords: Scene text recognition · Neural network · Feature learning

1 Introduction

Scene text recognition, where the task is aiming to recognize the text in the scene images, is a long-standing computer vision issue that could be widely used in the majority of applications like driverless vehicles, product recognition, handwriting recognition, and visual recognition. Different from the Optical Character

---

Y. Mou and L. Tan — Equal Contribution.

Electronic supplementary material The online version of this chapter (https://doi.org/10.1007/978-3-030-58555-6_10) contains supplementary material, which is available to authorized users.

© Springer Nature Switzerland AG 2020
A. Vedaldi et al. (Eds.): ECCV 2020, LNCS 12360, pp. 158–174, 2020.
https://doi.org/10.1007/978-3-030-58555-6_10

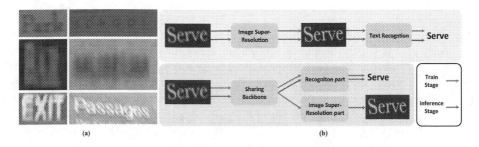

Fig. 1. (a)Low-quality images that are suffering from low-resolution, blur or shake will make a great challenge for text recognition. (b)Two types of strategies to solve the degrade images: the image-level solution and feature-level solution.

Recognition (OCR) which has been well-solved before, scene text recognition is still a challenging task owing to the variable scene conditions as occlusion, illumination, curvature distortions, perspective, etc.

Over recent years, inspired by the practical and research value, scene text recognition attracts growing attention [1,18,33,42,45]. We have witnessed the great improvement in this area due to the powerful feature extractor like deep convolution networks like Resnet [10] and more specific methods as Aster [33]. State-of-the-art scene text recognizers nowadays are base on two types of categories: the bottom-up approaches that recognize the text by each character and top-down approaches that recognize the text by the whole image. However, both of the categories are facing a condition that using the lexicon or not will make a great gap in the recognition result. Part of the above problem is caused by the low-quality images which are suffering the noising, blurred or low-resolution as shown in Fig. 1 (a). Due to the lack of sufficient details, those images easily cause the wrong result.

Generally, when facing low-quality images in other computer vision tasks, previous works prefer to solve this problem from the image-level. Embedding a super-resolution module in the ordinary model seems like a paradigm [3]. Following this trend, TextSR [39] training an ESRGAN [40]-Aster recognition network. Unfortunately, although this method shows better visual quality than original images, it improves limited in the recognition result. Especially when considers its efficiency, this structure is far from satisfied. For example, we train the ESRGAN-Aster with the Synth90K dataset [11] with a single NVIDIA 2080Ti, it needs nearly 30 days each epoch.

Motivated by this condition, we attempt to explore a more reasonable way to solve low-quality images. Different from general methods, we attempt to solve those degradation images from the feature-level as shown in Fig. 1 (b). Base on this idea, we proposed an end-to-end trainable scene text recognizer together with a pluggable super-resolution unit (PlugNet) for auxiliary training. As shown in Fig. 2, the PlugNet can be divided into four parts: rectification network, CNN backbone, recognition network, and pluggable super-resolution unit. Specifically, the PlugNet only takes a light-weight pluggable super-resolution unit (PSU) that constructed by upsampling layers and few convolution layers to improve the feature quality in the training stage. During the inference stage, the PSU will be removed which means no extra computation.

As the most popular text recognition framework, the CNN-LSTM shows a great performance nowadays. Owing to the special structure, the input for LSTM should be a one-dimension vector. So, in most the previous works tend to use deep CNN to squeeze the height-level features maps to generate one-dimension vectors. However, CNN shows limited performance to cope with spatial-level issues like rotation [14], shift [46]. Due to the loss of spatial acuity makes it difficult for both the recognition part and the rectified part to get effective learning. Therefore, in our work, we proposed a Feature Squeeze Module (FSM) trying to maintain more spatial information in the final one-dimension vectors. Specifically, we remove the down-sampling convolutional layers in the last three blocks to maintain the feature resolution and use a 1×1 convolution layer together with a reshape layer to generate the same one-dimension vectors straightly from the feature maps. Surprisingly, by adding more spatial information into the features, recognition performance improved significantly in all of the datasets. Additionally, maintain more feature-resolution helps the PSU could easily be attached to the CNN backbone.

Affected by the above observation and Feature Pyramid Networks [21], we suppose those low-level semantics will also enhance the final sharing feature maps. We designed a Feature Enhance Module (FEM) to further combine those semantics from low to high levels.

The proposed PlugNet is compared against several state-of-the-art scene text recognition method (as [19,25,42]) on various challenging text recognition datasets like SVT [37], ICDAR2015 [15], SVTP [29] and etc, to demonstrate its significant advantages.

In summary, the main contributions of this paper are as follows:

- We proposed an end-to-end trainable scene text recognizer (PlugNet), which combined with a light-weight pluggable super-resolution unit to handle degradation images from the feature-level with no extra computation during the inference stage.
- Observed the importance of feature resolution in the text recognition issue, we introduced a feature squeeze module (FSM) that offers a better way to connect the CNN-based backbone and the LSTM-based recognition model. It could also be used as a fresh baseline for top-down text recognition method.
- A feature enhance module is designed to combine those semantics from low to high levels which further strengthen the sharing feature maps.
- Experimental results have demonstrated the effectiveness of the proposed PlugNet, PlugNet achieves the state-of-the-art performance on several wildly-used text recognition benchmarks.

2   Related Works

Text recognition has made great progress in the last decades. Most of the text recognition methods can be divided into two categories: The bottom-up approaches which recognize the text by each character and top-down approaches which recognize the text by the whole image.

Traditional text recognition methods tend to use the bottom-up approach, in which the characters will be detected firstly by hand-crafted features and then follow with some subsequent steps like non-text component filtering, text line construction, and text line verification. These methods depend heavily on the result of character detection which using sliding-window [37,38], connected components [28] and Hough voting [43]. These methods often extract low-level hand-crafted features like HOG [43], SWT [44] to separate characters and backgrounds. Although those methods or relating occupy the major status before the deep learning era, it is still a challenge for text recognizers to reach satisfactory performance in the wild. As the most powerful extractor and classifier, we have witnessed lots of deep neural network-based frameworks further improved the performance. Bissacco *et al.* [4] utilizes the fully connected network with 5 hidden layers to extract the characters' features, then using the n-gram language model to recognition. Jaderberg *et al.* [12] introduced a CNN-based method to solve unconstrained. Charnet [22] trained a character-aware neural network for distorted scene text. Liao *et al.* [19] combined the text recognition methods with a semantic segmentation network for recognizing the text of arbitrary shapes.

On the other hand, the top-down approach has an advantage in character localization for using the whole image instead of individual characters. Jaderberg *et al.* [13] regards each word as an object and converts the text recognition to classification by training a classifier with a large number of classes. Inspired by speech recognition and Natural Language Processing [8,26], recurrent neural networks (RNN) are widely used in recent text recognition models in recent years. These methods solved this problem by converting the text recognition to sequence recognition which has high similarity. Su *et al.* [34] extracts the HOG features of the word and generates the character sequence with RNN. Busta *et al.* and Shi *et al.* [5,31] introduces an end to end model paradigm which using CNN for extracted the feature maps and using RNN to the decoder the feature maps. In recent years, the attention mechanism has inspired several works like Focus Attention [6], Moran [25].

Also, some special problems in text recognition have been proposed and solved well in the past three years. Aster, ESIR and Liao *et al.* [19,33,45] designed a rectification network to transform those irregular text. Liu *et al.* [23] proposed a data augmentation method to improve image feature learning for scene text recognition.

3   Approach

3.1   Overall Framework

The overall framework of our PlugNet is shown in Fig. 2. In order to solve the blur and low-resolution cases, we adopt the pluggable super-resolution unit (PSU) for auxiliary training to assist the recognition network. Hence, our PlugNet can be

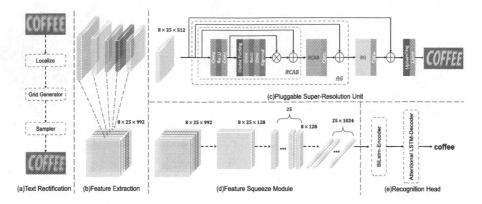

Fig. 2. Overall framework of our PlugNet. The pipeline of our method includes four parts: rectification network, sharing CNN backbone, recognition network and pluggable super-resolution unit.

divided into four parts: rectification network, sharing CNN backbone, PSU, and recognition part.

Rectification Network: The rectification network aims to rectify those irregular scene text. Our method employs the same strategy as Aster [33], which shows robust performance in irregular scene text recognition. The Rectification Network is composed of three parts: localization network, grid generator, and sampler. Localization network using a CNN-based network to localize the borders of text in the input image by n control points. The grid generator will utilize the localization result and compute the transformation matrix for each pixel by Thin-Plate-Spline (TPS). Finally, the sampler is attached to generate the rectified images.

Sharing CNN Backbone: As shown in Fig. 2, the sharing CNN backbone employed the Resnet-based structure to extract the feature maps. In this work, we keep a similar structure as Aster does. We describe the detailed structure in Table 1. Compared with the Aster, for achieving better expandability and retain more spatial information, we removed the down-sampling layers in last three CNN blocks so that the feature maps after the backbone have the dimension as $\frac{W}{4} \times \frac{H}{4} \times C$ where the W, H means the width and height of input image.

Recognition Part: Following the success of previous works like ESIR, Aster, we employed the LSTM-based method for text recognition for its advantage performance in solving whole sequences. The structure of the recognition part is shown in Fig. 2 (d)(e). Those features after the sharing CNN backbone will be used to the Feature Squeeze Module to generate the one-dimension vectors. After that, we employed a sequence-to-sequence model with attention mechanism which composed by two-layer Bidirectional LSTM (BiLSTM) with 256 hidden unit as an encoder and a two-layer attentional LSTM as a decoder. In detail, suppose the input sequence V_s whose shape can be denoted as $W \times (H \times C)$, a two-layer

Table 1. Structure of CNN Blocks in Fig. 2. Herein, the 's' means the stride of the first convolutional layer of each block.

Layers	Output size	Configurations
Block 0	32×100	$3 \times 3conv, s = 1$
Block 1	16×50	$\begin{bmatrix} 1 \times 1conv, 32 \\ 3 \times 3conv, 32 \end{bmatrix} \times 3, s = 2$
Block 2	8×25	$\begin{bmatrix} 1 \times 1conv, 64 \\ 3 \times 3conv, 64 \end{bmatrix} \times 3, s = 2$
Block 3	8×25	$\begin{bmatrix} 1 \times 1conv, 128 \\ 3 \times 3conv, 128 \end{bmatrix} \times 3, s = 1$
Block 4	8×25	$\begin{bmatrix} 1 \times 1conv, 256 \\ 3 \times 3conv, 256 \end{bmatrix} \times 3, s = 1$
Block 5	8×25	$\begin{bmatrix} 1 \times 1conv, 512 \\ 3 \times 3conv, 512 \end{bmatrix} \times 3, s = 1$

BiLSTM is attached to capture the long-range dependencies in both directions, obtaining a robust new sequence H_s of the same length as V_s. Next, a two-layer attentional LSTM is adopted to translate sequence H_s to a output sequence Y_s. Herein, to confirm the length of sequences, an end-of-sequence symbol (EOS) will be attached as a rest.

Pluggable SR Unit: Benefit from the FSM, the sharing CNN backbone could keep the image resolution. It helps the PSU which based-on the image super-resolution method can easily be attached in the whole network. In our work, the PSU is used to build the super-resolution images from the high-level features. This part will be detailed in Sect. 3.2.

3.2 Pluggable Super-Resolution Unit

As we have mentioned before, the PSU is designed to solve the degradation images from the feature level. Most of the recognition methods tend to embed the super-resolution network in the original recognition network. Limited by the efficiency, the SR-Recognition framework that solves those degradation images from the image-level shows an obvious bottleneck. Inspired by the success of multi-task learning, we utilized the PSU to help the sharing CNN backbone better represent the features of degradation images. We employed the RCAN [47] structure to build the PSU. As shown in Fig. 2 (c), we use two Residual Channel Attention Block (RCAB) to construct each Residual Group (RG). Then, two RGs are used to build the final PSU. After training, the PSU will be removed which means no extra computation in the inference stage.

3.3   Feature Enhancement

Feature Squeeze Module: As shown in Fig. 2 (d), we replaced those down-sampling convolution layers by the Feature Squeeze Module (FSM) to maintain more resolution information in the final one-dimension vectors. FSM only contains a 1×1 convolutional layer for the channel reduction and a reshape layer to generate the one-dimension vectors from the feature maps which means FSM adds few extra computations when compared to the baseline method. Based on the FSM, not only the CNN-LSTM text recognition framework has improved a lot, but also the PSU could benefit from high-resolution features which influenced a lot in the super-resolution issue.

Feature Enhance Module: Affected by the Feature Pyramid Networks [21] and the success of FSM, to further combine those semantics from low to high levels, we designed a Feature Enhance Module as shown in Fig. 2 (b). For the first two blocks, we use a down-sampling layer to transform their shape to $\frac{W}{4} \times \frac{H}{4}$. Then all of the features maps from low to high will be concatenated as the enhanced feature.

3.4   Training and Inference

Training Dataset. Our model is trained on the Synth90K (**90k**) [11] and SynthText (**ST**) [9]. The Synth90K includes 9 million synthetic text images generated from 90k words lexicon. Similarly, the synthetic is also synthetic dataset (SynthText). It is generated for text detection research, so the images should be cropped to a single text. We cropped 4 million text images for training our model which keeps the same size as [45] but less than [33] who cropped 7 million text images. When training our model, we do not separate the train data and test data, all images in these two datasets are used for training.

Training Super-Resolution Unit. Owing to the text recognition dataset has no separation of high-resolution and low-resolution images, training the super-resolution is not an easy task. To achieve this task, we adopted two strategies as Gaussian Blur and down-up sampling to generate low-quality images. Herein, we set a probability parameter α to ensure randomness.

$$I_{blur} = \begin{cases} f_{d-u}(f_{gau}(I)), & \text{if } p_1 >= \alpha; p_2 >= \alpha \\ f_{gau}(I), & \text{if } p_1 >= \alpha; p_2 < \alpha \\ f_{d-u}(I), & \text{if } p_1 < \alpha; p_2 >= \alpha \\ I, & \text{if } p_1 < \alpha; p_2 < \alpha \end{cases} \tag{1}$$

Herein, the f_{gau} refers to the Gaussian Blur and f_{d-u} refers to the down-up sampling. The random numbers $p_1, p_2 \in [0, 1]$ and we set $\alpha = 0.5$.

Nevertheless, another challenge exists in training the super-resolution branch. For the Rectification Network, it will change distribution of each pixel which makes a huge difference between the output image and input image. Following the original super-resolution methods and taking the Rectification Network into consideration, the loss L_{sr} can be described as:

$$L_{sr} = f_{loss}(f_{rn}(I), f_{blur}(f_{rn}(I))) \tag{2}$$

where the f_{loss} means loss function of super-resolution, the f_{rn} means the Rectification Network, and I refers to the input image and f_{blur} refers to the blur function as stated before. But, following this equation will cause a tricky problem. The $f_{blur}(f_{rn}(I))$ means the data generation strategies should take effect after the Rectification Network which means the input images are high-resolution. Therefore, we use the $f_{rn}(f_{blur}(I))$ to approximate $f_{blur}(f_{rn}(I))$ as:

$$L_{sr} = f_{loss}(f_{rn}(I), f_{rn}(f_{blur}(I))) \tag{3}$$

In this way, the Rectification Network can not only learn about solving low-quality images but also simplify the whole networks thus making it easy to achieve.

Loss Functions. Following the success of multi-task learning, recognition loss and super-resolution loss are combined to train our model end-to-end as:

$$L = L_{rec} + \lambda L_{sr} \tag{4}$$

where L_{rec} denotes recognition loss and L_{sr} denotes super-resolution loss. In order to balance the weight in two different tasks and keep the recognition performance, we add a parameter λ. In our method, we set the $\lambda = 0.01$.

In most of the time, the recognition problem could formulated as a classification problem [33,41], so we use a cross-entropy loss to describe L_{rec}:

$$L_{rec} = -\frac{1}{MN} \sum_{i=1}^{M} \sum_{j=1}^{N} y_{i,j} \log(s_{i,j}) \tag{5}$$

where i is the index of the sample in a batch and j is the index of the number in the label. In addition that, p is the ground truth label, s is the recognition result.

For the super-resolution branch, as has mentioned in [20] that L1 loss provides better convergence than L2 loss, we select L1 loss to train our network. So, for each pixel (i, j) in output O, the L_{sr} is employed as:

$$L_{sr} = \frac{1}{W \times H} \sum_{i=1}^{W} \sum_{j=1}^{H} \|O^{i,j} - I^{i,j}\| \tag{6}$$

Herein, the I means the ground truth of input image, W and H refers to the width and height of the input image.

4 Experiment

4.1 Datasets

We evaluate PlugNet over 7 widely used benchmarks as IIIT5K, ICDAR2003, ICDAR2013, ICDAR2015, SVTP, and CUTE80 to demonstrate its ability. Among these 7 datasets, SVT and SVTP are highly blurred and low-resolution which seems more typical. Herein, we evaluate PlugNet without any lexicon to show its robust performance.

IIIT5K [27] includes 3000 test images that are cropped from the website. Each image has a 50-word lexicon and a 1000-word lexicon in this dataset.

Street View Text (SVT) [37] contains 647 images, which are collected from the Google Street View. Many images in this dataset are suffering from noise, blur or having very low resolutions. Each image has a 50-word lexicon attached.

ICDAR 2003 (IC03) [24] contains 860 images after selection. Following [37], we discarded images that contain nonalphanumeric characters or have less than three characters.

ICDAR 2013 (IC13) [16] contains 1015 cropped text images. Most of text images inherit from IC03 and provide no lexicon.

ICDAR 2015 (IC15) [15] contains 2077 cropped text images collected by Google Glasses. IC15 is one of the most challenge datasets in text recognition in recent years. Same as IC13, no lexicon is attached to this dataset.

SVT-Perspective (SVT-P) [29] contains 645 cropped images from side-view angle snapshots in Google Street View. This dataset is not only suffering from noise, blur or having very low resolutions as SVT but also suffering from perspective distorted.

CUTE80 [30] contains 288 images cropped from the 80 high-resolution scene text images. This dataset focuses on the curved text and provides no lexicon.

4.2 Implementation Details

Our method implemented in Pytorch and trained end-to-end on the Synth90k and SynthText. The training images are all from these two datasets without any data augmentation or selection. The model is trained by batches of 128 examples. Each batch includes 64 samples from Synth90k and 64 samples from SynthText. During the training, the learning rate is initiated from the 1 and is decayed to 0.1 and 0.01 respectively after 0.6M and 0.8M iterations. We adopted the ADADELTA as an optimizer to minimize the objective function. In addition, all the experiments and training are accomplished on 2 NVIDIA GeForce GTX 2080Ti 11GB GPU.

Compared to the baseline method, both FSM and FEM needs very few computations when compared to the whole network. Therefore, in the inference stage, PSU is removing, the speed of PlugNet is 22 ms per image when the test batch size is 1 which is a little higher than the Aster (baseline method) as 20 ms. In the training stage, PSU is adding. The speed of the Plugnet is 0.97 s per batch (128), and the Aster is 0.63s per batch (128). The training process could be accelerated by a larger batch size.

Table 2. Scene text recognition performance among various widely used benchmarks under different feature resolutions.

Resolution	Data	SVT	SVTP	IIIT5K	IC03	IC13	IC15	CUTE80
1×25	90K	85.2	76.1	80.7	91.8	89.3	69.3	66.3
2×25	90K	$87.0_{\uparrow 1.8}$	$78.1_{\uparrow 2.0}$	$82.7_{\uparrow 2.0}$	$92.3_{\uparrow 1.5}$	$89.4_{\uparrow 0.1}$	$69.3_{\uparrow 0}$	$68.4_{\uparrow 2.1}$
4×25	90K	$87.9_{\uparrow 0.9}$	$79.5_{\uparrow 1.4}$	$82.2_{\downarrow 0.5}$	$92.7_{\uparrow 0.4}$	$89.8_{\uparrow 0.4}$	$71.4_{\uparrow 2.1}$	$69.1_{\uparrow 0.7}$
8×25	90K	$\mathbf{89.0}_{\uparrow 2.1}$	$\mathbf{82.0}_{\uparrow 2.5}$	$\mathbf{85.3}_{\uparrow 3.1}$	$\mathbf{94.3}_{\uparrow 1.6}$	$\mathbf{91.0}_{\uparrow 1.2}$	$\mathbf{73.6}_{\uparrow 2.2}$	$\mathbf{69.1}_{\uparrow 0}$

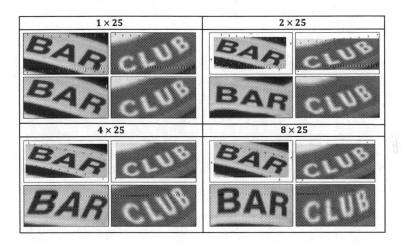

Fig. 3. Visualization of rectification results under different feature resolutions. Please zoom in to see the distribution of control points..

4.3   Ablation Study

Effectiveness of Feature Squeeze Module: The FSM is designed to offer better one-dimension vectors to connect the CNN part and the LSTM part. To further analyze the influence of the CNN feature resolution in the text recognition issue, we trained four networks with different CNN feature resolution under the 90 K dataset. For better comparison, we change the number of the channel of the 1×1 convolutional layer to keep the output of FSM has the same dimension as 25×1024.

Table 2 shows the result of four different networks in seven widely used text recognition datasets. With the broadening of the feature-resolution, the recognition accuracy gets increased in all of the datasets. It has already illustrated the importance of spatial information in text recognition tasks.

Additionally, we visualized the rectified results of the above four networks in Fig. 3. Clearly, decrease the loss spatial acuity helps the Rectification Network shows a much better location result of control points which helps the recognizer could overcome tougher irregular cases.

Effectiveness of Feature Enhance Module: As stated before, to obtain much robust feature maps for the recognition network, we designed a Feature

Table 3. Ablation study of PlugNet on several typical datasets. Herein, the SR-Plugnet is training following the SR-Recognition framework without PSU.

Methods	FSM	FEM	Data_Aug	ESRGAN	PSU	SVT	SVTP	IC15	CUTE80
Baseline(R) [33]	✗	✗	✗	✗	✗	89.5	78.5	76.1	79.5
PlugNet(R)	✓	✗	✗	✗	✗	$90.0_{\uparrow 0.5}$	$80.8_{\uparrow 2.3}$	$78.2_{\uparrow 2.1}$	$82.6_{\uparrow 3.1}$
PlugNet(R)	✓	✓	✗	✗	✗	$90.6_{\uparrow 0.6}$	$81.6_{\uparrow 0.8}$	$80.2_{\uparrow 2.0}$	$83.7_{\uparrow 1.1}$
PlugNet	✓	✓	✓	✗	✗	$89.8_{\downarrow 0.8}$	$82.2_{\uparrow 0.6}$	$79.8_{\downarrow 0.6}$	$81.6_{\downarrow 2.1}$
SR-PlugNet	✓	✓	✓	✓	✗	$90.6_{\uparrow 0.8}$	$80.8_{\downarrow 1.4}$	$79.4_{\downarrow 0.4}$	$82.6_{\uparrow 1.0}$
PlugNet	✓	✓	✓	✗	✓	$\mathbf{92.3}_{\uparrow 1.7}$	$\mathbf{84.3}_{\uparrow 3.5}$	$\mathbf{82.2}_{\uparrow 2.8}$	$\mathbf{85.0}_{\uparrow 2.4}$

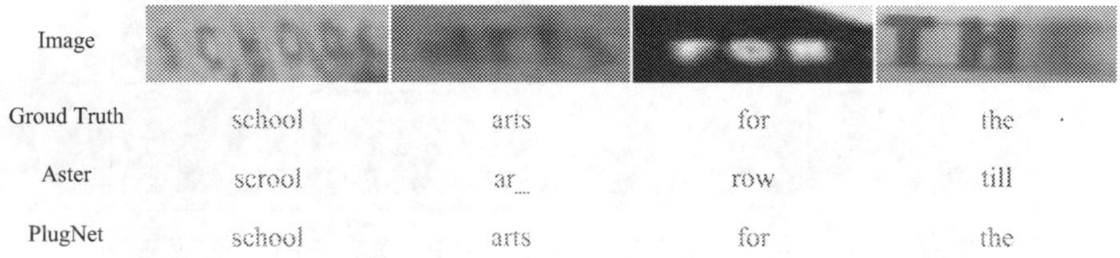

Image				
Groud Truth	school	arts	for	the
Aster	scrool	ar_	row	till
PlugNet	school	arts	for	the

Fig. 4. Several recognition results produced by our PlugNet and Baseline method Aster [33] in low-quality text images of the SVT dataset.

Enhance Module (FEM) that aims to combine the feature maps from low to high to provide diversity semantics.

To analyze the influence of FEM, we training the FSM enhanced model with and without FEM. Herein, we chose four typical datasets for evaluation: SVT that including many highly-blurred images, SVTP that suffering from blur and perspective, CUTE80 that contains many irregular cases, and the most widely used challenging dataset-IC15. The experimental result in Table 3 has illustrated the efficiency of FEM. We observe that adding the FEM, all of the results in these four datasets get improved as 0.6%, 0.8%, 2%, 1.1% in final recognition accuracy when compared to the model without FEM. It indicates that the feature enhancement module has improved feature quality by low-level semantics which in turn improved the recognition performance.

Effectiveness of Pluggable Super-Resolution Unit: So far, recognizing those scene texts with highly-blurred and low-resolution remains a challenging task, thus we employed the super-resolution method to better solve this problem.

As shown in Table 3, we conduct a set of ablation experiments by adding the PSU or not. Owing to PSU, we use the generated data rather than the raw data for training. Hence, to better compare the influence of PSU, we train the network without PSU both in raw data and generated data. The results show that the recognizer with PSU produced a much better performance in solving low-quality scene text images. Coupled with PSU, the recognition accuracy in SVT, SVTP, IC15 and CUTE80 has improved from 89.8%, 82.2%, 79.8%, 81.6% to 92.3%, 84.3%, 82.6%, 85.0%. In the visual level, we chose the recognition results of four

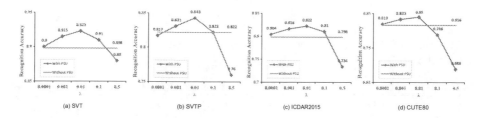

Fig. 5. Results of PlugNet recognition accuracy under different parameter λ value on various datasets.

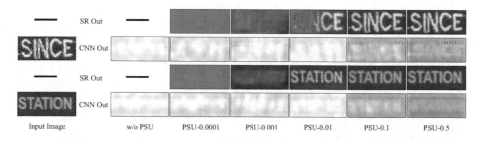

Fig. 6. Visualizing results of PlugNet under different parameter λ value.

low-quality images in the SVT dataset in Fig. 4 to show the improvement of our method when compared to the baseline method.

Herein, to make a comparison between the PSU method and general super-resolution combine with the recognizer method, we trained an end-to-end SR Plugnet (no PSU) model under the 90 K and ST datasets. The SR-Plugnet(no PSU) using the ESRGAN for the super-resolution part and Plugnet (no PSU) for the recognition part. Obviously, as shown in Table 3, the SR-Plugnet shows limited improvement in recognition results, which is a similar case to the TextSR [39]. Of course, we think SR-Plugnet may reach a higher performance when using a better training strategy, more proper parameters or adding more effective data. Like the TextSR using extra selected data for training to reach an even better performance. But consider the structure and efficiency, PSU is obviously a better choice to solve low-quality images. Finally, about the effect on the sharing feature maps will be discussed in the Sect. 4.4, in which we set different weights of PSU to show the changes of feature gradually.

4.4   Experiments on the Parameter λ

Proper parameters are necessary for training a multi-task network. In this work, we use the parameter λ in the formula 4 to balance the recognition branch and the super-resolution branch. To analyze the influence of the parameter λ, we train the model with the λ from 0.0001 to 0.5. We extensively evaluate these models on the challenging SVT, SVTP, ICDAR2015, and CUTE80 to demonstrate the influence of λ.

Table 4. Scene text recognition performance of PlugNet among various widely used benchmarks. The methods marked with * indicate they use the character-level annotations that will highly improve the performance in irregular text recognition datasets like CUTE80.

Methods	Data	IIIT5K	SVT	IC03	IC13	IC15	SVTP	CUTE80
Jaderberg et al.[12]	90K	–	71.7	89.6	81.8	-	-	-
Jaderberg et al.[13]	90K	–	80.7	93.1	90.8	-	-	-
Shi et al.[32]	90K	81.9	81.9	90.1	88.6	-	71.8	59.2
Lee et al.[17]	90K	78.4	80.7	88.7	90.0	-	-	-
Wang et al.[36]	90K	80.8	81.5	91.2	-	-	-	-
Cheng et al.[6]	90K, ST	87.4	85.9	94.2	93.3	70.6	-	-
Cheng et al.[7]	90K, ST	87.0	82.8	91.5	-	68.2	73.0	76.8
Liu et al.[22]	90K, ST	92.0	85.5	92.0	91.1	74.2	78.9	-
Bai et al.[2]	90K, ST	88.3	87.5	94.6	94.4	73.9	-	-
Liu et al.[23]	90K, ST	89.4	87.1	94.7	94.0	-	73.9	62.5
Luo et al.[25]	90K, ST	91.2	88.3	95.0	92.4	68.8	76.1	77.4
Liao et al.[19]	90K, ST	91.9	86.4	-	91.5	-	-	79.9
Zhan et al.[45]	90K, ST	93.3	90.2	-	91.3	76.9	79.6	83.3
Yang et al.[42]	90K, ST*	**94.4**	88.9	95.0	93.9	78.7	80.8	87.5
Wan et al.[35]	90K, ST*	93.9	90.1	-	92.9	79.4	84.3	85.2
Liao et al. -Seg [18]	90K	94.0	87.2	93.1	92.3	73.8	76.3	82.6
Liao et al. -SAM [18]	90K, ST*	93.9	90.6	95.2	**95.3**	77.3	82.2	**87.8**
Aster(Baseline)[33]	90K, ST	93.4	89.5	94.5	91.8	76.1	78.5	79.5
TextSR(SR-Aster)[39]	90K, ST	92.5	87.2	93.2	91.3	75.6	77.4	78.9
Ours	90K, ST	**94.4**	**92.3**	**95.7**	95.0	**82.2**	**84.3**	85.0

In Fig. 5, we set a baseline as training the whole network without the PSU. Obviously, adding the PSU improves the recognition performance in all of the datasets which also demonstrates the efficiency of PSU. From Fig. 5 (a)–(d), we can observe that the recognition accuracy improves monotonically when the λ is smaller than 0.01. After that, with the increase of the, the recognition accuracy decreased and will even have a negative impact. Obviously, 0.01 seems like a best choice of λ in most of the situation.

Based on this observation, we visualized the output feature maps of the sharing CNN backbone and the output of PSU in Fig. 6 to analyze the influence caused by the PSU. To visualize the feature maps, we calculate the average among the channels to generate one-channel images to represent the feature result. Since the Rectification Network will change the pixel distribution of each image, we removed this part when visualizing for better comparison. It is clear that by the increase of λ, the super-resolution result shows a growing visual quality. However, in the feature maps, there exist two types of effects: due to the increase of λ, the feature suffers much less noising, blur which helps the recognition part. Meanwhile, the increase of λ, let the sharing CNN backbone focus more on low-level images to rebuild the super-resolution images which makes a negative impact on the text recognition. These two types of effects make the PlugNet be sensitive to the λ in this work.

4.5   Comparison with State of the Art

We also compare our PlugNet with previous state-of-the-art text recognition methods on various widely used benchmarks to indicate the superiority of our method. Table 4 summarizes the recognition result among 7 widely used datasets including IIIT5K, SVT, IC03, IC13, IC15, SVTP, and CUTE80. Herein, we evaluated all the datasets without lexicon.

As Table 4 shows, our PlugNet outperforms all the previous state-of-the-art performance in 6 datasets and achieves competitive accuracy to the state-of-the-art techniques in the remain CUTE80 datasets. Especially in two low-quality text datasets as SVT and SVTP, our method shows a much robust performance. The CUTE80 dataset focuses on the high-resolution curved text images, so those methods that using the character level annotations to training the rectification part will perform much better.

Our method shows a significant improvement in most of the cases by using the combination of FSM, FEM and PSU. The PSU, FSM, and FEM are designed to obtaining more robust feature maps with high efficiency and performance. So, this constructure may also be useful for solving low-quality images in other computer vision tasks.

5   Conclusion

In this paper, we proposed an end-to-end trainable degradation aware scene text recognizer called PlugNet in short. The proposed method combined the pluggable super-resolution unit (PSU) to solve the low-quality text recognition from the feature-level. It only takes acceptable extra computation in the training stage and no additional computation in the inference stage. With PSU our method shows a significant improvement in a low-quality image feature representation that in turn improves the recognition accuracy. Moreover, in this paper, we further analyzed the important role of feature resolution in the text recognition issue and proposed the FSM for a better connection between CNN and LSTM for top-down recognition framework. Also, the FEM is attached to enhanced the backbone features by introducing those low-level semantics. Experiments show that FSM and FEM also improved performance markedly. Finally, our PlugNet achieves state-of-the-art performance on various widely used text recognition benchmark datasets, especially on SVT and SVTP which include many low-quality text images.

Acknowledgement. This work was supported by the Project of the National Natural Science Foundation of China Grant No. 61977027 and No. 61702208, the Hubei Province Technological Innovation Major Project Grant No. 2019AAA044 and the Colleges Basic Research and Operation of MOE Grant No. CCNU19Z02002, CCNU18KFY02.

References

1. Baek, J., et al.: What is wrong with scene text recognition model comparisons? dataset and model analysis. In: Proceedings of the IEEE International Conference on Computer Vision, pp. 4715–4723 (2019)
2. Bai, F., Cheng, Z., Niu, Y., Pu, S., Zhou, S.: Edit probability for scene text recognition. In: Proceedings of the IEEE Conference on Computer Vision and Pattern Recognition, pp. 1508–1516 (2018)
3. Bai, Y., Zhang, Y., Ding, M., Ghanem, B.: Finding tiny faces in the wild with generative adversarial network. In: Proceedings of the IEEE Conference on Computer Vision and Pattern Recognition, pp. 21–30 (2018)
4. Bissacco, A., Cummins, M., Netzer, Y., Neven, H.: Photoocr: reading text in uncontrolled conditions. In: Proceedings of the IEEE International Conference on Computer Vision, pp. 785–792 (2013)
5. Busta, M., Neumann, L., Matas, J.: Deep textspotter: an end-to-end trainable scene text localization and recognition framework. In: Proceedings of the IEEE International Conference on Computer Vision, pp. 2204–2212 (2017)
6. Cheng, Z., Bai, F., Xu, Y., Zheng, G., Pu, S., Zhou, S.: Focusing attention: towards accurate text recognition in natural images. In: Proceedings of the IEEE International Conference on Computer Vision, pp. 5076–5084 (2017)
7. Cheng, Z., Xu, Y., Bai, F., Niu, Y., Pu, S., Zhou, S.: Aon: towards arbitrarily-oriented text recognition. In: Proceedings of the IEEE Conference on Computer Vision and Pattern Recognition, pp. 5571–5579 (2018)
8. Cho, K., et al.: Learning phrase representations using rnn encoder-decoder for statistical machine translation. arXiv preprint arXiv:1406.1078 (2014)
9. Gupta, A., Vedaldi, A., Zisserman, A.: Synthetic data for text localisation in natural images. In: Proceedings of the IEEE Conference on Computer Vision and Pattern Recognition, pp. 2315–2324 (2016)
10. He, K., Zhang, X., Ren, S., Sun, J.: Deep residual learning for image recognition. In: Proceedings of the IEEE Conference on Computer Vision and Pattern Recognition, pp. 770–778 (2016)
11. Jaderberg, M., Simonyan, K., Vedaldi, A., Zisserman, A.: Synthetic data and artificial neural networks for natural scene text recognition. arXiv preprint arXiv:1406.2227 (2014)
12. Jaderberg, M., Simonyan, K., Vedaldi, A., Zisserman, A.: Deep structured output learning for unconstrained text recognition. arXiv preprint arXiv:1412.5903 (2015)
13. Jaderberg, M., Simonyan, K., Vedaldi, A., Zisserman, A.: Reading text in the wild with convolutional neural networks. Int. J. Comput. Vis. **116**(1), 1–20 (2015). https://doi.org/10.1007/s11263-015-0823-z
14. Jaderberg, M., et al.: Spatial transformer networks. In: Advances in Neural Information Processing Systems, pp. 2017–2025 (2015)
15. Karatzas, D., et al.: ICDAR 2015 competition on robust reading. In: 2015 13th International Conference on Document Analysis and Recognition (ICDAR), pp. 1156–1160. IEEE (2015)
16. Karatzas, D., et al.: ICDAR 2013 robust reading competition. In: 2013 12th International Conference on Document Analysis and Recognition, pp. 1484–1493. IEEE (2013)
17. Lee, C.Y., Osindero, S.: Recursive recurrent nets with attention modeling for ocr in the wild. In: Proceedings of the IEEE Conference on Computer Vision and Pattern Recognition, pp. 2231–2239 (2016)

18. Liao, M., Lyu, P., He, M., Yao, C., Wu, W., Bai, X.: Mask textspotter: an end-to-end trainable neural network for spotting text with arbitrary shapes. In: Proceedings of the European Conference on Computer Vision (ECCV), pp. 67–83 (2019)
19. Liao, M., et al.: Scene text recognition from two-dimensional perspective. Proc. AAAI Conf. Artif. Intell. **33**, 8714–8721 (2019)
20. Lim, B., Son, S., Kim, H., Nah, S., Mu Lee, K.: Enhanced deep residual networks for single image super-resolution. In: Proceedings of the IEEE Conference on Computer Vision and Pattern Recognition Workshops, pp. 136–144 (2017)
21. Lin, T.Y., Dollár, P., Girshick, R., He, K., Hariharan, B., Belongie, S.: Feature pyramid networks for object detection. In: Proceedings of the IEEE Conference on Computer Vision and Pattern Recognition, pp. 2117–2125 (2017)
22. Liu, W., Chen, C., Wong, K.Y.K.: Char-net: a character-aware neural network for distorted scene text recognition. Proc. AAAI **1**(2), 4 (2018)
23. Liu, Y., Wang, Z., Jin, H., Wassell, I.: Synthetically supervised feature learning for scene text recognition. In: Proceedings of the European Conference on Computer Vision (ECCV), pp. 435–451 (2018)
24. Lucas, S.M., et al.: ICDAR 2003 robust reading competitions: entries, results, and future directions. Int. J. Doc. Anal. Recogn. **7**(2–3), 105–122 (2005)
25. Luo, C., Jin, L., Sun, Z.: Moran: a multi-object rectified attention network for scene text recognition. Pattern Recogn. **90**, 109–118 (2019)
26. Mikolov, T., Chen, K., Corrado, G., Dean, J.: Efficient estimation of word representations in vector space. arXiv preprint arXiv:1301.3781 (2013)
27. Mishra, A., Alahari, K., Jawahar, C.: Top-down and bottom-up cues for scene text recognition. In: 2012 IEEE Conference on Computer Vision and Pattern Recognition, pp. 2687–2694. IEEE (2012)
28. Neumann, L., Matas, J.: Real-time scene text localization and recognition. In: 2012 IEEE Conference on Computer Vision and Pattern Recognition, pp. 3538–3545. IEEE (2012)
29. Quy Phan, T., Shivakumara, P., Tian, S., Lim Tan, C.: Recognizing text with perspective distortion in natural scenes. In: Proceedings of the IEEE International Conference on Computer Vision, pp. 569–576 (2013)
30. Risnumawan, A., Shivakumara, P., Chan, C.S., Tan, C.L.: A robust arbitrary text detection system for natural scene images. Expert Syst. Appl. **41**(18), 8027–8048 (2014)
31. Shi, B., Bai, X., Yao, C.: An end-to-end trainable neural network for image-based sequence recognition and its application to scene text recognition. IEEE Trans. Pattern Anal. Mach. Intell. **39**(11), 2298–2304 (2016)
32. Shi, B., Wang, X., Lyu, P., Yao, C., Bai, X.: Robust scene text recognition with automatic rectification. In: Proceedings of the IEEE Conference on Computer Vision and Pattern Recognition, pp. 4168–4176 (2016)
33. Shi, B., Yang, M., Wang, X., Lyu, P., Yao, C., Bai, X.: Aster: an attentional scene text recognizer with flexible rectification. IEEE Trans. Pattern Anal. Mach. Intell. **41**(9), 2035–2048 (2018)
34. Su, B., Lu, S.: Accurate scene text recognition based on recurrent neural network. In: Cremers, D., Reid, I., Saito, H., Yang, M.-H. (eds.) ACCV 2014. LNCS, vol. 9003, pp. 35–48. Springer, Cham (2015). https://doi.org/10.1007/978-3-319-16865-4_3
35. Wan, Z., He, M., Chen, H., Bai, X., Yao, C.: Textscanner: reading characters in order for robust scene text recognition. arXiv preprint arXiv:1912.12422 (2020)
36. Wang, J., Hu, X.: Gated recurrent convolution neural network for ocr. In: Advances in Neural Information Processing Systems, pp. 335–344 (2017)

37. Wang, K., Babenko, B., Belongie, S.: End-to-end scene text recognition. In: 2011 International Conference on Computer Vision, pp. 1457–1464. IEEE (2011)
38. Wang, K., Belongie, S.: Word spotting in the wild. In: Daniilidis, K., Maragos, P., Paragios, N. (eds.) ECCV 2010. LNCS, vol. 6311, pp. 591–604. Springer, Heidelberg (2010). https://doi.org/10.1007/978-3-642-15549-9_43
39. Wang, W., et al.: Textsr: content-aware text super-resolution guided by recognition. arXiv:1909.07113 (2019)
40. Wang, X., et al.: Esrgan: enhanced super-resolution generative adversarial networks. In: Proceedings of the European Conference on Computer Vision (ECCV) (2018)
41. Wei, K., Yang, M., Wang, H., Deng, C., Liu, X.: Adversarial fine-grained composition learning for unseen attribute-object recognition. In: Proceedings of the IEEE International Conference on Computer Vision, pp. 3741–3749 (2019)
42. Yang, M., et al.: Symmetry-constrained rectification network for scene text recognition. In: Proceedings of the IEEE International Conference on Computer Vision, pp. 9147–9156 (2019)
43. Yao, C., Bai, X., Shi, B., Liu, W.: Strokelets: a learned multi-scale representation for scene text recognition. In: Proceedings of the IEEE Conference on Computer Vision and Pattern Recognition, pp. 4042–4049 (2014)
44. Yin, X.C., Yin, X., Huang, K., Hao, H.W.: Robust text detection in natural scene images. IEEE Trans. Pattern Anal. Mach. Intell. **36**(5), 970–983 (2013)
45. Zhan, F., Lu, S.: Esir: end-to-end scene text recognition via iterative image rectification. In: Proceedings of the IEEE Conference on Computer Vision and Pattern Recognition, pp. 2059–2068 (2019)
46. Zhang, R.: Making convolutional networks shift-invariant again. arXiv preprint arXiv:1904.11486 (2019)
47. Zhang, Y., Li, K., Li, K., Wang, L., Zhong, B., Fu, Y.: Image super-resolution using very deep residual channel attention networks. In: Proceedings of the European Conference on Computer Vision (ECCV), pp. 286–301 (2018)

Memory Selection Network for Video Propagation

Ruizheng Wu[1(✉)], Huaijia Lin[1], Xiaojuan Qi[2], and Jiaya Jia[1,3]

[1] The Chinese University of Hong Kong, Hong Kong, China
{rzwu,linhj,leojia}@cse.cuhk.edu.hk
[2] University of Hong Kong, Hong Kong, China
xjqi@eee.hku.hk
[3] SmartMore, Shenzhen, China

Abstract. Video propagation is a fundamental problem in video processing where guidance frame predictions are propagated to guide predictions of the target frame. Previous research mainly treats the previous adjacent frame as guidance, which, however, could make the propagation vulnerable to occlusion, large motion and inaccurate information in the previous adjacent frame. To tackle this challenge, we propose a memory selection network, which learns to select suitable guidance from all previous frames for effective and robust propagation. Experimental results on video object segmentation and video colorization tasks show that our method consistently improves performance and can robustly handle challenging scenarios in video propagation.

1 Introduction

Video propagation is a fundamental technique in video processing tasks, including video colorization [18,46,47], video semantic segmentation [14,27], video object segmentation [5,16,20,21,29], to name a few. It aims at propagating information from an annotated or intermediate guidance frame to the entire video.

Prior work [5,16,20,21,29] mainly focused on propagating information in a frame-by-frame fashion as illustrated in Fig. 1(a) where adjacent frames are utilized to update the target one. This propagation pipeline is fragile due to accumulation of errors, since inaccurate predictions in previous frames inevitably influence target frame prediction. The influence is magnified especially when the target object disappears or is misclassified.

To address the error accumulation caused by frame-by-frame propagation, one feasible solution is to utilize the information from all previous frames to

---

R. Wu and H. Lin—Equal Contribution.

Electronic supplementary material The online version of this chapter (https://doi.org/10.1007/978-3-030-58555-6_11) contains supplementary material, which is available to authorized users.

A. Vedaldi et al. (Eds.): ECCV 2020, LNCS 12360, pp. 175–190, 2020.
https://doi.org/10.1007/978-3-030-58555-6_11

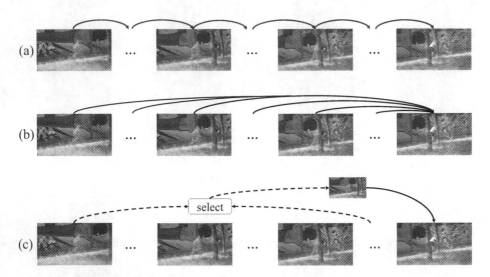

Fig. 1. Illustration of different strategies for propagating segmentation masks. (a) Frame-by-frame propagation. (b) Propagation with all previous frames. (c) Our proposed strategy for selecting the proper guidance frames for propagation.

propagate them to the current one, as illustrated in Fig. 1(b). Albeit reasonable, these frames contain a lot of redundant and cluttered information, and the problem becomes more serious as the number of previous frames increases. Thus, selecting the best frame for propagating information effectively and robustly in videos is a critical issue.

In this paper, we propose a memory selection network (MSN) to vastly benefit generic video propagation. To update information in the current frame, shown in Fig. 1(c), our MSN selects the most informative frames from a memory pool, which caches all previous frames as features. We note that this seemingly simple idea produces promising results. It effectively reduces error accumulation while not affecting computation cost much.

Our selection network serves as a generic and efficient component for video propagation to complement any propagation methods. Specifically, we apply our proposed memory selection network to different video propagation based approaches, including a classical temporal propagation network (TPN) built by us, and recent state-of-the-art propagation framework STM [28]. Their performance is boosted with incorporation of the memory selection network. Moreover, to further demonstrate the generality and usefulness, we conduct experiments on both video object segmentation and video colorization tasks. Our overall contributions are summarized below.

- We propose a memory selection network (MSN) to select suitable guidance frames for video propagation.
- MSN is generic to be integrated into any video propagation framework.

– Experimental results on video object segmentation and video colorization demonstrate that our approach boosts video propagation with limited computational cost.

2   Related Work

Video Propagation. Propagation across image and video pixels is a common technique in various computer vision tasks, such as image/video colorization [18,46,47], matting [19], object segmentation [14,28,29,39], and semantic segmentation [14]. Traditional priors for propagation are mainly optimization-based [18,19], which minimize the energy function on a graph. In addition, filtering-based approaches [10,32] propagate information using image or video filters, faster than the optimization-based method.

Recently, several methods model spatial or temporal pixel-pixel relationship with convolutional neural networks. Jampani et al. [14] used bilateral CNN to model the relationship between neighborhood pixels. Liu et al. [23,24] developed an affinity map for pixel propagation with CNN. In addition, there are a lot of object-level propagation approaches [4,14,21,28,29,39] using deep neural networks specifically designed for video object segmentation.

Most of the above propagation approaches treat adjacent previous frame as the guidance for propagation, allowing the system to easily accumulate errors through different propagation steps. Oh et al. [28] utilizes information from multiple previous frames and adaptively fuses them for propagation to the target.

In this paper, we design a generic module to select suitable frames for video propagation. It can be seamlessly inserted into these propagation approaches to improve stability, robustness and quality.

Semi-Supervised Video Object Segmentation. Semi-supervised object segmentation refers to the problem of segmenting all corresponding objects annotated in the first frame. A group of frameworks were proposed to tackle this problem [2,3,6,7,9,28,29,35,39,40]. Some of them [2,21,29] rely on the online learning technique, which requires time-consuming fine-tuning on the annotated frame for each testing sequence.

Among these approaches, one major stream contains propagation-based methods. MaskTrack [29] provides a classical propagation baseline method using the last frame mask or optical flow as guidance. Many following methods [16,20,28,39,43] are based on it and improve it with more components or better strategies. LucidTracker [16] incorporated additional data augmentation during online training. Li et al. [20] fixed long-term propagation errors by introducing a re-identification module to complement frame-by-frame propagation. The reference image is introduced as guidance for better propagation in the work of [39,43]. The network design is also improved correspondingly. These approaches utilize the previous adjacent frame for propagation, which makes the system easily fail in long-term propagation. STM [28] utilizes more previous frames in an effective way. Based on these propagation approaches, we propose a selection strategy for the guidance frame to improve performance from another perspective.

For high-quality long-term propagation, ConvGRU or ConvLSTM structures [34,41] were utilized to build an implicit memory module for long-term propagation. Such approaches may suffer from memory and optimization issues when capturing long-range dependency during the training stage. Our method differs from these RNN-based methods in that we build an external memory pool to select the appropriate guidance frame without memory constraints.

Apparently similar work to ours is BubbleNet [9] since both BubbleNet and our work design an additional network to help boost performance. The difference is also clear and fundamental: the BubbleNet network aims to find the best frame to be annotated by a human before applying any propagation methods, while our work determines which of the previous predictions would be most helpful for prediction of the current frame.

Video Colorization. Video colorization can also be addressed using video propagation approaches. Interactive colorization [44] propagates annotated strokes spatially across frames. The propagation procedure is guided by the matting Laplacian matrix and manually defined similarities. CNN-based methods [46,47] achieved colorization with fully-automatic or sparsely annotated color. Recently, Liu et al. [24] proposed a switchable temporal propagation network to colorize all frames in a video using a few color key-frames. Additionally, methods of [14,38] colorize the video sequence with the annotated first frame. To propagate annotated color information to the whole video, VPN [14] utilized a bilateral space to retrieve the pixel color and Vondrick et al. [38] leveraged pixel embedding for soft aggregation.

3   Proposed Method

3.1   Overview

We propose a generic memory selection network to select the appropriate guidance frame for general video propagation. In the following, we use the video object segmentation task as an example to illustrate our approach.

Formulation. We denote a video sequence with T frames as $\{x_t | t \in [0, T-1]\}$, where x_t refers to the raw frame at time step t. Given the annotation information y_0 of the first frame x_0, the goal of video propagation is to propagate the information to the whole video, i.e. to produce $\{y_t | t \in [1, T-1]\}$ from time step 1 to $T-1$ via a propagation module $\mathcal{P}$. For each target frame x_t, $\mathcal{P}$ utilizes the guidance image x_g, and the corresponding prediction result or annotation y_g, to obtain y_t. This can be formulated as

$$y_t = \mathcal{P}(x_t, x_g, y_g). \tag{1}$$

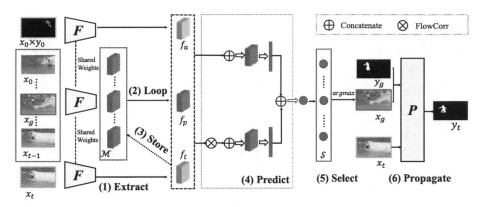

Fig. 2. Illustration of our framework. For each frame x_t in a video sequence, we first (1) **extract** feature f_t by a feature extractor $\mathcal{F}$, then we (2) **loop** for extracted features f_p in the memory pool $\mathcal{M}$, which is constructed from previous frames. At the same time, we also (3) **store** f_t back into $\mathcal{M}$ for later frames. For all f_p in $\mathcal{M}$, we (4) **predict** the selection score for each with the input of f_a, f_p and f_t, where f_a is the feature of the first frame masked with annotated objects. We then (5) **select** the frame x_g with the highest score S as guidance. Finally, x_g is utilized as the guidance frame for x_t to (6) **propagate** and obtain y_t. 'FlowCorr' is developed in FlowNet [13].

Previous frame-by-frame propagation is derived as $y_t - P(x_t, x_{t-1}, y_{t-1})$, where x_{t-1} serves as the guidance frame for frame x_t. In contrast, our approach aims to select the suitable guidance frame $x_g \in \{x_0, x_1, ..., x_{t-1}\}$ for propagation.

Workflow. The overall workflow of our framework is shown in Fig. 2. Our system builds a memory pool $\mathcal{M} = \{f_p | p = 0, 1, ..., T - 1\}$ by sequentially caching features of previous frames, where f_p represents the extracted representation of frames x_p with the feature extractor network $\mathcal{F}$. We also extract feature of the first frame masked with the annotated objects as f_a for subsequent selection score prediction. To select a proper guidance frame for x_t, we extract its feature f_t at first and estimate the selection score for all features $\{f_p | p = 0, 1, ..., t - 1\}$ cached in the memory pool via a light-weight selection network. It takes f_t, f_p and the feature of the annotated frame f_a as input, and outputs the corresponding selection scores. The frame with the highest score is selected as guidance for propagation, denoted as x_g. The propagation network $\mathcal{P}$ takes x_t, x_g and y_g as input to produce the final prediction y_t. f_t is cached back into $\mathcal{M}$ for subsequent frames.

It is worth noting that the feature extraction step takes much more time than the selection score estimation step, since the former is accomplished by a complicated network (i.e. VGG16 [33]) while the latter only uses a light-weight selection network consisting of only a few convolutional layers. Thus construction of the memory pool saves a lot of time by eliminating the feature extraction step of previous frames.

3.2   Memory Pool Construction

Representation. The memory pool $\mathcal{M}$ is a set of features $\{f_p\}$, where $f_p \in \mathbb{R}^{512 \times \frac{H}{32} \times \frac{W}{32}}$, where H and W are the original spatial sizes. f_p is extracted from the corresponding frame $x_p \in \mathbb{R}^{H \times W \times 3}$. We use a 2-D feature map instead of a vector to represent the memory because the spatial information is important in dense video propagation, e.g. video object segmentation and video colorization.

Construction Pipeline. To construct the memory pool $\mathcal{M}$, the feature extractor $\mathcal{F}$ first takes the first frame x_0 as input and obtains the feature f_0. Then f_0 is cached to initialize $\mathcal{M}$. Additionally, we need the feature f_a concerning only the annotated object to make the selection operation aware of the target object. To this end, f_a is extracted from the annotated object x_a, which is obtained from x_0, whose background is masked by the annotated mask y_0. For each time step of the video sequence, the extracted feature of the target frame f_t is also cached into $\mathcal{M}$, which guarantees the efficiency that each frame only needs to be processed once by $\mathcal{F}$.

3.3   Memory Selection Network

Observation. In frame-by-frame propagation, we have $y_t = \mathcal{P}(x_t, x_{t-1}, y_{t-1})$. Thus we can empirically infer that error accumulation of y_t stems from the prediction quality of y_{t-1} (the first factor) and the similarity between x_t and x_{t-1} (the second factor). We conduct experiments on YouTube-VOS [42] validation set[1] to verify the effect of these two factors (described below). For clarity, l_t indicates the ground-truth label of the tth frame and $\mathrm{IoU}(\cdot, \cdot)$ indicates the intersection over union between two masks in the following description.

The influence of the prediction quality of y_{t-1} is illustrated in Fig. 3(a). $\mathrm{IoU}(y_t, l_t)$ and $\mathrm{IoU}(y_{t-1}, l_{t-1})$ indicate the prediction quality of the last and target frames. As shown in Fig. 3 (a), prediction quality of the $(t-1)^{th}$ and t^{th} frames is positively related, i.e. low quality y_{t-1} degrades y_t.

As for the other factor, the relation of y_t and similarity between x_t and x_{t-1} are plotted in Fig. 3(b), where we use $\mathrm{IoU}(l_t, l_{t-1 \rightarrow t})$ to represent similarity between the two frames, where $l_{t-1 \rightarrow t}$ denotes the label warped from previous frame using optical flow [8]. It clearly draws the conclusion that the high similarity between x_t and x_{t-1} generally improves the propagation result y_t.

Selection Criterion. According to the observations above, error accumulation in the frame-by-frame propagation pipeline is mainly influenced by two factors, i.e., the prediction quality of the guidance frame and its similarity with the target frame. Intuitively, if segmentation of the previous frame prediction is erroneous, the inaccurate information can be propagated to the target frame and accumulate dramatically across frames. Moreover, frames with high similarity reduce errors in the propagation stage and can help robust propagation.

---

[1] YouTube-VOS online server returns a TEXT file containing the per frame IoU for each submission.

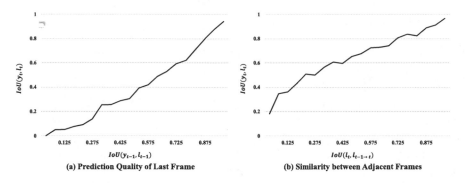

(a) Prediction Quality of Last Frame   (b) Similarity between Adjacent Frames

Fig. 3. Influence of two factors regarding error accumulation. Curves are plotted using TPN (Sect. 3.4).

Our selection network is designed to capture the above two factors. First, since the annotated frame is manually labeled by humans and is free of network prediction error, we adopt the feature map of the annotated frame f_a to help model the prediction quality of the guidance frame. We combine features from the annotated frame f_a and the guidance frame f_p with the FlowCorr operation $\text{FlowCorr}(f_p, f_a)$. The "FlowCorr" operation is developed in [13], which combines two feature maps by calculating the feature similarity between pixels. Next, to model similarity between f_p and target f_t, we further adopt the FlowCorr operation to combine their representation as $\text{FlowCorr}(f_p, f_t)$. The FlowCorr operation does not require computation of optical flow and is thus efficient.

Selection Network Design. The selection module is designed as a binary classification network with 'good' and 'bad' categories. Specifically, for each feature f_p in the memory pool $\mathcal{M}$, the goal of the selection network is to calculate a score to measure utility for selecting it as the guidance frame regarding the target frame t. To this end, the selection network takes f_p, the annotated object's feature f_a, and target feature f_t as input and outputs a selection score. The highest-score one is selected as the guidance frame.

Based on the two key factors above, our selection network shown in Fig. 2 adopts a two-stream structure. First, we use $\text{FlowCorr}(f_p, f_a)$, $\text{FlowCorr}(f_p, f_t)$ and $\text{FlowCorr}(f_t, f_a)$ to measure the relationship between guidance and annotated frames, guidance and target frames, and target and annotated frames, respectively. Then, a two stream network separately takes concatenation of $\{\text{FlowCorr}(f_p, f_a)$, $\text{FlowCorr}(f_p, f_t)$, $\text{FlowCorr}(f_t, f_a)\}$, and concatenation of $\{f_p, f_a, f_t\}$ as input and produces two feature vectors, which are further concatenated followed by a fully connected layer to generate the selection score. The detailed network structure is included in the supplementary material.

The memory selection network is light-weighted with only several convolution layers and fully-connected layers. The selection process can also be parallel for acceleration.

3.4   Video Propagation Frameworks

We select several video propagation based frameworks as baselines to verify the effectiveness of MSN.

Temporal Propagation Network (TPN). To verify the effectiveness of our proposed memory selection network, we build a classical temporal propagation network (TPN) as one baseline model. The design of TPN is similar to existing propagation-based frameworks [16,21,29]. It takes target frame x_t, selected guidance frame x_g, and corresponding predicted or annotated label y_g as input, and outputs the prediction label for the target frame y_t (i.e., Eq. (1)).

TPN consists of an appearance branch and a motion branch. The appearance branch takes x_t and y_g as input, while the motion branch takes optical flow $O_{t \to g}$ (between x_t and x_g) and y_g as input. Their output is further concatenated to obtain the final result. The detailed structure of TPN is described in our supplementary material. For each frame x_t, MSN selects proper x_g for TPN as input.

STM [28]. STM is a state-of-the-art semi-supervised video object segmentation network. It is composed of three modules: 1) memory encoder, 2) query encoder, and 3) query decoder. The memory encoder encodes previous masks as well as corresponding frames into the memorized features. The target frame is encoded by the query encoder into a new feature and is further fused with the propagated memorized features. The fused feature is utilized to decode the mask for the target frame. We also incorporate MSN into STM by selecting the suitable memorized feature for propagating mask information.

3.5   Training Pipeline

Two Stage Training. Since the *argmax* operation is non-differentiable, the whole system adopts two-stage training. In the first stage, different video propagation frameworks are trained to converge. For training TPN, we adopt IoU loss [22] for video object segmentation and $\mathcal{L}_1$ regression loss for video colorization. For STM, we adopt their official pre-trained model in our experiments.

In the second stage, video propagation networks are fixed. They are used to generate the training samples to train the memory selection network. MSN is a binary classification network to estimate the quality of the guidance frame for the current frame in propagation. We adopt binary cross-entropy loss for MSN training. The method to generate positive ('good' guidance frames) and negative ('bad' guidance frames) training samples is elaborated below.

Generating Training Samples for MSN. In the process of generating training samples for MSN, for each frame x_t in training sequences, we utilize the trained video propagation networks to propagate all previous frames $\{x_p | p = 0, 1, ..., t-1\}$ to x_t to obtain $t-1$ propagation results, denoted as

$\{y_{p \to t} | p = 0, 1, ..., t - 1\}$. With the label l_t of current frame, we obtain the IoU score of propagation results, i.e., $\{IoU(y_{p \to t}, l_t) | p = 0, 1, ..., t - 1]\}$.

To split $\{x_p | p = 0, 1, ..., t - 1\}$ into positive and negative samples, we first compute the highest IoU score IoU_{max} and the lowest score IoU_{min} among all these frames, and we set two thresholds σ_{pos} and σ_{neg} as hyper-parameters. The samples with IoU score in $[IoU_{max} - IoU_{max} * \sigma_{pos}, IoU_{max}]$ are split into positive samples, while those with score in $[IoU_{min}, IoU_{min} + IoU_{min} * \sigma_{neg}]$ are regarded as negative ones. The frames not belonging to either positive or negative samples are abandoned to avoid harming the classifier. These positive and negative samples are then used to train our memory selection network with binary cross-entropy loss. In our experiments, we empirically set the positive and negative thresholds as $0.05(\sigma_{pose})$ and $0.15(\sigma_{neg})$ respectively.

Implementation Details. We use Adam [17] stochastic optimization, with the initial learning rate as 1e-5 and polynomial learning policy. The input image is resized to a fixed-size 640×320. TPN/MSN are trained on YouTube-VOS for 30/6 epochs and fine-tuned on DAVIS for 50/10 epochs. TPN is trained by randomly sampling two frames in a video as the guidance and target frames.

4   Experiments

We evaluate our proposed memory selection network on two different video propagation tasks: video object segmentation and grayscale video colorization. We focus on their semi-supervised setting where only the first frame is annotated with segmented mask or color. For the video object segmentation task, we evaluate our method on YouTube-VOS [42], and DAVIS 2016 and 2017 datasets [30,31]. As for the video colorization dataset, following the work of [14], we conduct experiments on DAVIS 2016 dataset for evaluation.

The performance of the video object segmentation task is measured by region similarity $\mathcal{J}$ and contour accuracy $\mathcal{F}$ defined in [30]. Besides, For YouTube-VOS validation dataset, since there are 'seen' and 'unseen' categories, we provide $\mathcal{J}$ seen, $\mathcal{F}$ seen, $\mathcal{J}$ unseen and $\mathcal{F}$ unseen as corresponding metrics and *Overall* refers to the average score of them.

For grayscale video colorization, we evaluate the results with PSNR score and $\mathcal{L}_1$ distance between the generated results and its corresponding ground-truth.

4.1   Comparison with State-of-the-arts

Video Object Segmentation (VOS).

YouTube-VOS Dataset. YouTube-VOS [42] dataset is the largest video object segmentation dataset with diverse objects, which contains 3471 training videos and 474 validation ones. The validation videos contain totally 91 object categories, with 65 seen categories and 26 unseen ones.

Table 1. Results of Video Object Segmentation on YouTube-VOS validation set. 'OL' denotes online training. '*' denotes using pre-trained weights on DAIVS Dataset [30, 31]. For all propagation based methods, we consider one-frame propagation.

Methods	Seen		Unseen			OL
	$\mathcal{J}(\%)$	$\mathcal{F}(\%)$	$\mathcal{J}(\%)$	$\mathcal{F}(\%)$	$Overall(\%)$	
OSVOS [2]	59.8	60.5	54.2	60.7	58.8	✓
MaskTrack [29]	59.9	59.5	45.0	47.9	53.1	✓
OnAVOS [37]	60.1	62.7	46.6	51.4	55.2	✓
S2S [41]	71.0	70.0	55.5	61.2	64.4	✓
PReMVOS [25]	**71.4**	**75.9**	**56.5**	**63.7**	**66.9**	✓
OSMN [43]	60.0	60.1	40.6	44.0	51.2	
RVOS [35]	63.6	45.5	**67.2**	51.0	56.8	
DMM [45]	60.3	63.5	50.6	57.4	58.0	
RGMP [39]	59.5	–	45.2	–	53.8	
A-GAME [15]	66.9	–	61.2	–	66.0	
TPN	64.0	65.9	57.0	65.4	63.0	
TPN + MSN	65.7	68.0	58.0	66.3	64.5 (+1.5)	
*STM-1	71.1	74.4	64.0	69.7	69.9	
*STM-1 + MSN	**72.4**	**75.2**	65.4	**71.4**	**71.1** (+1.2)	

We compare our method with state-of-the-art methods. The quantitative results are presented in Table 1. We utilize STM with only one previous frame for propagation as baseline, denoted as 'STM-1', and we apply MSN to select one frame to replace the previous frame ('STM-1 + MSN'). We provide results with TPN and STM-1 as baseline video propagation frameworks, and incorporate our memory selection module (MSN) into them as 'TPN + MSN' and 'STM-1 + MSN'. For both video propagation networks, we achieve consistent improvement, in terms of *Overall* score, of 1.5% and 1.2% respectively. We note since the pre-trained model of STM on YouTube-VOS dataset is not provided, we here adopt their pre-trained model on DAVIS for inference.

DAVIS 2016 and 2017 Datasets. We further conduct experiments on DAVIS 2016 and 2017 datasets. DAVIS-2016 [30] is a popular single object segmentation benchmark, consisting of 30 training and 20 validation videos. DAVIS-2017 [31] is an extended version of DAVIS-2016 with multiple objects in a video sequence, consisting of 60 training and 30 validation videos.

We evaluate MSN with two baselines of TPN and STM [28] on the validation sets. Our memory selection module consistently benefits the baseline methods by choosing one suitable reference frame on both single-object- and multi-object-segmentation. MSN improves the baselines by 0.6% to 1.6% on both datasets, proving its effectiveness (Table 2).

Table 2. Comparison of video object segmentation methods on DAVIS 2016 and 2017 validation sets, where 'OL' indicates online learning techniques. For all propagation based methods, we consider one-frame propagation.

Methods	DAVIS-2016		DAVIS-2017			
	$\mathcal{J}(\%)$	$\mathcal{F}(\%)$	$\mathcal{J}(\%)$	$\mathcal{F}(\%)$	Runtime (s)	OL
OSVOS [2]	79.8	80.6	56.6	63.9	10	✓
PReMVOS [25]	84.9	**88.6**	**73.9**	**81.7**	–	✓
OSVOS-S [26]	85.6	86.4	64.7	71.3	4.5	✓
OnAVOS [37]	**86.1**	84.9	64.5	71.2	13	✓
CINM [1]	83.4	85.0	67.2	74.2	–	✓
MaskRNN [11]	80.7	80.9	60.5	–	–	✓
FAVOS [4]	82.4	79.5	54.6	61.8	**1.8**	✓
OSMN [43]	74.0	–	52.5	57.1	0.14	
VidMatch [12]	81.0	–	56.5	68.2	0.32	
FEELVOS [36]	81.1	82.2	69.1	74.0	0.51	
RGMP [39]	81.5	82.0	64.8	68.8	0.13	
A-GAME [15]	82.0	82.2	67.2	72.7	0.07	
DMM [45]	–	–	68.1	73.3	0.08	
TPN	75.8	74.2	58.9	62.7	0.17	
TPN+MSN	76.8	74.6	59.5	63.3	0.21	
STM-1 [28]	83.2	83.3	69.6	74.6	**0.06**	
STM-1 + MSN	**83.8**	**84.9**	**71.4**	**76.8**	0.10	

Visual Quality Results. A selected sequence is visualized in Fig. 4. For the results in TPN, the prediction error in the segmented mask accumulates and propagates along with the naive frame-by-frame propagation strategy. However, by selecting a proper guidance frame, we alleviate error accumulated and thus support high-quality long-term propagation.

Grayscale Video Colorization

Quantitative Results. We also evaluate our proposed memory selection network on the grayscale video colorization task using the same training and inference strategies as the video object segmentation task. To quantify the effectiveness of our memory selection network, following VPN [14], we evaluate our algorithm on DAVIS-2016 dataset. For each video sequence, we take the first frame as the annotated color frame and propagate color to the rest of grayscale frames. PSNR and $\mathcal{L}_1$ between predicted target frame and ground-truth one in RGB color space are adopted as the evaluation metrics. Table 3 gives comparison among our framework and others. 'TPN + MSN' achieves the best performance in terms of both PSNR and $\mathcal{L}_1$, and MSN improves results a lot on this task.

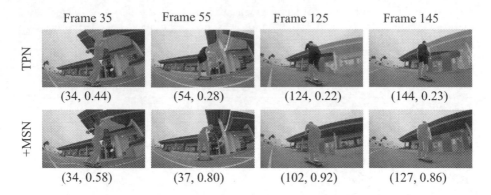

Fig. 4. Visualization of video object segmentation on YouTube-VOS validation set. For each target frame, $(\cdot, \cdot)$ denotes the index of selected guidance frame and the *Overall* score of propagated mask.

Table 3. Quantitative comparison of grayscale video colorization. ↑ means 'the higher the better'. ↓ means the opposite: 'the lower the better'.

Methods	PSNR ↑	$\mathcal{L}_1$ ↓	Runtime (s)
BNN-Identity [14]	27.89	13.51	0.29
VPN-Stage1 [14]	28.15	13.29	0.9
Levin et al. [18]	27.11	–	19
TPN	28.25	11.06	**0.23**
TPN+MSN	**28.57**	**10.76**	0.27

Visual Quality Results. Figure 5 shows visual results of a sample sequence. The color information reduces gradually by naive frame-by-frame propagation in TPN. TPN equipped with MSN preserves color information well since a better guidance frame is selected from the memory pool and propagated to each target frame. The whole framework propagates color information much longer than the baseline propagation network, which greatly helps colorization for its final quality.

4.2   Ablation Study

Comparison of Selection Strategies. In this section, we explore whether our designed selection network can be replaced by other simpler selection strategies.

- *VGG_select* : The guidance frame is selected by comparing its feature space distance with the target frame. The feature is extracted from pre-trained VGG [33] without fine-tuning.
- *VGG_mask_select* : To compare the distance, VGG feature distance of both the guidance frame and masked guidance frame with VGG feature of the target frame are separately computed and then added up.

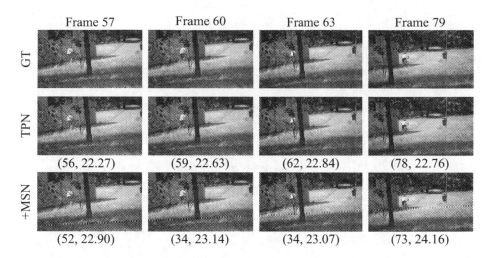

Fig. 5. Visualization of grayscale video colorization. $(\cdot,\cdot)$ below each frame denotes the index of selected guidance frame and the corresponding PSNR score with regard to ground-truth color image.

– *Time step gap* k : The guidance frame is selected by a fixed time-step gap k with the target frame. For each frame x_t, the prediction is calculated by $y_t = \mathcal{P}(x_t, x_{max(0,t-k)}, y_{max(0,t-k)})$.

We test the above selection strategies as well as our trained memory selection network for video object segmentation on YouTube-VOS validation set. The performance is demonstrated in Table 4. Simply selecting the most similar guidance frames in VGG feature space is insufficient for propagation since pre-trained VGG features are not aware of the propagation quality between frames. Moreover, the performance of simply selecting the frames with fixed time step gaps can be greatly erroneous since two frames far away may be significantly different in appearance. They increase difficulty of generating accurate motion information.

Table 4. Performance of different selection strategies on YouTube-VOS validation set. '*Overall*' metric defined in [42] measures the performance of different strategies.

Selection strategies	Overall
VGG_select	63.03
VGG_mask_select	63.08
Time step gap 1	63.04
Time step gap 5	63.13
Time step gap 10	59.67
Time step gap 20	55.61
MSN	**64.5**

Table 5. Performance and runtime for ensemble strategies. 'TPN-K' indicates ensembling predictions from the last K frames. '+MSN-K' indicates that the ensembled predictions are selected with the K highest selection scores.

Ensembles	Overall	Runtime
TPN-1	63.04	**0.09**
+MSN-1	64.54	0.13
TPN-3	65.27	0.29
+MSN-3	65.8	0.33
TPN-5	65.6	0.49
+MSN-5	**66.1**	0.53

Prediction Ensemble. Ensemble is an important means to improve propagation accuracy in the inference stage. In frame-by-frame propagation, the predictions from the last K frames are ensembled to produce the tth frame prediction. It is intriguing to investigate how to ensemble the predictions of selected guidance frames. Since MSN is trained as a binary classifier, the prediction score can be considered as the confidence of 'positive' for a guidance frame.

To ensemble of propagation for the memory selection network, the frames in the selection pool are ranked according to the scores obtained by the selection network. The top-K highest scoring frames are ensembled for the tth frame. We conduct experiments on K and test its performance and runtime. As shown in Table 5, ensembling multiple guidance frames consistently increases the accuracy on different K.

Oracle Results. We investigate the potential of MSN by applying ground-truth labels to select guidance frames. Specifically, for each target frame t with ground-truth label l_t, we compute the propagation results from all preceding frames, represented as $\{y_{p \to t} | p \in [0, t-1]\}$. The propagation mask with the highest accuracy $\mathrm{IoU}(y_{p \to t}, l_t)$ is selected as the prediction mask for the tth frame. As illustrated in Table 6, 'Oracle-MSN' achieves much better results than 'TPN' and 'TPN+MSN'. The results demonstrate that there is still much space to improve memory selection results.

Table 6. Oracle results in video object segmentation. '+MSN-Oracle' denotes selecting the guidance frame with the highest propagation accuracy. We report *Overall* and $\mathcal{J}$ scores for YouTube-VOS and DAVIS-2016, respectively.

Method	YouTube-VOS	DAVIS-2016
TPN	63.0	75.7
+MSN	64.5	76.4
+MSN-Oracle	75.3	82.4

5   Conclusion

We have presented a memory selection network for the robust video propagation by dynamically selecting the guidance frame to update information about the target frame. The memory selection network can select suitable guidance frames based on the quality of the guidance frame and its relationship with the target frame. Experimental results on video object segmentation and video colorization demonstrate that our method improves robustness of video propagation consistently.

References

1. Bao, L., Wu, B., Liu, W.: CNN in MRF: video object segmentation via inference in a CNN-based higher-order spatio-temporal MRF. In: CVPR (2018)
2. Caelles, S., Maninis, K.K., Pont-Tuset, J., Leal-Taixé, L., Cremers, D., Van Gool, L.: One-shot video object segmentation. In: CVPR (2017)
3. Chai, Y.: Patchwork: a patch-wise attention network for efficient object detection and segmentation in video streams. In: ICCV (2019)
4. Cheng, J., Tsai, Y.H., Hung, W.C., Wang, S., Yang, M.H.: Fast and accurate online video object segmentation via tracking parts. In: CVPR (2018)
5. Cheng, J., Tsai, Y.H., Wang, S., Yang, M.H.: Segflow: joint learning for video object segmentation and optical flow. In: ICCV (2017)
6. Ci, H., Wang, C., Wang, Y.: Video object segmentation by learning location-sensitive embeddings. In: ECCV (2018)
7. Duarte, K., Rawat, Y.S., Shah, M.: Capsulevos: semi-supervised video object segmentation using capsule routing. In: ICCV (2019)
8. Farnebäck, G.: Two-frame motion estimation based on polynomial expansion. In: Scandinavian conference on Image analysis (2003)
9. Griffin, B.A., Corso, J.J.: Bubblenets: learning to select the guidance frame in video object segmentation by deep sorting frames. In: CVPR (2019)
10. He, K., Sun, J., Tang, X.: Guided image filtering. TPAMI (2013)
11. Hu, Y.T., Huang, J.B., Schwing, A.: Maskrnn: instance level video object segmentation. In: NeurIPS (2017)
12. Hu, Y.T., Huang, J.B., Schwing, A.G.: Videomatch: matching based video object segmentation. In: ECCV (2018)
13. Ilg, E., Mayer, N., Saikia, T., Keuper, M., Dosovitskiy, A., Brox, T.: Flownet 2.0: evolution of optical flow estimation with deep networks. In: CVPR (2017)
14. Jampani, V., Gadde, R., Gehler, P.V.: Video propagation networks. In: CVPR (2017)
15. Johnander, J., Danelljan, M., Brissman, E., Khan, F.S., Felsberg, M.: A generative appearance model for end-to-end video object segmentation. In: CVPR (2019)
16. Khoreva, A., Benenson, R., Ilg, E., Brox, T., Schiele, B.: Lucid data dreaming for multiple object tracking. arXiv:1703.09554 (2017)
17. Kingma, D.P., Ba, J.: Adam: a method for stochastic optimization. arXiv preprint arXiv:1412.6980 (2014)
18. Levin, A., Lischinski, D., Weiss, Y.: Colorization using optimization. In: TOG (2004)
19. Levin, A., Lischinski, D., Weiss, Y.: A closed-form solution to natural image matting. TPAMI (2008)
20. Li, X., et al.: Video object segmentation with re-identification. In: The 2017 DAVIS Challenge on Video Object Segmentation-CVPR Workshops (2017)
21. Li, X., Change Loy, C.: Video object segmentation with joint re-identification and attention-aware mask propagation. In: ECCV (2018)
22. Li, Z., Chen, Q., Koltun, V.: Interactive image segmentation with latent diversity. In: CVPR (2018)
23. Liu, S., De Mello, S., Gu, J., Zhong, G., Yang, M.H., Kautz, J.: Learning affinity via spatial propagation networks. In: NeurIPS (2017)
24. Liu, S., Zhong, G., De Mello, S., Gu, J., Yang, M.H., Kautz, J.: Switchable temporal propagation network. arXiv:1804.08758 (2018)

25. Luiten, J., Voigtlaender, P., Leibe, B.: Premvos: proposal-generation, refinement and merging for video object segmentation. In: ACCV (2018)
26. Maninis, K.K., et al.: Video object segmentation without temporal information. TPAMI **41**, 1515–1530 (2018)
27. Miksik, O., Munoz, D., Bagnell, J.A., Hebert, M.: Efficient temporal consistency for streaming video scene analysis. In: ICRA (2013)
28. Oh, S.W., Lee, J.Y., Xu, N., Kim, S.J.: Video object segmentation using space-time memory networks. In: ICCV (2019)
29. Perazzi, F., Khoreva, A., Benenson, R., Schiele, B., Sorkine-Hornung, A.: Learning video object segmentation from static images. In: CVPR (2017)
30. Perazzi, F., Pont-Tuset, J., McWilliams, B., Van Gool, L., Gross, M., Sorkine-Hornung, A.: A benchmark dataset and evaluation methodology for video object segmentation. In: CVPR (2016)
31. Pont-Tuset, J., Perazzi, F., Caelles, S., Arbeláez, P., Sorkine-Hornung, A., Van Gool, L.: The 2017 davis challenge on video object segmentation. arXiv:1704.00675 (2017)
32. Rick Chang, J.H., Frank Wang, Y.C.: Propagated image filtering. In: CVPR (2015)
33. Simonyan, K., Zisserman, A.: Very deep convolutional networks for large-scale image recognition. arXiv:1409.1556 (2014)
34. Tokmakov, P., Alahari, K., Schmid, C.: Learning video object segmentation with visual memory. In: ICCV (2017)
35. Ventura, C., Bellver, M., Girbau, A., Salvador, A., Marques, F., Giro-i Nieto, X.: Rvos: end-to-end recurrent network for video object segmentation. In: CVPR (2019)
36. Voigtlaender, P., Chai, Y., Schroff, F., Adam, H., Leibe, B., Chen, L.C.: Feelvos: fast end-to-end embedding learning for video object segmentation. In: CVPR (2019)
37. Voigtlaender, P., Leibe, B.: Online adaptation of convolutional neural networks for video object segmentation. arXiv:1706.09364 (2017)
38. Vondrick, C., Shrivastava, A., Fathi, A., Guadarrama, S., Murphy, K.: Tracking emerges by colorizing videos. In: ECCV (2018)
39. Wug Oh, S., Lee, J.Y., Sunkavalli, K., Joo Kim, S.: Fast video object segmentation by reference-guided mask propagation. In: CVPR (2018)
40. Xu, K., Wen, L., Li, G., Bo, L., Huang, Q.: Spatiotemporal CNN for video object segmentation. In: CVPR (2019)
41. Xu, N., et al.: Youtube-VOS: Sequence-to-sequence video object segmentation. In: ECCV (2018)
42. Xu, N., et al.: Youtube-VOS: a large-scale video object segmentation benchmark. arXiv:1809.03327 (2018)
43. Yang, L., Wang, Y., Xiong, X., Yang, J., Katsaggelos, A.K.: Efficient video object segmentation via network modulation. In: CVPR (2018)
44. Yatziv, L., Sapiro, G.: Fast image and video colorization using chrominance blending. TIP **15**, 1120–1129 (2006)
45. Zeng, X., Liao, R., Gu, L., Xiong, Y., Fidler, S., Urtasun, R.: DMM-net: Differentiable mask-matching network for video object segmentation. In: ICCV (2019)
46. Zhang, R., Isola, P., Efros, A.A.: Colorful image colorization. In: ECCV (2016)
47. Zhang, R., et al.: Real-time user-guided image colorization with learned deep priors. TOG (2017)

Disentangled Non-local Neural Networks

Minghao Yin[1], Zhuliang Yao[1,2], Yue Cao[2], Xiu Li[1], Zheng Zhang[2], Stephen Lin[2], and Han Hu[2(✉)]

[1] Tsinghua University, Beijing, China
{yinmh17,yzl17}@mails.tsinghua.edu.cn, li.xiu@sz.tsinghua.edu.cn
[2] Microsoft Research Asia, Beijing, China
{yuecao,zhez,stevelin,hanhu}@microsoft.com

Abstract. The non-local block is a popular module for strengthening the context modeling ability of a regular convolutional neural network. This paper first studies the non-local block in depth, where we find that its attention computation can be split into two terms, a whitened pairwise term accounting for the relationship between two pixels and a unary term representing the saliency of every pixel. We also observe that the two terms trained alone tend to model different visual clues, e.g. the whitened pairwise term learns within-region relationships while the unary term learns salient boundaries. However, the two terms are tightly coupled in the non-local block, which hinders the learning of each. Based on these findings, we present the disentangled non-local block, where the two terms are decoupled to facilitate learning for both terms. We demonstrate the effectiveness of the decoupled design on various tasks, such as semantic segmentation on Cityscapes, ADE20K and PASCAL Context, object detection on COCO, and action recognition on Kinetics. Code is available at https://github.com/yinmh17/DNL-Semantic-Segmentation and https://github.com/Howal/DNL-Object-Detection

1 Introduction

The non-local block [26], which models long-range dependency between pixels, has been widely used for numerous visual recognition tasks, such as object detection, semantic segmentation, and video action recognition. Towards better understanding the non-local block's efficacy, we observe that it can be viewed as a self-attention mechanism for pixel-to-pixel modeling. This self-attention is modeled as the dot-product between the features of two pixels in the embedding space. At first glance, this dot-product formulation represents *pairwise* relationships. After further consideration, we find that it may encode *unary* information

---

M. Yin and Z. Yao—Equal contribution. This work is done when Minghao Yin and Zhuliang Yao are interns at MSRA.

Electronic supplementary material The online version of this chapter (https://doi.org/10.1007/978-3-030-58555-6_12) contains supplementary material, which is available to authorized users.

A. Vedaldi et al. (Eds.): ECCV 2020, LNCS 12360, pp. 191–207, 2020.
https://doi.org/10.1007/978-3-030-58555-6_12

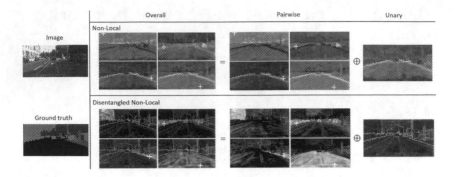

Fig. 1. Visualization of attention maps in the non-local block and our disentangled non-local block. With the disentanglement of our non-local block, the whitened pairwise term learns clear within-region clues while the unary term learns salient boundaries, which cannot be observed with the original non-local block

as well, in the sense that a pixel may have its own independent impact on all other pixels. Based on this perspective, we split the dot-product based attention into two terms: a whitened pairwise term that accounts for the impact of one pixel *specifically* on another pixel, and a unary term that represents the influence of one pixel *generally* over all the pixels.

We investigate the visual properties of each term without interference from the other. Specifically, we train two individual networks, with either the whitened pairwise term or the unary term removed in the standard attention formula of the non-local block. It is found that the non-local variant using the whitened pairwise term alone generally learns within-region relationships (the 2nd row of Fig. 3), while the variant using the unary term alone tends to model salient boundaries (the 3rd row of Fig. 3). However, the two terms do not learn such clear visual clues when they are both present within a non-local block, as illustrated in the top row of Fig. 1. This observation is verified via statistical analysis on the whole validation set. Also, the standard non-local block combining both terms performs even worse than the variant that includes only the unary term (shown in Table 2). This indicates that coupling the two terms together may be detrimental to the learning of these visual clues, and consequently affects the learning of discriminative features.

To address this problem, we present the disentangled non-local (DNL) block, where the whitened pairwise and unary terms are cleanly decoupled by using independent *Softmax* functions and embedding matrices. With this disentangled design, the difficulty in joint learning of the whitened pairwise and unary terms is greatly diminished. As shown in second row of Fig. 1, the whitened pairwise term learns clear within-region clues while the unary term learns salient boundaries, even more clearly than what is learned when each term is trained alone.

The disentangled non-local block is validated through various vision tasks. On semantic segmentation benchmarks, by replacing the standard non-local block with the proposed DNL block with all other settings unchanged, significantly greater accuracy is achieved, with a 2.0% mIoU gain on the Cityscapes validation set, 1.3% mIoU gain on ADE20k, and 3.4% on PASCAL-Context using a ResNet-101 backbone. With few bells and whistles, our DNL obtains state-of-the-art performance on the challenging ADE20K dataset. Also, with a task-specific DNL block, noticeable accuracy improvements are observed on both COCO object detection and Kinetics action recognition.

2   Related Works

Non-Local/Self-Attention. These terms may appear in different application domains, but they refer to the same modeling mechanism. This mechanism was first proposed and widely used in natural language processing [1,25] and physical system modeling [14,22,27]. The self-attention / relation module affects an individual element (e.g. a word in a sentence) by aggregating features from a set of elements (e.g. all the words in the sentence), where the aggregation weights are usually determined on embedded feature similarities among the elements. They are powerful in capturing long-range dependencies and contextual information.

In the computer vision, two pioneering works [15,26] first applied this kind of modeling mechanism to capture the relations between objects and pixels, respectively. Since then, such modeling methods have demonstrated great effectiveness in many vision tasks, such as image classification [16], object detection [10,15], semantic segmentation [30], video object detection [5,7,11,28] and tracking [29], and action recognition [26]. There are also works that propose improvements to self-attention modeling, e.g. an additional relative position term [15,16], an additional channel attention [8], simplification [2], and speed-up [17].

This paper also presents an improvement over the basic self-attention / non-local neural networks. However, our work goes beyond straightforward application or technical modification of non-local networks in that it also brings a new perspective for understanding this module.

Understanding Non-Local/Self-Attention Mechanisms. Our work is also related to several approaches that analyze the non-local/self-attention mechanism in depth, including the performance of individual terms [15,24,37] on various tasks. Also, there are studies which seek to uncover what is actually learnt by the non-local/self-attention mechanism in different tasks [2].

This work also targets a deeper understanding of the non-local mechanism, in a new perspective. Beyond improved understanding, our paper presents a more effective module, the disentangled non-local block, that is developed from this new understanding and is shown to be effective on multiple vision tasks.

3   Non-local Networks in Depth

3.1   Dividing Non-local Block into Pairwise and Unary Terms

Non-local block [26] computes pairwise relations between features of two positions to capture long-range dependencies. With $\mathbf{x}_i$ representing the input features at position i, the output features $\mathbf{y}_i$ of a non-local block are computed as

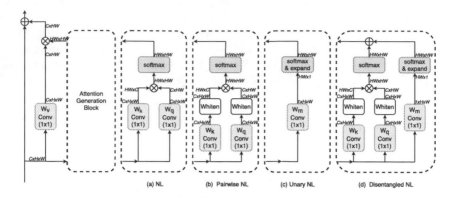

Fig. 2. Architectures of non-local block, disentangled non-local block, and other variants. The shapes of feature maps are indicated in gray, *e.g.*, C×H×W. "⊗" denotes matrix multiplication and "⊕" denotes element-wise addition. Blue boxes represent 1×1 convolution. *Softmax* is performed on the first dimension of feature maps (Color figure online)

$$\mathbf{y}_i = \sum_{j \in \Omega} \omega(\mathbf{x}_i, \mathbf{x}_j) g(\mathbf{x}_j), \tag{1}$$

where Ω denotes the set of all pixels on a feature map of size $H \times W$; $g(\cdot)$ is the *value* transformation function with parameter W_v; $\omega(\mathbf{x}_i, \mathbf{x}_j)$ is the embedded similarity function from pixel j (referred to as a *key* pixel) to pixel i (referred to as a *query* pixel), typically instantiated by an Embedded Gaussian as

$$\omega(\mathbf{x}_i, \mathbf{x}_j) = \sigma\left(\mathbf{q}_i^T \mathbf{k}_j\right) = \frac{\exp\left(\mathbf{q}_i^T \mathbf{k}_j\right)}{\sum_{t \in \Omega} \exp\left(\mathbf{q}_i^T \mathbf{k}_t\right)}, \tag{2}$$

where $\mathbf{q}_i = W_q \mathbf{x}_i$ and $\mathbf{k}_j = W_k \mathbf{x}_j$ denote the *query* and *key* embedding of pixel i and j, respectively, and $\sigma(\cdot)$ denotes the softmax function.

At first glance, $\omega(\mathbf{x}_i, \mathbf{x}_j)$ (defined in Eq. 2) appears to represent only a *pairwise* relationship in the non-local block, through a dot product operation. However, we find that it may encode some *unary* meaning as well. Considering a special case where the query vector is a constant over all image pixels, a *key* pixel will have global impact on all *query* pixels. In [2], it was found that non-local blocks frequently degenerate into a pure *unary* term in several image recognition tasks where each *key* pixel in the image has the same similarity with all *query* pixels. These findings indicate that the *unary* term does exist in the non-local block formulation. It also raises a question of how to divide Eq. (2) into *pairwise* and *unary* terms, which account for the impact of one *key* pixel specifically on another *query* pixel and the influence of one *key* pixel generally over all the *query* pixels, respectively.

To answer this question, we first present a *whitened* dot product between *key* and *query* to represent the *pure* pairwise term: $\left(\mathbf{q}_i - \boldsymbol{\mu}_q\right)^T \left(\mathbf{k}_j - \boldsymbol{\mu}_k\right)$, where $\boldsymbol{\mu}_q = \frac{1}{|\Omega|} \sum_{i \in \Omega} \mathbf{q}_i$ and $\boldsymbol{\mu}_k = \frac{1}{|\Omega|} \sum_{j \in \Omega} \mathbf{k}_j$ are the averaged *query* and *key* embedding over all pixels, respectively.

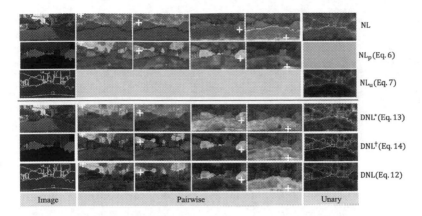

Fig. 3. Visualization of attention maps for all variants of the NL block mentioned in this paper. Column 1: image, ground truth and edges of ground truth. Columns 2–5: attention maps of pairwise terms. Column 6: attention maps of unary terms. As NL_u has no pairwise attention map, and NL_p has no unary attention map, we leave the corresponding spaces empty

To remove the *unary/global* component of *key* pixels, the *whitened* dot product is determined by maximizing the normalized differences between *query* and *key* pixels. In following proposition, we show how this can be achieved via an optimization objective, which allows for the whitened dot product to be computed.

Proposition 1. $\alpha^* = \frac{1}{|\Omega|} \sum_{i \in \Omega} \mathbf{q}_i$, $\beta^* = \frac{1}{|\Omega|} \sum_{m \in \Omega} \mathbf{k}_m$ is the optimal solution of the following optimization objective:

$$\underset{\alpha,\beta}{\arg\max} \quad \frac{\sum_{i,m,n \in \Omega} \left((\mathbf{q}_i - \alpha)^T (\mathbf{k}_m - \beta) - (\mathbf{q}_i - \alpha)^T (\mathbf{k}_n - \beta)\right)^2}{\sum_{i \in \Omega} \left((\mathbf{q}_i - \alpha)^T (\mathbf{q}_i - \alpha)\right) \cdot \sum_{m,n \in \Omega} \left((\mathbf{k}_m - \mathbf{k}_n)^T (\mathbf{k}_m - \mathbf{k}_n)\right)}$$
$$+ \frac{\sum_{m,i,j \in \Omega} \left((\mathbf{k}_m - \beta)^T (\mathbf{q}_i - \alpha) - (\mathbf{k}_m - \beta)^T (\mathbf{q}_j - \alpha)\right)^2}{\sum_{m \in \Omega} \left((\mathbf{k}_m - \beta)^T (\mathbf{k}_m - \beta)\right) \cdot \sum_{i,j \in \Omega} \left((\mathbf{q}_i - \mathbf{q}_j)^T (\mathbf{q}_i - \mathbf{q}_j)\right)} \tag{3}$$

Proof Sketch: The Hessian of the objective function O with respect to α and β is a non-positive definite matrix. The optimal α^* and β^* are thus the solutions of the following equations: $\frac{\partial O}{\partial \alpha} = 0$, $\frac{\partial O}{\partial \beta} = 0$. Solving this yields $\alpha^* = \frac{1}{|\Omega|} \sum_{i \in \Omega} \mathbf{q}_i$, $\beta^* = \frac{1}{|\Omega|} \sum_{m \in \Omega} \mathbf{k}_m$. Please see the appendix for a detailed proof.

By extracting the whitened dot product as the *pure* pairwise term, we can divide the dot product computation of the standard non-local block as

$$\mathbf{q}_i^T \mathbf{k}_j = \left(\mathbf{q}_i - \boldsymbol{\mu}_q\right)^T \left(\mathbf{k}_j - \boldsymbol{\mu}_k\right) + \boldsymbol{\mu}_q^T \mathbf{k}_j + \mathbf{q}_i^T \boldsymbol{\mu}_k + \boldsymbol{\mu}_q^T \boldsymbol{\mu}_k. \tag{4}$$

Note that the last two terms ($\mathbf{q}_i^T \boldsymbol{\mu}_k$ and $\boldsymbol{\mu}_q^T \boldsymbol{\mu}_k$) are factors that appear in both the numerator and denominator of Eq. (2). Hence, these two terms can

Table 1. Consistency statistics between attention maps of the non-local variants and the ground-truth within-category and boundary maps on the Cityscapes validation set

Method	pair ∩ within-category	pair ∩ boundary	unary ∩ boundary
random	0.259	0.132	0.135
pairwise NL (Eq. 6)	0.635	0.141	–
unary NL (Eq. 7)	–	–	0.460
NL (Eq. 2)	0.318	0.160	0.172
DNL* (Eq. 13)	0.446	0.146	0.305
DNL† (Eq. 14)	0.679	0.137	0.657
DNL (Eq. 12)	0.759	0.130	0.696

be eliminated (see proof in the Appendix). After this elimination, we reach the following *pairwise* and *unary* split of a standard non-local block:

$$\omega(\mathbf{x}_i, \mathbf{x}_j) = \sigma(\mathbf{q}_i^T \mathbf{k}_j) = \sigma(\underbrace{\left(\mathbf{q}_i - \boldsymbol{\mu}_q\right)^T \left(\mathbf{k}_j - \boldsymbol{\mu}_k\right)}_{\text{pairwise}} + \underbrace{\boldsymbol{\mu}_q^T \mathbf{k}_j}_{\text{unary}}), \quad (5)$$

where the first *whitened* dot product term represents the *pure* pairwise relation between a *query* pixel i and a *key* pixel j, and the second term represents the *unary* relation where a *key* pixel j has the same impact on all *query* pixels i.

3.2 What Visual Clues Are Expected to Be Learnt by Pairwise and Unary Terms?

To study what visual clues are expected to be learnt by the pairwise and unary terms, respectively, we construct two variants of the non-local block by using either the pairwise or unary term alone, such that the influence of the other term is eliminated. The two variants use the following similarity computation functions instead of the one in Eq. (2):

$$\omega_{\mathrm{p}}\left(\mathbf{x}_i, \mathbf{x}_j\right) = \sigma\left(\left(\mathbf{q}_i - \boldsymbol{\mu}_q\right)^T \left(\mathbf{k}_j - \boldsymbol{\mu}_k\right)\right), \quad (6)$$

$$\omega_{\mathrm{u}}\left(\mathbf{x}_i, \mathbf{x}_j\right) = \sigma(\boldsymbol{\mu}_q^T \mathbf{k}_j). \quad (7)$$

The two variants are denoted as "pairwise NL" and "unary NL", and illustrated in Fig. 2(b) and 2(c), respectively. We apply these two variants of non-local block to the Cityscapes semantic segmentation [6] (see Sect. 5.1 for detailed settings), and visualize their learnt attention (similarity) maps on several randomly selected validation images in Cityscapes, as shown in Fig. 3 (please see more examples in the Appendix). It can be seen that the pairwise NL block tends to learn pixel relationships within the same category region, while the unary NL block tends to learn the impact from boundary pixels to all image pixels.

This observation is further verified by quantitative analysis using the ground-truth region and boundary annotations in Cityscapes. Denote $P^{(i)} = \{\omega_{\mathrm{p}}(\mathbf{x}_i, \mathbf{x}_j)|$

$j \in \Omega\} \in \mathbb{R}^{H \times W}$ as the attention map of pixel i according to the pairwise term of Eq. (6), $U = \{\omega_u(\mathbf{x}_i, \mathbf{x}_j)|j \in \Omega\} \in \mathbb{R}^{H \times W}$ as the attention map for all query pixels according to the unary term of Eq. (7), $C^{(i)} \in \mathbb{R}^{H \times W}$ as the binary within-category region map of pixel i, and $E \in \mathbb{R}^{H \times W}$ as the binary boundary map indicating pixels with distance to ground truth contour of less than 5 pixels.

We evaluate the consistency between attention maps $A \in \{P^{(i)}, U\}$ and ground-truth boundary/same-category region $G \in \{C^{(i)}, E\}$ by their overlaps:

$$A \cap G = \sum_{j \in \Omega} A_j \odot G_j, \tag{8}$$

where A_j, G_j are the element values of the corresponding attention map and binary map at pixel j, respectively.

Table 1 shows the averaged consistency measures of the attention maps in Eq. (6) and Eq. (7) to ground-truth region maps (denoted as pairwise NL and unary NL) using all 500 validation images in the Cityscapes datasets. We also report the consistency measures by a random attention map for reference (denoted as random). The following can be seen:

- The attention map by the pairwise NL block of Eq. (6) has significantly larger overlap with the ground-truth same-category region than the random attention map (0.635 vs. 0.259), but has similar overlap with the ground-truth boundary region (0.141 vs. 0.132), indicating that *the pure pairwise term tends to learn relationship between pixels within same-category regions.*
- The attention map by the unary NL block of Eq. (7) has significantly larger overlap with the ground-truth boundary region than the random attention map (0.460 vs. 0.135), indicating that *the unary term tends to learn the impact of boundary pixels on all image pixels.* This is likely because the image boundary area provides the most informative cues when considering the general effect on all pixels.

3.3   Does the Non-local Block Learn Visual Clues Well?

We then study the learnt pairwise and unary terms by the non-local block. We follow Eq. (5) to split the standard similarity computation into the pairwise and unary terms, and normalize them by a softmax operation separately. After splitting and normalization, we can compute their overlaps with the ground-truth within-category region map and boundary region map, as shown in Table 1.

It can be seen that the pairwise term in the standard NL block which is jointly learnt with the unary term has significantly smaller overlap with the ground-truth within-category region than in the pairwise NL block where the pairwise term is learnt alone (0.318 vs. 0.635). It can be also seen that the unary term in the standard NL block which is jointly learnt with the pairwise term has significantly smaller overlap with the ground-truth boundary region than in the unary NL block where the unary term is learnt alone (0.172 vs. 0.460). These results indicate that neither of the pairwise and unary terms learn the visual clues of within-category regions and boundaries well, as also demonstrated in Fig. 1 (top).

3.4   Why the Non-Local Block Does Not Learn Visual Clues Well?

To understand why the non-local block does not learn the two visual clues well, while the two terms alone can clearly learn them, we rewrite Eq. (5) as:

$$
\sigma(\mathbf{q}_i \cdot \mathbf{k}_j) = \sigma\left(\left(\mathbf{q}_i - \boldsymbol{\mu}_q\right)^T \left(\mathbf{k}_j - \boldsymbol{\mu}_k\right) + \boldsymbol{\mu}_q^T \mathbf{k}_j \right)
$$

$$
= \frac{1}{\lambda_i} \sigma\left(\left(\mathbf{q}_i - \boldsymbol{\mu}_q\right)^T \left(\mathbf{k}_j - \boldsymbol{\mu}_k\right) \right) \cdot \sigma(\boldsymbol{\mu}_q^T \mathbf{k}_j) = \frac{1}{\lambda_i} \omega_{\mathrm{p}}(\mathbf{x}_i, \mathbf{x}_j) \cdot \omega_{\mathrm{u}}(\mathbf{x}_i, \mathbf{x}_j),
\tag{9}
$$

where λ_i is a normalization scalar such that the sum of attention map values over Ω is 1.

Consider the back-propagation of loss L to the pairwise and unary terms:

$$
\frac{\partial L}{\partial \sigma(\omega_{\mathrm{p}})} = \frac{\partial L}{\partial \sigma(\omega)} \cdot \frac{\partial \sigma(\omega)}{\partial \sigma(\omega_{\mathrm{p}})} = \frac{\partial L}{\partial \sigma(\omega)} \cdot \sigma(\omega_{\mathrm{u}}),
$$

$$
\frac{\partial L}{\partial \sigma(\omega_{\mathrm{u}})} = \frac{\partial L}{\partial \sigma(\omega)} \cdot \frac{\partial \sigma(\omega)}{\partial \sigma(\omega_{\mathrm{u}})} = \frac{\partial L}{\partial \sigma(\omega)} \cdot \sigma(\omega_{\mathrm{p}}).
$$

It can be seen that both gradients are determined by the value of the other term. When the value of the other term becomes very small (close to 0), the gradient of this term will be also very small, thus inhibiting the learning of this term. For example, if we learn the unary term to well represent the boundary area, the unary attention weights on the non-boundary area will be close to 0 and the pairwise term at the non-boundary area would thus be hard to learn well due to the vanishing gradient issue. On the other hand, if we learn the pairwise term to well represent the within-category area, the unary attention weights on the boundary area will be close to 0 and the pairwise term at the non-boundary area would also be hard to learn well due to the same vanishing gradient issue.

Another problem is the *shared* key transformation W_k used in both the pairwise and unary terms, causing the computation of the two terms to be coupled. Such coupling may introduce additional difficulties in learning the two terms.

4   Disentangled Non-local Neural Networks

In this section, we present a new non-local block, named disentangled non-local (DNL) block, which effectively disentangles the learning of pairwise and unary terms. In the following sections, we first describe how we modify the standard non-local (NL) block into a disentangled non-local (NL) block, such that the two visual clues described above can be learnt well. Then we analyze its actual behavior in learning visual clues using the method in Sect. 3.2.

4.1   Formulation

Our first modification is to change the *multiplication* in Eq. (9) to *addition*:

$$\omega(\mathbf{x}_i, \mathbf{x}_j) = \omega_{\mathrm{p}}(\mathbf{x}_i, \mathbf{x}_j) \cdot \omega_{\mathrm{u}}(\mathbf{x}_i, \mathbf{x}_j) \;\Rightarrow\; \omega(\mathbf{x}_i, \mathbf{x}_j) = \omega_{\mathrm{p}}(\mathbf{x}_i, \mathbf{x}_j) + \omega_{\mathrm{u}}(\mathbf{x}_i, \mathbf{x}_j). \tag{10}$$

The gradients of these two terms are

$$\frac{\partial L}{\partial \sigma(\omega_{\mathrm{p}})} = \frac{\partial L}{\partial \sigma(\omega)}, \frac{\partial L}{\partial \sigma(\omega_{\mathrm{u}})} = \frac{\partial L}{\partial \sigma(\omega)}.$$

So the gradients of each term will not be impacted by the other.

The second modification is to change the transformation W_k in unary term to be an independent linear transformation W_m with output dimension of 1:

$$\boldsymbol{\mu}_q^T \mathbf{k}_j = \boldsymbol{\mu}_q^T W_k \mathbf{x}_j \Rightarrow m_j = W_m \mathbf{x}_j. \tag{11}$$

After this modification, the pairwise and unary terms will no longer share the W_k transformation, which further decouples them.

DNL Formulation. With these two modifications, we obtain the following similarity computation for the disentangled non-local (DNL) block:

$$\omega^{\mathrm{D}}(\mathbf{x}_i, \mathbf{x}_j) = \sigma\left(\left(\mathbf{q}_i - \boldsymbol{\mu}_q\right)^T \left(\mathbf{k}_j - \boldsymbol{\mu}_k\right) \right) + \sigma(m_j). \tag{12}$$

The resulting DNL block is illustrated in Fig. 2 (d). Note that we adopt a single *value* transform for both pairwise and unary terms, which is similarly effective on benchmarks as using independent value transform but with reduced complexity.

Complexity. For an input feature map of $C \times H \times W$, we follow [26] by using $C/2$ dimensional *key* and *query* vectors. The space and time complexities are $\mathcal{O}^D(\text{space}) = (2C + 1)C$ and $\mathcal{O}^D(\text{time}) = \left((2C + 1)C + (\frac{3}{2}C + 2)HW\right)HW$, respectively. For reference, the space and time complexity of a standard non-local block are $\mathcal{O}(\text{space}) = 2C^2$ and $\mathcal{O}(\text{time}) = \left(2C^2 + (\frac{3}{2}C + 1)HW\right)HW$, respectively. The additional space and computational overhead of the disentangled non-local block over a standard non-local block is marginal, specifically 0.1% and 0.15% for $C = 512$ in our semantic segmentation experiments.

DNL Variants for Diagnostic Purposes. To diagnose the effects of the two decoupling modifications alone, we consider the following two variants:

$$\omega^{\mathrm{D}*}(\mathbf{x}_i, \mathbf{x}_j) = \sigma\left(\left(\mathbf{q}_i - \boldsymbol{\mu}_q\right)^T \left(\mathbf{k}_j - \boldsymbol{\mu}_k\right) + m_j \right), \tag{13}$$

$$\omega^{\mathrm{D}\dagger}(\mathbf{x}_i, \mathbf{x}_j) = \sigma\left(\left(\mathbf{q}_i - \boldsymbol{\mu}_q\right)^T \left(\mathbf{k}_j - \boldsymbol{\mu}_k\right) \right) + \sigma(\boldsymbol{\mu}_q^T \mathbf{k}_j), \tag{14}$$

which each involves only one of the two modifications.

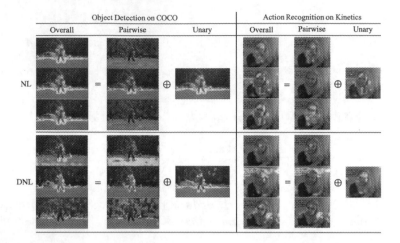

Fig. 4. Visualization of attention maps in NL and our DNL block on COCO object detection and Kinetics action recognition. The query points are marked in red. Please refer to appendix for more examples (Color figure online)

4.2 Behavior of DNL on Learning Visual Clues

We compute the overlaps of the pairwise and unary attention maps in DNL (Eq. 12) with the ground-truth within-category region map and boundary region map, as shown in Table 1.

It can be seen that the pairwise term in DNL has significantly larger overlap with the ground-truth within-category region than the one in the standard NL block (0.759 vs. 0.318), and the unary term has significantly larger overlap with the boundary region than that in the standard NL block (0.696 vs. 0.172). These results indicate better learning of the two visual clues by the DNL block in comparison to the standard NL block.

Compared with the blocks which learn the pairwise or unary terms alone (see the "pairwise NL" and "unary NL" rows), such measures are surprisingly 0.124 and 0.236 higher with DNL. We hypothesize that when one term is learned alone, it may encode some portion of the other clue, as it is also useful for inference. By explicitly learning both terms, our disentangling design can separate one from the other, allowing it to better extract these visual clues.

We then verify the effects of each disentangling modification by these measures. By incorporating the "disentangled transformation" modification alone (ω^*) as in Eq. (13), it achieves 0.446 and 0.305 on within-category modeling and boundary modeling, respectively, which is marginally better than the standard non-local block. By incorporating the "multiplication to addition" modification alone ($\omega^\dagger$) as in Eq. (14), it achieves 0.679 and 0.657 on within-category modeling and boundary modeling, respectively.

Table 2. Ablation study on the validation set of Cityscapes

(a) Decoupling strategy				(b) Pairwise and unary terms		
	mul → add	Non-shared W_k	mIoU		Pairwise term	Unary term	mIoU
Baseline	–	–	75.8	Baseline	–	–	75.8
NL	×	×	78.5	NL	✓	✓	78.5
DNL†(14)	✓	×	79.2	NL$_p$	✓	×	77.5
DNL*(13)	×	✓	79.0	NL$_u$	×	✓	79.3
DNL	✓	✓	80.5	DNL	✓	✓	80.5

The results indicate that the two modifications both benefit the learning of two visual clues and work better if combined together. The improvements in visual clue modeling by two disentangling strategies are also illustrated in Fig. 3.

Note such disentangling strategies also effect on other tasks beyond semantic segmentation. In object detection and action recognition tasks, we also observe clearer learnt visual clues by the DNL block than by the standard NL. As shown in Fig. 4, while in NL the pairwise term is almost hindered by the unary term (also observed by [2]), the pariwise term in DNL shows clear within-region meaning and appears significant in the final overall attention maps. The unary term in DNL also shows more focus to salient regions (not limited to boundaries which is different from that observed in the semantic segmentation task) than the one in an NL block. More examples will be shown in appendix.

5 Experiments

We evaluate the proposed DNL method on the recognition tasks of semantic segmentation, object detection/instance segmentation, and action recognition.

5.1 Semantic Segmentation

Datasets. We use three benchmarks for semantic segmentation evaluation.

Cityscapes [6] focuses on semantic understanding of urban street scenes. It provides a total of 5,000 finely annotated images, which is divided into 2,975/500/ 1,525 images for training, validation and testing. Additional 20,000 coarsely annotated images are also provided. The dataset contains annotations for over 30 classes, of which 19 classes are used in evaluation.

ADE20K [36] was used in the ImageNet Scene Parsing Challenge 2016 and covers a wide range of scenes and object categories. It contains 20 K images for training, 2K images for validation, and another 3 K images for testing. It includes 150 semantic categories for evaluation.

PASCAL-Context [21] is a set of additional annotations for PASCAL VOC 2010, which label more than 400 categories of 4,998 images for training and 5,105 images for validation. For semantic segmentation, 59 semantic classes and 1 background class are used in training and validation.

Table 3. Comparisons with state-of-the-art approaches on the Cityscapes test set

Method	Backbone	ASPP	Coarse	mIoU (%)
DANet [8]	ResNet-101			81.5
HRNet [23]	HRNetV2-W48			81.9
CCNet [17]	ResNet-101			81.4
ANN [38]	ResNet-101			81.3
OCNet [30]	ResNet-101	✓		81.7
ACFNet [31]	ResNet-101	✓		81.8
PSPNet [34]	ResNet-101		✓	81.2
PSANet [35]	ResNet-101		✓	81.4
DeepLabv3 [4]	ResNet-101	✓	✓	81.3
NL	ResNet-101		✓	80.8
DNL (ours)	ResNet-101		✓	82.0
NL	HRNetV2-W48		✓	82.5
DNL (ours)	HRNetV2-W48		✓	83.0

Architecture. We follow recent practice [17] by using dilated FCN [20] and a ResNet101 [13] backbone for our major segmentation experiments. The strides and dilations of 3×3 convolutions are set to 1 and 2 for stage4, and 1 and 4 for stage5. The baseline model uses a segmentation head consisting of a 3×3 convolution layer to reduce the channels to 512 and a subsequent classifier to predict the final segmentation results. For experiments with a non-local or a disentangled non-local block, the block is inserted right before the final classifier. The implementation and hyper-parameters mostly follow [17]. More details can be seen in appendix 4.

Ablation Study. We ablate several design components in the proposed disentangled non-local block on the Cityscapes validation set. A ResNet-101 backbone is adopted for all ablations.

DNL Variants. The disentangled non-local block has two decoupling modifications on the standard non-local block: multiplication to addition, and separate *key* transformations. In addition to comparing the full DNL model with the standard non-local model, we also conduct experiments for these two variants which include only one of the decoupling modifications. The results are shown in Table 2(a). While the standard non-local model brings 2.7% mIoU gains over a plain ResNet-101 model (78.5% vs. 75.8%), by replacing the standard non-local block by our disentangled non-local block, we achieve an additional 2.0% mIoU gain over the standard non-local block (80.5% vs. 78.5%), with almost no complexity increase. The variants that use each decoupling strategy alone achieve 0.5% and 0.7% mIoU gains over the standard non-local block (79.0 vs. 78.5 and 79.2 vs. 78.5), showing that both strategies are beneficial alone. They are also

Table 4. Comparisons with state-of-the-art approaches on the validation set and test set of ADE20K, and test set of PASCAL-Context

Method	Backbone	ADE20K		PASCAL-Context
		val mIoU (%)	test mIoU (%)	test mIoU (%)
CCNet [17]	ResNet-101	45.22	–	–
OCNet [30]	ResNet-101	45.45	–	–
EMANet [19]	ResNet-101	–	–	53.1
HRNetV2 [23]	HRNetV2-W48	42.99	–	54.0
EncNet [32]	ResNet-101	44.65	55.67	52.6
DANet [8]	ResNet-101	45.22	–	52.6
CFNet [33]	ResNet-101	44.89	–	54.0
ANN [38]	ResNet-101	45.24	–	52.8
DMNet [12]	ResNet-101	45.50	–	54.4
ACNet [9]	ResNet-101	45.90	55.84	54.1
NL	ResNet-101	44.67	55.58	50.6
DNL (ours)	ResNet-101	45.97	56.23	54.8
NL	HRNetV2-W48	44.82	55.60	54.2
DNL (ours)	HRNetV2-W48	45.82	55.98	55.3

both crucial, as combining them leads to significantly better performance than using each alone.

Effects of Pairwise and Unary Term Alone. Table 2(b) compares the methods using the pairwise term or unary term alone. Using the pairwise term alone achieves 77.5% mIoU, which is 1.7% better than the baseline plain network without it. Using the unary term alone achieves 79.3% mIoU, which is 3.5% better than the baseline plain network and even 0.8% mIoU better than the standard non-local network which models both pairwise and unary terms. These results indicate that the standard non-local block hinders the effect of the unary term, probably due to the coupling of two kinds of relationships. Our disentangled non-local networks effectively disentangle the two terms, and thus can better exploit their effects to achieve a higher accuracy of 80.5% mIoU.

Complexities. As discussed in Sect. 4.1, the time and space complexity of the DNL model over the NL model is tiny. Table 5 show the FLOPs and actual latency (single-scale inference using a single GPU) on semantic segmentation, using a ResNet-101 backbone and input resolution of 769 × 769.

Comparison with Other Methods
Results on Cityscapes. Table 3 shows comparison results for the proposed disentangled non-local network on the Cityscapes test set. Using a ResNet-101 backbone, the disentangled non-local network achieves 82.0% mIoU, 1.2% better than that of a standard non-local network. On a stronger backbone of HRNetV2-W48,

the disentangled non-local network achieves 0.5% better accuracy than a standard non-local network. Considering that the standard non-local network has 0.6% mIoU improvement over a plain HRNetV2-W48 network, such additional gains are significant.

Results on ADE20K. Table 4 shows comparison results of the proposed disentangled non-local network on the ADE20k benchmark. Using a ResNet-101 backbone, the disentangled non-local block achieves 45.97% and 56.23% on the validation and test sets, respectively, which are 1.30% and 0.65% better than the counterpart networks using a standard non-local block. Our result reveals a new SOTA on this benchmark. On a HRNetV2-W48 backbone, the DNL block is 1.0% and 0.38% better than a standard non-local block. Note on ADE20K, HRNetV2-W48 backbone does not perform better than a ResNet-101 backbone, which is different with the other datasets.

Results on PASCAL-Context. Table 3 shows comparison results of the proposed disentangled non-local network on the PASCAL-Context test set. On ResNet-101, our method improves the standard non-local method significantly, by 3.4% mIoU (53.7 vs. 50.3). On HRNetV2-W48, our DNL method is 1.1% mIoU better, which is significant considering that the NL method has 0.2% improvements over the plain counterpart.

Table 5. Complexity comparisons

	#param(M)	FLOPs(G)	latency(s/img)
Baseline	70.960	691.06	0.177
NL	71.484	765.07	0.192
DNL	71.485	765.16	0.194

5.2   Object Detection/Segmentation and Action Recognition

Object Detection and Instance Segmentation on COCO. We adopt the open source mmdetection [3] codebase for experiments. Following [26], the non-local variants are inserted right before the last residual block of c4. More details can be seen in appendix 4.

Table 6 shows comparisons of different methods. While the standard non-local block outperforms the baseline counterpart by 0.8% bbox mAP and 0.7% mask mAP, the proposed disentangled non-local block brings an additional 0.7% bbox mAP and 0.6% mask mAP in gains.

Action Recognition on Kinetics. We adopt the Kinetics [18] dataset for experiments, which includes ~240k training videos and 20k validation videos in 400 human action categories. We report the top-1 (%) and top-5 (%) accuracy on the validation set. More details can be seen in appendix 4.

Table 7 shows the comparison of different blocks. It can be seen that the disentangled design performs 0.36% better than using standard non-local block.

Table 6. Results based on Mask R-CNN, using R50 as backbone with FPN, for object detection and instance segmentation on COCO 2017 validation set

	AP^{bbox}	AP^{bbox}_{50}	AP^{bbox}_{75}	AP^{mask}	AP^{mask}_{50}	AP^{mask}_{75}
Baseline	38.8	59.3	42.5	35.1	56.2	37.9
NL	39.6	60.3	43.2	35.8	57.1	38.5
NL_p	39.8	60.4	43.7	35.9	57.3	38.4
NL_u	40.1	60.9	43.8	36.1	57.6	38.7
DNL	40.3	61.2	44.1	36.4	58.0	39.1

Table 7. Results based on Slow-only baseline using R50 as backbone on Kinetics validation set

	Top-1 Acc	Top-5 Acc
Baseline	74.94	91.90
NL	75.95	92.29
NL_p	76.01	92.28
NL_u	75.76	92.44
DNL	76.31	92.69

Discussion. For object detection and action recognition, similar to the semantic segmentation task, we observe that significantly clearer learnt visual clues by the proposed DNL model than by a standard NL model as shown in Fig. 4. But the accuracy improvement is not as large as in semantic segmentation. We hypothesize that it is probably because the semantic segmentation task aims at dense pixel-level prediction and may require more fine-grained relationship modeling of image pixels. In object detection and action recognition, the benefit of more fine-grained relationship modeling ability may not be as significant.

6 Conclusion

In this paper, we first study the non-local block in depth, where we find that its attention computation can be split into two terms, a whitened pairwise term and a unary term. Via both intuitive and statistical analysis, we find that the two terms are tightly coupled in the non-local block, which hinders the learning of each. Based on these findings, we present the disentangled non-local block, where the two terms are decoupled to facilitate learning for both terms. We demonstrate the effectiveness of the decoupled design for learning visual clues on various vision tasks, such as semantic segmentation, object detection and action recognition.

References

1. Britz, D., Goldie, A., Luong, M.T., Le, Q.: Massive exploration of neural machine translation architectures. arXiv preprint arXiv:1703.03906 (2017)
2. Cao, Y., Xu, J., Lin, S., Wei, F., Hu, H.: GCNET: non-local networks meet squeeze-excitation networks and beyond. arXiv preprint arXiv:1904.11492 (2019)
3. Chen, K., et al.: Mmdetection: open mmlab detection toolbox and benchmark. arXiv preprint arXiv:1906.07155 (2019)
4. Chen, L.C., Papandreou, G., Schroff, F., Adam, H.: Rethinking atrous convolution for semantic image segmentation. arXiv preprint arXiv:1706.05587 (2017)
5. Chen, Y., Cao, Y., Hu, H., Wang, L.: Memory enhanced global-local aggregation for video object detection. In: The Conference on Computer Vision and Pattern Recognition (CVPR), June 2020

6. Cordts, M., et al.: The cityscapes dataset for semantic urban scene understanding. In: Proceedings of the IEEE Conference on Computer Vision and Pattern Recognition, pp. 3213–3223 (2016)
7. Deng, J., Pan, Y., Yao, T., Zhou, W., Li, H., Mei, T.: Relation distillation networks for video object detection. In: The IEEE International Conference on Computer Vision (ICCV), October 2019
8. Fu, J., et al.: Dual attention network for scene segmentation. In: Proceedings of the IEEE Conference on Computer Vision and Pattern Recognition, pp. 3146–3154 (2019)
9. Fu, J., et al.: Adaptive context network for scene parsing. In: Proceedings of the IEEE International Conference on Computer Vision, pp. 6748–6757 (2019)
10. Gu, J., Hu, H., Wang, L., Wei, Y., Dai, J.: Learning region features for object detection. In: Proceedings of the European Conference on Computer Vision (ECCV), pp. 381–395 (2018)
11. Guo, C., et al.: Progressive sparse local attention for video object detection. In: The IEEE International Conference on Computer Vision (ICCV), October 2019
12. He, J., Deng, Z., Qiao, Y.: Dynamic multi-scale filters for semantic segmentation. In: Proceedings of the IEEE International Conference on Computer Vision, pp. 3562–3572 (2019)
13. He, K., Zhang, X., Ren, S., Sun, J.: Deep residual learning for image recognition. In: Proceedings of the IEEE Conference on Computer Vision and Pattern Recognition, pp. 770–778 (2016)
14. Hoshen, Y.: Vain: attentional multi-agent predictive modeling. In: Advances in Neural Information Processing Systems, pp. 2701–2711 (2017)
15. Hu, H., Gu, J., Zhang, Z., Dai, J., Wei, Y.: Relation networks for object detection (2017)
16. Hu, H., Zhang, Z., Xie, Z., Lin, S.: Local relation networks for image recognition (2019)
17. Huang, Z., Wang, X., Huang, L., Huang, C., Wei, Y., Liu, W.: Ccnet: Criss-cross attention for semantic segmentation. In: Proceedings of the IEEE International Conference on Computer Vision, pp. 603–612 (2019)
18. Kay, W., et al.: The kinetics human action video dataset. arXiv preprint arXiv:1705.06950 (2017)
19. Li, X., Zhong, Z., Wu, J., Yang, Y., Lin, Z., Liu, H.: Expectation-maximization attention networks for semantic segmentation. In: Proceedings of the IEEE International Conference on Computer Vision, pp. 9167–9176 (2019)
20. Long, J., Shelhamer, E., Darrell, T.: Fully convolutional networks for semantic segmentation. In: Proceedings of the IEEE Conference on Computer Vision and Pattern Recognition, pp. 3431–3440 (2015)
21. Mottaghi, R., et al.: The role of context for object detection and semantic segmentation in the wild. In: IEEE Conference on Computer Vision and Pattern Recognition (CVPR) (2014)
22. Santoro, A., et al.: A simple neural network module for relational reasoning. In: Advances in Neural Information Processing Systems, pp. 4967–4976 (2017)
23. Sun, K., et al.: High-resolution representations for labeling pixels and regions. arXiv preprint arXiv:1904.04514 (2019)
24. Tang, G., Sennrich, R., Nivre, J.: An analysis of attention mechanisms: the case of word sense disambiguation in neural machine translation. arXiv preprint arXiv:1810.07595 (2018)
25. Vaswani, A., et al.: Attention is all you need. In: Advances in Neural Information Processing Systems, pp. 5998–6008 (2017)

26. Wang, X., Girshick, R., Gupta, A., He, K.: Non-local neural networks. In: Proceedings of the IEEE Conference on Computer Vision and Pattern Recognition, pp. 7794–7803 (2018)
27. Watters, N., Zoran, D., Weber, T., Battaglia, P., Pascanu, R., Tacchetti, A.: Visual interaction networks: learning a physics simulator from video. In: Advances in Neural Information Processing Systems, pp. 4539–4547 (2017)
28. Wu, H., Chen, Y., Wang, N., Zhang, Z.: Sequence level semantics aggregation for video object detection. In: The IEEE International Conference on Computer Vision (ICCV), October 2019
29. Xu, J., Cao, Y., Zhang, Z., Hu, H.: Spatial-temporal relation networks for multi-object tracking. In: The IEEE International Conference on Computer Vision (ICCV), October 2019
30. Yuan, Y., Wang, J.: Ocnet: object context network for scene parsing. arXiv preprint arXiv:1809.00916 (2018)
31. Zhang, F., et al.: ACFNet: attentional class feature network for semantic segmentation. In: Proceedings of the IEEE International Conference on Computer Vision, pp. 6798–6807 (2019)
32. Zhang, H., et al.: Context encoding for semantic segmentation. In: Proceedings of the IEEE Conference on Computer Vision and Pattern Recognition, pp. 7151–7160 (2018)
33. Zhang, H., Zhang, H., Wang, C., Xie, J.: Co-occurrent features in semantic segmentation. In: Proceedings of the IEEE Conference on Computer Vision and Pattern Recognition, pp. 548–557 (2019)
34. Zhao, H., Shi, J., Qi, X., Wang, X., Jia, J.: Pyramid scene parsing network. In: Proceedings of the IEEE Conference on Computer Vision and Pattern Recognition, pp. 2881–2890 (2017)
35. Zhao, H., et al.: PSANet: point-wise spatial attention network for scene parsing. In: Proceedings of the European Conference on Computer Vision (ECCV), pp. 267–283 (2018)
36. Zhou, B., Zhao, H., Puig, X., Fidler, S., Barriuso, A., Torralba, A.: Scene parsing through ade20k dataset. In: Proceedings of the IEEE Conference on Computer Vision and Pattern Recognition, pp. 633–641 (2017)
37. Zhu, X., Cheng, D., Zhang, Z., Lin, S., Dai, J.: An empirical study of spatial attention mechanisms in deep networks. In: The IEEE International Conference on Computer Vision (ICCV), October 2019
38. Zhu, Z., Xu, M., Bai, S., Huang, T., Bai, X.: Asymmetric non-local neural networks for semantic segmentation. In: Proceedings of the IEEE International Conference on Computer Vision, pp. 593–602 (2019)

URVOS: Unified Referring Video Object Segmentation Network with a Large-Scale Benchmark

Seonguk Seo[1], Joon-Young Lee[2], and Bohyung Han[1(✉)]

[1] Seoul National University, Seoul, South Korea
bhhan@snu.ac.kr
[2] Adobe Research, San Jose, USA

Abstract. We propose a unified referring video object segmentation network (URVOS). URVOS takes a video and a referring expression as inputs, and estimates the object masks referred by the given language expression in the whole video frames. Our algorithm addresses the challenging problem by performing language-based object segmentation and mask propagation jointly using a single deep neural network with a proper combination of two attention models. In addition, we construct the first large-scale referring video object segmentation dataset called Refer-Youtube-VOS. We evaluate our model on two benchmark datasets including ours and demonstrate the effectiveness of the proposed approach. The dataset is released at https://github.com/skynbe/Refer-Youtube-VOS.

Keywords: Video object segmentation · Referring object segmentation

1 Introduction

Video object segmentation, which separates foreground objects from background in a video sequence, has attracted wide attention due to its applicability to many practical problems including video analysis and video editing. Typically, this task has been addressed in unsupervised or semi-supervised ways. Unsupervised techniques [7,30] perform segmentation without the guidance for foreground objects, and aim to estimate the object masks using salient features, independent motions, or known class labels automatically. Due to the ambiguity and the lack of flexibility in defining foreground objects, such approaches may be suitable for video analysis but not for video editing that requires to segment arbitrary objects

S. Seo—This work was done during an internship at Adobe Research.

Electronic supplementary material The online version of this chapter (https://doi.org/10.1007/978-3-030-58555-6_13) contains supplementary material, which is available to authorized users.

© Springer Nature Switzerland AG 2020
A. Vedaldi et al. (Eds.): ECCV 2020, LNCS 12360, pp. 208–223, 2020.
https://doi.org/10.1007/978-3-030-58555-6_13

or their parts flexibly. In the semi-supervised scenario, where the ground-truth mask is available at least in a single frame, existing methods [2, 22, 24, 28, 31, 34] propagate the ground-truth object mask to the rest of frames in a video. They fit well for interactive video editing but require tedious and time-consuming step to obtain ground-truth masks. To overcome such limitations, interactive approaches [1, 3, 4, 21] have recently been investigated to allow user interventions during inference.

Despite great progress in semi-supervised and interactive video object segmentation, pixel-level interactions are still challenging especially in mobile video editing and augmented reality use-cases. To address the challenge, we consider a different type of interaction, language expressions, and introduce a new task that segments an object referred by the given language expression in a video. We call this task as *referring video object segmentation.*

A naïve baseline for the task is applying referring image segmentation techniques [12, 16, 35, 36] to each input frame independently. However, it does not leverage temporal coherency of video frames and, consequently, may result in inconsistent object mask predictions across frames. Another option is a sequential integration of referring image segmentation and semi-supervised video object segmentation. In this case, a referring image segmentation method initializes an object mask at a certain frame and then a video object segmentation method propagates the mask to the rest of the frames. This would work well if the initialization is successful. However, it often overfits to the particular characteristics in the anchor frame, which may not be robust in practice in the presence of occlusions or background clutter. Recently, Khoreva *et al.* [10] tackle this task by generating a set of mask proposals and choosing the most temporally-consistent set of candidates, but such a post-selection approach has inevitable limitation in maintaining temporal coherence.

We propose URVOS, a unified referring video object segmentation network. URVOS is an end-to-end framework for referring video object segmentation, which performs referring image segmentation and semi-supervised video object segmentation jointly in a single model. In this unified network, we incorporate two attention modules, cross-modal attention and memory attention modules, where memory attention encourages temporal consistency while cross-modal attention prevents drift. In addition, we introduce a new large-scale benchmark dataset for referring video object segmentation task, called *Refer-Youtube-VOS*. Our dataset is one order of magnitude larger than the previous benchmark [10], which enables researchers to develop new models and validate their performance. We evaluate the proposed method extensively and observe that our approach achieves outstanding performance gain on the new large-scale dataset.

Our contributions are summarized below.

- We construct a large-scale referring video object segmentation dataset, which contains 27,000+ referring expressions for 3,900+ videos.
- We propose a unified end-to-end deep neural network that performs both language-based object segmentation and mask propagation in a single model.

Full : *"A person on the right dressed in blue black walking while holding a white bottle."*
First : *"A woman in a blue shirt and a black bag."*

Full : *"A person showing his skateboard skills on the road."*
First : *"A man wearing a white cap."*

Fig. 1. Annotation examples of Refer-Youtube-VOS dataset. "Full" denotes that annotators watch the entire video for annotation while "First" means that they are given only the first frame of each video.

- Our method achieves significant performance gains over previous methods in the referring video object segmentation task.

2   Related Work

Referring Image Segmentation. This task aims to produce a segmentation mask of an object in an input image given a natural language expression. Hu *et al.* [8] first propose the task with a baseline algorithm that relies on multimodal visual-and-linguistic features extracted from LSTM and CNN. RRN [12] utilizes the feature pyramid structures to take advantage of multi-scale semantics for referring image segmentation. MAttNet [36] introduces a modular attention network, which decomposes a multi-modal reasoning model into a subject, object and relationship modules, and exploits attention to focus on relevant modules. CMSA [35] employs cross-modal self-attentive features to bridge the attentions in language and vision domains and capture long-range correlations between visual and linguistic modalities effectively. Our model employs a variant of CMSA to obtain the cross-modal attentive features effectively.

Video Object Segmentation. Video object segmentation is categorized into two types. Unsupervised approaches do not allow user interactions during test time, and aim to segment the most salient spatio-temporal object tubes. They typically employ two-stream networks to fuse motion and appearance cues [13, 26,37] for learning spatio-temporal representations.

Semi-supervised video object segmentation tracks an object mask in a whole video, assuming that the ground-truth object mask is provided for the first frame. With the introduction of DAVIS [25] and Youtube-VOS [33] datasets, there has

Table 1. Datasets for referring video object segmentation. J-HMDB and A2D Sentences [6] focus on 'action' recognition along with 'actor' segmentation, which have different purposes than ours. Although Refer-DAVIS$_{16/17}$ are well-suited for our task, they are small datasets with limited diversity. Our dataset, Refer-Youtube-VOS, is the largest dataset containing objects in diverse categories.

Dataset	Target	Videos	Objects	Expressions
J-HMDB Sentences [6]	Actor	928	928	928
A2D Sentences [6]	Actor	3782	4825	6656
Refer-DAVIS$_{16}$ [10]	Object	50	50	100
Refer-DAVIS$_{17}$ [10]	Object	90	205	1544
Refer-Youtube-VOS (Ours)	Object	**3975**	**7451**	**27899**

been great progress in this task. There are two main categories, online learning and offline learning. Most approaches rely on online learning, which fine-tunes networks using the first-frame ground-truth at test-time [2,14,24]. While the online learning achieves outstanding results, its computational complexity at test-time limits its practical use. Offline methods alleviate this issue and reduce runtime [22,28,31,34]. STM [22] presents a space-time memory network by non-local matching between previous and current frames, which achieves state-of-the-art performance, even beating online learning methods. Our model also belongs to offline learning, which modifies the non-local module of STM for its integration into our memory attention network and exploits temporal coherence of segmentation results.

Multi-modal Video Understanding. The intersection of language and video understanding has been investigated in various areas including visual tracking [15], action segmentation [6,29], video captioning [19] and video question answering [5]. Gavrilyuk *et al.* [6] adopt a fully-convolutional model to segment an actor and its action in each frame of a video as specified by a language query. However, their method has been validated in the datasets with limited class diversities, A2D [32] and J-HMDB [9], which only have 8 and 21 predefined action classes, respectively. Khoreva *et al.* [10] have augmented the DAVIS dataset with language referring expressions and have proposed a way to transfer image-level grounding models to video domain. Although [10] is closely related to our work, it fails to exploit valuable temporal information in videos during training.

3 Refer-Youtube-VOS Dataset

There exist previous works [6,10] that constructed referring segmentation datasets for videos. Gavrilyuk *et al.* [6] extended the A2D [32] and J-HMDB [9] datasets with natural sentences; the datasets focus on describing the 'actors' and 'actions' appearing in videos, therefore the instance annotations are limited to only a few object categories corresponding to the dominant 'actors' performing a salient 'action'. Khoreva *et al.* [10] built a dataset based on DAVIS [25], but the scales are barely sufficient to learn an end-to-end model from scratch.

To facilitate referring video object segmentation, we have constructed a large-scale video object segmentation dataset, Youtube-VOS [33], with referring expressions. Youtube-VOS has 4,519 high-resolution videos with 94 common object categories. Each video has pixel-level instance segmentation annotation at every 5 frames in 30-fps videos, and their durations are around 3 to 6 seconds. We employed Amazon Mechanical Turk to annotate referring expressions. To ensure the quality of the annotations, we selected around 50 turkers after a validation test. Each turker was given a pair of videos, the original video and the mask-overlaid one with the target object highlighted, and was asked to provide a discriminative sentence within 20 words that describes the target object accurately. We collected two kinds of annotations, which describe the highlighted object (1) based on a whole video (Full-video expression) and (2) using only the first frame of the video (First-frame expression). After the initial annotation, we conducted verification and cleaning jobs for all annotations, and dropped objects if an object cannot be localized using language expressions only. The followings are the statistics and analysis of the two annotation types of the dataset after the verification.

Full-Video Expression. Youtube-VOS has 6,459 and 1,063 unique objects in train and validation split, respectively. Among them, we cover 6,388 unique objects in 3,471 videos ($6,388/6,459 = 98.9\%$) with 12,913 expressions in train split and 1,063 unique objects in 507 videos ($1,063/1,063 = 100\%$) with 2,096 expressions in validation split. On average, each video has 3.8 language expressions and each expression has 10.0 words.

First-Frame Expression. There are 6,006 unique objects in 3,412 videos ($6,006/6,459 = 93.0\%$) with 10,897 expressions in train split and 1,030 unique objects in 507 videos ($1,030/1,063 = 96.9\%$) with 1,993 expressions in validation split. The number of annotated objects is lower than that of the full-video expressions because using only the first frame makes annotation more ambiguous and inconsistent and we dropped more annotations during the verification. On average, each video has 3.2 language expressions and each expression has 7.5 words.

Dataset Analysis. Figure 1 illustrates examples of our dataset and shows the differences between two annotation types. The full-video expressions can use both static and dynamic information of a video while the first-frame expressions focus mostly on appearance information. We also provide the quantitative comparison of our dataset against the existing ones in Table 1, which presents that our dataset contains much more videos and language expressions.

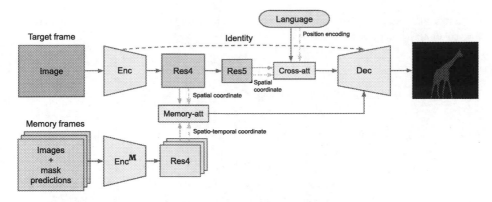

Fig. 2. The overall architecture of our framework. We employ ResNet-50 as our encoder and use the 4th and the 5th stage features (Res4 and Res5) to estimate memory and cross-modal attention, respectively. Both memory-attentive and cross-modal attentive features are progressively combined in the decoder.

4 Unified Referring VOS Network

Given a video with N frames and a language query Q, the goal of referring video object segmentation is to predict binary segmentation masks for the object(s) corresponding to the query Q in the N frames. As mentioned earlier, a naïve approach is to estimate the mask for each frame independently. However, a direct application of image-based solutions to referring object segmentation [12, 18,35,36] would fail to exploit valuable information, temporal coherence across the frames. Therefore, we cast the referring video object segmentation task as a joint problem of referring object segmentation in an image [12,18,35,36] and mask propagation in a video [22,31].

4.1 Our Framework

We propose a unified framework that performs referring image segmentation and video object segmentation jointly. Given a video and a referring expression, our network estimates an object mask in an input frame using the linguistic referring expression and the mask predictions in the previous frames. We iteratively process video frames until the mask predictions in all frames converge. Figure 2 illustrates the overall architecture of our network.

Visual Encoder. We employ ResNet-50 as our backbone network to extract visual features from an input frame. To include spatial information of the visual feature, we augment 8-dimensional spatial coordinates following [8]. Formally, let $\mathbf{F} \in \mathbb{R}^{H \times W \times C_{\mathbf{f}}}$ and $\mathbf{f}_p \in \mathbb{R}^{C_{\mathbf{f}}}$ denote a visual feature map[1] and a sliced visual feature at a certain spatial location p on $\mathbf{F}$, where $p \in \{1, 2, ..., H \times W\}$. We

---

[1] We use Res5 and Res4 feature maps in our model.

concatenate the spatial coordinates to the visual features $\mathbf{f}_p$ to obtain location-aware visual features $\bar{\mathbf{f}}_p$ as follows.

$$\bar{\mathbf{f}}_p = [\mathbf{f}_p; \mathbf{s}_p] \in \mathbb{R}^{C_f+8}, \tag{1}$$

where $\mathbf{s}_p$ is a 8-dimensional spatial coordinate features[2].

Language Encoder. Given a referral expression, a set of words in the expression is encoded as a multi-hot vector and projected onto an embedding space in C_e dimensions using a linear layer. To model the sequential nature of language expressions while maintaining the semantics in the expression, we add positional encoding [27] at each word position. Let $\mathbf{w}_l \in \mathbb{R}^{C_e}$ and $\mathbf{p}_l \in \mathbb{R}^{C_e}$ denote the embeddings for the l-th word and the position of the expression, respectively. Our lingual feature is obtained by the sum of the two embedding vectors, $i.e.$, $\mathbf{e}_l = \mathbf{w}_l + \mathbf{p}_l \in \mathbb{R}^{C_e}$.

Cross-Modal Attention Module. Using both visual and lingual features, we produce a joint cross-modal feature representation by concatenating the features in both the domains. Unlike [35], we first apply self-attention to each feature independently before producing a joint feature to capture complex alignments between both modalities effectively. Each self-attention module maps each feature to a C_a-dimensional space for both modalities as follows:

$$\widehat{\mathbf{f}}_p = \text{SA}^{\text{vis}}(\mathbf{f}_p) \in \mathbb{R}^{C_a}, \quad \widehat{\mathbf{e}}_l = \text{SA}^{\text{lang}}(\mathbf{e}_l) \in \mathbb{R}^{C_a} \tag{2}$$

where $\text{SA}^*(\cdot)$ ($* \in \{\text{vis}, \text{lang}\}$) denotes a self-attention module for each domain. Then a joint cross-modal feature at each spatial position p and each word position l is given by

$$\mathbf{c}_{pl} = [\widehat{\mathbf{f}}_p; \widehat{\mathbf{e}}_l] \in \mathbb{R}^{C_a+C_a}. \tag{3}$$

We collect all cross-model features $\mathbf{c}_{pl}$ and form a cross-modal feature map as $\mathbf{C} = \{\mathbf{c}_{pl} \mid \forall p, \forall l\} \in \mathbb{R}^{H \times W \times L \times (C_a+C_a)}$.

The next step is to apply self-attention to this cross-modal feature map $\mathbf{C}$. Figure 3(a) iluustrates our cross-modal attention module. We generate a set of (key, query, value) triplets, denoted by $(\mathbf{k}, \mathbf{q}, \mathbf{v})$, using 2D convolutions as follows:

$$\mathbf{k} = \text{Conv}_{\text{key}}(\mathbf{C}) \in \mathbb{R}^{L \times H \times W \times C_a} \tag{4}$$

$$\mathbf{q} = \text{Conv}_{\text{query}}(\mathbf{C}) \in \mathbb{R}^{L \times H \times W \times C_a} \tag{5}$$

$$\mathbf{v} = \text{Conv}_{\text{value}}(\mathbf{C}) \in \mathbb{R}^{L \times H \times W \times C_a} \tag{6}$$

---

[2] For each spatial grid (h, w), $\mathbf{s}_p = [h_{\min}, h_{\text{avg}}, h_{\max}, w_{\min}, w_{\text{avg}}, w_{\max}, \frac{1}{H}, \frac{1}{W}]$, where $h_*, w_* \in [-1, 1]$ are relative coordinates of the grid. H and W denotes the height and width of the whole spatial feature map.

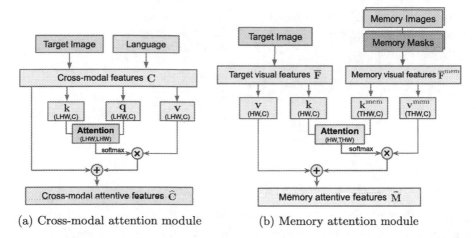

(a) Cross-modal attention module   (b) Memory attention module

Fig. 3. Detailed illustrations of cross-modal and memory attention modules. Each module retrieves relevant information from language and memory frames for the target image to obtain self-attentive features.

and we compute cross-modal attentive features by estimating the correlation between all combinations of pixels and words as

$$\widehat{\mathbf{c}}_{pl} = \mathbf{c}_{pl} + \sum_{\forall p', \forall l'} \text{Softmax}(\mathbf{q}_{pl} \cdot \mathbf{k}_{p'l'}) \mathbf{v}_{p'l'}, \tag{7}$$

where $\cdot$ denotes the dot-product operator. We average the self-attentive features over words and derive the final cross-modal feature as $\widehat{\mathbf{c}}_p = \frac{1}{L} \sum_l \mathbf{c}_{pl}$ and $\widehat{\mathbf{C}} = \{\widehat{\mathbf{c}}_p \mid \forall p\} \in \mathbb{R}^{H \times W \times C_b}$.

Memory Attention Module. To leverage information in the mask predictions at the frames processed earlier, we extend the idea introduced in [22] and design a memory attention module. This module retrieves the relevant information from the previous frame by computing the correlation between the visual feature map of the current frame and the mask-encoded visual feature map of the previous frame. Note that the mask-encoded visual features is obtained from another feature extractor that takes 4-channel inputs given by stacking an RGB image and its segmentation mask in the channel direction. We will call the current and previous frames as target and memory frames, respectively, hereafter.

Different from the previous method [22], we introduce a 12-dimensional spatio-temporal coordinate feature, $\widetilde{\mathbf{s}}_{tp}{}^3$, where the first 3 dimensions encode normalized temporal positions, the next 6 dimensions represent normalized vertical and horizontal positions, and the last 3 dimensions contain the information about duration and frame size of the whole video.

---

^{3} $\widetilde{\mathbf{s}}_{tp} = [t_{\min}, t_{\text{avg}}, t_{\max}, h_{\min}, h_{\text{avg}}, h_{\max}, w_{\min}, w_{\text{avg}}, w_{\max}, \frac{1}{T}, \frac{1}{H}, \frac{1}{W}]$.

Let T be the number of memory frames. For a target frame and T memory frames, we first compute key $(\mathbf{k}, \mathbf{k}^{\mathrm{mem}})$ and value $(\mathbf{v}, \mathbf{v}^{\mathrm{mem}})$ embeddings as follows:

$$\overline{\mathbf{F}} = \{[\mathbf{f}_p; \mathbf{s}_p] | \forall p\} \in \mathbb{R}^{H \times W \times (C_\mathbf{f}+8)} \tag{8}$$

$$\mathbf{k} = \mathrm{Conv}_{\mathrm{key}}(\overline{\mathbf{F}}) \in \mathbb{R}^{H \times W \times C_\mathrm{b}} \tag{9}$$

$$\mathbf{v} = \mathrm{Conv}_{\mathrm{value}}(\overline{\mathbf{F}}) \in \mathbb{R}^{H \times W \times C_\mathrm{b}} \tag{10}$$

$$\overline{\mathbf{F}}^{\mathrm{mem}} = \{[\mathbf{f}_{tp}^{\mathrm{mem}}; \widetilde{\mathbf{s}}_{tp}] | \forall t, \forall p\} \in \mathbb{R}^{T \times H \times W \times (C_\mathbf{f}+12)} \tag{11}$$

$$\mathbf{k}^{\mathrm{mem}} = \mathrm{Conv}_{\mathrm{key}}^{\mathrm{mem}}(\overline{\mathbf{F}}^{\mathrm{mem}}) \in \mathbb{R}^{T \times H \times W \times C_\mathrm{b}} \tag{12}$$

$$\mathbf{v}^{\mathrm{mem}} = \mathrm{Conv}_{\mathrm{value}}^{\mathrm{mem}}(\overline{\mathbf{F}}^{\mathrm{mem}}) \in \mathbb{R}^{T \times H \times W \times C_\mathrm{b}} \tag{13}$$

where $\mathbf{f}$ and $\mathbf{f}^{\mathrm{mem}}$ denotes target and memory visual features, and $\mathbf{s}$ and $\widetilde{\mathbf{s}}$ denotes spatial and spatio-temporal coordinate features, respectively. Then, the memory-attentive feature $\widehat{\mathbf{m}}_p$ at the spatial location p is given by

$$\widehat{\mathbf{m}}_p = \mathbf{v}_p + \sum_{\forall t', \forall p'} f(\mathbf{k}_p, \mathbf{k}_{t'p'}^{\mathrm{mem}}) \mathbf{v}_{t'p'}^{\mathrm{mem}} \tag{14}$$

and $\widehat{\mathbf{M}} = \{\widehat{\mathbf{m}}_p \mid \forall p\} \in \mathbb{R}^{H \times W \times C_\mathrm{b}}$. Figure 3(b) presents the detailed illustration of the memory attention module, which shows how it computes the relevance between target frame and memory frames using key-value structure. Since it attends the regions in the target frame that are relevant to previous predictions, our algorithm produces temporally coherent segmentation results. Note that we employ the 4th stage features (Res4) for both target and memory frames in this module because it requires more descriptive features to compute the correlation between local regions of the frames, while cross-modal attention module employs the 5th stage features (Res5) to exploit more semantic information.

Decoder with Feature Pyramid Network. We employ a coarse-to-fine hierarchical structure in our decoder to combine three kinds of semantic features; the cross-modal attentive feature map $\widehat{\mathbf{C}}$, the memory attentive feature map $\widehat{\mathbf{M}}$, and the original visual feature map $\mathbf{F}_l$ in different levels $l \in \{2, 3, 4, 5\}$. Figure 4 illustrates how our decoder combines those three features using a feature pyramid network in a progressive manner. Each layer in the feature pyramid network takes the output of the previous layer and the ResBlock-encoded visual feature in the same level $\mathbf{F}_l$. Additionally, its first and the second layers incorporate cross-attentive features $\widehat{\mathbf{C}}$ and memory-attentive features $\widehat{\mathbf{M}}$, respectively, to capture multi-modal and temporal information effectively. Note that each layer in the feature pyramid network is upsampled by the factor of 2 to match the feature map size to that of the subsequent level.

Instead of using the outputs from individual layers in the feature pyramid for mask generation, we employ an additional self-attention module following BFPN [23] to strengthen feature semantics of all levels. To this end, we first average the output features in all levels after normalizing their sizes and apply

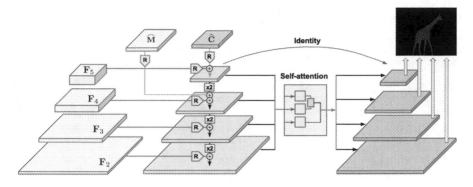

Fig. 4. Detailed illustration of our decoder. It first combines the cross-modal attentive feature map $\widehat{\mathbf{C}}$, the memory attentive feature map $\widehat{\mathbf{M}}$, and the original visual feature map $\mathbf{F}_l$ in multiple levels $l \in \{2,3,4,5\}$ in a progressive manner. 'R' denotes Res-Blocks and '×2' denotes upsampling layers in this figure. The multi-scale outputs are strengthened through a self-attention module, and then employed to estimate the final segmentation masks.

a self-attention to the combined feature map. The resulting map is rescaled to the original sizes, and the rescaled maps are aggregated to the original output feature maps forwarded through identity connections. Finally, these multi-scale outputs are employed to estimate segmentation masks in $1/4$ scale of the input image, following the same pipeline in [11].

Inference. Our network takes three kinds of inputs: a target image, memory images and their mask predictions, and a language expression. Since there is no predicted mask at the first frame, we introduce a novel two-stage procedure for its inference to fully exploit our two attention modules.

In the first stage, we run our network with no memory frame, which results in independent mask prediction at each frame based only on the language expression. After the initial per-frame mask estimation, we select an anchor frame, which has the most confident mask prediction for the language expression. To this end, we calculate the confidence score of each frame by averaging the pixelwsie final segmentation scores and select the frame with the highest one.

In the second stage, we update our initial segmentation results starting from the anchor to both ends using our full network. We first set the anchor frame as a memory frame, and re-estimate the object mask by sequentially propagating the mask prediction from anchor frame. After updating mask prediction at each frame, we add the image and its mask to the memory. In practice, however, cumulating all previous frames to the memory may cause memory overflow issues and slow down inference speed. To alleviate this problem, we set the maximum number of memory frames to T. If the number of memory frames reaches T, then we replace the least confident frame in the memory with the new prediction. Note that we leverage the previous mask predictions in the memory

Table 2. The quantitative evaluation of referring video object segmentation on the Refer-Youtube-VOS validation set.

Method	prec@0.5	prec@0.6	prec@0.7	prec@0.8	prec@0.9	$\mathcal{J}$	$\mathcal{F}$
Baseline (Image-based)	31.98	27.66	21.54	14.56	4.33	33.34	36.54
Baseline + RNN	40.24	35.90	30.34	22.26	9.35	34.79	38.08
Ours w/o cross-modal attention	41.58	36.12	28.50	20.13	8.37	38.88	42.82
Ours w/o memory attention	46.26	40.98	34.81	25.42	10.86	39.38	41.78
Ours	**52.19**	**46.77**	**40.16**	**27.68**	**14.11**	**45.27**	**49.19**

Table 3. The quantitative evaluation of referring video object segmentation on Refer-DAVIS$_{17}$ validation set.

Method	Pretrained	$\mathcal{J}$	$\mathcal{F}$
Khoreva et al. [10]	RefCOCO [20]	37.3	41.3
Ours	RefCOCO [20]	41.23	47.01
Baseline (frame-based)	Refer-YV (ours)	32.19	37.23
Basline + RNN	Refer-YV (ours)	36.94	43.45
Ours w/o cross-modal attention	Refer-YV (ours)	38.25	43.20
Ours w/o memory attention	Refer-YV (ours)	39.43	45.87
Ours (pretraining only)	Refer-YV (ours)	44.29	49.41
Ours	Refer-YV (ours)	**47.29**	**55.96**

frames and estimate the mask of the target frame. At the same time, we use a language expression for guidance during the second stage as well, which allows us to handle challenging scenarios like drift and occlusions. We iteratively refine segmentation by repeating the second stage based on the new anchor identified at each iteration.

5 Experiments

We first evaluate the proposed method on our Refer-Youtube-VOS dataset, and perform comparison to the existing work on the Refer-DAVIS$_{17}$ dataset [10]. We also provide diverse ablation studies to validate the effectiveness of our dataset and framework.

5.1 Implementation Details

We employ a pretrained ResNet-50 on the ImageNet dataset as our backbone network. Every frame of an input video is resized to 320×320. The maximum length of an expression, L, is 20 and the dimensionality of the word embedding space, C_e, is 1,000. We train our model using the Adam optimizer with a batch size 16. Our model is trained end-to-end for 120 epochs. The learning rate is initialized to 2×10^{-5} and decayed by the factor of 10 at every 80 epochs. We set the maximum number of memory frames, T, to 4.

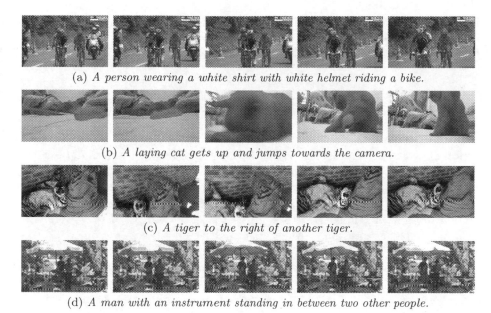

(a) *A person wearing a white shirt with white helmet riding a bike.*

(b) *A laying cat gets up and jumps towards the camera.*

(c) *A tiger to the right of another tiger.*

(d) *A man with an instrument standing in between two other people.*

Fig. 5. Qualitative results of our models on Refer-Youtube-VOS dataset.

5.2 Evaluation Metrics

We use two standard evaluation metrics, the region similarity ($\mathcal{J}$) and the contour accuracy ($\mathcal{F}$) following [25]. Additionally, we also measure prec@X, the percentage of correctly segmented frames in the whole dataset, given a predefined threshold X sampled from the range $[0.5, 0.9]$. Note that segmentation in a frame is regarded as successful if its $\mathcal{J}$ score is higher than a threshold.

5.3 Quantitative Results

Refer-Youtube-VOS. We present the experimental results of our framework on the Refer-Youtube-VOS dataset in Table 2. We follow the original Youtube-VOS dataset [33] to split the data into training and validation sets. In Table 2, 'Baseline' denotes a variant of the frame-based model [35] with our feature pyramid decoder while 'Baseline + RNN' is an extension of the baseline model, which applies a GRU layer to the visual features extracted from multiple input frames for sequential estimation of masks. 'Ours w/o cross-modal attention' and 'Ours w/o memory attention' are the ablative models without the cross-modal attention module and the memory attention module, respectively, for both training and inference. As shown in Table 2, our full model achieves remarkable performance gain over all the compared models on the Refer-Youtube-VOS dataset. The huge performance boost in our full model with respect to the ablative ones implies crucial role of the integrated attention modules in this referring video object segmentation task.

Table 4. The effects of dataset scale on our algorithm. We evaluate on the same validation set for each scale.

Dataset Scale	10%	20%	30%	50%	100%
$\mathcal{J}$		30.73	37.77	39.19	42.48	**45.27**
$\mathcal{F}$		32.92	41.02	43.15	46.05	**49.19**

Refer-DAVIS$_{17}$. DAVIS 2017 [25] is the most popular benchmark dataset for the video object segmentation task, which consists of 197 objects in 89 videos. Each video is composed of high-resolution frames with segmentation annotations, and involves various challenges including occlusions, multi-object interactions, camera motion, etc. Refer-DAVIS$_{17}$ [10] is the extension of DAVIS 2017 with natural language expressions. We evaluated all the models tested on the Refer-Youtube-VOS dataset. Because the number of videos in the DAVIS dataset is not sufficient to train the models for our task from scratch, we pretrain the models on Refer-Youtube-VOS and then fine-tune them on Refer-DAVIS$_{17}$. Table 3 shows the experimental results, where our model outperforms the existing method [10] and the ablative models. For the fair comparison with [10], we pretrained our model on a large-scale referring image segmentation benchmark [20]; our method turns out to be better than [10] under the same pretraining environment. Also, note that our model pretrained on Refer-Youtube-VOS with no fine-tuning on Refer-DAVIS$_{17}$ outperforms all other baselines while our full model further boosts accuracy significantly. This demonstrates the effectiveness of the new large-scale dataset and the proposed network.

5.4   Qualitative Results

Figure 5 illustrates the qualitative results of our method on the Refer-Youtube-VOS dataset. The proposed model segments the target objects successfully with sharp boundaries on many videos and queries. We observe that our framework handles occlusion, deformation, and target identification effectively. See our supplementary documents for more qualitative results.

5.5   Analysis

Dataset Scale. To investigate how the accuracy of a model changes depending on dataset sizes, we conduct experiments on four different subsets of the Refer-Youtube-VOS dataset, which contains 10%, 20%, 30%, and 50% of training examples, respectively. Table 4 presents the impact of dataset scale on model performance. As expected, the accuracy gradually improves upon the increase in the dataset size, which demonstrates the importance of a large-scale dataset on the referring video object segmentation task.

Table 5. Ablation study on the effects of inference procedures.

Inference scheme	$\mathcal{J}$	$\mathcal{F}$
Forward	43.13	49.07
Anchor + Previous	44.58	49.14
Ours	**45.27**	**49.19**

Table 6. Iteration of inference procedures in terms of region similarity ($\mathcal{J}$).

	Stage 1	Stage 2					
		Iter 1	Iter 2	Iter 3	Iter 4	Iter 5	Iter 10
$\mathcal{J}$	41.34	45.27	45.33	45.41	45.44	45.43	**45.46**

Inference Procedure. To validate the effectiveness of our inference scheme, we compare it with two other options for mask prediction. The baseline method, denoted by 'Forward', computes the mask at the first frame and propagates it in the forward direction until the end of video. We have also tested a variant ('Anchor + Previous') of the proposed two-stage inference method. 'Anchor + Previous' first estimates the masks in each frame independently and propagate an anchor frame in a sequential manner, where the previous T frames are used as memory frames during the second stage. Table 5 presents that our full inference technique gives the best performance, which implies that both use of anchor frames and memory frame selection by confidence contribute to improving segmentation results.

Iterative Inference. We study the benefit given by the multiple iterations of the second stage inference step. Table 6 illustrates that the iterative inference procedure gradually improves accuracy and tends to be saturated after 5 iterations.

6   Conclusion

We have proposed a unified referring video object segmentation network to exploit both language-based object segmentation and mask propagation in a single model. Our two attention modules, cross-modal attention and memory attention, collaborate to obtain accurate target object masks specified by language expressions and achieve temporally coherent segmentation results across frames. We also constructed the first large-scale referring video object segmentation dataset. Our framework accomplishes remarkable performance gain on our new dataset as well as the existing one. We believe the new dataset and our proposed method will foster the new direction in this line of research.

Acknowledgement. This work was supported by Institute for Information & Communications Technology Promotion (IITP) grant funded by the Korea government (MSIT) [2017-0-01779, 2017-0-01780].

References

1. Benard, A., Gygli, M.: Interactive video object segmentation in the wild. arXiv preprint arXiv:1801.00269 (2017)
2. Caelles, S., Maninis, K.K., Pont-Tuset, J., Leal-Taixé, L., Cremers, D., Van Gool, L.: One-shot video object segmentation. In: CVPR (2017)
3. Caelles, S., et al.: The 2018 davis challenge on video object segmentation. arXiv preprint arXiv:1803.00557 (2018)
4. Chen, Y., Pont-Tuset, J., Montes, A., Van Gool, L.: Blazingly fast video object segmentation with pixel-wise metric learning. In: CVPR (2018)
5. Fan, C., Zhang, X., Zhang, S., Wang, W., Zhang, C., Huang, H.: Heterogeneous memory enhanced multimodal attention model for video question answering. In: CVPR (2019)
6. Gavrilyuk, K., Ghodrati, A., Li, Z., Snoek, C.G.: Actor and action video segmentation from a sentence. In: CVPR (2018)
7. Goel, V., Weng, J., Poupart, P.: Unsupervised video object segmentation for deep reinforcement learning. In: NIPS (2018)
8. Hu, R., Rohrbach, M., Darrell, T.: Segmentation from natural language expressions. In: ECCV (2016)
9. Jhuang, H., Gall, J., Zuffi, S., Schmid, C., Black, M.J.: Towards understanding action recognition. In: CVPR (2013)
10. Khoreva, A., Rohrbach, A., Schiele, B.: Video object segmentation with language referring expressions. In: ACCV (2018)
11. Kirillov, A., Girshick, R., He, K., Dollár, P.: Panoptic feature pyramid networks. In: CVPR (2019)
12. Li, R., et al.: Referring image segmentation via recurrent refinement networks. In: CVPR (2018)
13. Li, S., Seybold, B., Vorobyov, A., Lei, X., Jay Kuo, C.C.: Unsupervised video object segmentation with motion-based bilateral networks. In: ECCV (2018)
14. Li, X., Change Loy, C.: Video object segmentation with joint re-identification and attention-aware mask propagation. In: ECCV (2018)
15. Li, Z., Tao, R., Gavves, E., Snoek, C.G., Smeulders, A.W.: Tracking by natural language specification. In: CVPR (2017)
16. Liu, C., Lin, Z., Shen, X., Yang, J., Lu, X., Yuille, A.: Recurrent multimodal interaction for referring image segmentation. In: ICCV (2017)
17. Maninis, K.K., Caelles, S., Pont-Tuset, J., Van Gool, L.: Deep extreme cut: From extreme points to object segmentation. In: CVPR (2018)
18. Margffoy-Tuay, E., Pérez, J.C., Botero, E., Arbeláez, P.: Dynamic multimodal instance segmentation guided by natural language queries. In: ECCV (2018)
19. Mun, J., Yang, L., Ren, Z., Xu, N., Han, B.: Streamlined dense video captioning. In: CVPR (2019)
20. Nagaraja, V.K., Morariu, V.I., Davis, L.S.: Modeling context between objects for referring expression understanding. In: Leibe, B., Matas, J., Sebe, N., Welling, M. (eds.) ECCV 2016. LNCS, vol. 9908, pp. 792–807. Springer, Cham (2016). https://doi.org/10.1007/978-3-319-46493-0_48

21. Oh, S.W., Lee, J.Y., Xu, N., Kim, S.J.: Fast user-guided video object segmentation by interaction-and-propagation networks. In: CVPR (2019)
22. Oh, S.W., Lee, J.Y., Xu, N., Kim, S.J.: Video object segmentation using space-time memory networks. In: ICCV (2019)
23. Pang, J., Chen, K., Shi, J., Feng, H., Ouyang, W., Lin, D.: Libra r-cnn: towards balanced learning for object detection. In: CVPR (2019)
24. Perazzi, F., Khoreva, A., Benenson, R., Schiele, B., Sorkine-Hornung, A.: Learning video object segmentation from static images. In: CVPR (2017)
25. Perazzi, F., Pont-Tuset, J., McWilliams, B., Van Gool, L., Gross, M., Sorkine-Hornung, A.: A benchmark dataset and evaluation methodology for video object segmentation. In: CVPR (2016)
26. Tokmakov, P., Alahari, K., Schmid, C.: Learning video object segmentation with visual memory. In: ICCV (2017)
27. Vaswani, A., et al.: Attention is all you need. In: NIPS (2017)
28. Voigtlaender, P., Chai, Y., Schroff, F., Adam, H., Leibe, B., Chen, L.C.: Feelvos: fast end-to-end embedding learning for video object segmentation. In: CVPR (2019)
29. Wang, H., Deng, C., Yan, J., Tao, D.: Asymmetric cross-guided attention network for actor and action video segmentation from natural language query. In: ICCV (2019)
30. Wang, W., et al.: Learning unsupervised video object segmentation through visual attention. In: CVPR (2019)
31. Wug Oh, S., Lee, J.Y., Sunkavalli, K., Joo Kim, S.: Fast video object segmentation by reference-guided mask propagation. In: CVPR (2018)
32. Xu, C., Hsieh, S.H., Xiong, C., Corso, J.J.: Can humans fly? action understanding with multiple classes of actors. In: CVPR (2015)
33. Xu, N., et al.: Youtube-vos: sequence-to-sequence video object segmentation. In: ECCV (2018)
34. Yang, L., Wang, Y., Xiong, X., Yang, J., Katsaggelos, A.K.: Efficient video object segmentation via network modulation. In: CVPR (2018)
35. Ye, L., Rochan, M., Liu, Z., Wang, Y.: Cross-modal self-attention network for referring image segmentation. In: CVPR (2019)
36. Yu, L., et al.: Mattnet: modular attention network for referring expression comprehension. In: CVPR (2018)
37. Zhou, T., Wang, S., Zhou, Y., Yao, Y., Li, J., Shao, L.: Motion-attentive transition for zero-shot video object segmentation. In: AAAI (2020)

Generalizing Person Re-Identification by Camera-Aware Invariance Learning and Cross-Domain Mixup

Chuanchen Luo[1,2], Chunfeng Song[1,2], and Zhaoxiang Zhang[1,2,3(✉)]

[1] University of Chinese Academy of Sciences, Beijing, China
[2] Center for Research on Intelligent Perception and Computing, NLPR, CASIA, Beijing, China
{luochuanchen2017,chunfeng.song,zhaoxiang.zhang}@ia.ac.cn
[3] Center for Excellence in Brain Science and Intelligence Technology, CAS, Shanghai, China

Abstract. Despite the impressive performance under the single-domain setup, current fully-supervised models for person re-identification (re-ID) degrade significantly when deployed to an unseen domain. According to the characteristics of cross-domain re-ID, such degradation is mainly attributed to the dramatic variation within the target domain and the severe shift between the source and target domain. To achieve a model that generalizes well to the target domain, it is desirable to take both issues into account. In terms of the former issue, one of the most successful solutions is to enforce consistency between nearest-neighbors in the embedding space. However, we find that the search of neighbors is highly biased due to the discrepancy across cameras. To this end, we improve the vanilla neighborhood invariance approach by imposing the constraint in a camera-aware manner. As for the latter issue, we propose a novel cross-domain mixup scheme. It alleviates the abrupt transfer by introducing the interpolation between the two domains as a transition state. Extensive experiments on three public benchmarks demonstrate the superiority of our method. Without any auxiliary data or models, it outperforms existing state-of-the-arts by a large margin. The code is available at https://github.com/LuckyDC/generalizing-reid.

Keywords: Domain adaptation · Person re-identification · Camera-aware invariance learning · Cross-domain mixup

1 Introduction

Person re-identification (re-ID) aims to associate images of the same person across non-overlapping camera views. As the fundamental component of intelligent surveillance systems, it has drawn wide attention both in the industry and

Electronic supplementary material The online version of this chapter (https://doi.org/10.1007/978-3-030-58555-6_14) contains supplementary material, which is available to authorized users.

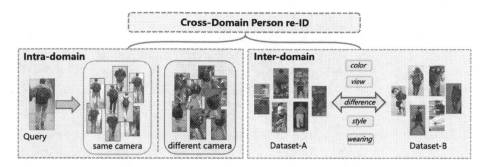

Fig. 1. The illustration cross-domain person re-ID. We consider intra-domain variation and inter-domain shift simultaneously. In terms of the former, cross-camera variations lead to a biased retrieval. As for the latter, the discrepancy between the source domain and the target domain hiders the effective adaptation.

academia. With the surge of deep learning techniques, recent years have witnessed great progress in fully-supervised person re-ID [13,20,33,34,38,51,52]. However, the success of this paradigm relies heavily on enormous annotated data in the target domain, which is usually prohibitive to acquire in practice. To bypass the scarcity of annotations, one can train the model with a relevant labeled dataset, a.k.a.the source domain. Unfortunately, due to the dramatic shift in data distribution, such a model would suffer a severe degradation in performance when directly deployed to the target domain. For this reason, it is desirable to investigate the problem of cross-domain person re-ID.

Given labeled source data and unlabeled target data, cross-domain re-ID dedicates to learn a model that generalizes well to the target domain. Compared with conventional unsupervised domain adaptation (UDA), it is characterized by the open-set setup and the domain hierarchy. The former implies the disjoint label space between the source domain and the target domain, which breaks the underlying assumption of most UDA methods. As for the latter, each domain can be further divided into multiple camera sub-domains, since the style of images is distinct across different cameras. According to such a hierarchy of domains, we impute the poor transfer performance to two factors, *intra-domain variation* and *inter-domain shift*. Wherein, the first factor is mainly derived from camera divergence. These issues are illustrated in Fig. 1. To achieve superior transfer performance, it is desirable to take both issues into account.

Recently, some studies [10,45,57,58] have verified the effectiveness of neighborhood invariance in coping with the intra-domain variation of the target domain. Equipped with a memory bank, these methods search neighbors of each probe throughout the whole dataset and impose a consistency constraint between them. However, due to the lack of supervision in the target domain, the model cannot suppress well the impact of inter-camera variation (including illumination, viewpoint, and background). In this case, the neighbor search is easily biased towards the candidates from the same camera as the probe. To be more specific, positive inter-camera matches are more likely to be arranged behind many negative intra-camera matches in the ranking list, which confuses the model learning. To address the issue, we improve the neighborhood invariance by imposing the constraint separately for intra-camera matching and inter-camera matching. Despite the simplicity, this proposal leads to considerable improvement over its vanilla counterpart.

To alleviate the adverse effect of inter-domain shift, early works [9, 40] employ extra generative models to transfer the image style across domains, which is essentially an advanced interpolation between the source and target manifold. By introducing stylized images as an intermediate domain, these methods expect to avoid the issues caused by the abrupt transfer between two very different domains. Along this insight, we explore to achieve the same goal by interpolating the samples from the two domains directly. Different from style transfer, the direct mixture in the pixel level leads to the change of content. Therefore, the identity label should also be mixed accordingly. This is exactly a mixup [49] process. However, it is nontrivial to employ vanilla mixup [49] in our case since it is initially customized for the *closed-set* classification problem. To make it applicable to *open-set* cross-domain re-ID, we augment mixup with a dynamic classifier. It can cover the label space of the input source-target pairs adaptively without the access to the exact label space of the target domain.

In summary, the contribution of this work is three-fold:

- To bypass the bias in the neighbor search, we impose the neighborhood invariance in a camera-aware way. Despite the simplicity, this approach leads to a significant improvement over its camera-agnostic counterpart.
- We propose a novel cross-domain mixup scheme to smooth the transition between the source domain and the target domain. It improves the transfer performance significantly with negligible overhead.
- Extensive experiments validate the effectiveness of our method. It achieves state-of-the-art performance on Market-1501, DukeMTMC-reID and MSMT17 datasets.

2   Related Work

Supervised person re-identification has made significant progress in recent years, thanks to the advent of deep neural networks [5,15,17] and large scale datasets [29,40,53,54]. The research in this field mainly focuses on the development of discriminative loss functions [16,22,48] or network architectures [28,34,37,61]. In term of the former direction, a series of deep metric learning methods [3,6,16,31,47] been proposed to enhance intra-class compactness and inter-class separability in the manifold. As for the customization of architecture, PCB [34] and its follow-ups [13,34,38,52] dominate the trend. Apart from the two directions mentioned above, some methods [20,33,51] attempt to involve auxiliary data for the fine-grained alignment of the human body. Despite their success in the single domain, these fully-supervised methods suffer from poor generalization ability, which prevents them from the practical application.

Cross-domain person re-identification pursues high performance in the target domain with the access to labeled source data and unlabeled target data. Early works [9,19,25,40] focus on reducing the domain gap between the two domains at the image level. They perform the image-to-image translation [7,62] from the source domain to the target domain and then train the model with translated images. Besides, some methods [18,39] attempt to connect the two domains with common auxiliary tasks. Wang et al. [39] share knowledge across domains through attribute classification. Huang et al. [18] perform human parsing and pose estimation on both domains simultaneously to enhance alignment and model generalization. Recently, some studies [10,45,46,57] recognize the importance of mining discriminative cues in the target domain. Yu et al. [46] mine underlying pairwise relationships according to the discrepancy between feature similarity and class probability. They then use a contrastive loss to enforce the mined relationships. Zhong et al. [57] investigate the impact of intra-domain variations and impose three types of invariance constraints on target samples, *i.e.*exemplar-invariance, camera-invariance [59], and neighborhood-invariance. Yang et al. [45] further introduce the idea of neighborhood-invariance to the patch level. Current leading methods [11,12,14,32,42,50] adopt a pseudo label estimation scheme. They label target samples by a clustering algorithm and then train the model accordingly. Such an operation will be performed repeatedly until the model converges, which results in a heavy computational burden.

MixUp [49] is a data augmentation technique initially proposed for the supervised classification problem. Afterwards, it was extended to random hidden layers by Verma et al. [35]. MixUp enhances the smoothness of the learned manifold by applying convex combinations of labeled samples for training. It has demonstrated its effectiveness on several classification benchmarks. Recently, MixUp has been successfully adapted to the field of semi-supervised learning [1,36] and domain adaptation [26,30]. Without the access to the ground-truths of unlabeled/target data, these methods conduct MixUp based on the prediction of original samples. Unfortunately, all of them focus on the closed-set scenario and cannot be applied to cross-domain re-ID directly. In parallel with our work, Zhong et al. [60] extend MixUp scheme to the open-set scenario where the number of target classes is given. They explain the insight from the viewpoint of the label reliability and achieve very positive results on CIFAR [21] as well as ImageNet [8].

3   Method

In the context of cross-domain person re-ID, we have access to a labeled source domain $\mathcal{S} = \{X_s, Y_s\}$ and an unlabeled target domain $\mathcal{T} = \{X_t\}$. The source domain contains N_s images of P persons. Each sample $x_i^s \in X_s$ is associated with an identity label y_i^s. The target domain consists of N_t images $\{x_i^t\}_{i=1}^{N_t}$ whose identity annotations are absent. In addition, the camera indices of images

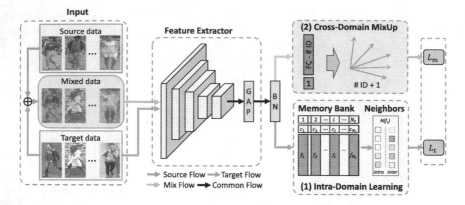

Fig. 2. The framework of our method. Firstly, mixed data is generated by the convex combination between source-target pairs. Then, it is fed into the network together with target data to acquire image embeddings. After the normalization by a BN layer, each type of embeddings is assigned to its corresponding component. (1) With the help of an augmented memory, the learning of target embeddings is supervised by intra-camera and inter-camera neighborhood consistency. (2) As for mixed embeddings, we maintain a dynamic classifier to cover the label space of each source-target pair adaptively.

($i.e. C_s = \{c_i^s\}_{i=1}^{N_s}$ and $C_t = \{c_i^t\}_{i=1}^{N_t}$) are also available in both domains. Given such information, the goal is to learn a model that generalizes well to the target domain.

3.1 Overview

As illustrated in Fig. 2, we feed-forward target samples and mixed samples into the network simultaneously. Wherein, each mixed sample is generated by interpolating between a source-target pair. For target data, we maintain a memory bank $M \in \mathbb{R}^{N_t \times d}$, where each slot $m_i \in \mathbb{R}^d$ stores the feature of the corresponding sample x_i^t. The memory is updated in a running-average manner during training:

$$m_i \leftarrow \sigma m_i + (1 - \sigma) f\left(x_i^t\right), \quad m_i \leftarrow m_i / \|m_i\|_2, \tag{1}$$

where σ denotes the momentum of the update, $f\left(x_i^t\right) \in \mathbb{R}^d$ represents the l_2-normalized feature of x_i^t extracted by the current model. In practice, the memory bank behaves as a non-parametric inner-product layer [44], by which we can obtain pairwise similarities between each input sample and all target instances on the fly. On the basis of such pairwise similarities, we can retrieve nearest-neighbors of each input image and impose a consistency constraint between them. As for mixed data, we compose a dynamic classifier to cover the label space of each source-target pair adaptively. It is built upon the source prototypes and the feature of the target instance. In the sequel, we will elaborate on the learning tasks customized for target data and mixed data.

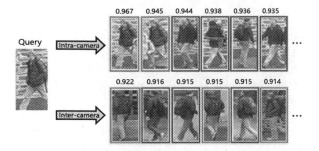

Fig. 3. The visualization of ranking lists in intra-camera matching and inter-camera matching. We perform retrieval on DukeMTMC-reID using a model pre-trained on Market-1501. The green frame indicates positive matches, while the red frame indicates negative matches. The score on the top of each gallery image represents its cosine similarity with the probe. (Color figure online)

3.2 Camera-Aware Neighborhood Invariance

Without the knowledge of the label space (*i.e.*identity annotations and the number of identities), it is infeasible to figure out the class assignment of target samples directly. In this case, the pairwise relationship is a potential cue to guide the feature learning in the target domain. In representation learning, it is generally assumed that each sample shares the same underlying label with its nearest-neighbors at a high probability. Equipped with the memory bank mentioned above, we can obtain the probability that x_i^t share the same identity with x_j^t on the fly:

$$p_{ij} = \frac{\exp\left(s \cdot m_j^T f\left(x_i^t\right)\right)}{\sum_{k=1}^{N_t} \exp\left(s \cdot m_k^T f\left(x_i^t\right)\right)}, \tag{2}$$

where s is a scaling factor that modulates the sharpness of the probability distribution. According to the above assumption, ECN [57] proposes to maximize such probabilities between each probe image and its nearest-neighbors in the whole dataset:

$$\mathcal{L}_{ag} = -\sum_j w_{i,j} \log p_{ij}, \qquad w_{i,j} = \begin{cases} \frac{1}{|\Omega_i|}, & j \neq i \\ 1, & j = i \end{cases}, \qquad \forall j \in \Omega_i, \tag{3}$$

where Ω_i represents the nearest-neighbors of x_i^t throughout the whole target domain. $|\Omega(x_i^t)|$ denotes the size of the neighbor set. For convenience, we term this loss function as camera-agnostic neighborhood loss, since it treats all candidates equally regardless of their camera indices while searching neighbors.

Due to the scene variation across cameras, there is a significant discrepancy in similarity distribution between inter-camera matching and intra-camera matching [41]. The average pairwise similarity of inter-camera matching is smaller than that of intra-camera matching. As a result, intra-camera candidates can easily dominate the top ranking list, whether or not they are positive matches. In this case, it is problematic to employ Eq. (3), since it would push inter-camera positive matches away from the probe. For clarity, we visualize an example in Fig. 3. From the figure, we observe that even the first positive inter-camera match has a lower similarity score than many negative intra-camera matches. When sorting all candidates in a camera-agnostic manner, positive inter-camera matches can be easily excluded from a pre-defined neighborhood range. An intuitive solution is to choose a larger neighborhood range. However, such a practice would involve more negative matches inevitably, which is detrimental to feature learning.

To bypass this dilemma, we propose to enforce neighborhood invariance separately for intra-camera matching and inter-camera matching. Suppose O_i^{intra} denotes the set of instances that share the same camera as x_i^t and O_i^{inter} represents the set of instances whose camera indexes are different from x_i^t. For sample x_i^t, intra-camera matching and inter-camera matching only have access to the instances in O_i^{intra} and O_i^{inter}, respectively. Therefore, the probability that x_i^t shares the same identity with an intra-camera candidate x_j^t is formulated as follows:

$$p_{i,j}^{intra} = \frac{\exp\left(s \cdot \boldsymbol{m}_j^{\mathrm{T}} f\left(x_i^t\right)\right)}{\sum_{k \in O_i^{intra}} \exp\left(s \cdot \boldsymbol{m}_k^{\mathrm{T}} f\left(x_i^t\right)\right)} \tag{4}$$

The definition of the probability that x_i^t shares the same identity with an inter-camera candidate is similar:

$$p_{i,j}^{inter} = \frac{\exp\left(s \cdot \boldsymbol{m}_j^{\mathrm{T}} f\left(x_i^t\right)\right)}{\sum_{k \in O_i^{inter}} \exp\left(s \cdot \boldsymbol{m}_k^{\mathrm{T}} f\left(x_i^t\right)\right)} \tag{5}$$

Accordingly, we replace the original camera-agnostic loss function Eq. (3) with the following two camera-aware loss functions:

$$\begin{aligned} \mathcal{L}_{intra} &= -\sum_j w_{i,j} \log p_{i,j}^{intra}, \quad \forall j \in \Omega_i^{intra}. \\ \mathcal{L}_{inter} &= -\sum_j w_{i,j} \log p_{i,j}^{inter}, \quad \forall j \in \Omega_i^{inter}. \end{aligned} \tag{6}$$

where Ω_i^{intra} and Ω_i^{inter} denote the neighbor sets of x_i^t throughout O_i^{intra} and O_i^{inter}, respectively. Different from ECN [57] that adopts fixed top-k nearest-neighbors, we define the neighborhood based on the relative similarity ratio to the top-1 nearest neighbors:

$$\Omega_i = \{j | \mathrm{sim}(x_i, x_j) > \epsilon \cdot \mathrm{sim}(x_i, \text{top-1 neighbor of } x_i)\} \tag{7}$$

Moreover, without the disturbance of cross-camera variations, the mined neighborhood for intra-camera matching is much more reliable than that for inter-camera matching. Thus, it is much easier to learn a discriminative intra-camera representation first, which can encourage accurate inter-camera matching. For this reason, we propose to employ $\mathcal{L}_{intra}$ before the involvement of $\mathcal{L}_{inter}$ in practice.

Remarks. Some related works [4, 23, 41, 43, 63] adopt a similar two-stage learning scheme. They focus on spreading the given local association (*i.e.* tracklet [4, 23] or identity [41, 63] within the same camera) to the global. They do not pay attention to the discrepancy in similarity distribution between the intra-camera matching and inter-camera matching.

3.3   Cross-Domain Mixup

In order to push the transfer performance ahead, it is desirable to handle the shift between the source domain and the target domain. Early efforts in this direction perform image-to-image translation from the source style to the target style. They introduce the stylized domain as an intermediate state to mitigate the performance loss of direct transfer. However, the style transfer process demands cumbersome generative models. Can we achieve the same goal in a more concise fashion?

Essentially, style transfe is an advanced interpolation between the source and target manifold. Considering that mixup [49] also conducts interpolation on the data manifold, we explore to employ it as the substitute for style transfer. According to the formulation of mixup [49], we mix samples and their labels simultaneously:

$$\lambda \sim \text{Beta}(\alpha, \alpha) \tag{8}$$

$$x_m = \lambda x_s + (1 - \lambda)x_t \qquad y_m = \lambda y_s + (1 - \lambda)y_t \tag{9}$$

where α is a hyper-parameter of Beta distribution. $(x_s, y_s) \in \mathcal{S}$ and $(x_t, y_t) \in \mathcal{T}$ denote samples from the source domain and the target domain, respectively. However, we have no access to the target annotation y_t in the context of cross-domain re-ID. Besides, the label space is disjoint between the two domains. Thus, it is infeasible to apply mixup operation directly. To address the issue, we propose to maintain a dynamic classifier that covers the label space of source-target pair adaptively. With the knowledge of the source label space, we can first define a classifier to identify P persons in the source domain. Then, we append a virtual prototype vector $\boldsymbol{w}_{virt} \in \mathbb{R}^d$ to the source classifier $\boldsymbol{W} \in \mathbb{R}^{P \times d}$:

$$\boldsymbol{W}' \leftarrow [\boldsymbol{W}, \boldsymbol{w}_{virt}] \qquad \boldsymbol{w}_{virt} = \frac{\|\boldsymbol{w}_{y_s}\|_2 \cdot f(x_t)}{\|f(x_t)\|_2}, \tag{10}$$

where $[\cdot]$ denotes the concatenate operation, $\boldsymbol{w}_{y_s}$ denotes the prototype vector of the y_s-th identity, $\boldsymbol{W}' \in \mathbb{R}^{(P+1) \times d}$ is the parameter matrix of the composed classifier. As expressed in the above equation, the dynamically created virtual prototype vector is derived from the feature of the target instance of the mixed pair. It has the same angular as the target feature and the same norm as the source prototype vector. The composed classifier can distinguish $(P+1)$ identities apart. Wherein, the $(P + 1)$-th identity corresponds to the target individual of the source-target pair. To make the labels compatible with the composed classifier, we pad one-hot labels of source samples to $(P + 1)$-d with the zero value. As for the labels of target samples, the final element of this $(P+1)$-d one-hot vector is always activated. Since without specific class assignment, target samples should always be identified as themselves. The feature learning of the mixed data is constrained by the cross-entropy loss between the prediction of the newly composed classifier and the mixed label:

$$\mathcal{L}_m = -\frac{1}{N_s} \sum_{i=1}^{N_s} \sum_{j=1}^{P+1} y_{i,j}^m \log p(j|x_i^m; \boldsymbol{W}'), \tag{11}$$

where $y_{i,j}^m$ denotes the j-th element of the mixed label y_i^m, $p(j|x_i^m; \boldsymbol{W}')$ denotes the probability predicted by the compose classifier that x_i^m belongs to the j-th identity. Empirically, we find that replacing the up-to-date feature $f(x^t)$ in Eq. (10) with its counterpart in the memory can benefit the stability of the training. Therefore, we adopt this practice in the following experiments. Our supplementary material provides detailed experimental results.

Remarks. Both our method and Virtual Softmax [2] introduce the concept of the virtual prototype. However, they are different in motivation and implementation. In terms of motivation, Virtual Softmax introduces the virtual prototype to enhance the discrimination of learned features under the fully-supervised setup. By contrast, our method uses it to adjust the label space of the classifier dynamically according to the input source-target pair. As for implementation, the direction of the virtual prototype is equal to that of the input feature in Virtual Softmax. Whereas in our method, the classifier operates on mixed samples. The virtual prototype has the same direction as the target instance of the input mixed pair.

3.4 Overall Loss Function

Neighborhood invariance for intra-camera matching and inter-camera-matching composes the supervision for the target domain, *i.e.*, $\mathcal{L}_t = \mathcal{L}_{intra} + \mathcal{L}_{inter}$. By combining it with the proposed constraint on the mixed data, we can obtain the final loss function for the model training:

$$\mathcal{L} = \mathcal{L}_t + \mathcal{L}_m. \tag{12}$$

One may ask why not impose a constraint (*e.g.*classification loss or triplet loss) on the source domain, just as other methods do. We remind that $\mathcal{L}_m$ already contains moderate supervision for the source domain. When the interpolation coefficient λ in Eq. (8) is sampled close to 1, $\mathcal{L}_m$ degrades to the classification loss on the source data. For experimental results, see our supplementary material.

4 Experiment

4.1 Dataset and Evaluation Protocol

We evaluate the performance of the proposed method on three public benchmarks, *i.e.*Market-1501 [53], DukeMTMC-reID [29,54], MSMT17 [40]. During training, we adopt two of the three datasets as the source domain and the target domain, respectively. During testing, we evaluate Cumulated Matching Characteristics (CMC) at rank-1, rank-5, rank-10 and mean average precision (mAP) in the testing set of the target domain.

4.2 Implementation Details

We adopt ResNet-50 [15] pre-trained on ImageNet [8] as the backbone of our model. The last downsampling operation of ResNet is discarded, which leads to an overall stride of 16. $\mathcal{L}_{intra}$ and $\mathcal{L}_{inter}$ are involved into the training at 10th and 30th epoch, respectively. In terms of optimizer, we employ Stochastic Gradient Descent (SGD) with a momentum of 0.9 and a weight decay of $1e-5$. The learning rate is set to 0.01 and 0.05 for the backbone layers and newly added layers, respectively. It is divided by 10 at 60th epoch. The whole training process lasts for 70 epochs. As for data, each mini-batch contains 128 source images and 128 target images. All input images are resized to 256×128. Random horizontal flip, random crop and random erasing [55] are utilized for data augmentation. Unless otherwise specified, we follow the setting of scaling factor $s = 10$, neighborhood range $\epsilon = 0.8$, momentum of memory updating $\sigma = 0.6$, and parameter of Beta distribution $\alpha = 0.6$. During testing, we adopt the output of the final Batch Normalization layer as the image embedding. Cosine similarity is used as the measure for retrieval. All experiments are conducted on two NVIDIA TITAN V GPUs using Pytorch [27] platform.

4.3 Parameter Analysis

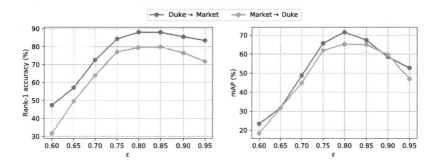

Fig. 4. Evaluation with different values of ϵ in Eq. (7).

Neighborhood Range ϵ. To analyze the effect of ϵ, we vary its value in a reasonable scope and evaluate the performance under these settings. As illustrated in Fig. 4, both rank-1 accuracy and mAP first improve as ϵ decreases. Assigning too small value to ϵ may introduce considerable false positives, which is harmful to the learning of discriminative features. We obtain the optimal performance around $\epsilon = 0.8$. Our method is somewhat sensitive to the setting of neighborhood range. For detailed analysis, see our supplementary material.

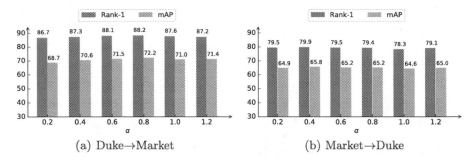

(a) Duke→Market   (b) Market→Duke

Fig. 5. Evaluation with different values of the Beta distribution parameter α in Eq. (8).

Beta Distribution Parameter α. The parameter α determines the distribution of interpolation coefficient λ. Assigning a larger value to α leads to an stronger regularization. To investigate its effect, we vary the parameter α to five different values and evaluate the performance under these settings. As shown in Fig. 5, both rank-1 accuracy and mAP fluctuate very slightly with the variation of α. This indicates that our method is relatively robust to the setting of cross-domain mixup.

Table 1. Evaluation with different values of the scaling factor s in Eq. (2).

s	Duke→ Market				Market→ Duke			
	Rank-1	Rank-5	Rank-10	mAP	Rank-1	Rank-5	Rank-10	mAP
6	80.3	90.1	92.9	60.1	75.3	84.0	86.7	58.5
8	86.4	93.3	95.4	67.5	78.9	88.0	90.9	64.1
10	**88.1**	**94.4**	**96.2**	**71.5**	79.5	88.3	91.4	**65.2**
12	84.6	92.7	94.9	67.8	**79.7**	**89.1**	**91.6**	**65.2**
14	74.4	89.1	93.0	57.1	76.1	87.0	89.6	61.0

Scaling Factor s. The scaling factor s in Eq. (2) is crucial to the final performance. Large s can sharpen the probability distribution and ease the optimization. However, assigning too large value may make the task too trivial to learn discriminative features. We train the model under five different values of s and report their results in Table 1. As shown in Table 1, we obtain the optimal performance at $s = 10$ on Market-1501 and $s = 12$ on DukeMTMC-reID. The performance degrades dramatically when s gets too large or too small.

Table 2. Ablation studies on Market-1501 and DukeMTMC-reID. **Supervised Learning**: Model trained with labeled target data. **Direct Transfer**: Model trained with only labeled source data.

Methods	Duke→Market				Market→Duke			
	R-1	R-5	R-10	mAP	R-1	R-5	R-1	mAP
Supervised Learning	90.7	96.6	98.0	74.8	82.7	91.0	93.7	66.4
Direct Transfer	48.9	65.1	71.8	19.8	30.0	44.9	50.9	15.0
$\mathcal{L}_s + \mathcal{L}_{ag}$	60.9	73.1	77.8	35.3	49.8	63.0	68.1	34.5
$\mathcal{L}_s + \mathcal{L}_{intra}$	70.6	83.9	88.6	44.6	70.0	82.4	86.0	52.1
$\mathcal{L}_m + \mathcal{L}_{intra}$	76.8	89.0	92.4	54.9	73.7	84.2	88.1	57.3
$\mathcal{L}_s + \mathcal{L}_{intra} + \mathcal{L}_{inter}$	81.2	91.7	94.2	59.2	76.2	87.5	90.4	59.6
$\mathcal{L}_m + \mathcal{L}_{intra} + \mathcal{L}_{inter}$	88.1	94.4	96.2	71.5	79.5	88.3	91.4	65.2

4.4   Ablation Study

In this section, we conduct extensive ablation studies on the adaptation between Market-1501 and DukeMTMC-reID. For the variants that do not involve cross-domain mixup loss $\mathcal{L}_m$, the supervision on the source data is necessary to ensure meaningful representations. Thus, we perform a classification task in the source domain just as ECN [57] and its follow-ups [10,58] do. Suppose $p(y_i^s|x_i^s)$ denotes the predicted probability that x_i^s belongs to the identity y_i^s. The loss function for the source data is defined as $\mathcal{L}_s = -\frac{1}{N_s}\sum_{i=1}^{N_s} \log p(y_i^s|x_i^s)$. See our supplementary material for ablation studies on other datasets.

Performance Bound. As reported in the first two rows of Table 2, the model achieves promising performance when trained and tested in the same domain (termed *Supervised Learning*). However, such a model performs poorly when directly deployed to an unseen domain (termed *Direct Transfer*). Specifically, the model trained on DukeMTMC-reID achieves only 48.9% rank-1 accuracy on Market-1501, which is 41.8% lower than its single-domain counterpart. *Supervised Learning* and *Direct Transfer* behave as the upper-bound and lower-bound of the transfer performance, respectively.

Effect of Camera-Aware Invariance Learning. To investigate the effect of camera-aware invariance learning, we impose neighborhood invariance separately for intra-camera matching and inter-camera matching. From Row 3–4 in Table 2, we observe a considerable improvement when replacing camera-agnostic neighborhood loss $\mathcal{L}_{ag}$ with its intra-camera counterpart $\mathcal{L}_{intra}$. To be specific, rank-1 accuracy improves from 60.9% to 70.6% and 49.8% to 70.0% on Market-1501 and DukeMTMC-reID, respectively. This is interesting since $\mathcal{L}_{intra}$ even omits massive inter-camera candidates during the optimization. Such a phenomenon verifies our hypothesis mentioned in Sect. 3.2. That is, the discrepancy between intra-camera matching and inter-camera matching makes camera-agnostic neighborhood constraint ambiguous for the optimization. Besides, a discriminative intra-camera representation is beneficial for the cross-camera association. Furthermore, we add inter-camera neighborhood loss $\mathcal{L}_{inter}$ to the supervisory signal to validate its effectiveness. As shown in Row 6 of Table 2, the injection of inter-camera neighborhood loss improves the performance significantly. It leads to a 10.6% and 6.2% gain in rank-1 accuracy on Market-1501 and DukeMTMC-reID, respectively. Without the proposed cross-domain mixup component, such a concise variant is already on par with existing state-of-the-art methods.

Effect of Cross-Domain MixUp. We further investigate the effect of cross-domain mixup by incorporating it into the training. As shown in Table 2, the involvement of $\mathcal{L}_m$ improves the variant "$\mathcal{L}_s + \mathcal{L}_{intra}$" by 6.2% and 3.7% in terms of rank-1 accuracy on Market-1501 and DukeMTMC-reID, respectively. The gain is 6.9% and 3.3% when applying $\mathcal{L}_m$ to the variant "$\mathcal{L}_s + \mathcal{L}_{intra} + \mathcal{L}_{inter}$". Such a significant improvement validates the effectiveness of cross-domain mixup. It is noteworthy that the rank-1 accuracy of our final variant is only **2.6%** and **3.2%** lower than the supervised counterpart on Market-1501 and DukeMTMC-reID, respectively.

4.5   Comparison with State-of-the-art Methods

Results on Market-1501 Dataset. We evaluate the performance of our method on Market-1501 using DukeMTMC-reID as the source domain. We compare the result with representative works of different directions, including the methods based on style transfer [9,25,40], the methods based on pseudo-label estimation [12,14,32], and those mining intra-domain cues [10,45,46,57]. As reported in Table 3, our method performs favorably against current leading methods in rank-1 accuracy and mAP. Note that both ECN [57] and its follow-ups [10,58] benefit a lot from CamStyle [59] augmentation, which requires an extra StarGAN [7]. MMT [14], the nearest rival, employs computationally intensive clustering operation and four ResNet-50 models in total (2 students and 2 teachers) to achieve the similar performance.

Results on DukeMTMC-reID Dataset. We adopt Market-1501 as the source domain and evaluate the performance of the proposed approach on DukeMTMC-reID. As shown in the right part of Table 3, our methods is competitive against other state-of-the-arts. Both SSG [12] and PAUL [45] mine discriminative cues at the part level, which is orthogonal to our concerns. It has been widely validated in the field of supervised re-ID that part models are more powerful than their vanilla counterparts in discrimination. Even so, our method achieves much higher performance than the two competitors. Compared with MMT [14], our method is superior in rank-1 and mAP with much less computational overhead.

Results on MSMT17 Dataset. We further evaluate the transfer performance of our method on MSMT17. MSMT17 is characterized by large scale and abundant variations, which makes it much more challenging. As shown in Table 4, the proposed approach outperforms MMT [14] while using DukeMTMC-reID as

Table 3. Comparison with state-of-the-art cross-domain methods on Market-1501 and DukeMTMC-reID. Red indicates the best and **Blue** the runner-up.

Methods	Market-1501				DukeMTMC-reID			
	R-1	R-5	R-10	mAP	R-1	R-5	R-10	mAP
PTGAN [40]	38.6	–	66.1	–	27.4	–	50.7	–
SPGAN [9]	51.5	70.1	76.8	22.8	41.1	56.6	63.0	22.3
TJ-AIDL [39]	58.2	74.8	81.1	26.5	44.3	59.6	65.0	23.0
CamStyle [59]	58.8	78.2	84.3	27.4	48.4	62.5	68.9	25.1
HHL [56]	62.2	78.8	84.0	31.4	46.9	61.0	66.7	27.2
MAR [46]	67.7	81.9	–	40.0	67.1	79.8	–	48.0
PAUL [45]	68.5	82.4	87.4	40.1	72.0	82.7	86.0	53.2
ARN [24]	70.3	80.4	86.3	39.4	60.2	73.9	79.5	33.4
ECN [57]	75.1	87.6	91.6	43.0	63.3	75.8	80.4	40.4
UDA [32]	75.8	89.5	93.2	53.7	68.4	80.1	83.5	49.0
PAST [50]	78.4	–	–	54.6	72.4	–	–	54.3
SSG [12]	80.0	90.0	92.4	58.3	73.0	80.6	83.2	53.4
AE [10]	81.6	91.9	94.6	58.0	67.9	79.2	83.6	46.7
ECN++ [58]	84.1	92.8	95.4	63.8	74.0	83.7	87.4	54.4
MMT [14]	87.7	94.9	96.9	71.2	78.0	88.8	92.5	65.1
Ours	88.1	94.4	96.2	71.5	79.5	88.3	91.4	65.2

the source domain. However, the performance is far inferior to MMT when the source domain is Market-1501. This is mainly attributed to the training instability induced by the unreliable neighborhood search at the early stage. We will investigate this issue in the future work.

Table 4. Comparison with state-of-the-art cross-domain methods on MSMT17. Red indicates the best and **Blue** the runner-up.

Methods	Market→MSMT				Duke→MSMT			
	R-1	R-5	R-10	mAP	R-1	R-5	R-10	mAP
PTGAN [40]	10.2	–	24.4	2.9	11.8	–	27.4	3.3
ECN [57]	25.3	36.3	42.1	8.5	30.2	41.5	46.8	10.2
AE [10]	25.5	37.3	42.6	9.2	32.3	44.4	50.1	11.7
SSG [12]	31.6	–	49.6	13.2	32.2	–	51.2	13.3
ECN++ [58]	40.4	53.1	58.7	15.2	42.5	55.9	61.5	16.0
MMT [14]	49.2	63.1	68.8	22.9	50.1	63.9	69.8	23.3
Ours	43.7	56.1	61.9	20.4	51.7	64.0	68.9	24.3

5   Conclusion

In this paper, we propose a superior model for cross-domain person re-identification that takes both intra-domain variation and inter-domain shift into account. We adopt a neighborhood invariance approach to supervise feature learning in the target domain. However, we find that the neighbor search is highly biased due to the dramatic discrepancy across cameras. To avoid this issue, we propose to impose the constraint in a camera-aware manner. Furthermore, we devise a novel cross-domain mixup scheme to bridge the gap between the source domain and the target domain. To be more specific, it introduces the interpolation between the two domains as an intermediate state of the transfer. Extensive experiments validate the effectiveness of each proposal. By taking the two proposals together, our method outperforms existing state-of-the-arts by a large margin.

Acknowledgement. This work was supported in part by the National Key R&D Program of China (No. 2018YFB1004602), the National Natural Science Foundation of China (No. 61836014, No. 61761146004, No. 61773375).

References

1. Berthelot, D., Carlini, N., Goodfellow, I., Papernot, N., Oliver, A., Raffel, C.: Mixmatch: a holistic approach to semi-supervised learning. arXiv:1905.02249 (2019)
2. Chen, B., Deng, W., Shen, H.: Virtual class enhanced discriminative embedding learning. In: NeurIPS (2018)
3. Chen, W., Chen, X., Zhang, J., Huang, K.: Beyond triplet loss: a deep quadruplet network for person re-identification. In: CVPR (2017)
4. Chen, Y., Zhu, X., Gong, S.: Deep association learning for unsupervised video person re-identification. In: BMVC (2018)
5. Chen, Y., Li, J., Xiao, H., Jin, X., Yan, S., Feng, J.: Dual path networks. In: NeurIPS (2017)
6. Cheng, D., Gong, Y., Zhou, S., Wang, J., Zheng, N.: Person re-identification by multi-channel parts-based CNN with improved triplet loss function. In: CVPR (2016)
7. Choi, Y., Choi, M., Kim, M., Ha, J.W., Kim, S., Choo, J.: StarGAN: unified generative adversarial networks for multi-domain image-to-image translation. In: CVPR (2018)
8. Deng, J., Dong, W., Socher, R., Li, L.J., Li, K., Fei-Fei, L.: ImageNet: a large-scale hierarchical image database. In: CVPR (2009)
9. Deng, W., Zheng, L., Ye, Q., Kang, G., Yang, Y., Jiao, J.: Image-image domain adaptation with preserved self-similarity and domain-dissimilarity for person re-identification. In: CVPR (2018)
10. Ding, Y., Fan, H., Xu, M., Yang, Y.: Adaptive exploration for unsupervised person re-identification. arXiv:1907.04194 (2019)
11. Fan, H., Zheng, L., Yan, C., Yang, Y.: Unsupervised person re-identification: clustering and fine-tuning. ACM Trans. Multimed. Comput. Commun. Appl. **14**, 83 (2018)

12. Fu, Y., Wei, Y., Wang, G., Zhou, Y., Shi, H., Huang, T.: Self-similarity grouping: a simple unsupervised cross domain adaptation approach for person re-identification. In: ICCV (2019)
13. Fu, Y., et al.: Horizontal pyramid matching for person re-identification. In: AAAI (2019)
14. Ge, Y., Chen, D., Li, H.: Mutual mean-teaching: Pseudo label refinery for unsupervised domain adaptation on person re-identification. In: ICLR (2020)
15. He, K., Zhang, X., Ren, S., Sun, J.: Deep residual learning for image recognition. In: CVPR (2016)
16. Hermans, A., Beyer, L., Leibe, B.: In defense of the triplet loss for person re-identification. arXiv:1703.07737 (2017)
17. Huang, G., Liu, Z., Van Der Maaten, L., Weinberger, K.Q.: Densely connected convolutional networks. In: CVPR (2017)
18. Huang, H., et al.: Eanet: enhancing alignment for cross-domain person re-identification. arXiv:1812.11369 (2018)
19. Huang, Y., Wu, Q., Xu, J., Zhong, Y.: Sbsgan: suppression of inter-domain background shift for person re-identification. In: ICCV (2019)
20. Kalayeh, M.M., Basaran, E., Gökmen, M., Kamasak, M.E., Shah, M.: Human semantic parsing for person re-identification. In: CVPR (2018)
21. Krizhevsky, A., Hinton, G., et al.: Learning multiple layers of features from tiny images (2009)
22. Li, K., Ding, Z., Li, K., Zhang, Y., Fu, Y.: Support neighbor loss for person re-identification. In: ACM MM (2018)
23. Li, M., Zhu, X., Gong, S.: Unsupervised person re-identification by deep learning tracklet association. In: ECCV (2018)
24. Li, Y.J., Yang, F.E., Liu, Y.C., Yeh, Y.Y., Du, X., Frank Wang, Y.C.: Adaptation and re-identification network: an unsupervised deep transfer learning approach to person re-identification. In: CVPR Workshop (2018)
25. Liu, J., Zha, Z.J., Chen, D., Hong, R., Wang, M.: Adaptive transfer network for cross-domain person re-identification. In: CVPR (2019)
26. Mao, X., Ma, Y., Yang, Z., Chen, Y., Li, Q.: Virtual mixup training for unsupervised domain adaptation. arXiv:1905.04215 (2019)
27. Paszke, A., et al.: Automatic differentiation in pytorch. In: NeurIPS Workshop (2017)
28. Quan, R., Dong, X., Wu, Y., Zhu, L., Yang, Y.: Auto-reid: searching for a part-aware convnet for person re-identification. In: ICCV (2019)
29. Ristani, E., Solera, F., Zou, R., Cucchiara, R., Tomasi, C.: Performance Measures and a Data Set for Multi-target, Multi-camera Tracking. In: Hua, G., Jégou, H. (eds.) ECCV 2016. LNCS, vol. 9914, pp. 17–35. Springer, Cham (2016). https://doi.org/10.1007/978-3-319-48881-3_2
30. Rukhovich, D., Galeev, D.: Mixmatch domain adaptaion: Prize-winning solution for both tracks of visda 2019 challenge. arXiv:1910.03903 (2019)
31. Schroff, F., Kalenichenko, D., Philbin, J.: Facenet: a unified embedding for face recognition and clustering. In: CVPR (2015)
32. Song, L., et al.: Unsupervised domain adaptive re-identification: theory and practice. arXiv:1807.11334 (2018)
33. Suh, Y., Wang, J., Tang, S., Mei, T., Mu Lee, K.: Part-aligned bilinear representations for person re-identification. In: ECCV (2018)
34. Sun, Y., Zheng, L., Yang, Y., Tian, Q., Wang, S.: Beyond part models: Person retrieval with refined part pooling (and a strong convolutional baseline). In: ECCV (2018)

35. Verma, V., et al.: Manifold mixup: better representations by interpolating hidden states. In: ICML (2019)
36. Verma, V., Lamb, A., Kannala, J., Bengio, Y., Lopez-Paz, D.: Interpolation consistency training for semi-supervised learning. arXiv:1903.03825 (2019)
37. Wang, C., Zhang, Q., Huang, C., Liu, W., Wang, X.: Mancs: a multi-task attentional network with curriculum sampling for person re-identification. In: ECCV (2018)
38. Wang, G., Yuan, Y., Chen, X., Li, J., Zhou, X.: Learning discriminative features with multiple granularities for person re-identification. In: ACM MM (2018)
39. Wang, J., Zhu, X., Gong, S., Li, W.: Transferable joint attribute-identity deep learning for unsupervised person re-identification. In: CVPR (2018)
40. Wei, L., Zhang, S., Gao, W., Tian, Q.: Person transfer GAN to bridge domain gap for person re-identification. In: CVPR (2018)
41. Wu, A., Zheng, W.S., Lai, J.H.: Unsupervised person re-identification by camera-aware similarity consistency learning. In: ICCV (2019)
42. Wu, J., et al.: Clustering and dynamic sampling based unsupervised domain adaptation for person re-identification. In: ICME (2019)
43. Wu, J., Yang, Y., Liu, H., Liao, S., Lei, Z., Li, S.Z.: Unsupervised graph association for person re-identification. In: ICCV (2019)
44. Wu, Z., Xiong, Y., Yu, S.X., Lin, D.: Unsupervised feature learning via non-parametric instance discrimination. In: CVPR (2018)
45. Yang, Q., Yu, H.X., Wu, A., Zheng, W.S.: Patch-based discriminative feature learning for unsupervised person re-identification. In: CVPR (2019)
46. Yu, H.X., Zheng, W.S., Wu, A., Guo, X., Gong, S., Lai, J.H.: Unsupervised person re-identification by soft multilabel learning. In: CVPR (2019)
47. Yu, R., Dou, Z., Bai, S., Zhang, Z., Xu, Y., Bai, X.: Hard-aware point-to-set deep metric for person re-identification. In: ECCV (2018)
48. Zhai, Y., Guo, X., Lu, Y., Li, H.: In defense of the classification loss for person re-identification. In: CVPR Workshops (2019)
49. Zhang, H., Cisse, M., Dauphin, Y.N., Lopez-Paz, D.: mixup: beyond empirical risk minimization. In: ICLR (2018)
50. Zhang, X., Cao, J., Shen, C., You, M.: Self-training with progressive augmentation for unsupervised cross-domain person re-identification. In: ICCV (2019)
51. Zhang, Z., Lan, C., Zeng, W., Chen, Z.: Densely semantically aligned person re-identification. In: CVPR (2019)
52. Zheng, F., et al.: Pyramidal person re-identification via multi-loss dynamic training. In: CVPR (2019)
53. Zheng, L., Shen, L., Tian, L., Wang, S., Wang, J., Tian, Q.: Scalable person re-identification: a benchmark. In: ICCV (2015)
54. Zheng, Z., Zheng, L., Yang, Y.: Unlabeled samples generated by GAN improve the person re-identification baseline in vitro. In: ICCV (2017)
55. Zhong, Z., Zheng, L., Kang, G., Li, S., Yang, Y.: Random erasing data augmentation. arXiv:1708.04896 (2017)
56. Zhong, Z., Zheng, L., Li, S., Yang, Y.: Generalizing a person retrieval model hetero- and homogeneously. In: ECCV (2018)
57. Zhong, Z., Zheng, L., Luo, Z., Li, S., Yang, Y.: Invariance matters: exemplar memory for domain adaptive person re-identification. In: CVPR (2019)
58. Zhong, Z., Zheng, L., Luo, Z., Li, S., Yang, Y.: Learning to adapt invariance in memory for person re-identification. arXiv:1908.00485 (2019)

59. Zhong, Z., Zheng, L., Zheng, Z., Li, S., Yang, Y.: Camstyle: a novel data augmentation method for person re-identification. IEEE Trans. Image Process. **28**, 1176–1190 (2019)
60. Zhong, Z., Zhu, L., Luo, Z., Li, S., Yang, Y., Sebe, N.: Openmix: reviving known knowledge for discovering novel visual categories in an open world. arXiv:2004.05551 (2020)
61. Zhou, K., Yang, Y., Cavallaro, A., Xiang, T.: Omni-scale feature learning for person re-identification. In: ICCV (2019)
62. Zhu, J.Y., Park, T., Isola, P., Efros, A.A.: Unpaired image-to-image translation using cycle-consistent adversarial networks. In: ICCV (2017)
63. Zhu, X., Zhu, X., Li, M., Murino, V., Gong, S.: Intra-camera supervised person re-identification: a new benchmark. In: ICCV Workshop (2019)

Semi-supervised Crowd Counting
via Self-training on Surrogate Tasks

Yan Liu[1], Lingqiao Liu[2], Peng Wang[3], Pingping Zhang[4], and Yinjie Lei[1($\boxtimes$)]

[1] College of Electronics and Information Engieering, Sichuan University,
Chengdu, China
`yanliu27@stu.scu.edu.cn`, `yinijie@scu.edu.cn`
[2] School of Computer Science, The University of Adelaide, Adelaide, Australia
`lingqiao.liu@adelaide.edu.au`
[3] School of Computing and Information Technology, University of Wollongong,
Wollongong, Australia
`pengw@uow.edu.au`
[4] School of Artificial Intelligence, Dalian University of Technology, Dalian, China
`jssxzhpp@mail.dlut.edu.cn`

Abstract. Most existing crowd counting systems rely on the availability of the object location annotation which can be expensive to obtain. To reduce the annotation cost, one attractive solution is to leverage a large number of unlabeled images to build a crowd counting model in semi-supervised fashion. This paper tackles the semi-supervised crowd counting problem from the perspective of feature learning. Our key idea is to leverage the unlabeled images to train a generic feature extractor rather than the entire network of a crowd counter. The rationale of this design is that learning the feature extractor can be more reliable and robust towards the inevitable noisy supervision generated from the unlabeled data. Also, on top of a good feature extractor, it is possible to build a density map regressor with much fewer density map annotations. Specifically, we proposed a novel semi-supervised crowd counting method which is built upon two innovative components: (1) a set of interrelated binary segmentation tasks are derived from the original density map regression task as the surrogate prediction target; (2) the surrogate target predictors are learned from both labeled and unlabeled data by utilizing a proposed self-training scheme which fully exploits the underlying constraints of these binary segmentation tasks. Through experiments, we show that the proposed method is superior over the existing semi-supervised crowd counting method and other representative baselines.

Keywords: Crowd counting · Surrogate tasks · Self-training · Semi-supervised learning

---

Y. Liu and L. Liu—Authors have equal contribution.

Electronic supplementary material The online version of this chapter (https://doi.org/10.1007/978-3-030-58555-6_15) contains supplementary material, which is available to authorized users.

© Springer Nature Switzerland AG 2020
A. Vedaldi et al. (Eds.): ECCV 2020, LNCS 12360, pp. 242–259, 2020.
https://doi.org/10.1007/978-3-030-58555-6_15

1   Introduction

Crowd counting is to estimate the number of people or objects from images or videos. Most existing methods formulate it as a density map regression problem [1–5], and solve it by using the pixel-to-pixel prediction networks [6–8]. Once the density map is estimated, the total object count can be trivially calculated. To train such a density map regression model, most existing crowd counting methods rely on a substantial amount of labeled images with the object location annotation, e.g., marking a dot at the center of corresponding persons. The annotation process can be labor-intensive and time-consuming. For example, to annotate the ShanghaiTech [3] dataset, 330,165 dots must be placed on corresponding persons carefully.

To reduce the annotation cost, an attractive solution is to learn the crowd counter in a semi-supervised setting which assumes availability of a small amount of labeled images and a large amount of unlabeled images. This is a realistic assumption since unlabeled images are much easier or effortlessly to obtain than labeled images. Then the research problem is how to leverage the unlabeled image to help train the crowd counter for achieving a reasonable performance.

To solve this problem, we propose a novel semi-supervised learning algorithm to obtain a crowd counting model. One key of our model is to use the unlabeled data to learn a generic feature extractor of the crowd counter instead of the entire network as most traditional methods do. The underlying motivations are threefold: (1) It is challenging to construct a robust semi-supervised learning loss term from unlabeled data for regression output. In contrast, learning a feature extractor is more robust and reliable towards the inevitable noisy supervision generated from unlabeled data; (2) the feature extractor often plays a critical role in a prediction model. If we have a good feature extractor, it is possible to learn a density map regressor, i.e., crowd counter, require much less ground-truth density map annotations; (3) there are a range of methods for learning feature extractor, and features can be even learned from other tasks rather than density map regression (i.e., surrogate tasks in this paper).

Inspired by those motivations, we propose to learn the feature extractor through a set of surrogate tasks: predicting whether the density of a pixel is above multiple predefined thresholds. Essentially, those surrogate tasks are binary segmentation tasks and we build multiple segmentation predictors for each of them. Since those tasks are derived from the density map regression, we expect that through training with these surrogate tasks the network can learn good features to benefit the density map estimation. For labeled images, we have ground-truth segmentation derived from the ground-truth density map. For unlabeled images, the ground-truth segmentation are not available. However, the unlabeled images can still be leveraged through a semi-supervised segmentation algorithm. Also, we notice that the correct predictions for the surrogate tasks should hold certain inter-relationship, e.g., if the density of a pixel is predicted to be higher than a high threshold, it should also be predicted higher than a low threshold. Such inter-relationships could serve as additional cues for jointly training segmentation predictors under the semi-supervised learning setting. Inspired by that, we developed a novel self-training algorithm to incorporate these inter-relationships to generate reliable pseudo-labels for semi-supervised learning. By conducting extensive experiments, we demonstrate the superior performance of the proposed method. To sum up, our main contributions are:

- We approach the problem of semi-supervised crowd counting from a novel perspective of feature learning. By introducing the surrogate tasks, we cast the original problem into a set of semi-supervised segmentation problem.
- We develop a novel self-training method which fully takes advantage of the inter-relationship between multiple binary segmentation tasks.

2 Related Works

Traditional Crowd Counting Methods include detection-based and regression-based methods. The detection-based methods use head or body detectors to obtain the total count in an image [9–11]. However, in extremely congested scenes with occlusions, detection-based methods can not produce satisfying predictions.

Regression-based methods are proposed [12,13] to tackle challenges in overcrowded scenes. In regression-based methods, feature extraction mechanisms [14,15] such as Fourier Analysis and Random Forest regression are widely used. However, traditional methods can not predict total counts accurately as they overlook the spatial distribution information of crowds.

CNN-based Crowd Counting Methods learn a mapping function from the semantic features to density map instead of total count [1]. Convolutional Neural Network (CNN) shows great potential in computer vision tasks. CNN-based methods are used to predict density maps. Recently, the mainstream idea is to leverage deep neural networks for density regression [2–4,16]. These methods construct multi-column structures to tackle scale variations. Then local or global contextual information is obtained for producing density maps.

Several works [5,17] combine the VGG [18] structure with dilated convolution to assemble the semantic features for density regression. While other works [19–22] introduce attention mechanisms to handle several challenges, e.g. background noise and various resolutions. Meanwhile works [23–27] leverage the multi-task frameworks, i.e., detection, segmentation or localization, which provide more accurate location information for density regression. Besides, the self-attention mechanism [28,29] and residual learning mechanism [30] are effective in regularizing the training of the feature extractor. Work [31] transforms the density value to the density level from close-set to open-set. Further, a Bayesian-based loss function [32] is proposed for density estimation. These above CNN-based methods require a large number of labeled images to train the crowd counter.

However, annotating the crowd counting dataset is a time-consuming and labor-intensive work.

Semi-/Weakly/Un-Supervised Crowd Counting Methods attempt to reduce the annotation burden by using semi-/weakly/un-supervised settings. In the semi-supervised setting, work [33] collects large unlabeled images as extra training data and constructs a rank loss based on the estimated density maps. Also, work in [34] leverages the total count as a weak supervision signal for density estimation. Besides, an auto-encoder structure [35] is proposed for crowd counting in an almost unsupervised setting. Another method for reducing the annotation burden is to use synthetic images [36]. For example, the GAN-based [37] and domain adaption based [38] frameworks combine the synthetic images and realistic images to train the crowd counter. These methods are effective in reducing the annotation burden. However, they can not obtain satisfying crowd counting performance because the inevitable noisy supervision may mislead the density regressor.

(a) Traditional semi-supervised methods.   (b) Proposed semi-supervised method.

Fig. 1. (a) Traditional semi-supervised methods use both labeled and unlabeled images to update the feature extractor and density regressor. (b) In the proposed method, the unlabeled images are only used for updating feature extractor.

3   Background: Crowd Counting as Density Estimation

Following the framework "learning to count" [1], crowd counting can be transformed into a density map regression problem. Once the density map is estimated, the total object count can be simply estimated by its summation, that is, $\hat{N} = \sum_{i,j} \hat{D}(i,j)$, where $\hat{D}(i,j)$ denotes the density value for pixel (i,j). The Mean Square Error (MSE) loss is commonly used in model training, that is,

$$\mathcal{L}_{MSE} = \sum_{(i,j)} |\hat{D}(i,j) - D(i,j)|^2, \tag{1}$$

where $\hat{D}$ is the estimated density map and D is the ground-truth density map.

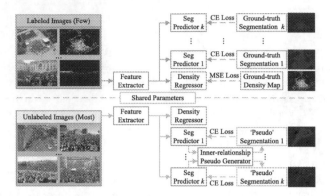

Fig. 2. The overview of our proposed method. We introduce a set of binary segmentation surrogate tasks. For labeled images, we construct loss terms on both original and surrogate tasks. For unlabeled images, we use the output of segmentation predictor and inter-relationship to generate "pseudo segmentation", which is shown in Fig. 3.

4   Methodology

In this paper, we are interested in learning a crowd counter based on the semi-supervised setting. Formally, we assume that we have a set of labeled images $L = \{I_i^l, D_i\}$, where D_i is the ground-truth density map, and a set of unlabeled images $U = \{I_i^u\}$. Our task is to learn a crowd counter by using both the labeled images and unlabeled images. In our setting, the unlabeled set contains much more images than the labeled set for training a crowd counting model.

4.1   Using Unlabeled Data for Feature Learning

Generally speaking, a network can be divided into two parts, a feature extractor and a task-specific predictor. The former converts the raw images into feature maps while the latter further transforms them to the desired output, e.g., density map, in the context of crowd counting. Most existing semi-supervised learning methods [33,39–41] learn those two parts simultaneously and seek to construct a loss term from unlabeled data applied to the entire network.

In contrast to the existing methods, we propose to learn the feature extractor and the task-specific predictor through different tasks and loss terms. In particular, in our method, the unlabeled data is only used for learning the feature extractor. This design is motivated by three considerations: (1) crowd counting is essentially a semi-supervised regression problem in our setting. Besides, it can be challenging to construct a robust semi-supervised regression loss term from unlabeled data (i.e., as most existing methods do). The noisy supervision generated from the loss term from unlabeled data may contaminate the task-specific predictor and lead to inferior performance. In our method, unlabeled data is only used to train the feature extractor as the noisy supervision will not directly affect the task-specific predictor; (2) feature extractor plays an important role in many fields like unsupervised feature learning [42,43], semi-supervised feature learning [44,45] and few-shot learning [46,47]. Indeed, with a good feature extractor, it is possible to reduce the need of a large amount of labeled data in training. In the context of crowd counting, this implies that much less ground-truth density map annotations are needed if we can obtain a robust feature extractor via other means; (3) feature extractor can be learned in various ways. In this way, we will have more freedom in designing semi-supervised learning algorithms for feature learning. Specifically, we propose to derive surrogate tasks from the original density map regression problem, and use those tasks for training the feature extractor. The schematic overview of this idea is shown in Fig. 1 (b). For labeled images, the target of surrogate task can be transformed from ground-truth annotation. For the unlabeled images, the ground-truth annotation becomes unavailable. However, the unlabeled images can still be leveraged to learn the surrogate task predictor and consequently the feature extractor in a semi-supervised learning manner. In the following sections, we first elaborate how to construct the surrogate loss and then describe the semi-supervised learning algorithm developed for the surrogate tasks.

4.2   Constructing Surrogate Tasks for Feature Learning

The surrogate task defined in this paper is to predict whether the density value of a pixel, $D(i,j)$, exceeds a given threshold. In other words, the prediction target of the surrogate task is defined as:

$$M(i,j) = \begin{cases} 1 & D(i,j) > \epsilon \\ 0 & D(i,j) <= \epsilon \end{cases}, \tag{2}$$

where (i,j) is the pixel coordinate, and ϵ is the predefined threshold. For labeled data, the ground-truth of D is known and thus M is known. For unlabeled data, no annotation of D is available and thus M is unknown. However, we can still use unlabeled data to construct loss term for indirectly supervising the prediction of M. Note that in this way, we essentially recast the original semi-supervised crowd counting problem into a semi-supervised segmentation problem since M only takes binary values.

In practice, we use multiple thresholds and generate multiple surrogate targets $\{M_k\}$ to consider the pixels with different density levels. To set these thresholds, we rank all non-zero density values from all the labeled images in ascending order and choose the thresholds as the value ranked at $r_k \times N$, where $r_k \in [0,1]$ $k = 1,..,c$, N is the total number of non-zero values and c indicates the number of surrogate tasks. Meanwhile, we create multiple segmentation predictor branches attached to the feature extractor. These surrogate tasks are parallel to the density map regressor, as shown in Fig. 2.

4.3 Inter-Relationship-Aware Self-Training (IRAST) for Semi-supervised Training on Surrogate Tasks

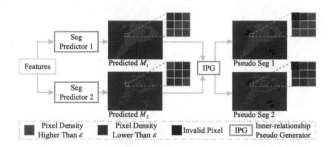

Fig. 3. The illustration of the inter-relationship between two segmentation predictors. We use a lower threshold ϵ_1 segmentation predictor to produce $\hat{M}_1$, and a higher threshold ϵ_2 segmentation predictor to produce $\hat{M}_2$. If a specific pixel in $\hat{M}_1$ is lower than ϵ_1, while in $\hat{M}_2$ is higher than ϵ_2, we can consider this pixel is invalid. The inter-relationship avoids such incorrect training signal flowing into the feature extractor.

To leverage the unlabeled data to train the surrogate task predictors and the feature extractor, a semi-supervised learning algorithm is needed. Self-training is one of the most commonly used semi-supervised learning algorithms in segmentation tasks [48,49]. It recursively generates pseudo-class-label for samples (pixels) with prediction confidence values higher than a given threshold $t_p = 0.9$. However, this straightforward solution largely ignores the underlying inter-relationship between multiple surrogate tasks. Recall that M takes binary values and $M(i,j) = 1$ if the density value of pixel (i,j) is greater than a given threshold. Suppose we have two segmentation results M_1 and M_2 estimated from two predictors corresponding to two thresholds ϵ_1 and ϵ_2 ($\epsilon_1 < \epsilon_2$), then there will be a conflict if one predictor gives the prediction $\hat{M}_1(i,j) = 0$ while the other gives the prediction $\hat{M}_2(i,j) = 1$. This is because $\hat{M}_1(i,j) = 0$ indicates the density value of the pixel is less than ϵ_1, but $\hat{M}_2(i,j) = 1$ implies the density value of pixel is larger than ϵ_2 and consequently larger than ϵ_1 since $\epsilon_1 < \epsilon_2$.

This inter-relationship could essentially act as an error correcting mechanism to test if the prediction made by surrogate predictors are likely to be accurate. Thus in our method, we incorporate it into the framework of self-training as an additional criterion for pseudo-label generation besides the commonly used thresholding criterion. Formally, we define the following rule for generating a pseudo label at the k-th predictor. Without loss of generality, we assume there are c predictors, ranking from 1 to c according to the descent order of their corresponding thresholds, that is, $\epsilon_a > \epsilon_b$ if $a > b$. The formal rule of generating pseudo-label is shown in Algorithm 1. The generation of pseudo-labels is online.

Algorithm 1: Pseudo-label Generation Rule

Input: Number of surrogate tasks c. Given threshold t_p, Predicted
    confidence value (posterior probability) $P(\hat{M}_k = 1)$ $k = 1, \cdots, c$;
    $P(\hat{M}_k = 0) = 1 - P(\hat{M}_k = 1)$

Output: A set of pseudo-label set $\{\mathcal{S}_k\}$, one for each k: $\mathcal{S}_k = \{(i, j, s_{ij})\}$,
    where s_{ij} is the generated pseudo-label for (i, j).

1 **for** $k \in [1, c]$ **do**
2   **for** *each location* (i, j) **do**
3   **if** $P(\hat{M}_k(i, j) = 1) > t_p$ *and* $P(\hat{M}_g(i, j) = 1) > t_p$ $\forall g < k$ **then**
4            $\mathcal{S}_k \leftarrow \mathcal{S}_k \cup (i, j, 1)$
5   **end**
6   **if** $P(\hat{M}_k(i, j) = 0) > t_p$ *and* $P(\hat{M}_h(i, j) = 0) > t_p$ $\forall h > k$ **then**
7            $\mathcal{S}_k \leftarrow \mathcal{S}_k \cup (i, j, 0)$
8   **end**
9   **end**
10 **end**

In nutshell, a pseudo label is generated in the surrogate binary segmentation task if its prediction confidence value for one class ("1" or "0" in our case) is greater than t_p and its prediction is not conflict with predictions of other predictors. An example of this scheme is illustrated in Fig. 3.

Discussion: The proposed method defines c binary segmentation tasks and one may wonder why not directly define a single c-way multi-class segmentation task. Then an standard multi-class self-training method can be used. We refer this method as **Multiple-class Segmentation Self-Training (MSST)**. Comparing with our approach, MSST has the following two disadvantages: (1) it does not have the "error correction" mechanism as described in the rule of generating pseudo label. The difference between MSST and IRAST is the standard one-vs-rest multi-class classification formulation and the error correcting output codes formulation [50]; (2) MSST may be overoptimistic towards the confidence score due to the softmax normalization of logits. Considering a three-way classification scenario for example, it is possible that the confidence for either class is low and the logits for all three classes are negative. But by chance, one class has relatively larger logits, say, $\{-100, -110$ and $-90\}$ for class 1, 2 and 3 respectively. After normalization, the posterior probability for the last class becomes near 1, and will exceed the threshold for generating pseudo labels. In contrast, the proposed IRAST does not have this issue since the confidence score will not be normalized across different classes (quantization level). We also conduct an ablation study in Sect. 6.3 to verify that MSST is inferior to IRAST.

5   Overall Training Process

In practice, we use the Stochastic Gradient Descent (SGD) to train the network[1]. For an labeled image, we construct supervised loss terms based on the density regression task and surrogate tasks, and the training loss is:

$$\mathcal{L}_L = \mathcal{L}_{MSE} + \lambda_1 \mathcal{L}_{SEG} = \sum_{(i,j)} \left(|\hat{D}(i,j) - D(i,j)|^2 + \lambda_1 \sum_{k=1}^{c} CE(M_k(i,j), \hat{M}_k(i,j)) \right),$$

where $CE()$ denotes the cross-entropy loss, $\hat{D}$ and $\hat{M}_k$ are the predicted density map and segmentation respectively; D and M_k are the ground-truth density map and segmentation respectively.

For an unlabeled image, we construct an unsupervised loss based on the surrogate tasks and use it to train the feature extractor:

$$\mathcal{L}_U = \lambda_2 \mathcal{L}_{SEG} = \lambda_2 \sum_{k=1}^{c} \sum_{(i,j,s_{ij}) \in \mathcal{S}_k} CE\left(\hat{M}_k(i,j), s_{ij} \right), \qquad (3)$$

where the $\mathcal{S}_k = \{(i,j,s_{ij})\}$ denotes the set of generated pseudo labels at the k-th segmentation predictor. Please refer to Algorithm 1 for the generation of $\mathcal{S}_k$.

6   Experimental Results

We conduct extensive experiments on three popular crowd-counting datasets. The purpose is to verify if the proposed methods can achieve superior performance over other alternatives in a **semi-supervised learning setting** and understand the impact of various components of our method. Note that works that methods in a fully-supervised setting or a unsupervised setting are **not directly comparable** to ours.

6.1   Experimental Settings

Datasets: ShanghaiTech [3], UCF-QNRF [27] and WorldExpo'10 [2] are used throughout our experiments. We modify the setting of each dataset to suit the need of semi-supervised learning evaluation. Specifically, the original training dataset is divided into labeled and unlabeled sets. The details about such partition are given as follows.

ShanghaiTech [3]: The ShanghaiTech dataset consists of 1,198 images with 330,165 annotated persons, which is divided into two parts: Part_A and Part_B. Part_A is composed of 482 images with 244,167 annotated persons; the training set includes 300 images; the remaining 182 images are used for testing. Part_B

---

[1] As the unlabeled set contains more images than the labeled set, we oversample labeled images to ensure the similar amount of labeled and unlabeled images occur in a single batch.

consists of 716 images with 88,498 annotated persons. The size of the training set is 400, and the testing set contains 316 images. In Part_A, we randomly pick up 210 images to consist the unlabeled set, 90 images to consist the labeled set (60 images for validation). Also, In Part_B, we randomly pick up 280 images to consist the unlabeled set, 120 images to consist the labeled set (80 images for validation).

UCF-QNRF [27]: The UCF-QNRF dataset contains 1,535 high-resolution images with 1,251,642 annotated persons. The training set includes 1,201 images, and the testing set contains 334 images. We randomly pick up 721 images to consist the unlabeled set, 480 images to consist the labeled set (240 images for validation).

World Expo'10 [2]: The World Expo'10 dataset includes 3980 frames from Shanghai 2010 WorldExpo. The training set contains 3380 images, and the testing set consists of 600 frames. Besides, the Region of Interest (ROI) is available in each scene. Each frame and the corresponding annotated person should be masked with ROI before training. We randomly pick up 2433 images to consist the unlabeled set, 947 images to consist the labeled set (271 images for validation).

Compared Methods: We compare the proposed IRAST method against four methods: (1) Label data only (Label-only): only use the labeled dataset to train the network. This is the baseline of all semi-supervised crowd counting approaches. (2) Learning to Rank (L2R): a semi-supervised crowd counting method proposed in [33]. As the unlabeled images used in this paper are not released, we re-implement it with the same backbone and test setting as our method to ensure a fair comparison. (3) Unsupervised Data Augmentation (UDA): UDA [39] is one of the state-of-the-art semi-supervised learning methods. It encourages the network to generate similar predictions for an unlabeled image and its augmented version. This method was developed for image classification. We modify it by using the estimated density map as the network output. (4) Mean teacher (MT): Mean teacher [40] is a classic consistency-based semi-supervised learning approach. Similar as UDA, it was originally developed for the classification task and we apply it to the regression task by changing the network work output as the estimated density maps. (5) Interpolation Consistency Training (ICT): ICT [41] is a recently developed semi-supervised learning approach. It is based on the mixup data augmentation [51] but performed on unlabeled data. Again, we tailor it for the density map regression task by changing the output as the density map. More details about the implementation of the compared methods can be found in the supplementary material.

Implementation Details: The feature extractor used in most of our experiment is based on the CSRNet [5]. We also conducted an ablation study in Sect. 6.3 to use Scale Pyramid Network (SPN) [17] as the feature extractor. Both CSRNet and SPN leverage VGG-16 [18] as the backbone. Also, three segmentation predictors are used by default unless specified. The thresholds for the corresponding surrogate tasks are selected as $\{0, 0.5N, 0.7N\}$ (please refer to Sect. 4.2 for the method of choosing thresholds). The segmentation predictors are attached to the 14-th layer of the CSRNet or the 13-th layer of SPN. The rest layers in those networks are viewed as the task specific predictor, i.e., the density map regressor. The segmentation predictors share the same network structure as the density map regressor. Please refer the supplementary material for the detailed structure of the network. In all experiments, we set the batch size as 1 and use Adam [52] as the optimizer. The learning rate is initially set to 1e-6 and halves per 30 epochs (120 epochs in total). Besides, we set t_p to 0.9 in experiments. Our implementation is based on PyTorch [53] and we will also release the code.

Evaluation Metrics: Following the previous works [2,3], the Mean Absolute Error (MAE) and Mean Squared Error (MSE) are adopted as the metrics to evaluate the performance of the compared crowd counting methods.

6.2   Datasets and Results

Evaluation on the ShanghaiTech Dataset: The experimental results on ShanghaiTech dataset are shown in Table 1. As seen, if we only use the labeled image, the network can only attain an MAE of 98.3 on Part A and 15.8 on Part B. In general, using a semi-supervised learning approach brings improvement. The L2R [33] shows an improvement around 8 people in the MAE of Part A but almost no improvement for Part B. Semi-supervised learning approaches modified from the classification task (UDA, MT, ICT) also lead to improved performance over Label-only on Part A. However, the improvement is not as large as L2R. Our approach, IRAST, clearly demonstrates the best performance. It leads to 11.4 MAE improvement over the Label-only on Part A (Fig. 4).

Table 1. The comparison on the ShanghaiTech dataset. The best results are in bold font.

	Method	Part_A		Part_B	
		MAE	MSE	MAE	MSE
Semi	Label-only	98.3	159.2	15.8	25.0
	L2R [33]	90.3	153.5	15.6	24.4
	UDA [39]	93.8	157.2	15.7	24.1
	MT [40]	94.5	156.1	15.6	24.5
	ICT [41]	92.5	156.8	15.4	23.8
	IRAST	**86.9**	**148.9**	**14.7**	**22.9**
	(Fully) CSRNet [5]	68.2	115.0	10.6	16.0

Table 2. The comparison on the UCF-QNRF dataset. The best results are in bold font.

	Method	UCF-QNRF	
		MAE	MSE
Semi	Label-only	147.7	253.1
	L2R [33]	148.9	249.8
	UDA [39]	144.7	255.9
	MT [40]	145.5	250.3
	ICT [41]	144.9	250.0
	IRAST	**135.6**	**233.4**
	(Fully) CSRNet [5]	119.2	211.4

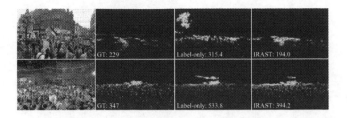

Fig. 4. A comparison of predicted density maps on the UCF-QNRF dataset.

Evaluation on the UCF-QNRF Dataset: The advantage of the proposed method is also well demonstrated on UCF-QNRF dataset, shown in Table 2. Again, the proposed method achieves the overall best performance, and exceeds the Label-only by around 12 MAE. The other semi-supervised learning approach does not work well on this dataset. In particular, L2R even achieves worse performance than the Label-only. This on the other hand clearly demonstrates the robustness of our approach. Also, from the results in both ShanghaiTech and UCF-QNRF, we can see that directly employing the semi-supervised learning approaches which were originally developed for classification may not achieve satisfying performance. It remains challenging for developing the semi-supervised crowd counting algorithm.

Table 3. The performance comparison in terms of MAE on the WorldExpo'10 dataset. The best results are in bold font.

	Method	Sce.1	Sce.2	Sce.3	Sce.4	Sce.5	Avg.
Semi	Label-only	2.4	16.9	9.7	41.3	3.1	14.7
	L2R [33]	2.4	20.9	9.8	31.9	4.4	13.9
	UDA [39]	**1.9**	20.3	10.9	34.5	3.6	14.2
	MT [40]	2.6	24.8	9.4	30.3	3.3	14.1
	ICT [41]	2.3	17.8	**8.3**	43.5	**2.8**	14.9
	IRAST	2.2	**12.3**	9.2	**27.8**	4.1	**11.1**
	(Fully) CSRNet [5]	2.9	11.5	8.6	16.6	3.4	8.6

Evaluation on the World Expo'10 Dataset: The results are shown in Table 3. As seen, IRAST again achieves the best MAE in 2 scenes and delivers the best MAE over other methods. The other semi-supervised learning methods achieve comparable performance and their performance gain over the Label-only is not significant.

6.3 Ablation Study

To understand the importance of various components in our algorithm, we conduct a serials of ablation studies.

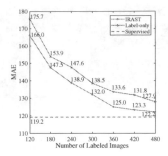

Fig. 5. The impact of the number of labeled images. Evaluated in terms of MAE on the UCF-QNRF dataset.

Varying the Number of Labeled Images: We first examine the performance gain over the Label-only under different amount of labeled images. We conduct experiments on the UCF-QNRF dataset. We vary the number of labeled image from 120 to 480 while fixing the amount of unlabeled images to be 481. The performance curves of the Label-only and IRAST are depicted in Fig. 5. As seen, IRAST achieves consistent performance gain over the Label-only, which is an evidence of the robustness of our method. Also, we can see that with IRAST, using 480 images can almost achieve comparable performance than the performance of a fully-supervised model which needs 961 training images.

IRAST on Labeled Set: The proposed method constructs an additional training task and one may suspect the good performance is benefited from the multi-task learning. To investigate this hypothesis, we also conduct an ablation study by learning the crowd counter on the labeled set only, but with both density map regression task and surrogate tasks. The results are shown in Table 4. As seen, using multiple-surrogate tasks for the labeled set does improve the performance to some extent, but still has a significant performance gap with the proposed method. This result clearly validates that our method can not be simply understood as a multi-task learning approach.

Other Alternative Surrogate Task: One alternative method is to use a multi-class segmentation predictor to train the feature extractor, namely MSST mentioned in Sect. 4.3. To compare MSST and IRAST, we conduct experiments on the ShanghaiTech Part_A and UCF-QNRF dataset. The results are shown in Table 5. As seen, MSST can achieve a better performance than Label-only method, which demonstrates the effectiveness of using a surrogate task for feature learning. However, MSST obtains a worse crowd counting performance than IRAST. Recall that MSST lacks an error correction mechanism to generate pseud-label, the superior performance of IRAST over MSST provides evidence to support the merit of our multiple surrogate binary-segmentation task modelling.

Table 4. Impact of the unlabeled images in the process of feature learning. Evaluated on the ShanghaiTech Part_A and UCF-QNRF dataset. The best results are in bold font.

Method	Part_A		UCF-QNRF	
	MAE	MSE	MAE	MSE
Label-only	98.3	159.2	147.7	253.1
IRAST on label	94.1	151.6	140.8	245.4
IRAST	**86.9**	**148.9**	**135.6**	**233.4**

Table 5. Comparison of IRAST and MSST. Evaluated on the ShanghaiTech Part_A and UCF-QNRF dataset. The best results are in bold font.

Method	Part_A		UCF-QNRF	
	MAE	MSE	MAE	MSE
Label-only	98.3	159.2	147.7	253.1
MSST	91.5	155.2	140.0	233.7
IRAST	**86.9**	**148.9**	**135.6**	**233.4**

The Importance of Considering the Inter-relationship: In IRAST, we leverage the Inter-Relationship (IR) between surrogate tasks to generate pseudo-labels. To verify the importance of this consideration, we conduct an ablation study by removing the inter-Relationship constraint for pseudo-label generation. The results are shown in Table 6. As seen, a decrease in performance is observed when the Inter-Relationship is not considered. This observation suggests that the Inter-Relationship awareness is essential to the proposed IRAST method.

Table 6. Impact of the inter-relationship. Evaluated on the ShanghaiTech Part_A and UCF-QNRF dataset. The best results are in bold font.

Method	Part_A		UCF-QNRF	
	MAE	MSE	MAE	MSE
Label only	98.3	159.2	147.7	253.1
IRAST w/o IR	93.5	155.5	139.8	240.3
IRAST	**86.9**	**148.9**	**135.6**	**233.4**

Table 7. Impact of the changing hyper-parameter t_p. Evaluated on the ShanghaiTech Part_A and UCF-QNRF dataset. The best results are in bold font.

Method	Part_A		UCF-QNRF	
	MAE	MSE	MAE	MSE
Label-only	98.3	159.2	147.7	253.1
$t_p = 0.6$	88.4	152.3	137.2	234.9
$t_p = 0.9$	**86.9**	**148.9**	**135.6**	**233.4**

The Impact of Changing the Prediction Confidence Threshold: We set hyper-parameter t_p to 0.9 in the previous experiments. To investigate the impact of t_p, we conduct experiments on ShanghaiTech part_A and UCF-QNRF dataset. The results are shown in Table 7. The results demonstrate setting diverse t_p does not impact crowd counting performance significantly. The crowd counting performance are comparable to our current results. It means the proposed IRAST method is robust.

Table 8. Impact of the feature extractor. Evaluated on the ShanghaiTech Part_A and UCF-QNRF dataset. The best results are in bold font.

Method	Part_A		UCF-QNRF	
	MAE	MSE	MAE	MSE
Label-only (CSRNet)	98.3	159.2	147.7	253.1
IRAST (CSRNet)	86.9	148.9	135.6	233.4
Label-only (SPN)	88.5	152.6	138.0	244.5
IRAST (SPN)	**83.9**	**140.1**	**128.4**	**225.3**

Table 9. Impact of the varing number of surrogate tasks. The best results are in bold font.

Tasks	Part_A		UCF-QNRF	
	MAE	MSE	MAE	MSE
1	89.8	149.8	142.8	236.5
2	88.9	149.6	139.1	237.8
3	**86.9**	**148.9**	**135.6**	**233.4**
4	90.1	150.2	137.5	236.8
5	90.3	150.9	137.8	234.4

Change of the Feature Extractor: So far, we conduct our experiment with the CSRNet [5] feature extractor. It is unclear if performance gain can still be achieved with other feature extractors. To investigate this, we conduct an experiment that uses SPN [17] as the feature extractor on the ShanghaiTech Part_A and UCF-QNRF dataset. Results are shown in Table 8. We can see that the significant performance gain can still be achieved. Also, we observe an improved overall performance by using SPN. This suggests that the advances in network architecture design for crowd counting can be readily incorporated into our method.

The Effect of Varying the Number of Surrogates Tasks: Finally, we test the impact of choosing the number of surrogate tasks. We incrementally adding more thresholds by following the threshold sequence $\{0, 0.5N, 0.7N, 0.8N, 0.9N\}$, e.g., $\{0, 0.5N, 0.7N\}$ is used for the three-task setting while $\{0, 0.5N, 0.7N, 0.8N\}$ is used for the four-task setting. The results are shown in Table 9. The results demonstrate setting three surrogate tasks for feature learning can achieve the best crowd counting performance. To have a finer grained partition of density value does not necessarily lead to improved performance.

7   Conclusions

In this paper, we proposed a semi-supervised crowd counting algorithm by creating a set of surrogate tasks for learning the feature extractor. A novel self-training strategy that can leverage the inter-relationship of different surrogate tasks is developed. Through extensive experiments, it is clear that the proposed method enjoys superior performance over other semi-supervised crowd counter learning approaches.

Acknowledgement. This work was supported by the Key Research and Development Program of Sichuan Province (2019YFG0409). Lingqiao Liu was in part supported by ARC DECRA Fellowship DE170101259.

References

1. Lempitsky, V., Zisserman, A.: Learning to count objects in images. In: Advances in Neural Information Processing Systems (NIPS), pp. 1324–1332 (2010)
2. Zhang, C., Li, H., Wang, X., Yang, X.: Cross-scene crowd counting via deep convolutional neural networks. In: IEEE Conference on Computer Vision and Pattern Recognition (CVPR), pp. 833–841 (2015)
3. Zhang, Y., Zhou, D., Chen, S., Gao, S., Ma, Y.: Single-image crowd counting via multi-column convolutional neural network. In: IEEE Conference on Computer Vision and Pattern Recognition (CVPR), pp. 589–597 (2016)
4. Sam, D.B., Surya, S., Babu, R.V.: Switching convolutional neural network for crowd counting. In: IEEE Conference on Computer Vision and Pattern Recognition (CVPR), pp. 5744–5752 (2017)

5. Li, Y., Zhang, X., Chen, D.: Csrnet: dilated convolutional neural networks for understanding the highly congested scenes. In: IEEE Conference on Computer Vision and Pattern Recognition (CVPR), pp. 1091–1100 (2018)
6. Kang, K., Wang, X.: Fully convolutional neural networks for crowd segmentation. arXiv preprint arXiv:1411.4464 (2014)
7. Peng, C., Zhang, X., Yu, G., Luo, G., Sun, J.: Large kernel matters-improve semantic segmentation by global convolutional network. In: IEEE Conference on Computer Vision and Pattern Recognition (CVPR), pp. 4353–4361 (2017)
8. Long, J., Shelhamer, E., Darrell, T.: Fully convolutional networks for semantic segmentation. In: IEEE Conference on Computer Vision and Pattern Recognition (CVPR), pp. 3431–3440 (2015)
9. Subburaman, V.B., Descamps, A., Carincotte, C.: Counting people in the crowd using a generic head detector. In: IEEE International Conference on Advanced Video and Signal-Based Surveillance (AVSS), pp. 470–475 (2012)
10. Dalal, N., Triggs, B.: Histograms of oriented gradients for human detection. In: IEEE Conference on Computer Vision and Pattern Recognition (CVPR), pp. 886–893 (2005)
11. Viola, P., Jones, M.: Robust real-time face detection. Int. J. Comput. Vis. (IJCV) **57**(2), 137–154 (2004)
12. Chen, K., Loy, C.C., Gong, S., Xiang, T., : Feature mining for localised crowd counting. In: British Machine Vision Conference (BMVC), pp. 21.1–21.11 (2012)
13. Chan, A.B., Vasconcelos, N.: Bayesian poisson regression for crowd counting. In: IEEE International Conference on Computer Vision (ICCV), pp. 545–551 (2009)
14. Idrees, H., Saleemi, I., Seibert, C., Shah, M.: Multi-source multi-scale counting in extremely dense crowd images. In: IEEE Conference on Computer Vision and Pattern Recognition (CVPR), pp. 2547–2554 (2013)
15. Fiaschi, L., Köthe, U., Nair, R., Hamprecht, F.A.: Learning to count with regression forest and structured labels. In: International Conference on Pattern Recognition (ICPR), pp. 2685–2688 (2012)
16. Sindagi, V.A., Patel, V.M.: Generating high-quality crowd density maps using contextual pyramid CNNs. In: IEEE International Conference on Computer Vision (ICCV), pp. 1861–1870 (2017)
17. Chen, X., Bin, Y., Sang, N., Gao, C.: Scale pyramid network for crowd counting. In: IEEE Winter Conference on Applications of Computer Vision (WACV), pp. 1941–1950 (2019)
18. Simonyan, K., Zisserman, A.: Very deep convolutional networks for large-scale image recognition. In: International Conference on Learning Representations (ICLR) (2015)
19. Jiang, X., et al.: Crowd counting and density estimation by trellis encoder-decoder networks. In: IEEE Conference on Computer Vision and Pattern Recognition (CVPR), pp. 6133–6142 (2019)
20. Zhao, M., Zhang, J., Zhang, C., Zhang, W.: Leveraging heterogeneous auxiliary tasks to assist crowd counting. In: IEEE Conference on Computer Vision and Pattern Recognition (CVPR), pp. 12736–12745(2019)
21. Ranjan, V., Le, H., Hoai, M.: Iterative crowd counting. In: European Conference on Computer Vision (ECCV), pp. 270–285 (2018)
22. Zhu, L., Zhao, Z., Lu, C., Lin, Y., Peng, Y., Yao, T.: Dual path multi-scale fusion networks with attention for crowd counting. arXiv preprint arXiv:1902.01115 (2019)

23. Liu, J., Gao, C., Meng, D., Hauptmann, A.G.: Decidenet: counting varying density crowds through attention guided detection and density estimation. In: IEEE Conference on Computer Vision and Pattern Recognition (CVPR), pp. 9175–9184 (2018)
24. Lian, D., Li, J., Zheng, J., Luo, W., Gao, S.: Density map regression guided detection network for RGB-D crowd counting and localization. In: IEEE Conference on Computer Vision and Pattern Recognition (CVPR), pp. 1821–1830 (2019)
25. Jiang, S., Lu, X., Lei, Y., Liu, L.: Mask-aware networks for crowd counting. IEEE Trans. Circ. Syst. Video Technol. (TCSVT) (2019)
26. Valloli, V.K. and Mehta, K.: W-net: reinforced u-net for density map estimation. arXiv preprint arXiv:1903.11249 (2019)
27. Idrees, H., et al.: Composition loss for counting, density map estimation and localization in dense crowds. In: European Conference on Computer Vision (ECCV), pp. 532–546 (2018)
28. Zhang, A., et al.: Relational attention network for crowd counting. In: IEEE International Conference on Computer Vision (ICCV), pp. 6788–6797 (2019)
29. Wan, J., Chan, A.: Adaptive density map generation for crowd counting. In: IEEE International Conference on Computer Vision (ICCV), pp. 1130–1139 (2019)
30. Sindagi, V.A., Yasarla, R., Patel, V.M.: Pushing the frontiers of unconstrained crowd counting: new dataset and benchmark method. In: IEEE International Conference on Computer Vision (ICCV), pp. 1221–1231 (2019)
31. Xiong, H., Lu, H., Liu, C., Liu, L., Cao, Z., Shen, C.: From open set to closed set: counting objects by spatial divide-and-conquer. In: IEEE International Conference on Computer Vision (ICCV), pp. 8362–8371 (2019)
32. Ma, Z., Wei, X., Hong, X., Gong, Y.: Bayesian loss for crowd count estimation with point supervision. In: IEEE International Conference on Computer Vision (ICCV), pp. 6142–6151 (2019)
33. Liu, X., Van De Weijer, J., Bagdanov, A.D.: Leveraging unlabeled data for crowd counting by learning to rank. In: IEEE Conference on Computer Vision and Pattern Recognition (CVPR), pp. 7661–7669 (2018)
34. von Borstel, M., Kandemir, M., Schmidt, P., Rao, M.K., Rajamani, K., Hamprecht, F.A.: Gaussian process density counting from weak supervision. In: Leibe, B., Matas, J., Sebe, N., Welling, M. (eds.) ECCV 2016. LNCS, vol. 9905, pp. 365–380. Springer, Cham (2016). https://doi.org/10.1007/978-3-319-46448-0_22
35. Sam, D.B., Sajjan, N.N., Maurya, H., Babu, R.V.: Almost unsupervised learning for dense crowd counting. In: AAAI Conference on Artificial Intelligence (AAAI), vol. 33, pp. 8868–8875 (2019)
36. Wang, Q., Gao, J., Lin, W., Yuan, Y.: Learning from synthetic data for crowd counting in the wild. In: IEEE Conference on Computer Vision and Pattern Recognition (CVPR), pp. 8198–8207 (2019)
37. Gao, J., Wang, Q., Yuan, Y.: Feature-aware adaptation and structured density alignment for crowd counting in video surveillance. arXiv preprint arXiv:1912.03672 (2019)
38. Gao, J., Han, T., Wang, Q., Yuan, Y.: Domain-adaptive crowd counting via inter-domain features segregation and gaussian-prior reconstruction. arXiv preprint arXiv:1912.03677 (2019)
39. Xie, Q., Dai, Z., Hovy, E., Luong, M.T., Le, Q.V.: Unsupervised data augmentation for consistency training. arXiv preprint arXiv:1904.12848 (2019)
40. Tarvainen, A., Valpola, H.: Mean teachers are better role models: weight-averaged consistency targets improve semi-supervised deep learning results. In: Advances in Neural Information Processing Systems (NIPS), pp. 1195–1204 (2017)

41. Verma, V., Lamb, A., Kannala, J., Bengio, Y., Lopez-Paz, D.: Interpolation consistency training for semi-supervised learning. In: International Joint Conference on Artificial Intelligence (IJCAI), pp. 3635–3641 (2019)
42. Zhang, F., Bo, D., Zhang, L.: Saliency-guided unsupervised feature learning for scene classification. IEEE Trans. Geosci. Remote Sensing (TGRS) **53**(4), 2175–2184 (2014)
43. Dosovitskiy, A., Springenberg, J.T., Riedmiller, M., Brox, T.: Discriminative unsupervised feature learning with convolutional neural networks. In: Advances in Neural Information Processing Systems (NIPS), pp. 766–774 (2014)
44. Yang, Y., Shu, G., Shah, M.: Semi-supervised learning of feature hierarchies for object detection in a video. In: IEEE Conference on Computer Vision and Pattern Recognition (CVPR), pp. 1650–1657 (2013)
45. Cheng, Y., Zhao, X., Huang, K. and Tan, T.: Semi-supervised learning for RGB-D object recognition. In: International Conference on Pattern Recognition (ICPR), pp. 2377–2382 (2014)
46. Gidaris, S., Bursuc, A., Komodakis, N., Pérez, P., Cord, M.: Boosting few-shot visual learning with self-supervision. In: IEEE International Conference on Computer Vision (ICCV), pp. 8059–8068 (2019)
47. Li, H., Eigen, D., Dodge, S., Zeiler, M., Wang, X.: Finding task-relevant features for few-shot learning by category traversal. In: IEEE Conference on Computer Vision and Pattern Recognition (CVPR), pp. 1–10 (2019)
48. Socher, R., Fei-Fei, L.: Connecting modalities: semi-supervised segmentation and annotation of images using unaligned text corpora. In: IEEE Computer Conference on Computer Vision and Pattern Recognition (CVPR), pp. 966–973 (2010)
49. Karnyaczki, S., Desrosiers, C.: A sparse coding method for semi-supervised segmentation with multi-class histogram constraints. In: IEEE International Conference on Image Processing (ICIP), pp. 3215–3219 (2015)
50. Dietterich, T.G., Bakiri, G.: Solving multiclass learning problems via error-correcting output codes. J. Artif. Intell. Res. (JAIR) **2**, 263–286 (1994)
51. Zhang, H., Cisse, M., Dauphin, Y.N.,Lopez-Paz, D.: mixup: beyond empirical risk minimization. In: International Conference on Learning Representations (ICLR) (2018)
52. Kingma, D.P., Ba, J.: A method for stochastic optimization. In: International Conference on Learning Representations (ICLR), pp. 1–13 (2014)
53. Paszke, A., et al.: Pytorch: an imperative style, high-performance deep learning library. In: Advances in Neural Information Processing Systems (NIPS), pp. 8024–8035 (2019)

Dynamic R-CNN: Towards High Quality Object Detection via Dynamic Training

Hongkai Zhang[1,2], Hong Chang[1,2(✉)], Bingpeng Ma[2], Naiyan Wang[3], and Xilin Chen[1,2]

[1] Key Laboratory of Intelligent Information Processing of Chinese Academy of Sciences (CAS), Institute of Computing Technology, Chinese Academy of Sciences, Beijing, China
hongkai.zhang@vipl.ict.ac.cn, {changhong,xlchen}@ict.ac.cn
[2] University of Chinese Academy of Sciences, Beijing, China
bpma@ucas.ac.cn
[3] TuSimple, San Diego, USA
winsty@gmail.com

Abstract. Although two-stage object detectors have continuously advanced the state-of-the-art performance in recent years, the training process itself is far from crystal. In this work, we first point out the inconsistency problem between the fixed network settings and the dynamic training procedure, which greatly affects the performance. For example, the fixed label assignment strategy and regression loss function cannot fit the distribution change of proposals and thus are harmful to training high quality detectors. Consequently, we propose *Dynamic R-CNN* to adjust the label assignment criteria (IoU threshold) and the shape of regression loss function (parameters of SmoothL1 Loss) automatically based on the statistics of proposals during training. This dynamic design makes better use of the training samples and pushes the detector to fit more high quality samples. Specifically, our method improves upon ResNet-50-FPN baseline with 1.9% AP and 5.5% AP_{90} on the MS COCO dataset with no extra overhead. Codes and models are available at https://github.com/hkzhang95/DynamicRCNN.

Keywords: Dynamic training · High quality object detection

1 Introduction

Benefiting from the advances in deep convolutional neural networks (CNNs) [13, 15,21,39], object detection has made remarkable progress in recent years. Modern detection frameworks can be divided into two major categories of one-stage detectors [28,31,36] and two-stage detectors [10,11,37]. And various

Electronic supplementary material The online version of this chapter (https://doi.org/10.1007/978-3-030-58555-6_16) contains supplementary material, which is available to authorized users.

© Springer Nature Switzerland AG 2020
A. Vedaldi et al. (Eds.): ECCV 2020, LNCS 12360, pp. 260–275, 2020.
https://doi.org/10.1007/978-3-030-58555-6_16

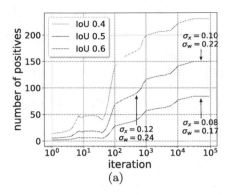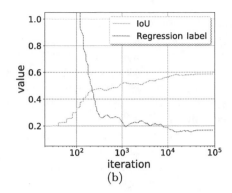

Fig. 1. (a) The number of positive proposals under different IoU thresholds during the training process. The curve shows the numbers of positives vary significantly during training, with corresponding changes in regression labels distribution (σ_x and σ_w stands for the standard deviation for x and w respectively). (b) The IoU and regression label of the 75th and 10th most accurate proposals respectively in the training procedure. These curves further show the improved quality of proposals.

improvements have been made in recent studies [5, 19, 23–25, 30, 42, 46, 47]. In the training procedure of both kinds of pipelines, a classifier and a regressor are adopted respectively to solve the *recognition* and *localization* tasks. Therefore, an effective training process plays a crucial role in achieving high quality object detection[1].

Different from the image classification task, the annotations for the classification task in object detection are the ground-truth boxes in the image. So it is not clear how to assign positive and negative labels for the proposals in classifier training since their separation may be ambiguous. The most widely used strategy is to set a threshold for the IoU of the proposal and corresponding ground-truth. As mentioned in Cascade R-CNN [3], training with a certain IoU threshold will lead to a classifier that degrades the performance at other IoUs. However, we cannot directly set a high IoU from the beginning of the training due to the scarcity of positive samples. The solution that Cascade R-CNN provides is to gradually refine the proposals by several stages, which are effective yet time-consuming. As for regressor, the problem is similar. During training, the quality of proposals is improved, however the parameter in SmoothL1 Loss is fixed. Thus it leads to insufficient training for the high quality proposals.

To solve this issue, we first examine an overlooked fact that the quality of proposals is indeed improved during training as shown in Fig. 1. We can find that even under different IoU thresholds, the number of positives still increases significantly. Inspired by the illuminating observations, we propose *Dynamic R-CNN*, a simple yet effective method to better exploit the dynamic quality of proposals for object detection. It consists of two components: *Dynamic Label Assignment* and *Dynamic SmoothL1 Loss*, which are designed for classification

---

[1] Specifically, high quality represents the results under high IoU.

and regression branches, respectively. First, to train a better classifier that is discriminative for high IoU proposals, we gradually adjust the IoU threshold for positive/negative samples based on the proposals distribution in the training procedure. Specifically, we set the threshold as the IoU of the proposal at a certain percentage since it can reflect the quality of the overall distribution. For regression, we choose to change the shape of the regression loss function to adaptively fit the distribution change of regression label and ensure the contribution of high quality samples to training. In particular, we adjust the β in SmoothL1 Loss based on the regression label distribution, since β actually controls the magnitude of the gradient of small errors (shown in Fig. 4).

By this dynamic scheme, we can not only alleviate the data scarcity issue at the beginning of the training, but also harvest the benefit of high IoU training. These two modules explore different parts of the detector, thus could work collaboratively towards high quality object detection. Furthermore, despite the simplicity of our proposed method, Dynamic R-CNN could bring consistent performance gains on MS COCO [29] with almost no extra computational complexity in training. *And during the inference phase, our method does not introduce any additional overhead.* Moreover, extensive experiments verify the proposed method could generalize to other baselines with stronger performance.

2   Related Work

Region-Based Object Detectors. The general practice of region-based object detectors is converting the object detection task into a bounding box classification and a regression problem. In recent years, region-based approaches have been the leading paradigm with top performance. For example, R-CNN [11], Fast R-CNN [10] and Faster R-CNN [37] first generate some candidate region proposals, then randomly sample a small batch with certain foreground-background ratio from all the proposals. These proposals will be fed into a second stage to classify the categories and refine the locations at the same time. Later, some works extended Faster R-CNN to address different problems. R-FCN [7] makes the whole network fully convolutional to improve the speed; and FPN [27] proposes a top-down pathway to combine multi-scale features. Besides, various improvements have been witnessed in recent studies [17,25,26,43,51].

Classification in Object Detection. Recent researches focus on improving object classifier from various perspectives [6,16,18,24,28,33,41,48]. The classification scores in detection not only determine the semantic category for each proposal, but also imply the localization accuracy, since Non-Maximum Suppression (NMS) suppresses less confident boxes using more reliable ones. It ranks the resultant boxes first using the classification scores. However, as mentioned in IoU-Net [18], the classification score has low correlation with localization accuracy, which leads to noisy ranking and limited performance. Therefore, IoU-Net [18] adopts an extra branch for predicting IoU scores and refining the classification confidence. Softer NMS [16] devises an KL loss to model the variance of bounding box regression directly, and uses that for voting in NMS. Another direction to improve is to raise the IoU threshold for training high quality classifiers, since training with different IoU thresholds will lead to classifiers with corresponding quality. However, as mentioned in Cascade R-CNN [3], directly raising the IoU threshold is impractical due to the vanishing positive samples. Therefore, to produce high quality training samples, some approaches [3,47] adopt sequential stages which are effective yet time-consuming. Essentially, it should be noted that these methods ignore the inherent dynamic property in training procedure which is useful for training high quality classifiers.

Bounding Box Regression. It has been proved that the performance of models is dependent on the relative weight between losses in multi-task learning [20]. Cascade R-CNN [3] also adopt different regression normalization factors to adjust the aptitude of regression term in different stages. Besides, Libra R-CNN [33] proposes to promote the regression gradients from the accurate samples; and SABL [44] localizes each side of the bounding box with a lightweight two step bucketing scheme for precise localization. However, they mainly focus on a fixed scheme ignoring the dynamic distribution of learning targets during training.

Dynamic Training. There are various researches following the idea of dynamic training. A widely used example is adjusting the learning rate based on the training iterations [32]. Besides, Curriculum Learning [1] and Self-paced Learning [22] focus on improving the training order of the examples. Moreover, for object detection, hard mining methods [28,33,38] can also be regarded as a dynamic way. However, they don't handle the core issues in object detection such as constant label assignment strategy. Our method is complementary to theirs.

3   Dynamic Quality in the Training Procedure

Generally speaking, Object detection is complex since it needs to solve two main tasks: *recognition* and *localization*. *Recognition* task needs to distinguish foreground objects from backgrounds and determine the semantic category for them. Besides, the *localization* task needs to find accurate bounding boxes for different objects. To achieve high-quality object detection, we need to further explore the training process of both two tasks as follows.

3.1   Proposal Classification

How to assign labels is an interesting question for the classifier in object detection. It is unique to other classification problems since the annotations are the ground-truth boxes in the image. Obviously, a proposal should be negative if it does not overlap with any ground-truth, and a proposal should be positive if its overlap with a ground-truth is 100%. However, it is a dilemma to define whether a proposal with IoU 0.5 should be labeled as positive or negative.

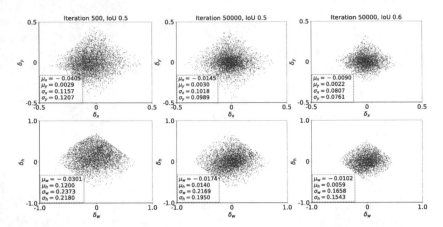

Fig. 2. Δ distribution at different iterations and IoU thresholds (we randomly select some points for simplicity). Column 1&2: under the same IoU threshold, the regression labels are more concentrated as the training goes. Column 2&3: at the same iteration, raising the IoU threshold will significantly change the distribution.

In Faster R-CNN [37], labels are assigned by comparing the box's highest IoU with ground-truths using a pre-defined IoU threshold. Formally, the paradigm can be formulated as follows (we take a binary classification loss for simplicity):

$$
\text{label} = \begin{cases} 1, & \text{if } \max IoU(b, G) \geq T_+ \\ 0, & \text{if } \max IoU(b, G) < T_- \\ -1, & \text{otherwise.} \end{cases} \tag{1}
$$

Here b stands for a bounding box, G represents for the set of ground-truths, T_+ and T_- are the positive and negative threshold for IoU. $1, 0, -1$ stand for positives, negatives and ignored samples, respectively. As for the second stage of Faster R-CNN, T_+ and T_- are set to 0.5 by default [12]. So the definition of positives and negatives is essentially hand-crafted.

Since the goal of classifier is to distinguish the positives and negatives, training with different IoU thresholds will lead to classifiers with corresponding quality [3].Therefore, to achieve high quality object detection, we need to train the classifier with a high IoU threshold. However, as mentioned in Cascade R-CNN, directly raising the IoU threshold is impractical due to the vanishing positive samples. Cascade R-CNN uses several sequential stages to lift the IoU of the proposals, which are effective yet time-consuming.

So is there a way to get the best of two worlds? As mentioned above, the quality of proposals actually improves along the training. This observation inspires us to take a progressive approach in training: At the beginning, the proposal network is not capable to produce enough high quality proposals, so we use a lower IoU threshold to better accommodate these imperfect proposals in second stage training. As training goes, the quality of proposals improves, we gradually have enough high quality proposals. As a result, we may increase the threshold to better utilize them to train a high quality detector that is more discriminative at higher IoU. We will formulate this process in the following section.

3.2   Bounding Box Regression

The task of bounding box regression is to regress the positive candidate bounding box b to a target ground-truth g. This is learned under the supervision of the regression loss function L_{reg}. To encourage the regression label invariant to scale and location, L_{reg} operates on the offset $\Delta = (\delta_x, \delta_y, \delta_w, \delta_h)$ defined by

$$
\begin{aligned}
\delta_x &= (g_x - b_x)/b_w, \quad \delta_y = (g_y - b_y)/b_h \\
\delta_w &= \log(g_w/b_w), \quad \delta_h = \log(g_h/b_h).
\end{aligned}
\tag{2}
$$

Since the bounding box regression performs on the offsets, the absolute values of Eq. (2) can be very small. To balance the different terms in multi-task learning, Δ is usually normalized by pre-defined *mean* and *stdev* (standard deviation) as widely used in many work [14,27,37].

However, we discover that the distribution of regression labels are shifting during training. As shown in Fig. 2, we calculate the statistics of the regression labels under different iterations and IoU thresholds. First, from the first two columns, we find that under the same IoU threshold for positives, the *mean* and *stdev* are decreasing as the training goes due to the improved quality of proposals. With the same normalization factors, the contributions of those high quality samples will be reduced based on the definition of SmoothL1 Loss function, which is harmful to the training of high quality regressors. Moreover, with a higher IoU threshold, the quality of positive samples is further enhanced, thus their contributions are reduced even more, which will greatly limit the overall performance. Therefore, to achieve high quality object detection, we need to fit the distribution change and adjust the shape of regression loss function to compensate for the increasing of high quality proposals.

4   Dynamic R-CNN

To better exploit the dynamic property of the training procedure, we propose Dynamic R-CNN which is shown in Fig. 3. Our key insight is **adjusting the second stage classifier and regressor to fit the distribution change of proposals**. The two components designed for the classification and localization branch will be elaborated in the following sections.

4.1   Dynamic Label Assignment

The Dynamic Label Assignment (DLA) process is illustrated in Fig. 3(a). Based on the common practice of label assignment in Eq. (1) in object detection, the DLA module can be formulated as follows:

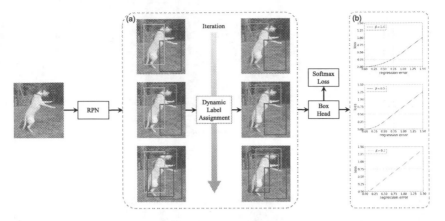

Fig. 3. The overall pipeline of the proposed Dynamic R-CNN. Considering the dynamic property of the training process, Dynamic R-CNN consists of two main components (a) Dynamic Label Assignment (DLA) process and (b) Dynamic SmoothL1 Loss (DSL) from different perspectives. From the left part of (a) we can find that there are more high quality proposals as the training goes. With the improved quality of proposals, DLA will automatically raise the IoU threshold based on the proposal distribution. Then positive (green) and negative (red) labels are assigned for the proposals by DLA which are shown in the right part of (a). Meanwhile, to fit the distribution change and compensate for the increasing of high quality proposals, the shape of regression loss function is also adjusted correspondingly in (b). Best viewed in color. (Color figure online)

$$
\text{label} = \begin{cases} 1, & \text{if } \max IoU(b,G) \geq T_{now} \\ 0, & \text{if } \max IoU(b,G) < T_{now}, \end{cases} \tag{3}
$$

where T_{now} stands for the current IoU threshold. Considering the dynamic property in training, the distribution of proposals is changing over time. Our DLA updates the T_{now} automatically based on the statistics of proposals to fit this distribution change. Specifically, we first calculate the IoUs I between proposals and their target ground-truths, and then select the K_I-th largest value from I as the threshold T_{now}. As the training goes, T_{now} will increase gradually which reflects the improved quality of proposals. In practice, we first calculate the K_I-th largest value in each batch, and then update T_{now} every C iterations using the mean of them to enhance the robustness of the training. It should be noted that the calculation of IoUs is already done by the original method, so there is almost no additional complexity in our method. The resultant IoU thresholds used in training are illustrated in Fig. 3(a).

4.2   Dynamic SmoothL1 Loss

The localization task for object detection is supervised by the commonly used SmoothL1 Loss, which can be formulated as follows:

$$
SmoothL1(x, \beta) = \begin{cases} 0.5|x|^2/\beta, & \text{if } |x| < \beta, \\ |x| - 0.5\beta, & \text{otherwise.} \end{cases} \tag{4}
$$

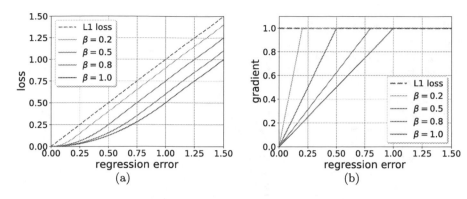

Fig. 4. We show curves for (a) loss and (b) gradient of SmoothL1 Loss with different β here. β is set default as 1.0 in the R-CNN part.

Here the x stands for the regression label. β is a hyper-parameter controlling in which range we should use a softer loss function like l_1 loss instead of the original l_2 loss. Considering the robustness of training, β is set default as 1.0 to prevent the exploding loss due to the poor trained network in the early stages. We also illustrate the impact of β in Fig. 4, in which changing β leads to different curves of loss and gradient. It is easy to find that a smaller β actually accelerate the saturation of the magnitude of gradient, thus it makes more accurate sample contributes more to the network training.

As analyzed in Sect. 3.2, we need to fit the distribution change and adjust the regression loss function to compensate for the high quality samples. So we propose Dynamic SmoothL1 Loss (DSL) to change the shape of loss function to gradually focus on high quality samples as follows:

$$DSL(x, \beta_{now}) = \begin{cases} 0.5|x|^2/\beta_{now}, & \text{if } |x| < \beta_{now}, \\ |x| - 0.5\beta_{now}, & \text{otherwise.} \end{cases} \tag{5}$$

Similar to DLA, DSL will change the value of β_{now} according to the statistics of regression labels which can reflect the localization accuracy. To be more specific, we first obtain the regression labels E between proposals and their target ground-truths, then select the K_β-th smallest value from E to update the β_{now} in Eq. (4). Similarly, we also update the β_{now} every C iterations using the median of the K_β-th smallest label in each batch. We choose median instead of mean as in the classification because we find more outliers in regression labels. Through this dynamic way, appropriate β_{now} will be adopted automatically as shown in Fig. 3(b), which will better exploit the training samples and lead to a high quality regressor.

To summarize the whole method, we describe the proposed Dynamic R-CNN in Algorithm 1. Besides the proposals P and ground-truths G, Dynamic R-CNN has three hyperparamters: IoU threshold top-k K_I, β top-k K_β and update iteration count C. Note that compared with baseline, we only introduce one

---

Algorithm 1. Dynamic R-CNN

Input:
    Proposal set P, ground-truth set G.
    IoU threshold top-k K_I, β top-k K_β, update iteration count C.
Output:
    Trained object detector D.
 1: Initialize IoU threshold and SmoothL1 β as T_{now}, β_{now}
 2: Build two empty sets $\mathcal{S}_I, \mathcal{S}_E$ for recording the IoUs and regression labels
 3: **for** $i = 0$ to max_iter **do**
 4:     Obtain matched IoUs I and regression labels E between P and G
 5:     Select thresholds I_k, E_k based on the K_I, K_β
 6:     Record corresponding values, add I_k to $\mathcal{S}_I$ and E_k to $\mathcal{S}_E$
 7:     **if** $i \% C == 0$ **then**
 8:         Update IoU threshold: $T_{now} = \text{Mean}(\mathcal{S}_I)$
 9:         Update SmoothL1 β: $\beta_{now} = \text{Median}(\mathcal{S}_E)$
10:         $\mathcal{S}_I = \emptyset, \mathcal{S}_E = \emptyset$
11:     Train the network with T_{now}, β_{now}
12: **return** Improved object detector D

---

additional hyperparameter. And we will show soon the results are actually quite robust to the choice of these hyperparameters.

5   Experiments

5.1   Dataset and Evaluation Metrics

Experimental results are mainly evaluated on the bounding box detection track of the challenging MS COCO [29] 2017 dataset. Following the common practice [14,28], we use the COCO `train` split (~118k images) for training and report the ablation studies on the `val` split (5k images). We also submit our main results to the evaluation server for the final performance on the `test-dev` split, *which has no disclosed labels*. The COCO-style Average Precision (AP) is chosen as the main evaluation metric which averages AP across IoU thresholds from 0.5 to 0.95 with an interval of 0.05. We also include other metrics to better understand the behavior of the proposed method.

5.2   Implementation Details

For fair comparisons, all experiments are implemented on PyTorch [34] and follow the settings in maskrcnn-benchmark[2] and SimpleDet [4]. We adopt FPN-based Faster R-CNN [27,37] with ResNet-50 [15] model pre-trained on ImageNet [9] as our baseline. All models are trained on the COCO 2017 `train` set and tested on `val` set with image short size at 800 pixels unless noted. Due to the scarcity of positives in the training procedure, we set the NMS threshold of RPN to 0.85 instead of 0.7 for all the experiments.

---

[2] https://github.com/facebookresearch/maskrcnn-benchmark.

Table 1. Comparisons with different baselines (our re-implementations) on COCO test-dev set. "MST" and "*" stand for multi-scale training and testing respectively. "2×" and "3×" are training schedules which extend the iterations by 2/3 times.

Method	Backbone	AP	AP_{50}	AP_{75}	AP_S	AP_M	AP_L
Faster R-CNN	ResNet-50	37.3	58.5	40.6	20.3	39.2	49.1
Faster R-CNN+2×	ResNet-50	38.1	58.9	41.5	20.5	40.0	50.0
Faster R-CNN	ResNet-101	39.3	60.5	42.7	21.3	41.8	51.7
Faster R-CNN+2×	ResNet-101	39.9	60.6	43.5	21.4	42.4	52.1
Faster R-CNN+3×+MST	ResNet-101	42.8	63.8	46.8	24.8	45.6	55.6
Faster R-CNN+3×+MST	ResNet-101-DCN	44.8	65.5	48.8	26.2	47.6	58.1
Faster R-CNN+3×+MST*	ResNet-101-DCN	46.9	68.1	51.4	30.6	49.6	58.1
Dynamic R-CNN	ResNet-50	39.1	58.0	42.8	21.3	40.9	50.3
Dynamic R-CNN+2×	ResNet-50	39.9	58.6	43.7	21.6	41.5	51.9
Dynamic R-CNN	ResNet-101	41.2	60.1	45.1	22.5	43.6	53.2
Dynamic R-CNN+2×	ResNet-101	42.0	60.7	45.9	22.7	44.3	54.3
Dynamic R-CNN+3×+MST	ResNet-101	44.7	63.6	49.1	26.0	47.4	57.2
Dynamic R-CNN+3×+MST	ResNet-101-DCN	46.9	65.9	51.3	28.1	49.6	60.0
Dynamic R-CNN+3×+MST*	ResNet-101-DCN	49.2	68.6	54.0	32.5	51.7	60.3

5.3   Main Results

We compare Dynamic R-CNN with corresponding baselines on COCO test-dev set in Table 1. For fair comparisons, We report our re-implemented results.

First, we prove that our method can work on different backbones. Dynamic R-CNN achieves 39.1% AP with ResNet-50 [15], which is 1.8 points higher than the FPN-based Faster R-CNN baseline. With a stronger backbone like ResNet-101, Dynamic R-CNN can also achieve consistent gains (+1.9 points).

Then, our dynamic design is also compatible with other training and testing skills. The results are consistently improved by progressively adding in 2× longer training schedule, multi-scale training (extra 1.5× longer training schedule), multi-scale testing and deformable convolution [51]. With the best combination, out Dynamic R-CNN achieves 49.2% AP, which is still 2.3 points higher than the Faster R-CNN baseline.

These results show the effectiveness and robustness of our method since it can work together with different backbones and multiple training and testing skills. It should also be noted that the performance gains are almost free.

5.4   Ablation Experiments

To show the effectiveness of each proposed component, we report the overall ablation studies in Table 2.

1) Dynamic Label Assignment (DLA). DLA brings 1.2 points higher box AP than the ResNet-50-FPN baseline. To be more specific, results in higher IoU metrics are consistently improved, especially for the 2.9 points gains in AP_{90}.

Table 2. Results of each component in Dynamic R-CNN on COCO `val` set.

Backbone	DLA	DSL	AP	ΔAP	AP_{50}	AP_{60}	AP_{70}	AP_{80}	AP_{90}
ResNet-50-FPN			37.0	–	58.0	53.5	46.0	32.6	9.7
ResNet-50-FPN		✓	38.0	+1.0	57.6	53.5	46.7	34.4	13.2
ResNet-50-FPN	✓		38.2	+1.2	57.5	53.6	47.1	35.2	12.6
ResNet-50-FPN	✓	✓	38.9	+1.9	57.3	53.6	47.4	36.3	15.2

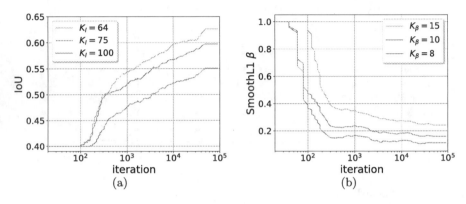

Fig. 5. Trends of (a) IoU threshold and (b) SmoothL1 β under different settings based on our method. Obviously the distribution has changed a lot during training.

It proves the effectiveness of our method for pushing the classifier to be more discriminative at higher IoU thresholds.

2) Dynamic SmoothL1 Loss (DSL). DSL improves the box AP from 37.0 to 38.0. Results in higher IoU metrics like AP_{80} and AP_{90} are hugely improved, which validates the effectiveness of changing the loss function to compensate for the high quality samples during training. Moreover, as analyzed in Sect. 3.2, with DLA the quality of positives is further improved thus their contributions are reduced even more. So applying DSL on DLA will also bring reasonable gains especially on high quality metrics. **To sum up, Dynamic R-CNN improves the baseline by 1.9 points AP and 5.5 points AP_{90}.**

3) Illustration of Dynamic Training. To further illustrate the dynamics in the training procedure, we show the trends of IoU threshold and SmoothL1 β under different settings based on our method in Fig. 5. Here we clip the values of IoU threshold and β to 0.4 and 1.0 respectively at the beginning of training. Regardless of the specific values of K_I and K_β, the overall trend of IoU threshold is increasing while that for SmoothL1 β is decreasing during training. These results again verify the proposed method work as expected.

Table 3. Ablation study on K_I.

K_I	AP	AP_{50}	AP_{60}	AP_{70}	AP_{80}	AP_{90}
–	37.0	58.0	53.5	46.0	32.6	9.7
64	38.1	57.2	53.3	46.8	35.1	**12.8**
75	**38.2**	57.5	53.6	**47.1**	**35.2**	12.6
100	37.9	**57.9**	**53.8**	46.9	34.2	11.6

Table 4. Ablation study on C.

C	AP	AP_{50}	AP_{60}	AP_{70}	AP_{80}	AP_{90}
–	37.0	58.0	53.5	46.0	32.6	9.7
20	38.0	57.4	53.5	47.0	35.0	12.5
100	38.2	57.5	53.6	47.1	35.2	12.6
500	38.1	57.6	53.5	47.2	34.8	12.6

Table 5. Ablation study on K_β.

Setting	AP	AP_{50}	AP_{60}	AP_{70}	AP_{80}	AP_{90}
$\beta = 1.0$	37.0	58.0	53.5	46.0	32.6	9.7
$\beta = 2.0$	35.9	57.7	53.2	45.1	30.1	8.3
$\beta = 0.5$	37.5	57.6	53.3	46.4	33.5	11.3
$K_\beta = 15$	37.6	57.3	53.1	46.0	33.9	12.5
$K_\beta = 10$	38.0	57.6	53.5	46.7	34.4	13.2
$K_\beta = 8$	37.6	57.5	53.3	45.9	33.9	12.4

Table 6. Inference speed comparisons using ResNet-50-FPN backbone on RTX 2080TI GPU.

Method	FPS
Dynamic R-CNN	13.9
Cascade R-CNN	11.2
Dynamic Mask R-CNN	11.3
Cascade Mask R-CNN	7.3

5.5 Studies on the Effect of Hyperparameters

Ablation Study on K_I in DLA. Experimental results on different K_I are shown in Table 3. Compared to the baseline, DLA can achieve consistent gains in AP regardless of the choice of K_I. These results prove the universality of K_I. Moreover, the performance on various metrics are changed under different K_I. Choosing K_I as 64/75/100 means that nearly 12.5%/15%/20% of the whole batch are selected as positives. Generally speaking, setting a smaller K_I will increase the quality of selected samples, which will lead to better accuracy under higher metrics like AP_{90}. On the contrary, adopting a larger K_I will be more helpful for the metrics at lower IoU. Finally, we find that setting K_I as 75 achieves the best trade-off and use it as the default value for further experiments. All these ablations prove the effectiveness and robustness of the DLA part.

Ablation Study on K_β in DSL. As shown in Table 5, we first try different β on Faster R-CNN and empirically find that a smaller β leads to better performance. Then, experiments under different K_β are provided to show the effects of K_β. Regardless of the certain value of K_β, DSL can achieve consistent improvements compared with various fine-tuned baselines. Specifically, with our best setting, DSL can bring 1.0 point higher AP than the baseline, and the improvement mainly lies in the high quality metrics like AP_{90} (+3.5 points). These experimental results prove that our DSL is effective in compensating for high quality samples and can lead to a better regressor due to the advanced dynamic design.

Ablation Study on Iteration Count C. Due to the concern of robustness, we update T_{now} and β_{now} every C iterations using the statistics in the last interval. To show the effects of different iteration count C, we try different values of C on

Table 7. The universality of Dynamic R-CNN. We apply the idea of dynamic training on Mask R-CNN under different backbones. "bbox" and "segm" stand for object detection and instance segmentation results on COCO `val` set, respectively.

Backbone	+Dynamic	AP^{bbox}	AP^{bbox}_{50}	AP^{bbox}_{75}	AP^{segm}	AP^{segm}_{50}	AP^{segm}_{75}
ResNet-50-FPN		37.5	58.0	40.7	33.8	54.6	36.0
	✓	39.4	57.6	43.3	34.8	55.0	37.5
ResNet-101-FPN		39.7	60.7	43.2	35.6	56.9	37.7
	✓	41.8	60.4	45.8	36.7	57.5	39.4

the proposed method. As shown in Table 4, setting C as 20, 100 and 500 leads to very similar results, which proves the robustness to this hyperparameter.

Complexity and Speed. As shown in Algorithm 1, the main computational complexity of our method lies in the calculations of IoUs and regression labels, which are already done by the original method. Thus the additional overhead only lies in calculating the mean or median of a short vector, which basically **does not increase the training time**. Moreover, since our method only changes the training procedure, obviously the inference speed will not be slowed down.

Our advantage compared to other high quality detectors like Cascade R-CNN is the efficiency. Cascade R-CNN increases the training time and slows down the inference speed while our method does not. Specifically, as shown in Table 6, Dynamic R-CNN achieves 13.9 FPS, which is ~1.25 times faster than Cascade R-CNN (11.2 FPS) under ResNet-50-FPN backbone. Moreover, with larger heads, the cascade manner will further slow down the speed. Dynamic Mask R-CNN runs ~1.5 times faster than Cascade Mask R-CNN. Note that the difference will be more apparent as the backbone gets smaller (~1.74 times faster, 13.6 FPS vs 7.8 FPS under ResNet-18 backbone with mask head), since the main overhead of Cascade R-CNN is the two additional headers.

5.6 Universality

Since the viewpoint of dynamic training is a general concept, we believe that it can be adopted in different methods. To validate the universality, we further apply the dynamic design on Mask R-CNN with different backbones. As shown in Table 7, adopting the dynamic design can not only bring ~2.0 points higher box AP but also improve the instance segmentation results regardless of backbones. Note that we only adopt the DLA and DSL which are designed for object detection, so these results further demonstrate the universality and effectiveness of our dynamic design on improving training procedure for current detectors.

5.7 Comparison with State-of-the-Arts

We compare Dynamic R-CNN with the state-of-the-art object detectors on COCO `test-dev` set in Table 8. Considering that various backbones and training/testing settings are adopted by different detectors (including deformable

Table 8. Comparisons of single-model results on COCO `test-dev` set.

Method	Backbone	AP	AP_{50}	AP_{75}	AP_S	AP_M	AP_L
RetinaNet [28]	ResNet-101	39.1	59.1	42.3	21.8	42.7	50.2
CornerNet [23]	Hourglass-104	40.5	56.5	43.1	19.4	42.7	53.9
FCOS [42]	ResNet-101	41.0	60.7	44.1	24.0	44.1	51.0
FreeAnchor [49]	ResNet-101	41.8	61.1	44.9	22.6	44.7	53.9
RepPoints [46]	ResNet-101-DCN	45.0	66.1	49.0	26.6	48.6	57.5
CenterNet [50]	Hourglass-104	45.1	63.9	49.3	26.6	47.1	57.7
ATSS [48]	ResNet-101-DCN	46.3	64.7	50.4	27.7	49.8	58.4
Faster R-CNN [27]	ResNet-101	36.2	59.1	39.0	18.2	39.0	48.2
Mask R-CNN [14]	ResNet-101	38.2	60.3	41.7	20.1	41.1	50.2
Regionlets [45]	ResNet-101	39.3	59.8	–	21.7	43.7	50.9
Libra R-CNN [33]	ResNet-101	41.1	62.1	44.7	23.4	43.7	52.5
Cascade R-CNN [3]	ResNet-101	42.8	62.1	46.3	23.7	45.5	55.2
SNIP [40]	ResNet-101-DCN	44.4	66.2	49.9	27.3	47.4	56.9
DCNv2 [51]	ResNet-101-DCN	46.0	67.9	50.8	27.8	49.1	59.5
TridentNet [25]	ResNet-101-DCN	48.4	69.7	53.5	31.8	51.3	60.3
Dynamic R-CNN	ResNet-101	42.0	60.7	45.9	22.7	44.3	54.3
Dynamic R-CNN*	ResNet-101-DCN	50.1	68.3	55.6	32.8	53.0	61.2

convolutions [8,51], image pyramid scheme [40], large-batch Batch Normalization [35] and Soft-NMS [2]), we report the results of our method with two types.

Dynamic R-CNN applies our method on FPN-based Faster R-CNN with ResNet-101 as backbone, and it can achieve 42.0% AP without bells and whistles. Dynamic R-CNN* adopts image pyramid scheme (multi-scale training and testing), deformable convolutions and Soft-NMS. It further improves the results to 50.1% AP, outperforming all the previous detectors.

6   Conclusion

In this paper, we take a thorough analysis of the training process of detectors and find that the fixed scheme limits the overall performance. Based on the advanced dynamic viewpoint, we propose Dynamic R-CNN to better exploit the training procedure. With the help of the simple but effective components like Dynamic Label Assignment and Dynamic SmoothL1 Loss, Dynamic R-CNN brings significant improvements on the challenging COCO dataset with no extra cost. Extensive experiments with various detectors and backbones validate the universality and effectiveness of Dynamic R-CNN. We hope that this dynamic viewpoint can inspire further researches in the future.

Acknowledgements. This work is partially supported by Natural Science Foundation of China (NSFC): 61876171 and 61976203, and Beijing Natural Science Foundation under Grant L182054.

References

1. Bengio, Y., Louradour, J., Collobert, R., Weston, J.: Curriculum learning. In: ICML (2009)
2. Bodla, N., Singh, B., Chellappa, R., Davis, L.S.: Soft-NMS - improving object detection with one line of code. In: ICCV (2017)
3. Cai, Z., Vasconcelos, N.: Cascade R-CNN: delving into high quality object detection. In: CVPR (2018)
4. Chen, Y., et al.: SimpleDet: a simple and versatile distributed framework for object detection and instance recognition. JMLR **20**(156), 1–8 (2019)
5. Chen, Y., Han, C., Wang, N., Zhang, Z.: Revisiting feature alignment for one-stage object detection. arXiv:1908.01570 (2019)
6. Cheng, B., Wei, Y., Shi, H., Feris, R., Xiong, J., Huang, T.: Revisiting RCNN: on awakening the classification power of faster RCNN. In: ECCV (2018)
7. Dai, J., Li, Y., He, K., Sun, J.: R-FCN: object detection via region-based fully convolutional networks. In: NIPS (2016)
8. Dai, J., Qi, H., Xiong, Y., Li, Y., Zhang, G., Hu, H., Wei, Y.: Deformable convolutional networks. In: ICCV (2017)
9. Deng, J., Dong, W., Socher, R., Li, L.J., Li, K., Fei-Fei, L.: ImageNet: a large-scale hierarchical image database. In: CVPR (2009)
10. Girshick, R.: Fast R-CNN. In: ICCV (2015)
11. Girshick, R., Donahue, J., Darrell, T., Malik, J.: Rich feature hierarchies for accurate object detection and semantic segmentation. In: CVPR (2014)
12. Girshick, R., Radosavovic, I., Gkioxari, G., Dollár, P., He, K.: Detectron (2018). https://github.com/facebookresearch/detectron
13. Gu, X., Chang, H., Ma, B., Zhang, H., Chen, X.: Appearance-preserving 3D convolution for video-based person re-identification. In: ECCV (2020)
14. He, K., Gkioxari, G., Dollar, P., Girshick, R.: Mask R-CNN. In: ICCV (2017)
15. He, K., Zhang, X., Ren, S., Sun, J.: Deep residual learning for image recognition. In: CVPR (2016)
16. He, Y., Zhu, C., Wang, J., Savvides, M., Zhang, X.: Bounding box regression with uncertainty for accurate object detection. In: CVPR (2019)
17. Huang, J., et al.: Speed/accuracy trade-offs for modern convolutional object detectors. In: CVPR (2017)
18. Jiang, B., Luo, R., Mao, J., Xiao, T., Jiang, Y.: Acquisition of localization confidence for accurate object detection. In: ECCV (2018)
19. Jiang, Z., Liu, Y., Yang, C., Liu, J., Gao, P., Zhang, Q., Xiang, S., Pan, C.: Learning where to focus for efficient video object detection. In: ECCV (2020). https://doi.org/10.1007/978-3-030-58517-4_2
20. Kendall, A., Gal, Y., Cipolla, R.: Multi-task learning using uncertainty to weigh losses for scene geometry and semantics. In: CVPR (2018)
21. Krizhevsky, A., Sutskever, I., Hinton, G.E.: ImageNet classification with deep convolutional neural networks. In: NIPS (2012)
22. Kumar, M.P., Packer, B., Koller, D.: Self-paced learning for latent variable models. In: NIPS (2010)
23. Law, H., Deng, J.: CornerNet: detecting objects as paired keypoints. In: ECCV (2018)
24. Li, H., Wu, Z., Zhu, C., Xiong, C., Socher, R., Davis, L.S.: Learning from noisy anchors for one-stage object detection. In: CVPR (2020)

25. Li, Y., Chen, Y., Wang, N., Zhang, Z.: Scale-aware trident networks for object detection. In: ICCV (2019)
26. Li, Z., Peng, C., Yu, G., Zhang, X., Deng, Y., Sun, J.: DetNet: design backbone for object detection. In: ECCV (2018)
27. Lin, T.Y., Dollar, P., Girshick, R., He, K., Hariharan, B., Belongie, S.: Feature pyramid networks for object detection. In: CVPR (2017)
28. Lin, T.Y., Goyal, P., Girshick, R., He, K., Dollar, P.: Focal loss for dense object detection. In: ICCV (2017)
29. Lin, T.Y., et al.: Microsoft COCO: common objects in context. In: ECCV (2014)
30. Liu, S., Huang, D., Wang, Y.: Receptive field block net for accurate and fast object detection. In: ECCV (2018)
31. Liu, W., et al.: SSD: Single shot multibox detector. In: ECCV (2016)
32. Loshchilov, I., Hutter, F.: SGDR: stochastic gradient descent with warm restarts. In: ICLR (2017)
33. Pang, J., Chen, K., Shi, J., Feng, H., Ouyang, W., Lin, D.: Libra R-CNN: towards balanced learning for object detection. In: CVPR (2019)
34. Paszke, A., et al.: Automatic differentiation in PyTorch. In: NIPS Workshop (2017)
35. Peng, C., et al.: MegDet: a large mini-batch object detector. In: CVPR (2018)
36. Redmon, J., Divvala, S., Girshick, R., Farhadi, A.: You only look once: unified, real-time object detection. In: CVPR (2016)
37. Ren, S., He, K., Girshick, R., Sun, J.: Faster R-CNN: towards real-time object detection with region proposal networks. In: NIPS (2015)
38. Shrivastava, A., Gupta, A., Girshick, R.: Training region-based object detectors with online hard example mining. In: CVPR (2016)
39. Simonyan, K., Zisserman, A.: Very deep convolutional networks for large-scale image recognition. In: ICLR (2015)
40. Singh, B., Davis, L.S.: An analysis of scale invariance in object detection - SNIP. In: CVPR (2018)
41. Tan, Z., Nie, X., Qian, Q., Li, N., Li, H.: Learning to rank proposals for object detection. In: ICCV (2019)
42. Tian, Z., Shen, C., Chen, H., He, T.: FCOS: fully convolutional one-stage object detection. In: ICCV (2019)
43. Wang, J., Chen, K., Yang, S., Loy, C.C., Lin, D.: Region proposal by guided anchoring. In: CVPR (2019)
44. Wang, J., et al.: Side-aware boundary localization for more precise object detection. In: ECCV (2020)
45. Xu, H., Lv, X., Wang, X., Ren, Z., Bodla, N., Chellappa, R.: Deep regionlets for object detection. In: ECCV (2018)
46. Yang, Z., Liu, S., Hu, H., Wang, L., Lin, S.: RepPoints: point set representation for object detection. In: ICCV (2019)
47. Zhang, H., Chang, H., Ma, B., Shan, S., Chen, X.: Cascade RetinaNet: maintaining consistency for single-stage object detection. In: BMVC (2019)
48. Zhang, S., Chi, C., Yao, Y., Lei, Z., Li, S.Z.: Bridging the gap between anchor-based and anchor-free detection via adaptive training sample selection. In: CVPR (2020)
49. Zhang, X., Wan, F., Liu, C., Ji, R., Ye, Q.: FreeAnchor: learning to match anchors for visual object detection. In: NeurIPS (2019)
50. Zhou, X., Wang, D., Krähenbühl, P.: Objects as points. arXiv:1904.07850 (2019)
51. Zhu, X., Hu, H., Lin, S., Dai, J.: Deformable convnets v2: more deformable, better results. In: CVPR (2019)

Boosting Decision-Based Black-Box Adversarial Attacks with Random Sign Flip

Weilun Chen[1,2], Zhaoxiang Zhang[1,2,3(✉)], Xiaolin Hu[4], and Baoyuan Wu[5,6]

[1] Center for Research on Intelligent Perception and Computing (CRIPAC), National Laboratory of Pattern Recognition (NLPR), Institute of Automation, Chinese Academy of Sciences (CASIA), Beijing, China
{chenweilun2018,zhaoxiang.zhang}@ia.ac.cn
[2] School of Artificial Intelligence, University of Chinese Academy of Sciences (UCAS), Beijing, China
[3] Center for Excellence in Brain Science and Intelligence Technology, CAS, Beijing, China
[4] Tsinghua University, Beijing, China
xlhu@mail.tsinghua.edu.cn
[5] The Chinese University of Hong Kong, Shenzhen, China
[6] Tencent AI Lab, Shenzhen, China
wubaoyuan1987@gmail.com

Abstract. Decision-based black-box adversarial attacks (decision-based attack) pose a severe threat to current deep neural networks, as they only need the predicted label of the target model to craft adversarial examples. However, existing decision-based attacks perform poorly on the l_∞ setting and the required enormous queries cast a shadow over the practicality. In this paper, we show that just randomly flipping the signs of a small number of entries in adversarial perturbations can significantly boost the attack performance. We name this simple and highly efficient decision-based l_∞ attack as Sign Flip Attack. Extensive experiments on CIFAR-10 and ImageNet show that the proposed method outperforms existing decision-based attacks by large margins and can serve as a strong baseline to evaluate the robustness of defensive models. We further demonstrate the applicability of the proposed method on real-world systems.

Keywords: Adversarial examples · Decision-based attacks

1 Introduction

Deep neural networks are susceptible to *adversarial examples* [6,48,51]. In terms of image classification, an imperceptible adversarial perturbation can alter the

Electronic supplementary material The online version of this chapter (https://doi.org/10.1007/978-3-030-58555-6_17) contains supplementary material, which is available to authorized users.

prediction of a well-trained model to any desired class [9, 44]. The effectiveness of these maliciously crafted perturbations has been further demonstrated in the physical world [4, 18, 30], leading to growing concerns about the security of widely deployed applications based on deep neural networks, especially in sensitive areas, *e.g.*, financial service, autonomous driving, and face verification. Developing adversarial attacks under various settings provides great insights to understand and resistant the vulnerability of deep neural networks.

A broad range of existing works on adversarial attacks [21, 37] mainly focus on the *white-box* setting, where the adversary is capable to access all the information about the target model. While white-box attacks serve as an important role to evaluate the robustness, nearly all the real-world models are not completely exposed to the adversary. Typically, at least the model structures and internal weights are concealed. We refer to this case as the *black-box* setting. One approach to make black-box attacks feasible is to utilize the transferability of adversarial examples [33, 34, 38, 39, 46]. Whereas, these *transfer-based black-box* attacks need a substitute model and suffer from a low attack success rate when conducting targeted attacks. Instead of relying on the substitute model, some methods successfully conduct both untargeted and targeted attacks via accessing the confidence scores or losses of the target model [11, 27, 28, 36]. Despite the efficiency of these *score-based black-box* attacks, their requirements are still too hard in some real-world cases. Perhaps the most practical setting is that only the final decision (top-1 label) can be observed, noted as the *decision-based black-box* setting [7]. Even under this rather restrictive setting, deep neural networks are extremely vulnerable [7, 10, 12, 27]. However, the considerable query numbers required by existing decision-based attacks diminish their applicability, not to mention the poor performance on undefended and weak defensive models under the l_∞ setting. Designing a versatile and efficient decision-based l_∞ adversarial attack is still an open problem.

In this paper, we focus on the challenging decision-based black-box l_∞ setting and consider deep neural networks used for image classification as the models being attacked. Inspired by the cruciality of the sign vector of adversarial perturbations, we propose a simple and highly efficient attack method and dub **Sign Flip Attack**. Our method does not need to estimate gradients and works in an iterative manner. In each iteration, we first reduce the l_∞ distance by projection, then randomly flip the signs of partial entries in the current adversarial perturbation. The core of our method is the novel random sign flip step, which iteratively adjusts the sign vector of the current perturbation to get closer to a good one. Extensive experiments on two standard datasets CIFAR-10 [29] and ImageNet [15] demonstrate the superiority of our method. Results on 7 defensive models indicate that our method can serve as a strong baseline to evaluate the robustness. We further apply our method on real-world systems to show its practical applicability. Examples of attacking a face verification API are shown in Fig. 1.

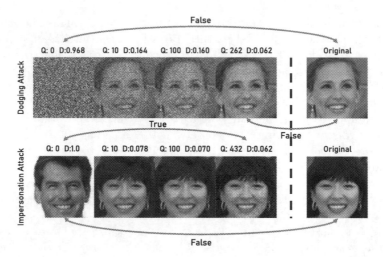

Fig. 1. Examples of attacking the face verification API in Tencent AI Open Platform. **Q** denotes the query number. **D** denotes the l_∞ distance towards the original image. True means the API classifies the two images as the same identity, and false means not the same. Best viewed in color with zoom in. (Color figure online)

2   Related Work

A variety of white-box attacks [9,19,21,37,40] have been developed since the vulnerability of deep neural networks discovered by Szegedy et al. [44]. The gradient-based nature of powerful white-box attacks is leveraged by quite a few defenses. However, Athalye et al. [3] has shown that most of them can be defeated. Defenses based on robust optimization [32,35,53] are the most effective, while still perform limited. In what follows, we will concentrate on the recent advances of black-box attacks.

Transfer-Based Black-Box Attacks. One intriguing property of adversarial examples is their good transferability [44]. Papernot et al. [39,40] constructed a local surrogate model by querying the target model, then used the resulting one to generate adversarial examples. Liu et al. [34] showed that adversarial examples crafted on an ensemble of models have better transferability. Dong et al. [16] integrated momentum into the iterative white-box attacks and achieved a superior black-box attack performance. However, transfer-based black-box attacks are not effective to generate targeted adversarial examples [11] and perform poorly on Ensemble Adversarial Training [45]. Besides, it is arduous to find or train an appropriate substitute model for a real-world deployed application.

Score-Based Black-Box Attacks. In the score-based black-box setting, the adversary can obtain the corresponding predicted probability or loss by querying the target model. Chen et al. [11] applied zeroth order optimization to craft adversarial examples. A similar technique was used in [5], different from [11] they used the estimated gradient to perform the fast gradient sign method

(FGSM) [21] and its iterative variant [30]. Ilyas et al. [27] used the natural evolutionary strategy (NES) for gradient estimation, then performed the projected gradient descent (PGD) [35]. The method can be further improved by exploiting two kinds of gradient priors [28]. Instead of gradient approximation, some attacks work in a gradient-free manner [1,2,22,36]. Recently, several works [14,20,23,31,42] consider additional models to improve query efficiency. Albeit effective, the applicability of these attacks is still restricted by their requirement of accessing the continuous outputs of the target model.

Decision-Based Black-Box Attacks. Brendel et al. [7] proposed the first effective decision-based black-box attack on deep neural networks, named the Boundary Attack, which starts from an adversarial point and performs random walks on the decision boundary while keeping adversarial. Ilyas et al. [27] extended their score-based method to this setting by estimating a proxy score. Cheng et al. [12] reformulated the original problem to a continuous version and applied zeroth order optimization. In [13], the same continuous problem was considered, however it computes the sign of the directional derivative instead of the magnitude, which leads to fast convergences. Recently, Chen et al. [10] proposed an unbiased estimate of the gradient direction at the decision boundary to improve the Boundary Attack. In each iteration, the adversarial example first approaches the boundary via a binary search, then moves along the estimated gradient direction to deviate from the decision boundary. In [17], an evolutionary attack method was proposed against face recognition systems. These methods generally require enormous queries and have poor performance under the l_∞ setting.

3   Approach

In this section, we first introduce preliminaries about adversarial examples and specify the threat model. Then, we present the proposed decision-based black-box l_∞ adversarial attack, which we dub **Sign Flip Attack** (SFA).

3.1   Preliminaries

We consider an image classifier $f : \mathbb{R}^d \to \mathbb{R}^k$ based on deep neural networks as the target model. For a given image $x \in [0,1]^d$ and its corresponding true label y, $f(x)_i$ denotes the probability that x belongs to class i, and $c(x) = \arg\max_{i \in \{1,...,k\}} f(x)_i$ refers to the predicted label. We only consider images that are correctly classified. The goal of the adversary is to find an adversarial perturbation $\delta \in \mathbb{R}^d$ such that $c(x + \delta) \neq y$ (untargeted attacks) or $c(x + \delta) = t$ ($t \neq y$, targeted attacks) , and $\|\delta\|_\infty \leq \epsilon$. Here, ϵ refers to the allowed maximum perturbation. We choose l_∞ distance to depict the perceptual similarity between the natural and adversarial images.

Suppose we have a suitable loss function $L(f(x), y)$, e.g., cross entropy loss, then we can formulate the task of generating untargeted adversarial examples as a constrained optimization problem:

$$\max_{\delta} L\big(f(\boldsymbol{x}+\boldsymbol{\delta}),y\big) \quad s.t. \ \|\boldsymbol{\delta}\|_{\infty} \leq \epsilon. \tag{1}$$

The constrained problem in Eq. 1 can be efficiently optimized by gradient-based methods under the white-box setting, such as PGD, which is an iterative method using the following update:

$$\boldsymbol{x}^n = \Pi_{B_{\infty}(\boldsymbol{x},\epsilon)}\big(\boldsymbol{x}^{n-1} + \eta \mathbf{sgn}(\nabla_{\boldsymbol{x}} L(f(\boldsymbol{x}^{n-1}),y)))\big). \tag{2}$$

Here, $B_{\infty}(\boldsymbol{x},\epsilon)$ refers to the l_{∞} ball around $\boldsymbol{x}$ with radius ϵ and Π is the projection operator. Specifically, in this case, the projection is an element-wise clip:

$$\big(\Pi_{B_{\infty}(\boldsymbol{x},\epsilon)}(\boldsymbol{x}')\big)_i = \min\{\max\{\boldsymbol{x}_i - \epsilon, \boldsymbol{x}_i'\}, \boldsymbol{x}_i + \epsilon\}. \tag{3}$$

3.2 Threat Models

We consider adversarial attacks under the decision-based black-box setting, with l_{∞} distance as the similarity constraint. That is, the adversary has no knowledge about the network architecture, internal weights, intermediate outputs or the predicted probability $f(\cdot)$, and can solely obtain the final decision $c(\cdot)$ of the target model by querying. As the value of $f(\cdot)$ can not be directly obtained, we consider the following constrained problem:

$$\min_{\delta} \|\boldsymbol{\delta}\|_{\infty} \quad s.t. \ \phi(\boldsymbol{x}+\boldsymbol{\delta}) = 1. \tag{4}$$

Here, $\phi : \mathbb{R}^d \to \{0,1\}$ is an indicator function, which takes 1 if the adversarial constraint is satisfied, that is, $c(\boldsymbol{x}+\boldsymbol{\delta}) \neq y$ in untargeted attacks and $c(\boldsymbol{x}+\boldsymbol{\delta}) = t$ in targeted attacks. $\phi(\boldsymbol{x})$ can be computed by querying the target model. The goal of the adversary is to find a successful adversarial perturbation in as few queries as possible. Thus in practice, we set a maximum distortion ϵ. Once $\|\boldsymbol{\delta}\|_{\infty} \leq \epsilon$, we stop the attack and report the query number.

3.3 Sign Flip Attack

The basic logic of our method is simple. For an image-label pair $\{\boldsymbol{x},y\}$, we start from a perturbation[1] $\boldsymbol{\delta}$ with a large l_{∞} norm such that $\phi(\boldsymbol{x}+\boldsymbol{\delta}) = 1$, then iteratively reduce $\|\boldsymbol{\delta}\|_{\infty}$ while keeping adversarial. Next, we will mainly discuss the targeted version of the proposed method, and t is the target label. The extension to the untargeted setting is straightforward.

In each iteration, we first add a random noise $\boldsymbol{\eta}$ to the current perturbation $\boldsymbol{\delta}$, then project the new perturbation onto a smaller l_{∞} ball, which can be formalized as:

$$\boldsymbol{\eta} \sim \{-\alpha, \alpha\}^d, \quad \boldsymbol{\delta}_p = \Pi_{B_{\infty}(0,\epsilon'-\alpha)}(\boldsymbol{\delta}+\boldsymbol{\eta}). \tag{5}$$

---

[1] Finding an initial perturbation is easy, any image which is from a different class (untargeted attacks) or a specific class (targeted class) can be taken as an initial adversarial example.

Here, $\epsilon' = \|\delta\|_\infty$ and $0 < \alpha < \epsilon'$. δ_p is the generated perturbation after the project step. α is an adjustable hyperparameter, which controls the shrinking magnitude of $\|\delta\|_\infty$. Intuitively, the project step often leads to decreases in $f(x+\delta)_t$ and increases in $f(x+\delta)_y$, as the adversarial example gets closer to a natural image classified with a high probability. In Fig. 2 (a), we plot the distribution of the probability increments $\Delta_p f_t = f(x+\delta_p)_t - f(x+\delta)_t$ and $\Delta_p f_y = f(x+\delta_p)_y - f(x+\delta)_y$, it shows that $\Delta_p f_t$ is always negative and $\Delta_p f_y$ is always positive. If $f(x+\delta_p)_t$ is less than $f(x+\delta_p)_y$ or any other entry in $f(x+\delta_p)$, we reject δ_p to hold the adversarial constraint.

As the project step always reduces $f(x+\delta)_t$ and increases $f(x+\delta)_y$, we expect to get a new perturbation using a single query, denoted by δ_s, which satisfies the following properties: 1) if δ_s does not violate the adversarial constraint, it should have a high probability to acquire a positive $\Delta_s f_t$ and a negative $\Delta_s f_y$; 2) the l_∞ norm of δ_s is equal to $\|\delta\|_\infty$. It is not desirable to alleviate the probability changes introduced by the project step at the cost of a greater l_∞ norm.

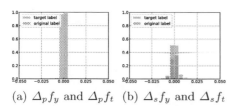 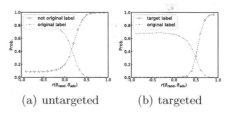

(a) $\Delta_p f_y$ and $\Delta_p f_t$ (b) $\Delta_s f_y$ and $\Delta_s f_t$ (a) untargeted   (b) targeted

Fig. 2. Distribution of the relative probability changes on the target and original label. We perform targeted SFA to 100 images from ImageNet on DenseNet-121. (a) the project step. (b) the random sign flip step. Note, we only consider the relative probability changes on successful trails, i.e., $\phi(x+\delta_p) = 1$ and $\phi(x+\delta_s) = 1$. For comparison, the success rates for the random sign flip step and the project step are 24.2% and 49.3% respectively.

Fig. 3. Relationship between the sign match rate and the predicted probability. The experiments are conducted on 1,000 images from ImageNet. δ_{adv} is an $\epsilon = 0.031$ l_∞ adversarial perturbation generated by 20-step PGD. δ_{rand} is randomly selected from $\{-\epsilon, \epsilon\}^d$. "original label", "target label" and "not original label" denote $f(x+\delta_{rand})_y$, $f(x+\delta_{rand})_t$ and $\max_{i\neq y} f(x+\delta_{rand})_i$, respectively.

Depicting the distribution of δ_s is arduous. Fortunately, adversarial examples found by PGD frequently lie on the vertices of l_∞ balls around natural images [36] and one can modify the sign vector of a given adversarial perturbation to generate a new one [28]. These discoveries suggest that searching among the vertices of l_∞ balls may have a higher success rate than searching in the l_∞ balls. Our experiments conducted on CIFAR-10 support this claim, the untargeted attack success rates for these two random sampling strategies are 43.0% and 18.2% respectively. Thus, we conjecture that one has a high probability to get a qualified δ_s by randomly changing the sign vector. Inspired by this, we propose the **random sign flip** step. In each iteration, we randomly select partial coordinates (*e.g.*, 0.1%), then flip the signs of the corresponding entries in $\boldsymbol{\delta}$. Suppose $\boldsymbol{s} \in \{0,1\}^d$ and $\boldsymbol{p} \in (0,1)^d$, the random sign flip step can be formulated as:

$$s_i \sim \textbf{Bernoulli}(p_i), \quad \boldsymbol{\delta}_s = \boldsymbol{\delta} \odot (1 - 2\boldsymbol{s}), \tag{6}$$

where s_i and p_i denote the i-th element of $\boldsymbol{s}$ and $\boldsymbol{p}$ respectively. $\odot$ is the Hadamard product. $\boldsymbol{p}$ is another crucial hyperparameter, which controls the sign flip probability of each entry in $\boldsymbol{\delta}$. We will discuss how to adjust these two hyperparameters later. Same as the project step, $\boldsymbol{\delta}_s$ which violates the adversarial constraint will be rejected. A simple illustration of the random sign flip step is shown in Fig. 4. Clearly, the random sign flip step does not alter the l_∞ norm, $\|\boldsymbol{\delta}_s\|_\infty = \|\boldsymbol{\delta}\|_\infty$. In practice, we plot the distribution of $\Delta_s f_t$ and $\Delta_s f_y$ in Fig. 2(b). It can be seen that $\boldsymbol{\delta}_s$ generated by the random sign flip step does have a certain probability to get a positive $\Delta_s f_t$ and a negative $\Delta_s f_y$.

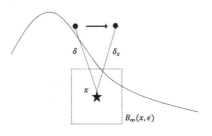

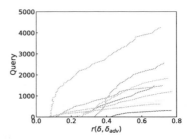

Fig. 4. Illustration of the random sign flip step. $\boldsymbol{\delta}_s$ is the generated perturbation of $\boldsymbol{\delta}$ after the random sign flip step, which also has a larger distance towards the decision boundary.

Fig. 5. The change of the sign match rate between $\boldsymbol{\delta}_{adv}$ and $\boldsymbol{\delta}$ during querying. $\boldsymbol{\delta}_{adv}$ is generated by targeted PGD with the last $\boldsymbol{\delta}$ as the initial start. Each curve represents an image chosen from CIFAR-10.

Why Does the Random Sign Flip Step Work? Let $r(\boldsymbol{\delta}_1, \boldsymbol{\delta}_2) = \frac{\text{sgn}(\boldsymbol{\delta}_1)^\top \text{sgn}(\boldsymbol{\delta}_2)}{d}$ be the sign match rate of $\boldsymbol{\delta}_1 \in \mathbb{R}^d$ and $\boldsymbol{\delta}_2 \in \mathbb{R}^d$. We find that,

1) There exist a large body of diverse adversarial sign vectors for most images. We choose 1,000 images from ImageNet. For each image-label pair $\{\boldsymbol{x} \in \mathbb{R}^d, y \in \mathbb{R}\}$, we generate 100 adversarial perturbations using 20-step targeted PGD with random starts and calculate the maximum sign match rate between any two of them. The results show that the maximum sign match rate is lower than 0.1 for 87.8% of images (82.5% for untargeted attacks).
2) For an image-label pair $\{\boldsymbol{x}, y\}$, consider a random vector $\boldsymbol{\delta}_{rand} \in \{-\epsilon, \epsilon\}^d$ and a "good" adversarial perturbation[2] $\boldsymbol{\delta}_{adv} \in [-\epsilon, \epsilon]^d$, $r(\boldsymbol{\delta}_{rand}, \boldsymbol{\delta}_{adv})$ has a negative correlation with $f(\boldsymbol{x} + \boldsymbol{\delta}_{rand})_y$ and a positive correlation with $f(\boldsymbol{x} + \boldsymbol{\delta}_{rand})_t$ (or $\max_{i \neq y} f(\boldsymbol{x} + \boldsymbol{\delta}_{rand})_i$), as presented in Fig. 3.

---

[2] $f(\boldsymbol{x} + \boldsymbol{\delta}_{adv})_t$ (or $\max_{i \neq y} f(\boldsymbol{x} + \boldsymbol{\delta}_{adv})_i$) is very close to 1.

As a large number of diverse adversarial sign vectors exist, we assume that there is a "good" adversarial perturbation $\delta_{adv} \in \mathbb{R}^d$ which has a relatively high sign match rate with the current perturbation $\delta \in \mathbb{R}^d$. Although we do not know the actual δ_{adv}, we can alter the sign match rate between δ_{adv} and δ through the random sign flip step. Then according to the second point mentioned above, what we prefer is to get a new perturbation with a higher sign match rate. In fact, once our method attacks successfully, we can use the resulting perturbation as an initial start and perform targeted PGD to construct such a δ_{adv}. We plot the change of the sign match rate between δ_{adv} and δ during querying in Fig. 5. It shows a clear uptrend. Thus, the random sign flip step serves as a tool to make the sign match rate become higher. But how large is the probability of acquiring a higher sign match rate after the random sign flip step?

Let $r(\delta, \delta_{adv}) = r\ (r \in [-1, 1])$ and $m = (1-r) \cdot d/2$ denote the total number of coordinates where δ and δ_{adv} have reverse signs. We perform one random sign flip step. That is, we randomly select $k\ (k \ll \min(m, d-m))$ coordinates and flip the corresponding signs in δ, resulting in a new vector δ_s. Then, we have

$$P(r(\delta_s, \delta_{adv}) > r) = \frac{\sum_{i=\lceil \frac{k+1}{2} \rceil}^{k} \binom{m}{i} \cdot \binom{d-m}{k-i}}{\binom{d}{k}}, \tag{7}$$

where $\binom{\cdot}{\cdot}$ is the binomial coefficient. It can be easily proven that $P(r(\delta_s, \delta_{adv}) > r)$ is smaller than $P(r(\delta_s, \delta_{adv}) < r)$ when r is larger than 0. Thus, it is no surprise that lots of δ_s are rejected during the optimization as mentioned in Fig. 2. Even though, our method performs much better than existing methods, see Sect. 4 for detailed information.

The complete Sign Flip Attack (SFA) is summarized in Algorithm 1. SFA works in an iterative manner. Given a correctly classified image, we first initialize the perturbation which satisfies the adversarial constraint, then push the initial adversarial example close to the original image through a binary search. In each iteration, there are two steps requiring a total of 2 queries. One is the project step described in Eq. 5, which reduces the l_∞ norm of the current perturbation. The other is the random sign flip step described in Eq. 6, which has a relatively high probability to generate a better adversarial perturbation. We clip adversarial examples to a legitimate range before querying the target model, and reject unqualified perturbations. Next, we will discuss several techniques to improve query efficiency.

Dimensionality Reduction. In general, attacking deep neural networks with high-dimensional inputs requires hundreds of thousands of queries under the black-box setting. Several works [2,11] have demonstrated that performing dimensionality reduction can boost attack efficiency. We adopt this strategy into our method. To be specific, we define p and η in the lower dimensionality, i.e., $p \in (0,1)^{d'}$, $\eta \in \{-\alpha, \alpha\}^{d'}$, $d' < d$, and choose bilinear interpolation $T : \mathbb{R}^{d'} \to \mathbb{R}^d$ as the upscaling function. Then,

$$\eta \leftarrow T(\eta), \quad s_i \sim \mathbf{Bernoulli}(p_i), \quad s \leftarrow \mathbf{sgn}(T(s)). \tag{8}$$

Algorithm 1. Sign Flip Attack

Input: indicator function ϕ, original image $\boldsymbol{x}$, threshold ϵ

1: Initialize $\boldsymbol{\delta} \in \mathbb{R}^d$, $\alpha \in \mathbb{R}_+$, $\boldsymbol{p} \in (0,1)^d$;
2: $\boldsymbol{\delta} \leftarrow \textbf{BinarySearch}(\boldsymbol{x}, \boldsymbol{\delta}, \phi)$
3: $\epsilon' = \|\boldsymbol{\delta}\|_\infty$
4: **while** $\epsilon' > \epsilon$ **do**
5:     $\boldsymbol{\eta} \sim \{-\alpha, \alpha\}^d$
6:     $\boldsymbol{\delta}_p = \Pi_{B_\infty(0,\epsilon'-\alpha)}(\boldsymbol{\delta} + \boldsymbol{\eta})$
7:     **if** $\phi(\boldsymbol{x} + \boldsymbol{\delta}_p) = 1$ **then**
8:         $\boldsymbol{\delta} \leftarrow \boldsymbol{\delta}_p$
9:     **end if**
10:    $s_i \sim \textbf{Bernoulli}(p_i)$
11:    $\boldsymbol{\delta}_s = \boldsymbol{\delta} \odot (1 - 2\boldsymbol{s})$
12:    **if** $\phi(\boldsymbol{x} + \boldsymbol{\delta}_s) = 1$ **then**
13:        $\boldsymbol{\delta} \leftarrow \boldsymbol{\delta}_s$
14:    **end if**
15:    Adjust $\boldsymbol{p}$, α
16:    $\epsilon' = \|\boldsymbol{\delta}\|_\infty$
17: **end while**
18: **return** $\boldsymbol{x} + \boldsymbol{\delta}$

With a suitable d', we can obtain higher attack success rates under limited queries. Note, our method works well even without dimensionality reduction.

Hyperparameter Adjustment. Our method has two hyperparameters α and $\boldsymbol{p}$ corresponding to two steps in each iteration. We dynamically adjust these two hyperparameters according to the success rate of several previous trails. For the project step, if the success rate is higher than 70%, we increase α by multiplying a fixed coefficient (*e.g.*, 1.5). If the success rate is lower than 30%, we reduce α by dividing the same coefficient. For the random sign flip step, if the success rate is higher than 70%, we increase $\boldsymbol{p}$ by adding a fixed increment (*e.g.*, 0.001). If the success rate is lower than 30%, we reduce $\boldsymbol{p}$ by subtracting the same increment. In practice, we adjust α and $\boldsymbol{p}$ once every 10 iterations. Besides, we also adjust p after each successful random sign flip step by $\boldsymbol{p} \leftarrow \boldsymbol{p} + 0.0001(1 - 2\boldsymbol{s})$. In this way, we roughly adjust each entry in $\boldsymbol{p}$. We set $\alpha = 0.004$ (around one pixel $1/255$), and set the initial flip probability for each coordinate as $p_i = 0.001$. We bound $\boldsymbol{p} \in [0.0001, 0.01]^d$ in each step, as we only want to flip a rather small number of entries.

4   Experiments

We compare Sign Flip Attack (SFA) with a comprehensive list of decision-based black-box attacks: Boundary Attack (BA) [7], Label-Only Attack (LO) [27], Hop-SkipJumpAttack (HSJA) [10], Evolutionary Attack (EA) [17] and Sign-OPT Attack (Sign-OPT) [13]. We use the l_∞ version of each baseline method if it exists. For BA and EA, we use their original versions. All experiments are conducted on two standard datasets, CIFAR-10 [29] and ImageNet [15].

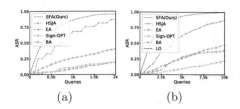

(a)   (b)

Fig. 6. Comparison of the attack success rates over the number of queries of various methods on **CIFAR-10** with $\epsilon = 0.031$. (a) untargeted attacks. (b) targeted attacks.

Table 1. Results on **CIFAR-10**. The query limits for untargeted and targeted attacks are 2,000 and 10,000, respectively. For Lable-Only Attack (LO), we only consider targeted attacks, since it is mainly designed for this setting. **SFA wo/SF** indicates SFA without Random Sign Flip.

Method	Untargeted			Targeted		
	ASR	AQ	MQ	ASR	AQ	MQ
BA [7]	40.2%	809	660	47.5%	4,629	4,338
LO [27]	–	–	–	0.2%	3,533	3,533
Sign-OPT [13]	19.7%	886	730	21.5%	5,515	5,438
EA [17]	19.4%	1,037	1,076	38.0%	4,139	3,611
HSJA [10]	87.1%	680	503	86.9%	3,731	3,197
SFA wo/SF	9.1%	–	–	0.2%	–	–
SFA (Ours)	**95.4%**	**409**	**282**	**99.4%**	**1,807**	**1,246**

The maximum l_∞ distortion and limited query numbers will be specified in each part. For our method, we use the same hyperparameters (described in Sect. 3.3) across all images. For baseline methods, we use their default settings. We quantify the performance in terms of three dimensions: attack success rate (ASR), average queries (AQ) and median queries (MQ). Average queries and median queries are calculated over successful trails. For some experiments, we also provide the results achieved by strong white-box attacks, *e.g.*, 100-step PGD. Additional results are provided in the supplementary material.

4.1   Attacks on Undefended Models

CIFAR-10 Setup. We use the first 1000 images of the validation set. 953 of them are correctly classified by our trained ResNet-18 [24]. For untargeted attacks, we set the maximum queries to 2,000. For targeted attacks, we set the maximum queries to 10,000. Following the protocol in [7], the target label is set to $t = (y + 1) \bmod 10$ for an image with label y. To ensure a fair comparison, we use the same initial perturbation for all methods in targeted attacks. As a convention, We bound the maximum l_∞ distortion to $\epsilon = 0.031(8/255)$.

ImageNet Setup. To verify the robustness of attack methods against different network architectures, we consider four prevailing model, ResNet-50 [24], VGG-16 [41], DenseNet-121 [25], and Inception-v3 [43]. For untargeted attacks, we randomly select 1,000 images and set the maximum queries to 20,000. For targeted attacks, we randomly select 200 images due to time concerns and set the maximum queries to 100,000. The target label is randomly chosen across 1,000 classes. Again, the same initial perturbation is applied for all methods for targeted attacks. We bound the maximum l_∞ distortion to $\epsilon = 0.031$. For our method, we apply dimensionality reduction described in Sect. 3.3. We simply set $d' = d/4$, *e.g.*, $d = 224 \times 224 \times 3$, $d' = 112 \times 112 \times 3$.

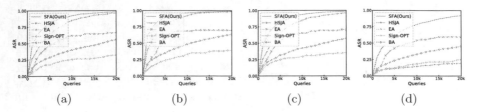

Fig. 7. Untargeted attack success rates versus numbers of queries on **ImageNet** with four different model architectures. (a) ResNet-50. (b) VGG-16. (c) DenseNet-121. (d)Inception-v3.

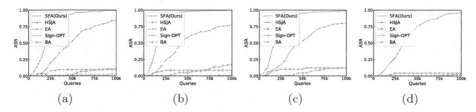

Fig. 8. Targeted attack success rates versus numbers of queries on **ImageNet** with four different model architectures. (a) ResNet-50. (b) VGG-16. (c) DenseNet-121. (d)Inception-v3.

We report the results for CIFAR-10 in Table 1. Untargeted and targeted attack results for ImageNet are shown in Table 2 and Table 3. The corresponding cumulative attack success rates of the number of queries are presented in Fig. 6, 7 and 8, respectively. Compared with existing methods, our method consistently achieves higher attack success rates across different datasets and various network architectures for both untargeted and targeted settings while requiring half or even fewer queries. In Table 3, we notice that the average and median queries of EA are better than ours. This is due to the average and median queries can be influenced by the attack success rates. As an instance, with Inception-v3 as the target model, our method achieves a 95.8% attack success rate, 91.5% higher than EA. On the 4.3% of images that EA attacks successfully, the average and median queries of our method are 6,657 and 5,604, respectively. Our method reduces more than 60% of queries compared to EA. A detailed comparison with EA and HSJA is provided in the supplementary material.

The Importance of Random Sign Flip. We study the effect of the random sign flip step. In Table 1, it can be seen that the random sign flip step boosts the attack success rate. With the random sign flip step, the success rates for untargeted and targeted attacks are 95.4% and 99.4%. However, without it, the attack success rates drop to 9.1% and 0.2%, even lower than random sampling.

Dimensionality Reduction Helps Attack Efficiency. We also study the effect of dimensionality reduction. We provide the results without dimensionality reduction in Table 2 and Table 3. The results show that, with the help of dimensionality reduction, it is able to achieve a higher attack success rate with

Table 2. Results of **untargeted** attacks on **ImageNet**. The maximum number of queries sets to 20,000. **SFA wo/DR** indicates SFA without dimensionality reduction.

Method	ResNet-50 [24]			VGG-16 [41]			DenseNet-121 [25]			Inception-v3 [43]		
	ASR	AQ	MQ	ASR	AQ	MQ	ASR	AQ	MQ	ASR	AQ	MQ
BA [7]	55.5%	6,547	4,764	63.7%	5,848	3,906	57.4%	6,285	4,170	45.3%	6,404	4,830
Sign-OPT [13]	31.8%	5,929	3,448	39.0%	4,984	3,624	35.7%	4,669	2,432	25.0%	6,548	5,153
EA [17]	65.7%	4,004	2,578	70.0%	3,016	2,886	70.6%	37,13	2,752	59.4%	4,247	2,836
HSJA [10]	96.1%	4,370	2,883	98.3%	3,044	1,554	96.3%	4,394	2,883	19.4%	5,401	2,538
SFA wo/DR	96.9%	3,193	1,570	**99.0%**	2,112	820	97.5%	2,972	1,652	91.1%	**3,759**	**2,134**
SFA(Ours)	**98.2%**	**2,712**	**1,288**	98.7%	**1,754**	**636**	**98.6%**	**2,613**	**1,200**	**92.1%**	4,501	2,602

Table 3. Results of **targeted** attacks on **ImageNet**. The maximum number of queries sets to 100,000. **SFA wo/DR** indicates SFA without dimensionality reduction.

Method	ResNet-50 [24]			VGG-16 [41]			DenseNet-121 [25]			Inception-v3 [43]		
	ASR	AQ	MQ	ASR	AQ	MQ	ASR	AQ	MQ	ASR	AQ	MQ
BA [7]	11.4%	62,358	57,336	17.1%	62,480	67,658	12.8%	58,879	56,646	3.4%	84,266	85,202
Sign-OPT [13]	2.7%	54,523	51,319	1.9%	91,172	82,492	0.0%	–	–	0.0%	–	–
EA [17]	9.6%	19,682	20,435	8.5%	12,126	8,534	12.2%	17,820	7,195	4.3%	18,164	15,362
HSJA [10]	84.5%	44,188	41,205	77.8%	39,400	36,172	79.7%	41,319	36,964	0.0%	–	–
SFA wo/DR	98.6%	29,440	26,208	98.5%	23,216	21,076	98.6%	28,151	25,824	**97.0%**	37,169	38,056
SFA(Ours)	**99.3%**	22,538	19,380	**99.2%**	16,627	15,008	**98.6%**	20,331	17,762	95.8%	36,681	32,210

fewer queries. Note that the results achieved by our method without dimensionality reduction are still much better than existing methods.

4.2 Attacks on Defensive Models

To investigate the effectiveness of decision-based black-box attacks against defenses, we conduct experiments on 7 defense mechanisms: Adversarial Training [35], Thermometer Encoding [8], Bit Depth Reduction [50], FeatDenoise [49], FeatScatter [52], KWTA [47] and TRADES [53]. For the first 3 defense mechanisms, we compare our method with BA [7], Sign-OPT [13], EA [17] and HSJA [10]. For other defense mechanisms, we only compare with the existing state-of-the-art method HSJA [10].

The overall results are presented in Table 4 and Table 5. Because the average and median queries can be affected by the attack success rate for an effective method, we also report the results achieved by our methods on the images that each baseline method successfully fools, please see the values in brackets. Our method significantly outperforms the existing decision-based attacks by large margins on nearly all evaluation metrics. In what follows, we will make a detailed discussion.

Firstly, our method performs much more stable than state-of-the-art decision-based attacks when confronting defenses that cause obfuscated gradients [3]. Thermometer Encoding [8] and Bit Depth Reduction [50] are two types of defenses that leverage per pixel quantization to resistant adversarial examples.

Table 4. Attack performance against Adversarial Training [35], Thermometer Encoding [8], and Bit Depth Reduction [50]. The limited query budgets set to 100,000, 50,000 and 50,000, respectively. The values in brackets denote results achieved by our method on the images that baseline methods successfully fool.

| Method | CIFAR-10 [29] | | | | | | ImageNet [15] | | |
| | Adv. Training [35] | | | Thermometer [8] | | | Bit Depth [50] | | |
	ASR	AQ	MQ	ASR	AQ	MQ	ASR	AQ	MQ
BA [7]	5.6%	1,811(1,294)	880(242)	11.0%	286(208)	134(122)	13.5%	165(173)	142(90)
Sign-OPT [13]	6.3%	5,908(324)	2,505(202)	8.0%	3,583(262)	1,285(122)	15.1%	2,284(154)	179(90)
EA [17]	12.1%	7,675(4,442)	1,594(1,462)	34.1%	5,254(752)	4,459(310)	75.3%	9,689(3,245)	5,626(858)
HSJA [10]	34.0%	14,829(7,694)	4,450(2,972)	13.1%	5,030(200)	310(132)	8.4%	6,664(243)	159(76)
SFA(Ours)	41.6%	15,512	5,486	92.3%	7,024	3,386	91.2%	7,227	2,100
BPDA [3] (white-box)	50.9%	–	–	100%	–	–	100%	–	–

Table 5. Attack performance against TRADES [53], FeatScatter [52], KWTA [47] and FeatDenoise [49]. The limited query budgets set to 100,000. The values in brackets denote results achieved by our method on the images that HSJA successfully fools.

| Method | | PGD(white-box) | SFA(Ours) | | | HSJA [10] | | |
		ASR	ASR	AQ	MQ	ASR	AQ	MQ
CIFAR-10 [29]	TRADES [53]	**34.0%**	29.8%	7,714	2,470	25.2%	13,115(4,372)	3,569(1,844)
	FeatScatter [52]	23.2%	**52.3%**	10,254	3,956	42.0%	14,393(4,441)	5,222(2,318)
	KWTA+AT [47]	33.6%	**74.7%**	16,935	3,660	35.4%	9,953(2,568)	2,187(688)
ImageNet [15]	FeatDenoise [49]	**88.0%**	51.3%	18,616	7,848	44.0%	23,866(13,620)	13,151(5,478)

Although these two defenses have been completely broken by current white-box attacks, it is still difficult to defeat them for decision-based attacks. For untargeted attacks on Thermometer Encoding, our method achieves a 92.3% attack success rate, while the highest achieved among existing methods is a mere 34.1%.

Secondly, our method can serve as a strong baseline to evaluate the robustness of defensive models. FeatScatter [52] and KWTA [47] are two recently published defenses. Both of them have shown superior resistance against white-box attacks than Adversarial Training [35] and done certain sanity checks such as transfer-based black-box attacks. However, according to our results, they reduce the performance of the original Adversarial Training. As presented in Table 5, our method has 52.3% and 74.7% attack success rates, around 29% and 41% higher than PGD, respectively. Their model accuracies are actually lower than ones in Adversarial Training.

Thirdly, for those defensive models which indeed increase the robustness, our method obtains better results than other decision-based attacks while still falls behind of white-box attacks. On CIFAR-10, for Adversarial Training [35] and TRADES [53], our method is slightly worse than PGD. Whereas, on ImageNet, the gap between our method and PGD is quite large. PGD achieves an 88.0% attack success rate against FeatDenoise [49], around 31% higher than our method. Attacking models with high dimensional inputs is arduous for *all* decision-based attacks. Our method takes a steady step towards closing the gap.

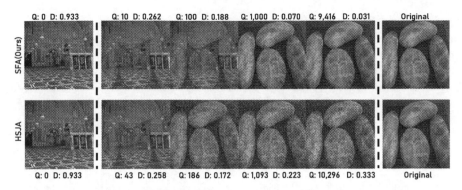

Fig. 9. An example of attacking the food API in Tencent AI Open Platform. **Q** denotes the query number. **D** denotes the l_∞ distance towards the original image. Green means the API classifies the image as food, red is otherwise. Best viewed in color with zoom in. (Color figure online)

4.3 Attacks on Real-World Applications

In this section, we investigate the applicability of decision-based attacks on real-world systems.

The target models are the face verification[3] API and food[4] API in Tencent AI Open Platform. For the face verification API, we set the similarity score threshold to 70. If the output is larger than 70, the decision is True—the two images are from the same identity, otherwise False. We choose 10 pairs of images from the Labeled Face in the Wild [26] dataset. We bound the maximum l_∞ distortion to $\epsilon = 0.062(16/255)$ and set the maximum query number to 5,000. The numbers of successfully attacked pairs of BA, Sign-OPT, EA, LO, HSJA, and our method are 1, 0, 5, 0, 2 and 9, respectively. The food API takes a single image as input and determines whether the input is about food. Our method successfully invades this API. We present an example in Fig. 9. The original and initial images are chosen from ImageNet.

5 Conclusion

In this paper, we proposed the Sign Flip Attack, a simple and highly efficient decision-based black-box l_∞ adversarial attack to craft adversarial examples. We introduced the novel random sign flip step to search for a better adversarial perturbation during the optimization, boosting the attack success rate and query efficiency. Comprehensive studies on CIFAR-10 and ImageNet demonstrate that our method has significantly outperformed existing methods. Experiments on several defensive models indicate the effectiveness of our method in evaluating

---

[3] https://ai.qq.com/product/face.shtml#compare.

[4] https://ai.qq.com/product/visionimgidy.shtml#food.

the robustness. Additionally, we applied our method to attack real-world applications successfully. These promising results suggest that our method can be viewed as a strong baseline to facilitate future research.

Acknowledgement. This work was supported in part by the Major Project for New Generation of AI under Grant No. 2018AAA0100400, the National Natural Science Foundation of China (No. 61836014, No. 61761146004, No. 61773375).

References

1. Al-Dujaili, A., O'Reilly, U.M.: Sign bits are all you need for black-box attacks. In: Proceedings of International Conference on Learning Representations (2020)
2. Alzantot, M., Sharma, Y., Chakraborty, S., Srivastava, M.: Genattack: practical black-box attacks with gradient-free optimization. arXiv preprint arXiv:1805.11090 (2018)
3. Athalye, A., Carlini, N., Wagner, D.A.: Obfuscated gradients give a false sense of security: circumventing defenses to adversarial examples. In: Proceedings of International Conference on Machine Learning (2018)
4. Athalye, A., Engstrom, L., Ilyas, A., Kwok, K.: Synthesizing robust adversarial examples. arXiv preprint arXiv:1707.07397 (2017)
5. Bhagoji, A.N., He, W., Li, B., Song, D.: Practical black-box attacks on deep neural networks using efficient query mechanisms. In: Ferrari, V., Hebert, M., Sminchisescu, C., Weiss, Y. (eds.) ECCV 2018. LNCS, vol. 11216, pp. 158–174. Springer, Cham (2018). https://doi.org/10.1007/978-3-030-01258-8_10
6. Biggio, B., et al.: Evasion attacks against machine learning at test time. In: Blockeel, H., Kersting, K., Nijssen, S., Železný, F. (eds.) ECML PKDD 2013. LNCS (LNAI), vol. 8190, pp. 387–402. Springer, Heidelberg (2013). https://doi.org/10.1007/978-3-642-40994-3_25
7. Brendel, W., Rauber, J., Bethge, M.: Decision-based adversarial attacks: reliable attacks against black-box machine learning models. In: Proceedings of International Conference on Learning Representations (2018)
8. Buckman, J., Roy, A., Raffel, C., Goodfellow, I.: Thermometer encoding: one hot way to resist adversarial examples. In: Proceedings of International Conference on Learning Representations (2018)
9. Carlini, N., Wagner, D.: Towards evaluating the robustness of neural networks. In: Proceedings of the IEEE Symposium on Security and Privacy (SP), pp. 39–57. IEEE (2017)
10. Chen, J., Jordan, M.I., Wainwright, M.: Hopskipjumpattack: a query-efficient decision-based attack. arXiv preprint arXiv:1904.02144 (2019)
11. Chen, P.Y., Zhang, H., Sharma, Y., Yi, J., Hsieh, C.J.: Zoo: zeroth order optimization based black-box attacks to deep neural networks without training substitute models. In: Proceedings of the 10th ACM Workshop on Artificial Intelligence and Security, pp. 15–26. ACM (2017)
12. Cheng, M., Le, T., Chen, P.Y., Yi, J., Zhang, H., Hsieh, C.J.: Query-efficient hard-label black-box attack: an optimization-based approach. In: Proceedings of International Conference on Learning Representations (2018)
13. Cheng, M., Singh, S., Chen, P.Y., Liu, S., Hsieh, C.J.: Sign-opt: a query-efficient hard-label adversarial attack. In: Proceedings of International Conference on Learning Representations (2020)

14. Cheng, S., Dong, Y., Pang, T., Su, H., Zhu, J.: Improving black-box adversarial attacks with a transfer-based prior. In: Advances in Neural Information Processing Systems, pp. 10934–10944 (2019)
15. Deng, J., Dong, W., Socher, R., Li, L.J., Li, K., Fei-Fei, L.: Imagenet: a large-scale hierarchical image database. In: Proceedings of the IEEE Conference on Computer Vision and Pattern Recognition, pp. 248–255 (2009)
16. Dong, Y., et al.: Boosting adversarial attacks with momentum. In: Proceedings of the IEEE Conference on Computer Vision and Pattern Recognition, pp. 9185–9193 (2018)
17. Dong, Y., et al.: Efficient decision-based black-box adversarial attacks on face recognition. In: Proceedings of the IEEE Conference on Computer Vision and Pattern Recognition, pp. 7714–7722 (2019)
18. Eykholt, K., et al.: Robust physical-world attacks on deep learning models. arXiv preprint arXiv:1707.08945 (2017)
19. Fan, Y., et al.: Sparse adversarial attack via perturbation factorization. In: Proceedings of European Conference on Computer Vision (2020)
20. Feng, Y., Wu, B., Fan, Y., Li, Z., Xia, S.: Efficient black-box adversarial attack guided by the distribution of adversarial perturbations. arXiv preprint arXiv:2006.08538 (2020)
21. Goodfellow, I.J., Shlens, J., Szegedy, C.: Explaining and harnessing adversarial examples. In: Proceedings of International Conference on Learning Representations (2014)
22. Guo, C., Gardner, J.R., You, Y., Wilson, A.G., Weinberger, K.Q.: Simple black-box adversarial attacks. arXiv preprint arXiv:1905.07121 (2019)
23. Guo, Y., Yan, Z., Zhang, C.: Subspace attack: Exploiting promising subspaces for query-efficient black-box attacks. In: Advances in Neural Information Processing Systems, pp. 3825–3834 (2019)
24. He, K., Zhang, X., Ren, S., Sun, J.: Deep residual learning for image recognition. In: Proceedings of the IEEE Conference on Computer Vision and Pattern Recognition, pp. 770–778 (2016)
25. Huang, G., Liu, Z., van der Maaten, L., Weinberger, K.Q.: Densely connected convolutional networks. In: Proceedings of the IEEE Conference on Computer Vision and Pattern Recognition, pp. 2261–2269 (2017)
26. Huang, G., Mattar, M., Berg, T., Learned-Miller, E.: Labeled faces in the wild: a database for studying face recognition in unconstrained environments. Technical report, October 2008
27. Ilyas, A., Engstrom, L., Athalye, A., Lin, J.: Black-box adversarial attacks with limited queries and information. In: Proceedings of International Conference on Machine Learning (2018)
28. Ilyas, A., Engstrom, L., Madry, A.: Prior convictions: black-box adversarial attacks with bandits and priors. In: Proceedings of International Conference on Learning Representations (2019)
29. Krizhevsky, A., et al.: Learning multiple layers of features from tiny images. Technical Report (2009)
30. Kurakin, A., Goodfellow, I., Bengio, S.: Adversarial examples in the physical world. In: Proceedings of International Conference on Learning Representations (2016)
31. Li, Y., Li, L., Wang, L., Zhang, T., Gong, B.: Nattack: learning the distributions of adversarial examples for an improved black-box attack on deep neural networks. In: Proceedings of International Conference on Machine Learning (2019)
32. Li, Y., et al.: Toward adversarial robustness via semi-supervised robust training. arXiv preprint arXiv:2003.06974 (2020)

33. Li, Y., Yang, X., Wu, B., Lyu, S.: Hiding faces in plain sight: disrupting AI face synthesis with adversarial perturbations. arXiv preprint arXiv:1906.09288 (2019)
34. Liu, Y., Chen, X., Liu, C., Song, D.: Delving into transferable adversarial examples and black-box attacks. In: Proceedings of International Conference on Learning Representations (2016)
35. Madry, A., Makelov, A., Schmidt, L., Tsipras, D., Vladu, A.: Towards deep learning models resistant to adversarial attacks. In: Proceedings of International Conference on Learning Representations (2017)
36. Moon, S., An, G., Song, H.O.: Parsimonious black-box adversarial attacks via efficient combinatorial optimization. In: Proceedings of International Conference on Machine Learning (2019)
37. Moosavi-Dezfooli, S.M., Fawzi, A., Frossard, P.: Deepfool: a simple and accurate method to fool deep neural networks. In: Proceedings of the IEEE Conference on Computer Vision and Pattern Recognition, pp. 2574–2582 (2016)
38. Papernot, N., McDaniel, P., Goodfellow, I.: Transferability in machine learning: from phenomena to black-box attacks using adversarial samples. arXiv preprint arXiv:1605.07277 (2016)
39. Papernot, N., McDaniel, P., Goodfellow, I., Jha, S., Celik, Z.B., Swami, A.: Practical black-box attacks against machine learning. In: Proceedings of the 2017 ACM on Asia Conference on Computer and Communications Security, pp. 506–519. ACM (2017)
40. Papernot, N., McDaniel, P., Jha, S., Fredrikson, M., Celik, Z.B., Swami, A.: The limitations of deep learning in adversarial settings. In: Proceedings of the IEEE European Symposium on Security and Privacy (EuroS&P), pp. 372–387. IEEE (2016)
41. Simonyan, K., Zisserman, A.: Very deep convolutional networks for large-scale image recognition. arXiv preprint arXiv:1409.1556 (2014)
42. Suya, F., Chi, J., Evans, D., Tian, Y.: Hybrid batch attacks: finding black-box adversarial examples with limited queries. In: USENIX Security Symposium (2020)
43. Szegedy, C., Vanhoucke, V., Ioffe, S., Shlens, J., Wojna, Z.: Rethinking the inception architecture for computer vision. In: Proceedings of the IEEE Conference on Computer Vision and Pattern Recognition, pp. 2818–2826 (2016)
44. Szegedy, C., et al.: Intriguing properties of neural networks. In: Proceedings of International Conference on Learning Representations (2013)
45. Tramèr, F., Kurakin, A., Papernot, N., Goodfellow, I., Boneh, D., McDaniel, P.: Ensemble adversarial training: attacks and defenses. In: Proceedings of International Conference on Learning Representations (2018)
46. Wu, D., Wang, Y., Xia, S.T., Bailey, J., Ma, X.: Skip connections matter: on the transferability of adversarial examples generated with resnets. In: Proceedings of International Conference on Learning Representations (2020)
47. Xiao, C., Zhong, P., Zheng, C.: Resisting adversarial attacks by k-winners-take-all. In: Proceedings of International Conference on Learning Representations (2020)
48. Xie, C., Wang, J., Zhang, Z., Zhou, Y., Xie, L., Yuille, A.: Adversarial examples for semantic segmentation and object detection. In: Proceedings of the IEEE International Conference on Computer Vision, pp. 1369–1378 (2017)
49. Xie, C., Wu, Y., Maaten, L.V.D., Yuille, A.L., He, K.: Feature denoising for improving adversarial robustness. In: Proceedings of the IEEE Conference on Computer Vision and Pattern Recognition (2019)
50. Xu, W., Evans, D., Qi, Y.: Feature squeezing: detecting adversarial examples in deep neural networks. In: Proceedings of Network and Distributed System Security Symposium. Internet Society (2018)

51. Xu, Y., et al.: Exact adversarial attack to image captioning via structured output learning with latent variables. In: Proceedings of the IEEE Conference on Computer Vision and Pattern Recognition, pp. 4135–4144 (2019)
52. Zhang, H., Wang, J.: Defense against adversarial attacks using feature scattering-based adversarial training. In: Advances in Neural Information Processing Systems (2019)
53. Zhang, H., Yu, Y., Jiao, J., Xing, E.P., Ghaoui, L.E., Jordan, M.I.: Theoretically principled trade-off between robustness and accuracy. In: Proceedings of International Conference on Machine Learning (2019)

Knowledge Transfer via Dense Cross-Layer Mutual-Distillation

Anbang Yao$^{(\boxtimes)}$ and Dawei Sun

Intel Labs, Beijing, China
{anbang.yao,dawei.sun}@intel.com

Abstract. Knowledge Distillation (KD) based methods adopt the one-way Knowledge Transfer (KT) scheme in which training a lower-capacity student network is guided by a pre-trained high-capacity teacher network. Recently, Deep Mutual Learning (DML) presented a two-way KT strategy, showing that the student network can be also helpful to improve the teacher network. In this paper, we propose Dense Cross-layer Mutual-distillation (DCM), an improved two-way KT method in which the teacher and student networks are trained collaboratively from scratch. To augment knowledge representation learning, well-designed auxiliary classifiers are added to certain hidden layers of both teacher and student networks. To boost KT performance, we introduce dense bidirectional KD operations between the layers appended with classifiers. After training, all auxiliary classifiers are discarded, and thus there are no extra parameters introduced to final models. We test our method on a variety of KT tasks, showing its superiorities over related methods. Code is available at https://github.com/sundw2014/DCM.

Keywords: Knowledge Distillation · Deep supervision · Convolutional Neural Network · Image classification

1 Introduction

In recent years, deep Convolutional Neural Networks (CNNs) have achieved remarkable success in many computer vision tasks such as image classification [24], object detection [9] and semantic segmentation [32]. However, along with the rapid advances on CNN architecture design, top-performing models [13,17,19,40,44,48,52] also pose intensive memory, compute and power costs, which limits their use in real applications, especially on resource-constrained devices.

A. Yao and D. Sun—Equal contribution.
Experiments were mostly done by Dawei Sun when he was an intern at Intel Labs China, supervised by Anbang Yao.

Electronic supplementary material The online version of this chapter (https://doi.org/10.1007/978-3-030-58555-6_18) contains supplementary material, which is available to authorized users.

© Springer Nature Switzerland AG 2020
A. Vedaldi et al. (Eds.): ECCV 2020, LNCS 12360, pp. 294–311, 2020.
https://doi.org/10.1007/978-3-030-58555-6_18

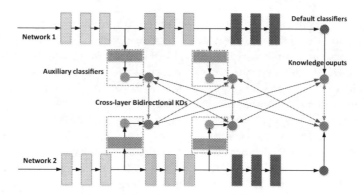

Fig. 1. Structure overview of the proposed method. For illustration, auxiliary classifiers are added to two hidden layers of each network, which are removed after training. Green/red arrows denote bidirectional knowledge distillation operations between the same-staged/different-staged layers of two networks. Best viewed in color.

To address this dilemma, Knowledge Transfer (KT) attracts great attentions among existing research efforts [43]. KT is typically treated as a problem of transferring learnt information from one neural network model to another. The first attempt of using KT to cope with model compression was made in [5] where Bucilă et al. used an ensemble of neural networks trained on a small annotated dataset to label a much larger unlabelled dataset. By doing this, a smaller and faster neural network can be trained using many more labeled samples, and thus the final model can have much better performance than that is trained solely on the original small dataset. [2] extended this idea to train a shallower and wider neural network. Hinton et al. advanced knowledge transfer research by introducing the well-known Knowledge Distillation (KD) [14] method adopting a teacher-student framework. In KD, a lower-capacity student network is enforced to mimic the probabilistic outputs by a pre-trained high-capacity teacher network as well as the one-hot ground-truth labels. Naturally, the teacher can also be an ensemble of multiple models. Since then, numerous KD variants [21,28,37,42,45,51,53] have been proposed, mostly focusing on using either feature maps or attention maps at the intermediate layers of the teacher network as the extra hints for improving KD designs. Following [14], these methods adopted the same teacher-student framework in which the teacher network is trained beforehand and fixed under the assumption that it always learns a better representation than the student network. Consequently, they all used the one-way KT strategy, where knowledge can only be transferred from a teacher network to a student network. Recently, Deep Mutual Learning (DML) [57] was proposed, which achieved superior performance by a powerful two-way KT design, showing that the probabilistic outputs from the last layer of both teacher and student networks can be beneficial to each other.

In this paper, we restrict our focus to advance two-way KT research in the perspective of promoting knowledge representation learning and transfer design.

Dense Cross-layer Mutual-distillation (DCM), an improved two-way KT method which is capable of collaboratively training the teacher and student networks from scratch, is the main contribution of this paper. Figure 1 shows the structure overview of DCM. Following the deep supervision methodology [18,27,42,44], we first add well-designed auxiliary classifiers to certain hidden layers of both teacher and student networks, allowing DCM to capture probabilistic predications not only from the last layer but also from the hidden layers of each network. To the best of our knowledge, deep supervision design is overlooked in the knowledge transfer field. Furthermore, we present dense cross-layer bidirectional KD operations to regularize the joint training of the teacher and student networks. On the one hand, knowledge is mutually transferred between the same-staged supervised layers. On the other hand, we find the bidirectional KD operations between the different-staged supervised layers can further improve KT performance, thanks to the well-designed auxiliary classifiers which alleviate semantic gaps of the knowledge learnt at different-staged layers of two networks. Note that there are no extra parameters added to final models as all auxiliary classifiers are discarded after training. Experiments are performed on image classification datasets with a variety of KT settings. Results show that our method outperforms related methods by noticeable margins, validating the importance of connecting knowledge representation learning with bidirectional KT design.

2 Related Work

In this section, we briefly summarize existing works related to our method.

Knowledge Distillation Applications. Although KD based methods were primarily proposed for model compression [14,21,26,28,37,45,51,53], there have been many attempts to extend them to other tasks recently. Two representative examples are lifelong learning and multi-modal visual recognition. In lifelong learning task, the combination of KD and other techniques such as fine-tuning and retrospection was applied, targeting to adapt a pre-trained model to new tasks while preserving the knowledge gained on old tasks [15,30,54]. When designing and training multiple-stream networks dedicated to action recognition [8], person re-identification [11], depth estimation and scene parsing [10,50], cross-modal distillation was used to facilitate the knowledge transfer between the network branches trained on different sources of data such as RGB and depth. Other KD extensions include but are not limited to efficient network design [6,47], style transfer [29], machine translation [1,22] and multi-task learning [25]. Our method differs from these approaches in task and formulation.

Co-training. Blum and Mitchell proposed a pioneering co-training framework [4] in which two models were trained separately on each of two views of labeled data first, and then more unlabelled samples as well as the predictions by each trained model were used to enlarge training data size. Recently, several deep co-training schemes have been proposed, mostly following the semi-supervised learning paradigm. [36] extended the idea of [4] via presenting a deep adversarial co-training method that uses adversarial samples to prevent multiple neural networks trained on different views from collapsing into each other. [3] proposed a cooperative learning mechanism in which two agents handling the same visual recognition task can transfer their current knowledge learnt on different sources of data to each other. [12] addressed multi-task machine translation problem with a dual co-training mechanism. [41] considered the co-training of several classifier heads of the same network. Unlike these methods, we aim to improve the two-way knowledge transfer design for supervised image classification task.

Deep Supervision. The basic idea of deep supervision is to add extra classifiers to the hidden layers of a deep CNN architecture, which will be removed after training. It was originally proposed in [27,44] to combat convergence issue when designing and training deep CNNs for image recognition task. Since then, it has been widely used in training deep CNN architectures specially designed to handle other visual recognition tasks such as edge detection [49], object detection [20, 31], semantic segmentation [58,59], human pose estimation [7,33] and anytime recognition [18,35]. In this paper, we extend the idea of deep supervision to promote the two-way knowledge transfer research.

3   Proposed Method

In this section, we detail the formulation and implementation of our method.

3.1   KD and DML

We first review the formulations of Knowledge Distillation (KD) [14] and Deep Mutual Learning (DML) [57]. For simplicity, we follow the teacher-student framework, and only consider the very basic case where there are one single teacher network and one single student network. Given the training data $X = \{x_n\}_{n=1}^N$ consisting of N samples collected from M image classes, the ground-truth labels are denoted as $Y = \{y_n\}_{n=1}^N$. Let W_t be a teacher network trained beforehand and fixed, and let W_s be a student network. In KD, the student network W_s is trained by minimizing

$$L_s = L_c(W_s, X, Y) + \lambda L_{kd}(\hat{P}_t, \hat{P}_s), \tag{1}$$

where L_c is the classification loss between the predications of the student network and the one-hot ground-truth labels, L_{kd} is the distillation loss, λ is a coefficient balancing these two loss terms. In [14], L_{kd} is defined as

$$L_{kd}(\hat{P}_t, \hat{P}_s) = -\frac{1}{N} \sum_{n=1}^N \sum_{m=1}^M \hat{P}_t^m(x_n) \log \hat{P}_s^m(x_n). \tag{2}$$

Given a training sample x_n, its probability of image class m is computed as

$$\hat{P}^m(x_n) = \frac{exp(z_n^m/T)}{\sum_{m=1}^M exp(z_n^m/T)}, \tag{3}$$

where z_n^m is the output logit for image class m obtained from the last layer (i.e., default classifier) of a neural network, and T is a temperature used to soften the probabilistic outputs. The distillation loss defined by Eq. 2 can be considered as a modified cross-entropy loss using the probabilistic outputs of the pre-trained teacher network as the soft labels instead of the one-hot ground-truth labels.

Now it is clear that KD encourages the student network to match the probabilistic outputs of the pre-trained teacher model via a one-way Knowledge Transfer (KT) scheme. *Two key factors to KD based methods are: the representation of knowledge and the strategy of knowledge transfer.* DML considers the latter one by presenting a two-way KT strategy in which the probabilistic outputs from both teacher and student networks can be used to guide the training of each other. DML can be viewed as a bidirectional KD method that jointly trains the teacher and student networks via interleavingly optimizing two objectives:

$$
\begin{aligned}
L_s &= L_c(W_s, X, Y) + \lambda L_{dml}(\hat{P}_t, \hat{P}_s) \\
L_t &= L_c(W_t, X, Y) + \lambda L_{dml}(\hat{P}_s, \hat{P}_t).
\end{aligned}
\tag{4}
$$

Here, λ is set to 1 and fixed [57]. As for the definition of the distillation loss L_{dml}, instead of using Eq. 2, DML uses Kullback-Leibler divergence:

$$
L_{dml}(\hat{P}_t, \hat{P}_s) = \frac{1}{N} \sum_{n=1}^{N} \sum_{m=1}^{M} \hat{P}_t^m(x_n) \log \frac{\hat{P}_t^m(x_n)}{\hat{P}_s^m(x_n)}.
\tag{5}
$$

The $\hat{P}^m(x_n)$ is the same as that in KD. KL divergence is equivalent to cross entropy from the perspective of gradients calculation. Unlike KD containing two separate training phases, DML can jointly train the teacher and student networks in an end-to-end manner, and shows much better performance. This is attributed to the two-way KT strategy. However, the information contained in the hidden layers of networks has not been explored by DML. Moreover, the problem of connecting more effective knowledge representation learning with bidirectional KT design has also not been studied by DML.

3.2   Dense Cross-Layer Mutual-Distillation

Our DCM promotes DML via jointly addressing two issues discussed above.

Knowledge Representation Learning with Deep Supervision. Ideally, the knowledge should contain rich and complementary information learnt by a network and can be easily understood by the other network. Recall that many KD variants [21,28,37,46,51,53] have validated that feature maps or attention maps extracted at the hidden layers of a pre-trained teacher network are beneficial to improve the training of a student network under the premise of using the one-way KT scheme. Being a two-way KT method, instead of using either intermediate feature maps or attention maps extracted in an unsupervised manner as the additional knowledge, our DCM adds relevant auxiliary classifiers to certain hidden layers of both teacher and student networks, aggregating probabilistic knowledge not only from the last layer but also from the hidden layers of each network. This is also inspired by the deep supervision methodology [27,44] which is overlooked in the knowledge transfer research. As showed in our experiments and [18,35,42,58], even adding well-designed auxiliary classifiers to the hidden layers of a modern CNN can only bring marginal or no accuracy improvement. This motivates us to present a more elaborate bidirectional KT strategy.

Cross-Layer Bidirectional KD. With default and well-designed auxiliary classifiers, rich probabilistic outputs learnt at the last and hidden layers of both teacher and student networks can be aggregated on the fly during the joint training. Moreover, these probabilistic outputs are in the same semantic space, and thus our DCM introduces dense cross-layer bidirectional KD operations to promote the two-way KT process, which are illustrated in Fig. 1.

Formulation. In the following, we detail the formulation of DCM. We follow the notations in the last sub-section. Let $Q = \{(t_k, s_k)\}_{k=1}^{K}$ be a set containing K pairs of the same-staged layer indices of the teacher network W_t and the student network W_s, indicating the locations where auxiliary classifiers are added. Let (t_{K+1}, s_{K+1}) be the last layer indices of W_t and W_s, indicating the locations of default classifier. DCM simultaneously minimizes the following two objectives:

$$L_s = L_c(W_s, X, Y) + \alpha L_{ds}(W_s, X, Y) + \beta L_{dcm_1}(\hat{P}_t, \hat{P}_s) + \gamma L_{dcm_2}(\hat{P}_t, \hat{P}_s)$$
$$L_t = L_c(W_t, X, Y) + \alpha L_{ds}(W_t, X, Y) + \beta L_{dcm_1}(\hat{P}_s, \hat{P}_t) + \gamma L_{dcm_2}(\hat{P}_s, \hat{P}_t), \tag{6}$$

where L_s/L_t is the loss of the student/teacher network. In this paper, we set α, β, γ and T to 1 and keep them fixed owing to easy implementation and satisfied results (*In fact, we tried the tedious manual tuning of these parameters, but just got marginal extra gains compared to this uniform setting*). Note that the teacher and student networks have the same loss definition. For simplicity, we take L_s as the reference and detail its definition in the following description. In L_s, L_c denotes the default loss which is the same as that in KD and DML. L_{ds} denotes the total cross-entropy loss over all auxiliary classifiers added to the different-staged layers of the student network, which is computed as

$$L_{ds}(W_s, X, Y) = \sum_{k=1}^{K} L_c(W_{s_k}, X, Y). \tag{7}$$

L_{dcm_1} denotes the total loss of the same-staged bidirectional KD operations, which is defined as

$$L_{dcm_1}(\hat{P}_t, \hat{P}_s) = \sum_{k=1}^{K+1} L_{kd}(\hat{P}_{t_k}, \hat{P}_{s_k}). \tag{8}$$

L_{dcm_2} denotes the total loss of the different-staged bidirectional KD operations, which is defined as

$$L_{dcm_2}(\hat{P}_t, \hat{P}_s) = \sum_{\{(i,j)|1\leq i,j\leq K+1, i\neq j\}} L_{kd}(\hat{P}_{t_i}, \hat{P}_{s_j}). \tag{9}$$

Algorithm 1. The DCM algorithm

Input   : Training data $\{X, Y\}$, two CNN models W_t and W_s, classifier
           locations $\{(t_k, s_k)\}_{k=1}^{K+1}$, learning rate γ_i

Initialise W_t and W_s, $i = 0$;

repeat

     $i \leftarrow i + 1$, update γ_i;

     1. Randomly sample a batch of data from $\{X, Y\}$;

     2. Compute knowledge set $\{(\hat{P}_{t_k}, \hat{P}_{s_k})\}_{k=1}^{K+1}$ at all supervised layers of
     two models by Eq. 3;

     3. Compute loss L_t and L_s by Eq. 6, Eq. 7, Eq. 8, and Eq. 9 ;

     4. Calculate gradients and update parameters:
        $W_t \leftarrow W_t - \gamma_i \frac{\partial L_t}{\partial W_t}$, $W_s \leftarrow W_s - \gamma_i \frac{\partial L_s}{\partial W_s}$

---

until *Converge*;

In Eq. 8 and Eq. 9, L_{kd} is computed with the modified cross-entropy loss defined by Eq. 2. It matches the probabilistic outputs from any pair of the supervised layers in the teacher and student networks. According to the above definitions, it can be seen: bidirectional KD operations are performed not only between the same-staged supervised layers but also between the different-staged supervised layers of the teacher and student networks. *Benefiting from the well-designed auxiliary classifiers, such two types of cross-layer bidirectional KD operations are complimentary to each other as validated in the experiments.* Enabling dense cross-layer bidirectional KD operations resembles a dynamic knowledge synergy process between two networks for the same task. The training algorithm of our DCM is summarized in Algorithm 1.

Connections to DML and KD. Regardless of the selection of measure function (Eq. 2 or Eq. 5) for matching probabilistic outputs, in the case where $Q = \emptyset$ meaning the supervision is only added to the last layer of the teacher and student networks, DCM becomes DML. In the extreme case where $Q = \emptyset$ and L_t is frozen, DCM becomes KD. Therefore, DML and KD are two special cases of DCM. Like KD and DML, DCM can be easily extended to handle more complex training scenarios where there are more than two neural networks. We leave this part as future research once a distributed system is available for training.

Setting of Q. In DCM, forming cross-layer bidirectional KD pairs to be connected depends on how to set Q. Setting Q needs to consider two basic questions: (1) Where to place auxiliary classifiers? (2) How to design their structures? Modern CNNs adopt a similar hierarchical structure consisting of several stages having different numbers of building blocks, where each stage has a down-sampling layer. In light of this, *to the first question, we use a practical principle, adding auxiliary classifiers merely to down-sampling layers of a backbone network* [18,35,42,58]. Existing works [18,42] showed that simple auxiliary classifiers usually worsen the training of modern CNNs as they have no convergence issues. Inspired by them, *to the second question, we use a heuris-*

tic principle, making the paths from the input to all auxiliary classifiers have the same number of down-sampling layers as the backbone network, and using backbone's building blocks to construct auxiliary classifiers with different numbers of building blocks and convolutional filters. Finally in DCM, we enable dense two-way KDs between all layers added with auxiliary classifiers. Although the aforementioned setting may not be the best, it enjoys easy implementation and satisfied results on many CNNs as validated in our experiments.

4   Experiments

In this section, we describe the experiments conducted to evaluate the performance of DCM. We first compare DCM with DML [57] which is closely related to our method. We then provide more comprehensive comparisons for a deep analysis of DCM. All methods are implemented with PyTorch [34]. For fair comparisons, the experiments with all methods are performed under the exactly same settings for the data pre-processing method, the batch size, the number of training epochs, the learning rate schedule, and the other hyper-parameters.

4.1   Experiments on CIFAR-100

First, we perform experiments on the CIFAR-100 dataset with a variety of knowledge transfer settings.

CIFAR-100 Dataset. It contains 50000 training samples and 10000 test samples, where samples are 32×32 color images collected from 100 object classes [23]. We use the same data pre-processing method as in [13,57]. For training, images are padded with 4 pixels to both sides, and 32×32 crops are randomly sampled from the padded images and their horizontal flips, which are finally normalized with the per-channel mean and std values. For evaluation, the original-sized test images are used.

Implementation Details. We consider 4 state-of-the-art CNN architectures including: (1) ResNets [13] with depth 110/164; (2) DenseNet [19] with depth 40 and growth rate 12; (3) WRNs [52] with depth 28 and widening factor 4/10; (4) MobileNet [16] as used in [57]. We use the code released by the authors to train each CNN backbone. In the experiments, we consider two training scenarios: (1) Two CNNs with the same backbone (e.g., WRN-28-10 & WRN-28-10); (2) Two CNNs with the different backbones (e.g., WRN-28-10 & ResNet-110). In the first training scenario, for ResNets, DenseNet and WRNs, we use the same settings as reported in the original papers [13,19,52]. For MobileNet, we use the same setting as ResNets, following DML [57]. *In the second training scenario, we use the training setting of the network having better capacity to train both networks.* In our method, we append two auxiliary classifiers to the different-staged layers of each CNN backbone. Specifically, we add each auxiliary classifier after the corresponding building block having a down-sampling layer. All auxiliary classifiers have the same building blocks as in the backbone network, a global

Table 1. Result comparison on the CIFAR-100 dataset. WRN-28-10(+) denotes the models trained with dropout. Bolded results show the accuracy margins of DCM compared to DML. *In this paper, for each joint training case on the CIFAR-100 dataset, we run each method 5 times and report "mean(std)" top-1 error rates (%). Results of all methods are obtained with the exactly same training hyper-parameters, and our CNN baselines mostly have better accuracies compared to the numbers reported in their original papers* [13,19,52,57].

Networks		Ind(baseline)		DML		DCM			
Net1	Net2	Net1	Net2	Net1	Net2	Net1	**DCM-DML**	Net2	**DCM-DML**
ResNet-164	ResNet-164	22.56(0.20)	22.56(0.20)	20.69(0.25)	20.72(0.14)	19.57(0.20)	**1.12**	19.59(0.15)	**1.13**
WRN-28-10	WRN-28-10	18.72(0.24)	18.72(0.24)	17.89(0.26)	17.95(0.07)	16.61(0.24)	**1.28**	16.65(0.22)	**1.30**
DenseNet-40-12	DenseNet-40-12	24.91(0.18)	24.91(0.18)	23.18(0.18)	23.15(0.20)	22.35(0.12)	**0.83**	22.41(0.17)	**0.74**
WRN-28-10	ResNet-110	18.72(0.24)	26.55(0.26)	17.99(0.24)	24.42(0.19)	17.82(0.14)	**0.17**	22.99(0.30)	**1.43**
WRN-28-10	WRN-28-4	18.72(0.24)	21.39(0.30)	17.80(0.11)	20.21(0.16)	16.84(0.08)	**0.96**	18.76(0.14)	**1.45**
WRN-28-10	MobileNet	18.72(0.24)	26.30(0.35)	17.24(0.13)	23.91(0.22)	16.83(0.07)	**0.41**	21.43(0.20)	**2.48**
WRN-28-10(+)	WRN-28-10(+)	18.64(0.19)	18.64(0.19)	17.62(0.12)	17.61(0.13)	16.57(0.12)	**1.05**	16.59(0.15)	**1.02**

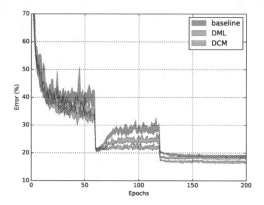

Fig. 2. Comparison of test curves at the different stages of jointly training two WRN-28-10 models. We show the range over 5 runs. Compared to the independent training method (baseline) and DML, DCM shows stably better performance during the whole training, and finally converges with the best accuracy on the test set.

average pooling layer and a fully connected layer. The differences are the number of building blocks and the number of convolutional filters. *Detailed designs of auxiliary classifiers and training hyper-parameter settings are provided in the supplementary material.* For each joint training case, we run each method 5 times and report "mean(std)" error rates. All models are trained on a server using 1/2 GPUs according to the GPU memory requirement.

First Training Scenario. Results of training two models with the same backbone are shown in the first part of Table 1 from which we can find: (1) Both DML and DCM obviously improve the model performance compared to the independent training method; (2) Generally, DCM performs better than DML. Taking the set of models having better mean accuracy as the example, the ResNet-164, WRN-28-10 and DenseNet-40-12 models trained with DCM show 1.12%, 1.28%

and 0.80% average margins to the models trained with DML respectively; (3) The accuracy gain of DCM against DML shows a trend: the higher the network capacity, the larger the accuracy gain.

Second Training Scenario. The second part of Table 1 provides the results of training two models with the different backbones, from which we can make similar observations as in the first training scenario. Besides, we can find another critical observation: Two networks with different capacities have different accuracy improvements. Comparatively, the lower-capacity ResNet-110/MobileNet/WRN-28-4 can benefit more from the high-capacity WRN-28-10 for both DML and DCM, and the corresponding accuracy improvement becomes much more large with DCM. For example, the WRN-28-4/MobileNet model trained with DCM shows 18.76%/21.43% mean error rate, outperforming the DML counterpart by 1.45%/2.48% margin.

The aforementioned experiments clearly validate the effectiveness of our method. Figure 2 shows an illustrative comparison of test curves at the different stages of training two WRN-28-10 jointly with three different methods.

4.2   Experiments on ImageNet

Next, we perform experiments to validate the generalization ability of our method to a much larger dataset.

ImageNet Classification Dataset. It has about 1.2 million training images and 50 thousand validation images including 1000 object classes [38]. For training, images are resized to 256×256, and 224×224 crops are randomly sampled from the resized images or their horizontal flips normalized with the per-channel mean and std values. For evaluation, we report top-1 and top-5 error rates using the center crops of resized validation data.

Implementation Details. On the ImageNet classification dataset, we use popular ResNet-18/50 [13] and MobileNetV2 [39] as the backbone networks, and consider the two same training scenarios as on the CIFAR-100 dataset. For all these CNN backbones, we use the same settings as reported in the original papers. In our method, we add two auxiliary classifiers to the different-staged layers of each CNN backbone. The auxiliary classifiers are constructed with the same building block as in the backbone network. The differences are the number of building blocks and the number of convolutional filters. *Detailed designs of auxiliary classifiers and training hyper-parameter settings are provided in the supplementary material.* For a concise comparison, we use the conventional data augmentation but not aggressive data augmentation methods. All models are trained on a server using 8 GPUs.

Results Comparison. The results are summarized in Table 2. It can be seen that both DML and DCM bring noticeable accuracy improvements to the baseline model in the first scenario of training two networks with the same structure jointly, and DCM is better than DML. Comparatively, the better one of two ResNet-18 models trained by DCM shows 28.67%/9.71% top-1/top-5 error rate

Table 2. Result comparison on the ImageNet classification dataset. For each network, we report top-1/top-5 error rate (%). Bolded results show the accuracy margins of DCM compared to the independent training method/DML.

Networks		Ind(baseline)		DML		DCM					
Net1	Net2	Net1	Net2	Net1	Net2	Net1	DCM-Ind	DCM-DML	Net2	DCM-Ind	DCM-DML
ResNet-18	ResNet-18	31.08/11.17	31.08/11.17	29.13/9.89	29.25/10.00	28.67/9.71	**2.41/1.46**	**0.46/0.18**	28.74/9.74	**2.34/1.43**	**0.51/0.26**
MobileNetV2	MobileNetV2	27.80/9.50	27.80/9.50	26.61/8.85	26.78/8.97	25.62/8.16	**2.18/1.34**	**0.99/0.69**	25.74/8.21	**2.06/1.29**	**1.04/0.76**
ResNet-50	ResNet-18	25.47/7.58	31.08/11.17	25.24/7.56	28.65/9.49	24.92/7.42	**0.55/0.16**	**0.32/0.14**	27.93/9.19	**3.15/1.98**	**0.72/0.30**

Table 3. Result comparison of jointly training two WRN-28-10 models on the CIFAR-100 dataset using different layer location settings for placing auxiliary classifiers. C1 denotes the default classifier over the last layer of the network, and C2, C3 and C4 denote 3 auxiliary classifiers with the increased layer distance to C1 (see supplementary material for details). We report "mean(std)" error rates (%) over 5 runs.

Classifier locations	WRN-28-10	
	Net1	Net2
baseline	18.72(0.24)	18.72(0.24)
C1+C4	17.16(0.14)	17.25(0.15)
C1+C3	16.89(0.21)	17.04(0.06)
C1+C2	17.40(0.20)	17.38(0.17)
C1+C2C3(default)	16.61(0.24)	16.65(0.22)
C1+C2C3C4	**16.59**(0.12)	16.73(0.17)

which outperforms the baseline model with a margin of 2.41%/1.46%. Impressively, our DCM shows at most 1.04%/0.76% accuracy improvement to DML on the MobileNetV2 model. These results are consistent with the results of training two networks with the same backbone on the CIFAR-100 dataset. In the second scenario of jointly training two different CNN backbones, the lower-capacity ResNet-18 benefits more from the high-capacity ResNet-50 than the reverse one for both DML and DCM, and the corresponding accuracy improvement becomes much larger by using DCM. Specifically, the ResNet-18 model trained by DCM can even reach 27.93%/9.19% top-1/top-5 error rate, showing 3.15%/1.98% and 0.72%/0.3% gain to the model trained with the independent training method and DML respectively. Although we use the conventional data augmentation, the best ResNet-18 model trained with our DCM shows 2.5% top-1 accuracy gain against the model (trained with aggressive data augmentation methods) released at the official GitHub page of Facebook[1].

---

[1] https://github.com/facebook/fb.resnet.torch.

Table 4. Result comparison of jointly training DenseNet-40-12 and WRN-28-10 on the CIFAR-100 dataset using different types of auxiliary classifiers. We report "mean(std)" error rates (%) over 5 runs.

Net1/Net2	Classifier type	DCM
DenseNet-40-12	APFC	25.10(0.25)
	Narrow	22.45(0.25)
	default	**22.35**(0.12)
WRN-28-10	APFC	18.23(0.10)
	Narrow	16.88(0.17)
	default	**16.61**(0.24)

Table 5. Result comparison of training two DenseNet-40-12 models jointly on the CIFAR-100 dataset using different settings of cross-layer bidirectional KD. DCM-1/DCM-2 performs KD operations between the same-staged/different-staged layers. We report "mean(std)" error rates (%) over 5 runs.

Method	Error (%)	
	Net1	Net2
baseline	24.91(0.18)	24.91(0.18)
DML	23.18(0.18)	23.15(0.20)
DML + DS	23.18(0.33)	23.08(0.28)
DCM-1	22.86(0.16)	22.89(0.14)
DCM-2	22.43(0.25)	22.51(0.18)
DCM	**22.35(0.12)**	**22.41(0.17)**

4.3 Deep Analysis of DCM

Finally, we conduct extensive ablative experiments on the CIFAR-100 dataset to better understand our method and show its capability to handle more challenging scenarios.

Setting of Q. Recall that the set Q plays a critical role in DCM. The setting of Q is closely related to two basic questions: (1) Where to place Auxiliary CLassiFiers (ACLFs)? (2) How to design the structure of ACLFs? As we discussed in the method section, we add ACLFs to the down-sampling layers of the network, following the common practices as used in [18,35,42,58]. However, a modern CNN architecture usually has several (e.g., 3/5) down-sampling layers, and thus there exist many layer location combinations for placing ACLFs. To this question, we conduct ablative experiments to jointly train two WRN-28-10 models considering different settings by adding ACLFs to at most three down-sampling layers. The results summarized in Table 3 show that the 2-ACLF model brings relatively large gain compared to the 1-ACLF model, while the 3-ACLF model gives negligible gain compared to the 2-ACLF model. Therefore, we add 2 ACLFs as the default setting of DCM for a good accuracy-efficiency trade-off. To the second question, we evaluate two additional kinds of ACLFs besides the default ACLFs used in DCM. Results are shown in Table 4 where "APFC" refers to a structure that uses an average pooling layer to down-sample input feature maps and then uses a fully connected layer to generate logits. "Narrow" refers to a narrower version (smaller width multipliers) compared to the default ACLF design. It can be seen that simple ACLFs may hurt performance sometimes (similar experiments are also provided in [18,42,58]), in which scenario our method has a small gain. Comparatively, large accuracy gains are obtained in the other two cases, therefore relatively strong ACLFs are required to make our method work properly. After training, ACLFs are discarded, and thus there are no extra parameters added to final models.

Analysis of Cross-Layer Bidirectional KDs. Recall that our DCM presents two cross-layer bidirectional KD designs (between either the same-staged or different-staged layers) to mutually transfer knowledge between two networks. In order to study their effects, we conduct two experiments in which we keep either the first or second bidirectional KD design. In the experiments, we consider the case of jointly training two DenseNet-40-12 models. Surprisingly, the results provided in Table 5 show that the second design brings larger performance gain than the first design, which means knowledge transfer between the different-staged layers is more effective. Because of the introduction of well-designed auxiliary classifiers, DCM enables much more diverse and effective bidirectional knowledge transfer which improves joint training performance considerably.

Accuracy Gain from Deep Supervision. There is a critical question to DCM: Is the performance margin between DCM and DML mostly owing to the auxiliary classifiers added to certain hidden layers of two networks as they have additional parameters? To examine this question, we conduct extensive experiments considering two different settings: (1) For DCM, we remove all the bidirectional KD connections except the one between the last layer of two DenseNet-40-12 backbones while retaining auxiliary classifiers. This configuration can be regarded as a straightforward combination of DML and Deep Supervision (DS); (2) Further, we add auxiliary classifiers to individual CNN backbones, and train each of them with DS independently while train two same backbones with DCM simultaneously. The results under the first setting are provided in Table 5, denoted as DML + DS. It can be observed that the combination of DML and DS only brings 0.07% average improvement to DenseNet-40-12, which only occupies 8.75% of the total margin brought by DCM. The results under the second setting are summarized in Table 6. It can be noticed that the average gain of DS to each baseline model is less than 0.85% in the most cases. Comparatively, DCM shows consistently large accuracy improvements over the baseline models, ranging from 2.07% to 2.99%.

Comparison with KD and its Variants. *Note that a fair comparison of DCM/DML with KD and its variants is impractical as the training paradigm*

Table 6. Comparison of Deep Supervision (DS) and DCM on the CIFAR-100 dataset. We report "mean(std)" error rates (%) over 5 runs.

Network	baseline	DS	DCM
ResNet-164	22.56(0.20)	21.38(0.32)	**19.57(0.20)**
DenseNet-40-12	24.91(0.18)	24.46(0.22)	**22.35(0.12)**
WRN-28-10	18.72(0.24)	18.32(0.13)	**16.61(0.24)**
WRN-28-10(+)	18.64(0.19)	17.80(0.29)	**16.57(0.12)**

is quite different. Here we illustratively study how much KD will work in our case, via using a pre-trained WRN-28-10 to guide the training of a ResNet-110. Surprisingly, KD shows a slightly worse mean error rate than baseline (26.66% vs. 26.55%). We noticed that during training, the soft labels generated by the teacher (WRN-28-10) are not so "soft" and the accuracy of these soft labels is very high (~99% on the CIFAR-100 dataset, meaning the model usually fits training data "perfectly"). These soft labels don't provide any more useful guidance than the hard labels and cause overfitting somehow. Using DML or DCM, the soft labels are generated dynamically as the teacher and student networks are jointly trained from the scratch, so they are comparatively softer and contain more useful guidance at every training iteration. *Besides, we provide horizontal comparisons of DCM with KD variants in the supplementary material.*

With Noisy Data. We also explore the capability of our method to handle noisy data. Following [55,56], we use CIFAR-10 dataset and jointly train two DenseNet-40-12 models as a test case. Before training, we randomly sample a fixed ratio of training data and replace their ground truth labels with randomly generated wrong labels. After training, we evaluate the models on the raw testing set. The results are summarized in Table 7. Three corruption ratios 0.2, 0.5, and 0.8 are considered. Compared to the case with 0.2 corruption ratio, the margin between DCM and baseline increases as the corruption ratio increases. One possible explanation of this phenomenon is that DCM behaves as a regularizer. When the training labels get corrupted, the baseline model will try to fit the training data and capture the wrong information, which causes severe overfitting. In the DCM configuration, things are different. Beyond the corrupted labels, the classifiers also get supervision from the soft labels generated by other different-staged or same-staged classifiers. These soft labels can prevent the classifiers from fitting the corrupted data and finally improve the generalization to a certain degree. In normal training without corrupted data, this can also happen. For example, if there is an image of a person with his or her dog, the human-annotated ground truth will be a 1-class label, either "dog" or "person", but not both. This kind of images can be seen as "noisy" data, and this is where soft-labels dynamically generated will kick in.

With Strong Regularization. The aforementioned experiments show DCM behaves as a strong regularizer which can improve the generalization of the mod-

Table 7. Result comparison on the CIFAR-10 dataset with noisy labels. We jointly train two DenseNet-40-12 models, and report "mean(std)" error rates (%) over 5 runs.

Corruption ratio	Method	Error (%)
0.2	baseline	9.85(0.24)
	DML	8.13(0.14)
	DCM	**7.11(0.11)**
0.5	baseline	17.93(0.39)
	DML	14.31(0.30)
	DCM	**12.08(0.34)**
0.8	baseline	35.32(0.42)
	DML	32.65(0.96)
	DCM	**31.26(0.94)**

els. In order to study the performance of DCM under the existence of other strong regularizations, we follow the dropout experiments in [52]. We add a dropout layer with $p = 0.3$ after the first layer of every building block of WRN-28-10. The results are shown in Table 1 as WRN-28-10(+). It can be seen that combining DCM with dropout achieves better performance than DCM, which means DCM is compatible with traditional regularization techniques like dropout.

Comparison of Efficiency. In average, DCM is about 1.5× slower than DML during the training phase due to the use of auxiliary classifiers. However, all auxiliary classifiers are discarded after training, so there is no extra computational cost to the resulting model during the inference phase compared with the independent training method and DML. With DCM, the lower-capacity model has similar accuracy but requires much less computational cost compared to high-capacity model. For example, as shown in Table 1, WRN-28-4 models trained with DCM show a mean error rate of 18.76% which is almost the same to that of WRN-28-10 models trained with the independent training method.

5   Conclusions

In this paper, we present DCM, an effective two-way knowledge transfer method for collaboratively training two networks from scratch. It connects knowledge representation learning with deep supervision methodology and introduces dense cross-layer bidirectional KD designs. Experiments on a variety of knowledge transfer tasks validate the effectiveness of our method.

References

1. Aguilar, G., Ling, Y., Zhang, Y., Yao, B., Fan, X., Guo, C.: Knowledge distillation from internal representations. In: AAAI (2020)

2. Ba, L.J., Caruana, R.: Do deep nets really need to be deep? In: NIPS (2014)
3. Batra, T., Parikh, D.: Cooperative learning with visual attributes. arXiv preprint arXiv:1705.05512 (2017)
4. Blum, A., Mitchell, T.: Combining labeled and unlabeled data with co-training. In: COLT (1998)
5. Bucilă, C., Caruana, R., Niculescu-Mizil, A.: Model compression. In: KDD (2006)
6. Chen, T., Goodfellow, I., Shlens, J.: Net2Net: accelerating learning via knowledge transfer. In: ICLR (2016)
7. Chen, Y., Wang, Z., Peng, Y., Zhang, Yu, G., Sun, J.: Cascaded pyramid network for multi-person pose estimation. In: CVPR (2018)
8. Garcia, N.C., Morerio, P., Murino, V.: Modality distillation with multiple stream networks for action recognition. In: Ferrari, V., Hebert, M., Sminchisescu, C., Weiss, Y. (eds.) ECCV 2018. LNCS, vol. 11212, pp. 106–121. Springer, Cham (2018). https://doi.org/10.1007/978-3-030-01237-3_7
9. Girshick, R., Donahue, J., Darrell, T., Malik, J.: Rich feature hierarchies for accurate object detection and semantic segmentation. In: CVPR (2014)
10. Guo, X., Li, H., Yi, S., Ren, J., Wang, X.: Learning monocular depth by distilling cross-domain stereo networks. In: Ferrari, V., Hebert, M., Sminchisescu, C., Weiss, Y. (eds.) ECCV 2018. LNCS, vol. 11215, pp. 506–523. Springer, Cham (2018). https://doi.org/10.1007/978-3-030-01252-6_30
11. Hafner, F., Bhuiyan, A., Kooij, J.F.P., Granger, E.: A cross-modal distillation network for person re-identification in RGB-depth. arXiv preprint arXiv:1810.11641 (2018)
12. He, D., et al.: Dual learning for machine translation. In: NIPS (2017)
13. He, K., Zhang, X., Ren, S., Sun, J.: Deep residual learning for image recognition. In: CVPR (2016)
14. Hinton, G., Vinyals, O., Dean, J.: Distilling the knowledge in a neural network. arXiv preprint arXiv:1503.02531 (2015)
15. Hou, S., Pan, X., Loy, C.C., Wang, Z., Lin, D.: Lifelong learning via progressive distillation and retrospection. In: Ferrari, V., Hebert, M., Sminchisescu, C., Weiss, Y. (eds.) ECCV 2018. LNCS, vol. 11207, pp. 452–467. Springer, Cham (2018). https://doi.org/10.1007/978-3-030-01219-9_27
16. Howard, A.G., et al.: MobileNets: efficient convolutional neural networks for mobile vision applications. arXiv preprint arXiv:1704.04861 (2017)
17. Hu, J., Shen, L., Sun, G.: Squeeze-and-excitation networks. In: CVPR (2018)
18. Huang, G., Chen, D., Li, T., Wu, F., van der Maaten, L., Weinberger, K.Q.: Multi-scale dense networks for resource efficient image classification. In: ICLR (2018)
19. Huang, G., Liu, Z., van der Maaten, L., Weinberger, K.Q.: Densely connected convolutional networks. In: CVPR (2017)
20. Jia, S., Bruce, N.D.B.: Richer and deeper supervision network for salient object detection. arXiv preprint arXiv:1901.02425 (2018)
21. Kim, J., Park, S., Kwak, N.: Paraphrasing complex network: network compression via factor transfer. In: NeurIPS (2018)
22. Kim, Y., Rush, A.M.: Sequence-level knowledge distillation. In: EMNLP (2016)
23. Krizhevsky, A., Hinton, G.: Learning multiple layers of features from tiny images. In: Tech Report (2009)
24. Krizhevsky, A., Sutskever, I., Hinton, G.E.: ImageNet classification with deep convolutional neural networks. In: NIPS (2012)
25. Kundu, J.N., Lakkakula, N., Babu, R.V.: UM-Adapt: unsupervised multi-task adaptation using adversarial cross-task distillation. In: ICCV (2019)

26. Lan, X., Zhu, X., Gong, S.: Knowledge distillation by on-the-fly native ensemble. In: NeurIPS (2018)
27. Lee, C.Y., Xie, S., Gallagher, P., Zhang, Z., Tu, Z.: Deeply-supervised nets. In: AISTATS (2015)
28. Lee, S.H., Kim, H.D., Song, B.C.: Self-supervised knowledge distillation using singular value decomposition. In: NeurIPS (2018)
29. Li, Y., Wang, N., Liu, J., Hou, X.: Demystifying neural style transfer. In: IJCAI (2016)
30. Li, Z., Hoiem, D.: Learning without forgetting. In: Leibe, B., Matas, J., Sebe, N., Welling, M. (eds.) ECCV 2016. LNCS, vol. 9908, pp. 614–629. Springer, Cham (2016). https://doi.org/10.1007/978-3-319-46493-0_37
31. Liu, W., et al.: SSD: single shot multibox detector. In: Leibe, B., Matas, J., Sebe, N., Welling, M. (eds.) ECCV 2016. LNCS, vol. 9905, pp. 21–37. Springer, Cham (2016). https://doi.org/10.1007/978-3-319-46448-0_2
32. Long, J., Shelhamer, E., Darrell, T.: Fully convolutional networks for semantic segmentation. In: CVPR (2015)
33. Newell, A., Yang, K., Deng, J.: Stacked hourglass networks for human pose estimation. In: Leibe, B., Matas, J., Sebe, N., Welling, M. (eds.) ECCV 2016. LNCS, vol. 9912, pp. 483–499. Springer, Cham (2016). https://doi.org/10.1007/978-3-319-46484-8_29
34. Paszke, A., et al.: Automatic differentiation in pytorch. In: NIPS Workshops (2017)
35. Phuong, M., Lampert, C.H.: Distillation-based training for multi-exit architectures. In: ICCV (2019)
36. Qiao, S., Shen, W., Zhang, Z., Wang, B., Yuille, A.: Deep co-training for semi-supervised image recognition. In: Ferrari, V., Hebert, M., Sminchisescu, C., Weiss, Y. (eds.) ECCV 2018. LNCS, vol. 11219, pp. 142–159. Springer, Cham (2018). https://doi.org/10.1007/978-3-030-01267-0_9
37. Romero, A., Ballas, N., Kahou, S.E., Chassang, A., Gatta, C., Bengio, Y.: FitNets: hints for thin deep nets. In: ICLR (2015)
38. Russakovsky, O., et al.: ImageNet large scale visual recognition challenge. Int. J. Comput. Vis. **115**(3), 211–252 (2015)
39. Sandler, M., Howard, A., Zhu, M., Zhmoginov, A., Chen, L.C.: MobileNetV2: inverted residuals and linear bottlenecks. In: CVPR (2018)
40. Simonyan, K., Zisserman, A.: Very deep convolutional networks for large-scale image recognition. In: ICLR (2015)
41. Song, G., Chai, W.: Collaborative learning for deep neural networks. In: NeurIPS (2018)
42. Sun, D., Yao, A., Zhou, A., Zhao, H.: Deeply-supervised knowledge synergy. In: CVPR (2019)
43. Sze, V., Chen, Y.H., Yang, T.J., Emer, J.S.: Efficient processing of deep neural networks: a tutorial and survey. Proc. IEEE **105**(12), 2295–2329 (2017)
44. Szegedy, C., et al.: Going deeper with convolutions. In: CVPR (2015)
45. Tian, Y., Krishnan, D., Isola, P.: Contrastive representation distillation. In: ICLR (2020)
46. Tian, Y., Krishnan, D., Isola, P.: Contrastive representation distillation. ICLR (2020)
47. Wang, Z., Deng, Z., Wang, S.: Accelerating convolutional neural networks with dominant convolutional kernel and knowledge pre-regression. In: Leibe, B., Matas, J., Sebe, N., Welling, M. (eds.) ECCV 2016. LNCS, vol. 9912, pp. 533–548. Springer, Cham (2016). https://doi.org/10.1007/978-3-319-46484-8_32

48. Xie, S., Girshick, R., Dollár, P., Tu, Z., He, K.: Aggregated residual transformations for deep neural networks. In: CVPR (2017)
49. Xie, S., Tu, Z.: Holistically-nested edge detection. In: ICCV (2015)
50. Xu, D., Ouyang, W., Wang, X., Nicu, S.: PAD-Net: multi-tasks guided prediction-and-distillation network for simultaneous depth estimation and scene parsing. In: CVPR (2018)
51. Yim, J., Joo, D., Bae, J., Kim, J.: A gift from knowledge distillation: fast optimization, network minimization and transfer learning. In: CVPR (2017)
52. Zagoruyko, S., Komodakis, N.: Wide residual networks. In: BMVC (2016)
53. Zagoruyko, S., Komodakis, N.: Paying more attention to attention: improving the performance of convolutional neural networks via attention transfer. In: ICLR (2017)
54. Zhai, M., Chen, L., Tung, F., He, J., Nawhal, M., Mori, G.: Lifelong GAN: continual learning for conditional image generation. In: ICCV (2019)
55. Zhang, C., Bengio, S., Hardt, M., Recht, B., Vinyals, O.: Understanding deep learning requires rethinking generalization. In: ICLR (2017)
56. Zhang, H., Cisse, M., Dauphin, Y.N., Lopez-Paz, D.: mixup: beyond empirical risk minimization. In: ICLR (2018)
57. Zhang, Y., Xiang, T., Hospedales, T.M., Lu, H.: Deep mutual learning. In: CVPR (2018)
58. Zhang, Z., Zhang, X., Peng, C., Xue, X., Sun, J.: ExFuse: enhancing feature fusion for semantic segmentation. In: Ferrari, V., Hebert, M., Sminchisescu, C., Weiss, Y. (eds.) ECCV 2018. LNCS, vol. 11214, pp. 273–288. Springer, Cham (2018). https://doi.org/10.1007/978-3-030-01249-6_17
59. Zhao, H., Shi, J., Qi, X., Wang, X., Jia, J.: Pyramid scene parsing network. In: CVPR (2017)

Matching Guided Distillation

Kaiyu Yue$^{(\boxtimes)}$, Jiangfan Deng, and Feng Zhou

Algorithm Research, Aibee Inc., Beijing, China
kaiyuyue@gmail.com

Abstract. Feature distillation is an effective way to improve the performance for a smaller student model, which has fewer parameters and lower computation cost compared to the larger teacher model. Unfortunately, there is a common obstacle—the gap in semantic feature structure between the intermediate features of teacher and student. The classic scheme prefers to transform intermediate features by adding the adaptation module, such as naive convolutional, attention-based or more complicated one. However, this introduces two problems: a) The adaptation module brings more parameters into training. b) The adaptation module with random initialization or special transformation isn't friendly for distilling a pre-trained student. In this paper, we present Matching Guided Distillation (MGD) as an efficient and parameter-free manner to solve these problems. The key idea of MGD is to pose matching the teacher channels with students' as an assignment problem. We compare three solutions of the assignment problem to reduce channels from teacher features with partial distillation loss. The overall training takes a coordinate-descent approach between two optimization objects— assignments update and parameters update. Since MGD only contains normalization or pooling operations with negligible computation cost, it is flexible to plug into network with other distillation methods. The project site is http://kaiyuyue.com/mgd.

1 Introduction

Deep networks [18,40] enjoy massive neuron parameters for achieving the state-of-the-art performances on lots of technique lines, such as visual recognition [9], image captioning [17], object detection system [36] and language understanding [5,37]. However, the industry prefers to carry out model inference on cheap devices, therefore the small model with few parameters is needed. The dilemma of achieving analogous performance to the large model in lightweight backbone recently motivates extensive research directions, such as channel pruning [10], lightweight model design [24,29], quantization [33] and neural architecture search (NAS) for efficient model [31,35]. Among them, model distillation is another active track, which aims to transfer knowledge or semantic feature information from a large teacher model into a light student model. Pioneered by dark

Electronic supplementary material The online version of this chapter (https://doi.org/10.1007/978-3-030-58555-6_19) contains supplementary material, which is available to authorized users.

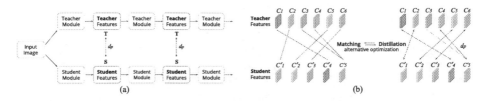

Fig. 1. Matching guided distillation. (a) MGD follows the general distillation paradigm. **T** and **S** are teacher and student feature tensors. d_p is the feature distance function. (b) MGD distills teacher channels (C_i) of intermediate features to student channels (C'_i) by solving a joint matching-distillation problem (alternative optimization).

knowledge [13], the main body of recent works [2,38] focuses on using distilling intermediate features to enrich the learnable information for guiding student in different tasks, such as classification [27] and detection [2].

However, a prominent challenging problem for these methods is on how to fill the gap in semantic feature structure between teacher and student. Roughly speaking, the contrast lies in two aspects: 1) The different channel dimensions of feature outputs used for distillation. 2) The different perceptual information or activations between two channel sets. Previous works [11,27,38] overcome these two obstacles by building an adaptive module between hidden layers of teacher and student. Whereas this manner can alleviate the issue, it still has two limitations: 1) The adaptation module introduces more parameters (including weights, gradients and optimizer states) into training [26]. These additional parameters induce the training harder despite the fact that they would be brushed off for model inference. 2) The adaptation module with random initialization or special non parameter transformation [30,41] isn't friendly for distilling a pre-trained student, because it would potentially disturb the student features. To avoid this break, [27,38] performs stage-wise training by separating optimization into multiple steps. But this way isn't perfect yet because it will plunge training into a cumbersome one.

To crack the limitations of former works, we propose a novel distillation method named Matching Guided Distillation (MGD). As shown in Fig. 1, MGD follows the general distillation paradigm. Given batches of data fed into teacher and student, MGD matches their intermediate channels from distillation position. The motivation is that whether the student has been pre-trained or not, each channel of it should be guided by its high related teacher channel directly to narrow the semantic feature gap. In order to implement element-wise losses, teacher channels would be reduced according to the matching graph. The method for channel reduction is flexible, we propose three manners: sparse matching, random drop and absolute max pooling. Experiments show that all three ways are effective on various tasks. For the whole training, we apply coordinate descent algorithm [34] to alternate between two optimizations for channels matching and weights updating. Furthermore, distilling a pre-trained student using MGD is efficient due to its parameter-free nature as same as training from-scratch.

2   Related Work

Correspondence Problem. Finding optimal correspondence between two sets of instances is a crucial step for a wide range of computer vision problems, such as shape retrieval [16], image retrieval [28], object categorization [6] and video action recognition [1]. Linear Assignment (LA) is the most classical correspondence problem that can be efficiently solved with Hungarian algorithm [20], unlike the NP-hard quadratic assignment problem [42]. Matching based training losses [7,15] contain the ideology of matching features, but they have a heavy computation. For example, the Wasserstein loss [7] uses the iterated optimization [3] to approximate the matching matrix, its computation cost will dramatically become large along with the growth of feature dimensions. It only can be used for the feature from the last fully-connected layer, so does [15]. In this paper, we treat relationships modeling between teacher and student channels as a LA problem, particularly the min cost assignment problem. The total matching cost function would be minimized by the Hungarian method to achieve a bipartite assignment graph. This graph represents the high related channel pairs between teacher and student feature sets.

Knowledge Distillation. Pioneered by [13], the classic method for knowledge distillation contains two constituents: logits from the last teacher layer used as the soft targets, and Kullback-Leibler divergence loss used to let student match these targets. However, the performance of output distillation is limited due to the very similar supervised signal from teacher model with ground-truth. More works [2,27,38] switch to feature distillation by combining intermediate features together to strongly supervise the student. All these works rely on certain adaptation modules between hidden layers, in order to solve the contrast of semantic feature structures. Feature correlation based methods provide more fine-grained recipes to perform knowledge distillation, such as attention transfer [21,41], neuron selectivity transfer [15]. These works focus on capturing and transferring the spatial information for intermediate feature maps. Another technique fashion for distillation is to design the loss function, including activation transfer loss with boundaries formed by hidden neurons [12], the loss for penalizing structural differences in relations [25]. In this paper, we propose a novel perception that performing distillation after matching intermediate channels between teacher and student. Our proposed approach is intuitive and lightweight. It introduces marginal computation costs during training. In the end, although the work of Jacobian matching-based distillation [30] seems related to ours, it still suffers from the problems discussed in Sect. 1. Because it not only uses adaption modules but also a specialized loss function.

Transfer Learning. Commonly fine-tuning is one of the effective methods for knowledge transfer learning. The student has been already pre-trained on a specific domain data, and then it's fine-tuned for another task with priori knowledge. The work [8] finds that the model with random initialization could be trained no worse than using pre-trained parameters. However, the model may suffer from the low capacity trained on a small dataset like Caltech-UCSD Birds 200 [32].

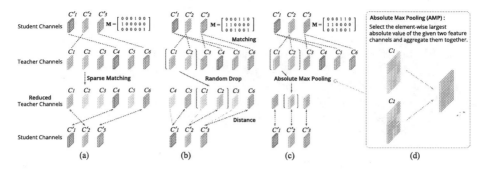

Fig. 2. Channel reduction methods. C_i indicates teacher channels of intermediate features as same as C'_i for student. $\mathbf{M}$ is the matching matrix. We propose three effective methods to play reduction for teacher channels: sparse matching, random drop and absolute max pooling. (a) Sparse Matching. Each student channel only matches one teacher channel. Unmatched teacher channels are directly ignored. (b) Random Drop. Each student channel matches more than one teacher channel. Matched teacher channels with the same student channel are *randomly* dropped to leave just one for guiding student. (c) Absolute Max Pooling (AMP). Each student channel matches more than one teacher channel. AMP picks out the element-wise largest absolute value along the channel axis over a group of matched teacher channels with the same student channel. (d) shows the detail about how AMP works on two channel tensors.

A number of previous works use ImageNet pre-trained models on different tasks such as detection and segmentation [36]. In this paper, we show that MGD can be also used for knowledge transfer learning with a pre-trained student for the further performance improvements in another task, particularly for the fine-grained categorization.

3 Methodology

In this section, we introduce a general formulation of the proposed Matching Guided Distillation (MGD). MGD consists in a parameter-free channel matching module that can guide to shave teacher channels in three effective manners: sparse matching, random drop and absolute max pooling, as shown in Fig. 2.

3.1 Feature Distillation Revisit

We begin by briefly reviewing feature distillation in general formulation. Suppose that 2D images[1] $\mathbf{X}$ are fed into the teacher f_T and student f_S networks that generate intermediate feature sets

$$\mathbf{T} = f_T(\mathbf{X}) \in \mathbb{R}^{C_T \times N}, \ \mathbf{S} = f_S(\mathbf{X}) \in \mathbb{R}^{C_S \times N}, \tag{1}$$

---

[1] Bold capital letters denote a matrix $\mathbf{X}$, bold lower-case letters a column vector $\mathbf{x}$. $\mathbf{x}_i$ and $\mathbf{x}^j$ represents the ith column and j^{th} row of the matrix $\mathbf{X}$ respectively. x_{ij} or $[\mathbf{X}]_{ij}$ denotes the scalar in the ith row and j^{th} column of the matrix $\mathbf{X}$. All non-bold letters represent scalars.

respectively at the target distillation positions. Without loss of generality, we assume the feature maps are of the same spatial size $N = HW$ (height H and width W) but could consist of different number of channels, e.g. $C_T = 512$ while $C_S = 128$. Given a teacher network f_T with frozen parameters, we wish to enhance the training of the student networks f_S using the hints from f_T. In a nutshell, the problem of feature distillation seeks for the optimum student network f_S that minimizes the loss of the main task together with feature discrepancy penalty:

$$Loss = L_{task} + \gamma L_{distill}, \tag{2}$$

where γ is a trade-off coefficient for distillation loss. The task loss L_{task}, for example, could be cross-entropy loss for classification or smooth-L_1 loss for object localization. The key to feature distillation is the design of the distillation loss, $L_{distill}$, which ensures the similarity between intermediate features $\mathbf{T}$ and $\mathbf{S}$:

$$L_{distill} = d_p(\sigma_T(\rho(\mathbf{T}, \mathbf{M})), \sigma_S(\mathbf{S})), \tag{3}$$

where σ_T, σ_S, d_p are the teacher and student feature transforms that convert raw feature into an easy-to-transfer form and distance functions respectively. In the past few years, various designs have been proposed to make better use of information contained in teacher networks. In MGD, σ_T is a marginal ReLU as same as [11]. Based on a recent comprehensive review [11] on these design aspects, we build the pipeline by employing the marginal ReLU for teacher transform σ_T:

$$\sigma_T(x) = \max(x, m), \tag{4}$$

where $m < 0$ is a margin value, computed as an expectation value over all training images. Following [11], we choose the partial L_2 loss function to calculate feature distance:

$$d_p(\mathbf{T}, \mathbf{S}) = \sum_i^C \sum_j^N \begin{cases} 0 & \text{if } s_{ij} \leq t_{ij} \leq 0 \\ (t_{ij} - s_{ij})^2 & \text{otherwise,} \end{cases} \tag{5}$$

for any pair of matrices $\mathbf{T}, \mathbf{S} \in \mathbb{R}^{C \times N}$ of the same dimension.

3.2 Channel Matching

The distillation loss (Eq. 3) plays a vital role in distilling the knowledge of a complex model into a simpler one. To achieve this goal, the design of distance function (d_p) and feature transforms (σ_T, σ_S) needs to ensure teacher's knowledge can be transferred to student with minimum loss. Despite that various choices have been proposed in the past few years (see [11] for an extensive review), it is still necessary to add an 1×1 convolutional layer or other module on student (σ_S) to bridge the semantic gap between $\mathbf{T}$ and $\mathbf{S}$. The presence of student transform not only adds burden on network complexity but also complicates the training procedure. We propose MGD by completely removing the student transform from the distillation pipeline, i.e. $\sigma_S(\mathbf{S}) = \mathbf{S}$. Instead, we directly match the channel via the reduction operation $\rho(\mathbf{T}, \mathbf{M})$ from teacher to student. This operation is parameter-free and efficient to optimize in training. Below we explain how to establish the correspondence $\mathbf{M}$ across channels and define the implementation of $\rho(\cdot, \cdot)$ in next section.

Given a pair of teacher feature $\mathbf{T} \in \mathbb{R}^{C_T \times N}$ and student one $\mathbf{S} \in \mathbb{R}^{C_S \times N}$, we first encode their pairwise relation in a distance matrix, $\mathbf{D} \in \mathbb{R}^{C_S \times C_T}$, whose element d_{ij} computes the Euclidean distance between i^{th} student and j^{th} teacher channels:

$$d_{ij} = \sum_{k=1}^{N} (s_{ik} - t_{jk})^2. \tag{6}$$

Our goal is to find a binary matrix $\mathbf{M} \in \{0, 1\}^{C_S \times C_T}$ that encodes the channel-wise correspondence, where $m_{ij} = 1$ if i-th student and j-th teacher channels are pertinent. In the special case when the teacher and student feature maps are of the same dimension (i.e., $C_T = C_S$), the matching is assumed to be one-to-one and the resulting matrix $\mathbf{M}'$ defines a permutation of C_T channels:

$$\Pi = \left\{ \mathbf{M}' \in \{0, 1\}^{C_T \times C_T} \mid \sum_{j=1}^{C_T} m'_{ij} = 1, \sum_{i=1}^{C_T} m'_{ij} = 1 \right\}. \tag{7}$$

In general, we resort to a many-to-one matching as the teacher channel number C_T is often several times more than the student one C_S. In order to make the distillation procedure evenly distributed over feature channels, we further constrain that each student channel has to be associated with $\alpha = \lfloor C_T / C_S \rfloor$ teacher channels[2]. More specifically, the many-to-one balanced matching $\mathbf{M}$ satisfies:

$$\Pi_b = \left\{ \mathbf{M} \in \{0, 1\}^{C_S \times C_T} \mid \sum_{i=1}^{C_S} m_{ij} = 1, \sum_{j=1}^{C_T} m_{ij} = \alpha \right\}. \tag{8}$$

This constraint enforces that $\mathbf{M}$ is a wide-shape matrix, where the sum of each column equals to one because each teacher channel can only be connected to one student channel. On the other hand, the sum of each row needs to be α. In another word, each student channel has to be associated with α teacher ones.

Given two sets of feature channels with the associated pairwise distance, the problem of channel matching consists in finding a balanced many-to-one mapping $\mathbf{M}$ such that the sum of matching cost is minimized:

$$\min_{\mathbf{M}} \text{trace}(\mathbf{D}^T \mathbf{M}) = \sum_{i=1}^{C_S} \sum_{j=1}^{C_T} d_{ij} m_{ij}, \text{ subject to } \mathbf{M} \in \Pi_b. \tag{9}$$

---

[2] Once teacher and student are decided, the factor α will be fixed. For the case when C_T is not divisible by C_S, we simply introduce $C_T - \alpha C_S$ dummy teacher channels that never contribute in the final operation. Although this solution is not optimal, we found the result is still promising on several datasets.

Although Eq. 9 is not a standard linear assignment problem, there still exists a globally optimal solution based on the Hungarian method [20]. Let $\mathbf{D}' = [\mathbf{D}; \cdots ; \mathbf{D}] \in \mathbb{R}^{C_T \times C_T}$ be a square matrix by concatenating α matrices $\mathbf{D}$ vertically. Our solution proceeds by first optimizing a standard linear assignment problem:

$$\min_{\mathbf{M}'} \operatorname{trace}(\mathbf{D}'^T \mathbf{M}') = \sum_{i=1}^{C_T} \sum_{j=1}^{C_T} d'_{ij} m'_{ij}, \text{ subject to } \mathbf{M}' \in \varPi, \qquad (10)$$

using Hungarian algorithm. We then evenly slice the resulting matrix $\mathbf{M}' = [\mathbf{M}'_1; \cdots ; \mathbf{M}'_\alpha]$ in row blocks, where each sub-matrix $\mathbf{M}'_i \in \{0,1\}^{C_S \times C_T}$ is of the same size. It's easy to prove that the optimal solution for Eq. 9 is:

$$\mathbf{M} = \sum_{i=1}^{\alpha} \mathbf{M}'_i \in \{0,1\}^{C_S \times C_T}. \qquad (11)$$

3.3   Channel Reduction

Once the channel-wise correspondence $\mathbf{M}$ is established, the distillation loss d_p in Eq. 3 would encourage the student channel to mimic the hidden feature of the related teacher channels. Because of the many-to-one nature for the mapping, we discuss below three parameter-free choices for reducing teacher feature via operation $\rho(\mathbf{T}, \mathbf{M})$ to match with student feature.

Sparse Matching. To match the C_T teacher channels with C_S student ones, the straightforward way is to pick an optimal subset of C_S teacher channels and construct a one-to-one matching with the C_S student channels. To do so, we simply formalize another linear assignment problem by introducing $C_T - C_S$ dummy student channels, each of which is put in an infinity distance from any teacher channel. This linear assignment problem is in the same form as Eq. 10 except the distance matrix $\mathbf{D}' \in \mathbb{R}^{C_T \times C_T}$ is constructed by appending $C_T - C_S$ rows of large constant (e.g. 1e10) to the end of the original $\mathbf{D} \in \mathbb{R}^{C_S \times C_T}$. After applying the Hungarian algorithm on Eq. 10, we could find for each student channel the most relevant teacher one, which are encoded in the first C_S rows of the resulting correspondence matrix, i.e., $\mathbf{M} = \mathbf{M}'_1 \in \{0,1\}^{C_S \times C_T}$. For instance, Fig. 2(a) illustrates an example of matching $C_T = 6$ teacher channels with $C_S = 3$ student ones, where the correspondence matrix $\mathbf{M} \in \{0,1\}^{3 \times 6}$ denotes three one-to-one matching pairs. In this case, the reduction operation can thus be defined as:

$$\rho_{\mathrm{SM}}(\mathbf{T}, \mathbf{M}) = \mathbf{M}\mathbf{T} \in \mathbb{R}^{C_S \times N}. \qquad (12)$$

Random Drop. The major limitation of the first sparse matching choice is that it only retains a small fraction (C_S / C_T) of information conveyed in the original teacher features. To reduce the information loss, our second choice for teacher reduction is to sample a random teacher channel from the ones associated with each student channel. More specifically, there are α non-zero elements in each row of $\mathbf{M}$ according to the constraint (Eq. 8). The random drop operation modifies the correspondence as $\mathbf{M}^{RD} \in \{0,1\}^{C_S \times C_T}$ by randomly keeping one non-zero element in each row, i.e., $\sum_{j=1}^{C_T} m_{ij}^{RD} = 1$ for any $i = 1, \cdots, C_S$. To have a better understanding, we visualize one case in Fig. 2(b), where the second student channel C_2' is associated with C_1 and C_2 teacher channels after the channel matching step. In random drop reduction, we randomly pick one of them (e.g. C_2) to match with the student. In order to maximize the randomness, we generate correspondence matrices $\mathbf{M}_i^{RD}$ independently for different spatial positions of the feature map. The overall reduction operation can be defined as:

$$\rho_{\mathrm{RD}}(\mathbf{T}, \mathbf{M}) = \left[\mathbf{M}_1^{RD} \mathbf{t}_1, \cdots, \mathbf{M}_N^{RD} \mathbf{t}_N \right] \in \mathbb{R}^{C_S \times N}, \tag{13}$$

where $\mathbf{t}_j \in \mathbb{R}^{C_T}$ denotes the jth column of the feature $\mathbf{T}$.

Absolute Max Pooling. Following [11], we place the distillation module before ReLU. Therefore both positive and negative values are transferred from teacher via the partial distance loss d_p to student. We hope the reduced teacher features still take the maximum activations in the same spatial position, including positive (usable) and negative (adverse) information. To reach this purpose, we propose a novel pooling mechanism, named Absolute Max Pooling (AMP), as shown in Fig. 2(c). Given a set of feature activations $\mathbf{x} = [x_1, \ldots, x_C]^T \in \mathbb{R}^C$, the AMP is designed to choose the element that yields the largest magnitude:

$$f_{\mathrm{AMP}}(\mathbf{x}) = \arg\max_{x_i} |x_i|. \tag{14}$$

Similar to the random drop idea, the AMP operation is performed independently for each spatial position $\mathbf{t}_j \in \mathbb{R}^{C_T}$ of the teacher feature map $\mathbf{T} \in \mathbb{R}^{C_T \times N}$. For each student channel $i = 1, \cdots, C_S$, AMP is used to select the most active teacher channel among the associated α ones. Because this teacher-student association has been encoded as the non-zero elements of ith row $\mathbf{m}^i \in \{0,1\}^{C_T}$ of matrix $\mathbf{M}$, we can write the overall reduction operation in matrix form as:

$$\rho_{\mathrm{AMP}}(\mathbf{T}, \mathbf{M}) = \left[f_{\mathrm{AMP}}(\mathbf{m}^i \circ \mathbf{t}_j) \right]_{ij} \in \mathbb{R}^{C_S \times N}, \tag{15}$$

where $\circ$ indicates the element-wise product between two vectors. As shown in Fig. 2(d), AMP pools these α feature nodes into single one along the channel axis.

Theoretical Summary. In the theoretical perspective, we describe the understanding on these three channel reduction methods as following. All these methods belong to variants of the LA problem. With the same matrix notation, their goal can be connected as seeking for the optimal matching matrix $\mathbf{M}$ by optimizing certain linear objective. The first sparse matching (Eq. 12) is a simple modification of the original Hungarian algorithm. The second random drop (Eq. 13) and the last absolute max pooling (Eq. 14) ideas progressively improve sparse matching by different approaches to reduce information loss from $\sigma_{\mathbf{T}}$. To better understand and verify this insight, we have done many experiments and ablation studies in Sect. 4.

3.4 Implementation Details

Distillation Position. In our experiments, we use contemporary models in recent years, including ResNet [9], MobileNet-V1 [14], -V2 [29] and ShuffleNet-V2 [24]. Commonly, these models contains four stages, each of which is composed of repeated unit blocks, as shown in Fig. 3. We apply distillation in the last unit block of each stage. In the cases of distilling MobileNet-V2 and ShuffleNet-V2, the first stage is skipped, because their first stage is only a convolutional layer and also its feature map size ($H \times W$) is distinct from teacher's.

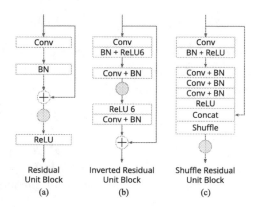

Fig. 3. The distillation position in a unit block is pinpointed by a color point. (a) For the standard residual unit in ResNet and MobileNet-V1, we use the output before ReLU. (b) For the inverted unit in MobileNet-V2, we use the intermediate feature before ReLU6. Although the output from the whole block isn't activated by ReLU-like operations too, the channel number is narrowed down into an unusable one for distillation. (c) The feature after channel shuffle is used for the shuffle unit. (Color figure online)

Coordinate Descent. To optimize the whole system (Eq. 3), we use Coordinate Descent algorithm [34] by alternating between solving the combinatorial matching problem and updating network weights. Postulating the matching is solved, we plug **M** in Eq. 3 and employ Stochastic Gradient Descent (SGD) to update student network weights f_S. After several epochs or iterations of training with SGD, the student is switched into evaluation mode without learning. Then we feed a dataset that is randomly sampled from *training* data into student and teacher, in order to update matching **M** that optimizes Eq. 9 for the next training rounds. Solving Eq. 9 takes the computational complexity of $O(C_T^3)$. This step introduces negligible cost in our implementation.

4   Experiments

Datasets. We run a number of major experiments to compare our MGD with other methods, using sparse matching (SM), random drop (RD) and absolute max pooling (AMP). We evaluate them in multiple tasks, including large-scale classification, fine-grained recognition with transfer learning, detection and segmentation. We have done on four popular open datasets. **CIFAR-100** [19] is composed of 50,000 images within 100 classes, and has a fixed input size of 32×32. ImageNet (**IN-1K**) [4] has 1.2 million training images and 50,000 validaton images in 1000 object categories. Birds-200-2011 (**CUB-200**) [32] is a dataset for categorizing the fine-grained objects, which contains 11,788 images of 200 bird classes. CUB-200 is used for testing MGD in transfer learning. **COCO** [23] is a standard object detection and instance segmentation benchmark.

Experimental Setting. Classification tasks use a standard training scheme. For the pre-trained student, we set init learning rate (LR) in 0.01, otherwise 0.1 for training from scratch. On CIFAR-100, the number of total epochs is 200, LR is dropped twice by 0.1 at 100th and 150th epoch. On IN-1K and CUB-200, total epochs number is 120, and LR is dropped by 0.1 after every 30 epochs. Momentum is 0.9 and weight decay is $5e-4$. We randomly crop 224 from 256 and then perform horizontal flip in IN-1K and CUB-200.

For object detection and instance segmentation, we follow the configurations in `Detectron2` [36]. The input size of training image is restricted in maximum size of 1333 and minimal size of 800. Horizontal flip is the only data augmentation. We use 8 GPUs and set mini-batch size of 2 images for each GPU. The total iterations number is 90k. The init LR is 0.01 and decreased by 0.1 at 60k and 80k iterations respectively.

The number of images for solving **M** depends on the dataset scale. We randomly sample 20k images on IN-1K and use all the training images on CIFAR-100 and CUB-200. In the detection and segmentation task, about 5k images are used for updating **M**.

4.1   Main Results

Classification. Following the competitor [11], student models are randomly initialized in tasks on CIFAR-100. We use WideResNet (WRN) [40] as the teacher, which is in the model setting of depth 28 and wide factor 4, indicated by s:28-4. The student has multiple settings in different compression aspects: depth, wide factor, and architecture. Results comparison is shown in Table 1. WRN with s:16-4 has the same wide factor but smaller depth, so its channels are in same number of teacher's at the same distillation positions. Since there is no need to reduce teacher channels, we only use MGD-SM to achieve the lowest error rate 20.53%. WRN with s:28-2 has the same depth but a larger wide factor, thus teacher channels need to be reduced. The results show that MGD-SM is better than [11] with +0.76%, MGD-RD is a little worse with +0.23%, and MGD-AMP achieves the lowest error of 21.12%. In experiments with s:16-2, whose both depth and wide factor are smaller than those of teacher, MGD-SM & RD are worse than the competitor and MGD-AMP is competitive. Another setting for student is using a different architecture, ResNet with s:56-2, which is deeper but with the same wide factor. MGD has the competitive results under this setting as well. The Table 1 shows that MGD can handle various compression types for student.

Table 1. Comparison of error rates (%) with various model settings on CIFAR-100. We average 5 runs to report final results.

Model Setting	Method	Error	Model Setting	Error	Model Setting	Error	Model Setting	Error
WRN w/ s:28-4	Teacher	21.09	–	21.09	–	21.09	–	21.09
WRN w/ s:16-4	Baseline	22.72	WRN w/ s:28-2	24.88	WRN w/ s:16-2	27.32	ResNet w/ s:56-2	27.68
	KD [13]	21.69		23.43		26.47		26.76
	FitNets [27]	21.85		23.94		26.30		26.35
	AT [41]	22.07		23.80		26.56		26.66
	Jacobian [30]	22.18		23.70		26.71		26.60
	AB [12]	21.36		23.19		26.02		26.04
	Overhaul [11]	20.89		21.98		24.08		**24.44**
	MGD - SM	**20.53**		21.22		24.72		25.20
	MGD - RD	–		22.21		24.64		26.01
	MGD - AMP	–		**21.12**		**24.06**		24.91

Table 2. Comparison of error rates (%) with MGD and previous works in large-scale classification on IN-1K.

Model	Method	Top-1 err.	Top-5 err.
ResNet-152	Teacher	21.69	5.95
ResNet-50	Baseline	23.85	7.13
	KD [13]	22.85	6.55
	AT [41]	22.75	6.35
	AB [12]	23.47	6.94
	Overhaul [11]	**21.65**	5.83
	MGD - SM	22.02	**5.68**
ResNet-50	Teacher	23.85	7.13
MobileNet-V1	Baseline	31.13	11.24
	KD [13]	31.42	11.02
	AT [41]	30.44	10.67
	AB [12]	31.11	11.29
	Overhaul [11]	28.75	9.66
	MGD - SM	28.79	9.65
	MGD - RD	29.55	10.02
	MGD - AMP	**28.53**	**9.65**

Table 3. Comparison of error rates (%) on CUB-200. The students are pre-trained on IN-1K. FT: fine-tune. FS: from-scratch.

Model	Method	Top-1 err.	Top-5 err.
ResNet-50	Teacher	20.02	6.06
MobileNet-V2	Baseline - FT	24.61	7.56
	Baseline - FS	54.97	27.0
	KD [13]	23.52	6.44
	AT [41]	23.14	6.97
	AB [12]	23.08	6.54
	Overhaul [11]	21.69	5.64
	MGD - SM	21.82	5.68
	MGD - RD	21.58	5.92
	MGD - AMP	**20.64**	**5.38**
ShuffleNet-V2	Baseline - FT	31.39	10.9
	Baseline - FS	66.28	35.7
	KD [13]	28.31	9.67
	AT [41]	28.58	9.29
	AB [12]	28.22	9.48
	Overhaul [11]	27.42	8.04
	MGD - SM	28.22	8.85
	MGD - RD	27.71	8.72
	MGD - AMP	25.95	**7.46**

In large-scale classification on IN-1K, as shown in Table 2, we use ResNet-152 to distill ResNet-50. Since the channel number is identical in each stage of them, we only investigate MGD-SM. The overall results from other works, except the Overhaul method, MGD beats other methods with maximum 1.45% improvement in top-1 accuracy. In the case of distilling MobileNet-V1 by ResNet-50, using MGD-SM has similar result with [11]. MGD w/ AMP is the best overall methods.

Table 4. Comparison of object detection and instance segmentation results. We distill lightweight backbones of RetinaNet for object detection (a, c, d) and EmbedMask for instance segmentation (b) on COCO. Here we experiment only with AMP because it's the best operation among three reduction manners.

backbone	method	AP^{bbox}	AP^{bbox}_{50}
ResNet-50	Teacher	36.37	55.37
ResNet-18	Baseline	30.30	47.53
	Overhaul [11]	30.02	46.95
	MGD - AMP	**31.15**	**48.60**
MobileNet-V2	Baseline	26.54	42.14
	Overhaul [11]	26.62	42.01
	MGD - AMP	**27.45**	**43.10**

(a) RetinaNet, 1x schedule + **single**-scale

backbone	method	AP^{mask}	AP^{mask}_{50}
ResNet-50	Teacher	33.5	54.1
ResNet-18	Baseline	26.1	43.7
	Overhaul [11]	26.3	43.9
	MGD - AMP	**26.9**	**44.2**
MobileNet-V2	Baseline	27.1	44.8
	Overhaul [11]	27.0	44.8
	MGD - AMP	**27.6**	**45.1**

(b) EmbedMask, 1x schedule + **single**-scale

backbone	method	AP^{bbox}	AP^{bbox}_{50}
ResNet-50	Teacher	37.01	56.03
ResNet-18	Baseline	30.78	47.88
	Overhaul [11]	30.26	47.22
	AT [41]	30.54	47.65
	AB [12]	31.32	48.70
	MGD - AMP	**31.38**	**48.79**

(c) RetinaNet, 1x schedule + **multi**-scale

backbone	method	AP^{bbox}	AP^{bbox}_{50}
ResNet-50	Teacher	38.73	56.72
ResNet-18	Baseline	34.63	53.08
	Overhaul [11]	34.42	52.90
	AT [41]	34.43	52.97
	AB [12]	34.92	53.50
	MGD - AMP	**35.10**	**53.76**

(d) RetinaNet, 2x schedule + **multi**-scale

Transfer Learning. We use fine-grained categorization on CUB-200 to investigate distillation for transfer learning. We implement MGD and our competitor Overhaul for using ResNet-50 to distill light students. The teacher ResNet-50 has been pre-trained on IN-1K and then trained on CUB-200. We use two prevailed lightweight models, MobileNet-V2 and ShuffleNet-V2. Conspicuously, they have fewer parameters than ResNet-50. The main results have been shown in Table 3. We have two baselines for each student: one is trained from scratch (Baseline-FS), and the other is fine-tuned (Baseline-FT) from IN-1K. We summarize the experimental results in two folds.

First, the results of two baselines show that transfer learning from a general data domain is helpful to the specific task. Students pre-trained on IN-1K could bring ~30% accuracy improvements at least.

Second, we adopt our three reduction methods of MGD to compare with Overhaul [11]. The experimental phenomenon of two students are same. MGD-AMP beats the Overhaul with 1.05% improvement and also makes MobileNet-V2 almost have the similar performance with teacher. ShuffleNet-V2 seems difficult to be distilled, there is a unignored gap with teacher. But MGD-AMP stably performs best to help ShuffleNet-V2 achieve the maximum top-1 error decrease from 31.39% to 25.95%.

No matter which reduction method we use to accomplish distillation for transfer learning, all the experimental results show that MGD is more friendly for distilling a pre-trained student.

Object Detection and Instance Segmentation. To verify the generalization of MGD, we extend it to object detection and instance segmentation. RetinaNet [22] is a modern one-stage detector, which has excellent performance in both precision and speed. EmbedMask [39] is a novel framework for instance segmentation, which utilizes embedding strategy to generate instance masks on

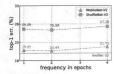
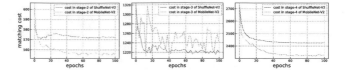

Fig. 4. Sensitivity to frequency of updating **M** on CUB-200.

Fig. 5. Curves of matching cost in three distillation stages of MobileNet-V2 and ShuffleNet-V2, which are distilled by ResNet-50 on CUB-200.

a unified one-stage structure. In this section, we experiment with RetinaNet and EmbedMask respectively, using three different backbones: ResNet-50 as teacher, ResNet-18 and MobileNet-V2 as students. All these backbones are pre-trained on IN-1K. We train the models on COCO `train2017` set and test them on `val2017`. Baselines are trained without distillation. As comparison, we also train with [11,12,41] under same configurations. The main results are presented in Table 4. For object detection, in both cases of distilling ResNet-18 and MobileNet-V2, MGD has stable improvements by 0.47–0.91 point. In segmentation, in the case of ResNet-18, we freeze the first two backbone stages to avoid OOM. MGD can bring about 0.8 mAP point for ResNet-18 and 0.6 point for MobileNet-v2, it outperforms the competitor. These results prove that MGD works more stable than previous works in object detection and instance segmentation.

4.2 Ablation Study

Frequency of Updating M. To find the best practices, we experiment on how intense the frequency of updating **M** could affect the MGD performance. Here all experiments adopt MGD-AMP. The baseline is updating **M** in the end of every training epoch (frequency = 1) as same as validation does. In Fig. 4, updating in every 2 training epochs achieves the best results both in MobileNet-V2 and ShuffleNet-V2 on CUB-200. If the frequency becomes larger than 2 in epochs (frequency = 4), it will produce higher error rates. This study suggests that MGD should not be updated either fast or lazily. On IN-1K, we update **M** after every epoch due to its large dataset scale.

Absolute Max Pooling. Absolute max pooling (AMP) leads to largest improvements compared to proposed SM and RD. Now we compare it with basic pooling operations, max pooling (MP) and average pooling (AvgP), to perform reduction along the channel axis. Table 5 shows that AMP behaves stably better than MP and AvgP. AvgP performs worst, because average pooling operation is easily to counteract the sharpness feature for a group of channel tensors. Albeit MP works closely with AMP, it's still not perfect because it would shave negative feature values by positive ones. This result shows the effect of AMP for preventing feature information loss from reduction. For a better illustration, we have a fundamental and intuitive experiment on these three pooling operations in Sect. A1.1.

With _vs._ Without Matching. This ablation checks the importance of channel math-
ing mechanism. We remove matching process
and simply use AMP as a feature reducer along
channel axis. The right table shows the results
of MGD with and without channels matching.
It proves the effectiveness of channels matching
in MGD. Both of distilling MobileNet-V2 and

Model	Top-1 err.	
With matching?		✓
ResNet-50	20.02	
MobileNet-V2	22.10	**20.64**
ShuffleNet-V2	27.62	**25.95**

ShuffleNet-V2 without matching are worse than that with matching about 1.5%
in top-1 error.

Capacity Analysis. Next, we illustrate MGD is more efficient for training than
other methods. We investigate the capacity of joint training with MGD and [11].
In experiments, we use four GeForce 1080Ti GPU cards to run training. Under
the same experimental settings, MobileNet-V2 has less parameters and memory
consumption than teacher without distillation. Table 6 shows [11] brings too
many parameters to cause OOM. As well known, additional distillation module
not only bring the learnable weights for training, but also additional gradients
and optimizer states into GPU memory [26]. In contrast, training with MGD
has little bit of additional parameters due to its basic nature of parameter-free.
Moreover it has better results.

Optimization Analysis. In this part, we analyze MGD in branches of visualization for understanding MGD with comprehensive vistas. We set our experiments
to check it in three aspects: matching cost, status of updating **M**, and features
matching & reduction.

First, Fig. 5 shows the descent curves of matching cost in three distillation
positions. We track the sum of matching costs in every 2 epochs when distilling
MobileNet-V2 and ShuffleNet-V2 with MGD-AMP. The curves show that all
the total costs are in the trend of descent during training. This phenomenon
is expected because the more related matched features are, the smaller their
matching cost becomes.

Second, Fig. 6 shows the updating status of **M** in distilling MobileNet-V2
on CUB-200. Due to the massive channel number of intermediate features, we
randomly select fifty student channels to visualize the updating status of **M**.
All the three sub-figures have a common view that at the beginning, most of
matching targets of each student channel change dramatically. Then they will
become stable after several training epochs. This result concludes that coordinate
descent is effective and friendly for the joint optimization with SGD.

Third, Fig. A1 in Appendix checks out the intermediate results of MGD in
multiple tasks. In order to check the rightness of matching status, we use the
intermediate features for visualization at the _earlier_ training iterations. We can
conclude the matching results can be trusted for guiding student to induce the
better results.

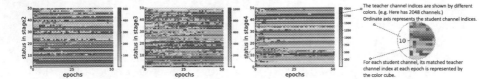

Fig. 6. Status of updating **M** in three distillation positions. Here we randomly select fifty student channels to check out their matched target channels, which are represented by the small cubes within different colors.

Table 5. Comparison of three pooling operations for channel reduction.

Model	Top-1 err.		
	MP	AvgP	AMP
ResNet-50	20.02		
MobileNet-V2	21.63	22.19	**20.64**
ShuffleNet-V2	26.20	27.27	**25.95**

Table 6. Capacity analysis. Memory consumption is measured by gigabyte. Parameters is in millions. Here *bs* indicates batch size.

Model	bs	Method	Memory	Parameters	Top-1 err.
ResNet-50	256	Teacher	8.10	25.1	20.02
MobileNet-V2	256	Student	5.01	3.5	24.61
	128	Overhaul [11]	OOM	31.5	21.69
	128	MGD - AMP	11.8	**29.1**	**20.64**

5   Discussion and Future Work

We have presented MGD as an effective distillation method within the parameter-free nature, and evaluated its three channel reduction ways in various tasks. We also experiment in multiple perspectives of ablation study to verify its effect. In the future, it's possible to supervise student in a dense manner, for example, using more than four positions to perform distillation with MGD.

References

1. Brendel, W., Todorovic, S.: Learning spatiotemporal graphs of human activities. In: ICCV (2011)
2. Chen, G., Choi, W., Yu, X., Han, T., Chandraker, M.: Learning efficient object detection models with knowledge distillation. In: NIPS (2017)
3. Cuturi, M.: Sinkhorn distances: lightspeed computation of optimal transport. In: NIPS (2013)
4. Deng, J., Dong, W., Socher, R., Li, L.J., Li, K., Fei-Fei, L.: ImageNet: a large-scale hierarchical image database. In: CVPR (2009)
5. Devlin, J., Chang, M.W., Lee, K., Toutanova, K.: Bert: pre-training of deep bidirectional transformers for language understanding. arXiv:1810.04805 (2018)
6. Duchenne, O., Joulin, A., Ponce, J.: A graph-matching kernel for object categorization. In: ICCV (2011)
7. Frogner, C., Zhang, C., Mobahi, H., Araya, M., Poggio, T.A.: Learning with a Wasserstein loss. In: NIPS (2015)
8. He, K., Girshick, R., Dollár, P.: Rethinking imagenet pre-training. In: ICCV (2019)
9. He, K., Zhang, X., Ren, S., Sun, J.: Deep residual learning for image recognition. In: CVPR (2016)

10. He, Y., Zhang, X., Sun, J.: Channel pruning for accelerating very deep neural networks. In: ICCV (2017)
11. Heo, B., Kim, J., Yun, S., Park, H., Kwak, N., Choi, J.Y.: A comprehensive overhaul of feature distillation. In: ICCV (2019)
12. Heo, B., Lee, M., Yun, S., Choi, J.Y.: Knowledge transfer via distillation of activation boundaries formed by hidden neurons. In: AAAI (2019)
13. Hinton, G., Vinyals, O., Dean, J.: Distilling the knowledge in a neural network. arXiv:1503.02531 (2015)
14. Howard, A.G., et al.: MobileNets: efficient convolutional neural networks for mobile vision applications. arXiv preprint arXiv:1704.04861 (2017)
15. Huang, Z., Wang, N.: Like what you like: knowledge distill via neuron selectivity transfer. arXiv:1707.01219 (2017)
16. Huet, B., Cross, A.D., Hancock, E.R.: Graph matching for shape retrieval. In: NIPS (1999)
17. Johnson, J., Karpathy, A., Fei-Fei, L.: DenseCap: fully convolutional localization networks for dense captioning. In: CVPR (2016)
18. Krizhevsky, A., Sutskever, I., Hinton, G.E.: ImageNet classification with deep convolutional neural networks. In: NIPS (2012)
19. Krizhevsky, A., et al.: Learning multiple layers of features from tiny images. Technical report, Citeseer (2009)
20. Kuhn, H.W.: The Hungarian method for the assignment problem. Nav. Res. Logist. Q. **2**, 83–97 (1955)
21. Lee, S., Song, B.C.: Graph-based knowledge distillation by multi-head self-attention network. arXiv:1907.02226 (2019)
22. Lin, T.Y., Goyal, P., Girshick, R., He, K., Dollár, P.: Focal loss for dense object detection. In: ICCV (2017)
23. Lin, T.-Y., et al.: Microsoft COCO: common objects in context. In: Fleet, D., Pajdla, T., Schiele, B., Tuytelaars, T. (eds.) ECCV 2014. LNCS, vol. 8693, pp. 740–755. Springer, Cham (2014). https://doi.org/10.1007/978-3-319-10602-1_48
24. Ma, N., Zhang, X., Zheng, H.-T., Sun, J.: ShuffleNet V2: practical guidelines for efficient CNN architecture design. In: Ferrari, V., Hebert, M., Sminchisescu, C., Weiss, Y. (eds.) Computer Vision – ECCV 2018. LNCS, vol. 11218, pp. 122–138. Springer, Cham (2018). https://doi.org/10.1007/978-3-030-01264-9_8
25. Park, W., Kim, D., Lu, Y., Cho, M.: Relational knowledge distillation. In: CVPR (2019)
26. Pudipeddi, B., Mesmakhosroshahi, M., Xi, J., Bharadwaj, S.: Training large neural networks with constant memory using a new execution algorithm. arXiv preprint arXiv:2002.05645 (2020)
27. Romero, A., Ballas, N., Kahou, S.E., Chassang, A., Gatta, C., Bengio, Y.: FitNets: hints for thin deep nets. arXiv:1412.6550 (2014)
28. Rubner, Y., Tomasi, C., Guibas, L.J.: The earth mover's distance as a metric for image retrieval. IJCV **40**, 99–121 (2000)
29. Sandler, M., Howard, A., Zhu, M., Zhmoginov, A., Chen, L.C.: MobileNet V2: inverted residuals and linear bottlenecks. In: CVPR (2018)
30. Srinivas, S., Fleuret, F.: Knowledge transfer with Jacobian matching. arXiv:1803.00443 (2018)
31. Tan, M., Le, Q.V.: EfficientNet: rethinking model scaling for convolutional neural networks. arXiv:1905.11946 (2019)
32. Wah, C., Branson, S., Welinder, P., Perona, P., Belongie, S.: The Caltech-UCSD Birds-200-2011 Dataset. Technical report, CNS-TR-2011-001, California Institute of Technology (2011)

33. Wang, K., Liu, Z., Lin, Y., Lin, J., Han, S.: HAQ: hardware-aware automated quantization with mixed precision. In: CVPR (2019)
34. Wright, S.J.: Coordinate descent algorithms. Math. Program. **151**(1), 3–34 (2015). https://doi.org/10.1007/s10107-015-0892-3
35. Wu, B., et al.: FBNet: hardware-aware efficient convnet design via differentiable neural architecture search. In: CVPR (2019)
36. Wu, Y., Kirillov, A., Massa, F., Lo, W.Y., Girshick, R.: Detectron2. https://github.com/facebookresearch/detectron2 (2019)
37. Yang, Z., Dai, Z., Yang, Y., Carbonell, J., Salakhutdinov, R., Le, Q.V.: XLNet: generalized autoregressive pretraining for language understanding. arXiv:1906.08237 (2019)
38. Yim, J., Joo, D., Bae, J., Kim, J.: A gift from knowledge distillation: fast optimization, network minimization and transfer learning. In: CVPR (2017)
39. Ying, H., Huang, Z., Liu, S., Shao, T., Zhou, K.: EmbedMask: embedding coupling for one-stage instance segmentation (2019)
40. Zagoruyko, S., Komodakis, N.: Wide residual networks. In: BMVC (2016)
41. Zagoruyko, S., Komodakis, N.: Paying more attention to attention: improving the performance of convolutional neural networks via attention transfer. In: ICLR (2017)
42. Zhou, F., De la Torre, F.: Factorized graph matching. In: CVPR (2012)

Clustering Driven Deep Autoencoder for Video Anomaly Detection

Yunpeng Chang[1], Zhigang Tu[1(✉)], Wei Xie[2], and Junsong Yuan[3]

[1] Wuhan University, Wuhan 430079, China
{changyunpeng,tuzhigang}@whu.edu.cn
[2] Central China Normal University, Wuhan 430079, China
xw@mail.ccnu.edu.cn
[3] State University of New York at Buffalo, Buffalo, NY 14260-2500, USA
jsyuan@buffalo.edu

Abstract. Because of the ambiguous definition of anomaly and the complexity of real data, video anomaly detection is one of the most challenging problems in intelligent video surveillance. Since the abnormal events are usually different from normal events in appearance and/or in motion behavior, we address this issue by designing a novel convolution autoencoder architecture to separately capture spatial and temporal informative representation. The spatial part reconstructs the last individual frame (LIF), while the temporal part takes consecutive frames as input and RGB difference as output to simulate the generation of optical flow. The abnormal events which are irregular in appearance or in motion behavior lead to a large reconstruction error. Besides, we design a deep k-means cluster to force the appearance and the motion encoder to extract common factors of variation within the dataset. Experiments on some publicly available datasets demonstrate the effectiveness of our method with the state-of-the-art performance.

Keywords: Video anomaly detection · Spatio-temporal dissociation · Deep k-means cluster

1 Introduction

Video anomaly detection refers to the identification of events which are deviated to the expected behavior. Due to the complexity of realistic data and the limited labelled effective data, a promising solution is to learn the regularity in normal videos with unsupervised setting. Methods based on autoencoder for abnormality detection [3,8,31,34,38,39], which focus on modeling only the normal pattern of the videos, have been proposed to address the issue of limited labelled data.

Since abnormal events can be detected by either appearance or motion, [23] uses two processing streams, where the first autoencoder learns common appearance spatial structures in normal events and the second stream learns its corresponding motion represented by an optical flow to learn a correspondence

© Springer Nature Switzerland AG 2020
A. Vedaldi et al. (Eds.): ECCV 2020, LNCS 12360, pp. 329–345, 2020.
https://doi.org/10.1007/978-3-030-58555-6_20

between appearances and their associated motions. However, optical flow may not be optimal for learning regularity as they are not specifically designed for this purpose [8,21]. Moreover, optical flow estimation has a high computational cost [33]. To overcome this drawback, we build a motion autoencoder with the stacked RGB difference [36] to learn motion information, where the RGB difference cue can be obtained much faster than the motion cue of optical flow.

In this paper, we decouple the spatial-temporal information into two sub-modules to learn regularity in both spatial and temporal feature spaces. Given the consecutive frames, the spatial autoencoder operates on the last individual frame (LIF) and the temporal autoencoder conducts on the rest of video frames. In our architecture, the temporal part produces the RGB difference between the rest of video frames and the LIF to get motion information. The spatial part, in the form of individual frame appearance, carries information about scenes and objects depicted in the video.

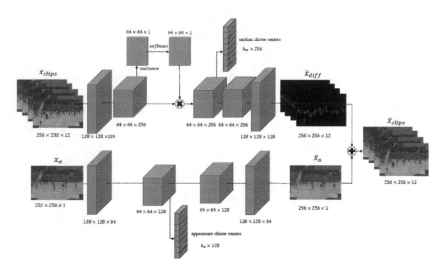

Fig. 1. Overview our video anomaly detection architecture. We dissociate the reconstruction of spatial-temporal information into two independent parts. The spatial part reconstructs the LIF, and the temporal part generates the RGB difference between the rest of video frames and the LIF. Two deep k-means clusters separately force the spatial encoder and the temporal encoder to obtain a more compressed data representation. The orange area represents our variance based attention module which can automatically assign an importance weight to the moving part of video clips in the motion autoencoder. (Color figure online)

As shown in Fig. 1, our two sub-modules can independently learn appearance and motion features, thus no matter the event is irregular in appearance feature space or motion feature space, the reconstruction of the input consecutive frames will get a large reconstruction error. Based on the characteristic that most part of the surveillance video is still and outliers have a high correlation to fast moving, we exploit a variance based attention module to automatically assign an importance weight to the moving part of video clips, which is helpful to accelerate the convergence of motion autoencoder.

In addition, we exploit two deep k-means clusters to separately force the spatial encoder and the temporal encoder to obtain a more compressed data representation and extract the common factors of variation within the normal dataset. By minimizing the distance between the data representation and cluster centers, normal examples are closely mapped to the cluster center while anomalous examples are mapped away from the cluster center.

In brief, our approach considers both appearance and motion features based on the perception that compared with normal behaviors, an abnormal behavior differs in their appearance and motion patterns. In summary, this paper makes the following contributions:

- We propose a novel autoencoder architecture to capture informative spatiotemporal representation to detect anomaly in videos by building a novel motion autoencoder, which takes consecutive frames as input and RGB difference as output to simulate the generation of optical flow. Hence the proposed method is much faster than the previous optical flow-based motion representation learning method, where the average running time of our approach is 32fps.
- We exploit a variance attention module to automatically assign an importance weight to the moving part of video clips, which is useful to improve the convergence performance of the motion autoencoder.
- We design a deep k-means cluster to force the autoencoder network to generate compact motion and appearance descriptors. Since the cluster is only trained on normal events, the distance between the cluster and the abnormal representations is much higher than between the normal patterns. The reconstruction error and the cluster distance are together used to assess the anomaly.

2   Related Work

2.1   Video Anomaly Detection with Two Stream Networks

Recently, many deep convolutional neural networks [10,25,27,35,40] have been proposed to extract high-level feature by learning temporal regularity on the video clips. To integrate spatial and temporal information together for video tasks, [30] firstly exploits a two-stream network, i.e. a separate RGB-stream and a optical flow-stream, in which the two streams are combined by late fusion for action classification. [38] introduces the two-stream architecture for anomaly detection. Still image patches and dynamic motion represented by optical flow are employed as input for two separate networks to respectively capture appearance and motion features, and the anomaly scores of these two streams are combined by late fusion for final evaluation. [26] utilizes two generator networks to learn the normal patterns of the crowd behavior, where a generator network takes the input frames to produce optical flow field images, and the other generator network reconstructs frames from the optical flow. However, the time cost of optical flow estimation is expensive [33]. In contrast, we used a RGB-difference strategy to simulate motion information, which is much faster than optical flow.

2.2   Data Representation and Data Clustering

Many anomaly detection methods [2,18,24,28,29] aim to find a "compact description" within normal events. Recently, several atuto-encoder based methods combine feature learning and clustering together. [5] jointly trains a CNN autoencoder and a multinomial logistic regression model to the autoencoder latent space. Similarly, [11] alternates the representation learning and clustering where a mini-batch k-Means is utilized as the clustering component. [37] proposes a Deep Embedded Clustering (DEC) method, which simultaneously updates the cluster centers and the data points' representations that are initialized from a pre-trained autoencoder. DEC uses soft assignments which are optimized to match stricter assignments through a Kullback-Leibler divergence loss. IDEC was subsequently proposed in [7] as an improvement of DEC by integrating the autoencoder's reconstruction error in the objective function. [13] proposes a supervised classification approach based on clustering the training samples into normality clusters. Based on this characteristic and inspired by the idea of [4], we design a deep k-means cluster to force the autoencoder network to generate compact feature representations for video anomaly detection.

3   Methods

To address the issues in video based anomaly detection, we introduce a clustering-driven autoencoder to map the normal data into a compact feature representation. Since the abnormal events are different from the normal events in appearance and/or in motion behavior, we decouple our model into two sub-modules, one for spatial part and one for temporal part.

Our proposed autoencoder is composed of three main components: (1) the appearance autoencoder network E_a and D_a, (2) the motion autoencoder network E_m and D_m, and (3) the deep k-means cluster. The spatial part, in the form of individual frame appearance, carries information about scenes and objects depicted in the video. The temporal part, feded the consecutive frames to generate the RGB difference, brings the movement information of the objects. The deep k-means cluster minimizes the distance between the data representation and cluster centers to force both the appearance encoder and the motion encoder networks to extract common factors within the training sets. The main structure of our network is shown in Fig. 1.

3.1   Spatial Autoencoder

Since some abnormal objects are partially associated with particular objects, the static appearance by itself is a useful clue [30]. To detect abnormal object with spatial features such as scenes and appearance, we feed the last frame of input video clips into the sptial autoencoder network. In our model, the appearance encoder is used to encode the input to a mid-level appearance representation from the original image pixels. The appearance autoencoder is trained with the goal of minimizing the reconstruction error between the input frame x_a and the output frame $\bar{x}_a$, therefore, the bottleneck latent-space z_a contains essential spatial information for frame reconstruction.

Given an individual frame, the appearance encoder converts it to appearance representation, denoted as z_a, and the appearance decoder reconstructs the input frame from the appearance representation, denoted as $\bar{x}_a$:

$$z_a = E_a(x_a; \theta_e^a) \tag{1}$$

$$\bar{x}_a = D_a(z_a; \theta_d^a) \tag{2}$$

where θ_e^a represents the set of the encoder's parameters, θ_d^a denotes the set of the decoder's parameters.

The loss function l_a for the appearance autoencoder is defined as Eq. (3):

$$l_a = \|x_a - \bar{x}_a\|_2 \tag{3}$$

3.2   Motion Autoencoder

Most two-stream based convolutional networks utilize warped optical flow as the source for motion modeling [30, 32]. Despite the motion feature is very useful, expensive computational cost of optical flow estimation impedes many real-time implementations. Inspired by [36], we build a motion representation without using optical flow, i.e., the stacked difference of RGB between consecutive frames and the target frame. As shown in Fig. 2, it is reasonable to hypothesize that the motion representation captured from optical flow could be learned from the simple cue of RGB difference [36]. Consequently, by learning temporal regularity and motion consistency, the motion autoencoder can learn to predict the RGB residual, and motion autoencoder can extract the data representation that contains essential motion information about the video frames.

We define x_{clips} to denote the consecutive frames, z_m to represent the motion representations, and x_{diff} to represent the RGB difference between the consecutive frames and the LIF, i.e., $x_{diff} = x_{clips} - x_a$. Given the consecutive frames, the motion encoder converts them to motion representations, and each motion representation is denoted as z_m. The motion decoder produces the RGB difference $\bar{x}_{diff}$ from the appearance representations:

$$z_m = E_m(x_{clips}; \theta_e^m) \tag{4}$$

$$\bar{x}_{diff} = D_m(z_m; \theta_d^m) \tag{5}$$

Fig. 2. Some examples of RGB video frames, RGB difference and optical flow.

where θ_e^m represents the set of the encoder's parameters, θ_d^m represents the set of the decoder's parameters. The loss function l_m for the motion autoencoder is given in Eq. (6):

$$l_m = \|x_{diff} - \bar{x}_{diff}\|_2 \tag{6}$$

3.3 Variance Attention Module

It is obvious that most part of the surveillance video is still, and the abnormal behaviors are more likely to have larger movement changes, thus we aim to learn a function to automatically assign the importance weight to the moving part of video clips. Based on this characteristic, we design a variance-based attention in temporal autoencoder to automatically assign the importance weight to the moving part of video clips. Accordingly, the abnormal object, e.g. pedestrian running fast at the subway entrance, will get larger motion loss which is helpful for fast moving abnormal events detection. Since input video clips contain irrelevant backgrounds, we utilize a temporal attention module to learn the importance of video clips. Given the representation of an input video clip x, the attention module feeds the embedded feature into a convolutional layer:

$$f_n(h, w) = W_g * x(h, w) \tag{7}$$

where $h \in (0, H]$ and $w \in (0, W]$. H and W denote the number of rows and columns of feature maps respectively. W_g represents the weight parameters of convolutional filter. We calculate the variance along the feature dimension followed by operating the l_2 normalization along spatial dimension to generate the corresponding attention map g_n:

$$v(h, w) = \frac{1}{D} \sum_{d=1}^{D} \left\| f_n(h, w, d) - \frac{1}{D} \sum_{d=1}^{D} f_n(h, w, d) \right\|_2 \tag{8}$$

$$att(h, w) = \left\| \frac{exp(v(h, w))}{\sum_{h=1, w=1}^{H, W} exp(v(h, w))} \right\|_2 \tag{9}$$

where $v(h, w)$ denotes the variance of feature maps at spatial location (h, w).

3.4   Clustering

The role of clustering is to force both the appearance encoder and motion encoder networks to extract the common factors of variation within the dataset. We utilize a deep k-means cluster method to minimize the distance between the data representation and the cluster centers. K is the number of clusters, c_k is the representation of cluster k, $1 < k < K$, and $C = \{c_1, ..., c_K\}$ is the set of representations.

For the motion representation $r_i \in R^D$ extracted from spatial location $i \in \{1, ..., N\}$, we first compute the Euclidean distance between the embeddings descriptors R^D and the corresponding cluster center. To constitute a continuous generalization of the clustering objective function, we adopt the soft-assignment to calculate the distance between the data representation r_i and the cluster centers C, where the distance is computed by Eq. (10):

$$D_m(r_i) = \sum_{k=1}^{K} \frac{e^{-\alpha\|r_i - c_k\|_2}}{\sum_{k=1}^{K} e^{-\alpha\|r_i - c_k\|_2}} \|r_i - c_k\|_2^2 \tag{10}$$

where the first part in Eq. (10) represents the soft-assignment of representation r_i to each cluster center c_k, α is a tunable hyper-parameter.

The cluster center matrix may suffer from redundancy problem if any two cluster centers getting too close. To address this issue, we introduce a penalization term to maxmimize the distance between each cluster. Inspired by [16], we construct a redundancy measure which is defined as dot product of the cluster center matrix C and its transpose C^T, and then subtracting the product by an identity matrix I:

$$R = \|CC^T - I\|_F \tag{11}$$

where $\|\|_F$ denotes the Frobenius norm of a matrix. This strategy encourages each cluster center to keep the distance from the other cluster centers and punish redundancy within the cluster centers. The objective function of our deep k-means cluster is defined as:

$$L_{cluster} = \sum_{i=1}^{N} D_m(z_i^m, C_m) + \sum_{i=1}^{N} D_a(z_i^a, C_a) + \lambda(R_m + R_a) \tag{12}$$

where D_m and D_a separately represents the distance between motion representations and their cluster centers, and the distance between appearance representations and their cluster centers. R_m and R_a respectively denotes the regularity on the motion cluster center matrix the and appearance cluster center matrix.

Since we optimize the deep k-means cluster on the training sets which contain only normal events, the anomaly events on the test set will not affect the cluster centers. During anomaly event detection, the cluster center will no longer be optimized. Hence the cluster centers can be deemed as a certain kind of normality within the training datasets.

3.5 Training Objective

To learn the model parameters, we combine all the loss functions into an objective function to train two autoencoders simultaneously: the spatial loss L_a constrains the model to produce the normal single frame; the motion loss L_m constrains the model to compute the RGB difference between the input video frames and the LIF; the cluster loss $L_{cluster}$ forces both motion and spatial autoencoder to minimize the distance between the data representation and the cluster centers:

$$Loss = L_a(x_a, \bar{x}_a) + L_m(x_{diff}, \bar{x}_{diff}) + \lambda_r * L_{cluster} \tag{13}$$

3.6 Anomaly Score

We train the model only in normal events, the reconstruction quality of video clips $\bar{x}_{clips}$ generated by $\bar{x}_a + x_{diff}$ can be used for anomaly detection, hence we compute the Euclidean distance between the x_{clips} and the $\bar{x}_{clips}$ of all pixels to measure the quality of reconstruction. The distance between data representation and the closest cluster center is another assessment to qualify the anomaly. For a given test video sequence, we define an anomaly score as:

$$s = \frac{1}{D_m * D_a * \|x_{clips} - \bar{x}_{clips}\|_2^2} \tag{14}$$

High score indicates the input video clips are more likely to be normal. Followed by [8], after calculating the score of each video over all spatial locations, we normalize the losses to get a score S(t) in the range of [0, 1] for each frame:

$$S(t) = \frac{s - min_t(s)}{max_t(s) - min_t(s)} \tag{15}$$

We use this normalized score S(t) to evaluate the probability of anomaly events contained in video clips.

4 Experiments

4.1 Video Anomaly Detection Datasets

We train our model on three publicly available datasets: the UCSD pedestrian [22], the Avenue [19], and the ShanghaiTech dataset [17]: (1) The UCSD Pedestrian 2 (Ped2) dataset contains 16 training videos and 12 testing videos with 12 abnormal events. All of these abnormal cases are about vehicles such as bicycles and cars. (2) The Avenue dataset contains 16 training videos and 21 testing videos in front of a subway station. All of these abnormal cases are about throwing objects, loitering and running. (3) The ShanghaiTech dataset contains 330 training videos and 107 testing ones with 130 abnormal events. All in all, it consists of 13 scenes and various anomaly types (Fig. 3).

Fig. 3. Some samples including normal and abnormal frames in the CUHK Avenue, the UCSD and the ShanghaiTech datasets are used for illustration. Red boxes denote anomalies in abnormal frames. (Color figure online)

4.2 Implementation Details

We resize all input video frames to 256×256 and use the Adam optimizer [15] to train our networks. To initialize the motion and spatial cluster centers, we jointly train the spatial and motion autoencoders in normal dataset without the cluster constraint at first by Eq. 3 and Eq. 6. At this stage, we set the learning rate as 1e−4, and train the spatial and motion autoencoders with 50 epochs for the UCSD Ped2 dataset, and 100 epochs for the Avenue dataset and the ShanghaiTech dataset. Then we freeze the spatial and motion autoencoders, and calculate the cluster centers via K-means to separately cluster the motion representation and spatial representation.

After initialization, the training process of our proposed model performs an alternate optimization. We first freeze the cluster centers and train the autoencoder parameters θ via Eq. 13. Then we freeze the spatial and motion autoencoders and optimize the cluster centers by Eq. 12. For the autoencoder part, we initialize the learning rate to 1e−4 and decrease it to 1e−5 at epoch 100. And we set the learning rate as 1e−5 to update the cluster centers. At this stage, we alternately train different part of our network with 100 epoch for the UCSD Ped2 dataset, and 200 epochs for the Avenue dataset and the ShanghaiTech dataset.

The final anomaly detection results are directly calculated based on both the reconstruction loss and the cluster distance according to Eq. 15.

4.3 Evaluation Metric

Following the prior works [17,19,21,22], we evaluate our method via the area under the ROC curve (AUC). The ROC curve is obtained by varying the threshold of the anomaly score. A higher AUC value represents a more accurate anomaly detection result. To ensure the comparability between different methods, we calculate AUC for the frame-level prediction [8,21,43].

4.4 Results

In this section, we compare the proposed method with different hand-crafted feature based methods [9,14,22] and deep feature based state-of-the-art meth-

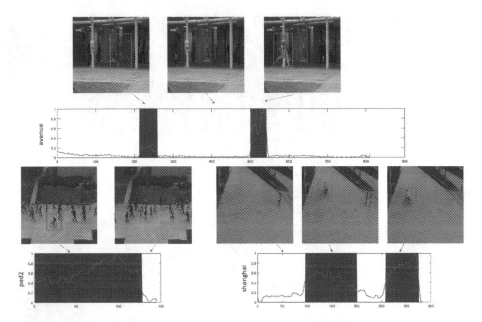

Fig. 4. Parts of the temporal regularity score of our method on the Avenue, UCSD Ped2 and ShanghaiTech datasets. The regularity score implies the possibility of normal, and the blue shaded regions are the anomaly in groundtruth.

ods including a 2D convolution autoencoder method (Conv2D-AE) [8], a 3D convolution autoencoder method (Conv3D-AE) [43], a convolution LSTM based autoencoder method (ConvLSTM-AE) [20], a stacked recurrent neural network (StackRNN) [21], and a prediction based method [17]. To be consistent with [17], we set T = 5. Specifically, our model takes 4 consecutive frames as the motion input and the last frame as the spatial autoencoder's input. We set both the motion cluster number and spatial cluster number to 32 for all datasets.

Table 1 shows the AUC results of our proposed method and the state-of-the-art approaches. We can see that our method outperforms all of them. In the upper part, compared to the hand-crafted feature based methods [14,22], the result of the proposed method is at least 4.3% more accurate (96.5% vs 92.2%) on the UCSD Ped2 dataset. In the below part, compared to the deep feature based approaches [6,8,17,20,21,43], our method also performs best on all the three datasets. Particularly, the performance of our algorithm is respectively 1.1%, 1.1%, and 0.5% better than [17] on the UCSD Ped2 dataset, the Avenue dataset, and the ShanghaiTech dataset. Besides, compared to the latest approach [23], the accuracy of our method is still 0.3% higher on the UCSD Ped2 dataset.

Figure 4 shows some qualitative examples of our method. We can find that for a normal frame, the predicted future frame tends to be close to the actual future prediction. For an abnormal frame, the predicted future frame tends to be blurry or distorted compared with the actual future frame.

Table 1. AUC of different methods on the Ped2, Avenue and ShanghaiTech datasets.

Algorithm	UCSD Ped2	Avenue	ShanghaiTech
MPPCA [14]	69.3%	–	–
MPPCA+SFA [22]	61.3%	–	–
MDT [22]	82.9%	–	–
MT-FRCN [9]	92.2%	–	–
Conv2D-AE [8]	85.0%	80.0%	60.9%
Conv3D-AE [43]	91.2%	77.1%	–
ConvLSTM-AE [20]	88.1%	77.0%	–
StackRNN [21]	92.2%	81.7%	68.0%
Abati [1]	95.41%	–%	72.5%
MemAE [6]	94.1%	83.3%	71.2%
Liu [17]	95.4%	84.9%	72.8%
Nguyen and Meunier [23]	96.2%	86.9%	–
Our method	96.5%	86.0%	73.3%

4.5 Ablation Study

In this subsection, we focus on investigating the effect of each component described in Sect. 3, including the variance attention mechanism, deep k-means clusters, and the combination of spatial information and temporal information. We combine different part of our components to conduct experiments on the Avenue dataset. For the first two parts, we consider only the motion loss and the spatial reconstruction loss. The anomaly score calculation is similar to Eq. 15. For the third part, we consider the reconstruction loss with the variance attention module. For the last part, we consider the full proposed model. Table 2 validates the effectiveness of each component. We can see that compared with the appearance information, the temporal regularity is more important for video anomaly detection. When combining the RGB difference with the spatial reconstruction, the performance improves by 2.9%. When the deep k-means cluster constraint is introduced, the spatiotemporal reconstruction multiplied by their cluster distance can further enhance the performance by 3.1%.

Table 2. Evaluation of different components of our architecture on the Avenue dataset. Results show that the combination of all components gives the best performance.

Motion	✓	–	✓	✓	✓	✓
Appearance	–	✓	✓	✓	✓	✓
Variance attention	–	–	–	✓	–	✓
Deep k-means	–	–	–	–	✓	✓
AUC	79.9%	71.2%	81.4%	82.8%	83.5%	86.0%

Table 3. AUC of the proposed method with different cluster numbers on the UCSD Ped2 dataset.

Algorithm	UCSD Ped2
without k-means	94.5%
4	95.6%
8	95.5%
16	96.0%
32	96.5%
64	96.4%

4.6 Exploration of Cluster Numbers

To evaluate the performance of the deep k-means cluster strategy on detecting abnormal events in videos, we conduct experiments on removing deep k-means cluster and changing the number of cluster centers. We use the UCSD-Ped2 datatset for testing and show the AUC results in Table 3. We separately set the number of the spatial cluster center and the motion cluster center to be 4, 8, 16, 32. Since the AUC value obtained by the autoencoder is already high at 94.5%, the cluster constraint can boost the performance by 1.1%. The AUC results of different size of cluster centers demonstrate the robustness of our method.

4.7 Attention Visualization

For a deeper understanding on the effect of our variance attention module, we visualize the motion encoder layer of the attention map. For comparison, we also show the input frames. Figure 5 shows two examples from the Avenue dataset. The left part of Fig. 5 is the normal example, where people walking normally. In the normal scene, the changing part of video sequence is relatively small, hence the attention weight of each location is quite consistent. On the other hand, the abnormal event contains a person throwing a bag, the variance attention module produces higher attention weight in areas where the movement is fast. The corresponding attention map shows that the value in the thrown bag area is much higher than the values in other areas. Since the variance attention module can automatically assign the importance weight to the moving part of video clips, the anomaly events such as running are more likely to cause higher reconstruction error. The experiments conducted in Sect. 4.5 demonstrate the effectiveness of the variance attention module.

4.8 Comparison with Optical Flow

We compare the performance and running time of RGB difference with the optical flow on the UCSD Ped2 dataset. One traditional optical flow algorithm TV-L1 [41] and one deep learning based optical flow method FlowNet2-SD [12]

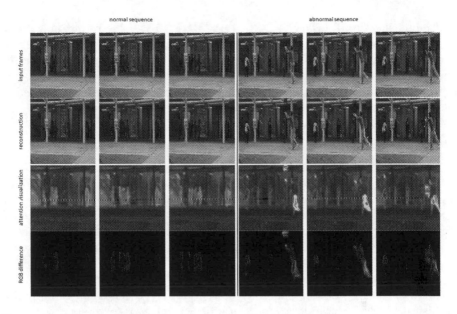

Fig. 5. The first row shows the input video frames, and the second row shows the reconstructed frames. The third row shows the visualization of the attention map in jet color map. The higher attention weight area is represented closer to red while the lower area is represented closer to blue. The forth row shows the RGB difference generated from the motion autoencoder. (Color figure online)

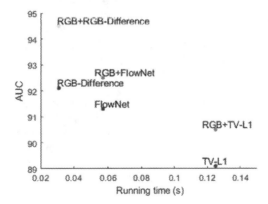

Fig. 6. AUC performance and running time on the UCSD Ped2 dataset. Compared with our "RGB+RGB difference" to the "RGB+FlowNet" method, the computational time of us is about 2 times faster, and the AUC performance is improved by 2.1%.

are selected for comparison. As shown in Fig. 6, our method is about 2.3 times faster than FlowNet2-SD [12]. Specifically, for one video frame, the FlowNet2-SD algorithm costs 0.071 s while our RGB difference strategy only needs 0.031 s. Furthermore, the accuracy of "RGB+RGB difference" is respectively 2.1% and

2.6% more than "RGB+FlowNet2-SD" and "RGB+TV-L1". We implement our method with an NVIDIA GeForce Titan Xp graphics card. It takes 0.0312 s to detect abnormal events per one video frame, i.e. 32fps, which is on par or faster than previous state-of-the-art deep learning based methods. For example, the fps of [17, 21], and [42] are respectively 25fps, 50fps, and 2fps (Where the results are copied from the original corresponding papers).

5   Conclusion

In this paper, we propose a novel clustering-driven deep autoencoder technique to generate the compact description within normal events. To learn regularity in both spatial and temporal feature spaces, we decouple the spatial-temporal information into two sub-modules. Given the consecutive frames, the spatial autoencoder operates on the last individual frame, and the temporal autoencoder processes on the rest of video frames to learn the temporal regularity by constructing the RGB difference. To force both the spatial encoder and the temporal encoder to obtain a more compact data representation, we minimize the distance between the data representation and cluster centers via two deep k-means clusters. Since the cluster is only trained on the normal events, the distance between the cluster and the representations of anomaly events is much higher than between the normal patterns. We use both the reconstruction error and the cluster distance to evaluate the anomaly. Extensive experiments on three datasets demonstrate that our method achieves the state-of-the-art performance.

Acknowledgements. This work was supported by the Fundamental Research Funds for the Central Universities (2042020KF0016 and CCNU20TS028). It was also supported by the Wuhan University-Huawei Company Project.

References

1. Abati, D., Porrello, A., Calderara, S., Cucchiara, R.: Latent space autoregression for novelty detection. In: IEEE Conference on Computer Vision and Pattern Recognition, pp. 481–490 (2019)
2. Blanchard, G., Lee, G., Scott, C.: Semi-supervised novelty detection. J. Mach. Learn. Res. **11**, 2973–3009 (2010)
3. Chang, Y., Tu, Z., Luo, B., Qin, Q.: Learning spatiotemporal representation based on 3D autoencoder for anomaly detection. In: Cree, M., Huang, F., Yuan, J., Yan, W.Q. (eds.) ACPR 2019. CCIS, vol. 1180, pp. 187–195. Springer, Singapore (2020). https://doi.org/10.1007/978-981-15-3651-9_17
4. Fard, M.M., Thonet, T., Gaussier, E.: Deep k-means: jointly clustering with k-means and learning representations. arXiv, Learning (2018)
5. Ghasedi Dizaji, K., Herandi, A., Deng, C., Cai, W., Huang, H.: Deep clustering via joint convolutional autoencoder embedding and relative entropy minimization. In: IEEE International Conference on Computer Vision (CVPR), pp. 5736–5745 (2017)

6. Gong, D., et al.: Memorizing normality to detect anomaly: memory-augmented deep autoencoder for unsupervised anomaly detection. In: IEEE International Conference on Computer Vision (ICCV), pp. 1705–1714 (2019)
7. Guo, X., Gao, L., Liu, X., Yin, J.: Improved deep embedded clustering with local structure preservation. In: International Joint Conferences on Artificial Intelligence (IJCAI), pp. 1753–1759 (2017)
8. Hasan, M., Choi, J., Neumann, J., Roy-Chowdhury, A.K., Davis, L.S.: Learning temporal regularity in video sequences. In: IEEE Conference on Computer Vision and Pattern Recognition, pp. 733–742 (2016)
9. Hinami, R., Mei, T., Satoh, S.: Joint detection and recounting of abnormal events by learning deep generic knowledge. In: IEEE International Conference on Computer Vision (ICCV), pp. 3619–3627 (2017)
10. Hinton, G.E., Salakhutdinov, R.R.: Reducing the dimensionality of data with neural networks. Science **313**(5786), 504–507 (2006)
11. Hsu, C., Lin, C.: CNN-based joint clustering and representation learning with feature drift compensation for large-scale image data. IEEE Trans. Multimed. **20**(2), 421–429 (2017)
12. Ilg, E., Mayer, N., Saikia, T., Keuper, M., Dosovitskiy, A., Brox, T.: FlowNet 2.0: evolution of optical flow estimation with deep networks. In: IEEE Conference on Computer Vision and Pattern Recognition (CVPR), pp. 2462–2470 (2017)
13. Ionescu, R.T., Khan, F.S., Georgescu, M.I., Shao, L.: Object-centric auto-encoders and dummy anomalies for abnormal event detection in video. In: IEEE Conference on Computer Vision and Pattern Recognition, pp. 7842–7851 (2019)
14. Kim, J., Grauman, K.: Observe locally, infer globally: a space-time MRF for detecting abnormal activities with incremental updates. In: IEEE Conference on Computer Vision and Pattern Recognition (CVPR), pp. 2921–2928 (2009)
15. Kingma, D.P., Ba, J.: Adam: a method for stochastic optimization. In: International Conference on Learning Representations (ICLR) (2015)
16. Lin, Z., et al.: A structured self-attentive sentence embedding. In: International Conference on Learning Representations (ICLR) (2017)
17. Liu, W., Luo, W., Lian, D., Gao, S.: Future frame prediction for anomaly detection-a new baseline. In: IEEE Conference on Computer Vision and Pattern Recognition, pp. 6536–6545 (2018)
18. Liu, Y., Zheng, Y.F.: Minimum enclosing and maximum excluding machine for pattern description and discrimination. In: International Conference on Pattern Recognition (ICPR), vol. 3, pp. 129–132 (2006)
19. Lu, C., Shi, J., Jia, J.: Abnormal event detection at 150 fps in matlab. In: IEEE International Conference on Computer Vision, pp. 2720–2727 (2013)
20. Luo, W., Liu, W., Gao, S.: Remembering history with convolutional LSTM for anomaly detection. In: International Conference on Multimedia and Expo (ICME), pp. 439–444 (2017)
21. Luo, W., Liu, W., Gao, S.: A revisit of sparse coding based anomaly detection in stacked RNN framework. In: IEEE International Conference on Computer Vision, pp. 341–349 (2017)
22. Mahadevan, V., Li, W., Bhalodia, V., Vasconcelos, N.: Anomaly detection in crowded scenes. In: 2010 IEEE Conference on Computer Vision and Pattern Recognition (CVPR), pp. 1975–1981. IEEE (2010)
23. Nguyen, T.N., Meunier, J.: Anomaly detection in video sequence with appearance-motion correspondence. In: IEEE International Conference on Computer Vision (ICCV), pp. 1273–1283 (2019)

24. Perera, P., Nallapati, R., Xiang, B.: OCGAN: one-class novelty detection using GANs with constrained latent representations. In: IEEE Conference on Computer Vision and Pattern Recognition, pp. 2898–2906 (2019)
25. Poultney, C., Chopra, S., Cun, Y.L., et al.: Efficient learning of sparse representations with an energy-based model. In: Advances in Neural Information Processing Systems, pp. 1137–1144 (2007)
26. Ravanbakhsh, M., Nabi, M., Sangineto, E., Marcenaro, L., Regazzoni, C., Sebe, N.: Abnormal event detection in videos using generative adversarial nets. In: IEEE International Conference on Image Processing (ICIP), pp. 1577–1581 (2017)
27. Rifai, S., Vincent, P., Muller, X., Glorot, X., Bengio, Y.: Contractive autoencoders: explicit invariance during feature extraction. In: International Conference on Machine Learning (ICML), pp. 833–840 (2011)
28. Ruff, L., et al.: Deep one-class classification. In: International Conference on Machine Learning, pp. 4393–4402 (2018)
29. Ruff, L., et al.: Deep semi-supervised anomaly detection. In: International Conference on Learning Representations (ICLR) (2020)
30. Simonyan, K., Zisserman, A.: Two-stream convolutional networks for action recognition in videos. In: Advances in Neural Information Processing Systems, pp. 568–576 (2014)
31. Srivastava, N., Mansimov, E., Salakhudinov, R.: Unsupervised learning of video representations using LSTMs. In: International Conference on Machine Learning, pp. 843–852 (2015)
32. Tu, Z., et al.: Multi-stream CNN: learning representations based on human-related regions for action recognition. Pattern Recogn. **79**, 32–43 (2018)
33. Tu, Z., et al.: A survey of variational and CNN-based optical flow techniques. Sig. Process. Image Commun. **72**, 9–24 (2019)
34. Tung, F., Zelek, J.S., Clausi, D.A.: Goal-based trajectory analysis for unusual behaviour detection in intelligent surveillance. Image Vis. Comput. **29**(4), 230–240 (2011)
35. Vincent, P., Larochelle, H., Bengio, Y., Manzagol, P.A.: Extracting and composing robust features with denoising autoencoders. In: International Conference on Machine Learning (ICML), pp. 1096–1103 (2008)
36. Wang, L., et al.: Temporal segment networks for action recognition in videos. IEEE Trans. Pattern Anal. Mach. Intell. **41**, 2740–2755 (2018)
37. Xie, J., Girshick, R., Farhadi, A.: Unsupervised deep embedding for clustering analysis. In: International Conference on Machine Learning, pp. 478–487 (2016)
38. Xu, D., Yan, Y., Ricci, E., Sebe, N.: Detecting anomalous events in videos by learning deep representations of appearance and motion. Comput. Vis. Image Underst. **156**, 117–127 (2017)
39. Yan, M., Meng, J., Zhou, C., Tu, Z., Tan, Y.P., Yuan, J.: Detecting spatiotemporal irregularities in videos via a 3D convolutional autoencoder. J. Vis. Commun. Image Represent. **67**, 102747 (2020)
40. Yu, T., Ren, Z., Li, Y., Yan, E., Xu, N., Yuan, J.: Temporal structure mining for weakly supervised action detection. In: IEEE International Conference on Computer Vision, pp. 5522–5531 (2019)
41. Zach, C., Pock, T., Bischof, H.: A duality based approach for realtime TV-L^1 optical flow. In: Hamprecht, F.A., Schnörr, C., Jähne, B. (eds.) DAGM 2007. LNCS, vol. 4713, pp. 214–223. Springer, Heidelberg (2007). https://doi.org/10.1007/978-3-540-74936-3_22

42. Zhao, B., Fei-Fei, L., Xing, E.P.: Online detection of unusual events in videos via dynamic sparse coding. In: IEEE Conference on Computer Vision and Pattern Recognition (CVPR), pp. 3313–3320 (2011)
43. Zimek, A., Schubert, E., Kriegel, H.P.: A survey on unsupervised outlier detection in high-dimensional numerical data. Stat. Anal. Data Min. 5(5), 363–387 (2012)

Learning to Compose Hypercolumns
for Visual Correspondence

Juhong Min[1,2], Jongmin Lee[1,2], Jean Ponce[3,4], and Minsu Cho[1,2(✉)]

[1] POSTECH, Pohang University of Science and Technology, Pohang, Korea
mscho@postech.ac.kr
[2] NPRC, The Neural Processing Research Center, Seoul, Korea
[3] Inria, Paris, France
[4] ENS, École normale supérieure, CNRS, PSL Research University,
75005 Paris, France
http://cvlab.postech.ac.kr/research/DHPF/

Abstract. Feature representation plays a crucial role in visual correspondence, and recent methods for image matching resort to deeply stacked convolutional layers. These models, however, are both monolithic and static in the sense that they typically use a specific level of features, *e.g.*, the output of the last layer, and adhere to it regardless of the images to match. In this work, we introduce a novel approach to visual correspondence that dynamically composes effective features by leveraging relevant layers conditioned on the images to match. Inspired by both multi-layer feature composition in object detection and adaptive inference architectures in classification, the proposed method, dubbed *Dynamic Hyperpixel Flow*, learns to compose hypercolumn features on the fly by selecting a small number of relevant layers from a deep convolutional neural network. We demonstrate the effectiveness on the task of semantic correspondence, *i.e.*, establishing correspondences between images depicting different instances of the same object or scene category. Experiments on standard benchmarks show that the proposed method greatly improves matching performance over the state of the art in an adaptive and efficient manner.

Keywords: Visual correspondence · Multi-layer features · Dynamic feature composition

1 Introduction

Visual correspondence is at the heart of image understanding with numerous applications such as object recognition, image retrieval, and 3D reconstruction [12]. With recent advances in neural networks [19,20,22,32,50], there has

---

Electronic supplementary material The online version of this chapter (https://doi.org/10.1007/978-3-030-58555-6_21) contains supplementary material, which is available to authorized users.

© Springer Nature Switzerland AG 2020
A. Vedaldi et al. (Eds.): ECCV 2020, LNCS 12360, pp. 346–363, 2020.
https://doi.org/10.1007/978-3-030-58555-6_21

been a significant progress in learning robust feature representation for establishing correspondences between images under illumination and viewpoint changes. Currently, the de facto standard is to use as feature representation the output of deeply stacked convolutional layers in a trainable architecture. Unlike in object classification and detection, however, such learned features have often achieved only modest performance gains over hand-crafted ones [6,40] in the task of visual correspondence [48]. In particular, correspondence between images under large intra-class variations still remains an extremely challenging problem [5,10,17,25,26,28–30,33,39,42,44–47,49,53,57] while modern neural networks are known to excel at classification [19,22]. What do we miss in using deep neural features for correspondence?

Most current approaches for correspondence build on monolithic and static feature representations in the sense that they use a specific feature layer, e.g., the last convolutional layer, and adhere to it regardless of the images to match. Correspondence, however, is all about precise localization of corresponding positions, which requires visual features at different levels, from local patterns to semantics and context; in order to disambiguate a match on similar patterns, it is necessary to analyze finer details and larger context in the image. Furthermore, relevant feature levels may vary with the images to match; the more we already know about images, the better we can decide which levels to use. In this aspect, conventional feature representations have fundamental limitations.

In this work, we introduce a novel approach to visual correspondence that dynamically composes effective features by leveraging relevant layers conditioned on the images to match. Inspired by both multi-layer feature composition, i.e., hypercolumn, in object detection [18,31,35,38] and adaptive inference architectures in classification [11,51,54], we combine the best of both worlds for visual correspondence. The proposed method learns to compose hypercolumn features on the fly by selecting a small number of relevant layers in a deep convolutional neural network. At inference time, this dynamic architecture greatly improves matching performance in an adaptive and efficient manner. We demonstrate the effectiveness of the proposed method on several benchmarks for semantic correspondence, i.e., establishing visual correspondences between images depicting different instances of the same object or scene categories, where due to large variations it may be crucial to use features at different levels.

2   Related Work

Feature Representation for Semantic Correspondence. Early approaches [3,4,15,27,37,52,55] tackle the problem of visual correspondence using hand-crafted descriptors such as HOG [6] and SIFT [40]. Since these lack high-level image semantics, the corresponding methods have difficulties with significant changes in background, view point, deformations, and instance-specific patterns. The advent of convolutional neural networks (CNN) [19,32] has led to a paradigm shift from this hand-crafted representations to deep features and boosted performance in visual correspondence [10,44,57]. Most approaches [5,17,29,47] learn to predict correlation scores between local regions in an input image pair, and some recent methods [25,26,28,45,46,49] cast this task as an image alignment problem in which a model learns to regress global geometric transformation parameters. All typically adopt a CNN pretrained on image classification as their backbone, and make predictions based on features from its final convolutional layer. While some methods [39,56] have demonstrated the advantage of using different CNN layers in capturing low-level to high-level patterns, leveraging multiple layers of deeply stacked layers has remained largely unexplored in correspondence problems.

Multi-layer Neural Features. To capture different levels of information distributed over all intermediate layers, Hariharan *et al.* propose the hypercolumn [18], a vector of multiple intermediate convolutional activations lying above a pixel for fine-grained localization. Attempts at integrating multi-level neural features have addressed object detection and segmentation [31,35,38]. In the area of visual correspondence, only a few methods [42,44,53] attempt to use multi-layer features. Unlike ours, however, these models use static features extracted from CNN layers that are chosen manually [44,53] or by greedy search [42]. While the use of hypercolumn features on the task of semantic visual correspondence has recently been explored by Min *et al.* [42], the method predefines hypercolumn layers by a greedy selection procedure, *i.e.*, beam search, using a validation dataset. In this work, we clearly demonstrate the benefit of a dynamic and learnable architecture both in strongly-supervised and weakly-supervised regimes and also outperform the work of [42] with a significant margin.

Dynamic Neural Architectures. Recently, dynamic neural architectures have been explored in different domains. In visual question answering, neural module networks [1,2] compose different answering networks conditioned on an input sentence. In image classification, adaptive inference networks [11,51,54] learn to decide whether to execute or bypass intermediate layers given an input image. Dynamic channel pruning methods [13,21] skip unimportant channels at runtime to accelerate inference. All these methods reveal the benefit of dynamic neural architectures in terms of either accuracy or speed, or both. To the best of our knowledge, our work is the first that explores a dynamic neural architecture for visual correspondence.

Our main contribution is threefold: (1) We introduce a novel dynamic feature composition approach to visual correspondence that composes features on the fly by selecting relevant layers conditioned on images to match. (2) We propose a trainable layer selection architecture for hypercolumn composition using Gumbel-softmax feature gating. (3) The proposed method outperforms recent state-of-the-art methods on standard benchmarks of semantic correspondence in terms of both accuracy and speed.

3   Dynamic Hyperpixel Flow

Given two input images to match, a pretrained convolutional network extracts a series of intermediate feature blocks for each image. The architecture we propose

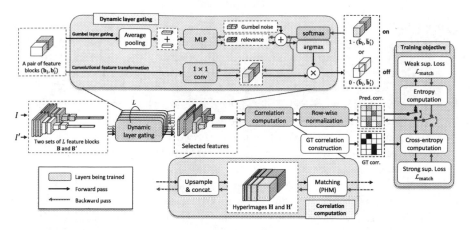

Fig. 1. The overall architecture of Dynamic Hyperpixel Flow (DHPF).

in this section, *dynamic hyperpixel flow*, learns to select a small number of layers (feature blocks) on the fly and composes effective features for reliable matching of the images. Figure 1 illustrates the overall architecture. In this section, we describe the proposed method in four steps: (i) multi-layer feature extraction, (ii) dynamic layer gating, (iii) correlation computation and matching, and (iv) training objective.

3.1 Multi-layer Feature Extraction

We adopt as a feature extractor a convolutional neural network pretrained on a large-scale classification dataset, *e.g.*, ImageNet [7], which is commonly used in most related methods [5,17,23,28,30,33,42,45–47,49]. Following the work on hypercolumns [18], however, we view the layers of the convolutional network as a non-linear counterpart of image pyramids and extract a series of multiple features along intermediate layers [42].

Let us assume the backbone network contains L feature extracting layers. Given two images I and I', source and target, the network generates two sets of L intermediate feature blocks. We denote the two sets of feature blocks by $\mathbf{B} = \{\mathbf{b}_l\}_{l=0}^{L-1}$ and $\mathbf{B}' = \{\mathbf{b}'_l\}_{l=0}^{L-1}$, respectively, and call the earliest blocks, $\mathbf{b}_0$ and $\mathbf{b}'_0$, *base* feature blocks. As in Fig. 1, each pair of source and target feature blocks at layer l is passed to the l-th layer gating module as explained next.

3.2 Dynamic Layer Gating

Given L feature block pairs $\{(\mathbf{b}_l, \mathbf{b}'_l)\}_{l=0}^{L-1}$, L layer gating modules learn to select relevant feature block pairs and transform them for establishing robust correspondences. As shown in the top of Fig. 1, the module has two branches, one for layer gating and the other for feature transformation.

Gumbel Layer Gating. The first branch of the l-th layer gating module takes the l-th pair of feature blocks $(\mathbf{b}_l, \mathbf{b}_l')$ as an input and performs global average pooling on two feature blocks to capture their channel-wise statistics. Two average pooled features of size $1 \times 1 \times c_l$ from $\mathbf{b}_l$ and $\mathbf{b}_l'$ are then added together to form a vector of size c_l. A multi-layer perceptron (MLP) composed of two fully-connected layers with ReLU non-linearity takes the vector and predicts a relevance vector $\mathbf{r}_l$ of size 2 for gating, whose entries indicate the scores for selecting or skipping ('on' or 'off') the l-th layer, respectively. We can simply obtain a gating decision using argmax over the entries, but this naïve gating precludes backpropagation since argmax is not differentiable.

To make the layer gating trainable and effective, we adopt the Gumbel-max trick [14] and its continuous relaxation [24,41]. Let $\mathbf{z}$ be a sequence of i.i.d. Gumbel random noise and let Y be a discrete random variable with K-class categorical distribution $\mathbf{u}$, i.e., $p(Y = y) \propto u_y$ and $y \in \{0, ..., K - 1\}$. Using the Gumbel-max trick [14], we can reparamaterize sampling Y to $y = \arg\max_{k \in \{0, ..., K-1\}} (\log u_k + z_k)$. To approximate the argmax in a differentiable manner, the continuous relaxation [24,41] of the Gumbel-max trick replaces the argmax operation with a softmax operation. By expressing a discrete random sample y as a one-hot vector $\mathbf{y}$, a sample from the Gumbel-softmax can be represented by $\hat{\mathbf{y}} = \text{softmax}((\log \mathbf{u} + \mathbf{z})/\tau)$, where τ denotes the temperature of the softmax. In our context, the discrete random variable obeys a Bernoulli distribution, i.e., $y \in \{0, 1\}$, and the predicted relevance scores represent the log probability distribution for 'on' and 'off', i.e., $\log \mathbf{u} = \mathbf{r}_l$. Our Gumbel-softmax gate thus has a form of

$$\hat{\mathbf{y}}_l = \text{softmax}(\mathbf{r}_l + \mathbf{z}_l), \tag{1}$$

where $\mathbf{z}_l$ is a pair of i.i.d. Gumbel random samples and the softmax temperature τ is set to 1.

Convolutional Feature Transformation. The second branch of the l-th layer gating module takes the l-th pair of feature blocks $(\mathbf{b}_l, \mathbf{b}_l')$ as an input and transforms each feature vector over all spatial positions while reducing its dimension by $\frac{1}{\rho}$; we implement it using 1×1 convolutions, i.e., position-wise linear transformations, followed by ReLU non-linearity. This branch is designed to transform the original feature block of size $h_l \times w_l \times c_l$ into a more compact and effective representation of size $h_l \times w_l \times \frac{c_l}{\rho}$ for our training objective. We denote the pair of transformed feature blocks by $(\bar{\mathbf{b}}_l, \bar{\mathbf{b}}_l')$. Note that if l-th Gumbel gate chooses to skip the layer, then the feature transformation of the layer can be also ignored thus reducing the computational cost.

Forward and Backward Propagations. During training, we use the *straight-through* version of the Gumbel-softmax estimator [24]: forward passes proceed with discrete samples by argmax whereas backward passes compute gradients of the softmax relaxation of Eq. (1). In the forward pass, the transformed feature pair $(\bar{\mathbf{b}}_l, \bar{\mathbf{b}}_l')$ is simply multiplied by 1 ('on') or 0 ('off') according to the gate's discrete decision $\mathbf{y}$. While the Gumbel gate always makes discrete decision $\mathbf{y}$ in the forward pass, the continuous relaxation in the backward pass allows gradients to propagate through softmax output $\hat{\mathbf{y}}$, effectively updating both branches, the feature transformation and the relevance estimation, regardless of the gate's decision. Note that this stochastic gate with random noise increases the diversity of samples and is thus crucial in preventing mode collapse in training. At test time, we simply use deterministic gating by argmax without Gumbel noise [24]. As discussed in Sect. 4:2, we found that the proposed hard gating trained with Gumbel softmax is superior to conventional soft gating with sigmoid in terms of both accuracy and speed.

3.3 Correlation Computation and Matching

The output of gating is a set of selected layer indices, $S = \{s_1, s_2, ..., s_N\}$. We construct a *hyperimage* $\mathbf{H}$ for each image by concatenating transformed feature blocks of the selected layers along channels with upsampling: $\mathbf{H} = [\zeta(\mathbf{b}_{s_1}), \zeta(\mathbf{b}_{s_2}), ..., \zeta(\mathbf{b}_{s_N})]$, where ζ denotes a function that spatially upsamples the input feature block to the size of $\mathbf{b}_0$, the *base* block. Note that the number of selected layers N is fully determined by the gating modules. If all layers are off, then we use the base feature block by setting $S = \{0\}$. We associate with each spatial position p of the hyperimage the corresponding image coordinates and hyperpixel feature [42]. Let us denote by $\mathbf{x}_p$ the image coordinate of position p, and by $\mathbf{f}_p$ the corresponding feature, *i.c.*, $\mathbf{f}_p = \mathbf{H}(\mathbf{x}_p)$. The hyperpixel at position p in the hyperimage is defined as $\mathbf{h}_p = (\mathbf{x}_p, \mathbf{f}_p)$. Given source and target images, we obtain two sets of hyperpixels, $\mathcal{H}$ and $\mathcal{H}'$. In order to reflect geometric consistency in matching, we adapt probablistic Hough matching (PHM) [4,17] to hyperpixels, similar to [42]. The key idea of PHM is to re-weight appearance similarity by Hough space voting to enforce geometric consistency. In our context, let $\mathcal{D} = (\mathcal{H}, \mathcal{H}')$ be two sets of hyperpixels, and $m = (\mathbf{h}, \mathbf{h}')$ be a match where $\mathbf{h}$ and $\mathbf{h}'$ are respectively elements of $\mathcal{H}$ and $\mathcal{H}'$. Given a Hough space $\mathcal{X}$ of possible offsets (image transformations) between the two hyperpixels, the confidence for match m, $p(m|\mathcal{D})$, is computed as $p(m|\mathcal{D}) \propto p(m_\mathrm{a}) \sum_{\mathbf{x} \in \mathcal{X}} p(m_\mathrm{g}|\mathbf{x}) \sum_{m \in \mathcal{H} \times \mathcal{H}'} p(m_\mathrm{a}) p(m_\mathrm{g}|\mathbf{x})$ where $p(m_\mathrm{a})$ represents the confidence for appearance matching and $p(m_\mathrm{g}|\mathbf{x})$ is the confidence for geometric matching with an offset $\mathbf{x}$, measuring how close the offset induced by m is to $\mathbf{x}$. By sharing the Hough space $\mathcal{X}$ for all matches, PHM efficiently computes match confidence with good empirical performance [4,15,17,42]. In this work, we compute appearance matching confidence using hyperpixel features by $p(m_\mathrm{a}) \propto \mathrm{ReLU}\left(\frac{\mathbf{f}_p \cdot \mathbf{f}'_p}{\|\mathbf{f}_p\| \|\mathbf{f}'_p\|}\right)^2$, where the squaring has the effect of suppressing smaller matching confidences. On the output $|\mathcal{H}| \times |\mathcal{H}'|$ correlation matrix of PHM, we perform soft mutual nearest neighbor filtering [47] to suppress noisy correlation values and denote the filtered matrix by $\mathbf{C}$.

Dense Matching and Keypoint Transfer. From the correlation matrix $\mathbf{C}$, we establish hyperpixel correspondences by assigning to each source hyperpixel $\mathbf{h}_i$ the target hyperpixel $\hat{\mathbf{h}}'_j$ with the highest correlation. Since the spatial resolutions

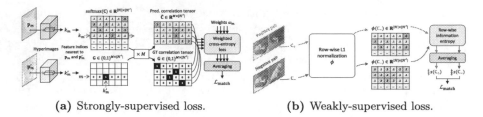

(a) Strongly-supervised loss.   (b) Weakly-supervised loss.

Fig. 2. Matching loss computation using (a) keypoint annotations (strong supervision) and (b) image pairs only (weak supervision). Best viewed in electronic form.

of the hyperimages are the same as those of base feature blocks, which are relatively high in most cases (*e.g.*, 1/4 of input image with ResNet-101 as the backbone), such hyperpixel correspondences produce quasi-dense matches.

Furthermore, given a keypoint $\mathbf{p}_m$ in the source image, we can easily predict its corresponding position $\hat{\mathbf{p}}'_m$ in the target image by transferring the keypoint using its nearest hyperpixel correspondence. In our experiments, we collect all correspondences of neighbor hyperpixels of keypoint $\mathbf{p}_m$ and use the geometric average of their individual transfers as the final prediction $\hat{\mathbf{p}}'_m$ [42]. This consensus keypoint transfer method improves accuracy by refining mis-localized predictions of individual transfers.

3.4 Training Objective

We propose two objectives to train our model using different degrees of supervision: strongly-supervised and weakly-supervised regimes.

Learning with Strong Supervision. In this setup, we assume that keypoint match annotations are given for each training image pair, as in [5,17,42]; each image pair is annotated with a set of coordinate pairs $\mathcal{M} = \{(\mathbf{p}_m, \mathbf{p}'_m)\}_{m=1}^{M}$, where M is the number of match annotations.

To compare the output of our network with ground-truth annotations, we convert the annotations into a form of discrete correlation matrix. First of all, for each coordinate pair $(\mathbf{p}_m, \mathbf{p}'_m)$, we identify their nearest position indices (k_m, k'_m) in hyperimages. On the one hand, given the set of identified match index pairs $\{(k_m, k'_m)\}_{m=1}^{M}$, we construct a ground-truth matrix $\mathbf{G} \in \{0,1\}^{M \times |\mathcal{H}'|}$ by assigning one-hot vector representation of k'_m to the m-th row of $\mathbf{G}$. On the other hand, we construct $\hat{\mathbf{C}} \in \mathbb{R}^{M \times |\mathcal{H}'|}$ by assigning the k_m-th row of $\mathbf{C}$ to the m-th row of $\hat{\mathbf{C}}$. We apply softmax to each row of the matrix $\hat{\mathbf{C}}$ after normalizing it to have zero mean and unit variance. Figure 2a illustrates the construction of $\hat{\mathbf{C}}$ and $\mathbf{G}$. Corresponding rows between $\hat{\mathbf{C}}$ and $\mathbf{G}$ can now be compared as categorical probability distributions. We thus define the strongly-supervised matching loss as the sum of cross-entropy values between them:

$$\mathcal{L}_{\text{match}} = -\frac{1}{M} \sum_{m=1}^{M} \omega_m \sum_{j=1}^{|\mathcal{H}'|} \mathbf{G}_{mj} \log \hat{\mathbf{C}}_{mj}, \tag{2}$$

where ω_m is an importance weight for the m-th keypoint. The keypoint weight ω_m helps training by reducing the effect of the corresponding cross-entropy term if the Eucliean distance between predicted keypoint $\hat{\mathbf{p}}'_m$ and target keypoint $\mathbf{p}'_m$ is smaller than some threshold distance δ_{thres}:

$$\omega_m = \begin{cases} (\|\hat{\mathbf{p}}'_m - \mathbf{p}'_m\| / \delta_{\text{thres}})^2 & \text{if } \|\hat{\mathbf{p}}'_m - \mathbf{p}'_m\| < \delta_{\text{thres}}, \\ 1 & \text{otherwise.} \end{cases} \tag{3}$$

The proposed objective for strongly-supervised learning can also be used for self-supervised learning with synthetic pairs [45,49][1], which typically results in trading off the cost of supervision against the generalization performance.

Learning with Weak Supervision. In this setup, we assume that only image-level labels are given for each image pair as either positive (the same class) or negative (different class), as in [23,47]. Let us denote the correlation matrix of a positive pair by $\mathbf{C}_+$ and that of a negative pair by $\mathbf{C}_-$. For $\mathbf{C} \in \mathbb{R}^{|\mathcal{H}| \times |\mathcal{H}'|}$, we define its correlation entropy as $s(\mathbf{C}) = -\frac{1}{|\mathcal{H}|} \sum_{i=1}^{|\mathcal{H}|} \sum_{j=1}^{|\mathcal{H}'|} \phi(\mathbf{C})_{ij} \log \phi(\mathbf{C})_{ij}$ where $\phi(\cdot)$ denotes row-wise L1-normalization. Higher correlation entropy indicates less distinctive correspondences between the two images. As illustrated in Fig. 2b, assuming that the positive images are likely to contain more distinctive correspondences, we encourage low entropy for positive pairs and high entropy for negative pairs. The weakly-supervised matching loss is formulated as

$$\mathcal{L}_{\text{match}} = \frac{s(\mathbf{C}_+) + s(\mathbf{C}_+^\top)}{s(\mathbf{C}_-) + s(\mathbf{C}_-^\top)}. \tag{4}$$

Layer Selection Loss. Following the work of [54], we add a soft constraint in our training objective to encourage the network to select each layer at a certain rate: $\mathcal{L}_{\text{sel}} = \sum_{l=0}^{L-1} (\bar{z}_l - \mu)^2$ where $\bar{z}_l$ is a fraction of image pairs within a mini-batch for which the l-th layer is selected and μ is a hyperparameter for the selection rate. This improves training by increasing diversity in layer selection and, as will be seen in our experiments, allows us to trade off between accuracy and speed in testing.

Finally, the training objective of our model is defined as the combination of the matching loss (either strong or weak) and the layer selection loss: $\mathcal{L} = \mathcal{L}_{\text{match}} + \mathcal{L}_{\text{sel}}$.

---

[1] For example, we can obtain keypoint annotations for free by forming a synthetic pair by applying random geometric transformation (*e.g.*, affine or TPS [8]) on an image and then sampling some corresponding points between the original image and the warped image using the transformation applied.

Table 1. Performance on standard benchmarks in accuracy and speed (avg. time per pair). The subscript of each method name denotes its feature extractor. Some results are from [25, 28, 33, 42]. Numbers in bold indicate the best performance and underlined ones are the second best. The average inference time (the last column) is measured on test split of PF-PASCAL [16] and includes all the pipelines of the models: from feature extraction to keypoint prediction.

Sup.	Sup. signal	Methods	PF-PASCAL				PF-WILLOW			Caltech-101		Time (ms)
			PCK @ α_{img}			α_{bbox}	PCK @ α_{bbox}			LT-ACC	IoU	
			0.05	0.1	0.15	0.1	0.05	0.1	0.15			
none	–	PF$_{\text{HOG}}$ [15]	31.4	62.5	79.5	45.0	28.4	56.8	68.2	0.78	0.50	>1000
self	synthetic pairs	CNNGeo$_{\text{res101}}$ [45]	41.0	69.5	80.4	68.0	36.9	69.2	77.8	0.79	0.56	40
		A2Net$_{\text{res101}}$ [49]	42.8	70.8	83.3	67.0	36.3	68.8	84.4	0.80	0.57	53
weak	bbox	SF-Net$_{\text{res101}}$ [33]	53.6	81.9	90.6	78.7	46.3	74.0	84.2	0.88	0.67	51
	image-level labels	Weakalign$_{\text{res101}}$ [46]	49.0	74.8	84.0	72.0	37.0	70.2	79.9	<u>0.85</u>	**0.63**	41
		RTNs$_{\text{res101}}$ [28]	55.2	75.9	85.2	–	41.3	71.9	86.2	–	–	376
		NC-Net$_{\text{res101}}$ [47]	54.3	78.9	86.0	70.0	33.8	67.0	83.7	<u>0.85</u>	0.60	261
		DCC-Net$_{\text{res101}}$ [23]	<u>55.6</u>	**82.3**	<u>90.5</u>	–	43.6	73.8	86.5	–	–	>261
		DHPF$^{\mu=0.4}_{\text{res50}}$ (ours)	54.8	79.0	89.8	<u>74.5</u>	48.7	75.7	87.3	<u>0.85</u>	0.59	**31**
		DHPF$_{\text{res50}}$ (ours)	54.7	79.0	89.7	<u>74.5</u>	**51.8**	<u>78.7</u>	<u>89.6</u>	<u>0.85</u>	0.59	<u>33</u>
		DHPF$_{\text{res101}}$ (ours)	**56.1**	<u>82.1</u>	**91.1**	**78.5**	<u>50.2</u>	**80.2**	**91.1**	**0.86**	<u>0.61</u>	56
strong	src & trg keypoint matches	SCNet$_{\text{vgg16}}$ [17]	36.2	72.2	82.0	48.2	38.6	70.4	85.3	0.79	0.51	>1000
		HPF$_{\text{res50}}$ [42]	60.5	83.4	92.1	76.5	46.5	72.4	84.7	**0.88**	**0.64**	34
		HPF$_{\text{res101}}$ [42]	60.1	84.8	92.7	78.5	45.9	74.4	85.6	<u>0.87</u>	<u>0.63</u>	63
		DHPF$^{\mu=0.4}_{\text{res50}}$ (ours)	70.2	<u>89.1</u>	94.0	85.0	45.8	73.3	86.6	0.86	0.60	**30**
		DHPF$_{\text{res50}}$ (ours)	<u>72.6</u>	88.9	<u>94.3</u>	<u>85.6</u>	<u>47.9</u>	<u>74.8</u>	<u>86.7</u>	0.86	0.61	<u>34</u>
		DHPF$_{\text{res101}}$ (ours)	**75.7**	**90.7**	**95.0**	**87.8**	**49.5**	**77.6**	**89.1**	<u>0.87</u>	0.62	58

4 Experiments

In this section we compare our method to the state of the art and discuss the results. The code and the trained model are available online at our project page.

Feature Extractor Networks. As the backbone networks for feature extraction, we use ResNet-50 and ResNet-101 [19], which contains 49 and 100 conv layers in total (excluding the last FC), respectively. Since features from adjacent layers are strongly correlated, we extract the base block from `conv1` maxpool and intermediate blocks from layers with residual connections (before ReLU). They amounts to 17 and 34 feature blocks (layers) in total, respectively, for ResNet-50 and ResNet-101. Following related work [5, 17, 23, 28, 33, 42, 45–47, 49], we freeze the backbone network parameters during training for fair comparison.

Datasets. Experiments are done on four benchmarks for semantic correspondence: PF-PASCAL [16], PF-WILLOW [15], Caltech-101 [34], and SPair-71k [43]. PF-PASCAL and PF-WILLOW consist of keypoint-annotated image pairs, 1,351

Table 2. Performance on SPair-71k dataset in accuracy (per-class PCK with $\alpha_{bbox} = 0.1$). TR represents transferred models trained on PF-PASCAL while FT denotes fine-tuned (trained) models on SPair-71k.

Sup.	Methods	aero	bike	bird	boat	bottle	bus	car	cat	chair	cow	dog	horse	mbike	person	plant	sheep	train	tv	all
self	TR CNNGeo$_{res101}$ [45]	21.3	15.1	34.6	12.8	31.2	26.3	24.0	30.6	11.6	24.3	20.4	12.2	19.7	15.6	14.3	9.6	28.5	28.8	18.1
	FT CNNGeo$_{res101}$ [45]	23.4	16.7	40.2	14.3	36.4	27.7	26.0	32.7	12.7	27.4	22.8	13.7	20.9	21.0	17.5	10.2	30.8	34.1	20.6
	TR A2Net$_{res101}$ [49]	20.8	17.1	37.4	13.9	33.6	29.4	26.5	34.9	12.0	26.5	22.5	13.3	21.3	20.0	16.9	11.5	28.9	31.6	20.1
	FT A2Net$_{res101}$ [49]	22.6	18.5	42.0	16.4	37.9	30.8	26.5	35.6	13.3	29.6	24.3	16.0	21.6	22.8	20.5	13.5	31.4	36.5	22.3
weak	TR WeakAlign$_{res101}$ [46]	23.4	17.0	41.6	14.6	37.6	28.1	26.6	32.6	12.6	27.9	23.0	13.6	21.3	22.2	17.9	10.9	31.5	34.8	21.1
	FT WeakAlign$_{res101}$ [46]	22.2	17.6	41.9	15.1	38.1	27.4	27.2	31.8	12.8	26.8	22.6	14.2	20.0	22.2	17.9	10.4	32.2	35.1	20.9
	TR NC-Net$_{res101}$ [47]	24.0	16.0	45.0	13.7	35.7	25.9	19.0	50.4	14.3	32.6	27.4	19.2	21.7	20.3	20.4	13.6	33.6	40.4	26.4
	FT NC-Net$_{res101}$ [47]	17.9	12.2	32.1	11.7	29.0	19.9	16.1	39.2	9.9	23.9	18.8	15.7	17.4	15.9	14.8	9.6	24.2	31.1	20.1
	TR DHPF$_{res101}$ (ours)	21.5	21.8	57.2	13.9	34.3	23.1	17.3	50.4	17.4	34.8	36.2	19.7	24.3	32.5	22.2	17.6	30.9	36.5	28.5
	FT DHPF$_{res101}$ (ours)	17.5	19.0	52.5	15.4	35.0	19.4	15.7	51.9	17.3	37.3	35.7	19.7	25.5	31.6	20.9	18.5	24.2	41.1	27.7
strong	FT HPF$_{res101}$ [42]	25.2	18.9	52.1	15.7	38.0	22.8	19.1	52.9	17.9	33.0	32.8	20.6	24.4	27.9	21.1	15.9	31.5	35.6	28.2
	TR DHPF$_{res101}$ (ours)	22.6	23.0	57.7	15.1	34.1	20.5	14.7	48.6	19.5	31.9	34.5	19.6	23.0	30.0	22.9	15.5	28.2	30.2	27.4
	FT DHPF$_{res101}$ (ours)	38.4	23.8	68.3	18.9	42.6	27.9	20.1	61.6	22.0	46.9	46.1	33.5	27.6	40.1	27.6	28.1	49.5	46.5	37.3

pairs from 20 categories, and 900 pairs from 4 categories, respectively. Caltech-101 [34] contains segmentation-annotated 1,515 pairs from 101 categories. SPair-71k [43] is a more challenging large-scale dataset recently introduced in [42], consisting of keypoint-annotated 70,958 image pairs from 18 categories with diverse view-point and scale variations.

Evaluation Metrics. As an evaluation metric for PF-PASCAL, PF-WILLOW, and SPair-71k, the probability of correct keypoints (PCK) is used. The PCK value given a set of predicted and ground-truth keypoint pairs $\mathcal{P} = \{(\hat{\mathbf{p}}'_m, \mathbf{p}'_m)\}_{m=1}^{M}$ is measured by $\mathrm{PCK}(\mathcal{P}) = \frac{1}{M}\sum_{m=1}^{M} \mathbb{1}[\|\hat{\mathbf{p}}'_m - \mathbf{p}'_m\| \leq \alpha_\tau \max(w_\tau, h_\tau)]$. As an evaluation metric for the Caltech-101 benchmark, the label transfer accuracy (LT-ACC) [36] and the intersection-over-union (IoU) [9] are used. Running time (average time per pair) for each method is measured using its authors' code on a machine with an Intel i7-7820X CPU and an NVIDIA Titan-XP GPU.

Hyperparameters. The layer selection rate μ and the channel reduction factor ρ are determined by grid search using the validation split of PF-PASCAL. As a result, we set $\mu = 0.5$ and $\rho = 8$ in our experiments if not specified otherwise. The threshold δ_{thres} in Eq. (3) is set to be $\max(w_\tau, h_\tau)/10$.

4.1 Results and Comparisons

First, we train both of our strongly and weakly-supervised models on the PF-PASCAL [16] dataset and test on three standard benchmarks of PF-PASCAL (test split), PF-WILLOW and Caltech-101. The evaluations on PF-WILLOW and Caltech-101 are to verify transferability. In training, we use the same splits of PF-PASCAL proposed in [17] where training, validation, and test sets respectively contain 700, 300, and 300 image pairs. Following [46,47], we augment the training pairs by horizontal flipping and swapping. Table 1 summarizes our result and those of recent methods [15,17,28,30,42,45–47,49]. Second, we train

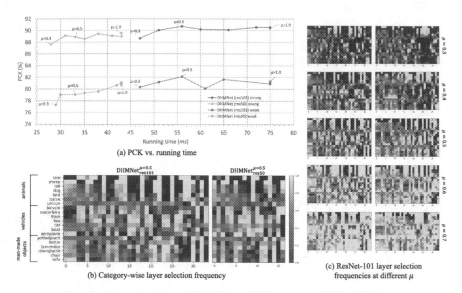

(a) PCK vs. running time

(b) Category-wise layer selection frequency

(c) ResNet-101 layer selection frequencies at different μ

Fig. 3. Analysis of layer selection on PF-PASCAL dataset (a) PCK vs. running time with varying selection rate μ (b) Category-wise layer selection frequencies (x-axis: candidate layer index, y-axis: category) of the strongly-supervised model with different backbones: ResNet-101 (left) and ResNet-50 (right) (c) ResNet-101 layer selection frequencies of strongly (left) and weakly (right) supervised models at different layer selection rates μ. Best viewed in electronic form.

our model on the SPair-71k dataset [43] and compare it to other recent methods [42,45–47,49]. Table 2 summarizes the results.

Strongly-Supervised Regime. As shown in the bottom sections of Table 1 and 2, our strongly-supervised model clearly outperforms the previous state of the art by a significant margin. It achieves 5.9%, 3.2%, and 9.1% points of PCK ($\alpha_{img} = 0.1$) improvement over the current state of the art [42] on PF-PASCAL, PF-WILLOW, and SPair-71k, respectively, and the improvement increases further with a more strict evaluation threshold, *e.g.*, more than 15% points of PCK with $\alpha_{img} = 0.05$ on PF-PASCAL. Even with a smaller backbone network (ResNet-50) and smaller selection rate ($\mu = 0.4$), our method achieves competitive performance with the smallest running time on the standard benchmarks of PF-PASCAL, PF-WILLOW, and Caltech-101.

Weakly-Supervised Regime. As shown in the middle sections of Table 1 and 2, our weakly-supervised model also achieves the state of the art in the weakly-supervised regime. In particular, our model shows more reliable transferablility compared to strongly-supervised models, outperforming both weakly [23] and strongly-supervised [42] state of the arts by 6.4% and 5.8% points of PCK respectively on PF-WILLOW. On the Caltech-101 benchmark, our method is comparable to the best among the recent methods. Note that unlike other benchmarks, the evaluation metric of Caltech-101 is indirect (*i.e.*, accuracy of mask transfer).

On the SPair-71k dataset, where image pairs have large view point and scale differences, the methods of [46,47] as well as ours do not successfully learn in the weakly-supervised regime; they (FT) all underperform transferred models (TR) trained on PF-PASCAL. This result reveals current weakly-supervised objectives are all prone to large variations, which requires further research in the future.

Effect of Layer Selection Rate μ [54]. The plot in Fig. 3a shows PCK and running time of our models trained with different layer selection rates μ. It shows that smaller selection rates in training lead to faster running time in testing, at the cost of some accuracy, by encouraging the model to select a smaller number of layers. The selection rate μ can thus be used for speed-accuracy trade-off.

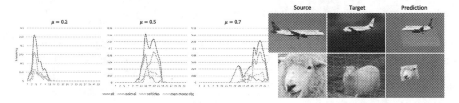

Fig. 4. Frequencies over the numbers of selected layers with different selection rates μ (x-axis: the number of selected layers, y-axis: frequency). Best viewed in electronics.

Fig. 5. Example results on SPair-71k dataset. The source images are warped to the target ones using resultant correspondences.

Analysis of Layer Selection Patterns. Category-wise layer selection patterns in Fig. 3b show that each group of animal, vehicle, and man-made object categories shares its own distinct selection patterns. The model with a small rate ($\mu = 0.3$) tends to select the most relevant layers only while the model with larger rates ($\mu > 0.3$) tends to select more complementary layers as seen in Fig. 3c. For each $\mu \in \{0.3, 0.4, 0.5\}$ in Fig. 3c, the network tends to select low-level features for vehicle and man-made object categories while it selects mostly high-level features for animal category. We conjecture that it is because low-level (geometric) features such as lines, corners and circles appear more often in the vehicle and man-made classes compared to the animal classes. Figure 4 plots the frequencies over the numbers of selected layers with different selection rate μ, where vehicles tend to require more layers than animals and man-made objects.

Qualitative Results. Some challenging examples on SPair-71k [43] and PF-PASCAL [16] are shown in Fig. 5 and 6 respectively: Using the keypoint correspondences, TPS transformation [8] is applied to source image to align target image. The object categories of the pairs in Fig. 6 are in order of table, potted plant, and tv. Alignment results of each pair demonstrate the robustness of our model against major challenges in semantic correspondences such as large changes in view-point and scale, occlusion, background clutters, and intra-class variation.

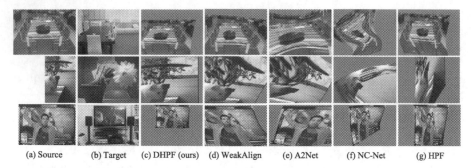

(a) Source   (b) Target   (c) DHPF (ours)   (d) WeakAlign   (e) A2Net   (f) NC-Net   (g) HPF

Fig. 6. Example results on PF-PASCAL [16]: (a) source image, (b) target image and (c) DHPF (ours), (d) WeakAlign [46], (e) A2Net [49], (f) NC-Net [47], and (g) HPF [42].

Ablation Study. We also conduct an ablation study to see the impacts of major components: Gumbel layer gating (GLG), conv feature transformation (CFT), probabilistic Hough matching (PHM), keypoint importance weight ω_m, and layer selection loss $\mathcal{L}_{\text{sel}}$. All the models are trained with strong supervision and evaluated on PF-PASCAL. Since the models with a PHM component have no training parameters, they are directly evaluated on the test split. Table 3 summarizes the results. It reveals that among others CFT in the dynamic gating module is the most significant component in boosting performance and speed; without the feature transformation along with channel reduction, our models do not successfully learn in our experiments and even fail to achieve faster per-pair inference time. The result of 'w/o ω_m' reveals the effect of the keypoint weight ω_m in Eq. (2) by replacing it with uniform weights for all m, *i.e.*, $\omega_m = 1$; putting less weights on easy examples helps in training the model by focusing on hard examples. The result of 'w/o $\mathcal{L}_{\text{sel}}$' shows the performance of the model using $\mathcal{L}_{\text{match}}$ only in training; performance drops with slower running time, demonstrating the effectiveness of the layer selection constraint in terms of both speed and accuracy. With all the components jointly used, our model achieves the highest PCK measure of 90.7%. Even with the smaller backbone network, ResNet-50, the model still outperforms previous state of the art and achieves real-time matching as well as described in Fig. 3 and Table 1.

Computational Complexity. The average feature dimensions of our model before correlation computation are 2089, 3080, and 3962 for each $\mu \in \{0.3, 0.4, 0.5\}$ while those of recent methods [23,33,42,47] are respectively 6400, 3072, 1024, 1024. The dimension of hyperimage is relatively small as GLG efficiently prunes irrelevant features and CFT effectively maps features onto smaller subspace, thus being more practical in terms of speed and accuracy as demonstrated in Table 1 and 3. Although [23,47] use lighter feature maps compared to ours, a series of 4D convolutions heavily increases time and memory complexity of the network, making them expensive for practical use (31 ms (ours) vs. 261 ms [23,47]).

Table 3. Ablation study on PF-PASCAL. (GLG: Gumbel layer gating with selection rates μ, CFT: conv feature transformation)

GLG	CFT	PHM	PCK (α_{img})			Time (ms)
			0.05	0.1	0.15	
0.5	✓	✓	75.7	90.7	95.0	58
0.4	✓	✓	73.6	90.4	95.3	51
0.3	✓	✓	73.1	88.7	94.4	47
	✓	✓	70.4	88.1	94.1	64
0.5		✓	43.6	74.7	87.5	176
0.5	✓		68.3	86.9	91.6	57
		✓	37.6	68.7	84.6	124
	✓		68.1	85.5	91.6	61
0.5			35.0	54.8	63.4	173
w/o ω_m			69.8	86.1	91.9	57
w/o $\mathcal{L}_{sel}$			68.1	89.2	93.5	56

Table 4. Comparison to soft layer gating on PF-PASCAL.

Gating function	PCK (α_{img})			Time (ms)
	0.05	0.1	0.15	
Gumbel$_{\mu=0.5}$	75.7	90.7	95.0	58
sigmoid	71.1	88.2	92.8	74
sigmoid$_{\mu=0.5}$	72.1	87.8	93.3	75
sigmoid + $\ell 1$	65.9	87.2	91.0	60

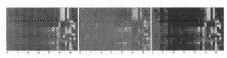

Fig. 7. ResNet-101 layer selection frequencies for 'sigmoid' (left), 'sigmoid$_{\mu=0.5}$' (middle), and 'sigmoid + $\ell 1$' (right) gating.

4.2 Comparison to Soft Layer Gating

The Gumbel gating function in our dynamic layer gating can be replaced with conventional soft gating using sigmoid. We have investigated different types of soft gating as follows: (1) 'sigmoid': The MLP of dynamic gating at each layer predicts a scalar input for sigmoid and the transformed feature block pairs are weighted by the sigmoid output. (2) 'sigmoid$_{\mu=0.5}$': In training the 'sigmoid' gating, the layer selection loss $\mathcal{L}_{sel}$ with $\mu = 0.5$ is used to encourage the model to increase diversity in layer selection. (3) 'sigmoid + $\ell 1$': In training the 'sigmoid' gating, the $\ell 1$ regularization on the sigmoid output is used to encourage the soft selection result to be sparse. Table 4 summarizes the results and Fig. 7 compares their layer selection frequencies.

While the soft gating modules provide decent results, all of them perform worse than the proposed Gumbel layer gating in both accuracy and speed. The slower per-pair inference time of 'sigmoid' and 'sigmoid$_{\mu=0.5}$' indicates that *soft* gating is not effective in skipping layers due to its non-zero gating values. We find that the sparse regularization of 'sigmoid + $\ell 1$' recovers the speed but only at the cost of significant accuracy points. Performance drop of soft gating in accuracy may result from the *deterministic* behavior of the soft gating during training that prohibits exploring diverse combinations of features at different levels. In contrast, the Gumbel gating during training enables the network to perform more comprehensive trials of a large number of different combinations of multi-level features, which help to learn better gating. Our experiments also show that *discrete* layer selection along with *stochastic* learning in searching the best combination is highly effective for learning to establish robust correspondences in terms of both accuracy and speed.

5   Conclusion

We have presented a dynamic matching network that predicts dense correspondences by composing hypercolumn features using a small set of relevant layers from a CNN. The state-of-the-art performance of the proposed method indicates that the use of dynamic multi-layer features in a trainable architecture is crucial for robust visual correspondence. We believe that our approach may prove useful for other domains involving correspondence such as image retrieval, object tracking, and action recognition. We leave this to future work.

Acknowledgements. This work is supported by Samsung Advanced Institute of Technology (SAIT) and also by Basic Science Research Program (NRF-2017R1E1A1A01077999) and Next-Generation Information Computing Development Program (NRF-2017M3C4A7069369) through the National Research Foundation of Korea (NRF) funded by the Ministry of Science, ICT, Korea. Jean Ponce was supported in part by the Louis Vuitton/ENS chair in artificial intelligence and the Inria/NYU collaboration and also by the French government under management of Agence Nationale de la Recherche as part of the "Investissements dâavenir" program, reference ANR-19-P3IA-0001 (PRAIRIE 3IA Institute).

References

1. Andreas, J., Rohrbach, M., Darrell, T., Klein, D.: Learning to compose neural networks for question answering. In: Proceedings of the Conference of the North American Chapter of the Association for Computational Linguistics: Human Language Technologies (NAACL) (2016)
2. Andreas, J., Rohrbach, M., Darrell, T., Klein, D.: Neural module networks. In: Proceedings of the IEEE Conference on Computer Vision and Pattern Recognition (CVPR) (2016)
3. Bristow, H., Valmadre, J., Lucey, S.: Dense semantic correspondence where every pixel is a classifier. In: Proceedings of the IEEE International Conference on Computer Vision (ICCV) (2015)
4. Cho, M., Kwak, S., Schmid, C., Ponce, J.: Unsupervised object discovery and localization in the wild: part-based matching with bottom-up region proposals. In: Proceedings of the IEEE Conference on Computer Vision and Pattern Recognition (CVPR) (2015)
5. Choy, C.B., Gwak, J., Savarese, S., Chandraker, M.: Universal correspondence network. In: Proceedings of the Neural Information Processing Systems (NeurIPS) (2016)
6. Dalal, N., Triggs, B.: Histograms of oriented gradients for human detection. In: Proceedings of the IEEE Conference on Computer Vision and Pattern Recognition (CVPR) (2005)
7. Deng, J., Dong, W., Socher, R., Li, L.J., Li, K., Fei-Fei, L.: ImageNet: a large-scale hierarchical image database. In: Proceedings of the IEEE Conference on Computer Vision and Pattern Recognition (CVPR) (2009)

8. Donato, G., Belongie, S.: Approximate thin plate spline mappings. In: Heyden, A., Sparr, G., Nielsen, M., Johansen, P. (eds.) ECCV 2002. LNCS, vol. 2352, pp. 21–31. Springer, Heidelberg (2002). https://doi.org/10.1007/3-540-47977-5_2

9. Everingham, M., Van Gool, L., Williams, C.K.I., Winn, J., Zisserman, A.: The pascal visual object classes (VOC) challenge. Int. J. Comput. Vis. (IJCV) **88**, 303–338 (2010)

10. Fathy, M.E., Tran, Q.-H., Zia, M.Z., Vernaza, P., Chandraker, M.: Hierarchical metric learning and matching for 2D and 3D geometric correspondences. In: Ferrari, V., Hebert, M., Sminchisescu, C., Weiss, Y. (eds.) ECCV 2018. LNCS, vol. 11219, pp. 832–850. Springer, Cham (2018). https://doi.org/10.1007/978-3-030-01267-0_49

11. Figurnov, M., et al.: Spatially adaptive computation time for residual networks. In: Proceedings of the IEEE Conference on Computer Vision and Pattern Recognition (CVPR) (2017)

12. Forsyth, D., Ponce, J.: Computer Vision: A Modern Approach, 2nd edn. Prentice Hall (2011)

13. Gao, X., Zhao, Y., Dudziak, L., Mullins, R., Xu, C.Z.: Dynamic channel pruning: feature boosting and suppression. In: Proceedings of the International Conference on Learning Representations (ICLR) (2019)

14. Gumbel, E.: Statistical theory of extreme values and some practical applications: a series of lectures. Applied mathematics series, U.S. Govt. Print. Office (1954)

15. Ham, B., Cho, M., Schmid, C., Ponce, J.: Proposal flow. In: Proceedings of the IEEE Conference on Computer Vision and Pattern Recognition (CVPR) (2016)

16. Ham, B., Cho, M., Schmid, C., Ponce, J.: Proposal flow: semantic correspondences from object proposals. IEEE Trans. Pattern Anal. Mach. Intell. (TPAMI) **40**, 1711–1725 (2018)

17. Han, K., et al.: SCNet: learning semantic correspondence. In: Proceedings of the IEEE International Conference on Computer Vision (ICCV) (2017)

18. Hariharan, B., Arbeláez, P., Girshick, R., Malik, J.: Hypercolumns for object segmentation and fine-grained localization. In: Proceedings of the IEEE Conference on Computer Vision and Pattern Recognition (CVPR) (2015)

19. He, K., Zhang, X., Ren, S., Sun, J.: Deep residual learning for image recognition. In: Proceedings of the IEEE Conference on Computer Vision and Pattern Recognition (CVPR) (2016)

20. Hu, J., Shen, L., Sun, G.: Squeeze-and-excitation networks. In: Proceedings of the IEEE Conference on Computer Vision and Pattern Recognition (CVPR) (2018)

21. Hua, W., De Sa, C., Zhang, Z., Suh, G.E.: Channel gating neural networks. arXiv preprint arXiv:1805.12549 (2018)

22. Huang*, G., Liu*, Z., van der Maaten, L., Weinberger, K.: Densely connected convolutional networks. In: Proceedings of the IEEE Conference on Computer Vision and Pattern Recognition (CVPR) (2017)

23. Huang, S., Wang, Q., Zhang, S., Yan, S., He, X.: Dynamic context correspondence network for semantic alignment. In: Proceedings of the IEEE International Conference on Computer Vision (ICCV) (2019)

24. Jang, E., Gu, S., Poole, B.: Categorical reparameterization with Gumbel-Softmax. In: Proceedings of the International Conference on Learning Representations (ICLR) (2017)

25. Jeon, S., Kim, S., Min, D., Sohn, K.: PARN: pyramidal affine regression networks for dense semantic correspondence. In: Ferrari, V., Hebert, M., Sminchisescu, C., Weiss, Y. (eds.) ECCV 2018. LNCS, vol. 11210, pp. 355–371. Springer, Cham (2018). https://doi.org/10.1007/978-3-030-01231-1_22

26. Kanazawa, A., Jacobs, D.W., Chandraker, M.: WarpNet: weakly supervised matching for single-view reconstruction. In: Proceedings of the IEEE Conference on Computer Vision and Pattern Recognition (CVPR) (2016)
27. Kim, J., Liu, C., Sha, F., Grauman, K.: Deformable spatial pyramid matching for fast dense correspondences. In: Proceedings of the IEEE Conference on Computer Vision and Pattern Recognition (CVPR) (2013)
28. Kim, S., Lin, S., Jeon, S.R., Min, D., Sohn, K.: Recurrent transformer networks for semantic correspondence. In: Proceedings of the Neural Information Processing Systems (NeurIPS) (2018)
29. Kim, S., Min, D., Ham, B., Jeon, S., Lin, S., Sohn, K.: FCSS: fully convolutional self-similarity for dense semantic correspondence. In: Proceedings of the IEEE Conference on Computer Vision and Pattern Recognition (CVPR) (2017)
30. Kim, S., Min, D., Lin, S., Sohn, K.: DCTM: discrete-continuous transformation matching for semantic flow. In: Proceedings of the IEEE International Conference on Computer Vision (ICCV) (2017)
31. Kong, T., Yao, A., Chen, Y., Sun, F.: HyperNet: towards accurate region proposal generation and joint object detection. In: Proceedings of the IEEE Conference on Computer Vision and Pattern Recognition (CVPR) (2016)
32. Krizhevsky, A., Sutskever, I., Hinton, G.E.: ImageNet classification with deep convolutional neural networks. In: Proceedings of the Neural Information Processing Systems (NeurIPS) (2012)
33. Lee, J., Kim, D., Ponce, J., Ham, B.: SFNet: learning object-aware semantic correspondence. In: Proceedings of the IEEE Conference on Computer Vision and Pattern Recognition (CVPR) (2019)
34. Li, F.F., Fergus, R., Perona, P.: One-shot learning of object categories. IEEE Trans. Pattern Anal. Mach. Intell. (TPAMI) **28**, 594–611 (2006)
35. Lin, T.Y., Dollár, P., Girshick, R., He, K., Hariharan, B., Belongie, S.: Feature pyramid networks for object detection. In: Proceedings of the IEEE Conference on Computer Vision and Pattern Recognition (CVPR) (2017)
36. Liu, C., Yuen, J., Torralba, A.: Nonparametric scene parsing: label transfer via dense scene alignment. In: Proceedings of the IEEE Conference on Computer Vision and Pattern Recognition (CVPR) (2009)
37. Liu, C., Yuen, J., Torralba, A.: Sift flow: dense correspondence across scenes and its applications. IEEE Trans. Pattern Anal. Mach. Intell. (TPAMI) **33**, 978–994 (2011)
38. Liu, S., Huang, D., Wang, Y.: Receptive field block net for accurate and fast object detection. In: Ferrari, V., Hebert, M., Sminchisescu, C., Weiss, Y. (eds.) ECCV 2018. LNCS, vol. 11215, pp. 404–419. Springer, Cham (2018). https://doi.org/10.1007/978-3-030-01252-6_24
39. Long, J.L., Zhang, N., Darrell, T.: Do convnets learn correspondence? In: Proceedings of the Neural Information Processing Systems (NeurIPS) (2014)
40. Lowe, D.G.: Distinctive image features from scale-invariant keypoints. Int. J. Comput. Vis. (IJCV) **60**, 91–110 (2004)
41. Maddison, C., Mnih, A., Whye Teh, Y.: The concrete distribution: a continuous relaxation of discrete random variables. In: Proceedings of the International Conference on Learning Representations (ICLR) (2017)
42. Min, J., Lee, J., Ponce, J., Cho, M.: Hyperpixel flow: semantic correspondence with multi-layer neural features. In: Proceedings of the IEEE International Conference on Computer Vision (ICCV) (2019)
43. Min, J., Lee, J., Ponce, J., Cho, M.: SPair-71k: a large-scale benchmark for semantic correspondence. arXiv preprint arXiv:1908.10543 (2019)

44. Novotny, D., Larlus, D., Vedaldi, A.: AnchorNet: a weakly supervised network to learn geometry-sensitive features for semantic matching. In: Proceedings of the IEEE Conference on Computer Vision and Pattern Recognition (CVPR) (2017)
45. Rocco, I., Arandjelovic, R., Sivic, J.: Convolutional neural network architecture for geometric matching. In: Proceedings of the IEEE Conference on Computer Vision and Pattern Recognition (CVPR) (2017)
46. Rocco, I., Arandjelović, R., Sivic, J.: End-to-end weakly-supervised semantic alignment. In: Proceedings of the IEEE Conference on Computer Vision and Pattern Recognition (CVPR) (2018)
47. Rocco, I., Cimpoi, M., Arandjelović, R., Torii, A., Pajdla, T., Sivic, J.: Neighbourhood consensus networks. In: Proceedings of the Neural Information Processing Systems (NeurIPS) (2018)
48. Schonberger, J.L., Hardmeier, H., Sattler, T., Pollefeys, M.: Comparative evaluation of hand-crafted and learned local features. In: Proceedings of the IEEE Conference on Computer Vision and Pattern Recognition (CVPR) (2017)
49. Seo, P.H., Lee, J., Jung, D., Han, B., Cho, M.: Attentive semantic alignment with offset-aware correlation kernels. In: Ferrari, V., Hebert, M., Sminchisescu, C., Weiss, Y. (eds.) ECCV 2018. LNCS, vol. 11208, pp. 367–383. Springer, Cham (2018). https://doi.org/10.1007/978-3-030-01225-0_22
50. Simonyan, K., Zisserman, A.: Very deep convolutional networks for large-scale image recognition. In: Proceedings of the International Conference on Learning Representations (ICLR) (2015)
51. Srivastava, R.K., Greff, K., Schmidhuber, J.: Highway networks. In: Proceedings of the International Conference on Machine Learning (ICML) (2015)
52. Taniai, T., Sinha, S.N., Sato, Y.: Joint recovery of dense correspondence and cosegmentation in two images. In: Proceedings of the IEEE Conference on Computer Vision and Pattern Recognition (CVPR) (2016)
53. Ufer, N., Ommer, B.: Deep semantic feature matching. In: Proceedings of the IEEE Conference on Computer Vision and Pattern Recognition (CVPR) (2017)
54. Veit, A., Belongie, S.: Convolutional networks with adaptive inference graphs. In: Ferrari, V., Hebert, M., Sminchisescu, C., Weiss, Y. (eds.) ECCV 2018. LNCS, vol. 11205, pp. 3–18. Springer, Cham (2018). https://doi.org/10.1007/978-3-030-01246-5_1
55. Yang, F., Li, X., Cheng, H., Li, J., Chen, L.: Object-aware dense semantic correspondence. In: Proceedings of the IEEE Conference on Computer Vision and Pattern Recognition (CVPR) (2017)
56. Zeiler, M.D., Fergus, R.: Visualizing and understanding convolutional networks. In: Fleet, D., Pajdla, T., Schiele, B., Tuytelaars, T. (eds.) ECCV 2014. LNCS, vol. 8689, pp. 818–833. Springer, Cham (2014). https://doi.org/10.1007/978-3-319-10590-1_53
57. Zhou, T., Krahenbuhl, P., Aubry, M., Huang, Q., Efros, A.A.: Learning dense correspondence via 3D-guided cycle consistency. In: Proceedings of the IEEE Conference on Computer Vision and Pattern Recognition (CVPR) (2016)

Stochastic Bundle Adjustment for Efficient and Scalable 3D Reconstruction

Lei Zhou[1]([⊠])(iD), Zixin Luo[1](iD), Mingmin Zhen[1](iD), Tianwei Shen[1](iD), Shiwei Li[2](iD), Zhuofei Huang[1](iD), Tian Fang[2](iD), and Long Quan[1](iD)

[1] Hong Kong University of Science and Technology, Hong Kong, China
{lzhouai,zluoag,mzhen,tshenaa,zhuangbr,quan}@cse.ust.hk
[2] Everest Innovation Technology, Hong Kong, China
{sli,fangtian}@altizure.com

Abstract. Current bundle adjustment solvers such as the Levenberg-Marquardt (LM) algorithm are limited by the bottleneck in solving the Reduced Camera System (RCS) whose dimension is proportional to the camera number. When the problem is scaled up, this step is neither efficient in computation nor manageable for a single compute node. In this work, we propose a stochastic bundle adjustment algorithm which seeks to decompose the RCS approximately inside the LM iterations to improve the efficiency and scalability. It first reformulates the quadratic programming problem of an LM iteration based on the clustering of the visibility graph by introducing the equality constraints across clusters. Then, we propose to relax it into a chance constrained problem and solve it through sampled convex program. The relaxation is intended to eliminate the interdependence between clusters embodied by the constraints, so that a large RCS can be decomposed into independent linear sub-problems. Numerical experiments on unordered Internet image sets and sequential SLAM image sets, as well as distributed experiments on large-scale datasets, have demonstrated the high efficiency and scalability of the proposed approach. Codes are released at https://github.com/zlthinker/STBA.

Keywords: Stochastic bundle adjustment · Clustering · 3D reconstruction

1 Introduction

Bundle Adjustment (BA) is typically formulated as a nonlinear least square problem to refine the parameters of cameras and 3D points. It is usually addressed by the Levenberg-Marquardt (LM) algorithm, where a linear equation system called *Reduced Camera System* (RCS) [15,25] must be solved in each iteration.

Electronic supplementary material The online version of this chapter (https://doi.org/10.1007/978-3-030-58555-6_22) contains supplementary material, which is available to authorized users.

© Springer Nature Switzerland AG 2020
A. Vedaldi et al. (Eds.): ECCV 2020, LNCS 12360, pp. 364–379, 2020.
https://doi.org/10.1007/978-3-030-58555-6_22

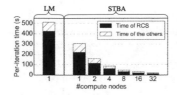

Fig. 1. Per-iteration time of bundle adjustment w.r.t. the compute node number. The Levenberg-Marquardt (LM) algorithm is limited by the bottleneck when solving the reduced camera system (RCS). Our STBA splits the RCS into independent sub-problems, which achieves a speedup on a single-threaded compute node. Besides, STBA allows parallel and distributed computing with multiple compute nodes which further improves the efficiency and scalability.

However, when the problem is scaled up, solving the RCS has been a bottleneck which takes a major portion of computation time (see the first bar of Fig. 1). The dimension of the RCS is proportional to the camera number, and thus the increase of cameras would ramp up the computation and memory consumption, although methods have been proposed to use efficient linear solvers [2, 13, 21, 23, 34] and economize on matrix manipulations [2, 22, 34]. Furthermore, different to other operations such as Jacobian or gradient evaluations, this step is indivisible, making it hard to fit BA for parallel and distributed computing.

In order to accomplish efficient and scalable reconstructions, clustering has been adopted as a useful practice to decompose a large problem into smaller, more manageable ones. For example, a number of SfM approaches have been developed in a divide and conquer fashion, which first reconstruct the partitioned sub-maps independently and then merge the partial reconstructions together [17, 29, 37, 38]. Although these methods are able to produce the initial sparse reconstructions in an efficient and scalable way, a full bundle adjustment is still indispensable to optimize the camera and point parameters globally. Therefore, in the context of BA, the methods [16, 36] proposed to distribute the objectives of BA to the split sub-models and optimize the sum of the objectives under the distributed optimization frameworks [5, 10], which, however, involves extra costly inner iterations and thus makes the optimization over-complicated.

In this work, we follow the direction of exploiting the clustering methods and push forward the investigation on how to integrate a clustering scheme into the BA problem systematically. Instead of applying a fixed, and one-time partition at the pre-processing step, we derive a stochastic clustering-based strategy within each LM iteration so as to decompose the RCS for efficiency and scalability.

- First, we reformulate the quadratic programming problem of an LM iteration based on the clustering of the visibility graph. Such a formulation splits the problem into the most elementary structures, but meanwhile introduces additional equality constraints and raises the computational cost.
- Second, in order to make the above problem efficiently solvable, we propose to relax the constraints into chance constraints [24] and then solve it with

sampled convex program [7]. The approach helps to eliminate the interdependence between different clusters by randomized constraint reduction, which hence decomposes the RCS into independent linear sub-problems related to the clusters. In this way, an approximate step can be achieved efficiently.
- Third, we present an add-on technique which helps to correct the approximate steps towards the steepest descent direction within a small trust region to improve the convergence.

Due to the stochastic process induced by the sampled convex program, we term our algorithm STochastic Bundle Adjustment (STBA), which brings the following tangible advantages. First, solving the split RCS in place of the original one has achieved a great speedup thanks to the reduced complexity. Second, the solving process can be made parallel and scalable to accommodate the growth of camera numbers, since all the sub-steps of a BA iteration can be decomposed. In Fig. 1, we visualize how the running time is reduced by distributing STBA over multiple compute nodes.

2   Related Works

Bundle adjustment (BA) is typically solved by the Levenberg-Marquardt (LM) algorithm [26], which approximately linearizes the error functions inside a local trust region and then solves a linear *normal equation* for an update step. SBA [25] first simplified the norm equation into a *reduced camera system (RCS)* through Schur complement by taking advantage of the special problem structure. After this, efforts were dedicated to solving the RCS faster in either exact or inexact ways. The exact solvers apply Cholesky factorization to the reduced camera matrix **S**, while exploiting variable ordering [3,12] and supernodal methods [12,30] for acceleration. The inexact solvers are based on the Conjugate Gradient (CG) method [19] coupled with various preconditioners [2,20,23], which attains inexact solutions with better efficiency. Apart from the algorithmic improvements, [22,34] presented well-optimized implementations of the LM solver to save the memory usage and exploit the CPU and GPU parallelism. However, despite the efforts above, solving a large and indivisible RCS will increasingly become the bottleneck of a BA solver when the problem is scaled up.

In order to make large-scale reconstruction tractable, the clustering methods are initially introduced into the structure from motion (SfM) domain. Basically, a divide-and-conquer strategy is applied, which first partitions a large scene into multiple sub-maps and then merges the partial reconstructions globally [17,29,37,38]. In the formulation of these approaches, a reduced optimization problem other than the original BA problem is addressed, thus leading to a sub-optimal result. For example, [29,38] factored out the internal variables inside the sub-maps and [37] registered all the cameras with motion averaging [9] without the involvement of points. In the realm of BA, [23] derived a block-diagonal preconditioner for the RCS from the clustering of cameras, but the clustering did not help to decompose the problem as it is done in the SfM algorithms [17,29,37,38]. Instead, [16,36] proposed to apply the distributed optimization frameworks like the Douglas-Rachford method [10] and ADMM [5] onto the empirically clusterized BA problems towards large scales. Although built upon a theoretical foundation, the methods required costly inner iterations and introduced a plethora of latent parameters during optimization.

3   Bundle Adjustment Revisited

In this section, we first revisit the bundle adjustment problem and its LM solution to give the necessary preliminaries and terminologies. Henceforth, vectors and matrices appear in boldface and $\|.\|$ denotes the L2 norm.

A bundle adjustment problem is built upon a bipartite visibility graph $\mathcal{G} = (\mathcal{C} \cup \mathcal{P}, \mathcal{E})$. Here, $\mathcal{C} = \{\mathbf{c}_i \in \mathbb{R}^d\}_{i=1}^m$ denotes the set of m cameras parameterized by d-dimensional vectors, $\mathcal{P} = \{\mathbf{p}_i \in \mathbb{R}^3\}_{i=1}^n$ denotes the set of n 3D points, and $\mathcal{E} = \{\mathbf{q}_i \in \mathbb{R}^2\}_{i=1}^q$ denotes the set of q projections. The objective is to minimize $F(\mathbf{x}) = \|\mathbf{f}(\mathbf{x})\|^2$, where $\mathbf{f}$ denotes a $2q$-dimensional vector of reprojection errors and $\mathbf{x}$ concatenates camera parameters $\mathbf{c} \in \mathbb{R}^{dm}$ and point parameters $\mathbf{p} \in \mathbb{R}^{3n}$, i.e., $\mathbf{x} = [\mathbf{c}^T, \mathbf{p}^T]^T$.

The LM algorithm achieves an update step $\Delta\mathbf{x}$ at each iteration by linearizing $\mathbf{f}(\mathbf{x})$ as $\mathbf{J}(\mathbf{x})\Delta\mathbf{x} + \mathbf{f}(\mathbf{x})$ in a trust region around $\mathbf{x}$, where $\mathbf{J}(\mathbf{x}) = \nabla_{\mathbf{x}}\mathbf{f}(\mathbf{x}) = [\mathbf{J}(\mathbf{c}), \mathbf{J}(\mathbf{p})]$ is the Jacobian matrix. Then the minimization of $F(\mathbf{x})$ is turned into

$$\min_{\Delta\mathbf{x}} \|\mathbf{J}(\mathbf{x})\Delta\mathbf{x} + \mathbf{f}(\mathbf{x})\|^2 + \lambda\|\mathbf{D}\Delta\mathbf{x}\|^2, \tag{1}$$

whose solution comes from the normal equation below

$$\begin{bmatrix} \mathbf{J}(\mathbf{x}) \\ \sqrt{\lambda}\mathbf{D} \end{bmatrix} \Delta\mathbf{x} = \begin{bmatrix} -\mathbf{f}(\mathbf{x}) \\ \mathbf{0} \end{bmatrix}, \tag{2}$$

where $\lambda > 0$ is the damping parameter and typically $\mathbf{D} = \mathrm{diag}(\mathbf{J}^T(\mathbf{x})\mathbf{J}(\mathbf{x}))^{\frac{1}{2}}$. For notational simplicity, we write $\mathbf{J} \triangleq \mathbf{J}(\mathbf{x})$, $\mathbf{J_c} \triangleq \mathbf{J}(\mathbf{c})$, $\mathbf{J_p} \triangleq \mathbf{J}(\mathbf{p})$ and $\mathbf{f} \triangleq \mathbf{f}(\mathbf{x})$. After multiplying $[\mathbf{J}^T, \sqrt{\lambda}\mathbf{D}^T]$ at both sides of Eq. 2, we have

$$(\mathbf{J}^T\mathbf{J} + \lambda\mathbf{D}^T\mathbf{D})\Delta\mathbf{x} = -\mathbf{J}^T\mathbf{f}, \tag{3}$$

which can be re-written in the form

$$\begin{bmatrix} \mathbf{B} & \mathbf{E} \\ \mathbf{E}^T & \mathbf{C} \end{bmatrix} \begin{bmatrix} \Delta\mathbf{c} \\ \Delta\mathbf{p} \end{bmatrix} = \begin{bmatrix} \mathbf{v} \\ \mathbf{w} \end{bmatrix}, \tag{4}$$

where $\mathbf{B} = \mathbf{J_c}^T\mathbf{J_c} + \lambda\mathrm{diag}(\mathbf{J_c}^T\mathbf{J_c})$, $\mathbf{C} = \mathbf{J_p}^T\mathbf{J_p} + \lambda\mathrm{diag}(\mathbf{J_p}^T\mathbf{J_p})$, $\mathbf{E} = \mathbf{J_c}^T\mathbf{J_p}$, $\mathbf{v} = -\mathbf{J_c}^T\mathbf{f}$, and $\mathbf{w} = -\mathbf{J_p}^T\mathbf{f}$. Equation 4 can be simplified by the Schur complement [25], which leads to

$$\mathbf{S}\Delta\mathbf{c} = \mathbf{v} - \mathbf{E}\mathbf{C}^{-1}\mathbf{w}, \tag{5}$$

$$\Delta\mathbf{p} = \mathbf{C}^{-1}(\mathbf{w} - \mathbf{E}^T\Delta\mathbf{c}), \tag{6}$$

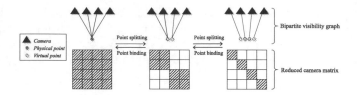

Fig. 2. Illustration of point splitting/binding over the visibility graph (top) and the corresponding structure of the reduced camera matrix (bottom). Point splitting helps to reshape the reduced camera matrix into the block-diagonal structure, while point binding does the inverse.

where $\mathbf{S} = \mathbf{B} - \mathbf{E}\mathbf{C}^{-1}\mathbf{E}^T$ is the Schur complement of $\mathbf{C}$. Here, $\mathbf{S}$, known as the *reduced camera matrix*, is a block structured symmetric positive definite matrix. The block $\mathbf{S}_{ij} \in \mathbb{R}^{d \times d}$ is nonzero iff cameras $\mathbf{c}_i$ and $\mathbf{c}_j$ observe at least one common point. Although a variety of sparse Cholesky factorization techniques [3,12,30] and preconditioned conjugate gradient methods [2,20,23] have been developed to solve the *reduced camera system (RCS)* of Eq. 5, it still can be prohibitive when the camera number m grows large.

4   Stochastic Bundle Adjustment

In this section, we present our stochastic bundle adjustment (STBA) method that decomposes the RCS into clusters inside the LM iterations. In Sect. 4.1, we first reformulate problem (1) based on the clustering of the visibility graph $\mathcal{G}$, yet subject to additional equality constraints. Next, in Sect. 4.2, we apply chance constrained relaxation to the reformulation and solve it by sampled convex program [7,8]. It manages to decompose the RCS into cluster-related linear sub-problems and yield an approximate STBA step efficiently. Third, a steepest correction step is proposed to remedy the approximation error of the STBA steps in Sect. 4.3. Finally, in Sect. 4.4, we present a practical implementation of the random constraint sampler required by the chance constrained relaxation.

4.1   Clustering Based Reformulation

In contrast to the previous methods [16,17,29,36–38] that partition the problem in the pre-processing stage, we present a reformulation of problem (1) to decompose the RCS into clusters inside the LM iterations.

Particularly, we consider the most general case that every single camera forms a cluster. In order to preserve all the projections $\mathcal{E}$, we apply **point splitting** to the physical points, as shown in Fig. 2. For a physical point $\mathbf{p}_i$ viewed by v_i cameras, we split it into v_i virtual points $\{\mathbf{p}_i^j\}_{j=1}^{v_i}$, each assigned to one cluster. Such a clustering will reformulate problem (1) equivalently as a new constrained quadratic programming (QP) problem as below

$$\min_{\Delta \mathbf{x}'} \; \|\mathbf{J}'\Delta\mathbf{x}' + \mathbf{f}\|^2 + \lambda\|\mathbf{D}'\Delta\mathbf{x}'\|^2, \tag{7}$$

$$\text{s.t.} \; \mathbf{A}\Delta\mathbf{x}' = \mathbf{0}. \tag{8}$$

Here, $\Delta\mathbf{x}' = [\Delta\mathbf{c}^T, \Delta\mathbf{p}'^T]^T$ is an expansion of $\Delta\mathbf{x}$ which considers the update steps for all the virtual points, and so is the Jacobian $\mathbf{J}' = [\mathbf{J_c}, \mathbf{J_p'}]$. Accordingly, $\mathbf{D}' = \mathrm{diag}(\mathbf{J}'^T\mathbf{J}')^{\frac{1}{2}}$. The noteworthy distinction between problems (7) and (1) is that the new equality constraints of Eq. 8 are imposed to enforce that the steps of the same points in different clusters are identical. For example, $\Delta\mathbf{p}_i^s = \Delta\mathbf{p}_i^t$ ($\forall s,t \in \{1,...,v_i\}$) for point $\mathbf{p}_i$ and the corresponding j-th row of $\mathbf{A}$ appears in a form as $\mathbf{a}_j = [0...,1,...,-1,...0]$. Since a point $\mathbf{p}_i$ introduces $v_i - 1$ equations, $\mathbf{A}$ has a row number of $r = \sum_{i=1}^{n}(v_i - 1)$. Besides, $\mathbf{A}$ is full row rank, because the rows each of which defines a unique equality constraint are linearly independent.

The constrained QP problem above can be easily solved by Lagrangian duality, which turns the problem into

$$\mathbf{H}_\lambda\Delta\mathbf{x}' = -(\mathbf{J}'^T\mathbf{f} + \mathbf{A}^T\boldsymbol{\nu}), \tag{9}$$

where $\mathbf{H}_\lambda = \mathbf{J}'^T\mathbf{J}' + \lambda\mathbf{D}'^T\mathbf{D}'$ and $\boldsymbol{\nu} = -(\mathbf{A}\mathbf{H}_\lambda^{-1}\mathbf{A}^T)^{-1}\mathbf{A}\mathbf{H}_\lambda^{-1}\mathbf{J}'^T\mathbf{f}$ are the Lagrangian multipliers. Equation 9 is in the similar format to Eq. 3, but has an additional term $\mathbf{A}^T\boldsymbol{\nu}$ on the right hand side compared with Eq. 3. While $\mathbf{J}'^T\mathbf{f}$ includes the gradients w.r.t. the independent virtual points, $\mathbf{A}^T\boldsymbol{\nu}$ acts as a **correction term** to ensure that the solution complies with the constraints of Eq. 8.

The ultimate benefit of the clustering-based reformulation is revealed below. By likewise applying Schur complement to Eq. 9, we have

$$\mathbf{S}'\Delta\mathbf{c} = \mathbf{v}' - \mathbf{E}'\mathbf{C}'^{-1}\mathbf{w}', \tag{10}$$

where $\mathbf{E}' = \mathbf{J_c}^T\mathbf{J_p'}$, $\mathbf{C}' = \mathbf{J_p'}^T\mathbf{J_p'} + \lambda\mathrm{diag}(\mathbf{J_p'}^T\mathbf{J_p'})$, $\mathbf{S}' = \mathbf{B} - \mathbf{E}'\mathbf{C}'^{-1}\mathbf{E}'^T$, and $[\mathbf{v}'^T, \mathbf{w}'^T]^T = -(\mathbf{J}'^T\mathbf{f} + \mathbf{A}^T\boldsymbol{\nu})$. Due to the fact that any two cameras do not share any common virtual points, $\mathbf{S}'$ now becomes a **block-diagonal** matrix, i.e., $\mathbf{S}'_{ij} = \mathbf{0}, \forall i \neq j$. Then Eq. 10 can be equivalently decomposed into m most elementary linear systems each corresponding to one camera.

4.2 Chance Constrained Relaxation

The major problem with the clustering based reformulation above is the excessive cost of evaluating the Lagrangian multipliers $\boldsymbol{\nu}$, because it requires the evaluation of $\mathbf{H}_\lambda^{-1}$. In order to make the problem practically solvable, we would like to eliminate the need to evaluate the correction term $\mathbf{A}^T\boldsymbol{\nu}$ of Eq. 9 by means of relaxation for problem (7).

We multiply a random binary variable θ_i with the i-th equality constraint of Eq. 8, which results in $\theta_i\mathbf{a}_i\Delta\mathbf{x}' = 0$. It could be interpreted that, if $\theta_i = 1$, the constraint $\mathbf{a}_i \Delta\mathbf{x}' = 0$ must be satisfied; otherwise, the constraint is allowed to be violated. In this way, Eq. 8 will be relaxed into chance constraints [24], which leads to

$$\min_{\Delta\mathbf{x}'} \|\mathbf{J}'\Delta\mathbf{x}' + \mathbf{f}\|^2 + \lambda\|\mathbf{D}'\Delta\mathbf{x}'\|^2, \tag{11}$$

$$\text{s.t. } \text{Prob}(\mathbf{a}_i\Delta\mathbf{x}' = 0) = \text{Prob}(\theta_i = 1) \geq \alpha, \quad i = 1, ..., r, \tag{12}$$

where $\alpha \in (0, 1]$ is a predefined confidence level. It means that, instead of enforcing the hard constraints, we allow them to be satisfied with a probability above α. The advantage of the chance constrained relaxation is that we can determine the reliability level of approximation by controlling α. The larger α is, the closer the chance constrained problem (11) will be to the original deterministic problem (7). One approach to problem (11) is called sampled convex program [7,8]. It extracts N independent samples $\Theta(\alpha) = \{\theta_i^{(n)} | i = 1, ..., r, n = 1, ..., N\}$ with a minimum sampling probability of α for each variable $\theta_i^{(n)}$ and replaces the chance constraints (12) with the sampled ones. Below we will elaborate on how problem (11) can be solved given the samples $\Theta(\alpha)$.

For a sample $\theta_i^{(n)} \in \Theta(\alpha)$, if $\theta_i^{(n)} = 0$, the equality constraint $\mathbf{a}_i\Delta\mathbf{x}' = 0$ is dropped; if $\theta_i^{(n)} = 1$, it enforces the equality of the steps of two virtual points, e.g., $\Delta\mathbf{p}_j^s = \Delta\mathbf{p}_j^t$. Here the virtual point $\mathbf{p}_j^s$ belongs to the single-camera cluster of camera $\mathbf{c}_s$ and similarly $\mathbf{p}_j^t$ to $\mathbf{c}_t$. Then we merge $\mathbf{p}_j^s$ and $\mathbf{p}_j^t$ into one point as shown in Fig. 2, and we call the operation **point binding** as opposed to **point splitting** introduced in Sect. 4.1. On the one hand, the point binding leads to the consequence that the Lagrangian multiplier $\nu_i = 0$, because the equality $\Delta\mathbf{p}_j^s = \Delta\mathbf{p}_j^t$ always holds. On the other hand, the block $\mathbf{S}'_{st}$ of $\mathbf{S}'$ (c.f. Eq. 10) becomes nonzero, since the merged point is now shared by cameras $\mathbf{c}_s$ and $\mathbf{c}_t$. After applying point binding to all the virtual points involved in the sampled constraints, i.e., $\{\theta_i^{(n)} \in \Theta(\alpha) | \theta_i^{(n)} = 1\}$, all the constraints will be eliminated and there is no need to evaluate $\boldsymbol{\nu}$ any more. Meanwhile, it will bring the cameras sharing common points into the same clusters.

Since the cameras in different clusters have no points in common after the point binding, the matrix $\mathbf{S}'$ will appear in a block-diagonal structure which we call **cluster-diagonal**, as illustrated in Fig. 2. It means that each diagonal block of $\mathbf{S}'$ corresponds to a camera cluster. In particular, we can intentionally design the sampler $\Theta(\alpha)$ in order to shape $\mathbf{S}'$ into the desired cluster-diagonal structures, as we will present in Sect. 4.4. As a result, this structure of $\mathbf{S}'$ still enables the decomposition of Eq. 10 into smaller independent linear systems each relating to one camera cluster. Provided that there are l clusters, Eq. 10 can be equivalently re-written as

$$\begin{cases} \mathbf{S}'_1\Delta\mathbf{c}_1 = \mathbf{b}_1, \\ \quad\quad ... \\ \mathbf{S}'_l\Delta\mathbf{c}_l = \mathbf{b}_l, \end{cases} \tag{13}$$

where $[\Delta\mathbf{c}_1^T, ..., \Delta\mathbf{c}_l^T]^T = \Delta\mathbf{c}$ and $[\mathbf{b}_1^T, ..., \mathbf{b}_l^T]^T = \mathbf{v}' - \mathbf{E}'\mathbf{C}'^{-1}\mathbf{w}'$. After Eqs. 13 are evaluated, we substitute $\Delta\mathbf{c}$ into Eq. 9 to give

$$\mathbf{J}_\mathbf{p}'\Delta\mathbf{p}' = \sum_{i=1}^{l} \mathbf{J}_{\mathbf{p}^i}\Delta\mathbf{p}^i = -\mathbf{f} - \mathbf{J}_\mathbf{c}\Delta\mathbf{c}, \tag{14}$$

where $\mathbf{J}_\mathbf{p}' = [\mathbf{J}_{\mathbf{p}^1}, ..., \mathbf{J}_{\mathbf{p}^l}]$ and $\Delta\mathbf{p}' = [\Delta\mathbf{p}^1, ..., \Delta\mathbf{p}^l]$ include the Jacobians and virtual point steps w.r.t. the l clusters respectively and we omit $\mathbf{D}'$ for ease of notation. To give a uniform step for a physical point, we equalize the steps of its virtual points in different clusters by solving the linear system below in place of Eq. 14:

$$\sum_{i=1}^{l} \mathbf{J}_{\mathbf{p}^i}\Delta\mathbf{p} = -\mathbf{f} - \mathbf{J}_\mathbf{c}\Delta\mathbf{c}. \tag{15}$$

Since $\sum_{i=1}^{l} \mathbf{J}_{\mathbf{p}^i} = \mathbf{J}_\mathbf{p}$, Eq. 15 gives the same solution as the point steps in Eq. 6: $\Delta\mathbf{p} = \mathbf{C}^{-1}(\mathbf{w} - \mathbf{E}^T\Delta\mathbf{c})$.

So far we have presented how an update step of the camera and point parameters is determined approximately by STBA. Besides this, we keep the other components of the LM algorithm unchanged [27]. For reference, we detail the full algorithm in the supplementary material.

4.3   Steepest Descent Correction

The chance constrained relaxation in the last section effectively decomposes the RCS, but leads to approximate solutions with decreased feasibility due to the random constraint sampling. Below, we provide an empirical analysis on the effect of the approximation and present a conditional correction step to remedy the approximation error.

The LM algorithm is known to be the interpolation of the Gauss-Newton and gradient descent methods, depending on the trust region radius controlled by the damping parameter λ. When λ is small and the LM algorithm behaves more like the Gauss-Newton method, the approximation induced by STBA in Sect. 4.2 is admissible, in that problem (1) itself is derived from the first order approximation of the error function $\mathbf{f}(\mathbf{x})$. And the LM algorithm can automatically contract the trust region when the approximation leads to the increase of the objective.

When λ is large, $i.e.$, the trust region is small, the LM algorithm is closer to the gradient descent method, which gives a step towards the steepest descent direction defined by the right hand side of Eq. 9, $i.e.$, $-(\mathbf{J}'^T\mathbf{f} + \mathbf{A}^T\boldsymbol{\nu})$. However, a problem with STBA is that the correction term $\mathbf{A}^T\boldsymbol{\nu}$ is eliminated approximately by the chance constrained relaxation. As a consequence, the derived step would deviate from the steepest descent direction and thus hamper the convergence. Therefore, we propose to recover $\mathbf{A}^T\boldsymbol{\nu}$ to remedy the deviation in such a case. Especially, when λ is large enough, the matrix $\mathbf{H}_\lambda$ in Eq. 9 will be dominated by the diagonal terms, so that we can approximate $\mathbf{H}_\lambda$ by $\mathrm{diag}(\mathbf{H}_\lambda)$. After that, $\mathbf{A}^T\boldsymbol{\nu} = -\mathbf{A}^T(\mathbf{A}\mathbf{H}_\lambda^{-1}\mathbf{A}^T)^{-1}\mathbf{A}\mathbf{H}_\lambda^{-1}\mathbf{J}'^T\mathbf{f}$ can be evaluated efficiently because of the

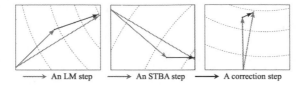

Fig. 3. **Visualization of the steepest descent correction steps** of sample camera parameters when $\lambda = 0.1$. It shows that the deviations of approximate STBA steps from the LM steps are effectively corrected.

sparsity of $\mathbf{A}$. Since the approximation $\mathbf{H}_\lambda \approx \mathrm{diag}(\mathbf{H}_\lambda)$ is not accurate unless λ is large, in practice, we enable the steepest descent correction particularly when $\lambda \geq 0.1$. Figure 3 visualizes the effect of the correction.

4.4   Stochastic Graph Clustering

The chance constrained relaxation in Sect. 4.2 necessitates an effective random constraint sampler $\Theta(\alpha)$. Among many of the possible designs, we propose a viable implementation named stochastic graph clustering in this section.

The design of the clustering method considers the following requirements. First, the sampler should be randomized with respect to the chance constraints (12). Since (12) indicates that the expectation $\mathrm{E}(\theta_i)$ should have $\mathrm{E}(\theta_i) \geq \alpha$ $(i = 1, ..., r)$, the upper bound of the confidence level α is defined as $\min_{i=1}^{r} \mathrm{E}(\theta_i)$. Therefore, the sampler should sample as many constraints as possible on average to increase the upper bound of α. Second, the random sampler is intended to partition the cameras into small independent clusters so that Eqs. 13 can be solved efficiently.

Concretely, the stochastic graph clustering operates over a camera graph $\mathcal{G}_c = (\mathcal{C}, \mathcal{E}_c)$, where the weight w_{ij} of an edge $e_{ij} \in \mathcal{E}_c$ between cameras $\mathbf{c}_i$ and $\mathbf{c}_j$ is equal to the number of points covisible by the two cameras. At the beginning, each camera forms an individual cluster as formulated in Sect. 4.1. Next, if $\mathbf{c}_i$ and $\mathbf{c}_j$ are joined, a number of w_{ij} pairs of virtual points viewed by $\mathbf{c}_i$ and $\mathbf{c}_j$ will be merged. Therefore, w_{ij} equality constraints will be satisfied. In order to join as many virtual points as possible while yielding a cluster structure of $\mathcal{G}_c$, we aim at finding a clustering that maximizes the modularity below inspired by [6]: $Q = \frac{1}{2s} \sum_{e_{ij} \in \mathcal{E}_c} \delta(\nu_i, \nu_j) \left(w_{ij} - \frac{k_i k_j}{2s} \right)$, where $s = \sum_{e_{ij} \in \mathcal{E}_c} w_{ij}$ is the total sum of edge weights, $k_i = \sum_j w_{ij}$ is the sum of weights of edges incident to camera $\mathbf{c}_i$, and ν_i denotes the cluster of $\mathbf{c}_i$. $\delta(\nu_i, \nu_j) = 1$ if $\nu_i = \nu_j$ and 0 otherwise. The modularity $Q \in [-1, 1]$ measures the density of connections inside clusters as opposed to those across clusters [6]. Therefore, a larger modularity generally indicates that more virtual points are merged inside clusters. Maximizing the modularity is NP-hard [31], but Louvain's algorithm [6] provides a greedy strategy which greedily joins the two clusters giving the maximum increase in modularity in a bottom-up manner. It can be efficiently applied to large graphs since its complexity is shown to be linear in the node number on sparse data [6].

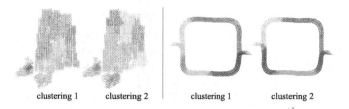

clustering 1   clustering 2   clustering 1   clustering 2

Fig. 4. Visualization of the random clustering results produced by stochastic graph clustering. Cameras of different clusters are in different colors. (Color figure online)

However, Louvain's algorithm [6] is deterministic due to its greedy nature. To ensure that every pair of virtual points is likely to be merged, we instead join clusters randomly according to a probability distribution defined based on the modularity increments [11], which is

$$\text{Prob}(N_x \cup N_y) = \frac{\exp(\beta \Delta Q(N_x, N_y))}{\sum_i \sum_j \exp(\beta \Delta Q(N_i, N_j))}, \tag{16}$$

where $\beta > 0$ is a scaling parameter. Two neighboring clusters N_x and N_y are more likely to join together if it leads to a larger modularity increment $\Delta Q(N_x, N_y)$. In order to limit the sizes of the sub-problems of STBA, we stop joining clusters if their sizes exceed Γ. In Fig. 4, we visualize the stochastic clustering results.

5   Experiments

5.1   Experiment Settings

Datasets. We run experiments on three different types of datasets: 1) 1DSfM dataset [33] which is composed of 14 sets of unordered Internet images; 2) KITTI dataset [18] containing 11 street-view image sequences; and 3) Large-Scale dataset which is collected by ourselves due to the absence of publicly available large-scale 3D datasets. It includes 4 image sets each comprising more than 30,000 images. The problem sizes all exceed the memory of a single compute node and thus we use them particularly for distributed experiments.

Comparisons. On 1DSfM and KITTI datasets, we compare our method with two standard trust region algorithms, Levenberg-Marquardt (LM) [25] and Dogleg (DL) [26]. For the LM algorithm, we use two variants: LM-sparse and LM-iterative, which exploit the exact sparse method and inexact iterative method [23] to solve the RCS (Eq. 5), respectively. For the distributed experiments on the Large-Scale dataset, we compare our distributed implementation of STBA against the state-of-the-art distributed solver DBACC [36]. The ablation studies on steepest descent correction and stochastic graph clustering are presented in Sect. 5.4 and the supplementary material, respectively.

Implementations. We implement LM, DL and STBA in C++, using Eigen for linear algebra computations. All the algorithms are implemented from the same code base, which means that they share the same elementary operations so that they can be compared equitably. For robustness, we use the Huber loss with a scale factor of 0.5 for the errors [35]. LM-sparse exploits the supernodal LL^T Cholesky factorization with COLAMD ordering [12] based on CHOLMOD, which is well suited for handling sparse data like KITTI [18]. LM-iterative uses the conjugate gradient method with the advanced cluster-jacobi preconditioner [23]. DL uses the same exact sparse solver as LM-sparse since it requires a reasonably good estimation of the Gauss-Newton step [26]. Dense LL^T factorization is used to solve the decomposed RCS (Eqs. 13) for STBA due to the dense connectivity inside camera clusters. Multi-threading is applied to the operations including the reprojection error and Jacobian computation, the preconditioner construction and the matrix-vector multiplications for all the methods as in [34].

Parameters. In the experiments, we assume that the camera intrinsics have been calibrated as in [25,26]. Camera extrinsics are parameterized with 6-d vectors, using axis-angle representations for rotations. We set the initial damping parameter λ to 1e−4 and the max iteration number to 100 for all the methods. The iterations could terminate early if the cost, gradient or parameter tolerance [1] drops below 1e−6. For STBA, we empirically set the scaling parameter β to 10 (Eq. 16) and the max cluster size Γ to 100.

Hardware. We use a compute node with an 8-core Intel i7-4790K CPU and a 32G RAM. The distributed experiments are deployed on a cluster with 6 compute nodes.

5.2   Performance Profiles

Following previous works [14,23], we evaluate the solvers with *Performance Profiles* over the total of 25 problems of 1DSfM [33] and KITTI [18]. We obtain the SfM results for 1DSfM by COLMAP [32][1] and the SLAM results for KITTI by stereo ORB-SLAM2 [28], while disabling the final bundle adjustment (BA). Since the SfM/SLAM results are generally accurate because the pipeline uses repeated BA for robust reconstruction, we make the problems more challenging by adding Gaussian noise to the points and camera centers following [16,20,29].

First of all, we give a brief introduction of performance profiles [14]. Given a problem $p \in \mathcal{P}$ and a solver $s \in \mathcal{S}$, let $F(p,s)$ denote the final objective the solver s attained when solving problem p. Then, for a number of solvers in $\mathcal{S}$, let $F^*(p) = \min_{s \in \mathcal{S}} F(p,s)$ denote the minimum objective the solvers $\mathcal{S}$ attained when solving problem p. Next, we define an objective threshold for problem p which is $F_\tau(p) = F^*(p) + \tau(F_0(p) - F^*(p))$, where $F_0(p)$ is the initial objective and $\tau \in (0,1)$ is the pre-defined tolerance determining how close the threshold is to the minimum objective. After this, we measure the efficiency of a solver s by

---

[1] Since one of the image sets *Union Square* has only 10 reconstructed images, we replace it with another public image set *ArtsQuad*.

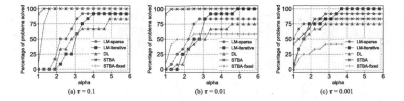

(a) $\tau = 0.1$ (b) $\tau = 0.01$ (c) $\tau = 0.001$

Fig. 5. Performance profiles [14] **of different solvers** when solving the total of 25 problems of 1DSfM [33] and KITTI [18].

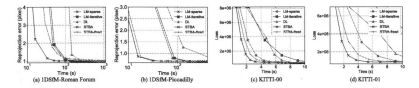

(a) 1DSfM-Roman Forum   (b) 1DSfM-Piccadilly   (c) KITTI-00   (d) KITTI-01

Fig. 6. The convergence curves of 4 scenes from 1DSfM and KITTI.

computing the time it takes to reduce the objective to $F_\tau(p)$, which is denoted by $T_\tau(p, s)$. And the most efficient solver is the one who takes the minimum time, *i.e.*, $\min_{s \in S} T_\tau(p, s)$.

The method Performance Profiles regards that the solver s solves the problem p if $T_\tau(p, s) \leq \alpha \min_{s \in S} T_\tau(p, s)$, where $\alpha \in [1, \infty)$. Therefore, if $\alpha = 1$, only the most efficient solver is thought to solve the problem, while if $\alpha \to \infty$, all the solvers can be seen to solve the problem. Finally, the performance profile of the solver s is defined w.r.t. α over the whole problem set $\mathcal{P}$ as $\rho(s, \alpha) = 100 * \frac{|\{p \in \mathcal{P} | T_\tau(p,s) \leq \alpha \min_{s \in S} T_\tau(p,s)\}|}{|\mathcal{P}|}$. It is basically the percentage of problems solved by s and is non-decreasing w.r.t. α.

We plot the performance profiles of the solvers in Fig. 5. To verify the benefits of using stochastic clustering, herein we also compare STBA with its variant STBA-fixed which uses a fixed clustering as previous methods [16,23,36]. When τ is equal to 0.1 and 0.01, our STBA is able to solve nearly 100% of the problems for any α, because it always reaches the objective threshold $F_\tau(p)$ with less time than LM and DL methods by a factor of more than 5. This is mainly attributed to the reduced per-iteration cost as we can see from the convergence curves in Fig. 6. When τ becomes 0.001 and the threshold $F_\tau(p)$ is harder to achieve, STBA is less efficient but still performs on par with LM-iterative and LM-sparse when $\alpha < 3$ and better than DL for any α. On the contrary, the performance of STBA-fixed drops drastically when τ decreases to 0.001. The performance change for STBA and STBA-fixed when τ decreases is mainly caused by the fact that the clustering methods come with the price of slower convergence near the stationary points [16,36]. Compared with the full second-order solvers such as LM and DL, clustering methods only utilize the second order information within clusters. However, as opposed to STBA-fixed which uses fixed clustering, STBA has mitigated the negative effect of clustering by introducing stochasticity

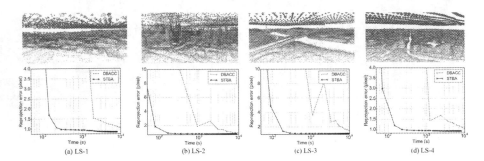

(a) LS-1   (b) LS-2   (c) LS-3   (d) LS-4

Fig. 7. Visualizations of SfM results (top) and convergence curves (bottom) of the Larse-Scale dataset. Cameras are drawn as blue pyramids. (Color figure online)

Table 1. Statistics of the distributed bundle adjustment solvers on the Large-Scale dataset. DBACC [36] consume many more Jacobian and RCS evaluations than STBA.

Data	#images	#clusters		RPE (pixel)		#Jacobian/RCS evaluations		Mean iteration time (s)	
		DBACC	STBA	DBACC	STBA	DBACC	STBA	DBACC	STBA
LS-1	29975	300	340	0.823	0.818	1011/1080	49/100	912.5	71.0
LS-2	33634	336	386	0.766	0.783	854/860	48/100	934.8	79.3
LS-3	33809	339	391	1.083	1.056	1025/1100	49/100	1107.0	89.9
LS-4	44276	444	505	0.909	0.882	877/900	49/100	988.1	71.2

so that different second-order information can be utilized to boost convergence as the clustering changes. Despite the slower convergence rate, the benefit of STBA that it can reduce most of the loss with the lowest time cost (*e.g.*, 99% loss reduction with only 1/5 time of the counterparts when $\tau = 0.01$ in Fig. 5(b)) is still supposed to be highlighted, especially for the real-time SLAM applications where bundle adjustment is called repeatedly to correct the drift.

5.3 Results on Large-Scale Dataset

To evaluate the scalability, we conduct distributed experiments on the Large-Scale (LS) dataset, which includes the urban scenes of four cities named *LS-1*, *LS-2*, *LS-3* and *LS-4*. We run a distributed SfM program of our own to produce initial sparse reconstructions of the four scenes and add Gaussian noise with a standard deviation of 3 meters to the camera centers and points. Then we compare our distributed STBA against the state-of-the-art distributed bundle adjustment framework DBACC [36].

We visualize the sparse reconstructions and the convergence curves of the four scenes in Fig. 7 and report the statistics in Table 1. As we can see from Fig. 7, STBA achieves faster convergence rates than DBACC by an order of magnitude. The main cause of the gap is that DBACC, which is based on the ADMM formulation [4], has to take inner iterations to solve a new minimization problem in every ADMM iteration. Although we have set the maximum inner

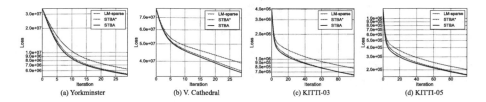

(a) Yorkminster	(b) V. Cathedral	(c) KITTI-03	(d) KITTI-05

Fig. 8. Convergence curves of LM-sparse, STBA and STBA* which does not use steepest descent correction (Sect. 4.3). Steepest descent correction helps to correct the deviations between the STBA and LM steps.

iteration number to merely 10, DBACC still takes many more Jacobian and RCS evaluations and thus has much longer iterations than STBA by an order of magnitude, as shown in Table 1. Besides, as opposed to DBACC, our STBA is free of too many hyper-parameters.

5.4   Ablation Study on Steepest Descent Correction

Here we perform an ablation study on steepest descent correction proposed in Sect. 4.3 to validate its efficacy. We compare our STBA with its variant called STBA* which does not use the correction. We run STBA, STBA* and LM-sparse on 1DSfM and KITTI. Since steepest descent correction is designed particularly for a small trust region, we set the lower bound of the damping parameter λ to 0.1 in the experiments. We observe that by using the correction, STBA consistently achieves a faster convergence than STBA* and performs on par with LM-sparse on all the scenes. Visualizations of the sample convergence curves w.r.t. the iterations are shown in Fig. 8, where STBA and LM-sparse have very close convergence curves. It manifests that steep descent correction indeed facilitates the correction of the approximation errors of the STBA steps and hence boosts the convergence.

6   Conclusion

In this paper, we rethink the proper way of integrating the clustering scheme into solving bundle adjustment by proposing STBA. First, STBA reformulates an LM iteration based on the clustering of the visibility graph, but meanwhile introduces additional equality constraints across the clusters. Second, we approximately relax the constraints as chance constraints and solve the problem by sampled convex program which randomly samples the chance constraints with the intention of splitting the large reduced camera system into small clusters. Not only does it reduce the per-iteration cost, but also allows parallel and distributed computing to accommodate the increase of the problem size. Moreover, we present a steepest descent correction technique to remedy the approximation errors of the STBA steps for a small trust region, and provide a practical implementation of stochastic graph clustering for constraint sampling. Extensive experiments on Internet SfM data, SLAM data and large-scale data demonstrate the efficiency and scalability of our approach.

Acknowledgement. This work is supported by Hong Kong RGC GRF16206819 & 16203518 and T22-603/15N.

References

1. Agarwal, S., Mierle, K., et al.: Ceres solver. http://ceres-solver.org
2. Agarwal, S., Snavely, N., Seitz, S.M., Szeliski, R.: Bundle adjustment in the large. In: Daniilidis, K., Maragos, P., Paragios, N. (eds.) ECCV 2010. LNCS, vol. 6312, pp. 29–42. Springer, Heidelberg (2010). https://doi.org/10.1007/978-3-642-15552-9_3
3. Amestoy, P.R., Davis, T.A., Duff, I.S.: An approximate minimum degree ordering algorithm. SIAM J. Matrix Anal. Appl. **17**(4), 886–905 (1996)
4. Bertsekas, D.P.: Parallel and Distributed Computation: Numerical Methods, vol. 3 (1989)
5. Bertsekas, D.P.: Constrained Optimization and Lagrange Multiplier Methods. Academic Press, Cambridge (2014)
6. Blondel, V.D., Guillaume, J.L., Lambiotte, R., Lefebvre, E.: Fast unfolding of communities in large networks. J. Stat. Mech. Theor. Exp. **2008**(10), P10008 (2008)
7. Calafiore, G., Campi, M.C.: Uncertain convex programs: randomized solutions and confidence levels. Math. Program. **102**(1), 25–46 (2005)
8. Campi, M.C., Garatti, S.: A sampling-and-discarding approach to chance-constrained optimization: feasibility and optimality. J. Optim. Theor. Appl. **148**(2), 257–280 (2011)
9. Chatterjee, A., Madhav Govindu, V.: Efficient and robust large-scale rotation averaging. In: ICCV (2013)
10. Combettes, P.L., Pesquet, J.C.: Proximal splitting methods in signal processing. In: Bauschke, H., Burachik, R., Combettes, P., Elser, V., Luke, D., Wolkowicz, H. (eds.) Fixed-Point Algorithms for Inverse Problems in Science and Engineering. Springer Optimization and Its Applications, vol. 49. Springer, New York (2011). https://doi.org/10.1007/978-1-4419-9569-8_10
11. Darmaillac, Y., Loustau, S.: MCMC Louvain for online community detection. arXiv preprint arXiv:1612.01489 (2016)
12. Davis, T.A., Gilbert, J.R., Larimore, S.I., Ng, E.G.: Algorithm 836: COLAMD, a column approximate minimum degree ordering algorithm. TOMS **30**(3), 377–380 (2004)
13. Dellaert, F., Carlson, J., Ila, V., Ni, K., Thorpe, C.E.: Subgraph-preconditioned conjugate gradients for large scale SLAM. In: IROS (2010)
14. Dolan, E.D., Moré, J.J.: Benchmarking optimization software with performance profiles. Math. Program. **91**(2), 201–213 (2002)
15. Engels, C., Stewénius, H., Nistér, D.: Bundle adjustment rules
16. Eriksson, A., Bastian, J., Chin, T.J., Isaksson, M.: A consensus-based framework for distributed bundle adjustment. In: CVPR (2016)
17. Fang, M., Pollok, T., Qu, C.: Merge-SfM: merging partial reconstructions. In: BMVC (2019)
18. Geiger, A., Lenz, P., Urtasun, R.: Are we ready for autonomous driving? The KITTI vision benchmark suite. In: CVPR (2012)

19. Hestenes, M.R., et al.: Methods of conjugate gradients for solving linear systems. J. Res. Natl. Bur. Stan. **49**(6), 409–436 (1952)
20. Jeong, Y., Nister, D., Steedly, D., Szeliski, R., Kweon, I.S.: Pushing the envelope of modern methods for bundle adjustment. PAMI **34**(8), 1605–1617 (2011)
21. Jian, Y.-D., Balcan, D.C., Dellaert, F.: Generalized subgraph preconditioners for large-scale bundle adjustment. In: Dellaert, F., Frahm, J.-M., Pollefeys, M., Leal-Taixé, L., Rosenhahn, B. (eds.) Outdoor and Large-Scale Real-World Scene Analysis. LNCS, vol. 7474, pp. 131–150. Springer, Heidelberg (2012). https://doi.org/10.1007/978-3-642-34091-8_6
22. Konolige, K., Garage, W.: Sparse sparse bundle adjustment. In: BMVC (2010)
23. Kushal, A., Agarwal, S.: Visibility based preconditioning for bundle adjustment. In: CVPR (2012)
24. Li, P., Arellano-Garcia, H., Wozny, G.: Chance constrained programming approach to process optimization under uncertainty. Comput. Chem. Eng. **32**(1–2), 25–45 (2008)
25. Lourakis, M.I., Argyros, A.A.: SBA: a software package for generic sparse bundle adjustment. TOMS **36**(1), 2 (2009)
26. Lourakis, M., Argyros, A.A.: Is Levenberg-Marquardt the most efficient optimization algorithm for implementing bundle adjustment? In: ICCV (2005)
27. Marquardt, D.W.: An algorithm for least-squares estimation of nonlinear parameters. J. Soc. Ind. Appl. Math. **11**(2), 431–441 (1963)
28. Mur-Artal, R., Tardós, J.D.: ORB-SLAM2: an open-source SLAM system for monocular, stereo and RGB-D cameras. IEEE Trans. Robot. **33**(5), 1255–1262 (2017)
29. Ni, K., Steedly, D., Dellaert, F.: Out-of-core bundle adjustment for large-scale 3D reconstruction. In: ICCV (2007)
30. Rotkin, V., Toledo, S.: The design and implementation of a new out-of-core sparse Cholesky factorization method. TOMS **30**(1), 19–46 (2004)
31. Schaeffer, S.E.: Survey: graph clustering. Comput. Sci. Rev. **1**(1), 27–64 (2007)
32. Schönberger, J.L., Frahm, J.M.: Structure-from-motion revisited. In: CVPR (2016)
33. Wilson, K., Snavely, N.: Robust global translations with 1DSfM. In: Fleet, D., Pajdla, T., Schiele, B., Tuytelaars, T. (eds.) ECCV 2014. LNCS, vol. 8691, pp. 61–75. Springer, Cham (2014). https://doi.org/10.1007/978-3-319-10578-9_5
34. Wu, C., Agarwal, S., Curless, B., Seitz, S.M.: Multicore bundle adjustment. In: CVPR (2011)
35. Zach, C.: Robust bundle adjustment revisited. In: Fleet, D., Pajdla, T., Schiele, B., Tuytelaars, T. (eds.) ECCV 2014. LNCS, vol. 8693, pp. 772–787. Springer, Cham (2014). https://doi.org/10.1007/978-3-319-10602-1_50
36. Zhang, R., Zhu, S., Fang, T., Quan, L.: Distributed very large scale bundle adjustment by global camera consensus. In: ICCV (2017)
37. Zhu, S., et al.: Parallel structure from motion from local increment to global averaging. arXiv preprint arXiv:1702.08601 (2017)
38. Zhu, S., et al.: Very large-scale global SfM by distributed motion averaging. In: CVPR (2018)

Object-Based Illumination Estimation with Rendering-Aware Neural Networks

Xin Wei[1,2], Guojun Chen[1], Yue Dong[1(✉)], Stephen Lin[1], and Xin Tong[1]

[1] Microsoft Research Asia, Beijing, China
yuedong@microsoft.com
[2] Zhejiang Univiersity, Hangzhou, China

Abstract. We present a scheme for fast environment light estimation from the RGBD appearance of individual objects and their local image areas. Conventional inverse rendering is too computationally demanding for real-time applications, and the performance of purely learning-based techniques may be limited by the meager input data available from individual objects. To address these issues, we propose an approach that takes advantage of physical principles from inverse rendering to constrain the solution, while also utilizing neural networks to expedite the more computationally expensive portions of its processing, to increase robustness to noisy input data as well as to improve temporal and spatial stability. This results in a rendering-aware system that estimates the local illumination distribution at an object with high accuracy and in real time. With the estimated lighting, virtual objects can be rendered in AR scenarios with shading that is consistent to the real scene, leading to improved realism.

1 Introduction

Consistent shading between a virtual object and its real-world surroundings is an essential element of realistic AR. To achieve this consistency, the illumination environment of the real scene needs to be estimated and used in rendering the virtual object. For practical purposes, the lighting estimation needs to be performed in real time, so that AR applications can accommodate changing illumination conditions that result from scene dynamics.

Traditionally, lighting estimation has been treated as an inverse rendering problem, where the illumination is inferred with respect to geometric and reflectance properties of the scene [4,29,31,32]. Solve the inverse rendering problem with a single image input is ill-conditioned, thus assumptions are made to simplify the lighting model or to limit the supported material type. Despite the ambiguity among shape, material, and lighting. Solving the optimization involved in inverse rendering entails a high computational cost that prevents

Electronic supplementary material The online version of this chapter (https://doi.org/10.1007/978-3-030-58555-6_23) contains supplementary material, which is available to authorized users.

real-time processing, since the forward rendering as a sub-step of such optimization already difficult to reach real-time performance without compromising the accuracy.

Recent methods based on end-to-end neural networks provide real-time performance [7,10,11,18,26,35], specifically, they regard the input image as containing partial content of the environment map and estimate high resolution environment light based on those contents, rich content in the input image is critical to infer the surrounding environment without ambiguity. However, many AR applications have interactions focused on a single object or a local region of a scene, where the partial content of the environment map in the input image is very limited. Thinking about guessing an environment map based on the image of a toy putting on the ground (like Fig. 3.h), although the ground is part of the environment map, it provides limited clues to rule out ambiguities when inferring the surroundings, yielding unstable estimation results.

On the contrary, given such a scenario, the appearance of the object and the shadow cast by the object provide strong cues for determining the environment light. Utilizing such cues with neural networks is however challenging, due to the complex relationship between the lighting, material, and appearance. First, the neural network needs to aware of the physical rules of the rendering. Second, physically based rendering is computationally intensive, simple analysis by synthesis is not suitable for a real-time application. Third, the diffuse and specular reflections follow different rules and the resulting appearance are mixed together in the input based on the unknown material property of the object. Previous methods already prove optimizing an end-to-end model for such a complex relationship is challenging and inefficient [35].

In this paper, we present a technique that can estimate illumination from the RGBD appearance of an individual object and its local image area. To make the most of the input data from a single object, our approach is to integrate physically-based principles of inverse rendering together with deep learning. An object with a range of known surface normals provides sufficient information for lighting estimation by inverse rendering. To deal with the computational inefficiency of inverse rendering, we employ neural networks to rapidly solve for certain intermediate steps that are slow to determine by optimization. Among these intermediate steps are the decomposition of object appearance into reflection components – namely albedo, diffuse shading, and specular highlights – and converting diffuse shading into a corresponding angular light distribution. On the other hand, steps such as projecting specular reflections to lighting directions can be efficiently computed based on physical laws without needing neural networks for speedup. Estimation results are obtained through a fusion of deep learning and physical reasoning, via a network that also accounts for temporal coherence of the lighting environment through the use of recurrent convolution layers.

In this way, our method takes advantage of physical knowledge to facilitate inference from limited input data, while making use of neural networks to achieve real-time performance. This rendering-aware approach moreover benefits from the robustness of neural networks to noisy input data. Our system is validated through experiments showing improved estimation accuracy from this use of inverse rendering concepts over a purely learning-based variant. The illumination estimation results also compare favorably to those of related techniques, and are demonstrated to bring high-quality AR effects.

2   Related Work

A large amount of literature exists on illumination estimation in computer graphics and vision. Here, we focus on the most recent methods and refer readers to the survey by Kronander et al. [24] for a more comprehensive review.

Scene-Based Lighting Estimation. Several methods estimate lighting from a partial view of the scene. In earlier works, a portion of the environment map is obtained by projecting the viewable scene area onto it, and the rest of the map is approximated through copying of the scene area [22] or by searching a panorama database for an environment map that closely matches the projected scene area [21].

Deep learning techniques have also been applied to this problem. Outdoor lighting estimation methods usually take advantage of the known prior distribution of the sky and predict a parametric sky model [18,44] or a sky model based on a learned latent space [17]. Another commonly used low-parametric light model is the spherical harmonic (SH) model, adopted by [7,12]. However, an SH model is generally inadequate for representing high frequency light sources. Recently, Gardner et al. [10] proposed a method estimating a 3D lighting model composed of up to 3–5 Gaussian lights and one global ambient term, improving the representation power of the parametric light model. Instead of depending on a low-order parametric model, our method estimates high resolution environment maps without the limitations of such models.

Recent scene-based methods [11,26,35] regard the input image as containing partial content of the environment map and estimate high resolution environment light based on those contents. The input image is usually regarded as a warped partial environment map that follows the spherical warping [11] or warping based on depth [35]. The quality their estimated light depends on the amount of content in the known partial environment, since fewer content in the partial environment leads to stronger ambiguity of the missing part, increasing the estimation difficulties. By contrast, our work seeks to estimate an environment map from only the shading information of an object, without requiring any content of the environment map, which is orthogonal to scene-based methods and could be a good complement.

Object-Based Lighting Estimation. Illumination has alternatively been estimated from the appearance of individual objects in a scene. A common technique is to include a mirrored sphere in the image and reconstruct the lighting environment through inverse rendering of the mirrored reflections [9,39,40]. This approach has

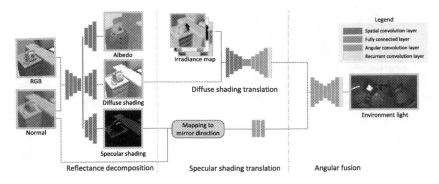

Fig. 1. Overview of our system. The input RGB image crop is first decomposed into albedo, diffuse shading and specular shading maps. This decomposition is assisted by shape information from a normal map, calculated from rough depth data. The diffuse shading is translated into the angular lighting domain with the help of auxiliary irradiance maps computed from the rough depth, and the specular shading is geometrically mapped to their mirror direction. The translated features are then processed in the angular fusion network to generate the final environment light estimate.

been employed with other highly reflective objects such as human eyes [30], and extended to recover multispectral lighting using a combination of light probes and color checker charts [27] (Fig. 1).

Avoiding the use of special light probes, several inverse rendering methods utilize general objects instead. Some estimate lighting from diffuse objects for which rough geometry is measured by a depth sensor [3,15,16], reconstructed by multi-view stereo [42], or jointly estimated with the lighting [4]. For homogeneous objects with a known shape, the BRDF has been recovered together with all-frequency illumination [29,31]. Cast shadows from objects with known geometry have also been analyzed to extract high-frequency lighting information [20,32]. With a complete scan and segmentation of a scene, inverse path-tracing has been used to estimate both the light and object BRDFs [2]. Inverse rendering approaches such as these have proven to be effective, but are too computationally slow for real-time AR applications due to iterative optimization. Recently, Sengupta et al. [33] trained a inverse rendering neural network; however, their direct rendering component assumes diffuse shading only and their method produces a low-resolution environment map limited by the direct renderer.

Deep learning has also been applied for illumination estimation from objects. Many of these works specifically address the case of human faces, for which prior knowledge about face shape and appearance can be incorporated [5,34,36–38,41,43,45]. For more general cases, neural networks have been designed to estimate an environment map from homogeneous specular objects of unknown shape [14] and for piecewise constant materials while considering background appearance [13]. By contrast, our approach does not make assumptions on the surface reflectance, and makes greater use of physical principles employed in inverse rendering.

3 Overview

Our method takes RGBD video frames from an AR sensor as input. For a cropped object area within a frame, an illumination environment map is estimated for the object's location. Illustrated in Fig. 2, the estimation process consists of three main components: reflectance decomposition, spatial-angular translation, and angular fusion.

Reflectance decomposition aims to separate object appearance into albedo, diffuse shading, and specular reflections. Such a decomposition facilitates inverse rendering, as diffuse and specular reflection arise from different physical mechanisms that provide separate cues about the lighting environment.

Spatial-angular translation then converts the computed diffuse and specular shading maps from the spatial domain into corresponding lighting distributions in the angular domain. Since the relationship between lighting and shading is different for diffuse and specular reflections, we perform the spatial-angular translation of diffuse and specular shading separately. For diffuse shading, a neural network is trained to infer a low-resolution environment map from it. On the other hand, the translation from the specular map to lighting angles is computed geometrically based on mirror reflection and a normal map of the object, calculated from its depth values.

The angular lighting maps from diffuse and specular reflections are then merged through an *angular fusion* network to produce a coherent environment map based on the different reflectance cues, as detailed in Sect. 4. To ensure temporal consistency of the estimated environment maps over consecutive video frames, we incorporate recurrent convolution layers [6,19] that account for feature maps from previous frames in determining the current frame's output.

4 Network Structures

Reflectance Decomposition Network. The input of the reflectance decomposition network is a cropped RGB input image and the corresponding normal map computed from the bilateral filtered input depth map as the cross-product of depths of neighboring pixels. After a series of convolutional, downsampling and upsampling layers, the network produces a set of decomposed maps including the diffuse albedo map, diffuse shading map, and specular shading map. Our reflectance decomposition is similar to traditional intrinsic image decomposition, but additionally separates shading into diffuse and specular components. Moreover, the normal map taken as input provides the network with shape information to use in decomposition, and the network also generates a refined normal map as output.

In practice, we follow the network structure and training loss of Li et al. [28] with the following modifications. We concatenate the normal map estimated from the rough depth input as an additional channel to the input. We also split the shading branch into two sibling branches, one for diffuse shading and the other for specular shading, each with the same loss function used for the shading component in [28]. The exact network structure is illustrated in the supplementary material.

Diffuse Shading Translation. The relationship between the angular light distribution and the spatial diffuse shading map is non-local, as light from one angular direction will influence the entire image according to the rendering equation:

$$R(x,y) = v(x,y)n(x,y) \cdot l \tag{1}$$

where $v(x,y)$ denotes the visibility between the surface point at pixel (x,y) and the light direction l, $n(x,y)$ represents the surface normal, and $R(x,y)$ is the irradiance map for lighting direction l.

Diffuse shading can be expressed as an integral of the radiance over all the angular directions. For discrete samples of the angular directions, the diffuse shading map will be a weighted sum of the irradiance maps for the sampled directions, with the light intensity as the weight. As a result, with an accurate shading map and irradiance maps, we can directly solve for the lighting via optimization [4,32]. However, optimization based solutions suffer from inaccurate shading and geometry and slow computation. Our translation also follows such a physically-based design. The neural network takes the same input as the numerical optimization, namely the diffuse shading map and the irradiance maps of sparsely sampled lighting directions. Then the network outputs the intensity values of those sampled lighting directions. In practice, we sample the pixel centers of a $8 \times 8 \times 6$ cube-map as the lighting directions to compute a 32×32 resolution auxiliary irradiance map. The diffuse shading map and each irradiance map is separately passed through one layer of convolution, then their feature maps are concatenated together. The concatenated features are sent through a series of convolution and pooling layers to produce a set of feature maps corresponding to $8 \times 8 \times 6$ angular directions. With the feature maps now defined in the angular domain, we rearrange them into a latitude and longitude representation and perform a series of angular domain convolutions to output an estimated environment map of 256×128 resolution.

Specular Shading Translation. Unlike the lowpass filtering effects of diffuse shading, specular reflections are generally of much higher frequency and cannot be adequately represented by the sparse low-resolution sampling used for diffuse shading translation, making the irradiance based solution inefficient. Since specular reflection has a strong response along the mirror direction, we thus map the decomposed specular shading to the angular domain according to the mirror direction, computed from the viewing direction and surface normal as $o = (2n - v)$. Each pixel in the specular shading map is translated individually, and the average value is computed among all the pixels that are mapped to the same angular map location, represented in terms of latitude and longitude.

In practice, depending on the scene geometry, some angular directions may not have any corresponding pixels in the specular shading map. To take this into account, we maintain an additional mask map that indicates whether an angular direction has a valid intensity value or not. After mapping to the 256×128 latitude and longitude map, the maps are processed by four convolution layers to produce angular feature maps.

Angular Fusion Network. After the diffuse and specular translation network, both the diffuse and specular feature maps are defined in the angular domain. Those feature maps will be concatenated together for the fusion network to determine the final lighting estimates.

The angular fusion network is a standard U-net structure, with a series of convolutions and downsampling for encoding, followed by convolution and upsampling layers for decoding. We also include skip links to map the feature maps from the encoder to the decoder, to better preserve angular details. Instead of having the estimated lighting from the diffuse spatial-angular translation network as input, we use the feature map (just before the output layer) of the translation network as the diffuse input to the fusion network. The feature map can preserve more information processed by the translation network (e.g. confidence of the current estimation/transformation), which may help with fusion but is lost in the lighting output.

Recurrent Convolution Layers. Due to the ill-conditioned nature of lighting estimation, there may exist ambiguity in the solution for a video frame. Moreover, this ambiguity may differ from frame to frame because of the different information that the frames contain (e.g., specular objects reflect light from different directions depending on the viewing angle, and cast shadows might be occluded by the objects in certain views). To help resolve these ambiguities and obtain a more coherent solution, we make use of temporal information from multiple frames in determining light predictions. This is done by introducing recurrent convolution layers into the diffuse shading translation network as well as the angular fusion network.

Depth Estimation for RGB Only Input. Our system is built with RGBD input in mind. However, scene depth could instead be estimated by another neural network, which would allow our system to run on RGB images without measured depth. We trained a depth estimation network adapted from [25] but with the viewpoint information as an additional input, as this may facilitate estimation over the large range of pitch directions encountered in AR scenarios. The estimated depth is then included as part of our input. Figure 7 presents real measured examples with light estimated using the predicted depth.

5   Training

5.1   Supervision and Training Losses

We train all the networks using synthetically generated data to provide supervision for each network individually. For the reflectance decomposition network, ground truth maps are used to supervise its training. Following previous networks for intrinsic image decomposition [28], we employ the L2 loss of the maps as well as the L2 loss of the gradients of the albedo and specular maps. The fusion network is trained with the ground truth full-resolution environment map, using the Huber loss.

Recurrent Convolution Training. We train the recurrent convolution layers with sequential data and expand the recurrent layers. In practice, we expand the recurrent layers to accommodate 10 consecutive frames during training. When training the recurrent layers, in addition to the Huber loss that minimizes the difference between the estimated lighting and the ground truth for each frame, we also include a temporal smoothness loss to promote smoothness between consecutively estimated lights L_i, L_{i+1}, defined as

$$\mathcal{L}_t(L_i, L_{i+1}) = ||L_i - \mathcal{W}_{i+1 \to i}(L_{i+1})||^2 \tag{2}$$

where $W_{i+1 \to i}$ is the ground truth optical flow from the lighting of the $(i+1)$-th frame to the i-th frame.

5.2   Training Data Preparation

The training data is composed of many elements such as reflectance maps and angular environment lighting which are difficult to collect in large volumes from real scenes. As a result, we choose to train our neural networks with synthetically generated data. For better network generality to real-world inputs, this synthetic data is prepared to better reflect the properties of real scenes. The environment lights consist of 500 HDR environment maps collected from the Internet, 500 HDR environment map captured by a 360 camera 14 K randomly generated multiple area light sources. We collected over 600 artist modeled objects with realistic texture details and material variations and organize those objects into local scenes that include about 1–6 objects. The scene is then lit by randomly selected environment lights. The viewpoint is uniformly sampled over the upper hemisphere to reflect most AR scenarios. Please refer to the supplementary materials for more details about training data preparation.

5.3   Implementation

We implement the neural networks in Tensorflow [1]. We train our system using the Adam [23] optimizer, with a 10^{-5} learning rate and the default settings for other hyper-parameters. The decomposition is separately trained using 2×10^6 iterations; the diffuse-translation network and the fusion network are first trained with 8×10^5 iterations without recurrent layers, and then finetuned over 10^5 iterations for the recurrent layer training. Finally, the whole system is finetuned end-to-end with 10^5 iterations. Training the decomposition network takes about 1 week on a single NVidia Titan X GPU. The diffuse-translation and fusion network training takes about 10 h, and the final end-to-end finetuning takes about 20 h.

6 Results

6.1 Validations

Single-Image Inputs. Beside video frames, our network does support single-image input by feeding ten copies of the single static image into the network to get the lighting estimation result. This allows us to perform validation and comparison to existing works on single-image inputs.

Error Metric. To measure the accuracy of the estimated light, we directly measure the RMSE compared to the reference environment map. In addition, since the goal of our lighting estimation is rendering virtual objects for AR applications, we also measure the accuracy of our system using the rendering error of a fixed set of objects with the estimated light. All the objects in this set are shown as virtual inserts in our real measured results.

Dataset. To evaluate the performance of our method on real inputs, we captured 180 real input images with the ground truth environment map, providing both numerical and visual results. To systematically analyze how our method works on inputs with different light, layout and materials, we also designed a comprehensive synthetic test set with full control of each individual factor. Specifically, we collected eight artist-designed scenes to ensure plausible layouts, various object shapes, and a range of objects. Ten random viewpoints are generated for each scene, under selected environment maps with random rotations.

Real Measured Environment Maps. For testing the performance of our method under different environment lights, we collected 20 indoor environment maps which are not used for neural network training. These environment maps can be classified into several categories that represent different kinds of lighting conditions. Specifically, we have five environment maps that contain a single dominant light source, five with multiple dominant lights, five with very large area light sources, and five environment maps with near ambient lighting.

Table 1 lists the average error for each category of environment light. Intuitively, lights with high frequency details, such as a dominant light, are more challenging than low frequency cases, like near-ambient lighting. Visual results for one typical example from each category are displayed in Fig. 2, which shows

Fig. 2. We test our system with different classes of environment maps. For each class, the left column shows the ground truth, and the right column shows our results.

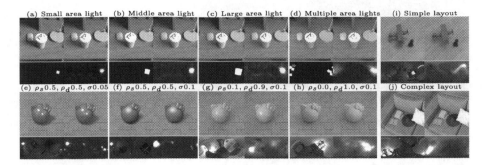

(a) Small area light (b) Middle area light (c) Large area light (d) Multiple area lights (i) Simple layout

(e) $\rho_s 0.5, \rho_d 0.5, \sigma 0.05$ (f) $\rho_s 0.5, \rho_d 0.5, \sigma 0.1$ (g) $\rho_s 0.1, \rho_d 0.9, \sigma 0.1$ (h) $\rho_s 0.0, \rho_d 1.0, \sigma 0.1$ (j) Complex layout

Fig. 3. (a–d) Light estimation results for the same scene lit by different-sized area light sources and multiple light sources. (e–h) Light estimation results for the same shape with varying material properties. (i, j) Light estimation results for scenes with simple or complex layouts. In each group, the left column shows the ground truth, and the right column shows our results.

Table 1. Average RMSE of estimated lights and re-rendered images.

	Render	Light
Environment light		
One dominant	0.057	2.163
Multiple lights	0.057	1.653
Large area light	0.045	1.222
Near ambient	0.038	0.994
Layout robustness		
Simple layout	0.046	0.915
Complex layout	0.049	1.508
Real captured test set		
Real captured	0.052	0.742

	Synthetic		Real	
Comparasion	Render	Light	Render	Light
Gardner et al. 17[11]	0.095	3.056	0.088	3.462
Gardner et al. 19[10]	0.100	1.614	0.093	1.303
Sato[32] + our decomposition	0.073	1.851	0.081	1.257
Sato[32] + ground truth shading	0.064	1.944	-	-
Ablation				
Diffuse only	0.050	1.447	0.055	0.768
Specualr only	0.056	1.440	0.061	0.756
Without decomposition	0.052	1.434	0.059	0.794
Direct regression	0.049	1.432	0.057	0.759
Without recurrent layer	0.048	1.429	0.056	0.760
Our results	**0.046**	**1.419**	**0.052**	**0.742**
Our results (estimated depth)	0.053	1.437	0.064	0.773

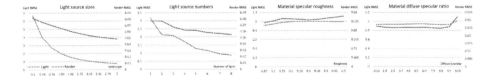

Fig. 4. Lighting estimation and re-rendering error with respect to area light source sizes, number of light sources and different surface materials.

that our method can correctly estimate environment lighting and produce consistent re-rendering results across all the types of environment lighting.

Synthetic Environment Maps. We also synthetically generate a set of lighting conditions that provide more continuous coverage over area light source size and number of light sources. For each set of lighting conditions, we render each scene

Fig. 5. Comparison to SH-based representation. Our method recovers an all-frequency environment map and produces sharp shadows similar to the ground truth. Fitting the ground truth light with 3- or 5-order SH still cannot reproduce such effects.

and each view under those lighting conditions with ten random rotations. We then plot the average error of our system on the different lighting conditions in the set.

Figure 4 plots the lighting estimates and re-rendering error for each dataset. Inputs with smaller area light sources are more challenging for lighting estimation, since small changes of light position leads to large error in both the angular lighting domain as well as the re-rendering results. Figure 3(a–d) exhibits a selected scene under various lighting conditions. For a near-diffuse scene lit by multiple area light sources, even the shadows of objects do not provide sufficient information to fully recover the rich details of each individual light source. However, the re-rendering results with our estimated lighting matches well with that rendered with ground truth light, making them plausible for most AR applications.

Object Materials. To analyze the how surface material affects our estimation results, we prepared a set of objects with the same shape but varying materials. We use a homogeneous specular BRDF with varying roughness and varying diffuse-specular ratios. The results, shown in Fig. 3(e–h), indicate that lighting estimation is stable to a wide range of object materials, from highly specular to purely diffuse. Since our method uses both shading and shadow cues in the scene for lighting estimation, the lighting estimation error is not sensitive to the material of the object, with only a slightly larger error for a purely diffuse scene.

Layouts. We test the performance of our method on scenes with different layouts by classifying our scenes based on complexity. The numerical results for the two layout categories are listed in Table 1. As expected, complex scene layouts lead to more difficult lighting estimation. However, as shown in Fig. 3(i, j), our method produces high quality results in both cases.

Spatially-Varying Illumination. Our object-based lighting estimation can be easily extended to support spatially-varying illumination effects, such as near-field illumination, by taking different local regions of the input image. Please find example results in the supplementary material.

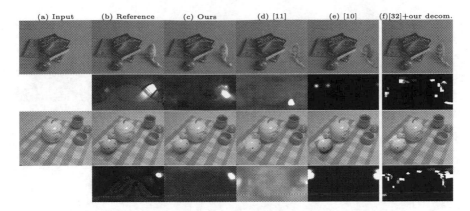

Fig. 6. We compare our method to existing lighting estimation methods on our real captured test set. (c) Our method correctly estimates the size and position of the light sources and produces correct shadow and specular highlights. (d) [11] overestimates the ambient light, the position of the dominant light is off. (e) [10] estimates a incorrect number of lights at incorrect locations. (f) [32] results in many incorrect light sources.

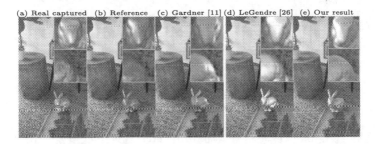

Fig. 7. We compare our method (with estimated depth) to existing lighting estimation methods on a real image from [26]. (a) The photograph of a real 3D-printed bunny placed in the scene. Rendering results of a virtual bunny under (b) captured ground truth environment map, (c) environment map estimated by [11], (d) [26] and (e) our method with estimated depth. Note that our result successfully reproduces specular highlight over the left side of the bunny (closed-up view inserted), similar to the ground-truth. [11] produces wrong specular highlights on the right ear and body of the bunny, and [26] results in wrong highlights all over the bunny.

6.2 Comparisons

Here, we compare our method to existing lighting estimation methods. For systematic evaluation with ground truth data, we compared with scene-based methods [10,11] as well as an inverse rendering method [32], on both synthetic and real measured test sets. Our method takes advantage of physical knowledge to facilitate inference from various cues in the input data, and outperforms all the existing methods, as shown in Table 1. In Figure 6, we show one example of a real captured example. Note that our method estimates an area light source with

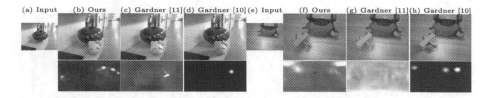

Fig. 8. Comparison on the redwood dataset [8]. Left: our method faithfully reproduces the four specular highlights (blue arrow) visible from the input object (green arrow), which are missing from other methods' results. Right: our method produces shadow with direction consistent to the input (blue and green arrows). Previous methods either fail to reproduce the shadow [11] or generate shadow in the wrong direction [10]. (Color figure online)

the right size at the correct location, while other methods either over-estimate the ambient light [11] or result in an incorrect light position [10,32].

We also compare our method on the redwood RGBD dataset [8], as illustrated in Fig. 8. Although without a ground truth reference, virtual objects rendered with our estimated lights produce consistent specular highlights and shadows, while existing methods fail to generate consistent renderings.

Scene based methods usually need input with a wider field of view, thus we compare our method with [11,26] on their ideal input. Their methods use the full image as input, while our method uses only a small local crop around the target object. The depth of the crop is estimated for input to our system. Fig. 7 illustrates one example result. Note that the rendering result with our estimated light matches well to the reference (the highlight should be at the left side of the bunny). More comparisons can be found in the supplementary material.

To compare with methods based on low-order spherical harmonics (SH) [12], we fit the ground-truth light with 3- and 5-order SH and compare the light and rendering results with our method. As shown in Fig. 5, a 5-order SH (as used by [12]) is incapable of representing small area light sources and generates blurred shadow; our method estimates a full environment map and produces consistent shadow in the rendering.

6.3   Ablation Studies

We conduct ablation studies to justify the components of our system, including the separation into diffuse and specular reflections, and to verify the effectiveness of the translation networks.

We first test our system without the diffuse or specular translation network, but with the remaining networks unchanged. Empirical comparisons are shown in Fig. 9. Without the specular translation network, the system fails to estimate the high-frequency lighting details and produces inconsistent specular highlights in the re-rendering results. Without the diffuse translation network, the system found difficulties estimating light from a diffuse scene.

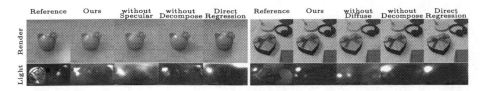

Fig. 9. Ablation study of our physically-based estimation. For a scene containing specular objects (left), our specular network can help to infer high-frequency lighting details from the decomposed specular reflections. For diffuse dominated scenes (right), the diffuse network plays an important role in estimating light from the diffuse shading and shadows. Without having the corresponding network or the decomposition, the system fails to produce accurate estimations. Direct regression also produces blurred light estimation. The inaccurate estimations lead to degraded rendering results, such as missing highlights (left) and shadows (right).

We then remove the decomposition network and feed the RGBD input directly to the spatial-angular translation networks, followed by the fusion network. It is also possible to train a neural network to regress the angular environment lighting directly from the RGBD input. Such a neural network structure shares a design similar to [14,26] but is trained on our dataset. As shown in Table 1 (numerical) and Fig. 9 (visual), compared to those alternative solutions, our method produces the best quality results by combining both diffuse and specular information with our physics-aware network design. With only RGB input, our method can also predict environment maps with estimated depth, and outperforms existing methods. Larger estimation error is found due to inaccuracy in the depth input, which could be improved by training our network together with the depth estimation network.

Please refer to our supplementary video to see how the recurrent convolution layers increase the temporal coherence of the estimated illumination.

Finally, we captured sequences of indoor scenes with dynamic lighting effects, and continuously changing viewpoints, demonstrating the advantages of having real-time lighting estimation for AR applications. Please refer to the supplementary material for the implementation details and real video sequence results. For all the real video results, we crop the central 384×384 region as the input of our method.

6.4 Performance

We test the runtime performance of our method on a workstation with an Intel 7920x CPU and a single NVidia RTX 2080 Ti GPU. For the full system, the processing time per frame is within 26 ms, of which 14 ms is needed for preparing the input and 12 ms is used for neural network inference. We regard adopting to mobile GPUs as future work.

7   Conclusion

We presented a scheme for realtime environment light estimation from the RGBD appearance of individual objects, rather than from broad scene views. By designing the neural networks to follow physical reflectance laws and infusing rendering knowledge with additional input, our method can robustly estimate environment lighting for scenes with arbitrary objects and various illumination conditions. Our recurrent convolution design also offers temporal and spatial smoothness which is critical for many AR applications.

Although our method supports near-field illumination effects by estimating light at different local regions of the input, a potential improvement would be to estimate a near-field illumination model, combining inferences from multiple different patches, to yield a more robust solution for near-field illumination.

Currently, our method estimates environment light only based on the shading information of objects and does not require the input contains contents of the environment map. Combining scene-based method [10,26,35] with our object-based method would be a potential future direction, yielding better quality results.

References

1. Abadi, M., et al.: TensorFlow: large-scale machine learning on heterogeneous systems (2015). https://www.tensorflow.org/, software available from tensorflow.org
2. Azinovic, D., Li, T.M., Kaplanyan, A., Niessner, M.: Inverse path tracing for joint material and lighting estimation. In: The IEEE Conference on Computer Vision and Pattern Recognition (CVPR), June 2019
3. Barron, J.T., Malik, J.: Intrinsic scene properties from a single RGB-D image. In: CVPR, pp. 17–24. IEEE, June 2013. https://doi.org/10.1109/CVPR.2013.10
4. Barron, J.T., Malik, J.: Shape, illumination, and reflectance from shading. IEEE Trans. Pattern Anal. Mach. Intell. **37**(8), 1670–1687 (2015). http://ieeexplore.ieee.org/document/6975182/
5. Calian, D.A., Lalonde, J.F., Gotardo, P., Simon, T., Matthews, I., Mitchell, K.: From faces to outdoor light probes. Comput. Graph. Forum **37**, 51–61 (2018)
6. Chaitanya, C.R.A., et al.: Interactive reconstruction of Monte Carlo image sequences using a recurrent denoising autoencoder. ACM Trans. Graph. **36**(4), 98:1–98:12 (2017)
7. Cheng, D., Shi, J., Chen, Y., Deng, X., Zhang, X.: Learning scene illumination by pairwise photos from rear and front mobile cameras. In: Computer Graphics Forum (2018)
8. Choi, S., Zhou, Q.Y., Miller, S., Koltun, V.: A large dataset of object scans. arXiv:1602.02481 (2016)
9. Debevec, P.: Rendering synthetic objects into real scenes: bridging traditional and image-based graphics with global illumination and high dynamic range photography. In: Proceedings of the 25th Annual Conference on Computer Graphics and Interactive Techniques, SIGGRAPH 1998, pp. 189–198. ACM, New York (1998)
10. Gardner, M.A., Hold-Geoffroy, Y., Sunkavalli, K., Gagne, C., Lalonde, J.F.: Deep parametric indoor lighting estimation. In: The IEEE International Conference on Computer Vision (ICCV), October 2019

11. Gardner, M.A., et al.: Learning to predict indoor illumination from a single image. ACM Trans. Graph. **36**(6), 1–14 (2017)

12. Garon, M., Sunkavalli, K., Hadap, S., Carr, N., Lalonde, J.F.: Fast spatially-varying indoor lighting estimation. In: The IEEE Conference on Computer Vision and Pattern Recognition (CVPR), June 2019

13. Georgoulis, S., Rematas, K., Ritschel, T., Fritz, M., Tuytelaars, T., Gool, L.V.: What is around the camera? In: ICCV (2017)

14. Georgoulis, S., et al.: Reflectance and natural illumination from single-material specular objects using deep learning. PAMI **40**, 1932–1947 (2017)

15. Gruber, L., Langlotz, T., Sen, P., Höherer, T., Schmalstieg, D.: Efficient and robust radiance transfer for probeless photorealistic augmented reality. In: 2014 IEEE Virtual Reality (VR), pp. 15–20, March 2014

16. Gruber, L., Richter-Trummer, T., Schmalstieg, D.: Real-time photometric registration from arbitrary geometry. In: 2012 IEEE International Symposium on Mixed and Augmented Reality (ISMAR), pp. 119–128, November 2012

17. Hold-Geoffroy, Y., Athawale, A., Lalonde, J.F.: Deep sky modeling for single image outdoor lighting estimation. In: The IEEE Conference on Computer Vision and Pattern Recognition (CVPR), June 2019

18. Hold-Geoffroy, Y., Sunkavalli, K., Hadap, S., Gambaretto, E., Lalonde, J.F.: Deep outdoor illumination estimation. In: IEEE International Conference on Computer Vision and Pattern Recognition (2017)

19. Huang, Y., Wang, W., Wang, L.: Bidirectional recurrent convolutional networks for multi-frame super-resolution. In: Advances in Neural Information Processing Systems, vol. 28 (2015)

20. Jiddi, S., Robert, P., Marchand, E.: Illumination estimation using cast shadows for realistic augmented reality applications. In: 2017 IEEE International Symposium on Mixed and Augmented Reality (ISMAR-Adjunct), pp. 192–193, October 2017

21. Karsch, K., et al.: Automatic scene inference for 3D object compositing. ACM Trans. Graph. **33**(3), 32:1–32:15 (2014)

22. Khan, E.A., Reinhard, E., Fleming, R.W., Bülthoff, H.H.: Image-based material editing. ACM Trans. Graph. **25**(3), 654–663 (2006). https://doi.org/10.1145/1141911.1141937

23. Kingma, D.P., Ba, J.: Adam: a method for stochastic optimization. In: ICLR, May 2015

24. Kronander, J., Banterle, F., Gardner, A., Miandji, E., Unger, J.: Photorealistic rendering of mixed reality scenes. Comput. Graph. Forum **34**(2), 643–665 (2015)

25. Laina, I., Rupprecht, C., Belagiannis, V., Tombari, F., Navab, N.: Deeper depth prediction with fully convolutional residual networks. In: 3DV (2016)

26. LeGendre, C., et al.: DeepLight: learning illumination for unconstrained mobile mixed reality. In: CVPR (2019)

27. LeGendre, C., et al.: Practical multispectral lighting reproduction. ACM Trans. Graph. **35**(4), 32:1–32:11 (2016)

28. Li, Z., Snavely, N.: CGintrinsics: better intrinsic image decomposition through physically-based rendering. In: European Conference on Computer Vision (ECCV) (2018)

29. Lombardi, S., Nishino, K.: Reflectance and illumination recovery in the wild. IEEE Trans. Pattern Anal. Mach. Intell. **38**(1), 129–141 (2016). https://doi.org/10.1109/TPAMI.2015.2430318

30. Nishino, K., Nayar, S.K.: Eyes for relighting. ACM Trans. Graph. **23**(3), 704–711 (2004)

31. Romeiro, F., Zickler, T.: Blind reflectometry. In: Daniilidis, K., Maragos, P., Paragios, N. (eds.) ECCV 2010. LNCS, vol. 6311, pp. 45–58. Springer, Heidelberg (2010). https://doi.org/10.1007/978-3-642-15549-9_4

32. Sato, I., Sato, Y., Ikeuchi, K.: Illumination from shadows. IEEE Trans. Pattern Anal. Mach. Intell. **25**(3), 290–300 (2003)

33. Sengupta, S., Gu, J., Kim, K., Liu, G., Jacobs, D.W., Kautz, J.: Neural inverse rendering of an indoor scene from a single image. In: Proceedings of the IEEE International Conference on Computer Vision, pp. 8598–8607 (2019)

34. Sengupta, S., Kanazawa, A., Castillo, C.D., Jacobs, D.W.: SfSNet: learning shape, reflectance and illuminance of faces 'in the wild'. In: CVPR (2018)

35. Song, S., Funkhouser, T.: Neural illumination: lighting prediction for indoor environments. In: The IEEE Conference on Computer Vision and Pattern Recognition (CVPR), June 2019

36. Sun, T., et al.: Single image portrait relighting. ACM Trans. Graph. **38**, 79-1 (2019)

37. Tewari, A., et al.: Self-supervised multi-level face model learning for monocular reconstruction at over 250 Hz. In: CVPR (2018)

38. Tewari, A., et al.: MoFA: model-based deep convolutional face autoencoder for unsupervised monocular reconstruction. In: ICCV (2017)

39. Unger, J., Gustavson, S., Ynnerman, A.: Densely sampled light probe sequences for spatially variant image based lighting. In: Proceedings of GRAPHITE, June 2006

40. Waese, J., Debevec, P.: A real-time high dynamic range light probe. In: Proceedings of the 27th Annual Conference on Computer Graphics and Interactive Techniques: Conference Abstracts and Applications (2002)

41. Weber, H., Prévost, D., Lalonde, J.F.: Learning to estimate indoor lighting from 3D objects. In: 2018 International Conference on 3D Vision (3DV), pp. 199–207. IEEE (2018)

42. Wu, C., Wilburn, B., Matsushita, Y., Theobalt, C.: High-quality shape from multi-view stereo and shading under general illumination. In: CVPR (2011)

43. Yi, R., Zhu, C., Tan, P., Lin, S.: Faces as lighting probes via unsupervised deep highlight extraction. In: ECCV (2018)

44. Zhang, J., et al.: All-weather deep outdoor lighting estimation. In: The IEEE Conference on Computer Vision and Pattern Recognition (CVPR), June 2019

45. Zhou, H., Sun, J., Yacoob, Y., Jacobs, D.W.: Label denoising adversarial network (LDAN) for inverse lighting of faces. In: CVPR (2018)

Progressive Point Cloud Deconvolution Generation Network

Le Hui, Rui Xu, Jin Xie$^{(\boxtimes)}$, Jianjun Qian, and Jian Yang$^{(\boxtimes)}$

Key Lab of Intelligent Perception and Systems for High-Dimensional Information of Ministry of Education, Jiangsu Key Lab of Image and Video Understanding for Social Security PCA Lab, School of Computer Science and Engineering, Nanjing University of Science and Technology, Nanjing, China
{le.hui,xu_ray,csjxie,csjqian,csjyang}@njust.edu.cn

Abstract. In this paper, we propose an effective point cloud generation method, which can generate multi-resolution point clouds of the same shape from a latent vector. Specifically, we develop a novel progressive deconvolution network with the learning-based bilateral interpolation. The learning-based bilateral interpolation is performed in the spatial and feature spaces of point clouds so that local geometric structure information of point clouds can be exploited. Starting from the low-resolution point clouds, with the bilateral interpolation and max-pooling operations, the deconvolution network can progressively output high-resolution local and global feature maps. By concatenating different resolutions of local and global feature maps, we employ the multi-layer perceptron as the generation network to generate multi-resolution point clouds. In order to keep the shapes of different resolutions of point clouds consistent, we propose a shape-preserving adversarial loss to train the point cloud deconvolution generation network. Experimental results on ShpaeNet and ModelNet datasets demonstrate that our proposed method can yield good performance. Our code is available at https://github.com/fpthink/PDGN.

Keywords: Point cloud generation · GAN · Deconvolution network · Bilateral interpolation

1 Introduction

With the development of 3D sensors such as LiDAR and Kinect, 3D geometric data are widely used in various kinds of computer vision tasks. Due to the great success of generative adversarial network (GAN) [10] in the 2D image domain, 3D data generation [5,7,11,16,36,38,45–47] has been receiving more and more attention. Point clouds, as an important 3D data type, can compactly

---

Electronic supplementary material The online version of this chapter (https://doi.org/10.1007/978-3-030-58555-6_24) contains supplementary material, which is available to authorized users.

A. Vedaldi et al. (Eds.): ECCV 2020, LNCS 12360, pp. 397–413, 2020.
https://doi.org/10.1007/978-3-030-58555-6_24

and flexibly characterize geometric structures of 3D models. Different from 2D image data, point clouds are unordered and irregular. 2D generative models cannot be directly extended to point clouds. Therefore, how to generate realistic point clouds in an unsupervised way is still a challenging and open problem.

Recent research efforts have been dedicated to 3D model generation. Based on the voxel representation of 3D models, 3D convolutional neural networks (3D CNNs) can be applied to form 3D GAN [40] for 3D model generation. Nonetheless, since 3D CNNs on the voxel representation require heavy computational and memory burdens, 3D GANs are limited to generate low-resolution 3D models. Different from the regular voxel representation, point clouds are spatially irregular. Therefore, CNNs cannot be directly applied on point clouds to form 3D generative models. Inspired by PointNet [26] that can learn compact representation of point clouds, Achlioptas et al. [1] proposed an auto-encoder based point cloud generation network in a supervised manner. Nonetheless, the generation model is not an end-to-end learning framework. Yang et al. [44] proposed the PointFlow generation model, which can learn a two-level hierarchical distribution with a continuous normalized flow. Based on graph convolution, Valsesia et al. [37] proposed a localized point cloud generation model. Shu et al. [31] developed a tree structured graph convolution network for point cloud generation. Due to the high computational complexity of the graph convolution operation, training the graph convolution based generation models is very time-consuming.

In this paper, we propose a simple yet efficient end-to-end generation model for point clouds. We develop a progressive deconvolution network to map the latent vector to the high-dimensional feature space. In the deconvolution network, the learning-based bilateral interpolation is adopted to enlarge the feature map, where the weights are learned from the spatial and feature spaces of point clouds simultaneously. It is desirable that the bilateral interpolation can capture the local geometric structures of point clouds well with the increase of the resolution of generated point clouds. Following the deconvolution network, we employ the multi-layer perceptron (MLP) to generate spatial coordinates of point clouds. By stacking multiple deconvolution networks with different resolutions of point clouds as the inputs, we can form a progressive deconvolution generation network to generate multi-resolution point clouds. Since the shapes of multi-resolution point clouds generated from the same latent vector should be consistent, we formulate a shape-preserving adversarial loss to train the point cloud deconvolution generation network. Extensive experiments are conducted on the ShapeNet [3] and ModelNet [41] datasets to demonstrate the effectiveness of our proposed method. The main contributions of our work are summarized as follows:

- We present a novel progressive point cloud generation framework in an end-to-end manner.
- We develop a new deconvolution network with the learning-based bilateral interpolation to generate high-resolution feature maps.

– We formulate a shape-preserving loss to train the progressive point cloud network so that the shapes of generated multi-resolution point clouds from the same latent vector are consistent.

The rest of the paper is organized as follows: Sect. 2 introduces related work. In Sect. 3, we present the progressive end-to-end point cloud generation model. Section 4 presents experimental results and Sect. 5 concludes the paper.

2 Related Work

2.1 Deep Learning on 3D Data

Existing 3D deep learning methods can be roughly divided into two classes. One class of 3D deep learning methods [22,27,33,41] convert the geometric data to the regular-structured data and apply existing deep learning algorithms to them. The other class of methods [17,24,26,28,32,35] mainly focus on constructing special operations that are suitable to the unstructured geometric data for 3D deep learning.

In the first class of 3D deep learning methods, view-based methods represent the 3D object as a collection of 2D views so that the standard CNN can be directly applied. Specifically, the max-pooling operation across views is used to obtain a compact 3D object descriptor [33]. Voxelization [22,41] is another way to represent the 3D geometric data with regular 3D grids. Based on the voxelization representation, the standard 3D convolution can be easily used to form the 3D CNNs. Nonetheless, the voxelization representation usually leads to the heavy burden of memory and high computational complexity because of the computation of the 3D convolution. Qi *et al.* [27] proposed to combine the view-based and voxelization-based deep learning methods for 3D shape classification.

In 3D deep learning, variants of deep neural networks are also developed to characterize the geometric structures of 3D data. [32,35] formulated the unstructured point clouds as the graph-structured data and employed the graph convolution to form the 3D deep learning representation. Qi *et al.* [26] proposed Point-Net that treats each point individually and aggregates point features through several MLPs followed by the max-pooling operation. Since PointNet cannot capture the local geometric structures of point clouds well, Qi *et al.* [28] proposed PointNet++ to learn the hierarchical feature representation of point clouds. By constructing the k-nearest neighbor graph, Wang *et al.* [39] proposed an edge convolution operation to form the dynamic graph CNN for point clouds. Li *et al.* [19] proposed PointCNN for feature learning from point clouds, where the χ-transform is learned to form the χ-convolution operation.

2.2 3D Point Cloud Generation

Variational auto-encoder (VAE) is an important type of generative model. Recently, VAE has been applied to point cloud generation. Gadelha *et al.* [8] proposed MRTNet to generate point clouds from a single image. Specifically, using a VAE framework, a 1D ordered list of points is fed to the multi-resolution encoder and decoder to perform point cloud generation in unsupervised learning. Zamorski *et al.* [46] applied the VAE and adversarial auto-encoder (AAE) to point cloud generation. Since the VAE model requires the particular prior distribution to make KL divergence tractable, the AAE is introduced to learn the prior distribution by utilizing adversarial training. Lately, Yang *et al.* [44] proposed a probabilistic framework (PointFlow) to generate point clouds by modeling them as a two-level hierarchical distribution. Nonetheless, as mentioned in PointFlow [44], it converges slowly and fails for the cases with many thin structures (like chairs).

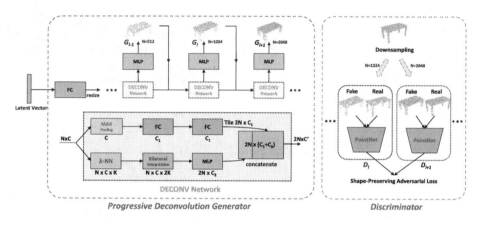

Fig. 1. The architecture of our progressive point cloud framework. The progressive deconvolution generator aims to generate point clouds, while the discriminator distinguishes it from the real point clouds.

Generative adversarial network (GAN) has also achieved great success in the field of image generation [2,6,21,23,29]. Recently, a series of attractive works [5,7,12,31,43] ignite a renewed interest in the 3D object generation task by adopting CNNs. Wu *et al.* [40] first proposed 3D-GAN, which can generate 3D objects from a probabilistic space by using the volumetric convolutional network and GAN. Zhu *et al.* [48] proposed a GAN-based neural network that can leverage information extracted from 2D images to improve the quality of generated 3D models. However, due to the sparsely occupied 3D grids of the 3D object, the volumetric representation approach usually faces a heavy memory burden, resulting in the high computational complexity of the volumetric convolutional network. To alleviate the memory burden, Achlioptas *et al.* [1] proposed a two-stage deep generative model with an auto-encoder for point clouds. It first maps data points into the latent representation and then trains a minimal GAN in the learned latent space to generate point clouds. However, the two-stage point cloud generation model cannot be trained in an end-to-end manner. Based on graph convolution, Valsesia *et al.* [37] focused on designing a graph-based generator

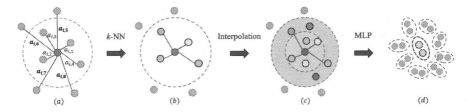

Fig. 2. The constructed deconvolution operation. First, we define the similarity between point pairs in the feature space (a). We choose the k nearest neighbor points (k-NN) in the feature space with the defined similarity in (b). Then we interpolate in the neighborhood to form an enlarged feature map in (c). Finally, we apply the MLP to generate new high-dimensional feature maps in (d). Note that we can obtain double numbers of points through the deconvolution operation.

that can learn the localized features of point clouds. Shu *et al.* [31] developed a tree structured graph convolution network for 3D point cloud generation.

3   Our Approach

In this section, we present our progressive generation model for 3D point clouds. The framework of our proposed generation model is illustrated in Fig. 1. In Sect. 3.1, we describe how to construct the proposed progressive deconvolution generation network. In Sect. 3.2, we present the details of the shape-preserving adversarial loss to train the progressive deconvolution generation network.

3.1   Progressive Deconvolution Generation Network

Given a latent vector, our goal is to generate high-quality 3D point clouds. One key problem in point cloud generation is how to utilize a one-dimensional vector to generate a set of 3D points consistent with the 3D object in geometry. To this end, we develop a special deconvolution network for 3D point clouds, where we first obtain the high-resolution feature map with the learning-based bilateral interpolation and then apply MLPs to generate the local and global feature maps. It is desirable that the fusion of the generated local and global feature maps can characterize the geometric structures of point clouds in the high-dimensional feature space.

Learning-Based Bilateral Interpolation. Due to the disordered and irregular structure of point clouds, we cannot directly perform the interpolation operation on the feature map. Therefore, we need to build a neighborhood for each point on the feature map to implement the interpolation operation. In this work, we simply employ the k-nearest neighbor (k-NN) to construct the neighborhood of each point in the feature space. Specifically, given an input with N feature vectors $\boldsymbol{x}_i \in \mathbb{R}^d$, the similarity between points i and j is defined as:

$$a_{i,j} = \exp\left(-\beta \|\boldsymbol{x}_i - \boldsymbol{x}_j\|_2^2\right) \tag{1}$$

where β is empirically set as $\beta = 1$ in our experiments. As shown in Figs. 2(a) and (b), we can choose k nearest neighbor points in the feature space with the defined similarity. And the parameter k is set as $k = 20$ in this paper.

Once we obtain the neighborhood of each point, we can perform the interpolation in it. As shown in Fig. 2 (c), with the interpolation, k points in the neighborhood can be generated to $2k$ points in the feature space. Classical interpolation methods such as linear and bilinear interpolations are non-learning interpolation methods, which cannot be adaptive to different classes of 3D models during the point cloud generation process. Moreover, the classical interpolation methods does not exploit neighborhood information of each point in the spatial and feature space simultaneously.

To this end, we propose a learning-based bilateral interpolation method that utilizes the spatial coordinates and features of the neighborhood of each point to generate the high-resolution feature map. Given the point $\boldsymbol{p}_i \in \mathbb{R}^3$ and k points in its neighborhood, we can formulate the bilateral interpolation as:

$$\tilde{\boldsymbol{x}}_{i,l} = \frac{\sum_{j=1}^{k} \varphi_l\left(\boldsymbol{p}_i, \boldsymbol{p}_j\right) \phi_l\left(\boldsymbol{x}_i, \boldsymbol{x}_j\right) \boldsymbol{x}_{j,l}}{\sum_{j=1}^{k} \varphi_l\left(\boldsymbol{p}_i, \boldsymbol{p}_j\right) \phi_l\left(\boldsymbol{x}_i, \boldsymbol{x}_j\right)} \tag{2}$$

where $\boldsymbol{p}_i$ and $\boldsymbol{p}_j$ are the 3D spatial coordinates, $\boldsymbol{x}_i$ and $\boldsymbol{x}_j$ are the d-dimensional feature vectors, $\varphi\left(\boldsymbol{p}_i, \boldsymbol{p}_j\right) \in \mathbb{R}^d$ and $\phi\left(\boldsymbol{x}_i, \boldsymbol{x}_j\right) \in \mathbb{R}^d$ are two embeddings in the spatial and feature spaces, $\tilde{\boldsymbol{x}}_{i,l}$ is the l-th element of the interpolated feature $\tilde{\boldsymbol{x}}_i$, $l = 1, 2, \cdots, d$. The embeddings $\varphi\left(\boldsymbol{p}_i, \boldsymbol{p}_j\right)$ and $\phi\left(\boldsymbol{x}_i, \boldsymbol{x}_j\right)$ can be defined as:

$$\varphi\left(\boldsymbol{p}_i, \boldsymbol{p}_j\right) = \text{ReLU}(\boldsymbol{W}_{\theta,j}^{\top}\left(\boldsymbol{p}_i - \boldsymbol{p}_j\right)), \quad \phi\left(\boldsymbol{x}_i, \boldsymbol{x}_j\right) = \text{ReLU}(\boldsymbol{W}_{\psi,j}^{\top}\left(\boldsymbol{x}_i - \boldsymbol{x}_j\right)) \tag{3}$$

where ReLU is the activation function, $\boldsymbol{W}_{\theta,j} \in \mathbb{R}^{3 \times d}$ and $\boldsymbol{W}_{\psi,j} \in \mathbb{R}^{d \times d}$ are the weights to be learned. Based on the differences between the points $\boldsymbol{p}_i$ and $\boldsymbol{p}_j$, $\boldsymbol{p}_i - \boldsymbol{p}_j$ and $\boldsymbol{x}_i - \boldsymbol{x}_j$, the embeddings $\varphi\left(\boldsymbol{p}_i, \boldsymbol{p}_j\right)$ and $\phi\left(\boldsymbol{x}_i, \boldsymbol{x}_j\right)$ can encode local structure information of the point $\boldsymbol{p}_i$ in the spatial and feature spaces, respectively. It is noted that in Eq. 2 the channel-wise bilateral interpolation is adopted. As shown in Fig. 3, the new interpolated feature $\tilde{\boldsymbol{x}}_i$ can be obtained from the neighborhood of $\boldsymbol{x}_i$ with the bilateral weight. For each point, we perform the bilateral interpolation in the k-neighborhood to generate new k points. Therefore, we can obtain a high-resolution feature map, where the neighborhood of each point contains $2k$ points.

After the interpolation, we then apply the convolution on the enlarged feature maps. For each point, we divide the neighborhood of $2k$ points into two regions according to the distance. As shown in Fig. 2 (c), the closest k points belong to the first region and the rest as the second region. Similar to PointNet [26], we first use the MLP to generate high-dimensional feature maps and then use the max-pooling operation to obtain the local features of the two interpolated points from two regions. As shown in Fig. 2 (d), we can double the number of points from the inputs through the deconvolution network to generate a high-resolution local feature map $\boldsymbol{X}_{local}$. We also use the max-pooling operation to extract the global feature of point clouds. By replicating the global feature for

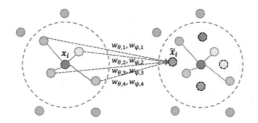

Fig. 3. The illustration of the learning-based bilateral interpolation method. The points in the neighborhood of the center point x_i are colored. We interpolate new points by considering the local geometric features of the points in the neighborhood. $W_{\theta,j}$ and $W_{\psi,j}$, $j = 1, 2, 3, 4$, are the weights in the spatial and feature spaces to be learned.

N times, where N is the number of points, we can obtain the high-resolution global feature map X_{global}. Then we concatenate the local feature map X_{local} and the global feature map X_{global} to obtain the output of the deconvolution network $X_c = [X_{local}; X_{global}]$. Thus, the output X_c can not only characterize the local geometric structures of point clouds, but also capture the global shape of point clouds during the point cloud generation process.

3D Point Cloud Generation. Our goal is to progressively generate 3D point clouds from the low resolution to the high resolution. Stacked deconvolution networks can progressively double the number of points and generate their high-dimensional feature maps. We use the MLP after each deconvolution network to generate the 3D coordinates of point clouds at each resolution. Note that two outputs of the DECONV block are the same, one for generating 3D coordinates of point clouds and the other as the features of the point clouds. We concatenate the generated 3D coordinates with the corresponding features as the input to the next DECONV block.

3.2 Shape-Preserving Adversarial Loss

Shape-Consistent Constraint. During the training process, different resolutions of 3D point clouds are generated. With the increase of the resolution of the output of the progressive deconvolution network, generated point clouds become denser. It is expected that the local geometric structures of the generated point clouds are as consistent as possible between different resolutions. Since our progressive deconvolution generation network is an unsupervised generation model, it is difficult to distinguish different shapes from the same class of 3D objects for the discriminator. Thus, for the specific class of 3D objects, the deconvolution generation networks at different resolutions might generate 3D point clouds with different shapes. Therefore, we encourage that the means and covariances of the neighborhoods of the corresponding points between different resolutions are as close as possible so that the corresponding parts of different resolutions of generated point clouds are consistent.

Shape-Preserving Adversarial Loss. We employ the mean and covariance of the neighborhoods of the corresponding points to characterize the consistency of the generated point clouds between different resolutions. We use the farthest point sampling (FPS) to choose centroid points from each resolution and find the k-neighborhoods for centroid points. The mean and covariance of the neighborhood of the i-th centroid point are represented as:

$$\boldsymbol{\mu}_i = \frac{\sum_{j \in \mathcal{N}_i} \boldsymbol{p}_j}{k}, \quad \boldsymbol{\sigma}_i = \frac{\sum_{j \in \mathcal{N}_i} (\boldsymbol{p}_j - \boldsymbol{\mu}_i)^\top (\boldsymbol{p}_j - \boldsymbol{\mu}_i)}{k-1} \tag{4}$$

where $\mathcal{N}_i$ is the neighborhood of the centroid point, $\boldsymbol{p}_j \in \mathbb{R}^3$ is the coordinates of the point cloud, $\boldsymbol{\mu}_i \in \mathbb{R}^3$ and $\boldsymbol{\sigma}_i \in \mathbb{R}^{3 \times 3}$ are the mean and covariance of the neighborhood, respectively.

Since the sampled centroid points are not completely matched between adjacent resolutions, we employ the Chamfer distances of the means and covariances to formulate the shape-preserving loss. We denote the centroid point sets at the resolutions l and $l+1$ by S_l and S_{l+1}, respectively. The Chamfer distance $d_1(S_l, S_{l+1})$ between the means of the neighborhoods from the adjacent resolutions is defined as:

$$d_1(S_l, S_{l+1}) = \max \left\{ \frac{1}{|S_l|} \sum_{i \in S_l} \min_{j \in S_{l+1}} \|\boldsymbol{\mu}_i - \boldsymbol{\mu}_j\|_2 , \quad \frac{1}{|S_{l+1}|} \sum_{j \in S_{l+1}} \min_{i \in S_l} \|\boldsymbol{\mu}_j - \boldsymbol{\mu}_i\|_2 \right\} \tag{5}$$

Similarly, the Chamfer distance $d_2(S_l, S_{l+1})$ between the covariances of the neighborhoods is defined as:

$$d_2(S_l, S_{l+1}) = \max \left\{ \frac{1}{|S_l|} \sum_{i \in S_l} \min_{j \in S_{l+1}} \|\boldsymbol{\sigma}_i - \boldsymbol{\sigma}_j\|_F , \quad \frac{1}{|S_{l+1}|} \sum_{j \in S_{l+1}} \min_{i \in S_l} \|\boldsymbol{\sigma}_j - \boldsymbol{\sigma}_i\|_F \right\} \tag{6}$$

The shape-preserving loss (SPL) for multi-resolution point clouds is defined as:

$$SPL(G_l, G_{l+1}) = \sum_{l=1}^{M-1} d_1(S_l, S_{l+1}) + d_2(S_l, S_{l+1}) \tag{7}$$

where M is the number of resolutions, G_l and G_{l+1} represents the l-th and $(l+1)$-th point cloud generators, respectively.

Based on Eq. 7, for the generator G_l and discriminator D_l, we define the following shape-preserving adversarial loss:

$$\begin{aligned} L(D_l) &= E_{s \sim p_{real}(s)}(\log D_l(\boldsymbol{s}) + \log(1 - D_l(G_l(\boldsymbol{z})))) \\ L(G_l) &= E_{z \sim p_z(z)}(\log(1 - D_l(G_l(\boldsymbol{z}))) + \lambda SPL(G_l(\boldsymbol{z}), G_{l+1}(\boldsymbol{z})) \end{aligned} \tag{8}$$

where $\boldsymbol{s}$ is the real point cloud sample, $\boldsymbol{z}$ is the randomly sampled latent vector from the distribution $p(\boldsymbol{z})$ and λ is the regularization parameter. Note that we ignore the SPL in $L(G_l)$ for $l = M$. Thus, multiple generators G and discriminators D can be trained with the following equation:

$$max_D \sum_{l=1}^{M} L(D_l), min_G \sum_{l=1}^{M} L(G_l) \tag{9}$$

where $D = \{D_1, D_2, \cdots, D_M\}$ and $G = \{G_1, G_2, \cdots, G_M\}$. During the training process, multiple generators G and discriminators D are alternatively optimized till convergence.

4   Experiments

4.1   Experimental Settings

We evaluate our proposed generation network on three popular datasets including ShapeNet [3], ModelNet10 and ModelNet40 [41]. ShapeNet is a richly annotated large-scale point cloud dataset containing 55 common object categories and 513,000 unique 3D models. In our experiments, we only use 16 categories of 3D objects. ModelNet10 and ModelNet40 are subsets of ModelNet, which contain 10 categories and 40 categories of CAD models, respectively.

Our proposed framework mainly consists of progressive deconvolution generator and shape-preserving discriminator. In this paper, we generate four resolutions of point clouds from a 128-dimensional latent vector. In the generator, the output size of 4 deconvolution networks are 256×32, 512×64, 1024×128 and 2048×256. We use MLPs to generate coordinates of point clouds. Note that MLPs are not shared for 4 resolutions. After the MLP, we adopt the $Tanh$ activation function. In the discriminator, we use 4 PointNet-like structures. For different resolutions, the network parameters of the discriminators are different. We use Leaky ReLU [42] and batch normalization [13] after every layer. The more detailed structure of our framework is shown in the supplementary material. In addition, we use the $k = 20$ nearest points as the neighborhood for the bilateral interpolation. During the training process, we adopt Adam [15] with the learning rate 10^{-4} for both generator and discriminator. We employ an alternative training strategy in [10] to train the generator and discriminator. Specifically, the discriminator is optimized for each generator step.

4.2   Evaluation of Point Cloud Generation

Visual results. As shown in Fig. 4, on the ShapeNet [3] dataset, we visualize the synthesized point clouds containing 4 categories, which are "Airplane", "Table", "Chair", and "Lamp", respectively. Due to our progressive generator, each category contains four resolutions of point clouds generated from the same latent vector. It can be observed that the geometric structures of different resolutions of generated point clouds are consistent. Note that the generated point clouds contain detailed structures, which are consistent with those of real 3D objects. More visualizations are shown in the supplementary material.

Quantitative evaluation. To conduct a quantitative evaluation of the generated point clouds, we adopt the evaluation metric proposed in [1,20], including Jensen-Shannon Divergence (JSD), Minimum Matching Distance (MMD), and Coverage (COV), the earth mover's distance (EMD), the chamfer distance (CD)

Fig. 4. Generated point clouds including "Airplane", "Table", "Chair" and "Lamp". Each category has four resolutions of point clouds (256, 512, 1024 and 2048).

and the 1-nearest neighbor accuracy (1-NNA). JSD measures the marginal distributions between the generated samples and real samples. MMD is the distance between one point in the real sample set and its nearest neighbors in the generation set. COV measures the fraction of point clouds in the real sample set that can be matched at least one point in the generation set. 1-NNA is used as a metric to evaluate whether two distributions are identical for two-sample tests. Table 1 lists our results with different criteria on the "Airplane" and "Chair" categories in the ShapeNet dataset. In Table 1, except for PointFlow [44] (VAE-

Table 1. The results on the "Airplane" and "Chair" categories. Note that JSD scores and MMD-EMD scores are multiplied by 10^2, while MMD-CD scores are multiplied by 10^3.

Category	Model	JSD (↓)	MMD (↓)		COV (%, ↑)		1-NNA (%, ↓)	
			CD	EMD	CD	EMD	CD	EMD
Airplane	r-GAN [1]	7.44	0.261	5.47	42.72	18.02	93.50	99.51
	l-GAN (CD) [1]	4.62	0.239	4.27	43.21	21.23	86.30	97.28
	l-GAN (EMD) [1]	3.61	0.269	3.29	47.90	50.62	87.65	85.68
	PC-GAN [18]	4.63	0.287	3.57	36.46	40.94	94.35	92.32
	GCN-GAN [37]	8.30	0.800	7.10	31.00	14.00	-	-
	tree-GAN [31]	9.70	0.400	6.80	61.00	20.00	-	-
	PointFlow [44]	4.92	**0.217**	3.24	46.91	48.40	75.68	75.06
	PDGN (ours)	**3.32**	0.281	**2.91**	**64.98**	**53.34**	**63.15**	**60.52**
Chair	r-GAN [1]	11.5	2.57	12.8	33.99	9.97	71.75	99.47
	l-GAN (CD) [1]	4.59	2.46	8.91	41.39	25.68	64.43	85.27
	l-GAN (EMD) [1]	2.27	2.61	7.85	40.79	41.69	64.73	65.56
	PC-GAN [18]	3.90	2.75	8.20	36.50	38.98	76.03	78.37
	GCN-GAN [37]	10.0	2.90	9.70	30.00	26.00	-	-
	tree-GAN [31]	11.9	**1.60**	10.1	58.00	30.00	-	-
	PointFlow [44]	1.74	2.24	7.87	46.83	46.98	60.88	59.89
	PDGN (ours)	**1.71**	1.93	**6.37**	**61.90**	**57.89**	**52.38**	**57.14**

based generation method), the others are GAN-based generation methods. For these evaluation metrics, in most cases, our point cloud deconvolution generation network (PDGN) outperforms other methods, demonstrating the effectiveness of the proposed method. Moreover, the metric results on the "Car" category and the mean result of all 16 categories are shown in the supplementary material.

Different from the existing GAN-based generation methods, we develop a progressive generation network to generate multi-resolution point clouds. In order to generate the high-resolution point clouds, we employ the bilateral interpolation in the spatial and feature spaces of the low-resolution point clouds to produce the geometric structures of the high-resolution point clouds. Thus, with the increase of resolutions, the structures of generated point clouds are more and more clear. Therefore, our PDGN can yield better performance in terms of these evaluation criteria. In addition, compared to PointFlow, our method can perform better on point clouds with thin structures. As shown in Fig. 5, it can be seen that our method can generate more complete point clouds. Since in PointFlow the VAE aims to minimize the lower bound of the log-likelihood of the latent vector, it may fail for point clouds with thin structures. Nonetheless, due to the bilateral deconvolution and progressive generation from the low resolution to the high resolution, our PDGN can still achieve good performance for point cloud generation with thin structures. For more visualization comparisons to PointFlow please refer to the supplementary material.

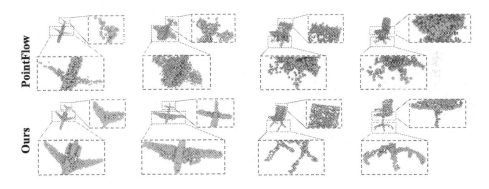

Fig. 5. Visualization results on the "Airplane" and "Chair" categories.

Classification Results. Following [40,44], we also conduct the classification experiments on ModelNet10 and ModelNet40 to evaluate our generated point clouds. We first use all samples from ModelNet40 to train our network with the iteration of 300 epochs. Then we feed all samples from ModelNet40 to the trained discriminator (PointNet) for feature extraction. With these features, we simply train a linear SVM to classify the generated point clouds. The settings of ModelNet10 are consistent with ModelNet40. The classification results are listed in Table 2. Note that for a fair comparison we only compare the point cloud

Table 2. Classification results on ModelNet10 (MN10) and ModelNet40 (MN40).

Model	MN10 (%)	MN40 (%)
SPH [14]	79.8	68.2
LFD [4]	79.9	75.5
T-L Network [9]	-	74.4
VConv-DAE [30]	80.5	75.5
3D-GAN [40]	91.0	83.3
PointGrow [34]	-	85.7
MRTNet [8]	91.7	86.4
PointFlow [44]	93.7	86.8
PDGN (ours)	**94.2**	**87.3**

generation methods in the classification experiment. It can be found that our PDGN outperforms the state-of-the-art point cloud generation methods on the ModelNet10 and ModelNet40 datasets. The results indicate that the generator in our framework can extract discriminative features. Thus, our generator can produce high-quality 3D point clouds.

Computational Cost. We compare our proposed method to PointFlow and tree-GAN in terms of the training time and GPU memory. We conduct point cloud generation experiments on the "Airplane" category in the ShapeNet dataset. For a fair comparison, both codes are run on a single Tesla P40 GPU using the PyTorch [25] framework. For training 1000 iterators with 2416 samples of the "Airplane" category, our proposed method costs about 1.9 days and 15G GPU memory, while PointFlow costs about 4.5 days and 7.9G GPU memory, and tree-GAN costs about 2.5 days and 9.2G GPU memory. Our GPU memory is larger than others due to the four discriminators.

4.3   Ablation Study and Analysis

Bilateral Interpolation. We conduct the experiments with different ways to generate the high-resolution feature maps, including the conventional reshape operation, bilinear interpolation and learning-based bilateral interpolation. In the conventional reshape operation, we resize the feature maps to generate new points. As shown in Fig. 6, we visualize the generated point clouds from different categories. One can see that the learning-based bilateral interpolation can generate more realistic objects than the other methods. For example, for the "Table" category, with the learning-based bilateral interpolation, the table legs are clearly generated. On the contrary, with the bilinear interpolation and reshape operation, the generated table legs are not complete. Besides, we also conduct a quantitative evaluation of generated point clouds. As shown in Table 3, on the "Chair" category, PDGN with the bilateral interpolation can obtain the best

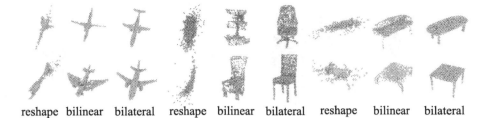

reshape bilinear bilateral   reshape bilinear bilateral   reshape   bilinear   bilateral

Fig. 6. Visualization results with different operations in the deconvolution network.

Table 3. The ablation study results on the "Chair" category.

Model	JSD (↓)	MMD (↓)		COV (%, ↑)		1-NNA (%, ↓)	
		CD	EMD	CD	EMD	CD	EMD
PDGN (reshape)	8.69	3.38	9.30	55.01	44.49	82.60	80.43
PDGN (bilinear interpolation)	5.02	3.31	8.83	53.84	48.35	69.23	68.18
PDGN (bilateral interpolation)	**1.71**	**1.93**	**6.37**	**61.90**	**57.89**	**52.38**	**57.14**
PDGN (adversarial loss)	3.28	3.00	8.82	56.15	53.84	57.14	66.07
PDGN (EMD loss)	3.35	3.03	8.80	53.84	53.34	60.89	68.18
PDGN (CD loss)	3.34	3.38	9.53	55.88	52.63	59.52	67.65
PDGN (shape-preserving loss)	**1.71**	**1.93**	**6.37**	**61.90**	**57.89**	**52.38**	**57.14**
PDGN (256 points)	5.57	5.12	9.69	39.47	42.85	67.56	70.27
PDGN (512 points)	4.67	4.89	9.67	47.82	51.17	71.42	67.86
PDGN (1024 points)	2.18	4.53	11.0	56.45	55.46	64.71	70.58
PDGN (2048 points)	**1.71**	**1.93**	**6.37**	**61.90**	**57.89**	**52.38**	**57.14**

metric results. In contrast to the bilinear interpolation and reshape operation, the learning-based bilateral interpolation exploits the spatial coordinates and high-dimensional features of the neighboring points to adaptively learn weights for different classes of 3D objects. Thus, the learned weights in the spatial and feature spaces can characterize the geometric structures of point clouds better. Therefore, the bilateral interpolation can yield good performance.

Shape-Preserving Adversarial Loss. To demonstrate the effectiveness of our shape-preserving adversarial loss, we train our generation model with the classical adversarial loss, EMD loss, CD loss and shape-preserving loss. It is noted that in the EMD loss and CD loss we replace the shape-preserving constraint (Eq. 7) with the Earth mover's distance and Chamfer distance of point clouds between the adjacent resolutions, respectively. We visualize the generated points with different loss functions in Fig. 7. One can see that the geometric structures of different resolutions of generated point clouds are consistent with the shape-preserving adversarial loss. Without the shape-preserving constraint on the multiple generators, the classical adversarial loss cannot guarantee the con-

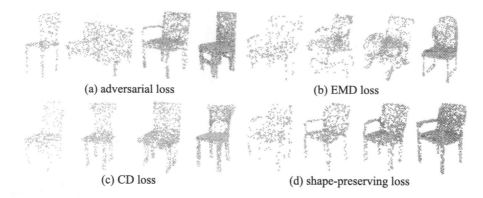

(a) adversarial loss   (b) EMD loss

(c) CD loss   (d) shape-preserving loss

Fig. 7. Visualization results of generated point clouds with different loss functions. For each loss, four resolutions of point clouds (256, 512, 1024 and 2048) are visualized.

sistency of generated points between different resolutions. Although the EMD and CD losses impose the constraint on different resolutions of point clouds, the loss can only make the global structures of point clouds consistent. On the contrary, the shape-preserving loss can keep the consistency of the local geometric structures of multi-resolution point clouds with the mean and covariance of the neighborhoods. Thus, our method with the shape-preserving loss can generate high-quality point clouds. Furthermore, we also conduct a quantitative evaluation of generated point clouds. As shown in Table 3, metric results show that the shape-preserving loss can obtain better results than the other three losses.

Point Cloud Generation with Different Resolutions. To verify the effectiveness of our progressive generation framework, we evaluate the metric results of generated point clouds in the cases of different resolutions. As shown in Table 3, for the "Chair" category, we report the results in the cases of four resolutions. One can see that as the resolution increases, the quality of the generated point clouds is gradually improved in terms of the evaluation criteria. As shown in Fig. 4, with the increase of resolutions, the local structures of point clouds become clearer. This is because our progressive generation framework can exploit the bilateral interpolation based deconvolution to generate the coarse-to-fine geometric structures of point clouds.

5   Conclusions

In this paper, we proposed a novel end-to-end generation model for point clouds. Specifically, we developed a progressive deconvolution network to generate multi-resolution point clouds from the latent vector. In the deconvolution network, we employed the learning-based bilateral interpolation to generate high-resolution feature maps so that the local structures of point clouds can be captured during the generation process. In order to keep the geometric structure of the generated point clouds at different resolutions consistent, we formulated the shape-preserving adversarial loss to train the point cloud deconvolution network. Experimental results on ShapeNet and ModelNet datasets verify the effectiveness of our proposed progressive point cloud deconvolution network.

Acknowledgments. This work was supported by the National Science Fund of China (Grant Nos. U1713208, 61876084, 61876083), Program for Changjiang Scholars.

References

1. Achlioptas, P., Diamanti, O., Mitliagkas, I., Guibas, L.: Learning representations and generative models for 3D point clouds. In: ICML (2018)
2. Arjovsky, M., Chintala, S., Bottou, L.: Wasserstein GAN. arXiv preprint arXiv:1701.07875 (2017)
3. Chang, A.X., et al.: ShapeNet: an information-rich 3D model repository. arXiv preprint arXiv:1512.03012 (2015)
4. Chen, D.Y., Tian, X.P., Shen, Y.T., Ouhyoung, M.: On visual similarity based 3D model retrieval. In: CGF (2003)
5. Choy, C.B., Xu, D., Gwak, J.Y., Chen, K., Savarese, S.: 3D-R2N2: a unified approach for single and multi-view 3D object reconstruction. In: Leibe, B., Matas, J., Sebe, N., Welling, M. (eds.) ECCV 2016. LNCS, vol. 9912, pp. 628–644. Springer, Cham (2016). https://doi.org/10.1007/978-3-319-46484-8_38
6. Denton, E.L., Chintala, S., Fergus, R., et al.: Deep generative image models using a Laplacian pyramid of adversarial networks. In: NeurIPS (2015)
7. Fan, H., Su, H., Guibas, L.J.: A point set generation network for 3D object reconstruction from a single image. In: CVPR (2017)
8. Gadelha, M., Wang, R., Maji, S.: Multiresolution tree networks for 3D point cloud processing. In: ECCV (2018)
9. Girdhar, R., Fouhey, D.F., Rodriguez, M., Gupta, A.: Learning a predictable and generative vector representation for objects. In: Leibe, B., Matas, J., Sebe, N., Welling, M. (eds.) ECCV 2016. LNCS, vol. 9910, pp. 484–499. Springer, Cham (2016). https://doi.org/10.1007/978-3-319-46466-4_29
10. Goodfellow, I., et al.: Generative adversarial nets. In: NeurIPS (2014)
11. Groueix, T., Fisher, M., Kim, V.G., Russell, B.C., Aubry, M.: A papier-mâché approach to learning 3D surface generation. In: CVPR (2018)
12. Gwak, J., Choy, C.B., Chandraker, M., Garg, A., Savarese, S.: Weakly supervised 3D reconstruction with adversarial constraint. In: 3DV (2017)
13. Ioffe, S., Szegedy, C.: Batch normalization: accelerating deep network training by reducing internal covariate shift. arXiv preprint arXiv:1502.03167 (2015)
14. Kazhdan, M., Funkhouser, T., Rusinkiewicz, S.: Rotation invariant spherical harmonic representation of 3D shape descriptors. In: SGP (2003)
15. Kingma, D.P., Ba, J.: Adam: a method for stochastic optimization. arXiv preprint arXiv:1412.6980 (2014)
16. Kulkarni, N., Misra, I., Tulsiani, S., Gupta, A.: 3D-RelNet: joint object and relational network for 3D prediction. In: ICCV (2019)
17. Landrieu, L., Simonovsky, M.: Large-scale point cloud semantic segmentation with superpoint graphs. In: CVPR (2018)
18. Li, C.L., Zaheer, M., Zhang, Y., Póczos, B., Salakhutdinov, R.: Point cloud GAN. arXiv preprint arXiv:1810.05795 (2018)

19. Li, Y., Bu, R., Sun, M., Wu, W., Di, X., Chen, B.: PointCNN: convolution on X-transformed points. In: NeurIPS (2018)
20. Lopez-Paz, D., Oquab, M.: Revisiting classifier two-sample tests. In: ICLR (2016)
21. Mao, X., Li, Q., Xie, H., Lau, R.Y., Wang, Z., Paul Smolley, S.: Least squares generative adversarial networks. In: ICCV (2017)
22. Maturana, D., Scherer, S.: VoxNet: a 3D convolutional neural network for real-time object recognition. In: IROS (2015)
23. Mirza, M., Osindero, S.: Conditional generative adversarial nets. arXiv preprint arXiv:1411.1784 (2014)
24. Monti, F., Boscaini, D., Masci, J., Rodola, E., Svoboda, J., Bronstein, M.M.: Geometric deep learning on graphs and manifolds using mixture model CNNs. In: CVPR (2017)
25. Paszke, A., et al.: Pytorch: an imperative style, high-performance deep learning library. In: NeurIPS (2019)
26. Qi, C.R., Su, H., Mo, K., Guibas, L.J.: PointNet: deep learning on point sets for 3D classification and segmentation. In: CVPR (2017)
27. Qi, C.R., Su, H., Nießner, M., Dai, A., Yan, M., Guibas, L.J.: Volumetric and multi-view CNNs for object classification on 3D data. In: CVPR (2016)
28. Qi, C.R., Yi, L., Su, H., Guibas, L.J.: PointNet++: deep hierarchical feature learning on point sets in a metric space. In: NeurIPS (2017)
29. Radford, A., Metz, L., Chintala, S.: Unsupervised representation learning with deep convolutional generative adversarial networks. arXiv preprint arXiv:1511.06434 (2015)
30. Sharma, A., Grau, O., Fritz, M.: VConv-DAE: deep volumetric shape learning without object labels. In: Hua, G., Jégou, H. (eds.) ECCV 2016. LNCS, vol. 9915, pp. 236–250. Springer, Cham (2016). https://doi.org/10.1007/978-3-319-49409-8_20
31. Shu, D.W., Park, S.W., Kwon, J.: 3D point cloud generative adversarial network based on tree structured graph convolutions. In: ICCV (2019)
32. Simonovsky, M., Komodakis, N.: Dynamic edge-conditioned filters in convolutional neural networks on graphs. In: CVPR (2017)
33. Su, H., Maji, S., Kalogerakis, E., Learned-Miller, E.: Multi-view convolutional neural networks for 3D shape recognition. In: ICCV (2015)
34. Sun, Y., Wang, Y., Liu, Z., Siegel, J.E., Sarma, S.E.: PointGrow: autoregressively learned point cloud generation with self-attention. arXiv preprint arXiv:1810.05591 (2018)
35. Te, G., Hu, W., Guo, Z., Zheng, A.: RGCNN: regularized graph CNN for point cloud segmentation. In: ACM MM (2018)
36. Tulsiani, S., Gupta, S., Fouhey, D.F., Efros, A.A., Malik, J.: Factoring shape, pose, and layout from the 2D image of a 3D scene. In: CVPR (2018)
37. Valsesia, D., Fracastoro, G., Magli, E.: Learning localized generative models for 3D point clouds via graph convolution. In: ICLR (2018)
38. Wang, N., Zhang, Y., Li, Z., Fu, Y., Liu, W., Jiang, Y.-G.: Pixel2Mesh: generating 3D mesh models from single RGB images. In: Ferrari, V., Hebert, M., Sminchisescu, C., Weiss, Y. (eds.) ECCV 2018. LNCS, vol. 11215, pp. 55–71. Springer, Cham (2018). https://doi.org/10.1007/978-3-030-01252-6_4
39. Wang, Y., Sun, Y., Liu, Z., Sarma, S.E., Bronstein, M.M., Solomon, J.M.: Dynamic graph CNN for learning on point clouds. arXiv preprint arXiv:1801.07829 (2018)
40. Wu, J., Zhang, C., Xue, T., Freeman, B., Tenenbaum, J.: Learning a probabilistic latent space of object shapes via 3D generative-adversarial modeling. In: NeurIPS (2016)

41. Wu, Z., et al.: 3D ShapeNets: a deep representation for volumetric shapes. In: CVPR (2015)
42. Xu, B., Wang, N., Chen, T., Li, M.: Empirical evaluation of rectified activations in convolutional network. arXiv preprint arXiv:1505.00853 (2015)
43. Yang, B., Wen, H., Wang, S., Clark, R., Markham, A., Trigoni, N.: 3D object reconstruction from a single depth view with adversarial learning. In: ICCV (2017)
44. Yang, G., Huang, X., Hao, Z., Liu, M.Y., Belongie, S., Hariharan, B.: PointFlow: 3D point cloud generation with continuous normalizing flows. In: ICCV (2019)
45. Yang, Y., Feng, C., Shen, Y., Tian, D.: FoldingNet: point cloud auto-encoder via deep grid deformation. In: CVPR (2018)
46. Zamorski, M., Zikeba, M., Nowak, R., Stokowiec, W., Trzcinski, T.: Adversarial autoencoders for compact representations of 3D point clouds. arXiv preprint arXiv:1811.07605 (2018)
47. Zhao, Y., Birdal, T., Deng, H., Tombari, F.: 3D point capsule networks. In: CVPR (2019)
48. Zhu, J., Xie, J., Fang, Y.: Learning adversarial 3D model generation with 2D image enhancer. In: AAAI (2018)

SSCGAN: Facial Attribute Editing via Style Skip Connections

Wenqing Chu[1], Ying Tai[1(✉)], Chengjie Wang[1], Jilin Li[1], Feiyue Huang[1], and Rongrong Ji[2]

[1] Youtu Lab Tencent, Shanghai, China
yingtai@tencent.com
[2] Xiamen University, Xiamen, China

Abstract. Existing facial attribute editing methods typically employ an encoder-decoder architecture where the attribute information is expressed as a conditional one-hot vector spatially concatenated with the image or intermediate feature maps. However, such operations only learn the local semantic mapping but ignore global facial statistics. In this work, we focus on solving this issue by editing the channel-wise global information denoted as the style feature. We develop a style skip connection based generative adversarial network, referred to as SSCGAN which enables accurate facial attribute manipulation. Specifically, we inject the target attribute information into multiple style skip connection paths between the encoder and decoder. Each connection extracts the style feature of the latent feature maps in the encoder and then performs a residual learning based mapping function in the global information space guided by the target attributes. In the following, the adjusted style feature will be utilized as the conditional information for instance normalization to transform the corresponding latent feature maps in the decoder. In addition, to avoid the vanishing of spatial details (*e.g.* hairstyle or pupil locations), we further introduce the skip connection based spatial information transfer module. Through the global-wise style and local-wise spatial information manipulation, the proposed method can produce better results in terms of attribute generation accuracy and image quality. Experimental results demonstrate the proposed algorithm performs favorably against the state-of-the-art methods.

Keywords: Facial attribute editing · Style feature · Skip connection

1 Introduction

Given a facial photo, attribute editing aims to translate the image to enable target attribute transfer while preserving the image content, i.e., the identity information, illumination, and other irrelevant attributes). During the past

© Springer Nature Switzerland AG 2020
A. Vedaldi et al. (Eds.): ECCV 2020, LNCS 12360, pp. 414–429, 2020.
https://doi.org/10.1007/978-3-030-58555-6_25

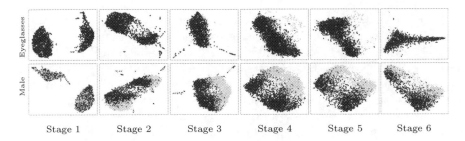

Fig. 1. Visualization of style features in different stages for attribute eyeglasses and male. We use t-SNE [27] to visualize the distributions of style features. Each point indicates a sample and different colors indicate whether this sample exhibits this attribute. It is observed that the style features at some specific stages are separated well, which means the style features could represent the attributes accurately. In addition, the style features in stage 4 and stage 3 for eyeglasses and male are the most discriminative, respectively. Therefore, the style features in different stages may capture different attribute information.

decades, attribute guided facial manipulation has drawn considerable attentions [6,12,23,33], and widely used in many real-world visual applications. However, it is very challenging to generate high-quality and accurate facial editing results due to the high-dimensional output space, complex facial structure, and vague attribute definitions. Besides, due to the absence of paired examples for training, it could only be tackled through an unpaired manner resulting in more difficulties.

With the rapid development of Generative Adversarial Networks (GANs) [8], there have been a large number of attempts [5,6,12,23,36] for facial editing. Most existing methods employ an encoder-decoder architecture, and rely on conditional GANs [28,29]. Specifically, the attribute information is represented as a one-hot vector [6], where each bit indicates a specific attribute. This vector is then expanded and concatenated with the input image or intermediate feature maps [6,23,36] to guide the feature transformation. Furthermore, the attribute information will be used as the supervised signals of the auxiliary loss combined with the cycle consistency loss [41] and adversarial loss [8], as to compose the overall objectives for stable and effective training.

However, providing the attributes in each spatial location and then manipulating the feature maps locally may ignore the global structure which leads to unsatisfactory performance. The channel-wise feature map manipulation is an important and effective technique for harvesting the global information in many visual tasks such as image classification [11] and semantic segmentation [40], which has not been well explored in facial attribute editing. That motivated us to perform attribute transfer via the manipulation of the global channel-wise statistics of the latent feature maps.

Following [13,18], we employ the channel-wise mean and variance of the feature maps as the global information and denote them as the *style feature*. Here, we take the advanced image generation method StyleGAN [18] as an example to verify the relationship between the style feature and different attributes. To be more specific, we leverage an efficient embedding algorithm [1] to compute the style feature of the well-annotated facial attribute dataset CelebA [26] and then employ the Neighborhood Components Analysis [31] to perform supervised dimensionality reduction for each attribute. Then we use t-SNE [27] to visualize the distributions of style features in different stages of the decoder. As shown in Fig. 1, we can observe that the style features at some specific stages are separated well, which means the style features could represent the attributes accurately. In addition, the style features in stage 4 and stage 3 for eyeglasses and male are the most discriminative, respectively. Therefore, the style features in different stages may control different attribute information.

Inspired by the good characteristic of the style feature on controlling facial attributes, we propose to edit the latent feature maps via style skip connections, which modify the global channel-wise statistics to achieve attribute transfer. Specifically, we leverage the style information in the encoder and target attributes to inference the desired statistic information of the latent feature maps in the decoder. Then the manipulated style information is utilized as the conditional input for instance normalization to adjust the distribution of the corresponding latent feature maps in the decoder. However, we find the style information is spatial invariant and may drop the spatial variations, which in some cases describe the local details like the pupil locations or hair texture. To address this issue, we further employ the spatial information based skip connections, which extract the spatial details from the latent feature maps and transfers them to the decoder. In summary, the global-wise style manipulation can handle the facial attribute transfer, and the local-wise spatial information transfer can make up the local finer details.

The main contributions of this work are as follows. First, we introduce a style skip connection based architecture to perform facial attribute editing which manipulates the latent feature maps in terms of global statistic information. Second, a spatial information transfer module is developed to avoid the vanishing of finer facial details. Third, the visual comparisons and quantitative analysis on the large-scale facial attribute benchmark CelebA [26] demonstrate that our framework achieves favorable performance against the state-of-the-art methods.

2   Related Work

Image-to-Image Translation. Recent years have seen tremendous progress in image-to-image translation, relying on generative adversarial networks [2,8]. To model the mapping from input to output images, Pix2pix [15] utilizes a patch-based adversarial loss which forces the generated images indistinguishable from target images and achieves reasonable results. However, the paired training is usually not available in real-world scenarios, CycleGAN [41], DiscoGAN [19], and UNIT [24] constrain the mapping through an additional cycle consistency loss. Furthermore, MUNIT [14] and DRIT [21] model the input images with disentangled content and attribute representations and thus generate diverse outputs. In addition, FUNIT [25] handles the few-shot image translation task which only provides a few target images for learning the mapping. However, these methods could only tackle image translation between two domains, thus they can not be applied to the facial attribute transfer task directly.

Facial Editing. Most facial editing methods are based on conditional GANs [28, 29]. The conditional information can be facial attributes [6, 23, 36, 39], expressions [33, 38], poses [3, 12] or reference images [5, 37]. Among them, facial attribute editing has caused great attentions due to its wide applications. IcGAN [32] generates the attribute-independent latent representation and then the target attribute information is combined as input to the conditional GANs. To achieve better multi-domain translation, StarGAN [6] and AttGAN [9] employ an additional attribute classifier to constrain the output image. Furthermore, STGAN [23] adopts a skip connection based architecture and transfers the feature maps selectively according to the desired attribute change which produces visually realistic editing. Similar to STGAN [23], RelGAN [36] also leverages the relative attribute differences for fine-grained control. Existing methods usually modify the entire feature maps locally according to the desired attributes which ignore the global information. Instead, we find that the statistical information like mean and variance of the feature maps are very informative.

Style-Based Face Generation. Recently, a number of improved GANs have been proposed [4, 17, 18, 35] which produce promising results with high resolution. StyleGAN [18] achieves impressive performance by adopting a style-based generator relying on the conditional adaptive instance normalization [13]. That has inspired several extensions [1, 34] to perform facial manipulation with Style-GAN. However, these methods [1, 34] employ an optimization-based embedding algorithm which uses 5000 gradient descent steps, taking about 7 minutes on an advanced V100 GPU device. Also, they are constrained by the pretrained StyleGAN model and could not be applied to other tasks flexibly.

3   Method

In this work, we introduce a facial attribute manipulation algorithm through editing the intermediate feature maps via style and spatial information guided skip connections. As shown in Fig. 1, the style features in multi-stages are responsible for different attributes, respectively. That inspired us to manipulate the facial image by adjusting the global statistic information in the feature maps. Different from existing methods that concatenate the target attribute information with the latent feature maps to achieve local feature transformation, our method aims to edit the facial attributes globally. As a result, the proposed approach can achieve more effective and accurate manipulation.

The overall framework is based on an encoder-decoder architecture shown in Fig. 2. Specifically, we leverage two kinds of skip connections between the

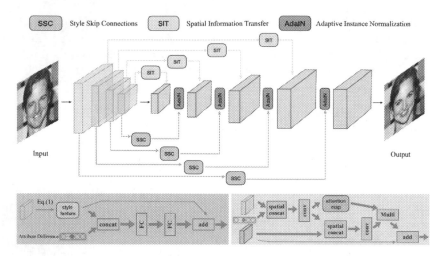

Fig. 2. Framework of the proposed algorithm. The overall framework is based on an encoder-decoder architecture. We combine the style skip connections and spatial information transfer to achieve facial attribute editing. Specifically, the style information in the encoder and target attributes are utilized to inference the desired statistic information for adjusting the latent feature maps in the decoder in a global way. Besides, we employ the spatial information based skip connections to transfer spatial details to the decoder, so that the proposed method can achieve local feature transformation and recovery for the pupil locations or hair texture.

encoder and decoder to incorporate the target attribute information. The goal of the first kind of skip connections is to obtain the style information of the latent feature maps in the encoder and then perform a residual learning based style transformation under the instruction of the target attributes. After that, we employ it as the conditional information for instance normalization on the corresponding latent feature maps in the decoder. To make up the vanishing of the facial details, we introduce a spatial information based skip connection which reserves the spatial variations in the latent feature maps. To be more specific, this information will be concatenated with the latent feature maps in the decoder to perform the local feature transformation. In the following, we first describe the network architecture. Next, we describe the style and spatial based image manipulation module, Finally, we present the loss functions for training and implementation details.

3.1 Multiple Skip Connections Architecture

Considering the style features in different stages control different attributes as shown in Fig. 1, a skip connection at some specific stage is not general for various attributes. Also, as demonstrated in StyleGAN [18], the style information in the low-resolution feature maps could represent coarse level attributes like pose, face shape and eyeglasses. Instead, the high-resolution feature maps could control finer details like eyes open/closed, color scheme and micro-structure. As a result, only using part of the skip connections may miss the control for some specific attributes. In addition, the spatial details are also sensitive to different resolutions. Therefore, we utilize multiple skip connections to manipulate both the coarse facial structure and finer local facial information. Specifically, we employ 6 stages for the encoder and decoder, respectively. Each stage in the encoder/decoder has one residual block and down-samples/up-samples the feature maps by 2 times, respectively. Besides, it has the style and spatial based skip connections sent to the corresponding stage in the decoder.

3.2 Style Skip Connections

In this section, we introduce the design methodology of the style manipulation part. The goal is to modify the latent feature maps in the decoder globally to enable accurate attribute transfer while preserving irrelevant facial information. Suppose the architectures for the encoder and decoder are symmetric and both have n stages. For simplicity, we denote the feature maps in the encoder and decoder as $\mathbf{f_{enc}^1}, \mathbf{f_{enc}^2}, \cdots, \mathbf{f_{enc}^n}$ and $\mathbf{f_{dec}^1}, \mathbf{f_{dec}^2}, \cdots, \mathbf{f_{dec}^n}$. For $t \in (1, 2, \cdots, n)$, the feature maps $\mathbf{f_{enc}^t}$ and $\mathbf{f_{enc}^{n+1-t}}$ have the same spatial and channel sizes.

We first describe how to represent the channel-wise global information for f_{dec}^t. Inspired by the neural style transfer approaches [7,22], we leverage the feature distributions in f_{dec}^t for representing the global facial statistics information. We find the prevalent Gram Matrices of feature maps used in [7] are too large and thus time-consuming compared to the statistics (i.e. mean and variance) of Batch Normalization (BN) layers used in [22]. Therefore, we consider using the mean and variance statistics in $\mathbf{f_{dec}^t}$ as the style features for efficiency. Suppose the size of the feature maps $\mathbf{f_{dec}^t}$ is $\mathbb{R}^{N_t \times M_t}$, where N_t is the number of the feature maps in the layer t and M_t is the result of the height multiplying with the width. Similar to [22], we adopt the mean μ_i^t and standard deviation σ_i^t of the i-th channel among all the positions of the feature map in the layer t to represent the style:

$$\begin{cases} \mu_i^t = \dfrac{1}{M_t} \sum_{j=1}^{M_t} (\mathbf{f_{dec}^t})_{i,j}. \\[2ex] \sigma_i^{t2} = \dfrac{1}{M_t} \sum_{j=1}^{M_t} ((\mathbf{f_{dec}^t})_{i,j} - \mu_i^t)^2. \end{cases} \tag{1}$$

Furthermore, we concatenate the μ^t and σ^t into a $(N_t \times 2)$-d vector as the style feature for feature maps $\mathbf{f_{dec}^t}$.

Next, a direct way is to utilize the attribute difference vector as input to generate the style information for adjusting the latent feature maps in the decoder. However, this solution may ignore the original image content and produce incorrect statistic information, which leads to a number of undesired changes in the generated images. To achieve accurate facial attribute editing, we employ the attribute difference vectors and the style information calculated from the latent feature maps in the encoder stage to produce the desired style information. Note that we find the style features at some specific stage are separated well, which means the style features in different stages could represent different attributes accurately as shown in Fig. 1. Therefore, we perform style information manipulation within the same stage of the encoder and decoder.

In the following, we describe how to perform style information based skip connections between $\mathbf{f}_{enc}^t$ and $\mathbf{f}_{dec}^{n+1-t}$. The style feature extracted in $\mathbf{f}_{enc}^t$ is concatenated with the attribute difference vector and fed into a 2 layer fully connected neural networks to predict the residual information as shown in Fig. 2. After that, we add it to the original style feature to obtain the desired style information which can be used as the conditional input to manipulate the corresponding $\mathbf{f}_{dec}^{n+1-t}$. Taking efficiency into consideration, we utilize the Adaptive Instance Normalization [13] (AdaIN) to manipulate the global statistic information of the latent feature maps. For all style based skip connections used in the proposed method, we adopt the same embedding way for the desired style feature and network structure.

3.3   Spatial Information Transfer

Although the style features could carry most facial information like coarse structure and facial components, the local information may be dropped due to the spatial invariant characteristic of style information and the low resolution of the last encoder stage. For example, the spatial details like the hair texture and pupil locations are very difficult to be embedded into the style features. Therefore, if only use the style feature based skip connections, the generated images may have accurate target attributes but look over smooth. As a result, they are not realistic enough and can not achieve satisfactory performance.

To address the above problem, we develop a spatial information transfer module to collect the spatial details and deliver them to the corresponding latent feature maps. Since the target attribute editing could only need part of the original facial image information, we also provide the attribute difference vector to extract the spatial information more accurately. Specifically, we expand the attribute difference vector spatially and concatenate it with the intermediate feature maps in the encoder, and then we adopt a convolution operation to generate a two-channel feature map. One of them is regarded as representing the spatial details. During the decoder stage, we combine the spatial map with the latent feature maps to predict the residual spatial information. The other one is processed by a sigmoid activation function and then used as an attention map because we want to avoid introducing noise from the residual information through the attention mechanism [33,39]. In the following, the attention map is leveraged to guide the fusion of the original intermediate feature maps and the residual one. Based on the dedicated design, the spatial information transfer module could benefit the editing and recovery of the local spatial details.

3.4   Loss Functions

We combine multiple loss functions to train the model in an unpaired manner. To better capture the translation information, we also employ the attribute difference vector $\mathbf{attr}_{diff}$ as the conditional information similar to STGAN [23] and RelGAN [36]. Given an input image $\mathbf{x}$, our framework can generate the output image $\mathbf{y}$ as below:

$$\mathbf{y} = \mathbf{G}(\mathbf{x}, \mathbf{attr}_{diff}). \tag{2}$$

To require the generated image $\mathbf{y}$ satisfying the objective of facial attribute editing, we utilize three constraints: 1) the generated facial image should be the same as input one when the attribute difference is none; 2) the generated facial image should be realistic and similar to the real facial images; 3) the generated image should exhibit the target attributes. Therefore, we employ three loss functions based on the above-mentioned constraints to train the network.

Reconstruction Loss. We set the attribute difference vector as $\mathbf{0}$ and fed it with $\mathbf{x}$ into the network to obtain $\mathbf{y}_{\mathbf{rec}}$:

$$\mathbf{y}_{\mathbf{rec}} = \mathbf{G}(\mathbf{x}, \mathbf{0}). \tag{3}$$

Then we combine the pixel and feature level reconstruction loss as below:

$$\mathcal{L}_{rec} = \mathbb{E}_{\mathbf{x}}\big[\mathcal{L}_1(\mathbf{y}_{\mathbf{rec}}, \mathbf{x}) + \mathcal{L}_{perceptual}(\mathbf{y}_{\mathbf{rec}}, \mathbf{x})\big], \tag{4}$$

where the perceptual loss $\mathcal{L}_{perceptual}$ introduced in [16] can improve the image quality as demonstrated in [14].

Adversarial Loss. In addition, we adopt the adversarial loss [8] which is effective in constraining the generated images looking realistic. The adversarial learning framework consists of two sub-networks, including a generator and a discriminator. Here we leverage the facial attribute editing network as the generator. Given an input image $\mathbf{x}$ and target attribute difference $\mathbf{attr}_{diff}$, our generator can produce the output image $\mathbf{y}$ according to Eq. 2. The discriminator is a fully convolutional neural network and required to distinguish the patches of the real ($\mathbf{x}$) and the generated images ($\mathbf{y}$). Then, the goal of the generator is to fool the discriminator via an adversarial loss denoted as $\mathcal{L}_{adv}$. We employ the same training scheme and loss functions as the Wasserstein GAN model [2] as below:

$$\begin{cases} \mathcal{L}_{\mathbf{dis}} = -\mathbb{E}_{\mathbf{x}, \mathbf{attr}_{diff}}\big[\log(1 - \mathbf{D}_{real}(\mathbf{y}))\big] - \mathbb{E}_{\mathbf{x}}\big[\log \mathbf{D}_{real}(\mathbf{x})\big], \\ \mathcal{L}_{\mathbf{adv}} = -\mathbb{E}_{\mathbf{x}, \mathbf{attr}_{diff}}\big[\log \mathbf{D}_{real}(\mathbf{y})\big], \end{cases} \tag{5}$$

where minimizing $\mathcal{L}_{dis}$ on the discriminator $\mathbf{D}_{real}$ tries to distinguish between the real and synthesized images. And optimizing $\mathcal{L}_{adv}$ leads to that the generator $\mathbf{G}$ produces visually realistic images.

Attribute Generation Loss. To achieve attribute transfer, we utilize an aux-iliary attribute generation loss similar to StarGAN [6]. It is achieved by an attribute classifier learned with the real images $\mathbf{x}$ and applied to the generated images $\mathbf{y}$ as a deep image prior. We denote the attribute classification and gen-eration loss functions as below:

$$\begin{cases} \mathcal{L}_{\mathbf{D_{attr}}} = -\sum_{i=1}^{n_{attr}} \left[\mathbf{attr}^i(\log \mathbf{D}_{attr}^i(\mathbf{x})) + (1 - \mathbf{attr}^i)\log(1 - \mathbf{D}_{attr}^i(\mathbf{x})) \right], \\ \mathcal{L}_{\mathbf{G_{attr}}} = -\sum_{i=1}^{n_{attr}} \left[\mathbf{attr}^i(\log \mathbf{D}_{attr}^i(\mathbf{y})) + (1 - \mathbf{attr}^i)\log(1 - \mathbf{D}_{attr}^i(\mathbf{y})) \right], \end{cases} \tag{6}$$

where the attribute classifiers $\mathbf{D}_{attr}$ are trained on the real images and optimizing $\mathcal{L}_{G_{attr}}$ aims to require the generated images to satisfy the target attributes.

Overall Objectives. The overall objective function for the proposed facial attribute editing network includes the reconstruction/adversarial loss to help generate high quality images, the attribute classification loss to ensure attribute transfer:

$$\mathcal{L}_{overall} = \lambda_r \mathcal{L}_{rec} + \mathcal{L}_{adv} + \mathcal{L}_{G_{attr}}. \tag{7}$$

where the hyper-parameters λ_r is set to 20.

Implementation Details. The proposed framework is implemented with PyTorch [30] and trained with 1 Nvidia V100 GPU. During training, we adopt the Adam [20] optimizer and set the batch size as 32. Similar to CycleGAN [41], we set the initial learning rate as 0.0002 and fix it for the first 100 epochs, and linearly decay the learning rate for another 100 epochs.

4   Results and Analysis

In this section, we first describe the basic experiment settings. Next, we per-form extensive ablation studies to evaluate different components of the pro-posed method, including the choices of style and spatial manipulation, embed-ding manner and multiple skip connections. Finally, we conduct both qualitative and quantitative experiments to compare the proposed algorithm with state-of-the-art methods.

Datasets. Following [6,23], we leverage the large scale facial attribution dataset Celeba [26] for evaluation. It contains around 200k facial images and annotates 40 attributes. We randomly select around 180k images for train and validation, and the rest is used as the test set. Besides, we choose 10 attributes to perform facial attribute transfer.

Table 1. Attribute generation accuracy for different skip connections.

Method	Bald	Bangs	Hair	Eyebrow	Glasses	Gender	Mouth	Mustache	Pale	Age	Average
Spatial	38.63	95.43	88.07	92.33	99.07	79.57	98.90	59.83	84.30	88.53	82.46
Style	69.60	**99.93**	99.83	**97.97**	**99.97**	98.20	99.87	61.83	97.13	98.30	92.26
SSCGAN	**85.40**	99.23	**99.30**	96.57	99.93	**99.10**	**99.90**	**65.73**	**98.03**	**99.00**	**94.21**

Evaluation Metrics. To evaluate the facial attribute editing performance, we take both the attribute generation accuracy and image quality into consideration. Similar to STGAN [23], we utilize the training data to train an attribute classifier and the average attribute classification accuracy on the test set is 95.55%. In all experiments, we use this pretrained classifier to verify the accuracy of facial editing results. In addition, we also follow ELEGANT [37] and RelGAN [36] to employ the Frechet Inception Distance (FID) [10] to demonstrate the image quality. FID aims to evaluate the distribution similarity between two datasets of images. As shown in [10,18], it correlates well to the human evaluation of image quality.

4.1 Ablation Study

Here, we investigate the effects of different algorithm designs by comparing the attribute generation accuracy and observing the qualitative results. First, we want to verify the effectiveness of style skip connections.

Style vs. Spatial. Based on the encoder-decoder architecture, we adopt the style and spatial skip connections separately to demonstrate their influence. Specifically, we have three settings, including SSCGAN-style, SSCGAN-spatial and SSCGAN (both style and spatial). From Table 1, we can find that SSCGAN-style achieves higher attribute generation accuracy compared with SSCGAN-spatial. Furthermore, employing both kinds of skip connections could obtain the best performance. We also present some qualitative results to demonstrate the editing results. As shown in Fig. 3, we can observe that SSCGAN-spatial does not change the lip color or eyebrow shape as it is not able to learn the global distribution for female appearance. Although SSCGAN-style could change the attributes well, the generated facial images are over smooth and can not maintain some input image information like pupil locations and background. That means the spatial skip connections are also very necessary.

Embedding Style Information. Different from the image generation method StyleGAN [18] which utilizes a random noise vector to generate the style information, the facial attribute editing task needs specific style information which combines the input image content and target attributes. Therefore, it is a key challenge to obtain plausible style information. We investigate 5 embedding ways to generate the style information.

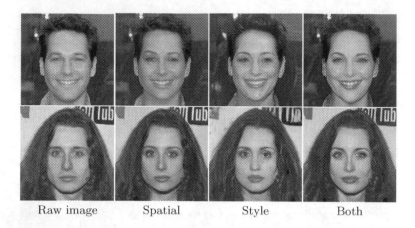

Raw image   Spatial   Style   Both

Fig. 3. An example of editing the attribute gender. We can observe that SSCGAN-spatial does not change the lip color or the eyebrow shape as it is not able to learn the global distribution for female appearance. In addition, SSCGAN-style does not maintain the original spatial information such as the pupil location and background.

- SSCGAN-att: directly leveraging the attribute difference vector to predict the style information through a 2 layer fully connected networks.
- SSCGAN-lm: computing the mean of the last stage in the encoder and concatenating it with attribute difference vector to obtain the style information.
- SSCGAN-lmv: calculating the mean and variance of the last stage and combining it with attribute difference to generate the style information.
- SSCGAN-mm: extracting means for each block in the encoder which are concatenated with attribute difference to predict the style information for the corresponding block in the decoder.
- SSCGAN: utilizing a residual learning based network to generate the style information for each block in the decoder.

Table 2. Attribute generation accuracy for different embedding settings.

Method	Bald	Bangs	Hair	Eyebrow	Glasses	Gender	Mouth	Mustache	Pale	Age	Average
SSCGAN-att	**88.40**	98.63	**99.43**	87.47	99.73	85.60	99.67	51.17	95.77	94.43	90.03
SSCGAN-lm	52.20	98.10	96.33	93.23	99.83	89.67	99.80	52.00	95.33	94.40	87.08
SSCGAN-lmv	61.40	98.57	96.60	93.33	99.53	93.07	99.60	47.90	94.60	96.17	88.07
SSCGAN-mm	55.73	98.63	98.77	95.37	99.63	92.17	99.57	49.13	93.40	93.63	87.60
SSCGAN	85.40	**99.23**	99.30	**96.57**	**99.93**	**99.10**	**99.90**	**65.73**	**98.03**	**99.00**	**94.21**

We use the same experiment setting to train these variants and show their performance in Table 2. We observe that the proposed SSCGAN achieves the best attribute generation accuracy. And embedding style information with the feature maps in different stages can surpass only using the last one. In addition, the

Table 3. Attribute generation accuracy for different layers.

Method	Bald	Bangs	Hair	Eyebrow	Glasses	Gender	Mouth	Mustache	Pale	Age	Average
SSCGAN-8	49.50	**99.47**	99.43	97.27	99.83	96.93	99.53	47.73	98.37	94.83	88.28
SSCGAN-16	71.57	99.47	**99.93**	**97.93**	99.97	97.77	99.90	74.27	**98.80**	96.77	93.63
SSCGAN-32	55.50	99.50	98.60	95.40	99.90	95.10	99.87	**68.70**	94.43	96.20	90.32
SSCGAN-64	34.60	97.93	95.33	93.83	99.83	87.37	99.47	45.57	90.43	88.70	83.30
SSCGAN-128	34.17	96.07	88.63	88.37	98.33	80.70	98.17	24.67	84.80	84.00	77.79
SSCGAN	**85.40**	99.23	99.30	96.57	**99.93**	**99.10**	**99.90**	65.73	98.03	**99.00**	**94.21**

Table 4. Comparisons of different methods on the attribute generation accuracy.

Method	Bald	Bangs	Hair	Eyebrow	Glasses	Gender	Mouth	Mustache	Pale	Age	Average
StarGAN	13.30	93.20	68.20	84.05	94.96	75.60	98.94	12.23	75.01	86.07	70.15
AttGAN	21.20	89.80	76.27	68.17	98.17	68.03	95.43	18.87	87.07	70.03	69.30
STGAN	58.93	99.23	87.27	95.07	99.37	73.34	98.70	45.20	96.89	78.13	83.21
RelGAN	51.39	96.50	98.33	72.33	99.10	99.60	85.57	45.37	91.97	95.83	83.59
SSCGAN	**85.40**	v99.23	**99.30**	**96.57**	**99.93**	**99.10**	**99.90**	**65.73**	**98.03**	**99.00**	**94.21**

Table 5. Comparisons of different methods on the FID scores.

Method	StarGAN	AttGAN	STGAN	RelGAN	Ours
FID	14.27	6.82	4.78	5.13	**4.69**

usage of both mean and variance information is helpful as SSCGAN-lmv obtains better results than SSCGAN-lm. In summary, generating style information in a residual learning manner for each style skip connection is the best way.

Multiple Skip Connections. Furthermore, we are interested in the influence of multiple skip connections. Specifically, we investigate to only use a single skip connection in the network architecture. Therefore, we can obtain 5 variants which only perform feature manipulation at 8×8, 16×16, 32×32, 64×64, 128×128 scale level which are denoted as SSCGAN-8, SSCGAN-16, SSCGAN-32, SSCGAN-64, SSCGAN-128. From Table 3, we can find that only using specific skip connection degrades the overall performance. In addition, the experimental results demonstrate that different scale level manipulations have different effects on the performance of attribute editing.

4.2 Comparisons with State-of-the-Arts

In the following, we compare the proposed framework with several state-of-the-art methods. We follow the pioneering STGAN [23] and RelGAN [36] to perform quantitative and qualitative experimental evaluations.

Baselines. The recently proposed StarGAN [6], AttGAN [9], STGAN [23] and RelGAN [36] are used as the competing approaches. They all use the encoder-decoder architecture and the overall objectives are also similar. To compare these

Raw image   StarGAN   AttGAN   STGAN   RelGAN   Ours

Fig. 4. An example of editing the attribute bangs. Existing methods all incorporate the attribute information through concatenating it with the feature maps. That may lead to inaccurate changes or appearance inconsistent. Our method based on global style manipulation could achieve better visual results.

existing methods under the same experimental setting including train/validation data split, image cropping manner, image resolution and, selected attributes, we use the official released codes and train these models under their default hyper-parameters. We find that the performance of the state-of-the-art methods AttGAN, STGAN and RelGAN on the attribute generation accuracy is close to those reported in the original paper. Therefore, the following comparisons are fair and convincing.

Quantitative Results. From Table 4, we can observe that the proposed method achieves the best average attribute generation accuracy (94.21%). STGAN [23] and RelGAN [36] leverages attribute difference vectors as conditional information, thus their results are better than StarGAN [6] and AttGAN [9]. However, they all introduce the attribute information locally by concatenating it with the intermediate feature maps in each spatial location, which leads to unsatisfactory editing performance. In contrast, our method is able to learn global appearances for different attributes which results in more accurate editing results. Furthermore, we compare the editing performance of these methods in terms of FID scores which can indicate the image quality well. Here, we provide FID scores for all generated images in Table 5. The experimental results demonstrate that our method performs favorable against existing facial attribute editing approaches.

Qualitative Results. In addition, we show an example to illustrate the facial editing performance for bangs of different methods in Fig. 4. The proposed style skip connections aim to manipulate the feature maps in a global channel-wise manner, and thus both input and output of the style skip connections are *compact vectors* which represent high-level semantics. In contrast, spatial concatenation learns the mapping on complex local regions which is a more difficult scenario than on the channel-wise vectors. As shown in the first row in Fig. 4, StarGAN modifies the irrelevant facial region and RelGAN produces inconsistent bangs compared with the hair. For the second row in Fig. 4, the results of AttGAN, STGAN and RelGAN are not correct around the hair. Furthermore, we show an example of the facial editing results for multiple attributes in Fig. 5.

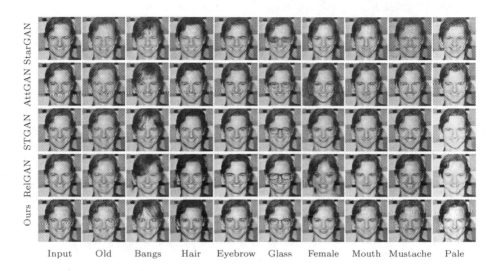

Input   Old   Bangs   Hair   Eyebrow   Glass   Female   Mouth   Mustache   Pale

Fig. 5. Results of different facial attribute editing methods. Existing methods all introduce the attribute information locally, which leads to unsatisfactory editing performance. Instead, through the global-wise style and local-wise spatial information manipulation, the proposed method can achieve favorable performance for most attributes.

5   Conclusions

In this work, we introduce a style skip connection based encoder-decoder architecture for facial attribute editing. To incorporate the target attributes with the image content, we propose to edit the statistics information of the intermediate feature maps in the decoder according to the attribute difference. The manipulation in the style space could translate the facial image in a global way which is more accurate and effective. Furthermore, a spatial information transfer module is developed to avoid the vanishing of the spatial details. In experiments, visual comparisons and quantitative results demonstrate that our method can generate accurate and high-quality facial results against state-of-the-art methods. In the future, we will investigate to apply the proposed algorithm to other visual tasks such as semantic segmentation, image colorization, to name a few.

References

1. Abdal, R., Qin, Y., Wonka, P.: Image2StyleGAN: how to embed images into the StyleGAN latent space? In: ICCV (2019)
2. Arjovsky, M., Chintala, S., Bottou, L.: Wasserstein generative adversarial networks. In: ICML (2017)
3. Bao, J., Chen, D., Wen, F., Li, H., Hua, G.: Towards open-set identity preserving face synthesis. In: CVPR (2018)
4. Brock, A., Donahue, J., Simonyan, K.: Large scale GAN training for high fidelity natural image synthesis. In: ICLR (2019)
5. Chang, H., Lu, J., Yu, F., Finkelstein, A.: Pairedcyclegan: asymmetric style transfer for applying and removing makeup. In: CVPR (2018)
6. Choi, Y., Choi, M., Kim, M., Ha, J.W., Kim, S., Choo, J.: StarGAN: unified generative adversarial networks for multi-domain image-to-image translation. In: CVPR (2018)
7. Gatys, L.A., Ecker, A.S., Bethge, M.: Image style transfer using convolutional neural networks. In: CVPR (2016)
8. Goodfellow, I., et al.: Generative adversarial nets. In: NIPS (2014)
9. He, Z., Zuo, W., Kan, M., Shan, S., Chen, X.: AttGAN: facial attribute editing by only changing what you want. IEEE Trans. Image Process. **28**(11), 5464–5478 (2019)
10. Heusel, M., Ramsauer, H., Unterthiner, T., Nessler, B., Hochreiter, S.: GANs trained by a two time-scale update rule converge to a local nash equilibrium. In: NIPS, pp. 6626–6637 (2017)
11. Hu, J., Shen, L., Sun, G.: Squeeze-and-excitation networks. In: CVPR (2018)
12. Huang, R., Zhang, S., Li, T., He, R.: Beyond face rotation: global and local perception gan for photorealistic and identity preserving frontal view synthesis. In: ICCV, pp. 2439–2448 (2017)
13. Huang, X., Belongie, S.: Arbitrary style transfer in real-time with adaptive instance normalization. In: ICCV (2017)
14. Huang, X., Liu, M.Y., Belongie, S., Kautz, J.: Multimodal unsupervised image-to-image translation. In: ECCV (2018)
15. Isola, P., Zhu, J.Y., Zhou, T., Efros, A.A.: Image-to-image translation with conditional adversarial networks. In: CVPR (2017)
16. Johnson, J., Alahi, A., Fei-Fei, L.: Perceptual losses for real-time style transfer and super-resolution. In: Leibe, B., Matas, J., Sebe, N., Welling, M. (eds.) ECCV 2016. LNCS, vol. 9906, pp. 694–711. Springer, Cham (2016). https://doi.org/10.1007/978-3-319-46475-6_43
17. Karras, T., Aila, T., Laine, S., Lehtinen, J.: Progressive growing of GANs for improved quality, stability, and variation. In: ICLR (2018)
18. Karras, T., Laine, S., Aila, T.: A style-based generator architecture for generative adversarial networks. In: CVPR (2019)
19. Kim, T., Cha, M., Kim, H., Lee, J.K., Kim, J.: Learning to discover cross-domain relations with generative adversarial networks. In: ICML (2017)
20. Kingma, D.P., Ba, J.: Adam: a method for stochastic optimization. In: ICLR (2015)
21. Lee, H.Y., Tseng, H.Y., Huang, J.B., Singh, M.K., Yang, M.H.: Diverse image-to-image translation via disentangled representations. In: ECCV (2018)
22. Li, Y., Wang, N., Liu, J., Hou, X.: Demystifying neural style transfer (2017)
23. Liu, M., et al.: STGAN: a unified selective transfer network for arbitrary image attribute editing. In: CVPR (2019)

24. Liu, M.Y., Breuel, T., Kautz, J.: Unsupervised image-to-image translation networks. In: NIPS (2017)
25. Liu, M.Y., et al.: Few-shot unsupervised image-to-image translation. In: ICCV (2019)
26. Liu, Z., Luo, P., Wang, X., Tang, X.: Deep learning face attributes in the wild. In: ICCV (2015)
27. Maaten, L.V.D., Hinton, G.: Visualizing data using t-SNE. J. Mach. Learn. Res. 9(Nov), 2579–2605 (2008)
28. Mirza, M., Osindero, S.: Conditional generative adversarial nets. arXiv preprint arXiv:1411.1784 (2014)
29. Odena, A., Olah, C., Shlens, J.: Conditional image synthesis with auxiliary classifier GANs. In: ICML (2017)
30. Paszke, A., et al.: Automatic differentiation in PyTorch (2017)
31. Pedregosa, F., et al.: Scikit-learn: machine learning in Python. J. Mach. Learn. Res. 12, 2825–2830 (2011)
32. Perarnau, G., Van De Weijer, J., Raducanu, B., Álvarez, J.M.: Invertible conditional GANs for image editing. arXiv preprint arXiv:1611.06355 (2016)
33. Pumarola, A., Agudo, A., Martinez, A.M., Sanfeliu, A., Moreno-Noguer, F.: Ganimation: anatomically-aware facial animation from a single image. In: ECCV (2018)
34. Shen, Y., Gu, J., Tang, X., Zhou, B.: Interpreting the latent space of GANs for semantic face editing. arXiv preprint arXiv:1907.10786 (2019)
35. Wang, T.C., Liu, M.Y., Zhu, J.Y., Tao, A., Kautz, J., Catanzaro, B.: High-resolution image synthesis and semantic manipulation with conditional GANs. In: CVPR (2018)
36. Wu, P.W., Lin, Y.J., Chang, C.H., Chang, E.Y., Liao, S.W.: ReLGAN: multi-domain image-to-image translation via relative attributes. In: ICCV (2019)
37. Xiao, T., Hong, J., Ma, J.: ELEGANT: exchanging latent encodings with GAN for transferring multiple face attributes. In: ECCV (2018)
38. Zakharov, E., Shysheya, A., Burkov, E., Lempitsky, V.: Few-shot adversarial learning of realistic neural talking head models. arXiv preprint arXiv:1905.08233 (2019)
39. Zhang, G., Kan, M., Shan, S., Chen, X.: Generative adversarial network with spatial attention for face attribute editing. In: ECCV (2018)
40. Zhang, H., et al.: Context encoding for semantic segmentation. In: CVPR (2018)
41. Zhu, J.Y., Park, T., Isola, P., Efros, A.A.: Unpaired image-to-image translation using cycle-consistent adversarial networks. In: ICCV (2017)

Negative Pseudo Labeling Using Class Proportion for Semantic Segmentation in Pathology

Hiroki Tokunaga[1], Brian Kenji Iwana[1], Yuki Teramoto[2], Akihiko Yoshizawa[2], and Ryoma Bise[1,3(✉)]

[1] Kyushu University, Fukuoka, Japan
{iwana,bise}@ait.kyushu-u.ac.jp
[2] Kyoto University Hospital, Kyoto, Japan
[3] Research Center for Medical Bigdata, National Institute of Informatics, Tokyo, Japan

Abstract. In pathological diagnosis, since the proportion of the adenocarcinoma subtypes is related to the recurrence rate and the survival time after surgery, the proportion of cancer subtypes for pathological images has been recorded as diagnostic information in some hospitals. In this paper, we propose a subtype segmentation method that uses such proportional labels as weakly supervised labels. If the estimated class rate is higher than that of the annotated class rate, we generate negative pseudo labels, which indicate, "input image does not belong to this negative label," in addition to standard pseudo labels. It can force out the low confidence samples and mitigate the problem of positive pseudo label learning which cannot label low confident unlabeled samples. Our method outperformed the state-of-the-art semi-supervised learning (SSL) methods.

Keywords: Pathological image · Semantic segmentation · Negative learning · Semi-supervised learning · Learning from label proportion

1 Introduction

Automated segmentation of cancer subtypes is an important task to help pathologists diagnose tumors since it is recently known that the proportion of the adenocarcinoma subtypes is related to the recurrence rate and the survival time after surgery [38]. Therefore, the proportional information of cancer subtypes has been recorded as the diagnostic information with pathological images in some advanced hospitals. To obtain an accurate proportion, in general, segmentation for each subtype region should be required. However, a pathologist does not segment the regions since it is time-consuming, and thus they roughly annotate the proportion. The ratio of the subtypes fluctuates depending on pathologists

---

H. Tokunaga and B. K. Iwana—Contributed Equally.

© Springer Nature Switzerland AG 2020
A. Vedaldi et al. (Eds.): ECCV 2020, LNCS 12360, pp. 430–446, 2020.
https://doi.org/10.1007/978-3-030-58555-6_26

due to subjective annotation and this proportional annotation also takes time. Therefore, automated semantic segmentation for cancer subtypes is required.

Many segmentation methods for segmenting tumor regions have been proposed. Although the state-of-the-art methods accurately distinguish regions in digital pathology images [12], most methods reported the results of binary segmentation (normal and tumor) but not for subtype segmentation. Although some methods [33] have tackled subtype segmentation tasks, the performance is not enough yet. We consider that this comes from the lack of sufficient training data; insufficient data cannot represent the complex patterns of the subtypes. To obtain sufficient training data, it requires the expert's annotations, and annotation for subtype segmentation takes much more time than binary segmentation. In addition, since the visual patterns of subtypes have various appearances depending on the tissues (*e.g..*, lung, colon), staining methods, and imaging devices, we usually have to prepare a training data-set for each individual case. Although it is considered that the pre-recorded proportional information will help to improve the segmentation performance, no methods have been proposed that use such information.

Let us clarify our problem setup with the proportional information that is labeled to each whole-slide images (WSI), which are widely used in digital pathology. To segment subtype regions, a WSI cannot be inputted to a CNN due to the huge size (*e.g..*, 100,000 × 50,000 pixels). Thus, most methods take a patch-based classification approach that first segments a large image into small patches and then classifies each patch [12]. In this case, the proportional rates of each class can be considered as the ratio between the number of patch images whose label is the same class and the total number of patches. This proportional label with a set (bag) of images (instances) can be considered as a weak-supervision. In our problem setup, a small amount of supervised data and a large amount of weakly-supervised data are given for improving patch-level subtype classification.

Semi-supervised learning (SSL) is a similar problem setup in which a small amount of supervised data and a large amount of unlabeled data are given. It is one of the most promising approaches to improve the performance by also using unlabeled data. One of the common approaches of SSL is a pseudo label learning that first trains with a small amount of labeled data, and then the confident unlabeled data, which is larger than a predefined threshold, is added to the training data and the classifier is trained iteratively [19]. This pseudo labeled learning assumes that the high confidence samples gradually increase with each iteration. However, the visual features in test data may be different from those in the training data due to their various appearance patterns, and thus many samples are estimated with low confidence. In this case, pseudo labels are not assigned to such low confidence samples and it does not improve the confidences (*i.e.*, the number of pseudo samples does not increase). Therefore, the iteration of pseudo labeling does not improve much the classification performance. If we set a lower confidence threshold to obtain enough amount of pseudo labels, the low confidence samples may contain the samples belong to the other class (noise samples), and it adversely affects the learning. If we naively use the proportional labels for this pseudo learning to optimize the threshold, this problem still remains since the order of confidence in low confidence is not accurate.

In this paper, we propose a negative pseudo labeling that generates negative pseudo labels in addition to positive pseudo labels (we call the standard pseudo label as a *positive pseudo label* in this paper) by using the class proportional information. A negative label indicates, "input image does not belong to this negative label." The method first estimates the subtype of each patch image (instance) in each WSI (bag), and then computes the estimated label proportion for each bag. If the estimated class rate is higher than that of the annotated class rate (*i.e.*, the samples of this class are over-estimated), we generate the negative pseudo labels for this class to the low confidence examples in addition to the standard pseudo labels. These positive and negative pseudo labels are added to the training data and the classifier is trained iteratively with increasing the pseudo labels. Our method can force out the low confidence samples in over-estimated classes on the basis of the proportional label, and it mitigates the problem of positive pseudo label learning that cannot label low confident unlabeled samples.

Our main contributions are summarized as follows:

- We propose a novel problem that uses proportional information of cancer subtypes as weak labels for semi-supervised learning in a multi-class segmentation task. This is a task that occurs in real applications.
- We propose Negative Pseudo Labeling that uses class proportional information in order to efficiently solve the weakly- and semi-supervised learning problem. Furthermore, Multi Negative Pseudo Labeling improves the robustness of the method in which it prevents the negative learning from getting hung up on obvious negative labels.
- We demonstrate the effectiveness of the proposed method on a challenging real-world task. Namely, we perform segmentation of subtype regions of lung adenocarcinomas. The proposed method outperformed other state-of-the-art SSL methods.

2   Related Works

Segmentation in Pathology: As data-sets for pattern recognition in pathology have opened, such as Camelyon 2016 [5], 2017 [4], many methods have been proposed for segmenting tumor regions from normal regions. As discussed above, most methods are based on a patch-based classification approach that segments a large image into small patches and then classifies each patch separately, since a WSI image is extremely large to be inputted into a CNN [3,7,9,13,22,35,37,41]. These methods use a fixed size patch image and thus they only use either the context information from a wide field of view or high resolution information but not both. To address this problem, multi-scale based methods have been proposed [2,18,28,29,33]. Tokunaga *et al.* [33] proposed an adaptively weighting multi-field-of-view CNN that can adaptively use image features from different-magnification images. Takahama *et al.* [28] proposed a two-stage segmentation method that first extracts the local features from patch images and then aggregates these features for the global-level segmentation using U-net. However, these supervised methods require a large amount of training data in order to represent the various patterns in a class, in particular, in the multi-class segmentation.

Negative Label Learning: Negative label learning [14,17], also called complementary label learning, is a new machine learning problem that was first proposed in 2017 [31]. In contrast to the standard machine learning that trains a classifier using positive labels that indicate "input image belongs to this label," negative label learning trains a classifier using a negative label that indicates "input image does not belong to this negative label." Ishida *et al.* [14,31] proposed loss functions for this negative label learning in general problem setup. Kim *et al.* [17] uses negative learning for filtering noisy data from training data. To the best of our knowledge, methods that introduce negative learning into semi- or weakly-supervised learning have not been proposed yet.

Semi-supervised Learning (SSL): SSL is one of the most promising approaches to improve the performance of classifiers using unlabeled data. Most SSLs take a pseudo label learning approach [6,16,19,30]. For example, pseudo labeling [19] (also called self-training [36]) first trains using labeled data and then confident unlabeled data, which is larger than a predefined threshold, is added to the training data and the classifier is trained iteratively. Consistency regularization [21] generates pseudo labeled samples on the basis of the idea that a classifier should output the same class distribution for an unlabeled example even after it has been augmented. These methods generate only positive pseudo labeled samples but not negative pseudo labels.

Learning from Label Proportions (LLP): LLP is the following problem setting: given the label proportion information of bags that contains a set of instances, estimating the class of each instance that is a member of a bag.

Rueping *et al.* [26] proposed a LLP that uses a large-margin regression by assuming the mean instance of each bag having a soft label corresponding to the label proportion. Kuck *et al.* [11] proposed a hierarchical probabilistic model that generates consistent label proportions, in which similar idea were also proposed in [8,23]. Yu *et al.* [39] proposed $\propto$SVM that models the latent unknown instance labels together with the known group label proportions in a large-margin framework. SVM-based method was also proposed in [25]. These methods assume that the features of instances are given *i.e.*, they cannot learn the feature representation. Recently, Liu *et al.* [15] proposed a deep learning-based method that leverages GANs to derive an effective algorithm LLP-GAN and the effectiveness was shown in using open dataset such as MNIST and CIFAR-10. The pathological image has more complex features compared with such open dataset and their method does not introduce the semi-supervised learning like fashion in order to represent such complex image features.

Unlike these current methods, we introduce negative labeling for semi-supervised learning using label proportion. It can effectively use the supervised data for LLP and mitigates the problem of the standard pseudo labeling that cannot label low confident unlabeled samples.

3   Negative Pseudo Labeling with Label Proportions

In the standard Pseudo Labeling, if the maximum prediction probability (Confidence) of an input image is higher than a set threshold, a pseudo label is assigned to the image. On the other hand, in the problem setting of this study, the class ratio (cancer type ratio) in the pathological image is known. In this section, we propose a new pseudo labeling method using this information.

3.1   Pseudo Labeling

Pseudo Labeling [19] is an SSL technique that assigns a pseudo label to an unlabeled input pattern if the maximum prediction probability (confidence) exceeds a threshold. This is based on the assumption that a high probability prediction is a correct classification result and can be used to augment the training patterns. It functions by repeated steps of learning from supervised data, classifying unlabeled data, and learning from the new pseudo labeled data. Notably, the advantage of using Pseudo Labeling is that it only carries the aforementioned assumption about the data and can be used with classifiers easily.

However, the downside of Pseudo Labeling is that it requires the unlabeled training data to be easily classified [24]. Specifically, difficult data that does not exceed the threshold is not assigned pseudo labels and data that is misclassified can actively harm the training. In addition, if the distributions of the supervised data and the unsupervised data are significantly different, Pseudo Labeling cannot perform accurate pseudo labeling and will not improve discrimination performance.

3.2   Negative Pseudo Labeling

Pathological images differ greatly in image units even in the same cancer cell class. Therefore, if there is only a small number of supervised data, the distribution of supervised data and unsupervised data is significantly different and the confidence is low, so Pseudo Labeling may not be performed. Therefore, we propose Negative Pseudo Labeling, which assigns a pseudo label to data with low confidence by utilizing the cancer type ratio.

In Negative Pseudo Labeling, in addition to normal pseudo labeling, negative pseudo labels are added to incorrectly predicted images using weak teacher information of cancer type ratio. Figure 1 shows the outline of Negative Pseudo Labeling. First, a training patch image is created from a small number of supervised data, and the CNN is trained (gray arrow, executed only the first time). Next, a tumor region is extracted from weakly supervised data, and a patch

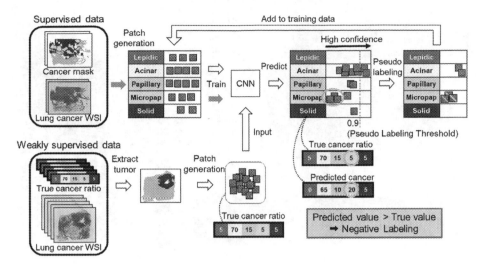

Fig. 1. Overview of negative pseudo labeling

image is created from that region. At this time, a set of patch images created from one pathological image has weak teacher information of the cancer type ratio. Next, the patch image is input to the CNN trained with supervised data, and the class probability is predicted. In the case of normal pseudo labeling, pseudo labels are given only to patches whose confidence is greater than or equal to the threshold. In the proposed method, negative pseudo labels are assigned to those with less than the threshold using cancer type ratio information.

By applying pseudo labeling not only for patch images but for whole pathological images, the predicted value of the cancer type ratio can be calculated. The correct answer and the predicted cancer type ratio are compared for each class. If the predicted value is larger than the correct answer value, a negative label is assigned to the corresponding image because the class is excessively incorrectly predicted. For example, in Fig. 1, the cyan class (Micro papillary) was predicted to be 20%. However, the correct ratio is 5%. Thus, the excess 15% in reverse order of per patch confidence is added as a negative label. This pseudo teacher is added to the training data, and the model is trained again. These steps are repeated until the loss of the validation data converges.

Specifically, the assignment of the actual negative pseudo label is performed according to the Negative Selectivity (NS) calculated by:

$$NS_c = SAE \times max(0, \ PCR_c - TCR_c) \times (1 - TCR_c/PCR_c), \qquad (1)$$

where c is the class, the sum absolute error SAE is the distance between the ratios, and TCR_c and PCR_c are the true cancer ratio and predicted cancer ratio, respectively. SAE is defined by the following equation and indicates the sum of errors between the ratios of one pathological image in the range of 0 to 1:

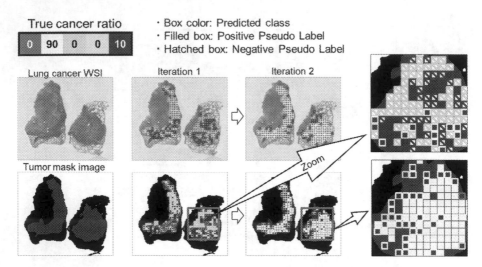

Fig. 2. Example of Negative Pseudo Label assignment. The color of the box is the predicted class, the filled boxes are patches with positive pseudo labels, and the hatched boxes are the patches with negative pseudo labels. By learning from the negative pseudo label, the prediction class changes to a different class from green. (Color figure online)

$$SAE = max(1, \sum_{c=1}^{C} |TCR_c - PCR_c|). \qquad (2)$$

Here, C represents the number of classification classes. The first term of Eq. (1) selects pathological images with many incorrect predictions, the second term gives negative pseudo labels to only excessively incorrect classes, and the third term gives negative pseudo labels to classes that are close to 0%.

In the proposed method, in addition to the positive pseudo label assigned by normal Pseudo Labeling, the new negative pseudo labels are added to enable effective learning. Figure 2 shows an example of a negative and positive pseudo label. From the ratio of correct cancer types, it can be seen that there should be none classified as green and 10% blue. However, when observing the predicted class in Iteration 1, it can be seen that blue and green occupy the majority classifications. Thus, they are incorrectly over-predicted. Therefore, a negative pseudo label (hatched box) is assigned to the patches in NS order. In Iteration 2, learning from the pseudo labels is performed and new predictions changed to the yellow class. In other words, negative pseudo labels are assigned to a patch that has been predicted as a class that should not exist which causes the model to change toward the correct class.

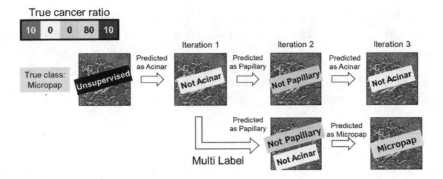

Fig. 3. Example of a negative label oscillating between wrong classes. By storing past label information, this can be prevented.

3.3   Multi Negative Pseudo Labeling

By giving a negative pseudo label, it is possible to change the prediction from one class to another. However, in negative learning, the new class is not necessarily the correct class. Figure 3 shows an example in which the learning does not progress because the negative label oscillates between the wrong classes. In the figure, the true correct class is Micro papillary, but only the weakly supervised ratio is known. When Negative Pseudo Labeling was performed, the patch was predicted to be Acinar, which was judged to be an incorrect prediction class based on the cancer type ratio. Therefore, a negative label of Acinar is assigned in Iteration 1. When the class was predicted again by Iteration 2, this time, the prediction was changed to another class which should not exist, Papillary. By repeating Negative Pseudo Labeling, this situation can oscillate between incorrect classes. In the case of Negative Pseudo Labeling, the loss function,

$$\mathcal{L}_{negative} = -\sum_{c=1}^{C} \boldsymbol{y}_c \log\left(1 - \log f(c|\boldsymbol{x}, \theta)\right), \tag{3}$$

can get stuck by only minimizing alternating predictions between erroneous classes.

Therefore, Negative Pseudo Labeling is extended to Multi Negative Pseudo Labeling to prevent these oscillations. In Negative Pseudo Labeling, positive and negative pseudo labels are added for each Iteration, and the past pseudo label information is not used. Therefore, we accumulate the pseudo label information given by each iteration and give a multi-negative pseudo label. However, simply adding negative labels increases the scale of the loss, therefore, we propose a loss function $\mathcal{L}_{multi-negative}$ that performs weighting based on the number of negative labels, or:

$$\mathcal{L}_{multi-negative} = -\frac{(C - |\boldsymbol{y}|)}{(C - 1)} \sum_{c=1}^{C} \boldsymbol{y}_c \log\left(1 - \log f(c|\boldsymbol{x}, \theta)\right), \tag{4}$$

where $\boldsymbol{y}_c$ is a negative label for a sample, $|\boldsymbol{y}|$ is the number of negative labels for a sample on the iteration, $\boldsymbol{x}$ is the input data, $f(c|\boldsymbol{x}, \theta)$ is the predicted probability of class c, and C is the number of classification classes. The loss $\mathcal{L}_{multi-negative}$ is the same with the loss (Eq. 3) for the single negative label when $|\boldsymbol{y}|$ is 1 (*i.e.,* a single negative pseudo label is given in the iteration), and the weight linearly decreases if the number of negative pseudo labels ($|\boldsymbol{y}|$) increases.

4   Adaptive Pseudo Labeling

Conventionally, Pseudo Labeling ignores the unlabeled data with low prediction probabilities and through the introduction of Negative Pseudo Labeling, it now is possible to use these inputs. However, in order to assign positive pseudo labels, it still is necessary to manually set a threshold value above which the positive pseudo label is assigned. In this section, we propose the use of Adaptive Pseudo Labeling, which adaptively determines the criteria for positive pseudo labeling based on the similarity of cancer types.

When an accurate prediction is made for a patch image and the predicted cancer type ratio is close to the correct cancer type ratio, we consider the similarity between the ratios as large. For example, in Fig. 4, the correct cancer type ratio is [5%, 70%, 15%, 0%, 10%] and the predicted cancer type ratio is [7%, 64%, 14%, 0%, 14%], the similarity between the ratios is high. However, in a single pathological image, it is not possible to judge that the class label of each patch image is accurate just because the similarity between the ratios is high. In Fig. 4, even if the green and blue predicted patch images are switched, the predicted cancer type ratio remains at 14% and the similarity does not change. However, if there are multiple pathological images and the ratio of cancer types differs for each pathological image, it is unlikely that the ratios will be the same when predictions between different classes are switched. Therefore, the prediction class of the patch image is considered reliable when the similarity of the cancer type ratio is high.

This is represented as the Positive Selectivity (PS) of the patch being assigned a pseudo label. PS is a parameter that determines how much of the images are given a positive pseudo label, and is defined by:

$$PS_c = (1 - SAE) \times min(TCR_c, PCR_c), \tag{5}$$

where SAE is the distance between the ratios defined by Eq. (2), and TCR_c and PCR_c are the true cancer ratio and predicted cancer ratio, respectively. Equation (5) is calculated for each class of pathological image, and positive pseudo labels are assigned to only PS_c total patch images in the order of confidence.

Figure 4 shows an example of Adaptive Pseudo Labeling. In Pseudo Labeling, pseudo labeling is performed with a fixed threshold regardless of the distance of the cancer type ratio, so it is typical of pseudo labeling being only applied to a small number of images. On the other hand, in Adaptive Pseudo Labeling, the selection rate is dynamically determined for each pathological image according to the distance of the ratio.

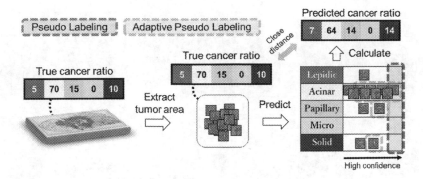

Fig. 4. Example of cancer type ratio information. When accurate prediction is performed, the similarity between the predicted cancer type ratio of the set of patch images and the correct cancer type ratio increases. (Color figure online)

5 Experimental Results

5.1 Dataset

In the experiment, we use a real-world dataset consisting of 42 supervised pathological images and 400 weakly supervised pathological images. Each pathological image contains a slide of a tumor that is up to $108,000 \times 54,000 \times 3$ pixels in size and was annotated by two pathologists. The supervised images have pixel-wise class labels and the weakly supervised images only have a general ratio of cancer types. As shown in Fig. 5, the supervised images are annotated into one non-cancer region and five cancer regions, Lepidic, Acinar, Papillary, Micro papillary, and Solid. In addition, there are vague areas in the tumor area, indicated in black, which were difficult for the pathologists to classify. These difficult regions were excluded when evaluating the accuracy of the cancer type segmentation. As for the weakly supervised images, the pathologists only provide an estimated ratio of cancer type in percent.

In order to construct patch images, the WSI images are broken up using a non-overlapping sliding window of size $224 \times 224 \times 3$ pixels at a magnification of $10\times$ (a typical magnification used for diagnosis). As a result, there are a total of 5,757, 2,399, 3,850, 3,343, and 4,541 labeled patches for Lepidic, Acinar, Papillary, Micro papillary, and Solid cancer cells, respectively. These patches are separated five pathological image-independent sets for 5-fold cross-validation. Specifically, each cross-validation labeled training set is created from patches of 33 to 34 of the original full-size images with 8 to 9 of the full-size images left out for the test sets. From each labeled training set, 25% random the patch images are taken and used as respective validation sets.

To extract patches for the pseudo labeled training set, patches of tumor regions were extracted from the weakly supervised data. Due to the weakly supervised data only containing the ratio of cancer types and not the full annotations, we used a Fix Weighting Multi-Field-of-View CNN [33] to extract tumor

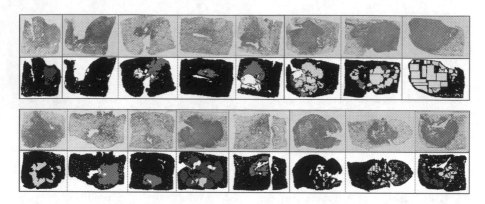

Fig. 5. Examples of 16 of the 42 pathological images and the ground truth regions. Red, yellow, green, cyan, and blue indicate Lepidic, Acinar, Papillary, Micro papillary, and Solid, respectively. (Color figure online)

proposal regions in order to create patches. The Multi-Field-of-View CNN combines multiple magnifications in order to accomplish segmentation. In this case, we used the magnifications of $10\times$, $5\times$, and $2.5\times$. In total, 52,714 patches from the weakly supervised data are used.

5.2   Experiment Settings

For the base CNN architecture, the proposed method uses a MobileNetV2 [27] with the weights pre-trained by ImageNet. During model training, the network was fine-tuned with the convolutional layers frozen [1]. The training was performed in two steps. First, the MobileNetV2 was trained using the supervised data for 100 epochs with early stopping based on a validation loss plateau of 10 epochs. Second, the network was fine-tuned using the Pseudo Labeling for a similar 100 epochs with early stopping. During the Pseudo Labeling step, the positive learning loss $\mathcal{L}_{positive}$,

$$\mathcal{L}_{positive} = -\sum_{c=1}^{C} \boldsymbol{y}_c \log f(c|\boldsymbol{x}, \theta), \tag{6}$$

where $\boldsymbol{y}_c$ is the correct label, $\boldsymbol{x}$ is input data, and $f(\cdot)$ is a classifier with parameter θ. For the Negative Pseudo Labeling, the multi-negative learning loss $\mathcal{L}_{multi-negative}$ from (4) is used. Furthermore, the network was trained using mini-batches of size 18 with an RAdam [20] optimizer with an initial learning rate of 10^{-5}.

5.3   Quantitative Evaluation

The prediction results were quantitatively evaluated using three evaluations recommended for region-based segmentation [10], Overall Pixel (OP),

Per-Class (PC), and the mean Intersection over Union (mIoU). These metrics are defined as:

$$OP = \frac{\sum_c TP_c}{\sum_c (TP_c + FP_c)}, \quad PC = \frac{1}{M} \sum_c \frac{TP_c}{TP_c + FP_c},$$

$$mIoU = \frac{1}{M} \sum_c \frac{TP_c}{TP_c + FP_c + FN_c}, \tag{7}$$

where M is the number of classes, and TP_c, FP_c, and FN_c are the numbers of true positives, false positives, and false negatives for class c, respectively.

Table 1. Quantitative evaluation results

Method	OP	PC	mIoU
Proposed	**0.636**	**0.588**	**0.435**
Supervised	0.576	0.527	0.373
Pseudo Labeling [19]	0.618	0.575	0.420
ICT [34]	0.461	0.430	0.289
Mean Teachers [32]	0.449	0.421	0.270
mixup [40]	0.529	0.476	0.324
VAT [21]	0.477	0.427	0.281

Table 2. Ablation results

Positive threshold	Negative labeling	OP	PC	mIoU
Fixed at 0.95	–	0.618	0.575	0.420
Adaptive	–	0.613	0.565	0.412
–	Single label	0.563	0.508	0.348
–	Multi label	0.551	0.478	0.336
Adaptive	Single label	0.621	0.585	0.426
Adaptive	Multi label	**0.636**	**0.588**	**0.435**

The results of the quantitative evaluations are shown in Table 1. In this table, we compare the results of using the proposed Multi Negative Pseudo Labeling with Adaptive Positive Pseudo Labeling (Proposed) with state-of-the-art SSL methods. As baselines, Supervised uses only the supervised training data and Pseudo Labeling [19] is the standard implementation of using positive pseudo labels with a threshold of 0.95. Interpolation Consistency Training (ICT) [34], Mean Teachers [32], mixup [40], Virtual Adversarial Training (VAT) [21] are other recent and popular SSL methods that we evaluated on our dataset. These methods were trained under similar conditions as Proposed. It should be noted that the SSL methods can only use the weakly supervised data as unlabeled data. From the quantitative evaluation results, it can be seen that the Proposed accuracy is better than the other SSL methods.

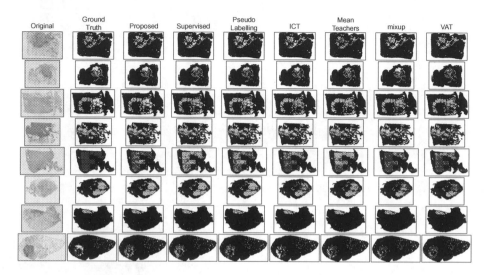

Fig. 6. Results from a sample test set.

Furthermore, we performed ablation experiments to demonstrate the importance of the Negative Pseudo Labeling. These results are shown in Table 2. In the result, the methods using only (Single/Multi) Negative Pseudo Labeling (NPL) did not work well due to class imbalance because NPL can not directly use an oversampling technique. When we use the Positive Labeling (PL) with NPL, the class imbalance problem can be mitigated by oversampling in PL, and thus the combination of PL and NPL improves the performance. In addition, Multi NPL further improved the performance compared with using NPL.

5.4   Qualitative Evaluation

In order to evaluate the results using qualitative analysis, the patches are recombined into the full pathological images. Figure 6 shows the results from the first cross-validation test set. The first two columns are the original image and ground truth and the subsequent eight columns are the synthesized images with the colors corresponding to the classes listed in Fig. 5. The figure is able to provide an understanding of how the proposed method was able to have higher quantitative results. For example, when comparing the fifth row of Fig. 6, the SSL methods, including the proposed method, generally had much better results than just using the labeled data in Supervised. Furthermore, the performance improvement from using Multi Negative Pseudo Labels can clearly be seen when comparing the results from Proposed and Pseudo Labeling. The figure also shows

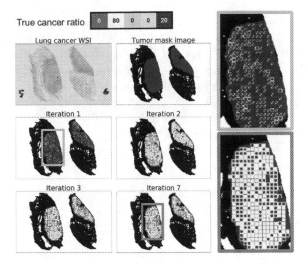

Fig. 7. Example of the proposed negative and positive pseudo labeling. The color of the box is the predicted class, the filled boxes are patches with positive pseudo labels, and the hatched boxes are the patches with negative pseudo labels. (Color figure online)

that many comparison methods were able to excel at particular images, even sometimes better than the proposed method. However, the proposed method had more consistently accurate results for all of the images.

In a specific instance, Fig. 7 shows the process of using positive and negative pseudo labels to direct the classifier to use the unsupervised data correctly. Before Iteration 1, the network is trained using only the supervised training set. According to the training, most of the weakly supervised patches were incorrectly classified as blue. Over the course of a few iterations, the pseudo labels of the patches shifted the distribution of classes to be more similar to the weakly supervised ratio. Before, the pseudo labels would have harmed the supervised training set. Now, instead, the pseudo labels are able to efficiently augment the supervised data.

6 Conclusion

This paper tackles cancer type classification using massive pathological images. A feature of using massive pathological images is that precise pixel-wise annotations are difficult and costly to acquire. Therefore, we proposed a new method of semi-supervised learning that incorporates positive and negative pseudo label learning. In this model, we introduce Negative Pseudo Labeling and its extension Multi Negative Pseudo Labeling. The Negative Pseudo Labeling assigns negative pseudo labels to unlabeled weakly supervised data in order to guide the pseudo label augmentation toward a ratio of classes provided by the weak supervision. Multi Negative Pseudo Labeling extends this idea and weights past negative pseudo labels in order to avoid getting stuck in alternating negative pseudo labels. Furthermore, we introduce Adaptive Pseudo Labeling for the positive pseudo labeling step, which dynamically selects the positive pseudo labels without the need of determining the threshold traditionally used in Pseudo Labeling. In order to evaluate the proposed method, we performed quantitative and qualitative analysis using pathological image-independent 5-fold cross-validation. We are able to demonstrate that the proposed method outperforms other state-of-the-art SSL methods. Through these promising results, we are able to show that segmentation via patch classification augmented with weak supervision is possible on real and large-scale pathological images.

Acknowledgment. This work was supported by JSPS KAKENHI Grant Number 20H04211.

References

1. Agrawal, P., Girshick, R., Malik, J.: Analyzing the performance of multilayer neural networks for object recognition. In: Fleet, D., Pajdla, T., Schiele, B., Tuytelaars, T. (eds.) ECCV 2014. LNCS, vol. 8695, pp. 329–344. Springer, Cham (2014). https://doi.org/10.1007/978-3-319-10584-0_22

2. Alsubaie, N., Shaban, M., Snead, D., Khurram, A., Rajpoot, N.: A multi-resolution deep learning framework for lung adenocarcinoma growth pattern classification. In: Nixon, M., Mahmoodi, S., Zwiggelaar, R. (eds.) MIUA 2018. CCIS, vol. 894, pp. 3–11. Springer, Cham (2018). https://doi.org/10.1007/978-3-319-95921-4_1

3. Altunbay, D., Cigir, C., Sokmensuer, C., GunduzDemi, C.: Color graphs for automated cancer diagnosis and grading. IEEE Trans. Biomed. Eng. **57**(3), 665–674 (2010)

4. Bandi, P., Geessink, O., Manson, Q., van Dijk, M., et al.: From detection of individual metastases to classification of lymph node status at the patient level: the camelyon17 challenge. IEEE Trans. Med. Imag. **38**(2), 550-560 (2018)

5. Bejnordi, B.E., Veta, M., van Diest, P.J., van Ginneken, B., et al.: Diagnostic assessment of deep learning algorithms for detection of lymph node metastases in women with breast cancer. Jama **318**(22), 2199–2210 (2017)

6. Berthelot, D., Carlini, N., Goodfellow, I., Papernot, N., Oliver, A., Raffel, C.A.: MixMatch: a holistic approach to semi-supervised learning. In: NeurIPS, pp. 5050–5060 (2019)

7. Chang, H., Zhou, Y., Borowsky, A., Barner, K., Spellman, P., Parvin, B.: Stacked predictive sparse decomposition for classification of histology sections. Int. J. Comput. Vis. **113**(1), 3–18 (2014)

8. Chen, B., Chen, L., Ramakrishnan, R., Musicant, D.: Learning from aggregate views. In: ICDE, p. 3 (2006)

9. Cruz-Roa, A., Basavanhally, A., Gonzalez, F., Gilmore, H., et al.: Automatic detection of invasive ductal carcinoma in whole slide images with convolutional neural networks. In: SPIE Medical Imaging (2014)

10. Csurka, G., Larlus, D., Perronnin, F.: What is a good evaluation measure for semantic segmentation? In: CVPR (2013)

11. Hendrik, K., de Nando, F.: SVM classifier estimation from group probabilities. In: CUAI, pp. 332–339 (2005)

12. Hou, L., Samaras, D., Kurc, T.M., Gao, Y., Davis, J.E., Saltz, J.H.: Patch-based convolutional neural network for whole slide tissue image classification. In: CVPR, pp. 2424–2433 (2016)
13. Hou, L., Samaras, D., Kurc, T.M., Gao, Y. et al.: Patch-based convolutional neural network for whole slide tissue image classification. In: CVPR, pp. 2424–2433 (2016)
14. Ishida, T., Niu, G., Menon, A.K., Sugiyama, M.: Complementary-label learning for arbitrary losses and models. arXiv preprint arXiv:1810.04327 (2018)
15. Liu, J., Wang, B., Qi, Z., Tian, Y., Shi, Y.: Learning from label proportions with generative adversarial networks. In: NeurIPS (2019)
16. Kikkawa, R., Sekiguchi, H., Tsuge, I., Saito, S., Bise, R.: Semi-supervised learning with structured knowledge for body hair detection in photoacoustic image. In: ISBI (2019)
17. Kim, Y., Yim, J., Yun, J., Kim, J.: NLNL: negative learning for noisy labels. In: ICCV, pp. 101–110 (2019)
18. Kong, B., Wang, X., Li, Z., Song, Q., Zhang, S.: Cancer metastasis detection via spatially structured deep network. In: Niethammer, M. (ed.) IPMI 2017. LNCS, vol. 10265, pp. 236–248. Springer, Cham (2017). https://doi.org/10.1007/978-3-319-59050-9 19
19. Lee, D.H.: Pseudo-label: the simple and efficient semi-supervised learning method for deep neural networks. In: ICML Workshops, vol. 3, p. 2 (2013)
20. Liu, L., et al.: On the variance of the adaptive learning rate and beyond. arXiv preprint arXiv:1908.03265 (2019)
21. Miyato, T., Maeda, S.I., Koyama, M., Ishii, S.: Virtual adversarial training: a regularization method for supervised and semi-supervised learning. IEEE Trans. Pattern Anal. Mach. Intell. **41**(8), 1979–1993 (2018)
22. Mousavi, H., Monga, V., Rao, G., Rao, A.U.: Automated discrimination of lower and higher grade gliomas based on histopathological image analysis. J. Pathol. Inform. **6**, 15 (2015)
23. Musicant, D., Christensen, J., Olson, J.: Supervised learning by training on aggregate outputs. In: ICDM, pp. 252–261 (2007)
24. Oliver, A., Odena, A., Raffel, C.A., Cubuk, E.D., Goodfellow, I.: Realistic evaluation of deep semi-supervised learning algorithms. In: NeurIPS, pp. 3235–3246 (2018)
25. Qi, Z., Wang, B., Meng, F.: Learning with label proportions via NPSVM. IEEE Trans. Cybern. **47**(10), 3293–3305 (2017)
26. Rueping, S.: SVM classifier estimation from group probabilities. In: ICML (2010)
27. Sandler, M., Howard, A., Zhu, M., Zhmoginov, A., Chen, L.C.: MobileNetV2: inverted residuals and linear bottlenecks. In: CVPR, pp. 4510–4520 (2018)
28. Shusuke, T., et al.: Multi-stage pathological image classification using semantic segmentation. In: ICCV (2019)
29. Sirinukunwattana, K., Alham, N.K., Verrill, C., Rittscher, J.: Improving whole slide segmentation through visual context - a systematic study. In: Frangi, A.F., Schnabel, J.A., Davatzikos, C., Alberola-López, C., Fichtinger, G. (eds.) MICCAI 2018. LNCS, vol. 11071, pp. 192–200. Springer, Cham (2018). https://doi.org/10.1007/978-3-030-00934-2_22
30. Sohn, K., et al.: Fixmatch: simplifying semi-supervised learning with consistency and confidence. arXiv preprint arXiv:2001.07685 (2020)
31. Takashi, I., Gang, N., Weihua, H., Masashi, S.: Learning from complementary labels. In: NeurIPS (2017)

32. Tarvainen, A., Valpola, H.: Mean teachers are better role models: weight-averaged consistency targets improve semi-supervised deep learning results. In: NeurIPS, pp. 1195–1204 (2017)

33. Tokunaga, H., Teramoto, Y., Yoshizawa, A., Bise, R.: Adaptive weighting multi-field-of-view CNN for semantic segmentation in pathology. In: CVPR (2019)

34. Verma, V., Lamb, A., Kannala, J., Bengio, Y., Lopez-Paz, D.: Interpolation consistency training for semi-supervised learning. In: IJCAI, pp. 3635–3641 (2019)

35. Wang, D., Khosla, A., Gargeya, R., Irshad, H., Beck, A.H.: Deep learning for identifying metastatic breast cancer. arXiv preprint arXiv:1606.05718 (2016)

36. Xie, Q., Hovy, E., Luong, M.T., Le, Q.V.: Self-training with noisy student improves ImageNet classification. arXiv preprint arXiv:1911.04252 (2019)

37. Xu, Y., Jia, Z., Ai, Y., Zhang, F., Lai, M., Chang, E.I.C.: Deep convolutional activation features for large scale brain tumor histopathology image classification and segmentation. In: ICASSP (2015)

38. Yoshizawa, A., Motoi, N., Riely, G.J., Sima, C., et al.: Impact of proposed IASLC/ATS/ERS classification of lung adenocarcinoma: prognostic subgroups and implications for further revision of staging based on analysis of 514 stage I cases. Mod. Pathol. **24**(5), 653 (2011)

39. Yu, F., Liu, D., Kumar, S., Tony, J., Chang, S.F.: $\propto$SVM for learning with label proportions. In: ICML, pp. 504–512 (2013)

40. Zhang, H., Cisse, M., Dauphin, Y.N., Lopez-Paz, D.: mixup: beyond empirical risk minimization. In: ICLR (2018)

41. Zhou, Y., Chang, H., Barner, K., Spellman, P., Parvin, B.: Classification of histology sections via multispectral convolutional sparse coding. In: CVPR Workshop, pp. 3081–3088 (2014)

Learn to Propagate Reliably on Noisy Affinity Graphs

Lei Yang[1(✉)] , Qingqiu Huang[1] , Huaiyi Huang[1] , Linning Xu[2] ,
and Dahua Lin[1]

[1] The Chinese University of Hong Kong, Shatin, Hong Kong
{yl016,hq016,hh016,dhlin}@ie.cuhk.edu.hk
[2] The Chinese University of Hong Kong, Shenzhen, China
linningxu@link.cuhk.edu.cn

Abstract. Recent works have shown that exploiting unlabeled data through label propagation can substantially reduce the labeling cost, which has been a critical issue in developing visual recognition models. Yet, how to propagate labels reliably, especially on a dataset with unknown outliers, remains an open question. Conventional methods such as linear diffusion lack the capability of handling complex graph structures and may perform poorly when the seeds are sparse. Latest methods based on graph neural networks would face difficulties on performance drop as they scale out to noisy graphs. To overcome these difficulties, we propose a new framework that allows labels to be propagated reliably on large-scale real-world data. This framework incorporates (1) a local graph neural network to predict accurately on varying local structures while maintaining high scalability, and (2) a confidence-based path scheduler that identifies outliers and moves forward the propagation frontier in a prudent way. Both components are learnable and closely coupled. Experiments on both ImageNet and Ms-Celeb-1M show that our confidence guided framework can significantly improve the overall accuracies of the propagated labels, especially when the graph is very noisy.

1 Introduction

The remarkable advances in visual recognition are built on top of large-scale annotated training data [6,7,11–13,15–17,28,29,33,34,40,41,41,42,46]. However, the ever increasing demand on annotated data has resulted in prohibitive annotation cost. Transductive learning, which aims to propagate labeled information to unlabeled samples, is a promising way to tackle this issue. Recent studies [18,21,25,26,38,50] show that transductive methods with an appropriate design can infer unknown labels accurately while dramatically reducing the annotation efforts.

Electronic supplementary material The online version of this chapter (https:// doi.org/10.1007/978-3-030-58555-6_27) contains supplementary material, which is available to authorized users.

A. Vedaldi et al. (Eds.): ECCV 2020, LNCS 12360, pp. 447–464, 2020.
https://doi.org/10.1007/978-3-030-58555-6_27

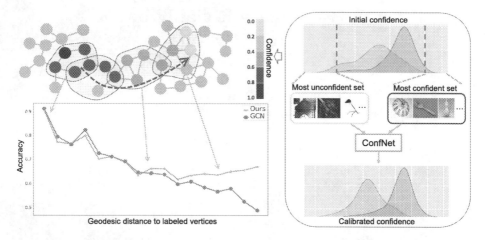

Fig. 1. In this paper, we propose a framework for transductive learning on noisy graphs, which contain a large number of outliers, *e.g.* out-of-class samples. The framework consists of a local predictor and a confidence-based path scheduler. The predictor updates local patches sequentially following a path driven by the estimated confidences. The path scheduler leverages both the confident and unconfident samples from the predictor to further calibrate the estimated confidence. The unconfident samples are usually images with low quality (*e.g.* the leftmost image is a clock with only top part), hard examples (*e.g.* the middle image is a spoon mixed with the background) or *out-of-class* samples (*e.g.* the rightmost image is a lamp but none of the labeled samples belong to this class). The lower left figure experimentally shows that the proposed method improves the reliability of propagation. When the distance from unlabeled samples to labeled ones increases, our method surpasses state-of-the-art by a significant margin

Many transductive methods adopt graph-based propagation [21,26,38,49] as a core component. Generally, these methods construct a graph among all samples, propagating labels or other relevant information from labeled samples to unlabeled ones. Early methods [1,47,49] often resort to a linear diffusion paradigm, where the class probabilities for each unlabeled sample are predicted as a linear combination of those for its neighbors. Relying on simplistic assumptions restricts their capability of dealing with complicated graph structures in real-world datasets. Recently, graph convolutional networks [21,38,39] have revealed its strong expressive power to process complex graph structures. Despite obtaining encouraging results, these GCN-based methods remain limited in an important aspect, namely the capability of coping with outliers in the graph. In real-world applications, unlabeled samples do not necessarily share the same classes with the labeled ones, leading to a large portion of *out-of-class* samples, which becomes the main source of outliers. Existing methods ignore the fact that the confidences of predictions on different samples can vary significantly, which may adversely influence the reliability of the predictions.

In this paper, we aim to explore a new framework that can propagate labels over noisy unlabeled data *reliably*. This framework is designed based on three principles: 1) *Local update:* each updating step can be carried out within a local part of the graph, such that the algorithm can be easily scaled out to a large-scale graph with millions of vertices. 2) *Learnable:* the graph structures over a real-world dataset are complex, and thus it is difficult to prescribe a rule that works well for all cases, especially for various unknown outliers. Hence, it is desirable to have a core operator with strong expressive power that can be learned from real data. 3) *Reliable path:* graph-based propagation is sensitive to noises – a noisy prediction can mislead other predictions downstream. To propagate reliably, it is crucial to choose a path such that most inferences are based on reliable sources.

Specifically, we propose a framework comprised of two learnable components, namely, a local predictor and a path scheduler. The local predictor is a lightweight graph neural network operating on local sub-graphs, which we refer to as *graph patches*, to predict the labels of unknown vertices. The path scheduler is driven by confidence estimates, ensuring that labels are gradually propagated from highly confident parts to the rest. The key challenge in designing the path scheduler is how to estimate the confidences effectively. We tackle this problem via a two-stage design. First, we adopt a *multi-view* strategy by exploiting the fact that a vertex is usually covered by multiple *graph patches*, where each patch may project a different prediction on it. Then the confidence can be evaluated on how consistent and certain the predictions are. Second, with the estimated confidence, we construct a candidate set by selecting the most confident samples and the most unconfident ones. As illustrated in Fig. 1, we devise a *ConfNet* to learn from the candidate set and calibrate the confidence estimated from the first stage. Highly confident samples are assumed to be labeled and used in later propagation, while highly unconfident samples are assumed to be outliers and excluded in later propagation. Both components work closely together to drive the propagation process. On one hand, the local predictor follows the scheduled path to update predictions; on the other hand, the path scheduler estimates confidences based on local predictions. Note that the training algorithm also follows the same coupled procedure, where the parameters of the local predictor and confidence estimator are learned end-to-end.

Our main contributions lie in three aspects: (1) A learnable framework that involves a local predictor and a path scheduler to drive propagation reliably on noisy large-scale graphs. (2) A novel scheme of exploiting both confident and unconfident samples for confidence estimation. (3) Experiments on ImageNet [6] and Ms-Celeb-1M [9] show that our proposed approach outperforms previous algorithms, especially when the graphs are noisy and the initial seeds are sparse.

2   Related Work

In this paper, we focus on graph-based transductive learning [14,21,26,38,49], which constructs a graph among all samples and propagates information from labeled samples to unlabeled ones. We summarize existing methods into three categories and briefly introduce other relevant techniques.

Early Methods. Conventional graph-based transductive learning [1,47,49] is mainly originated from smoothness assumption, which is formulated as a graph Laplacian regularization. They share the same paradigm to aggregate neighbors' information through linear combination. While relying on the simple assumption, these methods are limited by their capability of coping with complex graph structures in large-scale real-world datasets.

GCN-Based Methods. Graph Convolutional Network (GCN) and its variants [21,32,38,39] apply filters over the entire graph and achieve impressive performance on several tasks [48]. To extend the power of GCN to large-scale graphs, GraphSAGE [10] proposes to sample a fixed number of neighbors and apply aggregation functions thereon. FastGCN [4] further reduces memory demand by sampling vertices rather than neighbors in each graph convolution layer. However, they propagate labels in parallel across all vertices, regardless of the confidence difference among predictions. This ignorance on prediction confidence may adversely influence the reliability of propagation.

Confidence-Based Methods. Previous approaches either model the node label as a distribution along with uncertainty [2,31] or re-scale the weight of links by introducing attention to vertices [37,38]. Unlike these methods, which mainly focus on making use of confident samples, our approach learns from both confident and unconfident data for confidence estimation.

Inductive Learning. Inductive learning is closely related to transductive learning [36]. The key difference lies in that the former aims to learn a better model with unlabeled data [25,35,43–45], while the latter focuses on predicting labels for unlabeled data [49,50]. One important line of inductive learning is leveraging the predicted labels of unlabeled data in a supervised manner [19,23]. In this paper, we focus on transductive learning and show that the obtained pseudo labels can be applied to inductive learning as a downstream task.

Outlier Detection. Previous methods [3] either rely on crafted unsupervised rules [24] or employing a supervised method to learn from an extra labeled outlier dataset [5]. The unsupervised rules lack the capability of handling complex real-world dataset, while the supervised methods are easy to overfit to the labeled outliers and do not generalize well. In this paper, we *learn* to identify outliers from carefully selected confident and unconfident samples during propagation.

3   Propagation on Noisy Affinity Graphs

Our goal is to develop an effective method to propagate reliably over noisy affinity graphs, *e.g.* those containing lots of out-of-class samples, while maintaining reasonable runtime cost. This is challenging especially when the proportion of outliers are high and initial seeds are sparse. As real-world graphs often have complex and varying structures, noisy predictions can adversely affect these downstream along the propagation paths. We propose a novel framework for

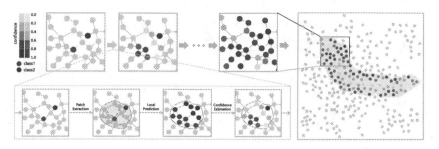

Fig. 2. Overview of our framework (better viewed in color). At each iteration, our approach consists of three steps: (1) Starting from the selected confident vertex, the patch extractor generates a graph patch. (2) Given the graph patch, the learned local predictor updates the predictions of all unlabeled vertices on the patch. (3) Given the updated predictions, the path scheduler estimates confidence for all unlabeled vertices. Over many iterations, labeled information are gradually propagated from highly confident parts to the rest (Color figure online)

graph-based propagation, which copes with the complexity in local graph structures via a light-weight graph convolutional network while improving the reliability via a confidence-based scheduler that chooses propagation paths prudently.

3.1   Problem Statement

Consider a dataset with $N = N_l + N_u$ samples, where N_l samples are labeled and N_u are unlabeled, and $N_l \ll N_u$. We denote the set of labeled samples as $\mathcal{D}_l = \{(\mathbf{x}_i, y_i)\}_{i=1}^{N_l}$, and that of unlabeled ones as $\mathcal{D}_u = \{\mathbf{x}_i\}_{i=N_l+1}^{N_l+N_u}$. Here, $\mathbf{x}_i \in \mathbb{R}^d$ is the feature for the i-th sample, which is often derived from a deep network in vision tasks, and $y_i \in \mathcal{Y}$ is its label, where $\mathcal{Y} = \{1, ..., m\}$. In our setting, $\mathcal{D}_u$ consists of two parts, namely in-class samples and out-of-class samples. For out-of-class data, their labels do not belong to $\mathcal{Y}$. The labeled set $\mathcal{D}_l$ only contains in-class labeled samples. The goal is to assign a label $\hat{y} \in \mathcal{Y} \bigcup \{-1\}$ to each unlabeled sample in $\mathcal{D}_u$, where $\hat{y} = -1$ indicates an unlabeled sample is identified as an outlier.

To construct an affinity graph $\mathcal{G} = (\mathcal{V}, \mathcal{E})$ on this dataset, we treat each sample as a vertex and connect it with its K nearest neighbors. The graph $\mathcal{G}$ can be expressed by an adjacency matrix $\mathbf{A} \in \mathbb{R}^{N \times N}$, where $a_{i,j} = \mathbf{A}(i,j)$ is the cosine similarity between $\mathbf{x}_i$ and $\mathbf{x}_j$ if $(i,j) \in \mathcal{E}$, and otherwise 0.

For label propagation, we associate each vertex with a probability vector $\mathbf{p}_i$, where $p_{ik} = \mathbf{p}_i(k)$ indicates the probability of the sample $\mathbf{x}_i$ belonging to the k-th class, and a confidence score $c_i \in [0, 1]$. For labeled samples, $\mathbf{p}_i$ is fixed to be a one-hot vector with $p_{ik} = 1$ for $k = y_i$. For unlabeled samples, $\mathbf{p}_i$ is initialized to be a uniform distribution over all classes and will be gradually updated as the propagation proceeds. We set a threshold o_τ to determine whether a sample is an outlier. After the propagation is completed, for those with confidence smaller than o_τ, the predicted label for each unlabeled sample $\mathbf{x}_i$ is set to $\hat{y}_i = -1$. For those with confidence larger than o_τ. the predicted label for each unlabeled sample $\mathbf{x}_i$ is set to be the class with highest probability in $\mathbf{p}_i$, as $\hat{y}_i = \mathrm{argmax}_k\, p_{ik}$.

3.2 Algorithm Overview

As shown in Fig. 2, our proposed propagation scheme is based on graph patches as the units for updating. Here, a *graph patch* is a sub-graph containing both labeled and unlabeled vertices. The algorithm performs updates over a graph patch in each step of propagation.

The propagation proceeds as follows. (1) At each iteration, we first randomly select a vertex from the *high-confidence vertex set S*, which contains both the initially labeled samples and those samples whose confidences are high enough to be considered as "labeled" as the propagation proceeds. (2) Starting from the selected *confident vertex*, we use a patch extractor to expand it into a graph patch, and then update the predictions on all unlabeled vertices in this patch, using a local predictor. (3) The path scheduler uses these predictions to re-estimate confidences for unlabeled vertices. In this work, both the local predictor and the path scheduler are formulated as a graph convolutional network (GCN) learned from the training data, in order to cope with the complexity of local graph structures. All the vertices whose confidence scores go beyond a threshold c_τ will be added into S and their predictions will not be updated again in future iterations. Note that the updated confidences would influence the choice of the next confident vertex and thus the propagation path. By iteratively updating predictions and confidences as above, the algorithm drives the propagation across the entire graph, gradually from high confident areas to the rest.

This propagation algorithm involves two components: a *local predictor* that generates confident graph patches and updates predictions thereon, and a *path scheduler* that estimates confidences and schedules the propagation path accordingly. Next, we will elaborate on these components in turn.

3.3 GCN-Based Local Predictor

Patch Extractor. A graph patch with the following properties is a good candidate for the next update. (1) *High confidence:* We define the *confidence of a graph patch* as the sum of its vertex confidences. A patch with high confidence is more likely to yield reliable predictions due to the availability of reliable information sources. (2) *Large expected confidence gain*: We define the *estimated confidence gain* of a patch $\mathcal{P}_i$ as $\sum_{v_j \in \mathcal{P}_i}(1 - c_j)$, *i.e.* the maximum possible improvement on the total confidence. Performing updates on those patches with large expected confidence gain can potentially speed up the propagation. To maintain sufficient confidence gain while avoiding excessive patch sizes, we consider a patch as *viable* for the next update if the expected gain is above a threshold Δc_τ and the size is below the maximum size s. Besides, to avoid selecting highly overlapped patches, once a vertex is taken as the start point, its m-hop neighbors will all be excluded from selecting as start points in later propagation.

To generate a graph patch $\mathcal{P}$, we start from the most confident vertex and add its immediate neighbors into a queue. Each vertex in the queue continues to search its unvisited neighbors until (1) the expected gain is above Δc_τ, which means that a viable patch is obtained; or (2) the size exceeds s, which means that no viable patch is found around the selected vertex and the algorithm randomly selects a new vertex from $\mathcal{S}$ to begin with. Note that our propagation can be parallelized by selecting multiple non-overlapped patches at the same time. We show the detailed algorithm in supplementary.

Graph patches are dynamically extracted along with the propagation. In early iterations, Δc_τ can often be achieved by a small number of unlabeled vertices, as most vertices are unlabeled and have low confidences. This results in more conservative exploration at the early stage. As the propagation proceeds, the number of confident vertices increases while the average expected confidence gain decreases, the algorithm encourages more aggressive updates over larger patches. Empirically, we found that on an affinity graph with $10K$ vertices with 1% of labeled seeds, it takes about 100 iterations to complete the propagation procedure, where the average size of graph patches is $1K$.

Design of Local Predictor. We introduce a graph convolutional network (GCN) to predict unknown labels for each graph patch. Given a graph patch $\mathcal{P}_i$ centered at $v_i \in \mathcal{V}$, the network takes as input the visual features $\mathbf{x}_i$, and the affinity sub-matrix restricted to $\mathcal{P}_i$, denoted as $\mathbf{A}(\mathcal{P}_i)$. Let $\mathbf{F}_0(\mathcal{P}_i)$ be a matrix of all vertex data for $\mathcal{P}_i$, where each row represents a vertex feature $\mathbf{x}_i$. The GCN takes $\mathbf{F}_0(\mathcal{P}_i)$ as the input to the bottom layer and carries out the computation through L blocks as follows:

$$\mathbf{F}_{l+1}(\mathcal{P}_i) = \sigma\left(\tilde{\mathbf{D}}(\mathcal{P}_i)^{-1}\tilde{\mathbf{A}}(\mathcal{P}_i)\mathbf{F}_l(\mathcal{P}_i)\mathbf{W}_l\right), \tag{1}$$

where $\tilde{\mathbf{A}}(\mathcal{P}_i) = \mathbf{A}(\mathcal{P}_i) + \mathbf{I}$; $\tilde{\mathbf{D}} = \sum_j \tilde{\mathbf{A}}_{ij}(\mathcal{P}_i)$ is a diagonal degree matrix; $\mathbf{F}_l(\mathcal{P}_i)$ contains the embeddings at the l-th layer; $\mathbf{W}_l$ is a matrix to transform the embeddings; σ is a nonlinear activation (*ReLU* in this work). Intuitively, this formula expresses a procedure of taking weighted average of the features of each vertex and its neighbors based on affinity weights, transforming them into a new space with $\mathbf{W}_l$, and then feeding them through a nonlinear activation. Note that this GCN operates locally within a graph patch and thus the demand on memory would not increase as the whole graph grows, which makes it easy to scale out to massive graphs with millions of vertices.

As the propagation proceeds, each vertex may be covered by multiple patches, including those constructed in previous steps. Each patch that covers a vertex v is called a *view* of v. We leverage the predictions from multiple views for higher reliability, and update the probability vector for each unlabeled vertex in $\mathcal{P}_i$ by averaging the predictions from all views, as

$$\mathbf{p}_i = \frac{1}{\sum \mathbb{1}_{v_i \in \mathcal{P}_j}} \sum_{v_i \in \mathcal{P}_j} \mathbf{F}_L(v_{i,j}). \tag{2}$$

3.4   Confidence-Based Path Scheduler

Confidence estimation is the core of the path scheduler. A good estimation of confidences is crucial for reliable propagation, as it allows unreliable sources to be suppressed. Our confidence estimator involves a *Multi-view* confidence estimator and a learnable *ConfNet*, to form a two-stage procedure. Specifically, the former generates an initial confidence estimation by aggregating predictions from multiple patches. Then ConfNet learns from the most confident samples and the most unconfident ones from the first stage, to further refine the confidence. The ultimate confidence is the average confidence of these two stages.

Multi-view Confidence Estimation. Previous studies [8, 22] have shown that neural networks usually yield over-confident predictions. In this work, we develop a simple but effective way to alleviate the over-confidence problem. We leverage the multiple views for each vertex v_i derived along the propagation process. Particularly, the confidence for v_i is defined as

$$c_i = \begin{cases} \max_k p_{ik}, & \text{if } v_i \text{ was visited multiple times} \\ \epsilon, & \text{if } v_i \text{ was visited only once} \end{cases} \tag{3}$$

where $\mathbf{p}_i$ is given in Eq. (2), and ϵ is a small positive value.

Here, we discuss why we use c_i as defined above to measure the confidence. When a vertex has only been visited once, it is difficult to assess the quality of the prediction, therefore it is safe to assume a low confidence. When a vertex has been visited multiple times, a high value of $\max_k p_{ik}$ suggests that the predictions from different views are consistent with each other. If not, *i.e.* different views vote for different classes, then the average probability for the best class would be significantly lower. We provide a proof in the supplementary showing that c_i takes a high value only when predictions are consistent and all with low entropy.

ConfNet. Among the initial confidence estimated from previous stage, the most confident samples are most likely to be genuine members while the most unconfident samples are most likely to be outliers, which can be regarded as positive samples and negative samples, respectively.

ConfNet is introduced to learn from the "discovered" genuine members and outliers. It aims to output a probability value for each vertex v to indicate how likely it is a genuine member instead of an outlier. Similar to the local predictor, we implement ConfNet as a graph convolutional network, following Eq. (1). Given a percentage η and sampled graph patches, we take the top-η confident vertices as the positive samples and the top-η unconfident vertices as the negative ones. Then we train the ConfNet using the vertex-wise binary cross-entropy as the loss function. The final confidence of a vertex is estimated as the average of multi-view confidence and the predicted confidence from the learned ConfNet.

3.5   Training of Local Predictor

Here we introduce how to train the local predictor. The training samples consist of graph patches with at least one labeled vertex. Instead of selecting graph patches consecutively during propagation, we sample a set of graph patches parallel for training. The sampling of graph patches follows the same principle, *i.e.* , selecting those with high confidence. Based on the sampled subgraphs, the local predictor predicts labels for all labeled vertices on sampled subgraphs. The cross-entropy error between predictions and ground-truth is then minimized over all labeled data to optimize the local predictor.

4   Experiments

4.1   Experimental Settings

Dataset. We conduct our experiments on two real-world datasets, namely, ImageNet [6] and Ms-Celeb-1M [9]. ImageNet compromises $1M$ images from $1,000$ classes, which is the most widely used image classification dataset. Ms-Celeb-1M is a large-scale face recognition dataset consisting of $100K$ identities, and each identity has about 100 facial images. Transductive learning in vision tasks considers a practical setting that obtains a pretrained model but its training data are unavailable. Given only the pretrained model and another unlabeled set with limited labeled data, it aims to predict labels for the unlabeled set. We simulate this setting with the following steps: (1) We randomly sample 10% data from ImageNet to train the feature extractor $\mathcal{F}$. (2) We use $\mathcal{F}$ to extract features for the rest 90% samples to construct $\mathcal{D}_{all}$. (3) We randomly sample 10 classes from $\mathcal{D}_{all}$ as $\mathcal{D}$, and randomly split 1% data from $\mathcal{D}$ as the labeled set $\mathcal{D}_l$. (4) With a noise ratio ρ, we construct the outlier set $\mathcal{D}_o$ by randomly sampling data from $\mathcal{D}_{all} \setminus \mathcal{D}$. (5) $\mathcal{D}_u$ is a union set of $\mathcal{D} \setminus \mathcal{D}_l$ and $\mathcal{D}_o$. Experiments on Ms-Celeb-1M follow the same setting except sampling 100 classes. We sample a small validation set $\mathcal{D}_v$ with the same size as $\mathcal{D}_l$, to determine the outlier threshold o_τ. To evaluate performance on graphs with different noise ratio, we set the noise ratio ρ to 0%, 10%, 30% and 50%.

Metrics. We assess the performance under the noisy transductive learning. Given the ground-truth of the unlabeled set, where the ground-truth of out-of-class outliers is set to -1, transductive learning aims to predict the label of each sample in $\mathcal{D}_u$, where the performance is measured by *top*-1 accuracy.

Implementation Details. We take ResNet-50 [11] as the feature extractor in our experiments. $K = 30$ is used to build the KNN affinity graph. c_τ is set to 0.9 as the threshold to fix high confident vertices. s and Δc_τ for generating graph patches is 3000 and 500. We use SGCs [39] for both local predictor and ConfNet. The depth of SGC is set to 2 and 1 for local predictor and ConfNet, respectively. The Adam optimizer is used with a start learning rate 0.01 and the training epoch is set to 200 and 100 for local predictor and ConfNet, respectively.

4.2   Method Comparison

We compare the proposed method with a series of transductive baselines. Since all these methods are not designed for noisy label propagation, we adapt them to this setting by adopting the same strategy as our method. Specifically, we first determine the outlier threshold o_τ on a validation set $\mathcal{D}_v$, and then take the samples whose confidence below the threshold o_τ as the noisy samples. The methods are briefly described below.

(1) LP [49] is the most widely used transductive learning approach, which aggregates the labels from the neighborhoods by linear combination. **(2) GCN** [21] is devised to capture complex graph structure, where each layer consists of a non-linear transformation and an aggregation function.
(3) GraphSAGE [10] is originally designed for node embedding, which applies trainable aggregation functions on sampled neighbors. We adapt it for transductive learning by replacing the unsupervised loss with the cross-entropy loss.
(4) GAT [38] introduces a self-attention mechanism to GCN, which enables specifying different weights to different nodes in a neighborhood.
(5) FastGCN [4] addresses the memory issue of GCN by a sampling scheme. Compared with GraphSAGE, it saves more memory by sampling vertices rather than neighbors at each layer. **(6) SGC** [39] simplifies the non-linear transformation of GCN, which comprises a linear local smooth filter followed by a standard linear classifier.
(7) Ours incorporates a local predictor and a confidence-based path scheduler. The two closely coupled components learn to propagate on noisy graphs reliably.

Table 1. Performance comparison of transductive methods on noisy affinity graphs. GraphSAGE† denotes using GCN as the aggregation function. For both ImageNet and Ms-Celeb-1M, 1% labeled images are randomly selected as seeds. We randomly select classes and initial seeds for 5 times and report the average results of 5 runs (see supplementary for the standard deviation of all experiments)

Noise ratio ρ	ImageNet				Ms-Celeb-1M			
	0%	10%	30%	50%	0%	10%	30%	50%
LP [49]	77.74	70.51	59.47	51.43	95.13	89.01	88.31	87.19
GCN [21]	83.17	75.37	66.28	64.09	99.6	99.6	96.37	96.3
GAT [38]	83.93	75.99	66.3	63.34	99.59	96.48	94.55	94.01
GraphSAGE [10]	82.42	73.42	63.84	59.12	99.57	95.68	92.21	91.06
GraphSAGE† [10]	81.39	73.53	63.42	58.99	99.59	95.62	92.38	91.19
FastGCN [4]	81.34	74.08	63.79	58.81	99.62	95.6	92.08	90.83
SGC [39]	84.78	76.71	67.97	65.63	99.63	97.43	96.71	96.5
Ours	**85.16**	**76.96**	**69.28**	**68.25**	**99.66**	**97.59**	**96.93**	**96.81**

Results. Table 1 shows that: (1) For LP, the performance is inferior to other learning-based approaches. (2) GCN shows competitive results under different settings, although it is not designed for the noisy scenario. (3) We employ Graph-SAGE with GCN aggregation and mean aggregation. Although it achieves a higher speedup than GCN, not considering the confidence of predictions makes the sampling-based method very sensitive to outliers. (4) Although GAT yields promising results when the graph size is $20K$, it incurs excessive memory demand when scaling to larger graphs, as shown in Fig. 3. Despite FastGCN is efficient, it suffers from the similar problem as GraphSAGE. (6) SGC, as a simplified version of GCN, achieves competitive results to GCN and GAT. As it has less training parameters, it may not easily overfit when the initial seeds are sparse. Figure 3 indicates that the performance of SGC becomes inferior to GCN when the graph size becomes large. (7) Table 1 illustrates that the noisy setting is very challenging, which deteriorates the performance of all algorithms marginally. The proposed method improves the accuracy under all noise ratios, with more significant improvement as the noise ratio becomes larger. It not only surpasses the sampling-based approaches by a large margin, but also outperforms the GNNs with the entire graph as inputs. Even in the well-learned face manifold, which is less sensitive to out-of-sample noise, our method still reduces the error rate from 3.5% to 3.19%. Note that the proposed method can be easily extended to the iterative scheme by using self-training [30]. As it can effectively estimate confidence, applying it iteratively can potentially lead to better results.

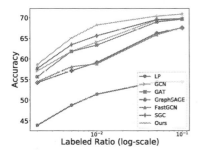
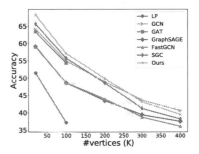

Fig. 3. Influence of labeled ratio   **Fig. 4.** Influence of graph size

Labeled Ratio. When the noise ratio ρ is 50%, we study the influence of different labeled ratios: 0.2%, 0.5%, 1%, 5% and 10%. Figure 3 shows that our method consistently outperforms other methods under all labeled ratios. When the initial seeds are very sparse, it becomes more challenging for both label propagation and confidence estimation. As our method learns from the discovered confident and unconfident samples along with the propagation, our method still performs well when there are a few initial seeds.

Graph Scale. The local update design makes the proposed method capable of scaling to large-scale graphs. As Fig 4 illustrates, LP suffers a severe performance drop when the graph size increases. GAT exceeds the memory limits when the number of vertices is beyond $100K$. Two sampling-based methods, GraphSAGE

Table 2. Comparison on local predictors and confidences. ConfNet† computes confidence as the average confidence from Multi-view and ConfNet

Confidence	GAT	GCN	SGC
Random	63.16	62.62	64.84
Multi-view	64.11	63.84	65.81
ConfNet	64.83	63.93	67.79
ConfNet†	65.95	65.21	68.25
GT	83.17	83.93	84.78

Table 3. Comparison on different source of initial confidence. FNR denotes *false noise ratio* of positive samples and TNR denotes *true noise ratio* of negative samples

Initial Confidence	Num	FNR	TNR	Acc
SGC($\eta=0.05$)	973	3.8%	66%	67.44
Multi-view($\eta=0.01$)	194	1.6%	70%	66.78
Multi-view($\eta=0.05$)	973	3.2%	65%	68.79
Multi-view($\eta=0.1$)	1947	4.1%	63%	67.81
GT($\eta=0.05$)	973	0%	100%	76.29

and FastGCN, are inferior to their counterparts operating on the entire graph. Although our method also operates on subgraphs, the reliable strategy enables it to perform well on noisy graphs under different scales. Note that when the graph size is $400K$, GCN performs better than ours. As we adopt SGC as the local predictor in our experiments, without non-linear transformation may limit its capability when graph scale is large. In real practice, we have the flexibility to select different local predictors according to the graph scale.

4.3   Ablation Study

We adopt a setting on ImageNet, where the labeled ratio is 1% and the noise ratio is 50%, to study some important designs in our framework.

Local Predictor. In our framework, the local predictor can be flexibly replaced with different graph-based algorithms. We compare the effectiveness of three learnable local predictors, namely GAT, GCN, and SGC. All three methods take the vertex features as input, and predict labels for unlabeled vertices. Comparing different columns in Table 2, all three local predictors outperforms LP (see Table 1) significantly, even using random confidence. The results demonstrate the advantage of learning-based approaches in handling complex graph structure.

Path Scheduler. As shown in different rows in Table 2, we study confidence choices with different local predictors. (1) *Random* refers to using random score between 0 and 1 as the confidence, which severely impairs the performance. (2) *Multi-view* denotes our first stage confidence estimation, *i.e.* , aggregating predictions from multiple graph patches, which provides a good initial confidence. (3) *ConfNet* indicates using the confidence predicted from ConfNet. Compared to Multi-view, the significant performance gain demonstrates the effectiveness of ConfNet. (4) *ConfNet†* is the ultimate confidence in our approach. It further increases the performance by averaging confidence from two previous stages, which shows that the confidence from Multi-view and ConfNet may be complementary to some extent. (5) *GT (Ground-truth)* denotes knowing all outliers in advance, which corresponds to the setting that noise ratio is 0 in Table 1. It indicates that the performance can be greatly boosted if identifying all outliers correctly.

Confidence Estimation. Table 3 analyzes the source of initial confidence for ConfNet training. η denotes the proportion of the most confident and unconfident samples, as defined in Sect. 3.4. SGC refers to using the prediction probabilities without Multi-view strategy. It shows that: (1) Comparison between SGC ($\eta = 0.05$) and Multi-view ($\eta = 0.05$) indicates that ConfNet is affected by the quality of initial confidence set. As Multi-view gives more precise confidence estimation, it provides more reliable samples for ConfNet training, leading to a better performance. (2) Comparison between GT ($\eta = 0.05$) and Multi-view ($\eta = 0.05$) further indicates that training on a reliable initial confidence set is a crucial design. (3) Comparison between Multi-view with three different η shows that choosing a proper proportion is important to ConfNet training. When η is small, although the positive and negative samples are more pure, training on a few samples impairs the final accuracy. When η is large, the introduction of noise in both positive and negative samples lead to the limited performance gain.

From another perspective, Fig 5 illustrates that the success of Multi-view and ConfNet is mainly due to altering the confidence distribution, where the gap between outliers and genuine members is enlarged and thus outliers can be identified more easily. Figure 7 shows that using ConfNet as a post-processing module in previous methods can also improve their capability of identifying outliers, leading to a significant accuracy gain with limited computational budget.

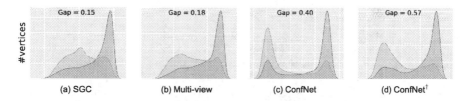

Fig. 5. Confidence distribution of outliers and genuine members. Orange represents the out-of-class noisy samples, while blue denotes the in-class unlabeled ones. Gap is computed as the difference between the mean of two distributions. It indicates that the proposed confidence estimation approach can enlarge the confidence gap between outliers and genuine ones, which is the key to our performance gain (Color figure online)

4.4 Further Analysis

Efficiency of Path Extraction. We refer to *visited times* of a vertex as the number of patches it belongs to. We conduct experiments on ImageNet with $10K$ vertices with $c_\tau = 0.9, \Delta c_\tau = 500$ and $s = 3000$. When propagating 100

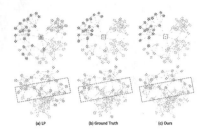

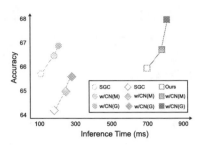

Fig. 6. Two graph patches with predictions from ImageNet, where different colors represent different classes (Color figure online)

Fig. 7. Apply ConfNet to different GNNs. CN(M) denotes ConfNet using MLP and CN(G) denotes ConfNet using GCN

iterations, the average visited times of vertices are about 6. Most samples are visited 2 times and only a very few samples are visited more than 10 times.

Conservative Prediction on Hard Cases. Except the out-of-sample noise, we also visualize the low confident samples when noise ratio is 0. As Fig 6 shows, when dealing with a hard case (the green box in the first row), our method gives the right prediction with very low confidence (small size of vertices) while LP gives a wrong prediction and misleads the predictions of downstream vertices. The second row shows that our confidence can identify inter-class boundaries, and remain conservative to these predictions, as highlighted in the green box.

Table 4. Two applications of our method in vision tasks. (a) We use the estimated confidence as indicators in active learning. (b) We apply the predicted labels to face recognition training in an inductive manner. (see supplementary for more details)

Labeled	Baseline	Random	GCN	Ours
1%	65.63	65.71	66.7	68.6

(a)

Test Protocol	Baseline	CDP	GCN	Ours
MegaFace [20]	58.21	59.15	59.33	60.02

(b)

4.5   Applications

Active Learning. Active learning desires an effective indicator to select representative unlabeled samples. Table 4(a) shows that our estimated confidence outperforms two widely used indicators. Specifically, the first one *randomly* selects unlabeled samples for annotation, while the second one applies a trained *GCN* to unlabeled samples and select those with large predicted entropy. *Baseline* refers to the accuracy before annotation. The result shows that our method brings larger accuracy gain by annotating the same number of unlabeled data.

Inductive Learning. The predicted labels from transductive learning can be used as "pseudo labels" in inductive learning. We randomly selects $1K$ person with $120K$ images from Ms-Celeb-1M, sampling 1% as the labeled data. We compare with CDP [45] and GCN [21] for generating "pseudo labels". Compared to these two methods, Table 4(b) shows our method brings larger performance gain on MegaFace [20], which demonstrates that the proposed method generates pseudo labels with higher quality.

5   Conclusion

In this paper, we propose a reliable label propagation approach to extend the transductive learning to a practical noisy setting. The proposed method consists of two learnable components, namely a GCN-based local predictor and a confidence-based path scheduler. Experiments on two real-world datasets show that the proposed approach outperforms previous state-of-the-art methods with reasonable computational cost. Ablation study shows that exploiting both confident and unconfident samples is a crucial design in our confidence estimation. Extending the proposed method to different kinds of noise, such as adversarial noise [27], is desired to be explored in the future.

Acknowledgment. This work is partially supported by the SenseTime Collaborative Grant on Large-scale Multi-modality Analysis (CUHK Agreement No. TS1610626 & No. TS1712093), the General Research Fund (GRF) of Hong Kong (No. 14203518 & No. 14205719).

References

1. Belkin, M., Niyogi, P., Sindhwani, V.: Manifold regularization: a geometric framework for learning from labeled and unlabeled examples. J. Mach. Learn. Res. **7**(Nov), 2399–2434 (2006)
2. Bojchevski, A., Günnemann, S.: Deep gaussian embedding of graphs: unsupervised inductive learning via ranking. arXiv preprint arXiv:1707.03815 (2017)
3. Chandola, V., Banerjee, A., Kumar, V.: Anomaly detection: a survey. ACM Comput. Surv. **41**(3), 1–58 (2009)
4. Chen, J., Ma, T., Xiao, C.: FastGCN: fast learning with graph convolutional networks via importance sampling. arXiv preprint arXiv:1801.10247 (2018)
5. Cheng, H.T., et al.: Wide and deep learning for recommender systems. In: Proceedings of the 1st Workshop on Deep Learning for Recommender Systems, pp. 7–10 (2016)
6. Deng, J., Dong, W., Socher, R., Li, L.J., Li, K., Fei-Fei, L.: ImageNet: a large-scale hierarchical image database. In: IEEE Conference on Computer Vision and Pattern Recognition, 2009, CVPR 2009, pp. 248–255. IEEE (2009)
7. Deng, J., Guo, J., Zafeiriou, S.: Arcface: additive angular margin loss for deep face recognition. arXiv preprint arXiv:1801.07698 (2018)
8. Gal, Y., Ghahramani, Z.: Dropout as a Bayesian approximation: representing model uncertainty in deep learning. In: International Conference on Machine Learning, pp. 1050–1059 (2016)

9. Guo, Y., Zhang, L., Hu, Y., He, X., Gao, J.: MS-Celeb-1M: a dataset and benchmark for large-scale face recognition. In: Leibe, B., Matas, J., Sebe, N., Welling, M. (eds.) ECCV 2016. LNCS, vol. 9907, pp. 87–102. Springer, Cham (2016). https://doi.org/10.1007/978-3-319-46487-9_6

10. Hamilton, W., Ying, Z., Leskovec, J.: Inductive representation learning on large graphs. In: Advances in Neural Information Processing Systems, pp. 1024–1034 (2017)

11. He, K., Zhang, X., Ren, S., Sun, J.: Deep residual learning for image recognition. In: Proceedings of the IEEE Conference on Computer Vision and Pattern Recognition, pp. 770–778 (2016)

12. Hu, J., Shen, L., Sun, G.: Squeeze-and-excitation networks. In: Proceedings of the IEEE Conference on Computer Vision and Pattern Recognition, pp. 7132–7141 (2018)

13. Huang, H., Zhang, Y., Huang, Q., Guo, Z., Liu, Z., Lin, D.: Placepedia: Comprehensive place understanding with multi-faceted annotations. In: Proceedings of the European Conference on Computer Vision (ECCV) (2020)

14. Huang, Q., Liu, W., Lin, D.: Person search in videos with one portrait through visual and temporal links. In: Proceedings of the European Conference on Computer Vision (ECCV), pp. 425–441 (2018)

15. Huang, Q., Xiong, Y., Lin, D.: Unifying identification and context learning for person recognition. In: The IEEE Conference on Computer Vision and Pattern Recognition (CVPR), June 2018

16. Huang, Q., Xiong, Y., Rao, A., Wang, J., Lin, D.: MovieNet: a holistic dataset for movie understanding. In: Proceedings of the European Conference on Computer Vision (ECCV) (2020)

17. Huang, Q., Xiong, Y., Xiong, Y., Zhang, Y., Lin, D.: From trailers to storylines: an efficient way to learn from movies. arXiv preprint arXiv:1806.05341 (2018)

18. Huang, Q., Yang, L., Huang, H., Wu, T., Lin, D.: Caption-supervised face recognition: training a state-of-the-art face model without manual annotation. In: Proceedings of the European Conference on Computer Vision (ECCV) (2020)

19. Iscen, A., Tolias, G., Avrithis, Y., Chum, O.: Label propagation for deep semi-supervised learning. In: Proceedings of the IEEE Conference on Computer Vision and Pattern Recognition, pp. 5070–5079 (2019)

20. Kemelmacher-Shlizerman, I., Seitz, S.M., Miller, D., Brossard, E.: The megaface benchmark: 1 million faces for recognition at scale. In: CVPR (2016)

21. Kipf, T.N., Welling, M.: Semi-supervised classification with graph convolutional networks. arXiv preprint arXiv:1609.02907 (2016)

22. Lakshminarayanan, B., Pritzel, A., Blundell, C.: Simple and scalable predictive uncertainty estimation using deep ensembles. In: Advances in Neural Information Processing Systems, pp. 6402–6413 (2017)

23. Lee, D.H.: Pseudo-label: The simple and efficient semi-supervised learning method for deep neural networks. In: Workshop on Challenges in Representation Learning, ICML. vol. 3, p. 2 (2013)

24. Noble, C.C., Cook, D.J.: Graph-based anomaly detection. In: Proceedings of the Ninth ACM SIGKDD International Conference on Knowledge Discovery and Data Mining, pp. 631–636 (2003)

25. Oliver, A., Odena, A., Raffel, C.A., Cubuk, E.D., Goodfellow, I.: Realistic evaluation of deep semi-supervised learning algorithms. In: Advances in Neural Information Processing Systems, pp. 3239–3250 (2018)

26. Perozzi, B., Al-Rfou, R., Skiena, S.: Deepwalk: Online learning of social representations. In: Proceedings of the 20th ACM SIGKDD International Conference on Knowledge Discovery and Data Mining, pp. 701–710. ACM (2014)
27. Qiu, H., Xiao, C., Yang, L., Yan, X., Lee, H., Li, B.: SemanticAdv: generating adversarial examples via attribute-conditional image editing. arXiv preprint arXiv:1906.07927 (2019)
28. Rao, A., et al.: A unified framework for shot type classification based on subject centric lens. In: Proceedings of the European Conference on Computer Vision (ECCV) (2020)
29. Rao, A., et al.: A local-to-global approach to multi-modal movie scene segmentation. In: Proceedings of the IEEE/CVF Conference on Computer Vision and Pattern Recognition, pp. 10146–10155 (2020)
30. Rosenberg, C., Hebert, M., Schneiderman, H.: Semi-supervised self-training of object detection models. WACV/MOTION 2 (2005)
31. Saunders, C., Gammerman, A., Vovk, V.: Transduction with confidence and credibility (1999)
32. Schlichtkrull, M., Kipf, T.N., Bloem, P., van den Berg, R., Titov, I., Welling, M.: Modeling relational data with graph convolutional networks. In: Gangemi, A., Navigli, R., Vidal, M.-E., Hitzler, P., Troncy, R., Hollink, L., Tordai, A., Alam, M. (eds.) ESWC 2018. LNCS, vol. 10843, pp. 593–607. Springer, Cham (2018). https://doi.org/10.1007/978-3-319-93417-4_38
33. Schroff, F., Kalenichenko, D., Philbin, J.: FaceNet: a unified embedding for face recognition and clustering. In: CVPR (2015)
34. Sun, Y., Chen, Y., Wang, X., Tang, X.: Deep learning face representation by joint identification-verification. In: NeurIPS (2014)
35. Tarvainen, A., Valpola, H.: Mean teachers are better role models: weight-averaged consistency targets improve semi-supervised deep learning results. In: Advances in Neural Information Processing Systems, pp. 1195–1204 (2017)
36. Vapnik, V.: 24 transductive inference and semi-supervised learning (2006)
37. Vashishth, S., Yadav, P., Bhandari, M., Talukdar, P.: Confidence-based graph convolutional networks for semi-supervised learning. arXiv preprint arXiv:1901.08255 (2019)
38. Veličković, P., Cucurull, G., Casanova, A., Romero, A., Lio, P., Bengio, Y.: Graph attention networks. arXiv preprint arXiv:1710.10903 (2017)
39. Wu, F., Zhang, T., Souza Jr, A.H.D., Fifty, C., Yu, T., Weinberger, K.Q.: Simplifying graph convolutional networks. arXiv preprint arXiv:1902.07153 (2019)
40. Wu, T., Huang, Q., Liu, Z., Wang, Y., Lin, D.: Distribution-balanced loss for multi-label classification in long-tailed datasets. In: Proceedings of the European Conference on Computer Vision (ECCV) (2020)
41. Xia, J., Rao, A., Xu, L., Huang, Q., Wen, J., Lin, D.: Online multi-modal person search in videos. In: Proceedings of the European Conference on Computer Vision (ECCV) (2020)
42. Xiong, Y., Huang, Q., Guo, L., Zhou, H., Zhou, B., Lin, D.: A graph-based framework to bridge movies and synopses. In: The IEEE International Conference on Computer Vision (ICCV), October 2019
43. Yang, L., Chen, D., Zhan, X., Zhao, R., Loy, C.C., Lin, D.: Learning to cluster faces via confidence and connectivity estimation. In: Proceedings of the IEEE Conference on Computer Vision and Pattern Recognition (2020)
44. Yang, L., Zhan, X., Chen, D., Yan, J., Loy, C.C., Lin, D.: Learning to cluster faces on an affinity graph. In: Proceedings of the IEEE Conference on Computer Vision and Pattern Recognition, pp. 2298–2306 (2019)

45. Zhan, X., Liu, Z., Yan, J., Lin, D., Loy, C.C.: Consensus-driven propagation in massive unlabeled data for face recognition. In: ECCV (2018)
46. Zhang, X., Yang, L., Yan, J., Lin, D.: Accelerated training for massive classification via dynamic class selection. In: Thirty-Second AAAI Conference on Artificial Intelligence (2018)
47. Zhou, D., Bousquet, O., Lal, T.N., Weston, J., Schölkopf, B.: Learning with local and global consistency. In: Advances in Neural Information Processing Systems, pp. 321–328 (2004)
48. Zhou, J., et al.: Graph neural networks: a review of methods and applications. arXiv preprint arXiv:1812.08434 (2018)
49. Zhu, X., Ghahramani, Z.: Learning from labeled and unlabeled data with label propagation. Technical report, Citeseer (2002)
50. Zhu, X.J.: Semi-supervised learning literature survey. University of Wisconsin-Madison Department of Computer Sciences, Technical report (2005)

Fair DARTS: Eliminating Unfair Advantages in Differentiable Architecture Search

Xiangxiang Chu[1]([⊠])[iD], Tianbao Zhou[2][iD], Bo Zhang[1][iD], and Jixiang Li[1][iD]

[1] Xiaomi AI Lab, Beijing, China
{chuxiangxiang,zhangbo11,lijixiang}@xiaomi.com
[2] Minzu University of China, Beijing, China
tianbaochou@163.com

Abstract. Differentiable Architecture Search (DARTS) is now a widely disseminated weight-sharing neural architecture search method. However, it suffers from well-known performance collapse due to an inevitable aggregation of skip connections. In this paper, we first disclose that its root cause lies in an **unfair advantage** in **exclusive competition**. Through experiments, we show that if either of two conditions is broken, the collapse disappears. Thereby, we present a novel approach called Fair DARTS where the exclusive competition is relaxed to be collaborative. Specifically, we let each operation's architectural weight be independent of others. Yet there is still an important issue of discretization discrepancy. We then propose a **zero-one** loss to push architectural weights towards zero or one, which approximates an expected multi-hot solution. Our experiments are performed on two mainstream search spaces, and we derive new state-of-the-art results on CIFAR-10 and ImageNet (Code is available here: https://github.com/xiaomi-automl/FairDARTS).

Keywords: Differentiable neural architecture search · Image classification · Failure of DARTS

1 Introduction

In the wake of the DARTS's open-sourcing [19], a diverse number of its variants emerge in the *neural architecture search* community. Some of them extend its use in higher-level architecture search spaces with performance awareness in mind [3,31], some learn a stochastic distribution instead of architectural parameters [9,10,31,32,35], and others offer remedies on discovering its lack of robustness [4,15,18,22,34].

T. Zhou and B. Zhang—Equal Contribution.

Electronic supplementary material The online version of this chapter (https://doi.org/10.1007/978-3-030-58555-6_28) contains supplementary material, which is available to authorized users.

© Springer Nature Switzerland AG 2020
A. Vedaldi et al. (Eds.): ECCV 2020, LNCS 12360, pp. 465–480, 2020.
https://doi.org/10.1007/978-3-030-58555-6_28

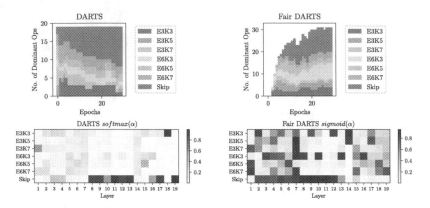

Fig. 1. Top: Stacked area plot of the number of dominant operations (In DARTS, it refers to the one with the highest architectural weight. In FairDARTS, it means the one whose $\sigma > \sigma_{threshold}$. Here we use $\sigma_{threshold} = 0.75$.) of DARTS and Fair DARTS when searching on ImageNet in search space S_2 (19 searchable layers). **Bottom:** Heatmaps of softmax (DARTS) and sigmoid (Fair DARTS) values in the last searching epoch. DARTS finally chooses a shallow model (11 layers removed by activating skip connections only) which obtains 66.4% top-1 accuracy. While in Fair DARTS, all operations develop independently that an excessive number of skip connections no longer leads to poor performance. Here it infers a deeper model (only one layer is removed) with 75.6% top-1 accuracy

In spite of these endeavors, the aggregation of skip connections in DARTS that noticed by [1,4,18,34] has not been solved with perfection. Observing that the aggregation leads to a dramatic **performance collapse** for the resulting architecture, P-DARTS [4] utilizes dropout as a workaround to restrict the number of skip connections during optimization. DARTS+ [18] directly puts a hard limit of two skip-connections per cell. RobustDARTS [34] finds out that these solutions coincide with high validation loss curvatures. To some extent, these approaches consider the poor-performing models as impurities from the solution set, for which they intervene in the training process to filter them out.

On the contrary, we extend the solution set and revise the optimization process so that aggregation of skip connections no longer causes the collapse. Moreover, there remains a **discrepancy problem** when discretizing continuous architecture encodings. DARTS [19] leaves it as future work, but till now it has not been deeply studied. We reiterate the basic premise of DARTS is that the continuous solution approximates a one-hot encoding. Intuitively, the smaller discrepancies are, the more consistent it will be when we transform a continuous solution back to a discrete one. We summarize our contributions as follows:

Firstly, we disclose the root cause that leads to the collapse of DARTS, which we later define as an *unfair advantage* that drives skip connections into a monopoly state in *exclusive competition*. These two indispensable factors work together to induce a performance collapse. Moreover, if either of the two conditions is broken, the collapse disappears.

Secondly, we propose the first **collaborative competition** approach by offering each operation an independent architectural weight. The unfair advantage no longer prevails as we break the second factor. Furthermore, to address the discrepancy between the continuous architecture encoding and the derived discrete one in our method, we propose a novel auxiliary loss, called *zero-one loss*, to steer architectural weights towards their extremities, that is, either completely enabled or disabled. The discrepancy thus decreases to its minimum.

Thirdly, based on the root cause of the collapse, we provide a unified perspective to view current DARTS cures for skip connections' aggregation. The majority of these works either make use of dropout [27] on skip connections [4,34], or play with the later termed *boundary epoch* by different early-stopping strategies [18,34]. They can all be regarded as preventing the first factor from taking effect. Moreover, as a direct application, we can derive a hypothesis that adding Gaussian noise also disrupts the unfairness, which is later proved to be effective.

Lastly, we conduct thorough experiments in two widely used search spaces in both proxy and proxyless ways. Results show that our method can escape from performance collapse. We also achieve state-of-the-art networks on CIFAR-10 and ImageNet.

2   Related Work

Lately, neural architecture search [36] has grown as a well-formed methodology to discover networks for various deep learning tasks. Endeavors have been made to reduce the enormous searching overhead with the weight-sharing mechanism [2,19,23]. Especially in DARTS [19], a nested gradient-descent algorithm is exploited to search for the graphical representation of architectures, which is born from gradient-based hyperparameter optimization [20].

Due to the limit of the DARTS search space, ProxylessNAS [3] and FBNet [31] apply DARTS in much larger search spaces based on MobileNetV2 [26]. ProxylessNAS also differs from DARTS in its supernet training process, where only two paths are activated, based on the assumption that one path is the best amongst all should be better than any single one. From a fairness point of view, as only two paths enhance their ability (get parameters updated) while others remain unchanged, it implicitly creates a bias. FBNet [31], SNAS [32] and GDAS [10] utilize the differentiable Gumbel Softmax [14,21] to mimic one-hot encoding. However, the one-hot nature implies an exclusive competition, which risks being exploited by unfair advantages.

Superficially, the most relevant work to ours is RobustDARTS [34]. Under several simplified search spaces, they state that the found solutions generalize poorly when they coincide with high validation loss curvature, where the supernet with an excessive number of skip connections happens to be such a solution. Based on this observation, they impose early-stop regularization by tracking the largest eigenvalue. Instead, our method doesn't need to perform early stopping.

3   The Downside of DARTS

In this section, we aim to excavate the disadvantages of DARTS that possibly impede the searching performance. We first prepare a minimum background.

3.1   Preliminary of Differentiable Architecture Search

For the case of convolutional neural networks, DARTS [19] searches for a *normal cell* and a *reduction cell* to build up the final architecture. A cell is represented as a directed acyclic graph (DAG) of N nodes in sequential order. Each node stands for a feature map. The edge $e_{i,j}$ from node i to j operates on the input feature x_i and its output is denoted as $o_{i,j}(x_i)$. The intermediate node j gathers all inputs from the incoming edges,

$$x_j = \sum_{i<j} o_{i,j}(x_i). \tag{1}$$

Let $\mathcal{O} = \{o_{i,j}^1, o_{i,j}^2, ..., o_{i,j}^M\}$ be the set of M candidate operations on edge $e_{i,j}$. DARTS relaxes this categorical choice to a softmax over all operations in $\mathcal{O}$ to form a mixed output:

$$\bar{o}_{i,j}(x) = \sum_{o\in\mathcal{O}} \frac{\exp(\alpha_{o_{i,j}})}{\sum_{o'\in\mathcal{O}} \exp(\alpha_{o'_{i,j}})} o(x), \tag{2}$$

where each operation $o_{i,j}$ is associated with a continuous coefficient $\alpha_{o_{i,j}}$. Regarding edge $e_{i,j}$, this softmax is utilized to approximate one-hot encoding $\beta_{i,j} = (\beta_{o_{i,j}^1}, \beta_{o_{i,j}^2}, ..., \beta_{o_{i,j}^M})$. Formally, let $\alpha_{o_{i,j}}$ denote the architectural weights vector $(\alpha_{o_{i,j}^1}, \alpha_{o_{i,j}^2}, ..., \alpha_{o_{i,j}^M})$. DARTS thus assumes the following as a valid approximation,

$$softmax(\alpha_{o_{i,j}}) \approx \beta_{i,j}. \tag{3}$$

The architecture search problem is reduced to learning α^* and network weights w^* that minimize the validation loss $\mathcal{L}_{val}(w^*, \alpha^*)$. DARTS resolves this problem with a bi-level optimization,

$$\min_{\alpha} \mathcal{L}_{val}(w^*(\alpha), \alpha)$$
$$\text{s.t. } w^*(\alpha) = \text{argmin}_w \mathcal{L}_{train}(w, \alpha). \tag{4}$$

We also adopt two common search spaces, the DARTS [19] search space (S_1) and the ProxylessNAS [3] search space (S_2) with minor modifications. More details are given in Sect. 2 (supplementary).

In S_2, the output of the l-th layer is a softmax-weighted summation of N choices. Formally, it can be written as

$$x_l = \sum_{k=1}^{N} \frac{\exp(\alpha_{l-1,l}^k)}{\sum_{j=1}^{N} \exp(\alpha_{l-1,l}^j)} o_{l-1,l}^k(x_{l-1}). \tag{5}$$

3.2   Performance Collapse Caused by Intractable Skip Connections

DARTS suffers from significant performance decay when *skip connections* become dominant [4, 18]. It was described as a competition-and-cooperation issue in the bi-level optimization [18]. Still, the reason behind this behavior is not clear, we hereby provide a different perspective.

First, to confirm this issue, we run DARTS $k = 4$ times with different random seeds. Following DARTS, we select 8 top-performing operations per cell (2 each for 4 intermediate nodes). Here we say one operation is *dominant* if it has top-2 $softmax(\alpha)$ among all incoming edges' candidates of a certain node. The results are shown in Fig. 2. In the beginning, all operations are given the same opportunity. As the over-parameterized network gradually converges, there is an evident aggregation of skip connections after 20 epochs (5 out of 8 in an extreme case).

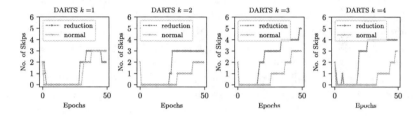

Fig. 2. The number of dominant skip connections continues to grow when searching with DARTS (run $k = 4$ times) on CIFAR-10 (in S_1)

When we utilize DARTS directly on ImageNet in S_2, which is a single branch architecture, the same phenomenon rigorously reappears. The number of dominant skip-connections (highest $softmax(\alpha)$ among all operations in that layer) steadily increases and reaches 11 out of 19 layers in the end, which is shown on the left of Fig. 1.

But why is this happening? The underlying reasons are rarely discussed in depth. A brief and superficial analysis regarding information flow is given in [4]. However, we claim that the reason for excessive skip connections is from **exclusive competition** among various operations. In Eq. 2 and Eq. 5, the skip connection is softmax-weighted and added to the output, which resembles a basic residual module as in ResNet [11]. While this module greatly benefits the training, the architectural weight of a skip connection increases much faster than its competitors. Moreover, *the softmax operation inherently provides an exclusive competition since increasing one is at the cost of suppressing others.* As a result, skip connections become gradually dominant during optimization. We have to keep in mind that skip connection works well because it is in cooperation with convolutions [11]. However, DARTS picks the top-performing one (skip connection here) and discards its collaborator (convolution), which results in a degenerate model.

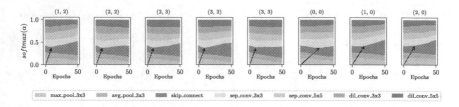

Fig. 3. The softmax evolution where skip connections gradually become dominant when running DARTS on CIFAR-10 (in S_1). Last two subplots of edge $(1, 0)$ and $(2,0)$ are from the normal cell, the rest are from the reduction cell. Black arrows point to boundary epochs where skip connections start to demonstrate its strength

We further study this effect from the experiments on CIFAR-10 by recording the competition progress in Fig. 3. The derived model has 8 skip connections in total[1]. ResNet [11] discovers that *skip connections begin to demonstrate power after a few epochs compared with models without them.* Interestingly, a similar phenomenon is also observed in our experiments. We term this tipping point a *boundary epoch*. The boundary epochs may vary from edge to edge, but are generally at the early stage. From Fig. 3, we observe that skip connections colored in red-orange progressively obtain higher architectural weights after some certain boundary epochs. Meantime, other operations are suppressed and steadily decline. We consider this benefit from the residual module as an unfair advantage by Definition 1.

Definition 1. *Unfair Advantage.* *Suppose that choosing one operation among others is a competition. This competition is deemed **exclusive** when only restricted operations can be selected. An operation in an exclusive competition is said to have an **unfair advantage** if this advantage contributes more to competition than to the performance of a resulted network.*

From the above discussion, we can draw **Insight 1: The root cause of excessive skip connections is the inherent unfair competition.** The skip connection has an unfair advantage by forming a residual module which is convenient for the supernet training, but not equally beneficial for the performance of the outcome network where the residual module is broken.

3.3   Non-negligible Discrepancy of Discretization

Apart from the above issue, DARTS reports that it suffers from discrepancies when discretizing continuous encodings [19]. To verify the problem, we run DARTS in S_1 on CIFAR-10, and in S_2 on ImageNet. The values of $softmax(\alpha)$ of the last iteration are displayed in Fig. 4 (S_1) and on the bottom left of Fig. 1 (S_2). For S_1, the largest value is about 0.3 while the smallest one is above 0.1[2].

---

[1] Corresponding to the experiment ($k = 3$) in Fig. 2.
[2] We run DARTS 4 times and it holds every time.

This range is somewhat too narrow to differentiate 'good' operations from 'bad'. For instance on edge 2 of the reduction cell, the values are very close to each other, [0.174, 0.170, 0.176, 0.112, 0.116, 0.132, 0.118], it's hard to say that an operation weighted by 0.176 is better than the other by 0.174. For S_2, the top-1 values are not so evidently particular from layer 2 to 7. Take the second layer for example, we have to use [0.235, 0.057, 0.17, 0.016, 0.187, 0.269, 0.066] to approximate [0, 0, 0, 0, 0, 1, 0]. This again confirms the existence of discrepancy.

In summary, DARTS is usually far from a good resemblance to a one-hot representation as required by its premise in Eq. 3. We often have to make ambiguous choices without high confidence. Hence, we learn **Insight 2: Relaxing from discrete categorical choices to continuous ones should make a close approximation**.

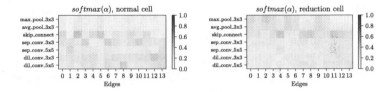

Fig. 4. Heatmap of softmax values in the normal cell and the reduction cell at the last searching epoch when running DARTS on CIFAR-10 (in search space S_1)

4   Fair DARTS

4.1   Stepping Out the Pitfalls of Skip Connections

Based on **Insight 1**, we propose a *cooperative mechanism* to eliminate the existing unfair advantage. Not only should we exploit skip connection for smoother information flow, but we also have to provide equal opportunities for other operations. In a word, they need to avoid being trapped by unfair advantage from skip connections. On this regard, we apply a *sigmoid activation* (σ) for each $\alpha_{i,j}$, so that each operation can be switched on or off independently without being suppressed. Formally, we replace Eq. 2 with the following,

$$\bar{o}_{i,j}(x) = \sum_{o \in \mathcal{O}} \sigma(\alpha_{o_{i,j}})o(x). \tag{6}$$

It's trivial to show that even if $\sigma(\alpha_{skip})$ saturates to 1, other operations still can be optimized cooperatively. Promising operations continue to grow their architectural weights to reduce $\mathcal{L}_{val}$, which leads to a **multi-hot** approximation. Instead, DARTS attempts to derive a one-hot estimation. The difference is that we have extended the solution set. Consequently, it allows us to tackle the discretization discrepancy. We are left to find out how to drive $\sigma(\alpha)$ towards each extremity (0 or 1). Next, we discuss it in greater detail.

4.2   Resolve Discrepancy from Continuous Representation to Discrete Encoding

To abide by **Insight 2**, we explicitly coerce an extra loss called *zero-one loss* to push the sigmoid value of architectural weights towards 0 or 1. Let $L_{0-1} = f(z)$ denote this loss component, where $z = \sigma(\alpha)$. To achieve our goal, the loss design must meet three basic criteria, a) It needs to have a global maximum at $z = 0.5$ (a fair starting point) and a global minimum at 0 and 1. b) The gradient magnitude $\frac{df}{dz}|_{z \approx 0.5}$ has to be adequately small to allow architectural weights to fluctuate, but large enough to attract z towards 0 or 1 when they are a bit far from 0.5. c) It should be differentiable for backpropagation.

According to the first requirement, we move $\sigma(\alpha)$ away from 0.5 towards 0 or 1 to minimize the discretization gap. The second one enacts explicit necessary constraints. Particularly, small gradients around the peak avoid stepping easily into two ends. Larger gradients around 0 and 1 instead help to quickly capture z nearby. Quite straightforward, we come up with a loss function to meet the above requirements, formally as,

$$L_{0-1} = -\frac{1}{N} \sum_{i}^{N} (\sigma(\alpha_i) - 0.5)^2 \tag{7}$$

In order to control its strength, we weight this loss by a coefficient w_{0-1}, thus the total loss for α is formulated as,

$$L_{total} = \mathcal{L}_{val}(w^*(\alpha), \alpha) + w_{0-1}L_{0-1}. \tag{8}$$

Like DARTS [19], the architectural weights can be optimized through back-propagation. From Eq. 8, the search objective is to find an architecture of high accuracy with a good approximation from a continuous encoding to a discrete one.

Moreover, the second requirement is indispensable, otherwise the gradient-based approach may step into local minimum too early. Here we design another loss as a negative example. Let L'_{0-1} be the following,

$$L'_{0-1} = -\frac{1}{N} \sum_{i}^{N} |(\sigma(\alpha_i) - 0.5)|. \tag{9}$$

It's trivial to see that $\frac{d|z-0.5|}{dz}|_{z>0.5} = 1$ and $\frac{d|z-0.5|}{dz}|_{z<0.5} = -1$. Once z stays away from 0.5, it may receive the same gradient (1 or -1) in the later iterations, thus rapidly pushing the architectural weights towards two ends. This phenomenon is illustrated in Fig. 5.

To conclude, by combining Eq. 4, 6 and 8, our method which we call Fair DARTS, can be now formally written as

$$\min_{\alpha} \mathcal{L}_{val}(w^*(\alpha), \alpha) + w_{0-1}L_{0-1}$$

$$\text{s.t. } w^*(\alpha) = \text{argmin}_w \mathcal{L}_{train}(w, \alpha).$$

$$\bar{o}_{i,j}(x) = \sum_{o \in \mathcal{O}} \sigma(\alpha_{o_{i,j}})o(x). \tag{10}$$

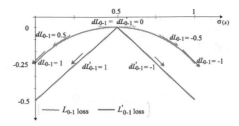

Fig. 5. Illustration about the auxiliary loss design: L_{0-1} (proposed) and L'_{0-1} (control)

It is also important to recognize that our zero-one loss is specially designed for Fair DARTS. Pushing $\sigma(\alpha)$ of one edge towards 0 or 1 is independent of others. It cannot be directly applied to DARTS given the exclusive competition by softmax. As the architectural weights converge to their extremities, it's natural to use a threshold value $\sigma_{threshold}$ in our approach to infer submodels instead of *argmax*.

5   Experiments and Results

5.1   Searching Architectures for CIFAR-10

At the search stage, we use similar hyperparameters and tricks as [19]. We apply the *first-order* optimization and it takes 10 GPU hours. All experiments are done on a Tesla V100. We select our target models with $\sigma_{threshold} = 0.85^3$. We use the same data processing and training trick as [4, 19].

Our collaborative approach performs well with skip connections aggregation. To verify this, we repetitively search 7 times on different random seeds and report the number of skip connections in Fig. 7 (see supplementary). Since the number of skip connections is more reasonable, we obtain an average top-1 accuracy 97.46%. Especially, the smallest FairDARTS-a reaches 97.46% accuracy on CIFAR-10 with reduced parameters and multiply-adds. A complete result of FairDARTS searched cells are shown in the supplementary (Fig. 5, Fig. 6 and Table 1).

5.2   Transferring to ImageNet

As a common practice, we transfer two searched cells (FairDARTS-a and b[4]) to ImageNet. We keep the same configurations and use the identical training tricks as DARTS [19]. Compared with SNAS [32] and DARTS, FairDARTS-A only uses 3.6M number of parameters and 417M multiply-adds to obtain 73.7% top-1 accuracy on ImageNet validation set. FairDARTS-B also achieves state-of-the-art 75.1% in S_1 with a smaller number of parameters than comparable counterparts (Table 2).

---

[3] The maximum number of edges for a node is also limited to 2 as in DARTS.

[4] Their architectures are given in Fig. 5 and 6 (supplementary).

Table 1. Comparison of architectures on CIFAR-10. [†]: MultAdds computed using the genotypes provided by the authors. [⋆]: Averaged on training the best model for several times . [‡]: Averaged on models from 7 runs of FairDARTS (Search + Full Train)

Models	Params (M)	×+ (M)	Top-1 (%)	Type
NASNet-A [36]	3.3	608[†]	97.35	RL
ENAS [23]	4.6	626[†]	97.11	RL
MdeNAS[35]	3.6	599[†]	97.45	MDL
DARTS(second order)[⋆][19]	3.3	528[†]	97.24 ± 0.09	GD
SNAS[⋆] [32]	2.8	422[†]	97.15 ± 0.02	GD
GDAS [10]	3.37	519[†]	97.07	GD
SGAS (Cri.2 avg.) [16]	3.9 ± 0.22[†]	640 ± 39[†]	97.33 ± 0.21	GD
P-DARTS [4]	3.4	532[†]	97.5	GD
PC-DARTS [33]	3.6	558[†]	97.43	GD
RDARTS [34]	-	-	97.05	GD
FairDARTS-a	**2.8**	**373**	97.46	GD
FairDARTS[‡]	3.32 ± 0.46	458 ± 61	97.46 ± 0.05	GD

5.3 Searching Proxylessly on ImageNet

Relaxing exclusive competition to collaboration greatly extends the size of the search space. In ProxylessNAS [3], there are 19 searchable layers and each layer contains 7 choices, consisting of 7^{19} possible models. In our approach, every choice can be activated independently, thus, S_2 contains $(2^7)^{19} = 128^{19}$ possible models. To our knowledge, this is a most gigantic search space ever proposed, about 18^{19} times that of [3].

For this search phase, we train for 30 epochs with a batch size of 1024, which takes about 3 GPU days. The final architectural weight matrix (after sigmoid activation) on the bottom right of Fig. 1 is used to derive target models. Under this cooperative setting, the skip connections and other inverted bottleneck blocks can be both chosen to work together, where the former facilitates the training and the latter learn the residual information [17]. In contrast, under the competitive setting of DARTS, it's impossible to achieve this, as shown in the bottom left of Fig. 1. Within 19 layers have 11 skip connection operation is preferred, which cuts down the overall depth of searchable layers to 8.

To be fair, we select at most two choices per layer if there are more than two above $\sigma_{threshold}$ (0.75) and use the same training tricks as [28]. We exclude squeeze and excitation [13] and refrain from using AutoAugment [8] tricks though they can boost the classification accuracy further. The searched model FairDARTS-D is shown in Fig. 6, which places the summation of two inverted bottleneck blocks nearby the down-sampling stage to keep more information. It also utilizes large kernels and big expansion blocks at the tail end. Further, We raise the $\sigma_{threshold}$ as 0.8 to get a more lightweight model FairDARTS-C. FairDARTS-C achieves 75.1% top-1 accuracy using only 4.2 M number of

Table 2. Comparison of architectures on ImageNet. *: Based on its published code. †: Searched on CIFAR-10. ††: Searched on CIFAR-100. ‡: Searched on ImageNet (cost more than those transferred). •: in GPU days. ◇: w/ SE and Swish

Models	×+ (M)	Params (M)	Top-1 (%)	Top-5 (%)	Cost•
MobileNetV2(1.4) [26]	585	6.9	74.7	92.2	-
NASNet-A [36]	564	5.3	74.0	91.6	2000
AmoebaNet-A [25]	555	5.1	74.5	92.0	3150
MnasNet-92 [28]	388	3.9	74.79	92.1	1667
DARTS [19]	574	4.7	73.3	91.3	4
FBNet-C [31]	375	5.5	74.9	92.3	9
Proxyless GPU‡ [3]	465*	7.1	75.1	92.4	8.3
FairNAS-C‡ [7]	321	4.4	74.7	92.1	10
SNAS [32]	522	4.3	72.7	90.8	1.5
GDAS [10]	581	5.3	74.0	91.5	0.2
P-DARTS†† [4]	577	5.1	74.9*	92.3*	0.3
PC-DARTS† [33]	586	5.3	74.9	92.2	3.8
FairDARTS-A†	417	3.6	73.7	91.7	0.4
FairDARTS-B‡	541	4.8	75.1	92.5	0.4
FairDARTS-C‡	380	4.2	75.1	92.4	3
FairDARTS-D‡	440	4.3	**75.6**	**92.6**	3
MobileNetV3 [12]	219	5.4	75.2	92.2	-
MoGA-A [6]	304	5.1	75.9	92.8	12
MixNet-M [30]	360	5.0	77.0	93.3	-
EfficientNet B0 [29]	390	5.3	76.3	93.2	-
SCARLET-A [5]	365	6.7	76.9	93.4	10
FairDARTS-C◇	386	5.3	**77.2**	**93.5**	3

parameters. To make comparisons with EfficientNetB0 [29], MobileNetV3 [12] and MixNet [30], FairDARTS-C obtains 77.2% top-1 accuracy with the same tricks such as squeeze-and-excitation [13], AutoAugment [8] and Swish [24].

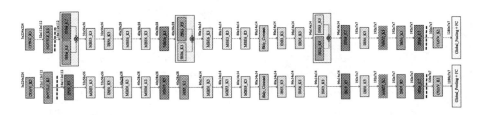

Fig. 6. The Architecture of Fair DARTS-D (top) and C (bottom). IBEx_Ky refers to an inverted bottleneck without an inset skip connection, while MBEx_Ky is the one with it. BOTTLE_K3 is the inverted bottleneck without expansion

6   Ablation Study and Analysis

6.1   Removing Skip Connections from S_1

As unfair advantages are mainly from skip connections, if we remove them from S_1 and get the reduced search space $S_1 \setminus \{skip\}$, we should expect a fair play even in an exclusive competition. Several runs of this experiment also show that there is indeed no more prevailing operations that suppress others, including other parameter-less ones like max-pooling and average pooling (Fig. 7). For $S_1 \setminus \{skip\}$, we run all the experiments with 7 different random seeds and we train the searched models from scratch. The best models $(96.88 \pm 0.18\%)$ are slightly higher than DARTS $(96.76 \pm 0.32\%)^5$, but lower than FairDARTS $(97.41 \pm 0.14\%)$ in S_1. The lowered accuracy indicates that adequate skip connections are indeed beneficial for accuracy.

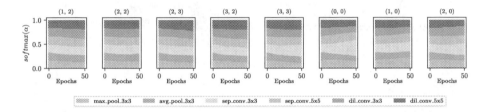

Fig. 7. Stacked area plot of the softmax evolution in $S_1 \setminus \{skip\}$ when running DARTS on CIFAR-10. With unfair advantages removed, all operations enjoy a fair treatment

6.2   How Does Zero-One Loss Matter?

Removing Zero-One Loss. We design two comparison groups for Fair DARTS with and without *zero-one loss*. Other settings are kept the same. We count the distribution of the sigmoid outputs from architectural weights and plot it on the left of Fig. 8. The one without zero-one loss covers a wide range between 0 and 0.6. So we have to make ambiguous choices again. Whereas the proposed loss has narrowed the distribution into two ends around 0 and 1. To further evaluate the influences of removing L_{0-1}, we repeat the searching for 7 times using different random seeds. The averaged top-1 accuracy for these models is 97.33 ± 0.15 (532M FLOPS on average, 74 M more than FairDARTS with L_{0-1}). Therefore, although the unfair advantage is balanced, making ambiguous choices still bring noises to the final search result, which is better solved by L_{0-1}. Discrepancy elimination seems to helps find more light-weight and accurate models.

Zero-One Loss Design. We run two experiments on CIFAR-10, one with L_{0-1} (proposed) and the other L'_{0-1} (control). To some extent, L_{0-1} allows stepping out of the local minimum while L'_{0-1} selects operations only at an early

---

^{5} This differs from DARTS' reported values as it trains one model for several times.

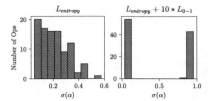
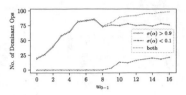

Fig. 8. Left: Histogram of sigmoid values in the last searching epoch without (left) and with L_{0-1} (right). On the right, this auxiliary loss has pushed the values towards 0 or 1. **Right:** Number of dominant operations in the last searching epoch running Fair DARTS on CIFAR-10 w.r.t the sensitivity weight w_{0-1} of auxiliary loss L_{0-1}

stage which depends greatly on the initialization. This matches our analysis in Sect. 4.2. The detailed results under both loss functions are shown in Fig. 2 (supplementary).

Loss Sensitivity. As the weight w_{0-1} of this auxiliary loss goes higher, it should squeeze more operations towards two ends, but it must not overshadow the main entropy loss. We perform several experiments where an integer w_{0-1} varies within $[0, 16]$. The right of Fig. 8 shows the final number of dominant operations for each. We select a reasonable $w_{0-1} = 10$ for the best trade-off.

6.3 Discussions from Fairness Perspective

We review the existing methods that seek to avoid the discussed weaknesses. In general, adding dropouts [4,34] to operations is similar to blending them with a simple additive Gaussian noise, both reduce the performance gain from unfair advantages. Early-stopping [18] avoids the case before unfairness prevails.

Adding Dropout to Skip Connections Reduces Unfairness. The operation-level dropout [27] inserted after skip connections by P-DARTS [4] can be viewed as an alleviation of unfair advantage. However, it comes with two obvious drawbacks. First, this dropout rate is hard to tune. Second, it is not so effective that they must involve another prior: setting the number of skip connections in the final cell to M. This is a very strong prior for searching good architectures [18].

Adding Dropout to All Operations Also Helps. Dropout troubles the training of skip connections and thus weakens the unfair advantage. Reasonably, higher dropout rates are more effective, especially for parameter-free operations. Therefore, RobustDARTS [34] adds dropout to all operations and obtains promising results.

Early Stopping Matters. DARTS+ [18] explicitly limits the maximum number of skip connections, which can be viewed as an early-stopping strategy nearby the previously mentioned *boundary epoch*, right before too many skip connections rise into power. RobustDARTS [34] also exploits early-stopping when the maximal Hessian eigenvalues change too fast.

Table 3. Experiment 1: Random sampling (7 models each, averaged) from regularized search space ($M = 2$). **Experiment 2:** Adding Gaussian noise to DARTS (repeated 4 times, averaged)

Methods	CIFAR-10 Top-1 Acc (%)
Random (M = 2)	97.01 ± 0.24
Random (M = 2, MultAdds $\geq$ 500M)	97.14 ± 0.28
DARTS + Gaussian (cosine decay)	97.12 ± 0.23

Limiting the Number of Skip Connections is a Strong Prior. In the regularized search space of P-DARTS [4] and DARTS+ [18], we find that simply by restricting $M = 2$, it is possible to generate competitive models even *without searching*. We randomly sample models from their search space and report the results in Table 3. In Experiment 1, the second group restricts the multiply-adds to be above 500M, to further leverage the average performance. Surprisingly, both groups outperform DARTS [19].

Random Noise Can Break Unfair Advantage. Based on our theory, we can boldly postulate that adding a random noise also disrupts the unfair advantage. Therefore, on top of DARTS [19], we mix the skip connections' architectural weights with a standard Gaussian noise $\mathcal{N}(0, 1)$, which has a cosine decay on 50 epochs. The results strongly confirm our hypothesis, as shown in Table 3. We repeat it 4 times to have similar results.

Remove Unfair Advantages or Destroy the Exclusive Competition? In principle, we can break either one of the indispensable factors to avoid collapse. However, FairDARTS breaks the latter which is simple and effective. Besides, it paves the way to eliminate the discrepancy by scheming an auxiliary loss L_{0-1}. Otherwise, the discrepancy issue remains hard to solve. However, to tackle the discrepancy issue, it's promising that the existing approaches might benefit from tricks like L_{0-1}. This remains to be our future work.

7   Conclusion

We unveil two indispensable factors of the DARTS's aggregation of excessive skip connections: **unfair advantages** and **exclusive competition**. We prove that breaking any one of them can improve the robustness. First, by allowing collaborative competition, each operation develops its architectural weight independently. Meanwhile, the non-negligible discrepancy of discretization is reduced at maximum by coercing a novel auxiliary loss which polarizes the architectural weights. In this regard, we achieve state-of-the-art performance both on CIFAR-10 and ImageNet. Second, disturbing the differentiable process with a Gaussian noise removes unfair advantage which leads to competitive results.

One of our future work is to make it more memory-friendly. As Gumbel softmax is used to replace categorical distribution [31], is there a similar way to our approach? More methods remain to be explored on our basis.

References

1. Bi, K., Hu, C., Xie, L., Chen, X., Wei, L., Tian, Q.: Stabilizing DARTS with amended gradient estimation on architectural parameters. arXiv preprint arXiv:1910.11831 (2019)
2. Brock, A., Lim, T., Ritchie, J.M., Weston, N.: SMASH: one-shot model architecture search through hypernetworks. In: International Conference on Learning Representations (2018)
3. Cai, H., Zhu, L., Han, S.: ProxylessNAS: direct neural architecture search on target task and hardware. In: International Conference on Learning Representations (2019)
4. Chen, X., Xie, L., Wu, J., Tian, Q.: Progressive differentiable architecture search: bridging the depth gap between search and evaluation. In: International Conference on Computer Vision (2019)
5. Chu, X., Zhang, B., Li, Q., Xu, R.: SCARLET-NAS: bridging the gap between scalability and fairness in neural architecture search. arXiv preprint arXiv:1908.06022 (2019)
6. Chu, X., Zhang, B., Xu, R.: MoGA: searching beyond MobileNetV3. In: International Conference on Acoustics, Speech, and Signal Processing (2020). https://arxiv.org/pdf/1908.01314.pdf
7. Chu, X., Zhang, B., Xu, R., Li, J.: FairNAS: rethinking evaluation fairness of weight sharing neural architecture search. arXiv preprint arXiv:1907.01845 (2019)
8. Cubuk, E.D., Zoph, B., Mane, D., Vasudevan, V., Le, Q.V.: AutoAugment: learning augmentation policies from data. In: Proceedings of the IEEE Conference on Computer Vision and Pattern Recognition (2019)
9. Dong, X., Yang, Y.: One-shot neural architecture search via self-evaluated template network. In: Proceedings of the IEEE International Conference on Computer Vision, pp. 3681–3690 (2019)
10. Dong, X., Yang, Y.: Searching for a robust neural architecture in four GPU hours. In: Proceedings of the IEEE Conference on Computer Vision and Pattern Recognition, pp. 1761–1770 (2019)
11. He, K., Zhang, X., Ren, S., Sun, J.: Deep residual learning for image recognition. In: Proceedings of the IEEE Conference on Computer Vision and Pattern Recognition, pp. 770–778 (2016)
12. Howard, A., et al.: Searching for MobileNetV3. In: International Conference on Computer Vision (2019)
13. Hu, J., Shen, L., Sun, G.: Squeeze-and-excitation networks. In: Proceedings of the IEEE Conference on Computer Vision and Pattern Recognition, pp. 7132–7141 (2018)
14. Jang, E., Gu, S., Poole, B.: Categorical reparameterization with gumbel-softmax. In: International Conference on Learning Representations (2017)
15. Li, G., Zhang, X., Wang, Z., Li, Z., Zhang, T.: StacNAS: towards stable and consistent optimization for differentiable Neural Architecture Search. arXiv preprint arXiv:1909.11926 (2019)
16. Li, G., Qian, G., Delgadillo, I.C., Müller, M., Thabet, A., Ghanem, B.: SGAS: sequential greedy architecture search. In: Proceedings of the IEEE Conference on Computer Vision and Pattern Recognition (2020)
17. Li, Y., Yuan, Y.: Convergence analysis of two-layer neural networks with ReLU activation. In: Advances in Neural Information Processing Systems, pp. 597–607 (2017)

18. Liang, H., et al.: DARTS+: Improved differentiable architecture search with early stopping. arXiv preprint arXiv:1909.06035 (2019)
19. Liu, H., Simonyan, K., Yang, Y.: DARTS: differentiable architecture search. In: International Conference on Learning Representations (2019)
20. Maclaurin, D., Duvenaud, D., Adams, R.: Gradient-based hyperparameter optimization through reversible learning. In: International Conference on Machine Learning (2015)
21. Maddison, C.J., Mnih, A., Teh, Y.W.: The concrete distribution: a continuous relaxation of discrete random variables. In: International Conference on Learning Representations (2017)
22. Nayman, N., Noy, A., Ridnik, T., Friedman, I., Jin, R., Zelnik-Manor, L.: XNAS: neural architecture search with expert advice. In: Advances in Neural Information Processing Systems (2019)
23. Pham, H., Guan, M.Y., Zoph, B., Le, Q.V., Dean, J.: Efficient neural architecture search via parameter sharing. In: International Conference on Machine Learning (2018)
24. Ramachandran, P., Zoph, B., Le, Q.V.: Searching for activation functions. arXiv preprint arXiv:1710.05941 (2017)
25. Real, E., Aggarwal, A., Huang, Y., Le, Q.V.: Regularized evolution for image classifier architecture search. In: International Conference on Machine Learning, AutoML Workshop (2018)
26. Sandler, M., Howard, A., Zhu, M., Zhmoginov, A., Chen, L.C.: MobileNetV2: inverted residuals and linear bottlenecks. In: Proceedings of the IEEE Conference on Computer Vision and Pattern Recognition, pp. 4510–4520 (2018)
27. Srivastava, N., Hinton, G., Krizhevsky, A., Sutskever, I., Salakhutdinov, R.: Dropout: a simple way to prevent neural networks from overfitting. J. Mach. Learn. Res. **15**(1), 1929–1958 (2014)
28. Tan, M., Chen, B., Pang, R., Vasudevan, V., Le, Q.V.: MnasNet: platform-aware neural architecture search for mobile. In: Proceedings of the IEEE Conference on Computer Vision and Pattern Recognition (2019)
29. Tan, M., Le, Q.V.: EfficientNet: rethinking model scaling for convolutional neural networks. In: International Conference on Machine Learning (2019)
30. Tan, M., Le., Q.V.: MixConv: mixed depthwise convolutional kernels. In: The British Machine Vision Conference (2019)
31. Wu, B., et al.: : FBNet: hardware-aware efficient convnet design via differentiable neural architecture search. In: Proceedings of the IEEE Conference on Computer Vision and Pattern Recognition (2019)
32. Xie, S., Zheng, H., Liu, C., Lin, L.: SNAS: stochastic neural architecture search. In: International Conference on Learning Representations (2019)
33. Xu, Y., et al.: PC-DARTS: partial channel connections for memory-efficient differentiable architecture search. In: International Conference on Learning Representations (2020)
34. Zela, A., Elsken, T., Saikia, T., Marrakchi, Y., Brox, T., Hutter, F.: Understanding and robustifying differentiable architecture search. In: International Conference on Learning Representations (2020). https://openreview.net/forum?id=H1gDNyrKDS
35. Zheng, X., Ji, R., Tang, L., Zhang, B., Liu, J., Tian, Q.: Multinomial distribution learning for effective neural architecture search. In: International Conference on Computer Vision (2019)
36. Zoph, B., Vasudevan, V., Shlens, J., Le, Q.V.: Learning transferable architectures for scalable image recognition. In: Proceedings of the IEEE Conference on Computer Vision and Pattern Recognition, vol. 2 (2018)

TANet: Towards Fully Automatic Tooth Arrangement

Guodong Wei[1,2], Zhiming Cui[2], Yumeng Liu[2], Nenglun Chen[2], Runnan Chen[2], Guiqing Li[1], and Wenping Wang[2(✉)]

[1] South China University of Technology, Guangzhou, China
csgdwei@mail.scut.edu.cn, ligq@scut.edu.cn
[2] The University of Hong Kong, Pok Fu Lam, Hong Kong
{zmcui,lym29,nolenc,rnchen2,wenping}@cs.hku.hk

Abstract. Determining optimal target tooth arrangements is a key step of treatment planning in digital orthodontics. Existing practice for specifying the target tooth arrangement involves tedious manual operations with the outcome quality depending heavily on the experience of individual specialists, leading to inefficiency and undesirable variations in treatment results. In this work, we proposed a learning-based method for fast and automatic tooth arrangement. To achieve this, we formulate the tooth arrangement task as a novel structured 6-DOF pose prediction problem and solve it by proposing a new neural network architecture to learn from a large set of clinical data that encode successful orthodontic treatment cases. Our method has been validated with extensive experiments and shows promising results both qualitatively and quantitatively.

Keywords: Deep learning · Orthodontics · Tooth arrangement · 6D pose prediction · Structure · Graph neural network

1 Introduction

Irregular tooth arrangements cause not only aesthetic issues but also compromised masticatory functions. Incorrect bite relationship, such as overjet or crowded teeth, may lead to disorders in chewing, which often induces other secondary diseases. With the growing concern of oral health, there is tremendous demand for orthodontic treatment. Although the number of people seeking orthodontic care is increasing rapidly, there is in general a severe lack of certified orthodontists to meet the demand. Currently, orthodontic treatment involves tedious manual operations and training professional orthodontists is a lengthy and costly process. Moreover, the quality of diagnosis and treatment depends in a large degree on the skills and experiences of individual orthodontists. Hence, it is imperative to develop a fully automated system for fast recommendation of

Electronic supplementary material The online version of this chapter (https://doi.org/10.1007/978-3-030-58555-6_29) contains supplementary material, which is available to authorized users.

A. Vedaldi et al. (Eds.): ECCV 2020, LNCS 12360, pp. 481–497, 2020.
https://doi.org/10.1007/978-3-030-58555-6_29

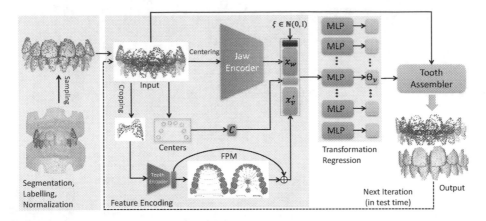

Fig. 1. The overall pipeline and the network architecture of our method. At the first stage, the input 3D dental model is automatically segmented to produce the label and point-cloud representation of each tooth. The point sets of each tooth is then normalized and sampled. The second stage involves a network consisting of four components: feature encoding, feature propagation, transformation (i.e. pose) regression, and tooth assembler modules. The final output is a rearranged dentition. For clarity, only one tooth-level encoder is illustrated here.

optimal tooth arrangements to improve the efficiency and quality of orthodontic treatment planning.

Tooth arrangement is an essential step of orthodontic treatment. Given a set of ill-positioned teeth of a patient, tooth arrangement aims to predict an ideal tooth layout that serves as the target arrangement to achieve through orthodontic treatment. In order to produce a satisfactory arrangement, multiple factors need to be taken into consideration. This makes tooth arrangement a complex task with its outcome quality heavily dependent on professional skills and subjective judgement of orthodontists.

Existing computer-aided systems used in orthodontic treatment planning provide a user interface for visualizing and manually editing individual teeth. As a related work in prosthodontics, Dai [8] performs complete denture tooth arrangement according to a set of heuristic rules, with teeth selected from a pre-specified set. In contrast we intend to solve a different and more challenging problem of tooth arrangement with patient-specific dentition for orthodontic treatment. The work in [6] automatically establishes proper dental occlusion by treating the upper teeth and lower teeth as two rigid objects, while our tooth arrangement problem requires pose adjustment of each individual tooth in a dentition.

To automatically determine the ideal positions of teeth for each specific patient is extremely challenging. Even though clinical rules like "Andrew's six keys" [1] suggest the necessary conditions for proper tooth alignment, the actual layout of patient's teeth may prevent the accessibility to the theoretically ideal poses. Therefore, a mathematical model developed by rule-based method can hardly lead to a clinically feasible outcome. Apart from this, detecting landmarks or other human-defined features on dental models is a tedious process and may also introduce errors at the very beginning of pose prediction. Moreover, dental models are texture-less and lack of sharp features, especially when we only have dental crowns, i.e. teeth outside gums. These characteristics make it hard to define orientation, position or other low-level features of a tooth precisely and consistently, while they are the prerequisites for a rule-based method.

We proposed a learning-based approach to predict an optimal treatment target from the initial irregular tooth positions of a patient before treatment, and thereby developed the first method for automatic tooth arrangement for orthodontic treatment. We formulate the tooth arrangement task as a structured 6D poses prediction problem, which has not been fully explored by the computer vision community. Our network aims to approximate the mapping from an input dental model representing the initial tooth arrangement to an ideal target poses via supervised learning. The network consists of four main components: a feature encoding module for information at the jaw-level and the tooth-level, a feature propagation module for information passing among teeth, a pose regression module for 6-DOF pose prediction and a differentiable tooth assembler module for rigid transformations. The loss function of the network is specially designed to capture intrinsic differences between different arrangements, enhance compact spatial relation and model the uncertainties in ground truth.

To summarize, the main contributions of this work are:

- We developed the first automatic tooth arrangement framework based on deep learning;
- We proposed the use of a graph-based feature propagation module to update features extracted by PointNet to provide crucial contextual information for successfully solving the structured poses prediction problem arising from the tooth arrangement task.
- We proposed a novel loss function that is able to provide effective supervision for aligning teeth by capturing intrinsic differences, spatial relations and uncertainties in the distribution of malaligned tooth layouts.

2   Related Work

6-DOF Pose Estimation Problem. The pose estimation problem has been extensively studied in recent years. It aims to infer the three-dimensional pose, which has six degrees of freedom, of an object present in an RGB image, [3, 5, 7, 18, 25, 33, 34, 45], RGB-D image [35, 39, 40], or point cloud data [26, 29, 30, 44]. Existing methods can be roughly categorized into the object coordinate regression approach and the template matching approach. The methods based on coordinate regression estimates the object's surface corresponding to each object at the pixel level, with the assumption that the corresponding 3D model is known for training [36]. The methods based on template matching perform alignment between known 3D models and image observations using various techniques, such as Iterative Closest Point (ICP) [4]. All these previous works do not consider multiple objects and their relative relationships, while the tooth arrangement problem that we face needs to predict 6-DOF poses of all the teeth (i.e. multiple objects) at the same time to form a regular layout, Most importantly, the 6-DOF pose estimation problem is concerned the relation between the pose of a known 3D shape and its image observation, In contrast, we aim to solve a more challenging problem of predicting the poses of regularly arranged teeth by learning from clinical data of orthodontic treatment.

Furniture Arrangement or Placement Problem. There have recently been many studies on how to automatically generate an optimized indoor scene composed of various furniture objects [10,14,21,37,38,42]. To simplify the problem, most of these methods use bounding boxes as proxies to roughly approximate the input objects, without taking into account the fine-grained geometric details of the objects. The work in [42] optimizes the configurations of given 3D models using learned priors. The core of their method is an energy function defined with a set of heuristic rules. The method in [31] addresses the problem of placing one 3D object with respect to others, assuming that all the objects are pre-aligned with the same orientation, thus only translation of the newly added component needs to be predicted. Since man-made furniture shapes usually have distinct sharp features, the orientations of these objects can easily be defined. As a comparison, we consider dental models which lack such distinct features, which makes it hard to precisely define orientations. Furthermore, most works on the furniture arrangement problem attempt to generate diverse arrangements for a given indoor scene, while the goal of the tooth arrangement problem is the best tooth arrangement for each specific patient.

3D Shape Generation Problem aims to generate realistic 3D shapes from user specifications or by inferring from images or partial models. Conditional generative methods [22] can generate realistic images based on the input condition. With the advance in geometric learning and 3D representation methods, various generative models have been proposed as powerful tools to process 3D shapes [24,27,28,32]. The problem of conditional 3D shape generation and the problem of automatic tooth arrangement both aim to generate 3D shapes according to given conditions. Recent works of conditional 3D shape generation [11,13,17,23] focus on generating realistic structured shapes that are able to adapt diverse shape variations. However they do not preserve the geometries of input objects, while this is a hard constraint in the tooth arrangement problem.

3 Method

3.1 Overview

An illustrated in Fig. 1, our proposed method contains two main stages. The first is a preprocessing stage that segments dental crowns from the whole model and then semantically labeling each individual tooth crown. The second stage uses a network with four main components to perform the following functions: a) a set of PointNet-based point feature encoders for jaw-level and tooth-level feature extraction; b) a graph-based feature propagation module that transfers information among teeth; c) the regressor for each tooth combines its corresponding tooth-level features, global features and a random conditional vector as input, and outputs the 6D transformation relative to the input position of this tooth; d) an assembler to map the 3D rotations represented in the axis-angle representation into rotation matrices for transforming the points, and output the rearranged point cloud. The details are described in the subsequent sections.

3.2 Preprocessing

Segmentation and labeling are critical as preprocessing operations for our tooth arrangement algorithm. There exist many off-the-shelf methods [20,41,43] for accurate automatic semantic segmentation and labeling on 3D dental meshes. We use the method in [41]. The tooth labels are assigned according to *FDI two-digit notation* for permanent teeth. Note that we only keep the crowns for all the teeth for use in our tooth arrangement computing. A local coordinate system is then defined for the model consisting of these crowns to normalize the position and orientation by coarsely aligning it with the world coordinate system. The resulting tooth set is denoted $X = \{X_v \subseteq \mathbb{R}^3 | v \in \mathcal{V}\}$, where $\mathcal{V}$ is the set of tooth crown labels and X_v is the point cloud of the crown with label v.

3.3 Network

Tooth Centering. Since the input teeth are sparsely distributed in the space, this may increase the difficulty in capturing the features among teeth that are far away from each other. So we first translate all the teeth to the origin so that $\widetilde{X}_v = \{p' = p - c_v | p \in X_v\}$ and $\widetilde{X} = \{\widetilde{X}_v | v \in \mathcal{V}\}$, where $c_v \in C$ is the geometric center of tooth v. This measure is key to decoupling the center positions of teeth from other features so as to enable the encoder to focus on extracting geometric features, such as shape details, orientation and size, in a translation-independent manner.

Feature Encoder. The feature encoders in our network are based on PointNet [27]. Using symmetric functions, PointNet achieves permutation invariance of point sets and is able to efficiently extract local features for each point and global features for the whole point cloud. Here, we will use its global features thus extracted. The quality of tooth arrangement is determined by the position and orientation of every tooth with respect to the others, and the information of each tooth and that of the whole dentition are equally important. We therefore extract jaw-level features $x_w = E_w(\widetilde{X}, C)$ and tooth-level features $x_v = E_v(\widetilde{X}_v)$, where E_w represents the encoder for the whole tooth crown set and E_v the encoder for individual teeth.

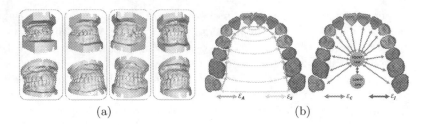

Fig. 2. (a) Representative examples of tooth arrangement in orthodontics. Each column contains dental models before (top) and after (bottom) the treatment; (b) The tooth graph used for feature propagation. We show the teeth connection in upper jaw here. The connection between jaws are also demonstrated.

Feature Propagation Module. Note that the jaw-level features are rather sparse and do not capture many geometric details, as shown in Fig. 8(b) and discussed in Sect. 4.5. The tooth-level features capture details, however, are encoded independently and so oblivious to the information from other teeth. So it is hard to achieve an accurate alignment of teeth by using these features. We introduce a graph-based feature propagation module (FPM) that allow geometric detail information to transfer among teeth via the connections of the graph.

Our feature propagation module G is based on the propagation model in [19]. First, we define a tooth graph as $\mathcal{G} = (\mathcal{N}, \mathcal{E}, \mathcal{H})$, where $\mathcal{N}$ is the set of nodes, each corresponding to a tooth v, $\mathcal{E}$ the set of undirected edges of the graph, and $\mathcal{H}$ the node embedding. In addition, two super nodes are created for the upper jaw and lower jaw. The node embedding of these two super nodes are set to zero vectors initially. The embedding h_v of any other node is initialized with its feature x_v. As illustrated in Fig. 2(b), $\mathcal{E}$ consists of four types of edges $\mathcal{E}_A, \mathcal{E}_S, \mathcal{E}_C, \mathcal{E}_J$, namely, $\mathcal{E} = \mathcal{E}_A \cup \mathcal{E}_S \cup \mathcal{E}_C \cup \mathcal{E}_J$, where $\mathcal{E}_A$ contains relationships between adjacent teeth in the same jaw, $\mathcal{E}_S$ connects left and right symmetric teeth in the same jaw, $\mathcal{E}_C$ consists of connections between each tooth node and its supper node of the corresponding jaw, and $\mathcal{E}_J$ is a set including single edge between two super nodes. Finally, local features x_v are updated in K iteration, each iteration with a fixed number of steps T, as follows:

$$m_v^{k,t+1} = \sum_{w \in N(v)} A_{e_{vw}}^k h_w^{k,t}, \tag{1}$$

$$h_v^{k+1,t+1} = \mathrm{GRU}(h_v^{k,t}, m_v^{k,t+1}), \tag{2}$$

where $N(v)$ denotes the set of neighboring nodes of v and $A_{e_{vw}}^k$ is a learned matrix for each type of edge in the graph $\mathcal{E}$. Both K and T are set to 3 in our experiments.

To further improve the network performance, we add a residual connection with the original feature. The final updated tooth feature x_v' is obtained after a residual operation,

$$x_v' = x_v + h_v^{K+1,T+1} \tag{3}$$

Pose Regressor. Considering that many other factors, such as the subjective judgment of clinical orthodontists or the age, gender and face appearance of patients, may also affect the layout of the optimal tooth arrangement, we generate a set of candidates instead of giving only one result. Inspired by the MoN loss [9], which is originally proposed to model the uncertainty in 3D recovery from a single image, we introduce a conditional weighting (CW) scheme. This scheme is designed to allow the network to generate multiple plausible arrangements and still be able to recommend a most appropriate one. To make the CW scheme work, here in the pose regressor, we only need to append a random vector $\xi \in \mathbb{N}(0, I)$ to the input features in training, where $\mathbb{N}(0, I)$ is a zero-mean Gaussian distribution. We set ξ to a zero vector in testing. Another part of the CW scheme lies in the loss function (Sect. 3.4). All the features are combined and fed into the corresponding pose regressors to predict 6D transformation parameters,

$$\Theta_v = \Psi_v \left(C, \mathbf{x}_w, \mathbf{x}_v^{'}, \xi \right) \tag{4}$$

where Θ_v consists of $\mathbf{r}_v = (r_v^x, r_v^y, r_v^z)$ in axis-angle representation for rotation and $\mathbf{t}_v \in \mathbb{R}^3$ for translation.

Tooth Assembler. The predicted transformation parameters are then passed to the assembler Φ to transform, assemble and generate the final output. This module maps the axis-angle representation of rotation $\mathbf{r}_v \in \mathcal{R}^3$ back into a rotation matrix $R_v \in SO(3)$ through an exponential map, which is differentiable. The exponential map $\exp : so(3) \rightarrow SO(3)$ connects the Lie algebra with the Lie group by

$$\exp(\mathbf{r}_\times) = I_{3\times3} + \frac{sin\theta}{\theta}\mathbf{r}_\times + \frac{1 - cos\theta}{\theta^2}\mathbf{r}_\times^2, \tag{5}$$

where $\theta = \|\mathbf{r}\|_2$ is the rotation angle. Let $\mathbf{r} = (r^x, r^y, r^z)$ be a rotation vector in axis-angle representation, with the associated skew-symmetric matrix

$$r_\times = \begin{bmatrix} 0 & -r^z & r^y \\ r^z & 0 & -r^x \\ -r^y & r^x & 0 \end{bmatrix}. \tag{6}$$

Then, given r_v for a tooth v, the assembler first maps it to a rotation matrix R_v using Eq. 5, and then applies the transformation to the input points to get the final output point cloud

$$X^* = \left\{ R_v p_v + c_v + t_v | v \in \mathcal{V}, p_v \in \widetilde{X}_v \right\}. \tag{7}$$

3.4   Loss Function

Geometric Reconstruction Loss. Based on the observation that teeth remain almost rigid during the treatment process and our network also keeps the shape of each tooth in input, we use iterative closest points method to align each pair of teeth in the prediction and ground truth (Fig. 4(c)). Then, for points in X_v^*, we find their correspondences $P_{\bar{X}}(X_v^*)$ by searching for the closest points in ground truth $\bar{X}_v$ based on the rigid alignment result. The function $P_{\bar{X}}(\cdot)$ represents this correspondence searching process. To eliminate the loss induced by global rigid transformation and reveal the intrinsic between two arrangements, we solve for a global rigid transformation Π to align the prediction and ground truth (Fig. 4(d)) by minimizing the following energy,

$$\underset{\Pi}{\text{argmin}} \sum_{v \in \mathcal{V}} \|[X_v^*|1]^\top - \Pi[P_{\bar{X}}(X_v^*)|1]^\top\|_2^2, \tag{8}$$

where X_v^* is the coordinate matrix of X_v^*. We solve the above problem by orthogonal Procrustes analysis. Finally, the reconstruction loss is calculated as

$$L_{recon}(X^*, \bar{X}) = \sum_{v \in \mathcal{V}} \|[X_v^*|1]^\top - \Pi[P_{\bar{X}}(X_v^*)|1]^\top\|_S, \tag{9}$$

where $\|\cdot\|_S$ represents the $Smooth_{L1}$ norm [12].

Geometric Spatial Relation Loss. To emphasize the fact that a good arrangement is mostly determined by the mutual spatial relation between all the teeth, we define the geometric spatial relation between two point sets $S_1, S_2 \subseteq \mathbb{R}^3$ as

$$V_{S_1, S_2} = \bigcup_{\substack{i \neq j \\ 1 \leq i,j \leq 2}} \left\{ x - y^* | y^* = \underset{y \in S_i}{\text{argmin}} \|x - y\|_2^2, x \in S_j \right\}. \tag{10}$$

Based on the simple observation that the distance between two teeth should not be larger than a threshold σ if the dentition is aesthetically and functionally satisfactory, we calculate the clamped V_{S_1, S_2} by clamping all elements into $[-\sigma, +\sigma]$ and denote as V_{S_1, S_2}^c. We empirically set $\sigma = 5.0$ in all our experiments. Finally, the geometric spatial relation loss is calculated as,

$$L_{spatial}(X^*, \bar{X}) = \sum_{q \in \mathcal{N}} \sum_{e \in \mathcal{P}(q)} \|V_{X_q^*, X_e^*}^c - V_{P_{\bar{X}}(X_q^*), P_{\bar{X}}(X_e^*)}^c\|_S, \tag{11}$$

where $\mathcal{P}(q) = \text{NBR}(q) \cup \text{OPS}(q)$. The functions $\text{NBR}(q)$ and $\text{OPS}(q)$ return neighboring nodes and the opposite jaw, respectively. If node q is a super node for a jaw, then X_q^* is the set of points of all teeth belongs to that jaw.

Conditional Weighting Loss. As discussed in Sect. 1, we introduce a mechanism that allows the network to model uncertainty and generate a distributional output for one input given a conditional vector ξ. Our approach is inspired by the MoN loss [9] with the following variation. We enable the network to recommend a most likely arrangement by using a conditional weighting loss, which is defined as follows

$$\sum_{1 \leq j \leq n} \underset{\xi_j \sim \mathbb{N}(0, I)}{\min} \left\{ \frac{1}{e^{\|\xi_j\|}} \cdot Loss(X^*, \bar{X}) \right\}, \tag{12}$$

where $Loss = L_{recon} + L_{spatial}$ and n is set to 2 in our experiments.

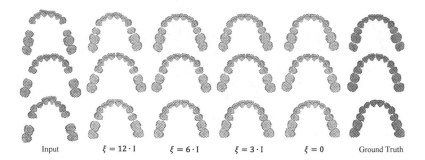

Input	$\xi = 12 \cdot I$	$\xi = 6 \cdot I$	$\xi = 3 \cdot I$	$\xi = 0$	Ground Truth

Fig. 3. For the same input model in test time, we give different values of ξ as conditions for the regressor, which result in predictions that have different distances with respect to the ground truth. It is observed that the prediction with $\xi = 0$ is often the most satisfactory arrangement.

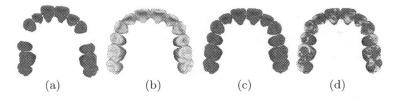

(a)	(b)	(c)	(d)

Fig. 4. (a) A pre-treatment model; (b) The corresponding post-treatment model (c) The aligned model with its tooth shapes from (a) and tooth poses from (b). It is used in the definition of our reconstruction loss (see Sect. 3.4); (d) Superposing (b) on (c) to visualize their differences.

3.5   Implementation and Training Details

Network Details. The dimensions of features encoded by the global and local PointNet encoders are 1024 and 512, respectively. The length of node embedding in FPM is set to 512. The random condition ξ is a 32-dimensional vector. The pose regressors consist of 3 linear layers with ReLU activator and dropout (0.3) in the first two layers. Only the Tanh activator is used in the last linear layer. The weights in the last layer of the regressors are initialized as zeros, as we assume that the teeth are more likely unmoved.

Training Details. Searching for corresponding point pairs is done before the training begins, since corresponding point pairs do not change due to rigid movement of teeth. Teeth that do not appear in both before and after treatment models are regarded as missing or extracted. We randomly sample 400 points on each tooth as the input. As for missing teeth, we set their positions with zeros. To augment the training data, all individual teeth of the input models, including pre-treatment and post-treatment models, are randomly rotated by an angle, within $[-30, +30]$ in our experiments, in a random direction and translated by a distance vector from the zero-mean Gaussian distribution $\mathbb{N}(0, 1^2)$. The complete set of teeth is also augmented by a random global rotation. Note that these augmented models are only used to enlarge the set of the simulated pre-treatment models. We assume that an augmented pre-treatment dental model $\mathcal{M}^*$ should be still mapped to the corresponding post-treatment model $\bar{\mathcal{M}}$, and an augmented treated model $\bar{\mathcal{M}}^*$ should also be mapped to its corresponding original post-treatment model before augmentation $\mathcal{M}$.

Our network is implemented with PyTorch and trained on a server using one 1080-Ti GPU. We use Adam optimizer. The batch size is set to be 16 with a learning rate initially equal to $1.0e-4$ and dropping down by 0.5 when the validation loss stops improving.

Table 1. Ablation study. The mean errors of translation ΔT_{avg}, rotation $\Delta \theta_{avg}$, ADD and PA-ADD together with their AUC scores are reported. The coordinate unit is *millimeter (mm)* except for $\Delta \theta_{avg}$, which is in *degree(°)*.

	ΔT_{avg}/AUC	$\Delta \theta_{avg}$/AUC	ADD/AUC	PA-ADD/AUC
NetBL+Lrecon	1.09/73.47	9.26/57.13	1.200/70.825	1.038/74.719
NetGL+Lchamfer	1.06/73.47	7.08/65.78	1.133/71.795	0.992/76.082
NetGL+Lrecon	1.03/74.43	6.70/67.30	1.096/72.821	0.957/77.032
NetGL+FPM+Lrecon	0.99/75.46	**6.64**/67.67	1.057/73.864	0.893/78.195
NetGL+FPM+Lrecon+CW	0.98/75.61	6.71/67.32	1.051/73.953	**0.886/78.512**
NetCom+Lcom	**0.97/76.00**	**6.64/67.71**	**1.036/74.362**	0.893/78.456

4   Experiments

4.1   Dataset

Our dataset consists of dental models of 300 patients, with males (47%) and females (53%) of age ranging from 6 to 18 years old. For each patient, there are two models scanned before and after treatment. All the three types of malocclusions (i.e., Class I, II, III) are observed in our dataset, according to Angle's classification [2]. Some examples are shown in Fig. 2(a). For network training, we randomly divide the 300 pairs of dental models of our dataset into three groups: 200 for training, 30 for validation, and 70 for testing.

4.2   Evaluation Metric

We evaluate the precision of our network prediction using the ADD metric [15], which is the mean point-wise distance between the predicted and ground truth models. We also report PA-ADD which is the ADD calculated after the rigid alignment between predicted jaw and ground truth jaw using Procrustes Analysis. In addition, we define PCT@K metric as the percentage of tooth predicted by the network with the error smaller than a threshold K. The error can be the shape reconstruction error, rotation or translation estimation error, etc. Similar to AUC [39] for 6-DOF pose estimation, we define PCT-AUC as the area

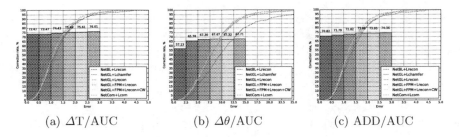

(a) ΔT/AUC   (b) $\Delta\theta$/AUC   (c) ADD/AUC

Fig. 5. The quantitative evaluation on the effectiveness of different components.

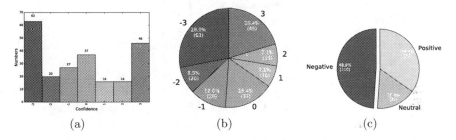

(a)   (b)   (c)

Fig. 6. Statistics of the user study.

under the PCT curve, which is the integral of PCT with respect to K. The PCT-AUC for shape reconstruction, rotation and translation errors are denoted as ADD-AUC, $\Delta\theta$-AUC and ΔT-AUC, respectively. We set the maximum K of PCT-AUC to be 5 mm for translation or reconstruction errors and 25° for rotation error. For ADD and PA-ADD metrics, the smaller values mean better precision. For the various AUC metrics, larger values indicate better precision.

4.3   Ablation Study

In this section, we will show the effectiveness of different components of our proposed network and the impact of different terms in our loss function.

The following three basic network architectures are used in the ablation study. They are the baseline network that has only the jaw-level global feature encoder in feature extracting stage (NetBL); the network with both jaw-level global feature encoder and tooth-level local feature encoders (NetGL); the complete network model proposed (NetCom), which contains all levels of feature encoders, the feature propagation module (FPM) and the conditional weighting mechanism (CW). Different losses are: our reconstruction loss L_{recon}(Lrecon); loss that replaces the $Smooth_{L1}$ with Chamfer distance (Lchamfer); the complete loss function we propose (Lcom). We have conducted six experiments with different combinations of the above networks with different loss functions. The results are reported in Table 1.

Global and Local Feature Integration. As shown in the 1st and 3rd rows in Table 1, introducing tooth-level local feature encoders to the network brings a significant improvement on the result, the PCT-AUC goes from 70.825 to 72.821. The improvement is almost completely caused by the growth in rotation estimation accuracy ($\Delta\theta$-AUC is increased by more than 10 points). As will be discussed in Sect. 4.5 later, this is mainly caused by the local feature extractors that help capture more details of the teeth, so that the rotations can be determined more accurately.

Feature Propagation. The feature propagation module (FPM) is introduced to make arrangements more compact. The 3rd and 4th rows of Table 1 validate the effectiveness of our feature propagation module. The improvement is mainly attributed to the translation estimation accuracy, which is increased by around 1 point in ΔT-AUC.

Conditional Weighting. The conditional weighting mechanism is designed to generate a distribution of predictions, so as to relieve the network from the ambiguities of ground truth due to subjective judgments of different dentists or insufficient input information. The 4th and 5th rows in Table 1 show that this mechanism have larger improvement on PA-AUC than ADD-AUC, because the CW may also mitigate ambiguities introduced by global rotations.

Reconstruction Loss. Based on the assumption that individual teeth have the same shape in each corresponding pair of pre-treatment model and the post-treatment model, we proposed to use MSE in reconstruction loss calculation. The 2nd and 3rd rows of Table 1 indicate that our loss is significantly better than the commonly used Chamfer Distance loss.

Spatial Relation Loss. The ablation study seems to suggest that the network learns better by emphasizing the reconstruction of the spatial relations between teeth. As can be seen in the last two rows of Table 1, the Add-AUC is increased by about 0.4 points. Although the improvement seems small in numerical value, we argue that this is significant in terms of shape variation because humans are visually sensitive to even slight misalignment of teeth.

Our complete model is able to achieve accurate tooth arrangement with around 0.97 mm translation error, 6.64° rotation error and 0.89 mm shape difference. A qualitative comparison between our complete method (NetCom+Lcom) and the baseline method (NetBL+Lrecon) is illustrated in Fig. 7. The results of our complete approach are significantly better than those of the baseline method. We show a more comprehensive comparison of these methods using different metrics in Fig. 5.

4.4   User Study

In order to evaluate the user perception of our results, we have conducted a user study. We randomly sampled 25 pairs of data from our test set, We recruited 9 students in dentistry and asked them to select the better one between the ground truth solutions and our predictions. The network predictions and ground truth were presented in random orders, with the original malaligned pre-treatment

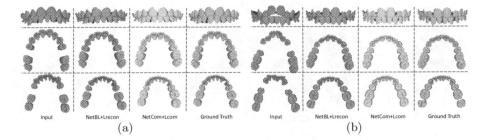

Fig. 7. A qualitative comparison between our complete method (NetCom+Lcom) and the baseline method (NetBL+Lrecon). Here we show two examples (a–b). Each example includes 3 rows and 4 columns. From top to bottom, the 3 rows are the complete dentition, the upper jaw and the lower jaw of a patient, respectively.

Fig. 8. (a) The critical points. Red: locally critical, Green: globally critical, Blue: both globally and locally critical; (b) The occlusion fields of the input, network output, and ground truth, respectively. Red: maximum distance; Green: minimum distance. (Color figure online)

models also presented as reference. Besides, they were asked to score their confidences for each of their selections with numbers between 0 to 3. The confidence score 3 indicates the selected one was much better than the other one, while score 0 indicates that they cannot tell which one is better.

As shown in Fig. 6(c), in 51.1% of totally $25 \cdot 9 = 225$ ratings, our network predictions are better than or equal to the post-treatment arrangements designed by dentists. In order to take the participants' confidences into account, we sum up these ratings weighted by their confidence scores, with the signs of the ratings set to negative if they prefer the arrangement by dentists (Fig. 6(a, b)). Normalized by $255 \cdot 3$, the final score is a weighted average of the ratings in the range of in $[-1, 1]$, where 1 indicates that our predictions are better and 0 indicates that our predictions and the ground truth judged to be of equal quality. The final score thus computed for our user study is -0.1037.

4.5 Visualization

Critical Points. To provide a better understanding of what our network has learned, we visualize the critical points related to local (tooth-level) and global

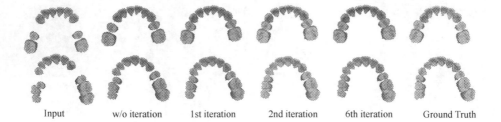

Input   w/o iteration   1st iteration   2nd iteration   6th iteration   Ground Truth

Fig. 9. By iteratively feeding the network output as input to the network, further improvement of arrangement is produced.

(jaw-level) features following the method in [27]. As shown in Fig. 8(a), the jaw-level feature extractor captures sparse features around the crown boundaries and the centers of teeth which can be helpful for the coarse arrangement of teeth, while the tooth-level local feature extractors capture denser features that describe the shape details of teeth much better and are beneficial for a more precise and compact arrangement.

Occlusion Field. The occlusion relationship is an important aspect to evaluate the quality of our network prediction. We visualize the occlusion relationship by displaying the minimal distance of every point in one jaw with respect to the opposite jaw, called the *occlusion field*. As illustrated in Fig. 8(b), the occlusion relationship in our prediction is improved significantly compared to the arrangement before treatment.

Distributional Output. We give different vectors ξ as input conditions to the network in test time to generate multiple predictions for an input. Interestingly, it turns out that the predictions are getting closer to the ground truth as the input condition vectors are closer to zero vectors (see Fig. 3).

5   Discussion

Failure Cases. Our method will fail if the input dental models deviate severely from the distribution of training data. To alleviate this problem, during testing, we feed the unsatisfactory output predictions as input back into the network again. As shown in Fig. 9, the arrangement is iteratively refined in this way.

Physical Constraints. Enforcing physical constraints in neural networks is an outstanding problem. Although we have encoded the left-right symmetry prior in FPM and propose $L_{spatial}$ for enhancing compact spatial relation, our network outputs do not guarantee to be physical feasible. Hence, a postprocessing procedure is needed to resolve these problems, such as penetration. See the supplementary materials for more details.

6   Conclusion

We present the first learning-based approach for automatic tooth arrangement in orthodontic treatment planning. By modeling the task as a structured 6-DOF poses prediction problem, we propose a network architecture composed of PointNet encoders and a graph-based feature propagation module, that is able to effectively capture crucial features for a compact alignment. Our novel loss function captures intrinsic geometric difference and uncertainties in ground truth. Extensive experiments validated that our method is able to achieve tooth alignments in quality comparable to those designed by orthodontists.

References

1. Andrews, L.F.: The six keys to normal occlusion. Am. J. Orthod. **62**(3), 296–309 (1972)
2. Angle, E.H.: Classification of malocclusion. Dent. Cosmos. **41**, 350–375 (1899)
3. Aubry, M., Maturana, D., Efros, A.A., Russell, B.C., Sivic, J.: Seeing 3D chairs: exemplar part-based 2D–3D alignment using a large dataset of cad models. In: Proceedings of the IEEE Conference on Computer Vision and Pattern Recognition, pp. 3762–3769 (2014)
4. Besl, P.J., McKay, N.D.: Method for registration of 3-D shapes. In: Sensor Fusion IV: Control Paradigms and Data Structures, vol. 1611, pp. 586–606. International Society for Optics and Photonics (1992)
5. Brachmann, E., Krull, A., Michel, F., Gumhold, S., Shotton, J., Rother, C.: Learning 6D object pose estimation using 3D object coordinates. In: Fleet, D., Pajdla, T., Schiele, B., Tuytelaars, T. (eds.) ECCV 2014. LNCS, vol. 8690, pp. 536–551. Springer, Cham (2014). https://doi.org/10.1007/978-3-319-10605-2_35
6. Chang, Y.B., Xia, J.J., Gateno, J., Xiong, Z., Zhou, X., Wong, S.T.: An automatic and robust algorithm of reestablishment of digital dental occlusion. IEEE Trans. Med. Imaging **29**(9), 1652–1663 (2010)
7. Collet, A., Martinez, M., Srinivasa, S.S.: The MOPED framework: object recognition and pose estimation for manipulation. Int. J. Robot. Res. **30**(10), 1284–1306 (2011)
8. Dai, N., Yu, X., Fan, Q., Yuan, F., Liu, L., Sun, Y.: Complete denture tooth arrangement technology driven by a reconfigurable rule. PLoS One **13**(6), e0198252 (2018)
9. Fan, H., Su, H., Guibas, L.J.: A point set generation network for 3D object reconstruction from a single image. In: Proceedings of the IEEE Conference on Computer Vision and Pattern Recognition, pp. 605–613 (2017)
10. Fisher, M., Ritchie, D., Savva, M., Funkhouser, T., Hanrahan, P.: Example-based synthesis of 3D object arrangements. ACM Trans. Graph. (TOG) **31**(6) (2012). Article no. 135
11. Gao, L., et al.: SDM-NET: deep generative network for structured deformable mesh. ACM Trans. Graph. (TOG) **38**(6) (2019). Article no. 243
12. Girshick, R.: Fast R-CNN. In: Proceedings of the IEEE International Conference on Computer Vision, pp. 1440–1448 (2015)
13. Groueix, T., Fisher, M., Kim, V.G., Russell, B.C., Aubry, M.: AtlasNet: Apapier-Mâché approach to learning 3D surfacegeneration. arXiv preprint arXiv:1802.05384 (2018)

14. Guerrero, P., Jeschke, S., Wimmer, M., Wonka, P.: Learning shape placements by example. ACM Trans. Graph. (TOG) **34**(4) (2015). Article no. 108
15. Hinterstoisser, S., et al.: Model based training, detection and pose estimation of texture-less 3D objects in heavily cluttered scenes. In: Lee, K.M., Matsushita, Y., Rehg, J.M., Hu, Z. (eds.) ACCV 2012. LNCS, vol. 7724, pp. 548–562. Springer, Heidelberg (2013). https://doi.org/10.1007/978-3-642-37331-2_42
16. Hwang, J.J., Azernikov, S., Efros, A.A., Yu, S.X.: Learning beyond human expertise with generative models for dental restorations. arXiv preprint arXiv:1804.00064 (2018)
17. Li, J., Xu, K., Chaudhuri, S., Yumer, E., Zhang, H., Guibas, L.: GRASS: generative recursive autoencoders for shape structures. ACM Trans. Graph. (TOG) **36**(4) (2017). Article no. 52
18. Li, Y., Wang, G., Ji, X., Xiang, Y., Fox, D.: DeepIM: deep iterative matching for 6d pose estimation. In: Proceedings of the European Conference on Computer Vision (ECCV), pp. 683–698 (2018)
19. Li, Y., Tarlow, D., Brockschmidt, M., Zemel, R.: Gated graph sequence neural networks. arXiv preprint arXiv:1511.05493 (2015)
20. Lian, C., et al.: MeshSNet: deep multi-scale mesh feature learning for end-to-end tooth labeling on 3D dental surfaces. In: Shen, D., et al. (eds.) MICCAI 2019. LNCS, vol. 11769, pp. 837–845. Springer, Cham (2019). https://doi.org/10.1007/978-3-030-32226-7_93
21. Majerowicz, L., Shamir, A., Sheffer, A., Hoos, H.H.: Filling your shelves: synthesizing diverse style-preserving artifact arrangements. IEEE Trans. Vis. Comput. Graph. **20**(11), 1507–1518 (2013)
22. Mirza, M., Osindero, S.: Conditional generative adversarial nets. arXiv preprint arXiv:1411.1784 (2014)
23. Mo, K., et al.: StructureNet: hierarchical graph networks for 3D shape generation. arXiv preprint arXiv:1908.00575 (2019)
24. Park, J.J., Florence, P., Straub, J., Newcombe, R., Lovegrove, S.: DeepSDF: learning continuous signed distance functions for shape representation. arXiv preprint arXiv:1901.05103 (2019)
25. Peng, S., Liu, Y., Huang, Q., Zhou, X., Bao, H.: PVNet: pixel-wise voting network for 6dof pose estimation. In: Proceedings of the IEEE Conference on Computer Vision and Pattern Recognition, pp. 4561–4570 (2019)
26. Qi, C.R., Liu, W., Wu, C., Su, H., Guibas, L.J.: Frustum pointnets for 3D object detection from RGB-D data. In: Proceedings of the IEEE Conference on Computer Vision and Pattern Recognition, pp. 918–927 (2018)
27. Qi, C.R., Su, H., Mo, K., Guibas, L.J.: PointNet: deep learning on point sets for 3D classification and segmentation. In: Proceedings of the IEEE Conference on Computer Vision and Pattern Recognition, pp. 652–660 (2017)
28. Qi, C.R., Yi, L., Su, H., Guibas, L.J.: PointNet++: deep hierarchical feature learning on point sets in a metric space. In: Advances in Neural Information Processing Systems, pp. 5099–5108 (2017)
29. Song, S., Xiao, J.: Sliding shapes for 3D object detection in depth images. In: Fleet, D., Pajdla, T., Schiele, B., Tuytelaars, T. (eds.) ECCV 2014. LNCS, vol. 8694, pp. 634–651. Springer, Cham (2014). https://doi.org/10.1007/978-3-319-10599-4_41
30. Song, S., Xiao, J.: Deep sliding shapes for amodal 3D object detection in RGB-D images. In: Proceedings of the IEEE Conference on Computer Vision and Pattern Recognition, pp. 808–816 (2016)

31. Sung, M., Su, H., Kim, V.G., Chaudhuri, S., Guibas, L.: ComplementMe: weakly-supervised component suggestions for 3D modeling. ACM Trans. Graph. (TOG) **36**(6) (2017). Article no. 226

32. Tatarchenko, M., Dosovitskiy, A., Brox, T.: Octree generating networks: efficient convolutional architectures for high-resolution 3D outputs. In: Proceedings of the IEEE International Conference on Computer Vision, pp. 2088–2096 (2017)

33. Tekin, B., Sinha, S.N., Fua, P.: Real-time seamless single shot 6D object pose prediction. In: Proceedings of the IEEE Conference on Computer Vision and Pattern Recognition, pp. 292–301 (2018)

34. Tremblay, J., To, T., Sundaralingam, B., Xiang, Y., Fox, D., Birchfield, S.: Deep object pose estimation for semantic robotic grasping of household objects. arXiv preprint arXiv:1809.10790 (2018)

35. Wang, C., et al.: DenseFusion: 6D object pose estimation by iterative dense fusion. In: Proceedings of the IEEE Conference on Computer Vision and Pattern Recognition, pp. 3343–3352 (2019)

36. Wang, H., Sridhar, S., Huang, J., Valentin, J., Song, S., Guibas, L.J.: Normalized object coordinate space for category-level 6D object pose and size estimation. In: Proceedings of the IEEE Conference on Computer Vision and Pattern Recognition, pp. 2642–2651 (2019)

37. Wang, K., Lin, Y.A., Weissmann, B., Savva, M., Chang, A.X., Ritchie, D.: PlanIT: planning and instantiating indoor scenes with relation graph and spatial prior networks. ACM Trans. Graph. (TOG) **38**(4) (2019). Article no. 132

38. Wang, K., Savva, M., Chang, A.X., Ritchie, D.: Deep convolutional priors for indoor scene synthesis. ACM Transactions on Graphics (TOG) **37**(4) (2018). Article no. 70

39. Xiang, Y., Schmidt, T., Narayanan, V., Fox, D.: PoseCNN: a convolutional neural network for 6D object pose estimation in cluttered scenes. arXiv preprint arXiv:1711.00199 (2017)

40. Xu, D., Anguelov, D., Jain, A.: PointFusion: deep sensor fusion for 3D bounding box estimation. In: Proceedings of the IEEE Conference on Computer Vision and Pattern Recognition, pp. 244–253 (2018)

41. Xu, X., Liu, C., Zheng, Y.: 3D tooth segmentation and labeling using deep convolutional neural networks. IEEE Trans. Vis. Comput. Graph. **25**(7), 2336–2348 (2018)

42. Yu, L.F., Yeung, S.K., Tang, C.K., Terzopoulos, D., Chan, T.F., Osher, S.: Make it home: automatic optimization of furniture arrangement. ACM Trans. Graph. **30**(4) (2011). Article no. 86

43. Zanjani, F.G., et al.: Mask-MCNet: instance segmentation in 3D point cloud of intra-oral scans. In: Shen, D., et al. (eds.) MICCAI 2019. LNCS, vol. 11768, pp. 128–136. Springer, Cham (2019). https://doi.org/10.1007/978-3-030-32254-0_15

44. Zhou, Y., Tuzel, O.: VoxelNet: end-to-end learning for point cloud based 3D object detection. In: Proceedings of the IEEE Conference on Computer Vision and Pattern Recognition, pp. 4490–4499 (2018)

45. Zhu, M., et al.: Single image 3D object detection and pose estimation for grasping. In: 2014 IEEE International Conference on Robotics and Automation (ICRA), pp. 3936–3943. IEEE (2014)

UnionDet: Union-Level Detector Towards Real-Time Human-Object Interaction Detection

Bumsoo Kim, Taeho Choi, Jaewoo Kang[✉], and Hyunwoo J. Kim[✉]

Korea University, Seoul 02841, Republic of Korea
{meliketoy,major1965,kangj,hyunwoojkim}@korea.ac.kr

Abstract. Recent advances in deep neural networks have achieved significant progress in detecting individual objects from an image. However, object detection is not sufficient to fully understand a visual scene. Towards a deeper visual understanding, the interactions between objects, especially humans and objects are essential. Most prior works have obtained this information with a bottom-up approach, where the objects are first detected and the interactions are predicted sequentially by pairing the objects. This is a major bottleneck in HOI detection inference time. To tackle this problem, we propose *UnionDet*, a one-stage meta-architecture for HOI detection powered by a novel union-level detector that eliminates this additional inference stage by directly capturing the region of interaction. Our one-stage detector for human-object interaction shows a significant reduction in interaction prediction time $(4\times\sim14\times)$ while outperforming state-of-the-art methods on two public datasets: V-COCO and HICO-DET.

Keywords: Visual relationships · Real-time detection · Human-object interaction detection · Object detection

1 Introduction

Recent advances in deep neural networks have achieved significant progress in detecting and recognizing individual objects from an image. However, to understand a scene, we need a deeper visual understanding that transcends the level of individual object detection. To understand what is happening in the image, not only do we have to accurately detect individual objects, but we also have to properly predict the interactions between the detected objects. Among the

---

B. Kim and T. Choi—Equal contribution.

Electronic supplementary material The online version of this chapter (https://doi.org/10.1007/978-3-030-58555-6_30) contains supplementary material, which is available to authorized users.

© Springer Nature Switzerland AG 2020
A. Vedaldi et al. (Eds.): ECCV 2020, LNCS 12360, pp. 498–514, 2020.
https://doi.org/10.1007/978-3-030-58555-6_30

interactions, in this paper, we focus on *human-object interaction (HOI) detection* that involves the localization and classification of interactions between humans and surrounding objects. HOI detection has been formally defined in [10] as the task to detect $\langle human, verb, object \rangle$ triplets within an image.

The main challenge of HOI detection boils down to a simple question: *"How can we localize* **interactions***?"*. When asked to localize the area of *"A person rides a horse."*, a human can naturally spot the tight area that covers both the person and the horse he/she is riding. This is the *union region* of the interacting objects that have been considered as a representation of visual relationships from previous works [21], and have been widely utilized in HOI detection [8,11,15,17, 29,30,35]. Ironically, no detector in the literature has been studied to *directly* capture the union region.

All the previous HOI detectors, therefore, incorporated a multi-stage and sequential pipeline that detects the individual objects first and 'associate' them to obtain the union region. This approach is far from intuitive and it makes HOI detectors inefficient. The sequential pipeline of object detection and interaction prediction makes end-to-end training impossible and creates a huge bottleneck in inference time for HOI detection. In standard object detection, one-stage detectors [19,20,26,32,33] were able to speed up two-stage detectors by eliminating the second stage while yielding a competitive performance. Yet in HOI detection, previous multi-stage models mainly focused on performance (e.g., average precision) leaving the large gap between high-performance and real-time detection unexplored. In this work, our goal is to fill the gap between the performance and inference time of HOI detection with a fast, single-stage model.

To this end, we propose *UnionDet:* a one-stage meta-architecture powered by a novel *union-level* detector that captures the union region of human-object interaction. Instead of associating the object detection results by feeding each object pair into a separate neural network afterward, we directly detect interacting $\langle human, object \rangle$ pairs with our novel union-level detection framework. This eliminates the need for heavy neural network inference after object detection and enables our model to detect interactions with minimal additional time on top of existing object detectors. Though the union-level detection sounds intuitive, detecting the union region is much more challenging than instance-level detection. In this paper, we study new challenges in union-level detection and address them by new techniques: (i) union anchor labeling, (ii) target object classification loss and (iii) union foreground focal loss. Based on these new methods, our proposed one-stage HOI detector achieves a $4\times \sim 14\times$ speed-up in additional inference time for interaction prediction while surpassing state-of-the-art performance on two HOI detection benchmark datasets: V-COCO (*Verbs in COCO*) and HICO-DET. The main **contributions** of our paper are threefold:

– We study new technical challenges in union-level detection, including bias toward human regions, inaccuracy of standard IoU-based matching and union regions containing multiple interactions and more than two objects.

- We propose a novel *union-level* detector that directly detects the interaction region. We study a new set of training techniques to address the new challenges of union-level detection.
- We propose a meta-architecture **UnionDet** equipped with our *union-level* detector. It is a single-stage HOI detector achieving 4×∼14× speed-up in interaction prediction and the *state-of-the-art* performance in two public datasets.

2 Related Work

2.1 One-Stage Object Detection

One-stage object detectors based on deep neural networks have formerly been proposed for faster detection [20,26,28]. These detectors have achieved a significant speed-up but they often come with a considerable loss of accuracy. One known problem is the class imbalance problem. Since one-stage object detectors densely sample anchor boxes, foreground anchor boxes are relatively much rarer than background anchor boxes (or negative samples), unlike two-stage methods that classify only a few anchor boxes after RPN. One common technique to resolve the class imbalance is hard negative mining which samples a few hard anchor boxes for training [27]. Later, RetinaNet [19] introduced focal loss to address the issue in a fundamental way by modifying the loss function to reduce the effect of easy negatives. Including these efforts, various techniques have been proposed to enhance one-stage object detection frameworks [32,33]. Recently, YOLACT [2] has expanded the capacity of one-stage networks to perform instance segmentation.

2.2 Human-Object Interactions

Human-Object Interaction (HOI) detection has been initially proposed in [10]. Later, human-object detectors have been improved using human body parts [6], human appearance [9], instance appearance [8] and spatial relationship of human-object pairs [8,15,17]. Especially, InteractNet [9] extended an existing object detector by introducing an action-specific density map to localize target objects based on the appearance of a detected human. Note that interaction detection based on visual cues from individual boxes often suffers from the lack of contextual information. So iCAN [8] proposed an instance-centric attention module that extracts contextual features complementary to the features from the localized objects/humans. GPNN [25] proposes a Graph Parsing Neural Network for HOI recognition—a general framework that explicitly represents HOI structures with graphs and automatically infers the optimal graph structures. Deep Contextual Attention [30] leverages contextual information by a contextual attention framework in HOI. Recent works in HOI have also explored external knowledge to improve the performance of HOI detection. Since the performance of HOI detection is dependent on how well we recognize the appearance of human actions, human pose information extracted from external models [3,5,7,13,16] shows meaningful improvement in performance [11,17,29,35]. Interactiveness Knowledge has also been implemented in previous works [17] by adding an additional inference stage where the model learns the probability of interactiveness by combining multiple HOI training datasets. Linguistic priors and knowledge graphs are also utilized to improve HOI detection performance. These sources are either used directly as an additional feature [12,24,31] or features to cluster the objects by their functions [1]. However, all the previous methods are multi-stage detectors focusing on accuracy and they are not suitable for real-time applications.

3   Method

We now introduce our method to detect human-object-interaction. To be specific, the goal is to capture $\langle human, verb, object \rangle$ triplets from an image without any external knowledge. The standard HOI detection benchmarks (e.g., V-COCO and HICO-DET) require the localization and classification of interactions. In this paper, we propose a one-stage HOI detector powered by our *union-level* detector, which directly detects the union region of an interacting pair. Since standard benchmarks require the instance-level localization of humans and objects, we *parallelly* combine the union-level detector and an instance-level detector, which allows more accurate instance-level localization. We name this meta-architecture **UnionDet** shown in Fig. 2. Our UnionDet is compatible with any one-stage object detectors such as SSD [20], RetinaNet [18], and STDN [34]. For a fair comparison with baseline HOI detectors, in this paper, we implement our model based on RetinaNet with ResNet50-FPN [14,18] since it's performance is comparably similar to Faster-RCNN—the dominant backbone network in previous works on HOI in literature.

We discuss new challenges in *union-level* detection and how to address them by the components in our union-level detector, which is the union branch in UnionDet in Fig. 2. We explain how to modify a standard instance-level detector in UnionDet for HOI detection and lastly, the details of training and inference are provided.

3.1   Challenges in Union-Level Detection

The union region of a pair of objects is an intuitive representation of visual relationships [21]. Union-level detection looks similar to instance-level detection. But standard object detectors are not directly applicable due to the following technical challenges. ***First***, a naive union-level detection often suffers from the large bias towards human regions since every union region of HOI has a human. The left figure in Fig. 1 shows that union predictions (green bboxes) by a vanilla detector are densely distributed around a human. ***Second***, the standard IoU is not an accurate metric for union bounding box matching. For instance, when one union region has two remote objects, a high IoU with the union region

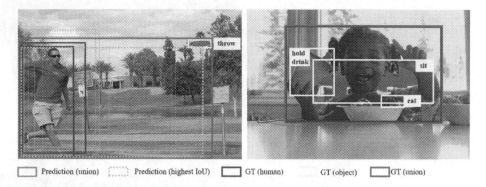

Prediction (union)   Prediction (highest IoU)   GT (human)   GT (object)   GT (union)

Fig. 1. Technical challenges in union detection. (Left) The box with the highest confidence for the ground-truth action 'throw' is highlighted in bold, and the region with the highest IoU with the ground-truth union region is dotted. As you can see, the highest confidence is biased towards the human region, and despite the high IoU, the dotted region failed to capture the target object (frisbee). (Right) Issues of overlapping union regions in *one vs many* relations. Even though the 3 different ⟨*human, object*⟩ pairs represent a different sets of interactions, all 3 pairs have an identical union region. (color figure online)

does not ensure that both human and target object is enclosed by the predicted region (see the left figure in Fig. 1). *Lastly*, one union region (or anchor box) may contain multiple interactions and more than two objects. These are often observed especially when a human bounding box contains multiple interacting objects. In the following explanation of Union Branch that performs union-level detection, we show in detail how we address these issues.

3.2   Union-Level Detector: Union Branch

Union Branch performs union-level detection which is the essence of our proposed meta-architecture, *UnionDet*. As in Fig. 2, Union Branch consists of three sub-branches that share the backbone Feature Pyramid Network. Out of the three sub-branches, the Action Classification sub-branch and the Union Box Regression sub-branch are the main sub-branches that contribute to the inference stage. Action Classification sub-branch performs multi-class classification for the interactions that are related to the union region, and the Union Box Regression sub-branch performs action-agnostic bounding box regression to predict the final union region with multiple actions. Vanilla detection results for union regions can be obtained through these two sub-branches. However, union regions inherently accompany several technical challenges as mentioned above. To address these challenges, Union Branch is trained by new techniques: i) union anchor labeling ii) target object classification loss iii) foreground focal loss. This provides accurate union-level detections even in various distances, see Fig. 4.

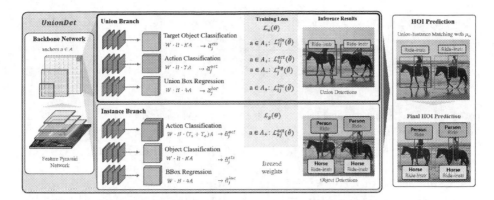

Fig. 2. The overall architecture of UnionDet. Our UnionDet is generally compatible with one-stage object detectors. The feature pyramid obtained from the backbone network is simultaneously fed to Union Branch and Instance Branch. While Union Branch directly captures the region of interaction, Instance Branch performs traditional object detection and action classification for more fine-grained HOI detection results.

Union Anchor Labeling. The standard IoU is not suitable for union-level detections. Especially, when objects are small and remote, an anchor box may fail to include either the subject or the object in interaction even when the ground-truth union bounding box and the anchor box has a high (e.g., 0.9) IoU. To address this, we propose a new labeling function to match union level labels to anchor boxes. Union Branch detects the union regions based on the set of anchors A generated from the backbone Feature Pyramid Network. During the forward propagation of Union Branch, each anchor $a_j \in A$ obtains a multi-label action prediction $\breve{a}_j^{act} \in \mathbb{R}^T$, target object class prediction $\breve{a}_j^{cls} \in \mathbb{R}^K$, and a location prediction $\breve{a}_j^{loc} \in \mathbb{R}^4$. T and K denote the number of interactions and the number of target object categories, respectively. $U_{ij} \in \{0,1\}$ indicates whether the i_{th} union-level ground-truth label $\breve{g}_i$ matches the j_{th} anchor a_j or not. We propose a new anchor labeling function, which is used during training. Let $\mathbb{1}(\cdot)$ as an indicator function. Given human box $\breve{h}_i$ and object box $\breve{o}_i$ of i-th ground truth union box $\breve{g}_i$, U_{ij} is calculated as:

$$U_{ij} = \mathbb{1}(\text{IoU}(a_j, \breve{g}_i^{loc}) > t_u) \cdot \mathbb{1}\left(\frac{a_j \cap \breve{h}_i^{loc}}{\breve{h}_i^{loc}} > t_h \right) \cdot \mathbb{1}\left(\frac{a_j \cap \breve{o}_i^{loc}}{\breve{o}_i^{loc}} > t_o \right), \quad (1)$$

where t_u, t_h, t_o indicate thresholds for union IoU, human inclusion ratio, and object inclusion ratio. They are set to 0.5 in our experiments. If multiple union-level ground truths are matched, the union with the largest IoU is associated with the anchor box so that an anchor box has at most one ground truth.

After labeling each anchor according to Eq. 1, we can build a basic loss function to train the Union Branch. Based on the positive anchor set $A_+ \subseteq A$

where $\{a_j| \sum_i U_{ij} = 1\}$ and the negative anchor samples $A_- \subseteq A$ where $\{a_j| \sum_i U_{ij} = 0\}$, the loss function $\mathcal{L}_u(\breve{\theta})$ is written as

$$\mathcal{L}_u(\breve{\theta}) = \sum_{a_j \in A_+} \sum_{\breve{g}_i \in \breve{\mathcal{G}}} U_{ij} \left[\mathcal{L}_{ij}^{act}(\breve{\theta}) + \mathcal{L}_{ij}^{loc}(\breve{\theta}) \right] + \sum_{a_j \in A_-} \mathcal{L}_j^{bg}(\breve{\theta}), \qquad (2)$$

where $\breve{\mathcal{G}}$ denotes the ground truth union box set and $\breve{\theta}$ denotes the Union Branch model parameters. $\mathcal{L}_{ij}^{act}(\breve{\theta}) = FL(\breve{a}_j^{act}, \breve{g}_i^{act}, \breve{\theta})$, $\mathcal{L}_{ij}^{loc}(\breve{\theta}) = smooth_{L1}(\breve{a}_j^{loc}, \breve{g}_i^{loc}, \breve{\theta})$, $\mathcal{L}_j^{bg} = FL(\breve{a}_j^{act}, \mathbf{0}, \breve{\theta})$, where FL and $smooth_{L1}$ each denotes focal loss [19] and Smooth L1 loss, respectively. After training the Union Branch with Eq. 2, a vanilla prediction of union regions can be obtained. However, it suffers from 1) the prediction being biased toward the human region, and 2) the noisy learning caused when multiple union regions overlap over each other.

Target Object Classification Loss. To address the first issue where the union prediction is biased toward the human region with a vanilla union-level detector, we design a pretext task, 'target object classification' from the detected union region. This encourages the union-level detector to focus more on target objects and helps the union-level detector to capture the region that encloses the target object. We add the target object classification loss to Eq. 2 and the loss function $\mathcal{L}_u$ of Union Branch is given as

$$\mathcal{L}_u(\breve{\theta}) = \sum_{a_j \in A_+} \sum_{\breve{g}_i \in \breve{\mathcal{G}}} U_{ij} \left[\mathcal{L}_{ij}^{act}(\breve{\theta}) + \mathcal{L}_{ij}^{loc}(\breve{\theta}) + \mathcal{L}_{ij}^{cls}(\breve{\theta}) \right] + \sum_{a_j \in A_-} \mathcal{L}_j^{bg}(\breve{\theta}), \qquad (3)$$

where $\mathcal{L}_{ij}^{cls}(\breve{\theta}) = BCE(\breve{a}_j^{cls}, \breve{g}_i^{cls}, \breve{\theta})$ is the Binary Cross Entropy loss. Though we do not use the target classification score at inference in the final HOI score function, we observed that learning to classify the target objects during training improves the union region detection as well as overall performance (see, Table 3).

Union Foreground Focal Loss. Union regions often overlap over each other when a single person interacts with multiple surrounding objects. The right subfigure in Fig. 1 shows an extreme example of overlapping union regions where different interaction pairs have the exactly same union region. In such cases where large portion of union regions overlap with each other, applying vanilla focal loss $\mathcal{L}_{ij}^{act}(\breve{\theta})$ as in Eq. 2 and Eq. 3 might mistakenly give negative loss to the overlapped union actions (more detailed explanation of such cases will be dealt in our supplement). To address this issue, we deployed a variation of focal loss where we selectively calculate losses for only positive labels for foreground regions. This is implemented by simply multiplying $\breve{g}_i^{act}$ to $\mathcal{L}_{ij}^{act}$, thus our final loss function is written as:

$$\mathcal{L}_u(\breve{\theta}) = \sum_{a_j \in A_+} \sum_{\breve{g}_i \in \breve{\mathcal{G}}} U_{ij} \left[\breve{g}_i^{act} \cdot \mathcal{L}_{ij}^{act}(\breve{\theta}) + \mathcal{L}_{ij}^{loc}(\breve{\theta}) + \mathcal{L}_{ij}^{cls}(\breve{\theta}) \right] + \sum_{a_j \in A_-} \mathcal{L}_j^{bg}(\breve{\theta}). \qquad (4)$$

3.3   Instance-Level Detector: Instance Branch

HOI detection benchmarks require the localization of instances in interactions. For more accurate instance localization, we added Instance Branch to our architecture, see Fig. 2. The Instance Branch parallelly performs instance-level HOI detection: object classification, bbox regression, and action (or *verb*) classification.

Object Detection. The instance-level detector was built based on a standard anchor-based single-stage object detector that performs object classification and bounding box regression. For training, we adopt the focal loss [19] to handle the class imbalance problem between the foreground and background anchors. The object detector is frozen for the V-COCO dataset and fine-tuned for the HICO-DET dataset. More discussion is available in the supplement.

Action Classification. The instance-level detector was extended by another sub-branch for action classification. We treat the action of subjects T_s and objects T_o as different types of actions. So, the action classification sub-branch predicts $(T_s + T_o)$ action types at every anchor. This helps to recognize the direction of interactions and can be combined with the interaction prediction from the Union Branch. For action classification, we only calculate the loss at the positive anchor boxes where an object is located at. This leads to more efficient loss calculation and improvement accuracy.

Training Loss $\mathcal{L}_g$ to Learn Instance-Level Actions. Instance Branch and Union Branch share the anchors A generated from the backbone Feature Pyramid Network. The set of instance-level ground-truth annotations for an input image is denoted as $\hat{\mathcal{G}}$. The ground-truth label $\hat{g}_i \in \hat{\mathcal{G}}$ at anchor box i consists of target class label $\hat{g}_i^{cls} \in \{0,1\}^K$, multi-label action types $\hat{g}_i^{act} \in \{0,1\}^{(T_s+T_o)}$ and a location $\hat{g}_i^{loc} \in \mathbb{R}^4$, i.e., $\hat{g}_i = (\hat{g}_i^{cls}, \hat{g}_i^{act}, \hat{g}_i^{loc}) \in \hat{\mathcal{G}}$. During the forward propagation of Instance Branch, each anchor $a_j \in A$ obtains a multi-label action prediction $\hat{a}_j^{act} \in \mathbb{R}^{T_s+T_o}$ and object class prediction $\hat{a}_j^{cls} \in \mathbb{R}^K$ after sigmoid activation, and a location prediction $\hat{a}_j^{loc} = \{x, y, w, h\}$ after bounding box regression. $I_{ij} \in \{0,1\}$ indicates whether object $\hat{g}_i$ matches anchor a_j or not, i.e., $I_{ij} = \mathbb{1}(IoU(a_j, \hat{g}_i) > t)$. We used threshold $t = 0.5$ in the experiments. The Object Classification and BBox Regression sub-branches are fixed with pre-trained weights of object detectors [19], and only the Action Classification sub-branch is trained. Given parameters of Action Classification sub-branch $\hat{\theta}$, the loss for the Instance Branch $\mathcal{L}_g(\hat{\theta})$ will be $\mathcal{L}_g(\hat{\theta}) = \mathcal{L}_{ij}^{act}(\hat{\theta}) = BCE(\hat{a}_j^{act}, \hat{g}_i^{act}, \hat{\theta})$.

3.4   Training UnionDet

The two branches of UnionDet shown in Fig. 2 (i.g., the Union Branch and Instance Branch) are trained jointly. Our overall loss is the sum of the losses of both branches, $\mathcal{L}_u$ and $\mathcal{L}_g$, where $\breve{\theta}$ is the parameters for Union Branch and $\hat{\theta}$ is the parameters for Instance Branch ($\theta = \breve{\theta} \cup \hat{\theta}$). The final loss becomes $\mathcal{L}(\theta) = \mathcal{L}_u(\breve{\theta}) + \mathcal{L}_g(\hat{\theta})$. For focal loss, we use $\alpha = 0.25$, $\gamma = 2.0$ as in [19]. Our model is trained with an Adam optimizer with a learning rate of 1e−5.

3.5   HOI Detection Inference

UnionDet at inference time parallelly performs the inference of Union Branch and Instance Branch and then seeks the highly-likely triplets using a summary score combining predictions from the subnetworks. Instance Branch performs object detection and action classification per anchor box. Non-maximum suppression with its object classification scores was performed. Union Branch directly detects the union region that covers the $\langle human, verb, object \rangle$ triplet. For Union Branch, non-maximum suppression was applied with union-level action classification scores. Instead of applying class-wise NMS as in ordinary object detection, we treated different action classes altogether to handle multi-label predictions of union regions.

Union-Instance Matching. As mentioned in Sect. 3.2, IoU is not an accurate measure for union regions, especially in the case where the target object of the interaction is remote and small. To search for a solid union region that covers the given human box b_h and object box b_o, we search for the union box b_u with our proposed *union-instance matching score* defined as

$$\mu_u = \frac{\text{IoU}(\lceil b_h \cup b_o \rfloor, b_u)}{2} + \frac{1}{2}\sqrt{\frac{(b_h \cap b_u)}{b_h} \cdot \frac{(b_o \cap b_u)}{b_o}}, \qquad (5)$$

where $\frac{b_1}{b_2}$ is the ratio of the areas of two bounding boxes b_1 and b_2 and $\lceil \cdot \rfloor$ stands for the tightest bounding box that covers the area. We use this union-instance matching score to calculate the HOI score instead of the standard IoU.

HOI Score. The detections from Union Branch and Instance Branch are integrated. This further improves the accuracy of the final HOI detection. Our HOI score function combines union-level action score s_u^a from Union Branch with the human category score s_h, human action score s_h^a, object class score s_o, instance-level action score s_o^a from Instance Branch. For each $\langle human, object \rangle$ pair, we first identify the best union area with the highest union-instance matching score μ_u in Eq. (5) and then calculate the HOI score $S_{h,o}^a$ as

$$S_{h,o}^a = (s_h \cdot s_h^a + s_o \cdot s_o^a) \cdot (1 + \mu_u \cdot s_u^a). \qquad (6)$$

When the action classes do not involve target objects, or no union region is predicted, the score will be $S_{h,o}^a = s_h \cdot s_h^a$ and $S_{h,o}^a = s_h \cdot s_h^a + s_o \cdot s_o^a$, respectively.

The calculation of Eq. (6) has in principle $O(n^3)$ complexity when the number of detections is n. However, our framework calculates the final triplet scores without any additional neural network inference after Union and Instance Branches.

The calculation time of Eq. (6) is negligible (<1 ms). The end-to-end inference time of our model is marginally increased (~9 ms) compared to the vanilla object detector (RetinaNet with ResNet50-FPN) thanks to the parallel architecture.

4   Experiments

In this section, we demonstrate the effectiveness of UnionDet in HOI detection. We first describe the two public datasets that we use as our benchmark: V-COCO and HICO-DET. Next, we perform various qualitative and quantitative analysis to show that our union-level detector successfully addresses the proposed technical challenges and captures quality union regions, leading to a fast and accurate one-stage HOI detector.

Fig. 3. Union-level detections (red) by Union Branch successfully group correct pairs of humans (blue), and target objects (yellow) among confusing cases caused by multiple triplets with the same action and target object types in an image. Best viewed in color. (Color figure online)

Datasets. To validate the performance of our model, we evaluate our model on two public benchmark datasets: the V-COCO (*Verbs in COCO*) dataset and HICO-DET dataset. ***V-COCO*** is a subset of COCO and has 5,400 `trainval` images and 4,946 `test` images. For V-COCO dataset, we report the AP_{role} over $T = 29$ interactions. Including the four interaction types that do not involve target objects, V-COCO has $T_s = 26$ active actions and $T_o = 25$ passive actions. As previous works, we exclude the interaction *point* during inference time, because only 31 instances appear in the test set. ***HICO-DET*** [4] is a subset of HICO dataset and has more 150K annotated instances of human-object pairs in 47,051 images (37,536 training and 9,515 testing) and is annotated with 600 ⟨*verb, object*⟩ interaction types. There are 80 unique object types, identical to the COCO object categories, and $T = 117$ unique verbs. In the HICO-DET dataset, we separate the 117 action classes into a_s, a_o, thus leading into a total action number of $T_s + T_o = 234$. For HICO-DET dataset, we follow the previous settings and report the mAP over three different category sets: (1) all 600 HOI categories in HICO (Full), (2) 138 HOI categories with less than 10 training instances (Rare), and (3) 462 HOI categories with 10 or more training instances (Non-Rare).

Union-Level Detection. Our union-level detector (Union Branch in Union-Det) directly detects union regions of HOI, see Fig. 3. Interestingly, the union-level detections are useful to disambiguate the confusing pairs with the same action and target object types (e.g., horse, or motorcycle) in an image. For example, when multiple people *ride* the same target objects as in Fig. 3, instance-level appearances are not sufficient to associate the correct pairs. Union-level detections successfully group them using the context in the union-region.

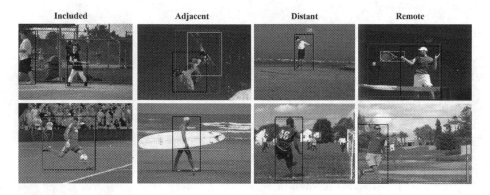

Fig. 4. Our union-level detector (Union Branch) successfully detects the union bounding boxes (red) in various distances: the interactions with included, adjacent, distant and remote target objects. Also, instance-level detections of human (blue), and target objects (yellow) by Instance Branch are visualized. Best viewed in color. (Color figure online)

Interactions in Various Distances. We discussed in Sect. 3.1 that a vanilla object detector is not directly applicable to union-level detection due to the bias toward human regions. This bias gets severer especially when a human interacts with remote target objects. Figure 4 shows that the bias is addressed by our pre-text task 'Target Object Classification' and UnionDet is able to detect target objects for various distances. We show four cases: included ($b_h \supset b_o$), adjacent ($\text{IoU}(b_h, b_o) > 0$), distant ($\text{IoU}(b_h, b_o) = 0$) and remote ($\text{IoU}(b_h, b_o) = 0$ and large distance), where b_h, and b_o are human and object bounding boxes. Especially the fourth column in Fig. 4 shows that UnionDet successfully captures the remote relation with small remote target objects (e.g., tennis ball and frisbee). Our ablation study in Table 3 provides that the 'Target Object Classification' improves HOI detection. Qualitative results of a vanilla union-level detector without the Target Object Classification sub-branch are provided in the supplement.

HOI Detection Results. In Fig. 5, we highlight the detected humans and objects by object detection with the blue and yellow boxes and the union region predicted by UnionDet with red boxes. The detected human-object interactions are visualized and given a pair of objects, the $\langle human, verb, object \rangle$ triplet with the highest HOI score is listed below each image. Note that our model can detect various types of interactions including one-to-one, many-to-one (multiple persons interacting with a single object), one-to-many (one person interacting with multiple objects), and many-to-many (multiple persons interacting with multiple objects) relationships.

Performance Analysis. We quantitatively evaluate our model on two datasets, followed by the ablation study of our proposed methods. We use the official evaluation code for computing the performance of both V-COCO and HICO-DET. In V-COCO, there are two versions of evaluation but most previous works have not explicitly stated which version was used for evaluation. We have specified the evaluation scenario if it has been referred in either the literature [35], authors' code or the reproduced code. We report our performance in both scenarios for a fair comparison with heterogeneous baselines. In both scenarios, our model outperforms state of the art methods [30]. Further, our model shows competitive

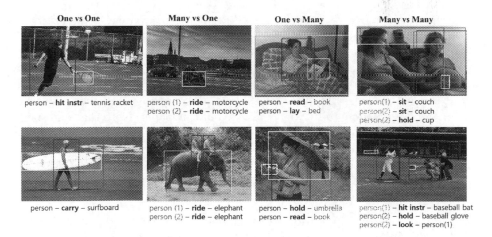

Fig. 5. The final HOI detection by our model combining the predictions of both branches of UnionDet. The columns from the left to the right show one-to-one, many-to-one (multiple persons interacting with a single object), one-to-many (one person interacting with multiple objects), and many-to-many (multiple persons interacting with multiple objects) relationships respectively.

Table 1. Comparison of performance and additional inference time on V-COCO test set. $\cdot_{\#1}$, $\cdot_{\#2}$ each refers to the performance with Scenario#1 and Scenario#2. While achieving $4\times\sim14\times$ speed-up in additional inference time for interaction prediction, our model surpasses all state-of-the-art performances in both Scenario#1 and Scenario#2. Our model also achieves competitive performance to those that deploy external knowledge or features. Note that our model does not use any external knowledge.

Method	Feature backbone	External Resources	AP_{role}	t(ms)
Models with external features				
Verb Embedding [31]	ResNet50	GloVe [23], VRD [21]	$45.9_{\#1}$	
RP_DC_D [17]	ResNet50	Pose [7,16]	$47.8_{\#1}$	
RPNN [35]	ResNet50	Keypoint [13]	$47.5_{\#2}$	
PMFNet [29]	ResNet50-FPN	Pose [5]	52.0	
Models with original comparison				
VSRL [10]	ResNet50-FPN	✗	31.8	-
InteractNet [9]	ResNet50-FPN	✗	$40.0_{\#2}$	55
BAR-CNN [15]	ResNet50-FPN	✗	43.6	125
GPNN [25]	**ResNet152**	✗	44.0	40
iCAN [8]	ResNet50	✗	$44.7_{\#1}$	75
TIN (RC$_D$) [17]	ResNet50	✗	$43.2_{\#1}$	70
DCA [30]	ResNet50	✗	47.3	130
UnionDet (Ours)	ResNet50-FPN	✗	$\mathbf{47.5}_{\#1}$ $\mathbf{56.2}_{\#2}$	**9.06**

performance compared to the baselines [17,31,35] that leverage heavy external features such as linguistic priors [22,23] or human pose features [3,5,7,16]. On HICO-DET, our model achieves state-of-the-art performance for both the official 'Default' setting and 'Known Object' setting. For a more comprehensive evaluation of HOI detectors, we also provide the performance of recent works leverage external knowledge [11,12,17,24,29,31,31,35], although the ***models with External Knowledge*** are beyond the scope of this paper. Note that our main focus is to build a fast single-stage HOI detector from visual features (Table 2).

Our **ablation study** in Table 3 shows that each component (foreground focal loss, target object classification loss $\mathcal{L}_{ij}^{cls}(\breve{\theta})$, union matching function μ_u) in our approach improves the overall performance of HOI detection.

Interaction Prediction Time. We measured inference time on a single Nvidia GTX1080Ti GPU. Our model achieved the fastest 'end-to-end' inference time (**77.6 ms**). However, the end-to-end inference time is not suitable for fair comparison since the end-to-end computation time of one approach may largely vary depending on the base networks or the backbone object detector. Therefore, we here compare the ***additional*** time for interaction prediction, excluding the time for object detection. The detailed analysis of end-to-end time will also be pro-

Table 2. Performance and additional inference time comparison in HICO-DET. Models with † used a heavier feature extraction backbone (i.e., ResNet152-FPN). Further experimental settings are discussed in detail in our supplement. Our model shows the fastest inference time while achieving state-of-the-art performance across the official evaluation metrics of HICO-DET.

Method	Ext src	Default			Known Object			t(ms)
		Full	Rare	Non Rare	Full	Rare	Non Rare	
Models with external features								
Verb Embedding [31]	[21,23]	14.70	13.26	15.13	-	-	-	
TIN (RP$_D$C$_D$) [17]	[7,16]	17.03	13.42	18.11	19.17	15.51	20.26	
Functional Gen. [1]	[22]	21.96	16.43	23.62	-	-	-	
RPNN [35]	[13]	17.35	12.78	18.71	-	-	-	
PMFNet [29]	[5]	17.46	15.65	18.00	20.34	17.47	21.20	
No-Frills HOI [11]†	[3]	17.18	12.17	18.68	-	-	-	
Analogies [24]	[22]	19.4	14.6	20.9	-	-	-	
Models with original comparison								
VSRL [10]	✗	9.09	7.02	9.71	-	-	-	-
HO-RCNN [15]	✗	7.81	5.37	8.54	10.41	8.94	10.85	-
InteractNet [9]	✗	9.94	7.16	10.77	-	-	-	55
GPNN [25]†	✗	13.11	9.41	14.23	-	-	-	40
iCAN [8]	✗	14.84	10.45	16.15	16.26	11.33	17.73	75
TIN (RC$_D$) [17]	✗	13.75	10.12	15.45	15.34	10.98	17.02	70
DCA [30]	✗	16.25	11.16	17.75	17.73	12.78	19.21	130
Ours	✗	**17.58**	**11.72**	**19.33**	**19.76**	**14.68**	**21.27**	**9.06**

vided in the supplement. Our approach increases the minimal inference time on top of a standard object detector by eliminating the additional pair-wise neural network inference on detected object pairs, which is commonly required in previous works. Table 1 compares the inference time of the HOI interaction prediction excluding the time of the object detection. Note that compared to other multi-stage pipelines that have heavy network structures after the object detection phase, our model additionally requires significantly less time 9.06 ms (11.7%) compared to the base object detector. Our approach achieves 4×∼14× **speed-up** compared to the baseline HOI detection models which require 40 ms ∼ 130 ms per image after the object detection phase. Since most multi-stage pipelines have extra overhead for switching heavy models between different stages and saving/loading intermediate results. In a real-world application on a single GPU, the gain from our approach is much bigger.

Table 3. Ablation Study on V-COCO test set of our model. The first row shows the performance with only the Instance Branch. It can be observed that our proposed Union Branch plays a significant role in HOI detection. The second~fourth row each shows the performance without the target object classification loss $\mathcal{L}_{ij}^{cls}(\check{\theta})$, μ_u substituted with standard IoU score in Eq. 6, foreground focal loss replaced with ordinary focal loss, respectively.

Union Branch	UnionDet components			Sce.#1	Sce.#2
	FFL	$\mathcal{L}_{ij}^{cls}(\check{\theta})$	μ_u		
-	-	-	-	38.4	51.0
✓	✓	-	✓	44.8	53.5
✓	✓	✓	-	45.0	53.6
✓	-	✓	✓	46.9	55.6
✓	✓	✓	✓	**47.5**	**56.2**

5   Conclusions

In this paper, we present a novel one-stage human-object interaction detector. By performing action classification and union region detection in parallel with object detection, we achieved the *fastest* inference time while maintaining comparable performance with state-of-the-art methods. Also, our architecture is generally compatible with existing one-stage object detectors and end-to-end trainable. Our model enables a unified HOI detection that performs object detection and human-object interaction prediction at near real-time frame rates. Compared to heavy multi-stage HOI detectors, our model does not need to switch models across different stages and save/load intermediate results. In the real-world scenario, our model will more beneficial.

Acknowledgement. This work was supported by the National Research Council of Science & Technology (NST) grant by the Korea government (MSIT)(No.CAP-18-03-ETRI), National Research Foundation of Korea (NRF-2017M3C4A7065887), and Samsung Electronics, Co. Ltd.

References

1. Bansal, A., Rambhatla, S.S., Shrivastava, A., Chellappa, R.: Detecting human-object interactions via functional generalization. In: AAAI, pp. 10460–10469 (2020)
2. Bolya, D., Zhou, C., Xiao, F., Lee, Y.J.: YOLACT: real-time instance segmentation. In: Proceedings of the IEEE International Conference on Computer Vision, pp. 9157–9166 (2019)
3. Cao, Z., Simon, T., Wei, S.E., Sheikh, Y.: Realtime multi-person 2D pose estimation using part affinity fields. In: Proceedings of the IEEE Conference on Computer Vision and Pattern Recognition, pp. 7291–7299 (2017)
4. Chao, Y.W., Liu, Y., Liu, X., Zeng, H., Deng, J.: Learning to detect human-object interactions. In: 2018 IEEE Winter Conference on Applications of Computer Vision (WACV), pp. 381–389. IEEE (2018)

5. Chen, Y., Wang, Z., Peng, Y., Zhang, Z., Yu, G., Sun, J.: Cascaded pyramid network for multi-person pose estimation. In: Proceedings of the IEEE Conference on Computer Vision and Pattern Recognition, pp. 7103–7112 (2018)

6. Fang, H.S., Cao, J., Tai, Y.W., Lu, C.: Pairwise body-part attention for recognizing human-object interactions. In: Proceedings of the European Conference on Computer Vision (ECCV), pp. 51–67 (2018)

7. Fang, H.S., Xie, S., Tai, Y.W., Lu, C.: RMPE: regional multi-person pose estimation. In: Proceedings of the IEEE International Conference on Computer Vision, pp. 2334–2343 (2017)

8. Gao, C., Zou, Y., Huang, J.B.: iCAN: instance-centric attention network for human-object interaction detection. arXiv preprint arXiv:1808.10437 (2018)

9. Gkioxari, G., Girshick, R., Dollár, P., He, K.: Detecting and recognizing human-object interactions. In: Proceedings of the IEEE Conference on Computer Vision and Pattern Recognition, pp. 8359–8367 (2018)

10. Gupta, S., Malik, J.: Visual semantic role labeling. arXiv preprint arXiv:1505.04474 (2015)

11. Gupta, T., Schwing, A., Hoiem, D.: No-frills human-object interaction detection: factorization, layout encodings, and training techniques. In: Proceedings of the IEEE International Conference on Computer Vision, pp. 9677–9685 (2019)

12. Gupta, T., Shih, K., Singh, S., Hoiem, D.: Aligned image-word representations improve inductive transfer across vision-language tasks. In: Proceedings of the IEEE International Conference on Computer Vision, pp. 4213–4222 (2017)

13. He, K., Gkioxari, G., Dollár, P., Girshick, R.: Mask R-CNN. In: Proceedings of the IEEE International Conference on Computer Vision, pp. 2961–2969 (2017)

14. He, K., Zhang, X., Ren, S., Sun, J.: Deep residual learning for image recognition. In: Proceedings of the IEEE Conference on Computer Vision and Pattern Recognition, pp. 770–778 (2016)

15. Kolesnikov, A., Kuznetsova, A., Lampert, C., Ferrari, V.: Detecting visual relationships using box attention. In: Proceedings of the IEEE International Conference on Computer Vision Workshops (2019)

16. Li, J., Wang, C., Zhu, H., Mao, Y., Fang, H.S., Lu, C.: CrowdPose: efficient crowded scenes pose estimation and a new benchmark. In: Proceedings of the IEEE Conference on Computer Vision and Pattern Recognition, pp. 10863–10872 (2019)

17. Li, Y.L., et al.: Transferable interactiveness knowledge for human-object interaction detection. In: Proceedings of the IEEE Conference on Computer Vision and Pattern Recognition, pp. 3585–3594 (2019)

18. Lin, T.Y., Dollár, P., Girshick, R., He, K., Hariharan, B., Belongie, S.: Feature pyramid networks for object detection. In: Proceedings of the IEEE Conference on Computer Vision and Pattern Recognition, pp. 2117–2125 (2017)

19. Lin, T.Y., Goyal, P., Girshick, R., He, K., Dollár, P.: Focal loss for dense object detection. In: Proceedings of the IEEE International Conference on Computer Vision, pp. 2980–2988 (2017)

20. Liu, W., et al.: SSD: single shot multibox detector. In: Leibe, B., Matas, J., Sebe, N., Welling, M. (eds.) ECCV 2016. LNCS, vol. 9905, pp. 21–37. Springer, Cham (2016). https://doi.org/10.1007/978-3-319-46448-0_2

21. Lu, C., Krishna, R., Bernstein, M., Fei-Fei, L.: Visual relationship detection with language priors. In: Leibe, B., Matas, J., Sebe, N., Welling, M. (eds.) ECCV 2016. LNCS, vol. 9905, pp. 852–869. Springer, Cham (2016). https://doi.org/10.1007/978-3-319-46448-0_51

22. Mikolov, T., Sutskever, I., Chen, K., Corrado, G.S., Dean, J.: Distributed representations of words and phrases and their compositionality. In: Advances in Neural Information Processing Systems, pp. 3111–3119 (2013)

23. Pennington, J., Socher, R., Manning, C.D.: GloVe: global vectors for word representation. In: Proceedings of the 2014 Conference on Empirical Methods in Natural Language Processing (EMNLP), pp. 1532–1543 (2014)

24. Peyre, J., Laptev, I., Schmid, C., Sivic, J.: Detecting unseen visual relations using analogies. In: Proceedings of the IEEE International Conference on Computer Vision, pp. 1981–1990 (2019)

25. Qi, S., Wang, W., Jia, B., Shen, J., Zhu, S.C.: Learning human-object interactions by graph parsing neural networks. In: Proceedings of the European Conference on Computer Vision (ECCV), pp. 401–417 (2018)

26. Redmon, J., Divvala, S., Girshick, R., Farhadi, A.: You only look once: unified, real-time object detection. In: Proceedings of the IEEE Conference on Computer Vision and Pattern Recognition, pp. 779–788 (2016)

27. Ren, S., He, K., Girshick, R., Sun, J.: Faster R-CNN: towards real-time object detection with region proposal networks. In: Advances in Neural Information Processing Systems, pp. 91–99 (2015)

28. Sermanet, P., Eigen, D., Zhang, X., Mathieu, M., Fergus, R., LeCun, Y.: OverFeat: integrated recognition, localization and detection using convolutional networks. arXiv preprint arXiv:1312.6229 (2013)

29. Wan, B., Zhou, D., Liu, Y., Li, R., He, X.: Pose-aware multi-level feature network for human object interaction detection. In: Proceedings of the IEEE International Conference on Computer Vision, pp. 9469–9478 (2019)

30. Wang, T., et al.: Deep contextual attention for human-object interaction detection. arXiv preprint arXiv:1910.07721 (2019)

31. Xu, B., Wong, Y., Li, J., Zhao, Q., Kankanhalli, M.S.: Learning to detect human-object interactions with knowledge. In: Proceedings of the IEEE Conference on Computer Vision and Pattern Recognition (2019)

32. Zhang, S., Wen, L., Bian, X., Lei, Z., Li, S.Z.: Single-shot refinement neural network for object detection. In: Proceedings of the IEEE Conference on Computer Vision and Pattern Recognition, pp. 4203–4212 (2018)

33. Zhao, Q., et al.: M2Det: a single-shot object detector based on multi-level feature pyramid network. In: Proceedings of the AAAI Conference on Artificial Intelligence, vol. 33, pp. 9259–9266 (2019)

34. Zhou, P., Ni, B., Geng, C., Hu, J., Xu, Y.: Scale-transferrable object detection. In: Proceedings of the IEEE Conference on Computer Vision and Pattern Recognition, pp. 528–537 (2018)

35. Zhou, P., Chi, M.: Relation parsing neural network for human-object interaction detection. In: Proceedings of the IEEE International Conference on Computer Vision, pp. 843–851 (2019)

GSNet: Joint Vehicle Pose and Shape Reconstruction with Geometrical and Scene-Aware Supervision

Lei Ke[1], Shichao Li[1], Yanan Sun[1], Yu-Wing Tai[1,2(✉)], and Chi-Keung Tang[1]

[1] The Hong Kong University of Science and Technology,
Clear Water Bay, Hong Kong
{lkeab,slicd,ysuncd,yuwing,cktang}@cse.ust.hk
[2] Kwai Inc., Shenzhen, China

Abstract. We present a novel end-to-end framework named as GSNet (**G**eometric and **S**cene-aware **Net**work), which jointly estimates 6DoF poses and reconstructs detailed 3D car shapes from single urban street view. GSNet utilizes a unique four-way feature extraction and fusion scheme and directly regresses 6DoF poses and shapes in a single forward pass. Extensive experiments show that our diverse feature extraction and fusion scheme can greatly improve model performance. Based on a divide-and-conquer 3D shape representation strategy, GSNet reconstructs 3D vehicle shape with great detail (1352 vertices and 2700 faces). This dense mesh representation further leads us to consider geometrical consistency and scene context, and inspires a new multi-objective loss function to regularize network training, which in turn improves the accuracy of 6D pose estimation and validates the merit of jointly performing both tasks. We evaluate GSNet on the largest multi-task ApolloCar3D benchmark and achieve state-of-the-art performance both quantitatively and qualitatively. Project page is available at https://lkeab.github.io/gsnet/.

Keywords: Vehicle pose and shape reconstruction · 3D traffic scene understanding

1 Introduction

Traffic scene understanding is an active area in autonomous driving, where one emerging and challenging task is to perceive 3D attributes (including translation, rotation and shape) of vehicle instances in a dynamic environment as Fig. 1 shows. Compared to other scene representations such as 2D/3D bounding boxes [5,27,37], semantic masks [7,40] and depth maps [60], representing traffic scene with 6D object pose and detailed 3D shape is more informative for spatial reasoning and motion planning of self-driving cars.

Electronic supplementary material The online version of this chapter (https://doi.org/10.1007/978-3-030-58555-6_31) contains supplementary material, which is available to authorized users.

ⓒ Springer Nature Switzerland AG 2020
A. Vedaldi et al. (Eds.): ECCV 2020, LNCS 12360, pp. 515–532, 2020.
https://doi.org/10.1007/978-3-030-58555-6_31

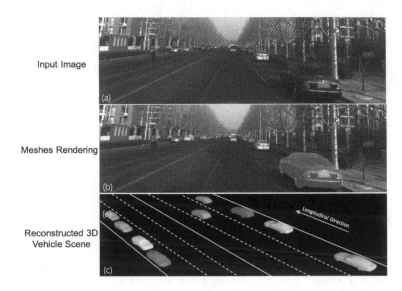

Input Image

Meshes Rendering

Reconstructed 3D
Vehicle Scene

Fig. 1. Joint vehicle pose and shape reconstruction results of our GSNet, where (a) is the input RGB image, (b) shows the reconstructed 3D car meshes projected onto the original image, (c) is a novel aerial view of the reconstructed 3D traffic scene. Corresponding car instances in (b) and (c) are depicted in the same color. (Color figure online)

Due to the lack of depth information in monocular RGB images, many existing works resort to stereo camera rigs [27,28] or expensive LiDAR [21,62,63]. However, they are limited by constrained perception range [27] or sparse 3D points for distant regions in the front view [48]. When using only a single RGB image, works that jointly reconstruct vehicle pose and shape can be classified into two categories: *fitting-based* and direct *regression-based*. *Fitting-based* methods [3,48,49] use a two-stage strategy where they first extract 2D image cues such as bounding boxes and keypoints and then fit a 3D template vehicle to best match its 2D image observations. The second stage is a post-processing step that is usually time-consuming due to iterative non-linear optimization, making it less applicable for real-time autonomous driving. On the contrary, *regression-based* methods [22,48] directly predict 3D pose/shape parameters with a single efficient forward pass of a deep network and is gaining increasing popularity with the growing scale of autonomous driving datasets.

Despite the recent regression-based methods having achieved remarkable performance for joint vehicle pose estimation and 3D shape reconstruction, we point out some unexplored yet valuable research questions: (1) Most regression-based networks [22,26,48] inherit classical 2D object detection architectures that solely use region of interest (ROI) features to regress 3D parameters. *How other potential feature representation can improve network performance* is less studied. (2) Deep networks require huge amounts of supervision [18], where useful supervisory signals other than manually annotated input-target pairs are favorable. Consistency brought by projective geometry is one possibility, yet the optimal design is still under-explored. Render-and-compare loss was used in [22] but it suffers from ambiguities where similar 2D projected masks can correspond to different 3D unknown parameters. For example, a mask similar to the ground truth mask is produced after changing the ground truth 3D pose by 180° around the symmetry axis, i.e., the prediction is not penalized enough despite being incorrect. (3) Previous regression-based works only penalize prediction error for single car instance and separate it from its environmental context, but a traffic scene includes the interaction between multiple instances and the relationship between instances with the physical world. We argue that considering these extra information can improve the training of a deep network.

We investigate these above questions and propose GSNet (**G**eometric and **S**cene-aware **Net**work), an end-to-end multi-task network that can estimate 6DoF car pose and reconstruct dense 3D shape simultaneously. We go beyond the ROI features and systematically study how other visual features that encode geometrical and visibility information can improve the network performance, where a simple yet effective four-way feature fusion scheme is adopted. Equipped with a dense 3D shape representation achieved by a *divide-and-conquer* strategy, we further design a multi-objective loss function to effectively improve network learning as validated by extensive experiments. This loss function considers geometric consistency using the projection of 66 semantic keypoints instead of masks which effectively reduces the ambiguity issue. It also incorporates a scene-aware term considering both inter-instance and instance-environment constraints.

In summary, our contributions are: (1) A novel end-to-end network that can jointly reconstruct 3D pose and dense shape of vehicles, achieving state-of-the-art performance on the largest multi-task ApolloCar3D benchmark [48]. (2) We propose an effective approach to extract and fuse diverse visual features, where systematic ablation study is shown to validate its effectiveness. (3) GSNet reconstructs fine-grained 3D meshes (1352 vertices) by our *divide-and-conquer* shape representation for vehicle instances rather than just 3D bounding boxes, wireframes [67] or retrieval [3,48]. (4) We design a new hybrid loss function to promote network performance, which considers both geometric consistency and scene constraints. This loss is made possible by the dense shape reconstruction, which in turn promotes the 6D pose estimation precision and sheds light on the benefit of jointly performing both tasks.

2   Related Work

Monocular 6DoF Pose Estimation. Traditionally, 6D object pose estimation is handled by creating correspondences between the object's known 3D model and 2D pixel locations, followed by Perspective-n-Point (PnP) algorithm [39,45,54]. For recent works, [2,13] construct templates and calculate the similarity score to obtain the best matching position on the image. In [38,45,59], 2D regional image features are extracted and matched with the features on 3D model to establish 2D-3D relation which thus require sufficient textures for matching. A single-shot deep CNN is proposed in [51] which regresses 6D object pose in one stage while in [50] a two-stage method is used: 1) SSD [35] for detecting bounding boxes and identities; 2) augmented autoencoder predicts object rotation using domain randomization [52]. Most recently, Hu et al. [14] introduces a segmentation-based method by combining local pose prediction from each visible part of the objects. Comparing to the cases in self-driving scenarios, these methods [16,42,59] are applied to indoor scenes with a small variance in translation especially along the longitudinal axis. Although using keypoints information, our model does *not* treat pose estimation as a PnP problem and is trained end-to-end.

Monocular 3D Shape Reconstruction. With the advent of large-scale shape datasets [4] and the progress of data-driven approaches, 3D shape reconstruction from a single image based on convolutional neural networks is drawing increasing interests. Most of these approaches [19,30,43,44,47,56,61,66] focus on general objects in the indoor scene or in the wild [15], where single object is shot in a close distance and occupies the majority of image area. Different from them, GSNet is targeted for more complicated traffic environment with far more vehicle instances to reconstruct per image, where some of them are even under occlusion at a long distance (over 50 m away).

Joint Vehicle Pose and Shape Reconstruction. 3D traffic scene understanding from a single RGB image is drawing increasing interests in recent years. However, many of these approaches only predict object orientation with 3D bounding boxes [1,6,29,33,46,60,62]. When it comes to 3D vehicle shape reconstruction, since the KITTI dataset [11] labels cars using 3D bounding boxes with no detailed 3D shape annotation, existing works mainly use wireframes [20,26,55,67] or retrieve from CAD objects [3,36,48,57]. In [64], the authors utilize 3D wireframe vehicle models to jointly estimate multiple objects in a scene and find that more detailed representations of object shape are highly beneficial to 3D scene understanding. DeepMANTA [3] adopts a coarse-to-fine refinement strategy to first regress 2D bounding box positions and generate 3D bounding boxes and finally obtain pose estimation results via 3D template fitting [25] by using the matched skeleton template to best fit its 2D image observations, which requires no image with 3D ground truth. *Most related to ours*, 3D-RCNN [22] regresses 3D poses and deformable shape parameters in a single forward pass, but it uses coarse voxel shape representation and the proposed render-and-compare loss causes ambiguity during training. Direct-based [48] further augments 3D-RCNN by adding mask pooling and offset flow. In contrast to these prior works, GSNet produces a more fine-grained 3D shape representation of vehicles by effective four-way feature fusion and *divide-and-conquer* shape reconstruction, which further inspires a geometrical scene aware loss to regularize network training with rich supervisory signals.

3   Pose and Shape Representation

6DoF Pose. The 6DoF pose for each instance consists of the 3D translation $\mathbf{T}$ and 3D rotation $\mathbf{R}$. $\mathbf{T}$ is represented by the object center coordinate $C_{obj} =$

$\{x, y, z\}$ in the camera coordinate system C_{cam}. Rotation $\mathbf{R}$ defines the rotation Euler angles about X, Y, Z axes of the object coordinate system C_{obj}.

Divide-and-Conquer Shape Representation. We represent vehicle shape with dense mesh consisting of 1352 vertices and 2700 faces, which is much more fine-grained compared to the volume representation used in [22]. We start with the CAD meshes provided by the ApolloCar3D database [48], which has different topology and vertex number for each car type. We convert them into the same topology with a fixed number of vertices by deforming a sphere using the SoftRas [34] method.

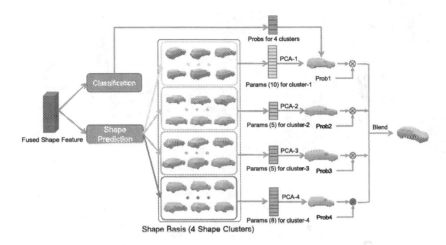

Fig. 2. Illustration of our divide-and-conquer 3D shape reconstruction module, where we obtain four independent PCA models for each shape cluster. Instance shape reconstruction is achieved by reconstructing shape in each cluster and blend them with the respective classification probabilities. This strategy achieves lower shape reconstruction error compared to other methods as shown in Table 4.

To ease the training of neural network for shape reconstruction, we reduce the shape representation dimension with principle component analysis (PCA) [41]. However, applying PCA to all available meshes directly [9,23,24] is sub-optimal due to the large variation of car types and shapes. We thus adopt a *divide-and-conquer* strategy as shown in Fig. 2. We first cluster a total of 79 CAD models into four subsets with K-Means algorithm utilizing the shape similarity between car meshes. For each subset, we separately learn a low dimensional shape basis with PCA. Denote a subset of k vehicle meshes as $M = \{m_1, m_2, ..., m_k\}$, we use PCA to find $n \leq 10$ dimensional shape basis, $\bar{\mathbf{S}} \in \mathbb{R}^{N \times n}$, where $N \gg n$. During inference, the network classifies the input instance into the 4 clusters and predicts the principle component coefficient for each cluster. The final shape is blended from the four meshes weighted by the classification score. With this strategy, we achieve lower shape reconstruction error than directly applying PCA to all meshes or retrieval which is detailed in our ablation study.

4   Network Architecture Design

Figure 3 shows the overall architecture of our GSNet for joint car pose and shape reconstruction. We design and extract four types of features from a complex traffic scene, after which a fusion scheme is proposed to aggregate them. Finally, multi-task prediction is done in parallel to estimate 3D translation, rotation and shape via the intermediate fused representations.

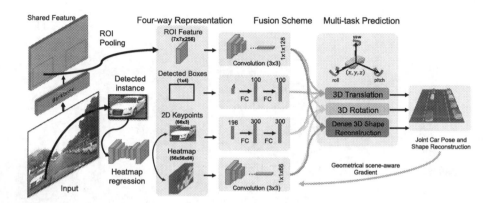

Fig. 3. Overview of our GSNet for joint vehicle pose and shape reconstruction. We use region-based 2D object detector [12] and a built-in heatmap regression branch to obtain ROI features, detected boxes, keypoint coordinates (global locations in the whole image) and corresponding heatmap (local positions and visibility in sub-region). GSNet performs an effective fusion of four-way input representations and builds three parallel branches respectively for 3D translation, rotation and shape estimation. 3D shape reconstruction is detailed in Fig. 2 and our hybrid loss function is illustrated in Sect. 5.

Diverse Feature Extraction and Representation. Existing methods [22,59] only use ROI features to regress 3D parameters, but we argue that using diverse features can better extract useful information in a complex traffic scene. Given an input image, we first use a region-based 2D object detector [12] to detect car instances and obtain its global location. Based on the bounding boxes, ROI pooling is used to extract appearance features for each instance. In a parallel branch, each detected instance is fed to a fully-convolutional sub-network to obtain 2D keypoint heatmaps and coordinates. The coordinates encode rich geometric information that can hardly be obtained with ROI features alone [65], while the heatmaps encode part visibility to help the network discriminate occluded instances.

Detected boxes are represented as 2D box center (b_x, b_y), width b_w and height b_h in pixel space. Camera intrinsic calibration matrix is $[f_x, 0, p_x; 0, f_y, p_y; 0, 0, 1]$ where f_x, f_y are focal lengths in pixel units and (p_x, p_y) is the principal point at the image center. We transform b_x, b_y, b_w, b_h from pixel space to the corresponding coordinates u_x, u_y, u_w, u_h in the world frame:

$$u_x = \frac{(b_x - p_x)z}{f_x}, u_y = \frac{(b_y - p_y)z}{f_y}, u_w = \frac{b_w}{f_x}, u_h = \frac{b_h}{f_y}, \tag{1}$$

where z is the fixed scale factor. For keypoint localization, we use the 66 semantic keypoints for cars defined in [48]. A 2D keypoint is represented as $\mathbf{p_k} = \{x_k, y_k, v_k\}$, where $\{x_k, y_k\}$ are the image coordinates and v_k denotes visibility. In implementation, we adapt [12] pre-trained for human pose estimation on COCO to initialize the keypoint localization branch. For extracting ROI features, we use FPN [31] as our backbone.

Fusion Scheme. We convert the extracted four-way inputs into 1D representation separately and decide which features to use for completing each task by prior knowledge. For global keypoint positions and detected boxes, we apply two fully-connected layers to convert them into higher level feature. For ROI feature maps and heatmaps, we adopt sequential convolutional operations with stride 2 to reduce their spatial size to 1×1 while keeping the channel number unchanged.

Instead of blindly using all features for prediction, we fuse different feature types that are most informative for each prediction branch. The translation $\mathbf{T}$ mainly affects the object location and scale during the imaging process, thus we concatenate the ROI feature, 2D keypoint feature and box position feature for translation regression. The rotation $\mathbf{R}$ determines the image appearance of the object given its 3D shape and texture, thus we utilize the fusion of ROI feature, heatmap feature and the keypoint feature as input. For estimating shape parameters $\mathbf{S}$, we aggregate the ROI and heatmap features.

Multi-task Prediction. We design three *parallel* estimation branches (translation, rotation and shape reconstruction) as shown in Fig. 3 since they are independent, where each branch directly regresses the targets with mutual benefits. Note that parts of input features such as ROI heatmap and keypoint positions are shared in different branches, which can be jointly optimized and is beneficial as shown by our experiments. In contrast to previous methods that predict translation or depth using a discretization policy [10], GSNet can directly regress the translation vector and achieve accurate result *without any post processing* or further refinement. For the shape reconstruction branch, the network classify the input instance and estimates the low-dimensional parameters (less than 30) for four clusters as described in Sect. 3.

5   Geometrical and Scene-Aware Supervision

To provide GSNet with rich supervisory signals, we design a composite loss functions consisting of multiple terms. Apart from ordinary regression losses, it also strives for geometrical consistency and considers scene-level constraints in both inter- and intra-instance manners.

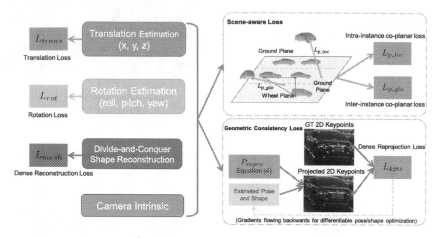

Fig. 4. The hybrid loss function for optimizing the GSNet. The scene-aware loss consists of two parts, L_{p_glo} for multiple car instances resting on common ground and L_{p_loc} for each single car at *a fine-grained level*. For geometrical consistency, camera intrinsics are used to project the predicted 3D semantic vertices on a car mesh to image and compared with the 2D detections.

Achieve Geometrical Consistency by Projecting Keypoints. With the rich geometric details of the 3D vehicles as shown in Fig. 4, we exploit the 2D-3D keypoints correspondence using a pinhole camera model to provide extra supervision signal. For a 3D semantic keypoint $\mathbf{p}_k = (x_0, y_0, z_0)$ on the predicted mesh with translation $\mathbf{T}_{pred}$ and rotation $\mathbf{R}_{pred}$, the reprojection equation is:

$$P_{repro} = s \begin{bmatrix} u_0 \\ v_0 \\ 1 \end{bmatrix} = \mathbf{k}[\mathbf{R}_{pred}|\mathbf{T}_{pred}]\mathbf{p}_k, \tag{2}$$

where $\mathbf{k}$ is the camera intrinsic matrix and (u_0, v_0) is the projection point in pixels units. For the ith projected keypoint $\mathbf{p}_i = (u_i, v_i)$, the reprojection loss is

$$L_{kpt\_i} = \|\mathbf{p}_i - \bar{\mathbf{p}}_i\|_2^2, \tag{3}$$

where $\bar{\mathbf{p}}_i = (\bar{u}_i, \bar{v}_i)$ is the corresponding image evidence given by our heatmap regression module. The total loss L_{kpts} for n semantic keypoints in a car instance is

$$L_{kpts} = \sum_{i=1}^{n} L_{kpt\_i} V_i, \tag{4}$$

where V_i is a boolean value indicating the visibility of ith keypoint in the image. This reprojection loss is *differentiable* and can be easily incorporated in the end-to-end training process. The correspondence of 2D keypoints and 3D mesh vertices is needed to compute the loss and we determine it by ourselves. We project each 3D vertex on the ground truth mesh to image plane and find its nearest neighboring 2D points. The 66 3D vertices whose 2D projections have the most 2D annotated neighbors are selected as the corresponding 3D landmarks. We also provide an ablation experiment on the influence of keypoints number in the supplementary file.

Scene-Aware Loss. Observe that most of cars rest on a common ground plane and the height of different instances is similar, thus the car centers are nearly co-planar. For each image, we locate mesh centers for four randomly-selected instances. Three of the centers define a plane $ax + by + cz + d = 0$ and denote the remaining car center coordinate as (x_1, y_1, z_1). As shown in Fig. 4, we introduce the *inter-instance co-planar loss* L_{p_glo} for multiple cars as:

$$L_{p\_glo} = \frac{|ax_1 + by_1 + cz_1 + d|}{\sqrt{a^2 + b^2 + c^2}}, \tag{5}$$

In addition, the centroids of the four wheels on a car should also lie in the same plane parallel to the ground. Thanks to the *dense* 3D mesh reconstructed by our multi-task network, we can readily obtain these four 3D coordinates. We thus propose the *intra-instance co-planar loss* L_{p_loc} to supplements L_{p_glo}. It is similar to Eq. 5 but the three points are chosen on the same instance.

Regression Losses. We use L2 loss L_{mesh} to penalize inaccurate 3D shape reconstruction as:

$$L_{mesh} = \frac{\sum_{j=1}^{m} \left\| \mathbf{M}_j - \bar{\mathbf{M}}_j \right\|_2^2}{m}, \tag{6}$$

where m is total number of vertices, $\mathbf{M}_j$ is the jth predicted vertex and $\bar{\mathbf{M}}_j$ is the ground truth vertex. For regression of 6DoF pose, we find that L1 loss performs better than L2 loss. The loss for translation regression is

$$L_{trans} = |\mathbf{T}_{pred} - \mathbf{T}_{gt}|, \tag{7}$$

where $\mathbf{T}_{gt}$ and $\mathbf{T}_{pred}$ are ground-truth and predicted translation vector, respectively. For regressing rotation in Euler angles, we restrict the range around each axis $[-\pi, \pi]$. Since this is a unimodal task, we define the regression loss as

$$L_{rot} = \begin{cases} |\mathbf{R}_{pred} - \mathbf{R}_{gt}| & \text{if } |\mathbf{R}_{pred} - \mathbf{R}_{gt}| \leq \pi, \\ 2\pi - |\mathbf{R}_{pred} - \mathbf{R}_{gt}| & \text{if } |\mathbf{R}_{pred} - \mathbf{R}_{gt}| > \pi, \end{cases} \tag{8}$$

where $\mathbf{R}_{pred}$ and $\mathbf{R}_{gt}$ are the predicted and ground truth rotation vector.

Sub-type Classification Loss. We also classify the car instance into 34 sub-types (sedan, minivan, SUV, etc.) and denote the classification loss as L_{cls}.

Final Objective Function. The final loss function L for training our GSNet is defined as:

$$\begin{aligned} L = {} & \lambda_{loc} L_{p\_loc} + \lambda_{glo} L_{p\_glo} + \lambda_{kpts} L_{kpts} \\ & + \lambda_{mesh} L_{mesh} + \lambda_{trans} L_{trans} + \lambda_{rot} L_{rot} + \lambda_{cls} L_{cls} \end{aligned} \tag{9}$$

where λs balance the above loss components. As validated by our experiments in Sect. 6.2, this hybrid loss function design significantly promotes the network's performance compared to using only regression losses alone.

6   Experiments

6.1   Datasets and Experimental Settings

ApolloCar3D. We use the most recent and largest multi-task ApolloCar3D dataset [48] to train and evaluate GSNet. This dataset contains 5,277 high-resolution $(2,710 \times 3384)$ images. We follow the official split where 4036 images are used for training, 200 for validation and the remaining 1041 for testing. Compared to KITTI [11], the instance count in ApolloCar3D is **20X** larger with far more cars per image (11.7 vs 4.8) where distant instances over 50 m are also annotated. In addition, ApolloCar3D provides 3D shape ground truth to evaluate shape reconstruction quantitatively, which is not available in KITTI.

Pascal3D+. We also train and evaluate GSNet on Pascal3D+ [58] dataset using its car category. There are totally 6704 in-the-wild images with 1.19 cars per image on average. It also provides both dense 3D shape and 6D pose annotation.

Evaluation Metrics. We follow the evaluation metrics in [48], which utilizes instance 3D average precision (A3DP) with 10 thresholds (criteria from loose to strict) for **jointly** measuring translation, rotation and 3D car shape reconstruction accuracy. The results on the loose and strict criterion are respectively denoted as *c-l* and *c-s*. During evaluation, Euclidean distance is used for 3D translation while arccos distance is used for 3D rotation. For 3D shape reconstruction, a predicted mesh is rendered into 100 views to compute IoU with the ground truth masks and the mean IoU is used. In addition to the absolute distance error, the relative error in translation is also evaluated to emphasize the model performance for nearby cars, which are more important for autonomous driving. We denote A3DP evaluated in relative and absolute version as *A3DP-Rel* and *A3DP-Abs* respectively.

Implementation Details. GSNet utilizes the Mask R-CNN [12] with ResNet-101 backbone pre-trained on the COCO 2017 dataset [32] for object detection and extracting ROI features (7×7). We discard detected objects with confidence score less than 0.3. The $\lambda_{loc}, \lambda_{glo}, \lambda_{kpts}, \lambda_{mesh}, \lambda_{trans}, \lambda_{rot}, \lambda_{cls}$ in Eq. 9 are set to 5.0, 5.0, 0.01, 10.0, 0.5, 1.0, 0.5 to balance the loss components. During training, we use Adam optimizer [17] with initial learning rate 0.0025 and reduce it by half every 10 epochs for total 30 epochs. The 2D keypoint localization branch is trained separately where we use 4,036 training images containing 40,000 labeled vehicles with 2D keypoints and set threshold 0.1 for deciding keypoint visibility. When building the dense shape representation, there are respectively 9, 24, 14, 32 meshes in the four clusters.

6.2   Ablation Study of Network Architecture and Loss Design

We conduct three ablation experiments on ApolloCar3D validation set to validate our network design, loss functions and dense shape representation strategy.

Table 1. Ablation study for GSNet on four-way feature fusion, which shows the relevant contribution of each representation with only regression losses. Performance is evaluated in terms of A3DP (*jointly* measuring translation, rotation and 3D car shape reconstruction accuracy), where *c-l* indicates results on loose criterion and *c-s* indicates strict criterion. GSNet exhibits a significant improvement compared to the baseline (with only ROI features), which promotes *A3DP-Rel* item *c-s* from 3.2 10.5. T and R in *6DoF Error* respectively represent 3D translation and rotation.

2D input representation				A3DP-Rel			A3DP-Abs			6DoF error	
ROI	Boxes	Heatmap	kpts	Mean	c-l	c-s	Mean	c-l	c-s	T	R
✓				6.8	20.1	3.2	7.0	17.7	5.1	2.41	0.33
✓	✓			$12.5_{\uparrow 5.7}$	$30.1_{\uparrow 10.4}$	$8.9_{\uparrow 5.7}$	$11.4_{\uparrow 4.4}$	$26.6_{\uparrow 8.9}$	$8.8_{\uparrow 3.7}$	$1.56_{\downarrow 0.85}$	$0.32_{\downarrow 0.01}$
✓	✓	✓		$13.7_{\uparrow 6.9}$	$32.5_{\uparrow 12.4}$	$9.2_{\uparrow 6.0}$	$12.4_{\uparrow 5.4}$	$29.2_{\uparrow 11.5}$	$9.2_{\uparrow 4.1}$	$1.53_{\downarrow 0.88}$	$0.24_{\downarrow 0.00}$
✓	✓	✓	✓	$14.1_{\uparrow 7.3}$	$32.9_{\uparrow 12.8}$	$10.5_{\uparrow 7.3}$	$12.8_{\uparrow 5.8}$	$29.3_{\uparrow 11.6}$	$9.9_{\uparrow 4.8}$	$1.50_{\downarrow 0.91}$	$0.24_{\downarrow 0.09}$

Is Extracting More Features Beneficial? We validate our four-way feature extraction fusion design by varying the number of used branches as: 1) Baseline: only using instance ROI features; 2) fusing transformed bounding box feature with the ROI feature; 3) combining predicted heatmap feature to the input; 4) further adding the 2D keypoint feature. The quantitative comparison is shown in Table 1. Compared to using ROI features alone, the injection of transformed detected boxes (center position, width and height) help provide geometric information, which help reduce translation error by 35.2% while improves *Rel-mAP* from 6.8 to 12.5 and *Abs-mAP* from 7.0 to 11.4. The introduction of keypoint heatmaps is beneficial especially for rotation estimation. This extra visibility information for the 2D keypoints reduces rotation error by 25.0% and further promoting *Rel-mAP* from 12.5 to 13.7 and *Abs-mAP* from 11.4 to 12.4. Finally, the 2D keypoint position branch complements the other three branches and improves model performance consistently for different evaluation metrics.

Effectiveness of the Hybrid Loss Function. Here we fix our network architecture while varying the components of loss function to validate our loss design. The experiments are designed as follows: 1) Baseline: adopt four-way feature fusion architecture, but only train the network with regression and classification losses without shape reconstruction; 2) adding 3D shape reconstruction loss; 3) incorporating geometrical consistency loss; 4) adding scene-aware loss but only use the inter-instance version; 5) adding the intra-instance scene-aware component to complete the multi-task loss function. As shown in Table 2, the reprojection consistency loss promotes 3D localization performance significantly, where the 3D translation error reduces over 10% and *Rel-mAp* increases from 15.1 to 17.6. The scene-aware loss brings obvious improvement compared to ignoring the traffic scene context, especially for the *A3DP-Rel* strict criterion *c-s* (increasing AP from 14.2 to 19.8). In addition, using both inter-instance and intra-instance loss components outperforms using inter-instance scene-aware loss alone. Compared to the baseline, our hybrid loss function significantly promotes the performance of *Rel-mAp* and *Abs-mAp* respectively to 20.2 and 18.9.

Table 2. Ablation study for GSNet using different loss components of the hybrid loss function, which shows the relevant contribution of each component. C0, C1, C2, C3, C4 respectively denote pose regression loss, 3D shape reconstruction loss, geometrical consistency loss, inter-instance scene-aware loss and intra-instance scene-aware loss. GSNet exhibits a significant improvement compared to the baseline (with only regression losses), especially in estimating the surrounding car instances as shown by *A3DP-Rel* (item *c-s* has been significantly boosted from 10.5 to 19.8).

Loss components					A3DP-Rel			A3DP-Abs			6DoF error	
C0	C1	C2	C3	C4	Mean	c-l	c-s	mean	c-l	c-s	T	R
✓					14.1	32.9	10.5	12.8	29.3	9.9	1.50	0.24
✓	✓				$15.1_{\uparrow 1.0}$	$34.8_{\uparrow 1.9}$	$11.3_{\uparrow 0.8}$	$15.0_{\uparrow 2.2}$	$32.0_{\uparrow 2.7}$	$13.0_{\uparrow 3.1}$	$1.44_{\downarrow 0.06}$	$0.23_{\downarrow 0.01}$
✓	✓	✓			$17.6_{\uparrow 3.5}$	$37.3_{\uparrow 4.4}$	$14.2_{\uparrow 3.7}$	$16.7_{\uparrow 3.9}$	$34.1_{\uparrow 4.8}$	$15.4_{\uparrow 5.5}$	$1.30_{\downarrow 0.20}$	$0.20_{\downarrow 0.04}$
✓	✓	✓	✓		$18.8_{\uparrow 4.7}$	$39.0_{\uparrow 6.1}$	$16.3_{\uparrow 5.8}$	$17.6_{\uparrow 4.8}$	$35.3_{\uparrow 6.0}$	$16.7_{\uparrow 6.8}$	$1.27_{\downarrow 0.23}$	$0.20_{\downarrow 0.04}$
✓	✓	✓	✓	✓	$\mathbf{20.2}_{\uparrow 6.1}$	$\mathbf{40.5}_{\uparrow 7.6}$	$\mathbf{19.8}_{\uparrow 9.3}$	$\mathbf{18.9}_{\uparrow 6.1}$	$\mathbf{37.4}_{\uparrow 8.1}$	$\mathbf{18.3}_{\uparrow 8.4}$	$\mathbf{1.23}_{\downarrow 0.27}$	$\mathbf{0.18}_{\downarrow 0.06}$

Is Jointly Performing Both Tasks Helpful? We argue that jointly performing dense shape reconstruction can in turn help 6D pose estimation. Without the introduction of the dense shape reconstruction task, we do not have access to the reconstruction loss C1 as well as the geometrical and scene-aware losses (C2, C3 and C4). Note that C1–C4 significantly improves estimation accuracy for translation and rotation.

Effectiveness of the Divide-and-Conquer Strategy. Table 4 compares model performance using different shape representations: retrieval, single PCA shape-space model and our divide-and-conquer strategy detailed in Sect. 3. Observe that our divide-and-conquer strategy not only reduces shape reconstruction error for around 10%, but also boosts the overall performance for traffic instance understanding. Also, we present shape reconstruction error distribution across different vehicle categories in our supplementary file.

6.3　Comparison with State-of-the-Art Methods

Quantitative Comparison on ApolloCar3D. We compare GSNet with state-of-the-art approaches that jointly reconstruct vehicle pose and shape on ApolloCar3D dataset as shown in Table 3. The most recent *regression-based* approaches are: 1) 3D-RCNN [22], which regress 3D instances from ROI features and add geometrical consistency by designing a differentiable render-and-compare mask loss; 2) Direct-based method in [48], which improves 3D-RCNN by adding mask pooling and offset flow. We can see that our GSNet achieves superior results among the existing *regression-based* methods across the evaluation metrics while being fast, nearly doubling the mAP performance of 3D-RCNN in *A3DP-Rel* entry. Compared to the *fitting-based* pose estimation methods using Epnp [25], which fit 3D template car model to 2D image observations in a time-consuming optimization process, GSNet performs comparably in *A3DP-Rel* and

A3DP-Abs metrics with a high-resolution shape reconstruction output not constrained by the existing CAD templates. Note that *fitting-based* methods consume long time and thus are not feasible for time-critical applications. Also note that *A3DP-Rel* is important since nearby cars are more relevant for self-driving car to make motion planning, where GSNet improves *c-l* AP performance by 15.75 compared to Kpts-based [48].

Table 3. Performance comparison with state-of-the-art 3D joint vehicle pose and shape reconstruction algorithms on ApolloCar3D dataset. Times is the average inference time for processing each image. GSNet achieves significantly better performance than state-of-the-art regression-based approaches (using a deep network to directly estimate the pose/shape from pixels) with both high precision and fast speed where inference time is *critical* in autonomous driving. * denotes fitting-based methods, which fits a 3D template car model to best match its 2D image observations (requires no image with 3D ground truth) and is time-consuming.

Model	Shape reconstruction	Regression-based	A3DP-Rel			A3DP-Abs			Times	Time-efficient
			Mean	c-l	c-s	Mean	c-l	c-s		
DeepMANTA (CVPR'17) [3]*	Retrieval	✗	16.04	23.76	19.80	20.10	30.69	23.76	3.38 s	✗
Kpts-based (CVPR'19) [48]*	Retrieval	✗	16.53	24.75	19.80	20.40	31.68	24.75	8.5 s	✗
3D-RCNN (CVPR'18) [22]	TSDF volume [8]	✓	10.79	17.82	11.88	16.44	29.70	**19.80**	0.29 s	✓
Direct-based (CVPR'19) [48]	Retrieval	✓	11.49	17.82	11.88	15.15	28.71	17.82	0.34 s	✓
Ours: GSNet	Detailed deformable mesh	✓	**20.21**	**40.50**	**19.85**	**18.91**	**37.42**	18.36	0.45 s	✓

Table 4. Results comparison between GSNet adopting retrieval, single PCA model and our divide-and-conquer shape module on Apollo-Car3D validation set.

Shape-space model	Shape reconstruction error	Rel-mAP
Retrieval	92.46	17.6
Single PCA	88.68	18.7
Divide-and-Conquer Shape Module	**81.33**	**20.2**

Table 5. Results on viewpoint estimation with annotated boxes on Pascal3D+ [58] for *Car*, where GSNet gets highest accuracy and lowest angular error.

Model	$Acc_{\pi/6}$ ↑	$MedErr$ ↓
RenderForCNN [49]	0.88	6.0°
Deep3DBox [37]	0.90	5.8°
3D-RCNN [22]	0.96	3.0°
Ours: GSNet	**0.98**	**2.4°**

Fig. 5. Qualitative comparison on the ApolloCar3D test set of different approaches by rendering 3D mesh output projected onto the input 2D image. The first row are the input images, the second row is the result of Direct-based [48] and the third row is predicted by our GSNet. The bottom row shows the reconstructed meshes in 3D space. Corresponding car instances are depicted in the same color. (Color figure online)

Fig. 6. Cross-dataset generalization of GSNet on KITTI [11] dataset. The first row are the input images and the second row are our reconstructed 3D car meshes projected onto the original image. Additional results are shown in our supplementary material.

Quantitative Comparison on Pascal3D+. To further validate our network, we evaluate GSNet on the Pascal3D+ [58] dataset using its car category. We follow the setting in [22,37] to evaluate the viewpoint and use $Acc_{\pi/6}$ and $MedErr$ adopted in [37,53] to report results in Table 5, where the median angular error improves by 20% from 3.0° to 2.4° compared to 3D-RCNN.

Qualitative Analysis. Figure 5 shows qualitative comparisons with other direct *regression-based* methods for joint vehicle pose and shape reconstruction. Compared with Direct-based [48], our GSNet produces more accurate 6DoF pose estimation and 3D shape reconstruction from monocular images due to the effective four-way feature fusion, the hybrid loss which considers both geometrical consistency and scene-level constraints and our divide-and-conquer shape reconstruction. Although directly regressing depth based on monocular images is considered as an ill-posed problem, our GSNet achieves high 3D estimation accuracy (our projected masks of car meshes on input images show an almost perfect match), particularly for instances in close proximity to the self-driving vehicle. The last column of the figure shows that the estimation of GSNet is still robust even in a relatively dark environment where the two left cars are heavily occluded. The last row visualizes the predicted 3D vehicle instances.

Figure 6 shows additional qualitative results on applying GSNet on KITTI [11]. Despite that GSNet is not trained on KITTI, the generalization ability of our model is validated as can be seen from the accurate 6D pose estimation and shape reconstruction of unseen vehicles. More results (including 3 temporally preceding frames of KITTI) are available in our supplementary material.

7   Conclusion

We present an end-to-end multi-task network GSNet, which jointly reconstructs 6DoF pose and 3D shape of vehicles from single urban street view. Compared to previous regression-based methods, GSNet not only explores more potential feature sources and uses an effective fusion scheme to supplement ROI features, but also provides richer supervisory signals from both geometric and scene-level perspectives. Vehicle pose estimation and shape reconstruction are tightly integrated in our system and benefit from each other, where 3D reconstruction delivers geometric scene context and greatly helps improve pose estimation precision. Extensive experiments conducted on ApolloCar3D and Pascal3D+ have demonstrated our state-of-the-art performance and validated the effectiveness of GSNet with both high accuracy and fast speed.

Acknowledgement. This research is supported in part by the Research Grant Council of the Hong Kong SAR under grant no. 1620818.

References

1. Brazil, G., Liu, X.: M3D-RPN: monocular 3D region proposal network for object detection. In: ICCV (2019)
2. Cao, Z., Sheikh, Y., Banerjee, N.K.: Real-time scalable 6DOF pose estimation for textureless objects. In: 2016 IEEE International Conference on Robotics and Automation (ICRA) (2016)
3. Chabot, F., Chaouch, M., Rabarisoa, J., Teulière, C., Chateau, T.: Deep MANTA: a coarse-to-fine many-task network for joint 2D and 3D vehicle analysis from monocular image. In: CVPR (2017)
4. Chang, A.X., et al.: ShapeNet: an information-rich 3D model repository. arXiv preprint arXiv:1512.03012 (2015)
5. Chen, X., Kundu, K., Zhang, Z., Ma, H., Fidler, S., Urtasun, R.: Monocular 3D object detection for autonomous driving. In: CVPR (2016)

6. Chen, X., Ma, H., Wan, J., Li, B., Xia, T.: Multi-view 3D object detection network for autonomous driving. In: CVPR (2017)
7. Cordts, M., et al.: The cityscapes dataset for semantic urban scene understanding. In: CVPR (2016)
8. Curless, B., Levoy, M.: A volumetric method for building complex models from range images. In: SIGGRAPH (1996)
9. Engelmann, F., Stückler, J., Leibe, B.: SAMP: shape and motion priors for 4D vehicle reconstruction. In: WACV (2017)
10. Fu, H., Gong, M., Wang, C., Batmanghelich, K., Tao, D.: Deep ordinal regression network for monocular depth estimation. In: CVPR (2018)
11. Geiger, A., Lenz, P., Urtasun, R.: Are we ready for autonomous driving? The KITTI vision benchmark suite. In: CVPR (2012)
12. He, K., Gkioxari, G., Dollár, P., Girshick, R.: Mask R-CNN. In: ICCV (2017)
13. Hinterstoisser, S., et al.: Gradient response maps for real-time detection of textureless objects. TPAMI **34**(5), 876–888 (2011)
14. Hu, Y., Hugonot, J., Fua, P., Salzmann, M.: Segmentation-driven 6D object pose estimation. In: CVPR (2019)
15. Kar, A., Tulsiani, S., Carreira, J., Malik, J.: Category-specific object reconstruction from a single image. In: CVPR (2015)
16. Kehl, W., Manhardt, F., Tombari, F., Ilic, S., Navab, N.: SSD-6D: Making RGB-based 3D detection and 6D pose estimation great again. In: ICCV (2017)
17. Kingma, D.P., Ba, J.: Adam: a method for stochastic optimization. In: ICLR (2015)
18. Kolotouros, N., Pavlakos, G., Black, M.J., Daniilidis, K.: Learning to reconstruct 3D human pose and shape via model-fitting in the loop. In: ICCV (2019)
19. Kong, C., Lin, C.H., Lucey, S.: Using locally corresponding cad models for dense 3D reconstructions from a single image. In: CVPR (2017)
20. Krishna Murthy, J., Sai Krishna, G., Chhaya, F., Madhava Krishna, K.: Reconstructing vehicles from a single image: shape priors for road scene understanding. In: 2017 IEEE International Conference on Robotics and Automation (ICRA) (2017)
21. Ku, J., Pon, A.D., Waslander, S.L.: Monocular 3D object detection leveraging accurate proposals and shape reconstruction. In: CVPR (2019)
22. Kundu, A., Li, Y., Rehg, J.M.: 3D-RCNN: instance-level 3D object reconstruction via render-and-compare. In: CVPR (2018)
23. Leotta, M.J., Mundy, J.L.: Predicting high resolution image edges with a generic, adaptive, 3-D vehicle model. In: CVPR (2009)
24. Leotta, M.J., Mundy, J.L.: Vehicle surveillance with a generic, adaptive, 3D vehicle model. TPAMI **33**(7), 1457–1469 (2010)
25. Lepetit, V., Moreno-Noguer, F., Fua, P.: EPnP: an accurate $o(n)$ solution to the PnP problem. IJCV **81**(2) (2009). Article number: 155. https://doi.org/10.1007/s11263-008-0152-6
26. Li, C., Zeeshan Zia, M., Tran, Q.H., Yu, X., Hager, G.D., Chandraker, M.: Deep supervision with shape concepts for occlusion-aware 3D object parsing. In: CVPR (2017)
27. Li, P., Chen, X., Shen, S.: Stereo R-CNN based 3D object detection for autonomous driving. In: CVPR (2019)
28. Li, P., Qin, T., Shen, S.: Stereo vision-based semantic 3D object and ego-motion tracking for autonomous driving. In: Ferrari, V., Hebert, M., Sminchisescu, C., Weiss, Y. (eds.) ECCV 2018. LNCS, vol. 11206, pp. 664–679. Springer, Cham (2018). https://doi.org/10.1007/978-3-030-01216-8_40

29. Liang, M., Yang, B., Wang, S., Urtasun, R.: Deep continuous fusion for multi-sensor 3D object detection. In: Ferrari, V., Hebert, M., Sminchisescu, C., Weiss, Y. (eds.) ECCV 2018. LNCS, vol. 11220, pp. 663–678. Springer, Cham (2018). https://doi.org/10.1007/978-3-030-01270-0_39

30. Lin, C.H., et al.: Photometric mesh optimization for video-aligned 3D object reconstruction. In: CVPR (2019)

31. Lin, T.Y., Dollár, P., Girshick, R., He, K., Hariharan, B., Belongie, S.: Feature pyramid networks for object detection. In: CVPR (2017)

32. Lin, T.-Y., et al.: Microsoft COCO: common objects in context. In: Fleet, D., Pajdla, T., Schiele, B., Tuytelaars, T. (eds.) ECCV 2014. LNCS, vol. 8693, pp. 740–755. Springer, Cham (2014). https://doi.org/10.1007/978-3-319-10602-1_48

33. Liu, L., Lu, J., Xu, C., Tian, Q., Zhou, J.: Deep fitting degree scoring network for monocular 3D object detection. In: CVPR (2019)

34. Liu, S., Li, T., Chen, W., Li, H.: Soft Rasterizer: a differentiable renderer for image-based 3D reasoning. In: ICCV (2019)

35. Liu, W., et al.: SSD: single shot multibox detector. In: Leibe, B., Matas, J., Sebe, N., Welling, M. (eds.) ECCV 2016. LNCS, vol. 9905, pp. 21–37. Springer, Cham (2016). https://doi.org/10.1007/978-3-319-46448-0_2

36. Mottaghi, R., Xiang, Y., Savarese, S.: A coarse-to-fine model for 3D pose estimation and sub-category recognition. In: CVPR (2015)

37. Mousavian, A., Anguelov, D., Flynn, J., Kosecka, J.: 3D bounding box estimation using deep learning and geometry. In: CVPR (2017)

38. Pavlakos, G., Zhou, X., Chan, A., Derpanis, K.G., Daniilidis, K.: 6-DoF object pose from semantic keypoints. In: 2017 IEEE International Conference on Robotics and Automation (ICRA) (2017)

39. Peng, S., Liu, Y., Huang, Q., Zhou, X., Bao, H.: PVNet: pixel-wise voting network for 6DoF pose estimation. In: CVPR (2019)

40. Pohlen, T., Hermans, A., Mathias, M., Leibe, B.: Full-resolution residual networks for semantic segmentation in street scenes. In: CVPR (2017)

41. Prisacariu, V.A., Reid, I.: Nonlinear shape manifolds as shape priors in level set segmentation and tracking. In: CVPR (2011)

42. Rad, M., Lepetit, V.: BB8: a scalable, accurate, robust to partial occlusion method for predicting the 3D poses of challenging objects without using depth. In: ICCV (2017)

43. Richter, S.R., Roth, S.: Matryoshka networks: predicting 3D geometry via nested shape layers. In: CVPR (2018)

44. Riegler, G., Osman Ulusoy, A., Geiger, A.: OctNet: learning deep 3D representations at high resolutions. In: CVPR (2017)

45. Rothganger, F., Lazebnik, S., Schmid, C., Ponce, J.: 3D object modeling and recognition using local affine-invariant image descriptors and multi-view spatial constraints. IJCV 66(3), 231–259 (2006). https://doi.org/10.1007/s11263-005-3674-1

46. Simonelli, A., Bulò, S.R.R., Porzi, L., López-Antequera, M., Kontschieder, P.: Disentangling monocular 3D object detection. In: ICCV (2019)

47. Sinha, A., Unmesh, A., Huang, Q., Ramani, K.: SurfNet: generating 3D shape surfaces using deep residual networks. In: CVPR (2017)

48. Song, X., et al.: ApolloCar3D: a large 3D car instance understanding benchmark for autonomous driving. In: CVPR (2019)

49. Su, H., Qi, C.R., Li, Y., Guibas, L.J.: Render for CNN: viewpoint estimation in images using CNNs trained with rendered 3D model views. In: ICCV (2015)

50. Sundermeyer, M., Marton, Z.-C., Durner, M., Brucker, M., Triebel, R.: Implicit 3D orientation learning for 6D object detection from RGB images. In: Ferrari, V., Hebert, M., Sminchisescu, C., Weiss, Y. (eds.) ECCV 2018. LNCS, vol. 11210, pp. 712–729. Springer, Cham (2018). https://doi.org/10.1007/978-3-030-01231-1_43

51. Tekin, B., Sinha, S.N., Fua, P.: Real-time seamless single shot 6D object pose prediction. In: CVPR (2018)

52. Tobin, J., Fong, R., Ray, A., Schneider, J., Zaremba, W., Abbeel, P.: Domain randomization for transferring deep neural networks from simulation to the real world. In: 2017 IEEE/RSJ International Conference on Intelligent Robots and Systems (IROS) (2017)

53. Tulsiani, S., Malik, J.: Viewpoints and keypoints. In: CVPR (2015)

54. Wagner, D., Reitmayr, G., Mulloni, A., Drummond, T., Schmalstieg, D.: Pose tracking from natural features on mobile phones. In: IEEE/ACM International Symposium on Mixed and Augmented Reality (2008)

55. Wu, J., et al.: Single image 3D interpreter network. In: Leibe, B., Matas, J., Sebe, N., Welling, M. (eds.) ECCV 2016. LNCS, vol. 9910, pp. 365–382. Springer, Cham (2016). https://doi.org/10.1007/978-3-319-46466-4_22

56. Wu, Z., et al.: 3D ShapeNets: a deep representation for volumetric shapes. In: CVPR (2015)

57. Xiang, Y., Choi, W., Lin, Y., Savarese, S.: Data-driven 3D voxel patterns for object category recognition. In: CVPR (2015)

58. Xiang, Y., Mottaghi, R., Savarese, S.: Beyond PASCAL: a benchmark for 3D object detection in the wild. In: WACV (2014)

59. Xiang, Y., Schmidt, T., Narayanan, V., Fox, D.: PoseCNN: a convolutional neural network for 6D object pose estimation in cluttered scenes. In: Robotics: Science and Systems (RSS) (2018)

60. Xu, B., Chen, Z.: Multi-level fusion based 3D object detection from monocular images. In: CVPR (2018)

61. Yan, X., Yang, J., Yumer, E., Guo, Y., Lee, H.: Perspective transformer nets: learning single-view 3D object reconstruction without 3D supervision. In: NIPS (2016)

62. Yang, B., Luo, W., Urtasun, R.: PIXOR: real-time 3D object detection from point clouds. In: CVPR (2018)

63. Yang, Z., Sun, Y., Liu, S., Shen, X., Jia, J.: STD: sparse-to-dense 3D object detector for point cloud. In: ICCV (2019)

64. Zeeshan Zia, M., Stark, M., Schindler, K.: Are cars just 3D boxes? Jointly estimating the 3D shape of multiple objects. In: CVPR (2014)

65. Zhao, R., Wang, Y., Martinez, A.M.: A simple, fast and highly-accurate algorithm to recover 3D shape from 2D landmarks on a single image. TPAMI **40**(12), 3059–3066 (2017)

66. Zhu, R., Kiani Galoogahi, H., Wang, C., Lucey, S.: Rethinking reprojection: closing the loop for pose-aware shape reconstruction from a single image. In: ICCV (2017)

67. Zia, M.Z., Stark, M., Schiele, B., Schindler, K.: Detailed 3D representations for object recognition and modeling. TPAMI **35**(11), 2608–2623 (2013)

Resolution Switchable Networks
for Runtime Efficient Image Recognition

Yikai Wang[1], Fuchun Sun[1], Duo Li[2], and Anbang Yao[2(✉)]

[1] Beijing National Research Center for Information Science and Technology
(BNRist), State Key Lab on Intelligent Technology and Systems,
Department of Computer Science and Technology,
Tsinghua University, Beijing, China
`wangyk17@mails.tsinghua.edu.cn, fcsun@tsinghua.edu.cn`
[2] Cognitive Computing Laboratory, Intel Labs China, Beijing, China
{`duo.li,anbang.yao`}`@intel.com`

Abstract. We propose a general method to train a single convolutional
neural network which is capable of switching image resolutions at inference. Thus the running speed can be selected to meet various computational resource limits. Networks trained with the proposed method are
named Resolution Switchable Networks (RS-Nets). The basic training
framework shares network parameters for handling images which differ
in resolution, yet keeps separate batch normalization layers. Though it is
parameter-efficient in design, it leads to inconsistent accuracy variations
at different resolutions, for which we provide a detailed analysis from
the aspect of the train-test recognition discrepancy. A multi-resolution
ensemble distillation is further designed, where a teacher is learnt on
the fly as a weighted ensemble over resolutions. Thanks to the ensemble and knowledge distillation, RS-Nets enjoy accuracy improvements
at a wide range of resolutions compared with individually trained models. Extensive experiments on the ImageNet dataset are provided, and
we additionally consider quantization problems. Code and models are
available at https://github.com/yikaiw/RS-Nets.

Keywords: Efficient design · Multi-resolution · Ensemble distillation

1 Introduction

Convolutional Neural Networks (CNNs) have achieved great success on image
recognition tasks [5,14], and well-trained recognition models usually need to be

Y. Wang—This work was done when Yikai Wang was an intern at Intel Labs China,
supervised by Anbang Yao who is responsible for correspondence.

Electronic supplementary material The online version of this chapter (https://
doi.org/10.1007/978-3-030-58555-6_32) contains supplementary material, which is
available to authorized users.

A. Vedaldi et al. (Eds.): ECCV 2020, LNCS 12360, pp. 533–549, 2020.
https://doi.org/10.1007/978-3-030-58555-6_32

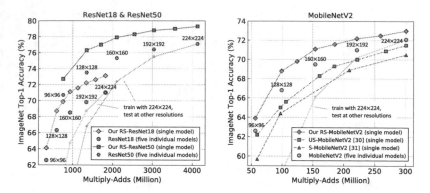

Fig. 1. ImageNet accuracy vs. FLOPs (Multiply-Adds) of our **single** models and the corresponding **sets** of individual models. A single RS-Net model is executable at each of the resolutions, and even achieves significantly higher accuracies than individual models. The results of two state-of-the-art switchable networks (switch by varying network widths) S-MobileNetV2 [31] and US-MobileNetV2 [30] are provided for comparison. Details are in Table 2 and Table 4.

deployed on mobile phones, robots or autonomous vehicles [1, 10]. To fit the resource constraints of devices, extensive research efforts have been devoted to balancing between accuracy and efficiency, by reducing computational complexities of models. Some of these methods adjust the structural configurations of networks, e.g., by adjusting the network depths [5], widths [10, 31] or the convolutional blocks [18, 33]. Besides that, adjusting the image resolution is another widely-used method for the accuracy-efficiency trade-off [9, 10, 16, 24]. If input images are downsized, all feature resolutions at different convolutional layers are reduced subsequently with the same ratio, and the computational cost of a model is nearly proportional to the image resolution ($H \times W$) [10]. However, for a common image recognition model, when the test image resolution differs from the resolution used for training, the accuracy quickly deteriorates [29]. To address this issue, existing works [9, 24, 28] train an individual model for each resolution. As a result, the total number of models to be trained and saved is proportional to the amount of resolutions considered at runtime. Besides the high storage costs, each time adjusting the resolution is accompanied with the additional latency to load another model which is trained with the target resolution.

The ability to switch the image resolution at inference meets a common need for real-life model deployments. By switching resolutions, the running speeds and costs are adjustable to flexibly handle the real-time latency and power requirements for different application scenarios or workloads. Besides, the flexible latency compatibility allows such model to be deployed on a wide range of resource-constrained platforms, which is friendly for application developers. In this paper, we focus on switching input resolutions for an image recognition model, and propose a general and economic method to improve overall accuracies. Models trained with our method are called **Resolution Switchable Networks (RS-Nets)**. Our contribution is composed of three parts.

First, we propose a parallel training framework where images with different resolutions are trained within a single model. As the resolution difference usually leads to the difference of activation statistics in a network [29], we adopt shared network parameters but privatized Batch Normalization layers (BNs) [12] for each resolution. Switching BNs enables the model to flexibly switch image resolutions, without needing to adjust other network parameters.

Second, we associate the multi-resolution interaction effects with a kind of train-test discrepancy (details in Sect. 3.2). Both our analysis and empirical results reach an interesting conclusion that the parallel training framework tends to enlarge the accuracy gaps over different resolutions. On the one hand, accuracy promotions at high resolutions make a stronger teacher potentially available. On the other hand, the accuracy drop at the lower resolution indicates that the benefits of parallel training itself are limited. Both reasons encourage us to further propose a design of ensemble distillation to improve overall performance.

Third, to the best of our knowledge, we are the first to propose that a data-driven ensemble distillation can be learnt on the fly for image recognition, based on the same image instances with different resolutions. Regarding the supervised image recognition, the structure of our design is also different from existing ensemble or knowledge distillation works, as they focus on the knowledge transfer among different models, e.g., by stacking multiple models [34], multiple branches [15], pre-training a teacher model, or splitting the model into sub-models [30], while our model is single and shared, with little extra parameters.

Extensive experiments on the ImageNet dataset validate that RS-Nets are executable given different image resolutions at runtime, and achieve significant accuracy improvements at a wide range of resolutions compared with individually trained models. Illustrative results are provided in Fig. 1, which also verify that our proposed method can be generally applied to modern recognition backbones.

2   Related Work

Image Recognition. Image recognition acts as a benchmark to evaluate models and is a core task for computer vision. Advances on the large-scale image recognition datasets, like ImageNet [3], can translate to improved results on a number of other applications [4,13,20]. In order to enhance the model generalization for image recognition, data augmentation strategies, e.g., random-size crop, are adopted during training [5,6,26]. Besides, common models are usually trained and tested with fixed-resolution inputs. [29] shows that for a model trained with the default 224×224 resolution and tested at lower resolutions, the accuracy quickly deteriorates (e.g., drops 11.6% at test resolution 128×128 on ResNet50).

Accuracy-Efficiency Trade-Off. There have been many attempts to balance accuracy and efficiency by model scaling. Some of them adjust the structural configurations of networks. For example, ResNets [5] provide several choices of network depths from shallower to deep. MobileNets [10,24] and ShuffleNets [33]

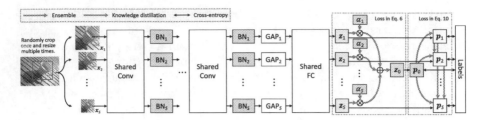

Fig. 2. Overall framework of training a RS-Net. Images with different resolutions are trained in parallel with shared Conv/FC layers and private BNs. The ensemble logit (z_0) is learnt on the fly as a weighted mean of logits ($z_1, z_2, \cdots, z_S$), shown as green arrows. Knowledge distillations are shown as red arrows. For inference, one of the S forward paths is selected (according to the image resolution), with its corresponding BNs, for obtaining its corresponding prediction $p_s, s \in \{1, 2, \cdots, S\}$. *The ensemble and knowledge distillation are not needed during inference.* (Color figure online)

can reduce network widths by using smaller width multipliers. While some other works [9,10,16,24] reduce the computational complexity by decreasing image resolutions at input, which is also our focus. Modifying the resolution usually does not make changes to the number of network parameters, but significantly affects the computational complexity [10].

Knowledge Distillation. A student network can be improved by imitating feature representations or soft targets of a larger teacher network [7,17,34]. The teacher is usually pre-trained beforehand and fixed, and the knowledge is transferred in one direction [22]. Yet [34] introduces a two-way transfer between two peer models. [25] performs mutual learning within one single network assisted by intermediate classifiers. [15] learns a native ensemble design based on multiple models for distillation. [30] conducts the knowledge distillation between the whole model and each split smaller model. Regarding the supervised image recognition, existing distillation works rely on different models, usually needing another teacher network with higher-capacity than low-capacity students. While our design is applied in a shared model, which is data-driven, collecting complementary knowledge from the same image instances with different resolutions.

3   Proposed Method

A schematic overview of our proposed method is shown in Fig. 2. In this section, we detail the insights and formulations of parallel training, interaction effects and ensemble distillation based on the multi-resolution setting.

3.1   Multi-resolution Parallel Training

To make the description self-contained, we begin with the basic training of a CNN model. Given training samples, we crop and resize each sample to a fixed-resolution image x^i. We denote network inputs as $\{(x^i, y^i) | i \in \{1, 2, \cdots, N\}\}$, where y^i is the ground truth which belongs to one of the C classes, and N is the amount of samples. Given network configurations with parameters $\boldsymbol{\theta}$, the predicted probability of the class c is denoted as $p(c|\boldsymbol{x}^i, \boldsymbol{\theta})$. The model is optimized with a cross-entropy loss defined as:

$$\mathcal{H}(\boldsymbol{x}, \boldsymbol{y}) = -\frac{1}{N} \sum_{i=1}^{N} \sum_{c=1}^{C} \delta(c, y^i) \log \left(p(c|\boldsymbol{x}^i, \boldsymbol{\theta}) \right), \tag{1}$$

where $\delta(c, y^i)$ equals to 1 when $c = y^i$, otherwise 0.

In this part, we propose **multi-resolution parallel training**, or called **parallel training** for brevity, to train a single model which can switch image resolutions at runtime. During training, each image sample is randomly cropped and resized to several duplicate images with different resolutions. Suppose that there are S resolutions in total, the inputs can be written as $\{(\boldsymbol{x}_1^i, \boldsymbol{x}_2^i, \cdots, \boldsymbol{x}_S^i, y^i) | i \in \{1, 2, \cdots, N\}\}$. Recent CNNs for image recognition follow similar structures that all stack Convolutional (Conv) layers, a Global Average Pooling (GAP) layer and a Fully-Connected (FC) layer. In CNNs, if input images have different resolutions, the corresponding feature maps in all Conv layers will also vary in resolution. Thanks to GAP, features are transformed to a unified spatial dimension (1×1) with equal amount of channels, making it possible to be followed by a same FC layer. During our parallel training, we share parameters of Conv layers and the FC layer, and therefore the training for multiple resolutions can be realized in a single network. The loss function for parallel training is calculated as a summation of the cross-entropy losses:

$$\mathcal{L}_{cls} = \sum_{s=1}^{S} \mathcal{H}(\boldsymbol{x}_s, \boldsymbol{y}). \tag{2}$$

Specializing Batch Normalization layers (BNs) [12] is proved to be effective for efficient model adaption [2, 19, 31, 35]. In image recognition tasks, resizing image results in different activation statistics in a network [29], including means and variances used in BNs. Thus during parallel training, we privatize BNs for each resolution. Results in the left panel of Fig. 4 verify the necessity of privatizing BNs. For the sth resolution, each corresponding BN layer normalizes the channel-wise feature as follows:

$$y_s' = \gamma_s \frac{y_s - \mu_s}{\sqrt{\sigma_s^2 + \epsilon}} + \beta_s, s \in \{1, 2, \cdots, S\}, \tag{3}$$

where μ_s and σ_s^2 are running mean and variance; γ_s and β_s are learnable scale and bias. Switching these parameters enables the model to switch resolutions.

3.2   Multi-resolution Interaction Effects

In this section, restricted to large-scale image datasets with fine-resolution images, we analyze the interaction effects of different resolutions under the parallel training framework. We start by posing a question: compared with individually trained models, how does parallel training affect test accuracies at different

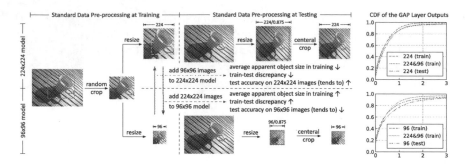

Fig. 3. Left: An illustration of interaction effects for the parallel training with two resolutions. Each red box indicates the region to be cropped, and the size of each blue dotted box is the apparent object (in this sample is a cup) size. For this example, in either one of the models, the apparent object size at testing is smaller than at training. [29] reveals that this relation still holds when averaging all data, which is called the train-test discrepancy. The data pre-processing for training or testing follows the standard image recognition method, which will be described in Sect. 4.1. Right: CDF curves for comparing the value distributions of feature activations. All curves are plotted on the validation dataset of ImageNet, but are based on different data pre-processing methods as annotated by (train) or (test). (Color figure online)

resolutions? As multi-resolution can be seen as a kind of data augmentation, we analyze from two aspects as follows.

The first aspect is straightforward. The model meets a wide range of image resolutions, which improves the generalization and reduces over-fitting. Thus if the setting of resolutions is suitable (e.g., not too diverse), the parallel training tends to bring overall accuracy gains at testing, especially for a high-capacity network such as ResNet50.

The second aspect is based on the specialty of large-scale image recognition tasks, where objects of interest randomly occupy different portions of image areas and thus the random-size crop augmentation is used during training. [29] reveals that for common recognition works, as the random-size crop is used for training but not for testing, there exists a train-test **discrepancy** that the average "apparent object size" at testing is smaller than that at training. Besides, [29] achieves accuracy improvements by alleviating such discrepancy, but is accompanied with the costs of test-time resolution augmentations and finetuning. Note that we do not aim to modify the data pre-processing method to alleviate the discrepancy. Instead, we are inspired to use the concept of the discrepancy to analyze multi-resolution interaction effects. We take the parallel training with two resolutions 224×224 and 96×96 for example. According to the left panel of Fig. 3 (the analysis in colors), compared with the model using only 224×224 images, the parallel training can be seen as augmenting 96×96 images, which reduces the average apparent object size at training and thus alleviates the discrepancy. On the contrary, compared with the individual model using 96×96 images, augmenting 224×224 images increases the discrepancy. Thus in this aspect, this parallel training tends to increase the test accuracy at 224×224 (actually $+1.6\%$ for ResNet18, as shown in Fig. 5), while tends to reduce the test accuracy at 96×96 (-0.8%). The right panel of Fig. 3 plots the Cumulative Distribution Function (CDF)[1] of output components of the Global Average Pooling (GAP) layer for a well-trained ResNet18 (as a ReLU layer is before the GAP, all components are nonnegative). We plot CDF to compare the value distributions of feature activations when using training or testing data pre-processing method. The parallel training seems to narrow the train-test gap for the 224×224 model, but widen the gap for the 96×96 model.

We take ResNet18 as an example and summarize the two aforementioned aspects. For the parallel training with two resolutions, the test accuracy at the high resolution increases compared with its individual model, as the two aspects reach an agreement to a large degree. As for the lower resolution, we find that the accuracy slightly increases if the two resolutions are close, otherwise decreases. Similarly, when multiple resolutions are used for parallel training, test accuracies increase at high resolutions but may decrease at lower resolutions, compared with individual models. Results in Table 2 show that for the parallel training with five resolutions, accuracies only decrease at 96×96 but increase at the other four resolutions. Detailed results in Fig. 5 also verify our analysis.

For image recognition, although testing at a high resolution already tends to achieve a good accuracy, using the parallel training makes it even better. This finding opens up a possibility that a stronger teacher may be available in this framework, and seeking a design based on such teacher could be highly effective.

3.3 Multi-resolution Ensemble Distillation

In this section, we propose a new design of ensemble distillation. Regarding the supervised image recognition, unlike conventional distillation works that rely on transferring knowledge among different models, ours is data-driven and can be applied in a shared model. Specifically, our design is learnt on the fly and the distillation is based on the same image instances with different resolutions.

As is commonly known, for image recognition tasks, models given a high resolution image are easy to capture fine-grained patterns, and thus achieve good performance [27,28]. However, according to the sample statistics in the middle column of Table 1, we find that there always exists a proportion of samples which are correctly classified at a low resolution but wrongly classified at another higher resolution. Such results indicate that model predictions at different image resolutions are complementary, and not always the higher resolution is better for each image sample. Therefore, we propose to learn a teacher on the fly as an ensemble of the predictions w.r.t. all resolutions, and conduct knowledge distillation to improve the overall performance. Our design is called **Multi-Resolution Ensemble Distillation (MRED)**.

---

[1] The CDF of a random variable X is defined as $F_X(x) = P(X \leq x)$, for all $x \in \mathbb{R}$.

Table 1. Proportions (%) of validation samples that are correctly classified ($\sqrt{}$) at a resolution but are wrongly classified ($\times$) at another, based on ResNet18. All models are well-trained. Lower numbers correspond to better performance, and numbers which are larger than the base are colored red, otherwise green. Benefited from MRED, all the proportions in the last column decrease.

$\times$ \ $\sqrt{}$	Individual training (base)					Parallel training					Parallel training + MRED				
	224	192	160	128	96	224	192	160	128	96	224	192	160	128	96
224	–	5.7	5.5	5.3	4.7	–	3.0↓	3.4↓	3.8↓	3.9↓	–	2.5↓	2.9↓	3.4↓	3.6↓
192	6.9	–	6.0	5.5	5.2	4.1↓	–	3.3↓	3.7↓	3.8↓	3.4↓	–	2.8↓	3.3↓	3.4↓
160	8.1	7.2	–	5.9	5.4	6.0↓	4.7↓	–	3.5↓	3.6↓	5.1↓	3.9↓	–	2.9↓	3.2↓
128	9.8	8.9	8.1	–	5.8	9.3↓	8.2↓	6.6↓	–	3.5↓	7.3↓	6.4↓	5.0↓	–	3.1↓
96	13.2	12.5	11.4	9.7	–	15.2↑	14.2↑	12.7↑	9.8↑	–	12.2↓	11.4↓	10.2↓	7.9↓	–

During the training process of image recognition, for each input image $\boldsymbol{x}^i$, the probability of the class c is calculated using a softmax function:

$$p(c|\boldsymbol{x}^i, \boldsymbol{\theta}) = p(c|\boldsymbol{z}^i) = \frac{\exp(z_c^i)}{\sum_{j=1}^{C} \exp(z_j^i)}, c \in \{1, 2, \cdots, C\}, \tag{4}$$

where $\boldsymbol{z}^i$ is the logit, the unnormalized log probability outputted by the network, and probabilities over all classes can be denoted as the model prediction $\boldsymbol{p}$.

In the parallel training framework, each image is randomly cropped and resized to S images with different resolutions. To better benefit from MRED, these S images need to be resized from a same random crop, as illustrated in the left-most part of Fig. 2. The necessity will be verified in Sect. 4.3.

As each image sample is resized to S resolutions, there are S corresponding logits $\boldsymbol{z}_1, \boldsymbol{z}_2, \cdots, \boldsymbol{z}_S$. We learn a group of importance scores $\boldsymbol{\alpha} = [\alpha_1\ \alpha_2\ \cdots\ \alpha_S]$, satisfying $\boldsymbol{\alpha} \geq 0$, $\sum_{s=1}^{S} \alpha_s = 1$, which can be easily implemented with a softmax function. We then calculate an ensemble logit $\boldsymbol{z}_0$ as the weighted summation of the S logits:

$$\boldsymbol{z}_0 = \sum_{s=1}^{S} \alpha_s \boldsymbol{z}_s. \tag{5}$$

To optimize $\boldsymbol{\alpha}$, we temporally froze the gradients of the logits $\boldsymbol{z}_1, \boldsymbol{z}_2, \cdots, \boldsymbol{z}_S$. Based on the ensemble logit $\boldsymbol{z}_0$, the corresponding prediction $\boldsymbol{p}_0$, called ensemble prediction, can be calculated via Eq. 4. Then $\boldsymbol{\alpha}$ is optimized using a cross-entropy loss between $\boldsymbol{p}_0$ and the ground truth, which we call the ensemble loss $\mathcal{L}_{ens}$:

$$\mathcal{L}_{ens} = -\frac{1}{N} \sum_{i=1}^{N} \sum_{c=1}^{C} \delta(c, y^i) \log\left(p(c|z_0^i)\right). \tag{6}$$

In knowledge distillation works, to quantify the alignment between a teacher prediction $\boldsymbol{p}_t$ and a student prediction $\boldsymbol{p}_s$, Kullback Leibler (KL) divergence is usually used:

$$\mathcal{D}_{kl}(\boldsymbol{p}_t\|\boldsymbol{p}_s) = \frac{1}{N} \sum_{i=1}^{N} \sum_{c=1}^{C} p(c|z_t^i) \log \frac{p(c|z_t^i)}{p(c|z_s^i)}. \tag{7}$$

We force predications at different resolutions to mimic the learnt ensemble prediction $\boldsymbol{p}_0$, and thus the distillation loss $\mathcal{L}_{dis}$ could be obtained as:

$$\mathcal{L}_{dis} = \sum_{s=1}^{S} \mathcal{D}_{kl}(\boldsymbol{p}_0 \| \boldsymbol{p}_s). \tag{8}$$

Finally, the overall loss function is a summation of the classification loss, the ensemble loss and the distillation loss, without needing to tune any extra weighted parameters:

$$\mathcal{L} = \mathcal{L}_{cls} + \mathcal{L}_{ens} + \mathcal{L}_{dis}, \tag{9}$$

where in practical, optimizing $\mathcal{L}_{ens}$ only updates $\boldsymbol{\alpha}$, with all network weights temporally frozen; optimizing $\mathcal{L}_{cls}$ and $\mathcal{L}_{dis}$ updates network weights.

We denote the method with Eq. 8 as our vanilla-version MRED. Under the parallel training framework, as the accuracy at a high resolution is usually better than at a lower resolution, accuracies can be further improved by offering dense guidance from predications at high resolutions toward predictions at lower resolutions. Thus the distillation loss can be extended to be a generalized one:

$$\mathcal{L}_{dis} = \frac{2}{S+1} \sum_{t=0}^{S-1} \sum_{s=t+1}^{S} \mathcal{D}_{kl}(\boldsymbol{p}_t \| \boldsymbol{p}_s), \tag{10}$$

where the index t starts from 0 referring to the ensemble term; as the summation results in $S(S+1)/2$ components in total, we multiply $\mathcal{L}_{dis}$ by a constant ratio $2/(S+1)$ to keep its range the same as $\mathcal{L}_{cls}$. We denote the method with Eq. 10 as our full-version MRED, which is used in our experiments by default.

The proposed design involves negligible extra parameters (only S scalars), without needing extra models. Models trained with the parallel training framework and MRED are named Resolution Switchable Networks (RS-Nets). An overall framework of training a RS-Net is illustrated in Fig. 2. **During inference**, the network only performs one forward calculation at a given resolution, without ensemble or distillation, and thus both the computational complexity and the amount of parameters equal to a conventional image recognition model.

4 Experiments

We perform experiments on ImageNet (ILSVRC12) [3,23], a widely-used image recognition dataset containing about 1.2 million training images and 50 thousand validation images, where each image is annotated as one of 1000 categories. Experiments are conducted with prevailing CNN architectures including a lightweight model MobileNetV2 [24] and ResNets [5], where a basic-block model ResNet18 and a bottleneck-block model ResNet50 are both considered. Besides, we also evaluate our method in handling network quantization problems, where we consider different kinds of bit-widths.

4.1 Implementation Details

Our basic experiments are implemented with PyTorch [21]. For quantization tasks, we apply our method to LQ-Nets [32] which show state-of-the-art performance in training CNNs with low-precision weights or both weights and activations.

We set $\mathbb{S} = \{224 \times 224, 192 \times 192, 160 \times 160, 128 \times 128, 96 \times 96\}$, as commonly adopted in a number of existing works [10,16,24]. During training, we pre-process the data for augmentation with an area ratio (cropped area/original area) uniformly sampled in $[0.08, 1.0]$, an aspect ratio $[3/4, 4/3]$ and a horizontal flipping. We resize images with the bilinear interpolation. Note that both $[0.08, 1.0]$ and $[3/4, 4/3]$ follow the standard data augmentation strategies for ImageNet [11,26,29], e.g., *RandomResizedCrop* in PyTorch uses such setting as default. During validation, we first resize images with the bilinear interpolation to every resolution in $\mathbb{S}$ divided by 0.875 [8,16], and then feed central regions to models.

Networks are trained from scratch with random initializations. We set the batch size to 256, and use a SGD optimizer with a momentum 0.9. For standard ResNets, we train 120 epochs and the learning rate is annealed from 0.1 to 0 with a cosine scheduling [6]. For MobileNetV2, we train 150 epochs and the learning rate is annealed from 0.05 to 0 with a cosine scheduling. For quantized ResNets, we follow the settings in LQ-Nets [32], which train 120 epochs and the learning rate is initialized to 0.1 and divided by 10 at 30, 60, 85, 95, 105 epochs. The weight decay rate is set to 1e−4 for all ResNets and 4e−5 for MobileNetV2.

4.2 Results

As mentioned in Sect. 1, common works with multi-resolution settings train and deploy multiple individual models separately for different resolutions. We denote these individual models as **I-Nets**, which are set as baselines. We use I-{resolution} to represent each individual model, e.g., I-224.

Basic Results. In Table 2, we report results on ResNet18, ResNet50 and MobileNetV2 (M-NetV2 for short). Besides I-Nets, we also report accuracies at five resolutions using the individual model which is trained with the largest resolution (I-224). For our proposed method, we provide separate results of the parallel training (parallel) and the overall design (RS-Net). As mentioned in Sect. 1, I-Nets need several times of parameter amount and high latencies for switching across models. We also cannot rely on an individual model to switch the image resolutions, as accuracies of I-224 are much lower than I-Nets at other resolutions (e.g., 15%–20% accuracy drop at the resolution 96×96). Similarly, each of the other individual models also suffers from serious accuracy drops, as can be seen in the right panel of Fig. 4. Our parallel training brings accuracy improvements at the four larger resolutions, while accuracies at 96×96 decrease, and the reason is previously analyzed in Sect. 3.2. Compared with I-Nets, our RS-Net achieves large improvements at all resolutions with only 1/5 parameters.

Table 2. Basic results comparison on ImageNet. We report top-1/top-5 accuracies (%), top-1 accuracy gains (%) over individual models (I-Nets), Multiply-Adds (MAdds) and total parameters (params). All experiments use the same data pre-processing methods. Our baseline results are slightly higher than the original papers [5, 24].

Network	Resolution	MAdds	I-Nets (base)	I-224	Our parallel	Our RS-Net
ResNet18	224 × 224	1.82G	71.0/90.0	71.0/90.0	73.0/90.9 $_{(+2.0)}$	**73.1/91.0** $_{(+2.1)}$
	192 × 192	1.34G	69.8/89.4	68.7/88.5 $_{(-1.1)}$	71.7/90.3 $_{(+1.9)}$	**72.2/90.6** $_{(+2.4)}$
	160 × 160	931M	68.5/88.2	64.7/85.9 $_{(-5.2)}$	70.4/89.6 $_{(+1.9)}$	**71.1/90.1** $_{(+2.6)}$
	128 × 128	596M	66.3/86.8	56.8/80.0 $_{(-9.5)}$	67.5/87.8 $_{(+1.2)}$	**68.7/88.5** $_{(+2.4)}$
	96 × 96	335M	62.6/84.1	42.5/67.9 $_{(-20.1)}$	61.5/83.5 $_{(-1.1)}$	**64.1/85.3** $_{(+1.5)}$
	Total params	55.74M	11.15M	11.18M	11.18M	
ResNet50	224 × 224	4.14G	77.1/93.4	77.1/93.4	78.9/94.4 $_{(+1.8)}$	**79.3/94.6** $_{(+2.2)}$
	192 × 192	3.04G	76.4/93.2	75.5/92.5 $_{(-0.9)}$	78.1/94.0 $_{(+1.7)}$	**78.8/94.4** $_{(+2.4)}$
	160 × 160	2.11G	75.3/92.4	72.4/90.7 $_{(-2.9)}$	76.9/93.1 $_{(+1.6)}$	**77.9/93.9** $_{(+2.6)}$
	128 × 128	1.35G	73.5/91.4	66.8/87.0 $_{(-6.7)}$	74.9/92.1 $_{(+1.4)}$	**76.3/93.0** $_{(+2.8)}$
	96 × 96	760M	70.7/89.8	54.9/78.2 $_{(-15.8)}$	70.2/89.4 $_{(-0.5)}$	**72.7/91.0** $_{(+2.0)}$
	Total params	121.87M	24.37M	24.58M	24.58M	
M-NetV2	224 × 224	301M	72.1/90.5	72.1/90.5	72.8/90.9 $_{(+0.7)}$	**73.0/90.8** $_{(+0.9)}$
	192 × 192	221M	71.0/89.8	70.2/89.1 $_{(-0.9)}$	71.7/90.2 $_{(+0.7)}$	**72.2/90.5** $_{(+1.2)}$
	160 × 160	154M	69.5/88.9	66.1/86.3 $_{(-3.2)}$	70.1/89.2 $_{(+0.6)}$	**71.1/90.2** $_{(+1.6)}$
	128 × 128	99M	66.8/87.0	58.3/81.2 $_{(-8.5)}$	67.3/87.2 $_{(+0.5)}$	**68.8/88.2** $_{(+2.0)}$
	96 × 96	56M	62.6/84.0	43.9/69.1 $_{(-18.7)}$	61.4/83.3 $_{(-1.2)}$	**63.9/84.9** $_{(+1.3)}$
	Total params	16.71M	3.34M	3.47M	3.47M	

Table 3. Results comparison for quantization tasks. We report top-1/top-5 accuracies (%) and top-1 accuracy gains (%) over individual models (I-Nets). All experiments are performed under the same training settings following LQ-Nets.

Network	Resolution	Bit-width (W/A): 2/32		Bit-width (W/A): 2/2	
		I-Nets (base)	Our RS-Net	I-Nets (base)	Our RS-Net
Quantized ResNet18	224 × 224	68.0/88.0	**68.8/88.4** $_{(+0.8)}$	64.9/86.0	**65.8/86.4** $_{(+0.9)}$
	192 × 192	66.4/86.9	**67.6/87.8** $_{(+1.2)}$	63.1/84.7	**64.8/85.8** $_{(+1.7)}$
	160 × 160	64.5/85.5	**66.0/86.5** $_{(+1.5)}$	61.1/83.3	**62.9/84.2** $_{(+1.8)}$
	128 × 128	61.5/83.4	**63.1/84.5** $_{(+1.6)}$	58.1/80.8	**59.3/81.9** $_{(+1.2)}$
	96 × 96	56.3/79.4	**56.6/79.9** $_{(+0.3)}$	52.3/76.4	**52.5/76.7** $_{(+0.2)}$
Quantized ResNet50	224 × 224	74.6/92.2	**76.0/92.8** $_{(+1.4)}$	72.2/90.8	**74.0/91.5** $_{(+1.8)}$
	192 × 192	73.5/91.3	**75.1/92.4** $_{(+1.6)}$	70.9/89.8	**73.1/91.0** $_{(+2.2)}$
	160 × 160	71.9/90.4	**73.8/91.6** $_{(+1.9)}$	69.0/88.5	**71.4/90.0** $_{(+2.4)}$
	128 × 128	69.6/88.9	**71.7/90.2** $_{(+2.1)}$	66.6/86.9	**68.9/88.3** $_{(+2.3)}$
	96 × 96	65.5/86.0	**67.3/87.4** $_{(+1.8)}$	61.7/83.4	**63.4/84.7** $_{(+1.7)}$

For example, the RS-Net with ResNet50 obtains about 2.4% absolute top-1 accuracy gains on average across five resolutions. Note that the number of FLOPs (Multiply-Adds) is nearly proportional to the image resolution [10]. Regarding ResNet18 and ResNet50, accuracies at 160 × 160 of our RS-Nets even surpass the accuracies of I-Nets at 224 × 224, significantly reducing about 49% FLOPs at runtime. Similarly, for MobileNetV2, the accuracy at 192 × 192 of RS-Net surpasses the accuracy of I-Nets at 224 × 224, reducing about 26% FLOPs.

Table 4. Top-1 accuracies (%) on MobileNetV2. Individual models (I-Nets-w, adjust via network width; I-Nets-r, adjust via resolution) and our RS-Net are trained under the same settings. Results of S-MobileNetV2 (S) [31] and US-MobileNetV2 (US) [30] are from the original papers.

Width	MAdds	I-Nets-w	S [31]	US [30]	Resolution	MAdds	I-Nets-r	Ours
1.0×	301M	72.1	70.5	71.5	224 × 224	301M	72.1	**73.0**
0.8×	222M	69.8	–	70.0	192 × 192	221M	71.0	**72.2**
0.65×	161M	68.0	–	68.3	160 × 160	154M	69.5	**71.1**
0.5×	97M	64.6	64.4	65.0	128 × 128	99M	66.8	**68.8**
0.35×	59M	60.1	59.7	62.2	96 × 96	56M	62.6	**63.9**
Model size		5×	1×	1×	Model size		5×	1×

Quantization. We further explore the generalization of our method to more challenging quantization problems, and we apply our method to LQ-Nets [32]. Experiments are performed under two typical kinds of quantization settings, including the quantization on weights (2/32) and the more extremely compressed quantization on both weights and activations (2/2). Results of I-Nets and each RS-Net based on quantized ResNets are reported in Table 3. Again, each RS-Net outperforms the corresponding I-Nets at all resolutions. For quantization problems, as I-Nets cannot force the quantized parameter values of each individual model to be the same, 2-bit weights in I-Nets are practically stored with more digits than a 4-bit model[2], while the RS-Net avoids such issue. For quantization, accuracy gains of RS-Net to ResNet50 are more obvious than those to ResNet18, we conjecture that under compressed conditions, ResNet50 better bridges the network capacity and the augmented data resolutions.

Switchable Models Comparison. Based on MobileNetV2, results in Table 4 indicate that under comparable FLOPs (Multiply-Adds), adjusting image resolutions (see I-Nets-r) achieves higher accuracies than adjusting network widths (see I-Nets-w). For example, adjusting resolution to 96 × 96 brings 2.5% higher absolute accuracy than adjusting width to 0.35×, with even lower FLOPs. Results of S-MobileNetV2 (S) [31] and US-MobileNetV2 (US) [30] are also provided for comparison. As we can see, our RS-Net significantly outperforms S and US at all given FLOPs, achieving 1.5%–4.4% absolute gains. Although both S and US can also adjust the accuracy-efficiency trade-off, they have marginal gains or even accuracy drops (e.g., at width 1.0×) compared with their baseline I-Nets-w. Our model shows large gains compared with our baseline I-Nets-r (even they are mostly stronger than I-Nets-w). Note that adjusting resolutions does not conflict with adjusting widths, and both methods can be potentially combined together.

---

[2] The total bit number of five models with individual 2-bit weights is $\log_2(2^2 \times 5) \approx 4.3$.

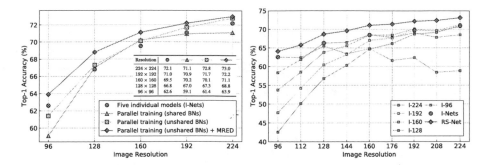

Fig. 4. Left: Comparison of parallel trainings with shared BNs and unshared (private) BNs, based on MobileNetV2. Individual models and Parallel (unshared BNs) + MRED (i.e., RS-Net) are provided for reference. Right: Comparison of our RS-Net and each individual model (from I-224 to I-96) tested at denser resolutions (the interval is 16), based on ResNet18. Each individual model suffers from serious accuracy drops at other resolutions, but our RS-Net avoids this issue.

4.3 Ablation Study

Importance of Using Private BNs. A quick question is that, why not share BNs as well? Based on MobileNetV2, the left panel of Fig. 4 shows that during the parallel training, privatizing BNs achieves higher accuracies than sharing BNs, especially at both the highest resolution (+1.7%) and the lowest (+2.3%). When BNs are shared, activation statistics of different image resolutions are averaged, which differ from the real statistics especially at two ends of resolutions.

Tested at New Resolutions. One may concern that if a RS-Net can be tested at a new resolution. In the right panel of Fig. 4, we test models at different resolutions with a smaller interval, using ResNet18. We follow a simple method to handle a resolution which is not involved in training. Suppose the resolution is sandwiched between two of the five training resolutions which correspond to two groups of BN parameters, we apply a linear interpolation on these two groups and obtain a new group of parameters, which are used for the given resolution. We observe that the RS-Net maintains high accuracies. RS-Net suppresses the serious accuracy drops which exist in every individual model.

Multi-resolution Interaction Effects. This part is for verifying the analysis in Sect. 3.2, and we do not apply MRED here. Parallel training results of ResNet18 are provided in Fig. 5, which show top-1 accuracy variations compared with each individual model. The left panel of Fig. 5 illustrates the parallel training with two resolutions, where all accuracies increase at 224 × 224. As the gap of the two resolutions increases, the accuracy variation at the lower resolution decreases. For example, compared with I-224 and I-96 respectively, the parallel training with 224 × 224 and 96 × 96 images increases the accuracy at 224 × 224

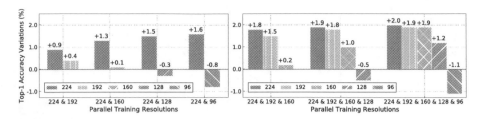

Fig. 5. Absolute top-1 accuracy variations (%) (compared with individual models) of parallel trainings, based on ResNet18. The result of each individual model (from I-96 to I-224) is used as the baseline. We use single numbers to represent the image resolutions.

Table 5. Top-1 accuracies (%) of different kinds of distillations and their accuracy variations compared with the full-version MRED, based on ResNet18. Definitions of $p_s, s \in \{1, 2, \cdots, S\}$ are referred to Sect. 3.3. The prediction p_1 corresponds to the largest resolution 224×224.

Resolution	Full-version MRED (base)	Vanilla-version MRED	Only distillations from $p_1, p_2, \cdots, p_{S-1}$	Only distillations from p_1
224×224	73.1	73.0 (-0.1)	72.3 (-0.8)	72.2 (-0.9)
192×192	72.2	72.1 (-0.1)	71.2 (-1.0)	71.7 (-0.5)
160×160	71.1	70.8 (-0.3)	70.2 (-0.9)	70.5 (-0.6)
128×128	68.7	68.3 (-0.4)	68.1 (-0.6)	68.1 (-0.6)
96×96	64.1	63.7 (-0.4)	63.5 (-0.6)	63.1 (-1.0)

from 71.0% to 72.6%, but decreases the accuracy at 96×96 from 62.6% to 61.8%. The right panel of Fig. 5 illustrates the parallel training with multiple resolutions, where we observe that most accuracies are improved and accuracies at the lowest resolutions may decrease. These results verify our analysis.

Verifying Each Element of Ensemble Distillation. In Sect. 3.3, we define two versions of MRED. The vanilla-version has only the distillation paths starting from the ensemble prediction p_0 toward all the other predictions, while the full-version has additional paths from predications at high resolutions toward predictions at lower resolutions. In Table 5, we compare the performance of the two versions as well as two other variants that omit the distillations from p_0, based on ResNet18. Results indicate that all kinds of the proposed distillations are indispensable. Besides, we also emphasize in Sect. 3.3 that for training a RS-Net, each image should be randomly cropped only once (called single-crop) and then resized to multiple resolutions (called multi-resolution). Results in Table 6 indicate that using multi-crop (applying a random crop individually for each resolution) will weaken the benefits of MRED, as accuracies at low resolutions are lower compared with using single-crop. We also verify the importance of multi-resolution by replacing the multi-resolution setting by five identical resolutions,

Table 6. Top-1 accuracies (%) comparison to verify the importance of using multi-resolution and single-crop, based on ResNet18. Multi-crop refers to applying the random crop individually for each resolution. Single-resolution refers to using five identical resolutions in a RS-Net instead of the original multi-resolution setting.

Resolution	Multi-resolution single-crop (base)	Multi-resolution multi-crop	Single-resolution multi-crop	Single-resolution single-crop
224 × 224	73.1	73.1 (−0.0)	72.1 (−1.0)	71.0 (−2.1)
192 × 192	72.2	72.0 (−0.2)	71.1 (−1.1)	69.8 (−2.4)
160 × 160	71.1	70.7 (−0.4)	69.6 (−1.5)	68.5 (−2.6)
128 × 128	68.7	68.0 (−0.7)	67.6 (−1.1)	66.3 (−2.4)
96 × 96	64.1	62.8 (−1.3)	63.3 (−0.8)	62.6 (−1.5)
Model size	1×	1×	5×	5×

called single-resolution. As resolutions are identical, an individual experiment is needed for each resolution. Results indicate that applying ensemble distillation to predictions of different crops has very limited benefits compared with applying it to predictions of different resolutions.

5 Conclusions

We introduce a general method to train a CNN which can switch the resolution during inference, and thus its running speed can be selected to fit different computational constraints. Specifically, we propose the parallel training to handle multi-resolution images, using shared parameters and private BNs. We analyze the interaction effects of resolutions from the perspective of train-test discrepancy. And we propose to learn an ensemble distillation based on the same image instances with different resolutions, which improves accuracies to a large extent.

Acknowledgement. This work is jointly supported by the National Science Foundation of China (NSFC) and the German Research Foundation (DFG) in project Cross Modal Learning, NSFC 61621136008/DFG TRR-169. We thank Aojun Zhou for the insightful discussions.

References

1. Cai, H., Zhu, L., Han, S.: ProxylessNAS: direct neural architecture search on target task and hardware. In: ICLR (2019)
2. Chang, W., You, T., Seo, S., Kwak, S., Han, B.: Domain-specific batch normalization for unsupervised domain adaptation. In: CVPR (2019)
3. Deng, J., Dong, W., Socher, R., Li, L., Li, K., Li, F.: ImageNet: a large-scale hierarchical image database. In: CVPR (2009)

4. Gordo, A., Almazán, J., Revaud, J., Larlus, D.: End-to-end learning of deep visual representations for image retrieval. IJCV **124**, 237–254 (2017). https://doi.org/10.1007/s11263-017-1016-8
5. He, K., Zhang, X., Ren, S., Sun, J.: Deep residual learning for image recognition. In: CVPR (2016)
6. He, T., Zhang, Z., Zhang, H., Zhang, Z., Xie, J., Li, M.: Bag of tricks for image classification with convolutional neural networks. In: CVPR (2019)
7. Hinton, G.E., Vinyals, O., Dean, J.: Distilling the knowledge in a neural network. arXiv preprint arXiv:1503.02531 (2015)
8. Hoffer, E., Weinstein, B., Hubara, I., Ben-Nun, T., Hoefler, T., Soudry, D.: Mix & match: training convnets with mixed image sizes for improved accuracy, speed and scale resiliency. arXiv preprint arXiv:1908.08986 (2019)
9. Howard, A., et al.: Searching for MobileNetV3. In: ICCV (2019)
10. Howard, A.G., et al.: MobileNets: efficient convolutional neural networks for mobile vision applications. arXiv preprint arXiv:1704.04861 (2017)
11. Huang, G., Liu, Z., van der Maaten, L., Weinberger, K.Q.: Densely connected convolutional networks. In: CVPR (2017)
12. Ioffe, S., Szegedy, C.: Batch normalization: accelerating deep network training by reducing internal covariate shift. In: ICML (2015)
13. Kornblith, S., Shlens, J., Le, Q.V.: Do better ImageNet models transfer better? In: CVPR (2019)
14. Krizhevsky, A., Sutskever, I., Hinton, G.E.: ImageNet classification with deep convolutional neural networks. In: NIPS (2012)
15. Lan, X., Zhu, X., Gong, S.: Knowledge distillation by on-the-fly native ensemble. In: NeurIPS (2018)
16. Li, D., Zhou, A., Yao, A.: HBONet: harmonious bottleneck on two orthogonal dimensions. In: ICCV (2019)
17. Li, Z., Hoiem, D.: Learning without forgetting. In: Leibe, B., Matas, J., Sebe, N., Welling, M. (eds.) ECCV 2016. LNCS, vol. 9908, pp. 614–629. Springer, Cham (2016). https://doi.org/10.1007/978-3-319-46493-0_37
18. Ma, N., Zhang, X., Zheng, H.-T., Sun, J.: ShuffleNet V2: practical guidelines for efficient CNN architecture design. In: Ferrari, V., Hebert, M., Sminchisescu, C., Weiss, Y. (eds.) Computer Vision – ECCV 2018. LNCS, vol. 11218, pp. 122–138. Springer, Cham (2018). https://doi.org/10.1007/978-3-030-01264-9_8
19. Mudrakarta, P.K., Sandler, M., Zhmoginov, A., Howard, A.G.: K for the price of 1: parameter-efficient multi-task and transfer learning. In: ICLR (2019)
20. Oquab, M., Bottou, L., Laptev, I., Sivic, J.: Learning and transferring mid-level image representations using convolutional neural networks. In: CVPR (2014)
21. Paszke, A., et al.: PyTorch: an imperative style, high-performance deep learning library. In: NeurIPS (2019)
22. Romero, A., Ballas, N., Kahou, S.E., Chassang, A., Gatta, C., Bengio, Y.: FitNets: hints for thin deep nets. In: ICLR (2015)
23. Russakovsky, O., et al.: ImageNet large scale visual recognition challenge. IJCV **115**, 211–252 (2015). https://doi.org/10.1007/s11263-015-0816-y
24. Sandler, M., Howard, A.G., Zhu, M., Zhmoginov, A., Chen, L.: MobileNetV2: inverted residuals and linear bottlenecks. In: CVPR (2018)
25. Sun, D., Yao, A., Zhou, A., Zhao, H.: Deeply-supervised knowledge synergy. In: CVPR (2019)
26. Szegedy, C., et al.: Going deeper with convolutions. In: CVPR (2015)
27. Szegedy, C., Vanhoucke, V., Ioffe, S., Shlens, J., Wojna, Z.: Rethinking the inception architecture for computer vision. In: CVPR (2016)

28. Tan, M., Le, Q.V.: EfficientNet: rethinking model scaling for convolutional neural networks. In: ICML (2019)
29. Touvron, H., Vedaldi, A., Douze, M., Jégou, H.: Fixing the train-test resolution discrepancy. In: NeurIPS (2019)
30. Yu, J., Huang, T.S.: Universally slimmable networks and improved training techniques. In: ICCV (2019)
31. Yu, J., Yang, L., Xu, N., Yang, J., Huang, T.S.: Slimmable neural networks. In: ICLR (2019)
32. Zhang, D., Yang, J., Ye, D., Hua, G.: LQ-Nets: learned quantization for highly accurate and compact deep neural networks. In: Ferrari, V., Hebert, M., Sminchisescu, C., Weiss, Y. (eds.) ECCV 2018. LNCS, vol. 11212, pp. 373–390. Springer, Cham (2018). https://doi.org/10.1007/978-3-030-01237-3_23
33. Zhang, X., Zhou, X., Lin, M., Sun, J.: ShuffleNet: an extremely efficient convolutional neural network for mobile devices. In: CVPR (2018)
34. Zhang, Y., Xiang, T., Hospedales, T.M., Lu, H.: Deep mutual learning. In: CVPR (2018)
35. Zhou, A., Ma, Y., Li, Y., Zhang, X., Luo, P.: Towards improving generalization of deep networks via consistent normalization. arXiv preprint arXiv:1909.00182 (2019)

SMAP: Single-Shot Multi-person Absolute 3D Pose Estimation

Jianan Zhen[1,2], Qi Fang[1], Jiaming Sun[1,2], Wentao Liu[2], Wei Jiang[1],
Hujun Bao[1], and Xiaowei Zhou[1(✉)]

[1] Zhejiang University, Hangzhou, China
xzhou@cad.zju.edu.cn
[2] SenseTime, Science Park, Hong Kong

Abstract. Recovering multi-person 3D poses with absolute scales from a single RGB image is a challenging problem due to the inherent depth and scale ambiguity from a single view. Addressing this ambiguity requires to aggregate various cues over the entire image, such as body sizes, scene layouts, and inter-person relationships. However, most previous methods adopt a top-down scheme that first performs 2D pose detection and then regresses the 3D pose and scale for each detected person individually, ignoring global contextual cues. In this paper, we propose a novel system that first regresses a set of 2.5D representations of body parts and then reconstructs the 3D absolute poses based on these 2.5D representations with a depth-aware part association algorithm. Such a single-shot bottom-up scheme allows the system to better learn and reason about the inter-person depth relationship, improving both 3D and 2D pose estimation. The experiments demonstrate that the proposed approach achieves the state-of-the-art performance on the CMU Panoptic and MuPoTS-3D datasets and is applicable to in-the-wild videos.

Keywords: Human pose estimation · 3D from a single image

1 Introduction

Recent years have witnessed an increasing trend of research on monocular 3D human pose estimation because of its wide applications in augmented reality, human-computer interaction, and video analysis. This paper aims to address the problem of estimating absolute 3D poses of multiple people simultaneously from a single RGB image. Compared to the single-person 3D pose estimation problem that focuses on recovering the root-relative pose, i.e., the 3D locations

---

J. Zhen and Q. Fang—Equal contribution.
X. Zhou—State Key Lab of CAD&CG.

Electronic supplementary material The online version of this chapter (https://doi.org/10.1007/978-3-030-58555-6_33) contains supplementary material, which is available to authorized users.

© Springer Nature Switzerland AG 2020
A. Vedaldi et al. (Eds.): ECCV 2020, LNCS 12360, pp. 550–566, 2020.
https://doi.org/10.1007/978-3-030-58555-6_33

Fig. 1. We propose a novel framework named SMAP to estimate absolute 3D poses of multiple people from a single RGB image. The figure visualizes the result of SMAP on an in-the-wild image. The proposed single-shot and bottom-up design allows SMAP to leverage the entire image to infer the absolute locations of multiple people consistently, especially in terms of the ordinal depth relations.

of human-body keypoints relative to the root of the skeleton, the task addressed here additionally needs to recover the 3D translation of each person in the camera coordinate system.

While there has been remarkable progress in recovering the root-relative 3D pose of a single person from an image [8,12,18,36], it was not until recently that more attention was paid to the multi-person case. Most existing methods for multi-person 3D pose estimation extend the single-person approach with a separate stage to recover the absolute position of each detected person separately. They either use another neural network to regress the 3D translation of the person from the cropped image [22] or compute it based on the prior about the body size [6,41,42], which ignore the global context of the whole image. Another line of work tries to recover body positions with a ground plane constraint [19], but this approach assumes that the feet are visible, which is not always true, and accurate estimation of ground plane geometry from a single image is still an open problem (Fig. 1).

We argue that the robust estimation of global positions of human bodies requires to aggregate the depth-related cues over the whole image, such as the 2D sizes of human bodies, the occlusion between them, and the layout of the scene. Recent advances in monocular depth estimation have shown that convolutional neural networks (CNNs) are able to predict the depth map from an RGB image [13,16], which is particularly successful on human images [15]. This observation motivates us to directly learn the depths of human bodies from the input image instead of recovering them in a post-processing stage.

To this end, we propose a novel single-shot bottom-up approach to multi-person 3D pose estimation, which predicts absolute 3D positions and poses of multiple people in a single forward pass. We regress the root depths of human bodies in the form of a novel root depth map, which only requires 3D pose annotations as supervision. We train a fully convolutional network to regress the root depth map, as well as 2D keypoint heatmaps, part affinity fields (PAFs), and part relative-depth maps that encode the relative depth between two joints of each body part. Then, the detected 2D keypoints are grouped into individuals based on PAFs using a part association algorithm, and absolute 3D poses are recovered with the root depth map and part relative-depth maps. The whole pipeline is illustrated in Fig. 2.

We also show that predicting depths of human bodies is beneficial for the part association and 2D pose estimation. We observe that many association errors occur when two human bodies overlap in the image. Knowing the depths of them allows us to reason about the occlusion between them when assigning the detected keypoints. Moreover, from the estimated depth, we can infer the spatial extent of each person in 2D and avoid linking two keypoints with an unreasonable distance. With these considerations, we propose a novel depth-aware part association algorithm and experimentally demonstrate its effectiveness.

To summarize, the contributions of this work are:

- A single-shot bottom-up framework for multi-person 3D pose estimation, which can reliably estimate absolute positions of multiple people by leveraging depth-relevant cues over the entire image.
- A depth-aware part association algorithm to reason about inter-person occlusion and bone-length constraints based on predicted body depths, which also benefits 2D pose estimation.
- The state-of-the-art performance on public benchmarks, with the generalization to in-the-wild images and the flexibility for both whole-body and half-body pose estimation. The code, demonstration videos and other supplementary material are available at https://zju3dv.github.io/SMAP.

2   Related Work

Multi-person 2D Pose. Existing methods for multi-person 2D pose estimation can be approximately divided into two classes. Top-down approaches detect human first and then estimate keypoints with a single person pose estimator [5,7,27,39]. Bottom-up approaches localize all keypoints in the image first and then group them to individuals [3,9,10,23,25,26,31]. Cao et al. [3] propose Open-Pose and use part affinity fields (PAFs) to represent the connection confidence between keypoints. They solve the part association problem with a greedy strategy. Newell et al. [23] propose an approach that simultaneously outputs detection and group assignments in the form of pixel-wise tags.

Single-Person 3D Pose. Researches on single-person 3D pose estimation from a single image have already achieved remarkable performances in recent years.

One-stage approaches directly regress 3D keypoints from images and can leverage shading and occlusion information to resolve the depth ambiguity. Most of them are learning-based [1, 8, 21, 28, 29, 35, 36, 40]. Two-stage approaches estimate the 2D pose first and then lift it to the 3D pose, including learning-based[18, 30, 43], optimization-based [38, 44] and exemplar-based [4] methods, which can benefit from the reliable result of 2D pose estimation.

Multi-person 3D Pose. For the task of multi-person 3D pose estimation, top-down approaches focus on how to integrate the pose estimation task with the detection framework. They crop the image first and then regress the 3D pose with a single-person 3D pose estimator. Most of them estimate the translation of each person with an optimization strategy that minimizes the reprojection error computed over sparse keypoints [6, 33, 34] or dense semantic correspondences [41]. Moon et al. [22] regard the area of 2D bounding box as a prior and adopt a neural network to learn a correction factor. In their framework, they regress the root-relative pose and the root depth separately. Informative cues for inferring the interaction between people may lose during the cropping operation. Another work [37] regresses the full depth map based on the existing depth estimation framework [16], but their 'read-out' strategy is not robust to 2D outliers. On the other hand, bottom-up approaches focus on how to represent pose annotations as several maps in a robust way [2, 19, 20, 42]. However, they either optimize the translation in a post-processing way or ignore the task of root localization. XNect [19] extends the 2D location map in [21] to 3D ones and estimates the translation with a calibrated camera and the ground plane constraint, but the feet may be invisible in crowded scenes and obtaining the extrinsic parameters of the camera is not practical in most applications. Another line of work tries to recover the SMPL model [41, 42], and their focus lies in using scene constraints and avoiding interpenetration, which is weakly related to our task. Taking these factors into consideration, a framework that both considers recovering the translation in a single forward pass and aggregating global features over the image will be helpful to this task.

Monocular Depth Estimation. Depth estimation from a single view suffers from inherent ambiguity. Nevertheless, several methods make remarkable advances in recent years [13, 16]. Li et al. [15] observe that Mannequin Challenge could be a good source for human depth datasets. They generate training data using multi-view stereo reconstruction and adopt a data-driven approach to recover a dense depth map, achieving good results. However, such a depth map lacks scale consistency and cannot reflect the real depth.

3   Methods

Figure 2 presents the pipeline of our approach, which consists of a single-shot bottom-up framework named SMAP. With a single RGB image as input, SMAP outputs 2D representations including keypoint heatmaps and part affinity fields [3]. Additionally, it also regresses 2.5D representations including root depth map

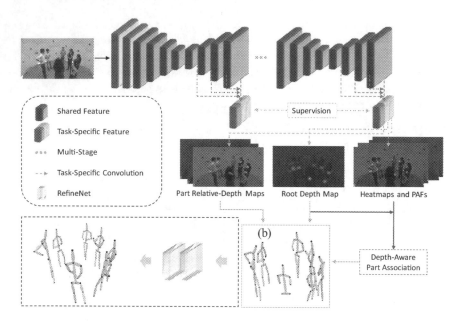

Fig. 2. Overview of the proposed approach. Given a single image, our single-shot network SMAP regresses several intermediate representations including 2D keypoint heatmaps, part affinity fields (PAFs), root depth map, and part relative-depth maps (Red means the child joint has a larger depth than its parent joint, and blue means the opposite). With a new depth-aware part association algorithm, body parts belonging to the same person are linked. With all these intermediate representations combined, absolute 3D poses of all people can be recovered. Finally, an optional RefineNet can be used to further refine the recovered 3D poses and complete invisible keypoints. (Color figure online)

and part relative-depth maps, which encode depth information of human bodies. Then, a depth-aware part association algorithm is proposed to assign detected 2D keypoints to individuals, depending on an ordinal prior and an adaptive bone-length constraint. Based on these results, the absolute 3D pose of each person can be reconstructed with a camera model. Individual modules of our system are introduced below.

3.1 Intermediate Representations

Given the input image, SMAP regresses the following intermediate representations, based on which 3D poses will be reconstructed:

Root Depth Map. As the number of people in an input image is unknown, we propose to represent the absolute depths for all human bodies in the image by a novel root depth map. The root depth map has the same size as the input image. The map values at the 2D locations of root joints of skeletons equal their absolute depths. An example is shown in Fig. 2. In this way, we are able to represent the depths of multiple people without predefining the number of people. During training, we only supervise the values of root locations. The proposed root depth map can be learned together with other cues within the same network as shown in Fig. 2 and only requires 3D poses (instead of full depth maps) as supervision, making our algorithm very efficient in terms of both model complexity and training data.

It is worth noting that visual perception of object scale and depth depends on the size of field of view (FoV), i.e., the ratio between the image size and the focal length. If two images are obtained with different FoVs, the same person at the same depth will occupy different proportions in these two images and seem to have different depths for the neural network, which may mislead the learning of depth. Thus, we normalize the root depth by the size of FoV as follows:

$$\widetilde{Z} = Z\frac{w}{f}, \tag{1}$$

where $\widetilde{Z}$ is the normalized depth, Z is the original depth, and f and w are the focal length and the image width both in pixels, respectively. So w/f is irrelevant to image resolution, but equals to the ratio between the physical size of image sensor and the focal length both in millimeters, i.e., FoV. The normalized depth values can be converted back to metric values in inference.

Keypoint Heatmaps. Each keypoint heatmap indicates the probable locations of a specific type of keypoints for all people in the image. Gaussian distribution is used to model uncertainties at the corresponding location.

Part Affinity Fields (PAFs). PAFs proposed in [3] include a set of 2D vector fields. Each vector field corresponds to a type of body part where the vector at each pixel represents the 2D orientation of the corresponding body part.

Part Relative-Depth Map. Besides the root depth, we also need depth values of other keypoints to reconstruct a 3D pose. Instead of predicting absolute depth values, for other keypoints we only regress their relative depths compared to their parent nodes in the skeleton, which are represented by part relative-depth maps. Similar to PAFs, each part relative-depth map corresponds to a type of body part, and every pixel that belongs to a body part encodes the relative depth between two joints of the corresponding body part. This dense representation provides rich information to reconstruct a 3D skeleton even if some keypoints are invisible.

Network Architecture. We use Hourglass [24] as our backbone and modify it to a multi-task structure with multiple branches that simultaneously output the above representations as illustrated in Fig. 2. Suppose the number of predefined joints is J. Then, there are $4J - 2$ channels in total (heatmaps and PAFs: $J + 2(J - 1)$, root depth map: 1, part relative-depth map: $J - 1$). Each output branch in our network only consists of two convolutional layers. Inspired by [14], we adopt multi-scale intermediate supervision. The L_1 loss is used to supervise the root depths and L_2 losses on other outputs. The effect of the network size and the multi-scale supervision will be validated in the experiments.

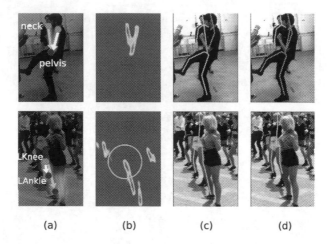

(a)   (b)   (c)   (d)

Fig. 3. Depth-aware part association. From left to right: candidate links, part affinity fields, pose estimation results from [3], and our results with depth-aware part association. The example in the first row shows the effect of ordinal prior. Under this circumstance, [3] assigns the pelvis to the occluded person while we give priority to the front person. The example in the second row shows the effect of the adaptive bone-length threshold, indicated by the green circle. As a noisy response occurs at the right ankle of the person it doesn't belong to, [3] will induce a false connection while our algorithm will not. (Color figure online)

3.2   Depth-Aware Part Association

Given 2D coordinates of keypoints from keypoint heatmaps after non-maximum suppression, we need to associate detected joints with corresponding individuals. Cao et al. [3] propose to link joints greedily based on the association scores given by PAFs. Basically, we follow their method to calculate association scores. However, PAFs scores might be unreliable due to occlusion between people. Figure 3 shows two examples where the above association strategy fails.

We propose to leverage the estimated depth maps to address the part association ambiguities caused by inter-person occlusion.

Ordinal Prior. A key insight to solve the occlusion issue is to give priority to the unoccluded person when assigning joints. The occlusion status can be inferred from the depth map. Therefore, we sort root joints from near to far according to the predicted root depth, instead of following the order of PAFs scores. Our association process starts with the root and proceeds along the skeleton tree successively.

Adaptive Bone-Length Constraint. To avoid linking two keypoints with an unreasonable distance, Cao et al. [3] constrain the association with a threshold defined as half of the image size. However, such a fixed threshold is ineffective as the 2D bone length depends on its depth. We adopt an adaptive bone-length threshold to penalize the unreasonable association, depending on predicted body depths. For each part, we compute the mean bone length in the training set in advance. Then, its maximal length in 2D is computed and used as the distance threshold for the corresponding part:

$$d_{cons} = \lambda \cdot \frac{D_{bone} \cdot f}{Z \cdot w} = \lambda \cdot \frac{D_{bone}}{\widetilde{Z}}, \qquad (2)$$

where D_{bone} is the 3D average length of a limb, $\widetilde{Z}$ is the normalized root depth predicted by our network, and λ is a relaxation factor. d_{cons} is used to filter unreasonable links. From Eq. 2 we can see that the adaptive threshold is not affected by camera intrinsic parameters or image resizing and is only related to the depth value estimated by our network and the statistical bone length.

As the association can benefit from depth information, we call it depth-aware part association. Figure 3(d) shows our qualitative results. The proposed scheme will also be validated by experiments.

3.3 3D Pose Reconstruction

Reconstruction. Following the connection relations obtained from part association, relative depths from child nodes to parent nodes can be read from the corresponding locations of part relative-depth maps. With the root depth and relative depths of body parts, we are able to compute the depth of each joint. Given 2D coordinates and joint depths, the 3D pose can be recovered through the perspective camera model:

$$\left[X, Y, Z \right]^T = Z K^{-1} \left[x, y, 1 \right]^T, \qquad (3)$$

where $[X, Y, Z]^T$ and $[x, y]^T$ are 3D and 2D coordinates of a joint respectively. K is the camera intrinsic matrix, which is available in most applications, e.g., from device specifications. In our experiments, we use the focal lengths provided by the datasets (same as [22]). For internet images with unknown focal lengths, we use a default value which equals the input image width in pixels. Note that the focal length will not affect the predicted ordinal depth relations between people.

Refinement. The above reconstruction procedure may introduce two types of errors. One is the cumulative error in the process of joint localization due to the hierarchical skeleton structure, and the other is caused by back projection when the depth is not accurate enough to calculate X and Y coordinates of 3D pose. Moreover, severe occlusion and truncation frequently occur in crowded scenes, which make some keypoints invisible. Therefore, we use an additional neural network named RefineNet to refine visible keypoints and complete invisible keypoints for each 3D pose. RefineNet consists of five fully connected layers. The inputs are 2D pose and 3D root-relative pose while the output is the refined 3D root-relative pose. The coordinates of invisible keypoints in the input are set to be zero. Note that RefineNet doesn't change the root depths.

4   Experiments

We evaluate the proposed approach on two widely-used datasets and compare it to previous approaches. Besides, we provide thorough ablation analysis to validate our designs.

4.1   Datasets

CMU Panoptic [11] is a large-scale dataset that contains various indoor social activities, captured by multiple cameras. Mutual occlusion between individuals and truncation makes it challenging to recover 3D poses. Following [41], we choose two cameras (16 and 30), 9600 images from four activities (Haggling, Mafia, Ultimatum, Pizza) as our test set, and 160k images from different sequences as our training set.

MuCo-3DHP and MuPoTS-3D [20]. MuCo-3DHP is an indoor multi-person dataset for training, which is composited from single-person datasets. MuPoTS-3D is a test set consisting of indoor and outdoor scenes with various camera poses, making it a convincing benchmark to test the generalization ability.

4.2   Implementation Details

We adopt Adam as optimizer with $2e-4$ learning rate, and train two models for 20 epochs on the CMU Panoptic and MuCo-3DHP datasets separately, mixed with COCO data [17]. The batch size is 32 and 50% data in each mini-batch is from COCO (same as [20,22]). Images are resized to a fixed size 832×512 as the input to the network. Note that resizing doesn't change FoV. Since the COCO dataset lacks 3D pose annotations, weights of 3D losses are set to zero when the COCO data is fed.

4.3   Evaluation Metrics

MPJPE. MPJPE measures the accuracy of the 3D root-relative pose. It calculates the Euclidean distance between the predicted and the groundtruth joint locations averaged over all joints.

RtError. Root Error (RtError) is defined as the Euclidean distance between the predicted and the groundtruth root locations.

3DPCK. 3DPCK is the percentage of correct keypoints. A keypoint is declared correct if the Euclidean distance between predicted and groundtruth coordinates is smaller than a threshold (15 cm in our experiments). PCK_{rel} measures relative pose accuracy with root alignment; PCK_{abs} measures absolute pose accuracy without root alignment; and PCK_{root} only measures the accuracy of root joints. AUC means the area under curve of 3DPCK over various thresholds.

PCOD. We propose a new metric named the percentage of correct ordinal depth (PCOD) relations between people. The insight is that predicting absolute depth from a single view is inherently ill-posed, while consistent ordinal relations between people are more meaningful and suffice many applications. For a pair of people (i, j), we compare their root depths and divide the ordinal depth relation into three classes: closer, farther, and roughly the same (within 30 cm). PCOD equals the classification accuracy of predicted ordinal depth relations.

Table 1. Results on the Panoptic dataset. For [22], we used the code provided by the authors and trained it on the Panoptic dataset. *The average of [42] is recalculated following the standard practice in [41], i.e., average over activities.

	Method	Haggling	Mafia	Ultim	Pizza	Average
MPJPE	PoPa et al. [32]	217.9	187.3	193.6	221.3	203.4
	Zanfir et al. [41]	140.0	165.9	150.7	156.0	153.4
	Moon et al. [22]	89.6	91.3	79.6	90.1	87.6
	Zanfir et al. [42]	72.4	78.8	66.8	94.3	78.1*
	Ours w/o Refine	71.8	72.5	65.9	82.1	73.1
	Ours	**63.1**	**60.3**	**56.6**	**67.1**	**61.8**
RtError	Zanfir et al. [41]	257.8	409.5	301.1	294.0	315.5
	Moon et al. [22]	160.2	151.9	177.5	127.7	154.3
	Ours	**84.7**	**87.7**	**91.2**	**78.5**	**85.5**
PCOD	Moon et al. [22]	92.3	93.7	95.2	94.2	93.9
	Ours	**97.8**	**98.5**	**97.6**	**99.6**	**98.4**

Table 2. Results on the MuPoTS-3D dataset. All numbers are average values over 20 activities.

		Matched people					All people	
	Method	PCK_{rel}	PCK_{abs}	PCK_{root}	AUC_{rel}	PCOD	PCK_{rel}	PCK_{abs}
Top down	Rogez. [33]	62.4	-	-	-	-	53.8	-
	Rogez. [34]	74.0	-	-	-	-	70.6	-
	Dabral. [6]	74.2	-	-	-	-	71.3	-
	Moon. [22]	**82.5**	31.8	31.0	40.9	92.6	**81.8**	31.5
Bottom up	Mehta. [19]	69.8	-	-	-	-	65.0	-
	Mehta. [20]	75.8	-	-	-	-	70.4	-
	Ours	80.5	**38.7**	**45.5**	**42.7**	**97.0**	73.5	**35.4**

4.4 Comparison with State-of-the-Art Methods

CMU Panoptic. Table 1 demonstrates quantitative comparison between state-of-the-art methods and our model. It indicates that our model outperforms previous methods in all metrics by a large margin. In particular, the error on the

Pizza sequence decreases significantly compared with the previous work. As the Pizza sequence shares no similarity with the training set, this improvement shows our generalization ability.

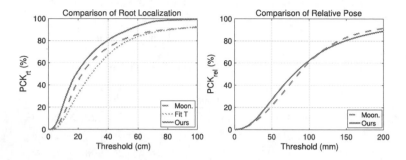

Fig. 4. Comparisons of root localization and relative pose. The curves are PCK_{root} and PCK_{rel} over different thresholds on the MuPoTS-3D dataset. Blue: result of [22]. Green: estimating the translation by minimizing the reprojection error. Red: our result. (Color figure online)

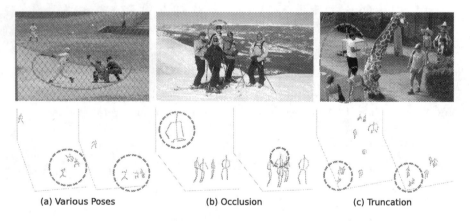

Fig. 5. Qualitative comparison. The results of three example images. For each example, the top row shows the input image, and the bottom row shows the results of [22] (left) and the proposed method (right), respectively. The red circles highlight the difference in localization of human bodies between two methods. (Color figure online)

MuPoTS-3D. We follow the protocol of [22]. Additionally, PCK_{root} and PCOD are used to evaluate the 3D localization of people. In terms of the absolute pose which we are more concerned with, it can be observed from Table 2 that our model is superior to [22] in relevant metrics including PCK_{abs}, PCK_{root} and PCOD by a large margin. It also demonstrates that our model has higher PCK_{rel} compared with all bottom-up methods and most top-down methods except [22].

Note that we achieve higher $\mathrm{AUC_{rel}}$ compared to [22] for the relative 3D pose of matched people.

Comparison with Top-Down Methods. We provide additional analysis to compare our single-shot bottom-up method to the state-of-the-art top-down method [22]. Figure 4 shows thorough comparisons in terms of $\mathrm{PCK_{root}}$ and $\mathrm{PCK_{rel}}$. For root localization, we compare to two methods: 1) regressing the scale from each cropped bounding box using a neural network as in [22] and 2) estimating the 3D translation by optimizing reprojection error with the groundtruth 2D pose and the estimated relative 3D pose from [22] ('FitT' in Fig. 4). We achieve better $\mathrm{PCK_{root}}$ over various thresholds than both of them. Notably, we achieve roughly 100% accuracy with a threshold 1m. As for relative pose estimation, [22] achieves higher $\mathrm{PCK_{rel}}$ (@15 cm) as it adopts a separate off-the-shelf network [36] that is particularly optimized for relative 3D pose estimation. Despite that, we obtain better $\mathrm{PCK_{rel}}$ when the threshold is smaller and higher $\mathrm{AUC_{rel}}$.

Figure 5 shows several scenarios (various poses, occlusion, and truncation) in which the top-down method [22] may fail as it predicts the scale for each detected person separately and ignore global context. Instead, the proposed bottom-up design is able to leverage features over the entire image instead of only using cropped features in individual bounding boxes.

Furthermore, our running time and memory remain almost unchanged with the number of people in the image while those of [22] grow faster with the number of people due to its top-down design, as shown in the supplementary material.

Depth Estimation. Apart from our method, there are two alternatives for depth estimation: 1) regressing the full depth map rather than the root depth map. 2) using the cropped image as the input to the network rather than the whole image. For the first alternative, since there is no depth map annotation in existing multi-person outdoor datasets, we use the released model of the state-of-the-art human depth estimator [15], which is particularly optimized for human depth estimation trained on a massive amount of in-the-wild 'frozen people' videos. For the second alternative, [22] is the state-of-the-art method that estimates root depth from the cropped image, so we compare with it. Figure 6 demonstrates scatter plots of the groundtruth root depth versus the predicted root depth of three methods on the MuPoTS-3D dataset. Ideally, the estimated depths should be linearly correlated to the ground truth, resulting a straight line in the scatter plot. Our model shows better consistency than baselines. Note that, while the compared methods are trained on different datasets, the images in the test set MuPoTS-3D are very different from the training images for all methods. Though not rigorous, this comparison is still reasonable to indicate the performance of these methods when applied to unseen images.

4.5   Ablation Analysis

Architecture. Table 3 shows how different designs of our framework affect the multi-person 3D pose estimation accuracy: 1) the performance of our model will degrade severely without depth normalization. As we discussed in Sect. 3.1,

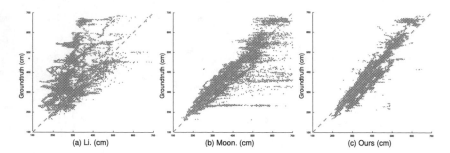

Fig. 6. Comparison with alternative depth estimation methods. The scatter plots show the consistency of root depth estimation on the MuPoTS-3D dataset. The X and Y axes are the predicted, groundtruth root depth, respectively. The dashed line means the ideal result, i.e., estimation equals ground truth. (a) 'Read-out' root depths from the full depth map estimated by [15]. (b) State-of-the-art top-down approach [22]. (c) Our approach.

Table 3. Ablation study of the structure design on the MuPoTS-3D dataset. The default backbone is Hourglass model with three stages, and 'Smaller Backbone' means one-stage model.

Design	Recall	$\mathrm{PCK_{root}}$	$\mathrm{PCK_{abs}}$	$\mathrm{PCK_{rel}}$	PCOD
Full model	**92.3**	**45.5**	**38.7**	**80.5**	**97.0**
No normalization	92.3	5.7	8.7	78.9	95.7
No multi-scale supervision	92.1	45.2	36.2	75.4	93.1
No RefineNet	92.3	45.5	34.7	70.9	97.0
Smaller backbone	91.1	43.8	35.1	75.7	96.4

Table 4. Ablation study of the part association. '2DPA' means the 2D part association proposed by [3]. 'DAPA' means the depth-aware part association we proposed. Both of them are based on the same heatmaps and PAFs results.

	Panoptic		MuPoTS-3D				
	Recall	2DPCK	Recall	$\mathrm{PCK_{root}}$	$\mathrm{PCK_{abs}}$	$\mathrm{PCK_{rel}}$	PCOD
2DPA	94.3	92.4	92.1	45.3	38.6	80.2	96.5
DAPA	**96.4**	**93.1**	**92.3**	**45.5**	**38.7**	**80.5**	**97.0**

normalizing depth values by the size of FoV makes depth learning easier. 2) Multi-scale supervision is beneficial. 3) To show that our performance gain in terms of the absolute 3D pose is mostly attributed to our single-shot bottom-up design rather than the network size, we test with a smaller backbone. The results show that, even with a one-stage hourglass network, our method still achieves higher $\mathrm{PCK_{root}}$ and $\mathrm{PCK_{abs}}$ than the top-down method [22].

Part Association. To compare the proposed depth-aware part association with the 2D part association in [3], we evaluate relevant metrics on Panoptic and

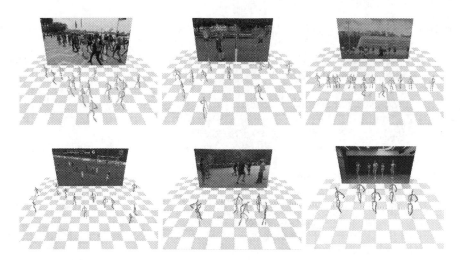

Fig. 7. Qualitative results on in-the-wild images from the Internet.

MuPoTS-3D datasets. Note that the threshold of 2DPCK is the half of the head size. Table 4 lists the results (2DPA vs. DAPA) and reveals that our depth-aware part association outperforms the 2D part association in all these metrics. Besides, Fig. 3 shows some qualitative examples.

RefineNet. Table 1 and 3 show that RefineNet is able to improve both relative and absolute pose estimation. It is able to complete invisible keypoints and refine visible keypoints with a learned 3D pose prior. The improvement is more significant on the Panoptic dataset since the training and test images are captured by cameras with similar views (Fig. 7).

Please refer to the supplementary material for more experimental details and results.

5 Conclusion

We proposed a novel single-shot bottom-up framework to estimate absolute multi-person 3D poses from a single RGB image. The proposed framework uses a fully convolutional network to regress a set of 2.5D representations for multiple people, from which the absolute 3D poses can be reconstructed. Additionally, benefited from the depth estimation of human bodies, a novel depth-aware part association algorithm was proposed and proven to benefit 2D pose estimation in crowd scenes. Experiments demonstrated state-of-the-art performance as well as generalization ability of the proposed approach.

Acknowledgements. The authors would like to acknowledge support from NSFC (No. 61806176), Fundamental Research Funds for the Central Universities (2019QNA5022) and ZJU-SenseTime Joint Lab of 3D Vision.

References

1. Alp Güler, R., Neverova, N., Kokkinos, I.: DensePose: dense human pose estimation in the wild. In: CVPR (2018)
2. Benzine, A., Luvison, B., Pham, Q.C., Achard, C.: Deep, robust and single shot 3D multi-person human pose estimation from monocular images. In: ICIP (2019)
3. Cao, Z., Simon, T., Wei, S.E., Sheikh, Y.: Realtime multi-person 2D pose estimation using part affinity fields. In: CVPR (2017)
4. Chen, C.H., Ramanan, D.: 3D human pose estimation= 2D pose estimation+ matching. In: CVPR (2017)
5. Chen, Y., Wang, Z., Peng, Y., Zhang, Z., Yu, G., Sun, J.: Cascaded pyramid network for multi-person pose estimation. In: CVPR (2018)
6. Dabral, R., Gundavarapu, N.B., Mitra, R., Sharma, A., Ramakrishnan, G., Jain, A.: Multi-person 3d human pose estimation from monocular images. In: 3DV (2019)
7. Fang, H.S., Xie, S., Tai, Y.W., Lu, C.: RMPE: regional multi-person pose estimation. In: ICCV (2017)
8. Guler, R.A., Kokkinos, I.: HoloPose: holistic 3D human reconstruction in-the-wild. In: CVPR (2019)
9. Hidalgo, G., et al.: Single-network whole-body pose estimation. In: ICCV (2019)
10. Insafutdinov, E., Pishchulin, L., Andres, B., Andriluka, M., Schiele, B.: DeeperCut: a deeper, stronger, and faster multi-person pose estimation model. In: Leibe, B., Matas, J., Sebe, N., Welling, M. (eds.) ECCV 2016. LNCS, vol. 9910, pp. 34–50. Springer, Cham (2016). https://doi.org/10.1007/978-3-319-46466-4_3
11. Joo, H., et al.: Panoptic studio: a massively multiview system for social interaction capture. TPAMI (2017)
12. Kanazawa, A., Black, M.J., Jacobs, D.W., Malik, J.: End-to-end recovery of human shape and pose. In: CVPR (2018)
13. Lee, J.H., Kim, C.S.: Monocular depth estimation using relative depth maps. In: CVPR (2019)
14. Li, W., et al.: Rethinking on multi-stage networks for human pose estimation. arXiv preprint arXiv:1901.00148 (2019)
15. Li, Z., et al.: Learning the depths of moving people by watching frozen people. In: CVPR (2019)
16. Li, Z., Snavely, N.: Megadepth: learning single-view depth prediction from Internet photos. In: CVPR (2018)
17. Lin, T.-Y., et al.: Microsoft COCO: common objects in context. In: Fleet, D., Pajdla, T., Schiele, B., Tuytelaars, T. (eds.) ECCV 2014. LNCS, vol. 8693, pp. 740–755. Springer, Cham (2014). https://doi.org/10.1007/978-3-319-10602-1_48
18. Martinez, J., Hossain, R., Romero, J., Little, J.J.: A simple yet effective baseline for 3D human pose estimation. In: ICCV (2017)
19. Mehta, D., et al.: XNect: real-time multi-person 3d human pose estimation with a single RGB camera. TOG (2020)
20. Mehta, D., et al.: Single-shot multi-person 3D pose estimation from monocular RGB. In: 3DV (2018)
21. Mehta, D., et al.: VNect: real-time 3D human pose estimation with a single RGB camera. TOG (2017)
22. Moon, G., Chang, J., Lee, K.M.: Camera distance-aware top-down approach for 3D multi-person pose estimation from a single RGB image. In: ICCV (2019)
23. Newell, A., Huang, Z., Deng, J.: Associative embedding: end-to-end learning for joint detection and grouping. In: NeurIPS (2017)

24. Newell, A., Yang, K., Deng, J.: Stacked hourglass networks for human pose esti-
    mation. In: Leibe, B., Matas, J., Sebe, N., Welling, M. (eds.) ECCV 2016. LNCS,
    vol. 9912, pp. 483–499. Springer, Cham (2016). https://doi.org/10.1007/978-3-319-
    46484-8_29
25. Nie, X., Zhang, J., Yan, S., Feng, J.: Single-stage multi-person pose machines. In:
    ICCV (2019)
26. Papandreou, G., Zhu, T., Chen, L.-C., Gidaris, S., Tompson, J., Murphy, K.: Per-
    sonLab: person pose estimation and instance segmentation with a bottom-up, part-
    based, geometric embedding model. In: Ferrari, V., Hebert, M., Sminchisescu, C.,
    Weiss, Y. (eds.) Computer Vision – ECCV 2018. LNCS, vol. 11218, pp. 282–299.
    Springer, Cham (2018). https://doi.org/10.1007/978-3-030-01264-9_17
27. Papandreou, G., et al.: Towards accurate multi-person pose estimation in the wild.
    In: CVPR (2017)
28. Pavlakos, G., Zhou, X., Daniilidis, K.: Ordinal depth supervision for 3D human
    pose estimation. In: CVPR (2018)
29. Pavlakos, G., Zhou, X., Derpanis, K.G., Daniilidis, K.: Coarse-to-fine volumetric
    prediction for single-image 3D human pose. In: CVPR (2017)
30. Pavllo, D., Feichtenhofer, C., Grangier, D., Auli, M.: 3D human pose estimation in
    video with temporal convolutions and semi-supervised training. In: CVPR (2019)
31. Pishchulin, L., et al.: Deepcut: joint subset partition and labeling for multi person
    pose estimation. In: CVPR (2016)
32. Popa, A.I., Zanfir, M., Sminchisescu, C.: Deep multitask architecture for integrated
    2D and 3D human sensing. In: CVPR (2017)
33. Rogez, G., Weinzaepfel, P., Schmid, C.: LCR-NET: localization-classification-
    regression for human pose. In: CVPR (2017)
34. Rogez, G., Weinzaepfel, P., Schmid, C.: LCR-Net++: multi-person 2D and 3D
    pose detection in natural images. TPAMI (2019)
35. Sun, X., Shang, J., Liang, S., Wei, Y.: Compositional human pose regression. In:
    ICCV (2017)
36. Sun, X., Xiao, B., Wei, F., Liang, S., Wei, Y.: Integral human pose regression. In:
    Ferrari, V., Hebert, M., Sminchisescu, C., Weiss, Y. (eds.) ECCV 2018. LNCS,
    vol. 11210, pp. 536–553. Springer, Cham (2018). https://doi.org/10.1007/978-3-
    030-01231-1_33
37. Véges, M., Lőrincz, A.: Absolute human pose estimation with depth prediction
    network. In: IJCNN (2019)
38. Xiang, D., Joo, H., Sheikh, Y.: Monocular total capture: posing face, body, and
    hands in the wild. In: CVPR (2019)
39. Xiao, B., Wu, H., Wei, Y.: Simple baselines for human pose estimation and tracking.
    In: Ferrari, V., Hebert, M., Sminchisescu, C., Weiss, Y. (eds.) ECCV 2018. LNCS,
    vol. 11210, pp. 472–487. Springer, Cham (2018). https://doi.org/10.1007/978-3-
    030-01231-1_29
40. Yang, W., Ouyang, W., Wang, X., Ren, J., Li, H., Wang, X.: 3D human pose
    estimation in the wild by adversarial learning. In: CVPR (2018)
41. Zanfir, A., Marinoiu, E., Sminchisescu, C.: Monocular 3D pose and shape estima-
    tion of multiple people in natural scenes-the importance of multiple scene con-
    straints. In: CVPR (2018)
42. Zanfir, A., Marinoiu, E., Zanfir, M., Popa, A.I., Sminchisescu, C.: Deep network
    for the integrated 3D sensing of multiple people in natural images. In: NeurIPS
    (2018)

43. Zhao, L., Peng, X., Tian, Y., Kapadia, M., Metaxas, D.N.: Semantic graph convolutional networks for 3D human pose regression. In: CVPR (2019)
44. Zhou, X., Zhu, M., Leonardos, S., Derpanis, K.G., Daniilidis, K.: Sparseness meets deepness: 3D human pose estimation from monocular video. In: CVPR (2016)

Learning to Detect Open Classes
for Universal Domain Adaptation

Bo Fu[1,2], Zhangjie Cao[1,2], Mingsheng Long[1,2(✉)], and Jianmin Wang[1,2]

[1] School of Software, BNRist, Tsinghua University, Beijing, China
microhhh9@gmail.com, caozhangjie14@gmail.com,
{mingsheng,jimwang}@tsinghua.edu.cn
[2] Research Center for Big Data, Tsinghua University, Beijing, China

Abstract. Universal domain adaptation (UniDA) transfers knowledge between domains without any constraint on the label sets, extending the applicability of domain adaptation in the wild. In UniDA, both the source and target label sets may hold individual labels not shared by the other domain. A *de facto* challenge of UniDA is to classify the target examples in the shared classes against the domain shift. A more prominent challenge of UniDA is to mark the target examples in the target-individual label set (open classes) as "unknown". These two entangled challenges make UniDA a highly under-explored problem. Previous work on UniDA focuses on the classification of data in the shared classes and uses per-class accuracy as the evaluation metric, which is badly biased to the accuracy of shared classes. However, accurately detecting open classes is the mission-critical task to enable real universal domain adaptation. It further turns UniDA problem into a well-established close-set domain adaptation problem. Towards accurate open class detection, we propose Calibrated Multiple Uncertainties (CMU) with a novel transferability measure estimated by a mixture of uncertainty quantities in complementation: entropy, confidence and consistency, defined on conditional probabilities calibrated by a multi-classifier ensemble model. The new transferability measure accurately quantifies the inclination of a target example to the open classes. We also propose a novel evaluation metric called H-score, which emphasizes the importance of both accuracies of the shared classes and the "unknown" class. Empirical results under the UniDA setting show that CMU outperforms the state-of-the-art domain adaptation methods on all the evaluation metrics, especially by a large margin on the H-score.

Keywords: Universal domain adaptation · Open class detection

B. Fu and Z. Cao—Equal contribution.

Electronic supplementary material The online version of this chapter (https://doi.org/10.1007/978-3-030-58555-6_34) contains supplementary material, which is available to authorized users.

A. Vedaldi et al. (Eds.): ECCV 2020, LNCS 12360, pp. 567–583, 2020.
https://doi.org/10.1007/978-3-030-58555-6_34

1   Introduction

Domain adaptation (DA) relieves the requirement of labeled data in deep learning by leveraging the labeled data from a related domain [28]. Most DA methods constrain the source and target label sets to some extent, which are easily violated in complicated practical scenarios. For example, we can access molecule datasets with annotated properties [39]. However, when predicting unknown molecules, we are exposed to two challenges: (**1**) The molecule structures such as scaffolds [13] may vary between training and testing sets, causing large *domain shift*; (**2**) Some molecules have property values never existing in our dataset such as unknown toxicity, which causes the *category shift*. To address the challenges, Universal Domain Adaptation (UniDA) [41] is raised to remove all label set constraints.

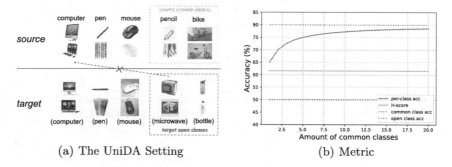

(a) The UniDA Setting   (b) Metric

Fig. 1. (a) The UniDA Setting. There are 3 common, 2 source private and 2 target private classes. The red cross means that the open class "microwave" is easily misclassified to "computer". (b) Comparison of per-class accuracy and H-score. Assuming that the amount of samples in each category is equal. The classification accuracy of common classes is 80%, and the accuracy of open classes is 50%. (Color figure online)

As shown in Fig. 1(a), in UniDA, given any labeled source domain and unlabeled target domain, we need to classify target data correctly if it belongs to the common label set or mark it as "unknown" otherwise. UniDA poses two technical challenges: (**1**) Distribution matching is still needed but should be constrained into the common label set; (**2**) As a new challenge, we need to detect data of the target open classes without any target labeled data or prior knowledge. Detecting open classes is the key to UniDA since it can directly solve the second challenge, and if it is solved, the first challenge can be easily addressed by remove the open class data and perform partial domain adaptation methods.

Universal Adaptation Network (UAN) [41] addresses the challenges by quantifying the transferability of each sample based on the uncertainty and the domain similarity. However, as we analyzed in Sect. 3.1, the transferability suffers from two shortcomings. First, they use entropy to measure uncertainty and auxiliary domain classifier to measure domain similarity. Entropy lacks discriminability for uncertain and sharp predictions, especially with a large number of classes. The predictions of the auxiliary domain classifier are mostly overconfident as shown in Fig. 4(b) in [41]. Second, the uncalibrated predictions make the transferability unreliable. Thus, UAN cannot detect open classes clearly. Such failure is hidden by the per-class accuracy used by UAN [41], which, as shown in Fig. 1(b), overly focuses on the common label set, especially under large-scale classes. How to detect open classes and how to evaluate UniDA are still unsolved problems.

In this paper, we propose **Calibrated Multiple Uncertainties (CMU)** with a novel measurement to quantify the transferability of each sample. We improve the quality of the transferability over the previous work in two aspects. 1) We design a new uncertainty measurement by compensating entropy with consistency and confidence for the lack of ability to tackle particular predictions; 2) The multi-classifier architecture for uncertainty computation naturally forms an ensemble, which is the most suitable calibration method for the domain shift setting. The new transferability can more accurately estimate the uncertainty and more clearly differentiate different samples by uncertainty, which improves the accuracy of open class detection. Furthermore, we propose a new evaluation metric called H-score as the harmonic mean of the accuracy on common label set and the accuracy of marking data in the target private label set as "unknown". As shown in Fig. 1(b), the new criterion is high only when target data in both common and private label sets are classified accurately.

The main contributions of this paper are:

(1) We emphasize the importance of detecting open classes for UniDA. We propose Calibrated Multiple Uncertainties (CMU) with a novel transferability composed of entropy, consistency, and confidence. The three uncertainties are complementary to discriminate different degrees of uncertainty clearly and are well-calibrated by multiple classifiers, which distinguish target samples from common classes and open classes more clearly.

(2) We point out that the evaluation metric: per-class accuracy, used by UAN highly biases to common classes but fails to test the ability to detect open classes, especially when the number of common classes is large. We design a new evaluation protocol: H-score, as the harmonic mean of target common data accuracy and private data accuracy. It evaluates a balance ability to classify common class samples and filter open class samples.

(3) We conduct experiments on UniDA benchmarks. Empirical results show that CMU outperforms UAN and methods of other DA settings on all evaluation metrics, especially on the H-score. Deeper analyses show that the proposed transferability can distinguish the common label set from the open classes effectively.

2 Related Work

Domain adaptation settings can be divided into closed set, partial, open set domain adaptation and universal domain adaptation based on the label set relationship. Universal domain adaptation removes all constraints on the label set and includes all other domain adaptation settings.

Closed Set Domain Adaptation assumes both domains share the same label set. Early deep closed set domain adaptation methods minimize Maximum Mean Discrepancy (MMD) on deep features [20,22,34]. Recently, methods based on adversarial learning [8,21,33] are proposed to play a two-player minimax game between the feature extractor and a domain discriminator. Adversarial learning methods achieves the state-of-the-art performance, which is further improved by recent works [5,12,14,15,19,23,24,29,31,37,40,45] with new architecture designs.

Partial Domain Adaptation requires that the source label set contains the target label set [2–4,11,44], which receives much more attention with access to large annotated dataset such as ImageNet [6] and Open Image [36]. To solve partial domain adaptation, one stream of works [2,3] uses target prediction to construct instance- and class-level weight to down-weight source private samples. Another stream [4,44] employs an auxiliary domain discriminator to quantify the domain similarity. Recent work [11] integrates the two weighting mechanisms.

Open Set Domain Adaptation (OSDA) is proposed by Busto *et al.* [25] to have private and shared classes in both domains but know shared labels. They use an Assign-and-Transform-Iteratively (ATI) algorithm to address the problem. Lian *et al.* [17] improves it by using entropy weight. Saito *et al.* [30] relaxed the problem by requiring no source private labels, so the target label set contains the source. Later OSDA methods [1,18,43] follow this more challenging setting and attack it by image translation [43] or a coarse-to-fine filtering process [18].

However, closed set, partial, open set domain adaptation are all restricted by label set assumptions. The latter two shed light on practical domain adaptation.

Universal Domain Adaptation (UniDA) [41] is the most general setting of domain adaptation, which removes all constraints and includes all the previous adaptation settings. It introduces new challenges to detect open classes in target data even with private classes in the source domain. UAN [41] evaluates the transferability of examples based on uncertainty and domain similarity. However, the uncertainty and domain similarity measurements, which are defined as prediction entropy and output of the auxiliary domain classifier, are not robust and discriminable enough. We propose a new uncertainty measurement as the mixture of entropy, consistency and confidence and design a deep ensemble model to calibrate the uncertainty, which characterizes different degrees of uncertainty and distinguishes target data in common label set from those in private label set.

3   Calibrated Multiple Uncertainties

In Universal Domain Adaptation (UniDA), a labeled source domain $\mathcal{D}^s = \{(\mathbf{x}^s, \mathbf{y}^s)\}$ and a unlabeled target domain $\mathcal{D}^t = \{(\mathbf{x}^t)\}$ are provided at training. Note that the source and target data are sampled from different distributions

p and q respectively. We use C^s and C^t to denote the label set of the source domain and the target domain. $C = C^s \cap C^t$ is the common label set shared by both domains while $\overline{C}^s = C^s \backslash C$ and $\overline{C}^t = C^t \backslash C$ are the label sets private to source and target respectively. p_{C^s} and p_C are used to denote the distributions of source data with labels in the label set C^s and C respectively, and q_{C^t}, q_C are defined similarly. Note that the target label set is not accessible at training and only used for defining the UniDA problem. UniDA requires a model to distinguish target data in C from those in $\overline{C}^t$, as well as predict accurate label for target data in C.

3.1 Limitations of Previous Works

The most important challenge for UniDA is detecting open classes. We compare several state-of-the-art domain adaptation methods with open class detection module in Table 1 including UniDA method, UAN [41], and open set DA methods, STA [18] and OSBP [30]. STA and OSBP both use the confidence for an extra class as the criterion to detect open classes. However, as stated below, confidence alone lacks discriminability for particular predictions. In UAN, transferability is derived from uncertainty and domain similarity. Optimally, uncertainty is a well-established measurement to distinguish samples from C and from $\overline{C}^s$ and $\overline{C}^t$. But the uncertainty is measured by entropy, which lacks discriminability for uncertain and extremely sharp predictions. For the domain similarity, the auxiliary domain classifier is trained with domain label by supervised learning. So the predictions are over-confident. All the open class detection criteria before are unilateral and lack the discriminability for particular predictions.

Furthermore, the confidence for STA and OSBP and the uncertainty and domain similarity for UAN are based on uncalibrated prediction, meaning the prediction does not reflect the exact confidence, uncertainty or domain similarity of the sample. So all the criteria before are not estimated accurately and thus fail to distinguish target data in the common label set from the private label set.

Table 1. Comparison of open class detection criterion for different methods

Criterion	Calibration	Entropy	Confidence	Consistency	Domain similarity
OSBP [30]	✗	✗	✓	✗	✗
STA [18]	✗	✗	✓	✗	✗
UAN [41]	✗	✓	✗	✗	✓
CMU	✓	✓	✓	✓	✗

3.2 Multiple Uncertainties

We design a novel transferability to detect open class. We adopt the assumption made by UAN: the target data in C have lower uncertainty than target data in

$\bar{\mathcal{C}}^t$. A well-defined uncertainty measurement should distinguish different degrees of uncertainty, e.g., distinguishing definitely uncertain predictions from slightly uncertain ones. Then we can rank the uncertainty of target samples and mark the most uncertain ones as open class data. We first analyze and compare different uncertainty measurements on the discriminability of various predictions.

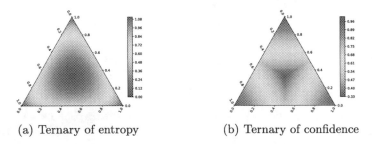

(a) Ternary of entropy   (b) Ternary of confidence

Fig. 2. Heatmap of entropy (a) and confidence (b) w.r.t. the probability values of three classes. Each edge is the value range $[0, 1]$. The corner area represents class distributions where one label is very likely, while the center area shows nearly uniform distribution.

Entropy measures the smoothness of the class distribution, which is higher for data in $\bar{\mathcal{C}}^t$ and lower for data in $\mathcal{C}$. We argue that *entropy exhibits low discriminability for highly uncertain and extremely sharp predictions*. Fig. 2(a) shows the value of entropy with respect to the probability of three classes. We can observe that when the probability distribution is close to uniform, i.e. very uncertain, the entropy is insensitive to probability changes. For sharp predictions, the entropy change in few classes is non-negligible. When there are a large number of classes, the relative difference of entropy values between sharp predictions is very small compared with the range of entropy values. For example, with m classes, the entropy values range is $[0, \log(m)]$ but the entropy difference between prediction $(1, 0, 0, ...)$ and $(0.5, 0.5, 0, ...)$ is $\log(2)$. When m is large, such difference can be ignored, but actually, the two predictions are quite different in terms of uncertainty. So estimating the uncertainty only by the entropy will fail to discriminate uncertain and extremely sharp predictions.

Confidence is higher for a more certain data point in $\mathcal{C}$. As shown in Fig. 2(b), confidence value shows ternary contour lines, where the confidence, i.e. the largest probabilities of three classes, is the same. We have the following statement on the length of the contour line: *The contour lines for extremely high and low confidence are short.* The proof is shown in the supplementary. On each contour line, even the confidence of different class distributions are the same, the degrees of uncertainty are different. For example, when the confidence is 0.5, the largest probability is 0.5, and the other two probabilities could be $(0.5, 0)$ or $(0.25, 0.25)$. It is obvious that $(0.5, 0.5, 0)$ is more uncertain than $(0.5, 0.25, 0.25)$. Therefore, confidence lacks discriminability in each contour line. The longer the contour line, the more class distributions in the contour line, the severer the problem of

Table 2. Comparison of calibration methods on out-of-distribution data

Method	Extra requirement	Extra computation
Temp scaling [9]	Target validation set	Training calibration parameters
Dropout [7]	Multiple dropout layers	Multiple full passes
Ensembles [16]	Multiple one-layer classifiers	Multiple one-layer passes
SVI [38]	Several times of model parameters	Several times of computation

confusing various class distributions. Thus, a shorter contour line exhibits higher discriminability for predictions, which, opposite and complementary to entropy, corresponds to extremely uncertain and confident predictions.

Based on the above analyses, confidence and entropy are complementary to cover both smooth and non-smooth class distributions. However, confidence suffers from prediction errors. If the classifier predicts an open class data as a class in $\mathcal{C}$ with high confidence, the confidence will mistakenly select the data as a common class sample. To compensate confidence, we employ **Consistency** built on multiple diverse classifiers $G_i|_{i=1}^m$, which reflects the agreement of different classifiers. The loss $\mathcal{E}(G_i)$ for the classifier G_i is defined as

$$\mathcal{E}(G_i) = \mathsf{E}_{(\mathbf{x},\mathbf{y})\sim p}L\left(\mathbf{y}, G_i(F(\mathbf{x}))\right) \tag{1}$$

where $i = 1, ..., m$ and L is the standard cross-entropy loss. To keep the diversity of different classifiers, we do not back-propagate gradients from $G_i|_{i=1}^m$ to the feature extractor F and initialize G_i with different random initialization. The lower the consistency value, the more likely the data is in $\mathcal{C}$. Consistency is more robust to prediction errors since the probability that all classifiers make the same mistake is low, which means all diverse classifiers predict a sample wrongly and coincidentally into the same class. Therefore, consistency compensates confidence for prediction errors. Confidence usually fails on smooth distribution because they are close to each other and show high consistency though they are uncertain.

Based on the above comparison, we can conclude that entropy, confidence and consistency all have their advantages and drawbacks and cannot individually represent the uncertainty. But they are complementary to each other and can collaborate to form an uncertainty measurement with high discriminability for all types of class distributions. Therefore, we choose the mixture of the three criteria. With each classifier G_i, $(i = 1, ..., m)$ predicting a probability $\hat{\mathbf{y}}_i^*$ for $\mathbf{x}^*$, $(* = s, t)$ over the source classes $\mathcal{C}^s$, we compute entropy w_{ent}, confidence w_{conf} and consistency w_{cons} as follows:

$$w_{\text{ent}}(\hat{\mathbf{y}}_i^t|_{i=1}^m) = \frac{1}{m}\sum_{i=1}^{m}\left(\sum_{j=1}^{|\mathcal{C}^s|} -\hat{y}_{ij}^t \log\left(\hat{y}_{ij}^t\right)\right), \tag{2}$$

$$w_{\text{conf}}(\hat{\mathbf{y}}_i^t|_{i=1}^m) = \frac{1}{m}\sum_{i=1}^{m}\max(\hat{\mathbf{y}}_i^t), \tag{3}$$

$$w_{\text{cons}}(\hat{\mathbf{y}}_i^t|_{i=1}^m) = \frac{1}{|\mathcal{C}^s|} \left\| \frac{1}{m} \sum_{i=1}^m \left(\hat{\mathbf{y}}_i^t - \frac{1}{m} \sum_{i=1}^m \hat{\mathbf{y}}_i^t \right)^2 \right\|_1, \tag{4}$$

where $\hat{y}_{ij}^t$ is the probability of j-th class and max take the maximum entry in $\hat{\mathbf{y}}_i^t$. w_{cons} is the standard deviation of all predictions. Multiple classifiers are employed to calibrate the entropy and the confidence.

We normalize the w_{ent} and w_{cons} by minmax normalization to unify them within $[0, 1]$. Then we compute w^t by aggregating the three uncertainties,

$$w^t = \frac{(1 - w_{\text{ent}}) + (1 - w_{\text{cons}}) + w_{\text{conf}}}{3}, \tag{5}$$

where the higher the $w_t(\mathbf{x}_0^t)$, the more likely $\mathbf{x}_0^t$ is in $\mathcal{C}$.

For source weight, since our novel w^t can distinguish target private data from common data more clearly and common class data should have high probability on one of the common classes, we sum the prediction of common data that are selected by w^t to compute weights $\mathbf{V}$ for source classes. Such class-level weight is only high for source classes in $\mathcal{C}$. Since source labels are available, the source weight w^s can be easily defined by taking the $\mathbf{y}^s$-th class weight:

$$\mathbf{V} = \text{avg}_{w^t(x^t) > w_0} \hat{\mathbf{y}}^t \text{ and } w^s(x^s) = V_{\mathbf{y}^s}, \tag{6}$$

where avg computes the average of $\hat{\mathbf{y}}^t$ and $\mathbf{V}_{\mathbf{y}^s}$ is the $\mathbf{y}^s$-th entry of $\mathbf{V}$.

3.3 Uncertainty Calibration

The transferability introduced above can estimate the uncertainty for all types of predictions to detect open classes. However, the criterion is still not reliable enough for UniDA. As shown in [16], overconfident predictions with low uncertainties exist among data of the "unknown" class, i.e., out-of-distribution data. So the uncertainty estimated from the prediction does not reflect the real uncertainty of the data samples, which deteriorates the reliability of the transferability.

Calibration is a widely-used approach to estimate the uncertainty more accurately, so we employ the most suitable calibration method for deep UniDA, where large-scale parameters and the domain gap need consideration. We compare existing calibration methods surveyed in [32]: Vanilla, Temp Scaling [9], Dropout [7], Ensemble [16], SVI [38], in terms of performance on out-of-distribution data. We do not include LL for the low performance in [32] and extra Bayesian Network.

As shown in Table 2, Temp Scaling requires a target validation set, which is not available in UniDA, or otherwise we know the components of the "unknown" class. Dropout and SVI require far more computation on deep networks. SVI can be embedded into particular network architectures. Ensemble naturally utilizes the current multi-classifier architecture in our framework and introduce no extra computation. From [32], we observe that when *testing on out-of-distribution data, Ensemble achieves the best performance on large-scale datasets.* Thus, Ensemble is the most suitable framework for UniDA and already embedded in our framework.

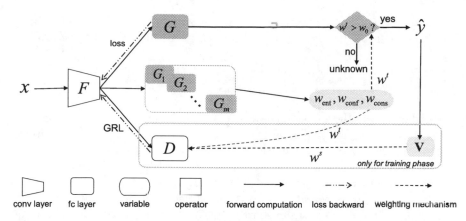

Fig. 3. The architecture of Calibrated Multiple Uncertainties (CMU). An input $\mathbf{x}$ is fed to F to output a feature, which is then input to a classifier G for prediction. The feature is also input to m classifiers $G_i|_{i=1}^m$ for ensemble. Three uncertainties: entropy, consistency and confidence are defined on the output of $G_i|_{i=1}^m$, to produce target weight w^t, which is used to decide a target data is "unknown" or not. The blue solid/dot lines represent the mechanism judging the "unknown" class. Target data with $w^t \geq w_0$ is common class data and is given a class prediction, and otherwise it is classified as "unknown". The source weight w^s is derived from class-level weight $\mathbf{V}$ based on the prediction $\hat{y}$ of target common class data. w^s and w^t are used to weighting samples in distribution matching. The part in the gray square is only used in the training phase.

3.4 Calibrated Multiple Uncertainties Framework

We first introduce the framework of CMU, which is shown in Fig. 3. CMU consists of a feature extractor F, a label classifier G, multiple classifiers G_1, G_2 ... G_m, and a domain discriminator D. For a data point $\mathbf{x}$, F extracts the feature $\mathbf{z} = F(\mathbf{x})$ and G predicts a probability $\hat{\mathbf{y}} = G(\mathbf{z})$ for $\mathbf{x}$. We derive the transferability measurement w^s and w^t for source and target data from the output of multiple classifiers as Eq. (5) and (6), which is used to weight each data sample in distribution matching. We train the F, G on D as [8] to enable classification and feature distribution matching where the losses are defined as:

$$\mathcal{E}(G) = \mathsf{E}_{(\mathbf{x},\mathbf{y})\sim p} L\left(\mathbf{y}, G(F(\mathbf{x}))\right) \tag{7}$$

$$\mathcal{E}(D) = -\,\mathsf{E}_{\mathbf{x}\sim p} w^s(\mathbf{x}) \log D\left(F\left(\mathbf{x}\right)\right) - \mathsf{E}_{\mathbf{x}\sim q} w^t(\mathbf{x}) \log\left(1 - D\left(F\left(\mathbf{x}\right)\right)\right) \tag{8}$$

where L is the cross-entropy loss. Combined with the loss for multiple classifiers in (1), the optimization of the new architecture can be defined as follows,

$$\max_{D}\min_{F,G} \mathcal{E}(G) - \lambda\mathcal{E}(D)$$

$$\min_{G_i|_{i=1}^m} \sum_{i=1}^m \mathcal{E}(G_i). \tag{9}$$

In the testing phase, given each input target sample $\mathbf{x}_0$, we first compute $w^t(\mathbf{x}_0)$ and then predict the class of $y(\mathbf{x})$ with a validated threshold w_0 as:

$$y(\mathbf{x}_0) = \begin{cases} \text{unknown} & w^t(\mathbf{x}_0) \leq w_0 \\ \text{argmax}\,(\hat{\mathbf{y}}_0) & w^t(\mathbf{x}_0) > w_0 \end{cases} \qquad (10)$$

which either rejects the $\mathbf{x}_0$ as "unknown" class or classifies it to a common class.

Our new transferability measurement consists of three complementary uncertainties covering all class distributions. We carefully compare and employ the most suitable calibration method to improve the quality of the uncertainty estimation. The proposed calibrated multiple uncertainties (CMU) can discriminate target data in $\bar{\mathcal{C}}^t$ from target data in $\mathcal{C}$ more clearly, which in turn helps discriminate source data in $\bar{\mathcal{C}}^s$ from source data in $\mathcal{C}$. Thus, CMU can simultaneously match the distributions of common classes and detect samples from open classes, which achieves high performance on both classifying common class and open class data.

4   Experiments

We conduct a thorough evaluation of CMU on universal domain adaptation benchmarks. Code is at https://github.com/thuml/Calibrated-Multiple-Uncertainties.

4.1   Setup

Datasets. We perform experiments on **Office-31** [28], **Office-Home** [35], **VisDA** [27] and **DomainNet** [26] datasets. For the first three datasets, we follow the same setup as [41]. **DomainNet** is by far the largest domain adaptation dataset, consists of six distinct domains: Clipart(C), Infograph(I), Painting(P), Quickdraw(Q), Real(R) and Sketch(S) across 345 classes. In the alphabet order, we use the first 150 classes as $\mathcal{C}$, the next 50 classes as $\bar{\mathcal{C}}^s$ and the rest as $\bar{\mathcal{C}}^t$. We choose 3 domains to transfer between each other due to the large amount of data.

Compared Methods. We compare the proposed CMU with (1) ResNet [10], (2) close-set domain adaptation: Domain-Adversarial Neural Networks (DANN) [8], Residual Transfer Networks (RTN) [22], (3) partial domain adaptation: Importance Weighted Adversarial Nets (IWAN) [44], Partial Adversarial Domain Adaptation (PADA) [3], (4) open set domain adaptation: Assign-and-Transform-Iteratively (ATI) [25], Open Set Back-Propagation (OSBP) [30]. (5) universal domain adaptation: Universal Adaptation Network (UAN) [41].

Evaluation Protocols. Previous work [41] uses the per-class accuracy as the evaluation metric, which calculates the instance accuracy of each class and then average. However, in per-class accuracy, the accuracy of each common class has the same contribution as the whole "unknown" class. So the influence of the

Table 3. Average class accuracy (%) and H-score (%) on **Office-31**

Method	Office-31												
	A → W		D → W		W → D		A → D		D → A		W → A		Avg	
	Acc	H-score	Acc	H-score	Acc	H-score	Acc	H-score	Acc	H-score	Acc	H-score	Acc	H-score
ResNet [10]	75.94	47.92	89.60	54.94	90.91	55.60	80.45	49.78	78.83	48.48	81.42	48.96	82.86	50.94
DANN [8]	80.65	48.82	80.94	52.73	88.07	54.87	82.67	50.18	74.82	47.69	83.54	49.33	81.78	50.60
RTN [22]	85.70	50.21	87.80	54.68	88.91	55.24	82.69	50.18	74.64	47.65	83.26	49.28	83.83	51.21
IWAN [44]	85.25	50.13	90.09	54.06	90.00	55.44	84.27	50.64	84.22	49.65	86.25	49.79	86.68	51.62
PADA [44]	85.37	49.65	79.26	52.62	90.91	55.60	81.68	50.00	55.32	42.87	82.61	49.17	79.19	49.98
ATI [25]	79.38	48.58	92.60	55.01	90.08	55.45	84.40	50.48	78.85	48.48	81.57	48.98	84.48	51.16
OSBP [30]	66.13	50.23	73.57	55.53	85.62	57.20	72.92	51.14	47.35	49.75	60.48	50.16	67.68	52.34
UAN [41]	85.62	58.61	94.77	70.62	97.99	71.42	86.50	59.68	85.45	60.11	85.12	60.34	89.24	63.46
CMU	**86.86**	**67.33**	**95.72**	**79.32**	**98.01**	**80.42**	**89.11**	**68.11**	**88.35**	**71.42**	**88.61**	**72.23**	**91.11**	**73.14**

"unknown" class is small, especially when the amount of common classes is large. As shown in Fig. 1(b), with a large number of classes, only classifying common class samples correctly can achieve fairly high per-class accuracy. Inspired by the F1-score, we propose the **H-score**: the harmonic mean of the instance accuracy on common class $a_{\mathcal{C}}$ and accuracy on the "unknown" class $a_{\overline{\mathcal{C}}^t}$ as:

$$h = 2 \cdot \frac{a_{\mathcal{C}} \cdot a_{\overline{\mathcal{C}}^t}}{a_{\mathcal{C}} + a_{\overline{\mathcal{C}}^t}}. \tag{11}$$

The new evaluation metric is high only when both the $a_{\mathcal{C}}$ and $a_{\overline{\mathcal{C}}^t}$ are high. So H-score emphasizes the importance of both abilities of UniDA methods.

Implementation Details. We implement our method in PyTorch framework with ResNet-50 [10] backbone pretrained on ImageNet [6]. The hyperparameters are tuned with cross-validation [42] and fixed for each dataset. To enable more diverse classifiers in deep ensemble, we use different data augmentations and randomly shuffled data in different orders for different classifiers.

4.2 Results

The classification results of Office-31, VisDA, Office-Home and DomainNet are shown in Table 3, 4 and 5. For a fair comparison with UAN, we compute per-class accuracy on Office-31 and VisDA. CMU outperforms UAN and all other methods. We compare H-score on all datasets and CMU consistently outperforms previous methods with a large margin on all datasets with various difficulties of detecting open classes. Some domain adaptation methods for other settings perform even worse than ResNet due to the violation of the label space assumption.

In particular, UAN performs well on per-class accuracy but not well on H-score, because the sub-optimal transferability measurement of UAN causes it unable to detect open classes clearly. The low accuracy of the "unknown" class pulls down the H-score. CMU outperforms UAN on H-score with a large margin, which demonstrates that CMU has higher-quality transferability measurement to more accurately detect target open classes $\overline{\mathcal{C}}^t$. This boosts the accuracy of the

Table 4. Tasks on **DomainNet** and **VisDA** dataset

Method	DomainNet (H-score)							VisDA	
	P → R	R → P	P → S	S → P	R → S	S → R	Avg	Acc	H-score
ResNet [10]	30.06	28.34	26.95	26.95	26.89	29.74	28.15	52.80	25.44
DANN [8]	31.18	29.33	27.84	27.84	27.77	30.84	29.13	52.94	25.65
RTN [22]	32.27	30.29	28.71	28.71	28.63	31.90	30.08	53.92	26.02
IWAN [44]	35.38	33.02	31.15	31.15	31.06	34.94	32.78	58.72	27.64
PADA [44]	28.92	27.32	26.03	26.03	25.97	28.62	27.15	44.98	23.05
ATI [25]	32.59	30.57	28.96	28.96	28.89	32.21	30.36	54.81	26.34
OSBP [30]	33.60	33.03	30.55	30.53	30.61	33.65	32.00	30.26	27.31
UAN [41]	41.85	43.59	39.06	38.95	38.73	43.69	40.98	60.83	30.47
CMU	**50.78**	**52.16**	**45.12**	**44.82**	**45.64**	**50.97**	**48.25**	**61.42**	**34.64**

Table 5. H-score (%) of tasks on **Office-Home** dataset

Method	Office-Home												
	A→C	A→P	A→R	C→A	C→P	C→R	P→A	P→C	P→R	R→A	R→C	R→P	Avg
ResNet [10]	44.65	48.04	50.13	46.64	46.91	48.96	47.47	43.17	50.23	48.45	44.76	48.43	47.32
DANN [8]	42.36	48.02	48.87	45.48	46.47	48.37	45.75	42.55	48.70	47.61	42.67	47.40	46.19
RTN [22]	38.41	44.65	45.70	42.64	44.06	45.48	42.56	36.79	45.50	44.56	39.79	44.53	42.89
IWAN [44]	40.54	46.96	47.78	44.97	45.06	47.59	45.81	41.43	47.55	46.29	42.49	46.54	45.25
PADA [44]	34.13	41.89	44.08	40.56	41.52	43.96	37.04	32.64	44.17	43.06	35.84	43.35	40.19
ATI [25]	39.88	45.77	46.63	44.13	44.39	46.63	44.73	41.20	46.59	45.05	41.78	45.45	44.35
OSBP [30]	39.59	45.09	46.17	45.70	45.24	46.75	45.26	40.54	45.75	45.08	41.64	46.90	44.48
UAN [41]	51.64	51.7	54.3	61.74	57.63	61.86	50.38	47.62	61.46	62.87	52.61	65.19	56.58
CMU	**56.02**	**56.93**	**59.15**	**66.95**	**64.27**	**67.82**	**54.72**	**51.09**	**66.39**	**68.24**	**57.89**	**69.73**	**61.60**

"unknown" class and further improves the quality of w^s, which further constrains feature distribution alignment within $\mathcal{C}$ and improves the common class accuracy.

4.3   Analysis

Varying Size of $\overline{\mathcal{C}}^s$ and $\overline{\mathcal{C}}^t$ Following UAN, with fixed $|\mathcal{C}^s \cup \mathcal{C}^t|$ and $|\mathcal{C}^s \cap \mathcal{C}^t|$, we explore the H-score with the various sizes of $\overline{\mathcal{C}}^t$ ($\overline{\mathcal{C}}^s$ also changes correspondingly) on task A → D in Office-31 dataset. As shown in Fig. 4(a), CMU outperforms all the compared methods consistently with different $\overline{\mathcal{C}}^t$, proving that CMU is effective and robust to diverse $\overline{\mathcal{C}}^s$ and $\overline{\mathcal{C}}^t$. In particular, when $\overline{\mathcal{C}}^t$ is large (over 10), meaning there are many open classes, CMU outperforms other methods with a large margin, demonstrating that CMU is superior in detecting open classes.

Varying Size of Common Label $\mathcal{C}$ Following UAN, we fix $|\mathcal{C}^s \cup \mathcal{C}^t|$ and varying $\mathcal{C}$ on task A → D in Office-31 dataset. We let $|\overline{\mathcal{C}}^s| + 1 = |\overline{\mathcal{C}}^t|$ to keep the relative size of $\overline{\mathcal{C}}^s$ and $\overline{\mathcal{C}}^t$ and vary $\mathcal{C}$ from 0 to 31. As shown in Fig. 4(b), CMU consistently outperforms previous methods on all size of $\mathcal{C}$. In particular, when the source domain and the target domain have no overlap on label sets, all the target data should be marked as "unknown". CMU achieves much higher H-score, indicating that CMU can detect open classes more effectively. When

$|\mathcal{C}| = 31$, the setting degrades to closed set domain adaptation, CMU and UAN perform similarly, because there is no open class to influence the adaptation.

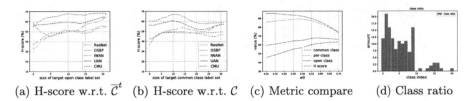

(a) H-score w.r.t. $\overline{\mathcal{C}}^t$ (b) H-score w.r.t. $\mathcal{C}$ (c) Metric compare   (d) Class ratio

Fig. 4. (a)(b) H-score with respect to $\overline{\mathcal{C}}^t$ and $\mathcal{C}$. In (a), we fix $|\mathcal{C}^s \cup \mathcal{C}^t|$ and $|\mathcal{C}^s \cap \mathcal{C}^t|$; In (b), we fix $|\mathcal{C}^s \cup \mathcal{C}^t|$. (c) Relationship between different metrics and w_0. (d) The class ratio of predicted labels (used to compute w^s) of target data in all source labels. Classes 0–9 are source commons and 10–19 are source privates.eps

Ablation Study. We go deeper into the efficacy of the proposed method by evaluating variants of CMU on Office-31. (1) CMU w/o cons is the variant without using the consistency component in the uncertainty in Eq. (4) but still using multiple classifiers to calibrate the entropy and confidence; (2) CMU w/o conf is the variant without integrating the average confidence of classifier in Eq. (3). (3) CMU w/o ent is the variant without integrating the average entropy of classifier into the criterion in Eq. (2). (4) CMU w/o ensemble is the variant without calibrating entropy and confidence but still using single classifier G to compute entropy and confidence while multiple classifiers are still used to compute consistency. (5) CMU w/ domain sim is the variant by adding the domain similarity as another component in the transferability like UAN [41].

Table 6. Ablation study tasks on **Office-31** dataset

Method	D → W		A → D		W → A		Avg (6 task)	
	Acc	H-score	Acc	H-score	Acc	H-score	Acc	H-score
CMU	95.72	79.32	89.11	68.11	88.61	72.23	91.11	73.14
w/o cons	95.01	78.65	88.74	67.25	87.82	71.44	90.43	72.23
w/o conf	95.23	78.84	88.92	67.48	88.04	71.71	90.62	72.52
w/o ent	94.11	75.68	86.81	63.97	87.24	68.66	89.07	69.78
w/o ensemble	93.68	74.43	86.39	63.81	88.67	72.26	88.93	69.50
w/domain sim	95.70	79.30	89.63	68.14	88.67	72.26	91.28	73.15

As shown in Table 6, CMU outperforms CMU w/o cons/conf/ent, especially w/o entropy, indicating the contribution of the multiple uncertainties is complementary to achieve a more complete and accurate uncertainty estimation. CMU outperforms CMU w/o ensemble, proving the calibration from the ensemble can more accurately estimate the uncertainty. CMU w/ domain sim performs similarly to CMU, indicating that domain similarity has little effect on detecting open classes, and thus we do not include it in the uncertainty estimation.

Comparison of Multiple Metrics. To justify our new H-score, we visualize the relationship between different metrics w.r.t. w_0 in Fig. 4(c). We can observe that the open class accuracy increases with w_0 increasing while the common class accuracy decreases with w_0 increasing. This is because, with higher w_0, more data are marked as "unknown" and more common data are misclassified to "unknown". Per-class accuracy varies in the same trend of common class accuracy, indicating that per-class accuracy bias common class accuracy while nearly neglect the open class accuracy. H-score is high only when both common class and "unknown" accuracies are high, which more comprehensively evaluates UniDA methods.

Threshold Sensitivity. We investigate the sensitivity of CMU with respect to threshold w_0 in task A → D. As shown in Fig. 4(c), with w_0 varying in a reasonable range $[0.45, 0.60]$, the H-score changes little, which proves that the performance is not very sensitive to the threshold w_0.

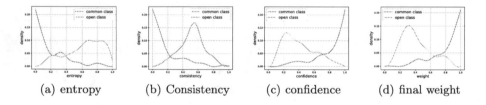

(a) entropy   (b) Consistency   (c) confidence   (d) final weight

Fig. 5. Density of each criterion within common and open class data.

Hypotheses Justification. To justify that our new transferability measurement distinguishes data in the common label set from those in the private label sets, we plot the estimated probability density function for different components of weights $w^s(x)$ in Eq. (6) and $w^t(x)$ in Eq. (5) on A → D task of Office-31. Figure 5(a)–5(d) show that the three uncertainties: entropy, consistency and confidence all distinguish target data in $\mathcal{C}$ and $\overline{\mathcal{C}}^t$ clearly, proving that the multi-classifier ensemble model can calibrate the uncertainty and estimate it more accurately. Figure 4(d) (0–9 is the common class) proves that the source class-level weight could assign high weights for common classes, which in turn demonstrates that the selected data to compute source weight are mostly common classes.

5   Conclusion

In this paper, we propose a novel approach: Calibrated Multiple Uncertainties (CMU) and a new evaluation metric: H-score for Universal Domain Adaptation

(UniDA). We design a novel transferability consisting of entropy, confidence and consistency, calibrated by a deep ensemble model. The new transferability exploits complementary characteristics of different uncertainties to cover all types of predictions. The calibration more accurately estimates the uncertainty and improves the quality of the transferability. The advanced transferability, in turn, improves the quality of source weight. CMU achieves a balanced ability to detect open classes and classify common class data correctly. We further propose a novel H-score to compensate for the previous per-class accuracy for ignorance of open classes. A thorough evaluation shows that CMU outperforms the state-of-the-art UniDA method on both the common set accuracy and the "unknown" class accuracy, especially with a large margin on detecting open classes.

Acknowledgement. This work was supported by the Natural Science Foundation of China (61772299, 71690231), and China University S&T Innovation Plan Guided by the Ministry of Education.

References

1. Busto, P.P., Iqbal, A., Gall, J.: Open set domain adaptation for image and action recognition. IEEE Trans. Pattern Anal. Mach. Intell. (2018)
2. Cao, Z., Long, M., Wang, J., Jordan, M.I.: Partial transfer learning with selective adversarial networks. In: CVPR, June 2018
3. Cao, Z., Ma, L., Long, M., Wang, J.: Partial adversarial domain adaptation. In: Ferrari, V., Hebert, M., Sminchisescu, C., Weiss, Y. (eds.) ECCV 2018. LNCS, vol. 11212, pp. 139–155. Springer, Cham (2018). https://doi.org/10.1007/978-3-030-01237-3_9
4. Cao, Z., You, K., Long, M., Wang, J., Yang, Q.: Learning to transfer examples for partial domain adaptation. In: Proceedings of the IEEE Conference on Computer Vision and Pattern Recognition, pp. 2985–2994 (2019)
5. Chen, Q., Liu, Y., Wang, Z., Wassell, I., Chetty, K.: Re-weighted adversarial adaptation network for unsupervised domain adaptation. In: CVPR, pp. 7976–7985 (2018)
6. Deng, J., Dong, W., Socher, R., Li, L.J., Li, K., Fei-Fei, L.: ImageNet: a large-scale hierarchical image database. In: CVPR 2009 (2009)
7. Gal, Y., Ghahramani, Z.: Dropout as a Bayesian approximation: representing model uncertainty in deep learning. In: International Conference on Machine Learning, pp. 1050–1059 (2016)
8. Ganin, Y., et al.: Domain-adversarial training of neural networks. JMLR **17**, 59:1–59:35 (2016)
9. Guo, C., Pleiss, G., Sun, Y., Weinberger, K.Q.: On calibration of modern neural networks. In: Proceedings of the 34th International Conference on Machine Learning, vol. 70, pp. 1321–1330. JMLR. org (2017)
10. He, K., Zhang, X., Ren, S., Sun, J.: Deep residual learning for image recognition. In: CVPR (2016)
11. Hu, J., Wang, C., Qiao, L., Zhong, H., Jing, Z.: Multi-weight partial domain adaptation. In: The British Machine Vision Conference (BMVC) (2019)
12. Hu, L., Kan, M., Shan, S., Chen, X.: Duplex generative adversarial network for unsupervised domain adaptation. In: CVPR, June 2018

13. Hu, Y., Stumpfe, D., Bajorath, J.: Computational exploration of molecular scaffolds in medicinal chemistry: miniperspective. J. Med. Chem. **59**(9), 4062–4076 (2016)
14. Kang, G., Zheng, L., Yan, Y., Yang, Y.: Deep adversarial attention alignment for unsupervised domain adaptation: the benefit of target Expectation maximization. In: Ferrari, V., Hebert, M., Sminchisescu, C., Weiss, Y. (eds.) ECCV 2018. LNCS, vol. 11215, pp. 420–436. Springer, Cham (2018). https://doi.org/10.1007/978-3-030-01252-6_25
15. Konstantinos, B., Nathan, S., David, D., Dumitru, E., Dilip, K.: Unsupervised pixel-level domain adaptation with generative adversarial networks. In: CVPR, pp. 95–104 (2017)
16. Lakshminarayanan, B., Pritzel, A., Blundell, C.: Simple and scalable predictive uncertainty estimation using deep ensembles. In: Advances in Neural Information Processing Systems, pp. 6402–6413 (2017)
17. Lian, Q., Li, W., Chen, L., Duan, L.: Known-class aware self-ensemble for open set domain adaptation. arXiv preprint arXiv:1905.01068 (2019)
18. Liu, H., Cao, Z., Long, M., Wang, J., Yang, Q.: Separate to adapt: open set domain adaptation via progressive separation. In: The IEEE Conference on Computer Vision and Pattern Recognition (CVPR), June 2019
19. Liu, Y.C., Yeh, Y.Y., Fu, T.C., Wang, S.D., Chiu, W.C., Frank Wang, Y.C.: Detach and adapt: learning cross-domain disentangled deep representation. In: CVPR, June 2018
20. Long, M., Cao, Y., Wang, J., Jordan, M.I.: Learning transferable features with deep adaptation networks. In: ICML (2015)
21. Long, M., Cao, Z., Wang, J., Jordan, M.I.: Conditional domain adversarial network. In: NeurIPS (2018)
22. Long, M., Zhu, H., Wang, J., Jordan, M.I.: Unsupervised domain adaptation with residual transfer networks. In: NeurIPS, pp. 136–144 (2016)
23. Maria Carlucci, F., Porzi, L., Caputo, B., Ricci, E., Rota Bulo, S.: Autodial: automatic domain alignment layers. In: ICCV, October 2017
24. Murez, Z., Kolouri, S., Kriegman, D., Ramamoorthi, R., Kim, K.: Image to image translation for domain adaptation. In: CVPR, June 2018
25. Panareda Busto, P., Gall, J.: Open set domain adaptation. In: ICCV, October 2017
26. Peng, X., Bai, Q., Xia, X., Huang, Z., Saenko, K., Wang, B.: Moment matching for multi-source domain adaptation. In: Proceedings of the IEEE International Conference on Computer Vision, pp. 1406–1415 (2019)
27. Peng, X., et al.: VisDA: a synthetic-to-real benchmark for visual domain adaptation. In: CVPR Workshops, pp. 2021–2026 (2018)
28. Saenko, K., Kulis, B., Fritz, M., Darrell, T.: Adapting visual category models to new domains. In: Daniilidis, K., Maragos, P., Paragios, N. (eds.) ECCV 2010. LNCS, vol. 6314, pp. 213–226. Springer, Heidelberg (2010). https://doi.org/10.1007/978-3-642-15561-1_16
29. Saito, K., Watanabe, K., Ushiku, Y., Harada, T.: Maximum classifier discrepancy for unsupervised domain adaptation. In: CVPR, June 2018
30. Saito, K., Yamamoto, S., Ushiku, Y., Harada, T.: Open set domain adaptation by backpropagation. In: Ferrari, V., Hebert, M., Sminchisescu, C., Weiss, Y. (eds.) ECCV 2018. LNCS, vol. 11209, pp. 156–171. Springer, Cham (2018). https://doi.org/10.1007/978-3-030-01228-1_10
31. Sankaranarayanan, S., Balaji, Y., Castillo, C.D., Chellappa, R.: Generate to adapt: aligning domains using generative adversarial networks. In: CVPR, June 2018

32. Snoek, J., et al.: Can you trust your model's uncertainty? Evaluating predictive uncertainty under dataset shift. In: Advances in Neural Information Processing Systems, pp. 13969–13980 (2019)
33. Tzeng, E., Hoffman, J., Saenko, K., Darrell, T.: Adversarial discriminative domain adaptation. In: CVPR (2017)
34. Tzeng, E., Hoffman, J., Zhang, N., Saenko, K., Darrell, T.: Deep domain confusion: Maximizing for domain invariance. arXiv preprint arXiv:1412.3474 (2014)
35. Venkateswara, H., Eusebio, J., Chakraborty, S., Panchanathan, S.: Deep hashing network for unsupervised domain adaptation. In: CVPR (2017)
36. Vittorio, F., Alina, K., Rodrigo, B., Victor, G., Matteo, M.: Open images challenge 2019 (2019). https://storage.googleapis.com/openimages/web/challenge2019.html
37. Volpi, R., Morerio, P., Savarese, S., Murino, V.: Adversarial feature augmentation for unsupervised domain adaptation. In: CVPR, June 2018
38. Wu, A., Nowozin, S., Meeds, E., Turner, R., Hernández-Lobato, J., Gaunt, A.: Deterministic variational inference for robust Bayesian neural networks. In: 7th International Conference on Learning Representations, ICLR 2019 (2019)
39. Wu, Z., et al.: Moleculenet: a benchmark for molecular machine learning. Chem. Sci. **9**(2), 513–530 (2018)
40. Xie, S., Zheng, Z., Chen, L., Chen, C.: Learning semantic representations for unsupervised domain adaptation. In: ICML, pp. 5423–5432 (2018)
41. You, K., Long, M., Cao, Z., Wang, J., Jordan, M.I.: Universal domain adaptation. In: The IEEE Conference on Computer Vision and Pattern Recognition (CVPR), June 2019
42. You, K., Wang, X., Long, M., Jordan, M.: Towards accurate model selection in deep unsupervised domain adaptation. In: ICML, pp. 7124–7133 (2019)
43. Zhang, H., Li, A., Han, X., Chen, Z., Zhang, Y., Guo, Y.: Improving open set domain adaptation using image-to-image translation. In: 2019 IEEE International Conference on Multimedia and Expo (ICME), pp. 1258–1263. IEEE (2019)
44. Zhang, J., Ding, Z., Li, W., Ogunbona, P.: Importance weighted adversarial nets for partial domain adaptation. In: CVPR, June 2018
45. Zhang, W., Ouyang, W., Li, W., Xu, D.: Collaborative and adversarial network for unsupervised domain adaptation. In: CVPR, June 2018

Visual Compositional Learning
for Human-Object Interaction Detection

Zhi Hou[1,2], Xiaojiang Peng[2], Yu Qiao[2(✉)], and Dacheng Tao[1]

[1] UBTECH Sydney AI Centre, School of Computer Science, Faculty of Engineering,
The University of Sydney, Darlington, NSW 2008, Australia
`zhou9878@uni.sydney.edu.au, dachengtao@sydney.edu.au`
[2] Shenzhen Key Lab of Computer Vision and Pattern Recognition, Shenzhen
Institutes of Advanced Technology, Chinese Academy of Sciences, Beijing, China
`{xj.peng,yu.qiao}@siat.ac.cn`

Abstract. Human-Object interaction (HOI) detection aims to localize
and infer relationships between human and objects in an image. It is
challenging because an enormous number of possible combinations of
objects and verbs types forms a long-tail distribution. We devise a deep
Visual Compositional Learning (VCL) framework, which is a simple yet
efficient framework to effectively address this problem. VCL first decomposes an HOI representation into object and verb specific features, and
then composes new interaction samples in the feature space via stitching
the decomposed features. The integration of decomposition and composition enables VCL to share object and verb features among different HOI
samples and images, and to generate new interaction samples and new
types of HOI, and thus largely alleviates the long-tail distribution problem and benefits low-shot or zero-shot HOI detection. Extensive experiments demonstrate that the proposed VCL can effectively improve the
generalization of HOI detection on HICO-DET and V-COCO and outperforms the recent state-of-the-art methods on HICO-DET. Code is
available at https://github.com/zhihou7/VCL.

Keywords: Human-object interaction · Compositional learning

1 Introduction

Human-Object interaction (HOI) detection [7,9,20,27] aims to localize and infer
relationships (verb-object pairs) between human and objects in images. The main
challenges of HOI come from the complexity of interaction and the long-tail distribution of possible verb-object combinations [7,29,36]. In practice, a few types
of interactions dominate the HOI samples, while a large number of interactions
are rare which are always difficult to obtain sufficient training samples.

Electronic supplementary material The online version of this chapter (https://
doi.org/10.1007/978-3-030-58555-6_35) contains supplementary material, which is
available to authorized users.

© Springer Nature Switzerland AG 2020
A. Vedaldi et al. (Eds.): ECCV 2020, LNCS 12360, pp. 584–600, 2020.
https://doi.org/10.1007/978-3-030-58555-6_35

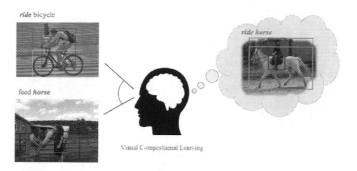

Fig. 1. An illustration of Visual Compositional Learning (VCL). VCL constructs the new concept of $\langle ride, horse \rangle$ from $\langle feed, horse \rangle$ and $\langle ride, bicycle \rangle$ via visual compositional learning

The visual scenes are composed of basic elements, such as objects, parts, and other semantic regions. It is well-acknowledged that humans perceive world in a compositional way in which visual scenes are treated as a layout of distinct semantic objects [16,30]. We can understand HOIs by decomposing them into objects and human interaction (verb) types. This decomposition helps to solve the rare Human-Object Interactions with long-tailed distribution. For example, in HICO-DET dataset [7], $\langle hug, suitcase \rangle$ is a rare case with only one example, while we have more than 1000 HOI samples including object "suitcase", and 500 samples including verb "hug". Obviously, object representations can be shared among different HOIs. And samples with the same verb usually exhibit similar human pose or action characteristics [36]. By compositing the concepts of "suitcase" and "hug" learned from these large number samples, one can handle the rare case $\langle hug, suitcase \rangle$. This inspires to reduce the complexity of HOI detection and handle unseen/rare categories via learning compositional components, *i.e.* human verb and objects from visual scenes. Note this idea is near but different from disentangling representation learning [5] (e.g. factors method [29] in HOI detection) which aims to separate the distinct and informative factors from input examples. Similar to disentangling, compositional learning of HOI also includes the decomposing step. Unlike disentangling, compositional learning further composes novel HOI examples with decomposed factors, which is helpful to address low-shot and zero-shot HOI detection.

Inspired by the above analysis, this paper proposes a deep Visual Compositional Learning (VCL) frame work for Human-Object Interaction Detection, which performs compositional learning on visual verb and object representations. VCL simultaneously encourages shared verb and object representation across images and HOI types. As illustrated in Fig. 1, with the semantic features of '*horse*' and '*ride*' in the images of $\langle feed, horse \rangle$ and $\langle ride, bicycle \rangle$, one may compose a new interaction feature $\langle ride, horse \rangle$ using off-the-shelf features.

To perform compositional learning for HOI detection, VCL faces three challenges. Firstly, verb features are usually highly correlated with human and object features. It is non-trivial to decouple the verb representations from those of human and objects in scenes. Unlike prior works [9,12], which extract the verb representations from human boxes, we build verb representation from the union box of human and object. Meanwhile, we share weights between verb and human stream in the mutli-branches structure to emphasize the verb representation on human box region. Our verb representation yields more discriminative cues for the final detection task. Secondly, the number of verbs and objects within a single image is limited for composition. We present a novel feature compositional learning approach by composing HOI samples with verb and object features from different images and different HOI types. In this way, our VCL encourages the model to learn the shared and distinctive verb and object representations that are insensitive to variations (*i.e.* the specific images and interactions). Thirdly, HOIs always exhibit long-tail distribution where a large number of categories have very few even zero training samples. VCL can compose new interactions and novel types of HOIs in the feature space (*i.e.* verb + object), e.g., the rare HOI $\langle wash, cat \rangle$ can be drawn from $\langle wash, dog \rangle$ and $\langle feed, cat \rangle$.

Overall, our main contributions can be summarized as follows,

- We creatively devise a deep Visual Compositional Learning framework to compose novel HOI samples from decomposed verbs and objects to relieve low-shot and zero-shot issues in HOI detection. Specifically, we propose to extract verb representations from union box of human and object and compose new HOI samples and new types of HOIs from pairwise images.
- Our VCL outperforms considerably previous state-of-the-art methods on the largest HOI Interaction detection dataset HICO-DET [7], particularly for rare categories and zero-shot.

2   Related Works

2.1   Human-Object Interaction Detection

Human-Object Interaction [7,8] is essential for deeper scene understanding. Different from **Visual Relationship Detection** [24], **HOI** is a kind of human-centric relation detection. Several large-scale datasets (V-COCO [11] and HICO-DET [7]) were released for the exploration of HOI detection. Chao *et al.* [7] introduced a multi-stream model combining visual features and spatial location features to help tackle this problem. Following the multi-stream structure in [7], Gao *et al.* [9] further exploited an instance centric attention module and Li *et al.* [20] utilized interactiveness to explicitly discriminate non-interactive pairs to improve HOI detection. Recently, Wang *et al.* [32] proposed a contextual attention framework for Human-Object Interaction detection. GPNN [27] and RPNN [38] were introduced to model the relationships with graph neural network among parts or/and objects. Pose-aware Multi-level Feature Network [31] aimed to generate robust predictions on fine-grained human object interaction. Different from the previous works [7,9,20,27,31,32,38] who mainly focus on learning better HOI features, we address the long-tailed and zero-shot issues in HOI detection via Visual Compositional Learning.

2.2 Low-Shot and Zero-Shot Learning

Our work also ties with low-shot learning [33] within long-tailed distribution [22] and zero-shot learning recognition [35]. Shen *et al.* [29] introduced a factor-ized model for HOI detection that disentangles reasoning on verbs and objects to tackle the challenge of scaling HOI recognition to the long tail of categories, which is similar to our work. But we design a compositional learning approach to compose HOI examples. Visual-language joint embedding models [26,36] enforce the representation of the same verb to be similar among different HOIs by the intrinsic semantic regularities. [26] further transferred HOIs embeddings from seen HOIs to unseen HOIs using analogies for zero-shot HOI detection. However, different from [26] who aims to zero-shot learning, VCL targets at Generalized Zero-Shot Learning [34]. In [3], a generic object detector was incorporated to generalize to interactions involving previously unseen objects. Also, Yang *et al.* [37] proposed to alleviate the predicate bias to objects for zero-shot visual rela-tionship detection. Similar to previous approaches [3,26,29,36], we also equally treat the same verb from different HOIs. However, all those works [3,26,29,36] largely ignore the composition of verbs and objects. In contrast, we propose the Visual Compositional Learning framework to relieve the long-tail and zero-shot issues of HOI detection jointly, and we demonstrate the efficiency by massive experiments, especially for rare/unseen data in HOI detection.

2.3 Feature Disentangling and Composing

Disentangled representation learning has attracted increasing attention in var-ious kinds of visual task [5,6,14,15,23] and the importance of Compositional Learning to build intelligent machines is acknowledged [4,5,10,19]. Higgins *et al.* [15] proposed Symbol-Concept Association Network (SCAN) to learn hierar-chical visual concepts. Recently, [6] proposed Multi-Object network (MONet) to decompose scenes by training a Variational Autoencoder together with a recur-rent attention network. However, both SCAN [15] and MONet [6] only validate their methods on the virtual datasets or simple scenes.

Besides, Compositional GAN [2] was introduced to generate new images from a pair of objects. Recently, Label-Set Operations network (LaSO) [1] combined features of image pairs to synthesize feature vectors of new label sets according to certain set operations on the label sets of image pairs for multi-label few-shot learning. Both Compositional GAN [2] and LaSO [1], however, compose the features from two whole images and depend on generative network or reconstruct loss. In addition, Kato *et al.* [18] introduced a compositional learning method for HOI classification [8] that utilizes the visual-language joint embedding model to the feature of the whole of image. But [18] did not involve multiple objects detection in the scene. Our visual compositional learning framework differs from them in following aspects: i) it composes interaction features from regions of images, ii) it simultaneously encourages *discriminative* and *shared* verb and object representations.

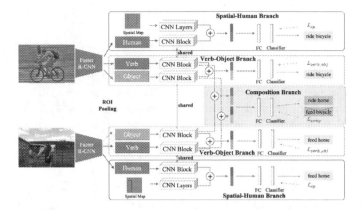

Fig. 2. Overview of the proposed Visual Compositional Learning (VCL) framework. Given two images, we first detect human and objects with Faster-RCNN [28]. Next, with ROI-Pooling and Residual CNN blocks, we extract human features, verb features (*i.e.* the union box of human and object), and object features. Then, these features are fed into the following branches: individual spatial-human branch, verb-object branch and composited branch. Finally, HOI representations from verb-object branch and composited branch are classified by a shared FC-Classifier, while HOI representations from spatial-human branch are classified by an individual FC-Classifier. *Note that all the parameters are shared across images and the newly composited HOI instances can be from a single image if the image includes multiple HOIs*

3   Visual Compositional Learning

In this section, we present our Visual Compositional Learning (VCL) framework for HOI detection. We first provide an overview of VCL and then detail the HOI branches of VCL. Last, we describe how we compose new HOIs and apply VCL to zero-shot detection.

3.1   Overview

To address the long-tail and zero-shot issues of HOI detection, we propose the Visual Compositional Learning (VCL) framework to learn shared object and verb features and compose novel HOI samples with these features. As shown in Fig. 2, to perform compositional learning, our VCL takes as input a randomly selected image pair. Then we employ a Faster R-CNN [28] with backbone ResNet-50 [13] to detect human and objects in images. Subsequently, we use ROI-Pooling and Residual CNN blocks to obtain features of human, verbs, and objects individually. Then, to obtain HOI detection, these features together with a human-object spatial map are fed into a spatial-human branch and a verb-object branch. Particularly, composited features are fed into the composited branch for compositional learning. It is worth noting that all the parameters are shared across images.

Fig. 3. Illustration of the process of composing new interactions. Given two images I_1 and I_2, we compose new interaction samples within single image and between them by first considering all possible verb-object pairs and then removing infeasible interactions

3.2 Multi-branch Network

Multi-branch architectures are usually used in previous HOI detection works [7,9,20] where each branch processes a kind of input information. Similarly, VCL includes a spatial-human branch and a verb-object branch. But unlike previous works, VCL has a composition branch which helps the training of HOI detection.

Spatial-Human Branch. The spatial-human branch processes human feature and the spatial map of human and object. Following [9], the input of spatial feature is a two-channel 64×64 tensor consisting of a person map and an object map. For the person map, the value of a position will be 1 if it is in in the person box otherwise 0. The object map is similar. We concatenate the spatial map with the human feature.

Verb-Object Branch. The verb-object branch in each image includes a verb stream and an object stream. Unlike prior works [9,12] which view the human features as verb representations, our newly introduced verb branch extracts *a verb representation from the union box of a human box and an object box*. Meanwhile, similar to [9,12], we share the weights between human stream in Spatial-Human branch and verb stream from the union box in Verb-Object branch. Our verb representation is more discriminative which contains more useful contextual information within the union box of human and object.

Composition Branch. We compose new verb-object interaction samples in the feature space from the verbs and objects between and within images and then *these synthesized samples are trained jointly with the interactions annotated in the dataset*. It in turn improves the generalization of HOI detection, particularly for rare and unseen HOIs.

3.3 Composing Interactions

The key idea of our proposed VCL framework is to compose new interaction samples within and between images. This composition process encourages the network to learn shared object and verb features across HOIs by composing massive diverse HOI samples. As shown in Fig. 3, the composition process mainly contains two stages: generating all possible interactions (*i.e.* verb-object pairs) and removing infeasible interactions. Given two images I_1 and I_2, we compose new interaction samples within single image and between images by first considering all possible verb-object pairs and then removing infeasible interactions in the HOI label space.

Existing HOI labels mainly contain one object and at least one verb, which set the HOI detection as a multi-label problem. To avoid frequently checking verb-object pairs, we design an efficient composing and removing strategy. First, we decouple the HOI label space into a verb-HOI matrix $\mathbf{A}_v \in R^{N_v \times C}$ and an object-HOI matrix $\mathbf{A}_o \in R^{N_o \times C}$, where N_v, N_o, and C denote the number of verbs, objects and HOI categories respectively. $\mathbf{A}_v$ ($\mathbf{A}_o$) can be viewed as the cooccurence matrix between verbs (objects) and HOIs. Then, given binary HOI label vectors $\mathbf{y} \in R^{N \times C}$, where N, C denote the number of interactions and HOI categories respectively. We can obtain the object label vector and verb label vector as follows,

$$\mathbf{l}_o = \mathbf{y}\mathbf{A}_o^\top, \ \mathbf{l}_v = \mathbf{y}\mathbf{A}_v^\top, \tag{1}$$

where $\mathbf{l}_o \in R^{N \times N_o}$ is usually one-hot vectors meaning one object of a HOI example, and $\mathbf{l}_v \in R^{N \times N_v}$ is possiblely multi-hot vectors meaning multiple verbs. *e.g.* $\langle \{hold, sip\}, cup \rangle$). Similarly, we can generate new interactions from arbitrary $\mathbf{l}_o$ and $\mathbf{l}_v$ as follows,

$$\hat{\mathbf{y}} = (\mathbf{l}_o \mathbf{A}_o) \& (\mathbf{l}_v \mathbf{A}_v), \tag{2}$$

where & denotes the "and" logical operation. The infeasible HOI labels that do not exist in the given label space are all-zero vectors after the logical operation. And then, we can filter out those inefasible HOIs. In implementation, we obtain verbs and objects from two images by ROI pooling and treat them within and between images as same. Therefore, we do not treat two levels of composition differently during composing HOIs.

Zero-Shot Detection. The composition process makes VCL handling zero-shot detection naturally. Specifically, with the above-mentioned operation, we can generate HOI samples for zero-shot (in the given HOI label space) between and within images which may not be annotated in the training set.

3.4   Training and Inference

Training. We train the proposed VCL in an end-to-end manner with Cross Entropy (CE) losses from multiple branches: $\mathcal{L}_{sp}$ from the spatial-human branch for original HOI instances, $\mathcal{L}_{verb\_obj}$ from verb-object branch for original HOI instances, and $\mathcal{L}_{comp}$ from verb-object branch for composited HOI instances. Formally, the total training loss is defined as follows,

$$\mathcal{L} = \mathcal{L}_{sp} + \lambda_1 \mathcal{L}_{verb\_obj} + \lambda_2 \mathcal{L}_{comp}, \tag{3}$$

where λ_1 and λ_2 are two hyper-parameters. We employ the composition process (*i.e.* composing new HOI instances) in each minibatch at training stage.

Inference. At test stage, we remove the composition operation and use the spatial-human branch and the verb-object branch to recognize interaction (*i.e.* human-object pair) of an input image. We predict HOI scores in a similar manner to [9]. For each human-object bounding box pair (b_h, b_o), we predict the score $S_{h,o}^c$ for each category $c \in 1, ..., C$, where C denotes the total number of possible HOI categories. The score $S_{h,o}^c$ depends on the confidence for the individual object detection scores (s_h and s_o) and the interaction prediction based on verb-object branch $s_{verb\_obj}^c$ and spatial-human branch s_{sp}^c. Specifically, Our final HOI score $S_{h,o}^c$ for the human-object bounding box pair (b_h, b_o) is:

$$S_{h,o}^c = s_h \cdot s_o \cdot s_{verb\_obj}^c \cdot s_{sp}^c \tag{4}$$

4 Experiment

In this section, we first introduce datasets and metrics and then provide the details of the implementation of our method. Next, we report our experimental results compared with state-of-the-art approaches and zero-shot results. Finally, we conduct ablation studies to validate the components in our framework.

4.1 Datasets and Metrics

Datasets. We adopt two HOI datasets HICO-DET [7] and V-COCO [11]. HICO-DET [7] contains 47,776 images (38,118 in train set and 9,658 in test set), 600 HOI categories constructed by 80 object categories and 117 verb classes. HICO-DET provides more than 150k annotated human-object pairs. V-COCO [11] provides 10,346 images (2,533 for training, 2,867 for validating and 4,946 for testing) and 16,199 person instances. Each person has annotations for 29 action categories and there are no interaction labels including objects.

Metrics. We follow the settings in [9], *i.e.* a prediction is a true positive only when the human and object bounding boxes both have IoUs larger than 0.5 with reference to ground truth and the HOI classification result is accurate. We use the role mean average precision to measure the performance on V-COCO.

4.2 Implementation Details

For HICO-DET dataset, we utilize the provided interaction labels to decompose and compose interactions labels. For V-COCO dataset, which only provides verb (*i.e.* action) annotations for COCO images [21], we obtain object labels (only 69 classes in V-COCO images) from the COCO annotations. Therefore, we have 69 classes of objects and 29 classes of actions in V-COCO that construct 238 classes of Human-Object pairs to facilitate the detection of 29 classes of actions with VCL. In addition, following the released code of [20], we also apply the same reweighting strategy in HICO-DET and V-COCO (See supplementary materials for comparison), and we keep it for the composited interactions. Besides, to prevent composited interactions from dominating the training of the model, we randomly select composited interactions in each minibatch to maintain the same number of composited interactions as we do of non-composited interactions.

For a fair comparison, we adopt the object detection results and pre-trained weights provided by authors of [9]. We apply two 1024-d fully-connected layers to classify the interaction feature concatenated by verb and object. We train our network for 1000k iterations on the HICO-DET dataset and 500k iterations on V-COCO dataset with an initial learning rate of 0.01, a weight decay of 0.0005, and a momentum of 0.9. We set λ_1 of 2 and λ_2 of 0.5 for HICO-DET dataset, while for V-COCO dataset, λ_1 is 0.5 and λ_2 is 0.1. In order to compose enough new interactions between images for training, we increase the number of interactions in each minibatch while reducing the number of augmentations for each interaction to keep the batch size unchanged. All experiments are conducted on a single Nvidia GeForce RTX 2080Ti GPU with Tensorflow.

4.3 Results and Comparisons

We compare our proposed Visual Compositional Learning with those state-of-the-art HOI detection approaches [3,9,12,20,26,27,31,32,36] on HICO-DET. The HOI detection result is evaluated with mean average precision (mAP) (%). We use the settings: Full (600 HOIs), Rare (138 HOIs), Non-Rare (462 HOIs) in Default mode and Known Object mode on HICO-DET.

Comparisons with State-of-the-Art. From Table 1, we can find that we respectively improve the performance of Default and Know modes with Full setting to 19.43% and 22.00% without external knowledge. In comparison with [31] who achieves the best mAP in Rare category on HICO-DET among the previous works, we improve the mAP by 1.97% in Full category and by 0.9% in Rare category. Particularly, we do not use pose information like [31]. We improve dramatically by over **3%** in Rare category compared to other visual methods [9,12,20,27,38]. Besides, we achieve 1.92% better result than [26] in the rare category although we only use the visual information.

Particularly, [3] incorporates ResNet101 as their backbone and finetune the object detector on HICO-DET. When we the same backbone and finetune the object detector on HICO-DET, we can largely improve the performance to **23.63%**. This also illustrate the great effect of object detector on current two-stage HOI detection method. *in the two-stage HOI detection, we should not only focus on HOI recognition such as [9,20,38], but also improve the object detector performance.*

Visualization. In Fig. 4, we qualitatively show that our proposed Visual Compositional Learning framework can detect those rare interactions correctly while the baseline model without VCL misclassifies on HICO-DET.

Table 1. Comparisons with the state-of-the-art approaches on HICO-DET dataset [7]. Xu *et al.* [36], Peyre *et al.* [26] and Bansal *et al.* [3] utilize language knowledge. We include the results of [3] with the same COCO detector as ours. * means we use the res101 backbone and finetune the object detector on HICO-DET dataset like [3]

Method	Default			Known Object		
	Full	Rare	NonRare	Full	Rare	NonRare
GPNN [27]	13.11	9.34	14.23	-	-	-
iCAN [9]	14.84	10.45	16.15	16.26	11.33	17.73
Li *et al.* [20]	17.03	13.42	18.11	19.17	15.51	20.26
Wang *et al.* [32]	16.24	11.16	17.75	17.73	12.78	19.21
Gupta *et al.* [12]	17.18	12.17	18.68	-	-	-
Zhou *et al.* [38]	17.35	12.78	18.71	-	-	-
PMFNet [31]	17.46	15.65	18.00	20.34	17.47	21.20
Xu *et al.* [36]	14.70	13.26	15.13	-	-	-
Peyre *et al.* [26]	19.40	14.63	**20.87**	-	-	-
Bansal *et al.* [3]	16.96	11.73	18.52	-	-	-
Bansal* *et al.* [3]	21.96	16.43	23.62	-	-	-
Ours (VCL)	**19.43**	**16.55**	20.29	**22.00**	**19.09**	**22.87**
Ours* (VCL)	**23.63**	**17.21**	**25.55**	**25.98**	**19.12**	**28.03**

4.4 Generalized Zero-Shot HOI Detection

We also evaluate our method in HOI zero-shot detection since our method can naturally be applied to zero-shot detection. Following [29], we select 120 HOIs from HOI label set and make sure that the remaining contains all objects and verbs. The annotations on HICO-DET, however, are long-tail distributed. The result of selecting preferentially rare labels is different from that of selecting non-rare labels and we do not know the specific partition of zero shot in [3,29]. For a fair comparison, we split the HOI zero-shot experiment into two groups: rare first selection and non-rare first selection. Rare first selection means we pick out as many rare labels as possible for unseen classes according to the number of instances for zero-shot (remain 92,705 training instances), while non-rare first selection is that we select as many non-rare labels as possible for unseen classes (remain 25,729 training instances). The unseen HOI detection result is evaluated with mean average precision (mAP) (%). We report our result in the settings: Unseen (120 HOIs), Seen (480 HOIs), Full (600 HOIs) in Default mode and Known Object mode on HICO-DET.

We can find in Table 2, both selection strategies witness a consistent increase in all categories compared to the corresponding baseline. Noticeably, both kinds of selecting strategies witness a surge of performance by **over 4%** than baseline in unseen category with VCL. Particularly, our baseline without VCL still achieves low results (3.30% and 5.06%) because we predict 600 classes (including 120 unseen categories) in baseline. The model could learn the verb and object

Table 2. Comparison of Zero Shot Detection results of our proposed Visual Compositional Learning framework. * means we uses the res101 backbone and finetune object detector on HICO-DET. Full means all categories including Seen and Unseen

Method	Default			Known Object		
	Unseen	Seen	Full	Unseen	Seen	Full
Shen *et al.* [29]	5.62	-	6.26	-	-	-
Bansal* *et al.* [3]	10.93	12.60	12.26	-	-	-
w/o VCL (rare first)	3.30	18.63	15.56	5.53	21.30	18.15
w/o VCL (non-rare first)	5.06	12.77	11.23	8.81	15.37	14.06
VCL (rare first)	**7.55**	**18.84**	**16.58**	**10.66**	**21.56**	**19.38**
VCL (non-rare first)	**9.13**	**13.67**	**12.76**	**12.97**	**16.31**	**15.64**
VCL* (rare first)	**10.06**	**24.28**	**21.43**	**12.12**	**26.71**	**23.79**
VCL* (non-rare first)	**16.22**	**18.52**	**18.06**	**20.93**	**21.02**	**20.90**

of unseen HOIs individually from seen HOIs. Meanwhile, non-rare first selection has more verbs and objects of unseen HOIs individually from seen HOIs than that of rare first selection. Therefore, non-rare first selection has better baseline than rare first selection. Besides, we improve [3,29] largely nearly among all categories. In detail, for a fair comparison to [3], we also use Resnet-101 and finetune the object detector on HICO-DET dataset, which can largely improve the performance. This demonstrates that our VCL effectively improves the detection of unseen interactions and maintains excellent HOI detection performance for seen interactions at the same time.

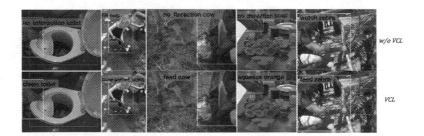

Fig. 4. Some rare HOI detections (Top 1 result) detected by the proposed Compositional Learning and the model without Compositional Learning. The first row is the results of baseline model without VCL. The second row is the results of VCL

4.5   Ablation Analysis

To evaluate the design of our VCL, we first conduct ablation studies about VCL and verb representation on HICO-DET dataset and V-COCO dataset. For V-COCO, we evaluate AP_{role} (24 actions with roles) following [7,11]. Besides, we evaluate two branches and composing interactions between images and/or within images on HICO-DET. See supplementary material for more analysis.

Table 3. Ablation study of the proposed Visual Compositional Learning framework on HICO-DET and V-COCO test set. VCL means Visual Compositional Learning and Union Verb means we learn verb feature from union box of human and object. Sharing W means sharing weights between human stream and verb stream for verb representation. Re-weighting means we use re-weighting strategy in the code of [20]

VCL	Union Verb	Sharing W	Re-weighting	HICO			V-COCO
				Full	Rare	NonRare	AP_{role}
-	-	✓	-	16.87	10.07	18.90	46.9
✓	-	✓	-	17.35	12.10	18.91	47.2
✓	✓	-	✓	18.93	15.68	19.90	47.8
-	-	✓	✓	18.03	13.62	19.35	47.4
✓	-	✓	✓	18.57	14.83	19.69	47.7
-	✓	✓	✓	18.43	14.14	19.71	47.5
✓	✓	✓	✓	**19.43**	**16.55**	**20.29**	**48.3**

Visual Compositional Learning is our core method. With VCL, we can respectively see an obvious increase by 1.00% in the Full category from Table 3 (row 6 vs row 7). Particularly, the improvement (2.41%) in Rare category is considerably better than that in Non-Rare category, which means the proposed VCL is more beneficial for rare data. Noticeably, the improvement of VCL with verb representation learning is better than that without verb representation, which implies that the more discriminative the verb representation is, the better improvement VCL obtains in HOI detection. We can see a similar trend of VCL on V-COCO in Table 3 where the performance decreases by 0.8% without VCL. Noticeably, V-COCO aims to recognize the actions (verbs) rather than the human-object pairs, which means the proposed VCL is also helpful to learn the discriminative action (verbs) representation. Besides, from row 1 and row 2 in Table 3, we can find the proposed VCL is pluggable to re-weighting strategy [17] (See details in supplementary materials).

Verb Representation, which we learn from the union box of human and object, is considerably helpful for the performance. We evaluate its efficiency by comparing human box and union box of human and object. From row 5 and row 7 in Table 3, we can find that with the proposed verb representation we improve the performance by 0.86% on HICO-DET and 0.6% in V-COCO respectively within VCL, which means learning verb representation from the union box of human and object is more discriminative and advantageous for HOI detection. Particularly, we *share the weights of resnet block between human and verb stream* after ROI pooling in the proposed verb representation method. Table 3 (row 3 vs row 7) shows sharing weights helps the model improve the performance largely from 18.93% to 19.43%. This may be explained by that the union region contains more noise and the model would emphasize the region of the human in the union box by sharing weights to obtain better verb representation.

Table 4. The branches ablation study of the model on HICO-DET test set. Verb-object branch only means we train the model without spatial-human branch.

Method	Full	Rare	NonRare
Two branches	19.43	16.55	20.29
Verb-Object branch only during inference	16.89	15.35	17.35
Spatial-Human branch only during inference	16.13	12.39	17.24
Verb-Object branch only	15.77	13.35	16.49
Verb-Object branch only (w/o VCL)	15.33	10.85	16.67

Branches. There are two branches in our method and we evaluate their contributions in Table 4. Noticeably, we apply VCL to Verb-Object branch during training, while we do not apply VCL to Spatial-Human. By keeping one branch each time on HICO-DET dataset during inference, we can find the verb-object branch makes the larger contribution, particularly for rare category (**3%**). This efficiently illustrates the advantage of VCL for rare categories. But we can improve the performance dramatically from 16.89% to 19.43% with Spatial-Human Branch. Meanwhile, we can find the proposed VCL is orthogonal to spatial-human branch from the last two rows in Table 4. Noticeably, *by comparing verb-object branch only during inference and verb-object branch only from training, we can find the spatial-human branch can facilitate the optimization of verb-object branch (improving the mAP from 15.77% to 16.89%).* See supplementary materials for more analysis of two branches in zero-shot detection where the performance of verb-object branch with VCL is over 3.5% better than spatial-human branch in unseen data.

Table 5. Composing Strategies study of VCL on HICO-DET test set

Method	Full (mAP %)	Rare (mAP %)	NonRare (mAP %)
Baseline (w/o VCL)	18.43	14.14	19.71
Within images	18.48	14.46	19.69
Between images	19.06	14.33	20.47
Between and within images	19.43	16.55	20.29

Composing Interactions Within and/or Between Images. In Table 5, we can find composing interaction samples between images is beneficial for HOI detection, whose performance in the Full category increases to 19.06% mAP, while composing interaction samples within images has similar results to baseline. It might be because the number of images including multiple interactions is few on HICO-DET dataset. Remarkably, composing interaction samples within and between images notably improves the performance up to 19.43% mAP in Full and **16.55%** mAP in Rare respectively. Those results mean composing interactions within images and between images is more beneficial for HOI detection.

4.6 Visualization of Features

We also illustrate the verb and object features by t-SNE visualization [25]. Figure 5 illustrates that VCL overall improves the discrimination of verb and object features. There are many noisy points (see black circle region) in Fig. 5 without VCL and verb presentation. Meanwhile, we can find the proposed verb representation learning is helpful for verb feature learning by comparing verb t-SNE graph between the left and the middle. Besides, the object features are more discriminative than verb. We think it is because the verb feature is more abstract and complex and the verb representation requires further exploration.

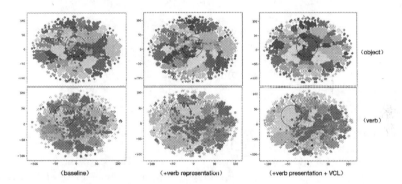

Fig. 5. Visual illustration of object features (80 classes) (up) and verb features (117 classes) (bottom) on HICO-DET dataset (20000 samples) via t-SNE visualization [25]. Left is the visual illustration of baseline, middle includes verb representation and the right uses both VCL and verb representation

5   Conclusion

In this paper, we propose a simple yet efficient deep Visual Compositional Learning framework, which composes the interactions from the shared verb and object latent representations between images and within images, to relieve the long-tail distribution issue and zero-shot learning in human-object interaction detection. Meanwhile, we extract a more discriminative verb representation from the union box of human and object rather than human box. Lastly, we evaluate the efficiency of our model on two HOI detection benchmarks, particularly for low-shot and zero-shot detection.

Acknowledgement. This work is partially supported by Science and Technology Service Network Initiative of Chinese Academy of Sciences (KFJ-STS-QYZX-092), Guangdong Special Support Program (2016TX03X276), National Natural Science Foundation of China (U1813218, U1713208), Shenzhen Basic Research Program (JCYJ20170818164704758, CXB201104220032A), the Joint Lab of CAS-HK, Australian Research Council Projects (FL-170100117).

References

1. Alfassy, A., et al.: LaSo: label-set operations networks for multi-label few-shot learning. In: Proceedings of the IEEE Conference on Computer Vision and Pattern Recognition, pp. 6548–6557 (2019)
2. Azadi, S., Pathak, D., Ebrahimi, S., Darrell, T.: Compositional GAN: Learning conditional image composition. arXiv preprint arXiv:1807.07560 (2018)
3. Bansal, A., Rambhatla, S.S., Shrivastava, A., Chellappa, R.: Detecting human-object interactions via functional generalization. arXiv preprint arXiv:1904.03181 (2019)
4. Battaglia, P.W., et al.: Relational inductive biases, deep learning, and graph networks. arXiv preprint arXiv:1806.01261 (2018)
5. Bengio, Y., Courville, A., Vincent, P.: Representation learning: a review and new perspectives. IEEE Trans. Pattern Anal. Mach. Intell. **35**(8), 1798–1828 (2013)
6. Burgess, C.P., et al.: Monet: Unsupervised scene decomposition and representation. arXiv preprint arXiv:1901.11390 (2019)
7. Chao, Y.W., Liu, Y., Liu, X., Zeng, H., Deng, J.: Learning to detect human-object interactions. In: 2018 IEEE Winter Conference on Applications of Computer Vision (WACV), pp. 381–389. IEEE (2018)
8. Chao, Y.W., Wang, Z., He, Y., Wang, J., Deng, J.: HICO: a benchmark for recognizing human-object interactions in images. In: Proceedings of the IEEE International Conference on Computer Vision, pp. 1017–1025 (2015)
9. Gao, C., Zou, Y., Huang, J.B.: iCAN: Instance-centric attention network for human-object interaction detection. arXiv preprint arXiv:1808.10437 (2018)
10. Garnelo, M., Shanahan, M.: Reconciling deep learning with symbolic artificial intelligence: representing objects and relations. Current Opin. Behav. Sci. **29**, 17–23 (2019)
11. Gupta, S., Malik, J.: Visual semantic role labeling. arXiv preprint arXiv:1505.04474 (2015)
12. Gupta, T., Schwing, A., Hoiem, D.: No-frills human-object interaction detection: Factorization, appearance and layout encodings, and training techniques. arXiv preprint arXiv:1811.05967 (2018)
13. He, K., Zhang, X., Ren, S., Sun, J.: Deep residual learning for image recognition. In: Proceedings of the IEEE Conference on Computer Vision and Pattern Recognition, pp. 770–778 (2016)
14. Higgins, I., et al.: beta-VAE: Learning basic visual concepts with a constrained variational framework. ICLR **2**(5), 6 (2017)

15. Higgins, I., et al.: Scan: Learning hierarchical compositional visual concepts. arXiv preprint arXiv:1707.03389 (2017)
16. Hoffman, D.D., Richards, W.: Parts of recognition (1983)
17. Japkowicz, N., Stephen, S.: The class imbalance problem: a systematic study. Intell. Data Anal. **6**(5), 429–449 (2002)
18. Kato, K., Li, Y., Gupta, A.: Compositional learning for human object interaction. In: Ferrari, V., Hebert, M., Sminchisescu, C., Weiss, Y. (eds.) Computer Vision – ECCV 2018. LNCS, vol. 11218, pp. 247–264. Springer, Cham (2018). https://doi.org/10.1007/978-3-030-01264-9_15
19. Lake, B.M., Ullman, T.D., Tenenbaum, J.B., Gershman, S.J.: Building machines that learn and think like people. Behav. Brain Sci. **40** (2017)
20. Li, Y.L., et al.: Transferable interactiveness prior for human-object interaction detection. arXiv preprint arXiv:1811.08264 (2018)
21. Lin, T.-Y., et al.: Microsoft COCO: common objects in context. In: Fleet, D., Pajdla, T., Schiele, B., Tuytelaars, T. (eds.) ECCV 2014. LNCS, vol. 8693, pp. 740–755. Springer, Cham (2014). https://doi.org/10.1007/978-3-319-10602-1_48
22. Liu, Z., Miao, Z., Zhan, X., Wang, J., Gong, B., Yu, S.X.: Large-scale long-tailed recognition in an open world. In: Proceedings of the IEEE Conference on Computer Vision and Pattern Recognition, pp. 2537–2546 (2019)
23. Locatello, F., Bauer, S., Lucic, M., Gelly, S., Schölkopf, B., Bachem, O.: Challenging common assumptions in the unsupervised learning of disentangled representations. arXiv preprint arXiv:1811.12359 (2018)
24. Lu, C., Krishna, R., Bernstein, M., Fei-Fei, L.: Visual relationship detection with language priors. In: Leibe, B., Matas, J., Sebe, N., Welling, M. (eds.) ECCV 2016. LNCS, vol. 9905, pp. 852–869. Springer, Cham (2016). https://doi.org/10.1007/978-3-319-46448-0_51
25. van den Maaten, L., Hinton, G.: Visualizing data using t-SNE. J. Mach. Learn. Res. **9**(Nov), 2579–2605 (2008)
26. Peyre, J., Laptev, I., Schmid, C., Sivic, J.: Detecting unseen visual relations using analogies. In: The IEEE International Conference on Computer Vision (ICCV), October 2019
27. Qi, S., Wang, W., Jia, B., Shen, J., Zhu, S.-C.: Learning human-object interactions by graph parsing neural networks. In: Ferrari, V., Hebert, M., Sminchisescu, C., Weiss, Y. (eds.) ECCV 2018. LNCS, vol. 11213, pp. 407–423. Springer, Cham (2018). https://doi.org/10.1007/978-3-030-01240-3_25
28. Ren, S., He, K., Girshick, R., Sun, J.: Faster R-CNN: towards real-time object detection with region proposal networks. In: Advances in Neural Information Processing Systems, pp. 91–99 (2015)
29. Shen, L., Yeung, S., Hoffman, J., Mori, G., Fei-Fei, L.: Scaling human-object interaction recognition through zero-shot learning. In: 2018 IEEE Winter Conference on Applications of Computer Vision (WACV), pp. 1568–1576. IEEE (2018)
30. Spelke, E.S.: Principles of object perception. Cogn. Sci. **14**(1), 29–56 (1990)
31. Wan, B., Zhou, D., Liu, Y., Li, R., He, X.: Pose-aware multi-level feature network for human object interaction detection. In: Proceedings of the IEEE International Conference on Computer Vision, pp. 9469–9478 (2019)
32. Wang, T., et al.: Deep contextual attention for human-object interaction detection. arXiv preprint arXiv:1910.07721 (2019)
33. Wang, Y.X., Girshick, R., Hebert, M., Hariharan, B.: Low-shot learning from imaginary data. In: Proceedings of the IEEE Conference on Computer Vision and Pattern Recognition, pp. 7278–7286 (2018)

34. Xian, Y., Lampert, C.H., Schiele, B., Akata, Z.: Zero-shot learning–a comprehensive evaluation of the good, the bad and the ugly. IEEE Trans. Pattern Anal. Mach. Intell. **41**(9), 2251–2265 (2018)
35. Xian, Y., Schiele, B., Akata, Z.: Zero-shot learning-the good, the bad and the ugly. In: Proceedings of the IEEE Conference on Computer Vision and Pattern Recognition, pp. 4582–4591 (2017)
36. Xu, B., Wong, Y., Li, J., Zhao, Q., Kankanhalli, M.S.: Learning to detect human-object interactions with knowledge. In: Proceedings of the IEEE Conference on Computer Vision and Pattern Recognition (2019)
37. Yang, X., Zhang, H., Cai, J.: Shuffle-then-assemble: learning object-agnostic visual relationship features. In: Ferrari, V., Hebert, M., Sminchisescu, C., Weiss, Y. (eds.) ECCV 2018. LNCS, vol. 11216, pp. 38–54. Springer, Cham (2018). https://doi.org/10.1007/978-3-030-01258-8_3
38. Zhou, P., Chi, M.: Relation parsing neural network for human-object interaction detection. In: The IEEE International Conference on Computer Vision (ICCV), October 2019

Deep Plastic Surgery: Robust and Controllable Image Editing with Human-Drawn Sketches

Shuai Yang[1], Zhangyang Wang[2], Jiaying Liu[1(✉)], and Zongming Guo[1]

[1] Wangxuan Institute of Computer Technology, Peking University, Beijing, China
{williamyang,liujiaying,guozongming}@pku.edu.cn
[2] Department of Electrical and Computer Engineering, University of Texas at Austin, Austin, USA
atlaswang@utexas.edu

Abstract. Sketch-based image editing aims to synthesize and modify photos based on the structural information provided by the human-drawn sketches. Since sketches are difficult to collect, previous methods mainly use edge maps instead of sketches to train models (referred to as edge-based models). However, human-drawn sketches display great structural discrepancy with edge maps, thus failing edge-based models. Moreover, sketches often demonstrate huge variety among different users, demanding even higher generalizability and robustness for the editing model to work. In this paper, we propose *Deep Plastic Surgery*, a novel, robust and controllable image editing framework that allows users to interactively edit images using hand-drawn sketch inputs. We present a sketch refinement strategy, as inspired by the coarse-to-fine drawing process of the artists, which we show can help our model well adapt to casual and varied sketches without the need for real sketch training data. Our model further provides a refinement level control parameter that enables users to flexibly define how "reliable" the input sketch should be considered for the final output, balancing between sketch faithfulness and output verisimilitude (as the two goals might contradict if the input sketch is drawn poorly). To achieve the multi-level refinement, we introduce a style-based module for level conditioning, which allows adaptive feature representations for different levels in a singe network. Extensive experimental results demonstrate the superiority of our approach in improving the visual quality and user controllablity of image editing over the state-of-the-art methods. Our project and code are available at https://github.com/TAMU-VITA/DeepPS.

Keywords: Image editing · Sketch-to-image translation · User control

Electronic supplementary material The online version of this chapter (https://doi.org/10.1007/978-3-030-58555-6_36) contains supplementary material, which is available to authorized users.

ⓒ Springer Nature Switzerland AG 2020
A. Vedaldi et al. (Eds.): ECCV 2020, LNCS 12360, pp. 601–617, 2020.
https://doi.org/10.1007/978-3-030-58555-6_36

1 Introduction

Human-drawn sketches reflect people's abstract expression of objects. They are highly concise yet expressive: usually several lines can reflect the important morphological features of an object, and even imply more semantic-level information. Meanwhile, sketches are easily editable: such an advantage is further amplified by the increasing popularity of touch-screen devices. Sketching thus becomes one of the most important ways that people illustrate their ideas and interact with devices. Motivated by the above, a series of sketch-based image synthesis and editing methods have been proposed in recent years. The common main idea underlying these methods is to train an image-to-image translation network to map a sketch to its corresponding color image. That can be extended to an image completion task where an additional mask is provided to specify the area for modification. These methods enable novice users to edit the photo by simply drawing lines, rather than resorting complicated tools to process the photo itself.

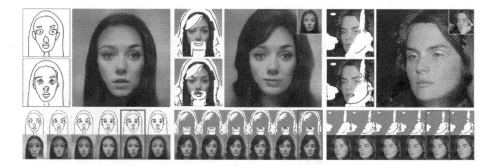

Fig. 1. Our Deep Plastic Surgery framework allows users to synthesize (left) and edit (middle, right) photos based on hand-drawn sketches. Our model is robust to tolerate the drawing errors and achieves the controllability on sketch faithfulness. For each group, we show the user input and our refined sketch in the left column, and the final output in the right column with the original photo in the upper right corner. The bottom row shows our results under an increasing refinement level, with a red box to indicate the user selection. Note that our model requires no real sketch for training. (Color figure online)

Due to the difficulty of collecting pairs of sketches and color images as training data, existing works [11,12,22] typically exploit edge maps (detected from color images) as "surrogates" for real sketches, and train their models on the paired edge-photo datasets. Despite certain success in shoe, handbag and face synthesis, edge maps look apparently different from the human drawings, the latter often being more causal, varied or even wild. As a result, those methods often generalize poorly when their inputs become human-drawn sketches, limiting their real-world usage. To resolve this bottleneck, researchers have studied edge pre-processing [22], yet with limited performance improvement gained so

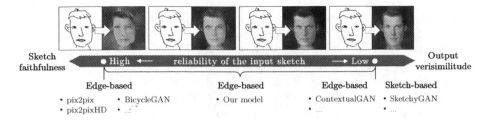

Sketch faithfulness ◄══════ ● High ◄── reliability of the input sketch ──► Low ● ══════► Output verisimilitude

Edge-based	Edge-based	Edge-based	Sketch-based
• pix2pix • BicycleGAN	• Our model	• ContextualGAN	• SketchyGAN
• pix2pixHD • ...		• ...	• ...

Fig. 2. Illustration of the sketch-to-image translation spectrum. Our model differs from existing models in that we allow users to define how "reliable" the input sketch should be considered for the final output, thus balancing between sketch faithfulness and output verisimilitude, which has not been well studied in previous approaches. As edge-based models, ContexualGAN [20] and our model realize verisimilitude without real sketch data for training, and our model further achieves controllability and efficiency.

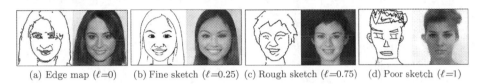

(a) Edge map (ℓ=0)   (b) Fine sketch (ℓ=0.25) (c) Rough sketch (ℓ=0.75)   (d) Poor sketch (ℓ=1)

Fig. 3. Our model works robustly on various sketches by setting refinement level ℓ adaptive to the quality of the input sketches, *i.e.*, higher ℓ for poorer sketches.

far. Some human-drawn sketch datasets have also been collected [24,39] to train sketch-based models [3,18]. However, the collection is too laborious to extend to larger scales or to meet all data subject needs.

As a compromise, it is valuable to study the adaption of edge-based models to the sketches. ContextualGAN [20] presents an intuitive solution. It retrieves the nearest neighbor of the input sketch from the learned generative edge-image manifolds, which relaxes the sketch constraint to trade for the image naturalness. However, neither edge-based models [11,12,22] nor ContextualGAN [20] allows for any user controllability on the *sketch faithfulness*, *i.e.*, to what extent we should stick to the given sketch? The former categories of methods completely hinge on the input sketch even it might yield highly unnatural outputs; while the latter mainly searches from natural manifolds and may produce visually disparate results from the sketch specification. That leaves little room for users to calibrate between freedom of sketching and the overall image verisimilitude: an important desirable feature for interactive photo editing.

In view of the above, we are motivated to investigate a new problem of controllable sketch-based image editing, that can work robustly on varied human-drawn sketches. Our main idea is to refine the sketches to approach the structural features of the edge maps, therefore avoiding the tedious collection of sketch data for training, while enabling users to control the refinement level freely. Figure 1 intuitively demonstrates our task: to improve the model's robustness to various sketch inputs by allowing users to navigate around editing results under different refinement levels and select the most desired one. The challenge of this problem lies in two aspects. First, in the absence of real sketches as reference, we have no paired or unpaired data to directly establish a mapping between sketches and edge maps. Second, in order to achieve controllability, it is necessary to extend the above mapping to a multi-level progress, which remains to be an open question. Please refer to Fig. 2 to get a sense of the difference between the proposed controllable model from common sketch-to-image translation models.

In this paper, we present *Deep Plastic Surgery*, a novel sketch-based image editing framework to achieve both **robustness** on hand-drawn sketch inputs, and the **controllability** on sketch faithfulness. Our key idea arises from our observation on the coarse-to-fine drawing process of the human artists: they first draw coarse outlines to specify the approximate areas where the final fine-level lines are located. Those are then gradually refined to converge to the final sharper lines. Inspired by so, we propose a dilation-based sketch refinement method. Instead of directly feeding the network with the sketch itself, we only specify the approximate region covering the final lines, created by edge dilation, which forces the network to find the mapping between the coarse-level sketches and fine-level edges. The level of coarseness can be specified and adjusted by setting the dilation radius. Finally, we treat sketches under different coarse levels as different stylized versions of the fine-level lines, and use the scale-aware style transfer to recover fine lines by removing their dilation-based styles. Our method only requires color images and their edge maps to train and can adapt to diversified sketch input. It can work as a plug-in for existing edge-based models, providing refinement for their inputs to boost their performance. Figure 3 shows an overall performance of our method on various sketches.

Our contributions are summarized as three-folds:

- We explore a new problem of controllable sketch-based image editing, to adapt edge-based models to human-drawn sketches, where the users have the freedom to balance the sketch faithfulness with the output verisimilitude.
- We propose a sketch refinement method using coarse-to-fine dilations, following the drawing process of artists in real world.
- We propose a style-based network architecture, which successfully learns to refine the input sketches into diverse and continuous levels.

2   Related Work

Sketch-Based Image Synthesis. Using the easily accessible edge maps to simulate sketches, edge-based models [6,11,25] are trained to map edges to their corresponding photos. By introducing masks, they are extended to image inpainting tasks to modify the specified photo areas [12,22,38] or provide users with sketch recommendations [7]. However, the drastic structural discrepancy between edges and human-drawn sketches makes these models less generalizable to sketches. As sketches draw increasing attentions and some datasets [24,39] are released, the discrepancy can be narrowed [3,18]. But existing datasets are far from enough and collecting sketches in large scale is still too expensive.

The most related method to our problem setting is ContextualGAN [20] that also aims to adapt edge-based models to sketches. It solves this problem by learning a generative edge-image manifold through GANs, and searching nearest neighbors to the input sketch in this manifold. As previously discussed, ContextualGAN offers no controllablity, and the influence of the sketch input might be limited for the final output. As can be seen in Fig. 10, ContextualGAN cannot well preseve some key sketch features. Besides, the nearest neighbor search costs time-consuming iterative back-propagation. It also relies on the generative manifolds provided by GANs, which can become hard to train as image resolution grows higher. Thus, results reported in [20] are of a limited 64×64 size. By comparison, our method is able to refine 256×256 sketches in a fast feed-forward way, with their refinement level controllable to better preserve the shape and details of the sketches and to facilitate flexible and user-friendly image editing.

Image-to-Image Translation. Image-to-image translation networks have been proposed to translate an image from a source domain into a target domain. Isola *et al.* [11] designed a general image-to-image translation framework named pix2pix to map semantic label maps or edge maps into photos. Follow-ups involve the diversification of the generated images [41], high-resolution translation [30], and multi-domain translation [4,34,35]. This framework requires that images in two domains exist as pairs for training. Zhu *et al.* [40] suggested a cycle consistency constraint to map the translated image back to its original version, which successfully trained CycleGAN on unpaired data. By assuming a shared latent space across two domains, UNIT [17] and MUNIT [10] are proposed upon CycleGAN to improve the translation quality and diversity.

Image Inpanting. Image inpanting aims to reconstruct the missing parts of an image. Early work [2] smoothly propagates pixel values from the known region to the missing region. To deal with large missing areas, examplar-based methods are proposed to synthesize textures by sampling pixels or patches from the known region in a greedy [5,28] or global [1,16,31] manner. However, the aforementioned methods only reuse information of known areas, but cannot create unseen content. In parallel, data-driven methods [8,26,29] are proposed to achieve creative image inpainting or extrapolation by retrieving, aligning and blending images of similar scenes from external data. Recent models such as Context Encoder [21] and DeepFill [37,38] build upon the powerful deep neural networks to leverage the extra data for semantic completion, which supports fast intelligent image editing for high-resolution images [33] and free-form masks [38].

3   The Deep Plastic Surgery Algorithm

As illustrated in Fig. 4, given an edge-based image editing model F trained on edge-image pairs $\{S_{gt}, I_{gt}\}$, our goal is to adapt F to human-draw sketches through a novel sketch refinement network G that can be trained without sketch data. G aims to refine the input sketch to match the fine edge maps S_{gt}. The output is then fed into F to obtain final editing results. Our model is further conditioned by a control parameter $\ell \in [0,1]$ indicating the refinement level, where larger ℓ corresponds to greater refinement.

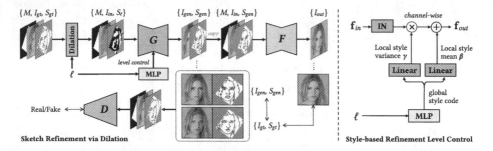

Fig. 4. Framework overview. A novel sketch refinement network G is proposed to refine the rough sketch S_ℓ modelled as dilated drawable regions to match the fine edges S_{gt}. The refined output S_{gen} is fed into a pretrained edge-based model F to obtain the final editing result I_{out}. A parameter ℓ is introduced to control the refinement level. It is realized by encoding ℓ into style codes and performing a style-based adjustment over the outputs $\mathbf{f}_{in}$ of the convolutional layers of G to remove the dilation-based styles.

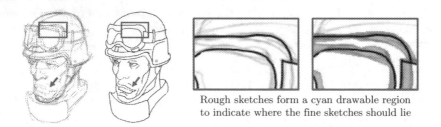

Rough sketches form a cyan drawable region to indicate where the fine sketches should lie

Fig. 5. Rough sketch (left) to fine sketch (middle). The sketches in the red boxes are enlarged and overlayed on the right. Image is copyrighted by Krenz Cushart [27]. (Color figure online)

3.1   Sketch Refinement via Dilation

Our sketch refinement method is inspired by the coarse-to-fine drawing process of human artists. As shown in Fig. 5, artists usually begin new illustrations with inaccurate rough sketches with many redundant lines to determine the shape of an object. These sketches are gradually fintuned by merge lines, tweaking details and fixing mistakes to obtain the final line drawings. When overlaying the final lines on the rough sketches, we find that the redundant lines in the rough sketches form a drawable region to indicate where the final lines should lie (tinted in cyan in Fig. 5). Thus *the coarse-to-fine drawing process is essentially a process of continuously reducing the drawable region.*

Based on the observation, we define our sketch refinement as an image-to-image translation task between rough and fine sketches, where in our problem,

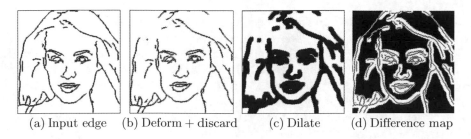

| (a) Input edge | (b) Deform + discard | (c) Dilate | (d) Difference map |

Fig. 6. Rough sketch synthesis. (a) S_{gt}. (b) Deformed edges with lines discarded. (c) $\Omega(S_{gt})$. (d) Overlay red S_{gt} above $\Omega(S_{gt})$ with discarded lines tinted in cyan.

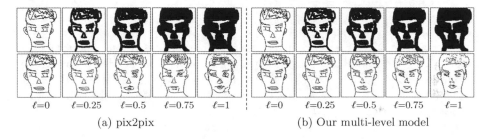

| $\ell=0$ | $\ell=0.25$ | $\ell=0.5$ | $\ell=0.75$ | $\ell=1$ | $\ell=0$ | $\ell=0.25$ | $\ell=0.5$ | $\ell=0.75$ | $\ell=1$ |

| (a) pix2pix | (b) Our multi-level model |

Fig. 7. Sketch refinement at different level ℓ. Top row: S_ℓ with different dilation radii. Bottom row: (a) Refinement results by pix2pix [11] trained separately for each level. (b) Refinement results by our proposed single model with multi-level control.

fine sketches S_{gt} are edge maps extracted from I_{gt} using HED edge detector [32] and *rough sketches are modelled as drawable regions* $\Omega(S_{gt})$ completely covering S_{gt}. In the following, we present our dilation-based drawable region generation method to automatically generate $\Omega(S_{gt})$ based on S_{gt} to form our training data.

Rough Sketch Data Generation. The pipeline of our drawable region generation is shown in Fig. 6. The main idea is to expand lines into areas by dilation operations used in mathematical morphology. However, directly learning to translate a dilated line back to itself will only make the network to simply extract the skeleton centered at the region without refining the sketches. Thus the fine lines are first randomly deformed before dilation. Supposing the radius of dilation is r, then we limit the offset of each pixel after deformation to no more than r, so that the ground truth fine lines are not centered at the drawable region but still fully covered by it, as shown in Fig. 6(d). In addition, noticing that artists will also infer new structures or details from the draft (see the upper lip pointed by the red arrow of Fig. 5), we further discard partial lines by removing random patches from the full sketches. By doing so, our network is motivated to learn to complete the incomplete structures such as the cyan lines in Fig. 6(d). Line deformation and discarding are only applied during the training phase.

Leveraging our dilation-based drawable region generation algorithm, sufficient paired data $\{\Omega(S_{gt}), S_{gt}\}$ is obtained. Intuitively, larger drawable regions provide more room for line-fintuning, which means a higher refinement level. To verify our idea of coarse-to-fine refinement, we train a basic image-to-image translation model of pix2pix [11] to map $\Omega(S_{gt})$ to S_{gt} and use a separate model for each dilation radius. As shown in Fig. 7(a), the rough facial structures are refined and a growing refinement is observed as the radius increases. This property makes it possible for convenient sketch editing control. In the next section, we will detail how we incorporate sketch refinement into one single model with effective level control, whose overall performance is illustrated in Fig. 7(b). The advantage is that coarse-level refinement can benefit from the learned robust fine-level features, thus achieving better performance.

3.2   Controllable Sketch-Based Image Editing

In our image editing task, we have a target photo I_{gt} as input, upon which a mask M is given to indicate the editing region. Users draw sketches S to serve as a shape guidance for the model to fill the masked region. The model will adjust S so that it better fits the contextual structures of I_{gt}, with the refinement level determined by a parameter ℓ.

Our training requires no human-drawn sketches. Instead, we use edge maps $S_{gt} = \text{HED}(I_{gt})$ [32] and generate their corresponding drawable regions $\Omega(S_{gt})$. As analyzed in Sect. 3.1, the refinement level is positively correlated with the dilation radius r. Therefore, we incorporate ℓ in the drawable region generation process (denoted as $\Omega_\ell(\cdot)$) to control r, where $r = \ell R$ with R the maximum allowable radius. The final drawable region with respect to ℓ takes the form of $S_\ell = \Omega_\ell(S_{gt}) \odot M$ where $\odot$ is the element-wise multiplication operator. Then we are going to train G to map S_ℓ back to the fine S_{gt} based on the contextual condition $I_{in} = I_{gt} \odot (1 - M)$, the spatial condition M and the level condition ℓ. Figure 4 shows an overview of our network architecture. G receives a concatenation of I_{in}, S_ℓ and M, with middle layers controlled by ℓ, and yields a four-channel tensor: the completed RGB channel image I_{gen} and the refined one channel sketch S_{gen}, i.e., $(I_{gen}, S_{gen}) = G(I_{in}, S_\ell, M, \ell)$. Here, we task the network with photo generation to enforce the perceptual guidance on the edge generation. It also enables our model to work independently if F is unavailable. Finally, a discriminator D is added to improve the results through adversarial learning.

Style-Based Refinement Level Control. As we will show later, conditioning by label concatenation or feature interpolation [36] fails to properly condition G about the refinement level. Inspired by AdaIN-based style transfer [9] and image generation [15], we propose an effective style-based control module to address this issue. Specifically, sketches at different coarse levels can be considered to have different styles. And G is tasked to destylize them to obtain the original S_{gt}. In AdaIN [9], styles are modelled as the mean and variance of the features and are transferred via distribution scaling and shifting (i.e., normalization+denormalization). Note that the same operation can also be used for its reverse process, i.e., destylization. To this end, as illustrated by Fig. 4, we propose to use a multi-layer perceptron to decode the condition ℓ into a global style code. For each convolution layer expect the first and the last ones in G, we have two affiliated linear layers to map the style code to the local style mean and variance for AdaIN-based destylization.

Loss Function. G is tasked to approach the ground truth photo and sketch:

$$\mathcal{L}_{\text{rec}} = \mathbb{E}_{I_{gt}, M, \ell}[\|I_{gen} - I_{gt}\|_1 + \|S_{gen} - S_{gt}\|_1 + \|I_{out} - I_{gt}\|_1], \tag{1}$$

where $I_{out} = F(I_{in}, S_{gen}, M)$ is the ultimate output in our problem. Here the quality of I_{out} is also considered to adapt G to the pretrained F in an end-to-end manner. Besides, perceptual loss $\mathcal{L}_{\text{perc}}$ [13] to measure the semantical similarity of the photos is computed as

$$\mathcal{L}_{\text{perc}} = \mathbb{E}_{I_{gt}, M, \ell}\Big[\sum_i \lambda_i\big(\|\Phi_i(I_{gen}) - \Phi_i(I_{gt})\|_2^2 + \|\Phi_i(I_{out}) - \Phi_i(I_{gt})\|_2^2\big)\Big], \tag{2}$$

where $\Phi_i(x)$ is the feature map of x in the i-th layer of VGG19 [23] and λ_i is the layer weight. Finally, we use hinge loss as our adversarial objective function:

$$\mathcal{L}_G = -\mathbb{E}_{I_{gt}, M, \ell}[D(I_{gen}, S_{gen}, M)], \tag{3}$$

$$\mathcal{L}_D = \mathbb{E}_{I_{gt}, M, \ell}[\sigma(\tau + D(I_{gen}, S_{gen}, M))] + \mathbb{E}_{I_{gt}, M}[\sigma(\tau - D(I_{gt}, S_{gt}, M))], \tag{4}$$

where τ is a margin parameter and σ is ReLU activation function.

Realistic Sketch-to-Image Translation. Under the extreme condition of $M = 1$, I_{gt} is fully masked out and our problem becomes a more challenging sketch-to-image translation problem. We experimentally find that the result will degrade without any contextual cues from I_{gt}. To solve this problem, we adapt our model by removing the I_{in} and M inputs, and train a separate model specifically for this task, which brings obvious quality improvement.

4   Experimental Results

4.1   Implementation Details

Dataset. We use CelebA-HQ dataset [14] with edge maps extracted by HED edge detector [32] to train our model. The masks are generated as the randomly rotated rectangular regions following [22]. To make a fair comparison with ContextualGAN [20], we also train our model on CelebA dataset [19].

Network Architecture. Our generator G utilizes the Encoder-ResBlocks-Decoder [13] with skip connections [11] to preserve the low-level information. Each convolutional layer is followed by AdaIN layer [9] except the first and the last layer. The discriminator D follows the SN-PatchGAN [38] for stable and fast training. Finally, we use pix2pix [11] as our edge-based baseline model F.

Network Training. We first train our network with $\ell = 1$ for 30 epochs, and then train with uniformly sampled $\ell \in [0, 1]$ for 200 epoches. The maximum

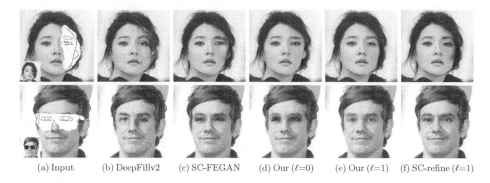

(a) Input   (b) DeepFillv2   (c) SC-FEGAN   (d) Our (ℓ=0)   (e) Our (ℓ=1)   (f) SC-refine (ℓ=1)

Fig. 8. Comparison with state-of-the-art methods on face edting. (a) Input photos, masks and sketches. (b) DeepFillv2 [38]. (c) SC-FEGAN [12]. (d) Our results with $\ell = 0$. (e) Our results with $\ell = 1$. (f) SC-FEGAN using our refined sketches as input.

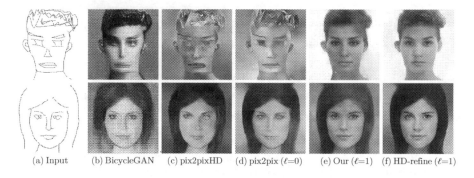

(a) Input   (b) BicycleGAN   (c) pix2pixHD   (d) pix2pix (ℓ=0)   (e) Our (ℓ=1)   (f) HD-refine (ℓ=1)

Fig. 9. Comparison with state-of-the-art methods on face synthesis. (a) Input human-drawn sketches. (b) BicycleGAN [41]. (c) pix2pixHD [30]. (d) pix2pix [11]. (e) Our results with $\ell = 1$. (f) pix2pixHD using our refined sketches as input.

allowable dilation radius is set to $R = 10$ for CelebA-HQ dataset [14] and $R = 4$ for CelebA dataset [19]. For all experiments, the weight for $\mathcal{L}_{\text{rec}}$, $\mathcal{L}_{\text{perc}}$, $\mathcal{L}_G$ and $\mathcal{L}_D$ are 100, 1, 1 and 1, respectively. To calculate $\mathcal{L}_{\text{perc}}$, we use the conv2_1 and conv3_1 layers of the VGG19 [23] weighted by 1 and 0.5, respectively. For hinge loss, we set τ to 10 and 1 for G and F, respectively.

Please refer to our supplementary material and project page for more details.

4.2   Comparisons with State-of-the-Art Methods

Face Editing and Synthesis. Figure 8 presents the qualitative comparison on face editing with two state-of-the-art inpainting models: DeepFillv2 [38] and SC-FEGAN [12]. The released DeepFillv2 uses no sketch guidance, which means the reliability of the input sketch is set to zero ($\ell = \infty$). Despite being one of the most advanced inpainting models, DeepFillv2 fails to repair the fine-scale facial structures well, indicating the necessity of user guidance. SC-FEGAN, on the

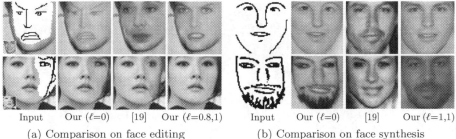

| Input | Our (ℓ=0) | [19] | Our (ℓ=0.8,1) | Input | Our (ℓ=0) | [19] | Our (ℓ=1,1) |

(a) Comparison on face editing   (b) Comparison on face synthesis

Fig. 10. Comparison with ContextualGAN [20] on face editing and face synthesis.

Table 1. User preference ratio of state-of-the-art methods.

Task	Face Editing			Face Synthesis			Face Synthesis*	
Method	DeepFillv2	SC-FEGAN	Ours	BicycleGAN	pix2pixHD	Ours	ContextualGAN	Ours
Score	0.032	0.238	**0.730**	0.024	0.031	**0.945**	0.094	**0.906**

* ContextualGAN is designed for image synthesis on 64 × 64 images. We have tried to extend ContextualGAN to 256 × 256. However, due to the inherent difficulty of training noise-to-image GAN on high resolution, ContextualGAN easily falls into model collapse with poor results. Therefore, we make a separate comparison with it on 64 × 64 images in the user study.

other hand, totally follows the inaccurate sketch and yields weird faces. Similar results can be found in the output of F when $\ell = 0$. By using a large refinement level ($\ell = 1$), the facial details become more natural and realistic. Finally, as an ablation study to indicate the importance of sketch-edge input adaption, we directly feed SC-FEGAN with our refined sketch (without fine-tuning upon SC-FEGAN), and observe improved results of SC-FEGAN.

Figure 9 shows the qualitative comparison on face synthesis with two state-of-the-art image-to-image translation models: BicycleGAN [41] and pix2pixHD [30]. As expected, both models as well as F (pix2pix [11]) synthesize facial structures that strictly match the inaccurate sketch inputs, producing poor results. Our model takes sketches as "useful yet flexible" constraints, and strikes a good balance between authenticity and consistency with the user guidance.

Comparison with ContextualGAN. As the most related work that accepts weak sketch constraint as our model, we further compare with it in this section. For face editing task, we implement ContextualGAN and adapt it to the completion task by additionally computing the appearance similarity between the known part of the photo during the nearest neighbor search. As shown in Fig. 10(a), the main downside of ContextualGAN is the distinct inpainting boundaries, likely due to that the learned generative manifold does not fully depict the real facial distribution. By comparison, our method produces more natural results. Figure 10(b) shows the sketch-to-image translation results, where the results of ContextualGAN are directly imported from the original paper. As can be seen, although realistic, the results of ContextualGAN lose certain

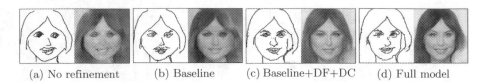

(a) No refinement   (b) Baseline   (c) Baseline+DF+DC   (d) Full model

Fig. 11. Effect of rough sketch models. (a) Input sketch and generated image without refinement. (b)–(d) Refinement results using different rough sketch models. (b) Baseline: edge dilation with a fixed single dilation radius. (c) Baseline + line deformation and discarding. (d) Edge dilation with multiple radii + line deformation and discarding.

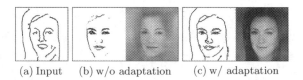

(a) Input   (b) w/o adaptation   (c) w/ adaptation

Fig. 12. Effect of adaptation to F.

attributes associated with the input such as the beard. It might be because the learned generative manifolds collapse for some uncommon attributes. As another possible cause, the nearest neighbor search might sometimes travel too far over the manifold, and results in found solutions less relevant to the initial points provided by user sketches. Our method preserves these attributes much better.

In terms of efficiency, for 64×64 images in Fig. 10, our implemented ContextualGAN requires about 7.89 s per image with a GeForce GTX 1080 Ti GPU, while the proposed feed-forward method only takes about **12 ms per image**.

Quantitative Evaluation. To better understand the performance of the compared methods, we perform user studies for quantitative evaluations. A total of 28 face editing and 38 face synthesis cases are used and participants are asked to select which result best balances the sketch faithfulness with the output verisimilitude. We finally collect totally 1,320 votes from 20 subjects and demonstrate the preference scores in Table 1. The study shows that our method receives most votes for both sketch detail preservation and output naturalness.

4.3   Ablation Study

In this section, we perform ablation studies to verify our model design. We test on the challenging sketch-to-image translation task for better comparison.

Rough Sketch Modelling. We first examine the effect of our dilation-based sketch modelling, which is the key of our sketch refinement. In Fig. 11, we perform a comparison between different rough sketch models. The dilation prompts the network to infer the facial details. Then the line deformation and discarding force the network to further infer and complete the accurate facial structures. In Fig. 11(d), we observe an improvement brought by learning multiple refinement

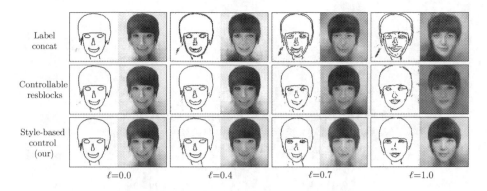

Fig. 13. Visual comparison on label conditioning.

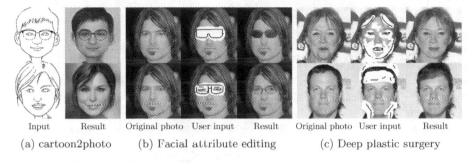

Fig. 14. Applications. More applications can be found in the supplemental material.

levels in one model over single level per model. The reason might be that coarse-level refinement can benefit from the learned more robust fine-level features.

Adaptation to F. Our generator G is trained together with a fixed F, which adapts G to F to improve the quality of the ultimate output. To verify the effect of the adaptation, we train a model without the loss terms related to I_{out} in Eqs. (1) and (2). Figure 12 presents the comparison of our model with and without adaptation. The sketch result without adaptation has its structure refined but some lines become indistinct. The reason might be the low proportion of the line region in the sketch. Through adaptation, G is motivated to generate sketches that are fully perceivable by F, which actually acts as a sketch-version perceptual loss [13], resulting in distinct lines and high-quality photos.

Refinement Level Control. We compare the proposed style-based conditioning with label concatenation and controllable resblock [36] in Fig. 13. Label concatenation yields stacking lines like those in draft sketches. Controllable resblock generates cleaner lines but still rough facial details. Our style-based conditioning surpasses controllable resblock in adaptive channel-wise control, which provides strongest results in both well-structured sketches and realistic photos.

User input Result   User input Result   User input Result   Original User input Result   Original User input Result

Fig. 15. Applications on handbag and shoe design.

Input   Result   User revision Refined result   User input I_{out} (ℓ=0)   New input I_{out} (ℓ=0.75)   level map

(a) User interactive revision   (b) Spatially non-uniform refinement

Fig. 16. User interaction for error revision and spatially non-uniform refinement.

4.4   Applications

Figure 14 shows various results with facial sketch inputs. Our model shows certain robustness on realistic photo rendering from cartoons. Our model can also edit facial attributes such as adding glasses. Finally, users can purposely perform "plastic surgery" digitally, such as removing wrinkles, lifting the eye corners. Alternatively, amateurs can intuitively edit the face with fairly coarse sketches to provide a general idea, such as face-lifting and bangs, and our model will tolerate the drawing errors and suggest a suitable "surgery" plan.

We further present our results on the handbag and shoe datasets [11] and Sketchy dataset [24]. The results are shown in Fig. 15, where our model can effectively design handbags and shoes.

4.5   Limitation and User Interaction

User Interactive Revision. While our approach has generated appealing results, limitations still exist. Our method cannot revise the structural error that exceeds the maximum allowable radius. This problem can be possibly solved by user interaction, where users can modify the input sketch when the output is still unsatisfactory under the maximum refinement level as shown in Fig. 16(a).

Spatially Non-uniform Refinement. Another limitation is that, when ℓ is large, the dilation operation will merge lines that are close to each other, which inevitably loses some structural details. One solution is to use adaptive spatially varied dilation radii. In addition, the accuracy of the structure can vary within one sketch, which also demands spatially non-uniform sketch refinement for more flexible controllability. Our model can be easily extended to spatially non-uniform refinement with user interaction. As shown in Fig. 16(b), the user first uses a low ℓ on the whole mask region to better comply with the structure guidance of the nose, mouth and stubble. Then, user can edit the mask and further improve the verisimilitude of the eye region with a high ℓ. It allows users to improve the overall facial structure while achieving better detail preservation.

5   Conclusion

In this paper, we raise a new a new problem of controllable sketch-based image editing, to adapt edge-based models to human-drawn sketches, and present a novel dilation-based sketch refinement method. Modelling the rough sketch as a drawable region via edge dilation, the network is effectively trained to infer accurate structural information. Leveraging the idea of style transfer, our network is able to undo the edge dilation of different levels in a destylization manner for multi-level refinement control. We validate by experiments the effectiveness and robustness of our method. Serving as a plug-in, our model can greatly improve the performance of edge-based models on the sketch inputs.

Acknowledgement. This work was supported in part by National Natural Science Foundation of China under contract No. 61772043, and in part by Beijing Natural Science Foundation under contract No. L182002 and No. 4192025. The research of Z. Wang was partially supported by NSF Award RI-1755701. This work was supported by China Scholarship Council.

References

1. Barnes, C., Shechtman, E., Finkelstein, A., Goldman, D.B.: PatchMatch: a randomized correspondence algorithm for structural image editing. ACM Trans. Graph. **28**(3), 24 (2009)
2. Bertalmio, M., Sapiro, G., Caselles, V., Ballester, C.: Image inpainting. In: Proceedings of ACM SIGGPRAH, pp. 417–424 (2000)
3. Chen, W., Hays, J.: SketchyGAN: towards diverse and realistic sketch to image synthesis. In: Proceedings of IEEE International Conference Computer Vision and Pattern Recognition, pp. 9416–9425 (2018)
4. Choi, Y., Choi, M., Kim, M., Ha, J.W., Kim, S., Choo, J.: StarGAN: unified generative adversarial networks for multi-domain image-to-image translation. In: Proceedings of IEEE International Conference Computer Vision and Pattern Recognition (2018)
5. Criminisi, A., Pérez, P., Toyama, K.: Region filling and object removal by exemplar-based image inpainting. IEEE Trans. Image Process. **13**(9), 1200–1212 (2004)
6. Dekel, T., Gan, C., Krishnan, D., Liu, C., Freeman, W.T.: Sparse, smart contours to represent and edit images. In: Proc. IEEE International Conference Computer Vision and Pattern Recognition, pp. 3511–3520 (2018)
7. Ghosh, A., et al.: Interactive sketch & fill: multiclass sketch-to-image translation. In: Proceedings of International Conference Computer Vision (2019)
8. Hays, J., Efros, A.A.: Scene completion using millions of photographs. ACM Trans. Graph. **26**(3), 4 (2007)
9. Huang, X., Belongie, S.: Arbitrary style transfer in real-time with adaptive instance normalization. In: roceedings of International Conference Computer Vision, pp. 1510–1519 (2017)
10. Huang, X., Liu, M.-Y., Belongie, S., Kautz, J.: Multimodal unsupervised image-to-image translation. In: Ferrari, V., Hebert, M., Sminchisescu, C., Weiss, Y. (eds.) ECCV 2018. LNCS, vol. 11207, pp. 179–196. Springer, Cham (2018). https://doi.org/10.1007/978-3-030-01219-9_11

11. Isola, P., Zhu, J.Y., Zhou, T., Efros, A.A.: Image-to-image translation with conditional adversarial networks. In: Proceedings of IEEE International Conference Computer Vision and Pattern Recognition, pp. 5967–5976 (2017)
12. Jo, Y., Park, J.: SC-FEGAN: face editing generative adversarial network with user's sketch and color. In: Proceedings of International Conference Computer Vision (2019)
13. Johnson, J., Alahi, A., Fei-Fei, L.: Perceptual losses for real-time style transfer and super-resolution. In: Leibe, B., Matas, J., Sebe, N., Welling, M. (eds.) ECCV 2016. LNCS, vol. 9906, pp. 694–711. Springer, Cham (2016). https://doi.org/10.1007/978-3-319-46475-6_43
14. Karras, T., Aila, T., Laine, S., Lehtinen, J.: Progressive growing of GANs for improved quality, stability, and variation. In: Proceedings of International Conference, Learning Representations (2018)
15. Karras, T., Laine, S., Aila, T.: A style-based generator architecture for generative adversarial networks. In: Proceedings of IEEE International Conference Computer Vision and Pattern Recognition, pp. 4401–4410 (2019)
16. Liu, J., Yang, S., Fang, Y., Guo, Z.: Structure-guided image inpainting using homography transformation. IEEE Trans. Multimedia 20(12), 3252–3265 (2018)
17. Liu, M.Y., Breuel, T., Kautz, J.: Unsupervised image-to-image translation networks. In: Advances in Neural Information Processing Systems, pp. 700–708 (2017)
18. Liu, R., Yu, Q., Yu, S.: An unpaired sketch-to-photo translation model (2019). arXiv:1909.08313
19. Liu, Z., Luo, P., Wang, X., Tang, X.: Deep learning face attributes in the wild. In: Proceedings of the IEEE International Conference on Computer Vision, pp. 3730–3738 (2015)
20. Lu, Y., Wu, S., Tai, Y.-W., Tang, C.-K.: Image generation from sketch constraint using contextual GAN. In: Ferrari, V., Hebert, M., Sminchisescu, C., Weiss, Y. (eds.) ECCV 2018. LNCS, vol. 11220, pp. 213–228. Springer, Cham (2018). https://doi.org/10.1007/978-3-030-01270-0_13
21. Pathak, D., Krähenbühl, P., Donahue, J., Darrell, T., Efros, A.A.: Context encoders: Feature learning by inpainting. In: Proceedings IEEE International Conference Computer Vision and Pattern Recognition, pp. 2536–2544 (2016)
22. Portenier, T., Hu, Q., Szabo, A., Bigdeli, S.A., Favaro, P., Zwicker, M.: Faceshop: deep sketch-based face image editing. ACM Trans. Graph. 37(4), 99 (2018)
23. Russakovsky, O., et al.: ImageNet large scale visual recognition challenge. Int. J. Comput. Vis. 115(3), 211–252 (2015)
24. Sangkloy, P., Burnell, N., Ham, C., Hays, J.: The sketchy database: learning to retrieve badly drawn bunnies. ACM Trans. Graph. 35(4), 119:1–119:12 (2016)
25. Sangkloy, P., Lu, J., Fang, C., Yu, F., Hays, J.: Scribbler: controlling deep image synthesis with sketch and color. In: Proceedings IEEE International Conference Computer Vision and Pattern Recognition, pp. 5400–5409 (2017)
26. Shan, Q., Curless, B., Furukawa, Y., Hernandez, C., Seitz, S.M.: Photo uncrop. In: Fleet, D., Pajdla, T., Schiele, B., Tuytelaars, T. (eds.) ECCV 2014. LNCS, vol. 8694, pp. 16–31. Springer, Cham (2014). https://doi.org/10.1007/978-3-319-10599-4_2
27. Simo-Serra, E., Iizuka, S., Ishikawa, H.: Real-time data-driven interactive rough sketch inking. ACM Trans. Graph. 37(4), 98 (2018)
28. Sun, J., Yuan, L., Jia, J., Shum, H.Y.: Image completion with structure propagation. ACM Trans. Graph. 24(3), 861–868 (2005)
29. Wang, M., Lai, Y., Liang, Y., Martin, R.R., Hu, S.M.: Biggerpicture: data-driven image extrapolation using graph matching. ACM Trans. Graph. 33(6) (2014)

30. Wang, T.C., Liu, M.Y., Zhu, J.Y., Tao, A., Kautz, J., Catanzaro, B.: High-resolution image synthesis and semantic manipulation with conditional GANs. In: Proceedings of IEEE International Conference Computer Vision and Pattern Recognition (2018)
31. Wexler, Y., Shechtman, E., Irani, M.: Space-time completion of video. IEEE Trans. Pattern Anal. Mach. Intell. **3**, 463–476 (2007)
32. Xie, S., Tu, Z.: Holistically-nested edge detection. In: Proceedings of IEEE International Conference Computer Vision and Pattern Recognition, pp. 1395–1403 (2015)
33. Yang, C., Lu, X., Lin, Z., Shechtman, E., Wang, O., Li, H.: High-resolution image inpainting using multi-scale neural patch synthesis. In: Proceedings of IEEE International Conference Computer Vision and Pattern Recognition (2017)
34. Yang, S., Liu, J., Wang, W., Guo, Z.: TET-GAN: text effects transfer via stylization and destylization. Proc. AAAI Conf. Artif. Intell. **33**, 1238–1245 (2019)
35. Yang, S., Wang, W., Liu, J.: TE141K: artistic text benchmark for text effect transfer. IEEE Trans. Pattern Anal. Mach. Intell. **PP**(99), 1–15 (2020). https://doi.org/10.1109/TPAMI.2020.2983697
36. Yang, S., Wang, Z., Wang, Z., Xu, N., Liu, J., Guo, Z.: Controllable artistic text style transfer via shape-matching GAN. In: Proceedings of International Conference Computer Vision, pp. 4442–4451 (2019)
37. Yu, J., Lin, Z., Yang, J., Shen, X., Lu, X., Huang, T.S.: Generative image inpainting with contextual attention. In: Proceedings of IEEE International Conference Computer Vision and Pattern Recognition, pp. 5505–5514 (2018)
38. Yu, J., Lin, Z., Yang, J., Shen, X., Lu, X., Huang, T.S.: Free-form image inpainting with gated convolution. In: Proceedings of International Conference Computer Vision (2019)
39. Yu, Q., Liu, F., Song, Y.Z., Xiang, T., Hospedales, T.M., Loy, C.C.: Sketch me that shoe. In: Proceedings of IEEE International Conference Computer Vision and Pattern Recognition, pp. 799–807 (2016)
40. Zhu, J.Y., Park, T., Isola, P., Efros, A.A.: Unpaired image-to-image translation using cycle-consistent adversarial networks. In: Proceedings of International Conference Computer Vision, pp. 2242–2251 (2017)
41. Zhu, J.Y., et al.: Toward multimodal image-to-image translation. In: Advances in Neural Information Processing Systems, pp. 465–476 (2017)

Rethinking Class Activation Mapping for Weakly Supervised Object Localization

Wonho Bae, Junhyug Noh, and Gunhee Kim[✉]

Department of Computer Science and Engineering,
Seoul National University, Seoul, Korea
bwh0324@gmail.com, jh.noh@vision.snu.ac.kr, gunhee@snu.ac.kr
http://vision.snu.ac.kr/projects/rethinking-cam-wsol

Abstract. Weakly supervised object localization (WSOL) is a task of localizing an object in an image only using image-level labels. To tackle the WSOL problem, most previous studies have followed the conventional class activation mapping (CAM) pipeline: (i) training CNNs for a classification objective, (ii) generating a class activation map via global average pooling (GAP) on feature maps, and (iii) extracting bounding boxes by thresholding based on the maximum value of the class activation map. In this work, we reveal the current CAM approach suffers from three fundamental issues: (i) the bias of GAP that assigns a higher weight to a channel with a small activation area, (ii) negatively weighted activations inside the object regions and (iii) instability from the use of the maximum value of a class activation map as a thresholding reference. They collectively cause the problem that the localization to be highly limited to small regions of an object. We propose three simple but robust techniques that alleviate the problems, including thresholded average pooling, negative weight clamping, and percentile as a standard for thresholding. Our solutions are universally applicable to any WSOL methods using CAM and improve their performance drastically. As a result, we achieve the new state-of-the-art performance on three benchmark datasets of CUB-200–2011, ImageNet-1K, and OpenImages30K.

Keywords: Weakly Supervised Object Localization (WSOL) · Class Activation Mapping (CAM)

1 Introduction

Many recent object detection algorithms such as Faster R-CNN [27], YOLO [25], SSD [22], R-FCN [7] and their variants [11, 20, 26] have been successful in

W. Bae and J. Noh—Equal contribution.

Electronic supplementary material The online version of this chapter (https:// doi.org/10.1007/978-3-030-58555-6_37) contains supplementary material, which is available to authorized users.

A. Vedaldi et al. (Eds.): ECCV 2020, LNCS 12360, pp. 618–634, 2020.
https://doi.org/10.1007/978-3-030-58555-6_37

challenging benchmarks of object detection [10,21]. However, due to the necessity of heavy manual labor for bounding box annotations, weakly supervised object localization (WSOL) has drawn great attention in computer vision research [5, 6,31,36,38–40]. Contrast to fully-supervised object detection, the models for WSOL are trained for the objective of classification solely relying on image-level labels. They then utilize the feature map activations from the last convolutional layer to generate class activation maps from which bounding boxes are estimated.

Since CAM approach [40] was initially introduced, most of previous studies on WSOL have followed its convention to first generate class activation maps and extract object locations out of them. However, this approach suffers from severe underestimation of an object region since the discriminative region activated through the classification training is often much smaller than the object's actual region. For instance, according to the class activation map $(\mathbf{M}_k)$ in Fig. 1, the classifier focuses on the *head* of the *monkey* rather than its whole *body*, since the activations of the *head* are enough to correctly classify the image as *monkey*. Thus, the bounding box reduces to delineate the small highly activated *head* region only. To resolve this problem, recent studies have devised architectures to obtain larger bounding boxes; for example, it erases the most discriminative region and trains a classifier only using the regions left, expecting the expansion of activation to the next most discriminative regions [1,5,6,13,17,18,31,34,35, 38]. These methods have significantly improved the performance of WSOL as well as other relevant tasks such as semantic segmentation.

In this work, however, we propose an approach different from the previous researches; instead of endeavoring to expand activations by devising a new architecture, we focus on correctly utilizing the information that already exists in the feature maps. The major contribution of our approach is three-fold.

1. We discover three underlying issues residing in the components of the CAM pipeline that hinder from properly utilizing the information from the feature maps for localization. Our thorough analysis on CAM reveals the mechanism of how each component of CAM negatively affects the localization to be limited to small discriminative regions of an object.
2. Based on the analysis, we propose three simple but robust techniques that significantly alleviate the problems. Since our solution does not introduce any new modules but replaces some of existing operations for pooling, weight averaging and thresholding with better ones, it is easily applicable to any CAM-based WSOL algorithms.
3. In our experiments, we show that our solutions significantly improve multiple state-of-the-art CAM-based WSOL models (*e.g.* HaS [31] and ADL [6]). More encouragingly, our approach achieves the new best performance on two representative benchmarks: CUB-200–2011 [33] and ImageNet-1K [28], and one recently proposed one: OpenImages30K [2,4].

2   Approach

In this section, we first outline three fundamental problems of CAM-based approach to WSOL that cause the localization to be limited to small discriminative

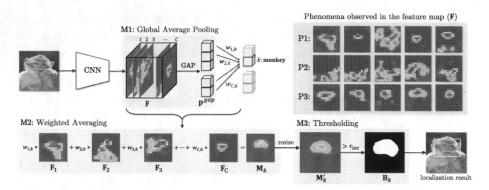

Fig. 1. Overview of the CAM pipeline. We investigate three phenomena of the feature maps (**F**). **P1.** The areas of the activated regions largely differ by channel. **P2.** The activated regions corresponding to the negative weights ($w_c < 0$) often cover large parts of the target object (*e.g. monkey*). **P3.** The most activated regions of each channel largely overlap at small regions. The three modules of CAM in gray boxes (**M1–M3**) do not take these phenomena into account correctly. It results in localization being limited to small discriminative regions.

regions of an object. To this end, three phenomena are visualized with feature maps in **P1**–**P3** of Fig. 1, and the corresponding modules where the problems occur related to the phenomena are described in **M1**–**M3**.

(i) **Global Average Pooing.** In practice, the areas of the activated regions largely differ by feature map channel. But, Global Average Pooling (GAP) is biased to assign a higher weight to a channel with small activated area. It results in the small region to be more focused when generating a class activation map.

(ii) **Weighted Averaging.** Ideally, the activated regions in the channel of a feature map corresponding to a negative weight are supposed to be *no-object regions* (*e.g.* background); however, they often occur inside the object, especially less import regions (*e.g. monkey*'s *body*). As a result, less important object regions are further suppressed in the class activation map.

(iii) **Thresholding.** The most activated regions largely overlap across different channels. Since a class activation map is generated by weighted-averaging all the channels and a bounding box is determined based on the threshold proportional to the maximum value of the class activation map, small overlapped regions with too high activations become overdominant to the localization.

Before presenting our solutions to the problems, we first review the class activation mapping (CAM) pipeline in Sect. 2.1. We then elaborate the problems and our solutions one by one in the following Sect. 2.2, 2.3 and 2.4.

2.1 Preliminary: Class Activation Mapping (CAM)

The current CAM approach based on the CNN trained for classification, generates a class activation map and localizes an object in the following way (Fig. 1).

Let the feature map be $\mathbf{F} \in \mathbb{R}_{\geq 0}^{H \times W \times C}$ where $\mathbb{R}_{\geq 0}$ is a non-negative real number. $\mathbf{F}_c \in \mathbb{R}_{\geq 0}^{H \times W}$ denotes c-th channel of $\mathbf{F}$ where $c = 1, \ldots, C$. First, $\mathbf{F}$ is passed into a global average pooling (GAP) layer that averages each $\mathbf{F}_c$ spatially and outputs a pooled feature vector, $\mathbf{p}^{\text{gap}} \in \mathbb{R}_{\geq 0}^{C}$ as

$$p_c^{\text{gap}} = \frac{1}{H \times W} \sum_{(h,w)} \mathbf{F}_c(h, w), \tag{1}$$

where p_c^{gap} denotes a scalar of $\mathbf{p}^{\text{gap}}$ at c-th channel, and $\mathbf{F}_c(h, w)$ is an activation of $\mathbf{F}_c$ at spatial position (h, w).

The pooled feature vector is then transformed into K-dim logits through an FC layer where K is the number of classes. We denote the weights of the FC layer as $\mathbf{W} \in \mathbb{R}^{C \times K}$. Hence, the class activation map $\mathbf{M}_k$ for class k becomes

$$\mathbf{M}_k - \sum_{c=1}^{C} w_{c,k} \cdot \mathbf{F}_c, \tag{2}$$

where $\mathbf{M}_k \subset \mathbb{R}^{H \times W}$ and $w_{c,k}$ is an (c, k) element of $\mathbf{W}$. For localization, $\mathbf{M}'_k$ is first generated by resizing $\mathbf{M}_k$ to the original image size. Then a localization threshold is computed as

$$\tau_{loc} = \theta_{loc} \cdot \max \mathbf{M}'_k, \tag{3}$$

where $\theta_{loc} \in [0, 1]$ is a hyperparameter. Next, a binary mask $\mathbf{B}_k$ identifies the regions where the activations of $\mathbf{M}'_k$ is greater than τ_{loc}: $\mathbf{B}_k = \mathbb{1}(\mathbf{M}'_k > \tau_{loc})$. Finally, the localization is predicted as the bounding box that circumscribes the contour of the regions with the largest positive area of $\mathbf{B}_k$.

2.2 Thresholded Average Pooling (TAP)

Problem. In a feature map ($\mathbf{F}$), the activated areas largely differ by channel as each channel captures different class information. The GAP layer, however, does not reflect this difference. It naively sums all the activations of each channel and divides them by $H \times W$ without considering the activated area in the channel as in Eq.(1). The difference in the activated area per channel is, however, not negligible. As an example in Fig. 2, suppose i-th channel $\mathbf{F}_i$ in (a) captures the *head* of a *bird* while j-th channel $\mathbf{F}_j$ captures its *body*. Although the area activated in $\mathbf{F}_i$ is much smaller than that in $\mathbf{F}_j$, the GAP layer divides both of them by $H \times W$, and thus the pooled feature value p_i^{gap} of $\mathbf{F}_i$ is also much smaller than p_j^{gap}. However, it does not mean the importance of $\mathbf{F}_i$ for classification is less than $\mathbf{F}_j$. For the GT class k (*bird*), to compensate this difference, the FC weight $w_{i,k}$ corresponding to $\mathbf{F}_i$ is trained to be higher than $w_{j,k}$. As a result,

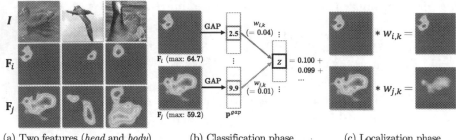

(a) Two features (*head* and *body*)   (b) Classification phase   (c) Localization phase

Fig. 2. An example illustrating a problem of the GAP layer. (a) $\mathbf{F}_i$ and $\mathbf{F}_j$ are the features capturing the *head* and *body* of a *bird*, respectively. (b) When the two features are passed to the GAP layer, although their max values are similar, the pooled feature values, p_i^{gap} and p_j^{gap}, are significantly different (2.5, 9.9). Despite the similar contributions of two features to the logit (z) as (0.100, 0.099), the FC weights, $w_{i,k}$ and $w_{j,k}$, are trained to be highly different to compensate the difference introduced by the GAP layer. (c) In the localization phase, the weighted feature with a small activated region, $w_{i,k} \cdot \mathbf{F}_i$, is highly overstated.

when generating a class activation map ($\mathbf{M}_k$), small activated regions of $\mathbf{F}_i$ are highly overstated due to the large value of $w_{i,k}$, which causes localization to be limited to small regions as localization depends on the maximum value of $\mathbf{M}_k$.

A batch normalization (BN) layer [16] can partially alleviate this issue through normalization as it forces the distributions of the activations to be similar by channel. However, it may also distort the activated area of a channel. For example, when a channel captures a small region like *ears* of a *monkey*, the BN layer expands its originally activated area through normalization, and as a result, localization can be expanded to the background if the channel is activated at the edge of the object. On the other hand, our proposed solution alleviates this problem without distorting the originally activated area.

Solution. To alleviate the problem of the GAP layer, we propose a *thresholded average pooling* (TAP) layer defined as

$$p_c^{\text{tap}} = \frac{\sum_{(h,w)} \mathbb{1}(\mathbf{F}_c(h,w) > \tau_{tap})\mathbf{F}_c(h,w)}{\sum_{(h,w)} \mathbb{1}(\mathbf{F}_c(h,w) > \tau_{tap})}, \tag{4}$$

where $\tau_{tap} = \theta_{tap} \cdot \max \mathbf{F}_c$ is a threshold value where $\theta_{tap} \in [0,1)$ is a hyperparameter. That is, our solution is to replace the GAP layer with the TAP layer (*i.e.* using Eq.(4) instead of Eq.(1)). The TAP layer can be regarded as a generalized pooling layers in between global max pooling (GMP) and global average pooling (GAP). Although GAP has an advantage over GMP for WSOL to expand the activation to broader regions, GMP also has an useful trait that it can precisely focus on the important activations of each channel for pooling.

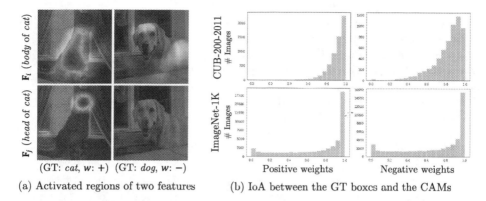

(a) Activated regions of two features   (b) IoA between the GT boxes and the CAMs

Fig. 3. An example illustrating the characteristics of the feature map channels with negative weights. (a) The activated regions of the i-th and j-th feature map channel for two different images. GT and w denote the ground truth class and sign of the corresponding weight. The less important regions (*e.g. body*) can be activated in multiple classes regardless of the weight sign due to their resemblance between different classes. (b) Intersection over Area (IoA) between the GT boxes and class activation maps generated from only the features corresponding to positive and negative weights, respectively, on both CUB-200–2011 and ImageNet-1K. It indicates how much activated regions and actual object regions are overlapped.

The TAP layer inherits the benefits of both GMP and GAP. By including much broader spatial areas than GMP, it can have the loss propagate to wider feature map activations than GMP which highlights only the most discriminative part [40]. Also, by excluding inactive or less active regions, the pooled channel value p_c^{tap} can better represent the core unique activations of each channel.

2.3   Negative Weight Clamping (NWC)

Problem. When CNNs are trained for classification, a large number of the weights from the FC layer are negative. Since feature map channels corresponding to negative weights only decrease the final logit of the GT class for classification, ideally they should not be activated for the sake of classification, which is not the case in general. According to the underlying assumption of the CAM method from Eq.(2), since only depreciating the values in a class activation map ($\mathbf{M}_k$), they should be activated in *no-object* regions like background. However, in reality, they are activated *inside* the object region, especially *less important* region for classification (**P2** in Fig. 1). As a result, it causes the localization to be limited to the discriminative region of an object further.

To better understand why this happens, we first take an example in Fig. 3(a). For the image of the first column whose GT class is *cat*, i-th and j-th channel of feature maps, $\mathbf{F}_i$ and $\mathbf{F}_j$, have positive weights for the class *cat* and they successfully capture the *body* and *head* of *cat*, respectively. Contrarily, for the image of the second column whose GT class is *dog*, the two channels have negative weights for the class *dog* and they are supposed to be activated in *no-dog* regions. However, the *body* of *dog* is activated in $\mathbf{F}_i$, because of the resemblance between the *body* of *cat* and *dog*. This phenomenon is very common in practice as *less important* object regions for classification are similar between different classes.

To make sure that this phenomenon commonly happens in WSOL benchmark datasets, we obtain the distributions of Intersection over Area (IoA) between the GT boxes and class activation maps generated from only the features corresponding to positive and negative weights, respectively, on CUB-200–2011 and ImageNet-1k, as described Fig. 3(b). Surprisingly, the distributions of the positive and negative weights are almost identical on both datasets, and it is highly prevalent that GT boxes are overlapped with the activated regions of negatively weighted features.

Solution. To mitigate the aforementioned problem, we simply clamp negative weights to zero to generate a class activation map. Hence, Eq.(2) is redefined as

$$\mathbf{M}_k = \sum_{c=1}^{C} \mathbb{1}(w_{c,k} > 0) \cdot w_{c,k} \cdot \mathbf{F}_c. \tag{5}$$

By doing this, we can secure the activations on *less important* object regions that are depreciated by the negative weights. One may think that this negative clamping may increase the chance of background selection if the feature with a negative weight correctly activates background. In our experiments, however, this method does not introduce further background noise, mainly because the features corresponding to positive weights are mostly activated in the object regions and their strengths are sufficiently large. Consequently, the activations in the background are still far below the threshold and can be easily filtered out.

2.4   Percentile as a Standard for Thresholding (PaS)

Problem. As shown in Eq.(3), the thresholding of CAM is simply based on the maximum value of a class activation map. If high activations largely overlap across feature map channels, due to the extremely high maximum value of $\mathbf{M}_k$, the region where activations are greater than the localization threshold is limited to a very small region. The top row in Fig. 4 shows such a case, where the values of a generated class activation map follow Zipf's law in Fig. 4(b) and the values in the discriminative region are exponentially larger than those in non-discriminative regions. On the other hand, when the activations are not solely concentrated in a small region as in the bottom case, the distribution of the activations follows a linearly decreasing pattern as shown in the bottom of (b), and the localization tends to cover the whole region of an object. While the thresholding of CAM works well with the bottom case but fails with the top one, our solution is designed to work robustly with both cases.

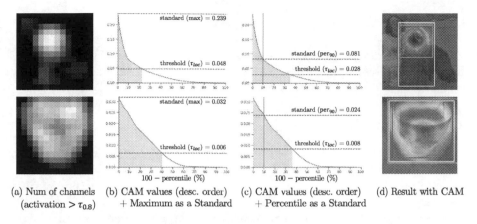

(a) Num of channels (b) CAM values (desc. order) (c) CAM values (desc. order) (d) Result with CAM
(activation > $\tau_{0.8}$)   + Maximum as a Standard   + Percentile as a Standard

Fig. 4. An example illustrating the problem of the overlap of high activations. (a) In the problematic case (top), when high activations (activation > $\tau_{0.8}$) are concentrated in the small discriminative region, the localization threshold τ_{loc} in Eq. (3) becomes too high due to the high maximum value of the class activation map, which results in localization being limited to a small region. (b) and (c) are the distributions of class activation map values in the descending order when using the maximum and percentile as thresholding standards, respectively. (d) illustrates the resulting class activation maps. The boxes in red, green and yellow represent the GT, prediction based on the maximum as a standard, and prediction based on the percentile as a standard, respectively. (Color figure online)

Solution. To alleviate the problem of having too large maximum value, a percentile can be employed as a substitute for the maximum value. The percentile is one of the simplest but the most robust metrics that are not sensitive to outliers nor exponential distributions of activations. Hence, Eq. (3) for the localization threshold τ_{loc} is redefined as

$$\tau_{loc} = \theta_{loc} \cdot \text{per}_i(\mathbf{M}'_k), \tag{6}$$

where per_i is an i-th percentile. Although any value in $[0, 1]$ is available for θ_{loc}, for percentile i, due to small object cases where even 80-th percentile of a class activation map is close to zero, we constraint the possible values for i to $[80, 100]$.

Figure 4(c) shows the same distributions as those in (b) except 90-th percentile is used as a standard. When using the maximum as a standard in Fig. 4(b), the thresholded percentiles (*i.e.* the value of x-axis at the threshold) in the top and the bottom are significantly different (*e.g.* 23% in top and 40% in bottom). Contrarily, by using 90-th percentile in Fig. 4(c), they are almost similar around 35%. Figure 4(d) shows the localization results are better when using the proposed percentile (*e.g.* yellow boxes) than using the maximum (*e.g.* green boxes).

3   Related Work

We review two directions of a related research: (i) CAM-based WSOL methods that attempt to resolve the problem of limited localization and (ii) spatial pooling methods for object localization in weakly supervised setting.

3.1   CAM-Based WSOL Methods

The major challenge of WSOL is to capture a whole object region rather than its most discriminative one. Since a CNN backbone is trained for classification, a class activation map is highly activated on the discriminative region not the whole region of an object. Hence, the expansion of the activation beyond the discriminative region in a feature map has been a major research topic for WSOL.

Image Masking. Bazzani *et al.* [1] improve localization by masking out the regions of which classification scores largely drop. Hide-and-Seek (HaS) [31] randomly hide patches in an image to make a classifier seek other regions. Choe *et al.* [5] improve HaS using GoogleNet Resize (GR) augmentation. Wei *et al.* [34] propose an adversarial erasing that progressively erases the most discriminative parts using multiple classifiers and combines them for final localization.

Feature Masking. Instead of masking in the image level, Kim *et al.* [17] and SeeNet [13] propose two-phase feature-level erasing methods, and ACoL [38] designs an end-to-end parallel adversarial architecture where two classifiers are learned to detect complementary regions via adversarially erasing feature maps. ADL [6] is an attention-based dropout layer that randomly generates masks on feature maps to expand the activations to less discriminative regions of an object.

Other Methods. SPG [39] uses highly discriminative regions from the latter layers as a mask supervision to the earlier layers. DANet [36] trains intermediate layers using the classes obtained from knowledge graphs of class hierarchy expecting the expansion of activations to the common object regions.

Unlike the previous methods, we aim at fully and correctly leveraging the information that already exists in the CAM pipeline. Thus, instead of endeavoring to devise a new architecture as done previously, we focus on discovering problematic steps of CAM and proposing simple but robust solutions.

3.2   Spatial Pooling Methods

Due to the absence of bounding box annotations, WSOL relies on the activations of feature maps to localize an object. Several approaches have been proposed to deal with how to extract the information from feature maps.

Representation Pooling. Oquab *et al.* [23] and Pinheiro *et al.* [24] respectively propose to use the max pooling and log-sum-exp layers as pooling methods for CAM. Although the max pooling accurately tells the most discriminative region of an object, its localization is highly limited to the small regions. Zhou *et al.*

[40] use a GAP layer proposed in Lin *et al.* [19] as a replacement for max pooling since the loss for average pooling benefits all the activated regions.

Gradient-Based Pooling. To utilize the GAP layer, the last layer of CNNs has to be converted to a FC layer following the GAP layer, which does not align with the structure of many of well-known classification CNNs [12,14,30,32]. Because of this limitation, GradCAM [29] and GradCAM+ [3] propose gradient-based methods to obtain a class activation map. Although gradient-based methods are applicable to any classification model with no modification of architecture, they are overwhelming in terms of the computation and memory cost without much improvement on the performance. Thus, using the GAP layer is still a de facto standard approach to WSOL, including recent works such as [6,13,36,39].

Score Pooling. Instead of pooling information only from the maximum scoring regions, WELDON [9] and WILDCAT [8] include the minimum scoring regions to regularize the class score. The score pooling (SP) proposed in WILDCAT is the closest idea to our TAP layer but they are fundamentally different in that SP is applied to a fixed number of activations on the class map to consider both positive and negative regions for classification, whereas TAP adaptively includes activations for pooling for every channel of the feature map to correctly estimate a weight for each channel in localization.

4   Experiments

We evaluate the proposed approach on two standard benchmarks for WSOL: CUB-200–2011 [33] and ImageNet-1K [28], and one recently proposed benchmark: OpenImages30K [2,4]. Our approach consistently improves the performance with various CNN backbones and WSOL methods; especially, we achieve the new state-of-the-art performance on all three datasets.

4.1   Experiment Setting

Datasets. CUB-200–2011 [33] consists of 200 bird species. The numbers of images in training and test sets are 6,033 and 5,755, respectively. ImageNet-1K [28] consists of 1,000 different categories; the numbers of images in training and validation sets are about 1.3 million and 50,000, respectively. We use bounding box annotations of the datasets only for the purpose of evaluation. OpenImages30K [2,4] consists of 29,819, 2,500 and 5,000 images for training, validation, and test sets, respectively, with binary mask annotations.

Implementation. To validate the robustness of our methods, we employ four different CNN backbones: VGG16 [30], ResNet50-SE [12,15], MobileNetV1 [14] and GoogleNet [32]. For VGG16, we replace the last pooling layer and two following FC layers with a GAP layer as done in [40]. We add SE blocks [15] on top of ResNet50 to build ResNet50-SE for CUB-200–2011 and ImageNet-1K following ADL [6], and leave ResNet50 as it is for OpenImages30K following Choe *et al.* [4]. For GoogleNet, we replace the last inception block with two

CONV layers based on SPG [39]. For the threshold τ_{tap} of TAP layer in Eq. (4), we set $\theta_{tap} = 0.1$ for VGG16 and MobileNetV1 and $\theta_{tap} = 0.0$ for ResNet50-SE and GoogleNet. Also, localization hyperparameters, i and θ_{loc}, in Eq. (6) are set to $\theta_{loc} = 0.35, i = 90$, which are fixed regardless of the backbones or datasets. The detailed hyperparameter tuning is described in the appendix.

Evaluation metrics. We report the performance of models using *Top-1 Cls*, *GT Loc*, and *Top-1 Loc* on CUB-200–2011 and ImageNet-1K, and PxAP on OpenImages30K. *Top-1 Cls* is the top-1 accuracy of classification, and *GT Loc* measures the localization accuracy with known ground truth classes. For *Top-1 Loc*, the prediction is counted as correct if the predictions on both classification and localization (*i.e.* IoU ≥ 0.5) are correct. Pixel Average Precision (PxAP) [4] is the area under a pixel precision and recall curve. As precision and recall are computed for all thresholds, PxAP is independent to the choice of a threshold.

4.2   Quantitative Results

Comparison with the State-of-the-Arts. As the proposed solutions are applicable to any CAM-based WSOL algorithms, we validate their compatibility with two recent state-of-the-art models. We select HaS [31] and ADL [6] as they are two of the best performing models for WSOL.

Table 1 provides the comparison of the proposed methods on HaS and ADL with various backbone structures and the state-of-the-art models: ACoL [38], SPG [39] and DANet [36]. We validate the proposed approaches further improve both HaS and ADL on CUB-200–2011 and ImageNet-1K. Especially, ADL with our approaches significantly outperforms all the state-of-the-art algorithms on CUB-200–2011, and obtain the comparable results on ImageNet-1K. To the best of our knowledge, Baseline + Ours with VGG16 and ResNet50-SE that are shown in Table 2 are the new state-of-the-art performance on CUB-200–2011 and ImageNet-1K, respectively.

Results with Different Backbones. To validate the robustness of our solutions, we experiment our approach with different backbones. Table 2 summarizes the results on CUB-200–2011 and ImageNet-1K. In terms of *Top-1 Loc* regarded as the most important metric for WSOL, our approach improves the *baseline*, which refers to Vanilla CAM [40], with significant margins (CUB: 14.18, ImageNet: 2.84 on average). The results are compatible or even better than the state-of-the-art methods on both datasets as shown in Table 1.

Results with Different Components. We further investigate the effectiveness of each of the proposed solutions using VGG16 on CUB-200–2011 and ImageNet-1K. Due to space constraint, we defer the results of the other backbones to the appendix. In Table 3, three leftmost columns denote whether each of our solutions is applied to the baseline, Vanilla CAM with VGG-16.

The TAP layer improves the performance of both classification (CUB: 69.95 → 74.91, ImageNet: 65.39 → 67.22) and localization (CUB: 37.05 → 48.53,

Table 1. Comparison of the proposed methods applied to ADL and HaS-32 with other state-of-the-art algorithms. The methods with * indicate the scores are referred from the original paper. – indicates no accuracy reported in the paper.

Backbone	Method	CUB-200–2011			ImageNet-1K		
		Top-1 Cls	GT Loc	Top-1 Loc	Top-1 Cls	GT Loc	Top-1 Loc
VGG16	ACoL*	71.90	–	45.92	67.50	–	**45.83**
	SPG*	**75.50**	–	48.93	–	–	–
	DANet*	75.40	–	52.52	–	–	–
	HaS-32	66.10	71.57	49.46	62.28	61.23	41.64
	HaS-32 + Ours	70.12	**78.58**	57.37	66.21	**61.48**	43.91
	ADL	69.05	73.96	53.40	68.03	59.24	42.96
	ADL + Ours	75.01	76.30	**58.96**	**68.67**	60.73	44.62
ResNet50	HaS-32	71.28	72.56	53.97	74.37	62.95	48.27
	HaS-32 + Ours	72.51	75.34	57.42	73.75	**63.84**	49.40
	ADL	**76.53**	71.99	57.40	75.06	61.04	48.23
	ADL + Ours	75.03	**77.58**	**59.53**	**75.82**	62.20	**49.42**
MobileNetV1	HaS-32	65.98	67.31	46.70	65.45	60.12	42.73
	Has-32 + Ours	71.16	75.04	55.56	65.60	**62.22**	44.31
	ADL	71.90	62.55	47.69	67.02	59.21	42.89
	ADL + Ours	**73.51**	**78.60**	**59.41**	**67.15**	61.69	**44.78**
GoogleNet	ACoL*	–	–	–	–	–	46.72
	SPG*	–	–	46.64	–	–	48.60
	DANet*	71.20	–	49.45	72.50	–	47.53
	Has-32	**75.35**	61.08	47.36	68.92	60.55	44.64
	Has-32 + Ours	74.25	67.03	50.64	67.86	62.36	45.36
	ADL	73.37	66.81	51.29	**74.38**	60.84	47.72
	ADL + Ours	73.65	**69.95**	**53.04**	74.25	**64.44**	**50.56**

Table 2. Performance of the proposed methods applied to Vanilla CAM (Baseline) with various backbone structures.

Backbone	Method	CUB-200–2011			ImageNet-1K		
		Top-1 Cls	GT Loc	Top-1 Loc	Top-1 Cls	GT Loc	Top-1 Loc
VGG16	Baseline	69.95	53.68	37.05	64.56	59.81	41.62
	+ Ours	**74.91**	**80.72**	**61.30**	**67.28**	**61.69**	**44.69**
ResNet50-SE	Baseline	**78.62**	56.49	43.29	77.22	58.21	46.64
	+ Ours	77.42	**74.51**	**58.39**	**77.25**	**64.40**	**51.96**
MobileNetV1	Baseline	72.09	58.92	44.46	67.34	59.45	43.29
	+ Ours	**75.82**	**74.28**	**57.63**	**68.07**	**61.85**	**45.55**
GoogleNet	Baseline	74.35	61.67	46.86	70.50	62.32	46.98
	+ Ours	**75.04**	**65.10**	**51.05**	**71.09**	**62.76**	**47.70**

ImageNet: 41.91 → 45.29). The weight clamping method as well as 90-th percentile standard also constantly improve the performance of localization regardless of datasets (CUB: 37.05 → 44.15, 48.45, ImageNet: 41.91 → 42.39, 44.04). With using all the solutions, the localization accuracies are maximized on both datasets.

Table 3. Performance variations of Vanilla CAM [40] with VGG16 according to different usage of our solutions. TAP, NWC and PaS refer to thresholded average pooling, negative weight clamping and percentile as a standard for thresholding.

Method	TAP	NWC	PaS	CUB-200–2011			ImageNet-1K		
				Top-1 Cls	GT Loc	Top-1 Loc	Top-1 Cls	GT Loc	Top-1 Loc
Baseline				69.95	53.68	37.05	65.39	59.65	41.91
+ Ours	✓			**74.91**	64.10	48.53	**67.22**	62.38	45.29
		✓		69.95	64.30	44.15	65.39	60.44	42.39
			✓	69.95	65.90	48.45	65.39	62.08	44.04
	✓	✓		**74.91**	73.58	54.41	**67.22**	62.48	45.24
	✓		✓	**74.91**	72.87	56.64	**67.22**	61.85	45.01
		✓	✓	69.95	76.42	54.30	65.39	**62.77**	44.40
	✓	✓	✓	**74.91**	**80.72**	**61.30**	**67.22**	62.68	**45.40**

Results on OpenImages30K. A drawback of *GT Loc* and *Top-1 Cls* is that they are sensitive to a localization threshold θ_{loc}. To validate that the robustness of our methods is not originated from a choice of the localization threshold, we compare the performance of our proposed solution applied to Vanilla CAM [40] and ADL [6] to other state-of-the-art algorithms on OpenImages30K using PxAP [4], which is independent to a threshold. Table 4 shows that CAM + Ours out-perform all the other methods of which performance is cited from [4]. Also, our proposed methods significantly improve ADL performance.

Table 4. Performance on OpenImages30K.

Method	VGG16	GoogleNet	ResNet50
HaS	56.9	58.5	58.2
ACoL	54.7	63.0	57.8
SPG	55.9	62.4	57.7
CutMix [37]	58.2	61.7	58.7
CAM	58.1	61.4	58.0
CAM + Ours	**59.6**	**63.3**	**60.9**
Δ	(+1.5)	(+1.9)	(+2.9)
ADL	58.3	62.1	54.3
ADL + Ours	59.3	**63.3**	55.7
Δ	(+1.0)	(+1.2)	(+1.4)

Discussion on Datasets. Interestingly, the improvement of localization performance by our methods is much higher on CUB-200–2011 than on ImageNet-1K and OpenImages30K. We conjecture the reasons are two-fold. First, our method works better on harder classification tasks such as CUB-200–2011 where more sophisticate part distinction is required. In other words, the discriminative regions of CUB-200–2011 are relatively smaller than those of the other datasets as many images in CUB-200–2011 share the common features such as *feathers* and *wings*. Since our proposed method focuses on expanding the localization to less-discriminative regions, it works better on such fine-grained classification problem. Second, negative weight clamping is more effective on single-object images such as CUB-200–2011. Contrary to the assumption of WSOL, ImageNet-1K and OpenImages30K contain multiple objects per image despite its single class labels. With an image of multiple objects, the features with negative weights tend to be activated in object regions of different classes. We elaborate it in the appendix more in detail.

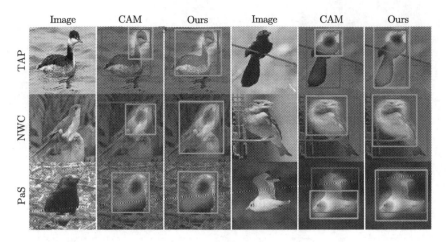

Fig. 5. Comparison of CAM and each of the proposed method. The boxes in red and green represent the ground truths and predictions of localization, respectively. (Color figure online)

4.3 Qualitative Results

Figure 5 provides some localization results that show the effectiveness of each of the proposed methods. All the methods contribute to expand localization from discriminative regions to the whole object regions through (i) TAP: balancing over/underestimated weights of features, (ii) NWC: securing depreciated activations due to negative weights, and (iii) PaS: lowering too high threshold due to the maximum standard. Note that PaS is robustly applicable whether the overlap of high activations is too severe (right) or not (left).

Figure 6 further provides localization results for the proposed methods on Vanilla CAM and ADL with VGG16 and ResNet50-SE. In general, the proposed methods help each model to utilize more activations in object regions, which results in the expansion of bounding boxes compared to the ones from CAM and ADL. We provide additional qualitative results on the other combination of backbones and modules in the appendix.

5   Conclusion

Class activation mapping (CAM), the foundation of WSOL algorithms, has three major problems which cause localization to be limited to small discriminative regions. Instead of devising a new architecture as done in most previous studies, we proposed three simple but robust methods to properly and efficiently utilize the information that already resides in feature maps. We validated the proposed method largely mitigated the problems, and as a result, achieved the new state-of-the-art performance on CUB-200–2011, ImageNet-1K, and OpenImages30K.

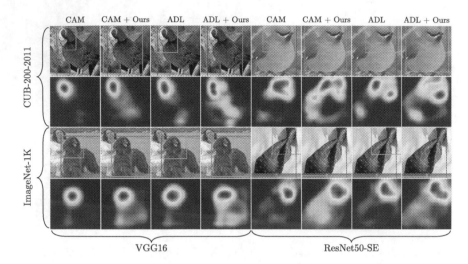

Fig. 6. Localization results of various models with and without our approach applied on CUB-200–2011 and ImageNet-1K datasets. The boxes in red and green represent the ground truths and predictions of localization, respectively. (Color figure online)

As a future work, we will investigate a more integrated algorithm to handle the aforementioned problems of the CAM method. Furthermore, instead of only using the information for a class as done in the current CAM method, using other external information such as weights of other classes may help to better localize an object by utilizing the relationship between different classes.

Acknowledgements. We appreciate Hyunwoo Kim and Jinhwan Seo for their valuable comments. This work was supported by AIR Lab (AI Research Lab) in Hyundai Motor Company through HMC-SNU AI Consortium Fund, and the ICT R&D program of MSIT/IITP (No. 2019-0-01309, Development of AI technology for guidance of a mobile robot to its goal with uncertain maps in indoor/outdoor environments and No.2019-0-01082, SW StarLab).

References

1. Bazzani, L., Bergamo, A., Anguelov, D., Torresani, L.: Self-Taught Object Localization With Deep Networks. In: WACV (2016)
2. Benenson, R., Popov, S., Ferrari, V.: Large-scale Interactive Object Segmentation with Human Annotators. In: CVPR (2019)
3. Chattopadhay, A., Sarkar, A., Howlader, P., Balasubramanian, V.N.: Grad-Cam++: Improved Visual Explanations for Deep Convolutional Networks. In: WACV (2018)
4. Choe, J., Oh, S.J., Lee, S., Chun, S., Akata, Z., Shim, H.: Evaluating Weakly Supervised Object Localization Methods Right. In: CVPR (2020)
5. Choe, J., Park, J.H., Shim, H.: Improved Techniques for Weakly-Supervised Object Localization. Arxiv:1802.07888 (2018)

6. Choe, J., Shim, H.: Attention-Based Dropout Layer for Weakly Supervised Object Localization. In: CVPR (2019)
7. Dai, J., Li, Y., He, K., Sun, J.: R-FCN: Object Detection via Region-Based Fully Convolutional Networks. In: NeurIPS (2016)
8. Durand, T., Mordan, T., Thome, N., Cord, M.: WILDCAT: Weakly Supervised Learning of Deep ConvNets for Image Classification, Pointwise Localization and Segmentation. In: CVPR (2017)
9. Durand, T., Thome, N., Cord, M.: WELDON: Weakly Supervised Learning of Deep Convolutional Neural Networks. In: CVPR (2016)
10. Everingham, M., Eslami, S.M.A., Van Gool, L., Williams, C.K.I., Winn, J., Zisserman, A.: The Pascal Visual Object Classes Challenge: A Retrospective. IJCV (2015)
11. He, K., Gkioxari, G., Dollár, P., Girshick, R.: Mask R-CNN. In: ICCV (2017)
12. He, K., Zhang, X., Ren, S., Sun, J.: Deep Residual Learning for Image Recognition. In: CVPR (2016)
13. Hou, Q., Jiang, P., Wei, Y., Cheng, M.M.: Self-Erasing Network for Integral Object Attention. In: NeurIPS (2018)
14. Howard, A.G., Zhu, M., Chen, B., Kalenichenko, D., Wang, W., Weyand, T., Andreetto, M., Adam, H.: MobileNets: Efficient Convolutional Neural Networks for Mobile Vision Applications. Arxiv:1704.04861 (2017)
15. Hu, J., Shen, L., Sun, G.: Squeeze-and-Excitation Networks. In: CVPR (2018)
16. Ioffe, S., Szegedy, C.: Batch Normalization: Accelerating Deep Network Training by Reducing Internal Covariate Shift. In: ICML (2015)
17. Kim, D., Cho, D., Yoo, D., Kweon, I.: Two-Phase Learning for Weakly Supervised Object Localization. In: ICCV (2017)
18. Li, K., Wu, Z., Peng, K.C., Ernst, J., Fu, Y.: Tell Me Where to Look: Guided Attention Inference Network. In: CVPR (2018)
19. Lin, M., Chen, Q., Yan, S.: Network in Network. In: ICLR (2014)
20. Lin, T.Y., Goyal, P., Girshick, R., He, K., Dollár, P.: Focal Loss for Dense Object Detection. In: ICCV (2017)
21. Lin, T.Y., Maire, M., Belongie, S., Hays, J., Perona, P., Ramanan, D., Dollár, P., Zitnick, C.L.: Microsoft COCO: Common Objects in Context. In: ECCV (2014)
22. Liu, W., Anguelov, D., Erhan, D., Szegedy, C., Reed, S., Fu, C.Y., Berg, A.C.: SSD: Single Shot Multibox Detector. In: ECCV (2016)
23. Oquab, M., Bottou, L., Laptev, I., Sivic, J.: Is Object Localization for Free? - Weakly-Supervised Learning With Convolutional Neural Networks. In: CVPR (2015)
24. Pinheiro, P.O., Collobert, R.: From Image-Level to Pixel-Level Labeling With Convolutional Networks. In: CVPR (2015)
25. Redmon, J., Divvala, S., Girshick, R., Farhadi, A.: You Only Look Once: Unified, Real-Time Object Detection. In: CVPR (2016)
26. Redmon, J., Farhadi, A.: YOLO9000: Better, Faster. CVPR, Stronger. In (2017)
27. Ren, S., He, K., Girshick, R., Sun, J.: Faster R-CNN: Towards Real-Time Object Detection with Region Proposal Networks. In: NeurIPS (2015)
28. Russakovsky, O., Deng, J., SU, H., Krause, J., Satheesh, S., Ma, S., Huang, Z., Karpathy, A., Khosla, A., Bernstein, M., Berg, A.C., Fei-Fei, L.: Imagenet Large Scale Visual Recognition Challenge. IJCV (2015)
29. Selvaraju, R.R., Cogswell, M., Das, A., Vedantam, R., Parikh, D., Batra, D.: Grad-Cam: Visual Explanations From Deep Networks via Gradient-Based Localization. In: ICCV (2017)

30. Simonyan, K., Zisserman, A.: Very Deep Convolutional Networks for Large-Scale Image Recognition. Arxiv:1409.1556 (2014)
31. Singh, K.K., Lee, Y.J.: Hide-And-Seek: Forcing a Network to Be Meticulous for Weakly-Supervised Object and Action Localization. In: ICCV (2017)
32. Szegedy, C., Liu, W., Jia, Y., Sermanet, P., Reed, S., Anguelov, D., Erhan, D., Vanhoucke, V., Rabinovich, A.: Going Deeper with Convolutions. In: CVPR (2015)
33. Wah, C., Branson, S., Welinder, P., Perona, P., Belongie, S.: The Caltech-UCSD Birds-200-2011 Dataset. Tech. Rep. Cns-Tr-2011-001, California Institute of Technology (2011)
34. Wei, Y., Feng, J., Liang, X., Cheng, M.M., Zhao, Y., Yan, S.: Object Region Mining With Adversarial Erasing: A Simple Classification to Semantic Segmentation Approach. In: CVPR (2017)
35. Wei, Y., Shen, Z., Cheng, B., Shi, H., Xiong, J., Feng, J., Huang, T.: TS2C: Tight Box Mining with Surrounding Segmentation Context for Weakly Supervised Object Detection. In: ECCV (2018)
36. Xue, H., Wan, F., Jiao, J., Ji, X., Qixiang, Y.: DANet: Divergent Activation for Weakly supervised Object Localization. In: ICCV (2019)
37. Yun, S., Han, D., Oh, S.J., Chun, S., Choe, J., Yoo, Y.: CutMix: Regularization Strategy to Train Strong Classifiers with Localizable Features. In: ICCV (2019)
38. Zhang, X., Wei, Y., Feng, J., Yang, Y., Huang, T.S.: Adversarial Complementary Learning for Weakly Supervised Object Localization. In: CVPR (2018)
39. Zhang, X., Wei, Y., Kang, G., Yang, Y., Huang, T.: Self-Produced Guidance for Weakly-Supervised Object Localization. In: ECCV (2018)
40. Zhou, B., Khosla, A., Lapedriza, A., Oliva, A., Torralba, A.: Learning Deep Features for Discriminative Localization. In: CVPR (2016)

OS2D: One-Stage One-Shot Object Detection by Matching Anchor Features

Anton Osokin[1,2(✉)] [iD], Denis Sumin[3], and Vasily Lomakin[3]

[1] National Research University Higher School of Economics, Moscow, Russia
aosokin@hse.ru
[2] Yandex, Moscow, Russia
[3] mirum.io, Moscow, Russia

Abstract. In this paper, we consider the task of one-shot object detection, which consists in detecting objects defined by a single demonstration. Differently from the standard object detection, the classes of objects used for training and testing do not overlap. We build the one-stage system that performs localization and recognition jointly. We use dense correlation matching of learned local features to find correspondences, a feed-forward geometric transformation model to align features and bilinear resampling of the correlation tensor to compute the detection score of the aligned features. All the components are differentiable, which allows end-to-end training. Experimental evaluation on several challenging domains (retail products, 3D objects, buildings and logos) shows that our method can detect unseen classes (e.g., toothpaste when trained on groceries) and outperforms several baselines by a significant margin. Our code is available online: https://github.com/aosokin/os2d.

Keywords: One-shot detection · Object detection · Few-shot learning

1 Introduction

The problem of detecting and classifying objects in images is often a necessary component of automatic image analysis. Currently, the state-of-the-art approach to this task consists in training convolutional neural networks (CNNs) on large annotated datasets. Collecting and annotating such datasets is often a major cost of deploying such systems and is the bottleneck when the list of classes of interest is large or changes over time. For example, in the domain of retail products on supermarket shelves, the assortment of available products and their appearance is gradually changing (e.g., 10% of products can change each month), which makes it hard to collect and maintain a dataset. Relaxing the requirement of a large annotated training set will make the technology easier to apply.

This work was done when Anton Osokin was with the Samsung-HSE lab.

Electronic supplementary material The online version of this chapter (https://doi.org/10.1007/978-3-030-58555-6_38) contains supplementary material, which is available to authorized users.

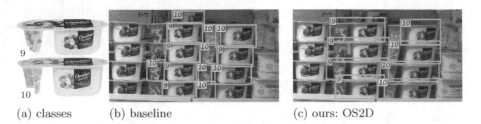

(a) classes   (b) baseline   (c) ours: OS2D

Fig. 1. Qualitative comparison of our OS2D model with the baseline (state-of-the-art object detection and image retrieval systems) when detecting unseen object classes. In this example, the goal is to detect objects that consist of several parts (a), which were not available in the training set. The baseline system has to finalize the object bounding box before knowing what class it is detecting (b), so as the object detector fails to merge parts, the retrieval system cannot fix the boxes and recognize correctly (c). Our model knows that the target objects consist of two parts, so it detects correctly.

In this paper, we consider the task of detecting objects defined by a single demonstration (one-shot object detection). In this formulation, a system at the detection (testing) stage has to "understand" what it needs to detect by a single exemplar of a target class. After the demonstration, the system should detect objects of the demonstrated target class on new images. Differently from the standard object detection systems, the classes of objects used for training and testing do not overlap. The task of one-shot detection is a step towards decreasing the complexity of collecting and maintaining data for object detection systems.

The computer vision community developed a few very successful methods for regular object detection, and currently, there are well-maintained implementations in all major deep learning libraries. One-shot detection is studied substantially less with one notable exception: human face detection combined with personality recognition. A significant simplification of this setting is that a human face is a well-defined object, and one can build a detector that works well on unseen faces. Defining a general object is harder [1], so such detector suffers from insufficient generalization. Several recent works tackled one-shot detection for general object classes (ImageNet [39] and COCO [20]). Usually, methods have two distinct stages: a detector of all objects (a region proposal network) and an embedding block for recognition. Works [17,42,44,47] linked the two stages by joint finetuning of network weights but, at the test stage, their models relied on class-agnostic object proposals (no dependence on the target class image). Differently, works [6,12] incorporated information about the target class into object proposals.

Contributions. First, we build a one-stage one-shot detector, OS2D, by bringing together the dense grid of anchor locations of the Faster R-CNN [34] and SSD [21] object detectors, the transformation model of Rocco et al. [35,37] designed for semantic alignment and the differentiable bilinear interpolation [15]. Our approach resembles the classical pipelines based on descriptor matching and subsequent geometric verification [22,23], but utilizes dense instead of sparse matches, learned instead of hand-crafted local features and a feed-forward geo-metric transformation model instead of RANSAC. The key feature of our model is that detection and recognition are performed jointly without the need to define a general object (instead, we need a good feature descriptor and transformation model). Figure 1 shows how detection and recognition can work jointly compared to the baseline consisting of separate detection and retrieval systems (see Sect. 6.2 for the baseline details).

Second, we design a training objective for our model that combines the ranked list loss [43] for deep metric learning and the standard robust L_1 loss for bounding box regression [8]. Our objective also includes remapping of the recognition targets based on the output of the forward pass, which allows us to use a relatively small number of anchors.

Finally, we apply our model to several domains (retail products based on the GroZi-3.2k dataset [7], everyday 3D objects, buildings and logos based on the INSTRE dataset [41]). For the retails products, we have created a new consistent annotation of all objects in GroZi-3.2k and collect extra test sets with new classes (e.g., toothpaste, when trained on groceries). Our method outperformed several baselines by a significant margin in all settings. Code of our method and the baselines together with all collected data is available online.

We proceed by reviewing the background works of Rocco et al. [35–37] in Sect. 2. We present our model and its training procedure in Sects. 3 and 4, respectively. Next, in Sect. 5, we review the related works. Finally, Scct. 6 contains the experimental evaluation, Sect. 7 – the conclusion.

2   Preliminaries: Matching Networks

We build on the works of Rocco et al. [35–37] that targeted the problem of semantic alignment where the goal was to align the source and target images. Their methods operate on dense feature maps $\{f_{kl}\}$, $\{g_{kl}\}$ of the spatial size $h^T \times w^T$ and feature dimensionality d. The feature maps are extracted from both images with a ConvNet, e.g., VGG or ResNet. To match the features, they compute ·a 4D tensor $c \in \mathbb{R}^{h^T \times w^T \times h^T \times w^T}$ containing correlations $c[k, l, p, q] = \frac{\langle f_{kl}, g_{pq} \rangle}{\|f_{kl}\|_2 \|g_{pq}\|_2}$ between all pairs of feature vectors from the two feature maps.

Next, they reshape the tensor c into a 3D tensor $\tilde{c} \in \mathbb{R}^{h^T \times w^T \times (h^T w^T)}$ and feed it into a ConvNet (with regular 2D convolutions) that outputs the parameters of the transformation aiming to map coordinates from the target image to the source image. Importantly, the kernel of the first convolution of such ConvNet has $h^T w^T$ input channels, and the overall network is designed to reduce the spatial size $h^T \times w^T$ to 1×1, thus providing a single vector of transformation parameters. This ConvNet will be the central element of our model, and hereinafter we will refer to it as TransformNet. Importantly, TransformNet has 3 conv2d layers with ReLU and BatchNorm [13] in-between and has receptive field of 15×15, i.e., $h^T = w^T = 15$ (corresponds to the image size 240×240 if using the features after the fourth block of ResNet). The full architecture is given in Appendix B.1.

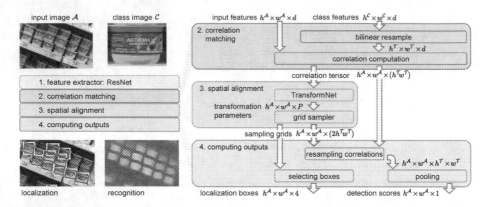

Fig. 2. (left) Inputs, outputs and main components of the OS2D model. Bottom left – the detection boxes (yellow) that correspond to the peaks in the score map, red parallelograms illustrate the corresponding affine transformations produced by TransformNet; and the detection score map (the lighter – the higher the score). **(right)** Details of the OS2D model. Boxes represent network layers, arrows – data flow (for convenience, we show the sizes of intermediate tensors). Best viewed in color.

TransformNet was first trained with full supervision on the level of points to be matched [35,36] (on real images, but synthetic transformations). Later, Rocco et al. [37, Section 3.2] proposed a *soft-inlier count* to finetune TransformNet without annotated point matches, i.e., with weak supervision. The soft-inlier count is a differentiable function that evaluates how well the current transformation aligns the two images. It operates by multiplying the tensor of correlations c by the soft inlier mask, which is obtained by warping the identity mask with a spatial transformer layer [15] consisting of grid generation and bilinear resampling.

3   The OS2D Model

We now describe our OS2D model for one-shot detection. The main idea behind OS2D is to apply TransformNet to a large feature map in a fully-convolutional way and to modify the soft-inlier count to work as the detection score. We will explain how to modify all the components of the model for our setting.

The OS2D model consists of the following steps (see Fig. 2): (1) extracting local features from both input and class images; (2) correlation matching of features; (3) spatially aligning features according to successful matches; (4) computing the localization bounding boxes and recognition score. We now describe these steps highlighting the technical differences to [35–37].

Feature Extraction. On an input image $\mathcal{A}$ and a class image $\mathcal{C}$, a feature extractor (same as in works [35–38]) computes dense feature maps $\{f_{kl}^{\mathcal{A}}\} \in \mathbb{R}^{h^{\mathcal{A}} \times w^{\mathcal{A}} \times d}$ and $\{f_{pq}^{\mathcal{C}}\} \in \mathbb{R}^{h^{\mathcal{C}} \times w^{\mathcal{C}} \times d}$, respectively. The spatial sizes of the two feature maps differ and depend linearly on the sizes of the respective images.

The architecture of TransformNet requires at least one of the extracted feature maps to be of the fixed size $h^T \times w^T$ (this size defines the number of the input channels of the first TransformNet convolution). As the input image $\mathcal{A}$ can be of a large resolution and thus need a large feature map, we chose to convert the class feature map $\{f_{pq}^{\mathcal{C}}\}$ to the fixed size to get $\{f_{pq}^{\mathcal{T}}\}$. Our experiments showed that resizing class images (as in [35–37]) significantly distorts the aspect ratio of some classes, e.g., a bottle becomes a can, which degrades the quality of feature matching. We found that it worked better to resize extracted class feature maps directly by applying the differentiable bilinear resampling [15].

We also found that it is very important to extract features from the input and class images in the Siamese way, i.e., with the networks with identical parameters, which ruled out more efficient (than an image pyramid) ways of computing multi-scale features maps, e.g., FPN [18]. We tried to untie the two branches and use FPN to extract features from the input image: such system worked well on the classes seen during training, but did not generalize to new classes.

Feature Matching and Alignment. Following Rocco et al. [35] we compute a 4D correlation tensor $c \in \mathbb{R}^{h^{\mathcal{A}} \times w^{\mathcal{A}} \times h^T \times w^T}$, $c[k, l, p, q] = \frac{\langle f_{kl}^{\mathcal{A}}, f_{pq}^{\mathcal{T}} \rangle}{\|f_{kl}^{\mathcal{A}}\|_2 \|f_{pq}^{\mathcal{T}}\|_2}$, between the input and resized class feature maps. After reshaping the tensor c to the 3D tensor $\tilde{c} \in \mathbb{R}^{h^{\mathcal{A}} \times w^{\mathcal{A}} \times (h^T w^T)}$, we apply TransformNet in a fully-convolutional way to obtain a $h^{\mathcal{A}} \times w^{\mathcal{A}} \times P$ tensor with parameters of transformations at each location of the input feature map. All transformations are defined w.r.t. local coordinates (different for all locations), which we then convert to the global w.r.t. the input feature map coordinates (we denote the transformations by G_{kl}).

The direction of the output transformations G_{kl} is defined by training the model. The TransformNet weights released by Rocco et al. [37] align the input image to the class image (according to their definition of the source and target), but we need the inverse of this to map everything to the coordinates w.r.t. the input image. Note that we cannot simply swap the two images because their feature maps are of different sizes. To use the released weights, we need to invert the direction, which, in the case of affine transformations, requires a batch inversion of 3×3 matrices. An alternative is to retrain TransformNet from scratch and to interpret its output transformations in the direction we need.

Finally, we feed all the transformations G_{kl} into a grid sampler to produce a grid of points aligning the class image at each location of the input image (we use the grids of the same size $h^T \times w^T$ as the TransformNet input). The output of the grid sampler is a tensor $g \in \mathbb{R}^{h^{\mathcal{A}} \times w^{\mathcal{A}} \times h^T \times w^T \times 2}$, $(g[k, l, p, q, 0], g[k, l, p, q, 1]) := G_{kl}(p, q)$ with coordinates w.r.t. the input feature map.

Recognition Scores. The next step is to use the grids of matching points to extract scores $s \in \mathbb{R}^{h^{\mathcal{A}} \times w^{\mathcal{A}} \times 1}$ indicating how likely the location has a detection. The soft-inlier count [37] is a natural candidate for this score, however computing it requires enormous amount of the device memory. Specifically, one needs to create a mask of the size $h^{\mathcal{A}} \times w^{\mathcal{A}} \times h^T \times w^T \times h^T \times w^T$, which is $h^T w^T = 225$ times larger than any tensor we have created so far. To circumvent the problem, we use a related but different quantity, which is more efficient to compute in a fully-convolutional way. We directly resample the correlation tensor c w.r.t. the computed grids, i.e., $\hat{s}[k,l,p,q] := c[g[k,l,p,q,0],g[k,l,p,q,1],p,q]$, $\hat{s} \in \mathbb{R}^{h^A \times w^A \times h^T \times w^T}$. The values $c[y,x,p,q]$ at non-integer points (y,x) are computed by differentiable bilinear resampling [15] on the 2D array $c[:,:,p,q]$. Note that this operation is not directly supported by the standard bilinear interpolation (different channels need to be resampled at different points). One can either loop over all the channels and resample them sequentially (can be slow) or create a specialized layer. The last step to get s is to pool $\hat{s}$ w.r.t. its two last dimensions. We use the average pooling and omit the boundary of the grids to reduce effect of background matches.

Localization Boxes. We extract the localization of the detections by taking max and min of grid tensor g w.r.t. its 3rd and 4th dimensions, i.e., output the tight bounding boxes around the transformed grid points.

4   Training the Model

Training Batches and Data Augmentation. In our datasets, input images are of high resolution and contain many objects (see Fig. 3 for examples). We can't downsample them to a small fixed size (as typical in object detection) because of the strong distortion of the aspect ratio, and each object might simply get too few pixels. Instead, when constructing a batch we randomly choose a scale and location at which the current image will be processed and resample it to provide a training image of the target size (random crop/scale data augmentation). For each batch, we collect a set of class images (we cannot use all classes, because there are too many) by taking the annotated classes of the batch and adding some random classes as negatives to fill the class batch size.

Objective Function. As in regular object detection we build the training objective from recognition and localization losses: the hinge-embedding loss with margins for recognition and the smoothed L_1 loss for localization:

$$\ell_{\text{rec}}^{\text{pos}}(s) = \max(m_{\text{pos}} - s, 0), \quad \ell_{\text{rec}}^{\text{neg}}(s) = \max(s - m_{\text{neg}}, 0), \tag{1}$$

$$\ell_{\text{loc}}(\boldsymbol{x}, \boldsymbol{y}) = \sum_{c=1}^{4} \begin{cases} \frac{1}{2}(x_c - y_c)^2, & \text{if } |x_c - y_c| < 1, \\ |x_c - y_c| - \frac{1}{2}, & \text{otherwise.} \end{cases} \tag{2}$$

Here $s \in [-1,1]$ is the recognition score (trained to be high for positives and low for negatives), m_{neg} and m_{pos} are the negative and positive margins, respectively, $\boldsymbol{x}, \boldsymbol{y} \in \mathbb{R}^4$ are the output and target encodings of bounding boxes (we use the standard encoding described, e.g., in [34, Eq. 2]).

As in detection and retrieval, our task has an inherent difficulty of non-balanced number of positives and negatives, so we need to balance summands in the objective. We started with the Faster R-CNN approach [34]: find a fixed number of hardest negatives per each positive within a batch (positive to negative ratio of 1:3). The localization loss is computed only for the positive objects. All the losses are normalized by the number of positives n_{pos} in the current batch. For recognition, we start with the contrastive loss from retrieval [26, 31, 40]: squared ℓ_{rec} and the positive margin m_{pos} set to 1 (never active as $s \in [-1, 1]$).

$$\mathcal{L}_{\mathrm{rec}}^{\mathrm{CL}} = \frac{1}{n_{\mathrm{pos}}} \sum_{i:t_i=1} \ell_{\mathrm{rec}}^{\mathrm{pos}}(s_i)^2 + \frac{1}{n_{\mathrm{pos}}} \sum_{i:t_i=0} \ell_{\mathrm{rec}}^{\mathrm{neg}}(s_i)^2, \tag{3}$$

$$\mathcal{L}_{\mathrm{loc}} = \frac{1}{n_{\mathrm{pos}}} \sum_{i:t_i=1} \ell_{\mathrm{loc}}(\boldsymbol{x}_i, \boldsymbol{y}_i). \tag{4}$$

Here, the index i loops over all anchor positions, $\{s_i, \boldsymbol{x}_i\}_i$ come from the network and $\{t_i, \boldsymbol{y}_i\}_i$ come from the annotation and assigned targets (see below).

We also tried the recently proposed ranked list loss (RLL) of Wang et al. [43], which builds on the ideas of Wu et al. [45] to better weight negatives. We have

$$\mathcal{L}_{\mathrm{rec}}^{\mathrm{RLL}} = \sum_{i:t_i=1} \frac{1}{\tilde{n}_{\mathrm{pos}}} \ell_{\mathrm{rec}}^{\mathrm{pos}}(s_i) + \sum_{i:t_i=0} w_i^{\mathrm{neg}} \ell_{\mathrm{rec}}^{\mathrm{neg}}(s_i), \tag{5}$$

$$w_i^{\mathrm{neg}} \propto \exp(T\ell_{\mathrm{rec}}(s_i, 0))[\ell_{\mathrm{rec}}^{\mathrm{neg}}(s_i) > 0]. \tag{6}$$

Here $\tilde{n}_{\mathrm{pos}}$ is the number of active positives, i.e., positives such that $\ell_{\mathrm{rec}}^{\mathrm{pos}}(s_i) > 0$. The weights w_i^{neg} are normalized in such a way that they sum to 1 over all the negatives for each image-class pair. The constant T controls how peaky the weights are. Wang et al. [43] fixed T in advance, but we found it hard to select this parameter. Instead, we chose it adaptively for each image-class pair in such a way that the weight for the negative with the highest loss is 10^3 times larger that the weights of the negatives at the margin boundary, i.e., $s_i = m_{\mathrm{neg}}$. Finally, we did not back-propagate gradients through the weights w_i^{neg} keeping those analogous to probabilities used for sampling negatives.

Target Assignment. For each position of the feature map extracted from the input image $\mathcal{A}$, we assign an anchor location w.r.t. which we decode from the output of the transformation net. At each location, as the anchor we use the rectangle corresponding to the receptive field of the transformation net. The next step is to assign targets (positive or negative) to all the anchor-class pairs and feed them into the loss functions as targets. First, we tried the standard object detection strategy, i.e., assign positive targets to the anchors with intersection over union (IoU) with a ground-truth object above 0.5 and negatives – to all anchors with IoU < 0.1. Note that we cannot force each ground-truth object to have at least one positive anchor, because we process an image in a training batch at only one scale, which might differ from the scale of the annotated objects. With the latter constraint, our set of positive anchors is sparser than typical in object detection. We tried to lower the IoU threshold for positives, but it never helped. To overcome this problem, we propose to use the standard IoU threshold only to determine localization targets (bounding boxes), and decide for classification targets *after* the output localization is computed. There, we set the high IoU thresholds of 0.8 for positives and 0.4 for negatives. We refer to this technique as *target remapping*. Target remapping is suitable for our model because we compute recognition scores at locations different from the anchors (due to the transformation model), which is not the case in regular object detection models.

5   Related Works

Object Detection. The standard formulation of object detection assumes a fixed list of target classes (usually 20–80) and an annotated dataset of images (preferably of a large size), where all the objects of the target classes are annotated with bounding boxes. Modern neural-network detectors are of two types: two-stage and one-stage. The two-stage detectors follow Faster R-CNN [8,34] and consist of two stages: region proposal (RPN) and detection networks. RPN generates bounding boxes that might contain objects, and the detection network classifies and refines them; a special pooling operation connects the two nets. In this approach, the RPN is, in fact, a detector of one joint *object* class. The one-stage methods like YOLO [32,33], SSD [21], RetinaNet [19] use only RPN-like networks and are often faster but less accurate than two-stage methods. Our OS2D architecture is a one-stage method and never computes the detections of a general object, which potentially implies better generalization to unseen classes.

One-shot Object Detection. Several recent works [4,6,12,16,17,25,29,42,44, 47] tackled the problems of one-shot and few-shot detection, and [6,12,17] are most relevant for us. Karlinsky et al. [17] built a two-stage model where RPN did not depend on the target class and recognition was done using the global representation of the target class image. Hsieh et al. [12] and Fan et al. [6] used attention mechanisms to influence the RPN object proposals (thus having non class-agnostic proposals) and later build context dependent global representations. Our model differs from the works [6,12,17] in the two key aspects: (1) we do not build one global representation of the image of the target classes, but work with lower-level features and match feature map to feature map instead of vector to vector; (2) our model is one-stage and never outputs non-class specific bounding boxes, which helps to generalize to unseen classes.

Image Retrieval. Image retrieval aims to search a large dataset of images for the ones most relevant to a query image. A specific example of such a formulation is to search for photos of a building given one photo of the same building. Differently from one-shot detection, the output is a list of images without object bounding boxes. Most modern methods for image retrieval model relevance via distances between global image representations obtained by pooling neural network features. These representations can rely on pre-trained networks [3] or, as common recently, be learned directly for image retrieval [2,10,31,43]. Differently, the work of Noh et al. [27] matches local image features similarly to OS2D. However, differently from OS2D, they do not match densely extracted features directly but subsample them with an attention mechanism, train the network not end-to-end (several steps), and the spatial verification of matches (RANSAC) is not a part of their network but a post-processing step.

Semantic Alignment. The works of Rocco et al. [35–37] studied the problem of semantic alignment, where the goal is to compute correspondence between two images with objects of the same class, and the primary output of their methods (what they evaluate) is the sparse set of matching keypoints. In one-shot detection, we are interested in the score representing similarity and the bounding box around the object, which is related, but different. Our model reuses the TransformNet architecture of [35–37] (network that takes dense correlation maps as input and outputs the transformation parameters). Differently from these works, we apply this network in a fully-convolutional way, which allows us to process input images of arbitrarily large resolution. As a consequence, we need to resample correlation maps in a fully-convolutional way, which makes the existing network head inapplicable. Another line of comparison lies in the way of training the models. Our approach is similar to [37], where the transformation is trained under weak supervision, whereas Rocco et al. [35,36] used strong supervision coming from synthetically generated transformations.

6   Experiments

Datasets and Evaluation. Most of prior works [4,12,16,17,25,44,47] constructed one-shot detection dataset by creating episodes from the standard object detection datasets (PASCAL [5] or COCO [20]) where the class image was randomly selected from one of the dataset images. However, we believe that this approach poses a task that requires either strong assumptions about what the target classes are (inevitably limit generalization to new classes, which is our main goal) or richer definitions of the target class (e.g., more than one shot). This happens because in the standard object detection the classes are very broad, e.g., contain both objects and their parts annotated as one class, or the objects look very different because of large viewpoint variation. A large number of training images allows to define such classes, which is not possible in the one-shot setting (e.g., detect a full dog given only a head as the class image or detect the front of a car given its side view). In other words, the difference of the previous settings to ours is similar to the difference of instance-level to semantic retrieval [11].

Having these points in mind, we chose to evaluate in domains where the one-shot setting is more natural. Specifically, we used the GroZi-3.2k dataset [7] for retail product detection as the development data. We create consistent bounding box annotations to train a detection system and collected extra test sets of retail products to evaluate on unseen classes. Additionally, we evaluate our methods on the INSTRE dataset [41] originally collected for large scale instance retrieval. The dataset is hard due to occlusions and large variations in scales and rotations [14]. The dataset has classes representing physical objects in the lab of the dataset creators and classes collected on-line (buildings, logos and common objects). We refer to the two parts of the datasets as INSTRE-S1 and INSTRE-S2, respectively. We provide the details about the dataset preparation in Appendix A.

In all experiments, we use the standard Pascal VOC metric [5], which is the mean average precision, mAP, at the intersection-over-union (IoU) threshold of 0.5. The metric uses the "difficult" flag to effectively ignore some detections.

For completeness, we report the results of our methods on the ImageNet-based setup of [17] in Appendix D. Confirming the difference of the tasks, OS2D

Table 1. Validation mAP of different OS2D training configs. The V1-train and V2-train configs selected for experiments in Table 2 are marked with "¶" and "§", respectively

	Config options	Training configs											
	TransformNet	V1			V2								
Training	Loss	CL	RLL	RLL	RLL	RLL	RLL	RLL	RLL	RLL	CL	CL	RLL
	Remap targets		+	+	+	+	+	+	+			+	
	Mine patches			+			+					+	
	Init transform							+	+	+	+	+	+
	Zero loc loss								+	+	+	+	+
	mAP	81.8	85.7	87.0¶	86.5	87.0	86.5	88.0	89.5§	**90.6**	87.4	88.1	87.1

is not competitive but our main baseline (see Sect. 6.2) slightly outperforms the method of [17].

6.1   Ablation Study

Model Architecture. The OS2D model contains two subnets with trainable parameters: TransformNet and feature extractor. For TransformNet, we use the network of the same architecture as Rocco et al. [35,37] and plug it into our model in the two ways: (V1) simplified affine transformations with $P = 4$ (only translation and scaling) applied in the direction natural to OS2D; (V2) full affine transformations with $P = 6$ applied in another direction (used in [35,37]). The V1 approach allows to have full supervision (the annotated bounding boxes define targets for the 4 parameters of the simplified transformations) and is more convenient to use (no need to invert transformations). The V2 approach as in [37] requires a form of weak supervision for training, but can be initialized from the weights released by Rocco et al. [35,37], and, surprisingly is fully-functional without any training on our data at all. Comparison of V1 vs. V2 in different settings is shown in Tables 1, 2, 3, 4 (see full descriptions below). For the features extractor, we consider the third block of ResNet-50 and ResNet-101. See Table 2 for the comparison (the description is below).

Training Configurations. We now evaluate multiple options of our OS2D training config (see Table 1 for the results). We experiment with the ResNet-50 feature extractor initialized from the weights trained on ImageNet [39] for image classification. For each run, we choose the best model by looking at the mAP on the validation set `val-new-cl` over the training iterations and report this quantity as mAP.

We start with the V1 approach and training with the standard contrastive loss for recognition (3) and smoothed L1 loss for localization (4) and gradually add more options. We observed that RLL (5) outperformed the contrastive loss

Table 2. Initializations for the OS2D feature extractor (mAP on the validation set). Symbol "†" marks the ImageNet models converted from Caffe, symbol "‡" marks the model with group norm [46] instead of batch norm

	Src task	Src data	V1-train	V2-init	V2-train
ResNet-50	—	—	59.6	2.0	67.9
	Classification	ImageNet	84.8	79.6	89.0
	Classification†	ImageNet	87.1	**86.1**	**89.5**
	Classification†‡	ImageNet	83.2	68.2	87.1
	Detection	COCO	81.9	77.5	88.1
ResNet-101	Classification	ImageNet	85.6	81.1	89.4
	Classification†	ImageNet	87.5	81.0	88.8
	Retrieval	landmarks	**88.7**	83.3	89.0
	Detection	COCO	85.6	73.1	86.8
	Alignment	PASCAL	85.7	77.2	88.7

(3), recognition target remapping (see Sect. 4) helped. In addition to the performance difference, these feature regularized the training process (the initial models were quickly starting to overfit). Finally, we implemented a computationally expensive feature of mining hard patches to feed into batches (like hard negative mining of images for image retrieval [2,31]). This step significantly slowed down training but did not improve V1 models.

Next, we switch to the V2 approach. The main benefit of V2 is to initialize TransformNet from the weights of Rocco et al. [35,37], but training runs without target remapping worsen the mAP at initialization. We also found that training signals coming to TransformNet from the localization and recognition losses were somewhat contradicting, which destabilized training. We were not able to balance the two, but found that simply zeroing out the localization loss helped (see Fig. 2 bottom-left for a visualization of the learned transformations). Finally, we re-evaluated the options, which helped V1, in the context of V2 and confirmed that they were important for successful training. Mining hard patches now improved results (see Table 1).

Finally, we selected the best config for V1 as V1-train and the best config overall as V2-train (we use V2-init to refer to its initialization). The additional implementation details are presented in Appendix B.

We also separately evaluated the effect of using the inverse and full affine transformations – they did not hurt mAP, but when visualizing the results we did not see the model learning transformations richer than translation and scaling, so we omitted the architectures in-between V1 and V2.

Initialization of the Feature Extractor. We observed that the size of the GroZi-3.2k dataset was not sufficient to train networks from scratch, so choosing the initialization for the feature extractor was important. In Table 2, we compared initializations from different pre-trained nets. The best performance was achieved when starting from networks trained on ImageNet [39] for image classification and the Google landmarks dataset [27] for image retrieval. Surprisingly, detection initializations did not work well, possibly due to the models largely ignoring color. Another surprising finding is that the V2-init models worked reasonably well even without any training for our task. The pre-trained weights of the affine transformation model [37] appeared to be compatible not only with the feature extractor trained with them, likely due to the correlation tensors that abstract away from the specific features.

Running Time. The running time of the feature extractor on the input image depends on the network size and is proportional to the input image size. The running time of the feature extractor on the class images can be shared across the whole dataset and is negligible. The running time of the network heads is in proportional to both the input image size and the number of classes to detect, thus in the case of a large number of classes dominates the process. On GTX 1080Ti in our evaluation regime (with image pyramid) of the `val-new-cl` subset, our PyTorch [28] code computed input features in 0.46 s per image and the heads took 0.052 s and 0.064 s per image per class for V1 and V2, respectively, out of which the transformation net itself took 0.020 s. At training, we chose the number of classes such that the time on the heads matched the time of feature extraction. Training on a batch of 4 patches of size 600×600 and 15 classes took 0.7 s.

6.2   Evaluation of OS2D Against Baselines

Class Detectors. We started with training regular detectors on the GroZi-3.2k dataset. We used the maskrcnn-benchmark system [24] to train the Faster R-CNN model with Resnet-50 and Resnet-101 backbones and the feature pyramid network to deal with multiple scales [18]. However, these systems can detect only the training classes, which is the `val-old-cl` subset.

Main Baseline: Object Detector + Retrieval. The natural baseline for one-shot detection consists in combining a regular detector of all objects merged into one class with the image retrieval system, which uses class objects as queries and the detections of the object detector as the database to search for relevant images. If both the detector and retrieval are perfect then this system solves the problem of one-shot detection. We trained object detectors with exactly same architectures as the class detectors above. The ResNet-50 and ResNet-101 versions (single class detection) delivered on the validation images (combined `val-old-cl` and `val-new-cl`) 96.42 and 96.36 mAP, respectively. For the retrieval, we used the software[1] of Radenović et al. [30,31]. We trained the models on the training subset of GroZi-3.2k and chose hyperparameters to maximize mAP on the `val-new-cl` subset. Specifically, GeM pooling (both on top of ResNet-50 and

---

[1] https://github.com/filipradenovic/cnnimageretrieval-pytorch.

Table 3. Comparison to baselines, mAP, on the task of retail product detection. *In this version of the main baseline, the detector is still trained on the training set of GroZi-3.2k, but the retrieval system uses the weights trained on ImageNet for classification

Method	Trained	val-old-cl	val-new-cl	dairy	paste-v	paste-f
Class detector	Yes	87.1	—	—	—	—
Main baseline	No*	72.0	69.1	46.6	34.8	31.2
Main baseline	Yes	87.6	86.8	70.0	44.3	40.0
Sliding window, square	No	57.6	58.8	33.9	8.0	7.0
Sliding window, target AR	No	72.5	71.3	63.0	65.1	45.9
Hsieh et al. [12]	Yes	**91.1**	88.2	57.2	32.6	27.6
Ours: OS2D V1-train	Yes	89.1	88.7	70.5	61.9	48.8
Ours: OS2D V2-init	No	79.7	86.1	65.4	68.2	48.4
Ours: OS2D V2-train	Yes	85.0	**90.6**	**71.8**	**73.3**	**54.5**

ResNet-101) and end-to-end trainable whitening worked best (see Appendix C for details).

Sliding Window Baselines. To evaluate the impact of our transformation model, we use the same feature extractor as in OS2D (paired with the same image pyramid to detect at multiple scales), but omit the transformation model and match the feature map of the class image directly with the feature map of the input image in the convolutional way. In the first version (denoted as "square"), we used the feature maps resized to squares identical to the input of the transformation model (equivalent to fixing the output of the transformation model to identity). In the second version (denoted as "target AR"), we did not resize the features and used them directly from the feature extractor, which kept the aspect ratio of the class image correct.

Extra Baselines. We have compared OS2D against the recently released code[2] of Hsieh et al. [12] (see Appendix C for details). We also compared against other available codes [17,25] with and without retraining the systems on our data but were not able to obtain more than 40 mAP on `val-new-cl`, which is very low, so we did not include these methods in the tables.

Comparison with Baselines. Tables 3 and 4 show quantitative comparison of our OS2D models to the baselines on the datasets of retail products and the INSTRE datasets, respectively. Figure 3 shows qualitative comparison to the main baseline (see Appendix E for more qualitative results). Note, that the datasets `paste-f`, INSTRE-S1 and INSTRE-S2 contain objects of all orientations. Features extracted by CNNs are not rotation invariant, so as is they do not match. To give matching-based methods a chance to work, we augment the class images with their 3 rotations (90, 180 and 270 °) for all the methods.

---

[2] https://github.com/timy90022/One-Shot-Object-Detection.

Table 4. Results on the INSTRE dataset, mAP

Method	ResNet-50		ResNet-101	
	INSTRE-S1	INSTRE-S2	INSTRE-S1	INSTRE-S2
Main baseline-train	72.2	64.4	79.0	53.9
Sliding window, AR	64.9	57.6	60.0	51.3
Hsieh et al. [12]	73.2	66.7	68.1	74.8
Ours: OS2D V1-train	83.9	73.8	87.1	76.0
Ours: OS2D V2-init	71.9	64.5	69.7	63.2
Ours: OS2D V2-train	**88.7**	**77.7**	88.7	**79.5**

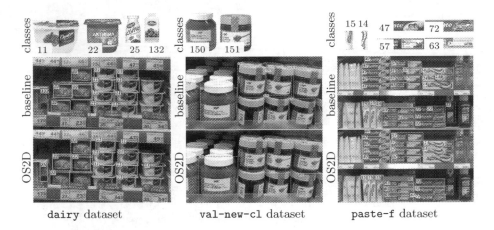

Fig. 3. Qualitative comparison of the best OS2D model vs. the best baseline. Correct detections – green boxes, incorrect – red boxes. (Color figure online)

First, note that when evaluated on the classes used for training some one-shot method outperformed the corresponding class detectors, which might be explained by having too little data to train regular detectors well enough. Second, in all the cases of classes unseen at the training time, the trained OS2D versions outperform the baselines. Qualitatively, there are at least two reasons for such significant difference w.r.t. our main baseline: the object detector of the baseline has to detect objects without knowing what class is has to detect, which implies that it is very easy to confuse object with its part (see examples in Fig. 1 and Fig. 3-right) or merge multiple objects together; the current retrieval system is not explicitly taking the geometry of matches into account, thus it is hard to distinguish different objects that consist of similar parts. Finally, note that the sliding window baseline with the correct aspect ratio performed surprisingly well and sometimes outperformed the main baseline, but the learned transformation model always brought improvements.

One particular difficult case for the matching-based OS2D models appears to be rotating in 3D objects (see wrong detection in Fig. 3-mid). The model matches the class image to only half of the object, which results in producing incorrect detection localization. Richer and maybe learnable way of producing bounding boxes from matched features is a promising direction for future work.

7 Conclusion

In this paper, we proposed the OS2D model for one-shot object detection. The model combines a deep feature extractor, correlation matching, feed-forward alignment network and bilinear interpolation in a differentiable way that allows end-to-end training. We trained our model with an objective function combining the recognition and localization losses. We applied our model to the challenging task of retail product recognition and construct a large dataset with consistent annotation for a large number of classes available (the recent SKU110k dataset [9] is of larger scale, but contains only object bounding boxes without the class labels). The OS2D model outperformed several strong baselines, which indicates the potential of the approach for practical usage.

Acknowledgments. We would like to personally thank Ignacio Rocco, Relja Arandjelović, Andrei Bursuc, Irina Saparina and Ekaterina Glazkova for amazing discussions and insightful comments, without which this project would not be possible. This research was partly supported by Samsung Research, Samsung Electronics, by the Russian Science Foundation grant 19-71-00082 and through computational resources of HPC facilities at NRU HSE.

References

1. Alexe, B., Deselaers, T., Ferrari, V.: What is an object? In: proceedings of the IEEE Conference on Computer Vision and Pattern Recognition (CVPR), pp. 73–80. IEEE (2010)
2. Arandjelović, R., Gronat, P., Torii, A., Pajdla, T., Sivic, J.: NetVLAD: CNN architecture for weakly supervised place recognition. IEEE Trans. Pattern Anal. Mach. Intell. (TPAMI) **40**(6), 1437–1451 (2018)
3. Babenko, A., Slesarev, A., Chigorin, A., Lempitsky, V.: Neural codes for image retrieval. In: Fleet, D., Pajdla, T., Schiele, B., Tuytelaars, T. (eds.) ECCV 2014. LNCS, vol. 8689, pp. 584–599. Springer, Cham (2014). https://doi.org/10.1007/978-3-319-10590-1_38
4. Chen, H., Wang, Y., Wang, G., Qiao, Y.: LSTD: a low-shot transfer detector for object detection. arXiv preprint arXiv:1803.01529 (2018)
5. Everingham, M., Van Gool, L., Williams, C.K.I., Winn, J., Zisserman, A.: The pascal visual object classes (VOC) challenge. Int. J. Comput. Vis. (IJCV) **88**(2), 303–338 (2010)
6. Fan, Q., Zhuo, W., Tang, C.K., Tai, Y.W.: Few-shot object detection with attention-RPN and multi-relation detector. In: proceedings of the IEEE Conference on Computer Vision and Pattern Recognition (CVPR), pp. 4013–4022 (2020)
7. George, M., Floerkemeier, C.: Recognizing products: a per-exemplar multi-label image classification approach. In: Fleet, D., Pajdla, T., Schiele, B., Tuytelaars, T. (eds.) ECCV 2014. LNCS, vol. 8690, pp. 440–455. Springer, Cham (2014). https://doi.org/10.1007/978-3-319-10605-2_29

8. Girshick, R.: Fast R-CNN. In: proceedings of the IEEE International Conference on Computer Vision (ICCV), pp. 1440–1448 (2015)
9. Goldman, E., et al.: Precise detection in densely packed scenes. In: proceedings of the IEEE Conference on Computer Vision and Pattern Recognition (CVPR), pp. 5227–5236 (2019)
10. Gordo, A., Almazán, J., Revaud, J., Larlus, D.: End-to-end learning of deep visual representations for image retrieval. Int. J. Comput. Vis. (IJCV) **12**, 237–254 (2017)
11. Gordo, A., Larlus, D.: Beyond instance-level image retrieval: Leveraging captions to learn a global visual representation for semantic retrieval. In: proceedings of the IEEE Conference on Computer Vision and Pattern Recognition (CVPR), pp. 6589–6598 (2017)
12. Hsieh, T.I., Lo, Y.C., Chen, H.T., Liu, T.L.: One-shot object detection with co-attention and co-excitation. In: Advances in Neural Information Processing Systems (NeurIPS), pp. 2725–2734 (2019)
13. Ioffe, S., Szegedy, C.: Batch normalization: accelerating deep network training by reducing internal covariate shift. arXiv preprint arXiv:1502.03167 (2015)
14. Iscen, A., Tolias, G., Avrithis, Y., Furon, T., Chum, O.: Efficient diffusion on region manifolds: recovering small objects with compact CNN representations. In: proceedings of the IEEE Conference on Computer Vision and Pattern Recognition (CVPR), pp. 2077–2086 (2017)
15. Jaderberg, M., Simonyan, K., Zisserman, A., Kavukcuoglu, K.: Spatial transformer networks. In: Advances in Neural Information Processing Systems (NIPS), pp. 2017–2025 (2015)
16. Kang, B., Liu, Z., Wang, X., Yu, F., Feng, J., Darrell, T.: Few-shot object detection via feature reweighting. In: proceedings of the IEEE International Conference on Computer Vision (ICCV), pp. 8420–8429 (2019)
17. Karlinsky, L., et al.: RepMet: representative-based metric learning for classification and one-shot object detection. arXiv preprint arXiv:1806.04728, 4323 (2019)
18. Lin, T.Y., Dollár, P., Girshick, R., He, K., Hariharan, B., Belongie, S.: Feature pyramid networks for object detection. In: proceedings of the IEEE Conference on Computer Vision and Pattern Recognition (CVPR), pp. 2117–2125 (2017)
19. Lin, T.Y., Goyal, P., Girshick, R., Kaiming He, P.D.: Focal loss for dense object detection. In: proceedings of the IEEE International Conference on Computer Vision (ICCV), pp. 2980–2988 (2017)
20. Lin, T.Y., et al.: Microsoft COCO: common objects in context. In: Fleet, D., Pajdla, T., Schiele, B., Tuytelaars, T. (eds.) ECCV 2014. LNCS, vol. 8693, pp. 740–755. Springer, Cham (2014). https://doi.org/10.1007/978-3-319-10602-1_48
21. Liu, W., et al.: SSD: single shot multibox detector. In: Leibe, B., Matas, J., Sebe, N., Welling, M. (eds.) ECCV 2016. LNCS, vol. 9905, pp. 21–37. Springer, Cham (2016). https://doi.org/10.1007/978-3-319-46448-0_2
22. Lowe, D.G.: Object recognition from local scale-invariant features. In: proceedings of the IEEE International Conference on Computer Vision (ICCV), **2**, pp. 1150–1157. IEEE (1999)
23. Lowe, D.G.: Distinctive image features from scale-invariant keypoints. Int. J. Comput. Vis. (IJCV) **60**(2), 91–110 (2004)
24. Massa, F., Girshick, R.: maskrcnn-benchmark: fast, modular reference implementation of instance segmentation and object detection algorithms in PyTorch. https://github.com/facebookresearch/maskrcnn-benchmark (2018). Accessed 01 March 2020
25. Michaelis, C., Ustyuzhaninov, I., Bethge, M., Ecker, A.S.: One-shot instance segmentation. arXiv:1811.11507v1 (2018)

26. Mobahi, H., Collobert, R., Weston, J.: Deep learning from temporal coherence in video. In: proceedings of the International Conference on Machine Learning (ICML), pp. 737–744 (2009)
27. Noh, H., Araujo, A., Sim, J., Weyand, T., Han, B.: Large-scale image retrieval with attentive deep local features. In: proceedings of the IEEE International Conference on Computer Vision (ICCV), pp. 3456–3465 (2017)
28. Paszke, A., et al.: PyTorch: an imperative style, high-performance deep learning library. In: Advances in Neural Information Processing Systems (NeurIPS), pp. 8026–8037 (2019)
29. Pérez-Rúa, J.M., Zhu, X., Hospedales, T., Xiang, T.: Incremental few-shot object detection. In: proceedings of the IEEE Conference on Computer Vision and Pattern Recognition (CVPR), pp. 13846–13855 (2020)
30. Radenović, F., Tolias, G., Chum, O.: CNN image retrieval learns from BoW: unsupervised fine-tuning with hard examples. In: Leibe, B., Matas, J., Sebe, N., Welling, M. (eds.) ECCV 2016. LNCS, vol. 9905, pp. 3–20. Springer, Cham (2016). https://doi.org/10.1007/978-3-319-46448-0_1
31. Radenović, F., Tolias, G., Chum, O.: Fine-tuning CNN image retrieval with no human annotation. IEEE Trans. Pattern Anal. Mach. Intell. (TPAMI) 41(6), 1655–1668 (2019)
32. Redmon, J., Divvala, S., Girshick, R., Farhadi, A.: You only look once: unified, real-time object detection. In: proceedings of the IEEE Conference on Computer Vision and Pattern Recognition (CVPR), pp. 779–788 (2016)
33. Redmon, J., Farhadi, A.: YOLO9000: better, faster, stronger. In: proceedings of the IEEE Conference on Computer Vision and Pattern Recognition (CVPR), pp. 7263–7271 (2017)
34. Ren, S., He, K., Girshick, R., Sun, J.: Faster R-CNN: towards real-time object detection with region proposal networks. IEEE Trans. Pattern Anal. Mach. Intell. (TPAMI) 39(6), 1137–1149 (2017)
35. Rocco, I., Arandjelović, R., Sivic, J.: Convolutional neural network architecture for geometric matching. In: proceedings of the IEEE Conference on Computer Vision and Pattern Recognition (CVPR), pp. 6148–6157 (2017)
36. Rocco, I., Arandjelović, R., Sivic, J.: Convolutional neural network architecture for geometric matching. IEEE Trans. Pattern Anal. Mach. Intell. (TPAMI) 41(11), 2553–2567 (2018)
37. Rocco, I., Arandjelovic, R., Sivic, J.: End-to-end weakly-supervised semantic alignment. In: proceedings of the IEEE Conference on Computer Vision and Pattern Recognition (CVPR), pp. 6917–6925 (2018)
38. Rocco, I., Cimpoi, M., Arandjelović, R., Torii, A., Pajdla, T., Sivic, J.: Neighbourhood consensus networks. In: Advances in Neural Information Processing Systems (NeurIPS), pp. 1651–1662 (2018)
39. Russakovsky, O., et al.: ImageNet large scale visual recognition challenge. Int. J. Comput. Vis. (IJCV) 115(3), 211–252 (2015)
40. Simo-Serra, E., Trulls, E., Ferraz, L., Kokkinos, I., Fua, P., Moreno-Noguer, F.: Discriminative learning of deep convolutional feature point descriptors. In: proceedings of the IEEE International Conference on Computer Vision (ICCV), pp. 118–126 (2015)
41. Wang, S., Jiang, S.: INSTRE: a new benchmark for instance-level object retrieval and recognition. ACM Trans. Multimedia Comput. Commun. Appl. (TOMM) 11(3), 1–21 (2015)
42. Wang, X., Huang, T.E., Darrell, T., Gonzalez, J.E., Yu, F.: Frustratingly simple few-shot object detection. arXiv preprint arXiv:2003.06957 (2020)

43. Wang, X., Hua, Y., Kodirov, E., Hu, G., Garnier, R., Robertson, N.M.: Ranked list loss for deep metric learning. In: proceedings of the IEEE Conference on Computer Vision and Pattern Recognition (CVPR), pp. 5207–5216 (2019)
44. Wang, Y.X., Ramanan, D., Hebert, M.: Meta-learning to detect rare objects. In: proceedings of the IEEE International Conference on Computer Vision (ICCV), pp. 9925–9934 (2019)
45. Wu, C.Y., Manmatha, R., Smola, A.J., Krähenbühl, P.: Sampling matters in deep embedding learning. In: proceedings of the IEEE International Conference on Computer Vision (ICCV), pp. 2840–2848 (2017)
46. Wu, Y., He, K.: Group Normalization. In: Ferrari, V., Hebert, M., Sminchisescu, C., Weiss, Y. (eds.) ECCV 2018. LNCS, vol. 11217, pp. 3–19. Springer, Cham (2018). https://doi.org/10.1007/978-3-030-01261-8_1
47. Yan, X., Chen, Z., Xu, A., Wang, X., Liang, X., Lin, L.: Meta R-CNN: towards general solver for instance-level low-shot learning. In: proceedings of the IEEE International Conference on Computer Vision (ICCV), pp. 9577–9586 (2019)

Interpretable Neural Network Decoupling

Yuchao Li[1], Rongrong Ji[1,2]([⊠]), Shaohui Lin[3], Baochang Zhang[4],
Chenqian Yan[1], Yongjian Wu[5], Feiyue Huang[5], and Ling Shao[6,7]

[1] Department of Artificial Intelligence, School of Informatics,
Xiamen University, Xiamen, China
rrji@xmu.edu.cn
[2] Peng Cheng Laboratory, Shenzhen, China
[3] National University of Singapore, Nanyang, Singapore
[4] Beihang University, Beijing, China
[5] BestImage, Tencent Technology (Shanghai) Co., Ltd., Shanghai, China
[6] Mohamed bin Zayed University of Artificial Intelligence, Abu Dhabi, UAE
[7] Inception Institute of Artificial Intelligence, Abu Dhabi, UAE

Abstract. The remarkable performance of convolutional neural networks (CNNs) is entangled with their huge number of uninterpretable parameters, which has become the bottleneck limiting the exploitation of their full potential. Towards network interpretation, previous endeavors mainly resort to the single filter analysis, which however ignores the relationship between filters. In this paper, we propose a novel architecture decoupling method to interpret the network from a perspective of investigating its calculation paths. More specifically, we introduce a novel architecture controlling module in each layer to encode the network architecture by a vector. By maximizing the mutual information between the vectors and input images, the module is trained to select specific filters to distill a unique calculation path for each input. Furthermore, to improve the interpretability and compactness of the decoupled network, the output of each layer is encoded to align the architecture encoding vector with the constraint of sparsity regularization. Unlike conventional pixel-level or filter-level network interpretation methods, we propose a path-level analysis to explore the relationship between the combination of filter and semantic concepts, which is more suitable to interpret the working rationale of the decoupled network. Extensive experiments show that the decoupled network achieves several applications, i.e., network interpretation, network acceleration, and adversarial samples detection.

Keywords: Network interpretation · Architecture decoupling

1 Introduction

Deep convolutional neural networks (CNNs) have dominated various computer vision tasks, such as object classification, detection and semantic segmentation.

Electronic supplementary material The online version of this chapter (https://doi.org/10.1007/978-3-030-58555-6_39) contains supplementary material, which is available to authorized users.

However, the superior performance of CNNs is rooted in their complex architectures and huge amounts of parameter, which thereby restrict the interpretation of their internal working mechanisms. Such a contradiction has become a key drawback when the network is used in task-critical applications such as medical diagnosis, automatic robots, and self-driving cars.

To this end, network interpretation have been explored to improve the understanding of the intrinsic structures and working mechanisms of neural networks [2,5,20,26,28,40,41]. Interpreting a neural network involves investigating the rationale behind the decision-making process and the roles of its parameters. For instance, some methods [5,22] view networks as a whole when explaining their working process. However, these approaches are too coarse-grained for exploring the intrinsic properties in the networks. In contrast, network visualization approaches [39,40] interpret the role of each parameter by analyzing the pixel-level feature representation, which always require complex trial-and-error experiments. Beyonds, Bau *et al.* [2] and Zhang *et al.* [41] explored the different roles of filters in the decision-making process of a network. Although these methods are more suitable for explaining the network, they characterize semantic concepts using only a single filter, which has been proven to be less effective than using a combination of multiple filters [10,34]. Under this situation, different combination

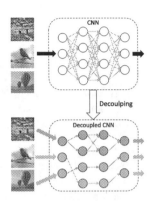

Fig. 1. An example of the neural network architecture decoupling. Each color represents a calculation path of specific input. (Color figure online)

of filters can be viewed as different calculation paths in the network, which inspires us to investigate the working process of networks based on a path-level analysis. The challenge, however, comes from the fact that each inference involves all filters in the network and has the same calculation process, making it difficult to interpret how each calculation path affects the final result. To overcome this problem, previous methods [36,37] explore the difference between the calculation paths of different inputs by reducing the number of parameters involved in the calculation process. For instance, Wang *et al.* [36] proposed a post-hoc analysis to obtain a unique calculation path of a specific input based on a pre-trained model, which however involves a huge number of complicated experiments. Moreover, Sun *et al.* [37] learned a network that generates a dynamic calculation path in the last layer by modifying the SGD algorithm. However, it ignores the fact that the responses of filters are also dynamic in the intermediate layers, and thus cannot interpret how the entire network works.

In this paper, we propose an interpretable network decoupling approach, which enables a network to adaptively select a suitable subset of filters to form a calculation path for each input, as shown in Fig. 1. In particular, Our design principle lies in a novel light-weight *architecture controlling module* as well as a novel learning process for network decoupling. Figure 2 depicts the framework of the proposed method. The architecture controlling module is first incorporated into each layer to dynamically select filters during network inference with a negligible computational burden. Then, we maximize the mutual information between the architecture encoding vector (*i.e.*, the output of the architecture controlling module) and the inherent attributes of the input images during training, which allows the network to dynamically generate the calculation path related to the input. In addition, to further improve the interpretability of decoupled networks, we increase the similarity between the architecture encoding vector of each convolutional layer and its output by minimizing the KL-divergence between them, making filter only respond to a specific object. Finally, we sparsify the architecture encoding vector to attenuate the calculation path and eliminate the effects of redundant filters for each input. We also introduce an improved semantic hashing scheme to make the discrete architecture encoding vector differentiable, which is therefore capable to be trained directly by stochastic gradient descent (SGD).

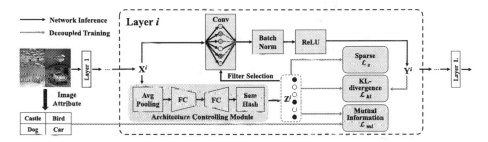

Fig. 2. The framework of the proposed interpretable neural network decoupling. The architecture encoding vector $\mathbf{z}^i$ is first constructed by the architecture controlling module, and then learned to determine the filter selection by Eq. 12. For network inference, we only use the selected filters based on each input. The mutual information loss $\mathcal{L}_{mi}$ is computed between the output of the architecture controlling module $\mathbf{z}^i$ and the attribute of the inputs to decouple the network architecture. The KL-divergence loss $\mathcal{L}_{kl}$ is computed by the output of convolutional layer $\mathcal{y}^i$ and $\mathbf{z}^i$ to disentangle the filters. The sparse loss $\mathcal{L}_s$ is used to sparsify the result of filter selection.

Correspondingly, the decoupled network becomes more interpretable, and one can trace the functional processing behavior layer-by-layer to form a hierarchical path towards understanding the working principle of the decoupled network. Meanwhile, each filter is only related to a set of similar input images after the decoupling, thus they also become more interpretable, and the combination of them forms a decoupled sub-architecture, which better characterizes the specific semantic concepts. Such a decoupled architecture further benefits from a low computational cost for network acceleration, as well as good hints for adversarial samples detection, which are subsequently validated in our experiments.

We summarize our three main contributions as follows:

- To interpret neural networks by dynamically selecting the filters for different inputs, we propose a lightweight architecture controlling module, which is differentiable and can be optimized by SGD based on the losses we propose.
- The decoupled network reserves similar performance of the original network and has better interpretable. Thus it enables the functional processing of each calculation path to be well interpreted, which helps better understand the rationale behind the network inference, as well as explore the relationship between filters and semantic concepts in the decoupled network.
- Our method is generic and flexible, which can be easily employed on the existing network architectures, such as VGGNets [32], ResNets [13], and Inceptions [33]. The decoupled architecture further benefits extensive applications, including network acceleration and adversarial samples detection.

2   Related Work

Network Interpretation. One way to interpret a network is to analyze how it responds to a specific input image for output prediction [5,20,22,26,42]. This strategy views the network as a whole to interpret the network prediction results by exploring the knowledge blind spots of neural networks [22], or by assigning each output feature an importance value for a particular prediction [26]. Moreover, a decision tree [42] or an explainer network [5] has been used to better understand the classification process. However, these methods only pay attention to the reason behind the network prediction result, and the roles of each parameter are ignored, making it difficult to understand their effects on the network.

To open the black-box of neural network and interpret the role of parameters, several methods [8,39,40] have been proposed to visualize the feature representations inside the network. For instance, Zeiler *et al.* [40] visualized the feature maps in the intermediate layers by establishing a deconvolutional network corresponding to the original one. Yoshinski *et al.* [39] proposed two visualization methods to explore the information contained in features: a respective post-hoc analysis on a pre-trained model and learning a network by regularized optimization. Visualizing feature representations is a very direct method to explain the role of parameters in a network, which however requires extensive experiments due to the enormous number of parameters.

In addition to the above methods, the functions of filters are also explored for interpreting networks [2,28,37,38,41]. They have evaluated the transferability of filters [38] or quantified the relationship between filters and categories [28] to explain their different roles. Compared with using a single filter to represent semantic concepts, methods in [10,34] have found that the semantic concepts can be better characterized by combining multiple filters. Wang *et al.* [34] further validated that clustering the activations of multiple filters can better represent semantic concepts than using a single filter. Fong *et al.* [10] mapped the semantic concepts into vectorial embeddings based on the responses of multiple filters and found that these embeddings can better characterize the features. Different from these methods, we interpret the working principle of a network based on a path-level analysis by decoupling the network, upon which we further disentangle each intra-layer filter to explore the interpretable semantic concepts across filters on the calculation path. Our method is more in line with the internal working mechanism of the network than these works, and has a better extension to other applications, such as network acceleration and adversarial samples detection.

Conditional Computation. Works on conditional computation tend to concentrate on the selection of model components when generating the calculation path. For instance, the work in [3] explored the influence of stochastic or nonsmooth neurons when estimating the gradient of the loss function. Later, an expert network was learned to find a suitable calculation path for each input by reinforcement learning [4] or SGD [6]. However, the requirement of a specific expert network makes these approaches cumbersome. Along another line, a halting score [9] or a differentiable directed acyclic graph [24] has been used to dynamically adjust the model components involved in the calculation process. Recently, a feature boosting and suppression method [11] was introduced to skip unimportant output channels of the convolutional layer for data-dependent inference, which is different from static pruning. However, it selects the same number of filters for each layer, without considering inter-layer differences. Different from the above works, we employ a novel architecture controlling module to decouple the network by fitting it to the data distribution. After decoupling, the network becomes interpretable, enabling us to visualize its intrinsic structure, accelerate the inference, and detect adversarial samples.

3   Architecture Decoupling

Formally speaking, the l-th convolutional layer in a network with a batch normalization (BN) [17] and a ReLU layer [29] transforms $\mathcal{X}^l \in \mathbb{R}^{C^l \times H_{in}^l \times W_{in}^l}$ to $\mathcal{Y}^l \in \mathbb{R}^{N^l \times H_{out}^l \times W_{out}^l}$ using the weight $\mathcal{W}^l \in \mathbb{R}^{N^l \times C^l \times D^l \times D^l}$, which is defined as:

$$\mathcal{Y}^l = \Big(BN\big(Conv(\mathcal{X}^l, \mathcal{W}^l)\big)\Big)_+, \tag{1}$$

where $(\cdot)_+$ represents the ReLU layer, and $Conv(\cdot, \cdot)$ denotes the standard convolution operator. (H_{in}^l, W_{in}^l) and (H_{out}^l, W_{out}^l) are the spatial size of the input and output in the l-th layer, respectively. D^l is the kernel size.

3.1   Architecture Controlling Module

For an input image, the proposed architecture controlling module selects the filters and generates the calculation path during network inference. In particular, we aim to predict which filters need to participate in the convolutional computation *before* the convolutional operation to accelerate network inference. Therefore, for the l-th convolutional layer, the architecture encoding vector $\mathbf{z}^l$ (*i.e.*, the output of the architecture controlling module) only relies on the input

$\mathcal{X}^l$ instead of the output $\mathcal{Y}^l$, which is defined as $\mathbf{z}^l = G^l(\mathcal{X}^l)$. Inspired by the effectiveness of the squeeze-and-excitation (SE) block [16], we select a similar SE-block to predict the importance of each filter. Thus, we first squeeze the global spatial information via global average pooling, which transforms each input channel $X_i^l \in \mathbb{R}^{H_{in}^l \times W_{in}^l}$ to a scalar s_i^l. We then design a sub-network structure $\bar{G}^l(\mathbf{s}^l)$ to determine the filter selection based on $\mathbf{s}^l \in \mathbb{R}^{C^l}$, which is formed by two fully connected layers, *i.e.*, a dimensionality-reduction layer with weights $\mathbf{W}_1^l$ and a dimensionality-increasing layer with weights $\mathbf{W}_2^l$:

$$\bar{G}^l(\mathbf{s}^l) = \mathbf{W}_2^l \cdot (\mathbf{W}_1^l \cdot \mathbf{s}^l)_+, \tag{2}$$

where $\mathbf{W}_1^l \in \mathbb{R}^{\frac{C^l}{\gamma} \times C^l}$, $\mathbf{W}_2^l \in \mathbb{R}^{N^l \times \frac{C^l}{\gamma}}$ and $\cdot$ represents the matrix multiplication. We ignore the bias for simplicity. To reduce the module complexity, we empirically set the reduction ratio γ to 4 in our experiments. The output of $\bar{G}^l(\mathbf{s}^l)$ is a real vector, while we need to binarize $\bar{G}^l(\mathbf{s}^l)$ to construct a binary vector $\mathbf{z}^l$, which represents the result of filter selection. However, a simple discretization using the sign function is not differentiable, which prevents the corresponding gradients from being directly obtained by back-propagation. Thus, we further employ an *Improved SemHash* method [19] to transform the real vector in $\bar{G}^l(\mathbf{s}^l)$ to a binary vector by a simple rounding bottleneck, which also makes the discretization become differentiable.

Improved SemHash. The proposed scheme is based on the different operations for training and testing. During training, we first sample a noise $\alpha \sim \mathcal{N}(0,1)^{N^l}$, which is added to $\bar{G}^l(\mathbf{s}^l)$, and then obtain $\widetilde{\mathbf{s}}^l = \bar{G}^l(\mathbf{s}^l) + \alpha$. After that, we compute a real vector and a binary vector by:

$$\mathbf{v}_1^l = \sigma'(\widetilde{\mathbf{s}}^l), \mathbf{v}_2^l = \mathbf{1}(\widetilde{\mathbf{s}}^l > 0), \tag{3}$$

where σ' is a saturating Sigmoid function [18] denoted as:

$$\sigma'(x) = \max\left(0, \min\left(1, 1.2\sigma(x) - 0.1\right)\right). \tag{4}$$

Here, σ is the Sigmoid function. $\mathbf{v}_1^l \in \mathbb{R}^{C^l}$ is a real vector with all elements falling in the interval $[0,1]$, and we calculate its gradient during back-propagation. $\mathbf{v}_2^l \in \mathbb{R}^{C^l}$ represents the discretized vector, which cannot be involved in the gradient calculation. Thus, we randomly use $\mathbf{z}^l = \mathbf{v}_1^l$ for half of the training samples and $\mathbf{z}^l = \mathbf{v}_2^l$ for the rest in the forward-propagation. We then mask the output channels using the architecture encoding vector (*i.e.*, $\mathcal{Y}^l * \mathbf{z}^l$) as the final output of this layer. In the backward-propagation, the gradient of $\mathbf{z}^l$ is the same as the gradient of $\mathbf{v}_1^l$.

During evaluation/testing, we directly use the sign function in the forward-propagation as:

$$\mathbf{z}^l = \mathbf{1}(\bar{G}^l(\mathbf{s}^l) > 0). \tag{5}$$

After that, we select suitable filters involved in the convolutional computation based on $\mathbf{z}^l$ to achieve fast inference.

3.2   Network Training

We expect the network architecture to be gradually decoupled during training, where the essential problem is how to learn an architecture encoding vector that fits the data distribution. To this end, we propose three loss functions for network decoupling.

Mutual Information Loss. When the network architecture is decoupled, different inputs should select their related sets of filters. We adopt mutual information $I(a; \mathbf{z}^l)$ between the result of filter selection $\mathbf{z}^l$ and the attribute of an input image a (*i.e.*, the unique information contained in the input image) to measure the correlation between the architecture encoding vector and its input image. $I(a; \mathbf{z}^l) = 0$ means that the result of filter selection is independent to the input image, *i.e.*, all the inputs share the same filter selection. In contrast, when $I(a; \mathbf{z}^l) \neq 0$, filter selection depends on the input image. Thus, we maximize the mutual information between a and $\mathbf{z}^l$ to achieve architecture decoupling. Formally speaking, we have:

$$
\begin{aligned}
I(a; \mathbf{z}^l) &= H(a) - H(a|\mathbf{z}^l) \\
&= \sum_a \sum_{\mathbf{z}^l} P(a, \mathbf{z}^l) log P(a|\mathbf{z}^l) + H(a) \\
&= \sum_a \sum_{\mathbf{z}^l} P(\mathbf{z}^l) P(a|\mathbf{z}^l) log P(a|\mathbf{z}^l) + H(a).
\end{aligned}
\tag{6}
$$

The mutual information $I(a; \mathbf{z}^l)$ is difficult to directly maximize, as it is hard to obtain $P(a|\mathbf{z}^l)$. Thus, we use $Q(a|\mathbf{z}^l)$ as a variational approximation to $P(a|\mathbf{z}^l)$ [1]. In fact, the KL-divergence is positive, so we have:

$$
\begin{aligned}
KL\big(P(a|\mathbf{z}^l), Q(a|\mathbf{z}^l)\big) \geq 0 &\Rightarrow \sum_a P(a|\mathbf{z}^l) log P(a|\mathbf{z}^l) \\
&\geq \sum_a P(a|\mathbf{z}^l) log Q(a|\mathbf{z}^l).
\end{aligned}
\tag{7}
$$

We then obtain the following equation:

$$
\begin{aligned}
I(a; \mathbf{z}^l) &\geq \sum_a \sum_{\mathbf{z}^l} P(\mathbf{z}^l) P(a|\mathbf{z}^l) log Q(a|\mathbf{z}^l) + H(a) \\
&\geq \sum_a \sum_{\mathbf{z}^l} P(\mathbf{z}^l) P(a|\mathbf{z}^l) log Q(a|\mathbf{z}^l) \\
&= \mathbb{E}_{\mathbf{z}^l \sim G^l(\mathcal{X}^l)} [\mathbb{E}_{a \sim P(a|\mathbf{z}^l)} [log Q(a|\mathbf{z}^l)]].
\end{aligned}
\tag{8}
$$

Equation 8 provides a lower bound for the mutual information $I(a; |\mathbf{z}^l)$. By maximizing this bound, the mutual information $I(a; \mathbf{z}^l)$ will also be maximized accordingly. In our paper, we use the class label as the attribute of the input image c in the classification task. Moreover, we reparametrize $Q(a|\mathbf{z}^l)$ as a neural network $\tilde{Q}(\mathbf{z}^l)$ that contains a fully connected layer and a softmax layer.

Thus, maximizing the mutual information in Eq. 8 is achieved by minimizing the following loss:

$$\mathcal{L}_{mi} = -\sum_{l=1}^{L} A_X * log\tilde{Q}(\mathbf{z}^l), \tag{9}$$

where A_X represents the label of the input image X. $\tilde{Q}(\mathbf{z}^l)$ is defined as $\mathbf{W}_{cla}^l \cdot \mathbf{z}^l$ with a fully connected weight $\mathbf{W}_{cla}^l \in \mathbb{R}^{K \times N^l}$, where K represents the number of categories in image classification.

KL-divergence Loss. After decoupling the network architecture, we guarantee that the filter selection depends on the input image. However, it is uncertain whether the filters become different (*i.e.*, detect different objects), which obstructs us from further interpreting the network. If a filter only responds to a specific semantic concept, it will not be activated when the input does not contain this feature. Thus, by limiting filters to only respond to specific category, they can be disentangled to detect different categories. To achieve this goal, we minimize the KL-divergence between the output of the current layer and its corresponding architecture encoding vector, which ensures that the overall responses of filters have a similar distribution to the responses of the selected subset. To align the dimension of the convolution output and architecture encoding vector, we further downsample $\mathcal{Y}^l$ to $\mathbf{y}^l \in \mathbb{R}^N$ using global average pooling. Then, the KL-divergence loss is defined as:

$$\mathcal{L}_{kl} = \sum_{l=1}^{L} KL(\mathbf{z}^l||\mathbf{y}^l). \tag{10}$$

As the output of filter is limited by the result of filter selection, it will be unique and only detects the specific object. Thus, all filters are different from each other, *i.e.*, each one performs its function.

Sparse Loss. An ℓ_1-regularization on $\mathbf{z}^l$ is further introduced to encourage the architecture encoding vector to be sparse, which makes the calculation path of each input becomes thinner. Thus, the sparse loss is defined as:

$$\mathcal{L}_s = \sum_{l=1}^{L} |\|\mathbf{z}^l\|_1 - R * N^l|, \tag{11}$$

where R represents the target compression ratio. Since z^l falls in the interval $[0, 1]$, the maximum value of $\|\mathbf{z}^l\|_1$ is N^l, and the minimum value is 0, where N^l is the number of filters. For example, we set R to 0.5 if activating only half of the filters.

Therefore, we obtain the overall loss function as follows:

$$\mathcal{L} = \mathcal{L}_{ce} + \lambda_m * \mathcal{L}_{mi} + \lambda_k * \mathcal{L}_{kl} + \lambda_s * \mathcal{L}_s, \tag{12}$$

where $\mathcal{L}_{ce}$ is the network classification loss. λ_m, λ_k and λ_s are the hyperparameters. Equation 12 can be effectively solved via SGD.

4   Experiments

We evaluate the effectiveness of the proposed neural network architecture decoupling scheme on three kinds of networks, *i.e.*, VGGNets [32], ResNets [13], and Inceptions [33]. For network acceleration, we conduct comprehensive experiments on three datasets, *i.e.*, CIFAR-10, CIFAR-100 [21] and ImageNet 2012 [31]. For quantifying the network interpretability, we use the interpretability of filters [41] and the representation ability of semantic features [10] on BRODEN dataset [2] to evaluate the original and our decoupled models.

4.1   Implementation Details

We implement our method using PyTorch [30]. The weights of decoupled networks are initialized using the weights from their corresponding pre-trained models. We add the architecture controlling module to all convolutional layers except the first and last ones. All networks are trained using stochastic gradient descent with a momentum of 0.9. For CIFAR-10 and CIFAR-100, we train all the networks over 200 epochs using a mini-batch size of 128. The learning rate is initialized by 0.1, which is divided by 10 at 50% and 75% of the total number of epochs. For ImageNet 2012, we train the networks over 120 epochs with a mini-batch size of 64 and 256 for VGG-16 and ResNet-18, respectively. The learning rate is initialized as 0.01 and is multiplied by 0.1 after the 30-th, 60-th and 90-th epoch. The real speed on the CPU is measured by a single-thread AMD Ryzen Threadripper 1900X. Except for the experiments on network acceleration, we automatically learn sparse filters by setting R to 0 in Eq. 11.

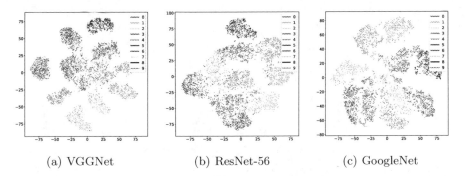

(a) VGGNet   (b) ResNet-56   (c) GoogleNet

Fig. 3. Visualization of the distribution of the integral calculation path in different networks on CIFAR-10.

4.2   Network Interpretability

Architecture Encoding. We collect the calculation paths from three different networks (*i.e.*, VGGNet, ResNet-56 and GoogleNet) to verify that the proposed network decoupling method can successfully decouple the network and ensure that it generates different calculation paths for different images. We first reduce the dimension of the calculation path (*i.e.*, the concatenation of architecture encoding vectors $\mathbf{z}^l$ across all layers) to 300 using Principal Component Analysis (PCA), and then visualize the calculation path by t-SNE [27]. As shown in Fig. 3, each color represents one category and each dot is a calculation path corresponding to an input. We can see that the network architecture is successfully decoupled after training by our method, where different categories of images have different calculation paths.

Filter State. After decoupling the network architecture, the state of a filter in the network has three possibilities: it responds to all the input samples, it does not respond to any input samples, or it responds to the specific inputs. These three possibilities are termed as *energetic filter*, *silent filter*, and *dynamic filter*, respectively. As shown in Fig. 4, we collect different states of filters in different layers. We can see that the proportion of dynamic filters increases with network depth increasing. This phenomenon

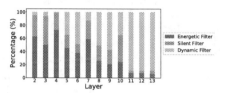

Fig. 4. The distribution of filters with different states in each layer of VGG-16 on ImageNet2012.

demonstrates that filters in the top layer tend to detect high-level semantic features, which are highly related to the input images. In contrast, filters in the bottom layer tend to detect low-level features, which are always shared across images. For more detailed analysis, refer to Section A.1 of the supplementary material.

Table 1. The average interpretability score of filters in the different layers of original networks and decoupled networks on BRODEN. The higher score is better.

Model	Top1-Acc	Top5-Acc	Conv2_2	Conv3_3	Conv4_3	Conv5_3
VGG-16	71.59	90.38	0.0637	0.0446	0.0627	0.0787
VGG-16$_{decoupled}$	71.51	90.32	**0.0750**	**0.0669**	**0.0643**	**0.0879**
Model	Top1-Acc	Top5-Acc	Block1	Block2	Block3	Block4
ResNet-18	69.76	89.08	0.0527	0.0212	0.0477	0.0521
ResNet-18$_{decoupled}$	67.62	87.78	**0.1062**	**0.0268**	**0.0580**	**0.0618**

Interpretable Quantitative Analysis. Following the works [2,10,41], we select the interpretability of filters and the representation ability of semantic features to measure the network interpretability. Specifically, we first select the original and our decoupled models which trained on ImageNet2012, and compute

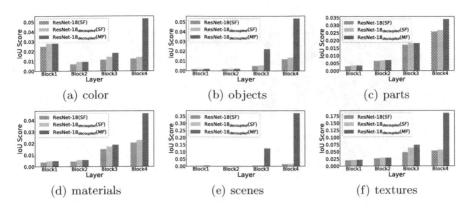

Fig. 5. Average representation ability of different concepts in ResNet-18 on BRODEN. SF/MF represents use single/multiple filters characterizing the semantic features.

the activation map of each filter/unit on BRODEN dataset. Then, the top quantile level threshold is determined over all spatial locations of feature maps. After that, low-resolution activation maps of all filters are scaled up to input-image resolution using bilinear interpolation and thresholded into a binary segmentation, so as to obtain the receptive fields of filters. The score of each filter f as segmentation for the semantic concept t in the input image I is reported as an intersection-over-union score $IoU^I_{f,t} = \frac{|S^I_f \cap S^I_t|}{|S^I_f \cup S^I_t|}$, where S^I_f and S^I_t denote the receptive field of filter f and the ground-truth mask of the semantic concept t in the input image, respectively. Given an image I, we associated filter f with the t-th part if $IoU^I_{f,t} > 0.01$. Finally, we measure the relationship between the filter f and concept t by $P_{f,t} = mean_I \mathbf{1}(IoU^I_{f,t} > 0.01)$ across all the input images. Based on [41], we can report the highest association between the filter and concept as the final interpretability score of filter f by $max_t P_{f,t}$. As shown in Table 1, the value in each layer is obtained by averaging the final interpretability score across all the corresponding filters. For ResNet-18, we collect the filters from the first convolutional layers in the last unit of each block. Compared to the original networks, our decoupled networks have the better interpretability under the similar classification accuracy. For instance, we achieve $1.2\times \sim 2\times$ score improvement of the filter interpretability than the original ResNet-18.

We further investigate the representation ability of network for specific semantic features before and after network decoupling. For the representation of semantic features from a single filter, we evaluate the highest association between each semantic feature in BRODEN (which has $1,197$ semantic features) and the filters using $max_{I,f} IoU^I_{f,t}$ as the representation ability of specific semantic features, based on [10]. For the representation of semantic features from multiple filters, we first occlude the semantic features in the original image and then collect the number of M filters by comparing the difference between the calculation path of the original image and the occluded image, where these filters are activated on the original image but inactivated due to the lack of specific semantic

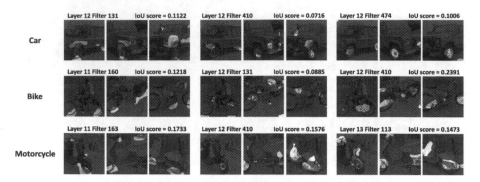

Fig. 6. Visualization of the receptive fields of filters which are inactivated because of the lack of semantic feature in images. We occlude the specific semantic feature (*i.e.,* wheel) in different images (*i.e.,* car, bike and motorcycle) on ImageNet and then collect the filters become inactivated due to the lack of the semantic feature.

features. After that, we merge their receptive field and calculate the value of IoU $IoU^I_{f \in M, t} = \frac{|S^I_{f \in M} \cap S^I_t|}{|S^I_{f \in M} \cup S^I_t|}$ as the representation ability of semantic feature t. As shown in Fig. 5, we average the representation ability of semantic features belonging to the same concepts in the different layers. The results demonstrates that our decoupled network has the better representation ability of semantic feature than the original ResNet-18. The combination of multiple filters, which collected by our path-level disentangling, achieves about $3\times$ improvement in the representation ability than the single ones. Moreover, we find that the bottom layers in the decoupled network always use the single filters to characterize the semantic features based on our path-level analysis, so the representation ability of semantic features in the bottom layers is similar in the single filter and multiple filters.

Semantic Concept Analysis. We further investigate the relationship between semantic concepts and calculation paths. To this end, we occlude the areas that contain similar semantic features (*i.e.,* wheels) in the images from different categories (*i.e,.* car, bike and motorcycle) to analyze the characterization of the same semantic concept in different categories. After that, we collect the filters which in the different parts of calculation path between the original images and the semantic lacked images. Our experiments only collect the three filters with highest IoU score in the last three convolutional layers of VGG-16. We find that the existence of a single semantic concept affects the state of multiple filters. For example, as shown in the first row of Fig. 6, when we only occlude the wheels of the car with black blocks, the 131-th, 410-th and 474-th filters in the 12-th convolutional layer become inactived, which makes the calculation path change. To further analyze the relationship between each filter and semantic concept, we visualize the receptive fields of filters on the input image to obtain the specific detection location of each one, and calculate the IoU score between the receptive fields of filters and the location area of the semantic concept. We find that dif-

Table 2. Results of the different networks on CIFAR-10 and CIFAR-100. * represents the result based on our implementation.

Model	CIFAR-10		CIFAR-100	
	FLOPs	Top-1 Acc(%)	FLOPs	Top-1 Acc(%)
ResNet-56	125M	93.17	125M	70.43
CP [15]	63M	91.80	–	–
L1 [23]*	90M	93.06	86M	69.38
Skip [35]*	103M	92.50	–	–
Ours	**63M**	**93.08**	**41M**	**69.72**
VGGNet	398M	93.75	398M	72.98
L1 [23]*	199M	93.69	194M	72.14
Slim [25]	196M	93.80	250M	73.48
Ours	**141M**	**93.82**	**191M**	**73.84**
GoogleNet	1.52B	95.11	1.52B	77.99
L1 [23]*	1.02B	94.54	0.87B	77.09
Ours	**0.39B**	**94.65**	**0.75B**	**77.28**

Table 3. Results of ResNet-18 on ImageNet2012. The baseline in our method has an 69.76% top-1 accuracy and 89.08% top-5 accuracy with 1.81B FLOPs and an average 180 ms testing on CPU based an image by running the whole of the validation dataset.

Model	Top-1 Acc↓ (%)	Top-5 Acc↓ (%)	FLOPs Reduction	CPU Time Reduction
SFP [14]	3.18	1.85	1.72×	1.38×
DCP [43]	2.29	1.38	1.89×	1.60×
LCL [7]	3.65	2.30	1.53×	1.25×
FBS [11]	2.54	1.46	1.98×	1.60×
Ours	**2.14**	**1.30**	**2.03×**	**1.64×**

ferent filters are responsible for different parts of the same semantic concept. For instance, the 131-th, 410-th and 474-th filters in the 12-th convolutional layer of VGG-16 are responsible for the features in the different parts of the wheel in "car" images, respectively. Therefore, the combination of these filters has the better representation ability of the wheel than the single ones.

4.3   Network Acceleration

In this subsection, we evaluate how our method can facilitate network acceleration. We decouple three different network architectures (*i.e.*, ResNet-56, VGGNet and GoogleNet) on CIFAR-10 and CIFAR-100, and set $R = 0$ to allow the networks to be learned automatically. The VGGNet in our experiments is the same as the network in [25]. As shown in Table 2, our method achieves the best trade-off between accuracy and speedup/compression rate, compared with static pruning [15,23,25] and dynamic pruning [35]. For instance, we achieve a $2\times$ FLOPs reduction with only a 0.09% drop in top-1 accuracy for ResNet-56 on CIFAR-10. For ImageNet 2012, the results of accelerating ResNet-18 are summarized in Table 3. When setting R to 0.6, our method also achieves the best performance with a 1.64× real CPU running speedup and 2.03× reduction in FLOPs compared with the static pruning [14,43] and dynamic pruning [7,11], while only decreasing by 1.30% in top-5 accuracy. The detail of hyper-parameter settings are presented in Section B of the supplementary material.

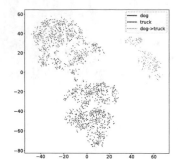

Fig. 7. The distribution of the integral calculation path of original images and adversarial samples in ResNet-56 on CIFAR-10.

Table 4. The Area-Under-Curve (AUC) score on adversarial samples detection. Higher is better.

Classifier	Method	Num. of samples		
		1	5	10
random forest	[36]	0.879	0.894	0.904
	Ours	**0.903**	**0.941**	**0.953**
adaboost	[36]	0.887	0.905	0.910
	Ours	**0.909**	**0.931**	**0.940**
gradient boosting	[36]	0.905	0.919	0.915
	Ours	**0.927**	**0.921**	**0.928**

4.4 Adversarial Samples Detection

We further demonstrate that the proposed architecture decoupling can help to detect the adversarial samples. Recently, several works [12] have concluded that neural networks are vulnerable to adversarial examples, where adding a slight amount of noise to an input image can disturb their robustness. We add noise to images belonging to the "dog" category to make the network predicts as "truck" and visualize the distribution of the calculation path between the original images and adversarial samples in ResNet-56 on CIFAR-10, as shown in Fig. 7. The result demonstrates that the calculation path of the adversarial samples "dog→truck" is different from that of the original "dog" and "truck" images. In other words, adversarial samples do not completely deceive our decoupled network, which can detect them by analyzing their calculation paths. More examples are given in Section C.1 of the supplementary material.

Based on the above observation, we use random forest, adaboost and gradient boosting as the binary classifier to determine whether the calculation paths are from real or adversarial samples. As shown in Table 4, we randomly select 1, 5 and 10 images from each class in the ImageNet 2012 training set to organize three different scales training datasets. The testing set is collected by selecting 1 image from each class in the ImageNet validation dataset. Each experiment is run five times independently. The results show that our method achieves an AUC score of 0.049 gain over Wang *et al.* [36] (*i.e.,* 0.953 *vs.* 0.904), when the number of training samples is 10 on random forest. It also demonstrates that the calculation paths obtained by our method are better than Wang *et al.* [36], with higher discriminability.

5 Conclusion

In this paper, we propose a novel architecture decoupling method to obtain an interpretable network and explore the rationale behind its overall working process based on a novel path-level analysis. In particular, an architecture controlling module is introduced and embedded into each layer to dynamically identify the activated filters. Then, by maximizing the mutual information between the architecture encoding vector and the input image, we decouple the network architecture to explore the functional processing behavior of each calculation path. Meanwhile, to further improve the interpretability of the network and inference, we limit the output of the convolutional layers and sparsifying the calculation path. Experiments show that our method can successfully decouple the network architecture with several merits, *i.e.,* network interpretation, network acceleration and adversarial samples detection.

Acknowledgements. This work is supported by the Nature Science Foundation of China (No. U1705262, No. 61772443, No. 61572410, No. 61802324 and No. 61702136), National Key R&D Program (No. 2017YFC0113000, and No. 2016Y FB1001503), Key R&D Program of Jiangxi Province (No. 20171ACH80022) and Natural Science Foundation of Guangdong Provice in China (No. 2019B1515120049).

References

1. Agakov, D.B.F.: The im algorithm: a variational approach to information maximization. In: NeurIPS (2004)
2. Bau, D., Zhou, B., Khosla, A., Oliva, A., Torralba, A.: Network dissection: quantifying interpretability of deep visual representations. In: CVPR (2017)
3. Bengio, Y., Léonard, N., Courville, A.: Estimating or propagating gradients through stochastic neurons for conditional computation. arXiv preprint arXiv:1308.3432 (2013)
4. Bolukbasi, T., Wang, J., Dekel, O., Saligrama, V.: Adaptive neural networks for efficient inference. In: ICML (2017)
5. Chen, R., Chen, H., Huang, G., Ren, J., Zhang, Q.: Explaining neural networks semantically and quantitatively. In: ICCV (2019)
6. Chen, Z., Li, Y., Bengio, S., Si, S.: You look twice: gaternet for dynamic filter selection in CNNS. In: CVPR (2019)
7. Dong, X., Huang, J., Yang, Y., Yan, S.: More is less: a more complicated network with less inference complexity. In: CVPR (2017)
8. Dosovitskiy, A., Brox, T.: Inverting visual representations with convolutional networks. In: CVPR (2016)
9. Figurnov, M., et al.: Spatially adaptive computation time for residual networks. In: CVPR (2017)
10. Fong, R., Vedaldi, A.: Net2vec: quantifying and explaining how concepts are encoded by filters in deep neural networks. In: CVPR (2018)
11. Gao, X., Zhao, Y., Dudziak, L., Mullins, R., Xu, C.Z.: Dynamic channel pruning: feature boosting and suppression. ICLR (2018)
12. Goodfellow, I.J., Shlens, J., Szegedy, C.: Explaining and harnessing adversarial examples. ICLR (2015)
13. He, K., Zhang, X., Ren, S., Sun, J.: Deep residual learning for image recognition. In: CVPR (2016)
14. He, Y., Kang, G., Dong, X., Fu, Y., Yang, Y.: Soft filter pruning for accelerating deep convolutional neural networks. IJCAI (2018)

15. He, Y., Zhang, X., Sun, J.: Channel pruning for accelerating very deep neural networks. In: ICCV (2017)
16. Hu, J., Shen, L., Sun, G.: Squeeze-and-excitation networks. In: CVPR (2018)
17. Ioffe, S., Szegedy, C.: Batch normalization: accelerating deep network training by reducing internal covariate shift. Int. Conf. Mach. Learn. (2015)
18. Kaiser, L., Bengio, S.: Can active memory replace attention? In: NeurIPS (2016)
19. Kaiser, Ł., et al.: Fast decoding in sequence models using discrete latent variables. ICML (2018)
20. Koh, P.W., Liang, P.: Understanding black-box predictions via influence functions. ICML (2017)
21. Krizhevsky, A., Hinton, G.: Learning multiple layers of features from tiny images. Tech. rep, Citeseer (2009)
22. Lakkaraju, H., Kamar, E., Caruana, R., Horvitz, E.: Identifying unknown unknowns in the open world: representations and policies for guided exploration. In: AAAI (2017)
23. Li, H., Kadav, A., Durdanovic, I., Samet, H., Graf, H.P.: Pruning filters for efficient convnets. ICLR (2016)
24. Liu, L., Deng, J.: Dynamic deep neural networks: optimizing accuracy-efficiency trade-offs by selective execution. In: AAAI (2018)
25. Liu, Z., Li, J., Shen, Z., Huang, G., Yan, S., Zhang, C.: Learning efficient convolutional networks through network slimming. In: ICCV (2017)
26. Lundberg, S.M., Lee, S.I.: A unified approach to interpreting model predictions. In: NeurIPS (2017)
27. Maaten, L., Hinton, G., Visualizing data using t-SNE: Visualizing data using t-SNE. J. Mach. Learn. Res. **9**, 2579–2605 (2008)
28. Morcos, A.S., Barrett, D.G., Rabinowitz, N.C., Botvinick, M.: On the importance of single directions for generalization. ICLR (2018)
29. Nair, V., Hinton, G.E.: Rectified linear units improve restricted boltzmann machines. In: International Conference on Machine Learning (2010)
30. Paszke, A., et al.: Automatic differentiation in pytorch. NeurIPS Workshop (2017)
31. Russakovsky, O., et al.: Imagenet large scale visual recognition challenge. IJCV **115**(3), 211–252 (2015)
32. Simonyan, K., Zisserman, A.: Very deep convolutional networks for large-scale image recognition. arXiv preprint arXiv:1409.1556 (2014)
33. Szegedy, C., et al.: Going deeper with convolutions. In: CVPR (2015)
34. Wang, J., Zhang, Z., Xie, C., Premachandran, V., Yuille, A.: Unsupervised learning of object semantic parts from internal states of cnns by population encoding. arXiv preprint arXiv:1511.06855 (2015)
35. Wang, X., Yu, F., Dou, Z.Y., Darrell, T., Gonzalez, J.E.: Skipnet: learning dynamic routing in convolutional networks. In: ECCV (2018)
36. Wang, Y., Su, H., Zhang, B., Hu, X.: Interpret neural networks by identifying critical data routing paths. In: CVPR (2018)
37. Yiyou, S., Sathya N., R., Vikas, S.: Adaptive activation thresholding: dynamic routing type behavior for interpretability in convolutional neural networks. In: ICCV (2019)
38. Yosinski, J., Clune, J., Bengio, Y., Lipson, H.: How transferable are features in deep neural networks? In: NeurIPS (2014)
39. Yosinski, J., Clune, J., Nguyen, A., Fuchs, T., Lipson, H.: Understanding neural networks through deep visualization. Int. Conf. Mach. Learn. Workshop (2015)

40. Zeiler, M.D., Fergus, R.: Visualizing and understanding convolutional networks. In: Fleet, D., Pajdla, T., Schiele, B., Tuytelaars, T. (eds.) ECCV 2014. LNCS, vol. 8689, pp. 818–833. Springer, Cham (2014). https://doi.org/10.1007/978-3-319-10590-1_53
41. Zhang, Q., Nian Wu, Y., Zhu, S.C.: Interpretable convolutional neural networks. In: CVPR (2018)
42. Zhang, Q., Yang, Y., Wu, Y.N., Zhu, S.C.: Interpreting cnns via decision trees. In: CVPR (2019)
43. Zhuang, Z., et al.: Discrimination-aware channel pruning for deep neural networks. In: NeurIPS (2018)

Omni-Sourced Webly-Supervised Learning for Video Recognition

Haodong Duan[1(✉)], Yue Zhao[1], Yuanjun Xiong[2], Wentao Liu[3], and Dahua Lin[1]

[1] The Chinese University of Hong Kong, Sha Tin, Hong Kong
dh019@ie.cuhk.edu.hk
[2] Amazon AI, Shanghai, China
[3] Sensetime Research, Tai Po, Hong Kong

Abstract. We introduce OmniSource, a novel framework for leveraging web data to train video recognition models. OmniSource overcomes the barriers between data formats, such as images, short videos, and long untrimmed videos for webly-supervised learning. First, data samples with multiple formats, curated by task-specific data collection and automatically filtered by a teacher model, are transformed into a unified form. Then a joint-training strategy is proposed to deal with the domain gaps between multiple data sources and formats in webly-supervised learning. Several good practices, including data balancing, resampling, and cross-dataset mixup are adopted in joint training. Experiments show that by utilizing data from multiple sources and formats, OmniSource is more data-efficient in training. With only 3.5M images and 800K min videos crawled from the internet without human labeling (less than 2% of prior works), our models learned with OmniSource improve Top-1 accuracy of 2D- and 3D-ConvNet baseline models by 3.0% and 3.9%, respectively, on the Kinetics-400 benchmark. With OmniSource, we establish new records with different pretraining strategies for video recognition. Our best models achieve **80.4%**, **80.5%**, and **83.6%** Top-1 accuracies on the Kinetics-400 benchmark respectively for training-from-scratch, ImageNet pre-training and IG-65M pre-training.

1 Introduction

Following the great success of representation learning in image recognition [15, 17,22,39], recent years have witnessed great progress in video classification thanks to the development of stronger models [3,38,43,47] as well as the collection of larger-scale datasets [3,31,32,56]. However, labelling large-scale image datasets [37,59] is well known to be costly and time-consuming. It is even more difficult to do so for trimmed video recognition. The reason is that most online videos are *untrimmed*, *i.e.* containing numerous shots with multiple concepts, making it unavoidable to first go through the entire video and then manually

Electronic supplementary material The online version of this chapter (https://doi.org/10.1007/978-3-030-58555-6_40) contains supplementary material, which is available to authorized users.

A. Vedaldi et al. (Eds.): ECCV 2020, LNCS 12360, pp. 670–688, 2020.
https://doi.org/10.1007/978-3-030-58555-6_40

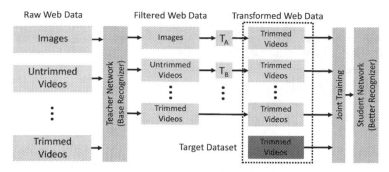

Fig. 1. OmniSource Framework. We first train a teacher network on the target dataset. Then, we use the teacher network to filter collected web data of different formats, to reduce noise and improve data quality. Specific transformations are conducted on the filtered out data corresponding to their formats. The target dataset and auxiliary web datasets are used for joint training of the student network

cut it into informative video clips based on a specific query. Such procedure requires far more efforts than image annotation where a simple glance and click is needed. As a result, while the quantity of web videos grows exponentially over the past 3 years, the Kinetics dataset merely grows from 300K videos in 400 classes [19] to 650K in 700 classes [2], partially limiting the scaling-up of video architectures [3] (Fig. 1).

Instead of confining ourselves to the well-annotated trimmed videos, we move beyond by exploring the abundant visual data that are publicly available on the Internet in a more labor-saving way. These visual data are in various formats, including images, short video clips, and long videos. They capture the same visual world while exhibiting different advantages: *e.g.* images may be of higher quality and focus on distinctive moments; short videos may be edited by the user, therefore contain denser information; long videos may depict an event in multiple views. We transform the data in different formats into a unified form so that a single model can combine the best of both worlds.

Recent works [12,29] explore the possibility of pre-training from massive unlabeled web images or videos only with hashtags. However, they restrict the scope to the data of a single format. Also, these methods usually require billions of images to obtain a pre-trained 2D CNN model that is resilient to noise, which poses great costs and restricts its practicability. Besides, to take advantage of representation learned from large-scale images for videos, we have to take extra steps to transfer the 2D ConvNets to the 3D counterparts, either by inflating [3] or distillation [13], and then perform fine-tuning on the target dataset, which is tedious and may be suboptimal.

In this work, we propose a simple and unified framework for video classification while utilizing multiple sources of web data in different formats simultaneously. To enhance data efficiency, we propose task-specific data collection, *i.e.* obtaining topmost results using class labels as keywords on search engines, making the supervision most informative. Our framework consists of three steps: (1) We train one (or more) teacher network on the labeled dataset; (2) For each source of data collected, we apply the corresponding teacher network to obtain pseudo-labels and filter out irrelevant samples with low confidence; (3) We apply different transforms to convert each type of web data (*e.g.* images) to the target input format (*e.g.* video clips) and train the student network.

There are two main obstacles during joint training with the labeled dataset and unlabeled web datasets. First, possible domain gaps occur. For example, web images may focus more on objects and contain less motion blur than videos. Second, teacher filtering may lead to unbalanced data distribution across different classes. To mitigate the domain gap, we propose to balance the size of training batches between the labeled dataset and unlabeled web datasets and apply cross-dataset *mixup*. To cope with data imbalance, we try several resampling strategies. All these techniques contribute to the success of our approach.

Compared to the previous methods, our method excels at the following aspects: (1) It leverages a mixture of web data forms, including images, trimmed videos and untrimmed videos into one student network, aiming at an *omni-sourced* fashion. (2) It is *data-efficient*. Empirical results show that only 2M images, a significantly smaller amount compared to the total frame number of Kinetics (240K videos, ~70M frames), are needed to produce notable improvements (about 1%). For trimmed videos, the required amount is around 0.5M. In stark contrast, 65M videos are collected to obtain a noise-resilient pre-trained model in [12,49]. It is also noteworthy that our framework can also benefit from the massively weakly-supervised pre-training from billions of images or videos.

To sum up, our contributions are as follows:

(1) We propose OmniSource, a simple and efficient framework for webly-supervised video classification, which can leverage web data in different formats.
(2) We propose good practices for problems during joint training with omni-sourced data, include source-target balancing, resampling and cross-dataset mixup.
(3) In experiments, our models trained by OmniSource achieve state-of-the-art performance on the Kinetics-400, for all pre-training strategies we tested.

2   Related Work

Webly-Supervised Learning. Leveraging information from the Internet, termed *webly-supervised learning*, has been extensively explored [11,14,25,50]. Divvala *et al.* in [7] proposes to automatically learn models from online resources for visual concept discovery and image annotation. Chen *et al.* reveals that images crawled from the Internet can yield superior results over the fully-supervised method [5]. For video classification, Ma *et al.* proposes to use web images to boost action recognition models in [28] at the cost of manually filtering web action images. To free from additional human labor, efforts have been made to learn video concept detectors [26,51] or to select relevant frames from videos [10,41,52]. These methods are based on frames thus fail to consider the rich temporal dynamics of videos. Recent works [12,29] show that webly-supervised learning can produce better pre-training models with very large scale noisy data ($\sim 10^9$ images and $\sim 10^7$ videos). Being orthogonal to the pre-training stage, our framework works in a joint-training paradigm and is complementary to large scale pre-training.

Semi-Supervised Learning. Our framework works under the semi-supervised setting where labeled and unlabeled(web) data co-exist. Representative classical approaches include label propagation [61], self-training [36], co-training [1], and graph networks [21]. Deep models make it possible to learn directly from unlabeled data via generative models [20], self-supervised learning [53], or consensus of multiple experts [54]. However, most existing methods are validated only on small scale datasets. One concurrent work [49] proposes to first train a student network with unlabeled data with pseudo-labels and then fine-tune it on the labeled dataset. Our framework, however, works on the two sources simultaneously, free from the pretrain-finetune paradigm and is more data-efficient.

Distillation. According to the setting of knowledge distillation [16] and data distillation [35], given a set of manually labeled data, we can train a base model in the manner of supervised learning. The model is then applied to the unlabeled data or its transforms. Most of the previous efforts [35] are confined to the domain of images. In [13], Rohit *et al.* proposes to distill spatial-temporal features from unlabeled videos with image-based teacher networks. Our framework is capable of distilling knowledge from multiple sources and formats within a single network.

Domain Adaptation. Since web data from multiple sources are taken as input, domain gaps inevitably exist. Previous efforts [4,6,45] in domain adaptation focus on mitigating the data shift [34] in terms of data distributions. On the contrary, our framework focuses on adapting visual information in different formats (*e.g.* still images, long videos) into the same format (*i.e.* trimmed video clips).

Video Classification. Video analysis has long been tackled using hand-crafted feature [24,46]. Following the success of deep learning for images, video classification architectures have been dominated by two families of models, *i.e.* two-stream [38,47] and 3D ConvNets [3,44]. The former uses 2D networks to extract image-level feature and performs temporal aggregation [18,47,58] on top while the latter learns spatial-temporal features directly from video clips [8,43,44].

3   Method

3.1   Overview

We propose a unified framework for omni-sourced webly-supervised video recognition, formulated in Sect. 3.2. The framework exploits web data of various forms (images, trimmed videos, untrimmed videos) from various sources (search engine, social media, video sharing platform) in an integrated way. Since web data can be very noisy, we use a teacher network to filter out samples with low confidence scores and obtain pseudo labels for the remaining ones (Sect. 3.4). We devise transformations for each form of data to make them applicable for the target task in Sect. 3.5. In addition, we explore several techniques to improve the robustness of joint training with web data in Sect. 3.6.

3.2   Framework Formulation

Given a target task (trimmed video recognition, e.g.) and its corresponding *target* dataset $\mathcal{D}_\mathcal{T} = \{(\mathbf{x}_i, \mathbf{y}_i)\}$, we aim to harness information from unlabeled web resources $\mathcal{U} = \mathcal{U}_1 \cup \cdots \cup \mathcal{U}_n$, where $\mathcal{U}_i$ refers to unlabeled data in a specific source or format. **First**, we construct the pseudo-labeled dataset $\widehat{\mathcal{D}}_i$ from $\mathcal{U}_i$. Samples with low confidence are dropped using a teacher model $\mathcal{M}$ trained on $\mathcal{D}_\mathcal{T}$, and the remaining data are assigned with pseudo-labels $\widehat{\mathbf{y}} = \text{PseudoLabel}(\mathcal{M}(\mathbf{x}))$. **Second**, we devise appropriate transforms $\mathcal{T}_i(\mathbf{x}) : \widehat{\mathcal{D}}_i \to \mathcal{D}_{\mathcal{A},i}$ to process data in a specific format (*e.g.* still images or long videos) into the data format (trimmed videos in our case) in the target task. We denote the union of $\mathcal{D}_{\mathcal{A},i}$ to be the *auxiliary* dataset $\mathcal{D}_\mathcal{A}$. **Finally**, a model $\mathcal{M}'$ (not necessarily the original $\mathcal{M}$), can be jointly trained on $\mathcal{D}_\mathcal{T}$ and $\mathcal{D}_\mathcal{A}$. In each iteration, we sample two mini-batches of data $\mathcal{B}_\mathcal{T}$, $\mathcal{B}_\mathcal{A}$ from $\mathcal{D}_\mathcal{T}$, $\mathcal{D}_\mathcal{A}$ respectively. The loss is a sum of cross entropy loss on both $\mathcal{B}_\mathcal{T}$ and $\mathcal{B}_\mathcal{A}$, indicated by Eq. 1.

$$\mathcal{L} = \sum_{\mathbf{x},\mathbf{y} \in \mathcal{B}_\mathcal{T}} \mathcal{L}(\mathcal{F}(\mathbf{x}; \mathcal{M}'), \mathbf{y}) + \sum_{\mathbf{x},\widehat{\mathbf{y}} \in \mathcal{B}_\mathcal{A}} \mathcal{L}(\mathcal{F}(\mathbf{x}; \mathcal{M}'), \widehat{\mathbf{y}}) \qquad (1)$$

For clarification, we compare our framework with some recent works on billion-scale webly-supervised learning in Table 1. OmniSource is capable of dealing with web data from multiple sources. It is designed to help a specific task, treats webly-supervision as co-training across multiple data sources instead of pre-training, thus is much more data-efficient. It is also noteworthy that our framework is orthogonal to webly-supervised pre-training [12].

Table 1. Difference to previous works. The notions follow Sect. 3.2: $\mathcal{U}$ is the unlabeled web data, $\mathcal{D}_T$ is the target dataset. $|\mathcal{U}|$, $|\mathcal{D}_A|$ denotes the scale of web data and filtered auxiliary dataset

	Webly-supervised pretrain [12,29]	Web-scale semi-supervised [49]	OmniSource (Ours)
Procedure	1. Train a model $\mathcal{M}$ on $\mathcal{U}$	1. Train a model $\mathcal{M}$ on $\mathcal{D}_T$	1. Train one (or more) model $\mathcal{M}$ on $\mathcal{D}_T$
	2. Fine-tune $\mathcal{M}$ on $\mathcal{D}_T$	2. Run $\mathcal{M}$ on $\mathcal{U}$ to pseudo-labeled $\hat{\mathcal{D}}$	2. Run $\mathcal{M}$ on $\bigcup_i \mathcal{U}_i$ to pseudo-labeled $\bigcup_i \hat{\mathcal{D}}_i$
		3. Train a student model $\mathcal{M}'$ on $\hat{\mathcal{D}}$	(Samples under certain threshold are dropped.)
		4. Fine-tune $\mathcal{M}'$ on $\mathcal{D}_T$	3. Apply transforms $\mathcal{T}_i : \hat{\mathcal{D}}_i \rightarrow \mathcal{D}_{A,i}$
			4. Train model $\mathcal{M}'$ (or $\mathcal{M}$) on $\mathcal{D}_T \cup \mathcal{D}_A$
$|\mathcal{U}|$	3.5B images **or** 65M videos	1B images **or** 65M videos	$|\mathcal{U}|$: 13M images **and** 1.4M videos (**0.4%–2%**)
			$|\mathcal{D}_A|$: 3.5M images **and** 0.8M videos (**0.1%–1%**)

3.3 Task-Specific Data Collection

We use class names as keywords for data **crawling**, with no extra query expansion. For tag-based system like Instagram, we use automatic permutation and stemming[1] to generate tags. We crawl web data from various sources, including search engine, social media and video sharing platform. Because Google restricts the number of results for each query, we conduct multiple queries, each of which is restricted by a specific period of time. Comparing with previous works [12,29] which rely on large-scale web data with hashtags, our task-specific collection uses keywords highly correlated with labels, making the supervision stronger. Moreover, it reduces the required amount of web data by 2 orders of magnitude (*e.g.* from 65M to 0.5M videos on Instagram).

After data collection, we first remove invalid or corrupted data. Since web data may contain samples very similar to validation data, data **de-duplication** is essential for a fair comparison. We perform content-based data de-duplication based on feature similarity. First, we extract frame-level features using an ImageNet-pretrained ResNet50. Then, we calculate the cosine similarity of features between the web data and target dataset and perform pairwise comparison after whitening. The average similarity among different crops of the same frame is used as the threshold. Similarity above it indicates suspicious duplicates. For Kinetics-400, we filter out 4,000 web images (out of 3.5M, 0.1%) and 400 web videos (out of 0.5M, 0.1%). We manually inspect a subset of them and find that less than 10% are real duplicates.

---

[1] For example, "beekeeping" can be transformed to "beekeep", and "keeping bee".

3.4   Teacher Filtering

Data crawled from the web are inevitably noisy. Directly using collected web data for joint training leads to a significant performance drop (over 3%). To prevent irrelevant data from polluting the training set, we first train a teacher network $\mathcal{M}$ on the target dataset and discard those web data with low confidence scores. For web images, we observe performance deterioration when deflating 3D teachers to 2D and therefore only use 2D teachers. For web videos, we find both applicable and 3D teachers outperform 2D counterparts consistently (Fig. 2).

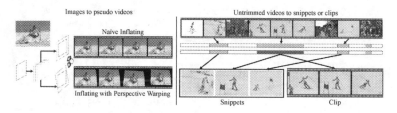

Fig. 2. Transformations. Left: Inflating images to clips, by replicating or inflating with perspective warping; Right: Extracting segments or clips from untrimmed videos, guided by confidence scores

3.5   Transforming to the Target Domain

Web Images. To prepare web images for video recognition training, we devise several ways to transform images into pseudo videos. The first naïve way is to replicate the image n times to form an n-frame clip. However, such clips may not be optimal since there is a visible gap between static clips and natural videos which visually change over time. Therefore, we propose to generate video clips from static images by viewing them with a moving camera. Given an image I, under the standard perspective projection model [9], an image with another perspective $\widetilde{I}$ can be generated by a homographic transform $\mathcal{H}$ which is induced by a homographic matrix $\mathbf{H} \in \mathbb{R}^{3 \times 3}$, $i.e.$, $\widetilde{I} = \mathcal{H}(I) = \mathcal{F}(I; \mathbf{H})$. To generate a clip $J = \{J_1, \cdots, J_N\}$ from I, starting from $J_1 = I$, we have

$$J_i = \mathcal{H}_i(J_{i-1}) = (\mathcal{H}_i \circ \mathcal{H}_{i-1} \circ \cdots \circ \mathcal{H}_1)(I) \tag{2}$$

Each matrix $\mathbf{H}_i$ is randomly sampled from a multivariate Gaussian distribution $\mathcal{N}(\mu, \Sigma)$, while the parameters μ and Σ are estimated using maximum likelihood estimation on the original video source. Once we get pseudo videos, we can leverage web images for joint training with trimmed video datasets.

Untrimmed Videos. Untrimmed videos form an important part of web data. To exploit web untrimmed videos for video recognition, we adopt different transformations respectively for 2D and 3D architectures.

For 2D TSN, *snippets* sparsely sampled from the entire video are used as input. We first extract frames from the entire video at a low frame rate (1 FPS). A 2D teacher is used to get the confidence score of each frame, which also divides frames into positive ones and negative ones. In practice, we find that only using positive frames to construct snippets is a sub-optimal choice. Instead, combining negative frames and positive frames can form harder examples, results in better recognition performance. In our experiments, we use 1 positive frame and 2 negative frames to construct a 3-snippet input.

For 3D ConvNets, video *clips* (densely sampled continuous frames) are used as input. We first cut untrimmed videos into 10-s clips, then use a 3D teacher to obtain confidence scores. Only positive clips are used for joint training.

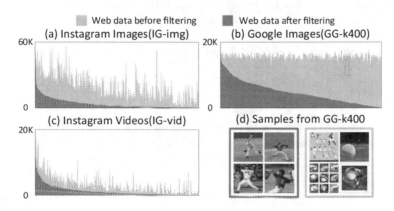

Fig. 3. Web Data Distribution. The inter-class distribution of three web datasets is visualized in (a, b, c), both before and after filtering. (d) gives out samples of filtered out images (cyan) and remained images (blue) for GG-K400. Teacher filtering successfully filters out lots of negative examples while making inter-class distribution more uneven (Color figure online)

3.6   Joint Training

Once web data are filtered and transformed into the same format of that in the target dataset $\mathcal{D}_\mathcal{T}$, we construct an *auxiliary* dataset $\mathcal{D}_\mathcal{A}$. A network can then be trained with both $\mathcal{D}_\mathcal{T}$ and $\mathcal{D}_\mathcal{A}$ using sum of cross-entropy loss in Eq. 1. As shown in Fig. 3, web data across classes are extremely unbalanced, especially after teacher filtering. Also there exists potential domain gap between $\mathcal{D}_\mathcal{T}$ and $\mathcal{D}_\mathcal{A}$. To mitigate these issues, we enumerate several good practices as follows.

Balance Between Target and Auxiliary Mini-Batches. Since the auxiliary dataset may be much larger than the target dataset and the domain gap may occur, the data ratio between target and auxiliary mini-batches is crucial for the final performance. Empirically, $|\mathcal{B}_\mathcal{T}| : |\mathcal{B}_\mathcal{A}| = 2{:}1{-}1{:}1$ works reasonably well.

Resampling Strategy. Web data are extremely unbalanced, especially after teacher filtering (see Fig. 3). To alleviate this, we explore several sampling policies: (1) sampling from a clipped distribution: classes whose samples exceeds threshold N_c are clipped; (2) sampling from distribution modified by a power law: the probability of choosing class with N samples is proportional to $N^p (p \in (0, 1))$. We find that (2) parameterized by $p = 0.2$ is generally a better practice.

Cross-Dataset *mixup*. Mixup [55] is a widely used strategy in image recognition. It uses convex combinations of pairs of examples and their labels for training, thus improving the generalization of deep neural networks. We find that technique also works for video recognition. When training teacher networks on $\mathcal{D}_\mathcal{T}$ only, we use the linear combination of two clip-label pairs as training data, termed as *intra-dataset* mixup. When both target and auxiliary datasets are used, the two pairs are samples randomly chosen from both datasets, termed as *cross-dataset* mixup. Mixup works fairly well when networks are trained from scratch. For fine-tuning, the performance gain is less noticeable.

4 Datasets

In this section, we introduce the datasets on which experiments will be conducted. Then we go through different sources from which web data are collected.

4.1 Target Datasets

Kinetics-400. The Kinetics dataset [3] is one of the largest video datasets. We use the version released in 2017 which contains 400 classes and each category has more than 400 videos. In total, it has around 240K, 19K, and 38K videos for training, validation and testing subset respectively. In each video, a 10 s clip is annotated and assigned a label. These 10 s clips constitute the data source for the default supervised learning setting, which we refer to **K400-tr**. The rest part of training videos is used to mimic untrimmed videos sourced from the Internet which we refer to **K400-untr**.

Youtube-car. Youtube-car [60] is a fine-grained video dataset with 10K training and 5K testing videos of 196 types of cars. The videos are untrimmed, last several minutes. Following [60], the frames are extracted from videos at 4 FPS.

UCF101. UCF101 [40] is a small scale video recognition dataset, which has 101 classes and each class has around 100 videos. We use the official split-1 in our experiments, which has about 10K and 3.6K videos for training and testing.

4.2 Web Sources

We collect web images and videos from various sources including search engines, social medias and video sharing platforms.

GoogleImage. GoogleImage is a search engine based web data source for Kinetics-400, Youtube-car and UCF101. We query each class name in the target dataset on Google to get related web images. We crawl 6M, 70K, 200K URLs for Kinetics-400, Youtube-car and UCF101 respectively. After data cleaning and teacher filtering, about 2M, 50K, 100K images are used for training on these three datasets. We denote the three datasets as **GG-k400**, **GG-car**, and **GG-UCF** respectively.

Instagram. Instagram is a social media based web data source for Kinetics-400. It consists of InstagramImage and InstagramVideo. We generate several tags for each class in Kinetics-400, resulting in 1,479 tags and 8.7M URLs. After removing corrupted data and teacher filtering, about 1.5M images and 500K videos are used for joint training, denoted as **IG-img** and **IG-vid**. As shown in Fig. 3, IG-img is significantly unbalanced after teacher filtering. Therefore, in the coming experiments, IG-img is used in combination with GG-k400.

YoutubeVideo. YoutubeVideo is a video sharing platform based web data source for Youtube-car. We crawl 28K videos from youtube by querying class names. After de-duplicating (remove videos in the original Youtube-car dataset) and teacher filtering, 17K videos remain, which we denote as **YT-car-17k**.

5   Experiments

5.1   Video Architectures

We mainly study two families of video classification architectures, namely Temporal Segment Networks [47] and 3D ConvNets [3], to verify the effectiveness of our design. Unless specified, we use ImageNet-pretrained models for initialization. We conduct all experiments using MMAction [57].

2D TSN. Different from the original setting in [47], we choose ResNet-50 [15] to be the backbone, unless otherwise specified. The number of segments is set to be 3 for Kinetics/UCF-101 and 4 for Youtube-car, respectively.

3D ConvNets. For 3D ConvNet, we use the SlowOnly architecture proposed in [8] in most of our experiments. It takes 64 consecutive frames as a video clip and sparsely samples 4/8 frames to form the network input. Different initialization strategies are explored, including training from scratch and fine-tuning from a pre-trained model. Besides, more advanced architecture like Channel Separable Network [43] and more powerful pre-training (IG-65M [12]) is also explored.

5.2   Verifying the Efficacy of OmniSource

We verify our framework's efficacy by examining several questions.

Why do We Need Teacher Filtering and are Search Results Good Enough? Some may question the necessity of a teacher network for filtering under the impression that a modern search engine might have internally utilized a visual recognition model, possibly trained on massively annotated data, to help generate the search results. However, we argue that web data are inherently noisy and we observe nearly half of the returned results are irrelevant. More quantitatively, 70%–80% of the web data are rejected by the teacher. On the other hand, we conduct an experiment without teacher filtering. Directly using collected web data for joint training leads to a significant (over 3%) performance drop on TSN. This reveals that teacher filtering is necessary to help retain the useful information from the crawled web data while eliminating the useless.

Does Every Data Source Contribute? We explore the contribution of different source types: images, trimmed videos and untrimmed videos. For each data source, we construct auxiliary dataset and use it for joint training with K400-tr. Results in Table 2 reveal that every source contributes to improving accuracy on the target task. When combined, the performance is further improved.

Table 2. Every source contributes. We find that each source contributes to the target task. With all sources combined (we intuitively set the ratio as: K400-tr: Web-img: IG-vid: K400-untr = 2: 1: 1: 1), the improvement can be more considerable. The conclusion holds for both 2D TSN and 3D ConvNets (Format: Top-1 Acc/ Top-5 Acc)

Arch/Dataset	K400-tr	+GG-k400	+GG& IG-img	+IG-vid	+K400-untr	+ All
TSN-3seg R50	70.6/89.4	71.5/89.5	72.0/90.0	72.0/90.3	71.7/89.6	73.6/91.0
SlowOnly4x16, R50	73.8/90.9	74.5/91.4	75.2/91.6	75.2/91.7	74.5/91.1	76.6/92.5

For images, when the combination of GG-k400 and IG-img is used, the Top-1 accuracy increases around 1.4%. For trimmed videos, we focus on IG-vid. Although being extremely unbalanced, IG-vid still improves Top-1 accuracy by over 1.0% in all settings. For untrimmed videos, we use the untrimmed version of Kinetics-400 (K400-untr) as the video source and find it also works well.

Do Multiple Sources Outperform a Single Source? Seeing that web data from multiple sources can jointly contribute to the target dataset, we wonder if multiple sources are still better than a single source with the same budget. To verify this, we consider the case of training TSN on both K400-tr and $\mathcal{D}_A$ = GG-k400 + IG-img. We fix the scale of auxiliary dataset to be that of GG-k400 and vary the ratio between GG-k400 and IG-img by replacing images from GG-k400 with those in IG-img. From Fig. 4, we observe an improvement of 0.3% without increasing $|\mathcal{D}_A|$, indicating that multiple sources provide complementary information by introducing diversity.

Does OmniSource Work with Different Architectures? We further conduct experiments on a wide range of architectures and obtain the results in Table 3. For TSN, we use EfficientNet-B4 [42] instead as the backbone, on which OmniSource improves Top-1 accuracy by 1.9%. For 3D-ConvNets, we conduct experiments on the SlowOnly-8x8-ResNet101 baseline, which takes longer input

Table 3. Improvement under various experiment configurations. OmniSource is extensively tested on various architectures with various pretraining strategies. The improvement is significant in **ALL** tested choices. Even for the SOTA setting, which uses 65M web videos for pretraining, OmniSource still improves the Top-1 accuracy by 1.0% (Format: Top-1/Top-5 Acc)

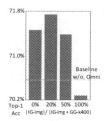

Fig. 4. Multi-source is better. Mixed sources lead to better performance with a constrained number of web images

Arch	Backbone	Pretrain	w/o. Omni	w/. Omni	Δ
TSN-3seg	ResNet50	ImageNet	70.6/89.4	73.6/91.0	+3.0/+1.6
TSN-3seg	ResNet50	IG-1B	73.1/90.4	75.7/91.9	+2.6/+1.5
TSN-3seg	Efficient-b4	ImageNet	73.3/91.0	75.2/92.0	+1.9/+1.0
SlowOnly-4x16	ResNet50	–	72.9/90.9	76.8/92.5	+3.9/+1.6
SlowOnly-4x16	ResNet50	ImageNet	73.8/90.9	76.6/92.5	+2.8/+1.6
SlowOnly-8x8	ResNet101	–	76.3/92.6	80.4/94.4	+4.1/+1.8
SlowOnly-8x8	ResNet101	ImageNet	76.8/92.8	80.5/94.4	+3.7/+1.6
irCSN-32x2	irCSN-152	IG-65M	82.6/95.3	83.6/96.0	+1.0/+0.7

and has a larger backbone. Our framework also works well in this case, improving the Top-1 accuracy from 76.3% to **80.4%** when training from scratch, from 76.8% to **80.5%** with ImageNet pretraining. The improvement on larger networks is higher, suggesting that deeper networks are more prone to suffering from the scarcity of video data and OmniSource can alleviate this.

Is OmniSource Compatible with Different Pre-Training Strategies?
As discussed, OmniSource alleviates the data-hungry issue by utilizing auxiliary data. One natural question is: how does it perform when training 3D networks from scratch? Can we simply drop ImageNet pretraining in pursuit of a more straightforward training policy? Indeed, we find that OmniSource works fairly well under this setting and interestingly the performance gain is more significant than fine-tuning. For example, SlowOnly-(4x16, R50) increases the Top-1 accuracy by 3.9% when training from scratch while fine-tuning only increases by 2.8%. The model trained from scratch beats the fine-tuned counterpart by 0.2% with OmniSource though being 0.9% lower with only K400-tr. Similar results can be observed for SlowOnly-(8x8, R101). With large-scale webly supervised pretraining, OmniSource still leads to significant performance improvement (+2.6% Top-1 for TSN-3seg-R50, +1.0% Top-1 for irCSN-32x2).

Do Features Learned by OmniSource Transfer to Other Tasks?
Although OmniSource is designed for a target video recognition task, the learned features also transfer well to other video recognition tasks. To evaluate the transfer capability, we finetune the learned model on two relatively smaller datasets: UCF101 [40] and HMDB51 [23]. Table 4 indicates that on both benchmarks, pretraining with OmniSource leads to significant performance improvements. Following standard evaluation protocol, SlowOnly-8x8-R101 achieves 97.3% Top-1 accuracy on UCF101, 79.0% Top-1 accuracy on HMDB51 with RGB input. When combined with optical flow, it achieves **98.6%** and **83.8%** Top-1 accuracy on UCF101 and HMDB51, which is the new state-of-the-art. More results on transfer learning are provided in the supplementary material.

Table 4. OmniSource features transfer well. We finetune on UCF101 and HMDB51 with `K400-tr` pretrained weight. Pretraining with OmniSource improves the performance significantly

Arch	UCF101-split1		HMDB51-split1	
	w.o. Omni	w/. Omni	w/o. Omni	w. Omni
TSN-3seg R50[FT]	91.5	93.3	63.5	65.9
SlowOnly 4x16, R50[FT]	94.7	96.0	69.4	70.7
SlowOnly4x16, R50[SC]	94.1	96.0	65.8	71.0

Drinking shots 16.3% + 18.4%   Drinking beer 38.0% - 2.0%   Rock scissors paper 10.0% + 12.0%   Shake hands 29.2% - 2.1%

Mopping floor 36.0% + 2.0%   Sweeping floor 54.0% + 10.0%   Eating doughnuts 18.4% + 8.2%   Eating burger 67.3% + 4.1%

Fig. 5. Confusing pairs improved by OmniSource. The original accuracy and change are denoted in black and in color (Color figure online)

Does OmniSource Work in Different Target Domains? Our framework is also effective and efficient in various domains. For a fine-grained recognition benchmark called **Youtube-car**, we collect 50K web images (`GG-car`) and 17K web videos (`YT-car-17k`) for training. Table 5 shows that the performance gain is significant: 5% in both Top-1 accuracy and mAP. On **UCF-101**, we train a two-stream TSN network with BNInception as the backbone. The RGB stream is trained either with or without `GG-UCF`. The results are listed in Table 6. The Top-1 accuracy of the RGB stream improves by 2.7%. When fused with the flow stream, there is still an improvement of 1.1%.

Where Does the Performance Gain Come From? To find out why web data help, we delve deeper into the collected web dataset and analyze the improvement on individual classes. We choose TSN-3seg-R50 trained either with or without `GG-k400`, where the improvement is 0.9% on average. We mainly focus on the confusion pairs that web images can improve. We define the *confusion score* of a class pair as $s_{ij} = (n_{ij} + n_{ji})/(n_{ij} + n_{ji} + n_{ii} + n_{jj})$, where n_{ij} denotes the number of images whose ground-truth are class i while being recognized as class j. Lower confusion score denotes better discriminating power between the two classes. We visualize some confusing pairs in Fig. 5. We find the improvement can be mainly attributed to two reasons: (1) Web data usually focus on key objects of action. For example, we find that in those pairs with the largest confusion score reduction, there exist pairs like "drinking beer" *vs.*"drinking shots", and "eating hotdog" *vs.*"eating chips". Training with web data leads to better object recognition ability in some confusing cases. (2) Web data usually include discriminative poses, especially for those actions which last for a short time. For example, "rock scissors paper" *vs.*"shaking hands" has the second-largest confusion score reduction. Other examples including "sniffing"-"headbutting", "break dancing"-"robot dancing", etc.

Table 5. Youtube-car

Setting	Top-1	mAP
Baseline	77.05	71.95
+GG-car	80.96	77.05
+YT-car-17k	81.68	78.61
+[GG-]+[YT-]	81.95	78.67

Table 6. UCF-101

Setting	+ Flow	Top-1
Baseline		86.04
+ GG-UCF		88.74
Baseline	✓	93.47
+ GG-UCF	✓	94.58

Table 7. Comparisons with Kinetics-400 state-of-the-art

Method	Backbone	Pretrain	Top-1	Top-5
TSN-7seg [47]	Inception-v3	ImageNet	73.9	91.1
TSM-8seg [27]	ResNet50	ImageNet	72.8	N/A
TSN-3seg (**Ours**)	ResNet50	ImageNet	73.6	91.0
TSN-3seg (**Ours**)	Efficient-b4	ImageNet	**75.2**	**92.0**
SlowOnly-8x8 [8]	ResNet101	–	75.9	N/A
SlowFast-8x8 [8]	ResNet101	–	77.9	93.2
SlowOnly-8x8 (**Ours**)	ResNet101	–	**80.4**	**94.4**
I3D-64x1 [3]	Inception-V1	ImageNet	72.1	90.3
NL-128x1 [48]	ResNet101	ImageNet	77.7	93.3
SlowFast-8x8 [8]	ResNet101	ImageNet	77.9	93.2
LGD-3D (RGB) [33]	ResNet101	ImageNet	79.4	94.4
STDFB [30]	ResNet152	ImageNet	78.8	93.6
SlowOnly-8x8 (**Ours**)	ResNet101	ImageNet	**80.5**	**94.4**
irCSN-32x2 [12]	irCSN-152	IG-65M	82.6	95.3
irCSN-32x2 (**Ours**)	irCSN-152	IG-65M	**83.6**	**96.0**

Table 8. Different ways to transform images into video clips. Still inflation is a strong baseline, while agnostic perspective warping performs best

Inflation	Top-1	Top-5
N/A	73.8	90.9
Replication (still)	74.1	91.2
Translation (random)	73.7	90.9
Translation (constant)	73.8	90.8
Perspective warp [spec]	74.4	91.3
Perspective warp [agno]	74.5	91.4

Table 9. Mixup technique can be beneficial to the model performance, both for intra- and cross-dataset cases. However, it works only when the model is trained from scratch

Pretraining	w. mixup	w.GG-img	Top-1	Top-5
ImageNet			73.8	90.9
ImageNet	✓		73.6	91.1
None			72.9	90.9
None	✓		73.3	90.9
None		✓	74.1	91.0
None	✓	✓	74.4	91.4

5.3 Comparisons with State-of-the-art

In Table 7, we compare OmniSource with current state-of-the-art on Kinetics-400. For 2D ConvNets, we obtain competitive performance with fewer segments and lighter backbones. For 3D ConvNets, considerable improvement is achieved for all pre-training settings with OmniSource applied. With IG-65M pre-trained irCSN-152, OmniSource achieves **83.6%** Top-1 accuracy, an absolute improvement of 1.0% with only 1.2% relatively more data, establishing a new record.

5.4 Validating the Good Practices in OmniSource

We conduct several ablation experiments on techniques we introduced. The target dataset is K400-tr and the auxiliary dataset is GG-k400 unless specified.

Transforming Images to Video Clips. We compare different ways to transform web images into clips in Table 8. Naïvely replicating still image brings limited improvement (0.3%). We then apply translation with randomized or constant speed to form pseudo clips. However, the performance deteriorates slightly, suggesting that translation cannot mimic the camera motion well. Finally, we resort to perspective warping to hallucinate camera motion. Estimating class-agnostic distribution parameters is slightly better, suggesting that all videos might share similar camera motion statistics.

Cross-Dataset Mixup. In Table 9, we find that mixup is effective for video recognition in both intra- and cross-dataset cases when the model is trained from scratch. The effect is unclear for fine-tuning. In particular, mixup can lead to 0.4% and 0.3% Top-1 accuracy improvement for intra- and inter-dataset cases.

Impact of Teacher Choice. Since both teacher and student networks can be 2D or 3D ConvNets, there are 4 possible combinations for teacher network choosing. For images, deflating 3D ConvNets to 2D yields a dramatic performance drop. Therefore, we do not use 3D ConvNet teachers for web images. For videos, however, 3D ConvNets lead to better filtering results comparing to its 2D counterpart. To examine the effect of different teachers, we fix the student model to be a ResNet-50 and vary the choices of teacher models (ResNet-50, EfficientNet-b4, and the ensemble of ResNet-152 and EfficientNet-b4). Consistent improvement is observed against the baseline (70.6%). The student accuracy increases when a better teacher network is used. It also holds for 3D ConvNets on web videos (Fig. 6).

Effectiveness When Labels Are Limited. To validate the effectiveness with limited labeled data, we construct 3 subsets of K400-tr with a proportion of 3%, 10%, and 30% respectively. We rerun the entire framework including data filtering with a weaker teacher. The final results on the validation set of K400-tr is shown in Fig. 7. Our framework consistently improves the performance as the percentage of labeled videos varies. Particularly, the gain is more significant when data are scarce, *e.g.* a relative increase of over 30% with 3% labeled data.

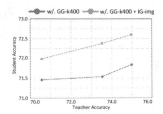
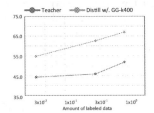
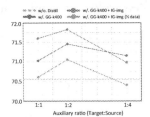

Fig. 6. Better teachers leads to better students

Fig. 7. OmniSource on subsets of Kinetics

Fig. 8. Accuracy with different ratios $|\mathcal{B}_{\mathcal{T}}| : |\mathcal{B}_{\mathcal{A}}|$

Balancing Between the Target and Auxiliary Dataset. We tune the ratio between the batch size of the target dataset $|\mathcal{B}_T|$ and the auxiliary dataset $|\mathcal{B}_A|$ and obtain the accuracy on Fig. 8. We test 3 scenarios: (1) the original GG-k400, clarified in Sect. 4.2; (2) [GG+IG]-k400, the union of GG-k400 and

Table 10. Resampling strategies. Simple resampling strategies lead to nontrivial improvement

Stategy	Top-1/5
None (original)	71.5/89.5
Clipped ($N_c = 5000$)	71.9/90.0
Power ($\sim N^p, p = 0.5$)	71.8/89.7
Power ($\sim N^p, p = 0.2$)	72.0/90.0

IG-img; (3) [GG+IG]-k400-half which is the half of (2). We observe that the performance gain is robust to the choice of $|\mathcal{B}_T|/|\mathcal{B}_A|$ in most cases. However, with less auxiliary data, the ratio has to be treated more carefully. For example, smaller $|\mathcal{D}_A|$ but larger $|\mathcal{B}_A|$ may cause overfitting auxiliary samples and hurt the overall result.

Resampling Strategies. The target dataset is usually balanced across classes. The nice property doesn't necessarily hold for the auxiliary dataset. Thus we propose several resampling strategies. From Table 10, we see that simple techniques to tailor the distribution into a more balanced one yield nontrivial improvements.

6   Conclusion

In this work, we propose OmniSource, a simple yet effective framework for webly-supervised video recognition. Our method can utilize web data from multiple sources and formats by transforming them into a same format. In addition, our task-specific data collection is more data-efficient. The framework is applicable to various video tasks. Under all settings of pretraining strategies, we obtain state-of-the-art performance on multiple benchmarks.

Acknowledgment. This work is partially supported by the SenseTime Collaborative Grant on Large-scale Multi-modality Analysis (CUHK Agreement No. TS1610626 & No. TS1712093), the General Research Fund (GRF) of Hong Kong (No. 14203518 & No. 14205719), and Innovation and Technology Support Program (ITSP) Tier 2, ITS/431/18F.

References

1. Blum, A., Mitchell, T.: Combining labeled and unlabeled data with co-training. In: COLT, pp. 92–100. ACM (1998)
2. Carreira, J., Noland, E., Hillier, C., Zisserman, A.: A short note on the kinetics-700 human action dataset. arXiv preprint arXiv:1907.06987 (2019)
3. Carreira, J., Zisserman, A.: Quo Vadis, action recognition? A new model and the kinetics dataset. In: CVPR, pp. 6299–6308 (2017)
4. Chen, M.-H., Kira, Z., AlRegib, G., Yoo, J., Chen, R., Zheng, J.: Temporal attentive alignment for large-scale video domain adaptation. In: ICCV, pp. 6321–6330 (2019)

5. Chen, X., Gupta, A.: Webly supervised learning of convolutional networks. In: ICCV, pp. 1431–1439 (2015)
6. Csurka, G.: A comprehensive survey on domain adaptation for visual applications. In: Csurka, G. (ed.) Domain Adaptation in Computer Vision Applications. ACVPR, pp. 1–35. Springer, Cham (2017). https://doi.org/10.1007/978-3-319-58347-1_1
7. Divvala, S.K., Farhadi, A., Guestrin, C.: Learning everything about anything: Webly-supervised visual concept learning. In: CVPR, pp. 3270–3277 (2014)
8. Feichtenhofer, C., Fan, H., Malik, J., He, K.: Slowfast networks for video recognition. In: ICCV, pp. 6202–6211 (2019)
9. Forsyth, D.A., Ponce, J.: Computer Vision: A Modern Approach. Prentice Hall Professional Technical Reference (2002)
10. Gan, C., Sun, C., Duan, L., Gong, B.: Webly-supervised video recognition by mutually voting for relevant web images and web video frames. In: Leibe, B., Matas, J., Sebe, N., Welling, M. (eds.) ECCV 2016. LNCS, vol. 9907, pp. 849–866. Springer, Cham (2016). https://doi.org/10.1007/978-3-319-46487-9_52
11. Gan, C., Yao, T., Yang, K., Yang, Y., Mei, T.: You lead, we exceed: labor-free video concept learning by jointly exploiting web videos and images. In: CVPR, pp. 923–932 (2016)
12. Ghadiyaram, D., Tran, D., Mahajan, D.: Large-scale weakly-supervised pre-training for video action recognition. In: CVPR, pp. 12046–12055 (2019)
13. Girdhar, R., Tran, D., Torresani, L., Ramanan, D.: Distinit: learning video representations without a single labeled video. In: ICCV, pp. 852–861 (2019)
14. Guo, S., et al.: CurriculumNet: weakly supervised learning from large-scale web images. In: Ferrari, V., Hebert, M., Sminchisescu, C., Weiss, Y. (eds.) ECCV 2018. LNCS, vol. 11214, pp. 139–154. Springer, Cham (2018). https://doi.org/10.1007/978-3-030-01249-6_9
15. He, K., Zhang, X., Ren, S., Sun, J.: Deep residual learning for image recognition. In: CVPR, pp. 770–778 (2016)
16. Hinton, G., Vinyals, O., Dean, J.: Distilling the knowledge in a neural network. In: NIPS Deep Learning and Representation Learning Workshop (2015)
17. Huang, G., Liu, Z., Van Der Maaten, L., Weinberger, K.Q.: Densely connected convolutional networks. In: CVPR, pp. 4700–4708 (2017)
18. Hussein, N., Gavves, E., Smeulders, A.W.M.: Timeception for complex action recognition. In: CVPR, pp. 254–263 (2019)
19. Kay, W., et al.: The kinetics human action video dataset. arXiv preprint arXiv:1705.06950 (2017)
20. Kingma, D.P., Mohamed, S., Rezende, D.J., Welling, M.: Semi-supervised learning with deep generative models. In: NeurIPS, pp. 3581–3589 (2014)
21. Kipf, T.N., Welling, M.: Semi-supervised classification with graph convolutional networks. arXiv preprint arXiv:1609.02907 (2016)
22. Krizhevsky, A., Sutskever, I., Hinton, G.E.: Imagenet classification with deep convolutional neural networks. In: NeurIPS, pp. 1097–1105 (2012)
23. Kuehne, H., Jhuang, H., Garrote, E., Poggio, T., Serre, T.: Hmdb: a large video database for human motion recognition. In: ICCV, pp. 2556–2563. IEEE (2011)
24. Laptev, I.: On space-time interest points. Int. J. Comput. Vis. **64**(2–3), 107–123 (2005). https://doi.org/10.1007/s11263-005-1838-7
25. Lee, K.-H., He, X., Zhang, L., Yang, L.: CleanNet: transfer learning for scalable image classifier training with label noise. In: CVPR, pp. 5447–5456 (2018)
26. Liang, J., Jiang, L., Meng, D., Hauptmann, A.G.: Learning to detect concepts from webly-labeled video data. In: IJCAI, pp. 1746–1752 (2016)

27. Lin, J., Gan, C., Han, S.: TSM: temporal shift module for efficient video understanding. In: ICCV, pp. 7083–7093 (2019)
28. Ma, S., Bargal, S.A., Zhang, J., Sigal, L., Sclaroff, S.: Do less and achieve more: training CNNS for action recognition utilizing action images from the web. Pattern Recognit. **68**, 334–345 (2017)
29. Mahajan, D., et al.: Exploring the limits of weakly supervised pretraining. In: Ferrari, V., Hebert, M., Sminchisescu, C., Weiss, Y. (eds.) ECCV 2018. LNCS, vol. 11206, pp. 185–201. Springer, Cham (2018). https://doi.org/10.1007/978-3-030-01216-8_12
30. Martinez, B., Modolo, D., Xiong, Y., Tighe, J.: Action recognition with spatial-temporal discriminative filter banks. In: ICCV, pp. 5482–5491 (2019)
31. Miech, A., Zhukov, D., Alayrac, J.-B., Tapaswi, M., Laptev, I., Sivic, J.: Howto100m: Learning a text-video embedding by watching hundred million narrated video clips. In: ICCV, pp. 2630–2640 (2019)
32. Monfort, M., et al.: Moments in time dataset: one million videos for event understanding. IEEE Trans. Pattern Anal. Mach. Intell. **42**, 502–508 (2019)
33. Qiu, Z., Yao, T., Ngo, C.-W., Tian, X., Mei, T.: Learning spatio-temporal representation with local and global diffusion. In: CVPR, pp. 12056–12065 (2019)
34. Quionero-Candela, J., Sugiyama, M., Schwaighofer, A., Lawrence, N.D.: Dataset Shift in Machine Learning. The MIT Press, Cambridge (2009)
35. Radosavovic, I., Dollár, P., Girshick, R., Gkioxari, G., He, K.: Data distillation: towards omni-supervised learning. In: CVPR, pp. 4119–4128 (2018)
36. Rosenberg, C., Hebert, M., Schneiderman, H.: Semi-supervised self-training of object detection models. WACV/MOTION, p. 2 (2005)
37. Russakovsky, O., et al.: Imagenet large scale visual recognition challenge. Int. J. Comput. Vis. **115**(3), 211–252 (2015). https://doi.org/10.1007/s11263-015-0816-y
38. Simonyan, K., Zisserman, A.: Two-stream convolutional networks for action recognition in videos. In: NeurIPS, pp. 568–576 (2014)
39. Simonyan, K., Zisserman, A.: Very deep convolutional networks for large-scale image recognition. arXiv preprint arXiv:1409.1556 (2014)
40. Soomro, K., Zamir, A.R., Shah, M.: UCF101: a dataset of 101 human actions classes from videos in the wild. arXiv preprint arXiv:1212.0402 (2012)
41. Sun, C., Shetty, S., Sukthankar, R., Nevatia, R.: Temporal localization of fine-grained actions in videos by domain transfer from web images. In: Proceedings of the 23rd ACM International Conference on Multimedia, pp. 371–380. ACM (2015)
42. Tan, M., Le, Q.: EfficientNet: rethinking model scaling for convolutional neural networks. In ICML, pp. 6105–6114 (2019)
43. Tran, D., Wang, H., Torresani, L., Feiszli, M.: Video classification with channel-separated convolutional networks. In: ICCV, pp. 5552–5561 (2019)
44. Tran, D., Wang, H., Torresani, L., Ray, J., LeCun, Y., Paluri, M.: A closer look at spatiotemporal convolutions for action recognition. In: CVPR, pp. 6450–6459 (2018)
45. Tzeng, E., Hoffman, J., Saenko, K., Darrell, T.: Adversarial discriminative domain adaptation. In CVPR, pp. 7167–7176 (2017)
46. Wang, H., Schmid, C.: Action recognition with improved trajectories. In: ICCV, pp. 3551–3558 (2013)
47. Wang, L., et al.: Temporal segment networks for action recognition in videos. IEEE Trans. Pattern Anal. Mach. Intell. **41**, 2740–2755 (2018)
48. Wang, X., Girshick, R., Gupta, A., He, K.: Non-local neural networks. In: CVPR, pp. 7794–7803 (2018)

49. Yalniz, I.Z., Jégou, H., Chen, K., Paluri, M., Mahajan, D.: Billion-scale semi-supervised learning for image classification. arXiv preprint arXiv:1905.00546 (2019)
50. Yang, J., Sun, X., Lai, Y.-K., Zheng, L., Cheng, M.-M.: Recognition from web data: a progressive filtering approach. IEEE Trans. Image Process. **27**(11), 5303–5315 (2018)
51. Ye, G., Li, Y., Xu, H., Liu, D., Chang, S.-F.: Eventnet: a large scale structured concept library for complex event detection in video. In: Proceedings of the 23rd ACM International Conference on Multimedia, pp. 471–480. ACM (2015)
52. Yeung, S., Ramanathan, V., Russakovsky, O., Shen, L., Mori, G., Fei-Fei, L.: Learning to learn from noisy web videos. In: CVPR, pp. 5154–5162 (2017)
53. Zhai, X., Oliver, A., Kolesnikov, A., Beyer, L.: S4l: self-supervised semi-supervised learning. In: ICCV, pp. 1476–1485 (2019)
54. Zhan, X., Liu, Z., Yan, J., Lin, D., Loy, C.C.: Consensus-driven propagation in massive unlabeled data for face recognition. In: Ferrari, V., Hebert, M., Sminchisescu, C., Weiss, Y. (eds.) ECCV 2018. LNCS, vol. 11213, pp. 576–592. Springer, Cham (2018). https://doi.org/10.1007/978-3-030-01240-3_35
55. Zhang, H., Cisse, M., Dauphin, Y.N., Lopez-Paz, D.: mixup: beyond empirical risk minimization. arXiv preprint arXiv:1710.09412 (2017)
56. Zhao, H., Torralba, A., Torresani, L., Yan, Z.: HACS: human action clips and segments dataset for recognition and temporal localization. In: ICCV, pp. 8668–8678 (2019)
57. Zhao, Y., Duan, H., Xiong, Y., Lin, D.: MMAction (2019). https://github.com/open-mmlab/mmaction
58. Zhou, B., Andonian, A., Oliva, A., Torralba, A.: Temporal relational reasoning in videos. In: Ferrari, V., Hebert, M., Sminchisescu, C., Weiss, Y. (eds.) ECCV 2018. LNCS, vol. 11205, pp. 831–846. Springer, Cham (2018). https://doi.org/10.1007/978-3-030-01246-5_49
59. Zhou, B., Lapedriza, A., Khosla, A., Oliva, A., Torralba, A.: Places: a 10 million image database for scene recognition. IEEE Trans. Pattern Anal. Mach. Intell. **40**(6), 1452–1464 (2017)
60. Zhu, C., et al.: Fine-grained video categorization with redundancy reduction attention. In: Ferrari, V., Hebert, M., Sminchisescu, C., Weiss, Y. (eds.) ECCV 2018. LNCS, vol. 11209, pp. 139–155. Springer, Cham (2018). https://doi.org/10.1007/978-3-030-01228-1_9
61. Zhu, X., Ghahramani, Z.: Learning from labeled and unlabeled data with label propagation. CMU CALD tech report CMU-CALD-02-107 (2002)

CurveLane-NAS: Unifying Lane-Sensitive Architecture Search and Adaptive Point Blending

Hang Xu[1], Shaoju Wang[2], Xinyue Cai[1], Wei Zhang[1], Xiaodan Liang[2(✉)], and Zhenguo Li[1]

[1] Huawei Noah's Ark Lab, Beijing, China
[2] Sun Yat-sen University, Guangzhou, China
xdliang328@gmail.com

Abstract. We address the curve lane detection problem which poses more realistic challenges than conventional lane detection for better facilitating modern assisted/autonomous driving systems. Current hand-designed lane detection methods are not robust enough to capture the curve lanes especially the remote parts due to the lack of modeling both long-range contextual information and detailed curve trajectory. In this paper, we propose a novel lane-sensitive architecture search framework named CurveLane-NAS to automatically capture both long-ranged coherent and accurate short-range curve information. It consists of three search modules: a) a feature fusion search module to find a better fusion of the local and global context for multi-level hierarchy features; b) an elastic backbone search module to explore an efficient feature extractor with good semantics and latency; c) an adaptive point blending module to search a multi-level post-processing refinement strategy to combine multi-scale head prediction. Furthermore, we also steer forward to release a more challenging benchmark named CurveLanes for addressing the most difficult curve lanes. It consists of 150K images with 680K labels (The new dataset can be downloaded at http://www.noahlab.com.hk/opensource/vega/#curvelanes). Experiments on the new CurveLanes show that the SOTA lane detection methods suffer substantial performance drop while our model can still reach an 80+% F1-score. Extensive experiments on traditional lane benchmarks such as CULane also demonstrate the superiority of our CurveLane-NAS, e.g. achieving a new SOTA 74.8% F1-score on CULane.

Keywords: Lane detection · Autonomous driving · Benchmark dataset · Neural architecture search · Curve lane

---

H. Xu and S. Wang—Equally Contributed.

Electronic supplementary material The online version of this chapter (https://doi.org/10.1007/978-3-030-58555-6_41) contains supplementary material, which is available to authorized users.

A. Vedaldi et al. (Eds.): ECCV 2020, LNCS 12360, pp. 689–704, 2020.
https://doi.org/10.1007/978-3-030-58555-6_41

(a) CULane Dataset (only ~2% images contains a curve lane)   (b) TuSimple Dataset (only ~30% images contains a curve lane)

(c) Our New CurveLanes Benchmark (More than 90% images contains a curve lane)

Fig. 1. Examples of curve lane detection. Comparing to straight lane, the detection of **curve lanes** is more crucial for trajectory planning in modern assisted and autonomous driving systems. However, the proportion of curve lanes images in current large-scale datasets is very limited, 2% in CULane Dataset (around 2.6K images) and 30% in TuSimple Dataset (around 3.9K images), which hinders the real-world applicability of the autonomous driving systems. Therefore, we establish a more challenging benchmark named CurveLanes for the community. It is the largest lane detection dataset so far (150K images) and over 90% images (around 135K images) contain curve lane.

1   Introduction

Lane detection is a core task in modern assisted and autonomous driving systems to localize the accurate shape of each lane in a traffic scene. Comparing to straight lane, the detection of curve lanes is more crucial for further down-streaming trajectory planning tasks to keep the car properly position itself within the road lanes during steering in complex road scenarios. As shown in Fig. 1, in real applications, curve lane detection could be very challenging considering the long varied shape of the curve lanes and likely occlusion by other traffic objects. Furthermore, the curvature of the curve lane is greatly increased for remote parts because of interpolation which makes those remote parts hard to be traced. Moreover, real-time hardware constraints and various harsh scenarios such as poor weather/light conditions [20] also limit the capacity of models.

Existing lane detection datasets such as TuSimple [28] and CULane [20] are not effective enough to measure the performance of curve lane detection. Because of the natural distribution of lanes in traffic scenes, most of the lanes in those datasets are straight lanes. Only about 2.1% of images in CULane (around 2.6K), and 30% in TuSimple contain curve lanes (around 3.9K). To better measure the challenging curve lane detection performance and facilitate the studies on the difficult road scenarios, we introduce a new large-scale lane detection dataset named CurveLanes consisting of 150K images with carefully annotated 680K curve lanes labels. All images are carefully picked so that almost all of the images contain at least one curve lane (more than 135K images). To our best knowledge, it is the largest lane detection dataset so far and establishes a more challenging benchmark for the community.

The most state-of-the-art lane detection methods are CNN-based methods. Dense prediction methods such as SCNN [20] and SAD [11] treat lane detection as a semantic segmentation task with a heavy encoder-decoder structure. However, those methods usually use a small input image which makes it hard to predict remote parts of curve lanes. Moreover, those methods are often limited to detect a pre-defined number of lanes. On the other hand, PointLaneNet [6] and Line-CNN [14] follow a proposal-based diagram which generates multiple point anchor or line proposals in the images thus getting rid of the inefficient decoder and pre-defined number of lanes. However, the line proposals are not flexible enough to capture variational curvature along the curve lane. Besides, PointLaneNet [6] prediction is based on one fixed single feature map and fails to capture both long-range and short-range contextual information at each proposal. They suffer from a great performance drop in predicting difficult scenarios such as the curve or remote lanes.

In this paper, we present a novel lane-sensitive architecture search framework named CurveLane-NAS to solve the above limitations of current models for curve lane detection. Inspired by recent advances in network architecture search (NAS) [4,15,16,18,22,27,35], we attempt to automatically explore and optimize current architectures to an efficient task-specific curve lane detector. A search space with a combination of multi-level prediction heads and a multi-level feature fusion is proposed to incorporate both long-ranged coherent lane information and accurate short-range curve information. Besides, since post-processing is crucial for the final result, we unify the architecture search with optimizing the post-processing step by adaptive point blending. The mutual guidance of the unified framework ensures a holistic optimization of the lane-sensitive model. Specifically, we design three search modules for the proposal-based lane detection: 1) an elastic backbone search module to allocate different computation across multi-size feature maps to explore an efficient feature extractor for a better trade-off with good semantics and latency; 2) a feature fusion search module is used to find a better fusion of the local and global context for multi-level hierarchy features; 3) an adaptive point blending module to search a novel multi-level post-processing refinement strategy to combine multi-level head prediction and allow more robust prediction over the shape variances and remote lanes. We consider a simple yet effective multi-objective search algorithm with the evolutionary algorithm to properly allocate computation with reasonable receptive fields and spatial resolution for each feature level thus reaching an optimal trade-off between efficiency and accuracy.

Experiments on the new CurveLanes show that the state-of-the-art lane detection methods suffer substantial performance drop (10%~20% in terms of F1 score) while our model remains resilient. Extensive experiments are also conducted on multiple existing lane detection benchmarks including TuSimple and CULane. The results demonstrate the effectiveness of our method, e.g. the searched model outperforms SCNN [20] and SAD [11] and achieves a new SOTA result 74.8% F1-score on the CULane dataset with a reduced FLOPS.

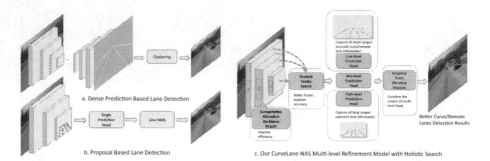

Fig. 2. Comparison of the Lane Detection frameworks: (a) dense prediction based (SCNN [20]), (b) proposal based (Line-CNN [14]), and (c) our CurveLane-NAS. CurveLane-NAS is a unified neural architecture search framework to discover novel holistic network architectures for more robust lane predictions.

2   Related Work

Lane Detection. Lane detection aims to detect the accurate location and shape of each lane on the road. It is the core problem in modern assisted and autonomous driving systems. Conventional methods usually are based on hand-crafted low-level features [7,8,13]. Deep learning has then been employed to extract features in an end-to-end manner. Most lane detection works follow pixel-level segmentation-based approach [10,19,21,36] as shown in Fig. 2 (a). These approaches usually adopt the dense prediction formulation, i.e., treat lane detection as a semantic segmentation task, where each pixel in an image is assigned with a label to indicate whether it belongs to a lane or not. [36] combines a semantic segmentation CNN and a recurrent neural network to enable a consistent lane detection. SCNN [20] generalizes traditional deep convolutions to slice-by-slice convolution, thus enabling message passing between pixels across rows and columns. However, pixel-wise prediction usually requires more computation and is limited to detect a pre-defined, fixed number of lanes. On the other hand, several works use proposal-based approaches for efficient lane detection as shown in Fig. 2 (b). These approaches generate multiple anchors or lines proposals in the images. PointLaneNet [6] finds the lane location by predicting the offsets of each anchor point. Line-CNN [14] introduces a line proposal network to propose a set of ray aiming to capture the actual lane. However, those methods predict on one single feature map and overlook the crucial semantic information for curve lanes and remote lanes in the feature hierarchy.

Neural Architecture Search. NAS aims at freeing expert's labor of designing a network by automatically finding an efficient neural network architecture for a certain task and dataset. Most works search a basic CNN architectures for a classification model [3,16,18,26,29] while a few of them focus on more complicated high-level vision tasks such as semantic segmentation and object detection [4,5,30,31]. Searching strategies in NAS area can be usually divided into three categories: 1) Reinforcement learning based methods [1,2,33,35] train a RNN policy controller to generate a sequence of actions to specify CNN architecture; Zoph et al. [34,35] apply reinforcement learning to search CNN, while the search cost is more than hundreds of GPU days. 2) Evolutionary Algorithms based methods and Network Morphism [12,17,23] try to "evolves" architectures by mutating the current best architectures; Real et al. [22] introduces an age property of the tournament selection evolutionary algorithm to favor the younger CNN candidates during the search. 3) Gradient-based methods [3,18,29] try to introduce an architecture parameter for continuous relaxation of the discrete search space, thus allowing weight-sharing and differentiable optimization of the architecture. SNAS [29] propose a stochastic differentiable sampling approach to improve [18]. Gradient-based methods are usually fast but not so reliable since weight-sharing makes a big gap between the searching and final training. RL methods usually require massive samples to converge thus a proxy task is usually required. In this paper, by considering the task-specific problems such as real-time requirement, severe road mark degradation, vehicle occlusion, we carefully design a search space and a sample-based multi-objective search algorithm to find an efficient but accurate architecture for the curve lane detection problem.

3   CurveLane-NAS Framework

In this paper, we present a novel lane-sensitive architecture search framework named CurveLane-NAS to solve the limitations of current models of curve lane detection. Figure 2 (a) and (b) show existing lane detection frameworks. We extend the diagram of (b) to our multi-level refinement model with a unified architecture search framework as shown in Fig. 2 (c). We propose a flexible model search space with multi-level prediction heads and multi-level feature fusion to incorporate both long-ranged coherent lane information and accurate short-range curve information. Furthermore, our search framework unifies NAS and optimizing the post-processing step in an end-to-end fashion.

The overview of our CurveLane-NAS framework can be found in Fig. 3. We design three search modules: 1) an elastic backbone search module to set up an efficient allocation of computation across stages, 2) a feature fusion search module to explore a better combination of local and global context; 3) an adaptive point blending module to search a novel multi-level post-processing refinement strategy and allow more robust prediction over the shape variances and remote lanes. We consider a simple yet effective multi-objective search algorithm to push the Pareto front towards an optimal trade-off between efficiency and accuracy while the post-processing search can be naturally fit in our NAS formulation.

3.1   Elastic Backbone Search Module

The common backbone of a lane detection is ImageNet pretrained ResNet [9], MoblieNet [24] and GoogLeNet [25], which is neither task-specific nor data-specific. The backbone is the most important part to extract relevant feature

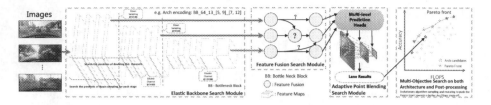

Fig. 3. An overview of our NAS for lane detection pipeline. Our unified search frameworks has three modules: 1) an elastic backbone search module to explore an efficient feature extractor with an optimal setting of network width, depth and when to raise channels/down-sampling, 2) a feature fusion search module to find a suitable fusion of several feature levels; 3) an adaptive point blending module to automatically highlight the most important regions by an adaptive masking and allow a more robust refinement over the shape variances and remote lanes. A unified multi-objective search algorithm is applied to generate a Pareto front with the optimal accuracy/FLOPS trade-off.

and handcrafting backbone may not be optimal for curve lane detection. Thus, we can resort to task-specific architecture search here to explore novel feature extractor for a better trade-off with good semantics and latency.

A common backbone aims to generate intermediate-level features with increasing down-sampling rates, which can be regarded as 4 stages. The blocks in the same stage share the same spatial resolution. Note that the early-stage usually has higher computational cost with more low-level features. The late-stage feature maps are smaller thus the computational cost is relatively smaller but losing a lot of spatial details. How to leverage the computation cost over different stages for an optimal lane network design? Inside the backbone, we design a flexible search space to find the optimal base channel size, when to down-sample and when to raise the channels as follows:

We build up the backbone by several stacked ResNet blocks [9]: basic residual block and bottleneck residual block. The backbone has the choice of 3 or 4 stages. We allow different base channel size $48, 64, 80, 96, 128$ and different number of blocks in each stage corresponding to different computational budget. The number of total blocks variates from 10 to 45. To further allow a flexible allocation of computation, we also search for where to raise the channels. Note that in the original ResNet18/50, the position of doubling the channel size block is fixed at the beginning of each stage. For example, as shown in Fig. 3, the backbone architecture encoding string looks like "BB_64_13_[5, 9]_[7, 12]" where the first placeholder encodes the block setting, 64 is the base channel size, 13 is the total number of blocks and [5, 9] are the position of down-sampling blocks and [7, 12] are the position of doubling channel size.

The total search space of the backbone search module has about 5×10^{12} possible choices. During searching, the models can be trained well from scratch with a large batch size without using the pre-trained ImageNet model.

3.2   Feature Fusion Search Module

As mentioned in DetNAS [5], neurons in the later stage of the backbone strongly respond to entire objects while other neurons are more likely to be activated by local textures and patterns. In the lane detection context, features in the later stage can capture long-range coherent lane information while the features in the early stage contain more accurate short-range curve information by its local patterns. In order to fuse different information across multi-level features, we propose a feature fusion search module to find a better fusion of the high-level and low-level features. We also allow predictions on different feature maps since the detailed lane shapes are usually captured by a large feature map. We consider the following search space for a lane-sensitive model:

Let $F_{1,...,t}$ denote the output feature maps from different stages of the backbone (t can be 3 or 4 depending on the choice of the backbone). From F_1 to F_t, the spatial size is gradually down-sampled with factor 2. Our feature fusion search module consists of M fusion layers $\{O_i\}$. For each fusion layer O_i, we pick two output feature levels with $\{F_1, ... , F_4\}$ as input features and one target output resolution. The two input features F_i first go through one 1x1 convolution layer to become one output channels c, then both features will do up-sampling/down-sampling to the target output resolution and be concatenated together. The output of each O_i will go through another 1x1 convolution with output channels c and to concatenate to the final output. Thus our search space is flexible enough to fuse different features and select the output feature layer to feed into the heads. For each level of the feature map, we also decide whether a prediction head should be added (at least one). The total search space of the feature fusion search module is relatively small (about 10^3 possible choices).

3.3   Adaptive Point Blending Search Module

Inspired by PointLaneNet [6], each head proposes many anchors on its feature map and predicts their corresponding offsets to generate line proposals. A lane line can be determined in the image by line points and one ending point. We first divide each feature map into a $w_f \times h_f$ grid G. If the center of the grid g_{ij} is near to a ground truth lane, g_{ij} is responsible for detecting that lane. Since a lane will go across several grids, multiple grids can be assigned to that lane and their confidence scores s_{ij} reflect how confident the grid contains a part of the lane. For each g_{ij}, the model will predict a set of offsets Δx_{ijz} and one ending point position, where Δx_{ijz} is the horizontal distance between the ground truth lane and a pre-defined vertical anchor points x_{ijz} as shown in Fig. 4. With the predicted Δx_{ijz} and the ending point position, each grid g_{ij} can forecast one potential lane l_{ij}.[1] Post-processing is required to summarize and filtering all line proposals and generate final results.

In PointLaneNet [6], a Line-NMS is adopted on one feature map to filter out lower confidence and a non-maximum suppression (NMS) algorithm is used to

---

[1] The description of the loss function can be found in the Appendix.

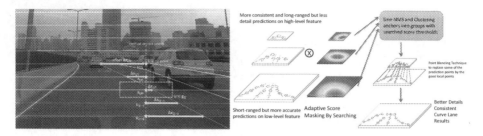

Fig. 4. (Left) For each prediction head, we predict the offsets Δx_{ijz} for each grid g_{ij} and their corresponding scores. The offset Δx_{ijz} is the horizontal distance between the ground truth and a pre-defined vertical anchor x_{ijz}. Each lane prediction can be recovered by the offset and the point anchors positions. (Right) We propose an adaptive Point Blending Search Module for a lane-sensitive detection. Adaptive score masking allows different regions of interest in multi-level prediction. A point blending technique is used to replace some of the prediction points in high confidence lines by the accurate local points. The remote and detailed curve shape can be amended by the local points.

filter out occluded lanes according to their confidence score.[2] However, it cannot fit in our multi-level situation. First, predictions from different levels of features cannot be treated equally. For example, predictions on the low-level feature map are more accurate in a short-range while predictions on high-level feature map are more consistent in a long-range but losing a lot of details. Moreover, we found that each grid can only predict the offsets precisely around its center and the Δx_{ijz} far away from the anchor is inaccurate. Using plain Line-NMS is not sensitive enough to capture the curve or remote part of the lanes.

An adaptive Point Blending Search Module is proposed to unify the post-processing into the NAS framework for lane-sensitive detection as shown in Fig. 4 (Right). In order to allow a different emphasize regions in multi-level prediction, we use an adaptive score masking m_f on the original score prediction for each feature map f. Let c_x and c_y denote the center position of each grid g_{ij} on certain feature map f. We consider a very simple score masking for each map as follow:

$$logit(m_f) = \alpha_{1f}(c_y) + \beta_{1f} + \alpha_{2f}\left[(c_x - u_{xf})^2 + (c_y - u_{yf})^2\right]^{\frac{1}{2}}. \quad (1)$$

There are two main terms in the Eq. (1): one term linearly related to the vertical position of each prediction, and another term is related to the distance from $[u_{xf}, u_{yf}]$. We use the above masking because we conjecture that low-level features may perform better in the remote part of the lane (near the center of the image) and such formulation allows flexible masking across feature maps.

Note that each grid with a good confidence score has precise local information about the lane near the center of the grid. Filtering out all other occluded lanes and only using the lane with the highest confidence score in Line-NMS might not capture the long range curvature for the remote part, since a low score may

---

[2] The detailed Line-NMS algorithm can be found in the Appendix.

Fig. 5. Examples of our new released CurveLanes dataset. All images are carefully selected so that each image contains at least one curve lane. It is the largest lane detection dataset so far and establishes a more challenging benchmark for the community. More difficult scenarios such as S-curves and Y-lanes can be found in this dataset.

be assigned. Thus, we further use a point blending technique for a lane-sensitive prediction. After modification of the original confidence score on each feature map, we first filter out those low score lanes by a suitable threshold and apply NMS to group the remaining lanes into several groups according to their mutual distance. In each group of lines, we iteratively swap the good local points in the lower score anchors with those remote points in the highest score anchors. For each high confidence anchor, some of its points are then replaced by the good local points to become the final prediction. The remote parts of lanes and curve shape can be amended by the local points and the details of lanes are better. The computation overhead is very small by adding a loop within each group of limited lines. The detailed algorithm can be found in the Appendix.

Thus, α_{1f}, β_{1f}, α_{2f}, $[u_{xf}, u_{yf}]$, the score thresholds and the mutual distance thresholds in NMS form the final search space of this module.

3.4 Unified Multi-objective Search

We consider a simple but effective multi-objective search algorithm that can generate a Pareto front with the optimal trade-off between accuracy and different computation constraints. Non-dominate sorting is used to determine whether one model dominates another model in terms of both FLOPS and accuracy. During the search, we sample a candidate architecture by mutating the best architecture along the current Pareto front. We considering following mutation: for the backbone search space, we randomly swap the position of downsampling and double-channel to their neighboring position; for the feature fusion search space, we randomly change the input feature level of each fusion layer; for the adaptive point blending search module, we disturb the best hyper-parameters. Note that the search of the post-processing parameters does not involve training thus the mutation can be more frequent on this module. Our algorithm can be run on multiple parallel computation nodes and can lift the Pareto front simultaneously. As a result, the search algorithm will automatically allocate

Table 1. Comparison of the three largest lane detection datasets and our new Curve-Lanes. Our new CurveLanes benchmark has substantially more images, bigger resolution, more average number of lanes and more curves lanes.

Datasets	Total amount of images	Resolution	Road type	# Lane > 5	Curves
TuSimple [28]	13.2K	1280 × 720	Highway	×	~30%
CULane [20]	133.2K	1640 × 590	Urban & Highway	×	~2%
BDD100K [32]	100K	1280 × 720	Urban & Highway	✓	~10%
Our CurveLanes	**150K**	**2650 × 1440**	Urban & Highway	✓	**>90%**

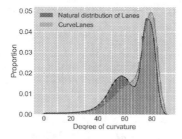 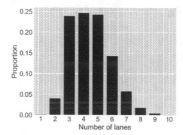

Fig. 6. (Left) Comparison of the distribution of the degree of curvature between common datasets and our CurveLanes. Ours has more proportion of various curvatures comparing to the natural distribution of lanes. (Right) The histogram of the average number of lanes per image in our CurveLanes. CurveLanes also has more number of lanes per image than CULane (<5) and TuSimple (<5) thus more challenging.

computation with reasonable receptive fields and spatial resolution towards an effecient and lane-sensitive detection.

4 Experiments

4.1 New CurveLanes Benchmark

We have released a new dataset named CurveLanes consisting of 150K lanes images with 650K carefully annotated lane labels for bench-marking difficult scenarios such as curves and multi-lanes in traffic lane detection. Table 1 shows a comparison between the existing lane detection datasets TuSimple [28], CULane [20] and BDD [32]. It can be found that our new CurveLanes benchmark has more images and a higher resolution. Moreover, our dataset has more lanes per image and more curves lanes. Thus, the new benchmark is suitable to compare the performance in the difficult situation for the community.

Figure 5 shows some typical examples of the CurveLanes. More difficult cases such as S-curves, Y-lanes can be found in this dataset. Figure 6 further shows the comparison of the distribution of the degree of curvature between common dataset and CurveLanes. The new CurveLanes has more turns and difficult curves CurveLanes also has more muti-lanes scenes thus more difficult. The whole dataset 150K is divided into: train:100K, val: 20K and testing 30K.

Table 2. Comparison of F1-measure of the state-of-the-art models on **CULane** test set. CurveLane-S, CurveLane-M, and CurveLane-L are the searched architectures of our method. Our method outperforms the SOTA models by a large margin with a small computational overhead.

Methods	SCNN [20]	SAD [11]	SAD	PointLane [6]	CurveLane-S	CurveLane-M	CurveLane-L
Backbone	SCNN	ENet	R101	R101	Searched	Searched	Searched
Normal	90.6	90.1	90.7	88.0	88.3	90.2	90.7
Crowded	69.7	68.8	70.0	68.1	68.6	70.5	72.3
Dazzle light	58.5	60.2	59.9	61.5	63.2	65.9	67.7
Shadow	66.9	65.9	67.0	63.3	68.0	69.3	70.1
No line	43.4	41.6	43.5	44.0	47.9	48.8	49.4
Arrow	84.1	84.0	84.4	80.9	82.5	85.7	85.8
Curve	64.4	65.7	65.7	65.2	66.0	67.5	68.4
Night	66.1	66.0	66.3	63.2	66.2	68.2	68.9
Crossroad	1990	1998	2052	1640	2817	2359	1746
FLOPS (G)	328.4	3.9	162.2	25.1	9.0	35.7	86.5
Total	71.6	70.8	71.8	70.2	**71.4**	**73.5**	**74.8**

4.2   Other Datasets and Evaluation Metrics

We conduct neural architecture search on two large lane detection datasets: the CULane [20], and the new CurveLanes dataset. We also transfer the searched architectures to the TuSimple [28] and test the generalization power of the proposed approach. **CULane** [20] is a large scale dataset on traffic lane detection which is collected by cameras in Beijing, China. The CULane dataset includes 88,880 training images, 9675 verification images, and 34,680 test images. The test dataset is divided into 1 normal and 8 challenging categories. **TuSimple** [28] is created by TuSimple specifically focuses on real highway scenarios. It includes 3626 training images and 2782 test images.

Evaluation Metrics. Evaluation metrics is important since it is also the target of our architecture search. We follow the literature [20] and use the corresponding evaluation metrics for each particular dataset. 1) CULane and CurveLanes. Following the official implementation of the evaluation [20], we compute the intersection-over-union (IoU) between GT labels and predictions, where each lane has 30 pixel width. Predictions whose IoUs are larger than 0.5 are considered as true positives (TP). The F1 measure is used as the evaluation metric: $F_1 = \frac{2 \times Precision \times Recall}{Precision + Recall}$, where $Precision = \frac{TP}{TP+FP}$ and $Recall = \frac{TP}{TP+FN}$. 2) TuSimple. We also use the official metric as the evaluation metrics: $Accuracy = \frac{N_{pred}}{N_{GT}}$, where N_{pred} is the number of correctly predicted lane points and N_{GT} is the number of ground-truth lane points.

NAS Implementation Details. During the search, we directly trained the model without ImageNet pretraining since the architecture of the backbone is changed. We found that for a large dataset like CULane and CurveLanes (more than 80K training images), ImageNet pretraining is not necessary and the resulting model converges well with only about 3~5% accuracy loss. We use FLOPS to

Table 3. Comparison of different algorithms on the new dataset **CurveLanes**. CurveLane-S, CurveLane-M, and CurveLane-L are the searched architectures of our method. The SOTA methods such as SCNN and SAD suffer substantial performance drop (20%~30% F1 score).

Method	F1	Precision	Recall	FLOPS(G)
SCNN [20]	65.02%	76.13%	56.74%	328.4
Enet-SAD [11]	50.31%	63.6%	41.6%	3.9
PointLaneNet [6]	78.47%	86.33%	72.91%	14.8
CurveLane-S	81.12%	93.58%	71.59%	7.4
CurveLane-M	81.80%	93.49%	72.71%	11.6
CurveLane-L	**82.29%**	91.11%	75.03%	20.7

Fig. 7. Qualitative result comparison on CurveLanes. Our method CurveLane-L performs better in the difficult scenarios such as curves, night and wet roads.

measure the computational complexity and construct the Pareto front. During search, we use SGD with cosine decay learning rate 0.04 to 0.0001, momentum 0.9. We train each candidate for 12 epochs. Empirically, we found that training with 12 epochs can well separate good models from bad models. We train and test the new architecture in parallel on four computation nodes, and each has 8 Nvidia V100 GPU cards. The batch size is 256 and the input size is 512×288. It takes about 1 h to complete evaluating one architecture for both datasets and it only takes 5 min to evaluate one setting of post-processing parameters. The total search cost is about 5000 GPU hours for one dataset. We set the number of blocks from 10–45 in order to get a complete Pareto front with different FLOPS. We set two random fusions ($M = 2$) with 128 output channels for the fusion search module.

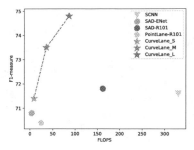

Fig. 8. (Left) Comparison between our methods and other SOTA methods on CULane. Although specifically designed for curve lane detection, our method still dominates most SOTA methods on the widely used CULane benchmark. (Right) Performance of the post-processing algorithms on TuSimple dataset: Ground-truth, Plain-NMS, and our Adaptive Point Blending. Our methods are more sensitive to the curve lanes and remote part of the lanes.

4.3   Lane Detection Results

After identifying the optimal architecture on each dataset, we fully train those models. We first pre-train those searched backbones on ImageNet following common practice [9] for fair comparison with other methods. We train 50 epochs with $bs = 256$, 40 epochs with $bs = 32$ and 30 epochs with $bs = 32$ for TuSimple, CULane and CurveLanes, respectively. SGD is used with initial learning rate 0.04 and a cosine decay learning rate 0.04 to 0.0001, momentum 0.9. The input size of both training and testing is 512×288 for three datasets. We consider three kinds of computational constraints thus the resulting models are denoted as CurveLane-S, CurveLane-M, and CurveLane-L picked from the Pareto front of each dataset. A detailed description of the hyper-parameter can be found in the supplementary materials. For other methods, we report the accuracy numbers of [11, 20] directly from the original papers for TuSimple and CULane. For the CurveLanes, we re-implement the official code from the [11, 20]. For Point-LaneNet [6], we use ResNet101 as the backbone.

Comparison with the State-of-the-Art on CULane. Table 2 shows the performance of the searched architectures on the CULane. Our CurveLane-L achieves a new SOTA result on CULane dataset with a 74.8 F1 score comparing to all the competing methods. Our CurveLane-M model is 1.9 higher F1-core, and 9x fewer FLOPS than SCNN; 1.5 higher F1-score, 4.5x fewer FLOPS than R101-SAD. Figure 8 (Left) shows a comparison between ours and other SOTA methods. Although our method is specifically designed for curve lane detection, it dominates most SOTA methods which proves the effectiveness of the proposed multi-objective search framework.

Results on new CurveLanes. The comparison is shown in Table 3. It can be found the SOTA methods such as SCNN and SAD suffer substantial performance drop (20%–30% F1 score). Our method is lane sensitive and performs far better

Table 4. Ablative study with the F1-measure on the CULane dataset. CurveLane_S to L denote our searched backbone architectures. The performance of models combined with all the modules are listed in the final column.

Backbone	Backbone Only	+Feature Fusion	+Multi Level Heads	+Adaptive Points Blending
ResNet101	70.2	$70.4^{+0.2}$	$71.9^{+1.5}$	$72.4^{+0.5}$
CurveLane_S	69.5	$70.1^{+0.6}$	$70.9^{+0.8}$	$71.5^{+0.6}$
CurveLane_M	71.7	$72.0^{+0.3}$	$72.2^{+0.2}$	$73.5^{+1.3}$
CurveLane_L	72.6	$73.1^{+0.5}$	$73.5^{+0.4}$	$74.8^{+1.3}$

than all the competing methods e.g. CurveLane-S can reach 81.12% with an F1 measure which is 16% higher than the SCNN. We also show some qualitative results on CurveLanes in Fig. 7. Better performance of CurveLane-NAS can be found in the difficult scenarios such as large-curves, night and wet roads. More qualitative comparisons can be found in the Appendix.

The searched architectures also show strong transferability in dealing with other lane detection tasks. We transfer the searched architecture to the TuSimple dataset. With the optimal architecture searched in CUlane, our method reached a comparable performance with SCNN but much faster. The detailed tables can be found in the Appendix.

Searched Architectures. The detailed searched architectures of CurveLane-NAS can be found in the supplementary materials. The architecture is quite different from the hand-craft design such as ResNet50. The backbone of the found architectures usually down-samples twice (only 3 stages). The positions of doubling channels are usually in the later stages to control the total FLOPS since larger channels in the early stage which will result in a great computational burden. Larger models use more heads. Most results of fusion module tend to select output features from the first stage and the second stage with the output one to enable more spatial information.

Ablative Study. We conduct ablation analysis of our proposed model modifications in Table 4. The study is based on different backbones (searched architectures and the ResNet101) on the CULane dataset. It can be found that multilevel heads are more useful for small models. Adaptive Points Blending modules can boost the performance more the larger models. Figure 8 (Right) further shows some qualitative comparisons, it can be found that our adaptive points blending can yield significantly better performance than Plain-NMS for the curve lanes and remote part of the lanes.

5   Conclusion

We propose CurveLane-NAS, a NAS pipeline unifying lane-sensitive architecture search and adaptive point blending for curve lane detection. The new framework can automatically fuse and capture both long-ranged coherent and accurate curve information and enable a more efficient computational allocation. The searched networks achieve state-of-the-art speed/FLOPS trade-off comparing to existing methods. Furthermore, we release a new largest lane detection dataset named CurveLanes for the community to establish a more challenging benchmark with more curve lanes/lanes per image.

References

1. Baker, B., Gupta, O., Naik, N., Raskar, R.: Designing neural network architectures using reinforcement learning. In: ICLR (2017)
2. Cai, H., Chen, T., Zhang, W., Yu, Y., Wang, J.: Efficient architecture search by network transformation. In: AAAI (2018)
3. Cai, H., Zhu, L., Han, S.: ProxylessNAS: direct neural architecture search on target task and hardware. In: ICLR 2019 (2019)
4. Chen, L.C., et al.: Searching for efficient multi-scale architectures for dense image prediction. In: NeurIPS (2018)
5. Chen, Y., Yang, T., Zhang, X., Meng, G., Pan, C., Sun, J.: DetNAS: neural architecture search on object detection. In: NeurIPS (2019)
6. Chen, Z., Liu, Q., Lian, C.: PointLaneNet: efficient end-to-end CNNs for accurate real-time lane detection. In: IV, pp. 2563–2568. IEEE (2019)
7. Chiu, K.Y., Lin, S.F.: Lane detection using color based segmentation. In: IV, pp. 706–711. IEEE (2005)
8. Gonzalez, J.P., Ozguner, U.: Lane detection using histogram-based segmentation and decision trees. In: 2000 IEEE Intelligent Transportation Systems. Proceedings (Cat. No. 00TH8493), ITSC 2000, pp. 346–351. IEEE (2000)
9. He, K., Zhang, X., Ren, S., Sun, J.: Deep residual learning for image recognition. In: CVPR (2016)
10. Hou, Y.: Agnostic lane detection. CoRR (2019). http://arxiv.org/abs/1905.03704
11. Hou, Y., Ma, Z., Liu, C., Loy, C.C.: Learning lightweight lane detection CNNs by self attention distillation. In: ICCV 2019 (2019)
12. Jiang, C., Xu, H., Zhang, W., Liang, X., Li, Z.: SP-NAS: serial-to-parallel backbone search for object detection. In: CVPR, pp. 11863–11872 (2020)
13. Lee, J.W., Cho, J.S.: Effective lane detection and tracking method using statistical modeling of color and lane edge-orientation. In: 2009 Fourth International Conference on Computer Sciences and Convergence Information Technology, pp. 1586–1591. IEEE (2009)
14. Li, X., Li, J., Hu, X., Yang, J.: Line-CNN: end-to-end traffic line detection with line proposal unit. IEEE Trans. Intell. Transp. Syst. 21, 248–258 (2019)
15. Liu, C., et al.: Auto-DeepLab: hierarchical neural architecture search for semantic image segmentation. In: CVPR (2019)
16. Liu, C., et al.: Progressive neural architecture search. In: Ferrari, V., Hebert, M., Sminchisescu, C., Weiss, Y. (eds.) ECCV 2018. LNCS, vol. 11205, pp. 19–35. Springer, Cham (2018). https://doi.org/10.1007/978-3-030-01246-5_2
17. Liu, H., Simonyan, K., Vinyals, O., Fernando, C., Kavukcuoglu, K.: Hierarchical representations for efficient architecture search. In: ICLR (2018)
18. Liu, H., Simonyan, K., Yang, Y.: Darts: differentiable architecture search. In: ICLR (2018)
19. Mamidala, R.S., Uthkota, U., Shankar, M.B., Antony, A.J., Narasimhadhan, A.V.: Dynamic approach for lane detection using google street view and CNN. CoRR (2019). http://arxiv.org/abs/1909.00798

20. Pan, X., Shi, J., Luo, P., Wang, X., Tang, X.: Spatial as deep: spatial CNN for traffic scene understanding. In: AAAI (2018)
21. Pizzati, F., Allodi, M., Barrera, A., García, F.: Lane detection and classification using cascaded CNNs. CoRR (2019). http://arxiv.org/abs/1907.01294
22. Real, E., Aggarwal, A., Huang, Y., Le, Q.V.: Regularized evolution for image classifier architecture search. In: AAAI, vol. 33, pp. 4780–4789 (2019)
23. Real, E., et al.: Large-scale evolution of image classifiers. In: ICML (2017)
24. Sandler, M., Howard, A., Zhu, M., Zhmoginov, A., Chen, L.C.: MobileNetV2: inverted residuals and linear bottlenecks. In: CVPR, pp. 4510–4520 (2018)
25. Szegedy, C., et al.: Going deeper with convolutions. In: CVPR, pp. 1–9 (2015)
26. Tan, M., Chen, B., Pang, R., Vasudevan, V., Le, Q.V.: MnasNet: platform-aware neural architecture search for mobile. In: CVPR (2019)
27. Tan, M., Le, Q.V.: EfficientNet: rethinking model scaling for convolutional neural networks. In: ICML (2019)
28. TuSimple: Tusimple lane detection challenge. In: CVPR Workshops (2017)
29. Xie, S., Zheng, H., Liu, C., Lin, L.: SNAS: stochastic neural architecture search. In: ICLR (2019)
30. Xu, H., Yao, L., Zhang, W., Liang, X., Li, Z.: Auto-FPN: automatic network architecture adaptation for object detection beyond classification. In: ICCV (2019)
31. Yao, L., Xu, H., Zhang, W., Liang, X., Li, Z.: SM-NAS: structural-to-modular neural architecture search for object detection. In: AAAI (2020)
32. Yu, F., et al.: BDD100K: a diverse driving video database with scalable annotation tooling. In: CVPR (2020)
33. Zhong, Z., Yan, J., Wu, W., Shao, J., Liu, C.L.: Practical block-wise neural network architecture generation. In: CVPR (2018)
34. Zoph, B., Le, Q.V.: Neural architecture search with reinforcement learning. In: ICLR (2017)
35. Zoph, B., Vasudevan, V., Shlens, J., Le, Q.V.: Learning transferable architectures for scalable image recognition. In: CVPR (2018)
36. Zou, Q., Jiang, H., Dai, Q., Yue, Y., Chen, L., Wang, Q.: Robust lane detection from continuous driving scenes using deep neural networks. arXiv preprint arXiv:1903.02193 (2019)

Contextual-Relation Consistent Domain Adaptation for Semantic Segmentation

Jiaxing Huang[1], Shijian Lu[1(✉)], Dayan Guan[1], and Xiaobing Zhang[2]

[1] Nanyang Technological University,
50 Nanyang Avenue, Singapore 639798, Singapore
{jiaxing.huang,shijian.lu,dayan.guan}@ntu.edu.sg
[2] University of Electronic Science and Technology of China, Chengdu, China
zhangxiaobing@std.uestc.edu.cn

Abstract. Recent advances in unsupervised domain adaptation for semantic segmentation have shown great potentials to relieve the demand of expensive per-pixel annotations. However, most existing works address the domain discrepancy by aligning the data distributions of two domains at a global image level whereas the local consistencies are largely neglected. This paper presents an innovative local contextual-relation consistent domain adaptation (CrCDA) technique that aims to achieve local-level consistencies during the global-level alignment. The idea is to take a closer look at region-wise feature representations and align them for local-level consistencies. Specifically, CrCDA learns and enforces the prototypical local contextual-relations explicitly in the feature space of a labelled source domain while transferring them to an unlabelled target domain via backpropagation-based adversarial learning. An adaptive entropy max-min adversarial learning scheme is designed to optimally align these hundreds of local contextual-relations across domain without requiring discriminator or extra computation overhead. The proposed CrCDA has been evaluated extensively over two challenging domain adaptive segmentation tasks (*e.g.*, GTA5 → Cityscapes and SYNTHIA → Cityscapes), and experiments demonstrate its superior segmentation performance as compared with state-of-the-art methods.

Keywords: Semantic segmentation · Unsupervised domain adaptation · Contextual-relation consistent

1 Introduction

Semantic segmentation has been a longstanding challenge in computer vision, which aims to assign class labels to every pixel of an image [59]. Deep learning based approaches have achieved great successes at the price of large-scale

Electronic supplementary material The online version of this chapter (https://doi.org/10.1007/978-3-030-58555-6_42) contains supplementary material, which is available to authorized users.

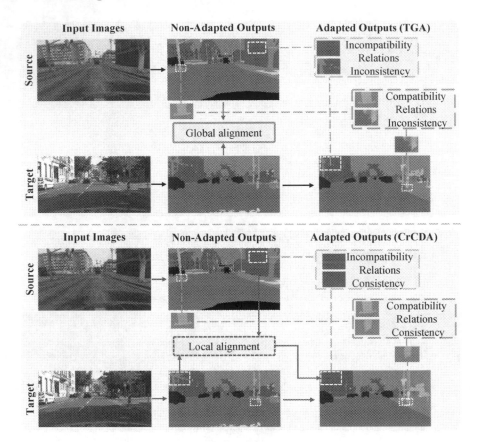

Fig. 1. Our contextual-relation consistent domain adaptation (CrCDA) improves domain adaptive semantic segmentation significantly: The traditional domain adaptive segmentation shown in the upper part employs discriminators for global alignment in the output space [44,45,49] (*e.g.*, probability-/entropy-/patch-represented output), which tends to introduce segmentation errors due to the neglect of local contextual consistency. Our CrCDA shown in the lower part adapts features at local level for contextual-relation consistency between the source and target domains which generates more accurate segmentation consistently. In the graph, "compatibility relations" refer to visual patterns with high co-occurrence frequency (*e.g.*, "pole" should be beside the "sidewalk"), and "incompatibility relations" refer to visual patterns with low co-occurrence frequency (*i.e.*, "sky" should not in the "building").

densely-annotated datasets [3,9,27] which are prohibitively expensive to collect [9]. One way of circumventing this constraint is to use synthesized images with automatically generated labels (*e.g.*, synthesized [36] or game-engine produced [35] data) in network training. Unfortunately, such models usually undergo a drastic performance drop when applied to real-world images [53] due to the domain bias and shift [30,32,39,40,44,48].

Unsupervised domain adaptation (UDA) has been introduced to address the domain bias/shift issue. To reduce the cross-domain discrepancy, most state-of-the-art UDA methods [17,31,44–46,49] exploit adversarial learning for distribution alignment in the intermediate feature [46], output [31,44] or latent [45,49] space. Among this cohort of adversarial-based methods, a common and pivotal step is the employment of a discriminator [16] that predicts a domain label for data being either source or target domain. However, the discriminator works only on image-level and merely achieves global consistency (*i.e.*, locational/spatial distributions consistency), where local contextual consistency (*i.e.*, region-wise contextual-relationships) is largely neglected.

Local contextual-relationships are ubiquitous and provide important cues for scene segmentation. They can be formulated in terms of semantic compatibility/incompatibility relations between one thing/stuff and its neighbouring things/stuff. Under this formulation, a compatibility relation is an indication of visual patterns with high co-occurrence frequency, e.g. a pole beside a sidewalk, and an incompatibility relation is an indication of visual patterns with low co-occurrence frequency, e.g. a person above a driving car. The contextual information has been extensively explored in supervised semantic segmentation, whereas the local contextual-relationships is largely neglected in unsupervised domain adaptive semantic segmentation though they're beneficial in addressing local contextual consistency and inconsistency in the target domain, as illustrated in Fig. 1.

To this end, we propose an unsupervised domain adaptation method for semantic segmentation that explicitly models the local contextual-relations in the feature space of source domain (with label) and then transfers this contextual information into the target domain (without label), ultimately improving target domain segmentation quality, as shown in Fig. 1. We first establish local contextual-relationships pseudo annotations in the source domain. This can be achieved by sampling regions from pixel-level ground-truth maps of source images and clustering the sampled regions to indexed N/M groups via Dbscan [12], as illustrated in Fig. 4. With the local contextual-relationships pseudo annotations in source domain, we can train a classifier C to explicitly models/learns the local contextual-relations in the feature space of source domain, and then transfers/enforces these local contextual-relations into target domain.

Following current discriminator-based global alignment methods [31,44,45, 49], a intuitive idea is to employ hundreds of discriminators to align hundreds of contextual-relations across domain where a single discriminator focuses on a single contextual-relation, or employ just one discriminator to align all contextual-relations across domains. Obviously, the former is cumbersome which requires much redundant computation, while the latter is not aware of a variety of contextual-relations in the data distribution and may end up biasing to low-level/simple difference. Therefore, different from current discriminator-based global alignment methods [31,44,45,49], we enforce these local contextual-relations on target domain via adaptive entropy max-minimizing (AEMM) between classifier C and feature extractor E that estimates prototypical feature representations of these local contextual-relations and congregates neighboring target incorrect samples/contextual-relations to the approximated correct source prototypes alternatively, ultimately leading to consistent local contextual-relations across domains. In this way, our method requires no discriminator which is normally used in UDA-based semantic segmentation and introduces training instability and extra components. In addition, this AEMM learning scheme can also be applied into pixel-/global-scale training.

The contributions of this work can be summarized in three aspects. First, we propose an unsupervised domain adaptation method for semantic segmentation that explicitly models the local contextual-relations in the feature space of source domain (with label) and then transfers this contextual information into target domain (without label). To the best of our knowledge, this is the first effort to explore contextual information for UDA-based semantic segmentation. Second, it introduces a novel adaptive entropy max-minimizing adversarial learning scheme to effectively align hundreds of local contextual-relations across domain, which requires no discriminator and adds no overhead. Third, it shows the proposed method can be seamlessly integrated into existing domain adaptation techniques without extra overhead except two classifiers and achieves consistent improvements on semantic segmentation. Fourth, extensive evaluations over two challenging UDA tasks GTA5 → Cityscapes and SYNTHIA → Cityscapes show that our method achieves superior semantic segmentation performance consistently.

2   Related Works

Current UDA-based semantic segmentation methods are threefold: adversarial learning based approach [5–7, 11, 13, 19, 23, 24, 28, 29, 31, 44, 46, 51], image translation based approach [2, 8, 18, 20, 25, 34, 42, 50, 52, 54], and pseudo-labels based approach [15, 21, 58, 61, 64].

Adversarial Learning Based Approach: Adversarial learning based UDA has been extensively explored for semantic segmentation, where a discriminator is employed to minimize the divergences between source and target domains in feature or output spaces. [19] first applies adversarial learning for UDA based semantic segmentation by aligning feature space at global scale. Curriculum domain adaptation [55] utilizes certain inferred properties (e.g., superpixel and global label distributions) as the guidance to train the segmentation network. In [44] and [7], the adversarial learning is used to align the global structure to benefit from the scene layout consistency across domains, where [7] integrates a target guided distillation module to achieve style adaptation. In addition, [38, 39] combines adversarial learning and co-training to achieve domain adaptation via maximizing the discrepancy between two classifiers' outputs.

Image Translation Based Approach: Inspired by the recent advances in image synthesis (e.g., CycleGAN [60]), a number of GAN-based methods are proposed to generate target images conditioned on the source, which can help reduce the domain discrepancy before training segmentation models. CyCADA

[18] uses CycleGAN to generate target images conditioned on the source images and achieves input space adaptation with a joint adversarial learning for feature alignment. A similar method, DCAN [50], implements channel-wise feature alignment to preserve spatial structures and semantic concepts in the generator and segmentation network. [42] transfers the information of the target domain to the learned embedding via the joint adversarial learning between generator and discriminator. Besides using GANs [16] to align the embedding across domains, [62] proposes a novel conservative loss to penalize the extremely easy and difficult cases while enhancing moderate examples.

Re-training Based Approach: Another approach of UDA based semantic segmentation is pseudo label re-training [26,63,64] that uses high-confident predictions as pseudo ground truth for the target unlabelled data to finetune the model trained on the source data. In [64], class balancing and spatial prior are included to guide the iterative re-training in target domain. [49] proposes a soft-assigned version of re-training, where it enforces the "most-confused" pixels (*e.g.*, with equal probabilities for all classes) to become more confident (*i.e.*, with either low or high probability for each class) by entropy minimization. [64] instead implements iterative learning on high-confident pixels.

Our method does not follow either global/class-wise feature space alignment using discriminators [7,19,28,29,31,46] or re-training on target data [41,64]. Instead, we enforce multi-scale feature space alignment via multi-scale entropy max-minimizing. To the best of our knowledge, this is the first end-to-end multi-scale UDA network that achieves competitive performance on two challenging UDA tasks.

3   Methods

In this section, we present our framework for contextual-relationships consistent domain adaptation (CrCDA): a discriminator-free adversarial training scheme between a feature extractor module and a classifier via adaptive entropy max-minimizing (AEMM) to align local contextual-relationships across domains. Figure 2 illustrates our network architecture.

3.1   Problem Definition

We focus on the problem of unsupervised domain adaptation (UDA) in semantic segmentation. Given the source data $X_s \subset \mathbb{R}^{H \times W \times 3}$ with C-class pixel-scale segmentation labels $Y_s \subset (1,C)^{H \times W}$ (*e.g.*, stimulated images from game engines) and the target data $X_t \subset \mathbb{R}^{H \times W \times 3}$ without labels (*i.e.*, real images), our goal is to learn a semantic segmentation model G that performs well on the target dataset X_t. Current adversarial learning methods rely heavily on discriminators to align the distributions of source and target domains via two loss functions: segmentation loss on source data and adversarial loss for alignment.

However, there exists a crucial limitation for these approaches: even if perfect adaptation is achieved through a discriminator, the alignment is implemented

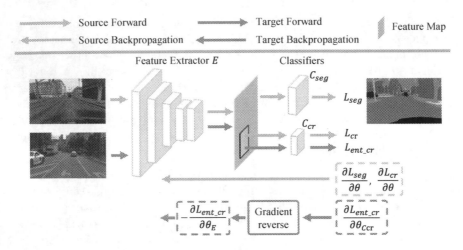

Fig. 2. Overview of our proposed contextual-relation consistent domain adaptation (CrCDA): Given images from source and target domains, the feature extractor E extracts features and feeds them to classifier C_{seg} and C_{cr} for classification at pixel and region scales. In the source flow (highlighted by arrows in blue), $\mathcal{L}_{seg}$ is computed based on the segmentation probability map from C_{seg}, $\mathcal{L}_{cr}$ is computed based on the classification probability maps from C_{cr}. In the target flow (highlighted by arrows in orange), $\mathcal{L}_{ent\_cr}$ is computed based on the classification probability maps from C_{cr}. The local-scale alignment is implemented in back-propagation by training the parts before and after the gradient reverse layer in adversarial scheme $w.r.t$ $\mathcal{L}_{ent\_cr}$. (Color figure online)

on global level (*i.e.*, image-level), where local contextual information may be lost/deconstructed. The reason lies in that the discriminator can only implement alignment at global level, which inputs the whole map but outputs a digit to represent domain labels (*e.g.*, 0 or 1). In some cases, parts of local regions (*i.e.*, local contextual-relations) have been well aligned across domains. However, the discriminator might deconstruct this existing local alignment during implementing the global marginal distribution alignment. In this paper, we define this phenomenon as "lack of local consistency (*i.e.*, local contextual inconsistency)", which is important to semantic segmentation in dense pixel-scale prediction.

3.2   Overview of Network Architecture

As shown in Fig. 2, our semantic segmentation model G consists of a feature extractor E and two classifiers (*i.e.*, C_{seg} and C_{cr}) where C_{seg} is for pixel-scale segmentation and C_{cr} is for local-scale contextual-relations learning/classification. E extracts features from input images. C_{seg} and C_{cr} classify features generated by E into pre-defined semantic classes. Specifically, C_{seg} processes features at pixel-scale, which aims to predict pixel-scale labels. The pre-defined semantic class domain for C_{seg} is the pixel-scale ground-truth, so there is

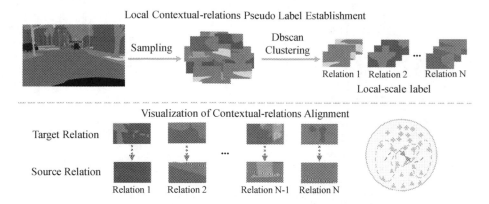

Fig. 3. Overview of local contextual-relation pseudo label establishment: "Dbscan clustering" means implementing Dbscan clustering based on the histogram of gradient. The effect of local contextual-relations alignment is shown at right-bottom part, with more visualization details provided in Fig. 4.

no difference between C_{seg} and traditional segmentation classifier. C_{cr} processes features at local scales, which aims to predict region-scale/contextual-relations labels. The pre-defined semantic class domain for C_{cr} is the clustered contextual-relations ground-truth. The establishment procedure of clustered contextual-relations labels is described in Sect. 3.3 and shown in Fig. 3. We train E and the classifiers (*i.e.*, C_{seg} and C_{cr}) in an adversarial scheme to reduce domain shifts at local scales to achieve local contextual-relation consistency.

3.3 Contextual-Relation Consistent Domain Adaptation

This subsection introduces our contextual-relation consistent domain adaptation at local scales, denoted as CrCDA*, via adaptive entropy max-minimizing, as shown in Fig. 2.

Contextual-Relation Pseudo Label Establishment. In order to implement local-scale task, we sample regions on the feature space and implement domain alignment at local scales to achieve local contextual-relation consistent domain adaptation, as shown in Fig. 3. Different from [22] that implements mode-agnostic patches alignment or [45] that aligns patch-indexed representation of the whole image only at global scales by a discriminator (*i.e.*, the probability distributions of patch index prediction of the whole images.), we aim to aligns inter-class relations within each single patch, *i.e.*, the probability distributions of pixel class prediction within each patch, *w.r.t* its mode via a classifier. Thus, the preliminary is to establish the region-scale label, where we first crop the pixel-scale ground-truth to many larger regions and then use Dbscan [12] to cluster them to assign each region a certain index label (*i.e.*, contextual-relation pseudo label). Specifically, we assign the index label to regions according to the clustering results based on the histogram of gradient.

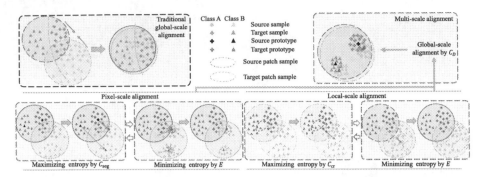

Fig. 4. Overview and comparison of the proposed AEMM at different scales: The mechanism of traditional global-scale domain adaptation is shown in the black box, where some samples are adapted into the wrong area due to the lack of local consistency (*i.e.*, local contextual-relation consistency). Our method is shown in the red boxes illustrating the alignment in pixel-scale, local-scale and global-scale. In pixel-scale alignment, C_{seg} firstly approximates the target prototypical features by maximizing entropy on target data and then E aims to congregate the features to the approximated prototypical features by minimizing entropy. Local-scale alignment works in the same scheme of pixel-scale adaptation while the only difference is the processing unit size (the former adapts a larger group of features; the latter adapts single pixel-scale features). As shown above, the global alignment is implemented by a domain classifier. Finally, the proposed AEMM can achieve feature alignment in different scales simultaneously.

For region-scale label (*i.e.*, contextual-relation pseudo label), we cluster regions into different groups based on the histogram of gradient and assign the index label. These region-scale/contextual-relation pseudo labels can assist our network to implement alignment at local scales. Detailed information about the region-scale/contextual-relation pseudo labels is in the supplementary materials.

Adaptive Entropy Max-Minimizing Adversarial Learning Scheme. In local-scale adaptation, C_{cr} aims to approximate the prototypical feature representations for each contextual-relation (*e.g.*, road-sidewalk, sky-building, pole-sidewalk, etc.) by implementing entropy maximization in target domain according to the source prototypical feature representations found via supervised learning in source domain. E focuses on extracting discriminative feature representations (near the approximated prototypical feature representations) by implementing entropy minimization. Specifically, the prototypical feature representations of source domain found with supervision are first utilized to estimate the prototypical feature representations for target data by entropy maximizing $w.r.t$ C_{cr}. E then adapts the extracted feature representations to the corresponding prototypical feature representations by minimizing the entropy. The overall unsupervised domain adaptation at local scales is achieved by the adversarial training between C_{cr} and E as illustrated in Fig. 4. Different from that applied in semi-supervised learning [37], our unsupervised domain adaptation training method, referred as adaptive entropy max-minimizing (AEMM) implements entropy max-min with a regularizer $\mathcal{R}(P) = ave\{P \log P\} \times \lambda_R$ (λ_R decreases with training iteration, details are shown in appendix) for better estimating the prototypes in the target domain where no labels are available.

Source Flow. In our local-scale adaptation setting, the source data contributes to L_{seg} and L_{cr}. Given a source image $x_s \subset X_s$, its corresponding segmentation label $y_s \subset Y_s$ and contextual-relation pseudo-label $y_{s\_cr} \subset Y_{s\_cr}$, $P_s^{(h,w,c)} = C_{seg}(E(x_s))$ is the predicted probability map $w.r.t$ each pixel over C classes; $P_{s\_cr}^{(i,j,n)} = C_{cr}(E(x_s))$ is the predicted probability map $w.r.t$ each region over N pre-defined contextual-relations classes. Therefore, it is a simple supervised learning objective to minimize L_{seg} and L_{cr}, which are expressed as:

$$\mathcal{L}_{seg}(E, C_{seg}) = \sum_{h,w}\sum_{c} -y_s^{(h,w,c)} \log P_s^{(h,w,c)} \tag{1}$$

$$\mathcal{L}_{cr}(E, C_{cr}) = \sum_{i,j}\sum_{n} -y_{s\_cr}^{(i,j,n)} \log P_{s\_cr}^{(i,j,n)} \tag{2}$$

Target Flow. As the target label is not accessible, we introduce the adversarial training scheme between feature extractor E and classifier C_{cr} to extract discriminative features for target data via adaptively max-minimizing entropy in target domain. Given a target image $x_t \subset X_t$, $P_{t\_cr}^{(i,j,n)} = C_{cr}(E(x_t))$ is the predicted probability map $w.r.t$ each region over N pre-defined contextual-relations classes. The entropy loss L_{ent_cr} is expressed as:

$$\mathcal{L}_{ent\_cr}(E, C_{cr}) = -\frac{1}{C}\sum_{i,j}\sum_{n} max\{P_{t\_cr}^{(i,j,n)} \log P_{t\_cr}^{(i,j,n)} - \mathcal{R}(P_{t\_cr}^{(i,j,n)}), 0\} \tag{3}$$

For local-scale adaptation, we use the same back-propagation optimizing scheme with the gradient reverse layer mentioned in [57]. The training objective can be express as:

$$\min_{\theta_E} \mathcal{L}_{seg} + \lambda_{cr}\mathcal{L}_{cr} + \lambda_{ent}\mathcal{L}_{ent\_cr}$$

$$\min_{\theta_{C_{seg}}} \mathcal{L}_{seg} \tag{4}$$

$$\min_{\theta_{C_{cr}}} \mathcal{L}_{cr} - \lambda_{ent}\mathcal{L}_{ent\_cr}$$

where λ_{ent} is a weight factor to control the balance of unsupervised adaptation on target data and supervised learning on source data.

3.4 CrCDA with Pixel-/Global-Scale

This subsection introduces our CrCDA with pixel-/global-scale, denoted as CrCDA, via adaptive entropy max-minimizing, as shown in Fig. 2. Our discriminator-free AEMM adversarial training scheme can also be extended into pixel-scale and global/image-scale to form multi-scale domain adaptation.

In multi-scale adaptation, for $\mathcal{L}_{seg}$, $\mathcal{L}_{cr}$ and $\mathcal{L}_{ent\_cr}$, the objectives are the same as that in local-scale adaptation. We extend the AEMM adversarial training scheme mentioned before into pixel-scale and global-scale adaptation. For pixel-scale adaptation, we implement pixel-scale entropy loss $\mathcal{L}_{ent}$ on target data to E and C_{seg}. For global-scale adaptation, we implement global-scale entropy loss $\mathcal{L}_{ent\_D}$ on target data to E and C_D, where C_D is a domain classifier. C_D takes the layout probability map concatenated by the two probability maps generated from C_{seg} and C_{cr} as input, and predicts domain label for it (e.g., 0 for source domain, 1 for target domain). The global-alignment is achieved by the adversarial training between C_D and (E, C_{seg}, C_{cr}). Finally, our multi-scale consistent domain adaptation network is able to align domain shift at global scales, local-scale and pixel-scale simultaneously.

Similar to local-scale adaptation, we formulate the pixel-scale entropy loss as:

$$\mathcal{L}_{ent\_pix}(E, C_{seg}) = -\frac{1}{C} \sum_{h,w} \sum_{c} max\{P_{t\_pix}^{(h,w,c)} \log P_{t\_pix}^{(h,w,c)} - \mathcal{R}(P_{t\_pix}^{(h,w,c)}), 0\} \quad (5)$$

For multi-scale adaptation, we also use the same back-propagation optimizing scheme with the gradient reverse layer mentioned in [13,14]. The training objective can be express as:

$$\min_{\theta_E} \mathcal{L}_{seg} + \lambda_{cr}\mathcal{L}_{C_{cr}} + \lambda_{ent}(\mathcal{L}_{ent\_pix} + \mathcal{L}_{ent\_cr}) + \lambda_D\mathcal{L}_D$$

$$\min_{\theta_{C_{seg}}} \mathcal{L}_{seg} - \lambda_{ent}\mathcal{L}_{ent\_pix} + \lambda_D\mathcal{L}_D$$

$$\min_{\theta_{C_{cr}}} \mathcal{L}_{C_{cr}} - \lambda_{ent}\mathcal{L}_{ent\_cr} + \lambda_D\mathcal{L}_D \quad (6)$$

$$\max_{\theta_{C_D}} \lambda_D\mathcal{L}_D$$

where $\mathcal{L}_D$ is provided in supplementary materials; λ_{cr}, λ_{ent} and λ_D are the weight factor to balance the unsupervised adaptation on target data and the task-specific objectives on source data.

4   Experiments

4.1   Datasets

We evaluate our unsupervised domain adaptation networks for semantic segmentation on two challenging synthesized-to-real tasks: GTA5 [35] → Cityscapes [9] and SYNTHIA [36] → Cityscapes. GTA5 contains $24,966$ synthesized images with high-resolution and we use the 19 common categories between GTA5 and Cityscapes in the same setting as in [44]. SYNTHIA contains $9,400$ synthetic images with 16 common categories in Cityscapes. We use either GTA5 or SYNTHIA as source domain. We use the unlabelled training set of Cityscapes as target domain, which includes 2975 real-world images.

Table 1. Results of domain adaptation task GTA5 → Cityscapes. "V" means the VGG16-based model and "R" means the ResNet101-based model.

Networks	Oracle	road	side.	build.	wall	fence	pole	light	sign	vege.	terr.	sky	pers.	rider	car	truck	bus	train	motor	bike	mIoU
FCN Wild [19]	V	70.4	32.4	62.1	14.9	5.4	10.9	14.2	2.7	79.2	21.3	64.6	44.1	4.2	70.4	8.0	7.3	0.0	3.5	0.0	27.1
CDA [55]	V	74.9	22.0	71.7	6.0	11.9	8.4	16.3	11.1	75.7	13.3	66.5	38.0	9.3	55.2	18.8	18.9	0.0	16.8	14.6	28.9
CyCADA [18]	V	83.5	**38.3**	76.4	20.6	16.5	22.2	26.2	**21.9**	80.4	28.7	65.7	49.4	4.2	74.6	16.0	26.6	2.0	8.0	0.0	34.8
AdaptSeg [44]	V	87.3	29.8	78.6	21.1	18.2	22.5	21.5	11.0	79.7	29.6	71.3	46.8	6.5	80.1	23.0	26.9	0.0	10.6	0.3	35.0
CBST [64]	V	66.7	26.8	73.7	14.8	9.5	**28.3**	25.9	10.1	75.5	15.7	51.6	47.2	6.2	71.9	3.7	2.2	5.4	**18.9**	**32.4**	30.9
CLAN [31]	V	**88.0**	30.6	79.2	23.4	20.5	26.1	23.0	14.8	**81.6**	**34.5**	72.0	45.8	7.9	80.5	**26.6**	**29.9**	0.0	10.7	0.0	36.6
AdvEnt [49]	V	86.9	28.7	78.7	28.5	**25.2**	17.1	20.3	10.9	80.0	26.4	70.2	47.1	8.4	81.5	26.0	17.2	**18.9**	11.7	1.6	36.1
PatAlign [45]	V	87.3	35.7	79.5	**32.0**	14.5	21.5	24.8	13.7	80.4	32.0	70.5	50.5	16.9	81.0	20.8	28.1	4.1	15.5	4.1	37.5
CrCDA (ours)	V	86.8	37.5	**80.4**	30.7	18.1	26.8	25.3	15.1	81.5	30.9	**72.1**	**52.8**	19.0	**82.1**	25.4	29.2	10.1	15.8	3.7	**39.1**
AdaptSeg [44]	R	86.5	36.0	79.9	23.4	23.3	23.9	35.2	14.8	83.4	33.3	75.6	58.5	27.6	73.7	32.5	35.4	3.9	30.1	28.1	42.4
CBST [64]	R	91.8	53.5	80.5	32.7	21.0	34.0	28.9	20.4	83.9	34.2	80.9	53.1	24.0	82.7	30.3	35.9	16.0	25.9	**42.8**	45.9
CLAN [31]	R	87.0	27.1	79.6	27.3	23.3	28.3	35.5	24.2	83.6	27.4	74.2	58.6	28.0	76.2	33.1	36.7	**6.7**	**31.9**	31.4	43.2
AdvEnt [49]	R	89.4	33.1	81.0	26.6	26.8	27.2	33.5	24.7	83.9	36.7	78.8	58.7	30.5	**84.8**	38.5	44.5	1.7	31.6	32.4	45.5
MaxSquare[4]	R	89.4	43.0	82.1	30.5	21.3	30.3	**34.7**	24.0	**85.3**	**39.4**	78.2	**63.0**	22.9	84.6	36.4	43.0	5.5	34.7	33.5	46.4
PatAlign [45]	R	92.3	51.9	82.1	29.2	25.1	24.5	33.8	33.0	82.4	32.8	82.2	58.6	27.2	84.3	33.4	**46.3**	2.2	29.5	32.3	46.5
CRST [63]	R	91.0	**55.4**	80.0	**33.7**	21.4	**37.3**	32.9	24.5	85.0	34.1	80.8	57.7	24.6	84.1	27.8	30.1	**26.9**	26.0	42.3	47.1
CrCDA (ours)	R	**92.4**	55.3	**82.3**	31.2	**29.1**	32.5	33.2	**35.6**	83.5	34.8	**84.2**	58.9	**32.2**	84.7	**40.6**	46.1	2.1	31.1	32.7	**48.6**

4.2   Implementation Details

For a fair comparison, similar to [31,44,49], we utilize Deeplab-V2 architecture [3] with ResNet-101 pretrained on ImageNet [10] as our single-scale semantic segmentation network ($E + C_{seg}$). To extend our model to multi-scale network, we simply copy and modify C_{seq} to create C_{cr} and C_D with different output channels (*e.g.*, N and 1) and different output sizes due to various scales (*i.e.*, region-size and global-size). We also apply our methods on VGG-16 [43] in the same way as employing ResNet-101. Following [13] [47], a gradient reverse layer is employed to reverse the entropy loss between E and (C_{seg}, C_{cr}) during pixel-/region-scale adaptation to achieve adversarial training. The domain classifier C_D works similar to a discriminator for global-scale alignment. During training, we utilize SGD [1] to optimize our networks with a momentum of 0.9 and a weight decay of $1e-4$. The initial learning rate is set as $2.5e-4$ and decayed by a polynomial policy with a power of 0.9, as illustrated in [3]. For all experiments, the hyper-parameters λ_{ent}, λ_D, λ_{cr} and N are set as $2.5e-5$, $2.5e-5$, $5e-3$ and 100, respectively.

4.3   Comparison with State-of-Art

We compare the experimental results of our method and state-of-the-art algorithms in two "Synthetic-to-real" UDA tasks with two different architectures: VGG-16 and ResNet-101. For "GTA5 → Cityscapes", we present the results in Table 1 with comparisons to the state-of-the-art domain adaptation methods [18,19,31,44,49,55–57]. Our contextual-relation consistent domain adaptation, expressed as CrCDA, achieves comparable performance to other state-of-the-art approaches on both architectures. Compared to Adapt-SegMap (output space global alignment) [44], category-level adversarial network (output

Table 2. Results of domain adaptation task SYNTHIA → Cityscapes. "V" means the VGG16-based model and "R" means the ResNet101-based model. "mIoU" and "mIoU*" are calculated over 16 and 13 classes, respectively.

Networks	Oracle	road	side.	build.	wall	fence	pole	light	sign	vege.	sky	pers.	rider	car	bus	motor	bike	mIoU	mIoU*
FCNs Wild [19]	V	11.5	19.6	30.8	4.4	0.0	20.3	0.1	11.7	42.3	68.7	51.2	3.8	54.0	3.2	0.2	0.6	20.2	22.1
CDA [55]	V	65.2	26.1	74.9	0.1	0.5	10.7	3.7	3.0	76.1	70.6	47.1	8.2	43.2	20.7	0.7	13.1	29.0	34.8
AdaptSeg [44]	V	78.9	29.2	75.5	-	-	-	0.1	4.8	72.6	76.7	43.4	8.8	71.1	16.0	3.6	8.4	-	37.6
CBST [64]	V	69.6	28.7	69.5	**12.1**	0.1	**25.4**	**11.9**	**13.6**	**82.0**	**81.9**	**49.1**	14.5	66.0	6.6	3.7	**32.4**	**35.4**	36.1
CLAN [31]	V	**80.4**	**30.7**	74.7	-	-	-	1.4	8.0	77.1	79.0	46.5	8.9	**73.8**	18.2	2.2	9.9	-	39.3
AdvEnt [49]	V	67.9	29.4	71.9	6.3	0.3	19.9	0.6	2.6	74.9	74.9	35.4	9.6	67.8	21.4	4.1	15.5	31.4	36.6
PatAlign [45]	V	72.6	29.5	77.2	3.5	0.4	21.0	1.4	7.9	73.3	79.0	45.7	14.5	69.4	19.6	7.4	16.5	33.7	39.6
CrCDA (ours)	V	74.5	30.5	**78.6**	6.6	**0.7**	21.2	2.3	8.4	77.4	79.1	45.9	**16.5**	73.1	**24.1**	**9.6**	14.2	35.2	**41.1**
AdaptSeg [44]	R	84.3	42.7	77.5	-	-	-	4.7	7.0	77.9	82.5	54.3	21.0	72.3	32.2	18.9	32.3	-	46.7
CLAN [31]	R	81.3	37.0	80.1	-	-	-	**16.1**	13.7	78.2	81.5	53.4	21.2	73.0	32.9	**22.6**	30.7	-	47.8
AdvEnt [49]	R	85.6	42.2	79.7	8.7	0.4	25.9	5.4	8.1	80.4	84.1	**57.9**	23.8	73.3	36.4	14.2	33.0	41.2	48.0
MaxSquare[4]	R	82.9	40.7	**80.3**	**10.2**	**0.8**	25.8	12.8	**18.2**	**82.5**	82.2	53.1	18.0	**79.0**	31.4	10.4	**35.6**	41.4	48.2
PatAlign [45]	R	82.4	38.0	78.6	8.7	0.6	26.0	3.9	11.1	75.5	84.6	53.5	21.6	71.4	32.6	19.3	31.7	40.0	46.5
CrCDA (ours)	R	**86.2**	**44.9**	79.5	8.3	0.7	**27.8**	9.4	11.8	78.6	**86.5**	57.2	**26.1**	76.8	**39.9**	21.5	32.1	**42.9**	**50.0**

space class-wise alignment) [31] and patch-represented global alignment [45] (patch-indexed latent space alignment), CrCDA consistently brings over +2.1% mIoU improvements on ResNet-101. We reckon this gain is from our end-to-end/concurrent multi-scale alignment, which indicates that local consistency (*i.e.*, local contextual-relation consistency) is very important as well as global consistency and they are complementary to each other. In Table 2, we present the adaptation result for the task "SYNTHIA → Cityscapes" and consistent improvements are observed *w.r.t* state-of-the-arts. Detailed analysis is included in next subsection.

4.4 Ablation Studies and Analysis

We analyze our proposed CrCDA with several state-of-the-art baselines. In general, both single-scale form (CrCDA*) and multi-scale form (CrCDA) achieve comparable results to all the baselines in all the settings.

As shown on the first three rows in Table 3, our pixel-scale AEMM adversarial network brings +1.4% improvements in terms of mIoU over MinEnt [49]. The reason lies in that direct entropy minimization does not take the domain gap into account while our AEMM training scheme pushes the source distribution closer to target distribution during maximizing entropy on target data.

For our CrCDA with single-scale form (CrCDA*) via AEMM, it outperforms MinEnt-based contextual-relations alignment by +1.6% on ResNet-101, as shown on the second block (row4-5) in Table 3. We reckon that these improvements are contributed by our adaptive entropy max-min training scheme which considers the domain mismatch/gap while MinEnt neglects.

Our CrCDA with multi-scale form integrating three scales' adaptation (pixel-, local- and global-scale), termed as CrCDA shown on the bottom block in Table 3, achieves state-of-the-art performances 48.6% mIoU on ResNet-101. Besides, CrCDA also outperforms all current methods by over +1.5%. Compared

Table 3. Ablation study of the proposed contextual-relation consistent domain adaptation on GTA5-to-Cityscapes using the ResNet-101 network. All settings/methods are with "L_{seg}" (bold texts represent our methods). CrCDA* represents the contextual-relation consistent domain adaptation with only single-scale (local scale).

Method	pixel-scale Ada.		local-scale Ada.		global-scale Ada.				mIoU
	L_{minent}	L_{ent}^{ours}	L_{minent}	L_{ent}^{ours}	L_{adv}	L_{patadv}	L_{advent}	L_D^{ours}	
Without Ada									36.6
MinEnt [49,61]	✓								42.4
Pixel-AEMM		✓							43.8
CrCDA*-MinEnt			✓						42.1
CrCDA*-AEMM				✓					43.7
AdaptSeg [44]					✓				41.4
PatAlign [45]						✓			41.3
AdvEnt [49]							✓		43.8
Global-AEMM								✓	44.3
Pixel+CrCDA*	✓		✓						45.6
Pixel+Global	✓							✓	46.0
CrCDA*+Global			✓					✓	46.1
CrCDA	✓		✓					✓	**48.6**

Table 4. Complementary study of the proposed contextual-relation consistent domain adaptation with local-scale to current global alignment UDA methods on GTA5-to-Cityscapes using the ResNet-101 network. All methods are default with "L_{seg}".

Method	local-level Ada.	global Ada.				mIoU
	L_{ent}^{ours}	L_{adv}	L_{puladv}	L_{advent}	L_D^{ours}	
CrCDA* (ours)	✓					43.7
AdaptSeg [44]		✓				41.4
PatAdv [45]			✓			41.3
AdvEnt [49]				✓		43.8
GlobalAlign (ours)					✓	44.3
AdaptSeg [44]+PatAlign [45]		✓	✓			43.2
CrCDA* + AdaptSeg [44]	✓	✓				44.8
CrCDA* + PatAlign [45]	✓		✓			44.7
CrCDA* + AdvEnt [49]	✓			✓		45.2
CrCDA*+GlobalAlign (ours)	✓				✓	**46.1**

to "Pixel+Global", CrCDA brings +2.6% improvement in mIoU, which demonstrates that local-scale alignment is essential as well as other scales (*e.g.*, pixel-scale and global-scale). In fact, the local contextual-relation consistent adaptation loss (*i.e.*, L_{ent_cr}) penalize groups of pixels predictions to achieve local-scale alignment, where global-scale adaptation loss operates more on image-scale (*e.g.*, scene layout) while that of pixel-scale works on the feature representation alignment of each independent pixels. The consistent results with different settings further confirm that complementary information has been learned in different

Target Image   Without Ada.   Ada.(AdvEnt)   Ada.(CrCDA)   Ground Truth

Fig. 5. Qualitative results for GTA5 → Cityscapes. Our approach (CrCDA) aligns low-level features (*e.g.*, boundaries of sidewalk, car and person *etc.*) as well as high-level features by multi-scale adversarial learning. In contrast, AdvEnt ignores low-level information because global alignment focuses more on high-level information. Thus, as shown above, CrCDA achieves both local and global consistencies while AdvEnt only achieves global consistency.

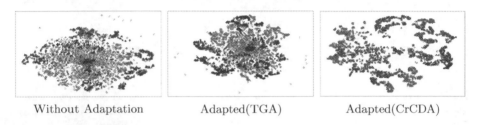

Without Adaptation   Adapted(TGA)   Adapted(CrCDA)

Fig. 6. Visualization of feature distributions via t-SNE [33]. "Blue": Source. "Red": Target. As shown in the first column, the feature distribution of source data is naturally more discriminative (discrete) than that of target data (uniformly distributed) due to only source supervision is available. Traditional global alignment (TGA) aligns them in global scale, where global consistency is achieved while local consistency is ignored. Thus the adapted target feature distribution is not discriminative. CrCDA aligns them with local-scale consistency (*i.e.*, local contextual-relation consistency), where both local and global consistencies are achieved. Thus the adapted target feature distribution is more discriminative and consistent with that of the source. (Color figure online)

scales' adaptation. The qualitative results and visualization of feature distributions are provided in Fig. 5 and 6, which further demonstrate our conjectures mentioned above. We also provide the complementary studies to demonstrate that our local contextual-relations alignment method is complementary to most existing global-scale alignment approaches, as shown in Table 4.

5   Conclusions

In this paper, we present the local contextual-relation consistent domain adaptation (CrCDA) to address the task of unsupervised domain adaptation for semantic segmentation. By taking a closer look at the local inconsistency (*i.e.*, local contextual-relations inconsistency) while implementing global adaptation, CrCDA is able to align the domain shift in local and global scales at the same time, where local semantic consistency is normally ignored by current approaches. The experimental results on the two challenging segmentation UDA tasks validate the state-of-the-art of CrCDA.

Acknowledgement. This research was conducted in collaboration with Singapore Telecommunications Limited and partially supported by the Singapore Government through the Industry Alignment Fund - Industry Collaboration Projects Grant.

References

1. Bottou, L.: Large-scale machine learning with stochastic gradient descent. In: Lechevallier, Y., Saporta, G. (eds.) Proceedings of COMPSTAT 2010, pp. 177–186. Springer, Heidelberg (2010). https://doi.org/10.1007/978-3-7908-2604-3_16
2. Bousmalis, K., Silberman, N., Dohan, D., Erhan, D., Krishnan, D.: Unsupervised pixel-level domain adaptation with generative adversarial networks. In: Proceedings of the IEEE Conference on Computer Vision and Pattern Recognition, pp. 3722–3731 (2017)
3. Chen, L.C., Papandreou, G., Kokkinos, I., Murphy, K., Yuille, A.L.: DeepLab: semantic image segmentation with deep convolutional nets, atrous convolution, and fully connected CRFs. IEEE Trans. Pattern Anal. Mach. Intell. **40**(4), 834–848 (2017)
4. Chen, M., Xue, H., Cai, D.: Domain adaptation for semantic segmentation with maximum squares loss. In: Proceedings of the IEEE International Conference on Computer Vision, pp. 2090–2099 (2019)
5. Chen, Q., Liu, Y., Wang, Z., Wassell, I., Chetty, K.: Re-weighted adversarial adaptation network for unsupervised domain adaptation. In: The IEEE Conference on Computer Vision and Pattern Recognition (CVPR), June 2018
6. Chen, Y., Li, W., Sakaridis, C., Dai, D., Van Gool, L.: Domain adaptive faster R-CNN for object detection in the wild. In: Proceedings of the IEEE Conference on Computer Vision and Pattern Recognition, pp. 3339–3348 (2018)
7. Chen, Y., Li, W., Van Gool, L.: Road: reality oriented adaptation for semantic segmentation of urban scenes. In: Proceedings of the IEEE Conference on Computer Vision and Pattern Recognition, pp. 7892–7901 (2018)
8. Choi, J., Kim, T., Kim, C.: Self-ensembling with GAN-based data augmentation for domain adaptation in semantic segmentation. In: Proceedings of the IEEE International Conference on Computer Vision, pp. 6830–6840 (2019)
9. Cordts, M., et al.: The cityscapes dataset for semantic urban scene understanding. In: Proceedings of the IEEE Conference on Computer Vision and Pattern Recognition, pp. 3213–3223 (2016)
10. Deng, J., Dong, W., Socher, R., Li, L.J., Li, K., Fei-Fei, L.: ImageNet: a large-scale hierarchical image database. In: 2009 IEEE Conference on Computer Vision and Pattern Recognition, pp. 248–255. IEEE (2009)
11. Du, L., et al.: SSF-DAN: separated semantic feature based domain adaptation network for semantic segmentation. In: Proceedings of the IEEE International Conference on Computer Vision, pp. 982–991 (2019)
12. Ester, M., Kriegel, H.P., Sander, J., Xu, X.: Density-based spatial clustering of applications with noise. In: International Conference on Knowledge Discovery and Data Mining, vol. 240, p. 6 (1996)
13. Ganin, Y., Lempitsky, V.: Unsupervised domain adaptation by backpropagation. arXiv preprint arXiv:1409.7495 (2014)

14. Ganin, Y., et al.: Domain-adversarial training of neural networks. J. Mach. Learn. Res. **17**(1), 2096–2030 (2016)
15. Gong, B., Shi, Y., Sha, F., Grauman, K.: Geodesic flow kernel for unsupervised domain adaptation. In: 2012 IEEE Conference on Computer Vision and Pattern Recognition, pp. 2066–2073. IEEE (2012)
16. Goodfellow, I., et al.: Generative adversarial nets. In: Advances in Neural Information Processing Systems, pp. 2672–2680 (2014)
17. Guan, D., et al.: Unsupervised domain adaptation for multispectral pedestrian detection. In: Proceedings of the IEEE Conference on Computer Vision and Pattern Recognition Workshops (2019)
18. Hoffman, J., et al.: CyCADA: cycle-consistent adversarial domain adaptation. arXiv preprint arXiv:1711.03213 (2017)
19. Hoffman, J., Wang, D., Yu, F., Darrell, T.: FCNs in the wild: Pixel-level adversarial and constraint-based adaptation. arXiv preprint arXiv:1612.02649 (2016)
20. Hong, W., Wang, Z., Yang, M., Yuan, J.: Conditional generative adversarial network for structured domain adaptation. In: Proceedings of the IEEE Conference on Computer Vision and Pattern Recognition, pp. 1335–1344 (2018)
21. Huang, J., Yuan, Z., Zhou, X.: A learning framework for target detection and human face recognition in real time. Int. J. Technol. Hum. Interact. (IJTHI) **15**(3), 63–76 (2019)
22. Isola, P., Zhu, J.Y., Zhou, T., Efros, A.A.: Image-to-image translation with conditional adversarial networks. In: Proceedings of the IEEE Conference on Computer Vision and Pattern Recognition, pp. 1125–1134 (2017)
23. Kang, G., Jiang, L., Yang, Y., Hauptmann, A.G.: Contrastive adaptation network for unsupervised domain adaptation. In: Proceedings of the IEEE Conference on Computer Vision and Pattern Recognition, pp. 4893–4902 (2019)
24. Kang, G., Zheng, L., Yan, Y., Yang, Y.: Deep adversarial attention alignment for unsupervised domain adaptation: the benefit of target expectation maximization. In: Proceedings of the European Conference on Computer Vision (ECCV), pp. 401–416 (2018)
25. Li, Y., Yuan, L., Vasconcelos, N.: Bidirectional learning for domain adaptation of semantic segmentation. In: Proceedings of the IEEE Conference on Computer Vision and Pattern Recognition, pp. 6936–6945 (2019)
26. Lian, Q., Lv, F., Duan, L., Gong, B.: Constructing self-motivated pyramid curriculums for cross-domain semantic segmentation: a non-adversarial approach. In: The IEEE International Conference on Computer Vision (ICCV), October 2019
27. Long, J., Shelhamer, E., Darrell, T.: Fully convolutional networks for semantic segmentation. In: Proceedings of the IEEE Conference on Computer Vision and Pattern Recognition, pp. 3431–3440 (2015)
28. Long, M., Cao, Y., Wang, J., Jordan, M.I.: Learning transferable features with deep adaptation networks. arXiv preprint arXiv:1502.02791 (2015)
29. Long, M., Zhu, H., Wang, J., Jordan, M.I.: Unsupervised domain adaptation with residual transfer networks. In: Advances in Neural Information Processing Systems, pp. 136–144 (2016)
30. Luo, Y., Liu, P., Guan, T., Yu, J., Yang, Y.: Significance-aware information bottleneck for domain adaptive semantic segmentation. arXiv preprint arXiv:1904.00876 (2019)
31. Luo, Y., Zheng, L., Guan, T., Yu, J., Yang, Y.: Taking a closer look at domain shift: category-level adversaries for semantics consistent domain adaptation. In: Proceedings of the IEEE Conference on Computer Vision and Pattern Recognition, pp. 2507–2516 (2019)

32. Luo, Y., Zheng, Z., Zheng, L., Guan, T., Yu, J., Yang, Y.: Macro-micro adversarial network for human parsing. In: Proceedings of the European Conference on Computer Vision (ECCV), pp. 418–434 (2018)
33. van der Maaten, L., Hinton, G.: Visualizing data using t-SNE. J. Mach. Learn. Res. 9(Nov), 2579–2605 (2008)
34. Murez, Z., Kolouri, S., Kriegman, D., Ramamoorthi, R., Kim, K.: Image to image translation for domain adaptation. In: Proceedings of the IEEE Conference on Computer Vision and Pattern Recognition, pp. 4500–4509 (2018)
35. Richter, S.R., Vineet, V., Roth, S., Koltun, V.: Playing for data: ground truth from computer games. In: Leibe, B., Matas, J., Sebe, N., Welling, M. (eds.) ECCV 2016. LNCS, vol. 9906, pp. 102–118. Springer, Cham (2016). https://doi.org/10.1007/978-3-319-46475-6_7
36. Ros, G., Sellart, L., Materzynska, J., Vazquez, D., Lopez, A.M.: The synthia dataset: a large collection of synthetic images for semantic segmentation of urban scenes. In: Proceedings of the IEEE Conference on Computer Vision and Pattern Recognition, pp. 3234–3243 (2016)
37. Saito, K., Kim, D., Sclaroff, S., Darrell, T., Saenko, K.: Semi-supervised domain adaptation via minimax entropy. In: Proceedings of the IEEE International Conference on Computer Vision, pp. 8050–8058 (2019)
38. Saito, K., Ushiku, Y., Harada, T., Saenko, K.: Adversarial dropout regularization. arXiv preprint arXiv:1711.01575 (2017)
39. Saito, K., Watanabe, K., Ushiku, Y., Harada, T.: Maximum classifier discrepancy for unsupervised domain adaptation. In: Proceedings of the IEEE Conference on Computer Vision and Pattern Recognition, pp. 3723–3732 (2018)
40. Saito, K., Yamamoto, S., Ushiku, Y., Harada, T.: Open set domain adaptation by backpropagation. In: Ferrari, V., Hebert, M., Sminchisescu, C., Weiss, Y. (eds.) ECCV 2018. LNCS, vol. 11209, pp. 156–171. Springer, Cham (2018). https://doi.org/10.1007/978-3-030-01228-1_10
41. Saleh, F.S., Aliakbarian, M.S., Salzmann, M., Petersson, L., Alvarez, J.M.: Effective use of synthetic data for urban scene semantic segmentation. In: Ferrari, V., Hebert, M., Sminchisescu, C., Weiss, Y. (eds.) ECCV 2018. LNCS, vol. 11206, pp. 86–103. Springer, Cham (2018). https://doi.org/10.1007/978-3-030-01216-8_6
42. Sankaranarayanan, S., Balaji, Y., Jain, A., Nam Lim, S., Chellappa, R.: Learning from synthetic data: addressing domain shift for semantic segmentation. In: Proceedings of the IEEE Conference on Computer Vision and Pattern Recognition, pp. 3752–3761 (2018)
43. Simonyan, K., Zisserman, A.: Very deep convolutional networks for large-scale image recognition. arXiv preprint arXiv:1409.1556 (2014)
44. Tsai, Y.H., Hung, W.C., Schulter, S., Sohn, K., Yang, M.H., Chandraker, M.: Learning to adapt structured output space for semantic segmentation. In: Proceedings of the IEEE Conference on Computer Vision and Pattern Recognition, pp. 7472–7481 (2018)
45. Tsai, Y.H., Sohn, K., Schulter, S., Chandraker, M.: Domain adaptation for structured output via discriminative patch representations. In: Proceedings of the IEEE International Conference on Computer Vision, pp. 1456–1465 (2019)
46. Tzeng, E., Hoffman, J., Saenko, K., Darrell, T.: Adversarial discriminative domain adaptation. In: Proceedings of the IEEE Conference on Computer Vision and Pattern Recognition, pp. 7167–7176 (2017)
47. Tzeng, E., Hoffman, J., Zhang, N., Saenko, K., Darrell, T.: Deep domain confusion: maximizing for domain invariance. arXiv preprint arXiv:1412.3474 (2014)

48. Vu, T.H., Choi, W., Schulter, S., Chandraker, M.: Memory warps for learning long-term online video representations. arXiv preprint arXiv:1803.10861 (2018)
49. Vu, T.H., Jain, H., Bucher, M., Cord, M., Pérez, P.: Advent: adversarial entropy minimization for domain adaptation in semantic segmentation. In: Proceedings of the IEEE Conference on Computer Vision and Pattern Recognition, pp. 2517–2526 (2019)
50. Wu, Z., et al.: DCAN: dual channel-wise alignment networks for unsupervised scene adaptation. In: Proceedings of the European Conference on Computer Vision (ECCV), pp. 518–534 (2018)
51. Yan, H., Ding, Y., Li, P., Wang, Q., Xu, Y., Zuo, W.: Mind the class weight bias: weighted maximum mean discrepancy for unsupervised domain adaptation. In: Proceedings of the IEEE Conference on Computer Vision and Pattern Recognition, pp. 2272–2281 (2017)
52. Zhan, F., Huang, J., Lu, S.: Adaptive composition GAN towards realistic image synthesis. arXiv preprint arXiv:1905.04693 (2019)
53. Zhang, C., Bengio, S., Hardt, M., Recht, B., Vinyals, O.: Understanding deep learning requires rethinking generalization. arXiv preprint arXiv:1611.03530 (2016)
54. Zhang, X., Gong, H., Dai, X., Yang, F., Liu, N., Liu, M.: Understanding pictograph with facial features: end-to-end sentence-level lip reading of Chinese. In: AAAI, pp. 9211–9218 (2019)
55. Zhang, Y., David, P., Gong, B.: Curriculum domain adaptation for semantic segmentation of urban scenes. In: Proceedings of the IEEE International Conference on Computer Vision, pp. 2020–2030 (2017)
56. Zhang, Y., Qiu, Z., Yao, T., Liu, D., Mei, T.: Fully convolutional adaptation networks for semantic segmentation. In: Proceedings of the IEEE Conference on Computer Vision and Pattern Recognition, pp. 6810–6818 (2018)
57. Zhao, H., Shi, J., Qi, X., Wang, X., Jia, J.: Pyramid scene parsing network. In: Proceedings of the IEEE Conference on Computer Vision and Pattern Recognition, pp. 2881–2890 (2017)
58. Zhong, Z., Zheng, L., Luo, Z., Li, S., Yang, Y.: Invariance matters: exemplar memory for domain adaptive person re-identification. In: Proceedings of the IEEE Conference on Computer Vision and Pattern Recognition, pp. 598–607 (2019)
59. Zhu, H., Meng, F., Cai, J., Lu, S.: Beyond pixels: a comprehensive survey from bottom-up to semantic image segmentation and cosegmentation. J. Vis. Commun. Image Represent. 34, 12–27 (2016)
60. Zhu, J.Y., Park, T., Isola, P., Efros, A.A.: Unpaired image-to-image translation using cycle-consistent adversarial networks. In: Proceedings of the IEEE International Conference on Computer Vision, pp. 2223–2232 (2017)
61. Zhu, X.J.: Semi-supervised learning literature survey. Technical report, University of Wisconsin-Madison Department of Computer Sciences (2005)
62. Zhu, X., Zhou, H., Yang, C., Shi, J., Lin, D.: Penalizing top performers: conservative loss for semantic segmentation adaptation. In: Proceedings of the European Conference on Computer Vision (ECCV), pp. 568–583 (2018)
63. Zou, Y., Yu, Z., Liu, X., Kumar, B.V., Wang, J.: Confidence regularized self-training. In: The IEEE International Conference on Computer Vision (ICCV), October 2019
64. Zou, Y., Yu, Z., Vijaya Kumar, B., Wang, J.: Unsupervised domain adaptation for semantic segmentation via class-balanced self-training. In: Proceedings of the European Conference on Computer Vision (ECCV), pp. 289–305 (2018)

Estimating People Flows to Better Count Them in Crowded Scenes

Weizhe Liu[1(✉)], Mathieu Salzmann[1,2], and Pascal Fua[1]

[1] CVLab, EPFL, Lausanne, Switzerland
{weizhe.liu,mathieu.salzmann,pascal.fua}@epfl.ch
[2] ClearSpace, Écublens, Switzerland

Abstract. Modern methods for counting people in crowded scenes rely on deep networks to estimate people densities in individual images. As such, only very few take advantage of temporal consistency in video sequences, and those that do only impose weak smoothness constraints across consecutive frames.

In this paper, we advocate estimating people flows across image locations between consecutive images and inferring the people densities from these flows instead of directly regressing. This enables us to impose much stronger constraints encoding the conservation of the number of people. As a result, it significantly boosts performance without requiring a more complex architecture. Furthermore, it also enables us to exploit the correlation between people flow and optical flow to further improve the results.

We will demonstrate that we consistently outperform state-of-the-art methods on five benchmark datasets.

Keywords: Crowd counting · Grid flow model · Temporal consistency

1 Introduction

Crowd counting is important for applications such as video surveillance and traffic control. Most state-of-the-art approaches rely on regressors to estimate the local crowd density in individual images, which they then proceed to integrate over portions of the images to produce people counts. The regressors typically use Random Forests [16], Gaussian Processes [7], or more recently Deep Nets [5, 14, 17, 22, 26, 30, 32–34, 36, 40, 41, 49, 55, 59].

When video sequences are available, some algorithms use temporal consistency to impose weak constraints on successive density estimates. One way is to use an LSTM to model the evolution of people densities from one frame to the next [49]. However, this does not explicitly enforce the fact that people numbers must be strictly conserved as they move about, except at very specific locations

Electronic supplementary material The online version of this chapter (https://doi.org/10.1007/978-3-030-58555-6_43) contains supplementary material, which is available to authorized users.

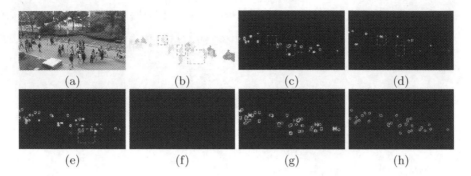

Fig. 1. From people flow to crowd density. (a) Original image. (b) Optical flow. Red denotes people moving right and blue moving left. The overlaid orange box encloses people moving slowly or not at all, the pink box people moving left, and the green box people moving right. (c) Estimated flow of people moving right. People moving left, such as those in the pink box, do not contribute to it, whereas those in the green box do. (d) Flow of people moving left. The situations within the pink and green box are reversed. (e) Estimated flow of people staying within the same grid location from one time instant to the next, such as those within the orange box. They are not necessarily static. They may simply not have had time to change location between the two time instants. (f) Estimated flow of people moving up. As no one does, it is almost zero everywhere. (g) Density map inferred by summing all the flows incident on a particular location. (h) Ground truth density map.

where they can move in or out of the field of view. Modeling this was attempted in [24] but, because expressing this constraint in terms of people densities is difficult, the constraints actually enforced were much weaker.

In this paper, we propose to regress people flows, that is, the number of people moving from one location to another in the image plane, instead of densities. To this end, we partition the image into a number of grid locations and, for each one, we define ten potential flows, one towards each neighboring location, one towards the location *itself*, and the last towards regions outside the image plane. In practice, the last one is only used at boundary locations. The flow towards the location itself enables us to account for people who stay in the same location from one instant to the next and the final flow to account for people who enter or exit the field of view. Figure 1 depicts some of the ten flows we compute. All the flows incident on a grid location are summed to yield an estimate of the people density in that location. The network can therefore be trained given ground-truth estimates only of the local people densities as opposed to people flows. In other words, even though we compute flows, our network only requires ground-truth density data for training purposes, like most others.

We will show that this formulation allows us to effectively impose people conservation constraints—people do not teleport from one region of the image to another—much more effectively than earlier approaches. This increases performance using network architectures that are neither deeper nor more complex than state-of-the-art ones. Furthermore, regressing people flows instead of densities provides a scene description that includes the motion direction and magnitude. This enables us to exploit the fact that people flow and optical flow should be highly correlated, as illustrated by Fig. 1, which provides an additional regularization constraint on the predicted flows and further enhances performance.

We will demonstrate on five benchmark datasets that our approach to enforcing temporal consistency brings a substantial performance boost compared to state-of-the-art approaches. Furthermore, if the cameras can be calibrated, we can apply our approach in the ground plane instead of the image plane, which further improves performance, as shown in the supplementary material. Our contribution is therefore a novel formulation of regressing people densities from video sequences that enforces strong temporal consistency constraints without requiring complex network architectures.

2   Related Work

Given a single image of a crowded scene, the currently dominant approach to counting people is to train a deep network to regress a people density estimate at every image location. This density is then integrated to deliver an actual count [8,15,18,19,21,23–25,27,29,37,39,42,46–48,50–54,57,60].

Enforcing Temporal Consistency. While most methods work on individual images, a few have nonetheless been extended to encode temporal consistency. Perhaps the most popular way to do so is to use an LSTM [13]. For example, in [49], the ConvLSTM architecture [38] is used for crowd counting purposes. It is trained to enforce consistency both in the forward and the backward direction. In [58], an LSTM is used in conjunction with an FCN [28] to count vehicles in video sequences. A Locality-constrained Spatial Transformer (LST) is introduced in [11]. It takes the current density map as input and outputs density maps in the next frames. The influence of these estimates on crowd density depends on the similarity between pixel values in pairs of neighboring frames.

While effective these approaches have two main limitations. First, at training time, they can only be used to impose consistency across annotated frames and cannot take advantage of unannotated ones to provide self-supervision. Second, they do not explicitly enforce the fact that people numbers must be conserved over time, except at the edges of the field of view. The recent method of [24] addresses both these issues. However, as will be discussed in more detail in Sect. 3.1, because the people conservation constraints are expressed in terms of numbers of people in neighboring image areas, they are much weaker than they should be.

Introducing Flow Variables. Imposing strong conservation constraints when tracking people has been a concern long before the advent of deep learning. For example, in [3], people tracking is formulated as multi-target tracking on a grid and gives rise to a linear program that can be solved efficiently using the K-Shortest Path algorithm [44]. The key to this formulation is the use as optimization variables of people flows from one grid location to another, instead of

Table 1. Notations.

T	number of time steps
K	number of locations in the image plane
I^t	image at t-th frame
m_j^t	number of people present at location j at time t
$f_{i,j}^{t-1,t}$	number of people moving from location i to location j between times $t-1$ and t
$N(j)$	neighborhood of location j that can be reached within a single time step

the actual number of people in each grid location. In [31], a people conservation constraint is enforced and the global solution is found by a greedy algorithm that sequentially instantiates tracks using shortest path computations on a flow network [56].

Such people conservation constraints have since been combined with additional ones to further boost performance. They include appearance constraints [1, 2,10] to prevent identity switches, spatio-temporal constraints to force the trajectories of different objects to be disjoint [12], and higher-order constraints [4,9].

However, all these works predate deep learning. These kind of flow constraints have never been used in a deep crowd counting context and are designed for scenarios in which people can still be tracked individually. In this paper, we demonstrate that this approach can also be brought to bear in a deep pipeline to handle dense crowds in which people cannot be tracked as individuals anymore.

3   Approach

We regress *people flows* from images. We take these flows to be counts between two consecutive time instants of people either moving from their current location to a neighboring one, staying at the same location, or moving in or out of the field of view. They are depicted by Fig. 2 and summarized in Table 1. People flows incident on a specific location are then summed to derive the number of people per location or *people count* per location. The *crowd density* then simply is the *people count* divided by the location area. Our key insight is that this formulation enables us to impose much tighter *people conservation constraints* than earlier approaches. By this, we mean that we can accurately model the fact that all people present in a location at a given instant either were already there at the previous one or came from a neighboring location. This assumes the image frequency to be high enough for people not being able to move beyond neighboring locations in the time that separates consecutive frames. This is a common assumption that has proved both valid and effective in many earlier works.

3.1   Formalization

Let us consider a video sequence $\mathbf{I} = \{\mathbf{I}^1, \ldots \mathbf{I}^T\}$ and three consecutive images $\mathbf{I}^{t-1}$, $\mathbf{I}^t$, and $\mathbf{I}^{t+1}$ from it. Let us assume that each image has been partitioned

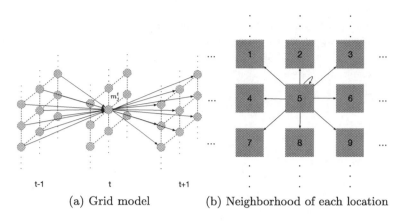

(a) Grid model   (b) Neighborhood of each location

Fig. 2. People flows. (a) The crowd density at time t at a given location can only come from neighboring grid locations at time $t-1$ and flow to neighboring grid locations at time $t+1$, in both cases including the location itself. (b) For each location not at the boundary of the image plane, there are nine locations reachable within a single time step, including the location itself. For locations at the edge of the image plane, we add a tenth location that represents the rest of the world. It allows for flows of people who either leave the image or enter it from outside.

into K rectangular grid locations. In our implementation, a location is one spatial position in the final convolutional feature map, corresponding to an 8×8 neighborhood in the image. However, other choices are possible.

The main constraint we want to enforce is that the number of people present at location j at time t is the number of people who were already there at time $t-1$ and stayed there plus the number of those who walked in from neighboring locations between $t-1$ and t. The number of people present at location j at time t also equals the sum of the number of people who stayed there until time $t+1$ and of people who went to a neighboring location between t and $t+1$.

Let m_j^t be the number of people present at location j at time t, or *people count* at that location. Let $f_{i,j}^{t-1,t}$ be the number of people who move from location i to location j between times $t-1$ and t, and $N(j)$ the neighborhood of location j that can be reached within a single time step. These notations are illustrated by Fig. 2(a) and summarized in Table 1. In practice, we take $N(j)$ to be the 8 neighbors of grid location j plus the grid location itself to account for people who remain at the same place, as depicted by Fig. 2(b). Our people conservation constraint can now be written as

$$\sum_{i \in N(j)} f_{i,j}^{t-1,t} = m_j^t = \sum_{k \in N(j)} f_{j,k}^{t,t+1} \ . \tag{1}$$

for all locations j that are *not* on the edge of the grid, that is, locations from which people cannot appear or disappear without being seen elsewhere in the image.

Most earlier approaches [5,17,20,23,25,30,59] regress the values of m_j^t, which makes it hard to impose the constraints of Eq. 1 because many different values of the flows $f_{i,j}^{t-1,t}$ can produce the same m_j^t values. For example, in [24], the equivalent constraint is

$$\forall j \quad m_j^t \leq \sum_{i \in N(j)} m_i^{t-1} \text{ and } m_j^t \leq \sum_{k \in N(j)} m_k^{t+1} . \tag{2}$$

It only states that the number of people at location j at time t is less than or equal to the total number of people at neighboring locations at time $t-1$ and that the same holds between times t and $t+1$. These are much looser constraints than the ones of Eq. 1. They guarantee that people cannot suddenly appear but do not account for the fact that people cannot suddenly disappear either. Our formulation lets us remedy this shortcoming. By regressing the $f_{i,j}^{t-1,t}$ from pairs consecutive images and computing the values of the m_j^t from these, we can impose the tighter constraints of Eq. 1.

3.2   Regressing the Flows

We now turn to the task of training a regressor that predicts flows that correspond to what is observed while obeying the above constraints and properly handling the boundary grid locations. Let us denote the regressor that predicts the flows from $\mathbf{I}^{t-1}$ and $\mathbf{I}^t$ as $\mathcal{F}$ with parameters Θ to be learned during training. In other words, $f^{t-1,t} = \mathcal{F}(I^{t-1}, I^t; \Theta)$ is the vector of predicted flows between all pairs of neighboring locations between times $t-1$ and t. In practice, $\mathcal{F}$ is implemented by a deep network. The predicted local people counts m_j^t, that is number of people per grid location j and at time t, are taken to be the sum of the incoming flows according to Eq. 1, and the predicted count for the whole image is the sum of all the m_j^t. As the flows are not directly observable, the training data comes in the form of *people counts* $\bar{m}_j^t$ per grid location j and at time t.

During training, our goal is therefore to find values of Θ such that

$$\bar{m}_j^t = \sum_{i \in N(j)} f_{i,j}^{t-1,t} = \sum_{k \in N(j)} f_{j,k}^{t,t+1} \text{ and } \quad f_{i,j}^{t-1,t} = f_{j,i}^{t,t-1} . \tag{3}$$

for all i, j, and t, except for locations at the edges of the image plane, where people can appear from and disappear to unseen parts of the scene. The first constraint is the people conservation constraint introduced in Sect. 3.1. The second accounts for the fact that, were we to play the video sequence in reverse, the flows should have the same magnitude but in the opposite direction. As will be discussed below, we enforce these constraints by incorporating them into the loss function we minimize to learn Θ. Finally, we impose that all the flows be non-negative by using ReLu normalization in the network that implements $\mathcal{F}$. Note that we only require the people flow to be non-negative, the fact that a location may contain less than 1 person simply means that the flow value will be less than 1.

Regressor Architecture. Recall that $f^{t-1,t} = \mathcal{F}(\mathbf{I}^{t-1}, \mathbf{I}^t; \Theta)$ is a vector of predicted flows from neighboring locations between times $t-1$ and t. In practice, $\mathcal{F}$ is implemented by the encoding/decoding architecture shown in Fig. 3, and $f^{t-1,t}$ has the same dimension as the image grid and 10 channels per location. The first are the flows to the 9 possible neighbors depicted by Fig. 2(b) and the tenth represents potential flows from outside the image and is therefore only meaningful at the edges. The fifth channel denotes the flow towards the location itself, which enables us to account for people who stay in the same location from one instant to the next.

To compute $f^{t-1,t}$, consecutive frames $\mathbf{I}^{t-1}$ and $\mathbf{I}^t$ are fed to the CAN encoder network of [25]. This yields deep features $s^{t-1} = \mathcal{E}_e(I^{t-1}; \Theta_e)$ and $s^t = \mathcal{E}_e(I^t; \Theta_e)$, where $\mathcal{E}_e$ denotes the encoder with weights Θ_e. These features are then concatenated and fed to a decoder network to output $f^{t-1,t} = \mathcal{D}(s^{t-1}, s^t; \Theta_d)$, where $\mathcal{D}$ is the decoder with weights Θ_d. $\mathcal{D}$ comprises the back-end decoder of CAN [25] with an additional final ReLU layer to guarantee that the output is always non-negative. The encoder and decoder specifications are given in the supplementary material.

Grid Size. In all our experiments, we treated each spatial location in the output people flow map as a separate location. Since our CAN [25] backbone outputs a down-sampled density map, each output grid location represent an 8×8 pixel block in input image. This down-sampling rate is common in crowd counting models [17,24,25] because it represents a good compromise between high-resolution of the density map and efficiency of the model. In the supplementary material, we will confirm this by showing that changing the down-sampling rate degrades performance.

Loss Function and Training. To obtain the ground-truth maps $\bar{m}^t$ of Eq. 3, we use the same approach as in most previous work [5,17,20,23,25,30,59]. In each image $\mathbf{I}^t$, we annotate a set of c^t 2D points $P^t = \{P_i^t\}_{1 \le i \le c^t}$ that denote the positions of the human heads in the scene. The corresponding ground-truth density map $\bar{m}^t$ is obtained by convolving an image containing ones at these locations and zeroes elsewhere with a Gaussian kernel $\mathcal{N}(\cdot|\mu, \sigma^2)$ with mean μ and standard deviation σ. We write

$$\bar{m}_j^t = \sum_{i=1}^{c^t} \mathcal{N}(p_j|\mu = P_i^t, \sigma^2) \, , \, \forall j \, . \tag{4}$$

where p_j denotes the center of location j. Note that this formulation preserves the constraints of Eq. 3 because we perform the same convolution across the whole image. In other words, if a person moves in a given direction by n pixels, the corresponding contribution to the density map will shift in the same direction and also by n pixels.

The final ReLU layer of the regressor guarantees that the estimated flows are non-negative. To enforce the constraints of Eq. 3, we define our combined loss function L_{combi} as the weighted sum of two loss terms. We write

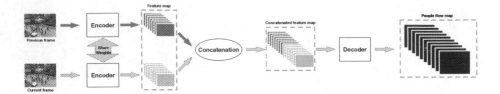

Fig. 3. Model Architecture: Two consecutive RGB image frames are fed to the same encoder network that relies on the CAN scale-aware feature extractor of [25]. These multi-scale features are further concatenated and fed to a decoder network to produce the final people flow maps.

$$L_{combi} = L_{flow} + \alpha L_{cycle} \,,$$

$$L_{flow} = \sum_{j \in I^t} \left[(\bar{m}_j^t - \sum_{i \in N(j)} f_{i,j}^{t-1,t})^2 + (\bar{m}_j^t - \sum_{k \in N(j)} f_{j,k}^{t,t+1})^2 \right],$$

$$L_{cycle} = \sum_{j \in I^t} \left[\sum_{i \in N(j)} (f_{i,j}^{t-1,t} - f_{j,i}^{t,t-1})^2 + \sum_{k \in N(j)} (f_{j,k}^{t,t+1} - f_{k,j}^{t+1,t})^2 \right]. \quad (5)$$

where $\bar{m}_j^t$ is the ground-truth crowd density value, that is, the *people count* at time t and location j of Eq. 4 and α is a scalar weight we set to 1 in all our experiments.

Although the variant of L_{combi} can be computed from only two consecutive frames, at training time we always use three to enforce the temporal consistency constraints of Eq. 1. Algorithm 1 describes our training scheme in more detail. Note that we do *not* assume all training frames to be annotated. Only frames V, $2V$, $3V$ need be with $V \geq 1$. To evaluate the loss function for frame kV, where k is an integer, we then use frames $kV - 1$, kV, and $kV + 1$, where one of the three is annotated. In practice, we could also use frames $kV - n$, kV, and $kV + n$ with $n \geq 1$.

3.3   Exploiting Optical Flow

When the camera is static, both the people flow discussed above and the optical flow that can be computed directly from the images stem for the motion of the people. They should therefore be highly correlated. In fact, this remains true even if the camera moves because its motion creates an apparent flow of people from one image location to another. However, there is no simple linear relationship between people flow and optical flow. To account for their correlation, we therefore introduce an additional loss function, which we define as

$$L_{optical} = \sum_j \delta(m_j > 0)(\mathbf{O}_j - \bar{o}_j^{t-1,t})^2 \,,$$

$$\text{where} \quad \mathbf{O} = \mathcal{F}_o(m^{t-1}, m^t; \Theta_o) \,. \quad (6)$$

Algorithm 1. Three-Frames Training Algorithm

Require: Training image sequence $\{\mathbf{I}^1, \ldots, \mathbf{I}^T\}$ with an interval V between annotated frames.

Require: Ground-truth density maps $\{\bar{m}^V, \bar{m}^{2V} \ldots, \bar{m}^{(T//V)V}\}$ computed by convolving the annotations according to Eq. 4.

    procedure TRAIN($\{\mathbf{I}^1, .., \mathbf{I}^T\}, \{\bar{m}^V, .., \bar{m}^{(T//V)V}\}$)
        Initialize the weights Θ of regressor network $\mathcal{F}$
        for # of gradient iterations **do**
          Pick 3 consecutive frames $(\mathbf{I}^{t-1}, \mathbf{I}^t, \mathbf{I}^{t+1})$, where t is a multiple of V, meaning
that only $\mathbf{I}^t$ is annotated
          Set $f^{t-1,t} = \mathcal{F}(I^{t-1}, I^t, \Theta)$
          Set $f^{t,t+1} = \mathcal{F}(I^t, I^{t+1}, \Theta)$
          Set $f^{t,t-1} = \mathcal{F}(I^t, I^{t-1}, \Theta)$
          Set $f^{t+1,t} = \mathcal{F}(I^{t+1}, I^t, \Theta)$
          Reconstruct density map m_1^t from $f^{t-1,t}$
          Reconstruct density map m_2^t from $f^{t,t+1}$
          Reconstruct density map m_3^t from $f^{t,t-1}$
          Reconstruct density map m_4^t from $f^{t+1,t}$
          Minimize L_{combi} of Eq. 5 w.r.t. Θ using Adam
        end for
    end procedure

---

where m^{t-1} and m^t are density maps inferred from our predicted flows using Eq. 1, $\mathbf{O}_j$ denotes the corresponding predicted optical flow at grid location j by a pre-trained regressor $\mathcal{F}_o$, $\bar{o}^{t-1,t}$ is the optical flow from frames $t-1$ to t computed by a state-of-the-art optical flow network [43], and the indicator function term $\delta(m_j > 0)$ ensures that the correlation is only enforced where there are people. This is especially useful when the camera moves to discount the optical flows generated by the changing background. We also use CAN [25] as the optical flow regressor $\mathcal{F}_o$ with 2 input channels, one for m^{t-1} and the other m^t. This network is pre-trained separately on the training data and then used to train the people flow regressor. We refer the reader to the supplementary material for implementation details.

Pre-training the regressor $\mathcal{F}_o$ requires annotations for consecutive frames, that is, $V = 1$ in the definition of Algorithm 1. When such annotations are available, we use this algorithm again but replace L_{combi} by

$$L_{all} = L_{combi} + \beta L_{optical}. \tag{7}$$

In all our experiments, we set β to 0.0001 to account for the fact that the optical flow values are around 4,000 times larger than the people flow values.

4 Experiments

In this section, we first introduce the evaluation metrics and benchmark datasets used in our experiments. We then compare our results to those of current state-of-the-art methods. Finally, we perform an ablation study to demonstrate the impact of individual constraints.

4.1   Evaluation Metrics

Previous works in crowd density estimation use the mean absolute error (MAE) and the root mean squared error ($RMSE$) as evaluation metrics [30,34,41,49, 55,59]. They are defined as

$$MAE = \frac{1}{N} \sum_{i=1}^{N} |z_i - \hat{z}_i| \text{ and } RMSE = \sqrt{\frac{1}{N} \sum_{i=1}^{N} (z_i - \hat{z}_i)^2} \, .$$

where N is the number of test images, z_i denotes the true number of people inside the ROI of the ith image and $\hat{z}_i$ the estimated number of people. In the benchmark datasets discussed below, the ROI is the whole image except when explicitly stated otherwise. In practice, $\hat{z}_i$ is taken to be $\sum_{p \in I_i} m_p$, that is, the sum over all locations or people counts obtained by summing the predicted people flows.

4.2   Benchmark Datasets and Ground-Truth Data

For evaluations purposes, we use five different datasets, for which the videos have been released along with recently published papers. The first one is a synthetic dataset with ground-truth optical flows. The other four are real world videos, with annotated people locations but without ground-truth optical flow. To use the optional optical flow constraints introduced in Sect. 3.3, we therefore use the pre-trained **PWC-Net** [43], as described in that section, to compute the loss function $L_{optical}$ of Eq. 6. Please refer to the supplementary material for additional details.

CrowdFlow [35]. This dataset consists of five synthetic sequences ranging from 300 to 450 frames each. Each one is rendered twice, once using a static camera and the other a moving one. The ground-truth optical flow is provided as shown in the supplementary material. As this dataset has not been used for crowd counting before, and the training and testing sets are not clearly described in [35], to verify the performance difference caused by using ground-truth optical flow vs. estimated one, we use the first three sequences of both the static and moving camera scenarios for training and validation, and the last two for testing.

FDST [11]. It comprises 100 videos captured from 13 different scenes with a total of 150,000 frames and 394,081 annotated heads. The training set consists of 60 videos, 9000 frames and the testing set contains the remaining 40 videos, 6000 frames. We use the same setting as in [11].

UCSD [6]. This dataset contains 2000 frames captured by surveillance cameras on the UCSD campus. The resolution of the frames is 238×158 pixels and the framerate is 10 fps. For each frame, the number of people varies from 11 to 46. We use the same setting as in [6], with frames 601 to 1400 used as training data and the remaining 1200 frames as testing data.

Venice [25]. It contains 4 different sequences and in total 167 annotated frames with fixed 1,280 × 720 resolution. As in [25], 80 images from a single long sequence are used as training data. The remaining 3 sequences are used for testing purposes.

WorldExpo'10 [55]. It comprises 1,132 annotated video sequences collected from 103 different scenes. There are 3,980 annotated frames, 3,380 of which are used for training purposes. Each scene contains a Region Of Interest (ROI) in which the people are counted. As in previous work [5,17,20,32–34,36,40,41,55,59] on this dataset, we report the *MAE* of each scene, as well as the average over all scenes.

4.3   Comparing Against Recent Techniques

Table 2. Comparative results on different datasets. (a) **CrowdFlow.** (b) **FDST.** (c) **Venice.** (d) **UCSD.** (e) **WorldExpo'10.**

Model	MAE	$RMSE$
MCNN [59]	172.8	216.0
CSRNet[17]	137.8	181.0
CAN[25]	124.3	160.2
OURS-COMBI	**97.8**	**112.1**
OURS-ALL-EST	**96.3**	**111.6**
OURS-ALL-GT	**90.9**	**110.3**

(a)

Model	MAE	$RMSE$
MCNN [59]	3.77	4.88
ConvLSTM [49]	4.48	5.82
WithoutLST [11]	3.87	5.16
LST [11]	3.35	4.45
OURS-COMBI	**2.17**	**2.62**
OURS-ALL-EST	**2.10**	**2.46**

(b)

Model	MAE	$RMSE$
MCNN [59]	145.4	147.3
Switch-CNN [34]	52.8	59.5
CSRNet[17]	35.8	50.0
CAN[25]	23.5	38.9
ECAN[25]	20.5	29.9
GPC[24]	18.2	26.6
OURS-COMBI	**15.0**	**19.6**

(c)

Model	MAE	RMSE
Zhang *et al.* [55]	1.60	3.31
Hydra-CNN [30]	1.07	1.35
CNN-Boosting [45]	1.10	-
MCNN [59]	1.07	1.35
Switch-CNN [34]	1.62	2.10
ConvLSTM [49]	1.30	1.79
Bi-ConvLSTM [49]	1.13	1.43
ACSCP [36]	1.04	1.35
CSRNet [17]	1.16	1.47
SANet [5]	1.02	1.29
ADCrowdNet [23]	0.98	1.25
PACNN [37]	0.89	1.18
SANet+SPANet [8]	1.00	1.28
OURS-COMBI	**0.86**	**1.13**
OURS-ALL-EST	**0.81**	**1.07**

(d)

Model	Scene1	Scene2	Scene3	Scene4	Scene5	**Average**
Zhang *et al.* [55]	9.8	14.1	14.3	22.2	3.7	12.9
MCNN [59]	3.4	20.6	12.9	13.0	8.1	11.6
Switch-CNN [34]	4.4	15.7	10.0	11.0	5.9	9.4
CP-CNN [41]	2.9	14.7	10.5	10.4	5.8	8.9
ACSCP [36]	2.8	14.05	9.6	8.1	2.9	7.5
IG-CNN [33]	2.6	16.1	10.15	20.2	7.6	11.3
ic-CNN[32]	17.0	12.3	9.2	8.1	4.7	10.3
D-ConvNet [40]	**1.9**	12.1	20.7	8.3	**2.6**	9.1
CSRNet [17]	2.9	11.5	8.6	16.6	3.4	8.6
SANet [5]	2.6	13.2	9.0	13.3	3.0	8.2
DecideNet [20]	2.0	13.14	8.9	17.4	4.75	9.23
CAN [25]	2.9	12.0	10.0	**7.9**	4.3	7.4
ECAN [25]	2.4	**9.4**	8.8	11.2	4.0	7.2
PGCNet [52]	2.5	12.7	8.4	13.7	3.2	8.1
OURS-COMBI	2.2	10.8	**8.0**	8.8	3.2	**6.6**

(e)

We denote our model trained using the combined loss function L_{combi} of Sect. 3.2 as **OURS-COMBI** and the one using the full loss function L_{all} of Sect. 3.3 with ground-truth optical flow as **OURS-ALL-GT**. In other words,

OURS-ALL-GT exploits the optical flow while **OURS-COMBI** does not. If the ground-truth optical flow is not available, we use the optical flow estimated by **PWC-Net** [43] and denote this model as **OURS-ALL-EST**.

Synthetic Data. Figure 4 depicts a qualitative result, and we report our quantitative results on the **CrowdFlow** dataset in Table 2 (a). **OURS-COMBI** outperforms the competing methods by a significant margin while **OURS-ALL-EST** delivers a further improvement. Using the ground-truth optical flow values in our L_{all} loss term yields yet another performance improvement, that points to the fact that using better optical flow estimation than **PWC-Net** [43] might help.

Real Data. Figure 5 depicts a qualitative result, and we report our quantitative results on the four real-world datasets in Tables 2(b), (c), (d) and (e). For **FDST** and **UCSD**, annotations in consecutive frames are available, which enabled us to pre-train the $\mathcal{F}_o$ regressor of Eq. 6. We therefore report results for both **OURS-COMBI** and **OURS-ALL-EST**. By contrast, for **Venice** and **WorldExpo'10**, only a sparse subset of frames are annotated, and we therefore only report results for **OURS-COMBI**.

For **FDST**, **UCSD**, and **Venice**, our approach again clearly outperforms the competing methods, with the optical flow constraint further boosting performance when applicable. For **WorldExpo'10**, the ranking of the methods depends on the scene being used, but ours still performs best on average and on Scene3. In short, when the crowd is dense, our approach dominates the others. By contrast, when the crowd becomes very sparse as in Scene1 and Scene5, models that comprise a pool of different regressors, such as [40], gain an advantage. This points to a potential way to further improve our own method, that is, to also use a pool of regressors to estimate the people flows.

4.4   Ablation Study

Table 3. Ablation study. (a) CrowdFlow. (b) FDST. (c) UCSD.

Model	MAE	$RMSE$	Model	MAE	$RMSE$	Model	MAE	$RMSE$
BASELINE	124.3	160.2	**BASELINE**	2.44	2.96	**BASELINE**	0.98	1.26
IMAGE-PAIR	125.7	164.1	**IMAGE-PAIR**	2.48	3.10	**IMAGE-PAIR**	1.02	1.40
AVERAGE	128.9	174.6	**AVERAGE**	2.52	3.14	**AVERAGE**	1.01	1.31
WEAK [24]	121.2	155.7	**WEAK** [24]	2.42	2.91	**WEAK** [24]	0.96	1.30
OURS-FLOW	113.3	140.3	**OURS-FLOW**	2.31	2.85	**OURS-FLOW**	0.94	1.21
OURS-COMBI	97.8	112.1	**OURS-COMBI**	2.17	2.62	**OURS-COMBI**	0.86	1.13
OURS-ALL-EST	**96.3**	**111.6**	**OURS-ALL-EST**	**2.10**	**2.46**	**OURS-ALL-EST**	**0.81**	**1.07**
(a)			(b)			(c)		

To confirm that the good performance we report really is attributable to our regressing flows instead of densities, we performed the following set of experiments. Recall from Sect. 3.2, that we use the CAN [25] architecture to regress

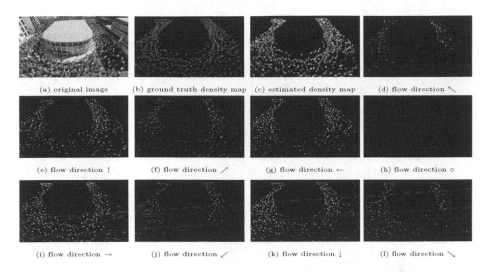

| (a) original image | (b) ground truth density map | (c) estimated density map | (d) flow direction ↖ |

| (e) flow direction ↑ | (f) flow direction ↗ | (g) flow direction ← | (h) flow direction ○ |

| (i) flow direction → | (j) flow direction ↙ | (k) flow direction ↓ | (l) flow direction ↘ |

Fig. 4. Density estimation in CrowdFlow. People are running counterclockwise. The estimated people density map is close to the ground-truth one. It was obtained by summing the flows towards the 9 neighbors of Fig. 2(b). They are denoted by the arrows and the circle. The latter corresponds to people not moving and is, correctly, empty. Note that the flow of people moving down is highest on the left of the building, moving right below the building, and moving up on the right of the building, which is also correct. Inevitably, there is also some noise in the estimated flow, some of which is attributable to body shaking while running.

the flows. Instead, we can use this network to directly regress the densities, as in the original paper. We will refer to this approach as **BASELINE**. In [24], it was suggested that people conservation constraints could be added by incorporating a loss term that enforces the conservation constraints of Eq. 2 that are weaker than those of Eq. 1, which are those we use in this paper. We will refer to this approach relying on weaker constraints while still using the CAN backbone as **WEAK**. As **OURS-COMBI**, it takes two consecutive images as input. For the sake of completeness, we implemented a simplified approach, **IMAGE-PAIR**, which takes the same two images as input and directly regresses the densities. To show that regressing flows does not simply smoothe the densities, we implement one further approach, **AVERAGE**, which takes three images as input, uses CAN to independently compute three density maps, and then averages them. To highlight the importance of the forward-backward constraints of Eq. 3, we also tested a simplified version of our approach in which we drop them and that we refer to **OURS-FLOW**.

We compare the performance of these five approaches on **CrowdFlow**, **FDST**, and **UCSD** in Table 3. Both **IMAGE-PAIR** and **AVERAGE** do worse than **BASELINE**, which confirms that temporal averaging of the densities is not the right thing to do. As reported in [24], **WEAK** delivers a small improvement. However, using our stronger constraints brings a much larger

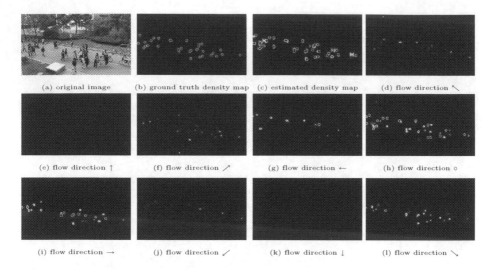

Fig. 5. Density estimation in FDST. People mostly move from left to right. The estimated people density map is close to the ground-truth one. It was obtained by summing the flows towards the 9 neighbors of Fig. 2(b). They are denoted by the arrows and the circle. Strong flows occur in (g), (h), and (i), that is, moving left, moving right, or not having moved. Note that the latter does not mean that the people are static but only that they have not had time to change grid location between the two time instants.

improvement, thereby confirming the importance of properly modeling the flows as we do here. As expected **OURS-FLOW** improves on **IMAGE-PAIR** in all three datasets, with further performance increase for **OURS-COMBI** and **OURS-ALL-EST**. This confirms that using people flows instead of densities is a win and the additional constraints we impose all make positive contributions.

5   Conclusion

We have shown that implementing a crowd counting algorithm in terms of estimating the people flows and then summing them to obtain people densities is more effective than attempting to directly estimate the densities. This is because it allows us to impose conservation constraints that make the estimates more robust. When optical flow data can be obtained, it also enables us to exploit the correlation between optical flow and people flow to further improve the results.

In this paper, we have focused on performing all the computations in image space, in large part so that we could compare our results to that of other recent algorithms that also work in image space. We have nonetheless shown in the supplementary material, that modeling the people flows in the ground plane yields even better performance. A promising application is to use drones for people counting because their internal sensors can be directly used to provide the camera registration parameters necessary to compute the homographies between the camera and the ground plane. In this scenario, the drone sensors also provide a motion estimate, which can be used to correct the optical flow measurements and therefore exploit the information they provide as effectively as if the camera was static.

Acknowledgments. This work was supported in part by the Swiss National Science Foundation.

References

1. BenShitrit, H., Berclaz, J., Fleuret, F., Fua, P.: Tracking multiple people under global apperance constraints. In: International Conference on Computer Vision (2011)
2. BenShitrit, H., Berclaz, J., Fleuret, F., Fua, P.: Multi-commodity network flow for tracking multiple people. IEEE Trans. Pattern Anal. Mach. Intell. **36**(8), 1614–1627 (2014)
3. Berclaz, J., Fleuret, F., Türetken, E., Fua, P.: Multiple object tracking using k-shortest paths optimization. IEEE Trans. Pattern Anal. Mach. Intell. **33**(11), 1806–1819 (2011)
4. Butt, A., Collins, R.: Multi-target tracking by Lagrangian relaxation to min-cost network flow. In: Conference on Computer Vision and Pattern Recognition, pp. 1846–1853 (2013)
5. Cao, X., Wang, Z., Zhao, Y., Su, F.: Scale aggregation network for accurate and efficient crowd counting. In: European Conference on Computer Vision (2018)
6. Chan, A., Liang, Z., Vasconcelos, N.: Privacy preserving crowd monitoring: counting people without people models or tracking. In: Conference on Computer Vision and Pattern Recognition (2008)
7. Chan, A., Vasconcelos, N.: Bayesian Poisson regression for crowd counting. In: International Conference on Computer Vision, pp. 545–551 (2009)
8. Cheng, Z., Li, J., Dai, Q., Wu, X., Hauptmann, A.G.: Learning spatial awareness to improve crowd counting. In: International Conference on Computer Vision (2019)
9. Collins, R.: Multitarget data association with higher-order motion models. In: Conference on Computer Vision and Pattern Recognition (2012)
10. Dicle, C., Camps, O.I., Sznaier, M.: The way they move: tracking multiple targets with similar appearance. In: International Conference on Computer Vision (2013)
11. Fang, Y., Zhan, B., Cai, W., Gao, S., Hu, B.: Locality-constrained spatial transformer network for video crowd counting. In: International Conference on Multimedia and Expo (2019)
12. He, Z., Li, X., You, X., Tao, D., Tang, Y.Y.: Connected component model for multi-object tracking. IEEE Trans. Image Process. **25**(8), 3698–3711 (2016)
13. Hochreiter, S., Schmidhuber, J.: Long short-term memory. Neural Comput. **9**(8), 1735–1780 (1997)
14. Idrees, H., et al.: Composition loss for counting, density map estimation and localization in dense crowds. In: European Conference on Computer Vision (2018)
15. Jiang, X., Xiao, Z., Zhang, B., Zhen, X.: Crowd counting and density estimation by trellis encoder-decoder networks. In: Conference on Computer Vision and Pattern Recognition (2019)
16. Lempitsky, V., Zisserman, A.: Learning to count objects in images. In: Advances in Neural Information Processing Systems (2010)

17. Li, Y., Zhang, X., Chen, D.: CSRNet: dilated convolutional neural networks for understanding the highly congested scenes. In: Conference on Computer Vision and Pattern Recognition (2018)
18. Lian, D., Li, J., Zheng, J., Luo, W., Gao, S.: Density map regression guided detection network for RGB-D crowd counting and localization. In: Conference on Computer Vision and Pattern Recognition (2019)
19. Liu, C., Weng, X., Mu, Y.: Recurrent attentive zooming for joint crowd counting and precise localization. In: Conference on Computer Vision and Pattern Recognition (2019)
20. Liu, J., Gao, C., Meng, D., Hauptmann, A.: DecideNet: counting varying density crowds through attention guided detection and density estimation. In: Conference on Computer Vision and Pattern Recognition (2018)
21. Liu, L., Qiu, Z., Li, G., Liu, S., Ouyang, W., Lin, L.: Crowd counting with deep structured scale integration network. In: International Conference on Computer Vision (2019)
22. Liu, L., Wang, H., Li, G., Ouyang, W., Lin, L.: Crowd counting using deep recurrent spatial-aware network. In: International Joint Conference on Artificial Intelligence (2018)
23. Liu, N., Long, Y., Zou, C., Niu, Q., Pan, L., Wu, H.: ADCrowdNet: an attention-injective deformable convolutional network for crowd understanding. In: Conference on Computer Vision and Pattern Recognition (2019)
24. Liu, W., Lis, K., Salzmann, M., Fua, P.: Geometric and physical constraints for drone-based head plane crowd density estimation. In: International Conference on Intelligent Robots and Systems (2019)
25. Liu, W., Salzmann, M., Fua, P.: Context-aware crowd counting. In: Conference on Computer Vision and Pattern Recognition (2019)
26. Liu, X., Weijer, J., Bagdanov, A.: Leveraging unlabeled data for crowd counting by learning to rank. In: Conference on Computer Vision and Pattern Recognition (2018)
27. Liu, Y., Shi, M., Zhao, Q., Wang, X.: Point in, box out: beyond counting persons in crowds. In: Conference on Computer Vision and Pattern Recognition (2019)
28. Long, J., Shelhamer, E., Darrell, T.: Fully convolutional networks for semantic segmentation. In: Conference on Computer Vision and Pattern Recognition (2015)
29. Ma, Z., Wei, X., Hong, X., Gong, Y.: Bayesian loss for crowd count estimation with point supervision. In: International Conference on Computer Vision (2019)
30. Oñoro-Rubio, D., López-Sastre, R.J.: Towards perspective-free object counting with deep learning. In: Leibe, B., Matas, J., Sebe, N., Welling, M. (eds.) ECCV 2016. LNCS, vol. 9911, pp. 615–629. Springer, Cham (2016). https://doi.org/10.1007/978-3-319-46478-7_38
31. Pirsiavash, H., Ramanan, D., Fowlkes, C.: Globally-optimal greedy algorithms for tracking a variable number of objects. In: Conference on Computer Vision and Pattern Recognition, pp. 1201–1208, June 2011
32. Ranjan, V., Le, H., Hoai, M.: Iterative crowd counting. In: European Conference on Computer Vision (2018)
33. Sam, D., Sajjan, N., Babu, R., Srinivasan, M.: Divide and grow: capturing huge diversity in crowd images with incrementally growing CNN. In: Conference on Computer Vision and Pattern Recognition (2018)
34. Sam, D., Surya, S., Babu, R.: Switching convolutional neural network for crowd counting. In: Conference on Computer Vision and Pattern Recognition, p. 6 (2017)

35. Schröder, G., Senst, T., Bochinski, E., Sikora, T.: Optical flow dataset and benchmark for visual crowd analysis. In: International Conference on Advanced Video and Signal Based Surveillance (2018)
36. Shen, Z., Xu, Y., Ni, B., Wang, M., Hu, J., Yang, X.: Crowd counting via adversarial cross-scale consistency pursuit. In: Conference on Computer Vision and Pattern Recognition (2018)
37. Shi, M., Yang, Z., Xu, C., Chen, Q.: Revisiting perspective information for efficient crowd counting. In: Conference on Computer Vision and Pattern Recognition (2019)
38. Shi, X., Chen, Z., Wang, H., Yeung, D., Wong, W., Woo, W.: Convolutional LSTM network: a machine learning approach for precipitation nowcasting. In: Advances in Neural Information Processing Systems, pp. 802–810 (2015)
39. Shi, Z., Mettes, P., Snoek, C.G.M.: Counting with focus for free. In: International Conference on Computer Vision (2019)
40. Shi, Z., Zhang, L., Liu, Y., Cao, X.: Crowd counting with deep negative correlation learning. In: Conference on Computer Vision and Pattern Recognition (2018)
41. Sindagi, V., Patel, V.: Generating high-quality crowd density maps using contextual pyramid CNNs. In: International Conference on Computer Vision, pp. 1879–1888 (2017)
42. Sindagi, V., Patel, V.: Multi-level bottom-top and top-bottom feature fusion for crowd counting. In: International Conference on Computer Vision (2019)
43. Sun, D., Yang, X., Liu, M., Kautz, J.: PWC-Net: CNNs for optical flow using pyramid, warping, and cost volume. In: Conference on Computer Vision and Pattern Recognition (2018)
44. Suurballe, J.: Disjoint paths in a network. Networks 4, 125–145 (1974)
45. Walach, E., Wolf, L.: Learning to count with CNN boosting. In: Leibe, B., Matas, J., Sebe, N., Welling, M. (eds.) ECCV 2016. LNCS, vol. 9906, pp. 660–676. Springer, Cham (2016). https://doi.org/10.1007/978-3-319 46475-6_41
46. Wan, J., Chan, A.B.: Adaptive density map generation for crowd counting. In: International Conference on Computer Vision (2019)
47. Wan, J., Luo, W., Wu, B., Chan, A.B., Liu, W.: Residual regression with semantic prior for crowd counting. In: Conference on Computer Vision and Pattern Recognition (2019)
48. Wang, Q., Gao, J., Lin, W., Yuan, Y.: Learning from synthetic data for crowd counting in the wild. In: Conference on Computer Vision and Pattern Recognition (2019)
49. Xiong, F., Shi, X., Yeung, D.: Spatiotemporal modeling for crowd counting in videos. In: International Conference on Computer Vision, pp. 5161–5169 (2017)
50. Xiong, H., Lu, H., Liu, C., Liu, L., Cao, Z., Shen, C.: From open set to closed set: counting objects by spatial divide-and-conquer. In: International Conference on Computer Vision (2019)
51. Xu, C., Qiu, K., Fu, J., Bai, S., Xu, Y., Bai, X.: Learn to scale: generating multipolar normalized density maps for crowd counting. In: International Conference on Computer Vision (2019)
52. Yan, Z., et al.: Perspective-guided convolution networks for crowd counting. In: International Conference on Computer Vision (2019)
53. Zhang, A., et al.: Relational attention network for crowd counting. In: International Conference on Computer Vision (2019)
54. Zhang, A., et al.: Attentional neural fields for crowd counting. In: International Conference on Computer Vision (2019)

55. Zhang, C., Li, H., Wang, X., Yang, X.: Cross-scene crowd counting via deep convolutional neural networks. In: Conference on Computer Vision and Pattern Recognition, pp. 833–841 (2015)
56. Zhang, L., Li, Y., Nevatia, R.: Global data association for multi-object tracking using network flows. In: Conference on Computer Vision and Pattern Recognition (2008)
57. Zhang, Q., Chan, A.B.: Wide-area crowd counting via ground-plane density maps and multi-view fusion CNNs. In: Conference on Computer Vision and Pattern Recognition (2019)
58. Zhang, S., Wu, G., Costeira, J., Moura, J.: FCN-rLSTM: deep spatio-temporal neural networks for vehicle counting in city cameras. In: International Conference on Computer Vision (2017)
59. Zhang, Y., Zhou, D., Chen, S., Gao, S., Ma, Y.: Single-image crowd counting via multi-column convolutional neural network. In: Conference on Computer Vision and Pattern Recognition, pp. 589–597 (2016)
60. Zhao, M., Zhang, J., Zhang, C., Zhang, W.: Leveraging heterogeneous auxiliary tasks to assist crowd counting. In: Conference on Computer Vision and Pattern Recognition (2019)

Generate to Adapt: Resolution Adaption Network for Surveillance Face Recognition

Han Fang, Weihong Deng$^{(\boxtimes)}$, Yaoyao Zhong, and Jiani Hu

Beijing University of Posts and Telecommunications, Beijing, China
{fanghan,whdeng,zhongyaoyao,jnhu}@bupt.edu.cn

Abstract. Although deep learning techniques have largely improved face recognition, unconstrained surveillance face recognition is still an unsolved challenge, due to the limited training data and the gap of domain distribution. Previous methods mostly match low-resolution and high-resolution faces in different domains, which tend to deteriorate the original feature space in the common recognition scenarios. To avoid this problem, we propose resolution adaption network (RAN) which contains Multi-Resolution Generative Adversarial Networks (MR-GAN) followed by a feature adaption network. MR-GAN learns multi-resolution representations and randomly selects one resolution to generate realistic low-resolution (LR) faces that can avoid the artifacts of down-sampled faces. A novel feature adaption network with translation gate is developed to fuse the discriminative information of LR faces into backbone network, while preserving the discrimination ability of original face representations. The experimental results on IJB-C TinyFace, SCface, QMUL-SurvFace datasets have demonstrated the superiority of our method compared with state-of-the-art surveillance face recognition methods, while showing stable performance on the common recognition scenarios.

Keywords: Surveillance face recognition · Generative adversarial networks · Feature adaption

1 Introduction

Surveillance face recognition is an important problem, which is widely existed in the real-world scenarios, *e.g.*, low-quality faces captured from surveillance cameras are used to match low-resolution (LR) faces or high-resolution (HR) faces. The performance on high-resolution testing sets such as LFW [21] has been greatly improved by SOTA face recognition methods [10,25,39] and the large-scale datasets [4,16,42]. However, due to the large distribution discrepancy between HR and LR faces, the performance of common recognition methods will deteriorate in surveillance face recognition significantly.

Electronic supplementary material The online version of this chapter (https://doi.org/10.1007/978-3-030-58555-6_44) contains supplementary material, which is available to authorized users.

A. Vedaldi et al. (Eds.): ECCV 2020, LNCS 12360, pp. 741–758, 2020.
https://doi.org/10.1007/978-3-030-58555-6_44

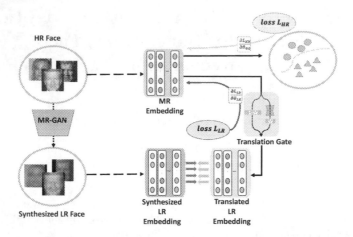

Fig. 1. In RAN, we follow the concept of *"generate to adapt"*. Unlike [32], MR-GAN is utilized to synthesize realistic resolution-degraded distribution as anchors. Then feature adaption network adopts translation gate to determine the source of translated LR features and minimizes the distance between translated and synthesized LR distribution. The embedding space is directly supervised by HR faces and indirectly supervised by synthesized LR faces, aiming to obtain robust multi-resolution embedding.

Since most of faces in existing datasets [4,16,42] are in high-quality, network will focus on learning more informative representation of high resolution such as eyebrows [5], while ignore the information of low-resolution, such as facial contour. When test in the surveillance face, the informative embedding can not catch the lost detail. One intuitive method is to employ face super-resolution [2, 8,43], and then apply synthesized faces for face recognition. Due to the inevitably introduced noise, the performance will be degraded with this method. The other approach translates the embedding of HR faces and down-sampled LR faces into a unified space to minimize the distance of same identity [18,44,54]. However, recent works [3,23] show down-sampling is not good for scale degradation. In this work, we aim to adopt MR-GAN based data argumentation and propose the progressive training procedure to fuse multi-resolution representations.

We propose a novel resolution adaption network (RAN) which includes multi-resolution generative adversarial networks (MR-GAN) to synthesize realistic LR faces. Feature adaption network is then included to progressively learn the multi-resolution (MR) knowledge. The framework is depicted in Fig. 1. Different from [3], which adopted GANs to generate LR images as an intermediate step to achieve image super-resolution, our MR-GAN aims to directly generate realistic LR faces that can be augmented in large-scale datasets and provide prior multi-resolution representations. The global and local mechanism is adopted in generator to pay attention to different areas. In the global stream of generator, input faces are down-sampled into three scales and passed to extract specific knowledge. Then multi-resolution representations are gradually combined and converged into stream of the lowest-resolution to obtain the refined global face by spatial attention. Multi-resolution fusion is conducted by connecting information from sub-encoders of higher-resolution repeatedly and one resolution can be selected randomly to refine realistic LR faces. Meanwhile, the local regions of lowest-scale face are employed to obtain the refined regions, aggregated with global face to generate realistic LR faces. So the coarse, but still discriminative faces can be employed to provide the low-resolution representations.

Following the concept of generating to adapt, we propose a novel feature adaption network to guide the HR model to fuse the discriminative information of the generated LR faces and maintain steady discrimination ability of the HR faces. So, the problem of domain shift by pulling features of different domains close to each other compulsively can be prevented. Specifically, translation gate is proposed to balance the source of translated embedding and preserve LR representations progressively. To minimize the distance between translated LR embedding and realistic LR embedding extracted by synthesized LR faces, HR model can be guaranteed to contain enough LR information and construct MR embedding, retaining both the information of facial details and contours.

In summary, this paper makes the following contributions:

- We propose multi-resolution GAN to synthesize realistic LR faces, which can avoid the artifacts of down-sampled faces. The representations of different resolutions are combined and injected into stream of the lowest resolution to refine LR faces. And the global and local architectures are both employed into generator and discriminator to reinforce the realism of generated faces.
- We propose feature adaption network to redirect HR model to focus on fusing LR information while preserving HR representations. This network employs translation gate to progressively extract LR knowledge from HR embedding, aiming to ensure that HR model contains enough LR information.
- We select small face from IJB-C [29] and construct testing set named IJB-C TinyFace to exploit unconstrained surveillance face recognition. Our method achieves state-of-the-art performance on surveillance datasets: SCface [15], QMUL-SurvFace [9], and IJB-C TinyFace and shows the stable performance on LFW [21], CALFW [53], CPLFW [52], AgeDB-30 [30] and CFP-FP [34].

2   Related Work

The method we proposed aims to learn and adapt embedding both in HR and LR domains. Therefore, we briefly review previous works from two aspects: common face recognition and surveillance face recognition.

Common Face Recognition. Face recognition [40] is a popular issue in computer vision. The performance has been greatly improved due to the development of discriminative loss functions [10,25,33,39,49,51] and deep architectures [17,19,20,35,36]. And the availability of large-scale datasets, such as CASIA-Webface [42], MS-Celeb-1M [16] and VGGFace2 [4] also contribute to the development of large-scale common face recognition. However, since most of faces in existing datasets are in high-quality, network will focus on learning more informative representations of high resolution, which fails to achieve satisfactory performance on low-resolution face recognition due to the large resolution gap.

Surveillance Face Recognition. There are two categories of method to resolve mismatch between HR and LR faces in surveillance face recognition. The most common studies have concentrated on face super-resolution. These hallucination based methods aim to obtain an identity preserved HR faces from the LR input and use synthesized HR faces for recognition. Bulat et al. proposed Super-FAN [2] to integrate a sub-network for facial landmark localization into a GAN-based super-resolution network. Chen et al. [8] suggested to employ facial prior knowledge, including facial landmark heatmaps and parsing maps to super-resolve LR faces. Zhang et al. [47] proposed a super-identity loss and presented domain integrated training approach to construct robust identity metric. Ataer-Cansizeoglu [1] proposed a framework which contains a super-resolution network and a feature extraction network for low-resolution face verification. The other category of works is to learn projection into a unified space and minimize the distances between LR and HR embedding. Zeng et al. [45] proposed to learn resolution-invariant features to preserve multi-resolution information and classify the identity. Lu et al. [27] proposed the deep coupled ResNet (DCR) model, consisting of one trunk network and two branch networks to extract discriminative features robust to the resolution. Yang et al. employed [41] multi-dimensional scaling method to learn a mapping matrix, projecting the HR and LR images into common space. Ge et al. proposed [13] selective knowledge distillation to selectively distill the most informative facial features from the teacher stream.

3   Methodology

3.1   Framework Overview

Instead of employing down-sampling and bicubic linear interpolation to obtain LR faces [5,27], our MR-GAN can generate LR faces to avoid artifacts, allowing us to leverage unpaired HR faces, which is crucial for tackling large-scale datasets where paired faces are unavailable. The proposed adaption network is adopted to improve performance on LR faces while still preserve the discrimination ability on HR faces. As shown in Fig. 1, our method consists of three steps: (i) Synthesize realistic LR faces; (ii) Employ HR faces and synthesized LR faces as training dataset to train HR and LR model respectively; (iii) Using feature adaption network to re-guide HR model to learn resolution-robust distribution.

3.2   Low-Resolution Face Synthesis

Resolution-Aggregated Generator. To minimize the distance between HR and LR domains, we first adopt simple down-sampling to obtain three inputs in three degrees of blur: x_{r_1}, x_{r_2} and x_{r_3}, where x_{r_1} maintains in the highest resolution and x_{r_3} is the face of the lowest resolution. Then we use generator to further refine the global and local information based on down-sampling.

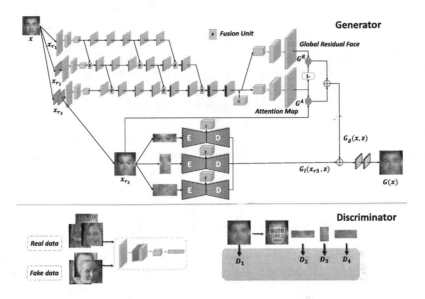

Fig. 2. The architecture of MR-GAN.

Inspired by HRNet [36], we introduce parallel sub-networks to repeatedly receive the information from sub-networks of higher-resolution and integrate the feature map in the global stream. The sub-networks adopt three strided convolutional layers to encode faces into feature maps. Then residual block is used to further deepen the network and make feature maps to maintain the same width and height. To aggregate the information from different streams, fusion units are adopted. We illustrate the details of fusion unit in Fig. 3, where all the operated feature maps are learned from residual blocks. The feature maps in the fusion unit can be denoted as $\{\mathbf{F}_{r_1}^1, \mathbf{F}_{r_2}^1, \mathbf{F}_{r_3}^1 ..., \mathbf{F}_{r_1}^k, \mathbf{F}_{r_2}^k, \mathbf{F}_{r_3}^k\}$, where superscript k indicates the feature map from k-th residual block and subscript r shows the feature map in the stream of resolution r. To fuse $\mathbf{F}_r$ from different resolutions, we concatenate two feature maps to deepen the channels. For instance, $F_{r_1}^a$ of $C_1 \times W \times H$ and $F_{r_2}^b$ of $C_2 \times W \times H$ could be integrated to get feature map of $(C_1 + C_2) \times W \times H$. To enhance resolution and

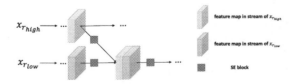

Fig. 3. Illustrate how the fusion unit connects the feature maps from different streams. The representations are selected by SE block before and after connection and flow into the stream of lower-resolution with the deeper channel.

identity-relevant information in the fusion unit, squeeze-and-excitation (SE) blocks [19] are aggregated before and after feature connection. With repeated feature fusion, the feature maps of higher-resolution are gradually injected into the stream of lower resolution. Meanwhile, since multi-resolution can be preserved at the most extent by connecting feature maps of different streams, we can inject the vector of random noise z to effectively select and simulate different degrees of resolution degradation randomly. To decode the low-resolution information and focus more on the resolution-relevant changes, we introduce spatial attention to ignore the background. So the output of global stream can be summarized as:

$$G_g(x, z) = G^A(x, z) \cdot x_{r_3} + (1 - G^A(x, z)) \cdot G^R(x, z), \tag{1}$$

where $G^R(x, z)$ is the output residual face and $G^A(x, z)$ represents the attention map to describe the contribution to output $G_g(x, z)$. So the important regions can be learned by the generator, and irrelevant pixels could be directly retained from x_{r_3}. The local generator G_l contains three identical sub-networks that learn separately to refine three center-cropped local patches: eyes, nose and mouth. These regions are obtained by the detected landmark and fixed. By passing encoder-decoder stream and injecting random vector z, three cropped local patches can be refined, which are further combined with global face $G_g(x, z)$ and then fed into two 1×1 strided convolutional layers to generate the faces $G(x)$.

Global-Local Focused Discriminator. We employ a series of discriminators to distinguish both global and local area, enhancing discrimination ability. Considering characteristics of LR faces, we adopt the same receptive regions as the local branch of generator, consisting of eyes, nose, and mouth to construct local discriminators, while a global discriminator receive the entire face. As shown in Fig. 2. These four discriminators ($D_k, k = 1, 2, 3, 4$) pay attention to discriminating different regions respectively. Compared with simple down-sampling and bicubic interpolation, MR-GAN attaches importance to guaranteeing the texture of local region keep fixed and naturally blurred with great visual quality.

3.3   Loss Function

The key objective of our MR-GAN is to generate LR face, while preserving the identity information to avoid artifacts. Several loss terms are proposed to learn realistic representations.

Perceptual Loss. To ensure the generated LR face preserve the same identity as input face, perceptual loss is introduced to reduce the differences in high-dimensional feature space. And the high-level feature representation $\boldsymbol{F}$ are extracted by the pre-trained expert network. So the loss can be formulated as:

$$L_{perceptual} = \sum \|\boldsymbol{F}(x) - \boldsymbol{F}(G(x))\|_1. \tag{2}$$

Adversarial Loss. Adversarial loss is employed for cross domain adaption from source to target distribution. The loss functions are presented as follows:

$$L_{adv}^{D} = \sum_{k=1}^{4} \boldsymbol{E}[(D_k(y) - 1)^2] + \sum_{k=1}^{4} \boldsymbol{E}[D_k(G(x))^2],$$

$$L_{adv}^{G} = \sum_{k=1}^{4} \boldsymbol{E}[(D_k(G(x)) - 1)^2], \qquad (3)$$

where x is the input HR face and y represents the realistic LR face. Subscript k points the discriminator of corresponding regions. Least square loss is adopted [28] to ensure the discriminator cannot distinguish the synthesized faces.

Pixel Loss. Besides the specially designed adversarial criticism and identity penalty, L_1 loss in the image space is also adopted for further refining the simple down-sampling and bridging the input-output gap, which is defined as follows:

$$L_{pixel} = \frac{1}{W \times H \times C} \|G(x) - x_{r_3}\|_1. \qquad (4)$$

As shown before, x_{r_3} is the input of the lowest resolution, which can be employed to accelerate the convergence speed and stabilize optimization.

Attention Activation Loss. As shown in Eq. 5, when all elements in $G^A(x)$ saturate to 0, all the output is treated as global output. To prevent learning identity-irrelevant information, attention activation loss is adopted to constrain the activation on the important mask and ignore the information around the background. So the loss function can be written as:

$$L_{att} = \|G^A(x, z)_{center} - 0\|_1 + \|G^A(x, z)_{edge} - 1\|_1, \qquad (5)$$

where $G^A(x, z)_{center}$ represents the 85×82 central patch of attention map and $G^A(x, z)_{edge}$ is the edge of attention map.

In summary, we have four loss functions for generating LR face and use hyper-parameters $\lambda_1, \lambda_2, \lambda_3$ and λ_4 to balance them. The overall objective is:

$$\begin{cases} L_D = \lambda_1 L_{adv}^{D}, \\ L_G = \lambda_1 L_{adv}^{G} + \lambda_2 L_{perceptual} + \lambda_3 L_{pixel} + \lambda_4 L_{att}. \end{cases} \qquad (6)$$

3.4   Feature Adaption Network

Due to the lack of enough LR faces in large-scale datasets, we propose to add generated target samples to balance the multi-resolution representations. However, due to the domain shift between HR and LR domains, it is hard to directly apply the method of simply minimizing distance with the same identities into surveillance face recognition. To overcome this issue, we propose the feature adaption network to preserve the discrimination ability in HR domain and apply it to improve competitiveness in LR domain dynamically.

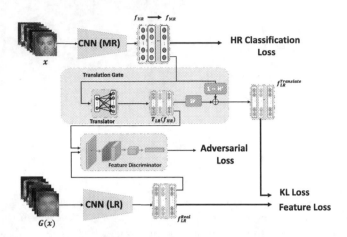

Fig. 4. The pipeline of feature adaption network.

The whole framework is shown in Fig. 4, which contains two streams. The stream at the bottom is trained by the generated LR face to offer the realistic LR representation and fixed in the following adaption learning. Stream at the top is used to learn the final multi-resolution embedding f_{MR}. To preserve discriminatory in HR faces, We employ ArcFace [10] as classification loss L_c^{HR}, making the model of top stream directly supervised by HR faces. Meanwhile, to improve the performance on LR face and avoid deteriorating the HR feature space by directly minimizing domain gap, we propose the translation gate. The translate gate employs translator to balance the LR component of f_{HR} and determine the source of $f_{LR}^{Translate}$. The translator consists of two batch normalization, ReLU and fully connected layers in sequence, which plays an intermediate role in amplifying the LR representations to obtain LR features $T_{LR}(f_{HR})$, making HR features f_{HR} focus on preserving LR information. By translating realistic LR features gradually, HR model at the top of stream can preserve more LR representations to obtain the multi-resolution embedding f_{MR}. To achieve this goal, we apply low-resolution adversarial network to ensure that translated LR embedding $T_{LR}(f_{HR})$ is realistic enough to confuse the discriminator. LSGAN [28] is adopted to pull them together. And the loss function can be seen as follows:

$$
\begin{aligned}
L_{feature}^D &= \boldsymbol{E}[(D(f_{LR}^{Real}) - 1)^2] + \boldsymbol{E}[D(T_{LR}(f_{HR}))^2], \\
L_{feature}^G &= \boldsymbol{E}[(D(T_{LR}(f_{HR})) - 1)^2].
\end{aligned}
\tag{7}
$$

By adopting LSGAN, $|D(T_{LR}(f_{HR})) - 0.5|$ is used to represent the confidence level of the translated LR features. The closer output of discriminator is to 0.5, the more realistic LR features are translated, representing that f^{HR} can preserve more LR information and obtain f^{MR} with balanced multi-resolution knowledge. With the increase of confidence, f_{HR} can also preserve and provide enough LR representations directly without translation. So, our translation gate adopts a weighted architecture to determine the final LR features:

$$
\begin{cases}
W = 1 - |D(T_{LR}(f_{HR})) - 0.5|, \\
f_{LR}^{Translate} = W \cdot T_{LR}(f_{HR}) + (1 - W) \cdot f_{HR},
\end{cases}
\tag{8}
$$

where W is the weight to balance $T_{LR}(f_{HR})$ and f_{HR}. After obtaining $f_{LR}^{Translate}$, we add L_1 loss and KL loss to learn the low-resolution face distribution in feature and probabilistic representation, further pulling translated embedding close to realistic embedding. The losses can be seen as follows:

$$
\begin{aligned}
L_f &= \| \frac{f_{LR}^{Translate}(x_{HR})}{\|f_{LR}^{Translate}(x_{HR})\|_2} - \frac{f_{LR}^{Real}(x_{LR})}{\|f_{LR}^{Real}(x_{LR})\|_2} \|, \\
L_p &= \sum p^{Real}(x_{LR}) \cdot \log \frac{p^{Real}(x_{LR})}{p^{Translate}(x_{HR})}.
\end{aligned}
\tag{9}
$$

Considering that f_{HR} contains the limited LR representations in the early stage of training, $T_{LR}(f_{HR})$ plays the dominant role in the feature and probabilistic supervision. Then as HR features can preserve and provide more realistic LR representations gradually, W will maintain within a stable range to balance two sources of low-resolution knowledge. With this weighted translation, f_{HR} can retain enough LR representation to construct resolution-robust embedding. So, total loss can be seen as:

$$
L_c = L_c^{HR} + \alpha L_{feature}^G + \beta L_p + \gamma L_f.
\tag{10}
$$

4   Experiments

4.1   Experiment Settings

In this section, we present results for proposed resolution adaption network. CASIA-WebFace [42] is used as HR faces to train both MR-GAN and feature

Table 1. Evaluation results on IJB-C TinyFace 1:1 covariate protocol. Results from row 2 to row 10 are implemented in the same ResNet-34 backbone network.

Method	10^{-7}	10^{-6}	10^{-5}	10^{-4}	10^{-3}	10^{-2}	10^{-1}
MS1Mv2 (ResNet100 + ArcFace) [10]	0.0300	0.0436	0.1002	0.2191	0.3842	0.5246	0.6948
CASIA-WebFace [42] (ResNet34 + ArcFace)	0.0261	0.0291	0.0420	0.0917	0.1961	0.3219	0.5409
Down-Sampling	0.0434	0.0629	0.1000	0.1486	0.2201	0.3510	0.5853
Cycle-GAN	0.0279	0.0468	0.0897	0.1399	0.2016	0.3065	0.5261
High-to-Low	0.0332	0.0454	0.0638	0.0916	0.1335	0.2113	0.3873
MR-GAN	0.0508	0.0715	0.1159	0.1736	0.2535	0.3861	0.6147
Down-Sampling + Adaption	0.0488	0.0764	0.1168	0.1890	0.2870	0.4452	0.6751
Cycle-GAN + Adaption	0.0524	0.1032	0.1508	0.2058	0.2819	0.4048	0.6254
High-to-Low + Adaption	0.0665	0.0940	0.1428	0.2132	0.2977	0.4281	0.6477
MR-GAN + Adaption (RAN)	**0.0699**	**0.1031**	**0.1616**	**0.2287**	**0.3273**	**0.4817**	**0.7095**

Fig. 5. Face images synthesized by different methods.

adaption network, which contains 494,414 images and 10,575 subjects. The realistic LR faces are selected from MillionCelebs [50]. We use MTCNN [48] for face detection and alignment. The detected landmarks are utilized to measure distance between the center point of eyes and mouth center. Faces whose distances less than 30 and more than 10 are selected as realistic LR faces. To evaluate the performance of feature adaption network, we utilize 34-layer deep residual architecture [17] as backbone and adopt SCface [15], QMUL-SurvFace [9] and low resolution subset of IJB-C [29] (IJB-C TinyFace) as test set. IJB-C [29] is a video-based face database which contains natural resolution variation. We follow the same rule to select realistic LR faces. All the detected LR faces are adopted and faces with same identity are selected for each anchor to construct the positive pairs, including 158,338 genuine comparisons. Following IJB-C 1:1 covariate verification protocol, the same 39,584,639 negative pairs are used in IJB-C TinyFace. SCface [15] consists of face images of 130 subjects. Following [27], 80 subjects are for testing and the other 50 subjects are used for fine-tuning. Face identification is conducted where HR faces are used as the gallery set and LR images captured at 4.2 m (d_1), 2.6 m (d_2) and 1.0 m (d_3) as the probe respectively. QMUL-SurvFace [9] consists of very-low resolution face images captured under surveillance cameras.

4.2   Implementation Details

All the training and testing faces are cropped and aligned into 112×112. In MR-GAN, We train the discriminator and generator by iteratively minimizing the discriminator and generator loss function with Adam optimization. Pixel-critic is employed at every 5 generator iterations. MobileFaceNets [6] has been adopted as the expert network and all the parameters are fixed. The hyper-parameters are empirically set as follows: $\lambda_1 = 2$, $\lambda_2 = 20$, $\lambda_3 = 20$, $\lambda_4 = 0.4$ and batch

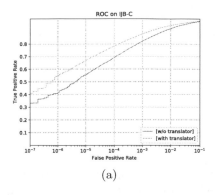

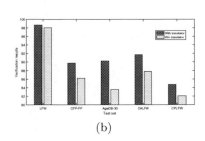

(a)   (b)

Fig. 6. (a). ROC curves depict the effectiveness of translator on IJB-C [29]. (b): The comparisons of "with/without translator" on HR domain is depicted to show discriminative recognition ability with translator. The results of LFW [21], CALFW [53], CPLFW [52], AgeDB-30 [30] and CFP-FP [34] are reported.

size = 16. And we set hyper-parameters of the optimizer as follows: $\beta_1 = 0.5$, $\beta_2 = 0.999$ and learning rate = 0.0002. ArcFace is adopted in feature adaption network as the classification loss. Following [10], the feature scale and angular margin m are set as 64 and 0.5 respectively. We set the batch size to 256 to train the pre-trained HR and LR model. There are three steps to obtain the MR embedding. First, we pre-train ResNet-34 by using CASIA-WebFace to obtain the HR model. The learning rate starts from 0.1 and is divided by 10 at 60,000, 100,000 and 140,000 iterations. Second, we finc-tunc HR model by adopting the generated LR CASIA-WebFace as training set to get the LR model. And the learning rate starts from 0.01 and is divided by 10 at 50,000 and 100,000 iterations. To simulate more changes of resolution, random Gaussian blur is added when training LR model. Finally, HR model continues to be finetuned by using HR faces with indirect supervision of fixed LR model to train MR model. The batch size is set to 200 in this step and learning rate starts from 0.01, which is divided by 10 at 50,000 iterations. The hyper-parameters can be set as follows: $\alpha = 0.05$, $\beta = 0.04$, $\gamma = 10$. We adopt SGD optimization for recognition and Adam optimization for adversarial learning. $L^G_{feature}$ is updated and utilized at every 4 discriminator iterations. Please refer to the supplementary material for full details on network architectures.

4.3   Ablation Study

Effects of LR Face Synthesis. Since existing large-scale datasets such as CASIA-Webface [42] and MS-Celeb-1M [16] contain a lot of HR faces, our method aims to generate LR faces with different resolutions to augment training set. However, most existing works adopt down-sampling to obtain LR face, which doesn't match the real environment. As shown in Fig. 5, faces generated by down-sampling are full of irregular twist and noise. The GAN-based synthesis

Table 2. Evaluation results on IJB-C TinyFace 1:1 covariate protocol. HR, LR and MR models trained on cleaned MS-Celeb-1M [16] are reported and compared.

Method	10^{-7}	10^{-6}	10^{-5}	10^{-4}	10^{-3}	10^{-2}	10^{-1}
HR model	0.0307	0.0418	0.0811	0.1801	0.3641	0.5591	0.7491
LR model	0.0643	0.0854	0.1356	0.2240	0.3479	0.5025	0.7033
MR model (RAN)	0.0664	0.1049	0.1678	0.2635	0.4135	0.5819	0.7597

method can keep the realism of faces when resolution is reduced. However, the faces generated by Cycle-GAN [55] are over-smoothed. Bulat et al. [3] aimed to adopt High-to-Low and Low-to-High for face super-resolution. They ignored to preserve the information around the facial details and employed the limited supervision in LR faces. So, the LR faces generated by High-to-Low generator can not be used for recognition directly. In contrast, our MR-GAN can integrate multi-resolution information to utilize the specific representation and focus more on the simulation of local region to obtain coarse, but discriminative details. More visualizations can be found in supplementary material.

To quantitatively compare with results on face recognition, we evaluate different methods on IJB-C TinyFace and report the results in Table 1. We translate all the faces of CASIA-WebFace to LR faces including: Down-sampling, Cycle-GAN [55], High-to-Low [3] and MR-GAN, and adopt the generated training set to fine-tune HR model. The results are depicted from row 3 to row 6. And with adaption, the performances are further improved. Since faces generated by High-to-Low [3] are very small which can not be recognized directly, the results are relatively low. However, High-to-Low still provides the coarse enough details during adaption learning, which shows the effectiveness. To better demonstrate the effect of RAN, we report the results of model [10] using larger datasets and more parameters, which is shown at the top. Our method utilizes the smaller model and training set to achieve the same performance and even far beyond them in some cases.

Effects of MR Feature Adaption. To prevent directly minimizing the distances of HR and LR domains due to the domain gap, translation gate is proposed to use translator to balance the source of translated LR features. Without translator, f_{HR} is directly adopted to minimize the distances between different domains. In Fig. 6(a), discrimination ability declines fast with the decrease of FAR by directly minimizing distance in feature and probabilistic distribution. In Fig. 6(b), the accuracy of LFW decreases to 97.7. However, with intermediate role of translator, translation gate can adopt weighted architecture to generate $T_{LR}(f_{HR})$ progressively. So, the accuracy of LFW can be kept into 98.7. The preserved results on IJB-C and high-resolution testing sets reveal that our MR embedding with translation gate can be adapted into two domains and shows significant effectiveness to handle difficult situations (Table 3).

Table 3. Rank-1 performance of face identification on SCface testing set. 'w/o FT' means testing with the trained model directly without fine-tuning.

Methods	d_1	d_2	d_3
RICNN [45]	23.00	66.00	74.00
LDMDS [41]	62.70	70.70	65.50
Coupled-ResNet [27]	73.30	93.50	98.00
TCN-ResNet [46]	74.60	94.90	98.60
Selective knowledge distillation [13]	43.50	48.00	53.50
Triplet Loss [26]	70.69	95.42	97.02
Quadruplet Loss [7]	74.00	96.57	98.41
DATL [14]	76.24	96.87	98.09
DAQL [14]	77.25	96.58	98.14
ArcFace [10](w/o FT)	35.00	85.80	98.00
MR-GAN (w/o FT)	65.00	91.50	86.50
RAN (w/o FT)	70.50	96.00	98.00
ArcFace [10]	56.80	91.00	97.50
MR-GAN	71.80	94.30	91.00
RAN	**81.30**	**97.80**	**98.80**

Performance on Large-scale Dataset. To show the effectiveness of our RAN on large-scale datasets, cleaned MS-Celeb-1M [16] which contains 5,084,598 faces and 97,099 subjects is used as training set. ResNet-50 and ArcFace are adopted as the basic training architecture and loss function. Same training steps are employed in this experiment. The results of HR, LR and MR models are depicted in Table 2. Since the large-scale datasets already contain a lot of low-resolution images, only adopting ArcFace loss for supervision can get high performance in HR model. By using MR-GAN to transform all the data set to the LR data set, LR model outperforms HR model where FAR is less than 10^{-4}. Furthermore, our RAN achieves the highest performance in all cases by integrating multi-resolution knowledge.

4.4 Compare with SOTA Methods

Comparisons on SCface. SCface defines face identification protocol. For each subject, there are 15 faces taken at three distances (five faces at each distance) by surveillance cameras, and one frontal mugshot image taken by a digital camera. For fair comparison, we implemented SOTA face recognition method Arc-Face [10] as HR model and follow [27] to fine-tune on SCface. The compared methods focus more on minimizing distance of intra-class in different resolutions. However, these methods directly minimize the distance of class, ignoring the resolution gap. And they simply adopt down-sampling to increase the diversity of resolutions and provide paired multi-resolution faces, which don't match

Table 4. Performance of face identification on QMUL-SurvFace. Most compared results are directly cited from [9] except ArcFace and RAN. In these face super-resolution methods including SRCNN [11], FSRCNN [12], VDSR [22], DRRN [38] and LapSRN [24], SphereFace [25] is used as recognition model.

Methods	TPIR20(%)@FPIR				AUC
	30%	20%	10%	1%	
DeepID2 [37]	12.8	8.1	3.4	0.8	20.8
VggFace [31]	5.1	2.6	0.8	0.1	14.0
FaceNet [33]	12.7	8.1	4.3	1.0	19.8
SphereFace [25]	21.3	15.7	8.3	1.0	28.1
SRCNN [11]	20.0	14.9	6.2	0.6	27.0
FSRCNN [12]	20.0	14.4	6.1	0.7	27.3
VDSR [22]	20.1	14.5	6.1	0.8	27.3
DRRN [38]	20.3	14.9	6.3	0.6	27.5
LapSRN [24]	20.2	14.7	6.3	0.7	27.4
ArcFace [10]	18.7	15.1	10.1	2.0	25.3
RAN	**26.5**	**21.6**	**14.9**	**3.8**	**32.3**

the real scenarios. Selective knowledge distillation [13] adopted HR model as teacher and LR model as student to try to restore LR model's ability to discriminate on facial details. Since high resolution information is already lost, sufficient representation cannot be recovered. Instead, our RAN focuses on retaining LR information from HR features through the resolution adaption, which can learn enough multi-resolution knowledge and achieve the best performance.

Comparisons on QMUL-SurvFace. QMUL-SurvFace contains very low LR faces which are drawn from real surveillance videos. We compare our RAN with face super-resolution (SR) methods and common recognition methods. As shown in Table 4, we conduct face identification. Large margin loss (ArcFace and SphereFace) have achieved the SOTA results in large-scale datasets. So, they improve the performance in HR domain, and also can be applied to LR domain. However, these face SR methods struggle to recover the identity information and focus more on the visual quality, inevitably degrading performance. By dynamically extracting MR knowledge in feature space from HR face, our method can perform better than face SR and common recognition methods.

5   Conclusion

This paper proposes Resolution Adaption Network (RAN) for realistic LR face synthesis and surveillance face recognition. We aim to generate LR faces for data augmentation and bridge the cross-resolution gap. In RAN, MR-GAN employs multi-resolution and global-local architecture, blurring face in random resolutions, to generate the identity-preserved and realistic LR faces. To use LR faces to better match with both LR faces and HR faces, feature adaption network is proposed to enhance LR knowledge and balance multi-resolution representations progressively. SOTA results are achieved for surveillance face recognition.

References

1. Ataer-Cansizoglu, E., Jones, M., Zhang, Z., Sullivan, A.: Verification of very low-resolution faces using an identity-preserving deep face super-resolution network. arXiv preprint arXiv:1903.10974 (2019)
2. Bulat, A., Tzimiropoulos, G.: Super-fan: integrated facial landmark localization and super-resolution of real-world low resolution faces in arbitrary poses with gans. In: Proceedings of the IEEE Conference on Computer Vision and Pattern Recognition, pp. 109–117 (2018)
3. Bulat, A., Yang, J., Tzimiropoulos, G.: To learn image super-resolution, use a gan to learn how to do image degradation first. In: Proceedings of the European Conference on Computer Vision (ECCV). pp. 185–200 (2018)
4. Cao, Q., Shen, L., Xie, W., Parkhi, O.M., Zisserman, A.: VGGFace2:: a dataset for recognising faces across pose and age. In: 2018 13th IEEE International Conference on Automatic Face & Gesture Recognition (FG 2018), pp. 67–74. IEEE (2018)
5. Chaitanya Mynepalli, S., Hu, P., Ramanan, D.: Recognizing tiny faces. In: Proceedings of the IEEE International Conference on Computer Vision Workshops. pp. 0–0 (2019)
6. Chen, S., Liu, Y., Gao, X., Han, Z.: MobileFaceNets: efficient CNNs for accurate real-time face verification on mobile devices. In: Zhou, J., et al. (eds.) CCBR 2018. LNCS, vol. 10996, pp. 428–438. Springer, Cham (2018). https://doi.org/10.1007/978-3-319-97909-0_46
7. Chen, W., Chen, X., Zhang, J., Huang, K.: Beyond triplet loss: a deep quadruplet network for person re-identification. In: Proceedings of the IEEE Conference on Computer Vision and Pattern Recognition, pp. 403–412 (2017)
8. Chen, Y., Tai, Y., Liu, X., Shen, C., Yang, J.: FSRNet: end-to-end learning face super-resolution with facial priors. In: Proceedings of the IEEE Conference on Computer Vision and Pattern Recognition, pp. 2492–2501 (2018)
9. Cheng, Z., Zhu, X., Gong, S.: Surveillance face recognition challenge. arXiv preprint arXiv:1804.09691 (2018)
10. Deng, J., Guo, J., Xue, N., Zafeiriou, S.: Arcface: additive angular margin loss for deep face recognition. In: Proceedings of the IEEE Conference on Computer Vision and Pattern Recognition, pp. 4690–4699 (2019)
11. Dong, C., Loy, C.C., He, K., Tang, X.: Learning a deep convolutional network for image super-resolution. In: Fleet, D., Pajdla, T., Schiele, B., Tuytelaars, T. (eds.) ECCV 2014. LNCS, vol. 8692, pp. 184–199. Springer, Cham (2014). https://doi.org/10.1007/978-3-319-10593-2_13
12. Dong, C., Loy, C.C., Tang, X.: Accelerating the super-resolution convolutional neural network. In: Leibe, B., Matas, J., Sebe, N., Welling, M. (eds.) ECCV 2016. LNCS, vol. 9906, pp. 391–407. Springer, Cham (2016). https://doi.org/10.1007/978-3-319-46475-6_25
13. Ge, S., Zhao, S., Li, C., Li, J.: Low-resolution face recognition in the wild via selective knowledge distillation. IEEE Trans. Image Process. **28**(4), 2051–2062 (2018)

14. Ghosh, S., Singh, R., Vatsa, M.: On learning density aware embeddings. In: Proceedings of the IEEE Conference on Computer Vision and Pattern Recognition, pp. 4884–4892 (2019)
15. Grgic, M., Delac, K., Grgic, S.: SCface-surveillance cameras face database. Multimed. Tools Appl. **51**(3), 863–879 (2011)
16. Guo, Y., Zhang, L., Hu, Y., He, X., Gao, J.: MS-Celeb-1M: a dataset and benchmark for large-scale face recognition. In: Leibe, B., Matas, J., Sebe, N., Welling, M. (eds.) ECCV 2016. LNCS, vol. 9907, pp. 87–102. Springer, Cham (2016). https://doi.org/10.1007/978-3-319-46487-9_6
17. He, K., Zhang, X., Ren, S., Sun, J.: Deep residual learning for image recognition. In: Proceedings of the IEEE conference on computer vision and pattern recognition, pp. 770–778 (2016)
18. Hennings-Yeomans, P.H., Baker, S., Kumar, B.V.: Simultaneous super-resolution and feature extraction for recognition of low-resolution faces. In: 2008 IEEE Conference on Computer Vision and Pattern Recognition, pp. 1–8. IEEE (2008)
19. Hu, J., Shen, L., Sun, G.: Squeeze-and-excitation networks. In: Proceedings of the IEEE Conference on Computer Vision and Pattern Recognition, pp. 7132–7141 (2018)
20. Huang, G., Liu, Z., Van Der Maaten, L., Weinberger, K.Q.: Densely connected convolutional networks. In: Proceedings of the IEEE Conference on Computer Vision and Pattern Recognition, pp. 4700–4708 (2017)
21. Huang, G.B., Mattar, M., Berg, T., Learned-Miller, E.: Labeled faces in the wild: a database for studying face recognition in unconstrained environments (2008)
22. Kim, J., Kwon Lee, J., Mu Lee, K.: Accurate image super-resolution using very deep convolutional networks. In: Proceedings of the IEEE Conference on Computer Vision and Pattern Recognition, pp. 1646–1654 (2016)
23. Kumar, A., Chellappa, R.: Landmark detection in low resolution faces with semi-supervised learning. arXiv preprint arXiv:1907.13255 (2019)
24. Lai, W.S., Huang, J.B., Ahuja, N., Yang, M.H.: Deep laplacian pyramid networks for fast and accurate super-resolution. In: Proceedings of the IEEE Conference on Computer Vision and Pattern Recognition, pp. 624–632 (2017)
25. Liu, W., Wen, Y., Yu, Z., Li, M., Raj, B., Song, L.: SphereFace: deep hypersphere embedding for face recognition. In: Proceedings of the IEEE Conference on Computer Vision and Pattern Recognition, pp. 212–220 (2017)
26. Liu, X., Song, L., Wu, X., Tan, T.: Transferring deep representation for nir-vis heterogeneous face recognition. In: 2016 International Conference on Biometrics (ICB), pp. 1–8. IEEE (2016)
27. Lu, Z., Jiang, X., Kot, A.: Deep coupled resnet for low-resolution face recognition. IEEE Signal Process. Lett. **25**(4), 526–530 (2018)
28. Mao, X., Li, Q., Xie, H., Lau, R.Y., Wang, Z., Paul Smolley, S.: Least squares generative adversarial networks. In: Proceedings of the IEEE International Conference on Computer Vision, pp. 2794–2802 (2017)
29. Maze, B., et al.: IARPA Janus benchmark-C: face dataset and protocol. In: 2018 International Conference on Biometrics (ICB), pp. 158–165. IEEE (2018)
30. Moschoglou, S., Papaioannou, A., Sagonas, C., Deng, J., Kotsia, I., Zafeiriou, S.: AgeDB: the first manually collected, in-the-wild age database. In: Proceedings of the IEEE Conference on Computer Vision and Pattern Recognition Workshops, pp. 51–59 (2017)
31. Parkhi, O.M., et al.: Deep face recognition. In: bmvc, vol. 1, p. 6 (2015)

32. Sankaranarayanan, S., Balaji, Y., Castillo, C.D., Chellappa, R.: Generate to adapt: aligning domains using generative adversarial networks. In: Proceedings of the IEEE Conference on Computer Vision and Pattern Recognition, pp. 8503–8512 (2018)

33. Schroff, F., Kalenichenko, D., Philbin, J.: FaceNet: a unified embedding for face recognition and clustering. In: Proceedings of the IEEE Conference on Computer Vision and Pattern Recognition, pp. 815–823 (2015)

34. Sengupta, S., Chen, J.C., Castillo, C., Patel, V.M., Chellappa, R., Jacobs, D.W.: Frontal to profile face verification in the wild. In: 2016 IEEE Winter Conference on Applications of Computer Vision (WACV), pp. 1–9. IEEE (2016)

35. Simonyan, K., Zisserman, A.: Very deep convolutional networks for large-scale image recognition. arXiv preprint arXiv:1409.1556 (2014)

36. Sun, K., Xiao, B., Liu, D., Wang, J.: Deep high-resolution representation learning for human pose estimation. arXiv preprint arXiv:1902.09212 (2019)

37. Sun, Y., Chen, Y., Wang, X., Tang, X.: Deep learning face representation by joint identification-verification. In: Advances in Neural Information Processing Systems, pp. 1988–1996 (2014)

38. Tai, Y., Yang, J., Liu, X.: Image super-resolution via deep recursive residual network. In: Proceedings of the IEEE Conference on Computer Vision and Pattern Recognition, pp. 3147–3155 (2017)

39. Wang, H., et al.: CosFace: large margin cosine loss for deep face recognition. In: Proceedings of the IEEE Conference on Computer Vision and Pattern Recognition, pp. 5265–5274 (2018)

40. Wang, M., Deng, W.: Deep visual domain adaptation: a survey. Neurocomputing 312, 135–153 (2018)

41. Yang, F., Yang, W., Gao, R., Liao, Q.: Discriminative multidimensional scaling for low-resolution face recognition. IEEE Signal Process. Lett. 25(3), 388–392 (2017)

42. Yi, D., Lei, Z., Liao, S., Li, S.Z.: Learning face representation from scratch. arXiv preprint arXiv:1411.7923 (2014)

43. Yu, X., Fernando, B., Hartley, R., Porikli, F.: Super-resolving very low-resolution face images with supplementary attributes. In: Proceedings of the IEEE Conference on Computer Vision and Pattern Recognition, pp. 908–917 (2018)

44. Zangeneh, E., Rahmati, M., Mohsenzadeh, Y.: Low resolution face recognition using a two-branch deep convolutional neural network architecture. Exp. Syst. Appl. 139, 112854 (2020)

45. Zeng, D., Chen, H., Zhao, Q.: Towards resolution invariant face recognition in uncontrolled scenarios. In: 2016 International Conference on Biometrics (ICB), pp. 1–8. IEEE (2016)

46. Zha, J., Chao, H.: TCN: Transferable coupled network for cross-resolution face recognition. In: ICASSP 2019–2019 IEEE International Conference on Acoustics, Speech and Signal Processing (ICASSP), pp. 3302–3306. IEEE (2019)

47. Zhang, K., et al.: Super-identity convolutional neural network for face hallucination. In: Proceedings of the European Conference on Computer Vision (ECCV), pp. 183–198 (2018)

48. Zhang, K., Zhang, Z., Li, Z., Qiao, Y.: Joint face detection and alignment using multitask cascaded convolutional networks. IEEE Signal Process. Lett. 23(10), 1499–1503 (2016)

49. Zhang, X., Zhao, R., Qiao, Y., Wang, X., Li, H.: AdaCos: adaptively scaling cosine logits for effectively learning deep face representations. In: Proceedings of the IEEE Conference on Computer Vision and Pattern Recognition, pp. 10823–10832 (2019)

50. Zhang, Y., et al.: Global-local GCN: Large-scale label noise cleansing for face recognition. In: Proceedings of the IEEE/CVF Conference on Computer Vision and Pattern Recognition, pp. 7731–7740 (2020)

51. Zhao, K., Xu, J., Cheng, M.M.: RegularFace: deep face recognition via exclusive regularization. In: Proceedings of the IEEE Conference on Computer Vision and Pattern Recognition, pp. 1136–1144 (2019)

52. Zheng, T., Deng, W.: Cross-pose LFW: a database for studying crosspose face recognition in unconstrained environments. Beijing University of Posts and Telecommunications, Technical Report, p. 18-01 (2018)

53. Zheng, T., Deng, W., Hu, J.: Cross-age LFW: a database for studying cross-age face recognition in unconstrained environments. arXiv preprint arXiv:1708.08197 (2017)

54. Zhou, C., Zhang, Z., Yi, D., Lei, Z., Li, S.Z.: Low-resolution face recognition via simultaneous discriminant analysis. In: 2011 International Joint Conference on Biometrics (IJCB), pp. 1–6. IEEE (2011)

55. Zhu, J.Y., Park, T., Isola, P., Efros, A.A.: Unpaired image-to-image translation using cycle-consistent adversarial networks. In: Proceedings of the IEEE International Conference on Computer Vision, pp. 2223–2232 (2017)

Learning Feature Embeddings
for Discriminant Model Based Tracking

Linyu Zheng[1,2]([⊠]), Ming Tang[1,3], Yingying Chen[1,2,4], Jinqiao Wang[1,2,4],
and Hanqing Lu[1,2]

[1] National Laboratory of Pattern Recognition, Institute of Automation,
Chinese Academy of Sciences, Beijing 100190, China
{linyu.zheng,tangm,yingying.chen,jqwang,luhq}@nlpr.ia.ac.cn
[2] School of Artificial Intelligence, University of Chinese Academy of Sciences,
Beijing 100049, China
[3] Shenzhen Infinova Limited, Shenzhen, China
[4] ObjectEye Inc., Beijing, China

Abstract. After observing that the features used in most online discriminatively trained trackers are not optimal, in this paper, we propose a novel and effective architecture to learn optimal feature embeddings for online discriminative tracking. Our method, called DCFST, integrates the solver of a discriminant model that is differentiable and has a closed-form solution into convolutional neural networks. Then, the resulting network can be trained in an end-to-end way, obtaining optimal feature embeddings for the discriminant model-based tracker. As an instance, we apply the popular ridge regression model in this work to demonstrate the power of DCFST. Extensive experiments on six public benchmarks, OTB2015, NFS, GOT10k, TrackingNet, VOT2018, and VOT2019, show that our approach is efficient and generalizes well to class-agnostic target objects in online tracking, thus achieves state-of-the-art accuracy, while running beyond the real-time speed. Code will be made available.

1 Introduction

Visual object tracking is one of the fundamental problems in computer vision. Given the initial state of a target object in the first frame, the goal of tracking is to estimate the states of the target in the subsequent frames [43,47,48]. Despite the significant progress in recent years, visual tracking remains challenging because the tracker has to learn a robust appearance model from very limited online training samples to resist many extremely challenging interferences, such as large appearance variation and heavy background clutters. In general, the key problem of visual tracking is how to construct a tracker which can not only tolerate the appearance variation of the target, but also exclude the background interference, while keeping the running speed that is as high as possible.

Electronic supplementary material The online version of this chapter (https://doi.org/10.1007/978-3-030-58555-6_45) contains supplementary material, which is available to authorized users.

There has been significant progress in deep convolutional neural networks (CNNs) based trackers in recent years. From a technical standpoint, existing state-of-the-art CNNs-based trackers mainly fall into two categories. (1) The one is to treat tracking as a problem of similarity learning and is only trained offline. Typically, SINT [40], SiamFC [2], and SiamRPN [29] belong to this category. Although these trackers achieve state-of-the-art performance on many challenging benchmarks, the lack of online learning prevents them from integrating background in an online and adaptive way to improve their discriminative power. Therefore, they are severely affected by heavy background clutters, hindering the further improvement of localization accuracy. (2) The other is to apply CNNs features to the trackers which are discriminatively trained online. Most of these trackers, such as HCF [32], ECO [5], LSART [38], and fdKCF* [46], extract features via the CNNs which are trained on ImageNet [9] for object classification task. Obviously, these features are not optimal for the visual tracking task. Therefore, such trackers are not able to make the visual tracking task sufficiently benefit from the powerful ability of CNNs in feature embedding learning. Even though CFNet [41] and CFCF [13] learnt feature embeddings for online discriminatively trained correlation filters-based trackers by integrating the KCF solver [16] into the training of CNNs, the negative boundary effect [23] in KCF severely degrades the quality of the feature embeddings they learn as well as their localization accuracy. Therefore, it is hard for their architectures to achieve high accuracy in online tracking.

To solve the above problem, in this paper, we propose a novel and effective architecture to learn optimal feature embeddings for online discriminative tracking. Our proposed network receives a pair of images, training image and test image, as its input in offline training[1]. First, an efficient sub-network is designed to extract the features of real and dense samples around the target object from each input image. Then, a discriminant model that is differentiable and has a closed-form solution is trained to fit the samples in the training image to their labels. Finally, the trained discriminant model predicts the labels of samples in the test image, and the predicted loss is calculated. In this way, the discriminant model is trained without circulant and synthetic samples like in KCF, avoiding the negative boundary effect naturally. On the other hand, because it is differentiable and has a closed-form solution, its solver can be integrated into CNNs as a layer with forward and backward processes during training. Therefore, the resulting network can be trained in an end-to-end way, obtaining optimal feature embeddings for the discriminant model-based tracker.

As an instance, we apply the popular ridge regression model in this work to demonstrate the power of the proposed architecture because ridge regression model not only is differentiable and has a closed-form solution, but also has been successfully applied by many modern online discriminatively trained

---

[1] In this paper, offline training refers to training deep convolutional neural networks, that is the process of learning feature embeddings, whereas discriminative training refers to training discriminant models, such as ridge regression and SVM. In our approach, each iteration of the offline training involves discriminative training.

trackers [5,7,16,22,38,46] due to its simplicity, efficiency, and effectiveness. In particular, we employ Woodbury identity [35] to ensure the efficient training of ridge regression model when high-dimensional features are used because it allows us to address the dependence of time complexity on the dimension of features. Moreover, we observed that the extreme imbalance of foreground-background samples encountered during network training slows down the convergence speed considerably and severely reduces the generalization ability of the learned feature embeddings if the commonly used mean square error loss is employed. In order to address this problem, we modify the original shrinkage loss [31] which is designed for deep regression learning and apply it to achieve efficient and effective training.

In online tracking, given the position and size of a target object in the current frame, we extract the features of real and dense samples around the target via the above trained network, *i.e.*, learned feature embeddings, and then train a ridge regression model with them. Finally, in the next frame, the target is first located by the trained model, and then its position and size are refined by ATOM [6].

It is worth mentioning that the core parts of our approach are easy-to-implement in a few lines of code using the popular deep learning packages. Extensive experiments on six public benchmarks, OTB2015, NFS, GOT10k, TrackingNet, VOT2018, and VOT2019, show that the proposed tracker DCFST, *i.e.*, learning feature embeddings with Differentiable and Closed-Form Solver for Tracking, is efficient and generalizes well to class-agnostic target objects in online tracking, thus achieves state-of-the-art accuracy on all six datasets, while running beyond the real-time speed. Figure 2b provides a glance of the comparison between DCFST and other state-of-the-art trackers on OTB2015. We hope that our simple and effective DCFST will serve as a solid baseline and help ease future research in visual tracking.

2   Related Work

In this section, we briefly introduce recent state-of-the-art trackers, with a special focus on the Siamese network based ones and the online discriminatively trained ones. In addition, we also shortly describe the recent advances in meta-learning for few-shot learning since our approach shares similar insights to theirs.

2.1   Siamese Network Based Trackers

Recently, Siamese network based trackers [2,40,45] have received much attention for their well-balanced accuracy and speed. These trackers treat visual tracking as a problem of similarity learning. By comparing the target image patch with the candidate patches in a search region, they consider the candidate with the highest similarity score as the target object. A notable characteristic of such trackers is that they do not perform online learning and update, achieving high FPS in online tracking. Typically, SiamFC [2] employed a fully-convolutional Siamese network to extract features and then used a simple cross-correlation layer to evaluate the candidates in a search region in a dense and efficient sliding-window way. GCT [12] introduced the graph convolution into SiamFC. SiamRPN [29] enhanced the accuracy of SiamFC by adding a region proposal sub-network after the Siamese network. CRPN [11] improved the accuracy of SiamRPN using a cascade structure. SiamDW [29] and SiamRPN++ [28] enabled SiamRPN to benefit from deeper and wider networks. Even though these trackers achieve state-of-the-art performance on many benchmarks, a key limitation they share is their inability to integrate background in an online and adaptive way to improve their discriminative power. Therefore, they are severely affected by heavy background clutters, hindering the further improvement of localization accuracy.

Different from Siamese trackers, our DCFST can integrate background in an online and adaptive way through training discriminant models. Therefore, it is more robust to background clutters than Siamese trackers.

2.2   Online Discriminatively Trained Trackers

In contrast to Siamese network based trackers, another family of trackers [7,16,46] train discriminant models online to distinguish the target object from its background. These trackers can effectively utilize the background of the target, achieving impressive discriminative power on multiple challenging benchmarks. The latest such trackers mainly focus on how to take advantage of CNNs features effectively. HCF [32], ECO [5], fdKCF* [46], and LSART [38] extracted features via the CNNs which are trained on ImageNet for object classification task and applied them to online discriminatively trained trackers. Obviously, these features are not optimal for the visual tracking task. Therefore, these trackers are not able to make the visual tracking task sufficiently benefit from the powerful ability of CNNs in feature embedding learning. In order to learn feature embeddings for online discriminatively trained correlation filters-based trackers, CFNet [41] and CFCF [13] integrated the KCF [16] solver into the training of CNNs. However, it is well known that KCF has to resort to circulant and synthetic samples to achieve the fast training of the filters, introducing the negative boundary effect and degrading the localization accuracy of trackers. Even though CFNet relaxed the boundary effect by cropping the trained filters, its experimental results show that this heuristic idea produces very little improvement. CFCF employed CCOT tracker [8], that is less affected by the boundary effect, in online tracking. However, its offline training does not aim to learn feature embeddings for CCOT, but for KCF, and its running speed is far away from real-time due to the low efficiency of CCOT. Therefore, we argue that it is hard for their architectures to achieve high accuracy and efficiency simultaneously in online tracking.

Different from CFNet and CFCF, our DCFST shares similar insights to DiMP's [3]. Both DCFST and DiMP propose an end-to-end trainable tracking architecture, capable of learning feature embeddings for online discriminatively trained trackers without circulant and synthetic samples. In addition, the training of their discriminant models in both offline training and online tracking are

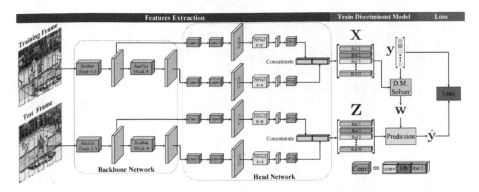

Fig. 1. Full architecture of the proposed network for learning feature embeddings. For each input image, N regions of interest (RoIs) with the target size are generated by uniform sampling. ResNet Block-3 and Block-4 backbone feature maps extracted from the input image are first passed through two convolutional layers to obtain two learned feature maps. Fixed-size feature maps of each RoI are then extracted using PrPool layers and further mapped to feature vectors using fully-connected layers. $\mathbf{X}$ and $\mathbf{Z}$ are data matrices composed of the learned feature vectors of all training and test samples, respectively. A discriminant model $\mathbf{w}$ is trained to fit the samples in $\mathbf{X}$ to their labels. Finally, $\mathbf{w}$ predicts the labels of the samples in $\mathbf{Z}$, and the predicted loss is calculated. Best viewed in color. (Color figure online)

identical. Therefore, they do not suffer from the negative boundary effect and can track target objects more accurately than CFNet and CFCF. The main differences between our DCFST and DiMP are as follows. (1) The architecture of DiMP forces it to use the square-shaped fragments to approximate the real foreground samples, ignoring the actual aspect ratio of the target object. In contrast, that of DCFST is more flexible, allowing us to sample identically to the actual size of the target object. Therefore, the foreground samples are approximate in DiMP, but relatively accurate in DCFST. (2) DiMP designed an iterative method to train its discriminant model, which cannot always guarantee an optimal solution. Whereas, DCFST uses a close-form solver which can always guarantee an optimal solution. (3) From the perspective of implementation, DCFST is much simpler than DiMP. The components and codes of the core parts of DCFST are much fewer than those of DiMP. Experiments show that DCFST outperforms CFNet and DiMP in tracking accuracy with large margins and it also outperforms CFCF in both tracking accuracy and speed with large margins.

2.3   Meta-Learning Based Few-Shot Learning

Meta-learning studies what aspects of the learner effect generalization across a distribution of tasks. Recently, differentiable convex optimization based meta-learning approaches greatly promote the development of few-shot learning. Instead of the nearest-neighbor based learners, MetaOptNet [27] used discriminatively trained linear predictors as base learners to learn representations for few-shot learning, and it aimed at learning feature embeddings that generalize well to novel categories under a linear classification rule. Bertinetto *et al.* [1] proposed both closed-form and iterative solvers, based on ridge regression and logistic regression components, to teach a CNN to use standard machine learning tools as part of its internal model, enabling it to adapt to novel data quickly.

To our best knowledge, the proposed DCFST is the first tracker to integrate the solver of a discriminant model that is differentiable and has a closed-form solution into the training of CNNs to learn optimal feature embeddings for online discriminative tracking without circulant and approximate samples. Experiments on multiple challenging benchmarks show that our approach achieves state-of-the-art accuracy at beyond the real-time speed and also sets a simple yet strong baseline for visual tracking. Therefore, we believe that it would promote the development of high-accuracy and real-time tracking.

3   Learning Feature Embeddings

The main task of an online discriminatively trained tracker is to train a discriminant model $\mathbf{w}$ which is able to not only fit the training samples well online, but also generalize well to the test samples. It is well known that not only different modeling methods, such as nearest neighbor and ridge regression, directly effect the generalization ability of $\mathbf{w}$, but features are also crucial to it. Therefore, our approach, DCFST, is developed by designing an architecture to learn optimal feature embeddings for discriminant model-based trackers, rather than more powerful discriminant models as most modern online discriminatively trained trackers did, to improve tracking accuracy.

As shown in Fig. 1, the proposed network receives a pair of 288×288 RGB images, training image and test image, as its input in offline training. It consists of the following three parts: features extraction network, discriminant model solver, and loss function. This section will present them one by one.

3.1   Features Extraction Network

For each input image, the features extraction consists of the following five steps.

(1) N RoIs with the target size are generated by uniform sampling across the whole image. In addition, the vector $\mathbf{y} \in \mathbb{R}^{N \times 1}$ containing their Gaussian labels is constructed as done in KCF [16] with standard deviation of 0.25.

(2) ResNet [14] Block-3 and Block-4 backbone feature maps are extracted from the input image and then passed through two convolutional layers to obtain two learned feature maps. Their strides are 8×8 and 16×16, respectively. Here, all convolutional kernels are 3×3 and all convolutional layers are followed by BatchNorm [18] and ReLU.

(3) Fixed-size feature maps of each RoI are respectively extracted from the above two learned feature maps using PrPool layers [19] and further mapped

to feature vectors using fully-connected layers. Specifically, the output sizes of two PrPool layers are 8×8 and 4×4, respectively, and both following fully-connected layers output a 512-dimensional feature vector.

(4) Two 512-dimensional feature vectors of each RoI are concatenated to produce the learned feature vector of it. Its dimension, denoted as D, is 1024.

(5) The learned feature vectors of all training RoIs form the training data matrix $\mathbf{X} \in \mathbb{R}^{N \times D}$. The test data matrix $\mathbf{Z} \in \mathbb{R}^{N \times D}$ is obtained in the same way.

It is worth noting that different from CFCF and CFNet whose training data matrices are circulant and most training samples are virtual ones, and different from DiMP whose training and test samples are always assumed to be square, in our DCFST, training data matrix is non-circulant and all training and test samples are real sampled identically to the actual size of the target object.

3.2   Discriminant Model Solver

We train a discriminant model that is differentiable and has a closed-form solution to fit the samples in $\mathbf{X}$ to their labels by integrating its solver into the proposed network. Because the discriminant model is differentiable and has a closed-form solution, its solver can be integrated into CNNs as a layer with forward and backward processes during training. As an instance, we apply the popular ridge regression model in this work to demonstrate the power of the proposed architecture. Ridge regression model has been confirmed to be simple, efficient and effective in the field of visual object tracking [5,22,38,39,46]. It can not only exploit all foreground and background samples to train a good regressor, but also effectively use high-dimensional features as the risk of over-fitting can be controlled by l2-norm regularization. Most importantly, it is differentiable and has a closed-form solution.

The optimization problem of ridge regression can be formulated as

$$\min_{\mathbf{w}} \|\mathbf{Xw} - \mathbf{y}\|_2^2 + \lambda \|\mathbf{w}\|_2^2, \tag{1}$$

where $\mathbf{X} \in \mathbb{R}^{N \times D}$ contains D-dimensional feature vectors of N training samples, $\mathbf{y} \in \mathbb{R}^{N \times 1}$ is the vector containing their Gaussian labels, and $\lambda > 0$ is the regularization parameter. Its optimal solution can be expressed as

$$\mathbf{w}^* = \left(\mathbf{X}^\top \mathbf{X} + \lambda \mathbf{I}\right)^{-1} \mathbf{X}^\top \mathbf{y}. \tag{2}$$

Solving for $\mathbf{w}^*$ by directly using Eq. 2 is time-consuming because $\mathbf{X}^\top \mathbf{X} \in \mathbb{R}^{D \times D}$ and the time complexity of matrix inversion is $O\left(D^3\right)$. Even if we obtain $\mathbf{w}^*$ by solving a system of linear equations with Gaussian elimination method, the time complexity is still $O\left(D^3/2\right)$, hindering the efficient training of the model when high-dimensional features are used. To address the dependence of the time complexity on the dimension of features, we employ the Woodbury formula [35]

$$\left(\mathbf{X}^\top \mathbf{X} + \lambda \mathbf{I}\right)^{-1} \mathbf{X}^\top \mathbf{y} = \mathbf{X}^\top \left(\mathbf{XX}^\top + \lambda \mathbf{I}\right)^{-1} \mathbf{y}, \tag{3}$$

where $\mathbf{X}\mathbf{X}^\top \in \mathbb{R}^{N \times N}$. It is easy to see that the right hand of Eq. 3 allows us to solve for $\mathbf{w}^*$ in time complexity $O\left(N^3/2\right)$. Usually, the number of online training samples is smaller than the dimension of features in tracking, $i.e.$, $N < D$. Therefore, in order to solve for $\mathbf{w}^*$ efficiently, we use the right hand of Eq. 3 if the dimension of the learned feature vectors is larger than the number of training samples, $i.e$, $D > N$. Otherwise, the left hand is used.

Last but not least, when we integrate the ridge regression solver into the training of CNNs, it is necessary to calculate $\partial \mathbf{w}^*/\partial \mathbf{X}$ in the backward process. Fortunately, $\partial \mathbf{w}^*/\partial \mathbf{X}$ is easy to be automatically obtained using the popular deep learning packages, such as TensorFlow and PyTorch, where automatic differentiation is supported. Specifically, during network training, given $\mathbf{X}$ and $\mathbf{y}$, only one line of code is necessary to solve for $\mathbf{w}^*$ in the forward process and there is no code needed in the backward process.

3.3   Fast Convergence with Shrinkage Loss

There exists extreme imbalance between foreground and background samples during network training. This problem slows down the convergence speed considerably and severely reduces the generalization ability of the learned feature embeddings if the commonly used mean square error loss $\mathcal{L}_{mse} = \|\mathbf{y} - \hat{\mathbf{y}}\|_2^2$ is employed, where $\mathbf{y}$ and $\hat{\mathbf{y}} = \mathbf{Z}\mathbf{w}^*$ are vectors containing the ground-truth labels and the predicted labels of the N test samples in $\mathbf{Z}$, respectively. In fact, this is because most background samples are easy ones and only a few hard samples provide useful supervision, making the network training difficult.

To address this problem and make the network training efficient and effective, we propose a new shrinkage loss

$$\mathcal{L} = \left\| \frac{\exp\left(\mathbf{y}\right) \odot \left(\mathbf{y} - \hat{\mathbf{y}}\right)}{1 + \exp\left(a \cdot \left(c - |\mathbf{y} - \hat{\mathbf{y}}|\right)\right)} \right\|_2^2, \tag{4}$$

where a and c are hyper-parameters controlling the shrinkage speed and location, respectively, and the absolute value and the fraction are element-wise. Specifically, $\mathcal{L}$ down-weights the losses assigned to easy samples and mainly focuses on a few hard ones, preventing the vast number of easy backgrounds from overwhelming the learning of feature embeddings. It is easy-to-implement in a few lines of code using the current deep learning packages.

In fact, Eq. 4 is a modified version of the original shrinkage loss [31]

$$\mathcal{L}_o = \left\| \frac{\exp\left(\mathbf{y}\right) \odot \left(\mathbf{y} - \hat{\mathbf{y}}\right)}{1 + \exp\left(a \cdot \left(c - \left(\mathbf{y} - \hat{\mathbf{y}}\right)\right)\right)} \right\|_2^2. \tag{5}$$

Mathematically, the main difference between Eq. 4 and Eq. 5 is that a sample is regarded as an easy one if its predicted value is larger than its label and their difference is less than c in Eq. 5, whereas in Eq. 4, we only consider the absolute difference to determine whether a sample is easy or not. In our experiments, we found that this modification can not only accelerate the convergence speed and reduce the validation loss in offline training, but also improve the accuracy in online tracking. Therefore, it may provide other researchers a better choice.

Moreover, it is worth noting that the motivations for using shrinkage loss in our approach and [31] are quite different. In [31], the purpose of using shrinkage loss is to prevent the discriminant model, *i.e.*, filters, from under-fitting to a few foreground samples after iterative training. However, there is no such concern in our approach because the discriminant model is trained directly by a close-form solution rather than an iterative method. The purpose of using shrinkage loss in our approach is similar to that of using focal loss [30] in training detectors, that is, preventing the vast number of easy backgrounds from overwhelming the learning of feature embeddings.

Based on the above design and discussions, the resulting network can be trained in an end-to-end way, obtaining optimal feature embeddings for the discriminant model-based tracker. Ideally, in online tracking, the learned feature embeddings should make the corresponding discriminant model, *e.g.*, ridge regression model in this work, trained with the features extracted via them robust not only to the large appearance variation of the target object, but also to the heavy background clutters, thereby improving tracking accuracy.

4   Online Tracking with Learned Feature Embeddings

4.1   Features Extraction

Suppose $(\mathbf{p}_t, \mathbf{s}_t)$ denotes the position and size of the target object in frame t. Given frame t and $(\mathbf{p}_t, \mathbf{s}_t)$, we sample a square patch centered at $\mathbf{p}_t$, with an area of 5^2 times the target area, and then resize it to 288×288. Finally, the training data matrix $\mathbf{X}_t$ is obtained using the approach presented in Sect. 3.1. Similarly, given frame $t + 1$ and $(\mathbf{p}_t, \mathbf{s}_t)$, the test data matrix $\mathbf{Z}_{t+1}$ can be obtained.

4.2   Online Learning and Update

In online tracking, according to Sect. 3.2, we can train a ridge regression model using the right hand or the left hand of Eq. 3 for online discriminative tracking. However, the time complexities of both sides are cubical with respect to D or N, hindering real-time performance. In order to solve for $\mathbf{w}^*$ more efficiently, we adopt the Gauss-Seidel based iterative approach [7]. Specifically, taking the left hand of Eq. 3 as an example, given $\mathbf{X}_t$ and $\mathbf{w}^*_{t-1}$, we decompose $\mathbf{X}_t^\top \mathbf{X}_t + \lambda \mathbf{I}$ into a lower triangular $\mathbf{L}_t$ and a strictly upper triangular $\mathbf{U}_t$, *i.e.*, $\mathbf{X}_t^\top \mathbf{X}_t + \lambda \mathbf{I} = \mathbf{L}_t + \mathbf{U}_t$. Then, $\mathbf{w}_t^*$ can be efficiently solved by iterative expressions:

$$\mathbf{w}_t^{*(j)} \leftarrow \mathbf{w}_{t-1}^*, \qquad\qquad\qquad j = 0, \qquad (6a)$$

$$\mathbf{w}_t^{*(j)} \leftarrow \mathbf{L}_t \setminus \left(\mathbf{X}_t^\top \mathbf{y} - \mathbf{U}_t \mathbf{w}_t^{*(j-1)} \right), \qquad j > 0, \qquad (6b)$$

where $\mathbf{w}_{t-1}^*$ is the trained model at frame $t - 1$, and j indicates the number of iterations. In practice, 5 iterations are enough for the satisfactory $\mathbf{w}_t^*$. Note that this iterative method is efficient because Eq. 6b can be solved efficiently with forward substitution, and the time complexity of each iteration is $O\left(D^2\right)$ instead of $O\left(D^3/2\right)$.

In order to locate the target object robustly, updating the appearance model is necessary. Following the popular updating method used in [16,22,41,46], we update $\mathbf{X}_t$ by means of the linear weighting approach, *i.e.*,

$$\begin{aligned} \widetilde{\mathbf{X}}_1 &= \mathbf{X}_1, \\ \widetilde{\mathbf{X}}_t &= (1 - \delta)\,\widetilde{\mathbf{X}}_{t-1} + \delta\mathbf{X}_t, \qquad t > 1, \end{aligned} \tag{7}$$

where δ is the learning rate. As a result, instead of $\mathbf{X}_t$, $\widetilde{\mathbf{X}}_t$ is used in Eq. 6 for solving for $\mathbf{w}_t^*$. In addition, we keep the weight of $\mathbf{X}_1$ not being less than 0.25 during updating because the initial target information is always reliable.

4.3　Localization and Refine

Given the trained model $\mathbf{w}_t^*$ and test data matrix $\mathbf{Z}_{t+1}$, we locate the target by $\hat{\mathbf{y}}_{t+1} = \mathbf{Z}_{t+1}\mathbf{w}_t^*$, and the sample corresponding to the maximum element of $\hat{\mathbf{y}}_{t+1}$ is regarded as the target object. After locating the target, we refine its bounding box by ATOM [6] for more accurate tracking, similar to DiMP [3].

5　Experiments

Our DCFST is implemented in Python using PyTorch. On a single TITAN X(Pascal) GPU, employing ResNet-18 and ResNet-50 respectively as the backbone network, our DCFST-18 and DCFST-50 achieve tracking speeds of 35 FPS and 25 FPS, respectively. Code will be made available.

5.1　Implementation Details

Training Data. To increase the generalization ability of the learned feature embeddings, we use the training splits of recent large-scale tracking datasets, including TrackingNet [33], GOT10k [17], and LaSOT [10], in offline training. During network training, each pair of training and test images is sampled from a video snippet within the nearest 50 frames. For training image, we sample a square patch centered at the target, with an area of 5^2 times the target area. For test image, we sample a similar patch, with a random translation and scale relative to the target's. These cropped patches are then resized to a fixed size 288×288. We use image flipping and color jittering for data augmentation.

Training Setting. We use the pre-trained model on ImageNet to initialize the weights of our backbone network and freeze them during network training. The weights of our head network are randomly initialized with zero-mean Gaussian distributions. We train the head network for 40 epochs with 1.5k iterations per epoch and 48 pairs of images per batch, giving a total training time of 30(40)

Table 1. Ablation studies on OTB2015. (a) The mean AUC scores of different loss functions. Our modified shrinkage loss achieves the best result. (b) The mean AUC scores and FPSs of different Ns. $N = 961$ achieves good balance.

(a) Loss Function.

Loss Function	Mean Square Error	Original Shrinkage	Modified Shrinkage
Mean AUC	0.682	0.698	0.709

(b) Number of Samples.

N	361 (19^2)	625 (25^2)	961 (31^2)	1369 (37^2)
Mean AUC	0.692	0.701	0.709	0.707
Mean FPS	46	40	35	32

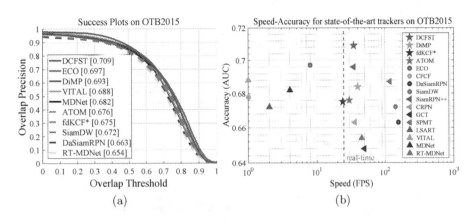

Fig. 2. State-of-the-art comparison on OTB2015. (a) The mean success plots of DCFST and nine state-of-the-art trackers. The mean AUC scores are reported in the legend. (b) Speed and accuracy plot of 16 state-of-the-art trackers. Our DCFST achieves the best accuracy, while running beyond the real-time speed.

hours for DCFST-18(50) on a single GPU. The ADAM [24] optimizer is employed with initial learning rate of 0.005, using a factor 0.2 decay every 15 epochs.

Parameters. We sample 961 RoIs in each input image, *e.g.*, $N = 961$. The regularization parameter λ in Eq. 1 is set to 0.1. Two hyper-parameters a and c in Eq. 4 are set to 10 and 0.2, respectively. The learning rate δ in Eq. 7 is 0.01.

5.2   Ablation Studies

In this section, we will investigate the effect of choosing the loss function and hyper-parameter in our approach. Our ablation studies are based on DCFST-18 and performed on OTB2015 [43].

Table 1a shows the mean AUC [43] scores of DCFST-18 with various loss functions. It can be seen that the tracking accuracy can be obviously improved (1.6% in AUC score) by using the original shrinkage loss instead of the mean square error one. This demonstrates that relaxing the imbalance of foreground-background samples in our approach is necessary and our choice of loss function is valid. Moreover, compared to the original shrinkage loss, our modified one can

Table 2. State-of-the-art comparison on OTB2015 in terms of mean overlap. The best three results are shown in red, blue, and magenta. Our DCFST outperforms its baseline tracker ATOM with a gain of 4%, using the same backbone network.

Tracker	DCFST	DiMP	fdKCF*	ATOM	SiamDW	DaSiamRPN	VITAL	ECO	MDNet	RT-MDNet
Mean Overlap	0.872	0.859	0.828	0.832	0.840	0.858	0.857	0.842	0.852	0.822

Table 3. State-of-the-art comparison on NFS in terms of mean AUC score. DiMP-18 and DiMP-50 are the DiMP tracker with ResNet-18 and ResNet-50 backbone network respectively. Both versions of our DCFST outperform other trackers.

Tracker	DCFST-50	DCFST-18	DiMP-50	DiMP-18	ATOM	UPDT	ECO	CCOT	MDNet	HDT	SFC
Mean AUC	0.641	0.634	0.620	0.610	0.584	0.537	0.466	0.488	0.429	0.403	0.401

Table 4. State-of-the-art comparison on the GOT10k test set in terms of average overlap (AO), and success rates (SR) at overlap thresholds 0.5 and 0.75. DaSiamRPN-18 is the DaSiamRPN tracker with ResNet-18 backbone network. Our DCFST-50 outperforms other trackers with large margins.

Tracker	DCFST-50	DCFST-18	DiMP-50	DiMP-18	ATOM	DaSiamRPN-18	CFNet	SiamFC	GOTURN
AO	0.638	0.610	0.611	0.579	0.556	0.483	0.374	0.348	0.347
SR(0.50)	0.753	0.716	0.717	0.672	0.634	0.581	0.404	0.353	0.375
SR(0.75)	0.498	0.463	0.492	0.446	0.402	0.270	0.144	0.098	0.124

Table 5. State-of-the-art comparison on the TrackingNet test set in terms of precision, normalized precision, and success. Our DCFST-50 achieves top results.

Tracker	DCFST-50	DCFST-18	DiMP-50	DiMP-18	ATOM	SiamRPN++	CRPN	SPMT	DaSiamRPN	CFNet
Precision	0.700	0.682	0.687	0.666	0.648	0.694	0.619	0.661	0.591	0.533
Norm. Prec.	0.809	0.797	0.801	0.785	0.771	0.800	0.746	0.778	0.733	0.654
Success (AUC)	0.752	0.739	0.740	0.723	0.703	0.733	0.669	0.712	0.638	0.578

further improve the tracking accuracy by a large margin (1.1% in AUC score). This confirms the effectiveness of our improvement.

Table 1b shows the mean AUC scores and FPSs of DCFST-18 with various Ns. It can be seen that $N = 961$ can not only provide beyond the real-time speed, but also high tracking accuracy. It is worth mentioning that the stride of ResNet Block-3 is 8 pixel, and when $N = 31^2$ and the size of input image is 288×288, the interval between two adjacent samples is 7.68 pixel. Therefore, the tracking accuracy will not be improved significantly when $N > 961$.

5.3 State-of-the-Art Comparisons

OTB2015 [43]. OTB2015 is the most popular benchmark for the evaluation of trackers, containing 100 videos with various challenges. On the OTB2015 experiment, we first compare our DCFST-18 against nine state-of-the-art trackers, DiMP [3], fdKCF* [46], ATOM [6], SiamDW [44], DaSiamRPN [49], VITAL [37],

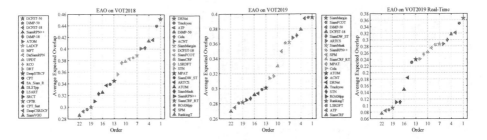

Fig. 3. State-of-the-art comparison on VOT2018, VOT2019, and VOT2019 real-time challenges in terms of expected average overlap (EAO).

ECO [5], MDNet [34], and RT-MDNet [20]. Success plot, mean overlap precision and AUC score [43] are employed to quantitatively evaluate all trackers. Figure 2a and Table 2 show the results. Our DCFST-18 obtains the mean AUC score and overlap precision of 70.9% and 87.2%, outperforming the second best trackers (ECO and DiMP) with significant gains of 1.2% and 1.3%, respectively. Additionally, Fig. 2b shows the comparison of DCFST-18 with the above trackers along with SiamRPN++ [28], CRPN [11], GCT [12], SPMT [42], LSART [38], and CFCF [13] in both mean AUC score and FPS. Our DCFST-18 achieves the best trade-off between accuracy and speed among all state-of-the-art trackers.

NFS [21]. We evaluate our DCFST on the 30 FPS version of NFS benchmark which contains 100 challenging videos with fast-moving objects. On the NFS experiment, we compare DCFST against DiMP, ATOM, UPDT [4], ECO, CCOT [8] along with the top-3 trackers, MDNet, HDT [36], and SFC [2], evaluated by NFS. All trackers are quantitatively evaluated by AUC score. Table 3 shows the results. Our DCFST-18 and DCFST-50 obtain the mean AUC scores of 63.4% and 64.1%, outperforming the latest state-of-the-art trackers DiMP-18 and DiMP-50 with gains of 2.4% and 2.1%, respectively. Additionally, DCFST-18 surpasses its baseline tracker ATOM[2] with a significant gain of 5.1%.

GOT10k [17]. We evaluate our DCFST on the test set of GOT10k which is a large-scale tracking benchmark and contains over 9000 training videos and 180 test videos. Here, the generalization capabilities of the tracker to unseen object classes is of major importance. Therefore, to ensure a fair comparision, on the GOT10k experiment, we retrain our DCFST and then compare it against DiMP, ATOM, DaSiamRPN along with the top-3 trackers, CFNet [41], SiamFC [2], and GOTURN [15], evaluated by GOT10k. Following the GOT10k challenge protocol, all trackers are quantitatively evaluated by average overlap, and success rates at overlap thresholds 0.5 and 0.75. Table 4 shows the results. In terms of success rate at overlap thresholds 0.5, our DCFST-18 and DCFST-50 obtain the scores of 71.6% and 75.3%, outperforming the latest state-of-the-art trackers DiMP-18 and DiMP-50 with gains of 4.4% and 3.6%, respectively. Additionally, DCFST-18 surpasses its baseline tracker ATOM with a significant gain of 8.2%.

---

[2] We state the relationship between DCFST and ATOM in supplementary materials.

TrackingNet [33]. We evaluate our DCFST on the test set of TrackingNet which is a large-scale tracking benchmark and provides 511 test videos to assess trackers. On the TrackingNet experiment, we compare DCFST against seven state-of-the-art trackers, DiMP, ATOM, SiamRPN++, DaSiamRPN, SPMT, CRPN, and CFNet. Following the TrackingNet challenge protocol, all trackers are quantitatively evaluated by precision, normalized precision, and AUC score. Table 5 shows the results. Our DCFST-18 and DCFST-50 obtain the mean AUC scores of 73.9% and 75.2%, outperforming the latest state-of-the-art trackers DiMP-18 and DiMP-50 with gains of 1.6% and 1.2%, respectively. Additionally, DCFST-18 surpasses its baseline tracker ATOM with a significant gain of 3.6%.

VOT2018/2019 [25,26]. We evaluate our DCFST on the 2018 and 2019 versions of Visual Object Tracking (VOT) challenge which contain 60 sequences, respectively. On the VOT2018 experiment, we compare DCFST against SiamRPN++, DiMP, ATOM along with the top-16 trackers on VOT2018 challenge. On the VOT2019 experiment, we compare DCFST-18 against the top-21 trackers on VOT2019 and VOT2019 real-time challenges, respectively. Following the VOT challenge protocol, all trackers are quantitatively evaluated by expected average overlap (EAO). Figure 3 shows the results. (1) On the VOT2018 challenge, our DCFST-18 and DCFST-50 obtain the EAO scores of 0.416 and 0.452, outperforming the latest state-of-the-art trackers DiMP-18 and DiMP-50 with gains of 1.4% and 1.2%, respectively. Additionally, DCFST-50 outperforms all other state-of-the-art trackers including SiamRPN++, and DCFST-18 surpasses its baseline tracker ATOM with a significant gain of 1.5%. (2) On the VOT2019 and VOT2019 real-time challenges, our DCFST-18 achieves the EAO scores of 0.361 and 0.317, respectively, surpassing its baseline tracker ATOM with significant gains of 6.9% and 7.7%, respectively. There are seven latest trackers perform well than DCFST-18 on VOT2019 challenge, however, five of them are obviously lower than DCFST-18 in terms of real-time performance. Although SiamRPN++ employs stronger backbone network (ResNet-50) and more training datas (ImageNet DET, ImageNet VID, YouTube, and COCO) than DCFST-18, DCFST-18 consistently outperforms it on all three challenges with large margins.

6   Conclusion

A novel and state-of-the-art tracker DCFST is proposed in this paper. By integrating the solver of a discriminant model that is differentiable and has a closed-form solution into convolutional neural networks, DCFST learns optimal feature embeddings for the discriminant model-based tracker in offline training, thus improves its accuracy and robustness in online tracking. Extensive experiments on multiple challenging benchmarks show that DCFST achieves state-of-the-art accuracy and beyond the real-time speed, and also sets a simple yet strong baseline for visual tracking due to its simplicity, efficiency and effectiveness.

Acknowledgements. This work was supported by the Research and Development Projects in the Key Areas of Guangdong Province (No. 2020B010165001). This work was also supported by National Natural Science Foundation of China under Grants 61772527, 61976210, 61806200, 61702510 and 61876086.

References

1. Bertinetto, L., Henriques, J.F., Torr, P., Vedaldi, A.: Meta-learning with differentiable closed-form solvers. In: International Conference on Learning Representations (2019). https://openreview.net/forum?id=HyxnZh0ct7
2. Bertinetto, L., Valmadre, J., Henriques, J.F., Vedaldi, A., Torr, P.H.S.: Fully-convolutional Siamese networks for object tracking. In: Hua, G., Jégou, H. (eds.) ECCV 2016. LNCS, vol. 9914, pp. 850–865. Springer, Cham (2016). https://doi.org/10.1007/978-3-319-48881-3_56
3. Bhat, G., Danelljan, M., Gool, L.V., Timofte, R.: Learning discriminative model prediction for tracking. In: The IEEE International Conference on Computer Vision (ICCV), October 2019
4. Bhat, G., Johnander, J., Danelljan, M., Khan, F.S., Felsberg, M.: Unveiling the power of deep tracking. In: Ferrari, V., Hebert, M., Sminchisescu, C., Weiss, Y. (eds.) ECCV 2018. LNCS, vol. 11206, pp. 493–509. Springer, Cham (2018). https://doi.org/10.1007/978-3-030-01216-8_30
5. Danelljan, M., Bhat, G., Shahbaz Khan, F., Felsberg, M.: Eco: efficient convolution operators for tracking. In: Proceedings of the IEEE Conference on Computer Vision and Pattern Recognition, pp. 6638–6646 (2017)
6. Danelljan, M., Bhat, G., Shahbaz Khan, F., Felsberg, M.: Atom: accurate tracking by overlap maximization. In: The IEEE Conference on Computer Vision and Pattern Recognition (CVPR), June 2019
7. Danelljan, M., Hager, G., Shahbaz Khan, F., Felsberg, M.: Learning spatially regularized correlation filters for visual tracking. In: Proceedings of the IEEE International Conference on Computer Vision, pp. 4310–4318 (2015)
8. Danelljan, M., Robinson, A., Shahbaz Khan, F., Felsberg, M.: Beyond correlation filters: learning continuous convolution operators for visual tracking. In: Leibe, B., Matas, J., Sebe, N., Welling, M. (eds.) ECCV 2016. LNCS, vol. 9909, pp. 472–488. Springer, Cham (2016). https://doi.org/10.1007/978-3-319-46454-1_29
9. Deng, J., Dong, W., Socher, R., Li, L.J., Li, K., Fei-Fei, L.: Imagenet: a large-scale hierarchical image database. In: 2009 IEEE Conference on Computer Vision and Pattern Recognition, pp. 248–255. IEEE (2009)
10. Fan, H., et al.: LaSoT: a high-quality benchmark for large-scale single object tracking. In: Proceedings of the IEEE Conference on Computer Vision and Pattern Recognition, pp. 5374–5383 (2019)
11. Fan, H., Ling, H.: Siamese cascaded region proposal networks for real-time visual tracking. In: The IEEE Conference on Computer Vision and Pattern Recognition (CVPR), June 2019
12. Gao, J., Zhang, T., Xu, C.: Graph convolutional tracking. In: The IEEE Conference on Computer Vision and Pattern Recognition (CVPR), June 2019
13. Gundogdu, E., Alatan, A.A.: Good features to correlate for visual tracking. IEEE Trans. Image Process. 27(5), 2526–2540 (2018)
14. He, K., Zhang, X., Ren, S., Sun, J.: Deep residual learning for image recognition. In: Proceedings of the IEEE Conference on Computer Vision and Pattern Recognition, pp. 770–778 (2016)
15. Held, D., Thrun, S., Savarese, S.: Learning to track at 100 FPS with deep regression networks. In: Leibe, B., Matas, J., Sebe, N., Welling, M. (eds.) ECCV 2016. LNCS, vol. 9905, pp. 749–765. Springer, Cham (2016). https://doi.org/10.1007/978-3-319-46448-0_45

16. Henriques, J.F., Caseiro, R., Martins, P., Batista, J.: High-speed tracking with kernelized correlation filters. IEEE Trans. Pattern Anal. Mach. Intell. **37**(3), 583–596 (2014)
17. Huang, L., Zhao, X., Huang, K.: Got-10k: A large high-diversity benchmark for generic object tracking in the wild. arXiv preprint arXiv:1810.11981 (2018)
18. Ioffe, S., Szegedy, C.: Batch normalization: Accelerating deep network training by reducing internal covariate shift. arXiv preprint arXiv:1502.03167 (2015)
19. Jiang, B., Luo, R., Mao, J., Xiao, T., Jiang, Y.: Acquisition of localization confidence for accurate object detection. In: Ferrari, V., Hebert, M., Sminchisescu, C., Weiss, Y. (eds.) Computer Vision – ECCV 2018. LNCS, vol. 11218, pp. 816–832. Springer, Cham (2018). https://doi.org/10.1007/978-3-030-01264-9_48
20. Jung, I., Son, J., Baek, M., Han, B.: Real-time MDNet. In: Ferrari, V., Hebert, M., Sminchisescu, C., Weiss, Y. (eds.) ECCV 2018. LNCS, vol. 11208, pp. 89–104. Springer, Cham (2018). https://doi.org/10.1007/978-3-030-01225-0_6
21. Kiani Galoogahi, H., Fagg, A., Huang, C., Ramanan, D., Lucey, S.: Need for speed: a benchmark for higher frame rate object tracking. In: Proceedings of the IEEE International Conference on Computer Vision, pp. 1125–1134 (2017)
22. Kiani Galoogahi, H., Fagg, A., Lucey, S.: Learning background-aware correlation filters for visual tracking. In: Proceedings of the IEEE International Conference on Computer Vision, pp. 1135–1143 (2017)
23. Kiani Galoogahi, H., Sim, T., Lucey, S.: Correlation filters with limited boundaries. In: Proceedings of the IEEE Conference on Computer Vision and Pattern Recognition, pp. 4630–4638 (2015)
24. Kingma, D.P., Ba, J.: Adam: A method for stochastic optimization. arXiv preprint arXiv:1412.6980 (2014)
25. Kristan, M., et al.: The sixth visual object tracking VOT2018 challenge results. In: Leal-Taixé, L., Roth, S. (eds.) ECCV 2018. LNCS, vol. 11129, pp. 3–53. Springer, Cham (2019). https://doi.org/10.1007/978-3-030-11009-3_1
26. Kristan, M., et al.: The seventh visual object tracking VOT2019 challenge results (2019)
27. Lee, K., Maji, S., Ravichandran, A., Soatto, S.: Meta-learning with differentiable convex optimization. In: The IEEE Conference on Computer Vision and Pattern Recognition (CVPR), June 2019
28. Li, B., Wu, W., Wang, Q., Zhang, F., Xing, J., Yan, J.: SiamRPN++: Evolution of siamese visual tracking with very deep networks. In: The IEEE Conference on Computer Vision and Pattern Recognition (CVPR), June 2019
29. Li, B., Yan, J., Wu, W., Zhu, Z., Hu, X.: High performance visual tracking with siamese region proposal network. In: Proceedings of the IEEE Conference on Computer Vision and Pattern Recognition, pp. 8971–8980 (2018)
30. Lin, T.Y., Goyal, P., Girshick, R., He, K., Dollár, P.: Focal loss for dense object detection. In: Proceedings of the IEEE International Conference on Computer Vision, pp. 2980–2988 (2017)
31. Lu, X., Ma, C., Ni, B., Yang, X., Reid, I., Yang, M.-H.: Deep regression tracking with shrinkage loss. In: Ferrari, V., Hebert, M., Sminchisescu, C., Weiss, Y. (eds.) Computer Vision – ECCV 2018. LNCS, vol. 11218, pp. 369–386. Springer, Cham (2018). https://doi.org/10.1007/978-3-030-01264-9_22
32. Ma, C., Huang, J.B., Yang, X., Yang, M.H.: Hierarchical convolutional features for visual tracking. In: Proceedings of the IEEE International Conference on Computer Vision, pp. 3074–3082 (2015)

33. Müller, M., Bibi, A., Giancola, S., Alsubaihi, S., Ghanem, B.: TrackingNet: a large-scale dataset and benchmark for object tracking in the wild. In: Ferrari, V., Hebert, M., Sminchisescu, C., Weiss, Y. (eds.) ECCV 2018. LNCS, vol. 11205, pp. 310–327. Springer, Cham (2018). https://doi.org/10.1007/978-3-030-01246-5_19

34. Nam, H., Han, B.: Learning multi-domain convolutional neural networks for visual tracking. In: Proceedings of the IEEE Conference on Computer Vision and Pattern Recognition, pp. 4293–4302 (2016)

35. Petersen, K.B., Pedersen, M.S., et al.: The matrix cookbook. Tech. Univ. Denmark **7**(15), 510 (2008)

36. Qi, Y., et al.: Hedged deep tracking. In: Proceedings of the IEEE Conference on Computer Vision and Pattern Recognition, pp. 4303–4311 (2016)

37. Song, Y., et al.: Vital: visual tracking via adversarial learning. In: Proceedings of the IEEE Conference on Computer Vision and Pattern Recognition, pp. 8990–8999 (2018)

38. Sun, C., Wang, D., Lu, H., Yang, M.H.: Learning spatial-aware regressions for visual tracking. In: Proceedings of the IEEE Conference on Computer Vision and Pattern Recognition, pp. 8962–8970 (2018)

39. Tang, M., Yu, B., Zhang, F., Wang, J.: High-speed tracking with multi-kernel correlation filters. In: The IEEE Conference on Computer Vision and Pattern Recognition (CVPR), June 2018

40. Tao, R., Gavves, E., Smeulders, A.W.: Siamese instance search for tracking. In: Proceedings of the IEEE Conference on Computer Vision and Pattern Recognition, pp. 1420–1429 (2016)

41. Valmadre, J., Bertinetto, L., Henriques, J., Vedaldi, A., Torr, P.H.: End-to-end representation learning for correlation filter based tracking. In: 2017 IEEE Conference on Computer Vision and Pattern Recognition (CVPR), pp. 5000–5008. IEEE (2017)

42. Wang, G., Luo, C., Xiong, Z., Zeng, W.: SPM-tracker: series-parallel matching for real-time visual object tracking. In: The IEEE Conference on Computer Vision and Pattern Recognition (CVPR), June 2019

43. Wu, Y., Lim, J., Yang, M.H.: Object tracking benchmark. IEEE Trans. Pattern Anal. Mach. Intell. **37**(9), 1834–1848 (2015)

44. Zhang, Z., Peng, H.: Deeper and wider siamese networks for real-time visual tracking. In: The IEEE Conference on Computer Vision and Pattern Recognition (CVPR), June 2019

45. Zheng, L., Chen, Y., Tang, M., Wang, J., Lu, H.: Siamese deformable cross-correlation network for real-time visual tracking. Neurocomputing **401**, 36–47 (2020)

46. Zheng, L., Tang, M., Chen, Y., Wang, J., Lu, H.: Fast-deepKCF without boundary effect. In: Proceedings of the IEEE International Conference on Computer Vision, pp. 4020–4029 (2019)

47. Zheng, L., Tang, M., Chen, Y., Wang, J., Lu, H.: High-speed and accurate scale estimation for visual tracking with Gaussian process regression. In: 2020 IEEE International Conference on Multimedia and Expo (ICME), pp. 1–6. IEEE (2020)

48. Zheng, L., Tang, M., Wang, J.: Learning robust Gaussian process regression for visual tracking. In: IJCAI, pp. 1219–1225 (2018)

49. Zhu, Z., Wang, Q., Li, B., Wu, W., Yan, J., Hu, W.: Distractor-aware siamese networks for visual object tracking. In: Ferrari, V., Hebert, M., Sminchisescu, C., Weiss, Y. (eds.) ECCV 2018. LNCS, vol. 11213, pp. 103–119. Springer, Cham (2018). https://doi.org/10.1007/978-3-030-01240-3_7

WeightNet: Revisiting the Design Space of Weight Networks

Ningning Ma[1], Xiangyu Zhang[2]([✉]), Jiawei Huang[2], and Jian Sun[2]

[1] Hong Kong University of Science and Technology, Sai Kung, Hong Kong
nmaac@cse.ust.hk
[2] MEGVII Technology, Beijing, China
{zhangxiangyu,huangjiawei,sunjian}@megvii.com

Abstract. We present a conceptually simple, flexible and effective framework for weight generating networks. Our approach is general that unifies two current distinct and extremely effective SENet and Cond-Conv into the same framework on weight space. The method, called *WeightNet*, generalizes the two methods by simply adding one more grouped fully-connected layer to the attention activation layer. We use the WeightNet, composed entirely of (grouped) fully-connected layers, to directly output the convolutional weight. WeightNet is easy and memory-conserving to train, on the kernel space instead of the feature space. Because of the flexibility, our method outperforms existing approaches on both ImageNet and COCO detection tasks, achieving better Accuracy-FLOPs and Accuracy-Parameter trade-offs. The framework on the flexible weight space has the potential to further improve the performance. Code is available at https://github.com/megvii-model/WeightNet.

Keywords: CNN · Weight network · Conditional kernel

1 Introduction

Designing convolution weight is a key issue in convolution networks (CNNs). The weight-generating methods [6,14,24] using a network, which we call weight networks, provide an insightful neural architecture design space. These approaches are conceptually intuitive, easy and efficient to train. Our goal in this work is to present a simple and effective framework, in the design space of weight networks, inspired by the rethinking of recent effective conditional networks.

Conditional networks (or dynamic networks) [2,11,38], which use extra sample-dependent modules to conditionally adjust the network, have achieved great success. SENet [11], an effective and robust attention module, helps many tasks achieve state-of-the-art results [9,31,32]. Conditionally Parameterized Convolution (CondConv) [38] uses over-parameterization to achieve great improvements but maintains the computational complexity at the inference phase.

© Springer Nature Switzerland AG 2020
A. Vedaldi et al. (Eds.): ECCV 2020, LNCS 12360, pp. 776–792, 2020.
https://doi.org/10.1007/978-3-030-58555-6_46

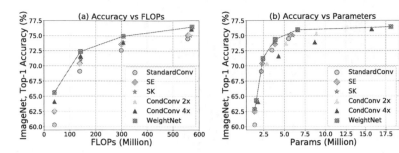

Fig. 1. Accuracy vs. FLOPs vs. Parameters comparisons on ImageNet, using ShuffleNetV2 [22]. (a) The trade-off between accuracy and FLOPs; (b) the trade-off between accuracy and number of parameters.

Both of the methods consist of two steps: first, they obtain an attention activation vector, then using the vector, SE scales the feature channels, while CondConv performs a mixture of expert weights. Despite they are usually treated as entirely distinct methods, they have some things in common. It is natural to ask: *do they have any correlations?* We show that we can link the two extremely effective approaches, by generalizing them in the weight network space.

Our methods, called *WeightNet*, extends the first step by simply adding one more layer for generating the convolutional weight directly (see Fig. 2). The layer is a grouped fully-connected layer applied to the attention vector, generating the weight in a group-wise manner. To achieve this, we rethink SENet and CondConv and discover that the subsequent operations after the first step can be cast to a *grouped fully-connected layer*, however, they are particular cases.

In that grouped layer, the output is direct the convolution weight, but the input size and the group number are variable. In CondConv the group number is discovered to be a minimum number of one and the input is small (4, 8, 16, etc.) to avoid the rapid growth of the model size. In SENet the group is discovered to be the maximum number equal to the input channel number.

Despite the two variants having seemingly minor differences, they have a large impact: they together control the parameter-FLOPs-accuracy tradeoff, leading to surprisingly different performance. Intuitively, we introduce two hyperparameters M and G, to control the input number and the group number, respectively. The two hyperparameters have not been observed and investigated before, in the additional grouped fully-connected layer. By simply adjusting them, we can strike a better trade-off between the representation capacity and the number of model parameters. We show by experiments on ImageNet classification and COCO detection the superiority of our method (Fig. 1).

Our main contributions include: 1) First, we rethink the weight generating manners in SENet and CondConv, for the first time, to be complete fully-connected networks; 2) Second, only from this new perspective can we revisit the novel network design space in the weight space, which provides more effective structures than those in convolution design space (group-wise, point-wise, and depth-wise convolutions, etc). In this new and rarely explored weight space, there could be new structures besides fully-connected layers, there could also be more kinds of sparse matrix besides those in Fig. 4. We believe this is a promising direction and hope it would have a broader impact on the vision community.

2   Related Work

Weight Generation Networks. Schmidhuber et al. [28] incorporate the "fast" weights into recurrent connections in RNN methods. Dynamic filter networks [14] use filter-generating networks on video and stereo prediction. HyperNetworks [6] decouple the neural networks according to the relationship in nature: a genotype (the hypernetwork), and a phenotype (the main network), that uses a small network to produce the weights for the main network, which reduces the number of parameters while achieving respectable results. Meta networks [24] generate weights using a meta learner for rapid generalization across tasks. The methods [6,14,24,25] provide a worthy design space in the weight-generating network, our method follows the spirits and uses a WeightNet to generate the weights.

Conditional CNNs. Different from standard CNNs [4,7,9,10,27,29,30,40], conditional (or dynamic) CNNs [15,17,20,37,39] use dynamic kernels, widths, or depths conditioned on the input samples, showing great improvement. Spatial Transform Networks [13] learns to transform to warp the feature map in a parametric way. Yang et al. [38] proposed conditional parameterized convolution to mix the experts voted by each sample's feature. The methods are extremely effective because they improve the Top-1 accuracy by more than 5% on the ImageNet dataset, which is a great improvement. Different from dynamic features or dynamic kernels, another series of work [12,35] focus on dynamic depths of the convolutional networks, that skip some layers for different samples.

Attention and Gating Mechanism. Attention mechanism [1,21,33,34,36] is also a kind of conditional network, that adjusts the networks dependent on the input. Recently the attention mechanism has shown its great improvement. Hu et al. [11] proposed a block-wise gating mechanism to enhance the representation ability, where they adopted a squeeze and excitation method to use global information and capture channel-wise dependencies. SENet achieves great success by not only winning the ImageNet challenge [5], but also helping many structures to achieve state-of-the-art performance [9,31,32]. In GaterNet [3], a gater network was used to predict binary masks for the backbone CNN, which can result in performance improvement. Besides, Li et al. [16] introduced a kernel selecting module, where they added attention to kernels with different sizes to enhance CNN's learning capability. In contrast, WeightNet is designed on kernel space which is more time-conserving and memory-conserving than feature space.

3   WeightNet

The WeightNet generalizes the current two extremely effective modules in weight space. Our method is conceptually simple: both SENet and CondConv generate

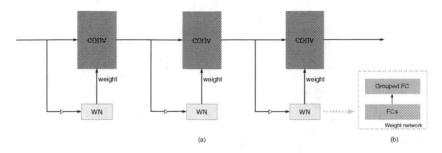

(a)   (b)

Fig. 2. The WeightNet structure. The convolutional weight is generated by Weight-Net that is comprised entirely of (grouped) fully-connected layers. The symbol (▷) represents the dimension reduction (global average pool) from feature space ($C \times H \times W$) to kernel space (C). The 'FC' denotes a fully-connected layer, 'conv' denotes convolution, and 'WN' denotes the WeightNet.

the *activation vector* using a global average pooling (GAP) and one or two fully-connected layers followed with a non-linear sigmoid operation; to this we simply add one more grouped fully-connected layer, to generate the weight *directly* (Fig. 2). This is different from common practice that applies the vector to feature space and we avoid the memory-consuming training period.

WeightNet is computationally efficient because of the dimension reduction (GAP) from $C \times H \times W$ dimension to a 1-D dimension C. Evidently, the Weight-Net only consists of (grouped) fully-connected layers. We begin by introducing the matrix multiplication behaviors of (grouped) fully-connected operations.

Grouped Fully-Connected Operation. Conceptually, neurons in a fully-connected layer have full connections and thus can be computed with a matrix multiplication, in the form $Y = WX$ (see Fig. 3 (a)). Further, neurons in a grouped fully-connected layer have group-wise sparse connections with activations in the previous layer.

Formally, in Fig. 3 (b), the neurons are divided exactly into g groups, each group (with i/g inputs and o/g outputs) performs a fully-connected operation (see the red box for example). One notable property of this operation, which can be easily seen in the graphic illustration, is that the weight matrix becomes a sparse, *block diagonal matrix*, with a size of ($o/g \times i/g$) in each block.

Grouped fully-connected operation is a general form of fully-connected operation where the group number is one. Next, we show how it generalizes Cond-Conv and SENet: use the grouped fully-connected layer to replace the subsequent operations after the activation vector and directly output the generated weight.

Denotation. We denote a convolution operation with the input feature map $\mathbf{X} \in \mathbb{R}^{C \times h \times w}$, the output feature map $\mathbf{Y} \in \mathbb{R}^{C \times h' \times w'}$, and the convolution weight $\mathbf{W}' \in \mathbb{R}^{C \times C \times k_h \times k_w}$. For simplicity, but without loss of generality, it is assumed that the number of the input channels equals to that of output channels,

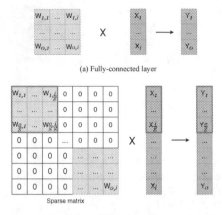

(a) Fully-connected layer

Sparse matrix

(b) Grouped fully-connected layer

Fig. 3. The **matrix multiplication** behaviors of the (grouped) fully-connected operations. Here i, o and g denote the numbers of the input channel, output channel and group number. (a) A standard matrix multiplication representing a fully-connected layer. (b) With the weight in a block diagonal sparse matrix, it becomes a general grouped fully-connected layer. Each group (red box) is exactly a standard matrix multiplication in (a), with i/g input channels and o/g output channels. Figure (a) is a special case of Fig. (b) where $g = 1$. (Color figure online)

here $(h, w), (h', w'), (k_h, k_w)$ denote the 2-D heights and the widths for the input, output, and kernel. Therefore, we denote the convolution operation using the symbol $(*)$: $\mathbf{Y}_c = \mathbf{W}'_c * \mathbf{X}$. We use $\boldsymbol{\alpha}$ to denote the attention activation vector in CondConv and SENet.

3.1 Rethinking CondConv

Conditionally parameterized convolution (CondConv) [38] is a mixture of m experts of weights, voted by a m-dimensional vector $\boldsymbol{\alpha}$, that is sample-dependent and makes each sample's weight dynamic.

Formally, we begin with reviewing the first step in CondConv, it gets $\boldsymbol{\alpha}$ by a global average pooling and a fully-connected layer $\mathbf{W}_{fc1}$, followed by a sigmoid $\sigma(\cdot)$: $\boldsymbol{\alpha} = \sigma(\mathbf{W}_{fc1} \times \frac{1}{hw} \sum_{i \in h, j \in w} \mathbf{X}_{c,i,j})$, here $(\times)$ denotes the matrix multiplication, $\mathbf{W}_{fc1} \in \mathbb{R}^{m \times C}$, $\boldsymbol{\alpha} \in \mathbb{R}^{m \times 1}$.

Next, we show the following mixture of expert operations in the original paper can essentially be replaced by a fully-connected layer. The weight is generated by multiple weights: $\mathbf{W}' = \alpha_1 \cdot \mathbf{W}_1 + \alpha_2 \cdot \mathbf{W}_2 + ... + \alpha_m \cdot \mathbf{W}_m$, here $\mathbf{W}_i \in \mathbb{R}^{C \times C \times k_h \times k_w}, (i \in \{1, 2, ..., m\})$. We rethink it as follows:

$$\mathbf{W}' = \mathbf{W}^T \times \boldsymbol{\alpha}$$
$$where \ \mathbf{W} = [\mathbf{W}_1 \mathbf{W}_2...\mathbf{W}_m]$$

(1)

Table 1. Summary of the configure in the grouped fully-connected layer. λ is the proportion of input size to group number, representing the major increased parameters.

Model	Input size	Group number	λ	Output size
CondConv	m	1	m	$C \times C \times k_h \times k_w$
SENet	C	C	1	$C \times C \times k_h \times k_w$
WeightNet	$M \times C$	$G \times C$	M/G	$C \times C \times k_h \times k_w$

Here $\mathbf{W} \in \mathbb{R}^{m \times CCk_hk_w}$ denotes the matrix concatenation result, $(\times)$ denotes the matrix multiplication (fully-connected in Fig. 3a). Therefore, the weight is generated by simply adding one more layer $(\mathbf{W})$ to the activation layer. That layer is a fully-connected layer with m inputs and $C \times C \times k_h \times k_w$ outputs.

This is different from the practice in the original paper in the training phase. In that case, it is memory-consuming and suffers from the batch problem when increasing m (batch size should be set to one when $m > 4$). In this case, we train with large batch sizes efficiently.

3.2 Rethinking SENet

Squeeze and Excitation (SE) [11] block is an extremely effective "plug-n-play" module that is acted on the feature map. We integrate the SE module into the convolution kernels and discover it can also be represented by adding one more grouped fully-connected layer to the activation vector $\boldsymbol{\alpha}$. We start from the reviewing of the $\boldsymbol{\alpha}$ generation process. It has a similar process with CondConv: a global average pool, two fully-connected layer with non-linear ReLU (δ) and sigmoid (σ): $\boldsymbol{\alpha} = \sigma(\mathbf{W}_{fc_2} \times \delta(\mathbf{W}_{fc1} \times \frac{1}{hw} \sum_{i \in h, j \in w} \mathbf{X}_{c,i,j}))$, here $\mathbf{W}_{fc1} \in \mathbb{R}^{C/r \times C}$, $\mathbf{W}_{fc2} \in \mathbb{R}^{C \times C/r}$, $(\times)$ in the equation denotes the matrix multiplication. The two fully-connected layers here are mainly used to reduce the number of parameters because $\boldsymbol{\alpha}$ here is a C-dimensional vector, a single layer is parameter-consuming.

Next, in common practice the block is used before or after a convolution layer, $\boldsymbol{\alpha}$ is computed right before a convolution (on the input feature $\mathbf{X}$): $\mathbf{Y}_c = \mathbf{W}'_c * (\mathbf{X} \cdot \boldsymbol{\alpha})$, or right after a convolution (on the output feature $\mathbf{Y}$): $\mathbf{Y}_c = (\mathbf{W}'_c * \mathbf{X}) \cdot \boldsymbol{\alpha}_c$, here ($\cdot$) denotes dot multiplication broadcasted along the C axis. In contrast, on kernel level, we analyze the case that SE is acted on $\mathbf{W}'$: $\mathbf{Y}_c = (\mathbf{W}'_c \cdot \boldsymbol{\alpha}_c) * \mathbf{X}$. Therefore we rewrite the weight to be $\mathbf{W}' \cdot \boldsymbol{\alpha}$, the ($\cdot$) here is different from the ($\times$) in Eq. 1. In that case, a dimension reduction is performed; in this case, no dimension reduction. Therefore, it is essentially a grouped sparse connected operation, that is a particular case of Fig. 3 (b), with C inputs, $C \times C \times k_h \times k_w$ outputs, and C groups.

3.3 WeightNet Structure

By far, we note that the group number in the general grouped fully-connected layer (Fig. 3 b) has values range from 1 to the channel number. That is, the group

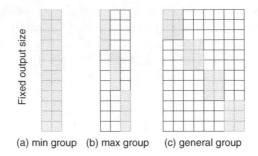

(a) min group   (b) max group   (c) general group

Fig. 4. The diagrams of the different cases in the **block diagonal matrix** (Fig. 3 b), that can represent the weights of the grouped fully-connected layer in CondConv, SENet and the general WeightNet. They output the same fixed size (convolution kernel's size $C \times C \times k_h \times k_w$), but have different group numbers: (a) the group number has a minimum number of one (CondConv), (b) the group number has a maximum number equals to the input size C (SENet), since (a) and (b) are extreme cases, (c) shows the general group number between 1 and the input size (WeightNet).

has a minimum number of one and has a maximum number of the input channel numbers. It, therefore, generalizes the CondConv, where the group number takes the minimum value (one), and the SENet, where it takes the maximum value (the input channel number). We conclude that they are two extreme cases of the general grouped fully-connected layer (Fig. 4).

We summarize the configure in the grouped fully-connected layer (in Table 1) and generalize them using two additional hyperparameters M and G. To make the group number more flexible, we set it by combining the channel C and a constant hyperparameter G, Moreover, another hyperparameter M is used to control the input number, thus M and G together to control the parameter-accuracy tradeoff. The layer in CondConv is a special case with $M = m/C, G = 1/C$, while for SENet $M = 1, G = 1$. We constrain $M \times C$ and $G \times C$ to be integers, M is divisible by G in this case. It is notable that the two hyperparameters are right there but have not been noticed and investigated.

Implementation Details. For the activation vector α's generating step, since α is a $(M \times C)$-dimensional vector, it may be large and parameter-consuming, therefore, we use two fully-connected layers with a reduction ratio r. It has a similar process with the two methods: a global average pool, two fully-connected layer with non-linear sigmoid (σ): $\alpha = \sigma(\mathbf{W}_{fc_2} \times \mathbf{W}_{fc1} \times \frac{1}{hw} \sum_{i \in h, j \in w} \mathbf{X}_{c,i,j})$, here $\mathbf{W}_{fc1} \in \mathbb{R}^{C/r \times C}$, $\mathbf{W}_{fc2} \in \mathbb{R}^{MC \times C/r}$, ($\times$) denotes the matrix multiplication, r has a default setting of 16.

In the second step, we adopt a grouped fully-connected layer with $M \times C$ input, $C \times C \times k_h \times k_w$ output, and $G \times C$ groups. We note that the structure is a straightforward design, and more complex structures have the potential to improve the performance further, but it is beyond the focus of this work.

Complexity Analysis. The structure of WeightNet decouples the convolution computation and the weight computation into two separate branches (see Fig. 2). Because the spatial dimensions $(h \times w)$ are reduced before feeding into the weight branch, the computational amount (FLOPs) is mainly in the convolution branch. The FLOPs complexities in the convolution and weight branches are $O(hwCCk_hk_w)$ and $O(MCCk_hk_w/G)$, the latter is relatively negligible. The parameter complexities for each branch are zero and $O(M/G \times C \times C \times k_h \times k_w)$, which is M/G times of normal convolution. We notate λ to represent it (Table 1).

Training with Batch Dimension. The weight generated by WeightNet has a dimension of batch size, here we briefly introduce the training method related to the batch dimension. We denote B as batch size and reshape the input $\mathbf{X}$ of the convolution layer to $(1, B \times C, h, w)$. Thus $\mathbf{X}$ has $B \times C$ channel numbers, which means we regard different samples in the same batch as different channels. Next, we reshape the generated weight $\mathbf{W}$ to (B, C, C, k_h, k_w). Then it becomes a group convolution, with a group number of B, the inputs and the outputs in each group are both equal to C. Therefore, we use the same memory-conserving method for both training and inference periods, and this is different from CondConv.

4   Experiments

In this section, we evaluate the WeightNet on classification and COCO detection tasks [19]. In classification task, we conduct experiments on a light-weight CNN model ShuffleNetV2 [22] and a deep model ResNet50 [7]. In the detection task, we evaluate our method's performance on distinct backbone models under RetinaNet. In the final analysis, we conduct ablation studies and investigate the properties of WeightNet in various aspects.

4.1   Classification

We conduct image classification experiments on ImageNet 2012 classification dataset, which includes 1000 classes [26]. Our models are first trained on the training dataset that consists of 1.28 million images and then evaluated over 50k images in the validation dataset. For the training settings, all the ShuffleNetV2 [22] models are trained with the same settings as [22]. For ResNet-50, we use a linear decay scheduled learning rate starting with 0.1, a batch size of 256, a weight decay of 1e−4, and 600k iterations.

ShuffleNetV2. To investigate the performance of our method on light-weight convolution networks, we construct experiments based on a recent effective network ShuffleNetV2 [22]. For a fair comparison, we retrain all the models by ourselves, using the same code base. We replace the standard convolution kernels in each bottleneck with our proposed WeightNet, and control FLOPs and the number of parameters for fairness comparison.

Table 2. ImageNet classification results of the WeightNet on ShuffleNetV2 [22]. For fair comparison, we control the values of λ to be 1×, to make sure that the experiments are under the same FLOPs and the same number of parameters.

Table 3. ImageNet classification results of the WeightNet on ShuffleNetV2 [22]. The comparison is under the same FLOPs and regardless of the number of parameters. To obtain the optimum performance, we set the λ to $\{8×, 4×, 4×, 4×\}$ respectively.

Model	# Params	FLOPs	Top-1 err.
ShuffleNetV2 (0.5×)	1.4M	41M	39.7
+ WeightNet (1×)	1.5M	41M	**36.7**
ShuffleNetV2 (1×)	2.2M	138M	30.9
+ WeightNet (1×)	2.4M	139M	**28.8**
ShuffleNetV2 (1.5×)	3.5M	299M	27.4
+ WeightNet (1×)	3.9M	301M	**25.6**
ShuffleNetV2 (2×)	5.5M	557M	25.5
+ WeightNet (1×)	6.1M	562M	**24.1**

Model	# Params	FLOPs	Top-1 err.
ShuffleNetV2 (0.5×)	1.4M	41M	39.7
+ WeightNet (8×)	2.7M	42M	**34.0**
ShuffleNetV2 (1×)	2.2M	138M	30.9
+ WeightNet (4×)	5.1M	141M	**27.6**
ShuffleNetV2 (1.5×)	3.5M	299M	27.4
+ WeightNet (4×)	9.6M	307M	**25.0**
ShuffleNetV2 (2×)	5.5M	557M	25.5
+ WeightNet (4×)	18.1M	573M	**23.5**

As shown in Table 1, λ is utilized to control the number of parameters in a convolution. For simplicity, we fix $G = 2$ when adjusting λ. In our experiments, λ has several sizes $\{1×, 2×, 4×, 8×\}$. To make the number of channels conveniently divisible by G when scaling the number of parameters, we slightly adjust the number of channels for ShuffleNetV2 1× and 2×.

We evaluate the WeightNet from two aspects. Table 2 reports the performance of our method considering parameters. The experiments illustrate that our method has significant advantages over the other counterparts under the same FLOPs and the same number of parameter constraints. ShuffleNetV2 0.5× gains 3% Top-1 accuracy without additional computation budgets.

In Table 3, we report the advantages after applying our method on ShuffleNetV2 with different sizes. Considering in practice, the storage space is sufficient. Therefore, without the loss of fairness, we only constrain the Flops to be the same and tolerate the increment of parameters.

ShuffleNet V2 (0.5×) gains 5.7% Top-1 accuracy which shows further significant improvements by adding a minority of parameters. ShuffleNet V2 (2×) gains 2.0% Top-1 accuracy.

To further investigate the improvement of our method, we compare our method with some recent effective conditional CNN methods under the same FLOPs and the same number of parameters. For the network settings of Cond-Conv [38], we replace standard convolutions in the bottlenecks with CondConv, and change the number of experts as described in CondConv to adjust parameters, as the number of experts grows, the number of parameters grows. To

Table 4. Comparison with recently effective attention modules on Shuf-fleNetV2 [22] and ResNet50 [7]. We show results on ImageNet.

Model	# Params	FLOPs	Top-1 err.
ShuffleNetV2 [22] (0.5×)	1.4M	41M	39.7
+ SE [11]	1.4M	41M	37.5
+ SK [16]	1.5M	42M	37.5
+ CondConv [38] (2×)	1.5M	41M	37.3
+ WeightNet (1×)	1.5M	41M	**36.7**
+ CondConv [38] (4×)	1.8M	41M	35.9
+ WeightNct (2×)	1.8M	41M	**35.5**
ShuffleNetV2 [22] (1.5×)	3.5M	299M	27.4
+ SE [11]	3.9M	299M	26.4
+ SK [16]	3.9M	306M	26.1
+ CondConv [38] (2×)	5.2M	303M	26.3
+ WeightNet (1×)	**3.9M**	301M	**25.6**
+ CondConv [38] (4×)	8.7M	306M	26.1
+ WeightNet (2×)	**5.9M**	**303M**	**25.2**
ShuffleNetV2 [22] (2.0×)	5.5M	557M	25.5
+ WeightNet (2×)	10.1M	565M	**23.7**
ResNet50 [7]	25.5M	3.86G	24.0
+ SE [11]	26.7M	3.86G	22.8
+ CondConv [38] (2×)	72.4M	3.90G	23.4
+ WeightNet (1×)	31.1M	3.89G	**22.5**

reveal the model capacity under the same number of parameters, for our proposed WeightNet, we control the number of parameters by changing λ. Table 4 describes the comparison between our method and other counterpart effective methods, from which we observe our method outperforms the other conditional CNN methods under the same budgets. The Accuracy-Parameters tradeoff and the Accuracy-FLOPs tradeoff are shown in Fig. 1.

From the results, we can see SE and CondConv boost the base models of all sizes significantly. However, CondConv has major improvements in smaller sizes especially, but as the model becomes larger, the smaller the advantage it has. For example, CondConv performs better than SE on ShuffleNetV2 0.5× but SE performs better on ShuffleNetV2 2×. In contrast, we find our method can be uniformly better than SE and CondConv.

To reduce the overfitting problem while increasing parameters, we add dropout [8] for models with more than 3M parameters. As we described in Sect. 3.3, λ represents the increase of parameters, so we measure the capacity of networks by changing parameter multiplier λ in $\{1\times, 2\times, 4\times, 8\times\}$. We further

Table 5. Object detection results comparing with baseline backbone. We show RetinaNet [18] results on COCO.

Backbone	# Params	FLOPs	mAP
ShuffleNetV2 [22] (0.5×)	1.4M	41M	22.5
+ WeightNet (4×)	2.0M	41M	**27.1**
ShuffleNetV2 [22] (1.0×)	2.2M	138M	29.2
+ WeightNet (4×)	4.8M	141M	**32.1**
ShuffleNetV2 [22] (1.5×)	3.5M	299M	30.8
+ WeightNet (2×)	5.7M	303M	**33.3**
ShuffleNetV2 [22] (2.0×)	5.5M	557M	33.0
+ WeightNet (2×)	9.7M	565M	**34.0**

Table 6. Object detection results comparing with other conditional CNN backbones. We show RetinaNet [18] results on COCO.

Backbone	# Params	FLOPs	mAP
ShuffleNetV2 [22] (0.5×)	1.4M	41M	22.5
+ SE [11]	1.4M	41M	25.0
+ SK [16]	1.5M	42M	24.5
+ CondConv [38] (2×)	1.5M	41M	25.8
+ CondConv [38] (4×)	1.8M	41M	25.0
+ CondConv [38] (8×)	2.3M	42M	26.4
+ WeightNet (4×)	2.0M	41M	**27.1**

analyze the effect of λ and the grouping hyperparameter G on each filter in the ablation study section.

ResNet50. For larger classification models, we conduct experiments on ResNet50 [7]. We use a similar way to replace the standard convolution kernels in ResNet50 bottlenecks with our proposed WeightNet. Besides, we train the conditional CNNs utilizing the same training settings with the base ResNet50 network.

In Table 4, based on ResNet50 model, we compare our method with SE [11] and CondConv [38] under the same computational budgets. It's shown that our method still performs better than other conditional convolution modules. We perform CondConv (2×) on ResNet50, the results reveal that it does not have further improvement comparing with SE, although CondConv has a larger number of parameters. We conduct our method (1×) by adding limited parameters and it also shows further improvement comparing with SE. Moreover, ShuffleNetV2 [22] (2×) with our method performs better than ResNet50, with only 40% parameters and 14.6% FLOPs.

4.2   Object Detection

We evaluate the performance of our method on COCO detection [19] task. The COCO dataset has 80 object categories. We use the *trainval35k* set for training and use the *minival* set for testing. For a fair comparison, we train all the models with the same settings. The batch size is set to 2, the weight decay is set to 1e-4 and the momentum is set to 0.9. We use anchors for 3 scales and 3 aspect ratios and use a 600-pixel train and test image scale. We conduct experiments

Table 7. Ablation on λ. The table shows the ImageNet Top-1 err. results. The experiments are conducted on ShuffleNetV2 [22]. By increasing λ in the range {1,2,4,8}, the FLOPs does not change and the number of parameters increases.

Model	λ			
	1	2	4	8
ShuffleNetV2 (0.5×)	36.7	35.5	34.4	**34.0**
ShuffleNetV2 (1.0×)	28.8	28.1	**27.6**	27.8
ShuffleNetV2 (1.5×)	25.6	25.2	**25.0**	25.3
ShuffleNetV2 (2.0×)	24.1	23.7	**23.5**	24.0

Table 8. Ablation on G. We tune the group hyperparameter G to {1, 2, 4}, we keep $\lambda = 1$. The results are ImageNet Top-1 err. The experiments are conducted on ShuffleNetV2 [22].

Model	G	# Params	FLOPs	Top-1 err.
ShuffleNetV2	$G = 1$	1.4M	41M	37.18
(0.5×)	$G = 2$	1.5M	41M	36.73
	$G = 4$	1.5M	41M	**36.37**
ShuffleNetV2	$G = 1$	2.3M	139M	29.09
(1.0×)	$G = 2$	2.4M	139M	28.77
	$G = 4$	2.6M	139M	**28.76**

Table 9. Ablation study on **different stages**. The ✓ means the convolutions in that stage is integrated with our proposed WeightNet.

Stage2	Stage3	Stage4	Top-1 err.
✓			39.13
	✓		36.82
		✓	36.43
✓	✓		35.73
✓		✓	36.44
	✓	✓	**35.30**
✓	✓	✓	35.47

Table 10. Ablation study on the number of the **global average pooling** operators in the whole network. We conduct ShuffleNetV2 0.5× experiments on ImageNet dataset. We compare the cases: one global average pooling in 1) each stage, 2) each block, and 3) each layer. GAP represents global average pooling in this table.

	Top-1 err.
Stage wise GAP	37.01
Block wise GAP	35.47
Layer wise GAP	35.04

on RetinaNet [18] using ShuffleNetV2 [22] as the backbone feature extractor. We compare the backbone models of our method with the standard CNN models.

Table 5 illustrates the improvement of our method over standard convolution on the RetinaNet framework. For simplicity we set $G = 1$ and adjust the size of WeightNet to {2×, 4×}. As we can see our method improves the mAP significantly by adding a minority of parameters. ShuffleNetV2 (0.5×) with WeightNet (4×) improves 4.6 mAP by adding few parameters under the same FLOPs.

To compare the performance between WeightNet and CondConv [38] under the same parameters, we utilize ShuffleNetV2 0.5× as the backbone and investigate the performances of all CondConv sizes. Table 6 reveals the clear advantage of our method over CondConv. Our method outperforms CondConv uniformly under the same computational budgets. As a result, our method is indeed robust and fundamental on different tasks.

4.3 Ablation Study and Analysis

The Influence of λ. By tuning λ, we control the number of parameters. We investigate the influence of λ on ImageNet Top-1 accuracy, conducting exper-

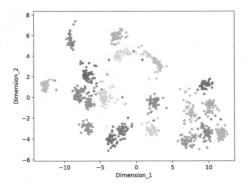

Fig. 5. Analysis for weights generated by our WeightNet. The figure illustrates of the weights of the 1,000 samples. The samples belong to 20 classes, which are represented by 20 different colors. Each point represents the weights of one sample. (Color figure online)

iments on the ShuffleNetV2 structure. Table 7 shows the results. We find that the optimal λ for ShuffleNetV2 $\{0.5\times, 1\times, 1.5\times, 2\times\}$ are $\{8,4,4,4\}$, respectively. The model capacity has an upper bounded as we increase λ, and there exists a choice of λ to achieve the optimal performance.

The Influence of G. To investigate the influence of G, we conduct experiments on ImageNet based on ShuffleNetV2. Table 8 illustrates the influence of G. We keep λ equals to 1, and change G to $\{1, 2, 4\}$. From the result we conclude that increasing G has a positive influence on model capacity.

WeightNet on Different Stages. As WeightNet makes the weights for each convolution layer changes dynamically for distinct samples, we investigate the influence of each stage. We change the static convolutions of each stage to our method respectively as Table 9 shows. From the result, we conclude that the last stage influences much larger than other stages, and the performance is best when we change the convolutions in the last two stages to our method.

The Number of the Global Average Pooling Operator. Sharing the global average pooling (GAP) operator contributes to improving the speed of the conditional convolution network. We compare the following three kinds of usages of GAP operator: using GAP for each layer, sharing GAPs in a block, sharing GAPs in a stage. We conduct experiments on ShuffleNetV2 [22] $(0.5\times)$ baseline with WeightNet $(2\times)$. Table 10 illustrates the comparison of these three kinds of usages. The results indicate that by adding the number of GAPs, the model capacity improves.

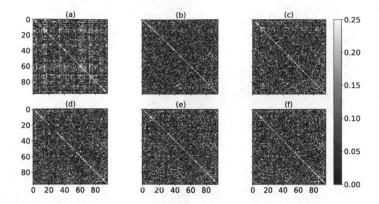

Fig. 6. Cosine similarity matrix. A 96 × 96 matrix represents 96 filters' pair-by-pair similarity, the smaller value (darker color) means the lower similarities. (a) Standard convolution kernel's similarity matrix, (b–f) WeightNet kernels' similarity matrixes. The colors in (b–f) are obviously much darker than (a), meaning lower similarity. (Color figure online)

Weight Similarity Among Distinct Samples. We randomly select 20 classes in the ImageNet validation set, which has 1,000 classes in total. Each class has 50 samples, and there are 1,000 samples in total. We extract the weights in the last convolution layer in Stage 4 from a well-trained ShuffleNetV2 (2×) with our WeightNet. We project the weights of each sample from a high dimensional space to a 2-dimension space by t-SNE [23], which is shown in Fig. 5. We use 20 different colors to distinguish samples from 20 distinct classes.

We observe two characteristics. First, different samples have distinct weights. Second, there are roughly 20 point clusters and the weights of samples in the same classes are closer to each other, which indicates that the weights of our method capture more class-specific information than static convolution.

Channel Similarity. We conduct the experiments to show the channel similarity of our method, we use the different filters' similarity in a convolution weight to represent the channel similarity. Lower channel similarity would improve the representative ability of CNN models and improve the channel representative capacities. Strong evidence was found to show that our method has a lower channel similarity.

We analyze the last convolution layer's kernel in the last stage of ShuffleNetV2 [22] (0.5×), where the channel number is 96, thus there are 96 filters in that convolution kernel. We compute the cosine similarities of the filters pair by pair, that comprise a 96×96 cosine similarity matrix.

In Fig. 6, we compare the channel similarity of WeightNet and standard convolution. We first compute the cosine similarity matrix of a standard convolution kernel and display it in Fig. 6-(a). Then for our method, because different samples do not share the same kernel, we randomly choose 5 samples in distinct classes from the ImageNet validation set and show the corresponding similarity matrix in Fig. 6-(b, c, d, e, f). The results clearly illustrate that our method has lower channel similarity.

5   Conclusion and Future Works

The study connects two distinct but extremely effective methods SENet and CondConv on weight space, and unifies them into the same framework we call WeightNet. In the simple WeightNet comprised entirely of (grouped) fully-connected layers, the grouping manners of SENet and CondConv are two extreme cases, thus involving two hyperparameters M and G that have not been observed and investigated. By simply adjusting them, we got a straightforward structure that achieves better tradeoff results. The more complex structures in the framework have the potential to further improve the performance, and we hope the simple framework in the weight space helps ease future research. Therefore, this would be a fruitful area for future work.

Acknowledgements. This work is supported by The National Key Research and Development Program of China (No. 2017YFA0700800) and Beijing Academy of Artificial Intelligence (BAAI).

References

1. Bahdanau, D., Cho, K., Bengio, Y.: Neural machine translation by jointly learning to align and translate. arXiv preprint arXiv:1409.0473 (2014)
2. Cao, Y., Xu, J., Lin, S., Wei, F., Hu, H.: GCNET: non-local networks meet squeeze-excitation networks and beyond. In: Proceedings of the IEEE International Conference on Computer Vision Workshops (2019)
3. Chen, Z., Li, Y., Bengio, S., Si, S.: GaterNet: dynamic filter selection in convolutional neural network via a dedicated global gating network. arXiv preprint arXiv:1811.11205 (2018)
4. Chollet, F.: Xception: deep learning with depthwise separable convolutions. In: Proceedings of the IEEE Conference on Computer Vision and Pattern Recognition, pp. 1251–1258 (2017)
5. Deng, J., Dong, W., Socher, R., Li, L.J., Li, K., Fei-Fei, L.: ImageNet: a large-scale hierarchical image database. In: 2009 IEEE Conference on Computer Vision and Pattern Recognition, pp. 248–255. IEEE (2009)
6. Ha, D., Dai, A., Le, Q.V.: Hypernetworks. arXiv preprint arXiv:1609.09106 (2016)
7. He, K., Zhang, X., Ren, S., Sun, J.: Deep residual learning for image recognition. In: Proceedings of the IEEE Conference on Computer Vision and Pattern Recognition, pp. 770–778 (2016)
8. Hinton, G.E., Srivastava, N., Krizhevsky, A., Sutskever, I., Salakhutdinov, R.R.: Improving neural networks by preventing co-adaptation of feature detectors. arXiv preprint arXiv:1207.0580 (2012)
9. Howard, A., et al.: Searching for mobilenetv3. In: Proceedings of the IEEE International Conference on Computer Vision, pp. 1314–1324 (2019)

10. Howard, A.G., et al.: MobileNets: efficient convolutional neural networks for mobile vision applications. arXiv preprint arXiv:1704.04861 (2017)
11. Hu, J., Shen, L., Sun, G.: Squeeze-and-excitation networks. In: Proceedings of the IEEE Conference on Computer Vision and Pattern Recognition, pp. 7132–7141 (2018)
12. Huang, G., Chen, D., Li, T., Wu, F., van der Maaten, L., Weinberger, K.Q.: Multi-scale dense networks for resource efficient image classification. arXiv preprint arXiv:1703.09844 (2017)
13. Jaderberg, M., Simonyan, K., Zisserman, A., et al.: Spatial transformer networks. In: Advances in Neural Information Processing Systems, pp. 2017–2025 (2015)
14. Jia, X., De Brabandere, B., Tuytelaars, T., Gool, L.V.: Dynamic filter networks. In: Advances in Neural Information Processing Systems, pp. 667–675 (2016)
15. Keskin, C., Izadi, S.: SplineNets: continuous neural decision graphs. In: Advances in Neural Information Processing Systems, pp. 1994–2004 (2018)
16. Li, X., Wang, W., Hu, X., Yang, J.: Selective kernel networks. In: Proceedings of the IEEE Conference on Computer Vision and Pattern Recognition, pp. 510–519 (2019)
17. Lin, J., Rao, Y., Lu, J., Zhou, J.: Runtime neural pruning. In: Advances in Neural Information Processing Systems, pp. 2181–2191 (2017)
18. Lin, T.Y., Goyal, P., Girshick, R., He, K., Dollár, P.: Focal loss for dense object detection. In: Proceedings of the IEEE International Conference on Computer Vision, pp. 2980–2988 (2017)
19. Lin, T.-Y., et al.: Microsoft COCO: common objects in context. In: Fleet, D., Pajdla, T., Schiele, B., Tuytelaars, T. (eds.) ECCV 2014. LNCS, vol. 8693, pp. 740–755. Springer, Cham (2014). https://doi.org/10.1007/978-3-319-10602-1_48
20. Liu, L., Deng, J.: Dynamic deep neural networks: optimizing accuracy-efficiency trade-offs by selective execution. In: Thirty-Second AAAI Conference on Artificial Intelligence (2018)
21. Luong, M.T., Pham, H., Manning, C.D.: Effective approaches to attention-based neural machine translation. arXiv preprint arXiv:1508.04025 (2015)
22. Ma, N., Zhang, X., Zheng, H.T., Sun, J.: Shufflenet v2: practical guidelines for efficient CNN architecture design. In: Proceedings of the European Conference on Computer Vision (ECCV), pp. 116–131 (2018)
23. Maaten, L.V.D., Hinton, G.: Visualizing data using t-SNE. J. Mach. Learn. Res. 9(Nov), 2579–2605 (2008)
24. Munkhdalai, T., Yu, H.: Meta networks. In: Proceedings of the 34th International Conference on Machine Learning, vol. 70, pp. 2554–2563. JMLR. org (2017)
25. Platanios, E.A., Sachan, M., Neubig, G., Mitchell, T.: Contextual parameter generation for universal neural machine translation. arXiv preprint arXiv:1808.08493 (2018)
26. Russakovsky, O., et al.: ImageNet large scale visual recognition challenge. Int. J. Comput. Vis. 115(3), 211–252 (2015)
27. Sandler, M., Howard, A., Zhu, M., Zhmoginov, A., Chen, L.C.: MobileNetV2: inverted residuals and linear bottlenecks. In: Proceedings of the IEEE Conference on Computer Vision and Pattern Recognition, pp. 4510–4520 (2018)
28. Schmidhuber, J.: Learning to control fast-weight memories: an alternative to dynamic recurrent networks. Neural Comput. 4(1), 131–139 (1992)
29. Simonyan, K., Zisserman, A.: Very deep convolutional networks for large-scale image recognition. arXiv preprint arXiv:1409.1556 (2014)
30. Szegedy, C., et al.: Going deeper with convolutions. In: Proceedings of the IEEE Conference on Computer Vision and Pattern Recognition, pp. 1–9 (2015)

31. Tan, M., et al.: MnasNet: platform-aware neural architecture search for mobile. In: Proceedings of the IEEE Conference on Computer Vision and Pattern Recognition, pp. 2820–2828 (2019)
32. Tan, M., Le, Q.V.: EfficientNet: rethinking model scaling for convolutional neural networks. arXiv preprint arXiv:1905.11946 (2019)
33. Vaswani, A., et al.: Attention is all you need. In: Advances in Neural Information Processing Systems, pp. 5998–6008 (2017)
34. Wang, F., et al.: Residual attention network for image classification. In: Proceedings of the IEEE Conference on Computer Vision and Pattern Recognition, pp. 3156–3164 (2017)
35. Wang, X., Yu, F., Dou, Z.Y., Darrell, T., Gonzalez, J.E.: SkipNet: learning dynamic routing in convolutional networks. In: Proceedings of the European Conference on Computer Vision (ECCV), pp. 409–424 (2018)
36. Woo, S., Park, J., Lee, J.Y., So Kweon, I.: CBAM: convolutional block attention module. In: Proceedings of the European Conference on Computer Vision (ECCV), pp. 3–19 (2018)
37. Wu, Z., et al.: BlockDrop: dynamic inference paths in residual networks. In: Proceedings of the IEEE Conference on Computer Vision and Pattern Recognition, pp. 8817–8826 (2018)
38. Yang, B., Bender, G., Le, Q.V., Ngiam, J.: CondConv: conditionally parameterized convolutions for efficient inference. In: Advances in Neural Information Processing Systems, pp. 1305–1316 (2019)
39. Yu, J., Yang, L., Xu, N., Yang, J., Huang, T.: Slimmable neural networks. arXiv preprint arXiv:1812.08928 (2018)
40. Zhang, X., Zhou, X., Lin, M., Sun, J.: ShuffleNet: an extremely efficient convolutional neural network for mobile devices. In: Proceedings of the IEEE Conference on Computer Vision and Pattern Recognition, pp. 6848–6856 (2018)

Author Index